Electrical Trade Principles

A practical approach

Steven Hanssen
Jeffery Hampson

6th Edition

Electrical Trade Principles
6th Edition
Steven Hanssen
Jeffery Hampson

Portfolio Manager: Sophie Kaliniecki
Product Manager: Sandy Jayadev
Content developer: Stephanie Davis
Senior project editor: Nathan Katz
Cover designer: Regine Abos (Studio Regina)
Text designer: Rina Gargano (Alba Design)
Permissions/Photo researcher: Liz McShane
Editor: Sylvia Marson
Proofreader: James Anderson
Indexer: Julie King
Art direction: Linda Davidson
Cover: Shutterstock.com/Roman Zaiets
Typeset by KnowledgeWorks Global Ltd.

Sixth edition published in 2023

For product information and technology assistance,
in Australia call **1300 790 853**;
in New Zealand call **0800 449 725**

For permission to use material from this text or product, please email
aust.permissions@cengage.com

National Library of Australia Cataloguing-in-Publication Data
ISBN: 9780170458856
A catalogue record for this book is available from the National Library of Australia.

Cengage Learning Australia
Level 7, 80 Dorcas Street
South Melbourne, Victoria Australia 3205

Cengage Learning New Zealand
Unit 4B Rosedale Office Park
331 Rosedale Road, Albany, North Shore 0632, NZ

For learning solutions, visit **cengage.com.au**

Printed in China by 1010 Printing International Limited.
3 4 5 6 7 26 25

Brief contents

Contents

Guide to the text

As you read this text you will find a number of features in every chapter to enhance your study of electrotechnology and help you understand how the theory is applied in the real world.

CHAPTER-OPENING FEATURES

Refer to the **Introduction** for a contextualised summary of the chapter.

Solving electrotechnology problems

This chapter provides electrotechnology workers with an introduction to engineering mathematical fundamentals required for industry participation. Mathematical skills provide people with thinking skills and techniques, which enable systematic or ordered problem solving. This chapter provides underpinning knowledge for the unit UEECD0038 from the UEE training package.

Identify the key concepts you will engage with through the **Learning Objectives** at the start of each chapter.

LEARNING OBJECTIVES

Basic units of measurement

- Recognise and use SI units, Greek alphabet, scientific and engineering notation, significant figures and mechanical quantities in calculations

Basic calculations

- Add and subtract, multiply and divide
- Simplify mathematical expressions involving square roots
- Understand exponent terms
- Use prefixes

Working with numbers using powers, exponents and indices

- Apply scientific notation
- Apply engineering notation
- Understand exponent terms
- Apply the laws of indices
- Perform transposition of equations

Algebra

- Perform substitution in algebraic equations
- Simplify algebraic equations
- Simplify mathematical expressions involving square roots

Interpreting data in graphical form

- Develop and interpret data in graphical form
- Develop and interpret vectors
- Solve right-angled triangles using Pythagoras' theorem and trigonometry ratios

Complete work calculations and report on solutions

- Understand the specific problem, solve the problem, check the solution and document the answer

FEATURES WITHIN CHAPTERS

All-important calculations are worked through step by step in easy-to-follow **Example boxes**. Worked examples show the process undertaken to solve a problem using the calculations.

EXAMPLE 1.7

Given the equation $I = \frac{V}{R}$, rewrite the equation with R as the subject.

Step 1 $I = \frac{V}{R}$ write the equation

Step 2 $I \times R = \frac{V}{R} \times R$ multiply both sides of the equation by R, thus $I \times R = V$

Step 3 $\frac{I \times R}{I} = \frac{V}{I}$ divide both sides of the equation by I, then

Step 4 $R = \frac{V}{I}$

Practice what you have learned in the text using the **Exercise boxes**.

EXERCISE 1.11

Simplify the following expressions.

a $(4a - 6b + 2c - 2d) + (5a - 4b + 2c - d) = ?$

b $(2x - 4y) + (6y - x) + (3z + 3x) = ?$

c $(4ab - 6a + 2b - 5) - (2ab - 2a - 10b + 12) = ?$

d $(x^2 + 2x + 4) + (2x^2 + 6x + 4) - (x^2 - 3x - 8) = ?$

e $n + (n + 2) = ?$

f $(6a - 4) - (4x - 10) = ?$

g $-(a + 8) + a = ?$

Switch On boxes highlight important tips and hints.

SWITCH ON

With knowledge of which direction the current is flowing in a conductor hold the conductor with the right hand so that the thumb points in the direction of conventional current flow; the curled fingers then point in the direction of the magnetic field around the conductor. Refer to Figure 4.22.

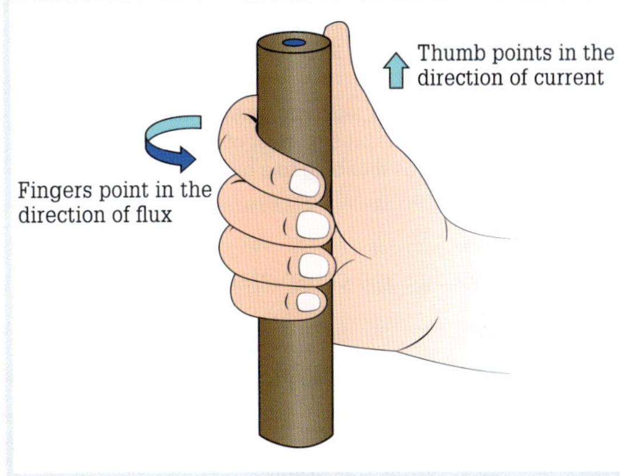

FIGURE 4.22 Right-hand grip rule

Check your understanding of the content by answering the **Review Questions** as you progress through the chapter.

REVIEW QUESTIONS

1 Name the three essential components in an electrical circuit.

2 What item is added to the basic electrical circuit to provide on-off control?

3 What is meant by the term 'closed circuit'?

4 What is meant by the term 'open circuit'?

5 What is the function of a circuit breaker in an electrical circuit?

6 What is the function of a battery in an electrical circuit?

7 Draw the circuit symbol for a battery.

8 Draw the circuit symbol for a filament lamp.

9 Express 1.5 kV in volts.

10 Express 0.015 A in milliampere (mA).

END-OF-CHAPTER FEATURES

At the end of each chapter you will find several tools to help you to review, practise and extend your knowledge of the key learning objectives.

Review your understanding of the key chapter topics with the **Chapter Review**.

CHAPTER REVIEW

1.1 Basic units of measurement

- There are seven SI fundamental units and 22 derived units that have been given specific names.
- A number of electrical quantities are represented by symbols from the Greek alphabet.

1.2 Basic calculations

- The acronym BODMAS is a mnemonic for Brackets, Orders, Division, Multiplication, Addition and Subtraction.
- The process of rounding a particular value by choosing a desired number of digits for measurement is known as rounding to a number of significant figures.
- Indices (the plural of index) provide a way of writing numbers in a more convenient form.
- Whatever you do to one side of an equation you must do the same to the other.

1.4 Algebra

- Algebra is a way of writing mathematics in a general form using algebraic expressions.
- When dividing algebraic terms you divide the numerical coefficients and then subtract exponents.
- To know the square root of a number, determine a number that, when multiplied by itself, provides the original number.
- To simplify a square root, display the number under

Test your knowledge and consolidate your learning through the **Trial Exam**.

TRIAL EXAM

For Chapter 1 knowledge assessment, please complete the following trial exam.

1 How many named derived units occur within the SI?
 a 3
 b 11
 c 22
 d 7

2 Which character set designates electrical quantities?
 a Latin
 b Arabic
 c Roman
 d Greek

3 Express the number 9 210 000 in engineering notation.

8 Simplify: $6a^2 \times 3a^2b$.
 a $18a^4b$
 b $3ab^{-1}$
 c $2a^4b$
 d $10a^{-2}b$

9 Simplify: $m^3n^5 \div m^6n^8$.
 a m^2n^{16}
 b $m^{-3}n^{-3}$
 c m^9n^{13}
 d $m^{18}n^{40}$

10 Simplify the following algebraic fraction: $8a^5 \div 4a$.
 a $2a^4$
 b $32a^4$
 c $4a^6$

END-OF-BOOK FEATURES

Check your work using the **Answers to the Exercises**.

Answers to the exercises

Chapter 1

Exercise 1.1

a 88
b 45
c 10

Exercise 1.2

a 4
b 1
c 4
d 4
e 3

h 6^5
i 3^6
j 4^2

Exercise 1.9

a $R_1 = R_T - (R_2 + R_3)$ or $R_1 = R_T - R_2 - R_3$
b $V_A = E_G - V_T$
c $L = \frac{F}{Bi}$
d $\rho = \frac{Ra}{l}$
e $f = \frac{X_L}{2\pi L}$

Guide to the online resources

FOR THE INSTRUCTOR

Cengage is pleased to provide you with a selection of resources that will help you prepare your learning and assessments. Contact your Cengage learning consultant for more information.

MINDTAP

Premium online teaching and learning tools are available on the MindTap platform – the personalised eLearning solution.

MindTap is a flexible and easy-to-use platform that helps build student confidence and gives you a clear picture of their progress. We partner with you to ease the transition to digital – we're with you every step of the way.

The *Cengage Mobile App* puts your course directly into students' hands with course materials available on their smartphone or tablet. Students can read on the go, complete practice quizzes or participate in interactive real-time activities.

A series *MindTap* for Hanssen and Hampson's Electrotechnology: Principles and Practice is full of innovative resources to support critical thinking, and help your students move from memorisation to mastery! Includes:

- Hanssen and Hampson's *Electrical Trade Principles and Electrotechnology Practice* eBooks
- Chapter reviews
- Concept checks
- Videos
- Trial exams
- Labelling activities
- Online chapters.

MindTap is a premium purchasable eLearning tool. Contact your Cengage learning consultant to find out how MindTap can transform your course.

INSTRUCTOR RESOURCES PACK

Premium resources that provide additional instructor support are available for this text including PowerPoints, TestBanks, Artwork from the text and Online-only chapters.

These resources save you time and are a convenient way to add more depth to your classes, covering additional content and with an exclusive selection of engaging features aligned with the text.

The Instructor Resource Pack is included for institutional adoptions of this text when certain conditions are met.

The pack is available to purchase for course-level adoptions of the text or as a standalone resource.

Contact your Cengage learning consultant for more information.

SOLUTIONS MANUAL

The **Solutions Manual** provides detailed answers to every question in the text.

MAPPING GRID

The **Mapping Grid** is a simple grid that shows how the content of this book relates to the units of competency needed to complete the Certificate III in Electrotechnology Electrician (UEE30811).

About the authors

Jeffery Hampson and Steven Hanssen each have more than 35 years of teaching experience in the VET sector. Jeff has taught electrical trade students at Skills Tech Australia (TAFE QLD) as a leading vocational teacher. Steven is a head teacher of electrical trades in TAFE NSW Sydney Region.

Acknowledgements

The authors wish to thank the many dedicated VET educators and industry experts who provided valuable feedback and willingly shared their knowledge in the preparation of this new edition of *Electrical Trade Principles*. Their helpful suggestions and robust critique, concerning many aspects of the technical content of this text, have ensured that the content remains up to date and reflects current industry best practice.

Cengage would like to thank the following lecturers and industry experts for their valuable feedback for this new edition of *Electrical Trade Principles*:

- Melissa Rawnsley, TAFE NSW
- Tristan McNaught, GO TAFE
- Paul Mansfield, TAFE SA.

Every effort has been made to trace and acknowledge copyright. However, if any infringement has occurred, the publishers tender their apologies and invite the copyright holders to contact them.

Solving electrotechnology problems

This chapter provides electrotechnology workers with an introduction to engineering mathematical fundamentals required for industry participation. Mathematical skills provide people with thinking skills and techniques, which enable systematic or ordered problem solving. This chapter provides underpinning knowledge for the unit UEECD0038 from the UEE training package.

LEARNING OBJECTIVES

Basic units of measurement

- Recognise and use SI units, Greek alphabet, scientific and engineering notation, significant figures and mechanical quantities in calculations

Basic calculations

- Add and subtract, multiply and divide
- Simplify mathematical expressions involving square roots
- Understand exponent terms
- Use prefixes

Working with numbers using powers, exponents and indices

- Apply scientific notation
- Apply engineering notation
- Understand exponent terms
- Apply the laws of indices
- Perform transposition of equations

Algebra

- Perform substitution in algebraic equations
- Simplify algebraic equations
- Simplify mathematical expressions involving square roots

Interpreting data in graphical form

- Develop and interpret data in graphical form
- Develop and interpret vectors
- Solve right-angled triangles using Pythagoras' theorem and trigonometry ratios

Complete work calculations and report on solutions

- Understand the specific problem, solve the problem, check the solution and document the answer

1.1 Basic units of measurement

The international system of measurement, called the SI, uses the first two letters from the French name, Système International d'Unités. In 1875 an international agreement, called the Treaty of the Metre (Convention du Metre), established the SI in Paris. The SI is maintained by the International Bureau of Weights and Measures. A feature of the SI is that it evolves to match the world's requirements for measurement. Moreover, there is an SI unit for each physical quantity.

SI fundamental units

At the centre of the SI is a small list of fundamental units. These units do not refer to any other units. There are seven base SI fundamental units as listed in **Table 1.1** and 22 named SI derived units (refer to **Table 1.2**). However, approved decimal prefixes, called SI prefixes, are used to create multiples or submultiples of SI units.

TABLE 1.1 SI base (fundamental) units

	SI base unit	
Base quantity	Name	Symbol
length	metre	m
mass	kilogram	kg
time	second	s
current	ampere	A
temperature	kelvin	K
amount of substance	mole	mol
luminous intensity	candela	Cd

Definitions of SI base units are as follows.

Metre

The length of the direction travelled by light in a vacuum during a time interval of 1/299 792 458 second.

Second

This is the time interval corresponding with 9 192 631 770 oscillations of a caesium-133 atom at 0 K.

Kilogram

The kilogram is the mass of platinum–iridium called the international prototype of the kilogram. Furthermore, the mass of 1 litre of water under standard conditions is one kilogram.

Ampere

The ampere is the unit of measure of electric current (I). The ampere is based on a fundamental physical constant, the elementary charge (e), which is the amount of electric charge in a single proton (positive) or electron (negative). It is a measure of the amount of electric charge in motion per unit time. The quantity of electric charge, whether or not in motion, is expressed by another SI unit, the coulomb (C). One coulomb is equal to about 6.241×10^{18} electric charges (e). One ampere therefore is the current in which one coulomb of charge travels across a given point in 1 second.

Kelvin

This is 1/273.16 of the triple point of water (ice point) (see 'Degree Celsius' below for explanation).

Mole

The Mole is the amount of material that contains many elementary entities (atoms, molecules, electrons, ions and the like) as there are atoms in 0.012 kilograms of carbon-12.

Candela

The candela is the luminous intensity in a particular direction from a source emitting a monochromatic radiation of 540×10^{12} Hz with a radian intensity of 1/683 watts per steradian.

Derived units

Other algebraically SI units are named SI derived units. At present, there are 22 named SI derived units, as listed in **Table 1.2**.

Definitions of some named SI derived units are as follows.

Degree Celsius (°C)

The degree Celsius is the SI derived unit of temperature. The unit honours the Swedish astronomer and physicist Anders Celsius (1701–44). Celsius scale indicates the freezing point of water (at one atmosphere of pressure) as 0 °C and the boiling point at 100 °C. In addition, the temperature of the triple point of water (where water exists in gaseous, liquid and solid states at the same time) is exactly 0.01 °C on the Celsius scale.

Newton (N)

The newton is the SI derived unit of force. A one-newton force accelerates a 1 kilogram mass at a rate of 1 metre per second per second. The unit honours Sir Isaac Newton (1642–1727), the British mathematician, physicist and natural philosopher.

Pascal (Pa)

The pascal is the SI derived unit of pressure. Pascal is often used to express the quantity of a container's internal pressure. The pascal is equal to one newton per square

TABLE 1.2 SI derived units

Derived quantity	Name	Symbol
absorbed dose (of ionising radiation)	gray	Gy
angle	radian	rad
catalytic activity	katal	kat
electric charge or quantity of electricity	coulomb	C
electrical capacitance	farad	F
electrical conductance	siemens	S
electrical inductance	henry	H
electrical resistance, impedance, reactance	ohm	Ω
energy, work, heat	joule	J
equivalent dose (of ionising radiation)	sievert	Sv
force, weight	newton	N
frequency	hertz	Hz
illuminance	lux	lx
luminous flux	lumen	lm
magnetic flux	weber	Wb
magnetic induction, magnetic flux density	tesla	T
power, radiant flux	watt	W
pressure, stress	pascal	Pa
radioactivity (decays per unit time)	becquerel	Bq
solid angle	steradian	sr
temperature relative to 273.15 K	degree Celsius	°C
voltage, electrical potential difference, electromotive force	volt	V

metre or 1 'kilogram per metre per second per second'. Pressure usually uses kilopascals (kPa). Air pressure is measured in hectopascals (hPa) and 1 hPa = 1 millibar. The pascal unit honours Blaise Pascal (1623–62).

Joule (J)

The joule is the SI derived unit of work or energy. Consequently, the joule is the work done by a force of one newton to move an object a metre in the direction of the applied force. The joule honours the British physicist James Prescott Joule (1818–89).

Watt (W)

The watt is the SI derived unit of power. Furthermore, power is the rate at which work occurs, or the rate at which energy disburses. In addition, one watt is equal to a power rate of 1 joule of work per second. Consequently, this unit represents and links both mechanical and electrical quantities. In electrical terms, 1 watt is the power produced by a current of 1 ampere flowing due to an electric potential of 1 volt. The unit honours James Watt (1736–1819), the British engineer who built practical steam engines.

Volt (V)

The volt is the SI derived unit of electric potential. In fact, separating electric charges creates potential energy, which uses energy units of measurement such as joules. Electric potential is the amount of potential energy present per unit of charge. Furthermore, the volt represents a potential of 1 joule per coulomb of charge. The voltage unit honours the Italian scientist Count Alessandro Volta (1745–1827).

Ohm (Ω)

The ohm is the SI derived unit of electric resistance. A conductor connected between two points of different potential allows current flow through the conductor. However, the value of the current depends on the potential difference across the two points, while the physical property of the conductor limits current flow. Finally, one ohm is the resistance that results from a potential difference of 1 volt per ampere of current. The unit honours the German physicist Georg Simon Ohm (1787–1854). The uppercase Greek letter omega is the symbol for the ohm (Ω).

Derived magnetic units

Weber (Wb)

The weber (Wb) is the SI derived unit of measurement for magnetic flux. Magnetic flux, when linking a circuit of one turn, produces an electromotive force (emf) of 1 volt when the magnetic flux reduces to zero in one second.

Tesla (T)

A tesla (T) is the SI derived unit of measurement for magnetic flux density and is one weber per square metre.

Henry (H)

A conductor or a component has an inductance of one henry (H) when a potential of 1 volt is induced across the circuit, conductor or part, when the current is changing at a rate of 1 ampere per second.

Greek alphabet

Historically, electrical theory has adopted some of the symbols of the Greek alphabet to designate a number of electrical quantities. Refer to **Table 1.3**.

TABLE 1.3 Greek symbols

Name	Greek symbols	Designates
alpha	α	temperature coefficient of resistance
beta	β	flux density
delta	Δ	variation of an inclusive quantity
zeta	Z	impedance
eta	η	efficiency or magnetising force
theta	θ	rotating angle
lambda	λ	wavelength

Name	Greek symbols	Designates
mu	μ	permeability or micro
pi	π	3.1416
rho	ρ	resistivity
psi	ψ	phase difference
tau	τ	time constant
phi	ϕ	phase angle or magnetic flux
omega	Ω	ohm

REVIEW QUESTIONS

1. What is the international system of measurement known as?
2. Name the seven SI base units.
3. Name the SI derived unit for force.
4. Name the SI derived unit of measurement for magnetic flux.
5. Which Greek letter represents the symbol for resistivity?

1.2 Basic calculations

Adding and subtracting, multiplying and dividing

The following equations are examples of basic calculations:

$$8 + 2 = \mathbf{10}$$
$$6 - 3 = \mathbf{3}$$
$$9 \times 2 = \mathbf{18}$$
$$40 \div 8 = \mathbf{5}$$

Exponents and roots (orders or indices)

An exponent term such as 10^3 has a base number (10) and an index number (3) called the base's power or exponent.

$$\text{Base}^{\text{exponent}} = 10^3$$

Order of operations

Operations mean mathematical expressions such as addition, subtraction, division, multiplication and the like. When an expression contains more than one operation, you can obtain different answers depending on the order in which you solve the expression. We have to follow rules for the order of operations so everyone arrives at the same answer. We often use grouping symbols, like brackets, to help us organise complex expressions into simpler ones. For example:

How do you calculate $3 + 4 \times 8$?

Is the answer 56 or is the answer 35?

To obtain the agreed answer we must apply the correct order of operations for addition, subtraction, multiplication and division. Multiplication and division must be solved before addition and subtraction. Therefore, for the expression above:

$3 + 4 \times 8 = 3 + (4 \times 8) = 3 + 32 =$ **35 is the agreed answer to the above question.**

Try entering $3 + 4 \times 8$ in your calculator as shown in **Figure 1.1**.

3	+	4	×	8	=

FIGURE 1.1 Calculator showing the calculation $3 + 4 \times 8$

The order of operations has a descending order that uses an acronym called BODMAS. BODMAS is a mnemonic for Brackets, Orders, Division, Multiplication, Addition and Subtraction. Note: A mnemonic is any learning technique that aids memory.

B Solve the calculations inside the *brackets* first. If there is more than one operation inside the brackets, then you must follow the rules of BODMAS.

O If the equation contains an order such as *power, indices (also called exponents)*, calculate these next.

D Solve the *division* calculations, working from left to right.

M Solve the *multiplication*.

A Solve the *addition*.

S Solve the *subtraction*, working from left to right.

However, using any of the above rules in the order addition first, subtraction afterwards, would give the wrong answer to:

$$11 - 4 + 2$$

The agreed answer is nine. This solution can be conceived by seeing the problem as the sum of positive eleven, negative four and positive two. The written answer is:

$$\sum 11, -4, 2 \text{ where } \sum \text{ means sum}$$

It is usual, wherever you need to calculate operations of equal precedence, to work from left to right. The following rules of thumb are useful:

1. Do any calculations inside parentheses (brackets).
2. Do all multiplication and division calculations by working from left to right.
3. Do all subtraction and addition calculations by working from left to right.

However, for example, the following multiplication:

$$11 \times \frac{12}{6}$$

is much easier when solved from right to left. That is:

$$\frac{12}{6} = 2$$

2 is then multiplied by 11 to provide the answer, which is 22:

$$2 \times 11 = 22$$

EXAMPLE 1.1

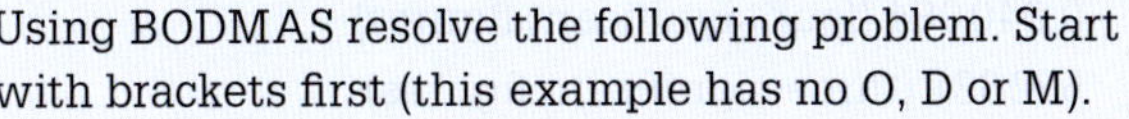

Using BODMAS resolve the following problem. Start with brackets first (this example has no O, D or M).

$$8 + (5 \times 12) - 6$$
$$= 8 + 60 - 6$$
$$= 68 - 6$$
$$= \mathbf{62}$$

This written result is:

$$\sum 8, 60, -6$$

EXERCISE 1.1

Use BODMAS to resolve the following problems.

a $10 + 5 \times 18 - 12$

b $15 \times \frac{18}{6}$

c $15 - \frac{20}{4}$

Significant figures

Significant figures are a way of rounding a particular value by choosing the desired number of figures for measurement. They are an expression of the accuracy of a number. 'Significant figures' refers to the number of important individual digits (0 through 9 inclusive) in the coefficient of an expression written in scientific notation. The number of significant figures in an expression indicates the precision of a measured quantity.

Rules to find the number of significant figures

Using the rules shown in **Figure 1.2**, 405 A has three significant figures, 22.00 Ω has four significant figures, however, each of the numbers 12 V, 12 000 A, and 0.0027 Ω, has only two significant figures.

All non-zero digits are significant
Zeros between non-zero digits are significant
Zeros at the end of a decimal are significant
All other zeros are not significant

FIGURE 1.2 Rules to find number of significant figures

In some numbers, zeros are significant. For example, in a voltage measurement of 205 volts the zero is significant.

EXAMPLE 1.2

A current of 1 A flowing in a conductor results in 6 250 000 000 000 000 000 electrons flowing past a point in the conductor in 1 second. As this is a very large number it is more convenient to display it in scientific notation. In order to accurately represent this number, we need to determine the significant figures, which are 6, 2, and 5.

The result is 6.25×10^{18} electrons per second (see Section 1.3 for scientific notation).

EXERCISE 1.2

Calculate the number of significant figures for the following measurements.

a 0.8354 A

b 20 V

c 67 480 Ω

d 0.0033 V

e 0.610 N

Rounding off significant figures

Rounding a number means to evaluate or approximate it. Rounding limits the digits in a number while preserving a similar value. The result is less precise, but simpler to use. For example, you have estimated that an electrical installation requires 235 m of 2.5 mm² twin and earth V90 TPS cable. Therefore, this cable length is a definite amount for ordering. However, the supplier provides cable in units of 100 m. It is necessary to round up the quantity of cable needed to the nearest 100 m; consequently, rounding the quantity of cable required up to 300 m. When using rounding, a judgement is required about the context for rounding. In the example of the 2.5 mm² cable, rounding down to 200 m would result in a shortfall.

Often with calculations, rounding can be either up or down depending on the last digit (last digit between 0–4 round down, 5–9 round up).

EXAMPLE 1.3

The current in a particular circuit was measured to be 18.678 A. This number can be expressed with various numbers of significant figures. It will be necessary to round off the value if insufficient precision is available. In this case the number is rounded in some manner to fit the available precision. Table 1.4 shows the results for various total precisions.

TABLE 1.4 Effect of rounding off

Precision (n)	Rounded to n significant figures
6	18.6780
5	18.678
4	18.68
3	18.7
2	19
1	20

This example shows how the level of precision affects the accuracy of the displayed value.

EXERCISE 1.3

- **a** Round 1238 to a precision of 3.
- **b** Round 639 to a precision of 1.
- **c** Round 1765 to a precision of 3.
- **d** Round 468 to a precision of 1.
- **e** Round 4131 to a precision of 5.
- **f** Round 295 to a precision of 2.

When rounding to a particular value of significant figures, the first significant figure is the first digit (not zero); beginning from left to right. For example, a measurement of 45.62 amperes shows that the first significant figure is 4. Therefore, the zeros left of the first non-zero are not significant. Again, given a measurement of 0.00457 volts, the zeros are not significant. The zeros show the location of the decimal point; 0.00457 (three significant figures).

All zeros between numbers are significant. In a measure of 15004 amperes, the zeros are significant. There are five significant figures.

Final zeros involving decimals are significant. Given a voltmeter reading of 30.0 volts, the last zero is significant and tells us that the reading uses three significant figures.

EXAMPLE 1.4

Express the resistance values of 396.6 Ω to three significant figures.

The result is **397 Ω**.

EXERCISE 1.4

Round the answers for the following significant figures as given in the brackets.

- **a** 26801 (two significant figures)
- **b** 0.008609 (three significant figures)
- **c** 85.372 (four significant figures)
- **d** 65674018 (five significant figures)
- **e** 0.040084 (four significant figures)
- **f** 582631 (three significant figures)
- **g** 5.004 (one significant figure)

REVIEW QUESTIONS

1. In the expression 10^N, what does the ^{N} represent?
2. What acronym expresses the order of operations in descending order?
3. Solve 20 + 5 × 18 − 11.
4. How many significant figures are in the value 0.0044 V?
5. Express 682631 to four significant figures.

1.3 Working with numbers using powers, exponents and indices

Scientific notation

Scientific notation is a useful system when dealing with very large or very small numbers. A number is in scientific notation when it is written as a number with only one digit to the left of the decimal point (the first element), then multiplied by a power of 10 (the second element).

EXAMPLE 1.5

A voltage reading of 875000 V is expressed in scientific notation as:

8.75	×	10^5 V
first element		second element

= **8.75 × 10^5 V**

Setting a number into scientific notation

1 To determine the first element, remove the decimal point from its original position and set the decimal point in between the first two non-zero digits in the number.

2 To determine the second element, count the number of places that the decimal point moves to its new position. The counting of places gives the power of 10 or index.

Note: If the decimal point moves to the left, the power of 10 is a positive index. However, if the decimal point moves to the right, the power of 10 is a negative index.

So, for example:

6255.0 A = 6.255 A (decimal place moved three places to the left)

0.007585 Ω = 0007.585 Ω (decimal place moved three places to the right)

3 Write the number as a product:

$$6255.0\text{ A} = \mathbf{6.2550 \times 10^{3}\text{ A}}$$

$$0.007585\ \Omega = \mathbf{7.585 \times 10^{-3}}\ \Omega$$

Refer to the Appendix 'Provide solutions to routine electrotechnology problems' for evaluating basic mathematical expressions using a calculator.

EXERCISE 1.5

Express the following numbers in scientific notation.

a 12845
b 2345000
c 0.010
d 0.000045

Engineering notation

Engineering notation is like scientific notation in that both notations use a numerical value and a power of 10. In addition, engineering notation is engineering format or E-format. However, engineering notation can have from one up to three numbers (999.9) to the left or preceding the decimal point. As such, this digit must not be less than one (0.1, 0.9 and so on). Furthermore, the power of 10 is expressed in multiples of three (3, 6, 9 and so on).

EXAMPLE 1.6

a Express the voltage reading 750000.0 V in engineering notation.

To determine the notation, move the decimal point from its original position and place it where the power of 10 is in multiples of three. In this example, moving the decimal point 3 places to the left produces the numerical value 750 with a 10^3 power. So:

$$750000.0\text{ V} = 750.0 \times 10^{3}\text{ V}$$

(decimal place moved three places to the left)

The prefix for 10^3 is 'kilo', symbol 'k', so the recorded voltage reading in engineering notation is **750 kV**.

In E-format 750 kV = 750E3 V

b Express the current reading 0.000989 A in engineering notation.

To determine the notation, move the decimal point from its original position and place it where the power of 10 is in multiples of three. In this example, moving the decimal point 6 places to the right produces the numerical value 989 with a 10^{-6} power. So:

$$0.000989\text{ A} = 989 \times 10^{-6}\text{ A}$$

(decimal place moved six places to the right)

The prefix for 10^{-6} is 'micro', symbol 'μ', so the recorded current reading in engineering notation is **989 μA**.

In E-format 0.000989 A = 989E-6 A.

Although the standard form of scientific notation is a very useful process, electricians use the engineering form to designate multiples and submultiples of measured values. Electronic multimeters have ranges that present an output in engineering notation.

EXERCISE 1.6

Express the following numbers in engineering notation.

a 15845
b 2365000
c 0.050
d 0.000065

Exponent terms

The exponent terms 10^2 and 5^3 have two similarities. They each have a base number (the 10 and the 5) and each have an index number (the 2 and the 3) called the base's power. The index number or power shows how many times the base number is multiplied by itself to provide the number represented by the full term. For example:

SWITCH ON

Positive indices

10^2 means $10 \times 10 = 100$

5^3 means $5 \times 5 \times 5 = 125$

Negative indices

Notice the minus sign before the indices.

$$10^{-2} \text{ means } \frac{1}{10^2} \text{ or } \frac{1}{10 \times 10} = 0.01$$

$$5^{-3} \text{ means } \frac{1}{5^3} \text{ or } \frac{1}{5 \times 5 \times 5} = 0.008$$

»

EXERCISE 1.7 

Evaluate the following numbers.

a 7^5

b 12^3

c 18^{-2}

The laws of indices

Indices (the plural of index) provide a way of writing numbers in a more convenient form. An index is a power or exponent. In fact, the management of powers, or indices or exponents is a vital skill to have in algebra.

What does x^4 or $x^4\ y^4$ mean? Well, they are an easier method for expressing the following:

x^4 means x times x times x times x

x^4y^4 means x times x times x times x times y times y times y times y

We also know that $64 = 4 \times 4 \times 4 = 4^3$. Here 4 is the base and 3 is the index. When termed as indices, we say '64 is equal to base 4 raised to the power 3'.

The laws of indices are:

1 (x^m) times $(x^n) = x^{m+n}$. For example: (x^2) times $(x^3) = x^5$

Note that $x^{-3} = \frac{1}{x^3}$ and $3^{-2} = \frac{1}{3^2} = \frac{1}{9}$

2 $\frac{x^m}{x^n} = x^{m-n}$. For example: $\frac{x^6}{x^2} = x^{6-2} = x^4$

Note that anything to the power zero is one.
For example: $x^0 = 1$

3 $(x^m)^n = x^{mn}$. For example: $(x^2)^4 = x^8$

Note that $(x^2y^3)^0 = 1$ and (x^m) times $x^{-m} = x^{m+(-m)} = x^0 = 1$ also.

$2^{-2} = \frac{1}{2^2} = \frac{1}{4}$ and $125^{1/3} = 5$ (cube root of $125 = 5 \times 5 \times 5$)

EXERCISE 1.8 

Express the following as Base and Index (that is x^y).

a $2^4 \times 2^3 = ?$

b $4^2 \times 4^3 = ?$

c $6^3 \times 6^6 = ?$

d $\frac{3^6}{3^2} = ?$

e $\frac{5^5}{5^3} = ?$

f $\frac{6^7}{6^9} = ?$

g $(4^2)^3 = ?$

h $(6^2)^3 = ?$

i $\frac{3^3 \times 3^5}{3^2} = ?$

j $\frac{4^6 \times 4^4}{4^3 \times 4^5} = ?$

Transposition of equations

Transposition is a method of isolating an algebraic variable to one side of an equation and everything else to the other side. This is also known as changing the subject of the equation. For example, if $E = V_1 + V_2$, then by changing the subject of this equation to V_2 we get $V_2 = E - V_1$. Algebraic equations can be solved using the Law of equations, which basically states:

Whatever you do to one side of the equals sign (=) in an equation you must do the same to the other side of the equals sign. Thus:

- If you multiply one side of the equals sign, you must multiply the other side of the equals sign by the same symbol or number.
- If you divide one side of the equals sign, you must divide the other side of the equals sign by the same symbol or number.
- If you add a number or symbol to one side of the equals sign, you must add the same number or symbol to the other side of the equals sign.
- If you subtract a number or symbol from one side of the equals sign, you must subtract the same number or symbol from the other side of the equals sign.
- If you take the square of one side of the equals sign, you must take the square of the other side of the equals sign.
- If you take the square root of one side of the equals sign, you must take the square root of the other side of the equals sign.

Refer to the Appendix 'Provide solutions to routine electrotechnology problems'.

EXAMPLE 1.7 

Given the equation $I = \frac{V}{R}$, rewrite the equation with R as the subject.

Step 1 $I = \frac{V}{R}$ write the equation

Step 2 $I \times R = \frac{V}{R} \times R$ multiply both sides of the equation by R, thus $I \times R = V$

Step 3 $\frac{I \times R}{I} = \frac{V}{I}$ divide both sides of the equation by I, then

Step 4 $R = \frac{V}{I}$

It is **crucial** when transposing equations to document each step fully. This process is necessary. It is the key to successful transposition.

EXAMPLE 1.8

$T = \frac{\pi}{4} \times \sqrt{2L}$ Transpose to find L.

Step 1 $4 \times T = 4 \times \frac{\pi}{4} \times \sqrt{2L}$ Multiply LHS and RHS by 4.

$4 \times T = \pi \times \sqrt{2L}$ This step removes the 4 from the RHS.

Step 2 $\frac{4 \times T}{\pi} = \pi \times \frac{\sqrt{2L}}{\pi}$ Divide LHS and RHS by π.

$\frac{4 \times T}{\pi} = 1 \times \sqrt{2L}$ This step removes the π from the RHS.

$\frac{4 \times T}{\pi} = \sqrt{2L}$

Step 3 $\left(\frac{4 \times T}{\pi}\right)^2 = \left(\sqrt{2L}\right)^2$ Square the LHS and the RHS.

$\left(\frac{4 \times T}{\pi}\right)^2 = 2L$ This step removes the square root from the RHS.

Step 4 $\frac{1}{2} \times \left(\frac{4 \times T}{\pi}\right)^2 = 2L \times \frac{1}{2}$ Divide the LHS and RHS by 2 (same as multiplying by ½).

$\frac{1}{2} \times \left(\frac{4 \times T}{\pi}\right)^2 = L$ This step removes the 2 from the RHS.

EXERCISE 1.9

Transpose the following equations.

a $R_T = R_1 + R_2 + R_3$ Find R_1

b $V_T = E_G - V_A$ Find V_A

c $F = B \times L \times i$ Find L

d $R = \frac{\rho \times I}{a}$ Find ρ

e $X_L = 2\pi fL$ Find f

f $N = \frac{60P}{2\pi T}$ Find T

g $I = \frac{V}{R}$ Find V

h $P = \frac{V^2}{R}$ Find V

i $R_X = \frac{R_A \times R}{R_B}$ Find R_B

j $N = \frac{120f}{\rho}$ Find f

REVIEW QUESTIONS

1 Express 7845 in scientific notation.

2 Name the notation that can have up to three numbers only in front of the decimal point, but never less than one number while the power of 10 is in multiples of three.

3 A number is expressed as x^y. What do the x and ^{y} represent?

4 What law provides a way of writing numbers in a more convenient form?

5 $R_T = R_1 + R_2 + R_3$ Find R_3

1.4 Algebra

Algebra is a way of writing mathematics in a general form using algebraic expressions. An algebraic expression uses letters or symbols rather than numbers. The general use of algebra is in writing equations.

Algebraic expressions

1 $x\,y$ means x multiplied by y. (With algebra, the multiplication sign is invisible. However, sometimes a dot (•) is used to replace a multiplication sign.)
2 $a \bullet a = a^2$ which means a multiplied by itself
3 $a \div b$ and a/b both mean a divided by b
4 $2a + b$ means b added to 2 a's
5 $3(a + b)$ means the sum of a and b multiplied by 3 (note that the brackets replace multiplication signs)
6 $x \div (a + b)$ means x divided by the sum of a and b
7 $(-x)\ y = -\ xy$ means $-x$ multiplied by y.

Like terms

Expressions such as $2a + 3b$ are not like terms – the factors are different. Like terms have the same factor or factors, for example $3a + 2b + 6a - 4b$. The like terms are $3a$ and $6a$, $2b$ and $-4b$.

What do we do with like terms? We add them:

$$3a + 2b + 6a - 4b$$
$$= 3a + 6a + 2b - 4b$$
$$= 9a - 2b$$

EXERCISE 1.10

Simplify the following expressions.

a $6x + 4x = ?$

b $5x - 3x = ?$

c $7a + 3a = ?$

d $-8a + 4a = ?$

e $7x - 10x = ?$

f $-6x - 4 + 5x + 8 = ?$

g $4a - 9 - 2a - 5 = ?$

h $6x + 2y - 4x + 3y = ?$

Terms that we cannot combine we simply rewrite. For example, $4a + 2b + 3a - 7ab = 7a + 2b - 7ab$.

Removal of brackets

When removing a bracket that has a sign outside the bracket, the signs of all the terms inside the brackets changes. For example:

$$-(x + y) = -x - y$$
$$-(x - y) = -x + y$$
$$-(-x - y) = x + y$$
$$(-x + y) = -x + y$$

To remove the brackets, combine the terms and rewrite them. For example:

$$(3x - 2y + z) + (4x - 3y + z)$$
$$= 3x - 2y + z + 4x - 3y + z$$
$$= 7x - 5y + 2z$$

EXERCISE 1.11

Simplify the following expressions.

a $(4a - 6b + 2c - 2d) + (5a - 4b + 2c - d) = ?$

b $(2x - 4y) + (6y - x) + (3z + 3x) = ?$

c $(4ab - 6a + 2b - 5) - (2ab - 2a - 10b + 12) = ?$

d $(x^2 + 2x + 4) + (2x^2 + 6x + 4) - (x^2 - 3x - 8) = ?$

e $n + (n + 2) = ?$

f $(6a - 4) - (4x - 10) = ?$

g $-(a + 8) + a = ?$

Multiplying terms

When multiplying algebraic terms, multiply numerical coefficients (1, 2 and so on) together, then list all the variables occurring for the terms being multiplied, and add the exponents (x^2) of like variables. For example:

$$x^2 \times 2x = 1 \times 2 \times x^{2+1} = 2x^3$$
$$4a^3 \times a^2 = 4 \times 1 \times a^{3+2} = 4a^5$$
$$5bc \times 3b^2 = 5b^1c^1 \times 3b^2 = 5 \times 3 \times b^{1+2} \times c^1 = 15b^3c$$

EXERCISE 1.12

Multiply the following terms.

a $3a^2 \times 8a^2b$

b $4xy^3 \times 4xy^2$

Dividing terms

Firstly, divide the numerical coefficients and then subtract the exponents.

For example:

$$\frac{6a^x}{3a^y} = \frac{6}{3}a^{x-y} = 2a^{x-y}$$
$$\frac{a^x b^y c}{a^w b^z} = a^{x-w} b^{y-z} c$$

EXERCISE 1.13

Divide the following terms.

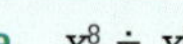

a $x^8 \div x^2$

b $m^4n^9 \div m^2n^3$

c $50u^2v^4 \div 5u^5v^7$

Algebraic fractions

To add or subtract algebraic fractions:

- Determine the lowest common multiple of the denominators.
- Express all fractions in terms of the lowest common denominator.
- Simplify the numerators to obtain the numerator of the answer.

For example:

$$\frac{2a}{4} + \frac{6a}{6}$$

The lowest common denominator (bottom line) for 4 and 6 is 12:

$$\frac{3 \times 2a}{12} + \frac{2 \times 6a}{12}$$

Always express denominators in their simplest form:

$$\frac{6a}{12} + \frac{12a}{12}$$

Simplify the numerator (top line):

$$\frac{6a + 12a}{12}$$

Add the numerators (top):

$$\frac{18a}{12}$$

Simplify:

$$\frac{3a}{2}$$

EXERCISE 1.14

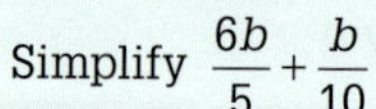

Simplify $\frac{6b}{5} + \frac{b}{10}$

Square roots

To determine the square root of a value, determine a number that when multiplied by itself gives the original value. For example, to solve the square root of 36 you want to find the number that multiplied by itself equals 36. The square root of 36 is six, which is a whole number.

Whole number square roots include:

$\sqrt{16} = 4$, $\sqrt{25} = 5$, $\sqrt{49} = 7$ and so on.

Note: When a negative sign comes before the square root sign then a negative answer is the result. For example:

$\sqrt{25} = 5$ and $-\sqrt{25} = -5$

Simplifying square roots

For example, when simplifying $\sqrt{24}$ it becomes $2\sqrt{6}$. To simplify the square root, make the factor ($\sqrt{24}$) into two factors, one must be the highest perfect square. (Perfect square numbers are 1, 4, 9, 16, 25, 49 and so on.)

EXAMPLE 1.9

$$\sqrt{24} = \sqrt{4 \times 6}$$
$$= \sqrt{4} \times \sqrt{6}$$

Resolving the square root of $\sqrt{4}$ gives $2 \times \sqrt{6}$

EXERCISE 1.15

a Show $\sqrt{96}$ as a simple expression.
b Show $\sqrt{250}$ as a simple expression.

To check your answer, square the number on the outside of the square root sign ($\sqrt{\ }$) and multiply it by the number on the inside.

However, most square roots cannot be reduced because they are already in their primary forms, such as $\sqrt{5}$, $\sqrt{11}$ and $\sqrt{15}$.

REVIEW QUESTIONS

1 Name the mathematical process that uses letters or symbols rather than numbers.
2 Simplify $10x + 4x = ?$
3 Simplify $10a - 4a = ?$
4 Simplify $n + (n + 5) = ?$
5 Simplify $-(a + 12) + a = ?$

1.5 Interpreting data in graphical form

Graphs play a significant role across the electrical industry by providing visual means of presenting information. The visual representations convey mathematical and pictorial data quickly and efficiently. The effectiveness, however, depends on the user's ability to understand the graphical form used and the conventions applied.

Graphs represent the relationship of data visually. A graph plots the relationship of one variable against another on two axes (called x and y) at right angles to each other. Electrotechnology makes use of line graphs to show how one quantity varies in relation to another. Line graphs consist of straight lines or curves.

As quantities may consist of different values, they use the general name of 'variables'. Usually when quantities vary, one quantity is dependent on the other. As shown in **Figure 1.3**, we plot the independent variable on the horizontal x-axis (also called the abscissa), and the dependent variable on the vertical y-axis (called the ordinate).

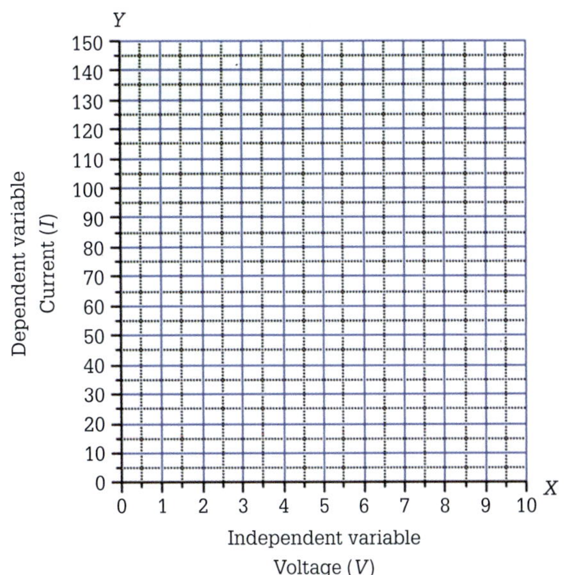

FIGURE 1.3 Independent and dependent variable

The strengths of line graphs are that:

- Graphs show definite magnitudes of data, meaning that given knowledge of one variable allows determination of the other.

- Graphs show movement in data clearly.
- Graphs display how one variable affects the other variable as it increases or decreases.
- Graphs allow predictions about the consequences of data.

Elements of a graph

Graph elements show the relationship between two variables: one variable is identified as *x* and the other *y*. The value of the variable *y* depends upon the value of the variable *x*. This is referred to as cause (*x*) and effect (*y*). We can express the relationship between variables using an equation. For example:

$$R = \frac{V}{I}$$

where *V* is the *x*-variable and *I* the *y*-variable. The value of *R* is how *y* relates to *x*. Graphs as shown in **Figure 1.4** provide a visual representation of the relationships between two variables, *x* and *y*.

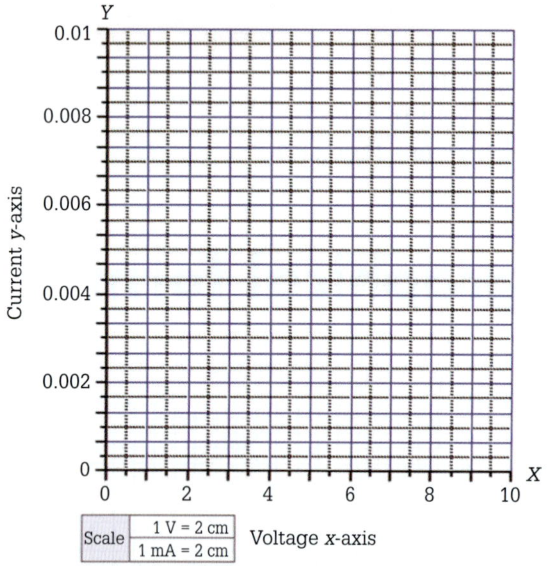

FIGURE 1.4 *x–y* graph

To correspond with the plotted variables the axes use a scale. In electrical graphs, we label the axes with the names of the two particular variables, such as voltage and current. The place where the two axes, x and y, intersect is called the 'point of origin' ($x = 0$ and $y = 0$).

Coordinates of points

The point is a simple relationship displayed on a graph. Pairs of numbers that contain two coordinates can be used to denote every point on the graph. The table of results contains the two coordinates obtained from the measured data, and are shown in **Figure 1.5**. Both an *x*- and a *y*-coordinate identify each pair of coordinates.

Table of results

Voltage	2	4	6	8	10
Current	0.002	0.004	0.006	0.008	0.01

FIGURE 1.5 Table of results

Scale

Before we identify the location of a point on a graph, we must have a suitable scale for the *x*-axis and *y*-axis. The scale is where we determine the maximum and minimum values of *x* and *y*. In our example we have chosen 2 centimetres = 1 V and 2 centimetres = 0.002 A. Note that the lengths of *x* and *y* can differ.

Identifying the *x*-coordinate

The *x*-coordinate of a point tells you how far from the point of origin the plotted value is on the horizontal axis, or *x*-axis. To plot the *x*-coordinate of a point on a graph:

1 Locate the point of origin and develop horizontally from that point.
2 Locate where on the *x*-axis the value from the results table is located.
3 Draw a straight line vertically from that point on the *x*-axis.

Figure 1.6 is a graph with two lines, *A* and *B*. In this figure, the *x*-coordinate of line *A* is 4 and the *x*-coordinate of line *B* is 8. These values correspond to values of voltage detailed in the table of results (see **Figure 1.5**).

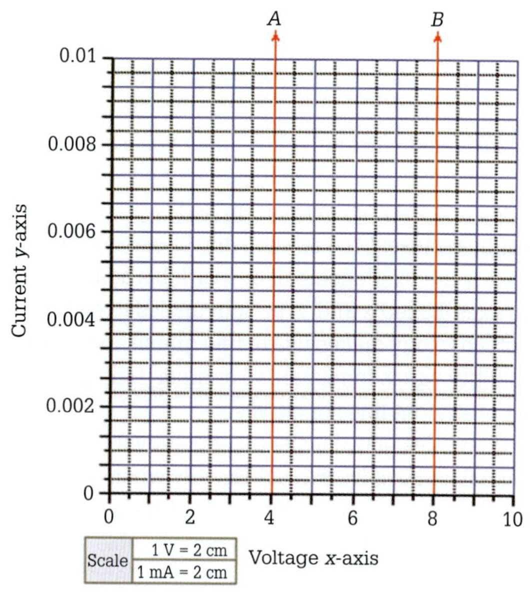

FIGURE 1.6 Identifying the *x*-coordinate

Identifying the *y*-coordinate

The *y*-coordinate of a point tells you how far from the point of origin the plotted value is on the vertical axis, or *y*-axis. To plot the *y*-coordinate of a point on a graph:

1 Locate the point of origin and develop vertically from that point.

2 Locate where on the y-axis the value from the results table is located.

3 Draw a straight line horizontally from that point on the y-axis.

Figure 1.7 is a graph with two lines, *C* and *D*. In this figure, the y-coordinate of line *C* is 0.004 and the y-coordinate of line *D* is 0.008. These values correspond to values of current detailed in the table of results (see **Figure 1.5**) for voltage values of 4 and 8 respectively.

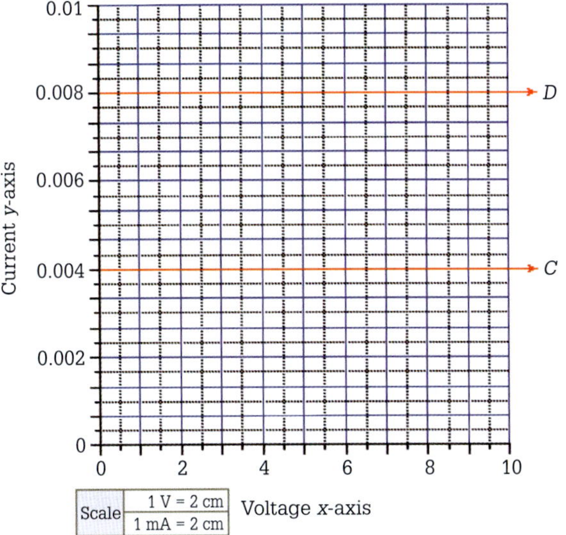

FIGURE 1.7 Identifying the y-coordinate

Plotting points on a graph

Once the coordinates of each *x* and *y* line are located on a graph, the point where the lines *A*, *B*, *C* and *D* intersect are the points we are plotting. **Figure 1.8** shows a method for plotting points on the graph. Note the coordinates are the x-value, y-value. In the example, the coordinates are (4, 0.004) and (8, 0.008).

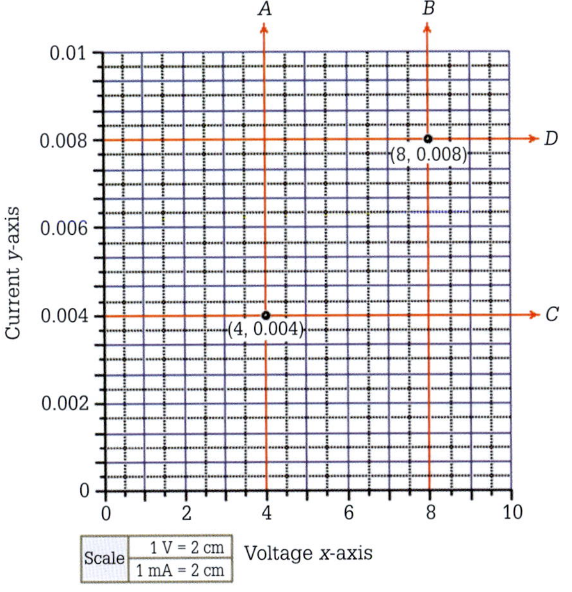

FIGURE 1.8 Plotting the points on a graph

The final graph in **Figure 1.9** is a straight line or linear graph that shows how *x* relates to *y*. Note that the graph starts at the point of origin.

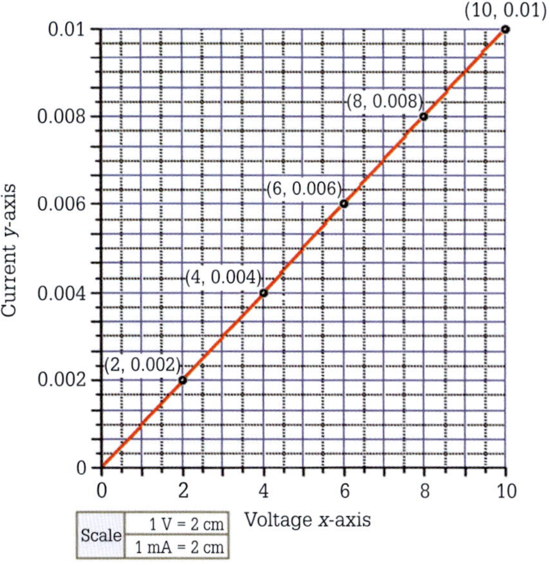

FIGURE 1.9 Final graph

Finally, the graph must have a name. As a result, the final graph's title is 'Ohm's law'. The graph illustrates the relationship between voltage and current in a direct current circuit containing only resistance.

Gradient

Gradient (slope), as shown in **Figure 1.10**, is a mathematical expression that informs how much a graphic line slants and in which direction.

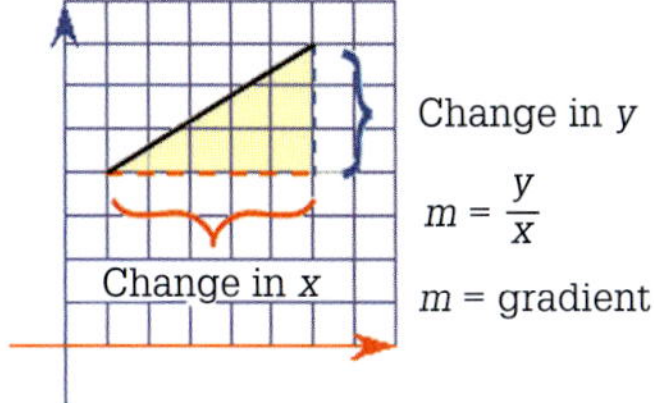

FIGURE 1.10 Gradient

Taking two points along the line, you divide how high the line rises on the y-axis by the distance the line extends along the x-axis. In fact, the gradient m represents the resulting value. If the gradient is positive, the line rises from left to right. However, if the gradient is negative, the line falls from left to right. The farther the value of the gradient m is from zero, then the steeper the line.

Types of graphs

The following types of graphs have different applications in industry.

Light distribution graph

These types of graphs show how light distributes from a particular lamp or luminaire. **Figure 1.11** shows an illustration of a light distribution polar diagram.

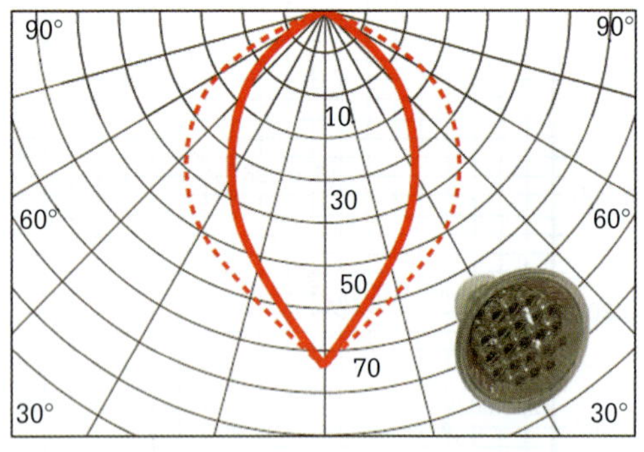

FIGURE 1.11 LED downlight 2 W light distribution curve

A polar diagram is a graphic representation of a luminaire's luminous intensity emitted in several directions as a relation of viewing angle in one or more planes. The polar graph uses an unbroken line to represent the light distribution perpendicular to the light source's longitudinal *y*-axis and by a dashed line that shows the light distribution (in yellow) in the direction of the longitudinal *x*-axis as shown in **Figure 1.12**.

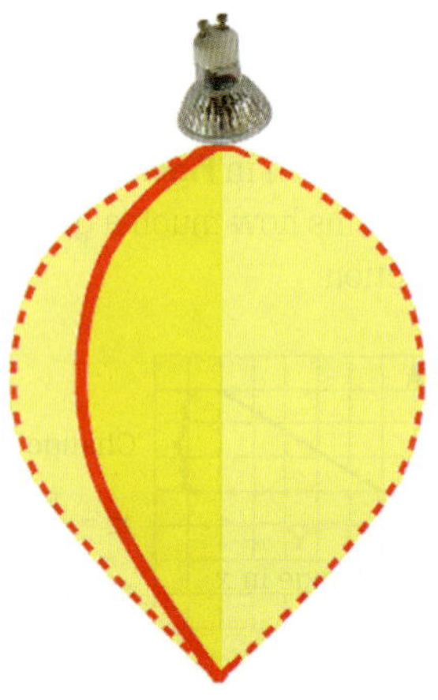

FIGURE 1.12 Light distribution perpendicular to the light source

The values of light distribution use standard scaling from the light source measured in candela/1000 lm. By using this standard, it becomes possible to show the different outputs by luminaires on a typical polar diagram.

Bar graph

A bar graph as illustrated in **Figure 1.13** is a chart with differently shaped bars each with lengths proportional to the values they represent. Plotting the differently shaped bars can occur either vertically or horizontally.

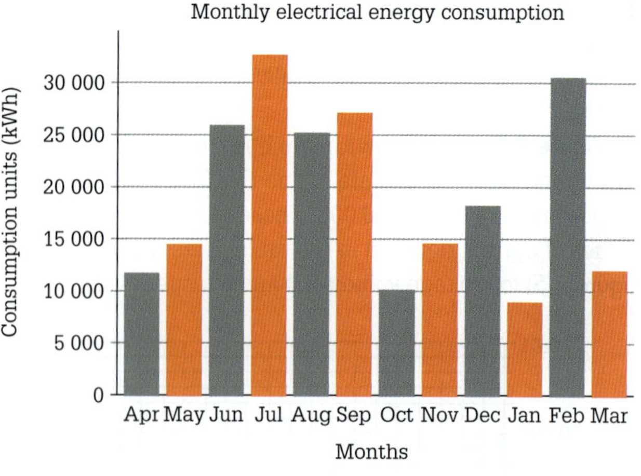

FIGURE 1.13 Bar graph: monthly electrical energy consumption

Bar charts are useful for plotting discrete (or irregular) data.

EXERCISE 1.16

Produce a bar graph using the following data in **Figure 1.14**. Include both supplied and self-generated units in each bar. (Note: Use a rectangle to represent data.)

ENERGY CONSUMPTION DATA FOR APR-MAR			
	Supplied units	Self-generated units	Total units
Apr	21802	628	22430
May	17792	1018	18810
Jun	21205	634	21839
Jul	17841	250	18091
Aug	13561	250	13811
Sep	17952	455	18407
Oct	13250	1416	14666
Nov	10853	667	11520
Dec	14994	7008	22002
Jan	11496	836	12332
Feb	11250	1984	13234
March	13409	1806	15215

FIGURE 1.14 Energy consumption data

Oscilloscope graphs

As a test instrument, the oscilloscope enables an electrician to look at the 'shape' of electrical signals in the form of graphs as shown in **Figure 1.15**.

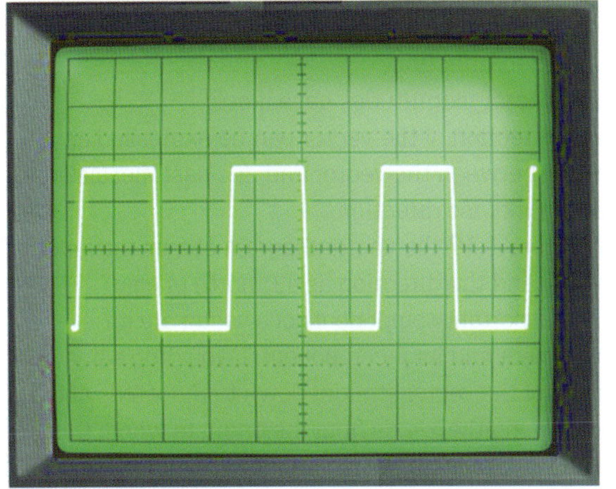

FIGURE 1.15 Oscilloscope graph

A graph of voltage against time has appeared as a trace on the screen. The nature of the input signal determines the type of oscilloscope graph displayed.

Solar energy graphs

Solar energy graphs (see **Figure 1.16**) display complicated data in a way that the properties of energy can be easily understood. These graphs show the effects of solar irradiation, wind speed, temperature, humidity, specific volume, entropy and enthalpy.

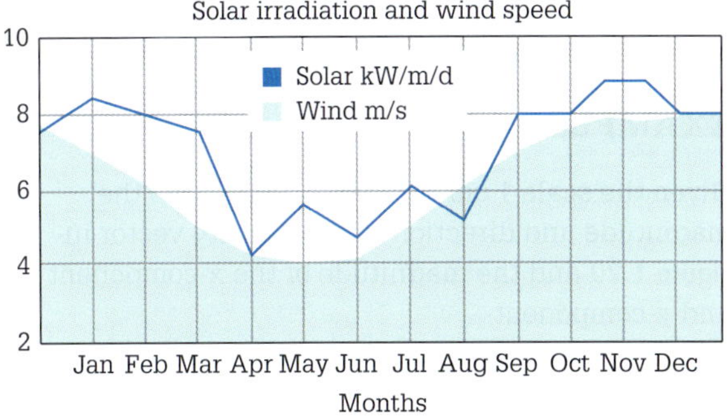

FIGURE 1.16 Solar energy graph

Time–current graph

A time–current graph, also called a time–current curve (TCC), shows the relationship between current (amperes) and response time (seconds) for the different types of over-current protection devices. In fact, most protective devices have an inverse time characteristic: as current increases response time decreases. **Figure 1.17** illustrates the TCC for three fuse ratings.

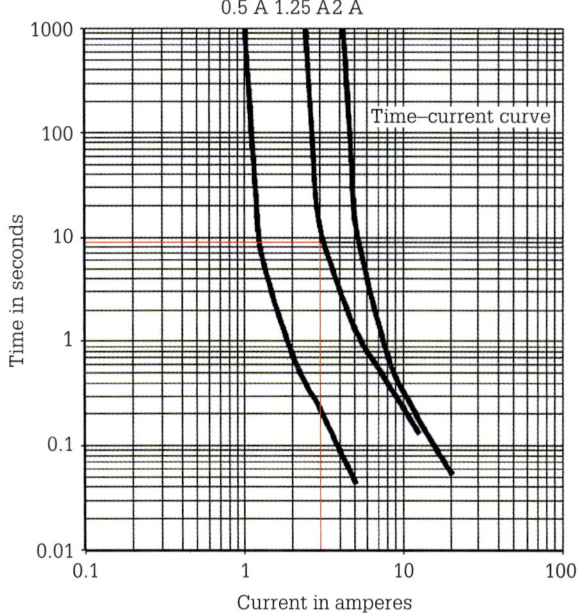

FIGURE 1.17 Time–current graph for fuses

Clearing time of an over-current fault for a fuse is determined from the graph by selecting the current at which the response time occurs. For example, select the 1.25 A fuse and from the graph we can see this fuse's response time when a fault current of 3 A flows. The vertical line at 3 A, is followed up until it intersects the 1.25 A fuse response curve. Clearing a 3 ampere fault occurs in nine seconds.

Electricians need to have knowledge of the different trip characteristics of circuit breakers. These characteristics are expressed as time–current graphs and are available for the different types of circuit breakers (a.c. and d.c.). The time–current graph as shown in **Figure 1.18** indicates how fast a circuit breaker will trip at any magnitude of current. The graph usually consists of a band with a lower and upper time limit. A particular breaker trips at any point within this region. Furthermore, the values on the *x*-axis represent multiples of the continuous current rating (I_N) for the breaker. The figures on the *y*-axis represent the time in seconds.

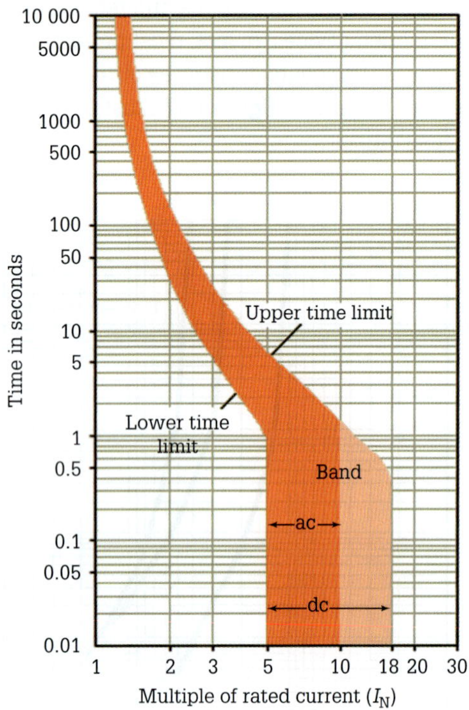

FIGURE 1.18 Time–current graph for a circuit breaker

To know the time it takes for a breaker to trip at a specific multiple of I_N, locate the required multiple on the *x*-axis. Then draw a vertical line, so it intersects the coloured band. Finally, draw a line to the *y*-axis and find the time to trip. From **Figure 1.18**, the circuit breaker trips in the time band interval between 6 seconds (min) and 27 seconds (max) when the current remains at three times I_N.

Vectors

A vector is a property that has two aspects. It has a magnitude (size) and direction. Examples of vectors include weight, momentum, position, displacement, velocity, acceleration and force. Special vectors called phasors demonstrate alternating current properties such as voltage and current. In contrast, there are scalars such as temperature and mass that have only size quantities. **Figure 1.19** is an example of a scaled vector diagram.

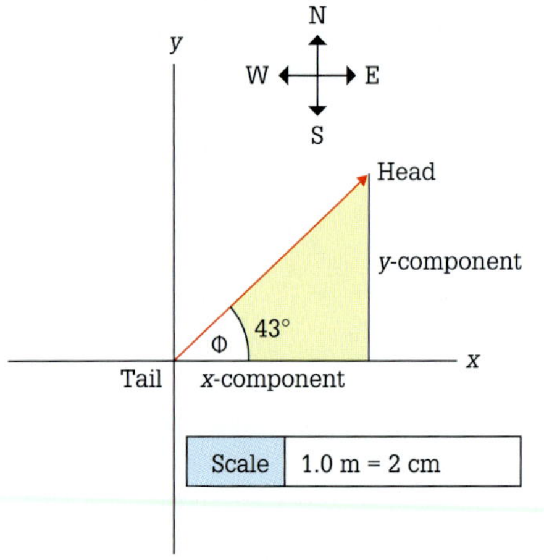

FIGURE 1.19 Scaled vector diagram

Vector diagrams show a line drawn to scale and an arrowhead pointing in a particular direction. The Greek letter theta (θ) designates angles with vector diagrams.

Note that there are several characteristics of a vector diagram:

- The diagram requires a scale.
- A vector has an arrow with head and tail drawn in a specified direction.
- The vector magnitudes have labels.
- The direction of the vector can be expressed as an angle between the vector and due east measured in an anticlockwise direction.

In **Figure 1.19** the vector diagram shows the magnitude as 1.75 m and the direction as 43° north of east.

The important components of a vector are its *x*-component and its *y*-component. If you drop a line from the head of the vector straight down to the *x*-axis and draw a vector along the *x*-axis from the origin to where this line hits the *x*-axis, then this line (in green) is the *x*-component of the vector. The vertical line from the head to the *x*-axis marks the vertical *y*-component of the vector. When both the *x*-component and the *y*-component are drawn, a right-angled triangle forms, with the vector being the hypotenuse.

The *x*- and *y*-components can also be viewed as vectors acting in various directions. The displacement vector illustrated in **Figure 1.20** is the vector sum of those two vectors. This sum is the result of adding the two vectors together, and the displacement vector is the resultant. Some vector diagrams have more than two vectors, with the resultant determined by the addition of those vectors.

The name given to vector diagrams with electrical values is 'phasor diagram'. Phasor diagrams are graphical representations of voltage or current magnitude in an alternating current circuit at any instant in time.

EXAMPLE 1.10

Given the scale 1 cm = 100 m/s, determine the magnitude and direction of the velocity vector in **Figure 1.20** and the magnitude of the *x*-component and *y*-component.

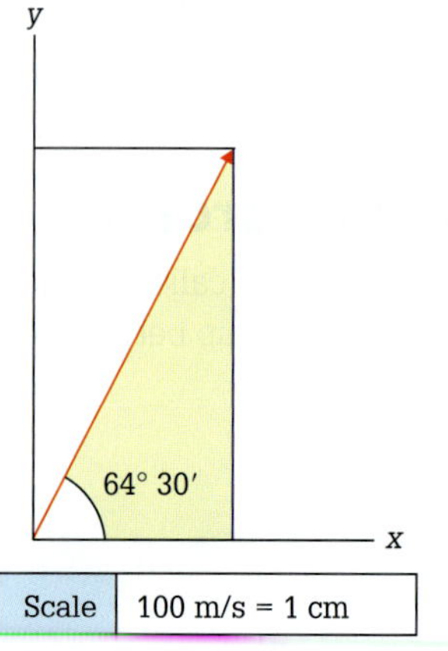

FIGURE 1.20 Velocity vector

»

The direction is 64° 30′ from due east. The magnitude is 516 m/s (5.16 cm in length multiplied by the factor 100 m/s).

Using a ruler, measure the lengths of the *x*- and *y*-components and multiply by 100 m/s:

$$x\text{-component} = 3 \text{ cm} \times 100 \text{ m/s} = \mathbf{300 \text{ m/s}}$$

$$y\text{-component} = 4.2 \text{ cm} \times 100 \text{ m/s} = \mathbf{420 \text{ m/s}}$$

EXERCISE 1.17

a A switch is relocated 3.0 m [east] then 4.0 m [north]. What is the switch's displacement?

b An electrician walked 100 m north, 200 m west, 40 m south and 20 m east. What is the electrician's displacement?

Trigonometric functions

A function is a quantity in mathematics so related with another quantity that if any alteration occurs to the latter there is a consequent change in the former. The dependent number is a function of the other (e.g. the circumference of a circle is a function of the diameter). There are six trigonometric functions. These functions are six different ratios that can be established from a Pythagorean right-angled triangle ($a^2 + b^2 = c^2$) as shown in **Figure 1.21**.

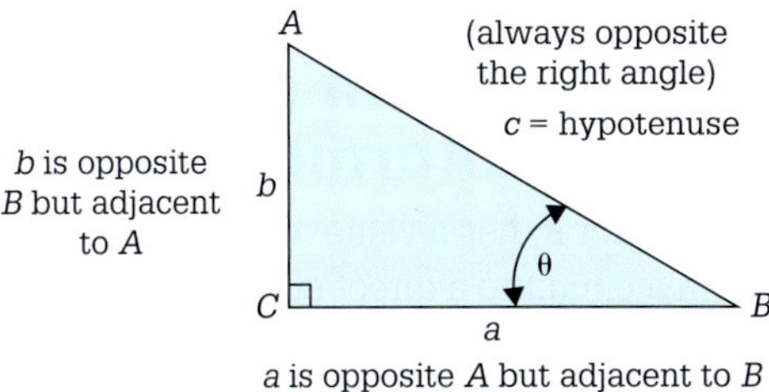

FIGURE 1.21 Right-angled triangle

The right angle in **Figure 1.21** is labelled with a letter *C* and the hypotenuse lower case *c*. In addition, using the letters *A* and *B* for the other two angles, name the sides opposite them as *a* and *b*. In a right-angled triangle, the trigonometric functions are ratios.

$$\text{sine } A = \frac{a}{c} \text{ and sine } B = \frac{b}{c} \left(\text{sine} = \frac{\text{opposite side}}{\text{hypotenuse}}\right)$$

$$\text{cosine } A = \frac{b}{c} \text{ and cosine } B = \frac{a}{c} \left(\text{cosine} = \frac{\text{adjacent side}}{\text{hypotenuse}}\right)$$

$$\text{tangent } A = \frac{a}{b} \text{ and tangent } B = \frac{b}{a} \left(\text{tangent} = \frac{\text{opposite side}}{\text{adjacent side}}\right)$$

$$\text{cotangent } A = \frac{b}{a} \text{ and cotangent } B = \frac{a}{b}$$

$$\text{secant } A = \frac{c}{b} \text{ and secant } B = \frac{c}{a}$$

$$\text{cosecant } A = \frac{c}{a} \text{ and cosecant } B = \frac{c}{b}$$

The cotangent, cosine and cosecant of the angle (θ) in **Figure 1.21**, are the sine, secant and tangent of the complement of the angle. For most purposes the three trigonometric functions, sine, cosine and tangent, are sufficient. All equations for the solution of triangles use the law of sines and cosines, together with the fact that the sum of the three angles in a triangle equals 180°. In addition, a tangent is an angle's sine divided by its cosine.

Change of sign

If the circle in **Figure 1.22** is divided into four quadrants the signs (+ or –) of the trigonometric functions are as follows.

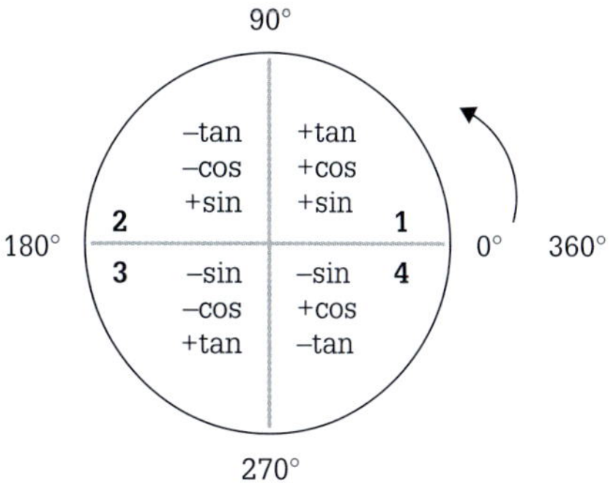

FIGURE 1.22 Four quadrants

A useful mnemonic for remembering which functions are positive is '**All S**tations **T**o **C**entral' (**All** are positive – **S**ine is positive – **T**angent is positive – **C**osine is positive).

EXERCISE 1.18

Using **Figure 1.23** and the given triangle sides, calculate the hypotenuse, the sine, cosine and tangent of the angles *A* and *B*.

a $a = 12, b = 30$

b $a = 45, b = 20$

c $a = 25, b = 15$

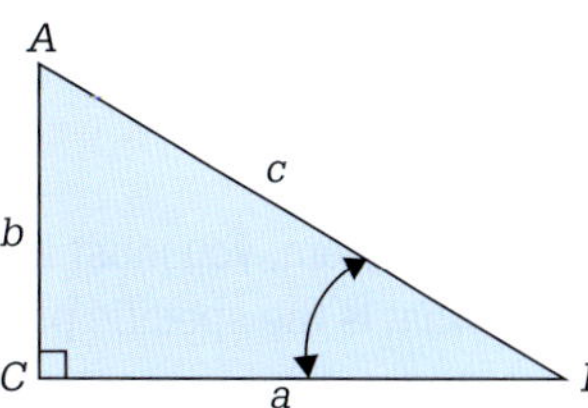

FIGURE 1.23 Exercise 1.18 triangle

REVIEW QUESTIONS

1 What kind of visual representations provide mathematical and pictorial data quickly and efficiently?
2 What is the strength of a line graph?
3 Define the term 'point' as it relates to graphs.
4 Describe the term 'gradient'.
5 What is a common use of a bar graph in the electrotechnology industry?
6 What is a vector?
7 For the triangle shown in Figure 1.24, if a = 10 and b = 5, calculate:
   a length of hypotenuse c.
   b sine of the angle B.
   c tangent of the angle B.
   d cosine of the angle A.

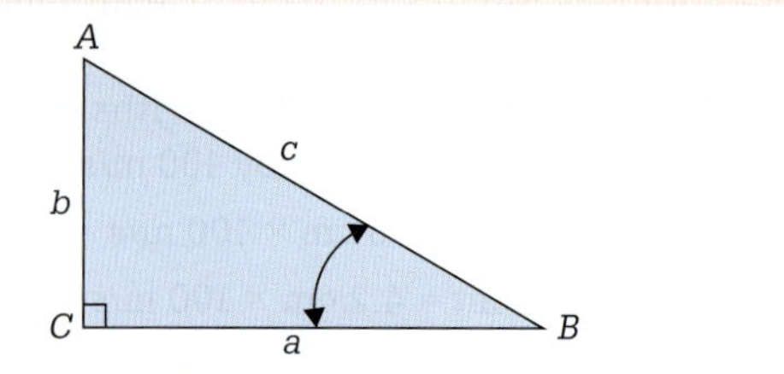

FIGURE 1.24 Trigonometric ratios

1.6 Complete work calculations and report on solutions

While being able to solve electrotechnology problems is a very important skill, being able to clearly document the problem and its solution is just as important. Often it is necessary to include supporting information, such as the reason for selecting a particular cable or particular motor, in a written report. These problems can be considered word problems, which are just a mathematical problem translated into words. When a problem is presented in this manner it requires a conscious decision to choose the appropriate strategy to solve the problem presented. This in itself is a skill, which can be learnt.

The four steps involved in the problem-solving process are:

1 Understanding the specific problem
2 Solving the problem
3 Checking the solution
4 Documenting the answer.

Understanding the specific problem

Before you can solve a problem, you must know what is being asked; that is, what is the specific problem. This is an important first step in solving problems that are presented as a string of words that do not appear to specify a particular mathematical operation. For this it is important to read and reread what is being asked. All too often the tendency is to skip ahead as soon as we see a familiar piece of information or give up completely if the problem does not make sense on the initial read.

Some useful strategies are:

- rereading a question more slowly and intently if it does not seem to make sense the first time
- seeking assistance from a colleague – the old adage states that two heads are better than one
- highlighting or underlining important information.

Identify important and extraneous information

Problem: Resistor R_1 has a value of 100 ohms and is connected in a circuit to a direct current source having a potential of 100 V. How much current is drawn from the source?

With experience you can look past the extraneous information to see an Ohm's law problem. Novices, however, can struggle to determine what is relevant in the information presented to them when it is in this format. The skill is in sifting through the information presented in a problem to locate the relevant parts. In the problem presented above, the relevant information is the resistance, potential difference and current.

A mathematical intervention strategy that can make problem solving easier is the schema approach. In this approach different word problems of the same type are compared and a mathematical sentence stem, which applies to all of them is developed. As an example, a simple addition problem can be expressed as:

[Quantity A] with [Quantity B] added becomes [end result].

This is the underlying procedure or schema required to solve the problem.

Solving the problem

Four useful strategies for problem solving are outlined below.

Visualising

An abstract written problem is often easier to solve when it is visualised. As an example, use the information presented in a written problem to draw a circuit diagram. Use a circuit diagram to write the known and unknown quantities. **Figure 1.25** shows how the sample problem from above can be drawn for visualisation.

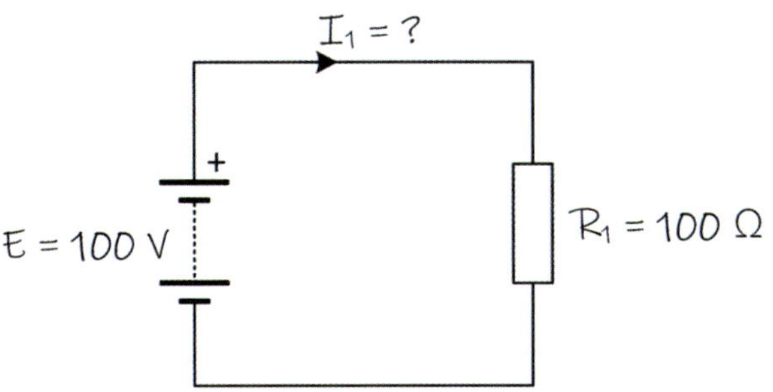

FIGURE 1.25 Visualising a problem from written description

Guess and check

Make a reasonable guess at the answer and then use this in the original problem. If this does not result in the correct answer, then adjust the initial guess higher or lower accordingly.

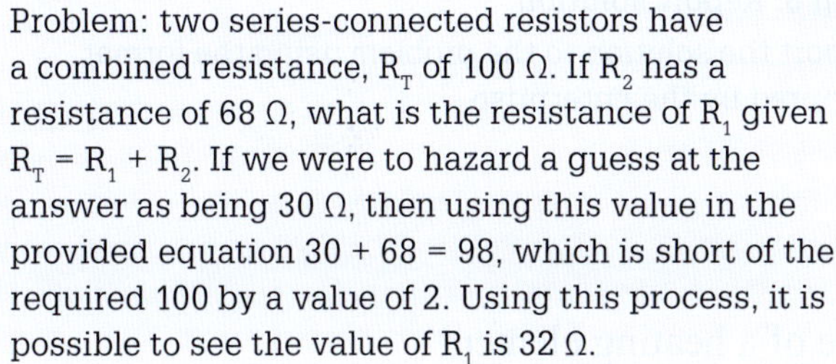

EXAMPLE 1.11

Problem: two series-connected resistors have a combined resistance, R_T of 100 Ω. If R_2 has a resistance of 68 Ω, what is the resistance of R_1 given $R_T = R_1 + R_2$. If we were to hazard a guess at the answer as being 30 Ω, then using this value in the provided equation 30 + 68 = 98, which is short of the required 100 by a value of 2. Using this process, it is possible to see the value of R_1 is 32 Ω.

This is **not** an efficient method for solving electrotechnology problems but is helpful in estimating (or guessing) the answer to see if the calculated value is feasible.

Find a pattern

To find patterns, extract and list all the relevant facts in a problem to allow for comparisons to be drawn. If we consider the information presented in **Figure 1.25** we can see:

$$E = 100\ V$$
$$R_1 = 100\ \Omega$$
$$I_1 = ?$$

Tabulating the pertinent information in this manner allows us to focus what we need to find based on what we know. In this case, find I_1 if E and R_1 are known quantities.

Work backward

Working backwards is useful if the problem is similar to that presented in **Example 1.11**. The steps are:

1 Starting with 100
2 Subtracting the 68 from the 100
3 Leaves a value of 32
4 Checking that 32 works when used instead of R_1 [32 + 68 does = 100]

Checking the solution

It is important to not rush to get an answer and move on without checking the answer. Checking that the answer is correct is a critical step in the process. Some strategies for checking answers include those outlined below.

Check with a colleague

Comparing answers with a colleague develops good reflective processes. If you and your colleague have different answers, it is important to discuss how each of you arrived at the answer and compare methods used in solving the problem.

Reread the problem along with the solution

Most of the time, it is possible to ascertain whether or not the answer is correct by substituting it into the original problem. If this solution does not work or it just 'looks wrong', then it is prudent to rework the solution.

Fixing mistakes

Here it is necessary to backtrack through the working out to find the exact point where the error occurred. This will be extremely challenging if every step to the solution is not documented. To this end, an answer without supporting working out is asking for failure.

Documenting working out

It is not possible to stress enough the importance of documenting every step taken to solve an electrotechnology problem. This facilitates being able to keep track of the thought processes and pick up errors before reaching a final solution.

Complete calculations and report solutions

If you follow a clear procedure, similar to that presented here, you will demonstrate that you have an obvious understanding of the problem. All answers must be noticeably discernible from the working, which must be complete. Reporting solutions to problems in this method allows a reviewer to follow your progress through the problem and see how you arrived at the documented solution.

Methods for recording and maintaining records

The methods of recording and the type of documentation produced should be appropriate to the nature of the enterprise and the purposes of the record. This means that the information needs to be captured, dealt with and safeguarded in an organised system which maintains its reliability and genuine nature over time.

Records can be created and stored using many different technologies, including paper-based files or electronic systems. Records can also be transferred from one medium to another through copying, imaging or digital transfer. In contrast to paper-based files, electronic records are easily updated, deleted, altered and manipulated. Basing records systems and filing practices on authorised retention periods can help to ensure that records are only kept for as long as needed and then promptly destroyed. Regardless of the technology used the objective remains the same: capture records so that they can be easily retrieved at a later date, understood and read as evidence of what occurred in an enterprise.

Note that an electronic record is a more fragile thing than paper, and can easily be overwritten, be lost or become inaccessible through technology change. Therefore a complete backup copy of such records should be stored in a separate safe location.

Communicating calculated solutions

When communicating calculated solutions, it is best to adopt a procedural approach as the following examples illustrate.

EXAMPLE 1.12

Problem:

Determine the expected resistance of a heating element that has a 4.8 kW rating at 240 V.

Step 1: Set out knowns and unknowns

Step 1 allows you to summarise the information you already know and what it is you need to calculate (the unknowns).

Step 2: Determine the correct equation to use

Step 2 allows you to use the information detailed in Step 1 to determine the appropriate equation to use.

Step 3: Perform the transposition

In Step 3, show the transposition of the equation documented in Step 2.

Step 4: Perform calculation

In Step 4, perform the calculations. Remember it is important to show all steps that are necessary to arrive at the answer. In this example, the steps are equation, transposition, substitute 'real values' in equation and solve the equation. Not only does this demonstrate that you clearly know what you are doing, it provides a blow-by-blow account for later revision.

Step 5: Highlight answer

Underline or highlight your answer.

These steps are summarised in Figure 1.26.

Step 6: Report solution

Report the solution to the problem using the format required by the enterprise.

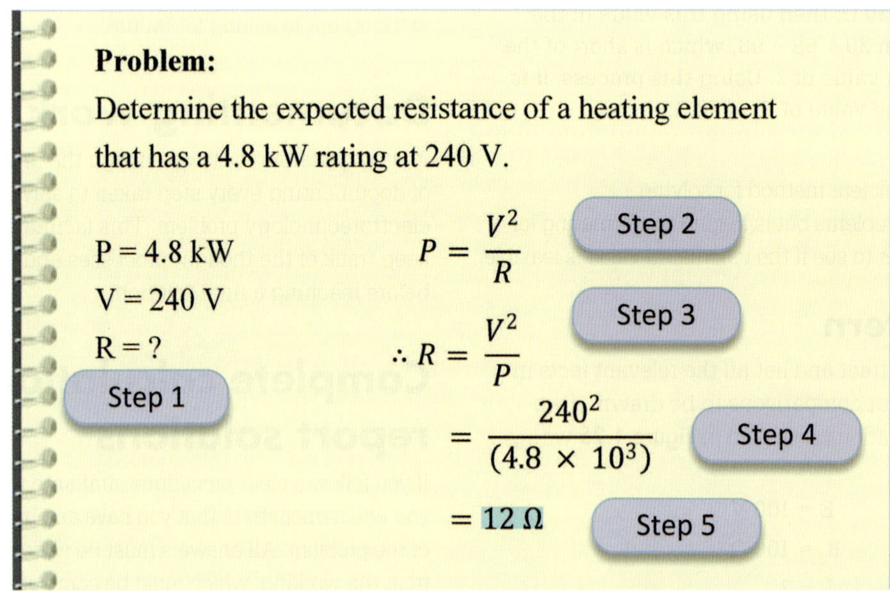

FIGURE 1.26 Documenting the problem and solution

EXAMPLE 1.13

Problem:

It is proposed to add a final sub-circuit in a three-phase, 230/400-volt non-domestic installation. Determine if the proposed sub-circuit is compliant with the requirements of AS/NZS 3000: 2018. The installation comprises the following:

Consumer main:

- Phases: 3
- Maximum demand: 55 A
- Route length: 15 m
- Cable size: 16 mm^2
- Cable configuration: V90 single-core thermoplastic and sheathed copper conductors
- Cable installation: the circuit is enclosed in heavy-duty rigid thermoplastic conduit with no other circuits. Conduit is buried in the ground having an ambient soil temperature of 25 °C and has a top cover of 0.65 m

Sub-main:

- Phases: 3
- Maximum demand: 40 A
- Route length: 45 m
- Cable size: 10 mm^2
- Cable configuration: V90 single-core thermoplastic and sheathed copper conductors
- Cable installation: the cables are clipped to the building structure in trefoil formation and installed in single circuit configuration, unenclosed in air

Proposed final sub-circuit:

- Phases: 1
- Maximum demand: 18 A
- Route length: 30 m
- Cable size: 4 mm^2
- Cable configuration: V90 two-core and earth thermoplastic and sheathed copper conductors
- Cable installation: the cables are clipped to the building structure and installed in single circuit configuration, unenclosed in air

Step 1: Draw the circuit diagram

Step 1 shows your ability to relate descriptive information to an accurate diagram. Make sure you show all relevant information on your diagram. Use the information presented in the problem to produce a diagram. Figure 1.27 shows what this looks like for **Example 1.13**.

Step 2: Determine consumer main parameters

Step 2 allows you to use the information detailed in Step 1 to determine the voltage drop in the consumer main segment of the installation.

Step 3: Determine sub-main parameters

Step 3 allows you to use the information detailed in Step 1 to determine the voltage drop in the sub-main segment of the installation.

Step 4: Determine final sub-circuit parameters

Step 4 allows you to use the information detailed in Step 1 to determine the voltage drop in the final sub-circuit segment of the installation.

Step 5: Determine suitability of proposed final sub-circuit

Step 5 allows you to determine the suitability of the proposed cable selection.

Step 6: Highlight answer

Underline or highlight your answer.

Step 7: Report solution

Report the solution to the problem using the format required by the enterprise.

Solution:

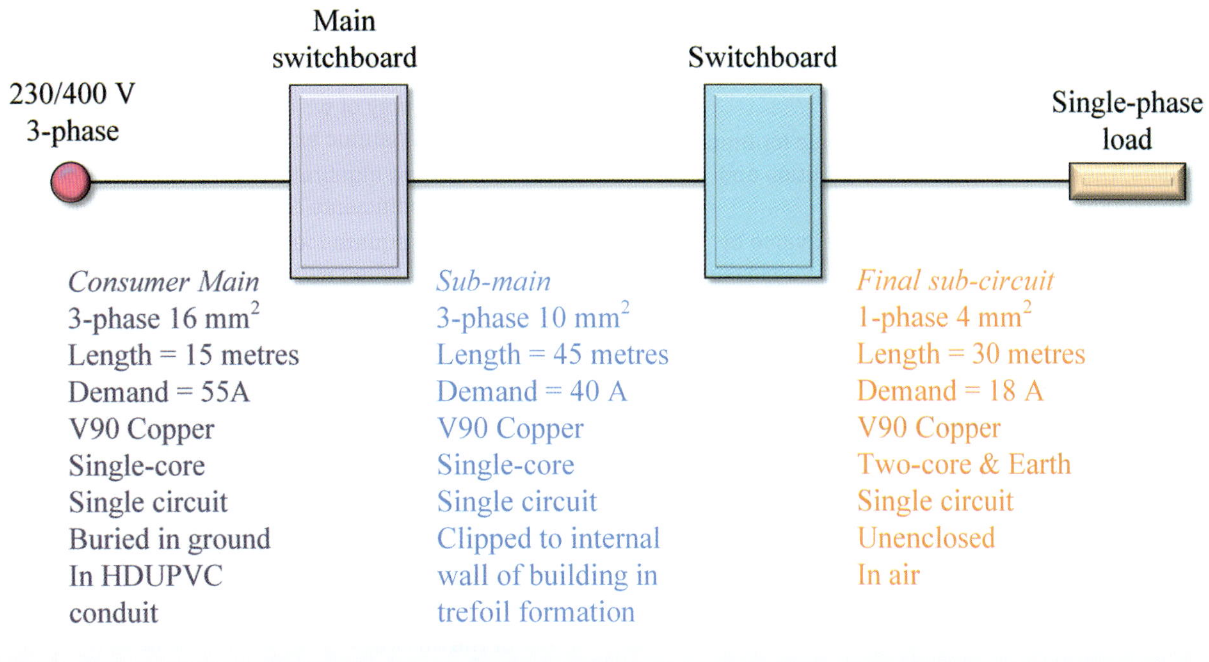

FIGURE 1.27 Summarising written information in a diagram

»

Consumer main:

From AS/NZS 3008.1 Table 41, 16 mm² V90 cable has a per unit voltage drop of 2.43 mV/A.m.

$$V_d = \frac{V_C \times L \times I}{1000} = \frac{2.43 \times 15 \times 55}{1000} = 2.0 \text{ volt}$$

$$V_d\% = \frac{V_d \times 100}{V} = \frac{2.0 \times 100}{400} = 0.5\%$$

The voltage drop in the consumer main represents 0.5% of available voltage.

Sub-main:

From AS/NZS 3008.1 Table 40, 10 mm² V90 cable has a per unit voltage drop of 3.86 mV/A.m.

$$V_d = \frac{V_C \times L \times I}{1000} = \frac{3.86 \times 45 \times 40}{1000} = 7.0 \text{ volt}$$

$$V_d\% = \frac{V_d \times 100}{V} = \frac{7 \times 100}{400} = 1.75\%$$

The voltage drop in the sub-main represents 1.75% of available voltage.

Final sub-circuit:

From AS/NZS 3008.1 Table 42, 4 mm² V90 cable has a per unit voltage drop of 9.71 mV/A.m. It is necessary to convert this three-phase value of per unit voltage drop to single-phase.

$$V_C \text{ (single-phase)} = V_C \text{ (three-phase)} \times 1.155 = 9.71 \times 1.155 = 11.2 \text{ mV/A.m}$$

So

$$V_d = \frac{V_C \times L \times I}{1000} = \frac{11.2 \times 30 \times 18}{1000} = 6.0 \text{ volt}$$

$$V_d\% = \frac{V_d \times 100}{V} = \frac{6.0 \times 100}{230} = 2.6\%$$

The voltage drop in the final sub-circuit represents 2.6% of available voltage.

This means the total per cent voltage drop in the installation is 4.85% (0.5 + 1.75 + 2.6). This is less than the 5 per cent voltage drop allowed by AS/NZS 3000.

REVIEW QUESTIONS

1 What are word problems?
2 List the four steps involved in the problem-solving process.
3 What are four useful strategies for problem solving?
4 Which of the strategies from Question 3 assists in making an abstract written problem easier to solve?
5 Why is documenting every step taken to solve an electrotechnology problem important?

CHAPTER REVIEW

1.1 Basic units of measurement

- There are seven SI fundamental units and 22 derived units that have been given specific names.
- A number of electrical quantities are represented by symbols from the Greek alphabet.

1.2 Basic calculations

- The acronym BODMAS is a mnemonic for Brackets, Orders, Division, Multiplication, Addition and Subtraction.
- The process of rounding a particular value by choosing a desired number of digits for measurement is known as rounding to a number of significant figures.
- Rounding a number means to assess or approximate it.

1.3 Working with numbers using powers, exponents and indices

- Scientific notation can only have one number in front of the decimal point.
- Engineering notation can have up to three numbers in front of the decimal point and the power of 10 must be in multiples of three.
- Indices (the plural of index) provide a way of writing numbers in a more convenient form.
- Whatever you do to one side of an equation you must do the same to the other.

1.4 Algebra

- Algebra is a way of writing mathematics in a general form using algebraic expressions.
- When dividing algebraic terms you divide the numerical coefficients and then subtract exponents.
- To know the square root of a number, determine a number that, when multiplied by itself, provides the original number.
- To simplify a square root, display the number under the square root symbol ($\sqrt{\ }$) in two factors, with one factor the largest possible perfect square.
- To add or subtract algebraic fractions:
  - determine the lowest common multiple of the denominators
  - express fractions in terms of the lowest common denominator
  - simplify the numerators to obtain the numerator of the answer.

1.5 Interpreting data in graphical form

- The graph consists of two labelled axes called the *x*-axis (horizontal) and the *y*-axis (vertical).
- The *x*-coordinate of a point is the value that tells you how far from the point of origin the value from the table of results is on the horizontal axis, or *x*-axis.
- The *y*-coordinate of a point is the value that indicates how far from the point of origin the value from the table of results is on the vertical axis, or *y*-axis.
- Gradient (slope) is a mathematical expression that informs how much a graphic line slants and in which direction.
- A vector is a quantity that has two aspects. It has a magnitude (size) and a direction.

1.6 Complete work calculations and report on solutions

- Often it is necessary to include supporting information, such as the reason for selecting a particular cable or particular motor, in a written report.
- The methods of recording and the type of documentation produced should be appropriate to the nature of the enterprise and the purposes of the record.

TRIAL EXAM

For Chapter 1 knowledge assessment, please complete the following trial exam.

1 How many named derived units occur within the SI?
  a 3
  b 11
  c 22
  d 7

2 Which character set designates electrical quantities?
  a Latin
  b Arabic
  c Roman
  d Greek

3 Express the number 9 210 000 in engineering notation.
  a 921×10^4
  b 92.1×10^5
  c 9.21×10^6
  d 92100×10^3

4 Express 4700 in scientific notation.
  a 0.00047
  b 4.7×10^3
  c 0.047×10^4
  d 0.0047

5 Round 46 421 to two significant figures.
  a 4.6
  b 47 000
  c 46
  d 46 000

6 Combine the terms $9x - 5x$.
  a $4x^2$
  b $-4x$
  c $-14x$
  d $4x$

7 Combine the terms: $(9ab - 7a + 7b - 2) - (6ab - 6a - 4b + 8)$.
  a $3ab - a + 11b - 10$
  b $15ab + a + 3b + 8$
  c $3ab - 13a - 3b - 6$
  d $15ab - 13a + 11b + 6$

8 Simplify: $6a^2 \times 3a^2b$.
  a $18a^4b$
  b $3ab^{-1}$
  c $2a^4b$
  d $10a^{-2}b$

9 Simplify: $m^3n^5 \div m^6n^8$.
  a m^2n^{16}
  b $m^{-3}n^{-3}$
  c m^9n^{13}
  d $m^{18}n^{40}$

10 Simplify the following algebraic fraction: $8a^5 \div 4a$.
  a $2a^4$
  b $32a^4$
  c $4a^6$
  d $4a^5$

11 Simplify the following square root: $\sqrt{24}$
  a $\sqrt{4} \times \sqrt{6}$
  b $6\sqrt{4}$
  c $2\sqrt{6}$
  d $\sqrt{2} \times \sqrt{6}$

12 A current of 135×10^{-6} A can also be written as:
  a 0.000 135 A
  b 1.35 A
  c 13.5 A
  d 0.135 A

13 A voltage of 200×10^3 V can also be written as:
  a 2 000 000 V
  b 200 000 V
  c 200 000 000 V
  d 20 000 000 V

14 A resistance of $72 \times 10^3\ \Omega$ can be expressed as:
  a 72 000 000 Ω
  b 72 000 Ω
  c 0.072 Ω
  d 0.000 072 Ω

15 A circuit with a current of 440×10^{-3} A is equivalent to a current of:
  a 0.44 A
  b 440 A
  c 4.4 A
  d 4400 A

16 Simplify: 4^3.

- a 12
- b 64
- c 4
- d 16

17 What is the SI symbol for resistance?

- a Ω
- b E
- c C
- d A

18 The SI unit of 'energy' is the:

- a joule
- b volt
- c farad
- d watt

19 Make T the subject of the following equation:

$$P = \frac{9.55N}{T}$$

20 Make R the subject of the following equation:

$$Z = \sqrt{R^2 + X^2}$$

21 Simplify the following:

- a $(15 \times 10^6) + (25 \times 10^6)$
- b $(64 \times 10^{-4}) \times (32 \times 10^6)$

22 Convert the following to engineering notation:

- a 0.0025 V
- b 0.0000045 A

23 The SI unit of pressure is the:

- a newton per square metre
- b coulomb
- c joule per second
- d Pascal

24 Transpose the equation $V = IR$ making I the subject of the equation.

25 The SI unit of mass is the:

- a gram
- b milligram
- c tonne
- d kilogram

26 The SI unit of current is the:

- a kilogram
- b kilowatt hour
- c ampere
- d joule

27 Plot and label an x-y graph in **Figure 1.28** using the following data.

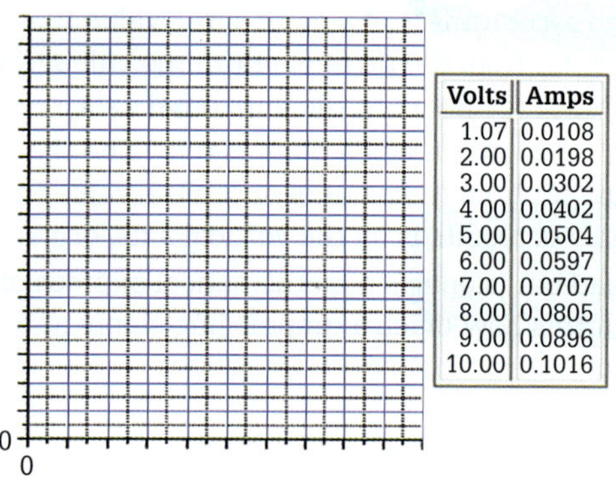

Volts	Amps
1.07	0.0108
2.00	0.0198
3.00	0.0302
4.00	0.0402
5.00	0.0504
6.00	0.0597
7.00	0.0707
8.00	0.0805
9.00	0.0896
10.00	0.1016

FIGURE 1.28 Graph grid exercise

28 An electrician walked 200 m north, 100 m west, 50 m south and 20 m east. What is the electrician's displacement?

29 Using the right-angled triangle in **Figure 1.29**, calculate the unknown side, the sine, cosine, and tangent of the angles A and B using the data provided in parts (a), (b) and (c).

- a $a = 3$, $b = 4$
- b $a = 6$, $b = 8$
- c $a = 14$, $b = 26$

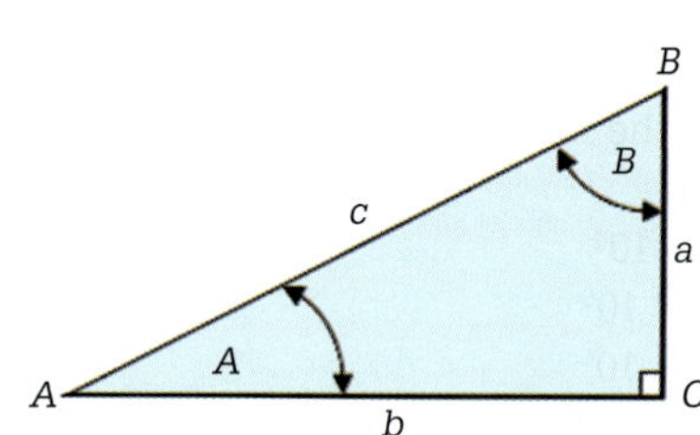

FIGURE 1.29 Right-angled triangle exercise

Single path d.c. circuits

This chapter provides essential knowledge required to solve problems in single path direct current circuits. It includes essential problem-solving procedures relating to measurements obtained from basic measuring devices including potential difference, current and resistance. This chapter provides underpinning knowledge for the unit UEECD0046 from the UEE training package.

LEARNING OBJECTIVES

Electrical concepts
- State the meaning of static and current electricity
- Outline how electricity is produced from renewable and non-renewable sources
- Explain how electricity is transported and utilised
- Perform basic calculations involving the quantity of electricity

Electrical circuit
- Show how a simple electrical circuit is represented using circuit symbols
- State the effect of various circuit conditions

Ohm's law
- Explain the relationship between voltage, current and resistance
- Calculate potential difference, current, and resistance

Electrical power
- Outline the relationship between power, work and energy
- Calculate power from potential difference, current, and resistance
- Explain why electrical components have power ratings
- Outline methods for measuring electrical power

Effects of an electric current
- Describe the effects of electrical current
- List the uses and disadvantages of the effects of electric current
- Describe how electrical energy is converted to other forms of energy
- Describe the physiological effects of current

EMF sources, energy sources and conversion
- Describe how other forms of energy are changed into electrical energy
- Calculate efficiency of electrical machines

Resistors
- Identify various types of fixed and variable resistors
- Determine the resistance value using resistance colour code
- Explain the importance of resistor power rating

Series circuits
- Identify series circuits
- Calculate currents, potential differences, resistances and power in series circuits

2.1 Electrical concepts

Static and current electricity

Electricity forms the basis of everyday life, from powering mobile phones, computers, lights, and air conditioners to nerves within the human body. The effect of electricity is obvious in the lightning created by a thunderstorm. Electricity is a general term used to describe the transference of electrical charges and is broadly categorised as being either static or current.

Static electricity

All matter is electrically neutral and can only become 'charged' when its atoms gain or lose electrons. In fact, electrically neutral matter has a 'zero charge'.

Electricity includes a variety of phenomena. One phenomenon is static which develops because of an imbalance in charges on the surface of the insulator. It follows that static electricity is an electric charge developed on the matter through friction. It is 'static' because although voltage pressure occurs only momentary current happens.

Friction between two objects of different charge is a process that causes electrons to leave the surface of one type of matter and relocate on the surface of another type of matter. The process is the triboelectric effect. The matter that loses electrons ends up with an excess of positive (+) charges on its surface. It follows that the material that gains electrons ends up with an excess of negative (–) charges on its surface. For this reason, objects with opposite charges attract each other thereby allowing the charges to enter into a state of balance. Dry human skin when rubbed has the best tendency to release electrons allowing the skin to become positively charged. In comparison, rubbed Teflon has the best tendency to become negatively charged. As a result, static electricity can cause materials to attract or repel each other. In addition, static electricity can produce a spark to jump from one type of matter to another.

Furthermore, a static charge is easy to obtain by rubbing a soft non-conductor against a hard insulator or a non-conducting material as illustrated in **Figure 2.1**.

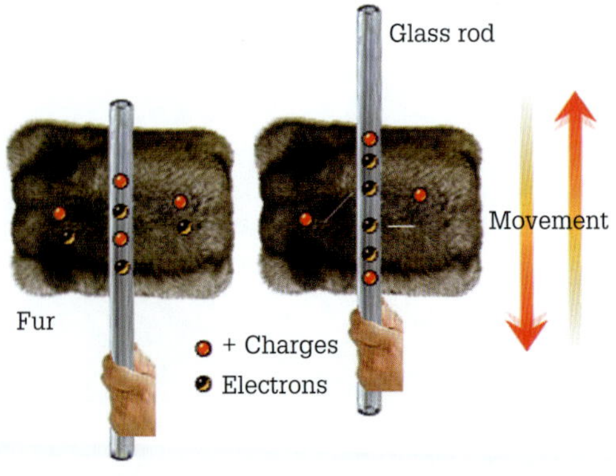

FIGURE 2.1 Static charge

Electrons are rubbed off the fur onto the glass rod causing the glass rod to accumulate electrons. A fundamental law of electrostatics is shown in **Figure 2.2**.

Like charges repel each other
Unlike charges attract each other

FIGURE 2.2 Fundamental law of electrostatics

The space around charged materials is an electrostatic field as shown in **Figure 2.3** and **Figure 2.4**.

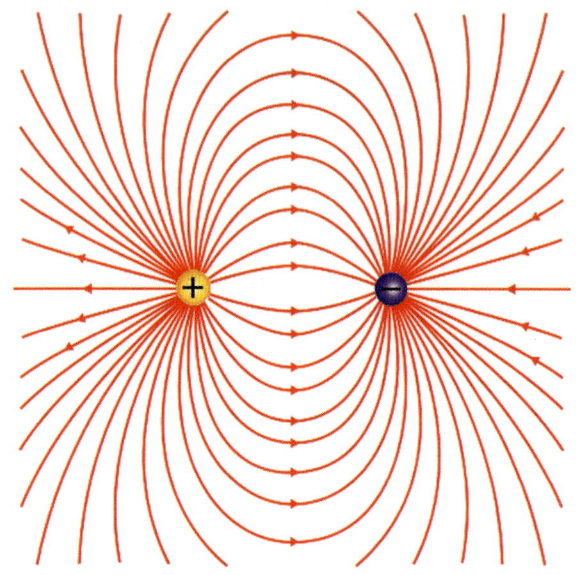

FIGURE 2.3 Electrostatic fields of force (unlike charges attracting each other)

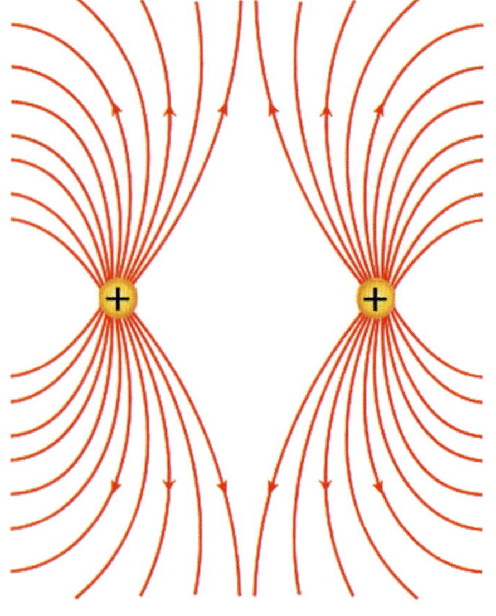

FIGURE 2.4 Electrostatic fields of force (like charges repelling each other)

For conventional reasons, the lines of force produced by a negatively charged material enter the charge while positive charges have the lines of force emitted from the charge. If a glass rod is rubbed with fur giving it a negative charge and then held against the grey pith ball (pith refers to the ball material, in this example Styrofoam) as illustrated in **Figure 2.5**, the glass rod imparts a negative charge to the grey ball.

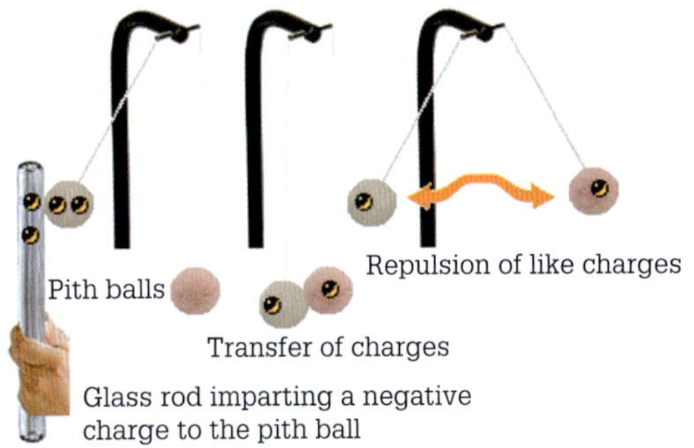

FIGURE 2.5 Reaction between charged materials

Consequently, the grey pith ball charges negatively while the pink pith ball remains neutral. When the grey pith ball is released, it will touch the pink ball. Upon touching, the pith balls remain in contact until the pink pith ball acquires a portion of the excess negative charge from the surface of the grey pith ball to make the negative charges on both pith balls equal. Consequently, following the law of charges, the balls repel each other.

Electrostatic discharges occur when a few electrons move across the space between objects of opposite charge. The electrons heat up the air allowing more electrons to jump across the space. The additional heat increases the air temperature even more. Because the movement of electrons happens very fast, the surrounding air gets so hot that it ionises as a spark for a short time. The same movement of charges occurs with lightning, as illustrated in **Figure 2.6**, except on a much larger scale with a massive electrostatic force field and immense electron transfer.

FIGURE 2.6 Lightning discharge

When in direct contact with integrated circuit technologies, or in the presence of explosive dusts or gases, avoid accumulating and discharging static electricity.

Current electricity

For current electricity to exist, as illustrated in **Figure 2.7**, there must be a closed conducting path for the charges to flow. Additionally, in order for the charges to flow they need a push, an electromotive force (emf) across the ends of the conductive path. Current electricity is the transference of a stream of charges through a conductor from a point of high emf to a point of low emf. A voltage source called a battery, or some other energy source establishes the emf.

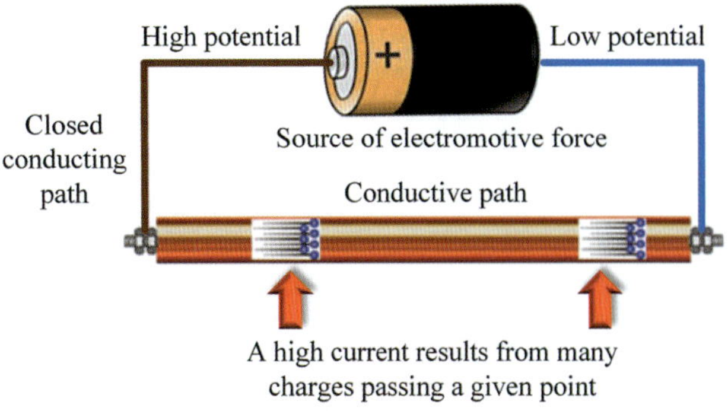

FIGURE 2.7 Current electricity

Types of electric current

There are two types of current: direct current (d.c.) and alternating current (a.c.). Direct current causes electrons to move in one direction through a closed circuit, while alternating current causes the electrons to regularly reverse their direction of travel through the closed circuit. In addition, alternating current occurs as single phase, two phases or three phases. Graphically, a horizontal straight line represents direct current while alternating current uses a sinusoidal or sine waveform as illustrated in **Figure 2.8**.

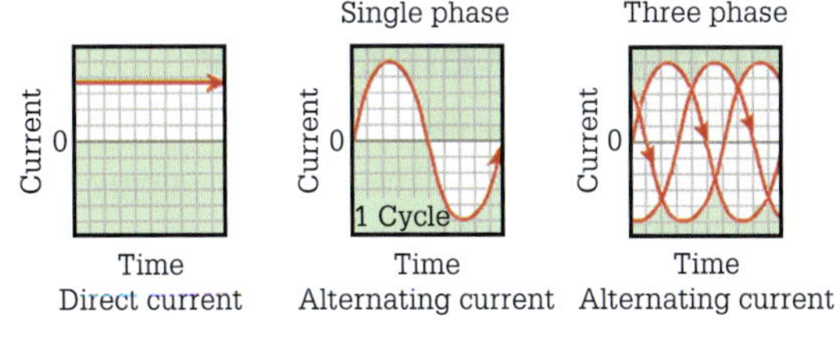

FIGURE 2.8 Graphical representations of d.c. and a.c.

In the 19th century, before the discovery of the electron, experimenters decided that current flowed from the positive terminal of the battery to the negative terminal, called the conventional flow of electric current. However, current consists of negatively charged particles called electrons flowing in the opposite direction – negative to positive – called electron flow. Despite this understanding, we still label diagrams in Australia with the conventional current direction.

Production of electricity

Power stations generate electricity and use different fuel resources such as coal, gas, biomass, nuclear energy, oil and water to power a prime mover to produce movement. In fact, steam turbine alternators, gas turbine alternators, diesel engine alternators, alternative energy systems such as wind, tidal and wave turbine alternators all operate on the same principle: conductors, movement and magnetic field. The exception is a solar cell. **Figure 2.9** shows the simplified process of generating electrical energy from fossil fuel resources.

Around 60 per cent of Australia's electricity comes from coal-fired power stations. Coal is a non-renewable energy source that will eventually run out. It is important for our future that we utilise more sustainable electricity generation methods.

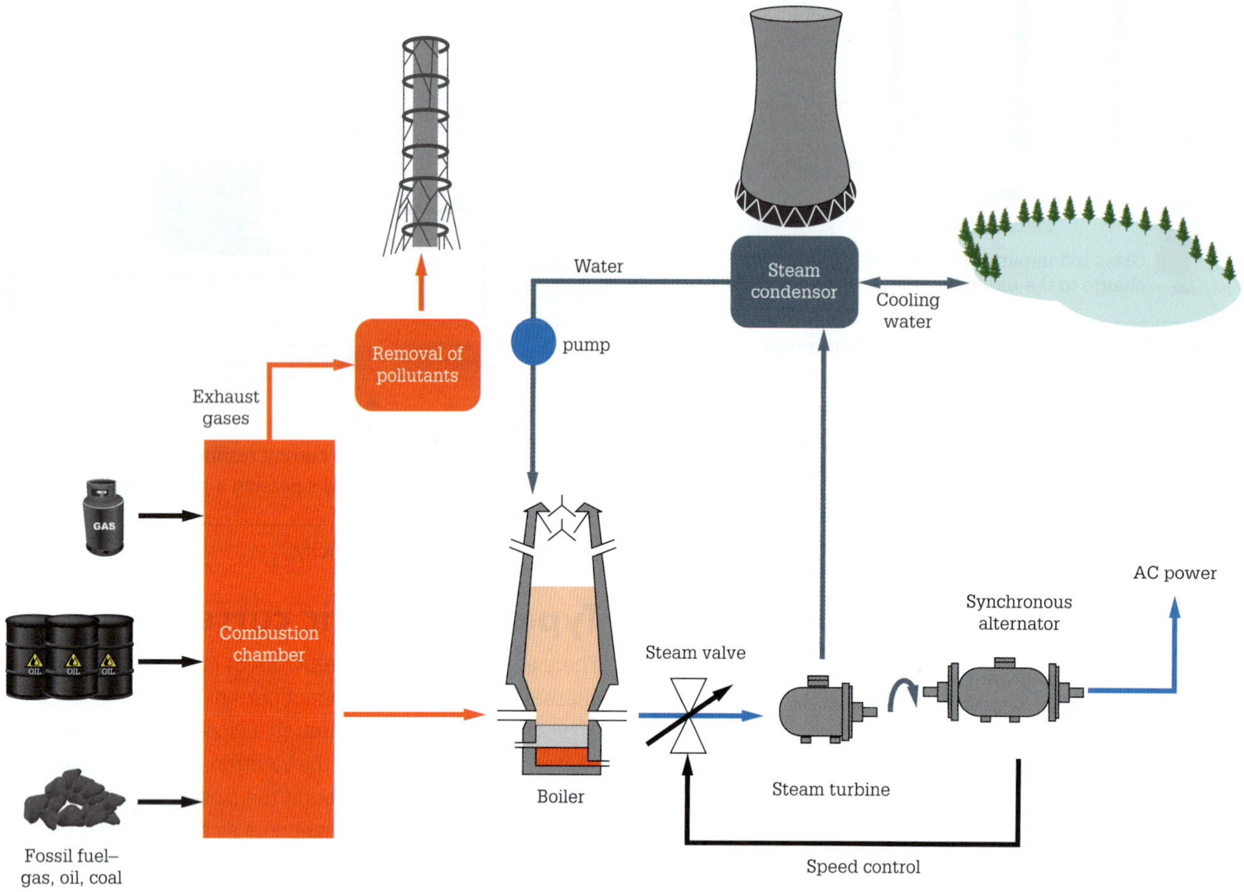

FIGURE 2.9 Burning fossil fuel to generate electricity

Generation of electricity

An alternator is a machine used to convert mechanical energy into electrical energy. The electrical energy takes the form of alternating current and voltage.

The elementary a.c. alternator as shown in **Figure 2.10** consists of a conductor or loop of wire in a magnetic field. The ends of the conductor loop connect to slip rings, which connect to an external circuit via two carbon brushes in contact with the rings. As the conductor loop rotates, it cuts magnetic lines of force. Induced across the ends of the loop is an emf at a particular frequency (complete cycles per second).

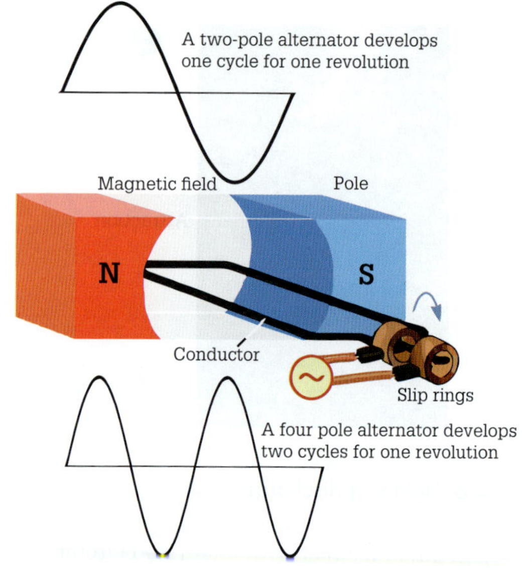

FIGURE 2.10 Elementary a.c. alternator

Effect of speed

The output voltage frequency of an alternator depends on rotor speed and the number of poles. Consequently, the faster the speed, the higher is the frequency; and the lower the speed, the lower is the frequency. The more poles there are, the higher the frequency is for a given speed.

The two-pole system of **Figure 2.10** must make one complete revolution to enable the conductor to pass both the north and south poles and thus to induce one cycle of emf. In addition, the frequency in hertz (cycles per second) of the induced emf is equal to the rotational speed of the two-pole alternator in revolutions per second (r/s). Furthermore, a four-pole alternator induces one cycle in half a revolution, or two cycles per revolution, while a six-pole alternator would induce three cycles per revolution.

$$N = \frac{120f}{p}$$

where N = rotational speed in revolutions per minute (rpm)
f = frequency in hertz (Hz)
p = number of poles

EXAMPLE 2.1

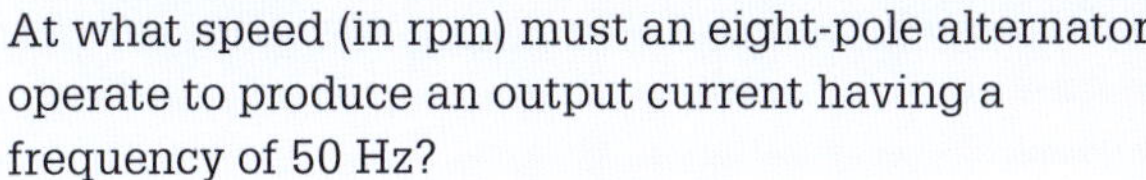

At what speed (in rpm) must an eight-pole alternator operate to produce an output current having a frequency of 50 Hz?

$$N = \frac{120f}{p}$$
$$= \frac{120 \times 50}{8}$$
$$= 750 \text{ rpm}$$

EXERCISE 2.1

a At what speed must a six-pole alternator run to give an output current having a frequency of 60 Hz?
b At what speed must a two-pole alternator run to give an output current having a frequency of 100 Hz?
c Determine the speed in rpm of an eight-pole alternator that produces an output current having a frequency of 25 Hz.
d Determine the number of poles required in an alternator running at 3600 rpm that produces an output current with a frequency of 120 Hz.
e Calculate the speed in rpm of an alternator with 12 poles that produces an output current having a frequency of 50 Hz.

Effect of velocity

An alternator produces a sinusoidal-shaped waveform as shown in **Figure 2.11**. The waveform shows that for given angular displacement, the rate of change in generated voltage varies. As the coil loop rotates between point A and point B, the generated voltage increases, but between point B and point C the generated voltage is more constant. This is due in part to the effects of angular velocity (ω in radians/s). Angular velocity is a vector measure of how fast the angular position of an object changes with time.

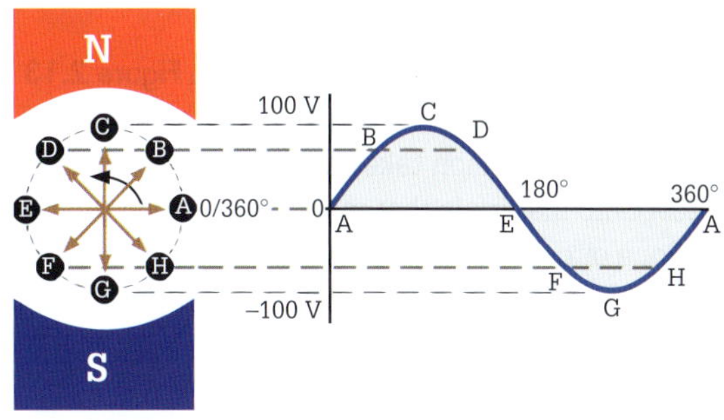

FIGURE 2.11 Sine wave

As a conductor loop revolves in a circular motion, it cuts the magnetic field at different angles and a particular voltage is induced in the conductor at different points in time resulting in a sine wave.

Renewable energy technologies

Australia's vision regarding energy sustainability is the use of renewable energy technology. Geothermal, solar, wind, tidal, biomass and wave are all examples of renewable energy technologies. These technologies save Australia tonnes of carbon dioxide emissions per year compared to coal-based power generation.

Turning wind into electricity

Wind turbines produce electricity. Wind turbines transform the kinetic energy in wind into mechanical energy to create electrical energy (electricity) through a generator (d.c. output) or alternator (a.c. output).

The turbine produces power as the wind turns a set of blades that rotates a generator or alternator, either directly or indirectly via a gearbox. The most common wind turbine is the horizontal axis turbine shown in **Figure 2.12**, which has blades like an aircraft propeller. A grid-connected wind turbine produces high levels of a.c. power, which feeds into transformers and then into the high-voltage transmission grid.

FIGURE 2.12 Wind turbine

Turning light into energy

Solar photovoltaic (PV) panels as shown in **Figure 2.13** generate clean electricity by converting the energy in sunlight. This conversion takes place within modules of photovoltaic materials that make up the solar panels. Stand-alone or grid-connected photovoltaic systems are available. The electricity generated by solar panels is direct current.

FIGURE 2.13 PV array

PV systems require an inverter to convert direct current into alternating current suitable for ordinary 230 V needs. For systems with battery backup (optional), the inverter regulates the charge of the batteries. Solar PV systems that feed into the grid require a grid interactive inverter.

Transmission and distribution systems

The intricate transmission and distribution network that carries electricity from power stations across Australia to your home or business is known as the electricity 'grid'. Australia's eastern and southern states have one of the largest interconnected electricity 'grids' or power systems in the world. This network services Queensland, New South Wales, the Australian Capital Territory, Victoria, South Australia and Tasmania.

Figure 2.14 illustrates a transmission and distribution system (or grid) from the alternator at the power station, through the various substations and feeders to the consumer. In addition, power stations produce three-phase alternating current at different voltages, depending upon the alternator design. For example, typical voltages are 11 kV and 22 kV. Transformers step up these voltages to 275 kV (high-tension feeder; HT) for long-distance transmission. Transformers at the bulk supply substations then step down this voltage to 66 kV, 33 kV or 11 kV. Consequently, energy is sold to local electricity distributors at these voltage levels.

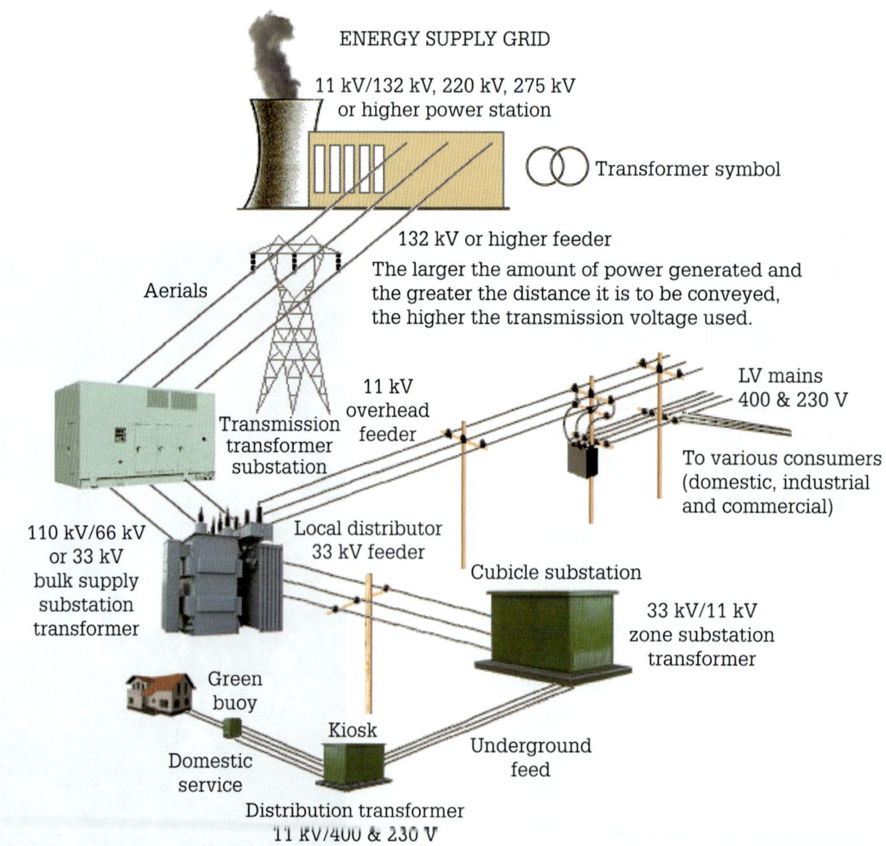

FIGURE 2.14 Transmission and distribution system

Besides changing the voltage, the purpose of the distribution substations is to isolate electrical faults in either the transmission system or the distribution system.

Local electricity distributors transmit the energy to their different consumers through their distribution lines. The three-phase voltage steps down from 33 kV to 11 kV. Finally, the consumers are supplied by low-tension (LT) feeders at low voltage, either 230 V (one phase) or 400 V multi-phase (two or three phase) as shown in **Figure 2.15**.

FIGURE 2.15 Low-tension feeder

A substation is an electrical transmission and distribution system facility where transformers change voltage from low to high (step-up) while decreasing current. Transformers can also change voltages from high to low (step-down) while increasing current. A transformer, as shown in **Figure 2.16**, is a static device used to change the values of alternating voltage to some different magnitude. In addition, transformers provide a means of isolating electrically one part of the transmission system from another.

FIGURE 2.16 Pole-mounted transformer

Transformation takes place at several points in the distribution system in succession, starting at the power station where the voltage increases for transmission purposes and then progressively reduces to the voltage required for domestic, commercial and industrial use. Modern electrical transmission and distribution systems receive energy contributions from a variety of sources including non-renewable and renewable energy technologies.

Utilisation of electricity by various loads

Storage of generated electricity cannot occur; instead it requires immediate use. It is important that the amount of electricity needed by the load at any point in time should match the generated output. A definition of a load is any device that uses the energy of the transmitted electricity. In fact, there are five types of loads driven by electricity; refer to **Figure 2.17**.

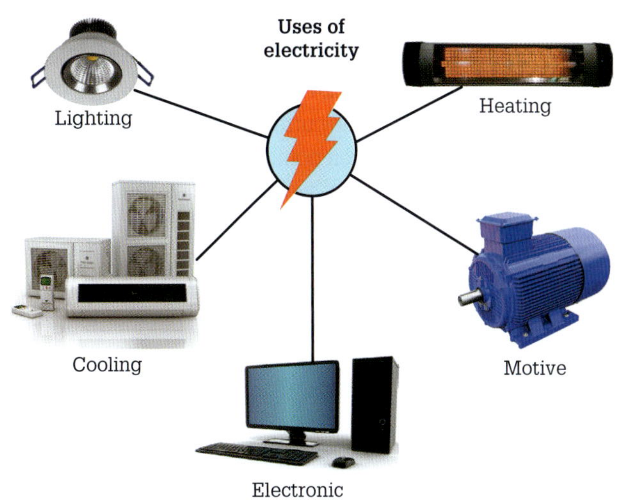

FIGURE 2.17 Utilisation of electricity

Source: Shutterstock.com/MAKFALI [downlight]; Shutterstock.com/diy13 [heater]; Shutterstock.com/Matveev Aleksandr [motive]; Shutterstock.com/Kateryna998 [computer]; Shutterstock.com/vipman [aircon]

Electrical quantities

The volt

The volt (V) is the electrical potential difference required between two conductors when one joule of energy transfers one coulomb of charge from one point to another as shown in **Figure 2.18**.

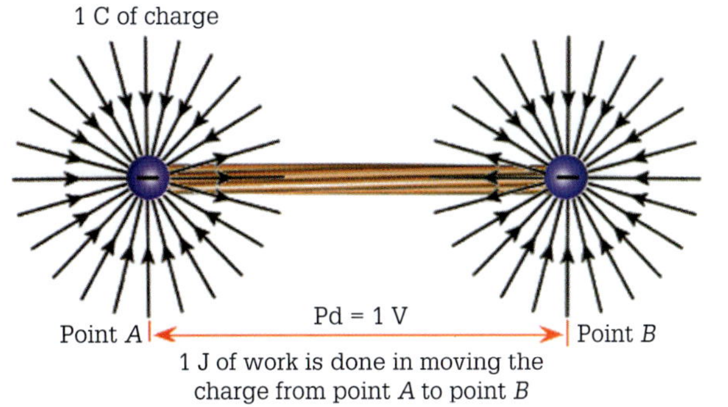

FIGURE 2.18 The volt

One volt (V) generates one joule of energy per coulomb.

Electromotive force (emf) is the term used to describe the voltage obtained from an energy source such as a generator or battery. The emf is energy per unit charge (voltage) and is unique to the generating system. In addition, the emf refers to an electrical force that can move electrical charges. When an emf is impressed across a conductor, then the conductor is 'live'.

A difference in electrical pressure or potential difference occurs across a load when a current flows.

The ampere

The ampere is the unit of measure of electric current (I). The ampere is based on a fundamental physical constant, the elementary charge (e), which is the amount of electric charge in a single proton (positive) or electron (negative). It is a measure of the amount of electric charge in motion per unit of time. The quantity of electric charge, whether or not in motion, is expressed by another SI unit, the coulomb (C). One coulomb is equal to about 6.241×10^{18} electric charges (e). One ampere therefore is the current in which one coulomb of charge travels across a given point in 1 second.

Prior to 2019, the ampere was defined as that constant current which, if maintained in two straight parallel conductors of infinite length, of negligible circular cross-section, and placed 1 metre apart in vacuum, would produce a force equal to 2×10^{-7} newtons per metre of length between these conductors as shown in **Figure 2.19**.

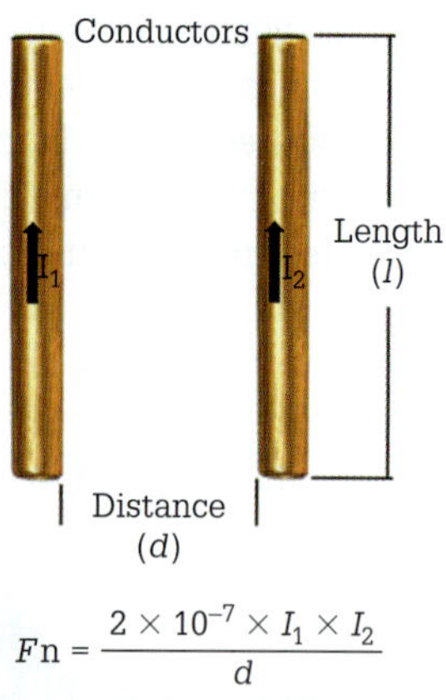

FIGURE 2.19 Defining electric current prior to 2019

Resistance

Where current flow meets opposition by a load (such as a light or appliance) or circuit conductors, the circuit has resistance (R), measured in ohms (Ω). A definition of resistance follows.

A water analogy in **Figure 2.20** demonstrates the principle of electric current. The force applied becomes the emf, and the pump is the battery, the valve is the switch, the quantity pumped is the charge, the difference in potential is the voltage, water flowing is the current, and the rough inner surface is the resistance.

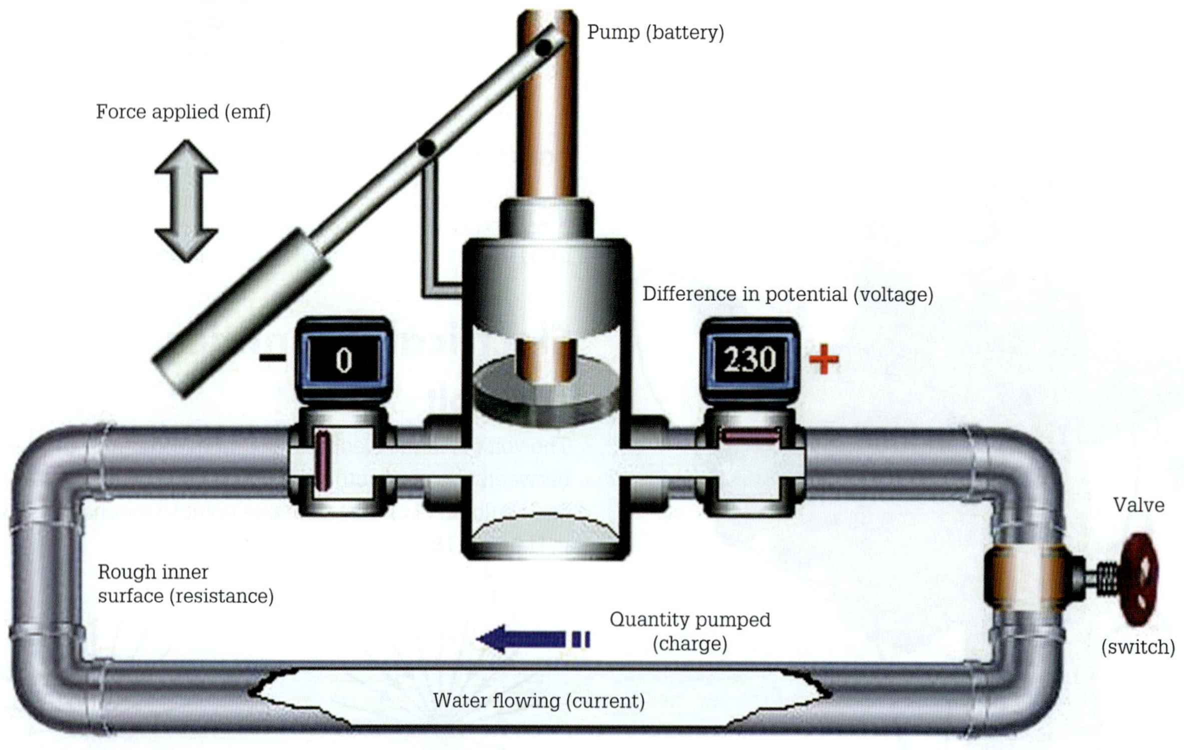

FIGURE 2.20 Water analogy of current electricity

Just as a difference in water pressure must exist for water to flow between two points, so a difference in electrical pressure (voltage) must exist to cause electrical current to flow. The higher the difference in pressure, the greater is the flow of current between the two points.

SWITCH ON

Potential difference is the difference in electrical potential between two conductors.

Current is a flow of charged electrons in a conductor.

Resistance is the opposition offered to the flow of current.

Charge

A charge (symbol Q) is measured using the SI unit of measurement called the coulomb, abbreviated to C. However, because one coulomb of charge is an enormous quantity of charge (1 C = 6.241×10^{18} electrons), values more commonly used are micro-coulombs (μC) and pico-coulombs (pC).

The coulomb is the unit of measurement for the quantity of charge (Q) that is moved when the flow of charges, called the current, flows past a point in a conductive path for 1 second. Calculate the amount of charge by applying the following equation:

$$\text{charge} = \text{current} \times \text{time}$$
$$Q = It$$

where Q = charge in coulombs (C)

I = current in amperes (A)

t = time in seconds (s)

Note: 1 ampere equates to 1 coulomb of charge per second. A definition of current is the rate of charge flowing past a set point in an electric circuit. Current is measured in coulombs per second (C/s), which is also called amperes (A).

EXAMPLE 2.2 

A copper conductor carries a current of 20 amperes to a load for 20 seconds. What is the quantity of charge transferred?

$$\begin{aligned} Q &= It \\ &= 20 \times 20 \\ &= \mathbf{400\ c} \end{aligned}$$

If a charge Q transfers through a conductive path in a certain time t then I is calculated as shown in Example 2.3.

EXAMPLE 2.3 

Calculate the current in amperes passing a given point in a conductor when a quantity of 10 000 coulombs passes that point in 10 seconds.

$$\begin{aligned} I &= \frac{Q}{t} \\ &= \frac{10000}{10} \\ &= \mathbf{1000\ A} \end{aligned}$$

REVIEW QUESTIONS

1. What is static electricity?
2. What is the direction for conventional current flow?
3. With the exception of photovoltaic, what is the general means of generating electricity?
4. Name three renewable energy sources.
5. What is the electricity 'grid'?
6. Name three categories of electrical load.
7. What is electrical current?
8. What is resistance?
9. A current of 50 amperes flows through a conductor for 15 seconds. What is the quantity of charge transferred?
10. Calculate the current in amperes passing a given point in a conductor when a quantity of 20 000 coulombs passes that point in 20 seconds.

2.2 Electrical circuit

An electrical circuit forms an unbroken loop within a conductive medium, allowing the movement of electrons from one point to another. The term describing the process is 'continuity'. A circuit is a complete uninterrupted loop with excellent continuity. In order to use electricity, we must have a power source that could be a battery, generator, alternator or solar panel. Conductors from the power source connect to the load device and for convenience a switch interrupts current to enable the load to be switched on or off.

Consequently, three items – a power source, load and conductors – form a simple circuit. **Figure 2.21** is a pictorial diagram of a simple circuit.

FIGURE 2.21 Simple electrical circuit

Anything else connected in the circuit such as a protection device or a control switch is an addition to the circuit. The control switch is a device designed to break the continuity under controlled conditions. The circuit illustrated in **Figure 2.22** shows the effect of an open circuit when the control switch is in the 'off' position. In this state current cannot flow from the battery to the lamp, which remains extinguished.

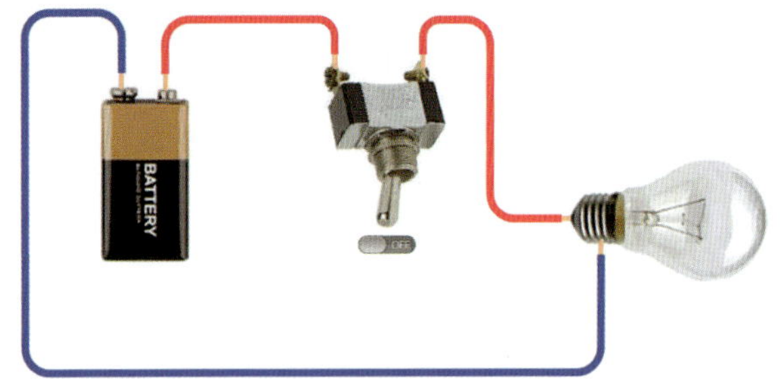

FIGURE 2.22 Simple electrical circuit with control switch – open circuit

When the control switch of the circuit of **Figure 2.22** is moved to the 'on' position it allows current to flow and the circuit is termed a closed circuit. Consequently, current flows from the positive to the negative terminal of the supply. The circuit illustrated in **Figure 2.23** shows the effect of a closed circuit when the control switch is in the 'on' position.

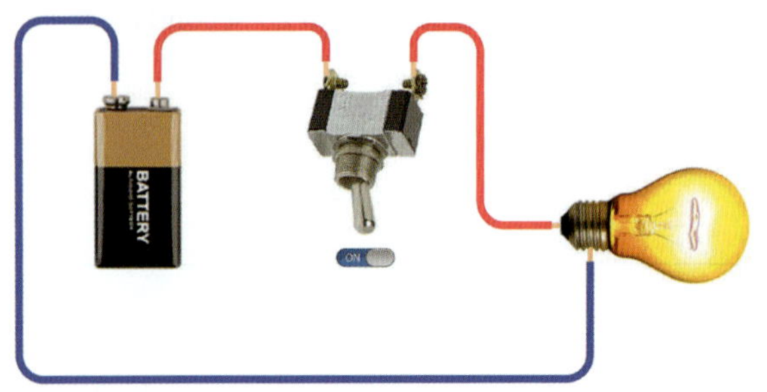

FIGURE 2.23 Simple electrical circuit with control switch – closed circuit

An open circuit can be the result of a broken conductor (no return path) as illustrated in **Figure 2.24**, an open switch, a weak conductor termination or a burnt-out load.

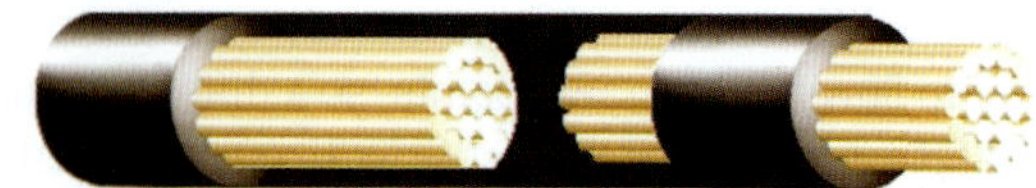

FIGURE 2.24 Open circuit conductor

In addition to the conditions of closed and open, another circuit condition is a short circuit. A short circuit with d.c. current is a condition whereby a positive conductor is connected to the negative conductor and bypasses the load or the conductor 'shorts' between conductors due to insulation failure or as a result of some object that causes the condition. In an alternating current system, short circuit can exist between the active conductor and the neutral or earth conductor. Short circuits of the power source are very dangerous because the high currents encountered cause the release of large amounts of heat energy as shown in **Figure 2.25**.

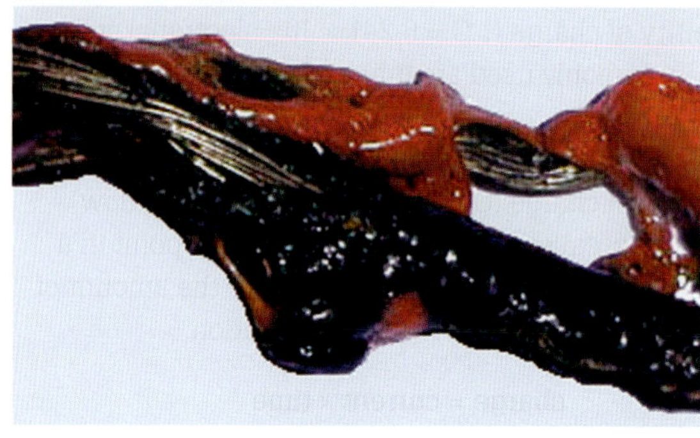

FIGURE 2.25 Effect of short circuit

Circuit protective devices such as fuses and circuit breakers protect the conductors from overheating in the event of a short circuit or overload. The circuit illustrated in **Figure 2.26** shows a circuit breaker added to our basic circuit to afford over-current protection, which will prevent the circuit conductors suffering the ill-effect of excessive heat as shown in **Figure 2.25**.

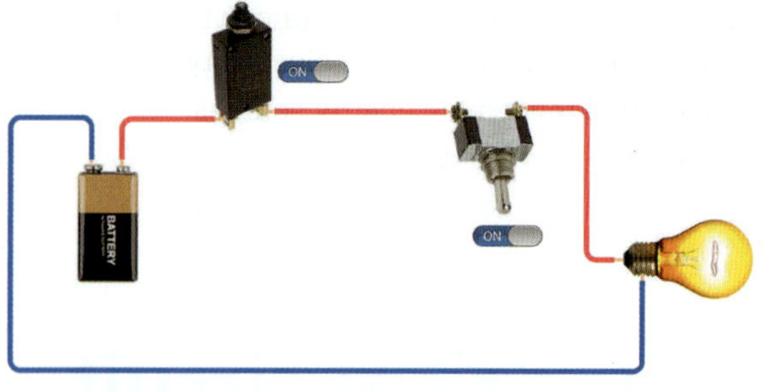

FIGURE 2.26 Simple electrical circuit with control switch and over-current protection

Circuit diagrams

The circuits represented so far used pictorial drawings to represent the physical components. While this is a suitable approach for a simple circuit and for someone who possesses reasonable drawing skills, it is not suitable for

circuits having many components and for the vast majority of electrical workers.

A better method is to use symbols to represent electrical components. There are many electrical and electronic components and each has its own unique symbol.

Circuit symbols

Figure 2.27 shows how a woman can be represented using pictorial representation and using a symbol.

Table 2.1 summarises common circuit symbols along with their function.

FIGURE 2.27 Representing a woman using pictorial representation and a symbol

Source: Shutterstock.com/Dean Drobot [left]; Shutterstock.com/ciputra [mid]; Shutterstock.com/N.Style [right]

TABLE 2.1 Circuit symbols and function

Component	Pictorial	Symbol	Function
Cell			To provide low voltage, low current electrical energy
Battery			To provide higher voltage and current electrical energy than a single cell
Conductor			To provide a conductive path from the source of electrical energy to the load device
Joined conductor			To provide a branching-off conductive path
Control switch			To provide on-off control of an electrical load in an electrical circuit
Filament lamp			To provide light (heat) energy from electrical energy
Fuse			To break the electric circuit should excessive current flow
Circuit breaker			To break the electric circuit should excessive current flow Has the advantage of being resettable

Source: [top to bottom] Shutterstock.com/Lipskiy; Shutterstock.com/Yoki5270; Shutterstock.com/nokkaew; Shutterstock.com/sockagphoto; Shutterstock.com/NsdPower; Shutterstock.com/Reshavskyi; Shutterstock.com/Mrs_ya; Shutterstock.com/Aleksei Golovanov

The circuit diagram

It is possible to combine these symbols to form what is known as a circuit (or schematic) diagram. A circuit diagram only shows connections and not the physical location of components.

The use of standard symbols makes reading and understanding the operation of circuits much simpler.

Figure 2.28 shows how circuit components can be represented using symbols in a circuit diagram.

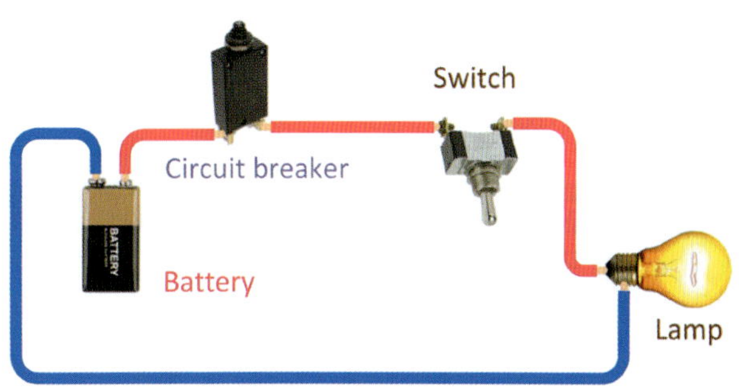

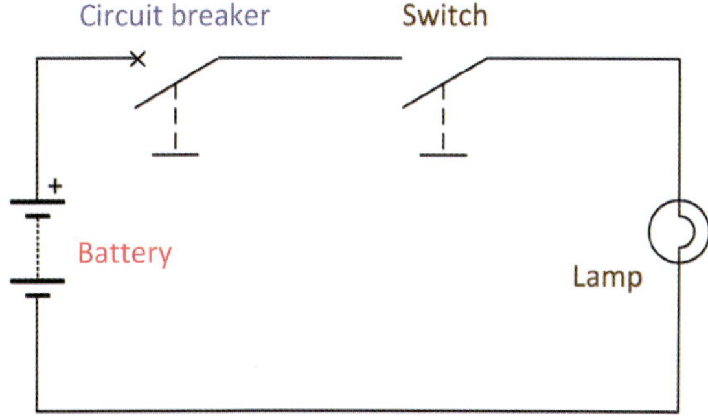

FIGURE 2.28 Using symbols (bottom) to represent components (top) in a circuit diagram

Note the orientation of the circuit symbols in the circuit diagram of **Figure 2.28**. By convention:

- the power source such as a battery is drawn to the left
- a load device such as a lamp is drawn to the right
- control and protective devices are drawn on the top line, which connects the positive of the power source to the load
- the return current path does not contain a control device and is drawn on the bottom
- control and protective devices such as switches and circuit breakers are orientated such that the 'hinge' side of the blade is closest to the load device and the switch blade operates in a clockwise direction.

Multiple and sub-multiple units

Units of measurement such as volt, ampere and ohm (to name a few) can have a prefix added, which multiplies or divides the base unit to create a more convenient unit. Prefixes align with engineering notation; therefore, the same prefixes occur for all units of measure.

The list in **Table 2.2** shows the most common prefixes. Each prefix is 1000 times greater than the prefix below it, or 1000 times less than the one above it. For example, one mA is the same value as 1000 μA or 0.001 A. **Figure 2.29** gives some examples.

TABLE 2.2 Common prefixes

Multiple or sub-multiple	Symbol	Prefix	Power
1 000 000 000 000	T	tera	10^{12}
1 000 000 000	G	giga	10^{9}
1 000 000	M	mega	10^{6}
1000	k	kilo	10^{3}
100	h	hecto	10^{2}
10	da	deca	10^{1}
0.1	d	deci	10^{-1}
0.01	e	centi	10^{-2}
0.001	m	milli	10^{-3}
0.000 001	μ	micro	10^{-6}
0.000 000 001	n	nano	10^{-9}
0.000 000 000 001	ρ	pico	10^{-12}

micro (μ)	means	$\frac{1}{1\,000\,000}$	or	one-millionth
milli (m)	means	$\frac{1}{1000}$	or	one-thousandth
kilo (k)	means	1000	or	one thousand
mega (M)	means	1 000 000	or	one million

FIGURE 2.29 Examples of prefixes and new symbols

Note: Realise the difference between the symbols M and m. Also, always convert a value into its basic unit before using it in an equation. As a result, any unknown value in the equation has expression in terms of its basic unit.

Converting units

A simple method of solving conversions from one prefix to another is to find the magnitude of one of these units. For example, 1 ampere (1 A) could be written as 1000 mA.

EXAMPLE 2.4 

Convert 100 μA to amperes.

$$1\ \mu A = 1 \times 10^{-6}\ A$$

Therefore:

$$100\ \mu A = 100 \times 10^{-6}\ A$$

$$= \mathbf{0.0001\ A}$$

Prefixes range from yotta (10^{24}), or one septillion, to yocto (10^{-24}), that is, one-septillionth. An advantage of the SI system of measurement is that the units for the same dimension relate to each other by 'powers of 10', which makes them easy to convert. For example, 0.1 A could be expressed as one hundred milliamperes (100 mA).

Manufacturer's notation for the electrical unit avoids the use of the decimal point. For example, the electrical part of the 'resistor' previously had its value indicated as 4.7 MΩ. Since decimal points could be lost due to poor print quality this resistor could be mistaken for 47 MΩ. So, in the new notation the resistor value is 4M7 Ω. That is, the symbol 'M' has taken the place of the decimal point.

REVIEW QUESTIONS

1. Name the three essential components in an electrical circuit.
2. What item is added to the basic electrical circuit to provide on-off control?
3. What is meant by the term 'closed circuit'?
4. What is meant by the term 'open circuit'?
5. What is the function of a circuit breaker in an electrical circuit?
6. What is the function of a battery in an electrical circuit?
7. Draw the circuit symbol for a battery.
8. Draw the circuit symbol for a filament lamp.
9. Express 1.5 kV in volts.
10. Express 0.015 A in milliampere (mA).

2.3 Ohm's law

Measuring circuit parameters

Before being able to apply Ohm's law it is necessary to know the potential difference across a resistance and the current flowing through the resistance. The instruments used to measure these quantities are the voltmeter and ammeter respectively.

The voltmeter

A voltmeter is used to measure potential difference (the voltage between any two points). A voltmeter is usually combined with the functions of other meters into an instrument called a multimeter. These instruments can be either digital or analogue.

As potential difference is measured between two points in a circuit, it is necessary for the voltmeter to have two test leads. For d.c. circuits it is important to observe correct polarity when connecting the voltmeter. The positive (red) terminal and lead connects to the point in the circuit closer to the positive terminal on the source of supply. The negative (black) terminal and lead connects to the point in the circuit closer to the negative terminal on the source of supply.

One advantage of the digital instrument is that if you get the polarity wrong, the meter displays the value preceded with a minus sign. For example, −10.0 V. The circuit symbol for a voltmeter is shown in **Figure 2.30**. Note the polarity mark on the voltmeter symbol.

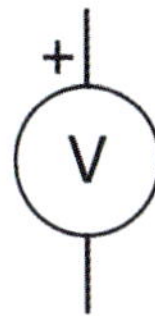

FIGURE 2.30 Voltmeter circuit symbol

The ammeter

An ammeter is used to measure current in an electrical circuit. An ammeter is usually combined with the functions of other meters into a multimeter.

As current is the flow of electrons, it is necessary to 'break' the circuit to install the ammeter to allow it to effectively count the number of electrons flowing through it and hence the circuit.

The ammeter has two test leads. For d.c. circuits it is important to observe correct polarity when connecting the ammeter. The positive (red) terminal and lead connects to the point in the circuit closer to the positive terminal on the source of supply. The negative (black) terminal and lead connects to the point in the circuit closer to the negative terminal on the source of supply.

The circuit symbol for an ammeter is shown in **Figure 2.31**. Note the polarity mark on the ammeter symbol.

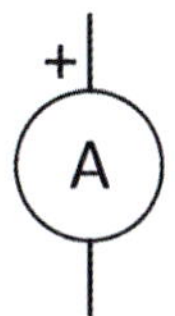

FIGURE 2.31 Ammeter circuit symbol

Measuring potential difference and current

Figure 2.32 shows how the voltmeter and ammeter connect in a circuit to measure the potential across the resistor and current through the resistor.

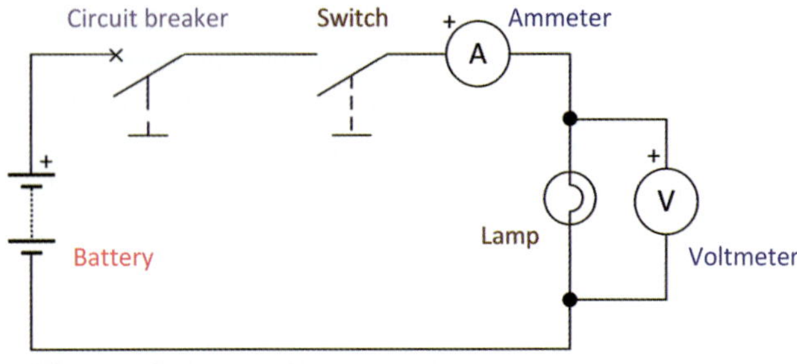

FIGURE 2.32 Using a voltmeter and ammeter in a circuit

Effect of open and closed circuit conditions

Figures 2.33 and **2.34** show the expected potential differences measured across various circuit components under closed and open circuit conditions respectively.

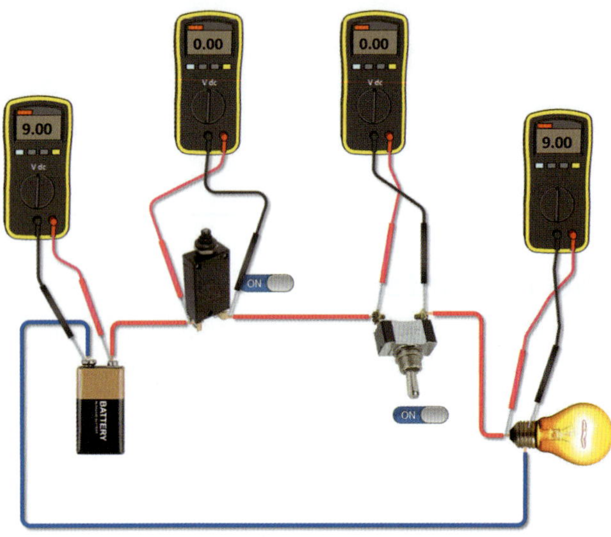

FIGURE 2.33 Measuring potential difference in a closed circuit

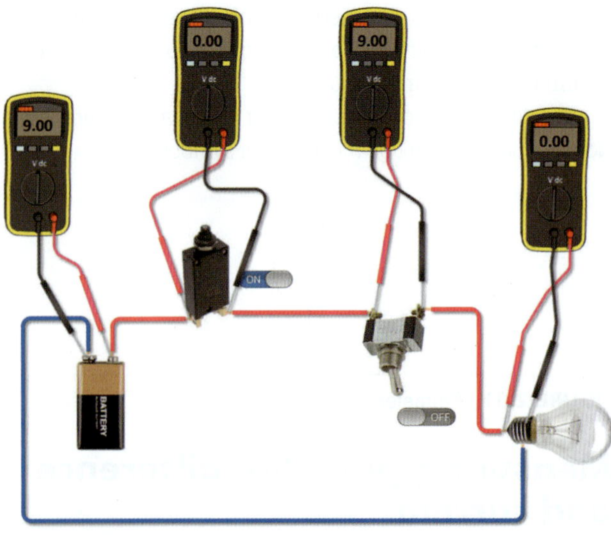

FIGURE 2.34 Measuring potential difference in an open circuit

Figures 2.35 and **2.36** show the expected circuit current measured under closed and open circuit conditions respectively.

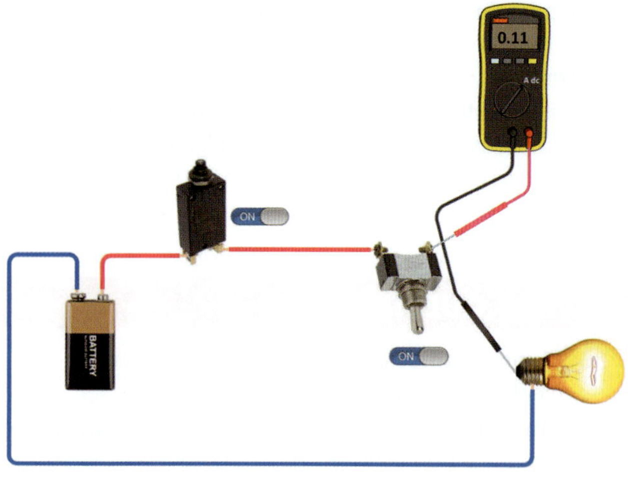

FIGURE 2.35 Measuring current in a closed circuit

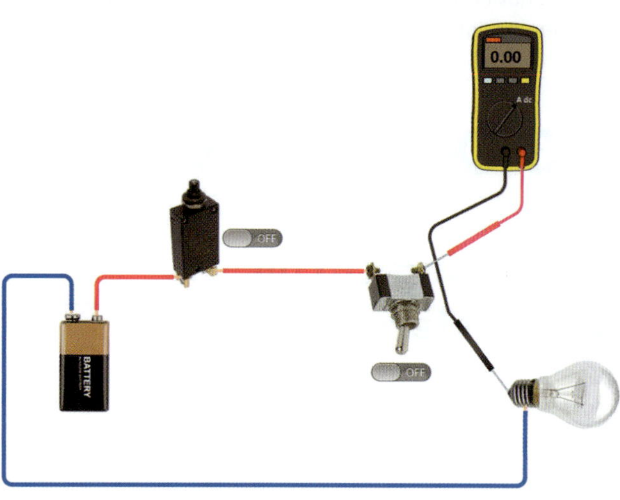

FIGURE 2.36 Measuring current in an open circuit

The resistor

Previously we saw how a filament can be used as a load device. Another common component in an electrical circuit is the resistor. Resistors are components that oppose current in an electrical circuit.

The circuit symbol for a resistor is shown in **Figure 2.37**. Note the absence of polarity markings on the resistor symbol. The resistor is a non-polarised, passive component. An instrument known as an ohmmeter is used to measure the resistance of components and circuits and like the voltmeter and ammeter is one of the functions included in a multimeter.

FIGURE 2.37 Resistor circuit symbol

Ohm's law explained

Georg Simon Ohm (1789–1854) was a German physicist and mathematician who undertook studies on resistance between 1825 and 1826. Ohm observed the effect on circuit current from varying the potential difference across a fixed resistance and varying the circuit resistance while maintaining potential difference constant. Ohm's law is a simple mathematical way for helping us to understand the relationship between potential difference, current, and resistance in electric circuits. This relationship is a law and is known as Ohm's law, which states:

SWITCH ON

The current flowing in a circuit is directly proportional to the applied potential difference and inversely proportional to the resistance in the circuit.

This observation gives rise to the equation:

$$I = \frac{V \text{ or } E \text{ or } U}{R}$$

where I = current through resistance in ampere (A)

V or E or U is potential difference across resistance in volts (V)

R is the value of the resistance in ohms (Ω)

Note: U is the American National Standards Institute (ANSI) reserve symbol; it is the chief symbol for the International Electrotechnical Commission (IEC) and the International Organization for Standardization (ISO). The reason behind U is that the measurement designate for voltage is V. It is nonsense to say that voltage (V) has a designate measurement symbol V. Consequently, the IEC and the ISO use voltage (U), which is measured in volts, V. Accepted in Australia the symbol 'U' occurs in the *Wiring Rules*.

EXAMPLE 2.5

What is the value of the current (I) in the circuit shown in **Figure 2.38** when the potential difference across the resistor (R) is 120 V and the resistor has resistance of 33 Ω?

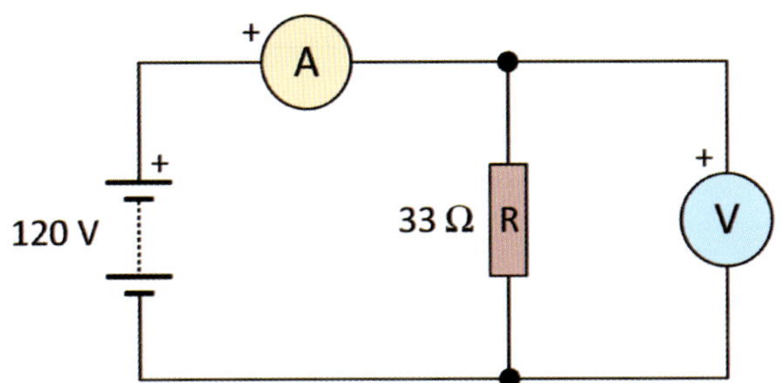

FIGURE 2.38 Ohm's law circuit (current)

$$I = \frac{V}{R} = \frac{120}{33}$$

= **3.64 A** (3 significant figures)

By transposing the Ohm's law equation it is possible to determine resistance from the potential difference and current values.

$$R = \frac{V}{I} \text{ in ohms}$$

EXAMPLE 2.6

Calculate the value of the resistance (R) in the circuit in **Figure 2.39** when the potential difference across the resistor (R) is 120 V and the circuit current is 3 A.

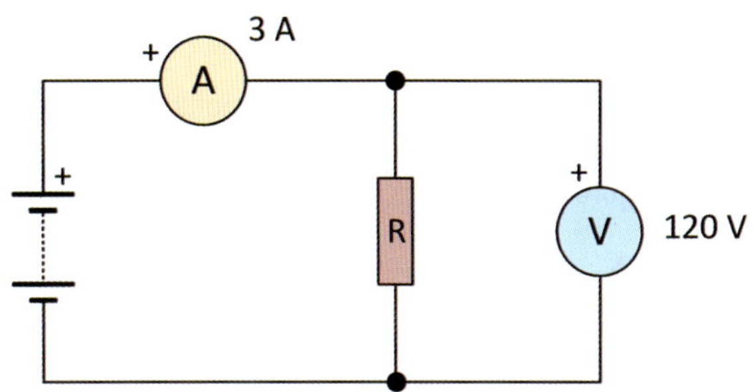

FIGURE 2.39 Ohm's law circuit (resistance)

$$R = \frac{V}{I} = \frac{120}{3} = \mathbf{40\,\Omega}$$

By transposing the Ohm's law equation, it is possible to determine potential difference from the resistance and current values.

$$V = I\,R$$

EXAMPLE 2.7

What is the magnitude of the potential difference (V) across a 33 Ω resistor carrying a current of 5 A as shown in **Figure 2.40**?

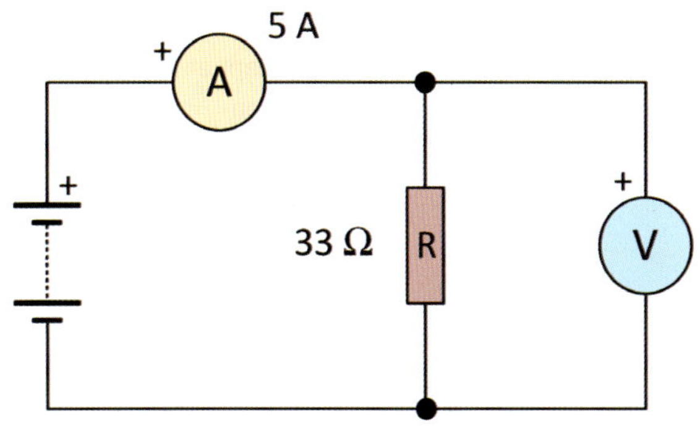

FIGURE 2.40 Ohm's law circuit (voltage)

$$V = I \times R = 5 \times 33 = \mathbf{165\,V}$$

Change in circuit parameters

If the circuit parameters change, the following effects occur.

1 **If the resistance remains unchanged:** Increasing the voltage causes the current to increase; decreasing the voltage causes the current to decrease.
2 **If the potential difference remains unchanged:** Increasing the value of the resistance decreases the current and decreasing the value of the resistance increases the current.
3 **For the current to remain unchanged:** Increasing the voltage means increasing the resistance. In contrast, decreasing the voltage means decreasing the resistance.

A simple method of showing the relationship between voltage, current and resistance is by using a triangle as shown in **Figure 2.41**.

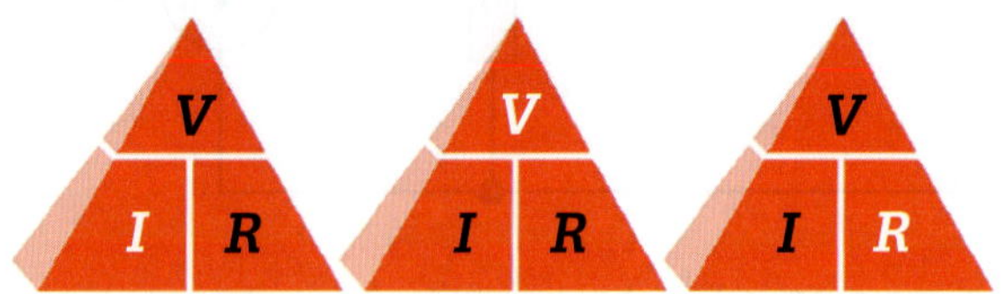

FIGURE 2.41 Ohm's law triangle

Cover up the quantity you wish to find with the thumb (white letters) and the remaining symbols give the formula required.

EXERCISE 2.2

a A potential difference of 60 V is measured across a 15 Ω resistor. Calculate the current flowing in mA through the resistor.

b A 48 Ω resistor has a current of 4 A flowing through it. Calculate the potential difference across the resistor.

c An electric toaster element when connected to a 200 V d.c. supply draws a current of 8 A. Calculate the resistance of the element.

Figure 2.42 shows typical results from applying Ohm's law in graphical form. The resistance in this case is a 1 Ω nichrome conductor which was kept at a constant temperature while readings were taken.

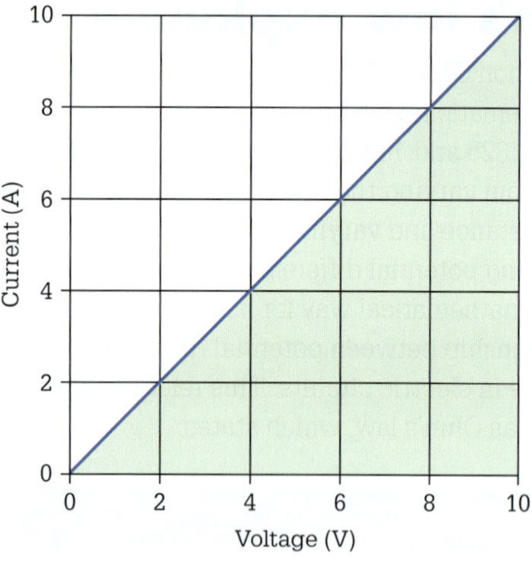

FIGURE 2.42 Graphical analysis of Ohm's law

1 The graph shown is a straight line passing through the origin.
2 Examination of the graph indicates that doubling the voltage doubles the current (a simple proportion exists).
3 Dividing the voltage by the current always gives the same value of resistance, 1 Ω in this case.

The graph shows the direct-proportion relationship between voltage and current when resistance is constant thereby verifying Ohm's law.

EXERCISE 2.3

Determine the value of the resistance of each resistor in **Figure 2.43** that shows the voltage and current relationship for two resistors A and B.

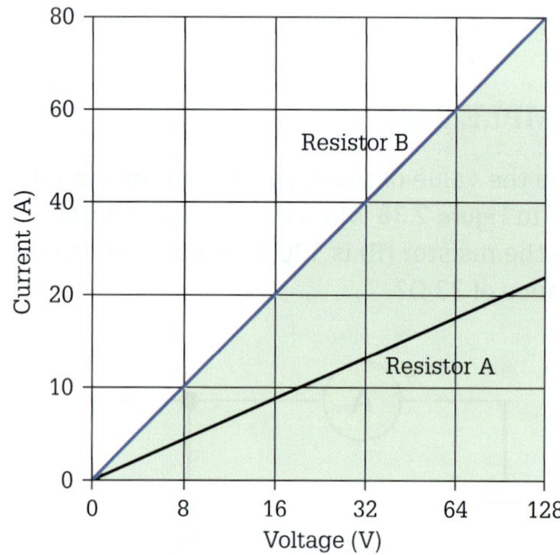

FIGURE 2.43 The relationship between voltage and current for two resistors

REVIEW QUESTIONS

1 What potential difference is required to enable a current of 0.5 A to flow through the 100 Ω resistor in Figure 2.44?

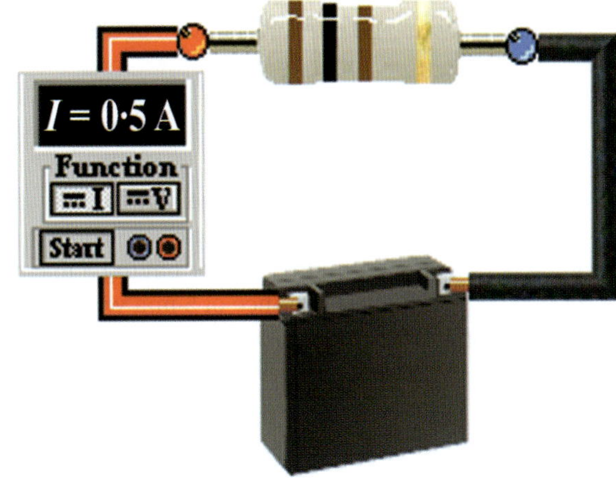

FIGURE 2.44 Resistor in circuit

2 What are the unknown values for Figure 2.45(a)–(h)?

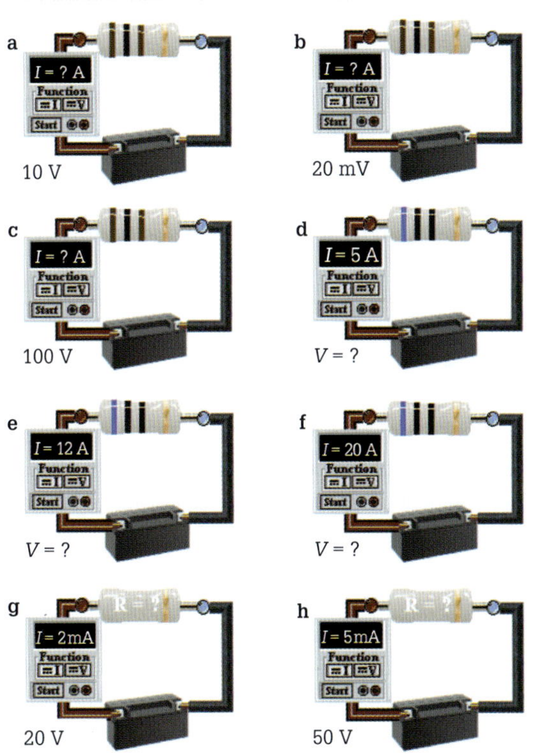

FIGURE 2.45 Unknown resistor values (a)–(h)

Note: 100 Ω = Brown Black Brown 60 Ω = Blue Black Black

3 A doorbell requires 150 mA of current in order to operate. If the potential difference across the bell is 12 V, what is the resistance of the bell?

4 The resistance of the d.c. motor windings of a series motor is 10 Ω. If the potential difference is 160 V, calculate the current drawn.

5 Find the current drawn by a 60 Ω toaster element when connected to a 200 V supply.

6 What is the resistance of an electric iron element that draws 3.8 A from a 200 V supply?

7 What is the potential difference across a lamp drawing a current of 3.5 A and having a resistance of 120 Ω?

8 A relay used to control a motor has a coil resistance of 150 Ω. What current will the relay draw from the supply if the potential difference across the coil is 100 V?

9 The resistance of the motor windings of a vacuum cleaner is 25 Ω. If the potential difference is 230 V, what current will it draw from the supply?

10 The coil of a relay draws a current of 65 mA when operating from a 160 V source. Find the resistance of the coil.

2.4 Electrical power

Force

Force is any push or pull on an object. Forces do not always give rise to motion. However, the object changes shape whether or not it moves when a force is exerted on it. A change of shape is obvious when you push on the side of a switchboard as you can observe the slight deformation of the metal.

If there is no force acting on a body, it remains at rest. However, what happens if a force causes the body to move? The body's position must change; therefore, force is an action capable of accelerating a body.

SWITCH ON

The unit of force is the newton (N). One newton is the force necessary to cause an acceleration of 1 ms^{-2} (metres per second every second) to a mass of 1 kg.

Note: Acceleration due to gravity on earth is 9.81 ms^{-2}, which may also be expressed as 9.81 m/s^2.

The force applied to a body is the product of the mass of a body (in kg) and its acceleration (in ms^{-2}).

$$F = ma$$

where F = force in newton (N)
m = mass in kilogram (kg)
a = acceleration in metres per second squared (ms^{-2})

The terms 'mass' and 'weight' are confusing. Mass is a measure of the inertia or state of rest of the body. The more mass a body has the harder it is to change its state of motion. In contrast, weight is a force.

The weight (W) of an object is the force of gravity on an object calculated as the mass (m) times the acceleration of gravity (g):

$$W = mg$$

where W = weight in newtons (N)
m = mass in kilograms (kg)
g = acceleration due to gravity in metres per second squared (ms^{-2})

Since the weight is a force, its SI unit is the newton (N).

To understand the difference, suppose we take an armature to the moon. The armature will weigh only about one-sixth as much as it did on earth since the force of gravity is weaker, but its mass will be the same. In fact, the armature still has the same amount of matter as shown in **Figure 2.46**.

FIGURE 2.46 Weight of an object on Earth and Moon

EXAMPLE 2.8

a Calculate the force required to move a 100 kg induction motor with an acceleration of 0.1 ms^{-2}.

$$F = ma$$
$$= 100 \times 0.1$$
$$= \mathbf{10\ newtons}$$

b An electrical fitter has to fit a steel key to the shaft of a rotor in order to fit a sprocket. What force is required to wedge the key having a mass of 0.050 kg through a distance of 5 mm with an acceleration of 0.5 ms^{-2} using a rawhide mallet?

$$F = ma$$
$$= 0.050 \times 0.5$$
$$= \mathbf{0.025\ newtons}$$

EXERCISE 2.4

a Calculate the force needed to draw a cable having a mass of 30 kilograms into a conduit with an acceleration of 0.2 ms^{-2}.

b Find the force acting downwards on a luminaire having a mass of 2 kg attached to the suspension cable if acceleration due to gravity is 9.81 ms^{-2}.

c A bundle of steel conduit has a mass of 250 kg. What value of force will be needed to raise it against gravity if acceleration due to gravity is 9.81 ms^{-2}?

Torque

In some cases, a force may be inclined to cause rotary movement. The turning effect is called the torque, or the turning moment of a force, and is measured as the force applied multiplied by the perpendicular distance between the direction of the force and the point about which rotation can occur. Imagine pushing a door to open it as illustrated in **Figure 2.47**. The force of your push causes the door to rotate about its hinges. How hard you need to push depends on the distance you are from the hinges. The closer you are to the hinges the harder it is to push. The torque you created on the door is smaller than it would have been had you pushed further away from the hinges. The magnitude of the force applied and the distance from the point of application to the hinge affect the tendency of the door to rotate. To obtain maximum torque for a given applied force, the force should always be in the direction of the resulting movement.

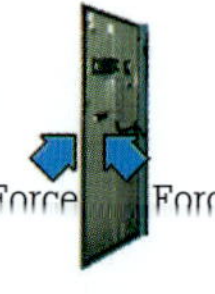

FIGURE 2.47 Forces acting at different angles

Force applied to the door can occur at an angle (sin θ) as well as straight on (90° and sin 90° = 1) as illustrated in **Figure 2.47**.

Torque, then, is proportional to the product of the force times the moment arm (r). In general, the torque about a given axis is:

$$T = Fr\sin\Theta$$

where T = torque in newton-metres (Nm)
r = moment arm in metre (m)
Θ = the angle of applied force

The moment arm length is r and θ is the angle by which the force acts, as shown in **Figure 2.48**.

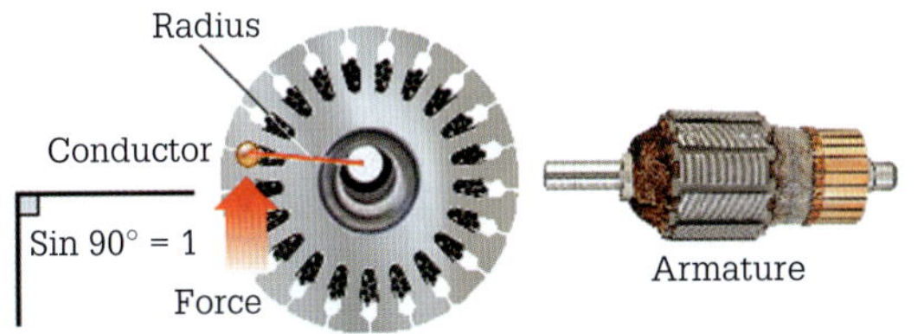

FIGURE 2.48 Moment arm length and sin θ

The newton-metre is also a way of expressing a joule (the unit for energy). However, torque is not energy because energy is a scalar quantity, whereas torque is a vector quantity.

EXAMPLE 2.9

The maximum force that an electrical fitter can use on the end of a 0.25 m long spanner is 100 newtons. What is the maximum torque that the fitter can apply to a bolt?

$$T = Fr\sin\Theta$$
$$= 100 \times 0.25 \times \sin 90°$$
$$= \mathbf{25\ Nm}$$

If a 1 m metal tube was placed over the handle of the spanner to increase the length, then the maximum torque that would be applied is 100 Nm.

Note: Tools use handles of such length as to prevent the application of too much torque. Too much torque can cause parts to bend or break under stress.

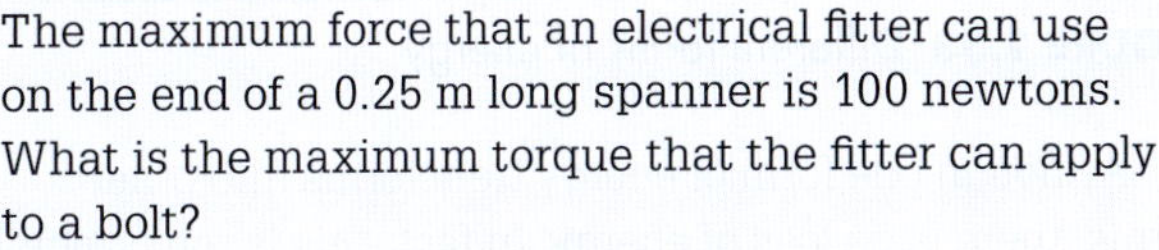

EXERCISE 2.5

A 50 mm steel tube requires the application of 40 Nm of torque for the threading operation. What minimum force acting at 90° occurs at the end of a conduit diestock of overall length 0.65 m?

Work

Work (unit = joule) is achieved whenever anything moves against a force or resistance.

There are two conditions that are involved: there must be movement, and this movement must be against some force or resistance as shown in **Figure 2.49**.

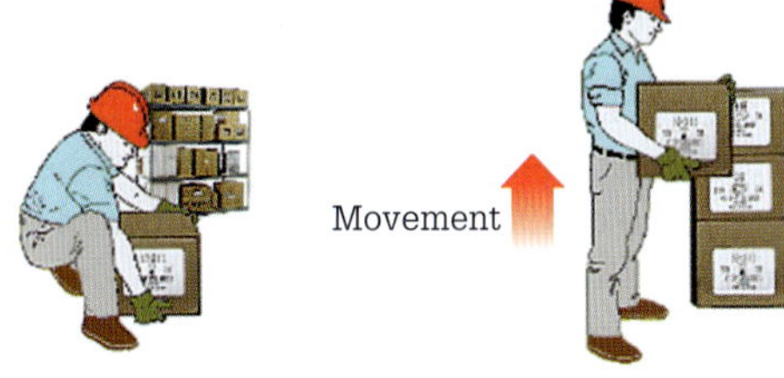

FIGURE 2.49 Work

The larger the force or resistance together with a bigger movement produced, the more work that is done. The amount of work done is measured by multiplying the size of the force or resistance by the distance the object is moved.

$$\text{work done} = \text{force} \times \text{distance (moved)}$$

This distance (measured in metres) must always be measured in the direction in which the force is acting. Gravity is one of the most common resistances against which work is done. In the case of lifting a mass against the force of gravity, the distance moved is measured vertically. In **Figure 2.50** the work done is the same whether the same cardboard box is hoisted vertically by the pulley or carried up the stairs by the electrician.

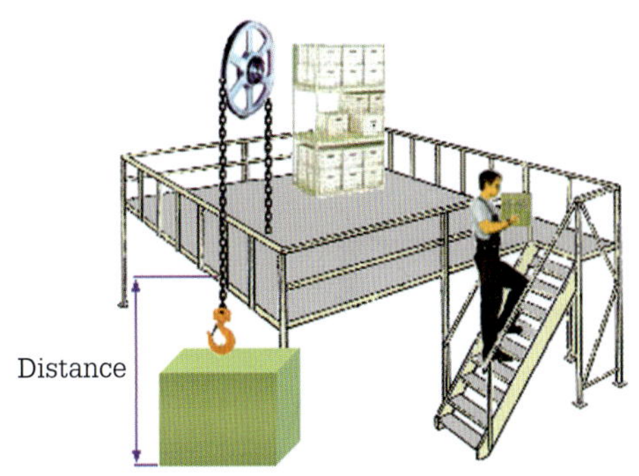

FIGURE 2.50 The same amount of work is being done

The actual distance travelled by the cardboard box may be different for the two methods of moving the cardboard box but the vertical distance is the same. Therefore, the work done is the same. In the SI system force is measured in newtons (N), distance is measured in metres (m), and the work done is calculated in joules (J).

SWITCH ON

One **joule** of work occurs when a force of 1 newton moves an object 1 metre in the direction of action of the force.

EXAMPLE 2.10

a The forklift truck in **Figure 2.51** can raise a box of electrical components to a height of 2 m when applying a force of 4900 N. How much work in joules occurs using the forklift?

$$\text{work done} = \text{force} \times \text{distance}$$
$$= 4900 \times 2$$
$$= \mathbf{9800\ joules}$$

FIGURE 2.51 Forklift truck

b How much work is done by the wire cable of the pile driver in **Figure 2.52** that draws the hammer mass through a height of 15 m while applying a force of 12 500 N?

$$\text{work done} = \text{force} \times \text{distance}$$
$$= 12500 \times 15$$
$$= \mathbf{187\,500\ joules}$$

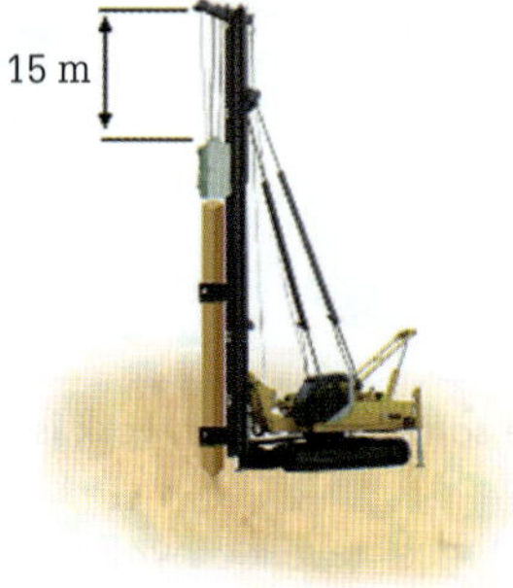

FIGURE 2.52 Pile driver

c A bundle of 20 mm conduit has a mass of 100 kg. What is the bundle's weight in newtons and the work done in joules when the conduit moves from ground level to a rack 1.8 m high?

$$\text{weight} = \text{mass} \times \text{force due to gravity}$$
$$= 100 \times 9.81$$
$$= \mathbf{981\ N}$$
$$\text{work done} = \text{force} \times \text{distance}$$
$$= 981 \times 1.8$$
$$= \mathbf{1765.8\ J}$$

Energy

Whenever work occurs, energy exists. The ability or capacity to do work defines energy. There are many different forms of energy. For example, electrical energy can be converted to light energy, magnetic energy and sound energy. Other forms of energy are chemical energy, heat energy and nuclear energy, as illustrated in **Figure 2.53**. Furthermore, energy and work have the same units, joules (J).

SWITCH ON

Energy is the ability or capacity to do work.

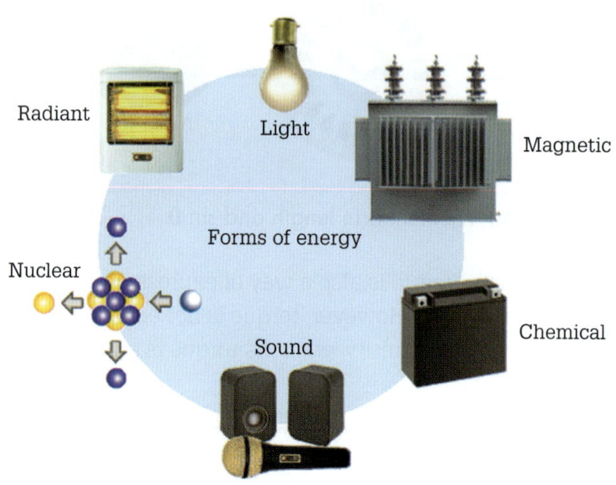

FIGURE 2.53 Different forms of energy

Although one form of energy transforms into other forms, there is no loss of energy. In fact, this is the principle of the conservation of energy. In other words, the total energy in the universe is constant.

SWITCH ON

Energy is not created or destroyed, it is merely transformed.

Potential energy

When anything rises against the force of gravity, work occurs and is stored as energy in the lifted object. Because the object is higher than it was originally, it possesses more energy. For example, when raising the dense mass of an electrode driver, work occurs. The work done is stored in the mass as energy of position, called potential energy.

However, when the electrode driver in **Figure 2.54** moves downwards, the driver's positional energy changes into the energy of motion. Therefore, motion energy drives the copper electrode into the ground, doing work against friction and pushing the ground out of the way.

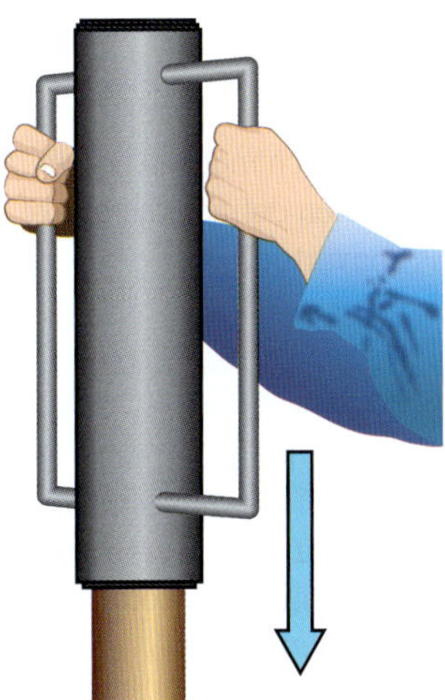

FIGURE 2.54 Work being done by an electrode driver

The potential energy of the electrode driver is equal to the work that is done if the driver were to fall to the ground. It is also equal (air resistance not considered) to the work done in lifting the driver up from the ground in the first place. In addition, the work done can be determined by considering the force needed to lift the electrode driver, and the distance moved.

An object that has energy stored in it because of its position has potential energy due to its position or state.

When a mass is lifted to a new position against gravitational force, the mass increases its potential energy (E_p). The following equation calculates potential energy:

$$E_p = mgh$$

where E_p = potential energy in joules (J)

m = mass in kilograms (kg)

g = acceleration due to gravity in metres per second squared (ms^{-2})

h = height of object in metres (m)

EXAMPLE 2.11

a Calculate the potential energy that an overhead line connection box and a riser bracket, as shown in Figure 2.55, possess when an electrician raises their combined 2.5 kg mass to a height of 3 m.

$$E_P = mgh$$
$$= 2.5 \times 9.81 \times 3$$
$$= 73.6 \text{ J (3 significant figures)}$$

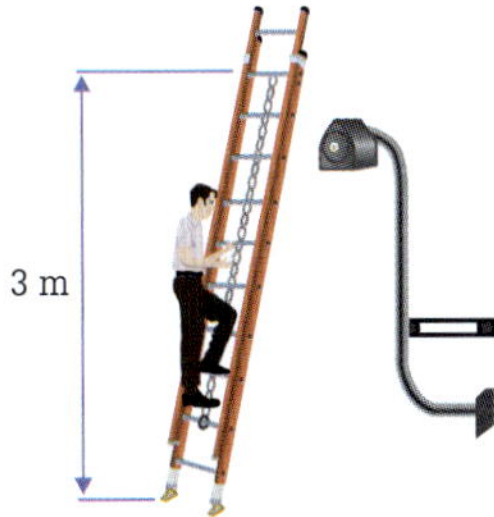

FIGURE 2.55 Potential energy

b Calculate the potential energy of an 80 kg three-phase motor lifted using a lifting trolley through a distance of 0.5 m, as shown in Figure 2.56.

$$E_P = mgh$$
$$= 80 \times 9.81 \times 0.5$$
$$= 392 \text{ J (3 significant figures)}$$

FIGURE 2.56 Lifting trolley

Kinetic energy

When force acting on a body produces movement, the force does work, and the body possesses kinetic energy (E_K) because of its motion. An object that has either vertical or horizontal motion has kinetic energy. There are many forms of kinetic energy such as vibrational, rotational and conveying. Conveying is kinetic energy due to the motion of a body from one location to another. In the hydroelectric dam shown in Figure 2.57, the water stored at a higher elevation is a source of potential energy. The potential energy converts to kinetic energy as it flows to the turbines.

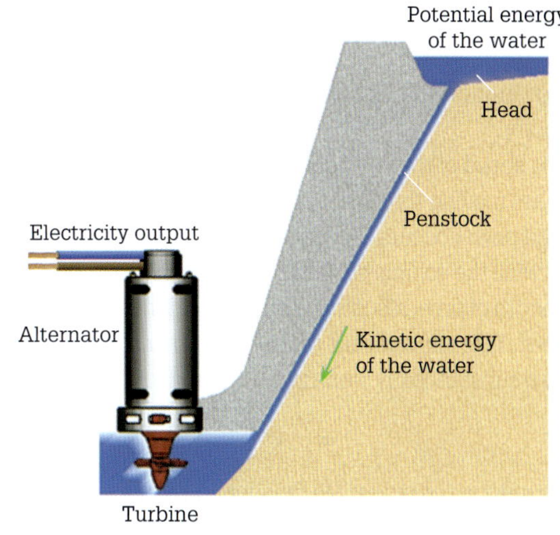

FIGURE 2.57 Potential energy of water transformed into kinetic energy

»

The gain in kinetic energy of the moving body is calculated from:

$$E_K = \frac{1}{2}mv^2$$

where E_K = kinetic energy in joules (J)
m = mass in kilograms (kg)
v = velocity of movement in metres per second (ms^{-1})

EXAMPLE 2.12

A conduit saddle as shown in **Figure 2.58** is about to be fixed to a brick wall using a wall plug. Calculate the kinetic energy of the hammer of 1 kg mass moving with a velocity of 10 ms^{-1} to insert the wall plug.

$$\begin{aligned}\text{kinetic energy} &= \frac{1}{2}mv^2\\ &= \frac{1}{2}\times 1\times 10^2\\ &= \mathbf{50\ joules}\end{aligned}$$

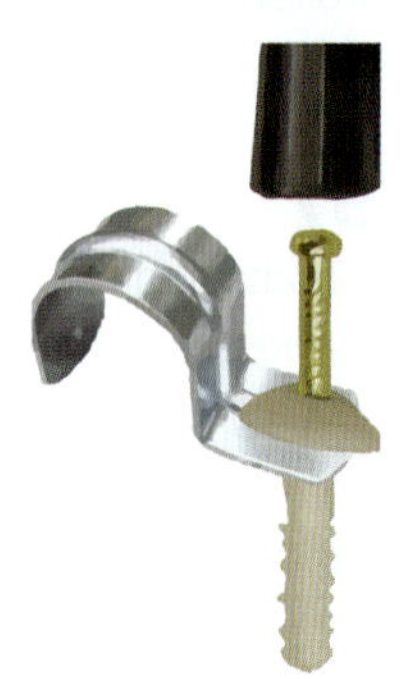

FIGURE 2.58 Wall plug

Kinetic energy in an air mass

A wind turbine converts the kinetic energy of the wind into electricity. In addition, the kinetic energy of the air mass passing through the turbine is proportional to the square of its velocity.

If the speed of the wind doubles with this same volume of air mass moving through the turbine, a fourfold increase in extracted power occurs. However, the doubling of wind speed allows twice the volume of air to pass through the turbine in a given length of time, resulting in an eightfold increase in power generated. This effect means that with a minor surge in wind velocity a substantial increase in power production occurs.

$$P \sim v^3$$

The amount of power (P) developed by the wind is proportional to the cube of its velocity (v).

Thermal energy

Thermal energy is energy related to temperature and implies that the higher the temperature, the more activity that transpires in the molecular movement of the body. If a body possesses more thermal energy than an adjacent body, then it will transfer thermal energy to the other body. Note that when energy moves from one place to another it is transient energy or heat.

Power

Up to now we have only considered how much work has occurred or how energy changes. Another important quantity is power, which defines the rate at which work is done or energy is converted.

EXAMPLE 2.13

To investigate power, let us suppose that there are 20 rolls of 1 mm^2 TPS twin plus earth cable each weighing 7 kg to be lifted 8 m from ground level to the third level. Determine the total work done in lifting this load against gravity.

$$\begin{aligned}\text{work done} &= ma(\text{force})\times d(\text{distance})\\ &= 7\times 20\ \text{rolls}\times 9.81\times 8\\ &= \mathbf{10\,987.2\ joules}\end{aligned}$$

Note: Acceleration due to gravity = 9.81 ms^{-2}.

We will now examine two possible scenarios as shown in **Figure 2.59**.

FIGURE 2.59 Two scenarios

An apprentice decides that he can take the rolls of cable separately, and he takes 90 seconds to carry each roll. The apprentice does 10 987.2 joules of work in 30 minutes. (We are not considering the work done by the apprentice in lifting his own weight.)

A tradesman aware of possible back injuries can obtain the crane that is on-site to raise all 20 rolls of cable at once in 15 seconds. The crane motor does, as the apprentice did, 10 987.2 joules of work; however, the crane does the work 120 times faster than the apprentice. So, the power of the crane is 120 times the power of the apprentice. When we speak of power, we mean how quickly work occurs.

Power, which is measured in watts, is the rate of doing work. Consequently, when 1 joule of work takes 1 second, the rate of doing work is 1 watt. It follows that power can be calculated from the following equation:

$$P = \frac{mad}{t}$$

where P = power in watts (W)

m = mass in kilograms (kg)

a = acceleration in metres per second squared (ms^{-2})

d = distance moved in metres (m)

t = time taken in seconds (s)

For the apprentice:

$$P = \frac{mad}{t}$$
$$\frac{(7 \times 20) \times 9.81 \times 8}{(30 \times 60)}$$
$$= \mathbf{6.104\ watts}$$

For the tradesman:

$$P = \frac{mad}{t}$$
$$\frac{(7 \times 20) \times 9.81 \times 8}{15}$$
$$= \mathbf{732.48\ watts}$$

In these scenarios, while the amount of work done is the same, the speed at which energy is converted is explained by the higher power for the tradesman.

EXAMPLE 2.14

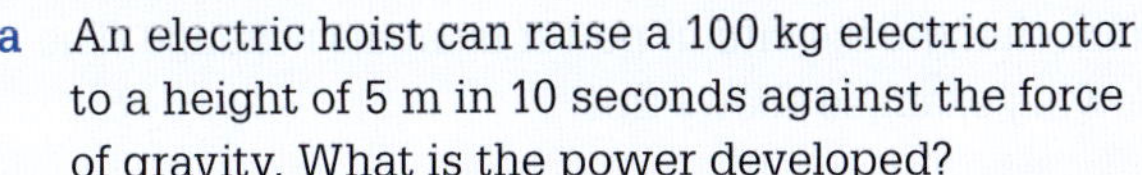

a An electric hoist can raise a 100 kg electric motor to a height of 5 m in 10 seconds against the force of gravity. What is the power developed?

$$P = \frac{mad}{t}$$
$$\frac{100 \times 9.81 \times 5}{10}$$
$$= \mathbf{490.5\ watts}$$

b A maintenance electrician weighing 55 kg has to attend to a bridge crane breakdown as shown in Figure 2.60. She has to install a motor contactor having a total mass of 1.5 kg. If she ascends the 25 m ladder in 60 seconds against the force of gravity, calculate the power developed by the maintenance electrician.

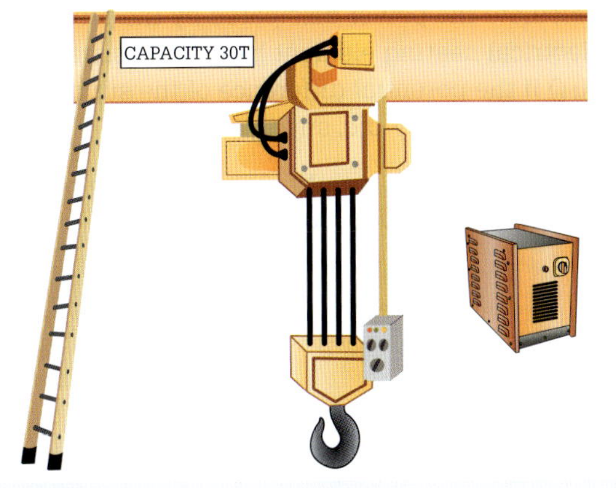

FIGURE 2.60 Bridge crane

$$P = \frac{mad}{t}$$
$$\frac{(55 + 1.5) \times 9.81 \times 25}{60}$$
$$= \mathbf{231\ watts}\ \text{(3 significant figures)}$$

c An electric motor requires a force of 400 N to move it to a new location. How much work occurs by moving the motor 10 m and what average power develops when the movement takes 120 s?

$$\text{work done} = \text{force} \times \text{distance}$$
$$= 400\ \text{N} \times 10\ \text{m}$$
$$= \mathbf{4.0\ kJ\ (2\ significant\ figures)}$$

$$\text{power} = \frac{\text{work done}}{\text{time}}$$
$$= \frac{4000}{120}$$
$$= \mathbf{33.33\ W\ (4\ significant\ figures)}$$

Mechanical power

The mechanical power required to drive a machine is the motor load. Mechanical power is the amount of work a machine, such as an electric motor, can do in a set time. The motor's mechanical power output is reliant on its rotational speed and the torque it develops. Determining the mechanical power developed by the motor uses the following equation:

$$P = \frac{2\pi NT}{60}$$

where P = mechanical power in watts (W)

N = speed of rotation in revolutions per minute (rpm)

T = torque in newton-metres (Nm)

EXAMPLE 2.15

Calculate the output power of an electric motor that maintains a speed of 1500 rpm against a load torque of 20 Nm.

$$P = \frac{2\pi NT}{60}$$
$$= \frac{2 \times \pi \times 1500 \times 20}{60}$$
$$= \mathbf{3142\ watts\ (4\ significant\ figures)}$$

EXERCISE 2.6

An electric motor develops a torque of 65 Nm at a speed of 1420 rpm. Calculate the power output in watts.

Power rating

Electrical equipment, such as an electric motor, has a rating determined by its power output. When the motor rating is 1.5 kW, it means that the output power of the motor is 1.5 kW. Note that the input power will be higher. Electrical components, such as resistors, have a rating according to the amount of heat they can dissipate without damage to themselves. For example, even though the resistance values may be the same, 10 Ω 1 W resistors can dissipate more heat than a 10 Ω 0.5 W resistor.

Electrical appliances usually have a label that specifies the power the appliance needs at the specified operating voltage. This is the power rating of the appliance. For example, the label shown in **Figure 2.61** specifies a power rating between 1100 W and 1200 W when operating at a voltage between 230 V and 240 V.

Model: HTR1200
230-240V ~50Hz
1100 – 1200W
Ser: xyz1234
Made in Australia

FIGURE 2.61 Appliance rating label

Resistors are rated in terms of the energy they can safely convert to heat. This is known as their power rating. Generally, the higher the power rating of a resistor, the larger the physical size of the resistor. **Figure 2.62** illustrates the effect physical size has on power rating.

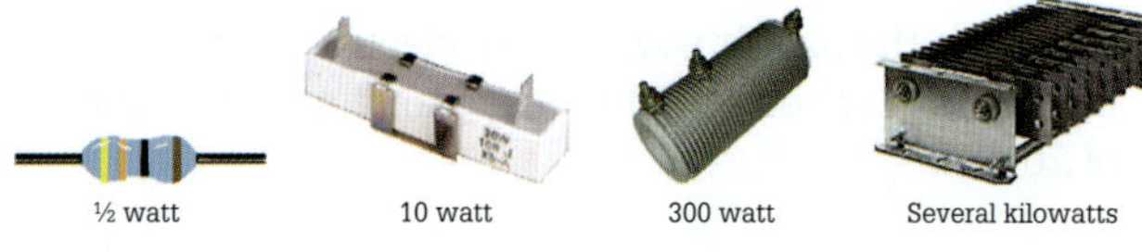

FIGURE 2.62 Resistor power handling capability

Power loss in electrical cables

When current flows through a conductor, heat is produced and this raises the temperature of the conductor. This temperature rise is due to conversion of electrical energy to heat in the conductor, which is now unable to deliver the required energy to the load and experiences a power loss. This loss can be reduced by effective design but it cannot be eliminated.

Except for bare aerials, all conductors are surrounded by electrical insulation. The insulation acts as a thermal insulation as well as being electrical insulation and reduces the rate at which heat is dissipated. However, this causes the conductor temperature to increase further and cause the resistance of the conductor to increase.

Electrical power

Electrical power represents the rate at which electrical energy converts to some other energy form. In a d.c. circuit, the following equation is used to calculate the power.

$$P = VI$$

where P = power in watts (W)
V = potential difference in volts (V)
I = current in amperes (A)

EXAMPLE 2.16

Calculate the power dissipated by an electrical heating element that draws a current of 5 A from a 200 V d.c. supply.

$$P = VI$$
$$= 200 \times 5$$
$$= \mathbf{1000\ W}$$

Change in power

If the voltage to a circuit increases then, according to Ohm's law, the current increases in the same proportion provided that the resistance remains unchanged. In addition, if the resistance of the circuit changes, then, according to Ohm's law, the current drawn by the circuit changes provided that the potential difference remains unchanged.

If the values of current or potential difference in a circuit change, the circuit power must also change. The following example shows this.

EXAMPLE 2.17

a Calculate the power dissipated by an electrical heating element having a resistance of 40 Ω and connected to a 200 V d.c. supply.

Firstly calculate the current drawn:

$$I = \frac{V}{R} = \frac{200}{40} = 5\,A$$

Next calculate the power dissipated:

$$P = VI = 200 \times 5 = \mathbf{1000\ W}$$

b Calculate the power dissipated by the previous electrical heating element having a resistance of 40 Ω if the potential difference of the supply is doubled to 400 V d.c. supply.

Firstly calculate the current drawn:

$$I = \frac{V}{R} = \frac{400}{40} = 10\,A$$

Next calculate the power dissipated:

$$P = VI = 400 \times 10 = \mathbf{4000\ W}$$

Notice that the power has increased. The reason is that power is the product of voltage and current. It follows that when the voltage and current doubled from their original values, then an increase in power by a factor of four occurs. Notice that the voltage increase creates a much larger change in the power taken by the circuit.

Using algebra and substituting different expressions for voltage and current from Ohm's law, the power equation becomes two new equations. These two equations relate the power to the current and the resistance, and to the voltage and the resistance.

SWITCH ON

Using Ohm's law equation for V

As $V = IR$ then by substituting for V:

$$P = IR \times I$$

This is simplified as:

$$P = I^2 R$$

Using Ohm's law equation for I

As $I = V/R$ then by substituting for I:

$$P = V \times \frac{V}{R}$$

This is simplified as:

$$P = \frac{V^2}{R}$$

The modifications together with the original equation, provide three equations for determining power in a circuit.

The equation $P = I^2R$ can be used to calculate the power. If either current or resistance is unknown, then use either of the other equations to calculate the power. In addition, use the Ohm's law equation to calculate any unknown quantity needed in order to use this power equation.

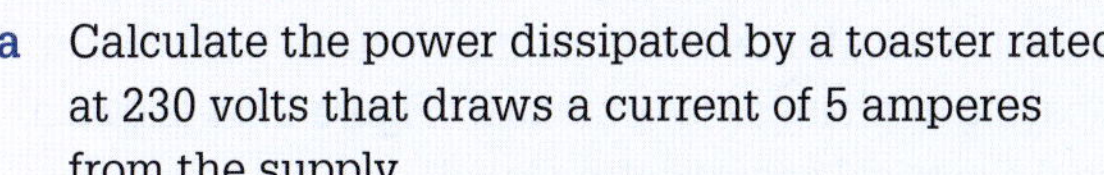

EXAMPLE 2.18

a Calculate the power dissipated by a toaster rated at 230 volts that draws a current of 5 amperes from the supply.

$$P = VI$$
$$= 230 \times 5$$
$$= \mathbf{1150\ W}$$

b The wiring in a circuit from its control device to the load terminals has a total resistance of 0.5 Ω. Calculate the power dissipated in the wiring when carrying a current of 20 A.

$$P = I^2 R$$
$$= 20^2 \times 0.5$$
$$= \mathbf{200\ W}$$

»

c Calculate the power that a 200 Ω resistor dissipates when connected across a 36 V supply.

$$P = \frac{V^2}{R}$$
$$= \frac{36^2}{200}$$
$$= \mathbf{6.48\ W}$$

If the voltage in this example remains unchanged, then doubling the resistance causes the power dissipated by the circuit to halve.

$$P = \frac{V^2}{R}$$
$$= \frac{36^2}{400}$$
$$= \mathbf{3.24\ W}$$

Therefore, if the resistance for the same circuit is halved (from 200 Ω to 100 Ω) while the voltage remains unchanged, the power dissipation doubles from the original value.

$$P = \frac{V^2}{R}$$
$$= \frac{36^2}{100}$$
$$= \mathbf{12.96\ W}$$

EXERCISE 2.7

a Calculate the power dissipated by a toaster rated at 230 volts that draws a current of 6.5 amperes from the supply.

b The wiring in a circuit from its control device to the load terminals has a total resistance of 0.25 Ω. Calculate the power dissipated in the wiring when carrying a current of 15 A.

c Calculate the power that a 36 Ω resistor dissipates when connected across a 36 V supply.

Electrical energy

Watt-second, joule, kWh

The unit of energy is watt-seconds or joules because power is measured in watts and the time in seconds. If the measured power uses kilowatts and the time measured is in hours, then the unit of energy is the kilowatt-hour, often called the 'unit of electricity'. For this reason, energy meters record the number of kilowatt-hours used.

$$E_E = VIt$$

where E_E = electrical energy in joules (J)

V = potential difference in volts (V)

I = current in amperes (A)

t = time in seconds (s)

The unit of energy is the joule (W.s); however, with large amounts of energy the unit used is the kilowatt-hour (kWh) where:

$$3600000 \text{ J} = 1000 \text{ W} \times 3600 \text{ seconds}$$
$$(1000 \text{ watts per hour} = 1 \text{ kWh})$$

As such we can calculate electrical energy in kilowatt-hours using the following equation.

$$E_E = \frac{VI}{1000} \times t$$

where E_E = electrical energy in kilowatt-hours (kWh)
V = potential difference in volts (V)
I = current in amperes (A)
t = time in hours (h)

EXAMPLE 2.19

a A 230 V supply delivers a current of 5 A to a load over a period of 30 minutes. Calculate the energy provided in kilowatt-hours (kWh).

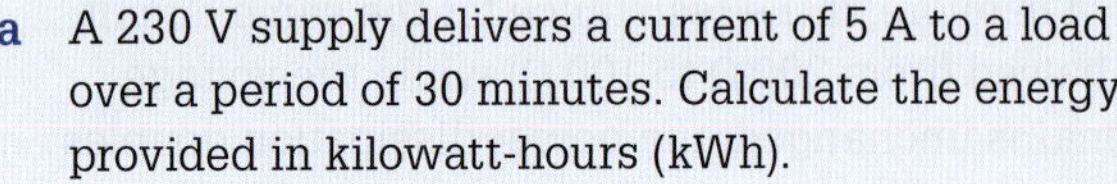

$$E_E = \frac{VI}{1000} \times t$$
$$= \frac{230 \times 5}{1000} \times 0.5$$
$$= \mathbf{0.575\ kWh}$$

b An electric heater converts 4.2 MJ of energy when connected to a 230 V supply for 60 minutes. Determine the power rating and the current drawn from the supply.

$$\text{power} = \frac{\text{energy}}{\text{time}}$$
$$= \frac{4.2 \times 10^6 \text{ J}}{60 \times 60 \text{ s}}$$
$$= 1167 \text{ J/s} = 1167 \text{ W (4 significant figures)}$$
$$\text{power } (P) = VI, \text{ therefore:}$$
$$I = \frac{P}{V}$$
$$= \frac{1167}{230}$$
$$= \mathbf{5.07\ A\ (3\ significant\ figures)}$$

c An installation has the following d.c. loads connected over a 24-hour period.

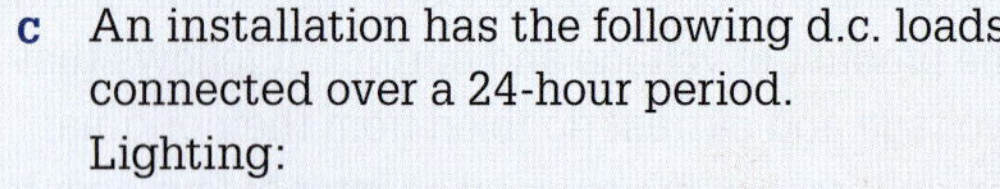

Lighting:

- 40 × 10 watts @ 24 V d.c. × six hours usage each

Other circuits:

- 249 litre d.c. compact fridge/freezer @ 24 V and 850 Wh/day
- hot water circulating pump d.c. @ 24 V and 1600 mA × three hours usage for gas hot water system with solar support
- 3500 watts continuous power @ 24 V from a d.c. inverter delivering 29.2 A continuously for six hours for 230 V a.c. appliances
- microwave oven d.c. @ 24 V with 660 watts input @ 27.5A over 30 minutes usage
- computer electrical systems outlet @ 19.5 V d.c. and 5 A each over four hours
- security system using 8A @ 24 V d.c. over 24 hours

Determine the energy in kWh over a 24-hour period.

$$40 \times 10 \times 6 \div 1000 = 2.4 \text{ kWh}$$
$$850 \text{ Wh} = 0.85 \text{ kWh}$$
$$1.6 \times 24 \times 3 \div 1000 = 0.1152 \text{ kWh}$$
$$3.5 \times 6 = 21 \text{ kWh}$$
$$660 \times 0.5 \div 1000 = 0.33 \text{ kWh}$$
$$19.5 \times 5 \times 4 \div 1000 = 0.39 \text{ kWh}$$
$$8 \times 24 \times 24 \div 1000 = 4.608 \text{ kWh}$$
$$\text{Total} = \mathbf{29.6932\ kWh}$$

Electrical energy is measured with a kilowatt-hour meter as shown in **Figure 2.63**, which can be either analogue (left) or digital (right). Electrical energy is costed per kilowatt-hour.

Energy rating label

Household appliances are required to display an energy rating label. This label specifies the amount of energy in kilowatt-hours (kWh) the appliance will use when operated for a year. This affords the consumer convenience when comparing appliances for purchase. **Figure 2.64** shows a typical electrical energy rating label.

FIGURE 2.63 Kilowatt-hour meters

Source: iStock.com/hbak; iStock.com/onurdongel

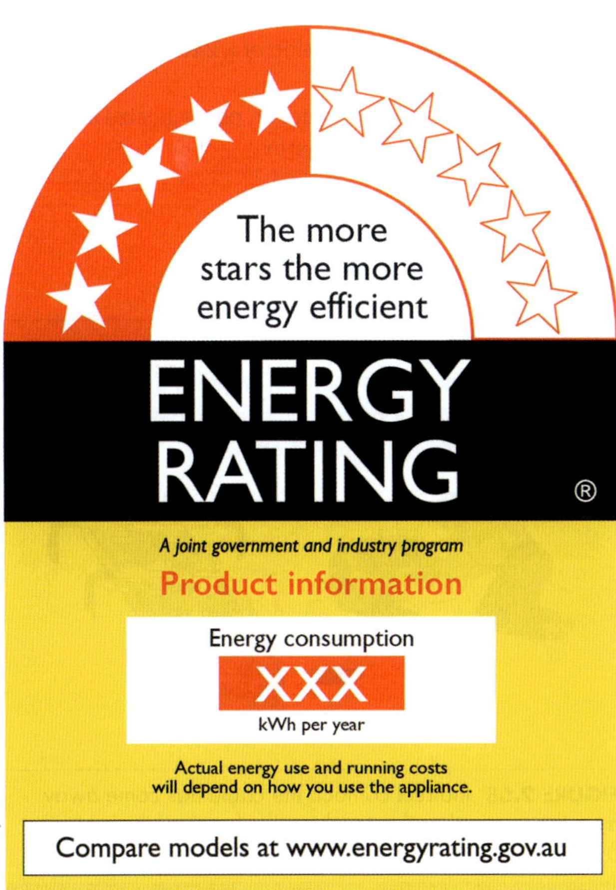

FIGURE 2.64 Energy rating label

Source: © Commonwealth of Australia

Measuring electrical power

Electrical power is measured with an instrument known as a wattmeter. Electrical power is the product of potential difference and current. As such, a wattmeter needs to measure both the current flowing in the circuit and the potential difference across the circuit. This means that the wattmeter is a four-terminal device; two terminals for the potential difference and two terminals for the current.

As with most measuring instruments, the wattmeter may be analogue or digital. Many digital wattmeters, such as the one shown below, have simplified circuit connections; two terminals identified as input or source and two terminals identified as output or load.

Figure 2.65 shows how an electronic wattmeter can connect into our previous circuit to measure the power dissipated by the lamp.

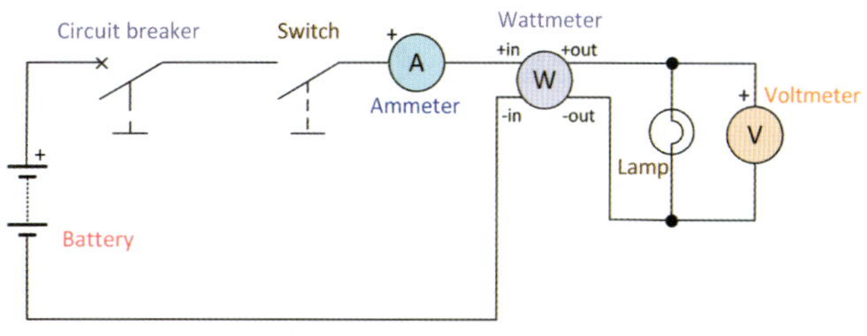

FIGURE 2.65 Measuring electrical power

Summary of electrical equations

Figure 2.66 is a summary of the 12 basic equations you should know. Adjacent to each quantity are three segments. In each segment, the basic quantity is expressed in terms of two other basic quantities.

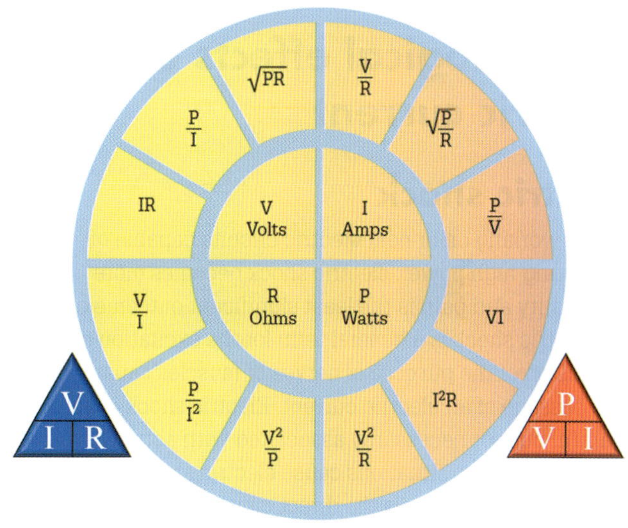

FIGURE 2.66 Summary of the 12 basic equations

REVIEW QUESTIONS

1. An electrician is using cable clips to fix TPS cable to a wall. What is the force the electrician is generating if a hammer of mass 1.5 kg accelerates towards the cable clip at 10 ms^{-2}?
2. A maintenance electrician is securing the frame of an electric motor to a concrete floor using a sleeve anchor. Calculate the torque required when a 0.45 m spanner applies a force of 150 N acting at right angles.
3. How much work is done when a 25 N force moves a motor a distance of 5 m?
4. Define 'energy'.
5. Define 'kinetic energy'.
6. Calculate the kinetic energy required to drive a 100 g steel slot wedge into an armature slot with a velocity of 40 ms^{-1}.
7. Define 'power'.
8. Calculate the power generated by a hoist electric motor that raises a 250 kg switchboard against the force of gravity a distance of 12 m in ten seconds.

»

9 An electric kettle draws a current of 9.5 A from a 230 V supply. How much power does it dissipate?

10 What instrument is used to measure electrical power?

11 Determine the power rating of a 240 V 16 A hot plate.

12 Comment on the difference in physical size of a 5 watt and a 300 watt resistor.

2.5 Effects of an electric current

When an electric current flows in a conducting medium, it creates change. These changes occur in the space around the conductor and within the conductor itself. The result is that an electric current has a:

- magnetic effect
- heating effect
- luminous effect
- chemical effect.

The magnetic and heating effects are always present whenever an electric current flows through a conductive medium. An electric current also has various effects on living tissue. Some of these effects are desirable and some are possibly fatal to life. These come under the category of physiological effects of electric current.

Physiological effects of electric current

Electric shock

When working on or near electric circuits capable of delivering high power, electric shock becomes more of a reality and pain is the least significant outcome of it. Electric shock is a general term for the excitation or confusion of the purpose of nerves or muscles caused by the passage of an electric current. Electric shock is usually painful but not necessarily associated with actual damage to a person. Two situations cause electric shock.

The first situation is the direct contact whereby a person contacts with a conductor that is live and becomes part of the fault path as shown in **Figure 2.67**.

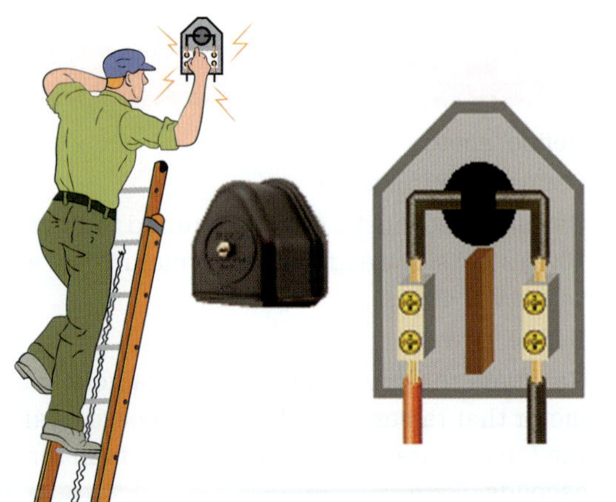

FIGURE 2.67 Direct contact: touching live mains in a connection box

The second is indirect contact whereby a person contacts material that has become live under fault conditions as shown in **Figure 2.68**.

FIGURE 2.68 Indirect contact: the cable has come away from the terminal and is touching the frame of the motor

SWITCH ON

Electrical injuries have three categories: nerve immobilisation, fibrillation and tissue destruction.

The effect that electric shock has on the body depends on the magnitude of the voltage, magnitude of the fault current, the body parts through which the current flows, the physical condition of the person and the duration of the shock. Serious electric shock is associated with alternating current and is rare with low-voltage direct currents.

The common feature of electric shock is a brutal piercing and deadening pain at the points of entry and exit. Muscular action is complex and under the control of various nerves. More to the point, neuromuscular coordination is conditioned by practice. Electric shock is an unintentional contraction of the muscles associated with the path of the current. This unconditioned response can tear tendons and the muscle tissue apart. Furthermore, nerves have neurons and stimulation of the neurons transpires in various ways: thermal, chemical, pressure or light stimulation, or a current entering the body. However, the current's effect on the nerves and the nerve centre of a person is one of prime control and restraint.

As a direct result of electric shock, a person may clench, and be unable to release, a conductor or an appliance. Alternatively, if a person has touched a live conductor without clenching it, the muscles of a person's back and legs may contract harshly so that the person moves backward rapidly. Another possible result of involuntary muscle contraction is that the muscles of the diaphragm and chest may contract and prevent breathing. Constriction could lead to death by suffocation.

Death may also occur as a result of the current flowing through the respiratory control centres of the central nervous system of a person. However, death is usually caused by direct interference with the action of the heart, called ventricular fibrillation.

The ventricles of the heart have three states: rest, regular beat and fibrillation. If the passage of current causes the heart to become violently disturbed, the heart may pass to the fibrillation state. Once in fibrillation death occurs unless action is taken to 'defibrillate' the heart. It is also possible that the experience of electric shock creates fear in a person, and the person dies of a heart attack rather than from the electric shock.

Body resistance

The resistance of the human body from hand-to-hand or from hand-to-foot is variable and depends upon the area of electric contact. Adding to the effect is whether that area is dry, moist or wet.

Refer to the Australian and New Zealand standards AS/NZS 60479.2: 2002 and AS/NZS 60479.1: 2010. In dry conditions, the body can offer some resistance to current, but above 30 V the body practically has no impeding effect at all. The body acts like a voltage-dependent resistor, meaning that as the voltage increases the resistance of the body decreases.

Physiological sensations of current

Two things determine the value of current flowing through a person experiencing an electric shock. These are:

1 the current path through the person
2 physical condition of the person.

The physiological values of current are suitable only for research on deciding on permissible leakage currents for electrical devices. At around 15–17 mA rms, the so-called 'grip' current, it becomes extremely difficult to release a conductor held in the hand. **Figure 2.69** shows reactions to 50 Hz current.

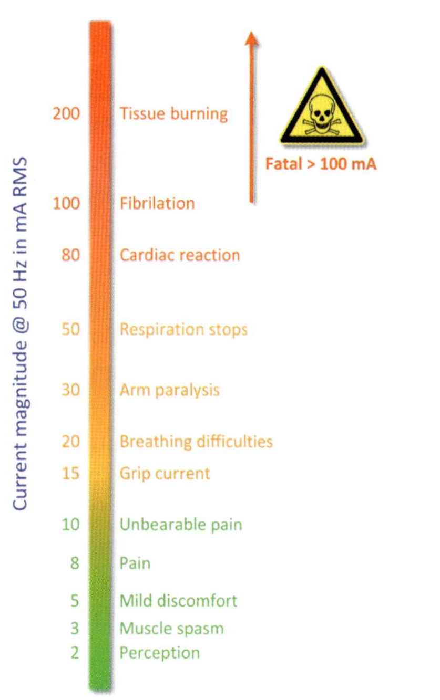

FIGURE 2.69 Possible reactions to 50 Hz current. Assumed conditions: hand grip of five seconds duration

Effect of voltage frequency

It is unfortunate that mains voltage occurs at a frequency of 50 Hz, because this voltage frequency has resonance with the nerve endings within the human body, producing maximum excitation. Lower or higher frequencies do not do this. However, if a person's body experiences high-frequency electric fields it will be heated up like any other conductor.

Note: This does not imply that voltages at frequencies other than 50 Hz are safe.

Electric shock is not an isolated event and many persons will experience this event at some point in their life. In Australia, the majority of electric shocks occur at the mains pressure of 230 V and 400 V respectively. Fatalities often occur at these voltages, while high-voltage injuries are mainly from burns. Because of so many material variables within the person, it is not possible to state the minimum current that kills. Muscle tissue does play a part and it appears that the current threshold is lower for women than for men.

Touch voltage

The touch voltage graph in **Figure 2.70** shows that a person in contact with 230 V requires releasing from this danger in 40 ms to avoid harmful effects. Similarly, a person in contact with 400 V requires releasing in 15 ms to avoid harmful effects.

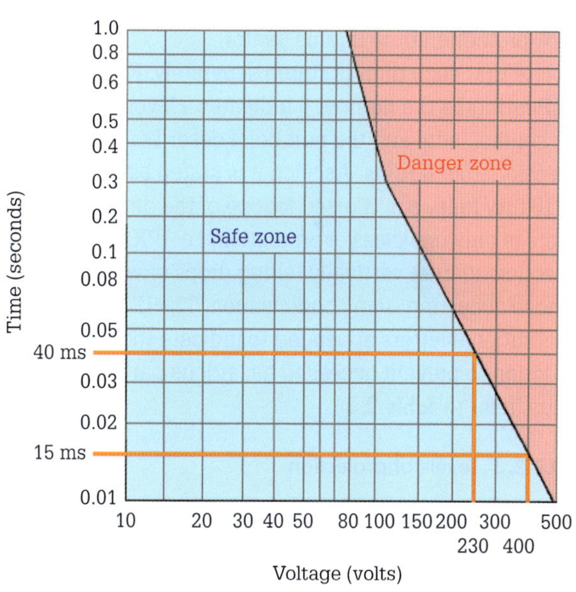

FIGURE 2.70 Touch voltage graph

Protection against the physiological effects of current

The fundamental principles

The Australian and New Zealand standard AS/NZS 3000: 2018 *Wiring Rules*, in section 1.7 'Protection for Safety' lists 10 fundamental principles for protection against the physiological effects of current. The carrying out of these 10 principles should ensure the safety of persons, livestock and property against hazards and harm that may arise in the sensible use of electrical installations. These are outlined below.

1 Protection against both direct and indirect contact by use of extra-low voltage. Protection can be effectively achieved by the use of PELV (protective extra-low voltage) to protect persons against electric shock, as a result of indirect contact or reduced area direct contact.

2 Protection against direct contact. Direct contact means contact by persons or livestock with exposed live parts that could result in electric shock. Protection against direct contact is called basic protection which is insulation of the live parts. Basic protection must be capable of withstanding the electrical, mechanical, chemical and thermal stresses under normal operating conditions. Because of this, basic insulation is only removable by destruction. In addition, protection can be achieved through one of the following methods. The residual current circuit breaker with overload protection (RCBO) provides protection against indirect contact and complementary protection against direct contact when protection such as enclosures or barriers (by placing them out of reach, using obstacles, construction or installation methods that prevent access) are used. Where electrical equipment is located in places open to all persons, a minimum degree of protection against direct contact corresponding to an appropriate ingress protection rating, IP4X or IPXXD, is necessary. The IP rating normally has two numbers:

- 1st protection from solid objects or materials
- 2nd protection from liquids (water).

Note that each digit is stated separately. For example, the IP rating 54 the number 5 indicates a level of protection from solid objects and the number 4 describes the level of protection from liquids. As such it is stated as 'IP 5-4' and not 'IP fifty four'. An 'X' used for one of the digits transpires if there is only one class of protection; i.e. IPX1 refers to protection against vertically falling drops of water such as those occurring from condensation. The IP standard has additional letters appended to indicate the level of protection to prevent access to hazardous parts by human beings (refer to **Table 2.3**).

TABLE 2.3 Levels of protection

Level	Protected against access to hazardous parts with
A	back of hand
B	finger
C	tool
D	wire

3 Protection against indirect contact. Protection against indirect contact intends to prevent hazardous situations due to an insulation fault between live parts and exposed conductive parts. In addition, indirect contact voltages are related to electrical faults that have not cleared. Measures to prevent the occurrence of a touch voltage include:

- earthed equipotential bonding
- provision of class II equipment or by equivalent insulation
- electrical separation
- automatic disconnection of the supply by using an RCBO
- use of isolation transformers.

In addition, enhanced protection arises by coordinating the disconnection times of protection devices within the installation.

4 Protection by use of residual current devices (RCDs). Safety switches are additional protection devices and must have an operating current of 30 mA or less. Safety switches are suitable for protection of persons against indirect contact.

5 Protection against thermal effects in normal service. Whenever current flows through a conductor, part of the electric energy changes into thermal energy, and this may not dissipate efficiently. Thermal protection refers to heat generated by the electrical equipment in normal use and under fault conditions. There are two types of thermal risks involving wiring:

1 internal overheating due to overloads or short circuits in cables
2 arcs caused by short circuits due to damaged insulation or from faulty, loose or broken conductors or switches.

Even if the incidents are of short duration, the effect is of thermal ageing on the dielectric performance of the insulation. The method used to verify these conditions is to carry out thermographic controls and install arc detectors. For protection against thermal effects in ordinary service, protective devices must break any overload current before it can cause a detrimental temperature rise to the conductor insulation.

6 Protection against unwanted voltages. Refer to Clause 1.5.11.4 of AS/NZS 3000: 2018 *Wiring Rules*. Surge protection devices offer protection against unwanted voltage by safeguarding all types of electric and electronic equipment from destructive high-voltage transients associated with lightning and high-voltage appliances. Surge protection devices operate instantaneously to divert surge current to ground. Inducement of unwanted voltages across equipment from power, telephone, telemetry or other cables, is limited.

7 Protection against over-current. Over-current means a current exceeding the rated value. For conductors the rated value is its current-carrying capacity. Over-currents in healthy circuits occur by motor starting currents, motor stalling and connection of excessive loads.

For the protection against overload current, protective devices (fuses, circuit breakers) must be provided in the circuit to break any overload current flowing in the circuit conductors. The protective devices can prevent dangerous events due to thermal and mechanical effects produced in the conductors and connections during an over-current condition. Protective devices such as fuses and circuit breakers provide a means to break any overload current flowing through the circuit conductors. They prevent dangerous events such as thermal and mechanical events developing in conductors and connections during an over-current condition.

8 Protection against fault currents. Fault currents occur when mechanical damage affects circuits and accessories causing insulation failure or breakdown leading to 'bridging' of live conductors. Consequently, the impedance falls and is assumed to be negligible. In these circumstances, the very high short-circuit current that flows is known as the 'prospective short-circuit current'. It follows that the protection device must be able to clear the fault at the installation point. Always install protection devices to control fault currents where conductor size changes.

9 Protection against over-voltage. Electrical installations require designing to withstand over-voltages caused by:

- lightning strikes
- switching surges
- transients
- noise
- incorrect connections
- other abnormal conditions or malfunctions.

Surge devices include metal oxide varistors (MOVs), silicon avalanche diodes and gas discharge tubes (GDT) and offer protection against over-voltage. In fact, some devices connect directly into the line and attempt to block incoming surges before they reach the equipment.

10 Protection against mechanical movement. A broad range of mechanical actions (cutting, bending) and movement (rotating, reciprocating) may present hazards to persons. Recognising the mechanical hazards is the first step towards protecting persons from the danger they present.

Magnetic effect of electric current

An electric current flowing in a conductor causes a magnetic field to surround the conductor as shown in **Figure 2.71**.

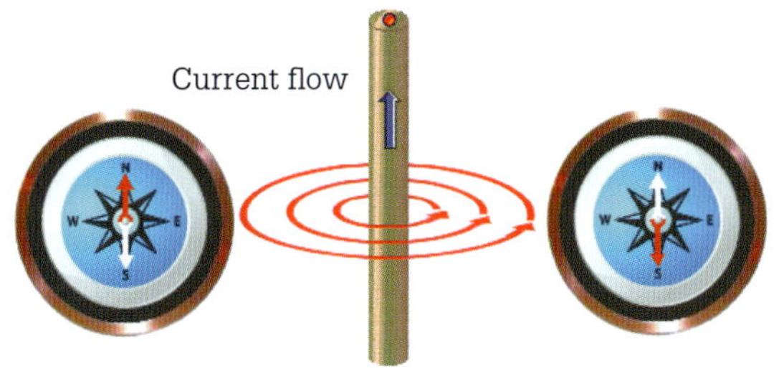

FIGURE 2.71 Magnetic field around a straight conductor

Placing a compass needle near the conductor and observing the deflection of the needle can reveal the direction of the magnetic lines of force.

When conductors carrying current are next to each other, they will exert a force on one another. This force may be attractive (**Figure 2.72**) or repulsive (**Figure 2.73**) depending on the direction of current flow in each conductor. The forces produced by the magnetic effect of an electric current form the principle of operation of various items of electrical equipment, which includes electric motors, solenoids, contactors and relays.

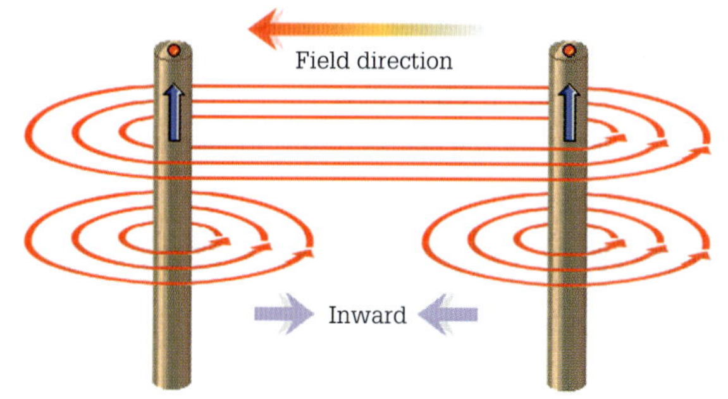

FIGURE 2.72 Attraction of like conductors

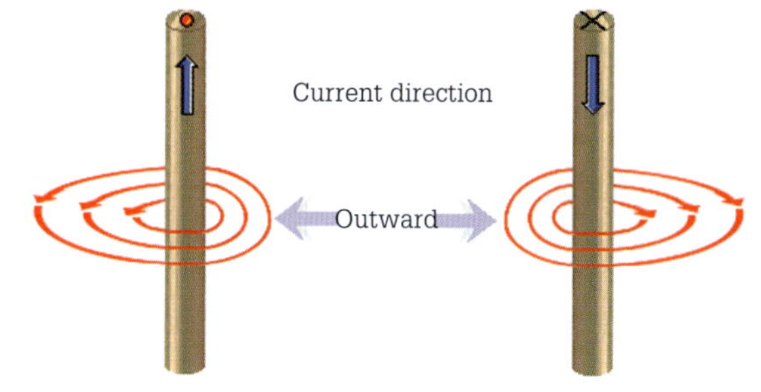

FIGURE 2.73 Repulsion of unlike conductors

Uses of the magnetic effect of electric current

Electromagnets are devices that take advantage of the magnetic effect of electric current and find use in many practical applications – from motors, generators and transformers producing large magnetic fields, to relay and solenoid switching devices and analogue-measuring instruments, microphones and computer memory. **Figure 2.74** shows various electromagnetic devices.

FIGURE 2.74 Electromagnetic devices

Unfavourable magnetic effects of electric current

The electromagnetic field can induce current in adjacent circuits. Conductors carrying different signals such as power and communication signals require separation from one another. The magnetic forces of attraction and repulsion that occur with high current flow mean that cables need securing. Furthermore, high-voltage transmission lines require rounding to reduce electric fields and prevent sparking (called the corona effect).

In the 2018 edition of AS/NZS 3000 *Wiring Rules* there are four clauses concerning magnetic effects: Clause 2.10.6, Clause 3.9.10, Clause 3.9.11 and Clause 4.1.4. Please refer to the *Wiring Rules* for the clause descriptions.

Heating (thermal) effect of electric current

The energy essential for current to overcome the resistance of a material dissipates as heat. The heat generated is the I^2R loss, or copper loss.

The outdated incandescent lamp used the thermal effect of current to heat a thin tungsten filament to incandescence – white hot. Light as an electromagnetic wave occurred because of the vibration of the charged particles within the atoms of the hot or incandescent filament.

Metal conductors convey high amounts of electrical energy from place to place. These conductors have resistance and, consequently, there is a waste of power in heating. Transmitting current long distances results in a high-energy loss. However, there are two ways to reduce this power waste:

1. Reduce the current.
2. Increase the diameter of the cable and thereby reduce the resistance. However, it is better to keep the current as low as possible because of the material cost of a larger-diameter cable.

When transmitting a given amount of power, doubling the voltage can halve the current.

Uses of the heating effect of electric current

Heating elements are devices that take advantage of the heating effect of electric current and find use in many practical applications in electrical appliances and apparatus – including water heaters, cooktops, ovens, kettles, electric blankets, toasters and electric grills to name a few. **Figure 2.75** shows various electrical heating devices.

FIGURE 2.75 Electrical heating devices.

Source: [L to R] Shutterstock.com/nikkytok; Shutterstock.com/Raf Quintero; Shutterstock.com/ppart; Shutterstock.com/MOAimage; Shutterstock.com/SeDmi

Unfavourable heating effects of electric current

The heating effect of current flow can cause degrading of the insulation of conductors enclosed in conduits and ducts because there may not be enough space available to dissipate heat to the surrounding atmosphere.

How efficiently a cable conducts current depends on its electrical resistance and temperature. When electric current flows through a cable, the temperature of the copper or aluminium conductor increases, up to an operating temperature of about 70 °C to 90 °C.

Heat dissipated by a cable in free air causes no problems. However, when cables are enclosed in confined spaces, such as ducts or conduits, their insulation overheats. In many practical situations, such as industrial and commercial installations, cable runs are exposed to the deposition of dusts over long periods. Dusts provide thermal insulation and can cause a self-heating effect when deposited on electric power cables. This condition can cause the onset of a runaway reaction where temperatures can reach the melting point of the cables' insulation.

Cables in proximity can also produce a thermal effect. Bunched cables can trap heat or prevent heat dissipation. In this situation, the cable amperage must be de-rated.

In the 2018 edition of AS/NZS 3000 *Wiring Rules* there are 10 clauses concerning heating/thermal effects: Clause 1.5.1b, Clause 2.10.6, Clause 3.7, Clause 3.7.2.1.1(e), Clause 4.1.2, Clause 4.2.3, Clause 4.5.2.1, Clause 4.5.2.2, Clause 4.10.6 and Clause 4.13.3. Please refer to the *Wiring Rules* for the clause descriptions.

Luminous effect

An electric current can also have a luminous effect, which means it is capable of producing light. Light is an electromagnetic wave. However, visible light is only one type of electromagnetic wave. The electromagnetic spectrum contains waves ranging from gamma, X-ray, ultraviolet (UV), infrared and microwave to radio waves. **Figure 2.76** shows visible light produced between 4.0×10^{14} Hz and 7.5×10^{14} Hz.

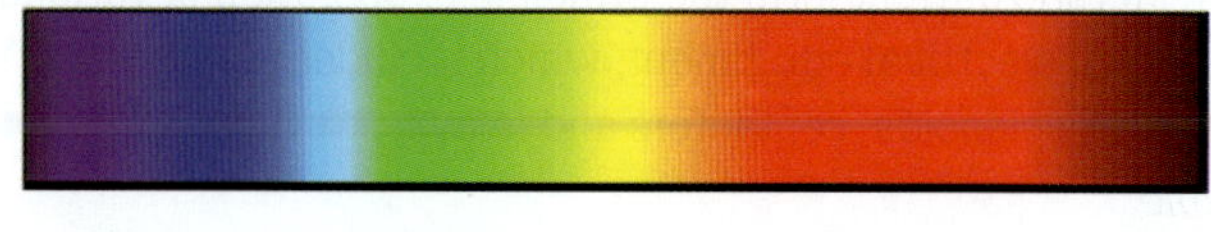

FIGURE 2.76 Visible light

Fluorescent lamp filled with mercury vapour

When high voltage occurs across the filaments of a fluorescent tube, the electrons emitted collide with the mercury atoms, causing them to emit UV photons. These photons strike the phosphor coating on the inner surface of the tube. The UV radiation emitted excites the phosphor atoms, raising their energy state, as a result of absorbing a photon. When these atoms return to their normal energy state the energy difference between the two states, excited and normal, is emitted in the form of photons. These photons display fluorescence in the visible range of the electromagnetic spectrum as shown in **Figure 2.77**.

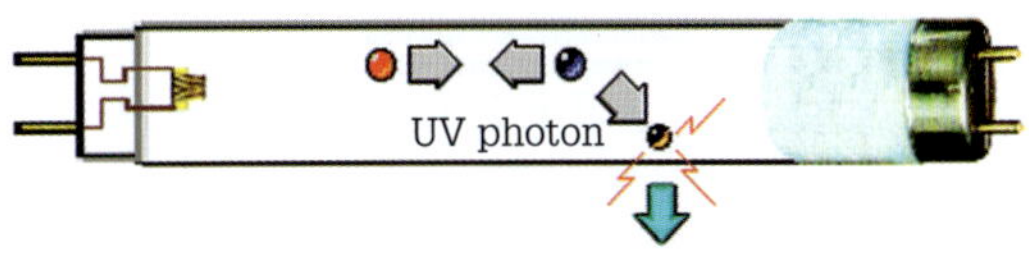

FIGURE 2.77 Fluorescent lamp ion collisions producing visible light

The electric arc

The electric arc lamp was one of the first uses for electricity. The operating principle was very simple – an arc was drawn between two carbon electrodes. This concept led to the development of the discharge lamp, which comprises a glass envelope filled with a gas and having two electrodes.

If two bare current carrying conductors are in contact and move apart (such as switch contacts opening), a stream of volatile particles called an arc happens. This arc can cause damage when an open circuit occurs. However, an arc can be useful, such as in the electric welder. The arc that results from welding demonstrates three effects of current – heating, magnetic and light (see **Figure 2.78**).

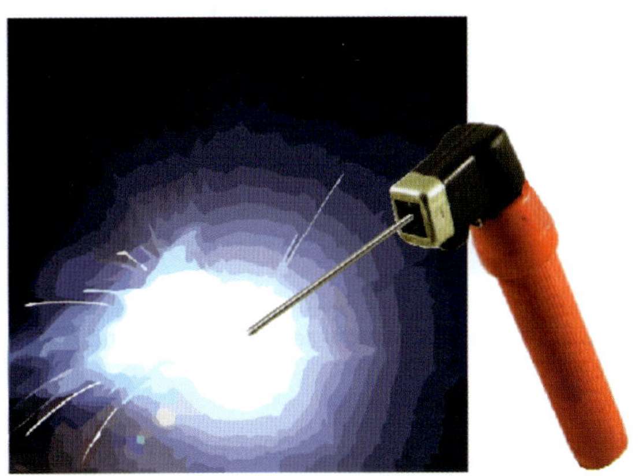

FIGURE 2.78 Electric arc

In the 2018 edition of AS/NZS 3000 *Wiring Rules* there are two clauses concerning arcs and sunlight effects: Clause 4.2.2.4 and Clause 3.10.3.7. Please refer to the *Wiring Rules* for the clause descriptions.

Chemical effect

There are two different chemical effects: the effect of electrolysis and the effect of the voltaic cell.

Electrolysis

Electrolysis is a process used in the refining of metals, battery charging and electroplating. The process uses an aqueous electrolyte that undergoes a chemical change. **Figure 2.79** shows a simple refinery process whereby an impure copper sheet connected to the anode slowly dissolves when a current passes through the anode and electrolyte to the cathode. The copper is carried through the electrolyte and deposited on a pure copper cathode while impurities in the anode fall to the bottom of the cell.

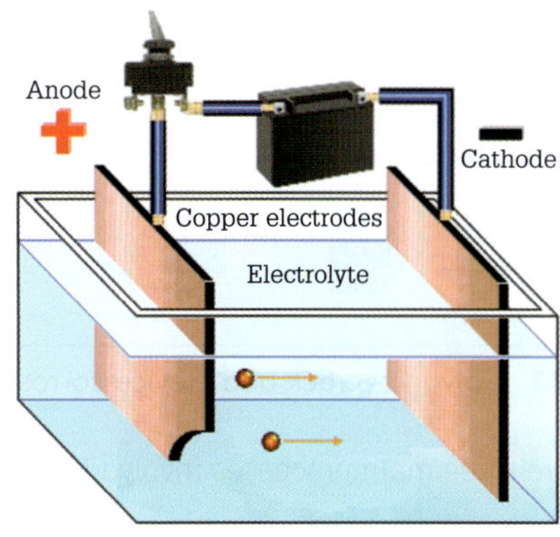

FIGURE 2.79 Electrolysis showing the movement of the copper to the cathode

Electroplating

Electroplating also uses the principle of electrolysis and is the process of coating metal objects with thin films of other metals obtained by electrolysis from a solution of their salts. The strip of metal acts as the anode. The anode and the cathode object requiring plating immerse in the electrolyte. When the anode and cathode connect to an external d.c. supply, the metal of the anode dissolves and passes over to the cathode where it is deposited as a thin coating as shown in **Figure 2.80**.

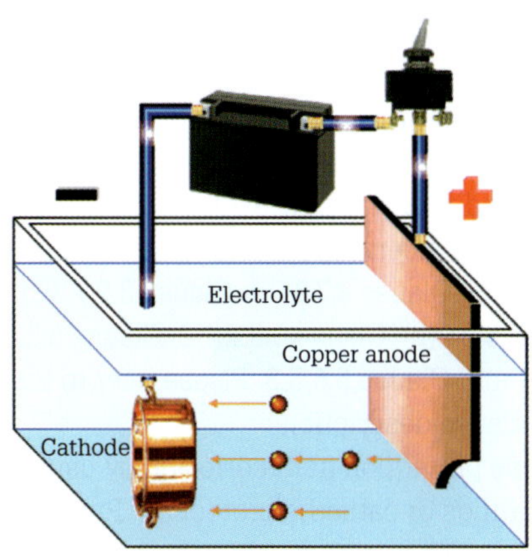

FIGURE 2.80 Electroplating converts voltage to chemical action

Electrolytic corrosion

Corrosion takes place when two different metals react in their ecosystem with corrosion substances forming. Examples of metal corrosion include the rusting of iron and the growth of patina on copper. Corrosion occurs in three types of environments: atmospheric, immersion and structural.

1 **Atmospheric** corrosion occurs owing to air pollution, salt-water spray and humidity.
2 **Immersion** corrosion takes place when metals are wet thereby allowing the creation of electrochemical cells.
3 **Structural** corrosion results from imperfections of a metal such as tiny structural cracks and crevices together with the presence of minute particles of other metals contained within the base metals' composition. **Figure 2.81** shows examples of structural corrosion between the copper and aluminium bar, which results in a high-resistance joint.

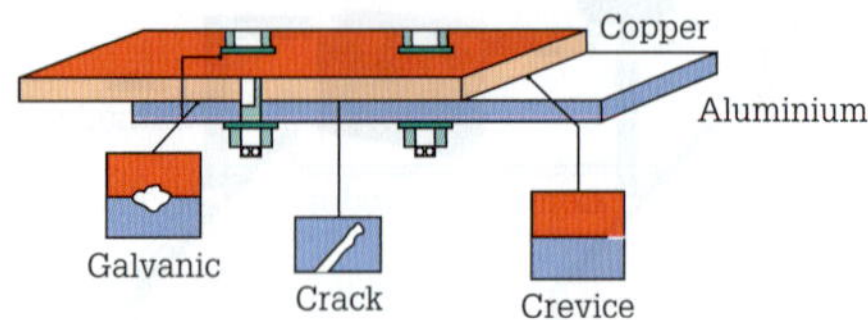

FIGURE 2.81 Galvanic, crack and crevice corrosion process

There are three main processes involved in corrosion of metal.

1 **Oxidation** is the giving up of electrons by the main metal, forming positive ions.
2 **Reduction** is the taking up of electrons by oxygen molecules and water to form negative ions.
3 **Rust formation** is the movement of the positive and negative ions towards each other to form a corrosion product.

These processes, illustrated in **Figure 2.82**, show that positive ions disperse through the electrolyte and the negative ions move through the metal.

FIGURE 2.82 Three processes of corrosion of metal

In the 2018 edition of AS/NZS 3000 *Wiring Rules* there are eight clauses concerning requirements for protection against corrosion: Clause 3.7.2.9.1, Clause 3.9.7.3(c), Clause 3.9.9, Clause 3.9.9.2(b), Clause 5.3.2.1.2, Clause 5.3.6.4, Clause 5.5.5.1 and Clause 5.5.5.3. Please refer to the *Wiring Rules* for the clause descriptions.

Metals have protection from corrosion by using protective coatings or cathodic protection. Protective coatings such as anticorrosive cold galvanising primers provide adequate results.

These coatings provide the degree of protection required by the Australian and New Zealand standard AS/NZS 3000: 2018 *Wiring Rules* concerning earth-fixing devices exposed to the weather.

Cathodic protection

There are two main systems of cathodic protection currently in use.

1 **Impressed current system** This system involves applying an external d.c. voltage to the metal requiring protection to keep it in the reduced or uncorroded condition. The metal needing protection maintains a lower potential than the attacking agent and forms the cathode of a d.c. power system as shown in **Figure 2.83**.

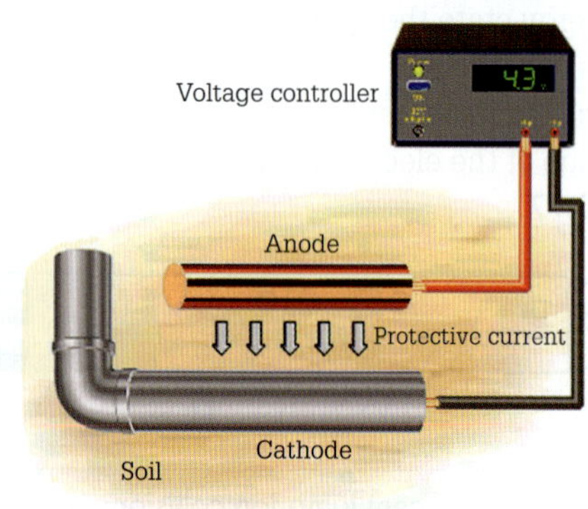

FIGURE 2.83 Impressed current system

Impressed current forces a reversal of the electrochemical current and causes the consumption of the anode rather than the metal. Currents from 50 mA up to 120 A are used depending on the surface area of the metal, soil and other factors. However, in remote areas, solar power systems provide the d.c. current.

2 **Sacrificial anode systems** A sacrificial anode is a metal or alloy that has a higher anodic potential than the metal that it is protecting. **Figure 2.84** shows the electrochemical series of metals.

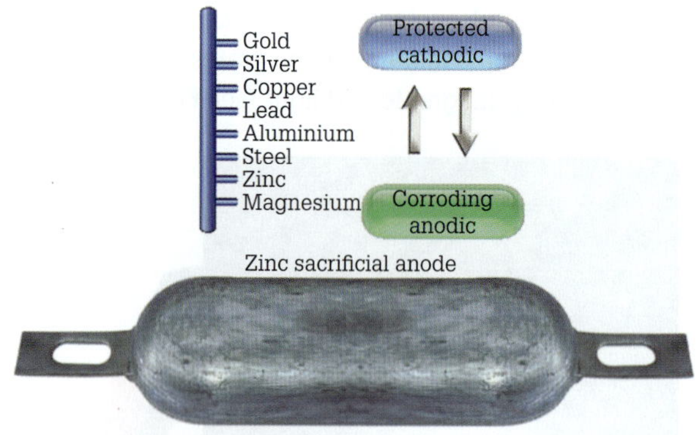

FIGURE 2.84 Electrochemical series of metals and a zinc sacrificial anode

Gas and oil come to us through pipes that are inclined to rust, but are protected by sacrificial anodes. The sacrificial anode oxidises away, thereby protecting the cathode metal. The protection of an oil pipeline with a magnesium sacrificial anode also works in this way. The anode corrodes, but the steel pipeline will not.

REVIEW QUESTIONS

1 What two effects of an electric current are always present when current flows through a conductive medium?
2 What do the terms 'direct' and 'indirect' contact mean?
3 Name the three categories of electrical injuries.
4 Why is 50 Hz current hazardous?
5 List three fundamental principles for protection against the physiological effects of current.
6 What is meant by the term 'grip current'?
7 How can the magnetic effect of an electric current be observed?
8 What is an undesirable consequence of the magnetic effect of an electric current?
9 What is an undesirable consequence of the heating effect of an electric current?
10 What was one of the first practical uses of electricity?
11 Which electrode dissolves in the process of electrolysis?
12 Name the three types of environment in which corrosion can occur.
13 What are the three main processes involved in corrosion?
14 Describe one system of cathodic protection.

2.6 EMF sources, energy sources and conversion

Energy losses and efficiency

Whenever a machine transforms energy from one form to another, there is always an energy loss. The energy loss takes place in the machine itself, causing:

- an increase in temperature
- a reduction in efficiency.

Electrical and mechanical energy losses develop in rotating machines such as electric motors while only electrical losses develop in stationary machines such as transformers.

Figure 2.85 shows that 3% of energy supplied to the machine occurs as various energy losses. Only 97% of the original energy turns the rotor.

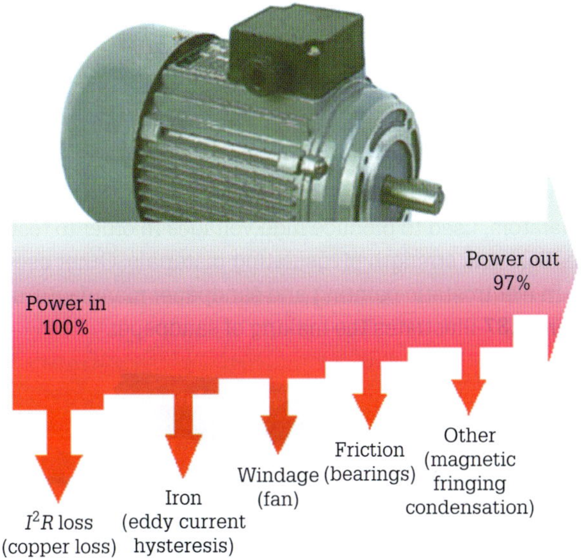

FIGURE 2.85 Power losses in an electric motor

Mechanical losses in an electric motor include bearing friction, brush friction and windage. It is evident that the friction losses depend upon the motor's speed. Bearing design and other friction devices such as brushes, commutator and slip rings add to friction losses. Windage losses occur because of the design and speed of the cooling fan. Turbulence produced by the rotating parts adds to the windage losses. Moisture condensation occurs when the motor is not operating and energy has to be expended to remove the moisture during motor operation. Electrical losses (I^2R) are due to current heating conductors while electromagnetic losses include magnetic fringing, hysteresis loss, and air gap electromagnetic losses between the rotor and stator.

Mechanical efficiency

Mechanical efficiency (η) is the ratio of the useful work expended (power output) to the power applied (power input) expressed as a percentage. Power output equals the power input less the sum of the power losses.

$$\eta = \frac{P_{out}}{P_{in}}$$

where η = efficiency expressed as a percentage
to express η as a percentage multiply the decimal value calculated by 100
P_{out} = output power in watts (W)
P_{in} = input power in watts (W)

Machine design takes into account the effect of friction. Consideration of the effect is necessary because work done against friction converts into heat energy and is wasted. The more work wasted – the less efficient is the machine.

The difference between input and output power defines the energy losses. The following equation shows variations of the basic efficiency equation.

We know that input power equals output power plus the sum of total losses:

$$P_{in} = P_{out} + P_{loss}$$

where P_{in} = input power in watts (W)
P_{out} = output power in watts (W)
P_{loss} = total power loss in watts (W)

therefore,

$$\eta = \frac{P_{in} - P_{loss}}{P_{in}}$$

or

$$\eta = \frac{P_{out}}{P_{out} + P_{loss}}$$

EXAMPLE 2.20

An electric motor rated at 150 kW has energy losses of 7.5 kW. Calculate the efficiency of the motor.

Note: The kW rating of a motor always refers to the mechanical power output of the motor.

$$\eta = \frac{P_{out}}{P_{out} + P_{loss}}$$

$$= \frac{150}{150+7.5}$$

$= 0.952$ or 95.2% (3 significant figures)

With mechanical efficiency this percentage is expressed as a number less than one, that is, 95.2% = 0.952. In practice, the efficiency of a machine is always less than one. The 0.048 (4.8%) difference in Example 2.20 represents the percentage power loss through the effects of condensation, friction and other factors.

EXAMPLE 2.21

A distribution transformer as shown in Figure 2.86 supplies a power output of 15 kW to the load. If the losses equal 1.5 kW, what is the efficiency of the transformer?

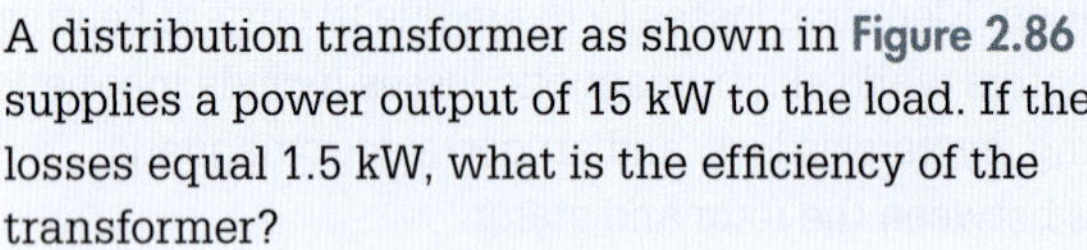

$$\eta = \frac{P_{out}}{P_{out} + P_{loss}}$$

$$= \frac{15}{15+1.5}$$

$= 0.909$ or 90.9% (3 significant figures)

FIGURE 2.86 Efficiency of a transformer

EXERCISE 2.8

a A d.c. generator is supplying 200 kW to a load. The prime mover is providing 220 kW to the generator. What is the efficiency of the generator?

b What are the total losses in this system?

Producing electrical energy

There are six primary methods of producing electrical energy as shown in Table 2.4.

TABLE 2.4 Primary methods of producing electrical energy

1 Friction	Energy produced by rubbing two materials together
2 Heat	Energy produced by heating the junction where two unlike metals are joined
3 Radiation	Energy produced by light (radiant energy) being absorbed by photoelectric cells
4 Chemical	Energy produced by chemical reaction in a voltaic cell
5 Pressure	Energy produced by compressing or decompressing specific crystals
6 Magnetism	Energy generated in a conductor passing through magnetic lines of force or the magnetic lines of force passing through the conductor.

Friction

If a cloth rubs an object, the object will display an effect called friction electricity. In fact, the object charges due to the rubbing process and is said to possess an electric charge.

Normally, static energy is an annoyance. For instance, a person walking across a dry carpet could maintain 35 000 electrostatic volts which could discharge when making contact with another object.

With aircraft, an electrostatic charge occurs due to the friction between the metal envelope of the aircraft and the passing air. This charge can interfere with radio communications. Static energy, though, does have several applications. Its main application is in Van de Graaff generators, used to produce high voltages in order to test the dielectric strength of insulating materials. Other uses are in electrostatic painting and sandpaper manufacturing. Figure 2.87 illustrates the making of sandpaper.

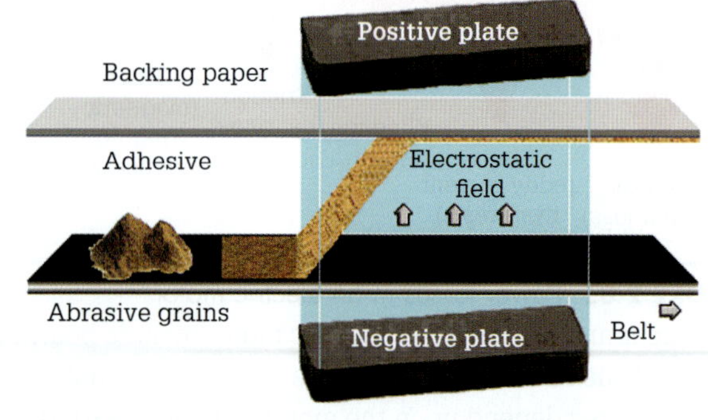

FIGURE 2.87 Making sandpaper by static means

The abrasive grains acquire a negative charge as they move across the negative plate. As unlike charges attract, the positive plate attracts the abrasive grains and their impact velocity enables them to be embedded into the adhesive.

Heat

In 1821, Thomas Seebeck discovered that the junction between two metals generates a voltage that is a function of temperature. So, if a closed circuit consists of conductors of two different metals, and if one junction of the two metals is at a higher temperature, an electromotive force develops with a particular polarity.

For example, in the case of copper and iron the electrons first flow along the iron from the hot junction to the cold one. In addition, electrons cross from the iron to the copper at the hot junction, and from the copper to the iron at the cold junction. The Seebeck effect has application in thermometry, namely the thermocouple.

Almost any two different types of metals make a thermocouple. However, a number of standard types are common because they maintain predictable output voltages and large temperature gradients. The ideal thermocouple consists of a pair of unbroken, uniform metal conductors of dissimilar material, joined together (forming a junction) by welding at one end. This end becomes the hot, or sensing, junction while the other end remains in the cold 'reference temperature' region (T_{ref}).

When maintained at a fixed cold temperature such as zero degrees Celsius the sensing junction becomes a probe to measure temperature.

Since the wire junctions are tiny, the thermometer measures temperatures at any selected point on various devices. The tiny voltage signal leaves the reference temperature region via copper conductors to a thermocouple thermometer at room temperature. Finally, the thermocouple thermometer measures the voltage and the temperature displayed as shown in **Figure 2.88**.

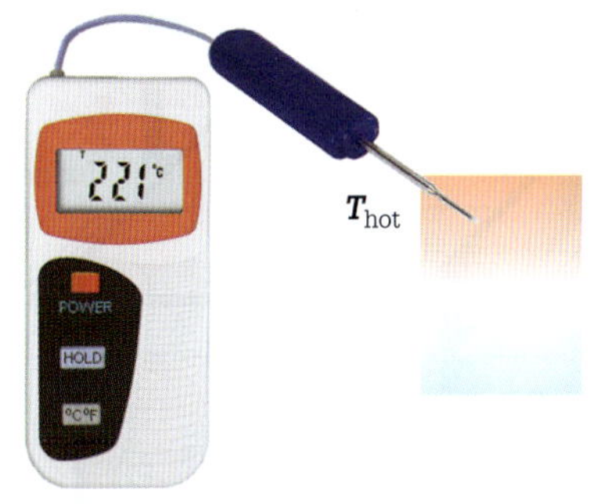

FIGURE 2.88 Thermocouple thermometer

Thermocouples are available as bare-wire bead thermocouples or as probes. The two most common thermocouple alloys for moderate temperatures are:

- **Iron–constantan** Its temperature range is 0–760 °C and it generates about 50 µV/°C.
- **Chromel–alumel** Chromel–alumel generates 40 µV/°C. The alumel wire is also magnetic.

Another material used is copper–constantan alloy.

Applications

Thermocouples sense the heat of furnaces and apply the voltage generated to a measuring device to display the voltage signal in terms of temperature. Thermocouples are also used as a detector for the pilot lights of gas water heaters. Referring to **Figure 2.89**, once the shut-off valve is not receiving a signal, the valve operates and turns off the gas.

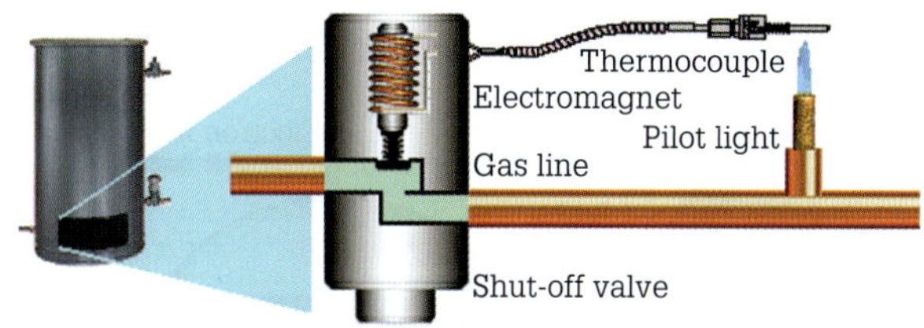

FIGURE 2.89 Water heater application for a thermocouple

Radiation

The sun's rays produce photon energy. The direct user of photon energy is the solar cell, or photovoltaic cell. A solar cell is formed by a light-sensitive P–N junction semiconductor, which, when exposed to sunlight, is excited to conduction by the photons in light as shown in **Figure 2.90**.

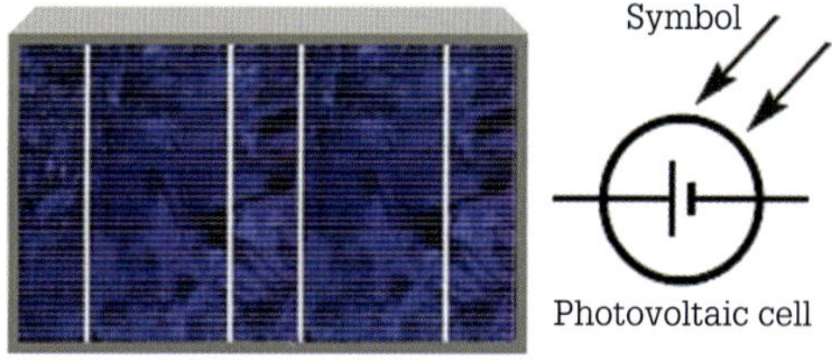

FIGURE 2.90 Solar cell

Silicon cells

Silicon, boron and phosphorus are used to make a silicon cell. Silicon atoms have four electrons in the outer shell, while phosphorus atoms have five electrons and boron atoms have three electrons. In addition, mixing boron atoms with silicon atoms creates a P-type material. However, adding phosphorus to silicon creates an N-type material. By joining the materials, a P–N junction diode forms. At the moment of joining, some of the electrons from the phosphorus-doped material cross over and fill the 'holes' present in the boron-doped layer, creating a depletion zone or barrier. The N-type material becomes positively charged, and the P-type material gains a negative charge. Eventually, balance is reached, and we have an electric field (0.55 eV to 0.66 eV) separating the N-type and P-type material. The consequence is that the displacement of electrons creates a fixed potential barrier or electrostatic field. The internal electrostatic field makes the solar cell work. In addition, the field acts as a diode, allowing electrons to flow from the N-type material to the P-type material, but not the other way around.

Solar cells

Solar cells are P–N junctions used in the reverse way. When light, in the form of photons, hits the cell and strikes an atom, photo-ionisation creates electron–hole pairs. The electrostatic field causes separation of these pairs, establishing an electromotive force in the process. As the electrons are concentrated in the N-type side of the junction and the holes are concentrated in the P-type side of the junction, the P-N junction behaves as a small battery. Consequently, if an external current path exists electrical energy is available to do work, as shown in **Figure 2.91**.

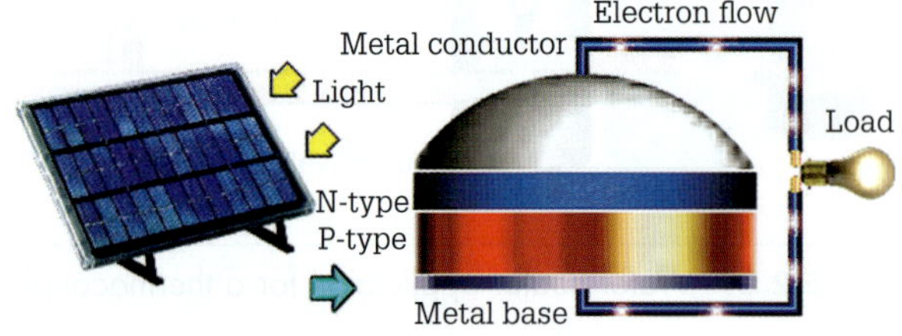

FIGURE 2.91 Solar cell

The electron flow provides the current, and the cell's electric field creates the voltage. The greater the amount of light falling on the cell's surface, the greater is the probability of photons releasing electrons and hence more electric energy is produced.

Forms of solar cells

Single-crystal silicon is not the only material used in photovoltaic cells. Other forms include polycrystalline silicon and amorphous silicon. Polycrystalline silicon consists of a myriad of microscopic crystals, while amorphous silicon has no crystal properties and its individual atoms bond together like glass. Other materials include cadmium sulphide, gallium arsenide and cadmium telluride. These materials have different barrier or depletion zones that respond to different wavelengths, or photons of different energies. **Figure 2.92** shows examples of different solar cells.

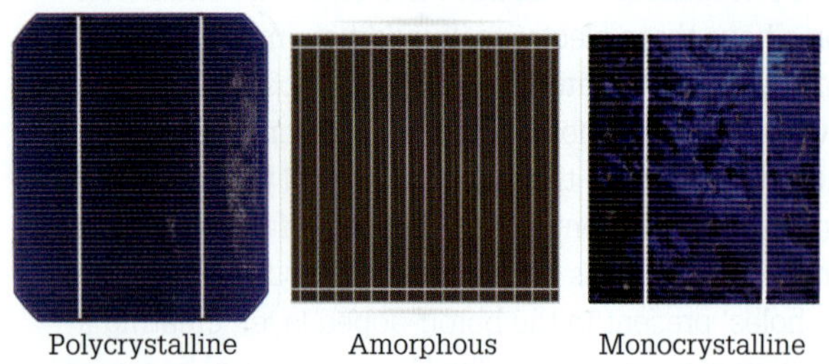

FIGURE 2.92 Forms of solar cells

Applications

Some solar panels, as shown in **Figure 2.93**, charge storage batteries in satellites, yachts and in locations where remote telecommunication installations exist. However, many solar panels using inverters feed energy into the grid.

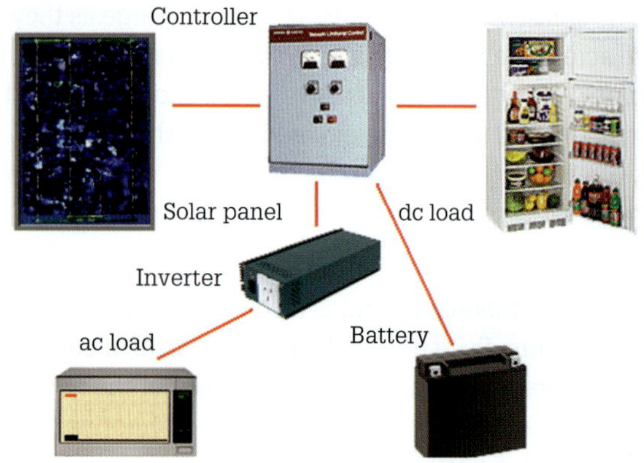

FIGURE 2.93 Solar panel and connected devices

Chemical

When a zinc electrode and copper electrode enter a dilute solution of sulphuric acid, the two metals react within the electrolyte and develop a potential difference of about 1 volt between them. Furthermore, when a conducting path joins the electrodes externally, the zinc electrode dissolves slowly into the acid electrolyte. In addition, the zinc goes into the electrolyte in the form of positive ions while its electrons remain on the electrode.

The copper electrode, on the other hand, does not dissolve in the electrolyte. Instead, it gives up its electrons to the positively charged ions of hydrogen in the electrolyte, turning them into molecules of hydrogen gas that bubble up around the electrode. The zinc ion combines with the sulphate ion to form zinc sulphate, and this salt falls to the bottom of the cell.

The effect of all this is that the dissolving zinc electrode becomes negatively charged, the copper electrode is left with a positive charge and electrons from the zinc pass through the external circuit to the copper electrode as shown in **Figure 2.94**.

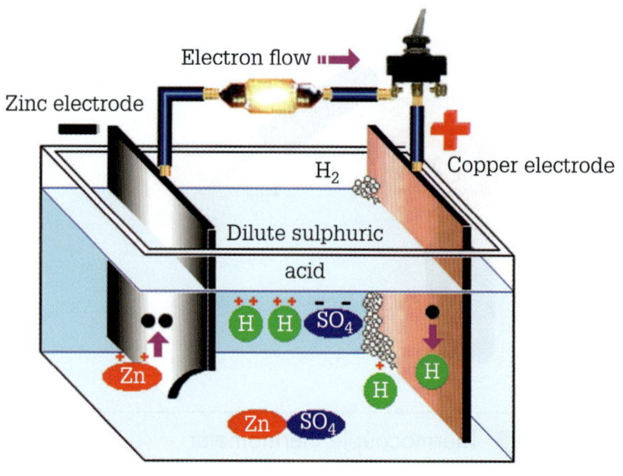

FIGURE 2.94 Voltaic cell

The dry cell

Instead of an acid electrolyte, a dry cell uses ammonium chloride in a jelly structure, so that the cell does not have to be vertical. A dry cell consists of a carbon rod that serves as the positive terminal and a zinc case that acts as the negative terminal. The voltage generated between the terminals is approximately 1.5 V.

Contained within the zinc case is the ammonium chloride together with a manganese dioxide depolariser mixed with carbon to improve its conducting ability. Refer to **Figure 2.95**.

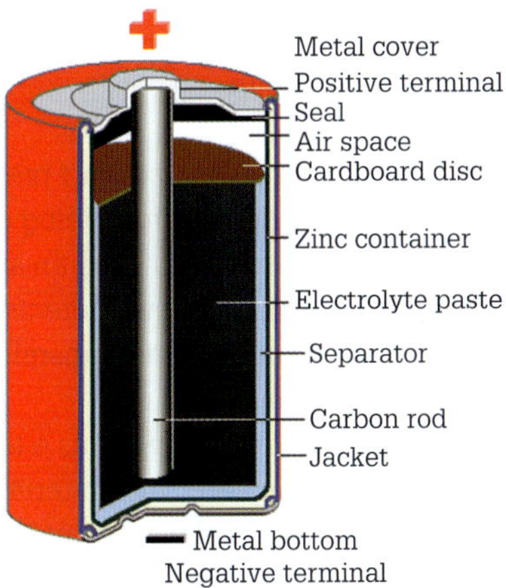

FIGURE 2.95 A carbon-zinc dry cell

The voltaic cell and the dry cell are primary cells. The chemical reactions within primary cells that cause production of electrical energy are not reversible because one electrode reduces.

However, secondary cells must charge before use by passing a current through them, and recharging can occur many times. Secondary cells are also known as accumulators or storage cells. Refer to **Figure 2.96**. **Note:** Some primary and secondary cells resemble each other in appearance only.

FIGURE 2.96 Examples of rechargeable 1.5 V, 9 V and 12 V secondary cells

Secondary cells provide a way of storing energy in a form that allows access to the electrical energy when needed. In fact, a grouping of cells produces the required voltage when connected together.

The fuel cell

Fuel cells are electrochemical devices similar to a battery. They have two electrodes, an anode catalyst and a cathode catalyst, separated by an electrolyte. However, these cells generate electrical energy, heat and pure water as long as fuel in the form of hydrogen and oxygen as the oxidant are used. At present, there are five types of fuel cells and the type of electrolyte categorises them. They are:

1. polymer
2. alkaline
3. solid oxide
4. phosphoric acid
5. molten carbonate.

An illustration of a polymer fuel cell is shown in **Figure 2.97**.

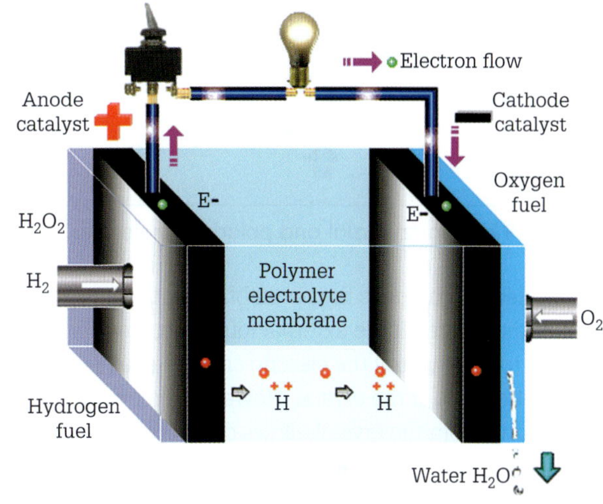

FIGURE 2.97 Polymer fuel cell

The electrolyte consists of a polymer foil coated on both sides with a catalyst containing platinum. The electrolyte provides only positively charged hydrogen ions (protons) to move between the anode and the cathode. This transference of the ions allows the ionisation of the hydrogen fuel and the reaction of the hydrogen ions with the oxygen oxidant. There are excess hydrogen electrons on the anode and an electron shortage on the oxygen cathode. When an external circuit occurs the hydrogen electrons complete the circuit at the cathode. They recombine with the hydrogen ions and the oxygen to form water. In addition, by regulating the hydrogen fuel the energy output from the cell is controlled.

Fuel cells demonstrate high-conversion efficiencies in relation to fuel input and electrical energy output when compared to other electrical energy generation systems. They have found applications in space exploration since the 1960s and have the potential to power motor vehicles and deliver emergency power.

Pressure

The molecules of some crystals and ceramics have the property of being permanently polarised. Permanently polarised means that some parts of the molecule become positively charged, while other parts are negatively charged. These materials produce an electric charge when the material changes dimension, as a result of an imposed external force. The charge produced is piezoelectricity. Many crystalline materials such as the natural crystals of quartz and Rochelle salted together with manufactured polycrystalline ceramics such as lead titanate zirconate and barium titanate exhibit piezoelectric effects.

A wafer of quartz cut from a crystal and shaped to the correct size constitutes a piezo element. Electrodes of silver or gold are added to opposite faces of the wafer to enable leads to be connected. When an external force compresses a quartz crystal, a displaced electric charge accumulates on the opposite faces of the crystal as shown in **Figure 2.98**.

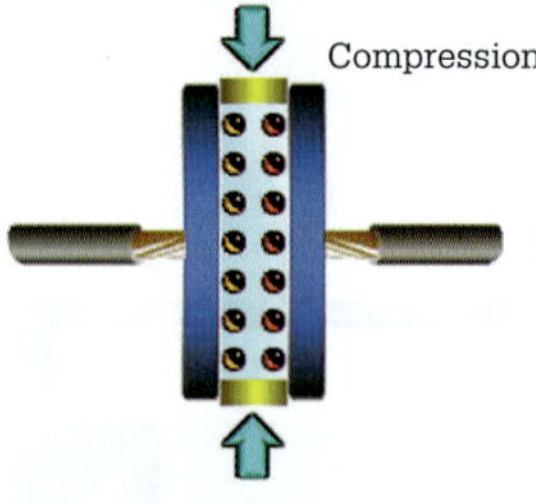

FIGURE 2.98 Quartz crystal and polarised molecules

This electric charge is the result of deflection by the lattice of the crystal. If the external force becomes constant, then electrons flow until the electric charges equalise on the opposite faces of the crystal. However, removing the external force from the crystal allows decompression and provides a small amount of electric energy to flow in the opposite direction as shown in **Figure 2.99**.

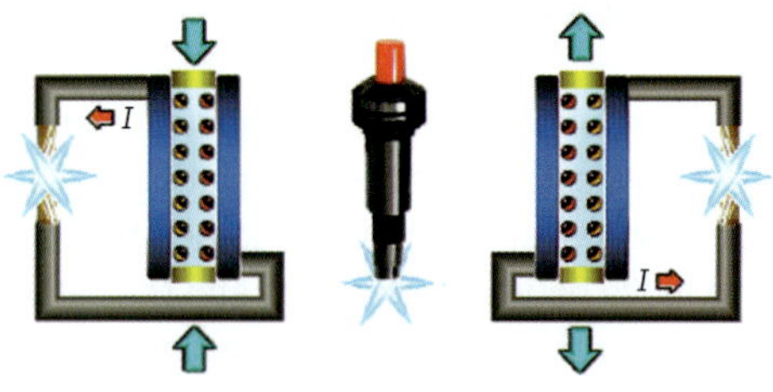

FIGURE 2.99 Compression and decompression of a quartz crystal

When squeezing particular types of crystals along specified directions, they develop an electric charge. This charge is proportional to the piezoelectric constant of the crystal and to the applied force. However, the dimensions of a piezo element change when a voltage or an applied electric field stresses a piezo element electrically. This phenomenon, which causes the element to vibrate, is electrostriction, or the reverse piezoelectric effect. **Figure 2.100** shows the flexural, shear and compression forms of a transducer.

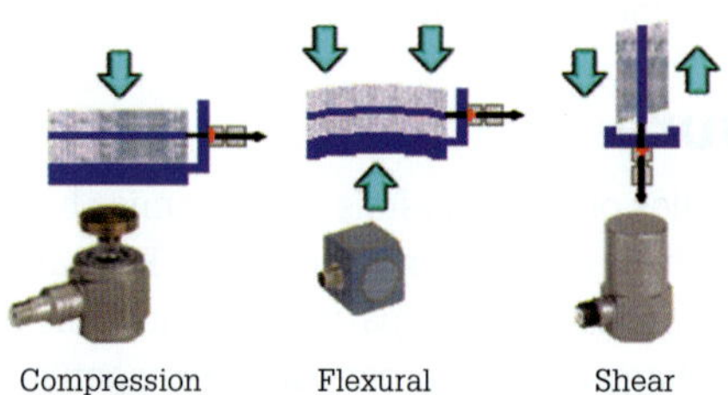

FIGURE 2.100 Compression, flexural and shear applications

Applications

Piezoelectric devices include buzzers inside pagers, ultrasonic cleaners and mobile phones, and in gas igniters as illustrated in **Figure 2.101**.

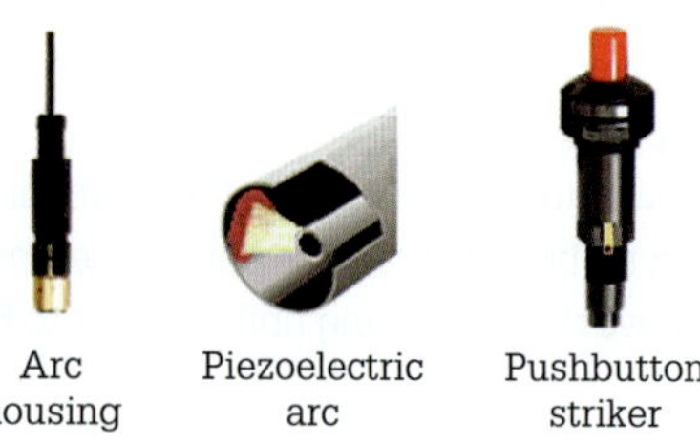

FIGURE 2.101 Piezo gas lighter

Piezoelectric sensors are able to convert pressure, force, vibration or shock into electrical energy. Being capable only of measuring active events, they have an application in flow meters, accelerometers and level detectors; and in motor vehicles to sense changes in the transmission, fuel injection and coolant pressure.

This effect enables the element to act as a translating device called an actuator. Piezoelectric materials used in power actuators convert electrical energy into mechanical energy, and in acoustic transducers the piezo converts electric fields into sound waves.

Telephones, microphones and musical instruments such as guitars use transducers. In addition, because the wafer of crystal has a resonant frequency dependent upon its thickness and metal electrodes, the piezo element makes an accurate clock.

Magnetism

The most useful and widely employed application of magnetism is in the production of electrical energy. Different sources provide the mechanical power needed to assist in this production.

These sources, called prime movers, include diesel, petrol or natural gas engines. Coal, oil, natural gas, biomass and nuclear energy are energy sources that are used to heat water to produce superheated steam.

Non-mechanical prime movers include water, steam, wind, wave motion and tidal current. Generators carry out the conversion of the energy sources into electric energy. However, in order to do this, three fundamental conditions – movement, conductors and a magnetic field – must exist before a voltage develops.

In accordance with these conditions, when the conductor or conductors move through a magnetic field cutting the lines of force, electrons enter the conduction band thereby developing an electric pressure for the production of alternating current in an external circuit as illustrated in **Figure 2.102**.

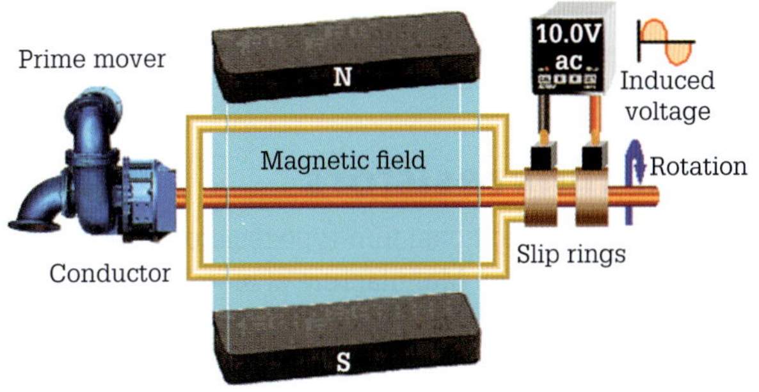

FIGURE 2.102 Alternator

This elementary alternator consists of a single wire loop called an armature with each end attached to slip rings and arranged to rotate midway between the magnetic poles. Two copper graphite brushes, which connect to the external circuit, make contact with the slip rings in order to collect the alternating current generated in the conductor when the alternator is in operation.

Another machine used for converting mechanical energy into electrical energy by means of electromagnetic induction is a dynamo or direct current generator, as illustrated in **Figure 2.103**.

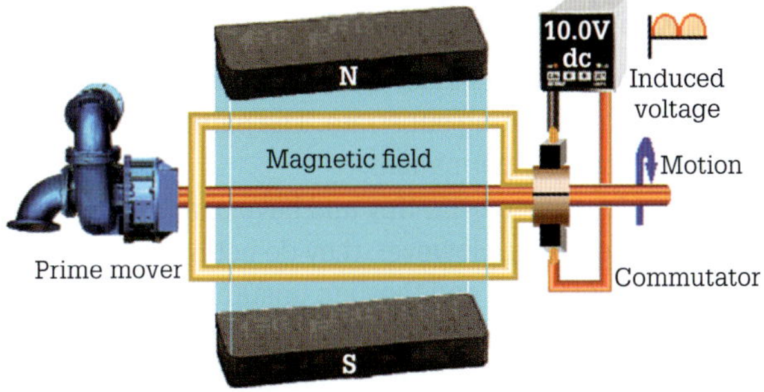

FIGURE 2.103 Generator

The fundamental difference between an alternator and a generator is that the alternator produces a.c. to the external circuit while the generator delivers d.c. In both machines, induced alternating current transpires in the armature, but the type of current delivered to an external circuit depends on the way in which the induced current is collected. In an alternator, the current feeds into an external circuit by means of brushes bearing against slip rings; in a generator, a form of rotating switch called the commutator is placed between the armature and the external circuits. The design of the commutator allows reversal of the connections with the external circuit at the instant of each reversal of induced current in the armature, producing rectified current or d.c. This rectified current is not pure like the current of a dry or secondary cell but a pulsating current constant in direction and uniformly varying in intensity.

Wind generation

This system utilises the power of the wind to drive the generator that provides power to the controller that feeds power to the various loads connected to the system as in **Figure 2.104**.

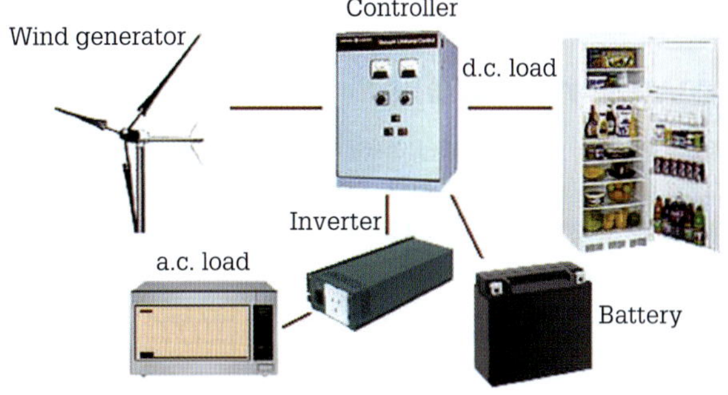

FIGURE 2.104 Wind power system

Wave and tidal generation

This system takes advantage of the oscillating effect of the ocean's movement. Wave action and tidal waters flowing in and out cause the turbo alternator to revolve and generate electrical energy as shown in **Figure 2.105**.

FIGURE 2.105 Wave and tidal generation

Hydroelectric power works on the same principle except that there is a constant flow of water in the same direction.

REVIEW QUESTIONS

1. What are two effects of the energy loss that takes place in machines?
2. Calculate the power loss in a 60 kW motor that has an input of 67 kW.
3. Calculate the efficiency of an 80 kW motor that has an input of 86 kW.
4. Name the six main methods of producing electricity.
5. What device utilises the Seebeck effect?
6. Name the device that produces electrical energy from radiant energy such as sunlight.
7. Name two categories of chemical cells.
8. Name the cell that produces electrical energy, heat and water.
9. What quantity is converted into electrical energy by a piezo-electric device?
10. What three fundamental conditions must exist in order to produce electrical energy using magnetism?

2.7 Resistors

Resistor types

There are various types of resistors. Resistors vary in resistance, their power rating in watts and the material of the resistive element. Resistors are designated as 'power' or 'precision'. Precision resistors have applications where exact resistance values and stability are fundamental considerations. They have restricted operating temperature limits and power dissipation ratings. Power resistors can also have exact resistance values and be quite stable, but their design emphasis is to optimise power dissipation.

Power resistors

Resistors, also known as 'power resistors', are ideal for severe-duty applications and they begin at 10 W and extend to megawatts. They have to be able to transfer large amounts of energy to their environment each second, and in order to achieve this, their temperature may reach 400 °C. **Figure 2.106** shows a grid resistor used for severe duty.

FIGURE 2.106 Grid resistor

Typical applications include load banks, primary resistance starters and secondary resistance starters, dynamic braking and plugging (braking of direct current motors), field discharge, harmonic filters and neutral earthing resistors.

Carbon resistors

The making of carbon resistors occurs by attaching connecting leads to the rod of a carbon composition material or by grinding a spiral path through pure crystalline carbon down to the ceramic base, as with carbon-film-type resistors. Film technology products have now replaced carbon composition resistors. Carbon film resistors have consistency and stability and their application is in the electronic, electrical and information industries. See **Figure 2.107** for the carbon resistor and the general symbol for a resistor.

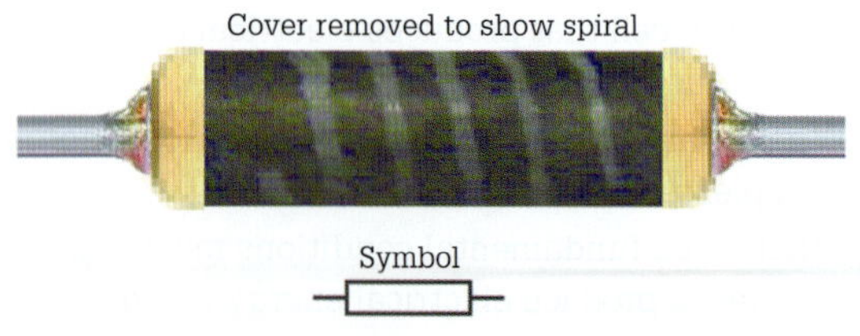

FIGURE 2.107 Carbon-film resistor and circuit symbol for fixed resistors

Metal-film resistors

There are two conventional film-type resistors, thick film and metal or thin film. Manufacture of metal-film resistors occurs using a vacuum sputtering system that creates multiple layers of mixed metals and passivate materials (glass) onto a ceramic substrate. Finally, the resistors have layers of coloured lacquer coated on them. In addition, these resistors can use metal films or oxides. Film-type resistors are small and demonstrate safe tolerances, but they become unstable at high-power levels.

Another type of film resistor uses conductive inks. They are unstable with limited power dissipation and poor resistance tolerances. See **Figure 2.108** for the metal-film resistor.

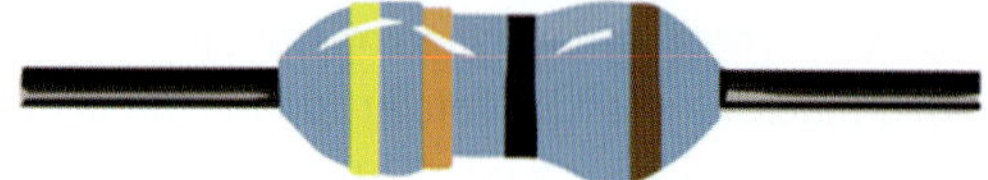

FIGURE 2.108 Metal-film resistor

Thick-film resistors

Thick-film resistors are made by silk-screening conductive paint onto an insulating substrate. The paint is 'fired' to make the assembly permanent.

Thick-film resistors are tiny and have characteristics like carbon resistors. However, they do not have tight resistance tolerances, and they are unstable. Thick-film technology is especially important in hybrid and integrated circuits because resistors can be 'printed' on to the substrate, eliminating board loading and soldering steps.

Tapped resistor

A tapped resistor as illustrated in **Figure 2.109** has two or more fixed taps that provide different resistance values.

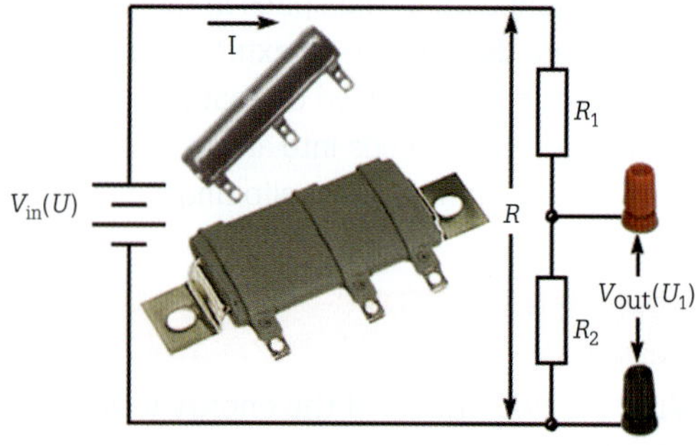

FIGURE 2.109 Tapped resistor

Many tapped resistors have application as voltage or potential dividers. The potential V_{out} at the output terminals will depend on the ratio of R_2 to the total resistance (R) and the current drawn by the load. Provided that the output current is very small, then:

$$V_{out} = V_{in} \times \frac{R_2}{R_1 + R_2}$$

This method of connection provides a variable-voltage source for low current loads that do not vary.

Wire-wound resistors

Winding a length of high-resistance wire on a heat-resisting ceramic insulating core makes wire-wound resistors. They can consume large power levels compared to other types of resistors. Wire-wound resistors demonstrate excellent resistance tolerances and controlled temperature characteristics. The precision resistance of wire-wound resistors determines their applications.

Wire-wound resistors have their limitations when used in high-frequency circuits. The limitation occurs due to their inductive and capacitive effects, and they are unsuited for use above 50 kHz even when specially wound. More to the point, wire-wounds usually exhibit an increase in resistance with high frequencies due to the 'skin' effect. The skin effect occurs at high frequencies of electron movement because the moving electrons travel close to the outer surface of the conductor.

Abrupt current or voltage surges are often hazardous to electronic products. In addition, transients or surges that enter microprocessor components can destroy those parts, and wipe out critical programming and data. Consequently, many electronic devices now include surge protection circuits. In these circuits the change in voltage drop across a low-value resistor, when the current changes due to a surge, can be used to control the power supply and prevent damage occurring.

Wire-wound resistors are a popular choice for these applications. They are inexpensive, available in low-resistance values, have low temperature coefficients and good surge-handling capability. Power supplies are a major application for wire-wound resistors, and they find use in miniature d.c.-to-d.c. converters, small battery-charging circuits, switching power supplies and uninterruptible power systems. Surge-resistor networks are necessary in telecommunication applications where voltage surges on telephone lines are perceived. Other applications for surge resistors include computers and peripherals, disk drives, microprocessor-based instrumentation, medical equipment, metering devices and ballasts. **Figure 2.110** shows two types of wire-wound resistors. The one on the top can easily dissipate up to 5 W of heat while the other can dissipate many times that amount.

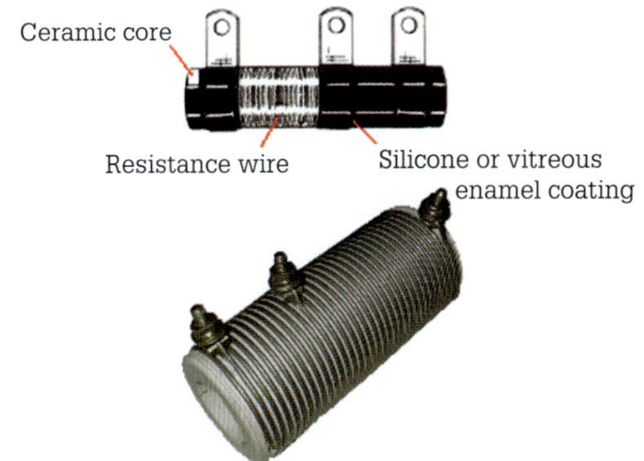

FIGURE 2.110 Wire-wound resistors

Variable resistors

Sometimes it is necessary to be able to vary the resistance in an electrical circuit in order to vary the speed of a fan or dim a light. A specific type of resistor, known as a variable resistor is ideal for these applications. Variable resistors are generally three-terminal devices that have a fixed value of resistance between the two outer terminals and a variable resistance between either of the outer terminals and a central moveable contact. **Figure 2.111** shows the terminal arrangement for this type of resistor along with its circuit symbol. The central terminal (T2) connects to a moving contact called the wiper. By moving the wiper in a clockwise direction, the resistance between T1 and the wiper increases while the resistance between the wiper and T3 decreases. These types of variable resistor may be classified as potentiometer – for varying potential difference or rheostat for varying current. When using potentiometers note the terminal arrangement and in which rotational direction the shaft turns to increase or decrease the resistance. **Figure 2.112** shows that the variable resistor is a potentiometer or rheostat.

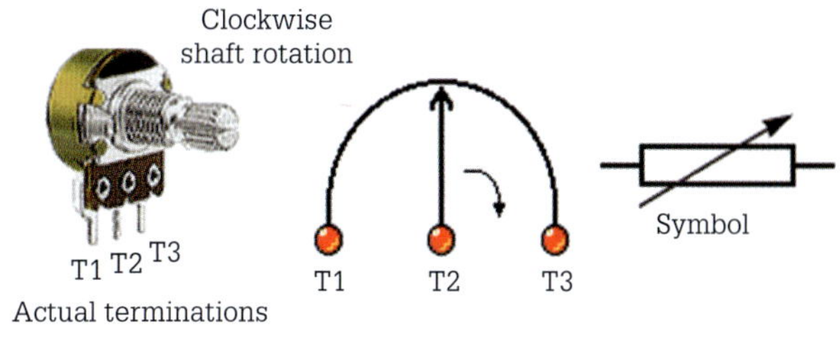

FIGURE 2.111 Potentiometer and circuit symbol

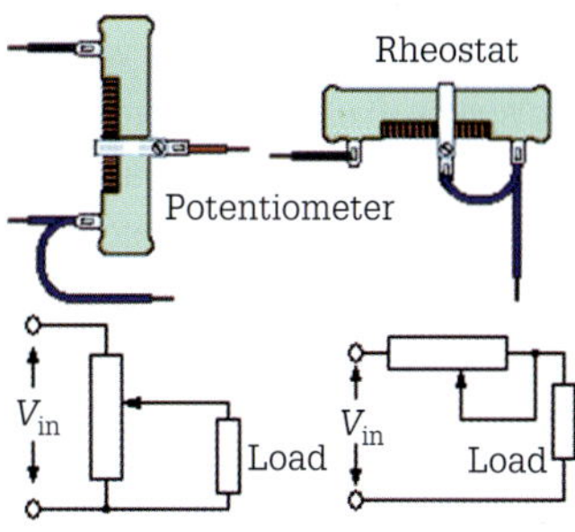

FIGURE 2.112 Potentiometer and rheostat

Potentiometers may be linear – where the resistance changes proportionally with movement or logarithmic – where the resistance changes disproportionately with movement. Logarithmic potentiometers find application in audio amplifiers for volume control. Linear potentiometers are often used for speed or light control.

Figure 2.113 shows a range of variable resistors.

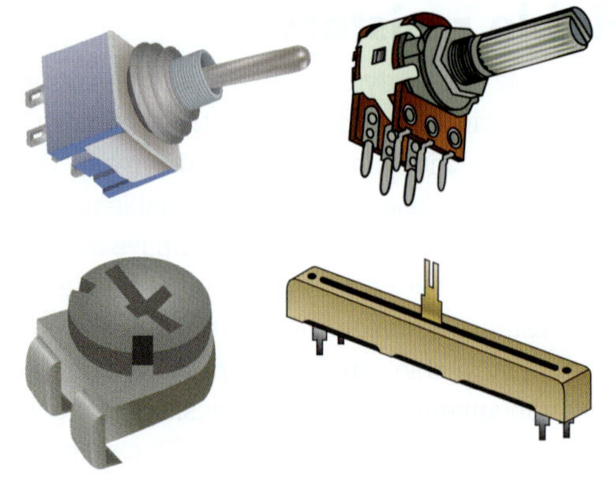

FIGURE 2.113 Variable resistors

Voltage-dependent resistors

A voltage-dependent resistor (VDR), also called a varistor, is a voltage surge protection device that connects in parallel across the a.c. input. Varistors protect various types of electronic devices and semiconductor elements from any switching transients and induced lightning voltage surges. These resistors use metal oxides like zinc oxide or titanium oxide as their resistor element. **Figure 2.114** shows VDRs and their circuit symbol.

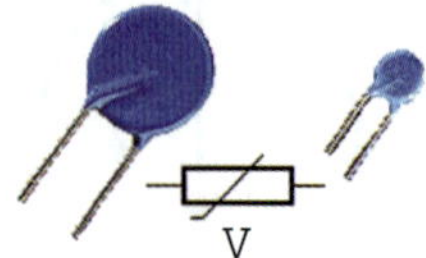

FIGURE 2.114 Metal oxide varistor and its circuit symbol

The zinc oxide varistor suppresses mains transients in domestic and industrial electronics as well as internally generated voltage spikes in electronic circuits. However, the titanium oxide varistor finds application as a spark-reduction device for miniaturised direct current motors and in relay circuits to protect the surface of the make-and-break contacts.

A varistor has a very high resistance at the supply or input voltage, but has a low resistance at the high voltage levels associated with surges that might damage the circuits the varistor is protecting.

The graph in **Figure 2.115** shows the conduction characteristic of a metal oxide varistor (MOV). From this graph it can be seen that voltages above or below a particular magnitude, either positive or negative, cause the varistor to conduct.

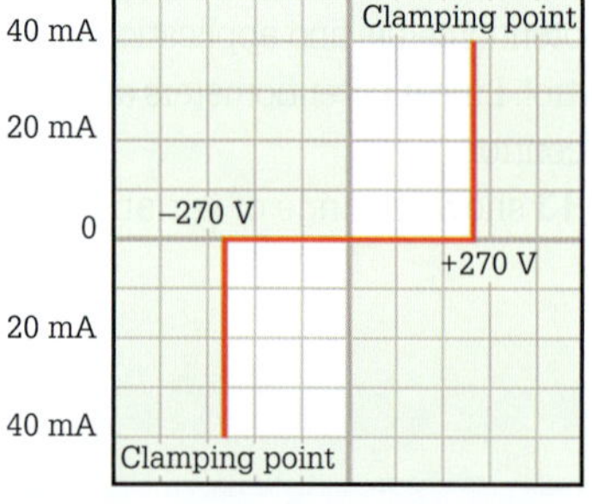

FIGURE 2.115 Metal oxide varistor conduction characteristic

In this case, the voltage surge clamps to 270 V, and when this peak voltage is achieved the MOV resistance changes from near-open circuit to highly conductive.

A VDR has several technical specifications such as voltage, frequency, current rating, efficiency, and duty cycle and noise rejection. However, the two essential specifications are the peak clamping voltage (in the case of 230 V mains the peak clamping voltage is usually 270 V rms) and the energy in joules it can absorb, for example 150 J. Sometimes, instead of energy, the peak current absorption specification; for example, 25000 A, is given. Surges are high-amplitude pulses of only a few millionths of a second in duration.

Surges are too fast for mechanical devices and circuit breakers. Surges can have a disruptive or destructive effect on all types of electrical equipment, especially high-speed, sensitive electronic equipment. Like other forms of energy, surges can change or be redirected.

The varistor's resistance rapidly decreases when a voltage spike or power surge occurs. The effect causes an instant shunt path for the over-voltage. This effect prevents the majority of the surge (a small remnant surge may enter) entering the protected circuit (see **Figure 2.116**). If the voltage surge is of sufficient time duration, a large current flowing in the VDR will cause the circuit-protection device to operate and thereby isolate the circuit from the supply.

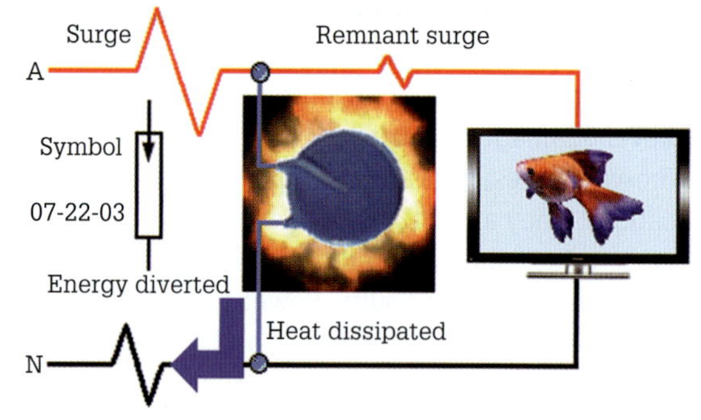

FIGURE 2.116 A VDR protecting a computer monitor from a voltage surge

Lightning arrestors

Lightning can distress a.c. power systems in two ways, either by direct or indirect strikes. A direct strike hits parts of the power supply system, injecting a very large amount of electrical current and creating voltage surges; while an indirect strike near the power supply system induces moderate currents and voltage in the power line.

Direct lightning strikes are the most destructive events, especially if they hit overhead power lines close to the structure's consumer terminals. When a hit occurs, the mains voltage is raised to many hundreds of thousands of volts, usually causing a flashover of insulation and fires. Furthermore, the lightning current can flow in the power line over long distances and surges have been estimated to reach as high as 50000 A.

Rural and urban areas serviced by overhead power lines are most susceptible to lightning strikes. These surge events occur in nanoseconds (billionths of a second) and events such as lightning cause significant damage immediately. One device used to control the action of lightning voltage surges is a silicon oxide varistor.

Silicon oxide varistor

Two metal electrodes separated by a silicon oxide compound constitute a silicon oxide varistor (SOV). Under normal conditions, the silicon oxide is an excellent insulator preventing line current flowing between the electrodes and the soil.

When an excessively high voltage occurs on the electrodes, the high-energy electrical field ionises the silicon oxide, releasing silicon ions.

Since the silicon ions are conductors, high-energy surge current conducts to ground. When the voltage falls towards normal, the silicon and oxygen elements recombine to form a silicon oxide thereby preventing conduction. Lightning arrestors such as these handle significant surges of up to 50 000 A, passing them harmlessly to earth. **Figure 2.117** shows an example of a lightning arrestor.

FIGURE 2.117 Lightning arrestor

Light-dependent resistors

A light-dependent resistor (LDR) is simply a device made using material such as cadmium sulphide (CdS) or selenium that is resistive and photosensitive. When light enters the semiconductor layer free charge carriers develop allowing current to flow more easily than when the LDR was in darkness.

LDRs have decreasing values of resistance with increasing illumination intensity. While values of over 100 kΩ appear in darkness, values of less than 100 Ω arise with high illumination intensity. **Figure 2.119** shows the characteristic curve for a typical LDR.

The LDR is one part of a voltage divider circuit that provides an output voltage that varies with illumination as shown in **Figure 2.118**.

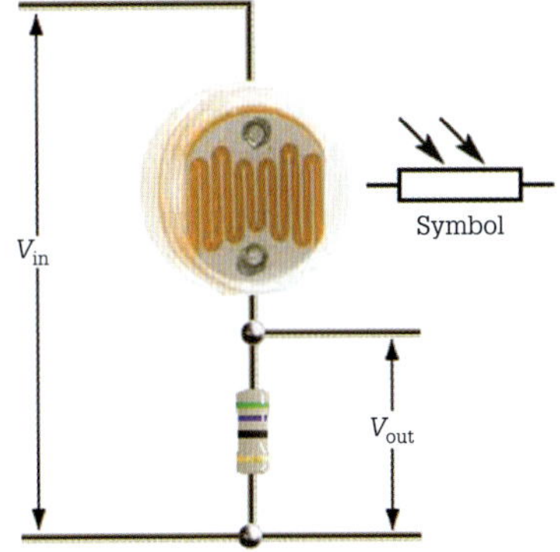

FIGURE 2.118 LDR and its circuit symbol

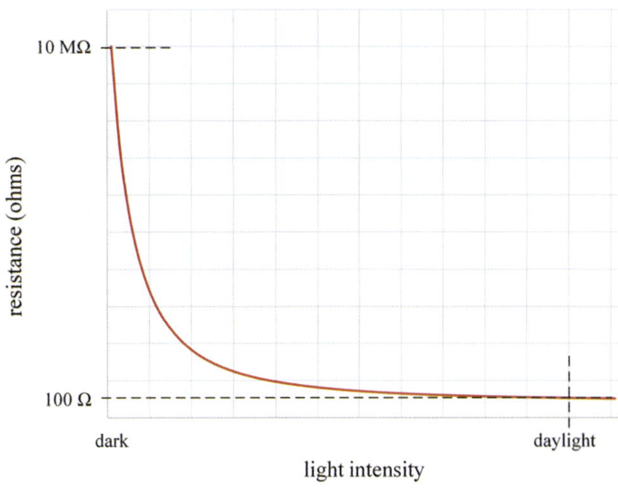

FIGURE 2.119 LDR characteristic

Thermistors

Thermistors use semiconducting metal oxides such as barium, titanium and strontium and are pressed into rods or discs and are fired like ceramics. Their resistance varies rapidly with increasing temperature. The thermistors in **Figure 2.120** are thermally sensitive resistors manufactured on a silicon base and can respond either positively (PTC) or negatively (NTC) to increasing temperature.

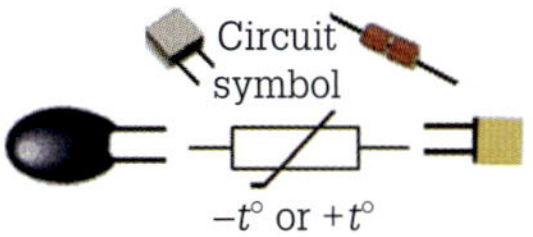

FIGURE 2.120 Thermistors

PTC thermistor – The temperature coefficients of these thermistors are positive only over a certain temperature range selected during manufacture. Most positive temperature coefficient thermistors consist of modified carbon resistors and are not silicon based. The graph in **Figure 2.121** shows the positive range of a PTC thermistor.

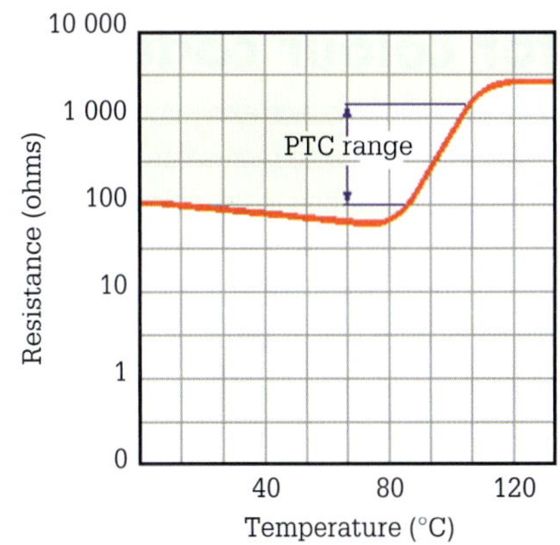

FIGURE 2.121 PTC thermistor characteristic

NTC thermistor – The resistance of negative temperature coefficient thermistors falls quickly with increasing temperature as shown in the graph in **Figure 2.122**.

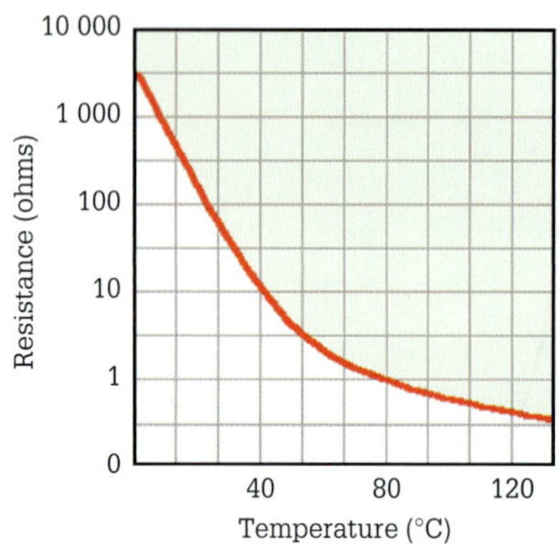

FIGURE 2.122 NTC thermistor characteristic

Applications

The NTC types have application for the suppression of high in-rush currents, such as at the switching on of power supplies, motor drives and audio amplifiers.

PTC thermistors have application in over-temperature and over-current protection devices for various types of motors, transformers and power machinery. As an over-current protection device it is able to switch from a low resistance value to a very high resistance value, reducing the current to a very low and safe level. The PTC types protect the windings of electric motors as shown in **Figure 2.123**. When used for motor winding protection, the thermistor should be placed in the connection end winding overhang and at least 8 mm below the surface of the winding using one thermistor per phase.

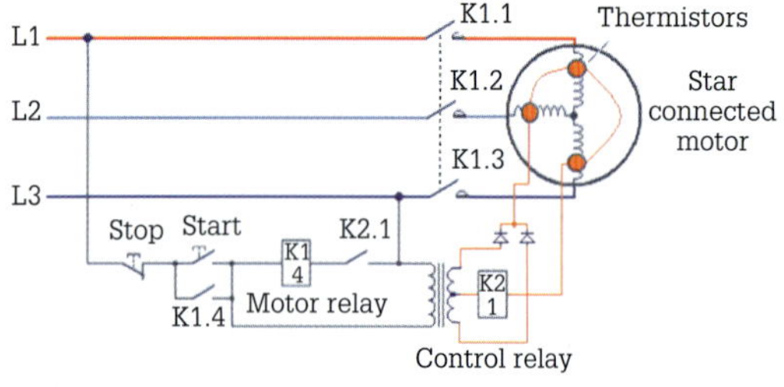

FIGURE 2.123 Three-phase motor protected by thermistors

Resistor colour code

A series of painted coloured bands exist around the body of low-power resistors. Each colour represents a number and reading the colours and referring to the colour code enables determination of the nominal resistance value. General-purpose resistors and precision resistors use the colour code. Resistors above a power capacity of two watts, such as the 20 W resistor illustrated in **Figure 2.124**, have a physical size that enables a recorded resistance value in numbers on the body of the resistor.

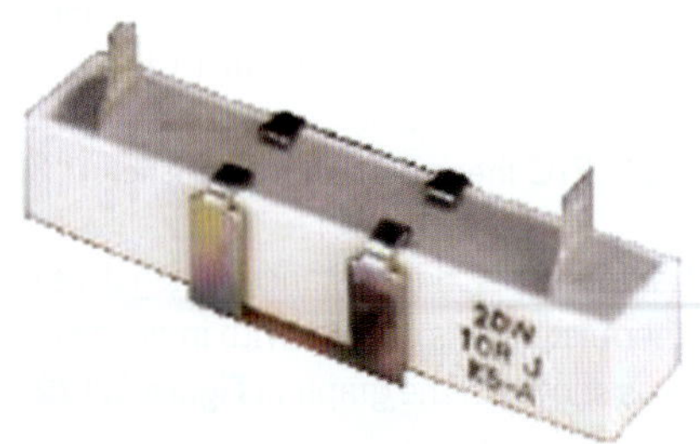

FIGURE 2.124 10 Ω, 20 W power resistor

Figure 2.125 shows the resistor colour code for identifying resistors.

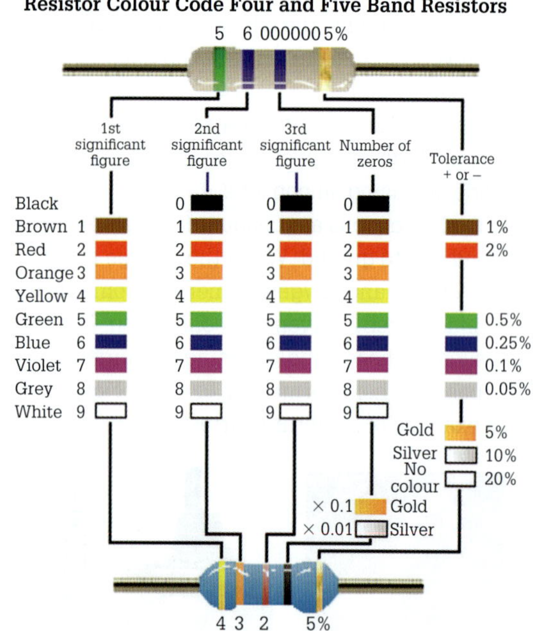

FIGURE 2.125 Resistor colour code

The coloured bands read from left to right with the tolerance band located to the right of the value bands. The first significant figure is never black, silver or gold.

General-purpose resistors have a tolerance of ±5% or higher than their nominal stated value and the resistance values are represented by three colour bands, plus another colour band for their tolerance. Precision resistors have a tolerance of less than ±5% and have their value represented by four colour bands and another band for their tolerance specification. In addition, tolerance bands are always located at a distance from the value-giving colours of the resistor.

Tolerance refers to the accuracy of the resistance. For example, a 100 Ω resistor with 10% tolerance indicates that the actual resistance value exists between 90 Ω and 110 Ω. A 100 Ω resistor with a 1% tolerance indicates that the real resistance could lie between 99 Ω and 101 Ω, making this resistor more precise than the 5% resistor.

In order to remember the colour code, different mnemonics exist. (Mnemonics are a way of remembering in rhyme form information; e.g. 30 days have September, April, June and so on…). In our mnemonic, the first letter of each colour indicates a word in rhyme. In addition, the first letter also symbolises a numerical value as shown in **Figure 2.126**.

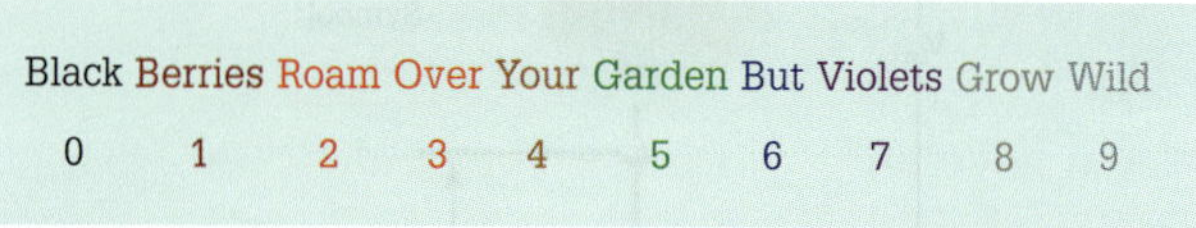

FIGURE 2.126 Resistor mnemonic code

Resistor power ratings

Resistors have a rating according to the amount of power that they can consume without experiencing a loss. This rating is in addition to the resistance rating of the resistor.

Some resistors such as the carbon-film and metal-film types demonstrate their power rating by their physical size. These resistors range in value from 0.25 W up to 1 W. Other resistors, such as wire-wound resistors, have their power rating printed directly on the body of the resistor.

Resistors come in standard values referred to as E ranges. The E value designates the number of resistor values in a decade (each 100 Ω). For example, in **Figure 2.127** there are 12 values between 10 Ω and 100 Ω in the E12 range. In addition, E12 resistors have a tolerance of 10% while the E24 has a 5% tolerance.

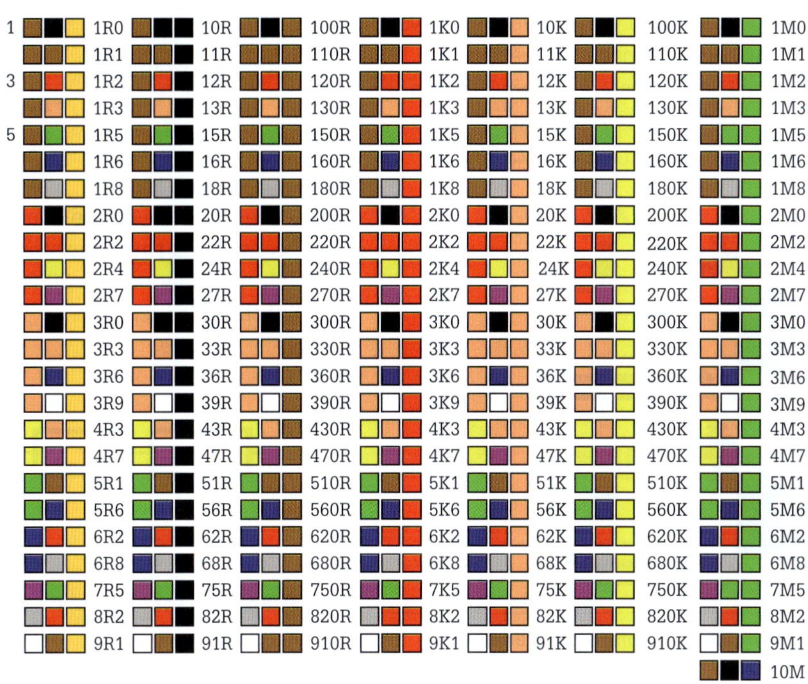

FIGURE 2.127 Colour code for the E12 to E24 range of resistors

REVIEW QUESTIONS

1. Name three applications for power resistors.
2. What are the two main classifications of film-type resistors?
3. What is a common application of a tapped resistor?
4. What limitation is placed on the use of wire-wound resistors?
5. Describe the difference between a potentiometer and a rheostat.
6. What is an MOV?
7. Name one application for a VDR.
8. What is the relationship between resistance and illumination for an LDR?
9. What is a common application for a PTC thermistor?
10. What coloured bands would identify the following resistors?
    - a 150 Ω ± 5%
    - b 120 Ω ± 10%
    - c 22 kΩ ± 2%
    - d 15 kΩ ± 10%
    - e 33 kΩ ± 10%
    - f 5M6 Ω ± 5%
    - g 470 kΩ ± 5%
    - h 1.8 Ω ± 1%
    - i 680 kΩ ± 10%
11. What are the values of the resistors in **Figure 2.128**?

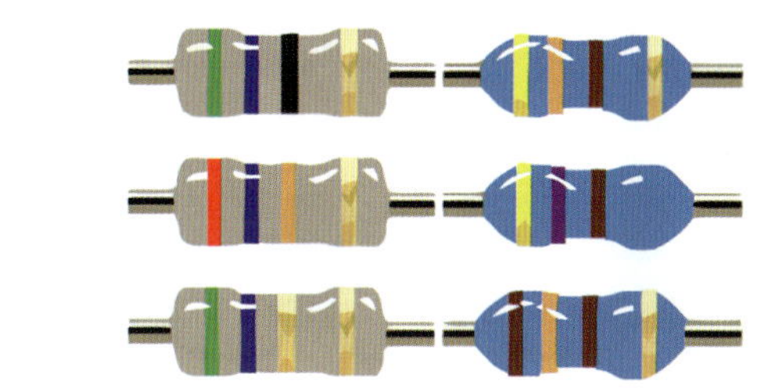

FIGURE 2.128 Resistor problems

12. What rating besides resistance value is critical when selecting a resistor?

2.8 Series circuits

Series circuit construction

An electric circuit is a closed conducting pathway in which current flows according to convention from positive to negative. When a constructed circuit occurs so that the movement of electrons in the conduction band has only one possible path, a series circuit exists. So, a series circuit is a circuit that presents only one path for current flow as shown in **Figure 2.129**.

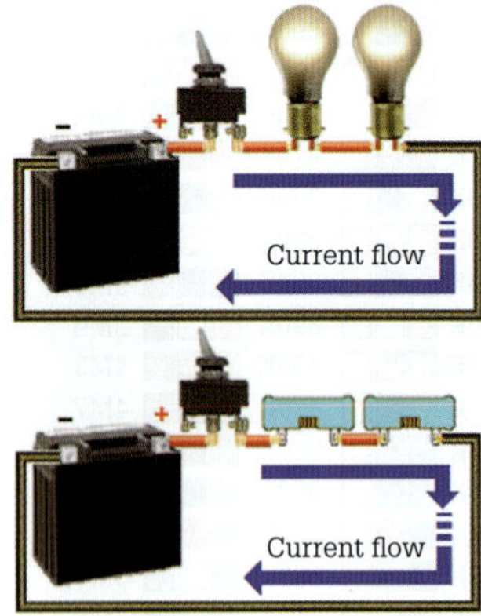

FIGURE 2.129 Series circuits with lamps and resistors

If the load devices connect so the total circuit current flows through each, then the loads are in series. Series circuits can consist of more than two loads. However, with circuits containing resistors most practical circuits do not have more than three series connected resistors. With regard to lighting circuits, lamps are rarely connected in series except with some lamp decoration circuits as in **Figure 2.130**.

FIGURE 2.130 Decoration lamps in series

The current flowing in a series circuit as shown in **Figure 2.130** must flow through each lamp inserted in the circuit in order for each lamp to be illuminated.

Each individual lamp offers added resistance, limiting the flow of current in the circuit. In a series circuit the total circuit resistance (R_T) is equal to the sum of the individual resistances, which has a similar effect to increasing the length of a conductor. As an equation:

$$R_{Total} = R_1 + R_2 + R_3 \text{ etc.}$$

In some circuits the total resistance is known and the value of a circuit resistor has to be determined. The equation can be transposed to ascertain the value of the unknown resistor.

EXAMPLE 2.22

a Three resistors – 15 Ω, 27 Ω and 39 Ω – are connected in series. What is the total resistance of the combination?

$$R_{Total} = R_1 + R_2 + R_3$$
$$= 15 + 27 + 39$$
$$= 81\ \Omega$$

b The total resistance of a circuit (see **Figure 2.131**) containing three resistors is 80 Ω. Two of the circuit resistors are 20 Ω each. Calculate the value of the third resistor.

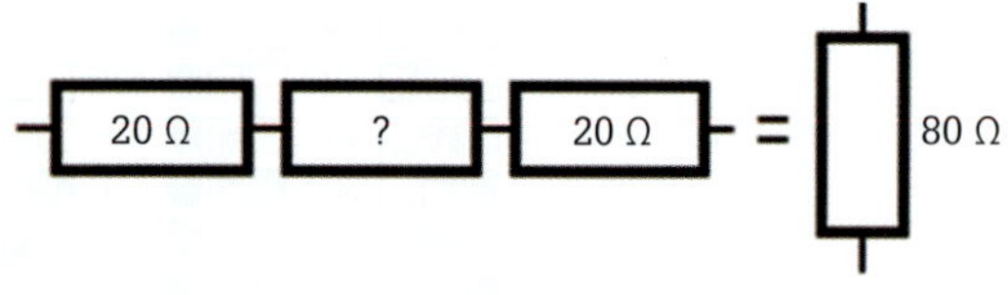

FIGURE 2.131 Resistors in series

$$R_2 = R_{Total} - (R_1 + R_3)$$
$$= 80 - (20 + 20)$$
$$= \mathbf{40}\ \Omega$$

c Due to unsatisfactory workmanship in a steel conduit run (**Figure 2.132**), there is high resistance at the conduit fittings. The readings are 0.003 Ω, 0.010 Ω and 0.006 Ω. Determine the total resistance of the fittings.

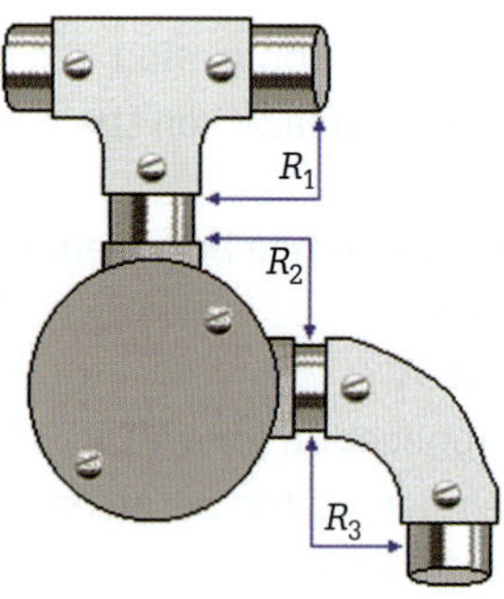

FIGURE 2.132 Metal conduit resistances

$$R_{Total} = R_1 + R_2 + R_3$$
$$= 0.003 + 0.010 + 0.006$$
$$= 0.019\ \Omega$$

EXERCISE 2.9

a Three resistors – 220 Ω, 270 Ω and 470 Ω – are connected in series. What is the total resistance of the combination?

b The total resistance of a circuit containing three resistors is 270 Ω. The resistance of one resistor is 120 Ω while another has a resistance of 56 Ω. Calculate the value of the third resistor.

Current in a series circuit

As there is only one path for current to flow, it must flow through each component. More to the point, to determine the current flowing throughout a series circuit, only the current flowing through one of the components is required. Ammeters connected into the circuit verify the fact that only one current flows through each component. Refer to **Figure 2.133**.

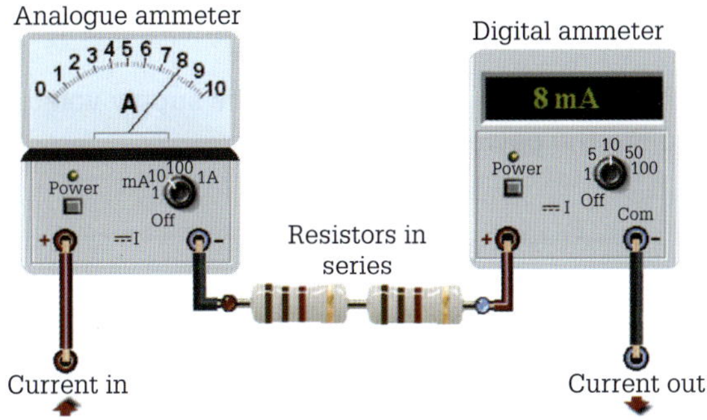

FIGURE 2.133 Ammeters measure the same current

Kirchhoff's voltage law

Gustav Kirchhoff extended the application of Ohm's law by formulating a simple law concerning the potential difference across components in a series circuit. His law states that the algebraic sum of the voltage drops in a series circuit is equal to the supply voltage. As an equation:

$$V_{Supply} = V_1 + V_2 + V_3 \text{ etc.}$$

In a series circuit, a portion of the supply voltage exists across each series load. The more loads there are in a series circuit the lower the potential difference across each load.

The potential difference across a load in a series circuit can be determined by using Ohm's law equation, $V = IR$, where I is the current and R is load resistance.

EXAMPLE 2.23

a A potential difference of 12 V appears across each resistor in a series circuit (see **Figure 2.134**) consisting of two 12 Ω resistors. Find the supply voltage across the series circuit.

$$V_{Supply} = V_1 + V_2$$
$$= 12 + 12$$
$$= 24\ V$$

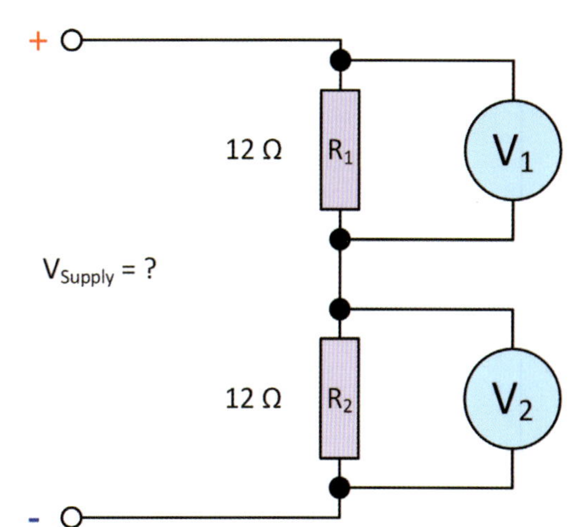

FIGURE 2.134 Example 2.23(a) series circuit

b The field circuit of a d.c. generator (see **Figure 2.135**) consists of four field coils, a rheostat and connecting leads all in series. If the resistance of each field coil is 0.008215 Ω, the rheostat 71.2 Ω and the wires 0.2 Ω, calculate the total resistance of the field circuit and the potential difference (V_d) across a field coil if the current through the field circuit is 1.25 A.

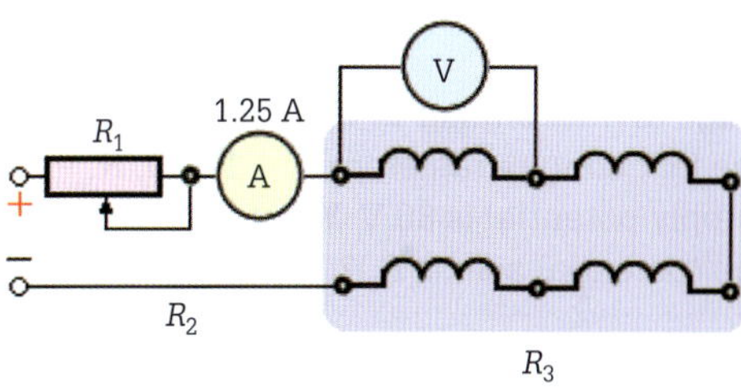

FIGURE 2.135 Example 2.23(b) components in series

$$R_{Total} = R_1 + R_2 + R_3$$
$$= 71.2 + 0.2 + (4 \times 0.008215)$$
$$= 71.43\ \Omega \text{ (4 significant figures)}$$
$$V_d \text{ across a coil} = IR$$
$$= 1.25 \times 0.008215$$
$$= 0.0103 \text{ volts (3 significant figures)}$$

Voltage divider

The circuit shown in **Figure 2.136** is often referred to as a voltage divider circuit.

»

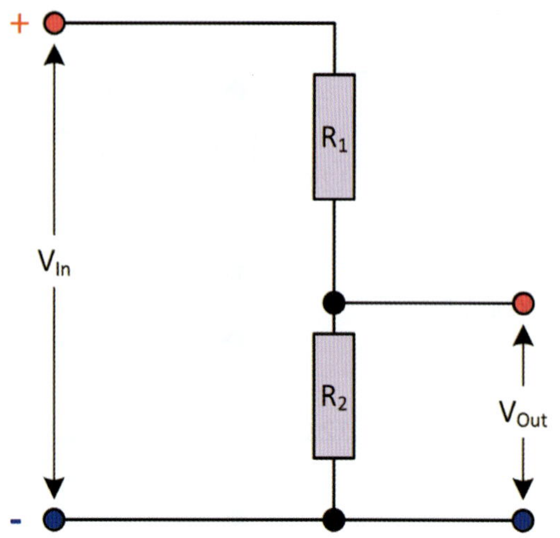

FIGURE 2.136 Voltage divider circuit

A voltage divider circuit consists of a number of elements connected in series across a voltage source. In addition, other voltages exist from connections between the elements. Usually the voltage divider consists of two resistors where:

$$V_{Out} = V_{in} \times \frac{R_2}{R_1 + R_2}$$

The voltage divider is a very simple method that produces a lower voltage from a higher voltage source.

EXERCISE 2.10

A voltage divider similar to that shown in **Figure 2.136** has the following parameters R_1 = 120 Ω, R_2 = 270 Ω, and supply potential is 80 V. What is the output voltage (potential across R_2)?

Power in a series circuit

In a series circuit, total power is equal to the sum of the power dissipated by each load. As an equation:

$$P_{Total} = P_1 + P_2 + P_3 \text{ etc.}$$

Any of the equations detailed in section 2.4 'Electrical power' are suitable for calculating power in a series circuit.

EXAMPLE 2.24

a A series circuit (see **Figure 2.137**) contains three resistors having values of 12 Ω, 22 Ω and 33 Ω. Find the total power dissipated when a potential of 120 V is applied across the combination.

The total resistance is determined first:

$$\begin{aligned} R_{Total} &= R_1 + R_2 + R_3 \\ &= 12 + 22 + 33 \\ &= 67\,\Omega \end{aligned}$$

»

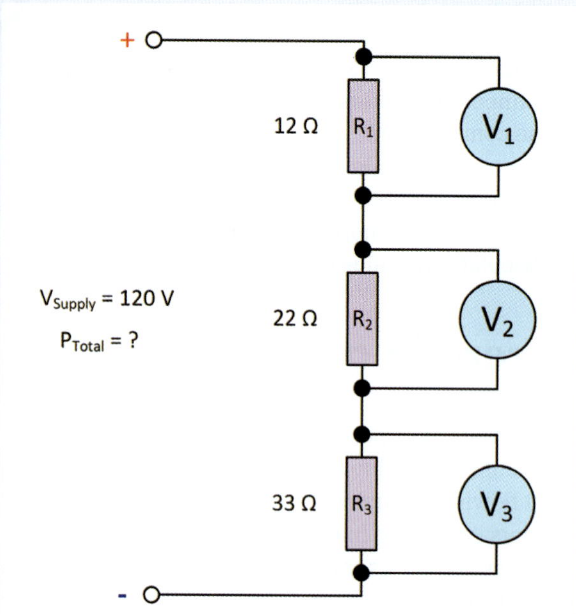

FIGURE 2.137 Example 2.24 three resistors in series

Using the total resistance and the supply voltage, calculate the power dissipated:

$$\begin{aligned} P_{Total} &= \frac{V^2}{R_{Total}} = \frac{120^2}{67} \\ &= 214.9\,\text{W (4 significant figures)} \end{aligned}$$

b For the series circuit of **Figure 2.137** calculate the power dissipated by each resistor.

Using the total resistance and the supply voltage, calculate the circuit current:

$$\begin{aligned} I &= \frac{V}{R_{Total}} = \frac{120}{67} \\ &= 1.79\,\text{A (3 significant figures)} \end{aligned}$$

The individual power dissipations are calculated using the $P = I^2R$ equation:

$$\begin{aligned} P_1 &= I^2 \times R_1 \\ &= 1.79^2 \times 12 \\ &= 38.4\,\text{W} \\ P_2 &= I^2 \times R_1 \\ &= 1.79^2 \times 22 \\ &= 70.5\,\text{W} \\ P_3 &= I^2 \times R_3 \\ &= 1.79^2 \times 33 \\ &= 105.7\,\text{W} \end{aligned}$$

EXERCISE 2.11

A series circuit comprises resistances of 22 Ω, 39 Ω and 56 Ω connected to a 90 V supply. Calculate the power dissipated by the circuit.

Practical series circuits

Series circuits connect coils in electric motors, generators and alternators. **Figure 2.138** shows direct current field coils connected in series.

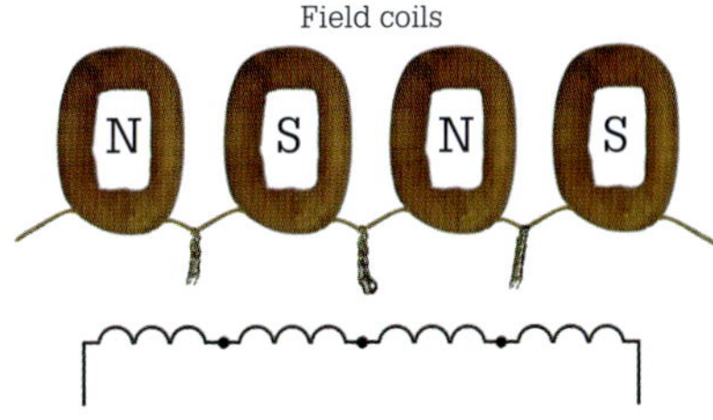

FIGURE 2.138 Series connected d.c. field coils

Control circuits as illustrated in **Figure 2.139** use the series method of connection.

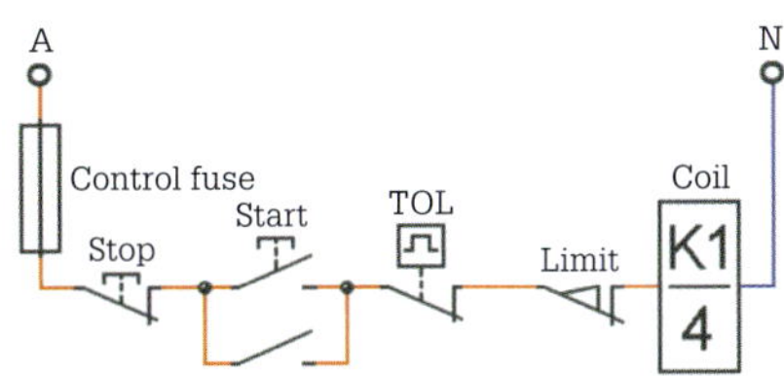

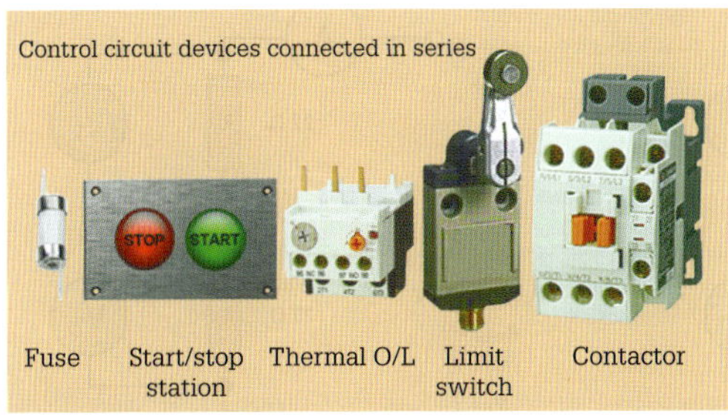

FIGURE 2.139 Control circuit devices connected in series

Batteries connect in series as shown in **Figure 2.140** in order to create a battery bank delivering a high voltage.

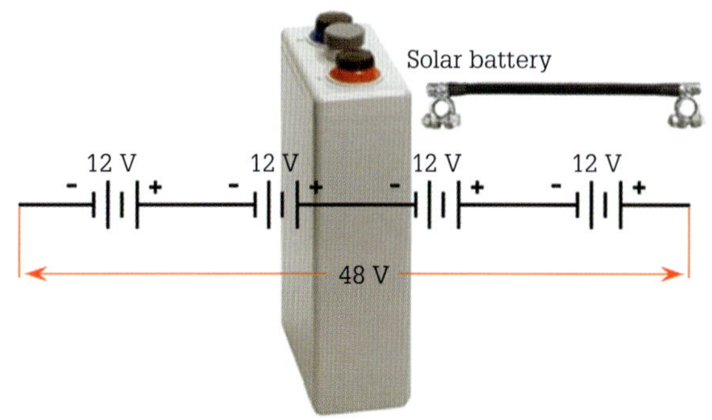

FIGURE 2.140 Series connected battery bank

Photovoltaic panels as illustrated in **Figure 2.141** connect in series as a 'series string'.

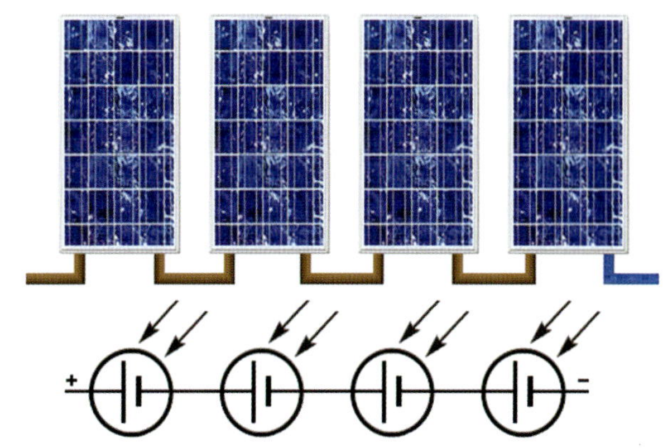

FIGURE 2.141 Photovoltaic panels in series

REVIEW QUESTIONS

1 In each circuit shown in **Figure 2.142**, calculate the value of the required resistance (R_{Total}, R_2, and R_{Total}).

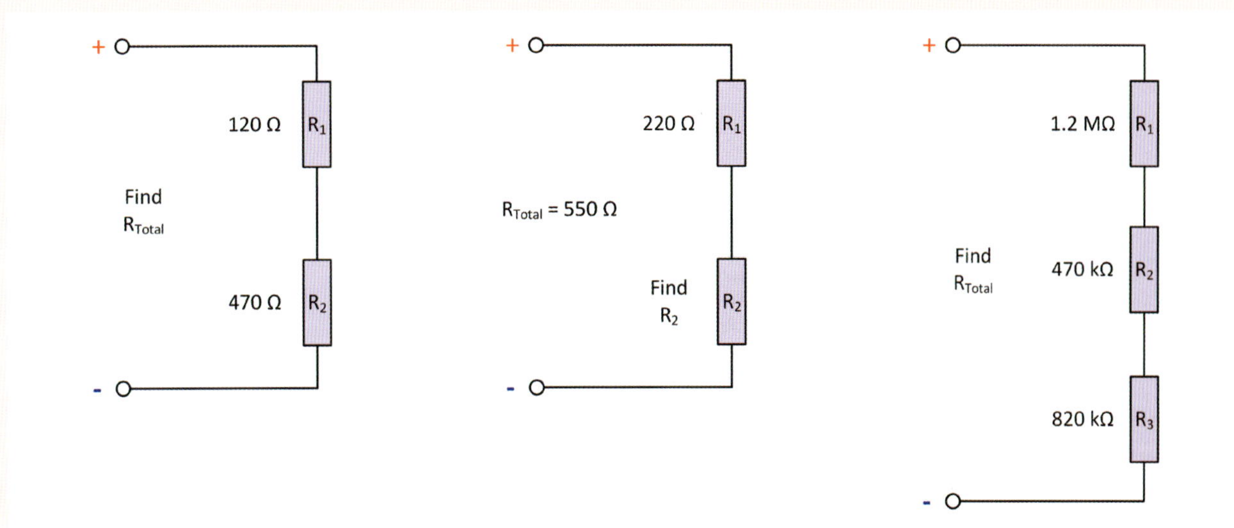

FIGURE 2.142 Unknown resistance

»

2 What are the values of current and total resistance in the following circuits of **Figure 2.143**?

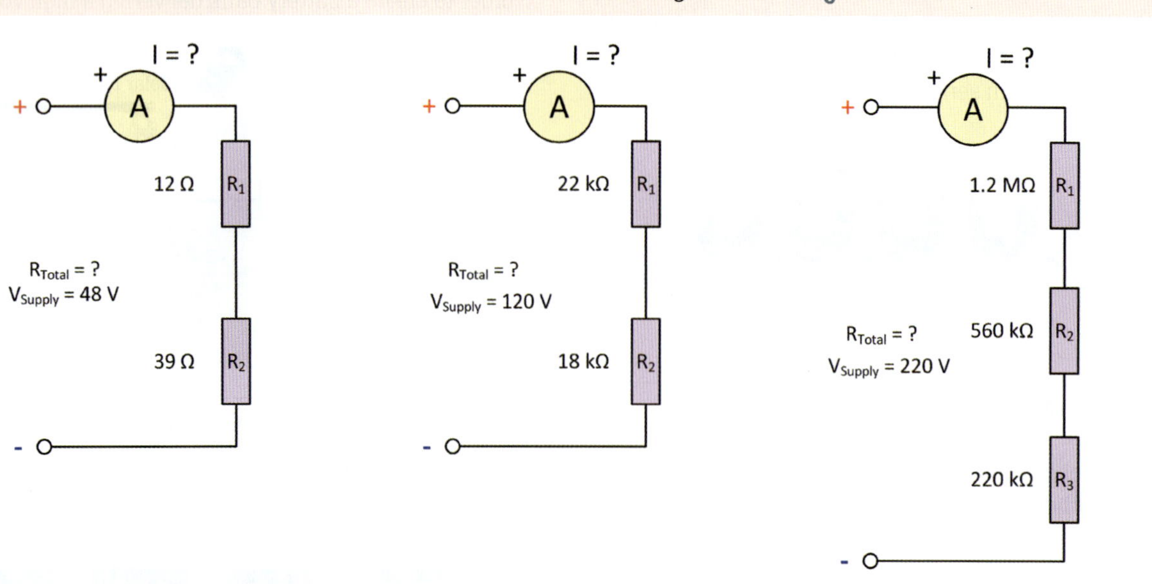

FIGURE 2.143 Current and total resistance

3 Calculate the total resistance, the total current, and potential across each resistor in the circuits of **Figure 2.144**.

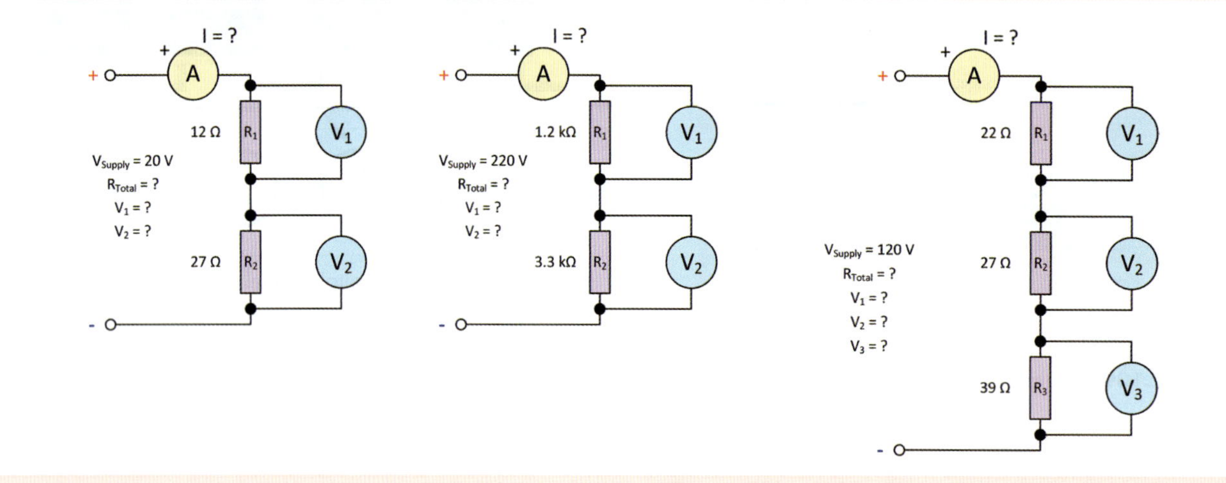

FIGURE 2.144 Current and resistance

4 Calculate the total current, total power dissipated by the circuit and power dissipated by each resistor in the circuits of **Figure 2.145**.

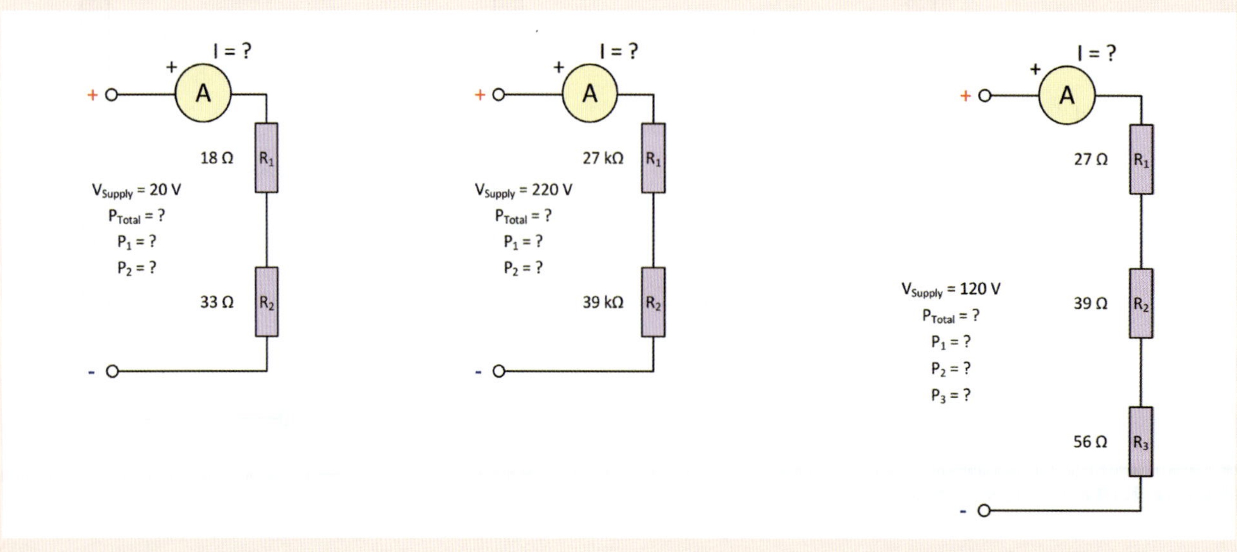

FIGURE 2.145 Power dissipated

CHAPTER REVIEW

2.1 Electrical concepts

- Static electricity refers to an electric charge created on persons or objects through friction.
- For current electricity to exist there must be a closed conducting path for the charges to flow.
- A fundamental law of electrostatics states that: Like charges repel one another and unlike charges attract one another.
- A difference in electrical pressure between the ends of a conductor or across an electrical component must exist when a current flows.
- Current is the flow of ions, electrons and holes carrying a charge through a continuous conducting path.
- Direct (d.c.) and alternating (a.c.) are two types of current.
- Low voltage is either 230 V or 400 V.
- Power stations generate electricity and use different energy resources such as coal, gas, biomass, nuclear energy, oil and water to power a prime mover to produce movement.
- Transmission networks move the electricity from the power stations to the distribution networks so it can be passed into your home or business.
- Electricity is transmitted at high voltages so that large amounts can travel efficiently over long distances.

2.2 Electrical circuit

- Power source, load, switch and conductors form a simple circuit.
- An open circuit is one where no circuit continuity occurs.
- Fuses and circuit breakers protect the conductors from over-temperature damage by short circuits and current surges.
- A circuit breaker trips a circuit while a fuse opens a circuit.
- In circuit diagrams, control and protective devices such as switches and circuit breakers are orientated such that the 'hinge' side of the blade is closest to the load device and the switch blade operates in a clockwise direction.
- Units of measurement such as volt, ampere, and ohm (to name a few) can have a prefix added, which multiplies or divides the base unit to create a more convenient unit.

2.3 Ohm's law

- Ohm's law provides the relationship between potential difference, current and resistance.
- Current is directly proportional to potential difference.
- Current is inversely proportional to resistance.
- An open circuit prevents current flow. A short circuit bypasses a circuit component. A closed circuit represents a condition where normal current flows.

2.4 Electrical power

- Force is any push or pull on an object.
- Torque is a measure of how much force acting on an object causes that object to rotate.
- The ability or capacity to do work defines energy.
- Kinetic energy is the energy of motion.
- Power is the rate of doing work.
- The mechanical power required to drive a machine is the motor load.

2.5 Effects of an electric current

- An electric current has magnetic, heating, luminous, chemical, and physiological effects.
- Electric shock is a general term for the excitation or confusion of the purpose of nerves or muscles caused by the passage of an electric current.
- A person can experience an electric shock by two methods: direct and indirect.
- At around 15–17 mA rms, the so-called 'grip' current, it becomes extremely difficult to release a conductor held in the hand.
- Parallel conductors that produce magnetic fields exert a force on each other.
- Light is an electromagnetic wave.
- Electrolysis is a process used in the refining of metals, battery charging and electroplating.
- Protect metals from corrosion by using protective coatings or cathodic protection.

2.6 EMF sources, energy sources and conversion

- Efficiency is the ratio of useful work expended to the power applied expressed as a percentage.
- There are six primary methods of producing electrical energy.
- Thermocouples are sensitive to temperature changes.
- The solar cell or photovoltaic cell converts sunlight directly into electrical energy.
- Fuel cells are electrochemical devices.
- Piezoelectric materials will produce an electric charge when the dimensions of the material change.
- When a conductor or conductors move through a magnetic field to cut the lines of force, electrical pressure is established.

2.7 Resistors

- Resistors have a designation as 'power' or 'precision'.
- Wire-wound resistors have application when precision regarding resistance is required.
- A variable resistor is either a potentiometer or rheostat.
- A voltage-dependent resistor (VDR), called a varistor, is a voltage surge protection device that connects in parallel across an a.c. input.
- Light-dependent resistors have decreasing values of resistance with increasing illumination intensity.
- Resistors are rated according to the amount of power they can dissipate.

2.8 Series circuits

- A series circuit provides only one path for current.
- The same amount of current flows through each component in a series circuit.

TRIAL EXAM

For Chapter 2 knowledge assessment, please complete the following trial exam.

1 Rubbing two surfaces together produces an electric charge. This electric charge is:
a electron electricity
b current electricity
c static electricity
d dynamic electricity

2 Which of the following electricity production methods is an example of a non-renewable energy source?
a wind powered turbine that rotates a generator to produce electricity
b coal fired boiler that converts water into steam, which drives a turbine that rotates a generator to produce electricity
c solar radiation on photovoltaic panels to produce electricity
d dammed water being gravity fed to a turbine that rotates a generator to produce electricity

3 The sector of the electrical power delivery system responsible for delivering electrical power to domestic residences is:
a transmission
b generation
c distribution
d feeding

4 Which of the following electrical loads utilises motive power?
a water heater
b radiant heater
c lamp
d electric motor

5 The coulomb is:
a the SI unit for voltage
b the SI unit for inductance
c the SI unit for charge
d the SI unit for capacitance

6 A current of 15 A flows through a conductor for 90 seconds. What quantity of electricity is transferred?
a 6 C
b 22.5 C
c 135 C
d 1350 C

FIGURE 2.146 Circuit symbol – 1

7 The symbol of **Figure 2.146** represents:
a switch
b lamp
c battery
d circuit breaker

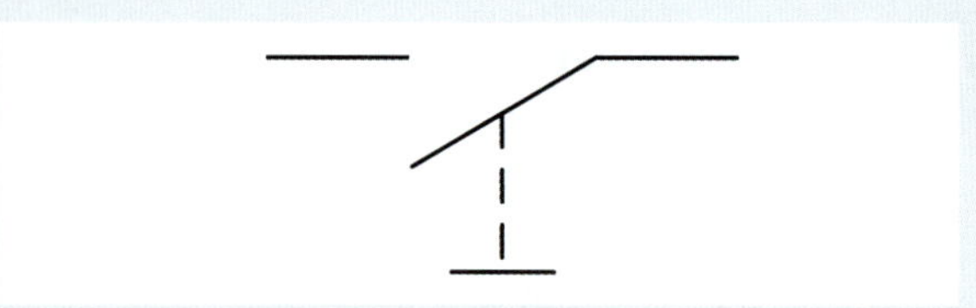

FIGURE 2.147 Circuit symbol – 2

8 The symbol of **Figure 2.147** represents:
a switch
b lamp
c battery
d circuit breaker

9 A battery provides:
a a source of electrical pressure
b opposition to an electric current
c constant resistance in a circuit
d kinetic energy

10 An open circuit is a condition where:
a an avalanche current occurs
b electrical conduction occurs
c no current flows
d internal short circuiting happens

11 Mega is the prefix used when a unit is multiplied by:
a 10^{6}
b 10^{3}
c 10^{-3}
d 10^{-6}

12 Milli is the prefix used when a unit is multiplied by:
a 10^{6}
b 10^{3}
c 10^{-3}
d 10^{-6}

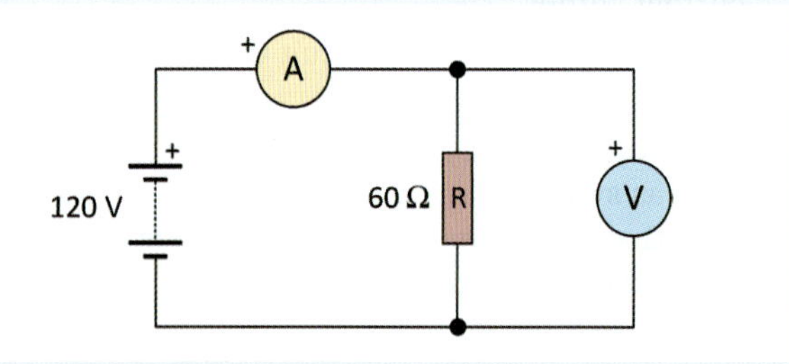

FIGURE 2.148 Circuit

13 What is the reading on the ammeter for the circuit shown in **Figure 2.148**?
a 2 A
b 5 A
c 0.5 A
d 50 A

14 What is the reading on the voltmeter for the circuit shown in **Figure 2.148**?
a 60 V
b 120 V
c 30 V
d 2 V

15 Calculate the current drawn by a 1 kΩ load when connected to a 50 V d.c. supply.
a 200 A
b 5 MA
c 1000 A
d 50 mA

16 Calculate the potential across a 150 Ω resistor drawing a current of 100 mA d.c.
a 150 mV
b 1.5 V
c 15 V
d 150 V

17 Calculate the resistance of a load drawing a current of 200 mA when connected across a 200 V d.c. supply.
a 1 kΩ
b 1 Ω
c 200 Ω
d 100 Ω

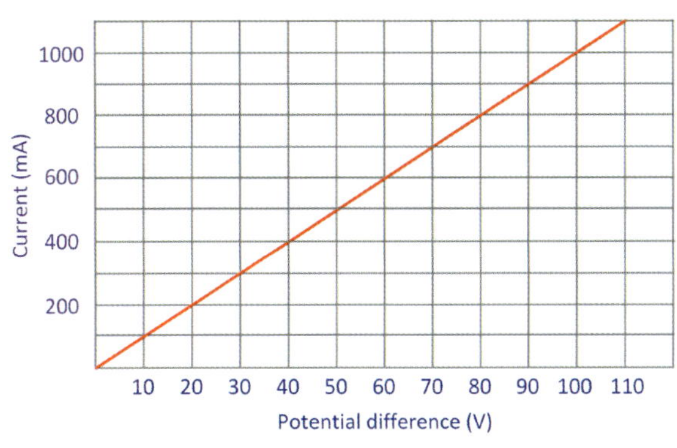

FIGURE 2.149 Graph

18 The graph shown in **Figure 2.149** shows the current that will flow through a resistive load for varying values of potential difference. What is the value of the resistance?
a 10 Ω
b 20 Ω
c 50 Ω
d 100 Ω

19 Calculate the power dissipated by a load drawing 100 mA when connected across a 20 V d.c. supply.
a 10 W
b 24 W
c 4 W
d 2 W

20 What is the power rating of a 230 V, 48 Ω industrial heating element?
a 4.8 W
b 1100 W
c 2.3 W
d 11 kW

21 A load dissipates 30 W when drawing 1.5 A from a d.c. supply. As the supply voltage is increased, what power develops when the current doubles?
a 60 W
b 300 W
c 120 W
d 150 W

22 Calculate how much force pushes a 60 kg switchboard with an acceleration of 0.03 m/s^2.
a 2000 N
b 1.8 N
c 0.0005 N
d 0.054 N

23 What is the maximum torque that an electrician can apply to a conduit threader of 0.5 m in length while applying a force of 50 N?
a 25 Nm
b 100 Nm
c 0.005 Nm
d 250 Nm

24 An electrician raises a consumer's switchboard to a height of 2 m while applying a force of 20 N. How much work occurs?
a 0.2 J
b 10 J
c 40 J
d 0.1 J

25 Calculate the potential energy of a 60 kg air-conditioning unit that is lifted using a lifting trolley through a distance of 0.75 m.
a 441 J
b 106.67 J
c 150 J
d 300 J

26 Calculate the kinetic energy of a rawhide hammer of 2 kg mass moving at a velocity of 3 m/s when installing a bearing onto an electric motor shaft.
a 18 J
b 6 J
c 9 J
d 3 J

27 A scissor lift can raise a 150 kg three-phase motor to a height of 2 m against the force of gravity in ten seconds. What is the power developed?
a 294 W
b 2943 W
c 900 W
d 8829 W

28 Whenever a machine transforms energy from one form to another, there is always:
a an energy gain
b power stability
c an energy loss
d mechanical power developed

29 Calculate the output power of an electric motor that has a constant speed of 1450 rpm when the load torque is 20 Nm.
a 3037 W
b 182.2 kW
c 71.5 W
d 29 kW

30 If the voltage in a circuit is held constant while the circuit resistance is halved then the power consumption will:
a stay the same
b double
c halve
d increase by four times

31 The instrument used to measure power in an electrical circuit is the:
a ammeter
b ohmmeter
c voltmeter
d wattmeter

32 Typical physiological effects of an electric current include:
a magnetism, pressure and radiation
b electrolysis and corrosion
c ventricular fibrillation, asphyxia and muscle spasms
d illumination and arcing

33 What is the mA range for the term 'grip current'?
a 250–1000 mA
b 75–100 mA
c 15–17 mA
d 5–8 mA

34 Name the device used to protect a person from an earth fault current.
a main switch
b fuse
c residual current device
d circuit breaker

35 What may occur to an electrical connection because of electrolysis between a copper busbar and an aluminium busbar?
a ionisation
b an electrostatic charge
c high-resistance joint
d corrosion

36 What type of electrical effect develops when a current flows through a copper conductor?
a piezoelectric effect
b ionic effect
c magnetic effect
d photovoltaic effect

37 Which of the following effects of electric current does an Automatic External Defibrillator utilise?
a heating
b magnetic
c physiological
d chemical

38 Which of the following effects of electric current does the electroplating process utilise?
a heating
b magnetic
c physiological
d chemical

39 Which of the following effects of electric current does an electric kettle utilise?
a heating
b magnetic
c physiological
d chemical

40 Which of the following effects of electric current does an electric motor utilise?
a heating
b magnetic
c physiological
d chemical

41 An electric motor rated at 300 kW has energy losses of 5 kW. What is the efficiency of the motor?
a 100.1%
b 1.6%
c 67.2%
d 98.36%

42 A practical example of electric current generated by chemical means is a:
a welding machine
b thermocouple
c battery
d transformer

43 A practical example of a device that generates electric current by thermal means is:
a electroplating
b a solar panel
c a thermocouple
d an electric motor

44 A practical example of a device that generates electric current from a conductor moving through a magnetic field is a:
a solar panel
b generator
c fuel cell
d solenoid

45 A practical example of a device that generates electric current from solar radiation is a:
a solar panel
b generator
c fuel cell
d solenoid

46 Which of the following resistor types is most suitable for high power applications?
a carbon film
b metal film
c carbon composite
d wire wound

47 The most suitable application for a rheostat is:
a field current control in a d.c. motor
b volume control in an audio amplifier
c light intensity controller
d variable voltage control

48 Which of the following resistors is most suitable for use as a surge protection device?
a LDR
b VDR
c PTC thermistor
d potentiometer

49 Which of the following resistors is most suitable for use in controlling outdoor lighting?
a LDR
b VDR
c PTC thermistor
d potentiometer

50 Which of the following resistors is most suitable for use as an over-temperature sensor in electric motors?
a LDR
b VDR
c PTC thermistor
d potentiometer

51 When a light dependent resistor is exposed to increasing illumination, its resistance:
a increases
b decreases
c does not change
d drops to zero

52 A resistor has the coloured bands orange–orange–red–gold. What is its value?
a 330 Ω ± 10%
b 33 kΩ ± 5%
c 3.3 kΩ ± 5%
d 3.3 kΩ ± 10%

53 Power loss in a cable is due to:
a failure of cable insulation
b conductor resistance
c reduced supply voltage
d reduced current demand

54 A series circuit has:
a large power consumption
b no voltage drop
c only one component connected
d one path for current

55 If the voltage supplied to a series-connected load decreased by 50% and the circuit resistance remains the same, the total circuit current will:
a increase
b decrease in the same proportion as the voltage
c it would remain the same
d burn out the series-connected load

56 The total resistance of a series circuit is equal to the:
a product of the circuit resistances
b sum of the circuit resistances
c square of the current times the voltage
d largest load resistance

57 A series circuit contains three loads each having the same resistance. What would the equivalent resistance of the circuit be if one load is shorted?
a twice the resistance of one load
b infinity resistance
c zero resistance
d remain the same resistance

58 Three lamps are connected in series across a 50 V d.c. supply. If the circuit current is 5 A and the resistance of two of the lamps totals 5 Ω, calculate the resistance of the third lamp.
a 25 Ω
b 245 Ω
c 10 Ω
d 5 Ω

3 Multiple path d.c. circuits

This chapter provides electrotechnology workers with additional knowledge and problem-solving skills that relate to the relationship between current, voltage, resistance and power in parallel and combination-type d.c. circuits. Electrotechnology workers will gain knowledge of various types of measuring instruments, resistance measurement and testing and capacitors. This chapter provides underpinning knowledge for the unit UEECD0044 from the UEE training package.

LEARNING OBJECTIVES

Resistance

- Explain how various quantities affect resistance

Parallel circuits

- Identify parallel circuits
- Calculate currents, voltages, resistances and power in parallel circuits

Series–parallel circuits

- Identify series–parallel circuits
- Calculate currents, voltages, resistances and power in series–parallel circuits

Meters in a circuit

- Explain the basic principles of using d.c. ammeters, voltmeters and ohmmeters
- Use analogue and digital meters

Resistance measurement

- Make accurate measurements of voltage, current and resistance
- Identify various electrical instruments

Capacitors and capacitance

- Describe a simple capacitor
- Define capacitance
- Explain how a capacitor stores energy
- Explain how a capacitor charges and discharges
- Test a capacitor

Capacitor circuits

- Perform basic calculations involving capacitive circuits

3.1 Resistance

Factors affecting resistance

An electrotechnology worker needs knowledge of the essential physical features and electrical characteristics of the common types of conductors. Knowledge is essential because electrical circuits make use of various types of conductors.

The physical features and electrical characteristics that affect resistance are length, cross-sectional area, resistivity and temperature.

Length

If we consider two conductors of the same cross-sectional area and material construction, with one conductor being three times the length of the other, then that conductor must have three times the resistance of the other. A shorter conductor allows electrons to move through at a higher rate rather than a longer one. **Figure 3.1** illustrates the effect of length.

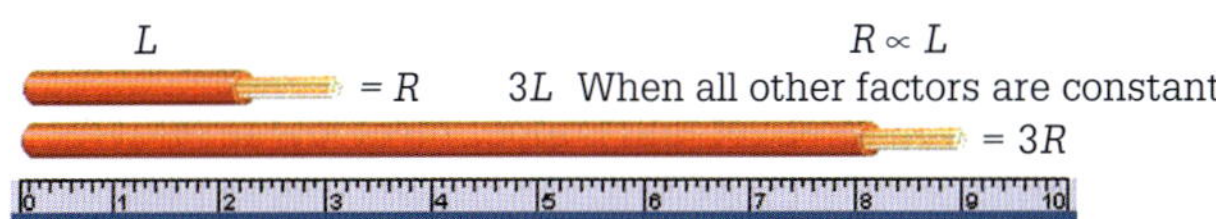

FIGURE 3.1 Resistance is directly proportional to length

As resistance is proportional to length, considering the effect length has on two otherwise similar conductors, we can say that the ratio of R_1:R_2 equals L_1:L_2. This relationship gives rise to the equation:

$$R_2 = \frac{R_1 L_2}{L_1}$$

where R_1 = is the resistance of conductor 1 in ohms (Ω)

R_2 = is the resistance of conductor 2 in ohms (Ω)

L_1 = is the length of conductor 1 in metres (m)

L_2 = is the length of conductor 2 in metres (m)

EXAMPLE 3.1 

The resistance of a 50 m length of copper wire is 1.2 Ω. Determine:

a the resistance of an 80 m length of the same wire

$$R_2 = \frac{R_1 L_2}{L_1}$$

$$= \frac{1.2 \times 80}{50}$$

$$= 1.92\ \Omega$$

b the length of the same wire when the resistance is 1.5 Ω

$$R_2 = \frac{R_1 L_2}{L_1}$$

$$L_2 = \frac{R_2 L_1}{R_1}$$

$$= \frac{1.5 \times 50}{1.2}$$

$$= 62.5\ \text{m}$$

Cross-sectional area

The cross-sectional area (CSA) of a circular conductor relates to its diameter by the equation:

$$A = \frac{\pi d^2}{4} \text{ in mm}^2$$

EXAMPLE 3.2 

a What is the cross-sectional area of a copper conductor having a nominal diameter of 1.13 mm?

$$A = \frac{\pi d^2}{4}$$

$$= \frac{\pi \times 1.13^2}{4}$$

$$= 1.00\ \text{mm}^2 \text{ (3 significant figures)}$$

b What is the cross-sectional area of a copper conductor having a nominal diameter of 1.78 mm?

$$A = \frac{\pi d^2}{4}$$

$$= \frac{\pi \times 1.78^2}{4}$$

$$= 2.49\ \text{mm}^2 \text{ (3 significant figures)}$$

If we consider two circular conductors of the same length and material, with one conductor being twice the diameter of the other, then the smaller conductor will have four times the resistance of the larger conductor. In each conductor electrons move at the same speed but there are many more electrons in the larger CSA conductor. This results in a larger current that means the resistance is less in a conductor with a larger CSA. **Figure 3.2** illustrates that resistance is inversely proportional to CSA.

$$R \propto \frac{1}{A}$$

When all other factors are constant

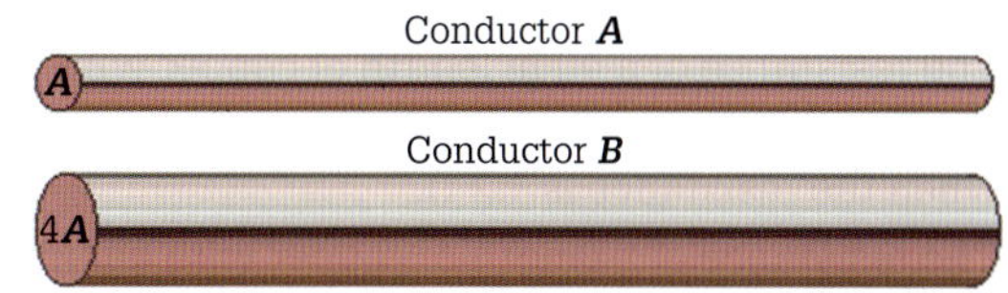

FIGURE 3.2 Resistance is inversely proportional to CSA

In comparing conductor *A* with conductor *B* in **Figure 3.2**, the ratio of the area *A* to area *B* is 1:4. However, the ratio of the resistance of conductor *A* to conductor *B* is 4:1. The difference in resistance occurs because an increase in the CSA reduces the electron density for a given current. As a result, the number of electron and atom collisions that can occur decreases, which decreases resistance to the movement of electrons flowing through the conductor. (Analogy: it is more difficult to force water (current) through a small-diameter pipe than a large-diameter pipe.) Therefore, the resistance of a conductor is inversely proportional to the CSA. It is conductor CSA that determines the current-carrying capacity.

As resistance is inversely proportional to area, considering the effect area has on two otherwise similar conductors, we can say that the ratio of R_1:R_2 equals A_2:A_1. This relationship gives rise to the equation:

$$R_2 = \frac{R_1 A_1}{A_2}$$

where R_1 = is the resistance of conductor 1 in ohms (Ω)

R_2 = is the resistance of conductor 2 in ohms (Ω)

A_1 = is the area of conductor 1 in square metres (m^2) or square millimetres (mm^2)

A_2 = is the area of conductor 2 in square metres (m^2) or square millimetres (mm^2) [this must be the same unit of measure as A_1]

EXAMPLE 3.3

The resistance of a length of copper wire having an area of 4 mm^2 is 1.2 Ω. Determine:

a the resistance of a length of the same wire having an area of 10 mm^2

$$R_2 = \frac{R_1 A_1}{A_2}$$
$$= \frac{1.2 \times 4}{10}$$
$$= 0.48\ \Omega$$

b the area of the same wire when the resistance is 1.92 Ω

$$R_2 = \frac{R_1 A_1}{A_2}$$
$$\therefore A_2 = \frac{R_1 A_1}{R_2}$$
$$= \frac{1.2 \times 4}{1.92}$$
$$= 2.5\ mm^2$$

EXERCISE 3.1

a The resistance of a 40 m length of copper wire is 1.8 Ω. Determine the resistance of a 100 m length of the same wire.

b The resistance of a 40 m length of copper wire is 1.8 Ω. Determine the length of the same wire when the resistance is 1.2 Ω.

c The resistance of a length of copper wire having an area of 4 mm^2 is 0.98 Ω. Determine the resistance of a length of the same wire having an area of 6 mm^2.

d The resistance of a length of copper wire having an area of 4 mm^2 is 0.98 Ω. Determine the area of the same wire when the resistance is 0.392 Ω.

Resistivity

The resistivity (symbol ρ (Greek letter rho)) of a material is a characteristic of the atomic lattice of the material. Different materials have different resistivity while alloying alters the resistivity. Resistivity refers to the resistance measured between opposite faces of a unit cube of a material expressed as the ohm-metre (Ωm) at a specified temperature. **Figure 3.3** illustrates the concept of the ohm-metre.

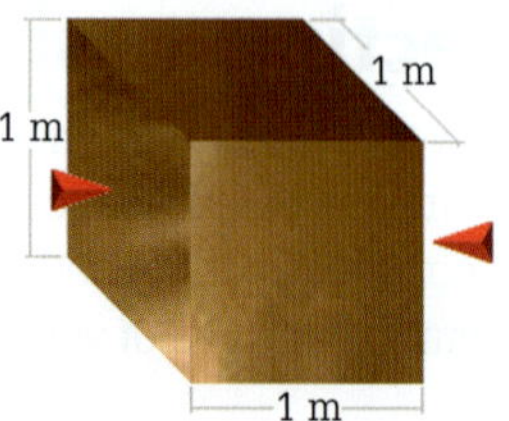

Resistance between opposite sides

FIGURE 3.3 Ohm-metre

The resistivity of a material depends upon the chemical composition of the material together with the physical treatment the material has received. For example, impurities and the addition of an alloying metal alter the atomic lattice of the material, while heat treatment and cold working modify the grain structure. Each of these treatments affects the resistivity.

The resistivity of several conductive materials at 20 °C can be found in **Table 3.1**.

An inspection of **Table 3.1** will show that good conductors such as silver and copper have low resistivity, while insulators like cross-linked polyethylene (XLPE) have a high resistivity. In between these two are the semiconductors silicon and germanium. In addition, resistivity is the proportionality constant in the following equation to determine resistance:

$$R = \frac{\rho l}{A}$$

where R = is the resistance of material in ohms (Ω)

ρ = is the resistivity of material in ohm-metres (Ωm)

A = is the area of material in square metres (m^2)

l = is the length of the material in metres (m)

TABLE 3.1 Conductor resistivity

Conductor resistivity (ρ) at 20 °C in ohm-metres			
Pure metals used for conductors		Alloys used as resistance wire	
aluminium	2.83×10^{-8}	German silver	33×10^{-8}
copper	1.725×10^{-8}	advance	49×10^{-8}
gold	2.32×10^{-8}	manganin	48×10^{-8}
lead	2.04×10^{-8}	nichrome	112×10^{-8}
platinum	10.09×10^{-8}	constantan	47×10^{-8}
silver	1.62×10^{-8}	Insulators	
steel	16.6×10^{-8}	paper	1×10^{10} Ωm
Semiconductors		mica	2×10^{14} Ωm
carbon	5×10^{-5} Ωm	Teflon	1×10^{15} Ωm
germanium	5.5×10^{-1} Ωm	porcelain	1×10^{16} Ωm
silicon	5.5×10^{-2} Ωm	glass	8×10^{16} Ωm
		PVC	1×10^{12} Ωm
		XLPE	1×10^{14} Ωm

Measurement of resistivity and the establishment of resistivity tables are essential in the electrical industry because they provide information on the suitability of materials for a specified electrical design.

EXAMPLE 3.4

a Find the resistance of a copper conductor 100 m in length if it has a CSA of 2.5 mm². The resistivity of copper is 1.72×10^{-8} Ωm.

Length $(l) = 100$ m

Area $(A) = 2.5\ \text{mm}^2 = 2.5 \times 10^{-6}\ \text{m}^2$

Resistivity $(\rho) = 1.72 \times 10^{-8}\ \Omega\text{m}$

$$R = \frac{\rho l}{A}$$

$$= \frac{1.72 \times 10^{-8} \times 100}{2.5 \times 10^{-6}}$$

$$= \mathbf{0.688\ \Omega}$$

b Calculate the resistance of one core of a 3 km length of aluminium XPLE X 90 aerial two-core 25 mm² aerial bundle. Resistivity of aluminium is 2.82×10^{-8} Ωm.

Length $(l) = 3\ \text{km} = 3000$ m

Area $(A) = 25\ \text{mm}^2 = 25 \times 10^{-6}\ \text{m}^2$

Resistivity $(\rho) = 2.82 \times 10^{-8}\ \Omega\text{m}$

$$R = \frac{\rho l}{A}$$

$$= \frac{2.82 \times 10^{-8} \times 3000}{25 \times 10^{-6}}$$

$$= \mathbf{3.384\ \Omega}$$

c Calculate the CSA of a nichrome conductor 25 m long and with a resistance of 60 Ω. The resistivity of nichrome is 110×10^{-8} Ωm.

$$R = \frac{\rho l}{A}$$

$$\therefore A = \frac{\rho l}{R}$$

$$= \frac{(110 \times 10^{-8}) \times 25}{60}$$

$$= 0.458\ \text{mm}^2\ \text{(3 significant figures)}\ [458 \times 10^{-9}\ \text{m}^2]$$

EXERCISE 3.2

a Find the resistance of a copper conductor that is 200 m in length if it has a CSA of 1.5 mm². The resistivity of copper is 1.72×10^{-8} Ωm.

b Calculate the CSA of a nichrome conductor that is 50 m long and has a resistance of 110 Ω. The resistivity of nichrome is 110×10^{-8} Ωm.

c Determine the resistance of 100 m of copper cable having a diameter of 4 mm if the resistivity of copper is 1.72×10^{-8} Ωm.

Temperature

The electrical resistance of a material also depends upon its temperature. Temperature affects conductors, insulators and semiconductors in the following manner.

SWITCH ON

The resistance of pure metallic conductors and alloys increases with increasing temperature.

The resistance of insulating materials decreases with increasing temperature.

In semiconductors the resistance decreases with increasing temperature.

Electrical equipment created at a particular ambient temperature may function at a temperature very different from that at which it was made, and therefore, its electrical resistance will change. Transformers, electric motors, alternators and lamp filaments all have a current flowing through them that produces heat. Consequently, the coil windings of such devices will have a lower resistance value when cold than when operating.

It is possible that due to temperature rise, the resistance of the coil conductors could increase by as much as 45 per cent. This means that the current through the cold devices is much greater than when the devices are in operation. Protection devices require careful design to withstand this initial current surge. Lamps, for example, nearly always burn out when being switched on.

Changes in resistance caused by both ambient temperature change and operating temperature are factors considered in the design of the electrical device. For example, power lines are subject to seasonal temperature variations that affect the line resistance. Consequently, temperature variation affects seasonal loading.

It is important to be able to calculate the resistance of a conductor at any given temperature. To do this, it is necessary to consider the temperature coefficient of resistance for the material concerned. The symbol for temperature coefficient of resistance is α (Greek alpha).

The temperature coefficient of resistance of a material at 0 °C can be defined as the change in resistance of a 1 ohm sample of a given material, when its temperature is increased from 0 °C to 1 °C. For example, a copper conductor that has a resistance of 1 Ω at 0 °C has a resistance of 1.004 27 Ω at 1 °C. Therefore, it can be said that the temperature coefficient of resistance of copper at 0 °C is 0.004 27 ohm per ohm per degree Celsius, or 0.004 27 Ω/Ω/°C.

If the resistance of a material at 0 °C is available, then using the following equation allows the resistance of the material at any other temperature to be determined.

$$R_\theta = R_0(1 + \alpha_0 \theta)$$

where: R_Θ = conductor resistance in ohms (Ω) at Θ °C

R_0 = conductor resistance in ohms (Ω) at 0 °C

α_0 = temperature coefficient of resistance at 0 °C

Θ = conductor temperature in °C

EXAMPLE 3.5

A length of copper wire has a resistance of 100 Ω when its temperature is 0 °C. Determine its resistance at 60 °C if the temperature coefficient of resistance of copper at 0 °C is 0.00427 °C.

$$R_\theta = R_0(1+\alpha_0\theta)$$

Therefore at 60°C

$$R_{60} = 100[1+(0.00427)(60)]$$
$$= 100[1+0.2562]$$
$$= \mathbf{125.62\ \Omega}$$

In practice, it is often difficult to measure resistances at 0 °C, so temperature coefficients of resistance use other temperatures, such as 20 °C.

Table 3.2 gives the temperature coefficients of resistance at 20 °C for several conductors, both pure and alloy.

TABLE 3.2 Temperature coefficients of resistance

Material	Element/Alloy	Alpha (α) at 20 °C
nickel	element	0.0059
iron	element	0.0057
molybdenum	element	0.0046
tungsten	element	0.0044
aluminium	element	0.0039
copper	element	0.00393
silver	element	0.0038
platinum	element	0.0037
gold	element	0.0037
zinc	element	0.0038
steel	alloy	0.003
nichrome	alloy	0.00017
manganin	alloy	0.000015
constantan	alloy	0.000074

The α (alpha) temperature constant refers to the temperature coefficient of resistance and represents the change in resistance per ohm per degree change in temperature. In other words, for each degree rise above 20 °C each ohm of resistance will increase by the constant alpha. Many materials that convey electrical energy experience a change in specific resistance when their temperature varies. Note that values of specific resistance use a standard temperature of 20 °C. Consequently, resistance values for conductors at any temperature other than the standard temperature of 20 °C must be determined through another equation:

$$R = R_{ref}[1 + \alpha(T - T_{ref})]$$

where: R = conductor resistance in ohms (Ω) at temperature T °C

R_{ref} = conductor resistance in ohms (Ω) at reference temperature, usually 20 °C

α = temperature coefficient of resistance at 20 °C

T_{ref} = reference temperature for which α is specified for the conductor itself in °C

T = the conductor temperature in °C

EXAMPLE 3.6

Use **Table 3.2** for temperature coefficient values for these calculations.

a The copper stator windings of an electric motor have a resistance of 120 Ω at a temperature of 20 °C. Determine the resistance of the stator windings when the motor temperature is 75 °C when operating at full load.

$$R = R_{ref}\,[1 + \alpha\,(T - T_{ref})] = 120\,[1 + 0.00393\,(75 - 20)]$$
$$= 145.9\ \Omega \text{ (4 significant figures)}$$

b A 16 mm^2 copper consumer mains can carry a current of 100 A if protected by a circuit breaker when contained within an underground enclosure. With this current, the copper conductors insulated with V-90 insulation may operate at a temperature of 75 °C. Calculate the resistance of 20 m of the consumer mains at this conductor temperature.

Firstly calculate the resistance:

$$R = \frac{\rho l}{A} = \frac{(1.72 \times 10^{-8}) \times 20}{16 \times 10^{-6}} = 0.0215\ \Omega$$

Now calculate resistance at conductor temperature of 75 °C

$$R = R_{ref}\,[1 + \alpha\,(T - T_{ref})] = 0.0215\,[1 + 0.00393\,(75 - 20)]$$
$$= 0.0261\ \Omega \text{ (3 significant figures) per conductor}$$

c The resistance of 100 m of 2.5 mm^2 plain annealed copper conductor is 0.5375 Ω at 20 °C. Find the resistance at 90 °C.

$$R = R_{ref}\,[1 + \alpha\,(T - T_{ref})]$$
$$= 0.5375[1 + 0.00393\,(90 - 20)]$$
$$= \mathbf{0.685\ \Omega} \text{ (3 significant figures)}$$

The positive temperature coefficient of resistance for a material means that its resistance increases with an increase in temperature. Intrinsic metals have positive coefficients of resistance. By contrast, a negative coefficient of resistance for a material means that the resistance decreases as temperature increases. Materials such as germanium and silicon semiconductors have a negative temperature coefficient of resistance. Many industrial processes that rely on the accurate measurement and control of temperature use these semiconductors.

Effect of temperature on current-carrying capacity

Heat (Q) occurs when current flows through a conductor. More to the point, the amount of heat energy produced is dependent upon the resistance of the material, the square of the current that the material carries and the time in seconds that the current is present.

$$\text{heat energy} = P \times t = I^2 R \times t \text{ in joules}$$

The heat produced by current flowing in a conductor depends upon the load. To reduce the heat energy produced, the resistance of the conductor must reduce.

The installation method also affects the temperature and, therefore, the current-carrying capacity of the insulated conductor. For example, if the insulated conductor is unenclosed in free air, as shown in **Figure 3.4**, the heat generated dissipates quickly into the surrounding air. The same insulated conductor, when enclosed in conduit, will be unable to dissipate the heat because the heat stays within the conduit. It follows that the trapped heat increases the internal temperature.

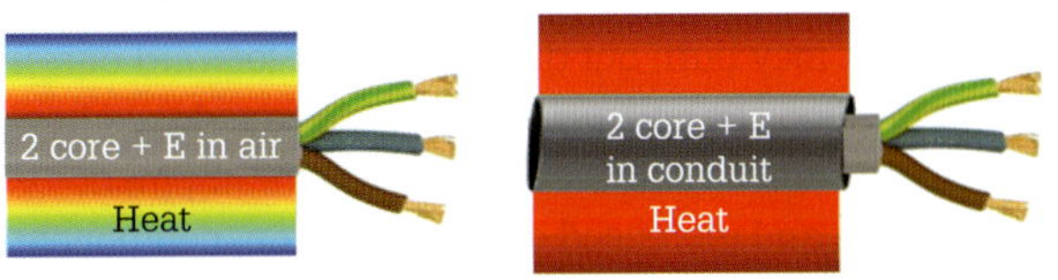

FIGURE 3.4 Effect of temperature on cables

It is not the conductor material that is affected, but the insulation covering the conductor. A typical insulation rating for conductors is V-90, which means that the insulation can withstand a maximum conductor temperature of 90 °C but in normal use the conductor temperature is 75 °C. As different installation methods can apply in the field, AS/NZS 3008.1.1: 2017 'Electrical installations, Selection of cables', has a series of tables that indicate the maximum current-carrying capacities for commonly used cables. For example, Table 10 of AS/NZS 3008.1.1 indicates that a 2.5 mm^2 two core + E 0.6/1 kV flat insulated and sheathed V-90 cable can carry 26 A if unenclosed and touching the surface [column 5] and only 23 A if enclosed in a non-metallic conduit in air [column 11]. For this reason, the cable has a lower current-carrying capacity due to the decreased ability to radiate heat into the environment.

Resistivity of soils and earth electrodes

The resistivity of soils is due to the complex interaction of permeability, porosity, ionic content of the soil fluids and a clay mineralisation. During resistivity studies, current is injected into the soil using a pair of electrodes and the potential difference is measured between them. The apparent resistivity calculated from the data is the average resistivity of all soils and rock that influences the flow of current. The reason for measuring soil resistivity is to find a location that has the lowest possible resistance. Resistivity of soils can vary between 100 Ωm for loams and up to 200 000 Ωm for sandstone. Lower resistivity provides a greater margin of protection for personnel and equipment. AS/NZS 3000: 2018 *Wiring Rules*, does not refer to a minimum soil resistance, only a minimum depth for the electrode.

Figure 3.5 shows this depth. The Rules also state that in general the earth electrode should be located in a position exposed to the weather and outside the building. For many Australian conditions, the minimum requirements of depth are all that is needed. However, there are factors that could change the earth electrode requirements. For example, an industrial or commercial facility could increase in size, thus demanding extra earth electrodes or an electrode driven to a greater depth. In addition, in some areas the water table is falling and the electrode may exist in dry soil of high resistance. Furthermore, effective earthing is mainly for the safety of persons, but it also provides for the protection of plant and equipment.

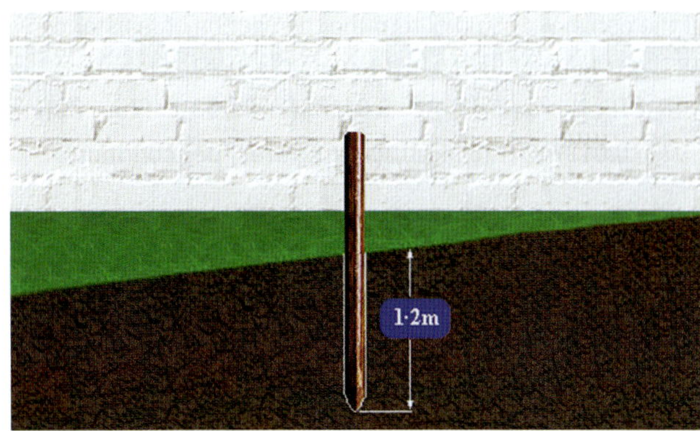

FIGURE 3.5 Earth electrode

Resistance of a cable under load

When a current flows through a conductor, heat is produced and this raises the temperature of the conductor. This temperature rise is due to power loss in the conductor which is now unable to deliver the required energy to the load. The power loss can be reduced by effective design but it cannot be eliminated.

Except for bare aerials, all conductors are surrounded by electrical insulation. The insulation acts as thermal insulation and reduces the rate at which heat is dissipated. Consequently, this causes the conductor temperature to increase further and as a result the resistance of the conductor increases.

A 16 mm^2 copper cable insulated with V-90 insulation buried in an underground enclosure and protected by a circuit breaker is capable of delivering 100 A. While the copper itself can withstand a high temperature, the V-90 refers to the maximum operating temperature of the conductors and hence what the insulation will withstand without deterioration and insulation failure.

If the ambient temperature is 30 °C the copper conductor may experience a 60 °C temperature rise due to the current flow. **Table 3.3** illustrates how temperature can affect conductor resistance and consequent circuit performance when supplying a 3.6697 Ω resistive load. This circuit has a total consumer mains resistance of 0.0226 Ω (two cables × 10 m) at 20 °C.

TABLE 3.3 Voltage, current and resistance

	Active	Neutral	Load	Total
Resistance	0.0113 Ω	0.0113 Ω	3.6697 Ω	3.6923 Ω
Current	62.29 A	62.29 A	62.29 A	62.29 A
Voltage	0.704 V	0.704 V	228.592 V	230 V

Note: 16 mm² Cu = 1.13 Ω/1000m.

At 20 °C, we have 228.592 V across the domestic load and a 1.408 V drop across the consumer mains. If the temperature were to rise to 75 °C (the operating temperature of the cable) we could determine the change of resistance for the consumer mains (note: α of copper = 0.003 93).

$$R = R_{\text{ref}}[1 + \alpha(T - T_{\text{ref}})] = 0.0113\,[1 + 0.00393\,(75 - 20)]$$
$$= 0.013742\ \Omega \text{ (5 significant figures) per conductor}$$

TABLE 3.4 Voltage across the load

	Active	Neutral	Load	Total
Resistance	0.013 742 Ω	0.013 742 Ω	3.6697 Ω	3.697 Ω
Current	62.21 A	62.21 A	62.21 A	62.21 A
Voltage	0.855 V	0.855 V	228.29 V	230 V

The voltage across the load went to 228.29 V as shown in **Table 3.4** and the voltage drop across the consumer mains increased from 1.408 V to 1.71 V, as a result of the temperature increasing. The overall resistance of the circuit increased due to increased temperature, causing a reduction in the supply current. Even though the changes seem small, they can be significant for power lines between a power station and the substation, and between the substation and the consumer.

Voltage drop across a conductor

If a power tool connects to a 230 V supply via an extension lead the voltage available for driving the tool may be only 228 V. This voltage drop could be the result of a low supply voltage from the energy authority or it could be caused by high ambient temperatures or excessive extension cord length. If the voltage is below the operating voltage of the tool, then its performance falls and the tool may suffer damage as a result.

Installation circuits exceeding 25 m in length need examination to see if the voltage drop is greater than the 5% allowed, calculated in accordance with Clause 3.6.2, AS/NZS 3000: 2018 *Wiring Rules*.

Other electrical equipment affected by a reduced voltage includes resistive loads such as heaters. Furthermore, devices – because of the drop in voltage – are unable to reach the required temperature and so are unable to provide the expected power output.

Inductive loads such as electric motors may run slower (torque ∝ voltage²), therefore 5% reduction in motor voltage results in a 10% reduction in the motor's torque output and it may burn out due to a higher forward current passing through the coil windings owing to reduced back electromotive force. This results in shorter equipment operating life and increased cost. In addition, components on printed circuit boards may not have sufficient voltage across them to perform correctly. Under-voltage for sensitive electronic equipment, such as computers, laser printers and copy machines can cause the equipment to lock up or suddenly power down. The result is data loss, increased cost and possible equipment failure. Long consumer mains, sub-mains or final sub-circuit conductors often cause reduced voltage, or under-voltage. The correct sizing of these conductors ensures proper operating voltage for a safe and efficient electrical system.

EXAMPLE 3.7

A single-phase motor takes 35 A from a 230 V supply. The motor connects to the supply by a twin + E 25 mm² TPS cable 30 m long. Using this data, calculate the voltage at the motor terminals.

The first step is to find the cable resistance. The total length of conductors (active and neutral) is 2 × 30 m = 60 m, and the resistivity of plain annealed copper is 1.72 × 10⁻⁸ Ωm. **Note:** 25 mm² = 25 × 10⁻⁶ m².

$$R = \frac{\rho l}{A}$$
$$= \frac{1.72 \times 10^{-8} \times 60}{25 \times 10^{-6}}$$
$$= \mathbf{0.04128\ \Omega}$$
$$V_d = I \times R$$
$$= 35 \times 0.04128$$
$$= \mathbf{1.4448\ V}$$
$$\text{motor voltage} = \text{supply voltage} - \text{voltage drop}$$
$$= 230 - 1.4448$$
$$= \mathbf{228.5552\ V}$$

REVIEW QUESTIONS

1 What is the relationship between the area of a conductor and its resistance?

2 What is the relationship between the length of a conductor and its resistance?

3 What is the relationship between the operating temperature of a metal conductor and its resistance?

4 The resistance of a 50 m length of copper wire is 1.8 Ω. Determine the length of the same wire when the resistance is 1.2 Ω.

5 The resistance of a length of copper wire having an area of 2.5 mm² is 0.92 Ω. Determine the area of the same wire when the resistance is 0.230 Ω.

»

6 Find the resistance of a copper conductor 150 m in length if it has a CSA of 1.5 mm^2. The resistivity of copper is 1.72×10^{-8} Ωm.

7 Will the resistance of insulators increase or decrease with an increase in temperature?

8 The copper armature windings of a d.c. electric motor have a resistance of 15 Ω at a temperature of 20 °C. Determine the resistance of the armature coils when the temperature of the motor reaches 120 °C.

9 The copper windings of a transformer have a resistance of 150 Ω at a temperature of 20 °C. In operation, the winding resistance increases to 180 Ω. Determine the operating temperature of the transformer.

10 Why is the current rating of a particular cable higher when installed unenclosed than when installed enclosed in a conduit?

3.2 Parallel circuits

Loads

Energy-transforming devices called loads may connect from a standard point (node) in a circuit. The branching from the standard point is a parallel connection whereby the current from the voltage source splits into separate branches as shown in **Figure 3.6**.

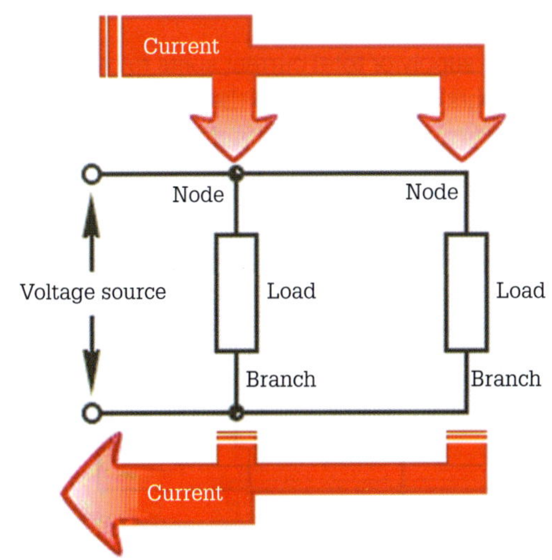

FIGURE 3.6 Parallel connection

A node is a standard point in a circuit where two or more circuit loads connect. In addition, a branch is that part of a circuit that lies between two nodes.

Practical parallel circuits

Parallel circuits (see **Figure 3.7**) connect lights, socket outlets, residential loads, sub-circuits and various other types of loads.

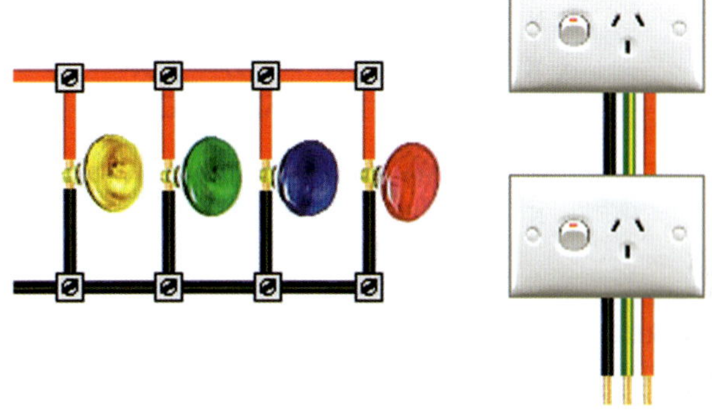

FIGURE 3.7 Decoration lights and socket outlets connected in parallel

Parallel circuits connect residential loads in parallel across supply mains as shown in **Figure 3.8**.

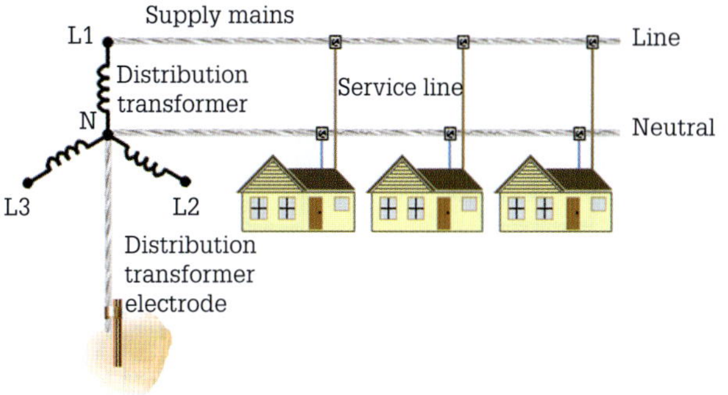

FIGURE 3.8 Residential loads connected in parallel

Parallel circuits connect sub-circuits to the feed. **Figure 3.9** shows fuse and circuit breaker/safety switch protected sub-circuits.

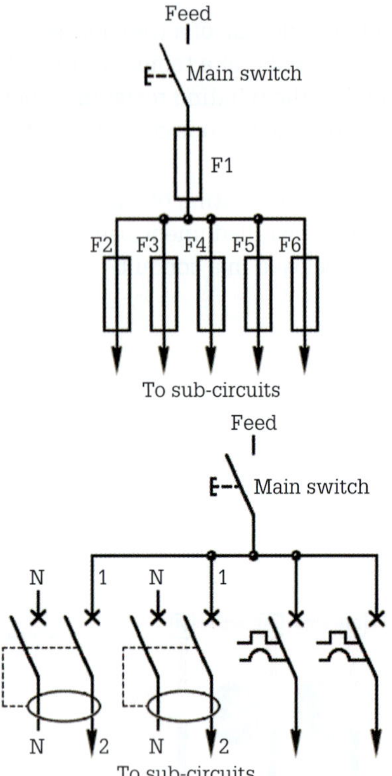

FIGURE 3.9 Fuse and circuit breaker/safety switch protected sub-circuits

Each resistor in the circuits of **Figure 3.10** offers another path for current to travel from positive to negative.

The node does not physically have to be a single point. It follows that as long as the current has several alternative paths the circuit is parallel. The circuits in **Figure 3.10** are identical even though they appear different.

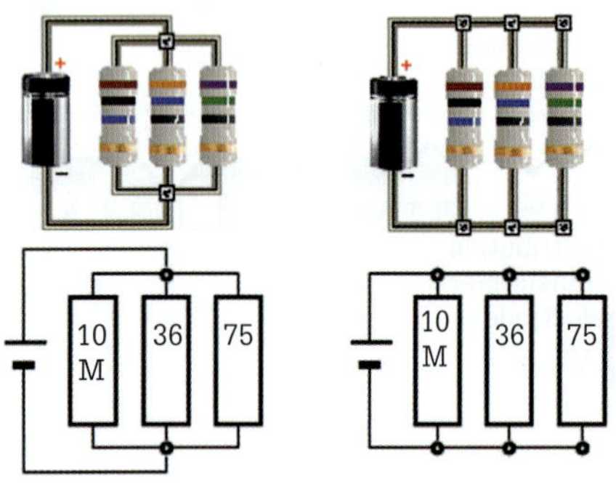

FIGURE 3.10 Resistors connected in parallel

Potential difference in a parallel circuit

The potential difference across each resistor in **Figure 3.11** is the same for each resistor and it is evident that this is so. Each connects directly to the supply voltage. The following equation is distinct:

$$V_{Supply} = V_1 = V_2 = V_3 \text{ etc.}$$

When voltmeters as shown in **Figure 3.11** are connected across resistors connected in parallel, the potential difference across each resistor is the same as the supply potential.

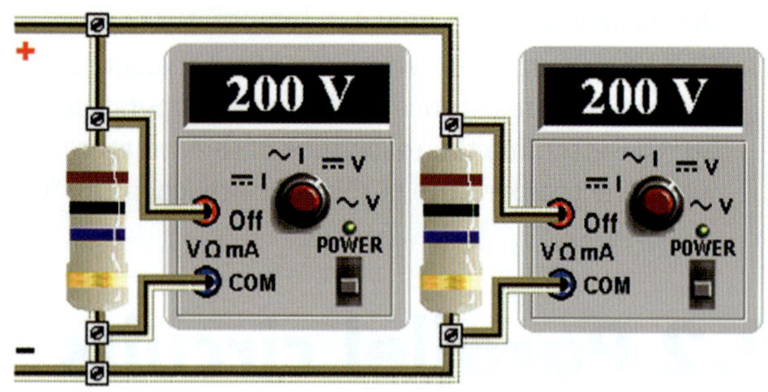

FIGURE 3.11 Voltmeters connected in parallel

Kirchhoff's current law

Kirchhoff's current law states that current leaving the voltage source equals the current returning to the voltage source. In a parallel circuit such as in **Figure 3.12** the current leaves the voltage source, divides at a node, and flows through the branches to then recombine and return to the voltage source.

Since the supply or total current splits to become branch currents as it travels from the source, the sum of the branch currents must equal the total current:

$$I_{Supply} = I_1 + I_2 + I_3 \text{ etc.}$$

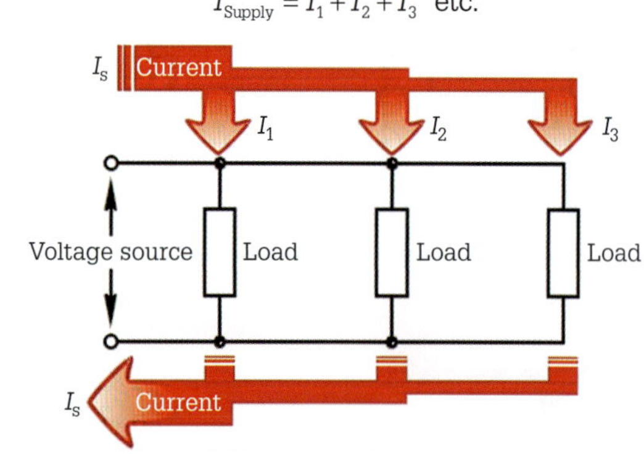

FIGURE 3.12 Kirchhoff's current law

If we connect ammeters into a parallel circuit we will see that the total current is the sum of the branch currents as shown in **Figure 3.13**.

$$\begin{aligned} I_{Supply} &= I_1 + I_2 \\ &= 1.5 + 0.75 \\ &= 2.25 \text{ A} \end{aligned}$$

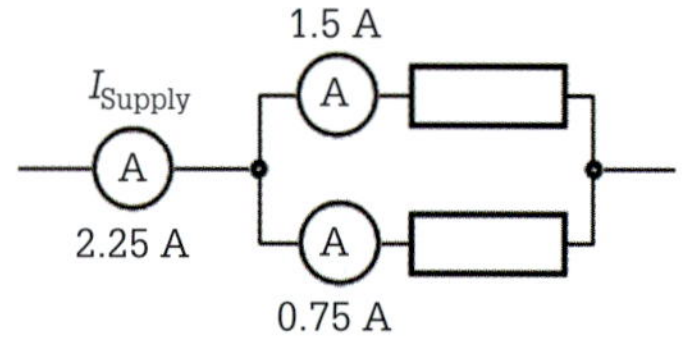

FIGURE 3.13 Current measurement

EXAMPLE 3.8

Two resistors, one 30 Ω and the other 50 Ω, connect in parallel across a 300 V supply. Calculate the current flowing through each resistor and the total current supplied to the circuit in **Figure 3.14**.

$$\begin{aligned} I_1 &= \frac{V}{R_1} \\ &= \frac{300}{30} \\ &= \mathbf{10\ A} \\ I_2 &= \frac{V}{R_2} \\ &= \frac{300}{50} \\ &= \mathbf{6\ A} \\ I_T &= I_1 + I_2 \\ &= 10 + 6 \\ &= \mathbf{16\ A} \end{aligned}$$

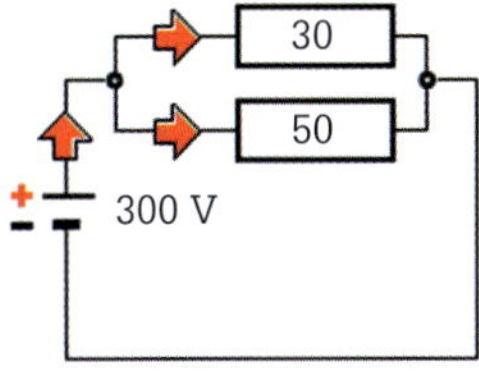

FIGURE 3.14 Resistors in parallel

Current divider

A current divider uses two or more resistors connected in parallel. **Figure 3.15** illustrates a two-resistor current divider.

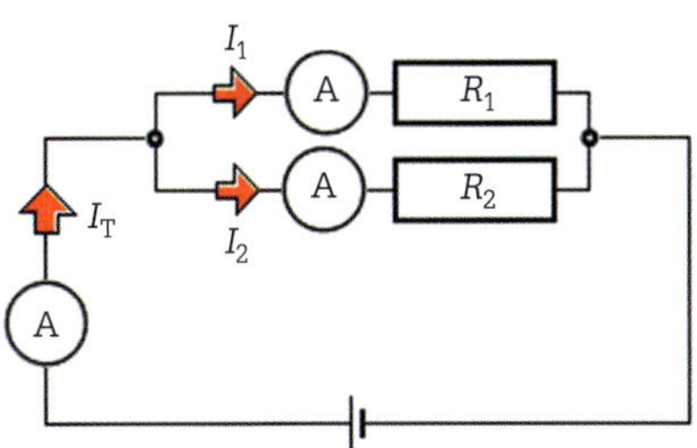

FIGURE 3.15 Current divider

If there are only two individual resistors in parallel then a simple equation can determine current value through each resistor.

$$I_1 = I_T \frac{R_2}{R_1 + R_2} \quad I_2 = I_T \frac{R_1}{R_1 + R_2}$$

For this reason a parallel circuit is called a 'current divider' because of its ability to proportion the total current into fractional parts.

Equivalent resistance in a parallel circuit

In a parallel circuit, the equivalent resistance is the total opposition to current presented by all the resistances or loads. Connecting more resistances in parallel has the effect of increasing the total current drawn from the supply. If current increases then there must be a subsequent decrease in opposition to current. Another way of putting it is to say that the circuit is becoming more conductive (less resistance).

Conductance

Resistance is an opposition to current flow; the opposite of resistance is conductance (symbol G, measured in Siemens, symbol S). Conductance is a measure of the ability of a material (solid or flowing medium) to conduct an electric current (i.e. to pass electrons). A reciprocal relationship exists between resistance and conductance. The reciprocal of a number (x) arises by dividing the number into one.

$$\text{reciprocal} = \frac{1}{X}$$

Modern scientific calculators perform this function with a key labelled [x^{-1}] as shown in **Figure 3.16**.

FIGURE 3.16 Locating x^{-1} key on a scientific calculator

Mathematically, in terms of resistance and conductance:

$$\text{resistance} = \frac{1}{G} \text{ and conductance} = \frac{1}{R}$$

When the resistance of a material has a value, dividing the number one by the value provides the conductance. For example, a resistance of 0.12 Ω has a conductance of 8.33 Siemens (3 significant figures).

$$\begin{aligned} G &= \frac{1}{R} \\ &= \frac{1}{0.12} \\ &= 8.33\ \text{S (3 significant figures)} \end{aligned}$$

In terms of resistance, adding more resistance in parallel decreases the total or circuit resistance. In terms of conductance, however, more resistance in parallel results in a higher total conductance (electrons flow with greater conductance).

Reciprocal method of finding equivalent resistance

In a parallel circuit the total circuit conductance equals the sum of the branch conductance:

$$G_{equivalent} = G_1 + G_2 + G_3 \text{ etc.}$$

Given that resistance and conductance are reciprocals of one another, we can express the previous equation in terms of resistance:

$$\frac{1}{R_{equivalent}} = \frac{1}{R_1} + \frac{1}{R_2} + \frac{1}{R_3} \text{ etc.}$$

To solve for $R_{equivalent}$, the equation becomes:

$$R_{equivalent} = \frac{1}{\frac{1}{R_1} + \frac{1}{R_2} + \frac{1}{R_3} \text{ etc.}}$$

This equation is known as the reciprocal of the sum of reciprocals and applies for any number of parallel-connected resistances and is valid whether the resistances are of equal or unequal value.

You will see from this exercise that the equivalent resistance of a number of parallel-connected resistors is always less than the smallest value of individual branch resistance.

There are a number of methods to determine the equivalent resistance of parallel resistances. Choosing a method depends on how many resistances there are, and whether or not their values of resistance are the same.

EXAMPLE 3.9

Three resistors consisting of 20 Ω, 30 Ω and 40 Ω are in parallel as shown in **Figure 3.18**. Determine the equivalent resistance of the circuit.

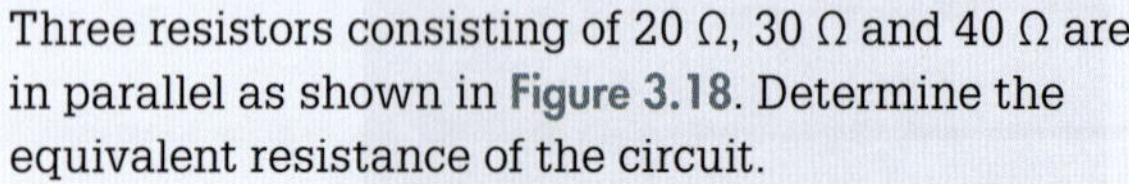

$$R_{equivalent} = \frac{1}{\frac{1}{R_1} + \frac{1}{R_2} + \frac{1}{R_3}} = \frac{1}{\frac{1}{20} + \frac{1}{30} + \frac{1}{40}}$$

$= 9.23\ \Omega$ (3 significant figures)

Fortunately, the scientific calculator makes solving this equation relatively straightforward:

FIGURE 3.17 Example 3.9 calculator operation

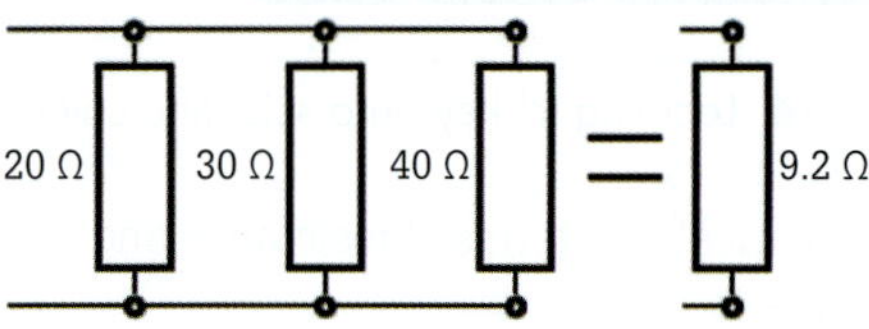

FIGURE 3.18 Example 3.9 parallel resistance

EXERCISE 3.3

a Three resistors consisting of 15 Ω, 33 Ω and 64 Ω connect in parallel. Determine the equivalent resistance of the circuit.

b Four 100 Ω resistors connect in parallel. Determine the equivalent resistance of the circuit.

c Three resistors connect in parallel to present an equivalent of resistance of 33 Ω. If two of the resistors have resistance values of 68 Ω and 100 Ω respectively, what is the resistance of the third resistor?

Equal resistances

The easiest method of determining the total resistance is when the circuit under study concerns two resistors of equal value connected in parallel. It is a fact that when the cross-sectional area of the material doubles, with all other factors remaining constant, the resistance of the material halves. So, the total resistance of two equal resistances connected in parallel is one-half the value of either resistance. In addition, the total resistance of three equal-value resistors connected in parallel is one-third the value of any of the resistances as demonstrated in **Figure 3.19**. Therefore:

$$R_T = \frac{R}{n}$$

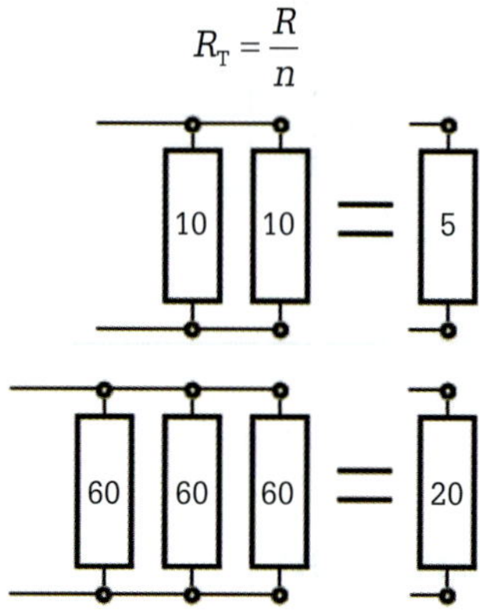

FIGURE 3.19 Total resistance

Resistances that are multiples of one another

If the different resistance values are multiples of one another, a variation of the equal-resistance method can apply because any one resistance acts as two or more other resistors in parallel.

For example, the six Ω resistor in **Figure 3.20** can be considered as being two 12 Ω resistors in parallel. When the circuit resistors under study have values that are multiples of each other, the lower resistance values are resistor combinations producing the same useful circuit.

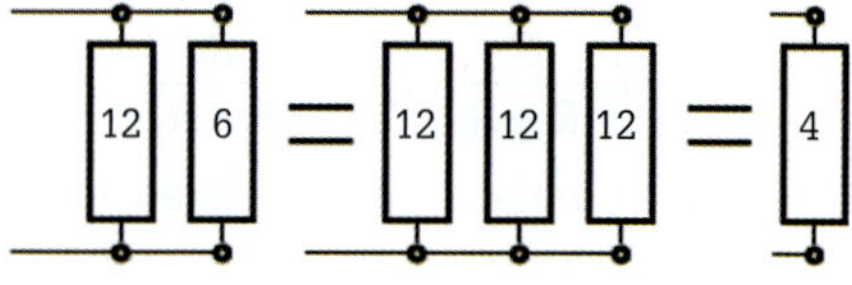

FIGURE 3.20 Each resistance group equals 4 Ω

Product/sum method of finding resistance

This method applies when there are two resistances whose values are not the same and where they cannot convert to the same multiple. Consequently, the values of the two resistances must be multiplied together to find their product. Then the two resistances are added together to get their sum. Finally, the product divided by the sum gives the total resistance as shown by the following equation.

$$R_t = \frac{product}{sum} \text{ thus } R_{equivalent} = \frac{R_1 \times R_2}{R_1 + R_2}$$

EXAMPLE 3.10

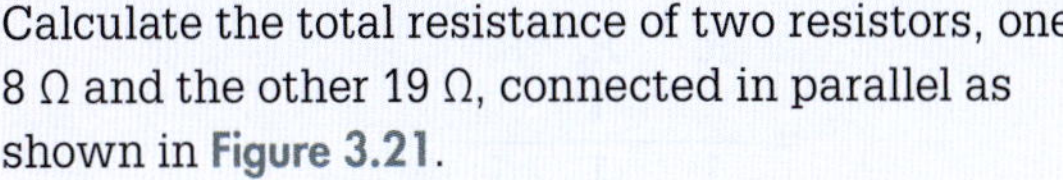

Calculate the total resistance of two resistors, one 8 Ω and the other 19 Ω, connected in parallel as shown in Figure 3.21.

$$R_{equivalent} = \frac{R_1 \times R_2}{R_1 + R_2}$$
$$= \frac{8 \times 19}{8 + 19}$$
$$= 5.63\ \Omega \text{ (3 significant figures)}$$

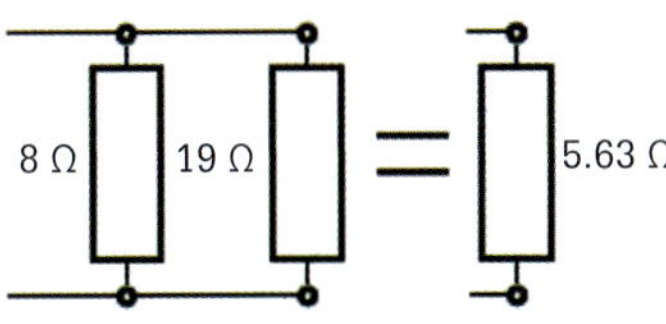

FIGURE 3.21 Example 3.10 total resistance

Ohm's law method of finding resistance

If the supply voltage and the total circuit current are available, calculation of total resistance uses Ohm's law.

$$R_{equivalent} = \frac{V_T}{I_T}$$

EXAMPLE 3.11

Three resistors of 1 kΩ, 2 kΩ and 3 kΩ connect across a 12 V supply as shown in Figure 3.22. Calculate the total resistance.

$$I_1 = \frac{E}{R_1} = \frac{12}{(1 \times 10^3)} = 12 \text{ mA}$$

$$I_2 = \frac{E}{R_2} = \frac{12}{(2 \times 10^3)} = 6 \text{ mA}$$

$$I_3 = \frac{E}{R_3} = \frac{12}{(3 \times 10^3)} = 4 \text{ mA}$$

$$I_T = I_1 + I_2 + I_3 = 12 + 6 + 4 \text{ mA} = 22 \text{ mA}$$

$$R_{equivalent} = \frac{V_T}{I_T}$$
$$= \frac{12}{(22 \times 10^{-3})}$$
$$= 545.5\ \Omega \text{ (4 significant figures)}$$

»

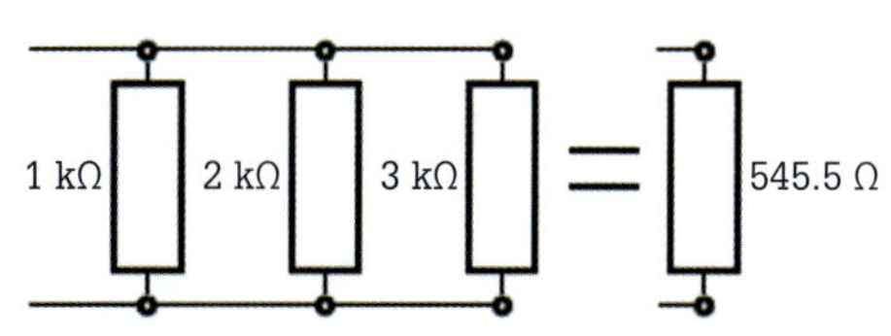

FIGURE 3. 22 Example 3.11 total resistance

Power dissipation

Power distribution in a parallel circuit follows the various rules used for series circuits. In a series circuit, the total power dissipated is equal to the sum of the power dissipated by each individual load. Furthermore, the total power can also be determined directly using the values of total circuit current, source voltage and total circuit resistance. These same relationships are true for power in a parallel circuit:

$$P_{Total} = P_1 + P_2 + P_3 \text{ etc.}$$

$$P = I^2R \quad P = VI \quad P = \frac{V^2}{R}$$

The following examples demonstrate the application of the equations to determine total power in various practical situations.

EXAMPLE 3.12

a A heating element draws 2.5 A from a 240 V supply. Determine the total power rating of the element shown in Figure 3.23.

$$P = VI$$
$$= 240 \times 2.5$$
$$= \mathbf{600\ W}$$

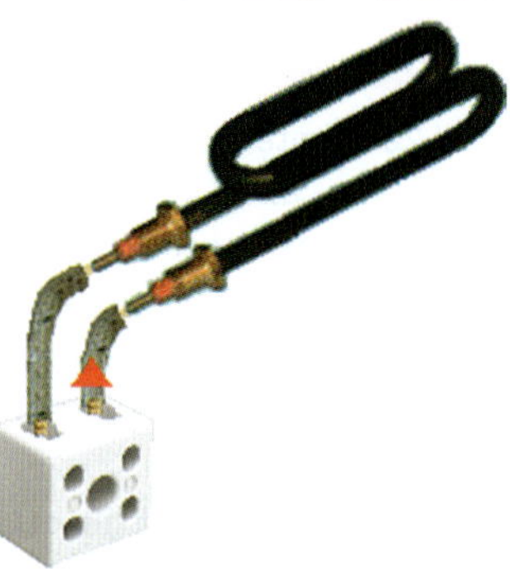

FIGURE 3.23 Example 3.12(a) power rating

b A two-element flat heating array (Figure 3.24) draws 5 A from the supply. If each element has a resistance of 96 Ω, determine the total power of the heating array.

$$P = I^2R$$
$$= 5^2 \times 48$$
$$= 1200 \text{ W}$$

»

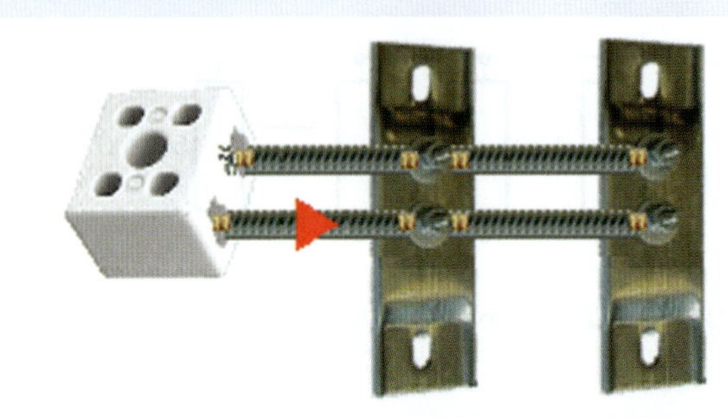

FIGURE 3.24 Example 3.12(b) heating array

c The parallel lighting circuit of **Figure 3.25** consists of three identical incandescent lamps connected across 230 V. If the resistance of each lamp when hot is 2116 Ω, calculate the power dissipated by each lamp.

$$P = \frac{V^2}{R} = \frac{230^2}{2116} = 25\text{ W}$$

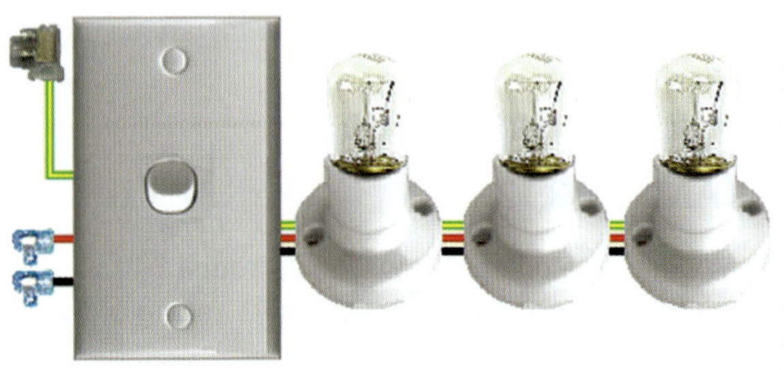

FIGURE 3.25 Example 3.12(c) lighting circuit

Circuit conditions

An open circuit occurring in a branch of a parallel circuit prevents current flowing in that branch only. This causes a reduction in the total current drawn from the supply. As such the equivalent circuit resistance increases as there is less current flowing. The following **Example 3.13** highlights this effect.

EXAMPLE 3.13

a Three resistors consisting of 33 Ω, 56 Ω and 100 Ω connect in parallel. Determine the equivalent resistance of the circuit.

$$R_{equivalent} = \frac{1}{\frac{1}{R_1}+\frac{1}{R_2}+\frac{1}{R_3}} = \frac{1}{\frac{1}{33}+\frac{1}{56}+\frac{1}{100}} = 17.2\ \Omega \text{ (3 significant figures)}$$

b Determine the equivalent resistance of the circuit above if the 33 Ω resistor becomes open circuit.

$$R_{equivalent} = \frac{1}{\frac{1}{R_1}+\frac{1}{R_2}} = \frac{1}{\frac{1}{56}+\frac{1}{100}} = 35.9\ \Omega \text{ (3 significant figures)}$$

A short circuit occurring across a branch has a similar effect as placing a short circuit across the supply. The excessive current flow would cause any over-current protective device to operate to protect the supply cables from over-heating.

REVIEW QUESTIONS

1 If a current of 20 A enters a node with two paths leading away from it and 12 A flows in one path, how much current flows through the other path?

2 Two lamps having an operating resistance of 150 Ω each connect in parallel across a 120 V supply. Determine the potential difference across each lamp.

3 Two resistors, one of 20 Ω and the other of 50 Ω, connect in parallel across a 200 V supply. Determine the current through each resistor and the total current drawn by the supply.

4 Calculate the total resistance of a 22 Ω and 33 Ω resistor connected in parallel.

5 Two resistors, one of 15 Ω and the other of 30 Ω, connect in parallel to a 120 V supply. Determine the total circuit current.

6 Three resistors consisting of 10 Ω, 22 Ω and 82 Ω connect in parallel. Determine the total resistance using the reciprocal method.

7 Two resistors drawing a total current of 150 mA connect in parallel across a 120 V supply. Find their total resistance.

8 A lamp draws a current of 0.65 A from a 230 V supply. Calculate the power dissipated by the lamp.

9 A load bank consisting of three 30 Ω resistor elements in parallel draws 15 A from the supply. What is the total power dissipated by the load bank?

10 A lighting circuit consists of three lamps. The power dissipated by the lamps are 60 W, 25 W and 100 W respectively. If the lamps connect across a 230 V supply, find:

- a the current in each lamp
- b the resistance of each lamp
- c the total circuit current
- d the total circuit resistance
- e the total power supplied to the circuit.

3.3 Series–parallel circuits

Combinational networks

Previously, series direct current circuits (**Chapter 2**) and parallel direct current circuits (Section 3.2) were considered separately. However, many circuits consist of both series and parallel parts. A circuit of this type is a series–parallel circuit or a combinational network. At least three elements are required to create a series–parallel circuit as shown in **Figure 3.26**.

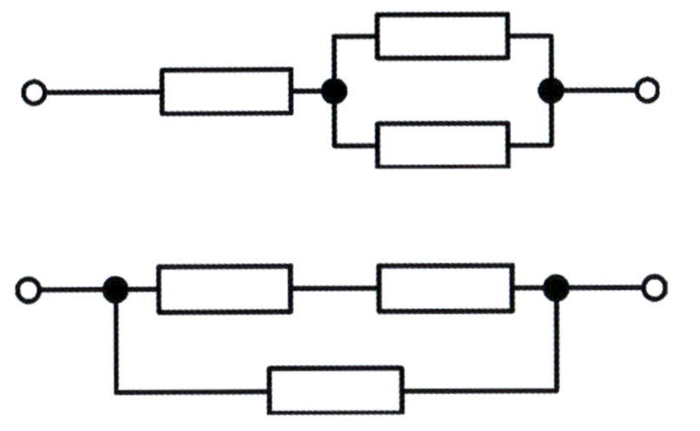

FIGURE 3.26 Two examples of a simple series–parallel circuit

Practical series–parallel circuits

Field coils for electric motors, generators and alternators may connect in series–parallel as illustrated in **Figure 3.27**.

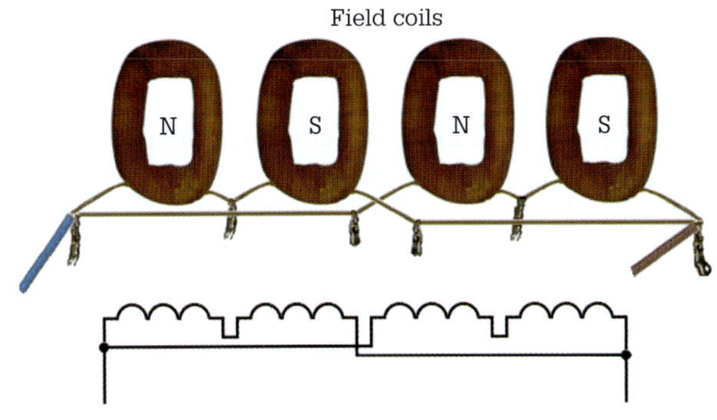

FIGURE 3.27 Series–parallel-connected d.c. field coils

Several batteries may connect in a series–parallel combination as shown in **Figure 3.28** to form a battery bank capable of delivering higher voltage and current than an individual battery.

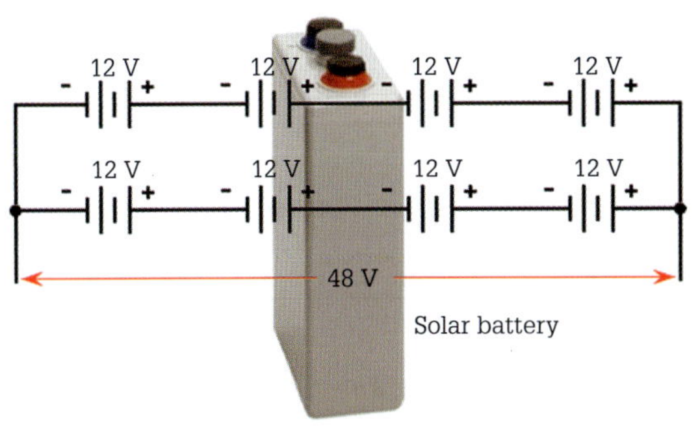

FIGURE 3.28 Series–parallel-connected battery bank

Photovoltaic panels as shown in **Figure 3.29** also connect in a series–parallel arrangement to provide a higher voltage and current output.

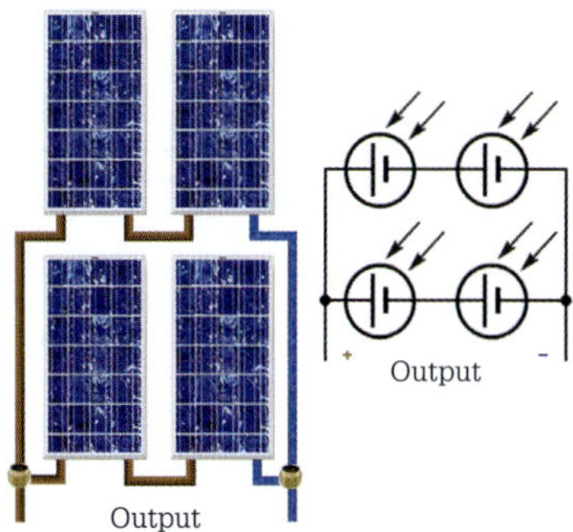

FIGURE 3.29 Photovoltaic panels in series–parallel

Redrawing combinational networks

It is easy to find the various voltages, currents and resistances in series and parallel circuits as the rules applying to series and parallel circuits are specific. Furthermore, in a combinational network, some parts of the circuit are series connected, while other parts connect in parallel.

Thus, in some sections of a combinational network, the rules for series circuits apply, and in other sections the rules for parallel circuits apply. However, before solving a problem or analysing a combinational network, recognition is required of which parts are series connected and which connect in parallel. Admittedly, this is a simple process. However, at times it is necessary to redraw circuits so that the parts and their relationships are evident as in **Figures 3.30** and **3.31**.

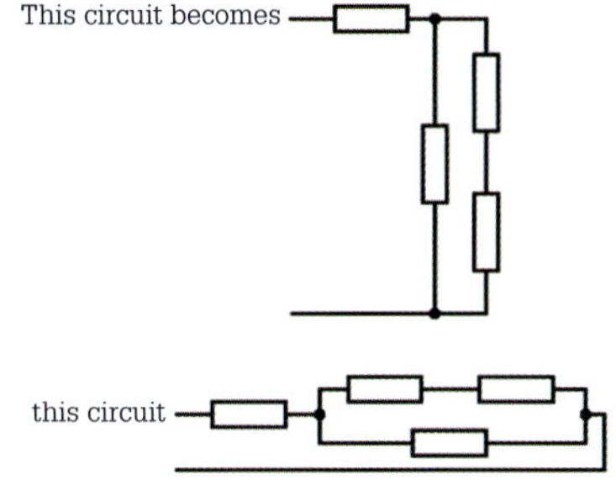

FIGURE 3.30 Redrawing combinational networks

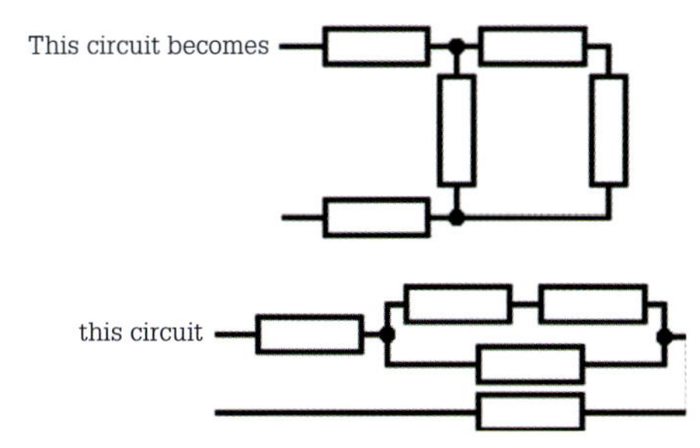

FIGURE 3.31 Redrawing combinational networks

Reducing combinational networks

With many combinational networks, only the supply voltage and the values of the different resistances are obvious. To determine the equivalent resistance it is necessary to reduce the circuit to its simplest form (see **Figure 3.32**).

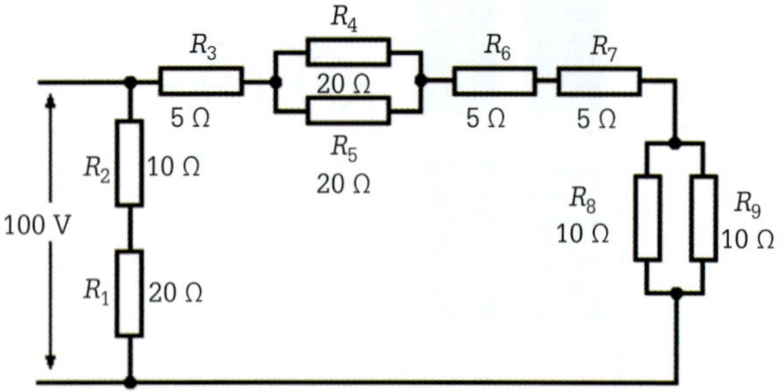

FIGURE 3.32 Combinational networks

A method used for simplifying a circuit starts with calculation of the total circuit current. However, in order to determine the total current, the value of the equivalent resistance of the circuit must be determined. Those values enable the calculation of the voltage drop across any of the resistances or the current through any of the circuit branches to happen.

In examining the circuit in **Figure 3.32**, R_1 is in series with R_2 while R_3 is in series with the R_4–R_5 parallel combinations, R_6 and R_7, and with the R_8–R_9 parallel combination. In addition, the series combination R_1–R_2 is in parallel with the R_3–R_9 network. All complex resistive circuits like this comprise a number of simpler series or parallel systems obeying all the laws and rules that apply to such circuits. In order to resolve the circuit in **Figure 3.32** to one equivalent resistance, the circuit needs clarification. **Figure 3.33** demonstrates this process.

- **Step One:** The most logical point to do this is by first simplifying the series resistors
- **Step Two:** Now simplify the parallel systems, R_4–R_5 and R_8–R_9
- **Step Three:** This simplification has created a series system, R_3–R_9
- **Step Four:** Finally, this parallel combination is reduced to produce a single equivalent resistor with a value of 15 Ω.

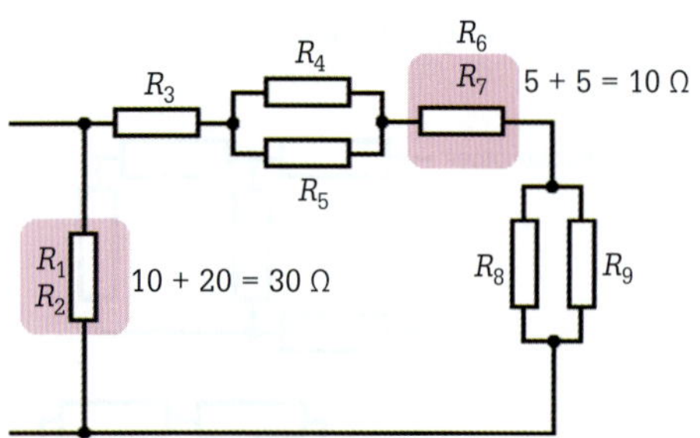

Step One: Simplifying the series resistors

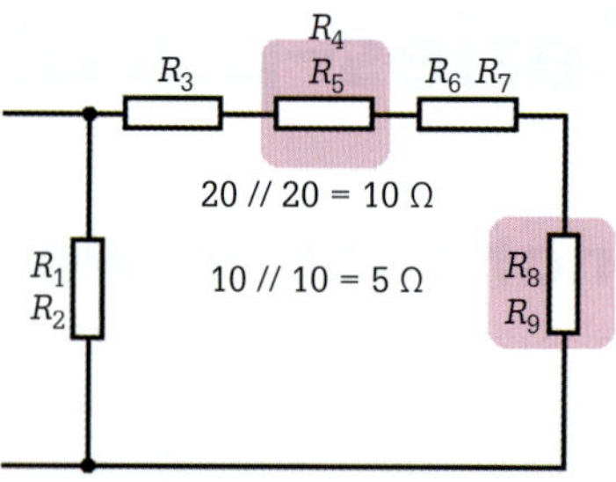

Step Two: Simplifying the parallel networks

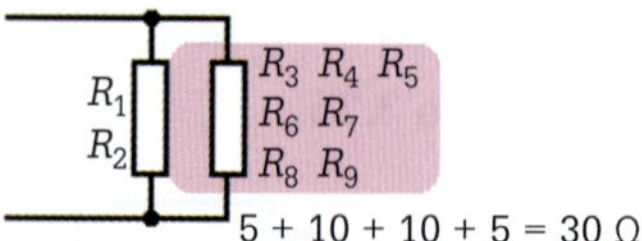

Step Three: Simplifying the series network

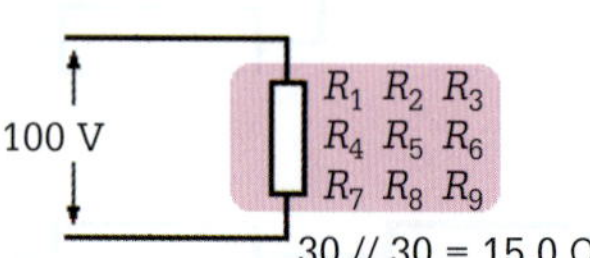

Step Four: Final simplification of the combinational network

FIGURE 3.33 Circuit clarification

Referring back to **Figure 3.32**, the current flowing through each resistor, the potential difference across them and their power dissipation, as well as total current and power, is determined by using the value of the supply voltage.

EXERCISE 3.4

Reduce each of the following combinational networks to a single equivalent resistance.

a

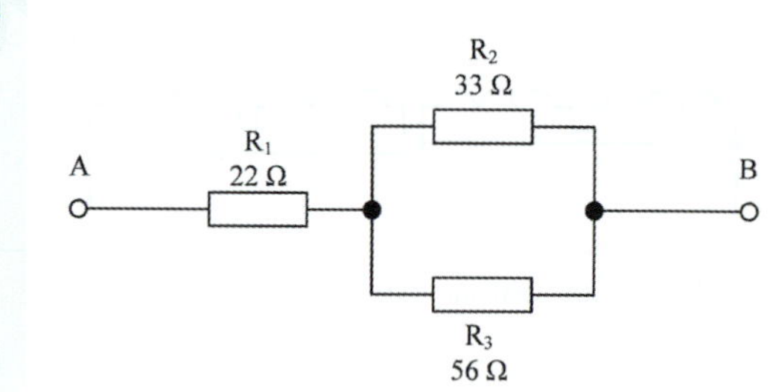

FIGURE 3.34 Exercise 3.4(a) combinational network

b

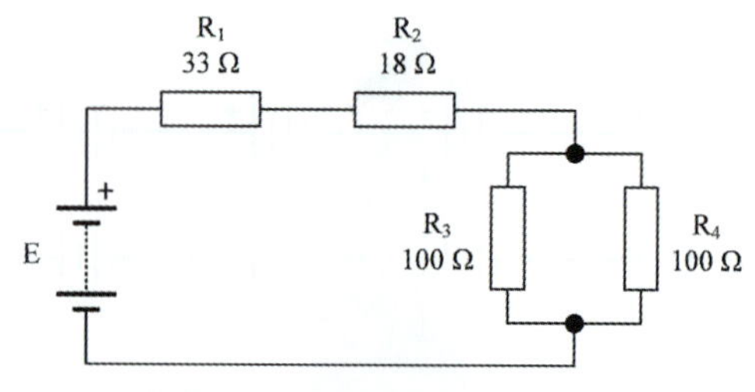

FIGURE 3.35 Exercise 3.4(b) combinational network

Current values

In any direct current circuit, the total current is determined by dividing the supply voltage by the total resistance. In a series circuit, the total current is the same current flowing through the parts. However, by contrast, the current in a parallel circuit divides and flows through more than one branch. Consequently, branch currents are determined by applying Ohm's law. For example, using the values of supply emf (100 V) and the total resistance determined from reducing the combinational network of **Figure 3.32** we will determine the following.

EXAMPLE 3.14

What is the value of I_{Total} for the circuit of **Figure 3.32**?

$$I_T = \frac{V}{R} = \frac{100}{15} = \mathbf{6.67\ A} \text{ (3 significant figures)}$$

What is the value of current that flows through resistors R_1 and R_2?

$$I_{R_1} = \frac{V}{R_1+R_2} = \frac{100}{20+10} = \mathbf{3.33\ A} \text{ (3 significant figures)}$$

and

$$I_{R_2} = \frac{V}{R_1+R_2} = \frac{100}{20+10} = \mathbf{3.33\ A} \text{ (3 significant figures)}$$

Because R_1 and R_2 are in series the current passing through them is the same. Therefore, the calculated current value is the result of the supply voltage divided by their combined resistance.

What is the current through the series resistors, R_3, R_6 and R_7?

$$I_{R_3} = \frac{V}{R_3 \text{ to } R_9} = \frac{100}{30} = \mathbf{3.33\ A} \text{ (3 significant figures)}$$

or

$$I_{R_3} = I_T - I_{R_1} = 6.67 - 3.33 = \mathbf{3.33\ A} \text{ (3 significant figures)}$$

These resistors also form a series circuit; therefore, the same current flows through them.

What is the current through R_4 and R_5?

As voltage across the parallel branches of **Figure 3.32** is unknown, the current through each resistor must be determined by applying the current–divider rule.

The symbol '//' means in parallel.

$$I \text{ through each } R = \frac{R_T}{R \text{ of resistor being measured}} \times I \text{ entering the branch}$$

$$I_{R_4} = \frac{R_4 // R_5}{R_4} \times I_{R_3} = \frac{20//20}{20} \times 3.33 = \frac{10}{20} \times 3.33 = \mathbf{1.665\ A}$$

or

$$I_{R_4} = \frac{I_{R_3}}{2} = \frac{3.33}{2} = \mathbf{1.665\ A}$$

$$I_{R_5} = \frac{R_4 // R_5}{R_5} \times I_{R_3} = \frac{20//20}{20} \times 3.33 = \frac{10}{20} \times 3.33 = \mathbf{1.665\ A}$$

or

$$I_{R_5} = \frac{I_{R_3}}{2} = \frac{3.33}{2} = \mathbf{1.665\ A}$$

What is the current through R_8 and R_9?

$$I_{R_8} = \frac{R_8 // R_9}{R_8} \times 3.33 = \frac{10//10}{10} \times 3.33 = \frac{5}{10} \times 3.33 = \mathbf{1.665\ A}$$

or

$$I_{R_8} = \frac{I_{R_3}}{2} = \frac{3.33}{2} = \mathbf{1.665\ A}$$

$$I_{R_9} = \frac{R_4 // R_9}{R_8} \times I_{R_3} = \frac{10//10}{10} \times 3.33 = \frac{5}{10} \times 3.33 = \mathbf{1.665\ A}$$

or

$$I_{R_9} = \frac{I_{R_3}}{2} = \frac{3.33}{2} = \mathbf{1.665\ A}$$

Now that all the currents throughout the circuit have a value, the voltages developed across individual resistors in **Figure 3.32** require calculation.

Voltage values

Now that we know the resistance values and have just calculated the current flowing through each resistor, we can determine the voltage across each resistor by directly using Ohm's law and Kirchhoff's voltage law.

$$V_{R_1} = IR = 3.33 \times 20 = \mathbf{66.6\ V} \qquad V_{R_2} = IR = 3.33 \times 10 = \mathbf{33.3\ V} \qquad V_{R_3} = IR = 3.33 \times 5 = \mathbf{16.65\ V}$$

$$V_{R_4} = IR = 1.665 \times 20 = \mathbf{33.3\ V} \qquad V_{R_5} = IR = 1.665 \times 20 = \mathbf{33.3\ V} \qquad V_{R_6} = IR = 3.33 \times 5 = \mathbf{16.65\ V}$$

$$V_{R_7} = IR = 3.33 \times 5 = \mathbf{16.65\ V} \qquad V_{R_8} = IR = 1.665 \times 10 = \mathbf{16.65\ V} \qquad V_{R_9} = IR = 1.665 \times 10 = \mathbf{16.65\ V}$$

$V_{R1} + V_{R2} = V_s$; also

$$V_{R_3} + V_{R_4} // V_{R_5} + V_{R_6} + V_{R_7} + V_{R_8} // V_{R_9} = V_S$$

»

Power values

Calculation of power dissipated by individual resistors can use any power equation. Total power for the circuit in **Figure 3.32** can be determined using total resistance of the circuit, supply voltage and total circuit current.

Method 1

$$P = \frac{V^2}{R}$$
$$= \frac{100 \times 100}{15}$$
$$= \mathbf{667\ W}\ \text{(3 significant figures)}$$

Method 2

$$P = I_{\text{Total}}^2 \times R_{\text{Total}}$$
$$= 6.67 \times 6.67 \times 15$$
$$= \mathbf{667\ W}\ \text{(3 significant figures)}$$

Method 3

$$P = I_{\text{Total}} \times V_{\text{Supply}}$$
$$= 6.67 \times 100$$
$$= \mathbf{667\ W}\ \text{(3 significant figures)}$$

EXERCISE 3.5

For the circuit shown in **Figure 3.36**, determine:

a circuit equivalent resistance

b circuit current

c current through each resistor

d potential difference across each resistor

e power dissipated by each resistor.

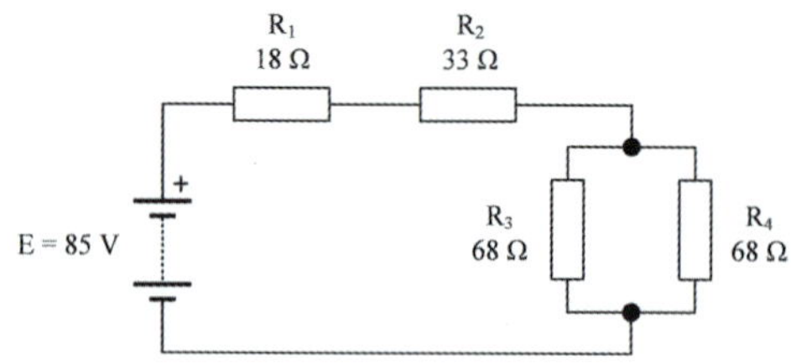

FIGURE 3.36 Exercise 3.5 combinational network

REVIEW QUESTIONS

1 Redraw the circuits of **Figure 3.37** using symbols (values are not necessary) so that the relationships between the various elements are evident.

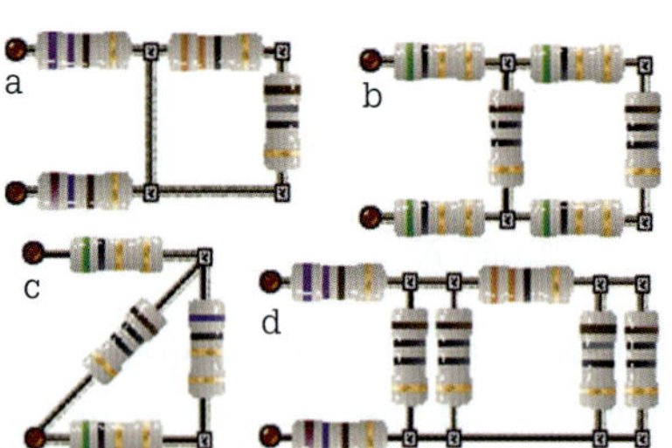

FIGURE 3.37 Pictorial circuit 1

2 What is the simplest circuit value for each of the circuits in **Figure 3.38**?

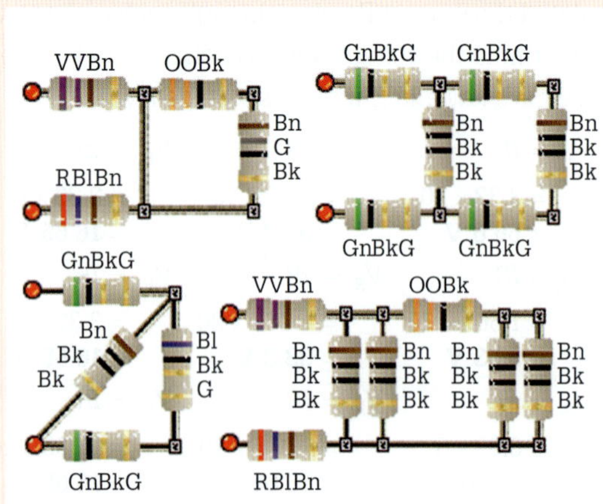

FIGURE 3.38 Pictorial circuit 2

3 Determine the value of the total current in each circuit of **Figure 3.39** when each circuit connects to 100 V.

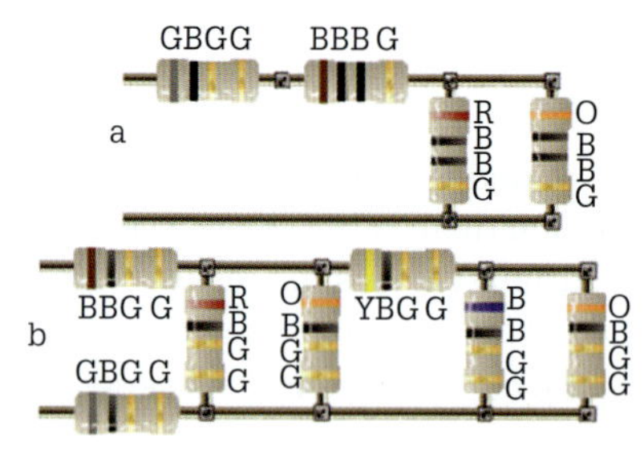

FIGURE 3.39 Combinational networks – current

4 What value of potential is developed across each resistor in the following circuits in **Figure 3.40** if connected across a 100 V supply?

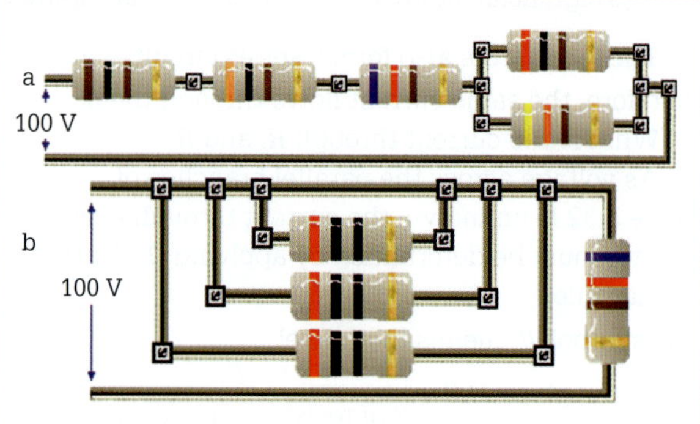

FIGURE 3.40 Combinational networks – potential

»

5 The 200 V d.c. four-pole shunt motor in **Figure 3.41** has an armature resistance of 0.68 Ω, a total interpole resistance of 0.32 Ω and a total shunt winding resistance of 120 Ω. Determine the total motor resistance, the current through the interpoles and the total power dissipated by the motor.

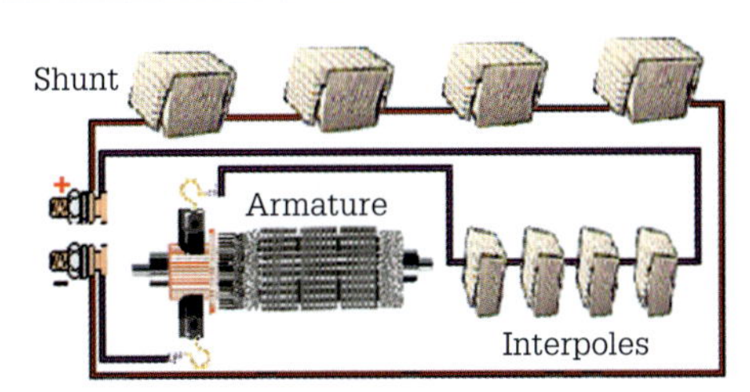

FIGURE 3.41 Four-pole shunt motor with interpoles

6 Determine the total motor resistance of the 200 V d.c. four-pole shunt motor in **Figure 3.41** if one of the interpoles develops an open circuit.

7 Determine the total motor resistance of the 200 V d.c. four-pole shunt motor in **Figure 3.41** if the armature develops a short circuit.

8 A toaster has three 40 Ω heating elements connected in series–parallel (two in series connected in parallel across a third). Determine the equivalent resistance of the combined heating elements, the total power dissipated by the toaster, and the voltage across each heating element when connected across a 200 V d.c. supply.

3.4 Meters in a circuit

Measuring instruments

A meter is an instrument designed to measure accurately and display an electrical quantity such as current, voltage, resistance and power in a readable form. There are two types of meter instruments: analogue and digital. Analogue meters are electromechanical in design and use the motion of a pointer to register readable data on a graduated scale. However, the digital meter is electronic in design and uses a numerical display to register the value of the measured quantity. Many electronic meters use analogue-to-digital converters (ADCs) to convert the analogue voltage and current input waveforms that are present at the meter terminals into digital values that represent the waveform magnitudes at a given point in time.

The term 'analogue' means changing continuously. For example, an analogue clock displays the time using non-stop movement of hands around the clock face. The display method of an analogue meter is a movement (to move the pointer across a scale). There are two types of analogue movements: moving coil and moving iron. See **Figure 3.42** for an example of the moving coil meter movement.

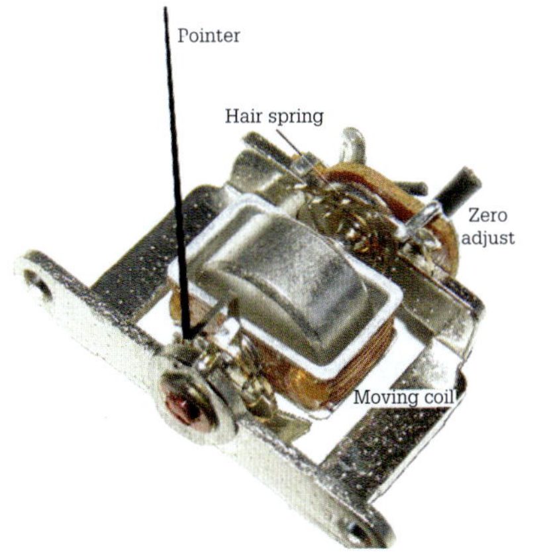

FIGURE 3.42 Moving-coil meter movement

Moving-coil meter

The moving-coil meter is the most commonly used analogue meter for measurements in d.c. circuits. The meter contains a moving coil mounted on jewel bearings between the poles of a permanent magnet (refer to **Figure 3.43**). This type of coil movement is a d'Arsonval movement. The coil has fine magnet wire wound on an aluminium former. In operation, current is required to deflect the meter pointer to the end of the full-scale deflection (FSD). However, due to the fine magnet wire the maximum current for this meter movement is limited from 3 µA to 100 µA while the coil resistance can vary from 1 kΩ to 20 kΩ depending on wire gauge and number of turns.

When direct current flows through the coil, a magnetic field occurs around each leg of the coil. This magnetic field reacts to the field developed by permanent magnets. As a result, the coil pivots by developing forces acting in opposite directions. This action results in the pointer moving across a calibrated scale. The degree of motion relates to the magnitude of the coil current. Control or restoring springs restrain the motion by providing an increasing retarding torque as the coil pivots. Finally, once the coil is de-energised the springs return the coil to its zero position. In addition, adjustment of one of these springs allows the pointer to be directly on zero when no current flows through the coil.

If the moving-coil meter connects through a rectifier (changes a.c. to d.c.) it can measure alternating current. When incorporated in a multimeter this meter can measure current, voltage and resistance separately.

Calibrated scale

The forces that produce the pivoting of the coil in a moving-coil meter are directly proportional to the magnitude of the current flowing through the coil. If the current flowing through the meter movement is 50 µA and the current is doubled (100 µA), then the distance that the needle will move across the scale will double. Consequently, the scale of a moving-coil meter is linear (uniform) as shown in **Figure 3.43**.

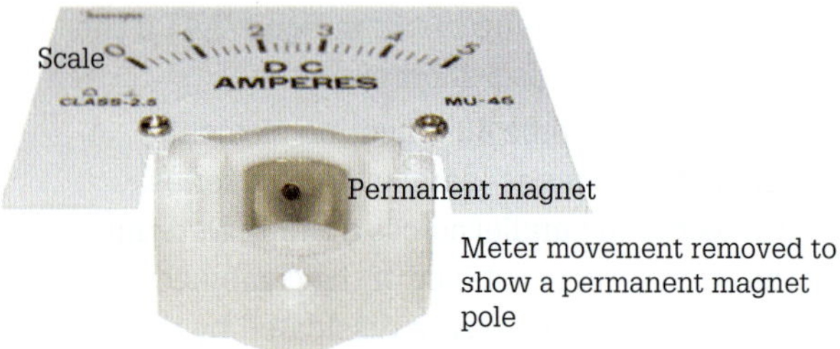

FIGURE 3.43 Linear scale and permanent magnetic pole

Loading effect

As a meter movement needs current to operate, it acts as another load attached to the circuit to which it connects. Thus, the lower the meter-movement current, the less will be the additional meter-load current taken from the circuit and the more accurate the measurement.

Measuring potential difference

A potential difference across the meter coil will cause current to be driven through the coil. The magnitude of this current is dependent on the coil resistance for a given potential difference. It is possible to determine the magnitude of this current using Ohm's law. If, for example, the meter coil resistance is 20 kΩ, and it requires a current of 10 µA to deflect the pointer to full-scale, then it will require the following potential difference to achieve FSD.

$$\begin{aligned} V &= IR \\ &= 10\times10^{-6}\times20\times10^{3} \\ &= \mathbf{0.2\ V\ or\ 200\ mV} \end{aligned}$$

This implies that any voltage greater than 200 mV will cause the pointer to move past the FSD value, and this may damage the movement. To allow the movement to measure higher potentials it is necessary to connect a resistor in series with the coil of the meter movement. This resistor is known as a multiplier. The purpose of the multiplier is to limit the meter current at higher potential to the FSD current of the meter movement. We can calculate the value of multiplier resistance from the following equation.

$$R_{Multiplier} = \frac{E-(I_{FSD}\times R_{meter})}{I_{FSD}}$$

where: $R_{\text{Multiplier}}$ = the value of the multiplier resistance in ohms (Ω)

I_{FSD} = the full-scale deflection current in amperes (A)

R_{meter} = the resistance of the meter in ohms (Ω)

E = the maximum potential to measure in volts (V)

EXAMPLE 3.15

Find the resistance of the multiplier resistor that will allow a meter having a full-scale deflection current of 10 µA and a resistance of 20 kΩ to read full scale when measuring a voltage of 10 V.

$$\begin{aligned} R_{Multiplier} &= \frac{E-(I_{FSD}\times R_{meter})}{I_{FSD}} \\ &= \frac{10-\left((10\times10^{-6})\times(20\times10^{3})\right)}{(10\times10^{-6})} \\ &= 980\ \text{k}\Omega \end{aligned}$$

Using a similar process, calculation of additional multiplier resistance values can occur.

For every 1 V of FSD range required, 98 kΩ must be added to the series multiplier. Refer to **Figure 3.44**.

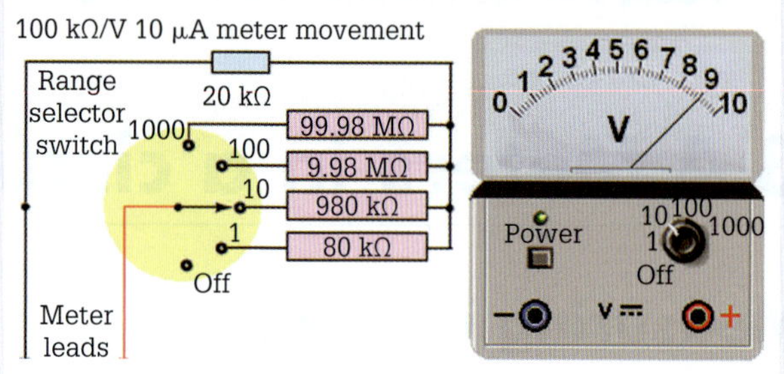

FIGURE 3.44 Multiplier resistors

Sensitivity

Sensitivity is an important parameter concerning circuit loading, and analogue meters such as voltmeters have a sensitivity specified in ohms per volt Ω/V. Sensitivity relates to the resistance of the meter, which includes the meter movement plus any multiplier resistance. It follows that high-sensitivity meters are able to respond to low levels of input – levels that a low-sensitivity meter cannot detect. A meter with high sensitivity is preferred. It is possible to calculate meter sensitivity from:

$$\text{sensitivity} = \frac{1}{I_{FSD}}$$

Where I_{FSD} = the full-scale deflection current in amperes (A)

EXAMPLE 3.16

Determine the sensitivity of two analogue meters where one meter movement has a full-scale deflection current of 20 µA, and the other meter movement has a full-scale deflection current of 200 µA.

Meter 1

$$\begin{aligned} \text{Sensitivity} &= \frac{1}{I_{FSD}} \\ &= \frac{1}{20\times10^{-6}} \\ &= 50\ \text{k}\Omega/\text{V} \end{aligned}$$

Meter 2

$$\begin{aligned} \text{Sensitivity} &= \frac{1}{I_{FSD}} \\ &= \frac{1}{200\times10^{-6}} \\ &= 5\ \text{k}\Omega/\text{V} \end{aligned}$$

Meter 1 has the highest sensitivity so it will have less circuit loading than meter 2.

Note that multiplying the sensitivity of a voltmeter by the FSD range in use will give the resistance of the meter-movement circuit, which connects in parallel to the circuit requiring measurement.

Field-effect transistor (FET) voltmeter

The ideal voltmeter would have an infinite resistance and draw zero current from the circuit under test. Field-effect transistor (FET) voltmeters come close to this ideal. These voltmeters use an FET voltage amplifier to minimise the current drawn from the circuit under test. However, the current drawn from the circuit under test is far too small for any analogue meter to indicate, so these meters amplify this current so that the analogue movement can display the reading. In addition, an FET voltmeter has very high input resistance and excellent sensitivity.

SWITCH ON

Safety precautions with voltmeters

- Visually inspect the meter and leads for damage.
- Ensure the meter and leads are of the appropriate Category rating for the measurement situation.
- Always connect a voltmeter in parallel (or across loads and feeds).
- If the meter has a range selector then always start with the highest range.
- Always observe the correct polarity.
- Never use the d.c. range to measure a.c.

Measuring current

Ammeter

An ammeter measures current flow and analogue meters present this information via a graduated scale using a d'Arsonval meter movement as shown in **Figure 3.45**. This type of movement contains a moving coil mounted on jewel bearings between the poles of a permanent magnet.

FIGURE 3.45 d'Arsonval meter movement

An ammeter connects in series with the load in a circuit and as such it must offer very little opposition to the current so that it does not significantly place another load in the circuit it is measuring.

The ammeter's sensitivity is a measure of the amount of current that causes full-scale deflection of the ammeter's internal meter movement. The lower the current for the meter movement, the more sensitive is the meter and the smaller the loading effect by the meter on the circuit. However, a small loading effect is desirable, since a measurement should try to have the least possible effect on the quantity being measured.

When measuring current, and to prevent the meter movement from burning out, a resistor placed parallel with the meter movement shunts (diverts) most of the current around the meter movement. The value of the shunt resistor does not affect the current flowing in the meter movement at a full-scale reading. **Figure 3.46** illustrates an analogue d.c. ammeter with external shunt resistors.

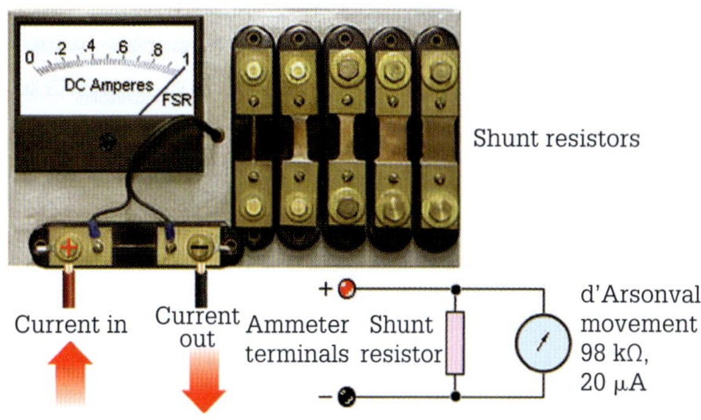

FIGURE 3.46 Analogue ammeter with external shunt resistors

The same 98 kΩ, 20 μA meter movement as shown in **Figure 3.46** can measure larger values of current. When resistors connect in parallel, the smaller resistance value carries more current than the larger resistance value. Calculation of the shunt resistance occurs using the following equation:

$$R_{Shunt} = \frac{I_{FSD} \times R_{meter}}{I - I_{FSD}}$$

where: R_{Shunt} = the value of shunt resistance in ohms (Ω)

I_{FSD} = the full-scale deflection current in amperes (A)

R_{meter} = the resistance of the meter in ohms (Ω)

I = the maximum current to measure in amperes (A)

EXAMPLE 3.17

Find the resistance of the shunt resistor that will allow a meter having a full-scale deflection current of 10 mA and a resistance of 10 Ω to read full scale when measuring a current of 100 mA.

$$R_{Shunt} = \frac{I_{FSD} \times R_{meter}}{I - I_{FSD}}$$

$$= \frac{0.01 \times 10}{0.1 - 0.01}$$

$$= 1.11\ \Omega \text{ (3 significant figures)}$$

An illustration of the various shunt resistors occurs in **Figure 3.47**.

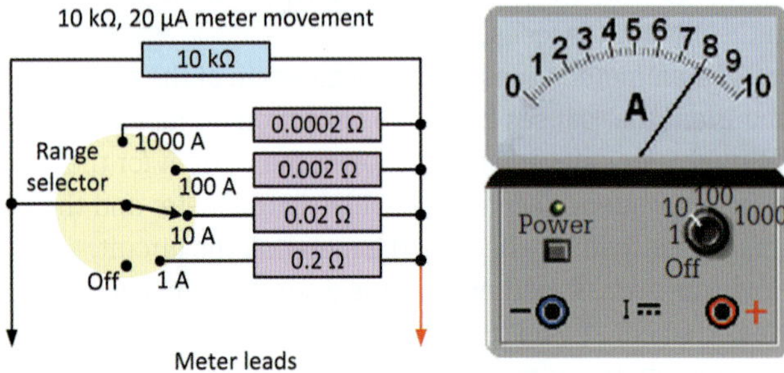

FIGURE 3.47 Ammeter with a 10 kΩ, 20 μA meter movement and shunt resistors

Remember that $P = I^2R$ and this is dissipated as heat, so the chosen resistors must be rated in terms of the wattage they must dissipate, otherwise they will burn out.

Note how the shunt lowers the total effective resistance of the meter as the current range rises. If a mistake occurs and an ammeter is used to measure voltage, it will effectively short the circuit out, causing heavy currents to flow. For example, the one-ampere range has a 0.2 Ω shunt, which when placed across 230 V will cause the following current to flow:

$$I_{\text{short}} = \frac{V}{R} = \frac{230}{0.2} = \mathbf{1150\,A}$$

A fault current of this magnitude, if the circuit protective device allows it, would melt the meter immediately.

SWITCH ON

Safety precautions with ammeters

- Visually inspect the meter and leads for damage.
- Ensure the meter and leads are of the appropriate Category rating for the measurement situation.
- De-energise a circuit before connecting or disconnecting an ammeter.
- Always connect an ammeter in series.
- If the meter has a range selector then always start with the highest range.
- Always observe the correct polarity.
- Never use the d.c. range to measure a.c.
- When measuring d.c. observe the meter's polarity.

EXERCISE 3.6

A moving coil meter movement has a coil resistance of 5 kΩ and a full-scale deflection current of 200 μA. Determine:

a the value of a multiplier resistance to enable the movement to read 50 V full-scale

b the value of a shunt resistance to enable the movement to read 500 mA full-scale

c the sensitivity of the movement.

Moving-iron meter

The moving-iron meter has a moving-iron core attached to the pointer and a fixed soft-iron armature mounted within a coil. The current passing through the coil causes the moving-iron core and the armature to magnetise with the same polarity. Consequently, they repel one another. However, because the core is fixed, the armature moves; this in turn moves the pointer to register a value on the meter scale. The distance the pointer moves is in proportion to the size of the current producing the magnetic field.

A moving-iron meter can measure both d.c. (by changing the meter scale calibration) and root-mean-squared (rms) values of alternating current.

Alternating currents with frequencies of up to 300 Hz require a moving-iron meter for measurement. This is because when a.c. passes through the coil the magnetic field alternates at the same frequency as the current. So, the moving-iron core and the armature become magnetised with the same polarity. In most modern meters, shaping of the fixed and armature irons enables a near-linear measurement scale. In addition, some moving-iron ammeters have 100% over-scale capability as shown in **Figure 3.48**. Consequently, this feature is useful for measuring the starting current of induction motors (a.c. motors).

FIGURE 3.48 Moving-iron meter scale

The coil with these meters has many turns of fine magnet wire to produce a strong magnetic field with only a small current flow.

Taut-band movements

Some analogue movements use a taut band. Taut-band suspensions are suited to high-vibration applications. A taut band is a means of suspending the moving mechanism (coil or moving iron) by soldered or welded bands of metal at the top and bottom of the coil. More to the point, the bands keep the coil, or moving iron, in a state of almost frictionless suspension. The advantages of taut-band movements include high torque and low internal friction and high sensitivity. Finally, taut-band suspension allows the meter to read accurately at all points on the scale.

Reading analogue scales

Analogue meter scales have major and minor graduations. Examples of major graduations are 10, 20, 30, while minor graduations might be 12, 21.

Each scale of an analogue meter has a division of smaller parts called minor graduations. In addition, each minor graduation has a definite value. To determine this definite value, divide the full-scale reading by the number of minor graduations. In **Figure 3.49**, the full-scale reading is 30 volts, and there are 30 minor graduations. **Figure 3.50** shows a scale and various scale readings on a multimeter.

$$\text{value of each graduation} = \frac{\text{full scale reading}}{\text{no. of graduations}}$$

$$= \frac{30}{30}$$

$$= \mathbf{1\,V\ per\ graduation}$$

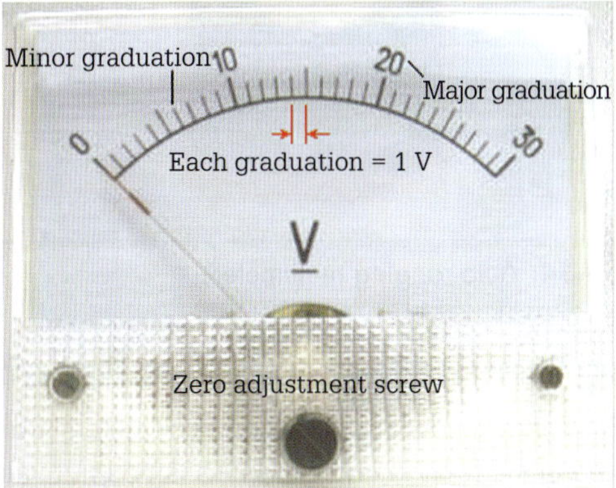

FIGURE 3.49 Graduation values

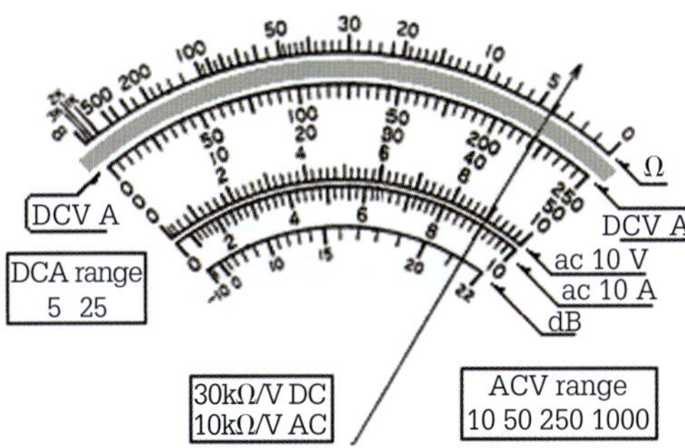

FIGURE 3.50 Various scale readings on an analogue multimeter

Ohmmeters

Ohmmeters as shown in **Figure 3.51** allow users to solve quickly and easily a variety of different measurement problems. These problems include measuring the resistance of conductors, components, printed circuit board tracks, electrical connectors, switch and relay contacts and connections. In addition, an ohmmeter can detect the quality of conductor continuity. An ohmmeter connects in parallel with the unknown resistance measured, but internally connects in series. Ohmmeters operate by passing a low magnitude current through the component measured, and then measure the current required. This current value is converted into resistance through the application of Ohm's law and displayed as ohms by the meter scale. Finally, when measuring resistance, always select the ohmmeter range that allows the largest scale deflection.

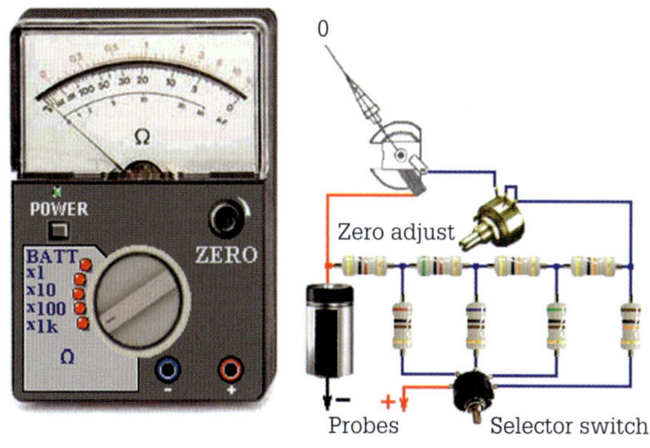

FIGURE 3.51 Analogue ohmmeter showing zero adjust and selector switch

SWITCH ON

Safety precautions with ohmmeters

- De-energise the circuit before connecting an ohmmeter.
- Always start with the highest range.
- Adjust the ohmmeter to 'zero' before measuring resistance.
- Switch off the ohmmeter when measuring is finished.

Digital multimeters

Digital meters carry out the same function as analogue meters; however, there are major differences between the two types of meters. Digital meters are easier to read because of the numerical display, which includes the placing of the decimal point to enable an exact display of the reading (see **Figure 3.52**). Furthermore, digital meters are more sensitive than analogue meters, except for field-effect transistor (FET) voltmeters. A digital meter is not necessarily more accurate than an analogue meter as accuracy of a measurement defines how near the indicated measurement is to the actual value of the value.

FIGURE 3.52 Digital multimeter

Digital ohmmeters

Digital ohmmeters use a low-noise, high-resolution analogue-to-digital converter. The measurement taken shows on a high-intensity-digit LCD readout called an alphanumeric display. Some digital ohmmeters use a four-terminal output configuration to reduce test lead wire and contact resistance errors. In fact, a four-terminal measurement technique provides for stable and repeatable low-resistance readings, and it has high accuracy, typically ± 0.3%. The size range can be from 0.1 μΩ to 1999 Ω. Finally, a four-terminal measurement system is one in which a constant current source is applied to the terminals of a resistor via the current terminals on the ohmmeter.

Digital ohmmeters have an auto-ranging and auto-decimal point, and some have an auto-power down after three minutes. In addition, a nine-volt battery powers them. Refer to **Figure 3.53**.

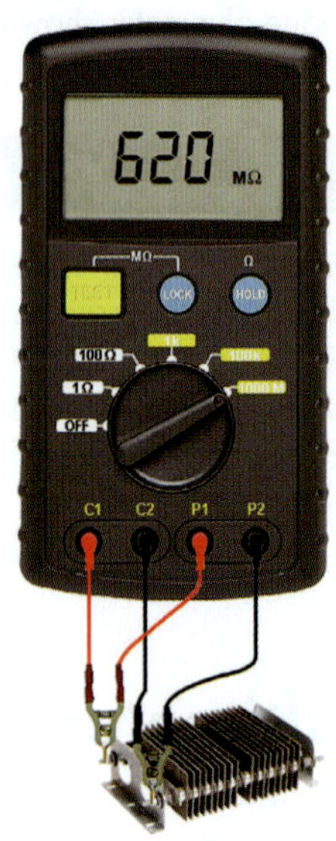

FIGURE 3.53 Four-terminal digital ohmmeter

Auto-ranging multimeter

An auto-ranging multimeter has far fewer selection places on its function pad – often just amps, volts, ohms and frequency. When the probes of this meter connect across a resistance, the meter automatically selects the proper range to show the measure with appropriate accuracy. An auto-ranging multimeter looks simple to use: press the voltage button and the meter does the rest. However, the meter can be slow to read the measured value because it needs to discern in which range to operate. In addition, if the display moves around for a long time before settling, it can be a problem when needing quick measures. More to the point, this indecision with the correct number display can cause difficulties, especially if the measured factor changes at the same time. However, some auto-ranging multimeters have an option of fixing the range to speed up readings. Refer to **Figure 3.54**.

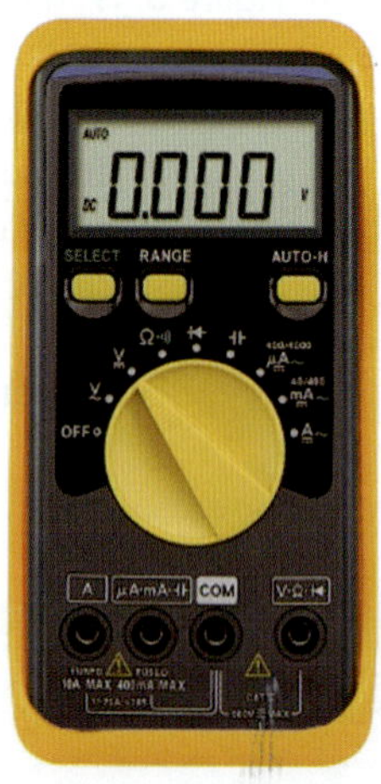

FIGURE 3.54 Auto-ranging multimeter

Using ammeters and voltmeters

Figure 3.55 shows how ammeters and voltmeters are connected in a circuit in order to establish circuit values.

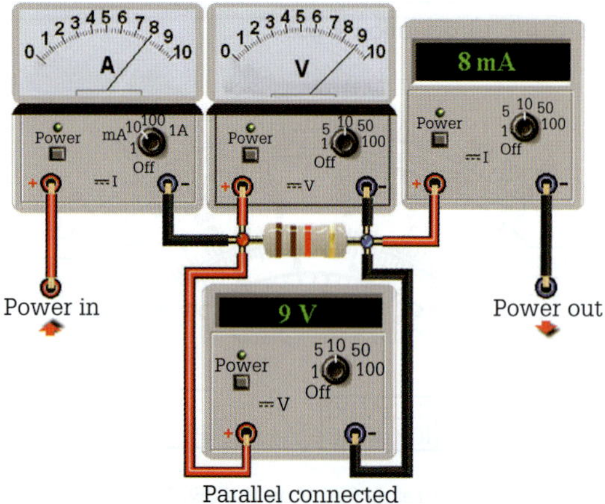

FIGURE 3.55 Ammeters are connected in series and voltmeters are connected in parallel

EXAMPLE 3.18

Determine the resistance of the test circuit in **Figure 3.55**.

Use the values displayed on the voltmeter and ammeter in the Ohm's law equation for resistance.

$$R = \frac{E}{I}$$

$$= \frac{9}{8 \times 10^{-3}}$$

$$= 1.125\ \text{k}\Omega$$

Comparison between digital and analogue measurement

TABLE 3.5 Definitions of variables

Variable	Definition
Resolution	The smallest value of the input (or output) signal that meters can measure. In analogue meters, resolution is the difference between adjacent scale divisions. In contrast, resolution in digital meters is the number shown by a one-digit change in the least-significant digit.
Accuracy	How near the indicated measurement is to the actual value of the signal. As for an error, this is '% error'.
Sensitivity	A measurement of the meter's input resistance per volt. In other words, the amount of power the meter takes from the circuit to function.
Noise	An undesirable signal from an external source. Equipment such as a.c. power lines, motors, generators, transformers, fluorescent lights, CRT displays, computers and radio transmitters create noise.
Damping	The technique that allows the analogue pointer to settle due to the measured quantity's change in value.
Stability	The instrument design which allows the instrument to deliver a constant reading in relation to a fixed input.
Taut-band	A mechanism whereby the moving element in an analogue meter is suspended.

Refer to **Table 3.6** for a more detailed comparison between digital and analogue meters, depending on which variable you are measuring. In general, expanded capabilities of digital meters include:

- better overload protection
- automatic as well as manual range selection options
- memory
- frequency measurement sensing
- peak hold and data hold
- capacitance measurement
- analogue bar graphs
- logic probes.

TABLE 3.6 Comparison illustration

Characteristic	Digital meters	Analogue meters
Display	Easy to read. Present measurement data in a direct format on a digital display. Are free from the parallax error of analogue meters.	Use a needle and a calibrated scale to indicate values. Can be difficult to read when scales are non-linear. Parallax error is possible.
Resolution	Often specified in 'digits' of resolution. For example, the term 4 digits refers to the number of digits displayed on the display of a multimeter. A 4-digit monitor has a resolution to 9999 counts. 0.01% for a 4-digit display.	Limited by the width of the scale pointer, vibration of the pointer, the accuracy of printing of scales, zero calibration, number of ranges and errors due to non-horizontal use of the mechanical display. About 1%, rarely better than 0.25%.
Accuracy	DC volts and amps 0.05% of full scale +1 digit AC volts and amps 0.2% of full scale +1 digit	Analogue meters usually have their accuracy listed as a percentage of the full-scale deflection reading. About 2% with some models achieving 0.1% are available. Meter movements because of their moving parts are subject to wear and shock failures and this affects accuracy.
Stability	Can be ultra-stable depending on manufacturer. Uses stable on-board references such as a thermally stabilised voltage reference. Resistance functions are referenced to a highly stabilised metal-foil resistor. Some meters may display the least significant digit changing intermittently.	Suspension (taut band)/Air damper
Polarity	Indicates negative input signals by turning 'on' a minus sign display-segment.	Pointer deflects in the wrong direction – may cause damage to meter movement.
Warm-up time (response time)	Usually < 1 sec	Has the ability to read subtle changes in current and see cycles if there are any.

»

Characteristic	Digital meters	Analogue meters
Input impedance (sensitivity)	Usually 10 MΩ, 50 µA	Most analogue multimeters of the moving-point type are unbuffered, and draw current from the circuit under test to deflect the meter pointer. The impedance of the meter varies depending on the basic sensitivity of the meter movement and the range which is selected. Typical sensitivity is 20 000 Ω/V (50 µA).
Noise	Electrical noise can cause inaccuracy. Shielded twin lead may be required if meter reading is erratic due to noise pickup.	Slight effect from electrical noise.

Testing equipment

It is often necessary for testing to be carried out live (for example, when testing meters, voltage, load and phase sequence). However, Work Health and Safety Regulations require employers to ensure that persons conducting tests for electrical system integrity and operability conduct the tests in a safe manner using a safe work system, appropriate personal protective equipment (PPE) and appropriate test equipment.

Before measuring, the electrician should revise the safety basics. For example, is the testing equipment suitable and approved for the measurement?

Live testing requires awareness of impulse faults (also called transients). These faults travel on power source waves as spikes and can cause severe injury. Note that insulation that has the dielectric strength to withstand the normal working voltage of the circuit will break down under large-impulse voltage spikes.

Test instruments need to be compliant for use in a defined energy-level measurement category. There are four energy-level measurement categories, I to IV, and they describe how much electrical energy could be present during an impulse voltage fault.

Finally, all testing instruments should be stored and maintained according to manufacturer's instructions.

Multimeters, testing and other measurement devices

The most versatile item of electrical test equipment is an instrument called the multimeter. The analogue versions are called VOMs (Volt-Ohm-Meter) while those with a digital readout are DMMs (digital multimeters). These meters have the ability to measure a range of electrical variables: current, voltage, capacitance and resistance. In addition, they have other functions such as diode testing, transistor testing and continuity checking. **Figure 3.56** shows two different types of multimeter.

FIGURE 3.56 Analogue and digital multimeters

Clamp tester

The clamp tester, also known as a tong meter or a clip-on ammeter, can be either analogue or digital. An important feature of these meters is that they can measure current by clamping around a live conductor. These meters are able to record readings in hard-to-reach or poorly lit areas due to the meter pointer or digital reading locking feature.

This type of meter is an average current sensing instrument and has an rms calibration. In addition, analogue clamp meters have taut-band moving-iron meter movement, and manual zero adjustment, and they can read both a.c. and d.c. Refer to **Figure 3.57**.

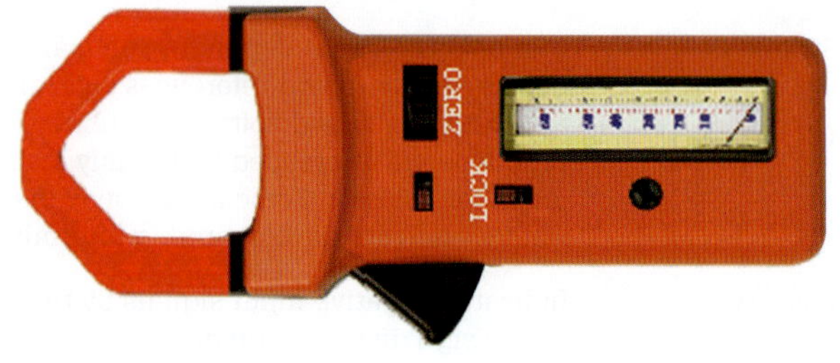

FIGURE 3.57 Analogue clamp meter

Hall Effect

The Hall Effect occurs when the current carriers moving through a material experience a deflection towards the edge of the material because of an applied magnetic field. This deflection provides a measurable potential difference across the sides of the material.

When a current-carrying conductor loop enters a magnetic field, a voltage develops between one side of the conductor and the other. For voltage to happen, the magnetic lines of force must be at right angles to the line containing the conductor. Consequently, the voltage appears at the sides of the conductor that are at right angles to the magnetic lines of force.

If the conductor has a rectangle shape, and the magnetic lines of force are perpendicular, then an uneven distribution of charges creates a potential difference. In addition, this voltage appears between the opposite edges of the conductor. Furthermore, this transverse voltage (shown in **Figure 3.58**) is a Hall Effect and its magnitude is equal to the interrelationship between current, the elementary charge, the magnetic field and the bulk density of the material.

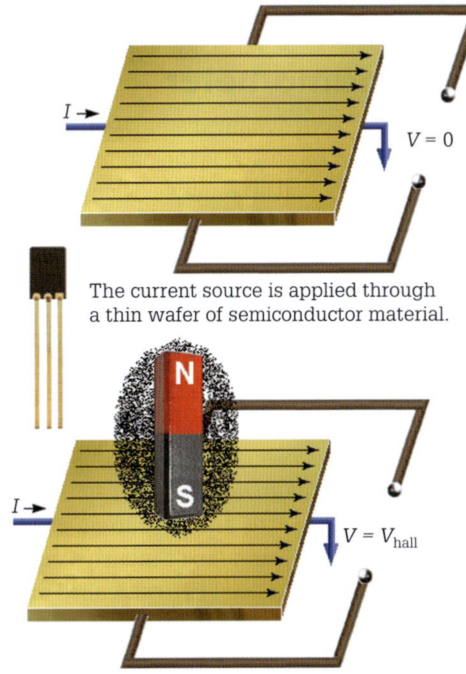

FIGURE 3.58 Hall Effect and sensor

Hall Effect sensors occur in many types of sensing devices. If the quantity measured includes a magnetic field, a Hall sensor will perform the task. Applications of Hall Effect sensors are:

- direction detection
- linear sensing
- speed sensing
- position sensing
- gear-tooth sensing
- contactless switching.

Hall Effect clamp meter

The Hall sensor has a location in the air gap of the magnetic core of the clamp meter. To operate the clamp, place a current-carrying conductor through the aperture of the core. The conductor produces a magnetic field proportional to the level of current. The metal core concentrates this magnetic field around the Hall sensor. Consequently, the Hall sensor develops a signal voltage sending it to an amplifying circuit. Once amplified the voltage signal indicates on the meter display the current flowing through the conductor. The Hall device responds to both varying and stationary magnetic fields; therefore, the meter can read both a.c. and d.c. However, with d.c. measurement, the meter requires zeroing to remove residual magnetism that may be present in the metal of the clamp. Refer to **Figure 3.59**.

FIGURE 3.59 Auto-ranging clamp meter and application

Voltage detectors

The voltage detector's pencil-like probe via a series of LEDs checks both a.c. and d.c. voltages within a range of 12 to 690 V a.c. and 6 to 220 V d.c. The LEDs will energise according to the voltage that is present across the testing points. For example, if the voltage was 110 V d.c, all the LEDs up to and including the 110 V LED would be lit up. Some of these voltage indicators have audible continuity test capability and automatic power on and off. It is an excellent device for checking for 230 V and a 400 V potential presence. Refer to **Figure 3.60**.

FIGURE 3.60 LED voltage detector

AC voltage finder

The a.c. voltage finder is a very safe device to use. It requires no contact with live conductors. This device detects the electromagnetic field due to current flowing in the conductor. The a.c. voltage finder in **Figure 3.61** senses any a.c. voltage above 120 V and indicates this voltage presence with a flashing LED and audible sound.

FIGURE 3.61 a.c. voltage finder

Series test lamp

These represent the simplest tool for low-voltage mains testing and proving dead. A series test lamp as illustrated in **Figure 3.62** is rugged and reliable and meets the requirements of Australian and New Zealand standards. Furthermore, test lamps are used throughout the electricity supply and electrical industries.

FIGURE 3.62 Series test lamps

The series test lamp consists of insulated fixed probes, a 2 m lead, lamp holder, pilot lamp, switch and fuse. However, the hazard associated with this tester when in service is the lamp current.

Continuity tester

If the continuity of a conductor is unbroken, the continuity tester will send out a loud sound, light an indicator lamp or display a combination of both. **Figure 3.63** shows an acoustical continuity tester.

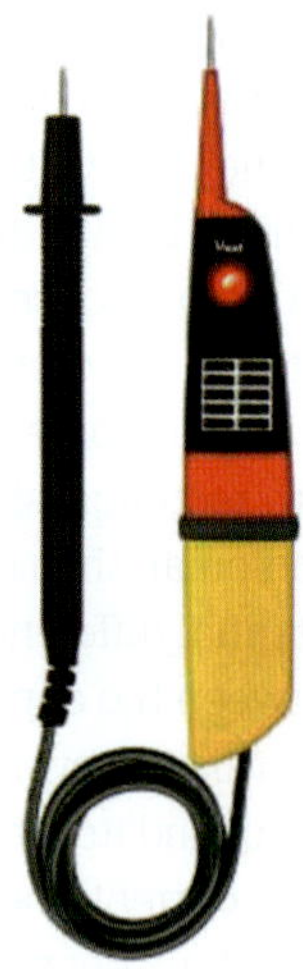

FIGURE 3.63 Acoustical continuity tester

Electrodynamometer

The purpose of an electrodynamometer (watt-hour meter) is to measure energy. As energy is the product of power and time, the watt-hour meter requires designing around these two parameters. The analogue electrodynamometer is an induction motor whose speed is directly proportional to the power moving through it. The total number of revolutions over a given time is comparable to the total watt-hours dissipated. These meters consist of a current coil wound with thick wire to carry the line current and two voltage coils wound with a large number of turns of fine wire. Located between the voltage coils and the current loop is an aluminium disc rotor. When the meter operates, the two voltage coils and the current coil create their own magnetic flux. These fluxes induce a current in the aluminium disc. The induced current, called an eddy current, creates a magnetic field and the relationship of the three magnetic fields enables the aluminium disc to rotate. Refer to **Figure 3.64**.

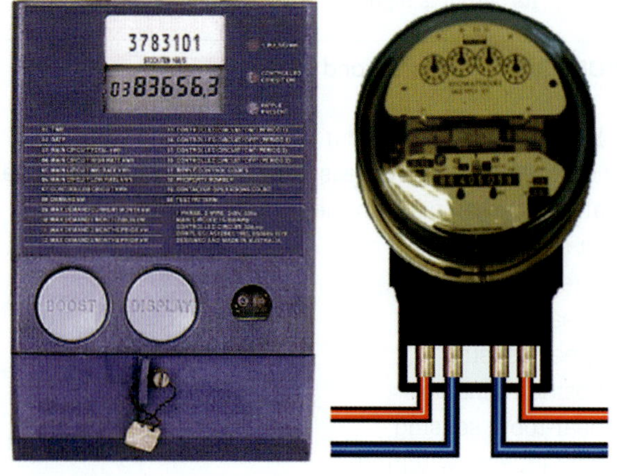

FIGURE 3.64 Digital and analogue watt-hour meters

Digital wattmeters

Digital wattmeters provide direct displays of watts, kilowatts, kilovolt amps, kilovolt amps reactive and the power factor. They are simple, easy to use power-metering devices with a variety of current transformers. In addition, they can measure power directly at the consumer's box or monitor the power consumed by appliances as shown in Figure 3.65.

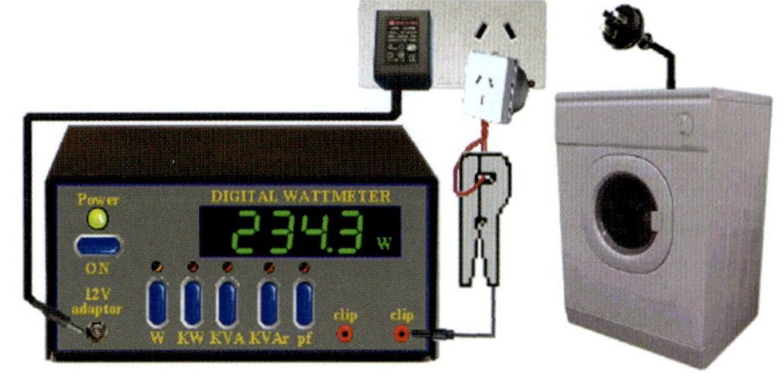

FIGURE 3.65 Appliance under test (plugs removed to show sequence)

REVIEW QUESTIONS

1. How does an ammeter connect in a circuit?
2. How does a voltmeter connected in a circuit?
3. Which analogue meter movement responds only to direct current?
4. Which analogue meter movement responds to direct current and alternating current?
5. What is the type of meter that can measure current without physical connection to the test circuit?
6. Figure 3.66 shows the meter scale on a moving-coil meter. Determine the meter readings indicated by arrows A to E on the 30 A range.
7. Figure 3.66 shows the meter scale on a moving-coil meter. Determine the meter readings indicated by arrows A to E on the 250 V range.
8. A moving-coil meter movement has a coil resistance of 10 kΩ and a full-scale deflection current of 100 μA. Determine:
   a. the value of a multiplier resistance to enable the movement to read 150 V full-scale
   b. the value of a shunt resistance to enable the movement to read 200 mA full-scale
   c. the sensitivity of the movement.

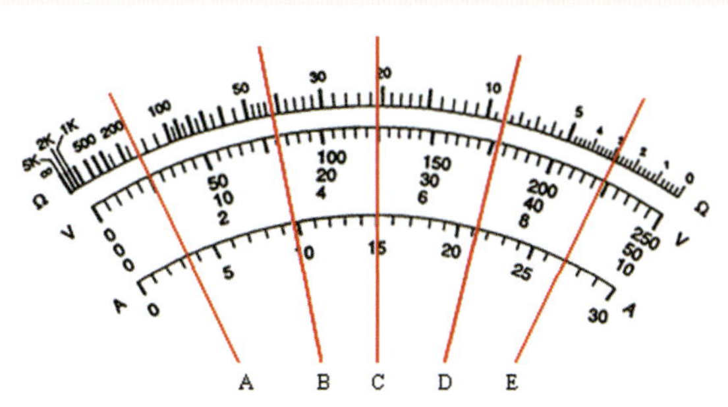

FIGURE 3.66

3.5 Resistance measurement

There are many reasons why the resistance of materials is measured. The following list details some of those reasons.

- Component testing and sorting items such as resistors and inductors require verification of their specified resistance tolerance.
- Contactors, relays and switches all require testing to verify that their contact resistance is below specified design limits.
- Cable and conductor manufacturers must measure the resistance of the wires they produce. A high resistance means that the current-carrying capacity of the cable reduces; on the other hand, too low a resistance means that the manufacturer is using more material in relation to the diameter of the conductor than necessary, and this can be very expensive.
- Bonding resistance measurement means that any metal-to-metal connection, weld joints, crimp lug connections for cables, connector integrity, switch gear to power lines connections and bolted joint resistance on busbars, all need to be proven for contact reliability.
- Fuse manufacturers need to measure the resistance of the fuse elements for quality control with respect to the dimensions of the part.
- Electrical motor and generator manufacturers need to determine the maximum temperature that motors and generators reach under full load. To determine maximum temperature, manufacturers use the temperature coefficient of the metal used in the winding. First, measurement of the winding resistance of the motor or generator at ambient temperature occurs. Second, the motor or generator runs at full load for a specified time, and the resistance is measured again. Finally, using the change in resistance value, calculation of the internal motor or generator temperature transpires.
- Welding shops need to measure the resistance of welding cables to ensure that welding quality does not deteriorate.
- Supply authorities need to measure distribution joint resistance and the soil resistivity for both the supply network earthing point and for lightning arrestors.

- Wiring and maintenance of power cables and switchgear require switch contacts and cable joints to have low resistance. More to the point, low resistance prevents a joint or contact from becoming excessively hot. Consequently, a weak cable joint or switch contact through heat will fail. It follows that routine preventive maintenance with regular resistance checks ensures the best possible life performance of power cables and switchgear.

SWITCH ON

Due to the electrical hazard, before taking resistance measurements prove that the component or device is de-energised.

Substitution method

One method used to determine the value of an unknown resistance is to compare it with a known standard resistance. This process, called the substitution method, works on the principle that if the unknown and known resistances are of the same value, they will allow the same magnitude of current to flow when connected in turn in the same circuit, as shown in **Figure 3.67**.

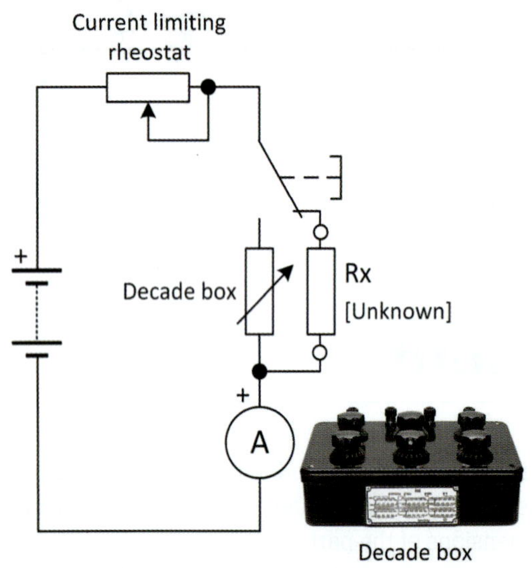

FIGURE 3.67 Substitution method

Source: Shutterstock.com/Shiyan Sergiy

The substitution method uses a resistance or decade box to indicate the value of the unknown resistance (R_X) being measured. Resistance boxes are often of the 'dial' type, in which the sum of the different readings on the various dials shows the total resistance measured. These decade boxes have very sensitive resistors and draw low current. For accurate measurement, the wattage dissipation for the resistors is 10 mW.

With the substitution process, known and unknown resistors connect into a circuit containing a power source, a two-way switch, a rheostat and an ammeter. The unknown resistor connects in the circuit, and the reading on the ammeter can be observed. Then the calibrated variable resistors dial into the circuit. Adjustment of these resistors occurs until the same reading shows on the ammeter. The value of the resistor is then equal to the value recorded on the various dials. The current limiter rheostat prevents a high current from flowing in the circuit. The power source must provide a constant voltage to deliver a constant current. Finally, to facilitate accurate resistance measurement, the ammeter should have excellent sensitivity to prevent additional loading of the circuit.

Modern decade resistance boxes are variable resistance output devices with several advanced features. This means connecting resistances into the measuring circuit via a keypad or incrementally by a rotary switch. Measured values display, in ohms, on a large liquid crystal screen (LCD). In addition, modern decade boxes are capable of storing values in internal memory. These resistance boxes can have a measuring range from zero to 24 MΩ with an efficiency of ±0.1 per cent.

Typical applications of decade boxes include measurements on coils, motor and transformer windings, contact resistance on switches, relays, pushbutton contacts, checking and calibration of insulation testers and the calibration of micro-ohmmeters.

Voltmeter/ammeter method

This system uses a voltmeter and ammeter to measure the potential difference across, and the current through, a resistor. Once these values are known it is possible to calculate the value of the resistance using Ohm's law. As the ammeter and the voltmeter both load the circuit, their positioning in the circuit should provide the minimum loading.

There are two approaches employed with this process. The first involves knowing the sensitivity of the meters. The sensitivity in ohms per volt, refers to the resistance of the meter as seen by the load under test. What the load sees is relative to the full-scale value of the voltage range selected. For example, with a 1000 Ω/V meter the 10 V range would have a resistance of 10 000 Ω and the 50 V range 50 kΩ (50 000 Ω), and so on. The second process involves the inclusion of a rheostat to enable a number of readings, so determination of the average value of the ratio of voltage to current occurs. In addition, there are two connection methods for the meters. The first method is suitable when the unknown resistor has a low value. Correspondingly, the second method is more accurate when the unknown resistor has a high value.

Low-resistance method (also called short-shunt method)

Using the meter sensitivity approach

The voltmeter in **Figure 3.68** has a sensitivity of 1000 Ω/V. Its resistance (R_{meter}) equals 10 000 Ω on the 10 V FSD range. The voltmeter indicates a reading of 10 V. In addition, the ammeter shows a reading of 0.80 A, or 800 mA. Using these values, the calculation of the unknown resistor R_X can occur.

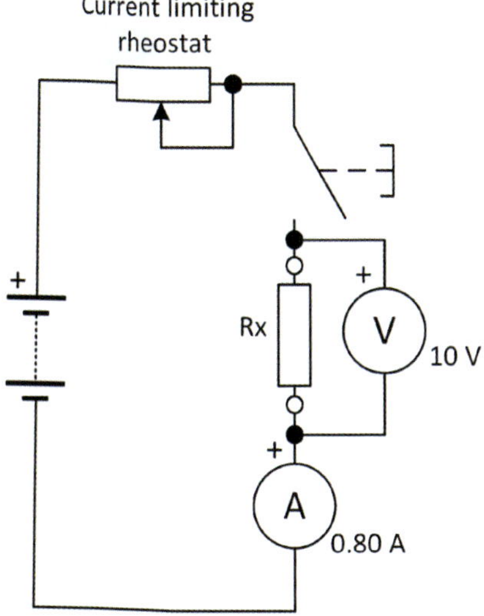

FIGURE 3.68 Low-resistance method

EXAMPLE 3.19

Ignoring the voltmeter resistance:

$$R_{Apparent} = \frac{V}{I} = \frac{10}{0.80} = 12.5\ \Omega$$

Including the voltmeter resistance, the current through the voltmeter is:

$$I_{Meter} = \frac{V}{R_{Meter}} = \frac{10}{10\,000} = 1\ \text{mA}$$

Therefore, current through the resistor is:

$$I_R = I - I_{Meter} = 0.8 - 0.001 = 0.799\ \text{A}$$

The actual value of resistance is:

$$R_X = \frac{V}{I_R} = \frac{10}{0.799} = 12.516\ \Omega \text{ (5 significant figures)}$$

The error incurred by including the current through the voltmeter in the first method is:

$$Error = \frac{R_X - R_{Apparent}}{R_X} = \frac{12.516 - 12.5}{12.516} = 0.128\ \% \text{ (3 significant figures)}$$

The percentage error is not significant for low-value resistance. With this circuit it should be noted that the ammeter measures the current flowing through the resistor and the voltmeter. To reduce the loading effect of the voltmeter and to increase efficiency, use of a meter with high sensitivity (in the order of 20000 Ω/V) is required.

The inclusion of a rheostat in the circuit (see **Figure 3.68**) enables a number of readings to be taken by adjusting the rheostat so that the average value of the ratio of voltage to current can be determined. Alternatively, a graph of voltage against current can be plotted and the slope of the graph will give the resistance.

High-resistance method (also called the long-shunt method)

The high-resistance method places the voltmeter across both the ammeter and the unknown resistance. In this method, shown in **Figure 3.69**, the reading on the voltmeter includes the voltage drop across the ammeter.

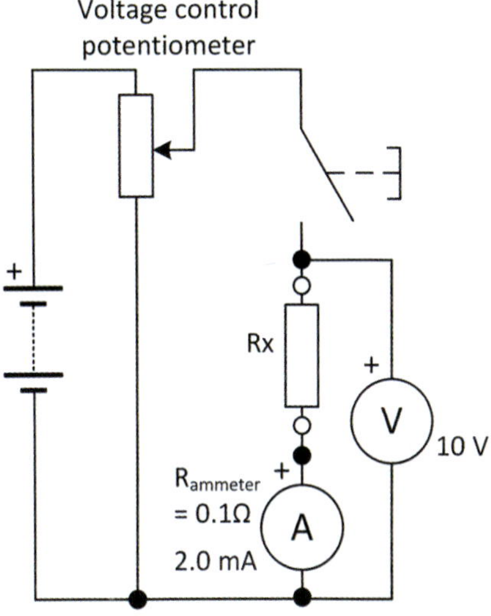

FIGURE 3.69 High-resistance method

The ammeter in this circuit has a resistance of 0.1 Ω and a current reading of 2 mA, while the voltmeter reads 10 V. Using these values it is possible to calculate the resistance of R_X.

EXAMPLE 3.20

Ignoring the ammeter resistance:

$$R_{Apparent} = \frac{V}{I} = \frac{10}{(2 \times 10^{-3})} = 5000\ \Omega\ (5.00\ \text{k}\Omega)$$

Including the ammeter resistance, the voltage across the ammeter is:

$$V_{Ammeter} = I \times R_{Ammeter} = (2 \times 10^{-3}) \times 0.1 = 0.000\,2\ \text{V}\ (200\ \mu\text{V})$$

Therefore, the potential across the resistor is:

$$V_R = V - V_{Ammeter} = 10 - 0.000\,2 = 9.999\,8\ \text{V}$$

The actual value of resistance is:

$$R_X = \frac{V_R}{I} = \frac{9.999\,8}{(2 \times 10^{-3})} = 4999.9\ \Omega$$

The error incurred by including the V_d across the ammeter in the first method is:

$$Error = \frac{R_X - R_{Apparent}}{R_X} = \frac{4999.9 - 5000}{4999.9} = -0.00200\ \% \text{ (3 significant figures)}$$

High-resistance method with potentiometer

The addition of a potentiometer in the circuit of **Figure 3.69** allows a number of readings by adjusting the potentiometer. Furthermore, this process enables the determination of the average value of the ratio of voltage to current.

Analogue ohmmeter

A simple analogue ohmmeter consists of a battery, a selection of resistors, a variable resistor and a moving-coil meter movement. In fact, ohmmeters use a d.c. measuring system to ensure true resistance is measured.

The ohmmeter is a portable instrument for measuring relatively low values of resistance. In addition, the range of resistance measures available varies according to the instrument. In order to use a standard series-type ohmmeter correctly, short the lead terminals by putting the leads together. After this occurs, adjust the variable resistor to zero the pointer at FSD to compensate for variations in battery voltage. This zero-resistance adjustment will allow the circuit to produce a full-scale deflection. The meter is now ready for use.

The scale of ohmmeters is non-linear and reads from right to left. When in use the unknown resistor (R_X) connects between the terminals via meter leads. Because a zero adjustment occurred, the value of current exists. Consequently, the connection of the resistance causes the flow of current to decrease. The result is that this lower value of current translates into ohms on the ohm-calibrated scale. Furthermore, because of the scale's non-linear design, accurate readings become progressively more difficult towards the high-value end of the scale.

Digital ohmmeter

Digital ohmmeters display the measured resistance via a digital readout. The readout presents in ohms and multiples of ohms. These devices are small, portable, easy to operate, highly reliable and the digital readout is easy to read. In addition, the device makes use of integrated circuits and is shock resistant.

Under field conditions, the digital ohmmeter gives accurate test results and is unaffected by the presence of nearby magnetic fields, short circuiting or power surges. The instrument in **Figure 3.70** is equipped with a microprocessor to perform the Ohm's law calculation.

FIGURE 3.70 Digital ohmmeter

The resistance of the main earthing conductor must not be higher than 0.5 Ω as shown in **Figure 3.71**.

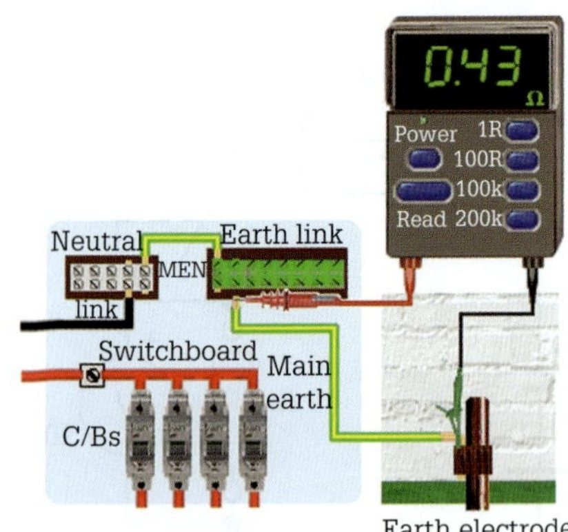

FIGURE 3.71 Testing main earth resistance (main earth disconnected from earth link)

SWITCH ON

To measure the resistance of the main earth conductor, disconnect main earth from the earthing link.

Kelvin ohmmeter

A Kelvin ohmmeter uses a four-terminal measurement method in which a constant current passes through the measured material via the current terminals on the ohmmeter. These ohmmeters use special measuring probes called Kelvin clips. Kelvin clips comprise a set of leads with combined current input probes and potential measuring probes. It follows that these ohmmeters have optimal measurement accuracy at very low values of resistance.

Use of highly stable semiconductor reference elements, a microprocessor and a comparator, together with precision stabilised resistors, allows long-term accuracy without the need for regular recalibration. Because of these design parameters, they are excellent instruments for measuring low values of resistance. Errors normally caused by test lead and contact resistance are virtually eliminated. It is evident that in many applications the contact resistance from leads can exceed the value of the load. Moreover, Kelvin ohmmeters bypass this potential error source by providing two terminals of constant current and an additional two terminals for high-impedance voltage measurement.

Figure 3.72 shows that there are virtually no contact or lead resistance errors created by the voltage measurement. The fact is that minuscule current flows in the voltage sense leads.

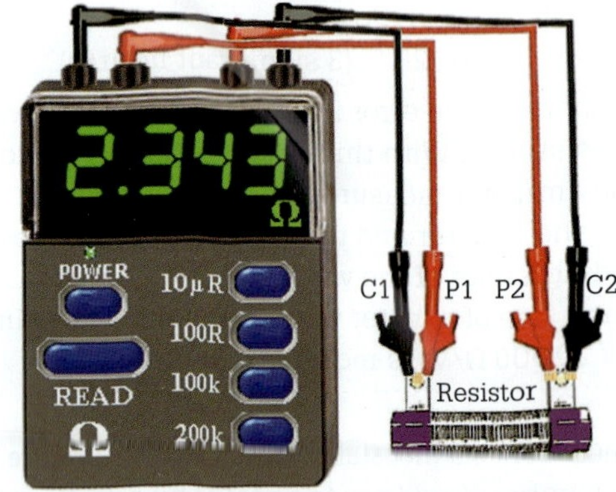

FIGURE 3.72 Kelvin ohmmeter

Test current with other ohmmeters heats up the unknown resistance and changes its value by the introduction of thermal emf. When two dissimilar metals join, an emf can be generated (thermocouple effect) which affects the resistance being measured. The measurement process is fast and gives accurate resistance measurement of an unknown resistance. The pictorial diagram of **Figure 3.72** illustrates how the four-wire system helps to eliminate lead wire and contact resistance as potential error sources. The measuring current passes through the unknown resistance using the C1 and C2 leads for connection. The placing of these leads should be outside the P1 and P2 leads. In addition, the voltage drop across the unknown resistance using the P1 and P2 leads is measured. Finally, the measured current passes to an internal reference resistance. A comparison occurs between the voltages dropped across the unknown resistance with the volt drop across the internal standard resistance. From the ratio of these two voltages, the resistance value of the unknown resistance is calculated and displayed.

Additional features

On some Kelvin ohmmeters, there exists a data output option that permits use of the Kelvin with a printer, data logger or data acquisition system. In addition, some of these ohmmeters have protection against damage by voltages of up to 480 V rms (root-mean-squared) by internal fuses, spark gaps and clamping diodes. They also have a microprocessor and comparator that can be a very useful option for batch measurement of components.

The cause of an error when making a low-resistance measurement is weak or inappropriate connection to the unknown resistance. Lead connections should be clean, mechanically firm and free from oxides that can cause an insulating barrier. Typical users of Kelvin ohmmeters are manufacturers of cable, wire, switchgear, electric motors and small transformers where the accuracy and reliability of low-resistance readings are of the utmost importance.

SWITCH ON

Safety precautions with resistance tests

- Always disconnect the power supply on a circuit before making any resistance tests and bleed any capacitors that are in the circuits under test.
- Use extreme care in testing solid-state components, as the voltage from the internal batteries of the ohmmeter can damage many of these components.
- Turn off meters in order to lengthen the life of the batteries.

Bridge circuits

The simplest form of a bridge circuit consists of two identical resistors connected in parallel across a voltage power source. This arrangement becomes a bridge circuit when a connection called a 'bridge' connects between the two resistors. Refer to **Figure 3.73**.

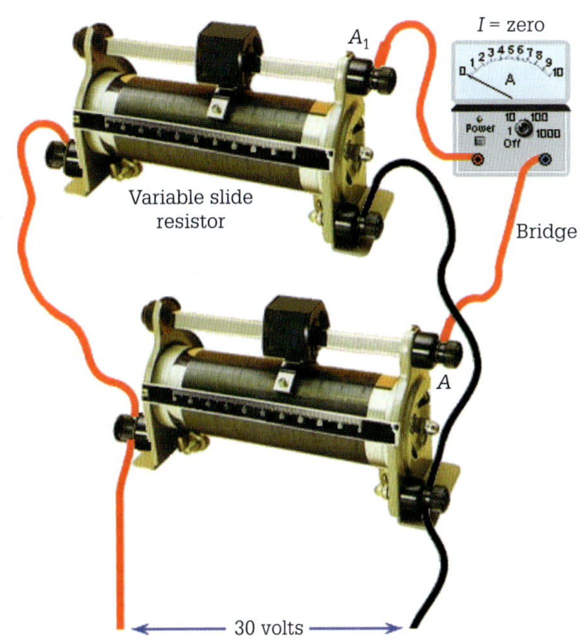

FIGURE 3.73 Balanced bridge

The potential across both resistors is the same value because the resistors are identical. So, points A–A_1 are at equal potential because both are at the mid-points of their resistors (that is the bridge is balanced). If a bridge connects between points of equal potential no current will flow through the bridge. The current, however, flows from a more positive to a less positive end as depicted in **Figure 3.74**, if the bridge connects between points of unequal potential.

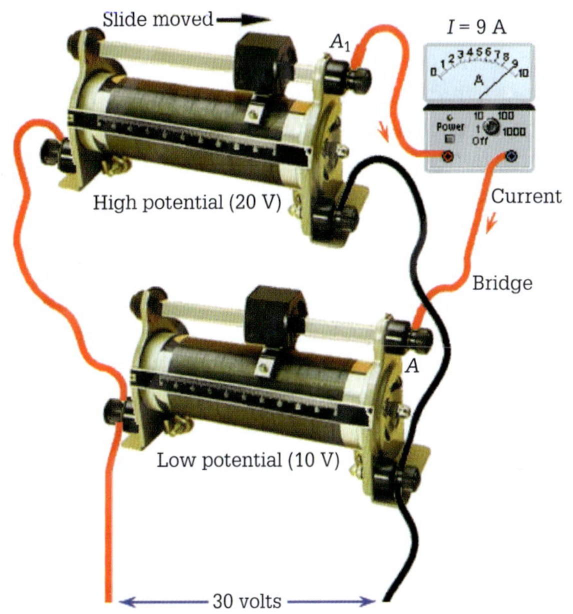

FIGURE 3.74 Unbalanced bridge – unequal potential

The direction of the current flow depends upon which point is more positive. If point A_1 is more positive than point A, current will flow from point A_1 to point A. As such, the potential at the two ends of the bridge controls the direction of the bridge current. When the bridge is across corresponding potentials, no current flows and the bridge is balanced. However, when the bridge connects across points of unequal potentials, current flows and the bridge is unbalanced.

Wheatstone bridge

The basic Wheatstone bridge circuit consists of four resistors, R_1, R_2, R_3 and R_x, which connect to form the sides of a square. A centre-zero galvanometer connects across two opposite corners while a direct current voltage source connects to the other two corners. The galvanometer is a sensitive milli-ammeter or micro-ammeter with centre-zero position setting. The null method of measurement depends on the bridge being adjusted so that the galvanometer reads zero current. Refer to **Figure 3.75** for the arrangement of the Wheatstone bridge.

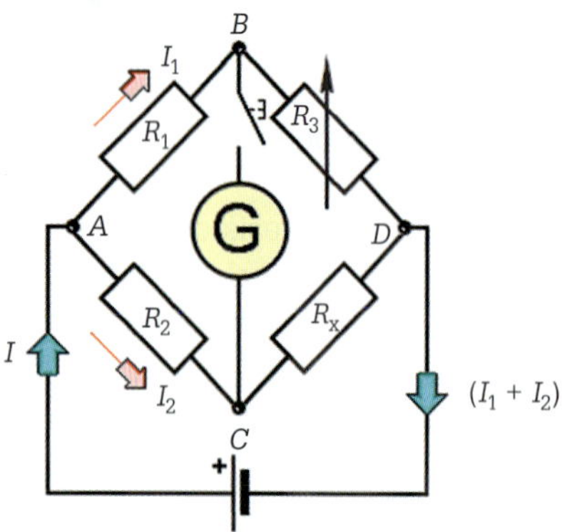

FIGURE 3.75 Wheatstone bridge

The current (I) entering the bridge circuit at point A divides. The current I_1 flows through resistor R_1 and a current I_2 through resistor R_2. Since no current flows through the galvanometer, current I_1 continues through R_3, while the current I_2 flows through R_x.

All these currents recombine at point D creating the total circuit current '$I_1 + I_2$' from the battery. Since no current flows through the galvanometer, points B and C are at the same potential. Therefore:

Potential difference across ABD = potential difference across ACD

As potential difference equals the product of current and resistance ($V = I.R$), the preceding statements have a new expression.

$$I_1 R_1 = I_2 R_2 \text{ and } I_1 R_3 = I_2 R_x$$

Then by dividing the first equation by the second we obtain:

$$\frac{I_1 R_1}{I_1 R_3} = \frac{I_2 R_2}{I_2 R_x}$$

The currents cancel and the equation becomes:

$$\frac{R_1}{R_3} = \frac{R_2}{R_x}$$

Only the resistance values, R_1, R_2 and R_3 are known. Consequently, calculation of the fourth unknown resistance, R_x, transpires from the balanced bridge relationship:

$$R_x = \frac{R_3 \times R_2}{R_1}$$

The Wheatstone bridge compares an unknown resistance R_x with others of known values, that is, R_1 and R_2, which have fixed values, and R_3, which is variable. Consequently, when R_3 varies until no deflection occurs on the galvanometer the bridge is 'balanced'. Essentially, we do not need to know the actual values of R_1 and R_2 – we need only their ratio. Wheatstone bridge type circuits occur in a.c. circuits to enable the determination of unknown values of inductance and capacitance, as well as resistance.

EXAMPLE 3.21

The bridge shown in **Figure 3.75** is balanced when $R_1 = 24\ \Omega$, $R_2 = 120\ \Omega$ and $R_3 = 10\ \Omega$. Determine the value of the unknown resistor R_x?

$$R_X = \frac{R_3 \times R_2}{R_1} = \frac{10 \times 120}{24} = 50\ \Omega$$

Figure 3.76 shows a practical form of the Wheatstone bridge. The slide wire bridge is a board using wire of uniform resistance.

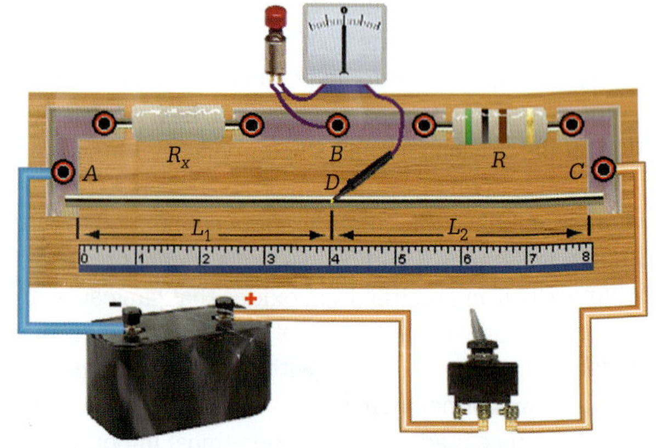

FIGURE 3.76 Slide wire bridge

The unknown resistance (R_x) connects in series across the left-hand element terminals, and a known-value resistor connects across the right-hand element terminals. When the power supply switch closes, a position is located for the slider on the wire at which the galvanometer reads zero. In addition, the ruler provides the measured lengths, L_1 and L_2. Previously, we saw that the resistance of a conductor is directly proportional to its length provided all other factors remain unchanged. Consequently, the ratio of the lengths L_1 to L_2 can replace the ratio of the resistances R_2 to R_1 in the previous equation.

$$R_x = R\frac{L_1}{L_2}$$

Choose the known-value resistor (R) so that the balance point (zero on the galvanometer) occurs near the centre of the uniform resistance wire.

EXAMPLE 3.22

By using the slide wire bridge of Figure 3.76 determine the value of the unknown resistance when the standard known resistance used is 50 Ω and the ratio of the lengths L_1 to L_2 is 3:5.

$$R_X = R\,\frac{L_1}{L_2}$$

$$= 50 \times \frac{3}{5}$$

$$= 30\ \Omega$$

EXERCISE 3.7

a The bridge shown in Figure 3.75 is balanced when $R_1 = 36\ \Omega$, $R_2 = 180\ \Omega$ and $R_3 = 1000\ \Omega$. Determine the value of the unknown resistor R_x.

b By using the slide wire bridge of Figure 3.76, determine the value of the unknown resistance when the standard known resistance used is 200 Ω and the ratio of the lengths L_1 to L_2 is 3.5:4.5.

Bridge circuits with strain gauges

The resistance of a strain gauge varies with the strain the device is under. Strain is deformation of a material due to an applied force. More precisely, strain is the fractional change in length. At a constant temperature, and when the device is strained, its resistance changes. In addition, strain occurs both parallel and perpendicular to the applied force.

Strain gauges like those in Figure 3.77 consist of a metal foil, photo-etched to form a desired pattern supported by a resin backing film. They are used in Wheatstone bridge configurations due to the high accuracy of measurement.

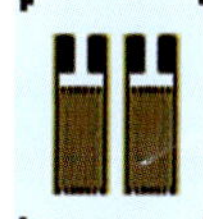
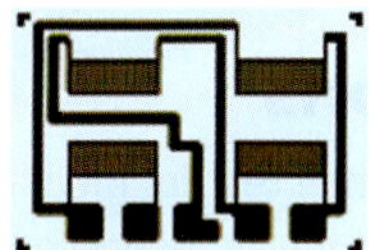

FIGURE 3.77 Strain gauges

Strain gauges typically provide small signal levels and it is important to have accurate instrumentation to amplify the signal. Figure 3.78 shows the quarter-bridge method for measuring deformation of a material.

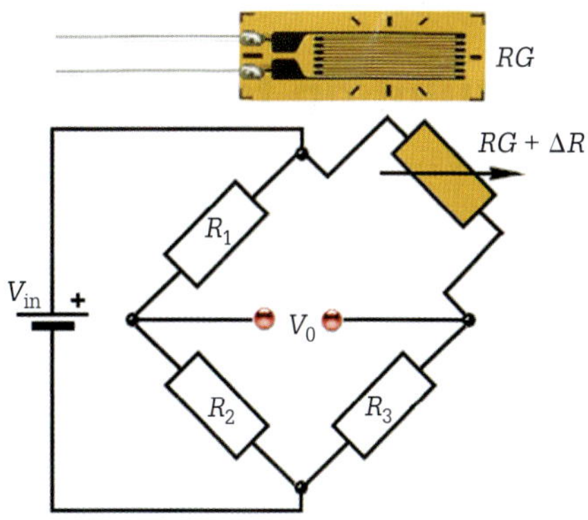

FIGURE 3.78 Quarter-bridge circuit

When the unknown resistance is supplied by a strain gauge, the Wheatstone bridge is used to measure the voltage difference called the output voltage (V_0). The other legs are simply completion resistors with resistance equal to that of the gauge. The output potential in most cases linearly relates to the deformation of the strain gauge. As the resistance of the strain gauge varies, the bridge becomes unbalanced. This unbalance causes a potential difference across the middle of the bridge, which can be measured with a voltmeter. Consequently, by using this potential difference and a calibration factor, calculation of a strain dimension can be made. There are a number of different variables for strain gauges, vibration, torque, bending and deflection.

Calculation of a strain dimension uses the following equation:

$$V_0 = V_2 - V_1 = \left[\frac{R_3}{RG + \Delta R + R_3} - \frac{R_2}{R_1 + R_2}\right] V_{IN}$$

EXAMPLE 3.23

Determine the voltage output of a quarter-bridge strain gauge which has the following circuit values:

- $R_1 = R_2 = 1\ \Omega$
- $R_3 = 100\ \Omega$
- $RG = 100\ \Omega$
- $\Delta R = 1\ \Omega$
- $V_{IN} = 5$ V
- $V_0 = ?$

$$V_0 = \left[\frac{100}{100 + 1 + 100} - \frac{1}{1 + 1}\right] \times 5$$

$$= -0.0124\ \text{V (3 significant figures)}$$

Insulation resistance testers

A major cause of equipment failure is the breakdown of the insulation. The breakdown of insulation has a number of interlinked causes and an insulation test checks the condition of the insulation. Furthermore, insulation resistance values can vary between zero and infinity. The insulation test injects a d.c. voltage at twice the working voltage between earth and the insulation. For example, the insulation of a 230 V system will be tested at 500 V d.c., and a 400 V system will also be tested at 500 V d.c. for 60 seconds (400 V 3Ø = 230 V to earth). However, some manufacturers and repair facilities use 1000 V d.c. as the test voltage. In addition, the output current of insulation testers varies according to whoever has manufactured it, but is between 50 μA and 1.5 mA.

Insulation breakdown

There are several basic causes of the electrical breakdown of insulation, including electrical stress, mechanical stress, thermal stress and the deposition of contaminants from the environment that attack the insulation chemically. Normal periods of operation will lead to 'ageing' of the insulation through these mechanisms. Air will gradually oxidise organic insulation materials while the effect of moisture, oil and salt will break down the efficiency of the insulation even more rapidly. Electrical stresses, particularly sustained over-voltage or voltage impulses caused by lightning strikes or the switching of high-inductive loads such as electric motors, will create current discharges in voids within the insulation that will expand and find other parallel paths through-and-over the surface of the insulation. The ageing of insulation is a slow process leading to insulation failure. However, if the operating conditions change, the breakdown process will increase. Furthermore, the voltage impulse diagnostic technique monitors insulation condition in order to undertake any required preventive maintenance action.

Insulation resistance testing

As insulation ages, leakage currents within the insulation may increase. Leakage currents cause the dielectric characteristic of the insulation to change, which alters the degree of electrical protection of the material. The insulation resistance test is the quickest method to gain proof of the fitness of the insulation. Refer to **Figure 3.79**.

FIGURE 3.79 Analogue and digital insulation resistance testers

AS/NZS 3000: 2018 *Wiring Rules* requirements

Refer to the following clauses:

- 8.3.3 Mandatory tests
- 8.3.5 Continuity of the earthing system
- 8.3.6 Insulation resistance

The AS/NZS 3000: 2018 *Wiring Rules* requires tests be undertaken to prove the continuity of the earthing system and measure the resistance of the main earthing conductor, any bonding conductors and protective earthing conductors. These tests ensure that the earthing system will cause protective devices to operate if there is a fault between live parts and the general mass of earth. A correctly installed and tested earthing system makes certain that exposed conductive parts of electrical equipment do not reach hazardous voltages when these faults occur.

An insulation resistance test guarantees the reliability of the insulation at the time of testing. More to the point, effective insulation resistance prevents:

- electric shock hazards for persons from unintentional contact
- fire risks from short circuits
- equipment damage.

SWITCH ON

Safety precautions

Insulation resistance testers similar to those shown in **Figure 3.79** are suitable for use on de-energised electrical cables and wiring systems. Testing insulation resistance of motors, transformers and similar equipment is undertaken using a Hi-Pot tester, which is similar to the insulation resistance tester.

Always ensure that there are no electronic devices or components connected in the circuit under test. This implies that damage to the devices will occur.

- Tests performed on single phase are between active and earth, neutral and earth and start-to-run windings, and with three phase between each phase and each phase to earth.
- The minimum insulation resistance allowed for wiring is 1 MΩ.

Applications

Insulation resistance testers measure the insulation resistance of conductors, power cables and similar wiring systems. Applications include:

- testing at completion of an installation to verify conformance to 'rule' specification
- routine preventive maintenance
- diagnostic testing enabling isolation of faulty components for repair.

In addition, an insulation resistance tester detects earth faults, as well as short circuits.

Crank insulation tester

The generator insulation resistance tester obtains its energy from a small built-in hand-cranked generator. Insulation resistance measurements occur while the handle turns. Refer to **Figure 3.80**.

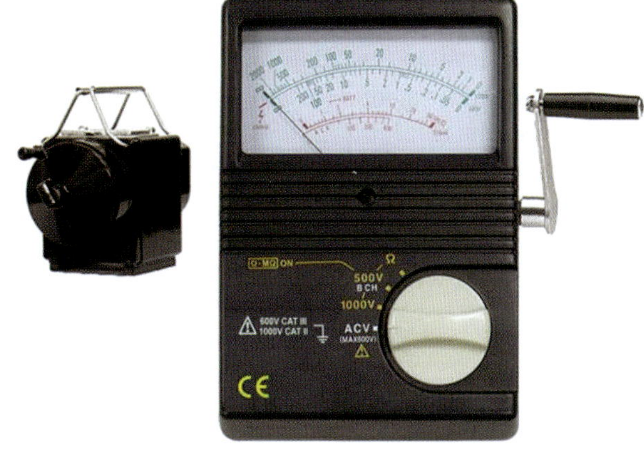

FIGURE 3.80 Hand crank insulation resistance testers

If insulation measurements are taken and recorded over time, the series of results will show a gradual decline that the insulation experiences. The recording of insulation resistance tests enables implementation of active maintenance plans.

Testing of the insulation resistance tester

Employers of persons performing tests on works or safety equipment must implement three specific requirements. These are:

1. Ensure that the instrument used is appropriate for the test.
2. Ensure the testing of the instrument every six months.
3. Keep a record of tests on the instrument for two years.

An insulation resistance tester requires verification for use by testing it with all the following resistors with a 1% tolerance rating.

0.5 Ω, 1 Ω, 2 Ω, 20 Ω, 10 kΩ, 1 MΩ

As the insulation resistance tester and the resistors possess accuracy limits, the insulation resistance tester ought not to read in excess of or less than the calculated results for overall accuracy of the instrument. As such, the minimum and maximum readings that should be realised for the 10 kΩ resistor are:

Minimum reading

10 000 Ω less the 1% tolerance of the resistor
= 10 000 − 100 = **9900 Ω**.

The meter has a tolerance of 2%
= 9900 − 198 Ω = **9702 Ω**.

Maximum reading

10 000 Ω plus the tolerance
= 10 000 + 100 = **10 100 Ω**.

The meter has a tolerance of 2%
= 10 100 + 202 = **10 302 Ω**.

Checking the output voltage of the insulation resistance tester

Checking occurs when the instrument connects across a 1 MΩ resistor with a voltmeter connected. The voltmeter should read within +20% (600 V) or −10% (450 V) on the 500 V range. In addition, checking the 1000 V range requires a 10 MΩ resistor with the same reading parameters as the one mega-ohm resistor maintained. Refer to **Figures 3.81** and **3.82**.

Testing insulation resistance between active and neutral and the testing of MIMS cable prior to installing pot seals

Again, the requirement is 1 MΩ or greater.

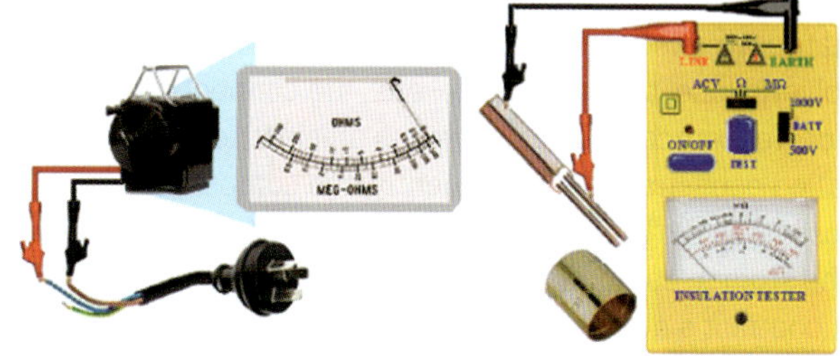

FIGURE 3.81 Insulation resistance testing of MIMS cable

Testing the equipotential bonding of the pool reinforcing metal using the ohm's range

This test confirms if the bonding resistance value meets the Standard – to be low enough so that the circuit's protective device will operate.

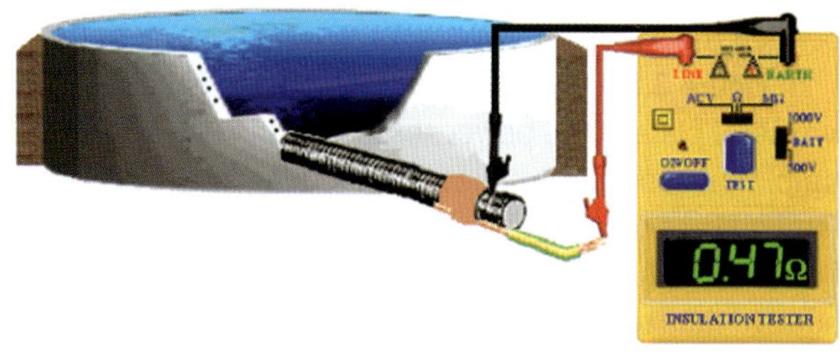

FIGURE 3.82 Equipotential bonding testing

REVIEW QUESTIONS

1. What is an important safety requirement to undertake before measuring circuit resistance?
2. What is the resistance measurement method that compares an unknown resistance with a known standard resistance?
3. What are typical applications of decade boxes?
4. Which voltmeter/ammeter method is more accurate when the measured value is low?
5. Why is it important to short out the leads of an analogue ohmmeter before taking a reading?
6. What type of scale does an analogue ohmmeter have?

»

7 Explain why a Kelvin ohmmeter is an excellent resistance-measuring instrument for low values of resistance.

8 Which resistance-measuring meter provides optimal measurement accuracy at very low values of resistance?

9 The ammeter in a long-shunt resistance-measuring circuit has a resistance of 0.1 Ω and a current reading of 10 mA, while the voltmeter reads 8 V. Using these values calculate the resistance of R_x.

10 Insulation must meet a certain standard. What test confirms this?

11 A 400 V circuit requires an insulation test. At what voltage must the circuit be tested?

12 What is the minimum allowed insulation resistance for wiring?

3.6 Capacitors and capacitance

Capacitors

If a potential difference exists between two conductive surfaces, then an electric field is the result of the separation of unlike charges. The strength of the electric field depends on the distance the charges separate. When two conductors (called plates), separated by an insulating material (called a dielectric – symbol ε_r), such as air, mica, polyethylene, tantalum oxide, etc., are connected across a power source, the structure has an electrical property called capacitance. Consequently, the composition possesses the ability to store a quantity of electric charge. The storage contains excess electrons on one plate and in contrast a deficiency of electrons on the other. By connecting a conductor across the capacitor terminals, the charged capacitor can regain electron balance by discharging its stored energy.

Electrical and electronic circuits use capacitors. For example, capacitors are used with split-phase motors to enable the motor to produce a starting torque or a starting and running torque beyond the capacity of the motor (e.g. compressors). Capacitors smooth rectified a.c. outputs and they are a component in time-delay circuits. Sometimes capacitance is undesirable: such as the capacitance between the conductors of overhead transmission lines.

In **Figure 3.83**, a potential difference exists between two metal plates. The power source, acting like an electron pump, transfers some of the free electrons from the positive metal plate to the negative metal plate. This transfer of electrons causes one plate to become increasingly negative thus receiving a charge of –Q, while the other plate becomes increasingly positive, receiving a charge of +Q. The build-up of opposite charges on the plates gradually establishes an electrostatic field between the plates. As the charging process continues, the negative plate saturates with negative charges that prevent any additional transfer of electrons. When saturation occurs, the potential difference across the two metal plates is equal and opposite of the driving source voltage (in **Figure 3.83** a battery).

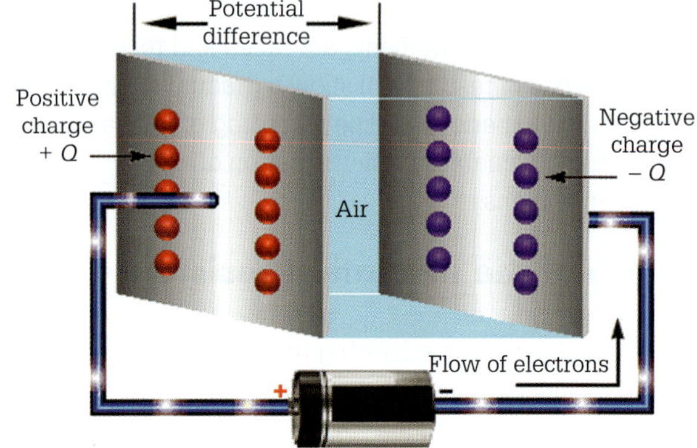

FIGURE 3.83 Two charged metal plates separated by an insulator

This saturation process causes polarisation. Consequently, the electric charges cannot move across the capacitor and current ceases to flow.

Capacitance

The capacitance (symbol C) is the amount of charge collected for a set potential difference across the capacitor plates. In addition, the coulomb/volt or farad represents the unit of capacitance.

A farad is a considerable value for practical purposes and in practical capacitors the capacitance is typically in the order of microfarads (1 µF = 10^{-6} F), nanofarads (1 nF = 10^{-9} F), or picofarads (1 pF = 10^{-12} F).

The capacitance is the ratio of accumulated charge divided by the supply voltage. This is expressed as an equation:

$$C = \frac{Q}{V}$$

where C = the capacitance in farads (F) (1 farad is 1 coulomb per volt)

Q = the stored electric charge in coulombs (C)

V = the potential difference in volts (V)

Additionally, capacitance is that property of a capacitor that delays the change of potential difference across it.

EXAMPLE 3.24

Two conductors are separated by air and have an accumulated charge of 10 milli-coulomb when connected to a 200 V d.c. supply. Calculate the capacitance of the capacitor.

$$C = \frac{Q}{V} = \frac{10 \times 10^{-3}}{200} = 50\ \mu F$$

Factors affecting capacitance

There are three factors affecting the quantity of capacitance: plate area, plate spacing and the type of dielectric. These factors affect the density of the electrostatic field that develops between the plates for a given potential across the plates.

The capacitance of an energy-storage system relates to its physical dimensions and to the nature of the material (called a dielectric) between the plates as shown in **Figure 3.84**. Larger plates have an increased capacity for electron acceptance. Consequently, more free electrons can transfer from the positive plate before the accumulated charge can build up an equal and opposite voltage. However, although a greater charge collects, it takes longer to achieve the same voltage.

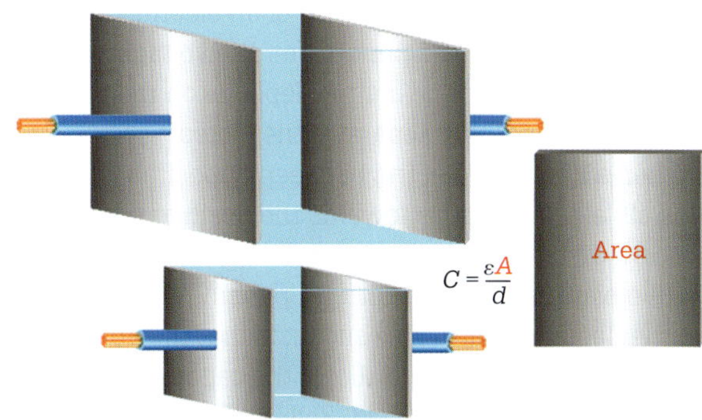

FIGURE 3.84 Plate area affects capacitance

In contrast, small plates will have less capacity for electron storage and hence the same voltage will occur with less electron transfer and in less time.

Plate spacing as shown in **Figure 3.85** also changes capacitance. Close spacing results in a denser electrostatic field while plates spaced further apart result in a reduced field.

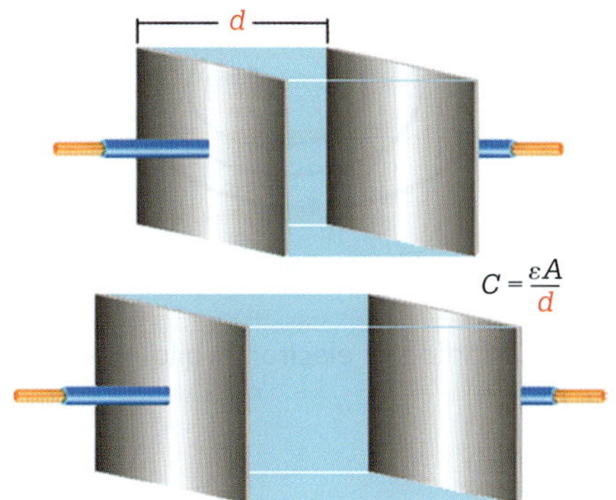

FIGURE 3.85 Plate spacing affects capacitance

Dielectric material

A dielectric is the insulating material located between the plates of a capacitor. The purpose of dielectric materials is to concentrate electrostatic lines of force between the plates. The ease by which an electrostatic flux forms in a dielectric is a measure of its permittivity. Permittivity has the Greek symbol epsilon ε. However, the permittivity of the material is rated in terms of its dielectric constant (K).

The increase in capacitance caused by different dielectrics is the relative permittivity (ε_r) or dielectric constant of the dielectric. The increase suggests that a material's dielectric constant determines the electrostatic energy stored in that material per unit volume for a given voltage. The reference unit for dielectric constant is air that has a value of one. When a dielectric constant varies, the capacitance will also vary.

It is evident that the greater the constant, the greater is the permittivity of the dielectric. Glass in **Figure 3.86**, for instance, with relative permittivity of 7.5, has seven and one-half times the permittivity of free space (air), and consequently allows the establishment of an electrostatic field 7.5 times stronger than that of a vacuum, all other factors being equal.

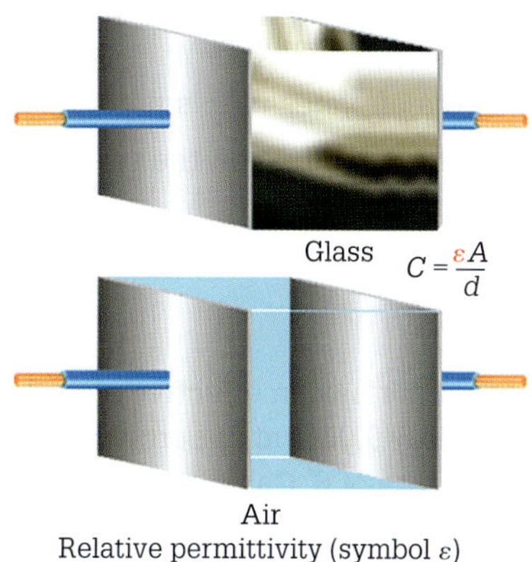

FIGURE 3.86 Glass and air dielectric material

Table 3.7 lists the values of relative permittivity (ε_r) for some dielectric materials.

TABLE 3.7 Relative permittivity of various materials

Material	Minimum	Maximum
air	1	1
epoxy resin	3.4	3.7
glass	3.8	14.5
mica	5.4	8.7
neoprene	4.0	6.7
oil, mineral	2.2	4
paper	1.5	3.0
polystyrene	2.4	3.0
porcelain	6.0	8.0
polyester	3.1	3.3
quartz	3.8	4.4
Teflon	2.1	2.1

The dielectric constant of a material determines the electrostatic energy which may be stored in that material per unit volume for a given voltage.

Calculating capacitance

The simplest capacitor is a pair of parallel conducting plates separated by a dielectric. For this capacitor the equation for capacitance is:

$$C = \frac{\varepsilon A}{d}$$

where C = the capacitance in farads (F)

ε = absolute permittivity of the dielectric

A = the active area of the plates in square metres (m^2)

d = the distance between the plates in metres (m)

However, the absolute permittivity of the dielectric is rarely used, and since $\varepsilon = \varepsilon_r \varepsilon_0$ the equation becomes:

$$C = \frac{\varepsilon_r \varepsilon_0 A}{d}$$

where C = the capacitance in farads (F)

ε_0 = the absolute permittivity (= 8.85×10^{-12} C^2/Nm^2)

ε_r = the relative permittivity

A = the area of plates in in square metres (m^2)

d = the distance between the two opposite plates in metres (m)

EXAMPLE 3.25

A ceramic capacitor has an effective plate area of 5 cm^2 separated by 0.05 mm of porcelain, which has a relative permittivity of 6. Calculate the capacitance of the capacitor in picofarads.

$$C = \frac{\varepsilon_0 \; \varepsilon_r \; A}{d}$$

$$= \frac{(8.85 \times 10^{-12}) \times 6 \times (5 \times 10^{-4})}{0.05 \times 10^{-3}}$$

$$= 531 \text{ pF}$$

EXERCISE 3.8

A capacitor has an effective plate area of 10 cm^2 separated by 0.05 mm of paper, which has a relative permittivity of 1.5. Calculate the capacitance of the capacitor in picofarads.

EXAMPLE 3.26

Two 50 mm square metal plates have a 0.5 mm thick mica dielectric. Assuming a relative permittivity of 5 for the mica determine the capacitance.

Firstly determine the active plate area:

$$A = 50 \times 50$$

$$= 2500 \text{ mm}^2$$

»

Next calculate the capacitance:

$$C = \frac{\varepsilon_0 \; \varepsilon_r \; A}{d}$$

$$= \frac{(8.85 \times 10^{-12}) \times 5 \times (2500 \times 10^{-6})}{0.5 \times 10^{-3}}$$

$$= 221 \text{ pF (3 significant figures)}$$

EXERCISE 3.9

A capacitor consists of two parallel plates having a surface area of 2 cm^2 separated by a distance of 0.2 mm. Determine the capacitance of the capacitor if the dielectric material separating the plates is tantalum oxide: ε_r for tantalum oxide is 26.0.

Electrostatic field

If a potential difference develops between two plates, an electric field results because of displaced electrons or charges. The strength of the field depends on the amount of charges displaced. Because of this, the process of transferring electron charges from one plate to the other results in the storage of energy within the space between the plates. This energy, in the form of an electrostatic field, remains stored for a period after the voltage stops. In addition, the storage time mainly depends on the type of insulating material that separates the plates.

When air is the dielectric between oppositely charged plates, its electrons move towards the positive plate. The result is a distorted polarised atom shape. By contrast, the negative plate repels the electrons, pushing them further from the negative plate. This suggests that the ability to attract and to repel charges allows the capacitor to store energy as an electrostatic field. Refer to **Figure 3.87**.

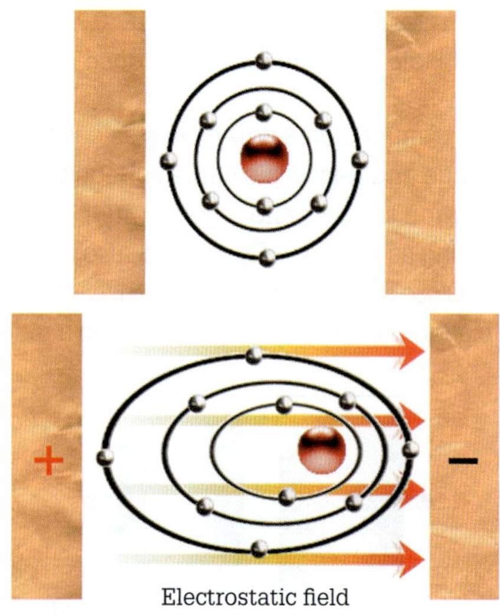

FIGURE 3.87 Distortion of electron orbital paths due to electrostatic force

Electrostatic field strength

A dielectric cannot withstand an indefinitely large voltage difference across it. If this voltage exceeds an absolute value, the dielectric will decompose and a discharge will take place through it. In addition, the dielectric strength is the voltage needed to make a particular thickness of a dielectric conduct. **Note:** The SI unit for dielectric strength is Volt per metre (Vm^{-1}) but the practical unit is kV/mm (or kV mm^{-1}).

$$\text{voltage to conduct} = \frac{V}{d} \text{ in } Vm^{-1}$$

Table 3.4 shows the dielectric strengths (kV/mm) of some materials.

The values given in Table 3.8, show that two plates 1 mm apart with mica in between them can withstand a voltage up to 150 kV before the mica starts to deteriorate.

The voltage gradient is an important electrical parameter for the dielectric because it enables manufacturers to design the dielectric to withstand the applied voltage. This implies that the applied voltage, divided by the dielectric thickness, determines the voltage gradient.

TABLE 3.8 Dielectric strength of various materials (kV/mm)

Material	Minimum	Maximum
air	5	
epoxy resin	19.7	
glass	30	100
mica	100	150
neoprene	15.7	27.6
oil, mineral	110.7	
paper	1.5	3.0
polystyrene	40	50

EXAMPLE 3.27

A dielectric operates with a potential difference of 200 V across it. Calculate the voltage gradient through the dielectric if the dielectric has a thickness of 40 µm.

$$V_{grad} = \frac{V}{d} = \frac{200}{40 \times 10^{-6}} = \mathbf{5 \times 10^6\ Vm^{-1}\ or\ 5\ kV/mm}$$

EXERCISE 3.10

Two parallel rectangular plates are spaced 5 mm apart in air and the voltage between them is 400 V. Determine the voltage gradient across the dielectric.

Energy stored in a capacitor

The amount of energy stored relies on the capacitance and the square of the applied voltage.

$$W = \tfrac{1}{2} CV^2 \text{ in joules}$$

where W = the energy in joules (J)
C = the capacitance in farads (F)
V = the potential difference across the plates in volts (V)

EXAMPLE 3.28

A 50 µF capacitor connects across a 100 V d.c. supply. Find the energy stored when disconnection occurs.

$$W = \tfrac{1}{2} CV^2 = \tfrac{1}{2}(50 \times 10^{-6}) \times 100^2 = \mathbf{0.25\ J}$$

EXERCISE 3.11

Determine the energy stored in a 100 µF capacitor when charged to 200 V.

Types of capacitors

A variable capacitor uses air, a dielectric film or an electrolyte paste as the dielectric. The movement of one plate via rotation away from the fixed plate varies the capacitance. Another technique used to vary a capacitance is the compression method whereby one plate moves towards the fixed plate. Because of this, the capacitance varies by altering the dielectric thickness. Variable capacitors can tune radio receivers or finely trim the capacitance in a circuit. Figure 3.88 shows different variable capacitors.

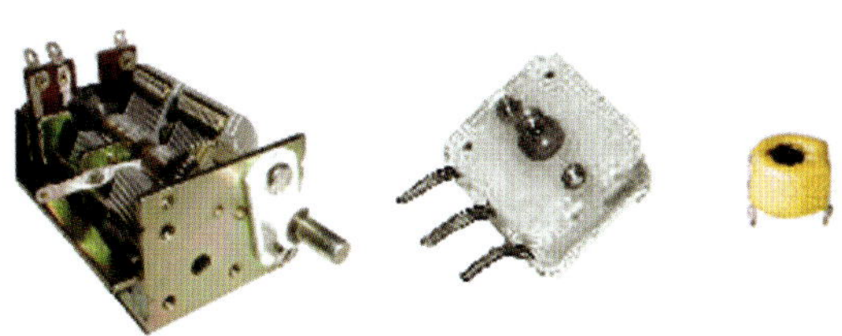

FIGURE 3.88 Tuning and trimmer capacitors

The **polyester (metallised) capacitors** in Figure 3.89 are widely used in a large range of circuits.

FIGURE 3.89 Polyester capacitors

Polystyrene capacitors (see **Figure 3.90**), have a low temperature coefficient. As such, they suit applications where circuit balance is critical. In addition, they are high-voltage capacitors.

FIGURE 3.90 Polystyrene capacitors

Ceramic capacitors (see **Figure 3.91**) have low values of capacitance but maintain high stability with low loss at a wide frequency range and have a linear temperature coefficient of capacitance.

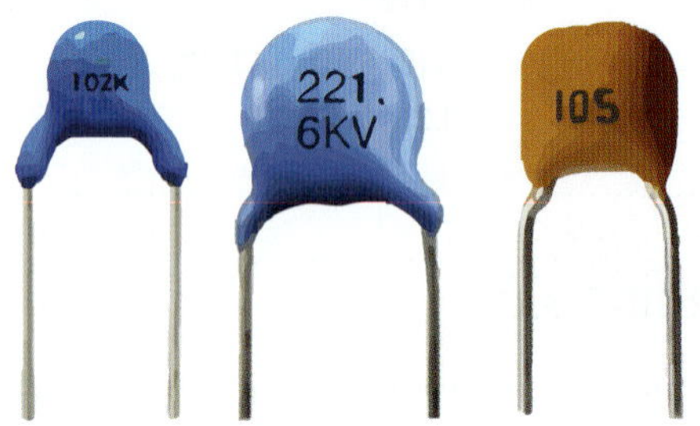

FIGURE 3.91 Ceramic capacitors

Monolithic multilayer ceramic capacitors (see **Figure 3.92**) have excellent stability and include a high capacitance in a small package.

FIGURE 3.92 Monolithic capacitors

A capacitor with a very thin dielectric will have a high capacitance for its size, and this applies to electrolytic capacitors. These capacitors contain an electrolyte in the form of a paste such as aluminium borate and are foil-rolled. The electrolytic capacitor is polarised with only the negative lead marked on the surface of the capacitor with an arrow or a series of negative symbols. Polarisation implies that these capacitors have a definite positive and negative terminal. It is essential that they correctly connect into a circuit. Incorrect polarity will result in an increasing leakage current that will cause the electrolyte to boil, building up pressure until the capacitor explodes. It follows that electrolytic capacitors must never connect across alternating current.

Electrolytic capacitors have their d.c. working voltage (VDCW) indicated on them. However, some electrolytes use a code to show the maximum safe voltage allowed. Refer to **Figure 3.93**. Note that on printed-circuit board (pcb) mount-type capacitors the longer lead indicates the positive side of the capacitor. Most electrolytic capacitors are clearly marked with the value of the capacitor in microfarads (μF), the polarity of the leads and the working voltage (for example 220 μF, 50 volts).

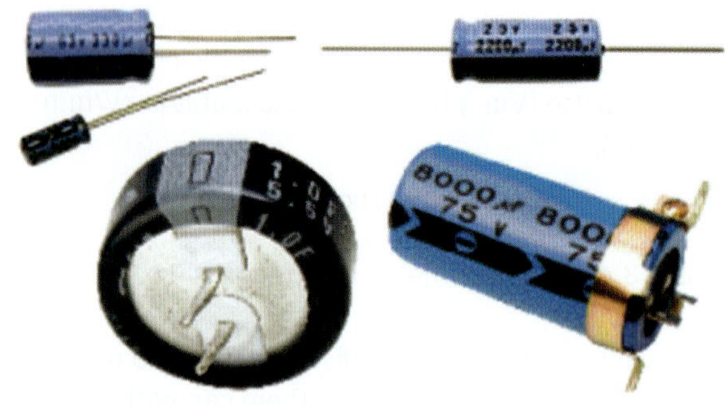

FIGURE 3.93 Electrolytic capacitors

The tantalum capacitors shown in **Figure 3.94** are also electrolytic capacitors. However, they contain a very high capacitance in a small package.

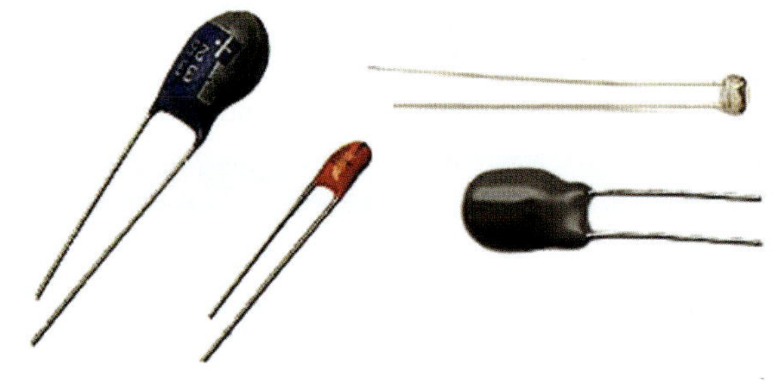

FIGURE 3.94 Tantalum capacitors

Capacitor identification codes

Most electrolytic capacitors are clearly marked with the value of the capacitor in microfarads (μF), the polarity of the leads and the working voltage (e.g. 220 μF, 50 volts).

Tantalum bead capacitors have the capacitance value, the voltage and polarity marking printed on the casing.

A capacitor marked 0.22 means 0.22 μF. In addition, if 2n2 is a marking, then this means a 2.2 nF capacitor. Other capacitors have a two-digit number written on the body; for example, 10. This number indicates that it is a 10 pF capacitor. Furthermore, other capacitors have a three-digit code; the number 103, for example, means 0.01 μF capacitor. However, some three-digit-code capacitors have a capital letter such as a K, which signifies the value of one thousand pF (picofarads).

Finally, an international colour-coding scheme identifies capacitor values and tolerances. However, there are no international agreements in place to standardise this capacitor identification. **Table 3.9** shows the capacitor colour code and **Table 3.10** the voltage colour code.

TABLE 3.9 Capacitor colour codes

Colour	Digit A	Digit B	Multiplier D
black	0	0	×1
brown	1	1	×10
red	2	2	×100
orange	3	3	×1000
yellow	4	4	×10 000
green	5	5	×100 000
blue	6	6	×1 000 000
violet	7	7	
grey	8	8	×0.01
white	9	9	×0.1
gold			×0.1
silver			×0.01

TABLE 3.10 Capacitor voltage colour codes

Colour	Voltage rating				
	Type J	Type K	Type L	Type M	Type N
black	4	100		10	10
brown	6	200	100	1.6	
red	10	300	250	4	35
orange	15	400		40	
yellow	20	500	400	6.3	6
green	25	600		16	15
blue	35	700	630		20
violet	50	800			
grey		900		25	25
white	3	1000		2.5	3
gold		2000			
silver					

Capacitor voltage reference:
Type J – Dipped tantalum capacitors
Type K – Mica capacitors
Type L – Polyester/polystyrene capacitors
Type M – Electrolytic 4 band capacitors
Type N – Electrolytic 3 band capacitors.

Application of capacitors

Power capacitors are used to store and release very high amounts of electrical energy. Their primary areas of application are in the supply authority's distribution system, induction motors, fluorescent lighting and for power factor correction. Notice that all these applications involve alternating current. Where pulses of high energy are required; for example, in the testing of high-voltage transformers, direct current power capacitors are used.

Motor starting power capacitors connect in series with the starting and running winding of a split-phase capacitor start/run motor. The result is that these capacitors enable the motor to develop high starting torque. A high starting torque is necessary when motors start refrigerators and compressors against the compression of their systems. The capacitors in **Figure 3.95** when connected across the input terminals of an induction motor improve the efficiency of the power generation and distribution system. In addition, a power capacitor connected in series with the run winding of a motor improves its running torque.

FIGURE 3.95 Motor starting and running capacitors

Inductive loads such as motors, transformers and the ballasts in fluorescent lamp circuits produce low power factor (low PF). Installing a capacitor like the one in **Figure 3.96** provides a leading current to counterbalance the lagging current drawn by the inductive loads.

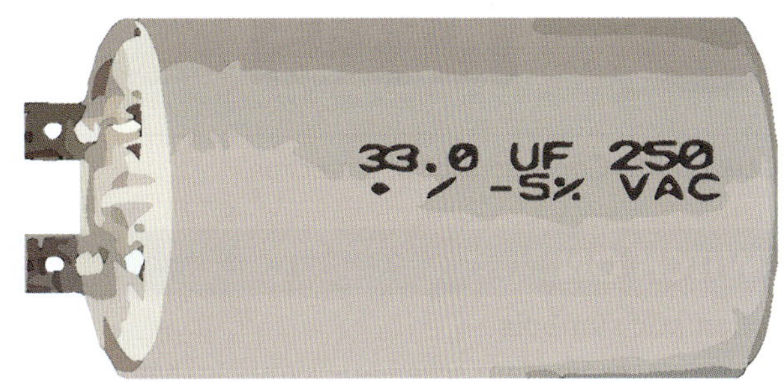

FIGURE 3.96 Capacitor to compensate for low pf

The high-voltage polyethylene (MKT) capacitor of **Figure 3.97** suppresses voltage spikes that can occur on 230 V mains.

FIGURE 3.97 0.01 μF Mains suppressor capacitor

In the supply authority's grid system, there are numbers of capacitors parallel connected to form capacitor stacks. For example, 50 capacitors each of 100 μF connect to create a stack with a capacitance of 5000 μF. A number of these stacks (about 70) might be series connected to create a bank of capacitors in order to operate at a grid voltage of 132 kV. The bank of capacitors has several purposes. For example, capacitor banks damp out the higher harmonics in voltage waves. They also compensate for transmission line inductive reactance and transformer inductive reactance in the transmission circuit. Finally, capacitors in series eliminate some of the inductance of a long transmission line. Series inductance occurs in long transmission lines when a large current flows and causes the production of heat. In addition, the current causes a large voltage drop. However, when capacitors connect in series with a transmission line, the capacitance effectively compensates for the characteristic inductance in the line to lower the total impedance. By injecting reactive power, the voltage drop across the line improves and heat as energy wasted falls. For this reason, the transmission circuit can transfer more energy.

Correcting the power factor reduces the total current flowing in the transmission lines and hence the associated line voltage losses and I^2R losses. Refer to **Figure 3.98**.

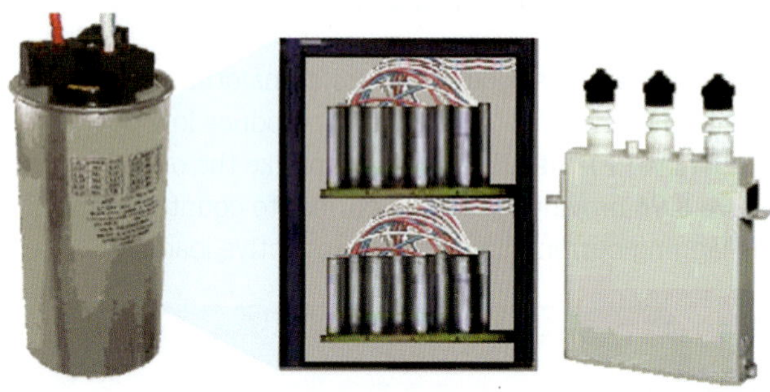

FIGURE 3.98 Capacitors for pf correction

SWITCH ON

Schematic symbols

Figure 3.99 shows four schematic symbols that relate to capacitors.

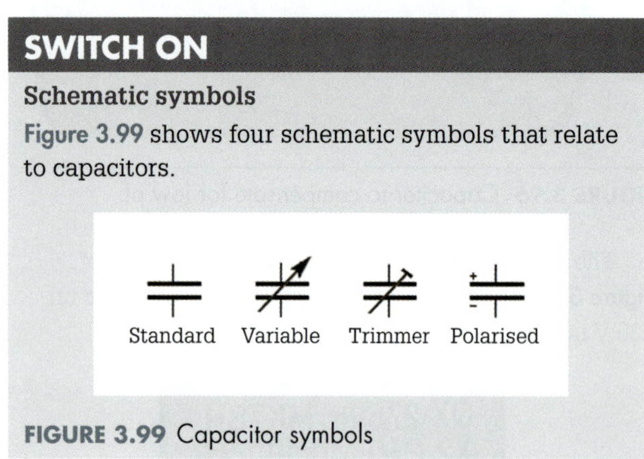

FIGURE 3.99 Capacitor symbols

The capacitor on d.c.

Referring to **Figure 3.100**, at the instant the switch closes to position A there is no potential difference across the capacitor, the supply potential is impressed across the resistor only and maximum current flows through the circuit. Although the capacitance has no effect on the initial value of the current, as the charging progresses the increase of an opposing potential across the capacitor reduces the current to zero. Occurring at the same time, the potential across the series resistor falls to zero. Consequently, no current passes through it. **Figure 3.101** shows graphs indicating the increase of potential and the decay of current with time.

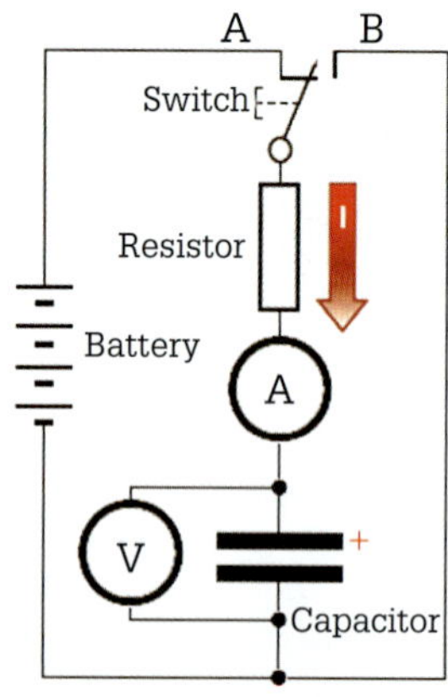

FIGURE 3.100 Capacitor charging

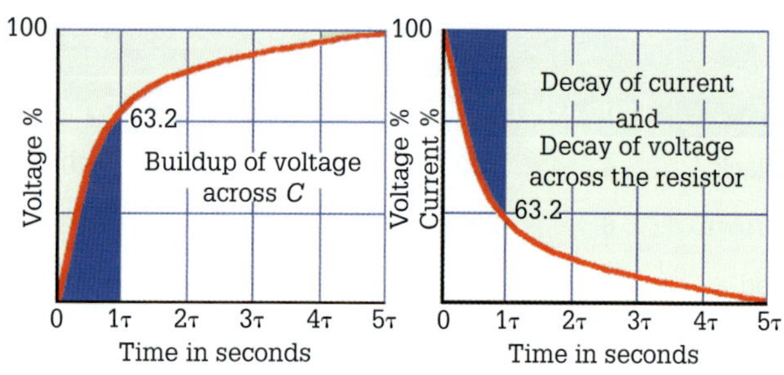

FIGURE 3.101 Voltage and current transients

The time delays experienced in the increase of voltage and the decay of current relate to the values of resistance and capacitance in the circuit and may be in the order of milliseconds, seconds, minutes or even hours. The time constant equation approximates the time delay period caused by the capacitive effect of the circuit.

Time constant

The time constant for a resistive-capacitive (RC) circuit is the time taken for current or potential to change by 63.2%. The Greek letter Tau (τ) is used to represent time constant in an equation. After five time constants (5), the change will be 99.3% complete.

$$\tau = RC$$

where τ = the time constant in seconds (s)

R = the resistance in ohms (Ω)

C = the capacitance in farads (F)

As shown in **Figure 3.101** the capacitance does not prevent the voltage from charging, it merely restricts the rate at which it can charge the capacitor. After a time interval equal to five time constants the potential difference across the capacitor will be approximately equal to the supply potential and the capacitor is fully charged.

Discharging

When the capacitor is disconnected from the supply, the stored energy starts to collapse. This decay is caused by leakage current paths within the dielectric that allows the capacitor to discharge. Electrolytic capacitors, because of their unique design characteristics, may hold this charge for hours. When the switch in the circuit of **Figure 3.102** changes to position B, a current flows through the resistor and discharges the capacitor. Because the current has now reversed, the resistor has a polarity opposite to the polarity it had while charging. However, the voltage now holds the same magnitude. At the instant the switch closes to position B, both the current and the voltage are at their maximum value. As the capacitor discharges its stored energy into the circuit, both the voltage and the current fall at the same rate finally reaching zero after five time constants. **Figure 3.103** shows the current and voltages against time for C and R.

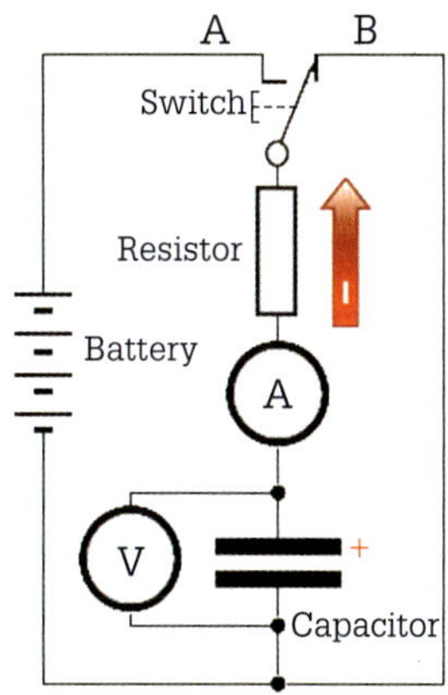

FIGURE 3.102 Capacitor discharging

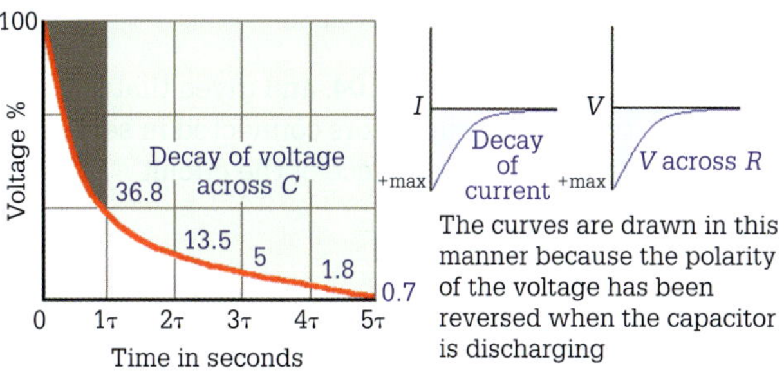

FIGURE 3.103 Voltage transients

EXAMPLE 3.29

A circuit contains a 0.5 µF capacitor in series with a 1M Ω resistor connected across a 100 V d.c. supply.

Determine:

a the time constant

b the time it will take for the capacitor to attain almost its full charge

c the charging current at the instant the switch is closed

d the energy stored in the capacitor 60 minutes after closing the switch.

The τ of the circuit is the product of the capacitance and the series resistance.

$$\tau = CR = (0.5 \times 10^{-6}) \times (1 \times 10^{6})$$
$$= 0.5 \text{ s}$$

The time for fully charging the capacitor takes five time constants so,

$$\text{time taken} = 5 \times 0.5$$
$$= 2.5 \text{ seconds}$$

After five time constants, the voltage reaches its approximate full value (assumed) of 100 V. Therefore, for all time after 2.5 seconds the voltage will be equal to 100 V.

The charging current at the instant the switch is closed is:

$$I_{Inst} = \frac{E}{R} = \frac{100}{(1 \times 10^{6})} = 100 \ \mu\text{A}$$

$$W = \tfrac{1}{2}CV^2$$
$$= \tfrac{1}{2}(0.5 \times 10^{-6}) \times 100^2$$
$$= \mathbf{0.0025 \text{ joules or } 2.5 \text{ mJ}}$$

Note: Provided at least 5τ have elapsed, the energy will be constant.

The capacitive effect exists whether the supply voltage is direct or alternating. However, with alternating current the voltage is continually reversing, causing the voltage across the capacitor to be in a continuous state of delay. As build-up of the potential across the capacitor is delayed, the current in the circuit is considered to be ahead of, or leading, the voltage.

EXERCISE 3.12

A circuit contains a 10 µF capacitor in series with a 1M Ω resistor connected across a 200 V d.c. supply.

Determine:

a the time constant

b the time it will take for the capacitor to attain almost its full charge

c the charging current five time constants after the switch is closed.

REVIEW QUESTIONS

1 What is a capacitor?

2 What is the capacitance of two conductors separated by 10 mm of air if 0.1 coulomb of charge is stored when connected across 200 V d.c.?

3 What factors determine the amount of capacitance of a capacitor?

4 Describe the term 'permittivity'.

5 What is another term for 'voltage gradient' in relation to a dielectric?

6 Determine the maximum voltage that two conductors 5 mm apart with the dielectric polystyrene (assume 50 kV mm^{-1}) between them can withstand.

7 How is energy stored in a capacitor?

8 Calculate the amount of energy stored within a 200 µF capacitor when connected across a 200 V d.c. supply.

9 What type of capacitor has small values of capacitance but maintains high stability with low loss over a wide frequency range?

10 What percentage of a capacitor's final voltage does the capacitor potential at one time constant equal?

11 A circuit contains a 100 µF capacitor in series with a 2.2 kΩ resistor connected across a 48 V d.c. supply. Determine:

- a the time constant
- b the time it will take for the capacitor to be fully charged
- c the voltage across the resistor at the instant the switch is closed
- d the charging current at the instant the switch is closed.

3.7 Capacitor circuits

Capacitors connect in series or parallel to produce resultant values, which may be the sum of the individual capacitance or the reciprocal of the capacitance values. Capacitors connected in series as in **Figure 3.104** have the same effect as increasing the thickness of the dielectric between the plates of a single capacitor.

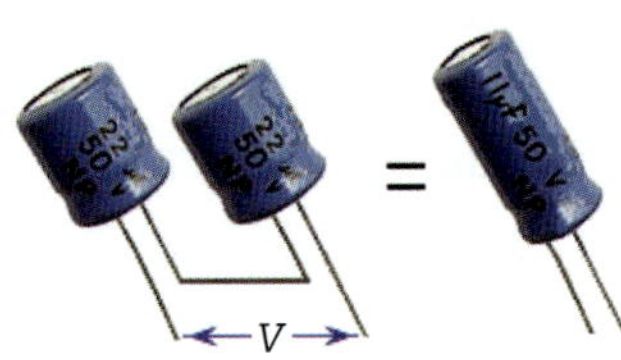

FIGURE 3.104 Capacitors in series across a voltage source

Total capacitance in series circuit

Figure 3.104 shows two series-connected capacitors. Since there is only one path for current flow, then as charge (Q) equals the product of current and time, it must be the same in all parts of the circuit. This is because the same magnitude flows through each series-connected capacitor for the same period of time. The current flows through the circuit until the sum of the potentials across the capacitors equals the source potential. Since the total charge (Q) is the same in all parts of the circuit:

$$Q_T = Q_1 = Q_2 \text{ etc.}$$

Also,

$$C = \frac{Q}{V} \text{ in farads and by transposing } Q = CV$$

where C = the capacitance in farads (F)

Q = the quantity of charge in coulombs (C)

V = the supply voltage in volts (V)

Since the sum of the potential differences across each capacitor must equal the source potential:

$$V_T = V_1 + V_2 \text{ etc.}$$

Substituting Q/C for the voltages in the above equation we obtain:

$$\frac{Q_T}{C_T} = \frac{Q_1}{C_1} + \frac{Q_2}{C_2} \text{ etc.}$$

Since all the charges (Q) are the same we can divide each term by Q and eliminate all Qs:

$$\frac{1}{C_T} = \frac{1}{C_1} + \frac{1}{C_2} \text{ etc.}$$

Taking the reciprocal of both sides of the equation we obtain the equation used to determine the total capacitance of series-connected capacitors:

$$C_T = \frac{1}{\frac{1}{C_1} + \frac{1}{C_2}} \text{ etc.}$$

EXAMPLE 3.30

Using the circuit of **Figure 3.104**, and given that there are two 100 µF capacitors connected in series, determine the total capacitance of the circuit.

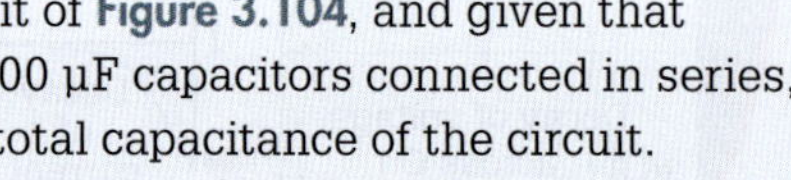

$$C_T = \frac{1}{\frac{1}{C_1} + \frac{1}{C_2}}$$

$$= \frac{1}{\frac{1}{100} + \frac{1}{100}} \quad [\mu F]$$

$$= 50 \text{ µF}$$

Total capacitance in a parallel circuit

Capacitors connected in parallel as in **Figure 3.105** have the effect of increasing the plate area of the capacitors. When capacitors connect in parallel the plate of each capacitor relates directly to each other and to the same voltage polarity. So, the sum of all the capacitors provides the total capacitance for the circuit.

FIGURE 3.105 Capacitors in parallel

For capacitors connected in parallel the total charge (Q) is the sum of all the individual charges.

$$Q_T = Q_1 + Q_2 \text{ etc.}$$

We already know that:

$$C = \frac{Q}{V} \text{ in farads and by transposing } Q = CV$$

Substituting CV for Q in our first equation we obtain:

$$C_T V = C_1 V + C_2 V \text{ etc.}$$

As potential difference in a parallel circuit is the same across each component we can divide both sides of the above equation by V and obtain the equation used to determine the total capacitance of parallel-connected capacitors.

$$C_T = C_1 + C_2 \text{ etc.}$$

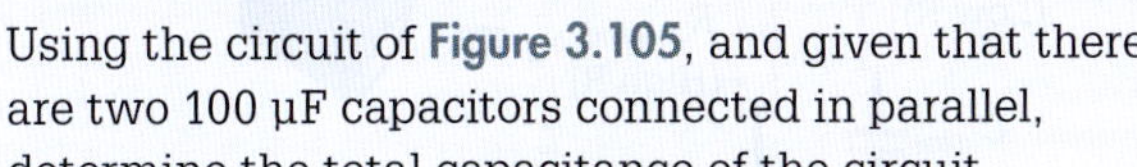

EXAMPLE 3.31

Using the circuit of **Figure 3.105**, and given that there are two 100 µF capacitors connected in parallel, determine the total capacitance of the circuit.

$$C_T = C_1 + C_2 = 100 + 100 \; [\mu F] = 200 \; \mu\text{F}$$

EXERCISE 3.13

- **a** Determine the equivalent capacitance of three series-connected capacitors having values of 100 µF, 50 µF and 33 µF.
- **b** Determine the equivalent capacitance of three parallel-connected capacitors having values of 220 µF, 150 µF and 33 µF.
- **c** A 120 µF and a 150 µF capacitor connect in parallel with a third capacitor. The equivalent capacitance of the combination is 470 µF. Determine the capacitance of the third capacitor.
- **d** A 120 µF and a 150 µF capacitor connect in series with a third capacitor. The equivalent capacitance of the combination is 47 µF. Determine the capacitance of the third capacitor.

Faults in capacitors

Faults can occur in capacitors by three means. The first is an open circuit whereby a lead connecting to the plates of the capacitor has disconnected. The second fault concerns short circuiting because of dielectric breakdown. Finally, a faulty capacitor may produce capacitance loss due to dielectric current leakage between the plates. If this leakage is high, there is a corresponding loss of capacitance and a heating of the capacitor itself. Testing of capacitors requires an ohmmeter that should be set to Ω × 10k or higher.

Discharge resistors

When a capacitor disconnects from the live circuit, say by opening a switch, the capacitor remains charged. The charge can remain stored for a long time, especially with power capacitors. Therefore, there is a very high risk of electric shock for anyone working on the circuit or touching the capacitor.

SWITCH ON

Before cleaning or touching capacitors they should be discharged through a bleeder resistor, as shown in **Figure 3.106**, to remove any charge still stored on the plates.

FIGURE 3.106 Capacitor and its bleeder resistor

Discharging should occur between terminals and with power capacitors between both terminals and earth. Short circuiting their leads with a copper conductor can discharge small capacitors, but this practice may damage power capacitors. Under no circumstances should short circuiting be undertaken. This implies that the sudden release of energy could vaporise or explode the shorting device.

A capacitor connected with electrical equipment that has no uninterruptible discharge path requires a discharge device to reduce the maximum voltage across the capacitor. For example, for capacitors rated at or below 650 V they must drop to a voltage below 50 V within 60 seconds. For capacitors whose rated voltage is above 650 V the time is five minutes. The fact that the capacitor has a bleeding device fitted does not ensure that the discharge of the capacitor has taken place. Bleeder resistor values can be determined by applying the following exponential equation:

$$R = \frac{t}{\ln\left(\frac{V_0}{V}\right)C}$$

where R = resistance of the bleeder resistor in ohms (Ω)

t = time in seconds (s)

ln = calculator function – returns the natural logarithm of a real or complex number, expression or list (e.g. ln 10 = 2.3)

V_0 = capacitor working voltage in volts (V)

V = voltage the capacitor must be discharged to in volts (V)

C = capacitance of the capacitor in farads (F)

EXAMPLE 3.32

What bleeder resistance value would a 100 µF capacitor charged to 500 V require if it discharges to 50 volts in one minute or less?

$$R = \frac{t}{\ln\left(\frac{V_0}{V}\right)C}$$

$$= \frac{60}{\ln\left(\frac{500}{50}\right)\times 100\times 10^{-6}}$$

$$= \frac{60}{2.3\times 100\times 10^{-6}}$$

$$= \mathbf{260\,576.69\ \Omega}$$

SWITCH ON

Capacitor safety precautions

Capacitors can contain significant energy that can cause electric shock and arc burns if touched. It is essential to de-energise and discharge capacitors and their circuits before work occurs. A voltage tester will determine if there is any charge on the capacitor.

Due to the chemical nature of capacitors, it is possible for these devices to develop a charge while isolated from the circuit. It is important to connect an approved discharging device across the capacitor terminals. When in storage, capacitors should have their terminals shorted out with a shorting cable in order to prevent chemical action charging the capacitor.

A capacitor should have discharged in its work environment. If this has not occurred, then the requirements of Australian and New Zealand standard AS/NZS 4836: 2010 'Safe working on low-voltage electrical installations', Clause 3.2.5, 'Work on or near exposed energised conductors', must be consulted. In addition, have recourse to 'Codes of practice' for safe electrical work before starting work on capacitors.

Testing capacitors

Capacitors require discharging before and after testing. A residual charge left in the capacitor will cause an error with measurement and may damage the testing instrument. In addition, you should make sure that the instrument's battery voltage is less than the direct current working voltage (VDCW) of the capacitor. An ohmmeter can quickly test small capacitors. When an analogue ohmmeter connects to a suitable capacitor, the pointer first deflects towards zero ohms and then gradually drops back as the capacitor charges (due to the battery voltage within the ohmmeter). If the pointer deflects to zero ohms and does not drop back, the capacitor is defective. Slight kicks downscale as the pointer climbs show leakage current through or across the dielectric.

Electrolytic capacitors only act as capacitors when the ohmmeter connects in one direction, but act as a short circuit when the leads from the ohmmeter to the capacitor interchange. These capacitors will indicate zero ohms in one direction and will climb to a high value of MΩ in the opposite direction. Testing of large capacitors for power factor improvement is difficult with an ohmmeter. Tests on such capacitors require power factor or current measurements, or a capacitance meter. A capacitance meter can perform open-circuit tests, short-circuit tests, and test for current leakage from the dielectric.

Equivalent circuit

The material used for the plates in a capacitor has resistance. So, just as there is no perfect dielectric, there is no such thing as an intrinsic (pure) capacitor. In reality, a capacitor contains both a series resistance and a parallel (leakage current paths) resistance interacting with its capacitive characteristics. **Figure 3.107** shows the equivalent circuit for a capacitor bringing together all these different factors.

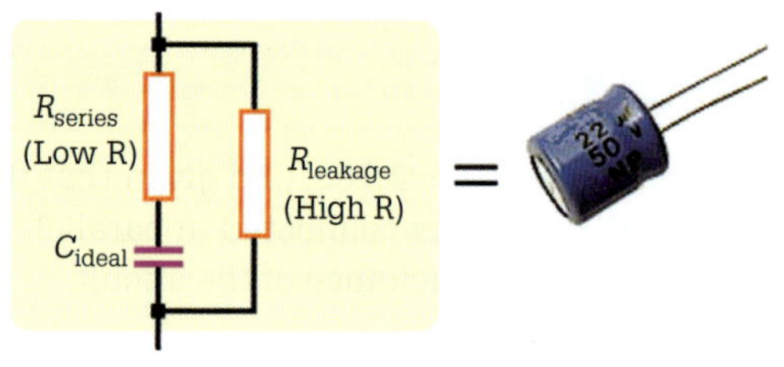

FIGURE 3.107 Equivalent circuit of a capacitor

Capacitor designs

The goal in component engineering is minimum size. The smaller the components, the more varied the control circuitry and features. The result is a smaller overall device package. With capacitors there are two limiting factors: working-voltage and capacitance. These two factors tend to oppose one another. However, given a choice of any dielectric materials, the only way to increase the voltage capacity of the capacitor is to increase the thickness of the dielectric.

This has the effect, as we have seen, of reducing the capacitance. To increase capacitance the plate area has to be increased – but this goes against the goal of minimum size. However, there are surface-mount and single-in-line (SIL) package capacitors which meet the minimum size parameters (2 mm to 5 mm thick) but have very small values of capacitance; that is, in the picofarad measurement range. Furthermore, in order to obtain the desired capacitance in a small volume, most capacitors are rolled up like a Swiss roll or layered like the fingers of a comb, with both combs meshed together.

REVIEW QUESTIONS

1 A 10 µF and a 47 µF capacitor connect in series. What is the equivalent capacitance of the combination?

2 Connecting capacitors in series has a similar effect to modifying which capacitor parameter?

3 Connecting capacitors in parallel has a similar effect to modifying which capacitor parameter?

4 A 33 µF capacitor connects in parallel with a 100 µF capacitor and a 22 µF capacitor. What is the total capacitance of the combination?

5 Name two faults that can occur in capacitors.

6 What is the purpose of the capacitor discharge resistor?

7 A power capacitor bank connected across 400 V consists of two stacks of capacitors. Each stack consists of 10 capacitors, each of 22 µF, connected in parallel. The two stacks connect in series with each other. Determine:

a the capacitance of a stack of capacitors

b the capacitance of the bank

c the potential difference across an individual stack.

8 An analogue ohmmeter is used to test a discharged small capacitor. How is a healthy capacitor indicated?

CHAPTER REVIEW

3.1 Resistance

- The resistance of a material is affected by its length, CSA, type of material and its temperature.
- The resistivity of a material is measured in ohmmetres.

3.2 Parallel circuits

- The same potential difference is impressed across each branch in a parallel circuit.
- Current leaving the voltage source equals the current returning to the voltage source.
- The total current in a parallel circuit divides into branch currents.
- In parallel circuits the more resistances there are the smaller is the total resistance. In addition, the total resistance value is less than the lowest valued resistance.

$$R_{equivalent} = \frac{1}{\frac{1}{R_1} + \frac{1}{R_2} + \frac{1}{R_3} \text{ etc.}}$$

- Total power in a parallel circuit is the sum of the power occurring in the individual branches.

$$P_{Total} = P_1 + P_2 + P_3 \text{ etc.}$$

3.3 Series–parallel circuits

- At least three elements are required to form a series–parallel circuit.
- To determine the potential difference across any of the resistances or the current through any of the branches, the magnitude of the total circuit current is required.

3.4 Meters in a circuit

- A meter is designed to measure accurately and display a quantity, such as voltage, current, resistance or power in a readable form.
- There are two analogue meter movements used: moving coil and moving iron.
- The purpose of the multiplier is to limit the movement current, at the new maximum voltage, to the FSD of the meter movement.
- A moving-iron meter can measure both d.c. (by changing the meter scale calibration) and root-mean-squared (rms) values of alternating current.
- Digital ohmmeters are built around a low-noise, high-resolution analogue-to-digital converter.
- Some digital ohmmeters use a four-terminal output configuration to reduce test lead wire and contact resistance errors.
- A tong tester is a clamp meter or a clip-on ammeter and can be either analogue or digital.
- The voltage detector's pencil-like probe via a series of LEDs checks both a.c. and d.c. voltages.
- The a.c. voltage finder requires no contact with live conductors.
- Electrodynamometers measure energy.

3.5 Resistance measurement

- The substitution method works on the principle that if the unknown and known resistances are of the same value, they will allow the same size of current to flow when connected in turn in the same circuit.
- A voltmeter and an ammeter measure the voltage difference across, and the current through, a resistor and the resistance value can be determined by Ohm's law.
- The ohmmeter is a portable instrument for measuring relatively low values of resistance.
- The power supply on a circuit requires disconnection before making any resistance tests.
- The insulation of a 230 V and a 400 V system requires testing at 500 V d.c. to earth.
- The minimum insulation resistance allowed is 1 MΩ.
- When a Wheatstone bridge balances, the output voltage is zero.
- Strain gauges are resistive devices.

3.6 Capacitors and capacitance

- Capacitors store an electric charge.
- There are three primary factors that determine the amount of capacitance that an energy-storage system possesses: plate area, plate spacing and type of dielectric.
- Electrolytic capacitors occur where a high amount of capacitance is required.
- The electrolytic capacitor is polarised.
- Power capacitors are used to store and release electrical energy.
- The time constant (τ) of a capacitive circuit is the time it takes for the transient voltage and current to make a 63.2% change in value.
- The time constant (τ) of the circuit is the product of the capacitance and the series resistance.

3.7 Capacitor circuits

- Capacitors may be connected in series or parallel to produce resultant values.
- The fact that a capacitor has a bleeding device fitted does not ensure that the discharge of the capacitor has taken place.
- Capacitors can contain significant energy to cause electric shock and arc burns if touched.

TRIAL EXAM

For Chapter 3 knowledge assessment, please complete the following trial exam.

1 If the length of a conductor is decreased while the CSA remains constant, the resistance of the conductor:
  a is inversely proportional to the length
  b remains the same
  c increases
  d decreases

2 When comparing two conductors of the same length and material, with one conductor having twice the diameter of the other, what value statement describes the resistance of the smaller conductor compared to the larger?
  a double the resistance
  b four times the resistance
  c smaller resistance
  d same resistance

3 If the temperature of a pure metal conductor decreases, then the resistance of the conductor:
  a gradually prevents electron flow
  b remains the same
  c decreases
  d increases

4 Determine the resistance of a 40 m length of aluminium conductor that has a CSA of 10 mm^2. The resistivity of aluminium is 2.83×10^{-8} Ωm.
  a 0.1132 Ω
  b 0.0377 Ω
  c 0.021 225 Ω
  d 0.000 44 Ω

5 A parallel circuit has:
  a a single current path
  b more than one current path
  c different voltages across each element
  d a total resistance equal to the sum of the individual elements

6 The supply current in a parallel circuit equals the:
  a square of the voltage divided by the resistance
  b branch resistance divided by the voltage
  c sum of the branch currents
  d product of the branch currents

7 The potential difference across each of two elements connected in parallel is:
  a dependent on the resistance value of the individual element
  b twice the supply voltage
  c the same as the supply voltage
  d half the supply voltage

8 Connecting resistors in parallel has a similar effect to:
  a reducing the length of a wire conductor
  b increasing the diameter of a wire conductor
  c increasing the length of a wire conductor
  d reducing the diameter of a wire conductor

9 Calculate the total resistance of five 100 Ω resistors connected in parallel:
  a 20 Ω
  b 100 Ω
  c 5 Ω
  d 4 Ω

10 Determine the total current supplied to a parallel circuit containing a 180 Ω resistor and a 120 Ω resistor when connected across a 90 V supply.
  a 1 A
  b 20 mA
  c 1.25 A
  d 2 A

11 Determine the resistance of an unknown resistor in a circuit if it connects in parallel with a 2 kΩ resistor when the total resistance is 800 Ω.
  a 1000 Ω
  b 500 Ω
  c 8000 Ω
  d 1333 Ω

12 Determine the power dissipated by a parallel circuit containing three resistors of 180 W each when connected across a supply voltage of 100 V.

a 167 Ω
b 80 Ω
c 40 Ω
d 3 Ω

13 A current divider circuit consists of two resistors, 1800 Ω and 600 Ω. Determine how much current flows through the 1800 Ω resistor when the total circuit current is 4 A.

a 2.666 A
b 1 A
c 3 A
d 6 A

14 The series–parallel circuit in **Figure 3.108** consists of a 50 Ω resistor connected in series with a parallel combination of two resistors of 600 Ω and 200 Ω. Calculate the total resistance of the circuit.

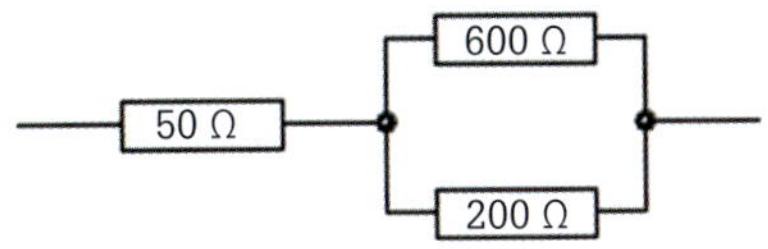

FIGURE 3.108 Series–parallel problem

a 37.5 Ω
b 850 Ω
c 450 Ω
d 200 Ω

15 Using **Figure 3.109**, determine the potential difference across the 20 Ω resistor at a supply voltage of 60 V.

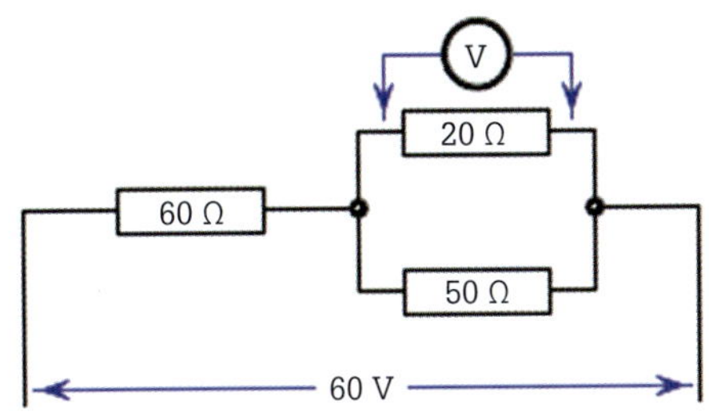

FIGURE 3.109 Series–parallel problem

a 30 V
b 11.54 V
c 9.23 V
d 20 V

16 Using **Figure 3.110**, determine the current through the 40 Ω resistor when the circuit voltage is 24 V.

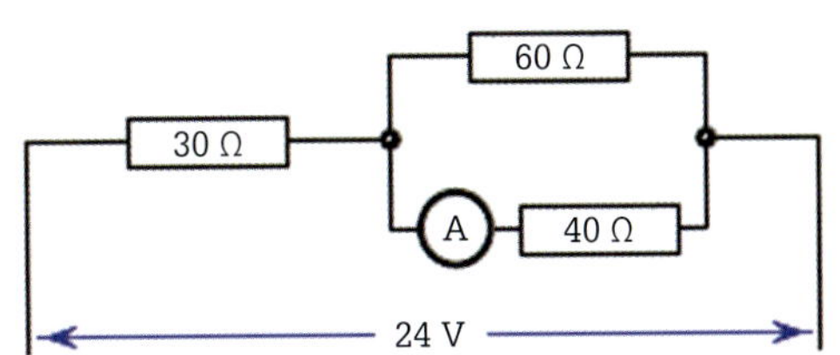

FIGURE 3.110 Series–parallel problem

a 0.185 A
b 0.444 A
c 0.267 A
d 0.288 A

17 Using **Figure 3.111**, determine the resistance of the series resistance when the equivalent resistance is 1 kΩ.

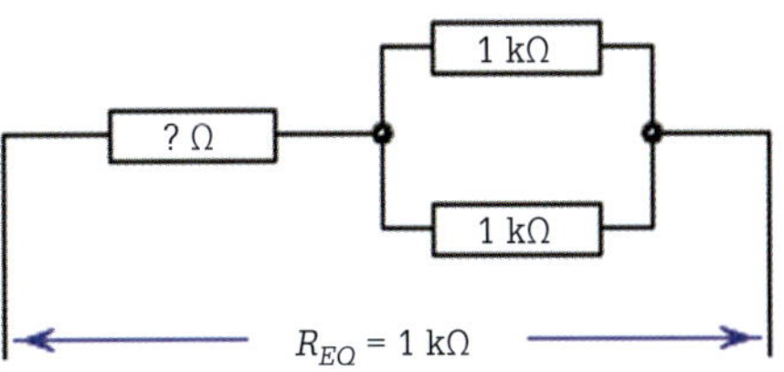

FIGURE 3.111 Series–parallel problem

a 1 kΩ
b 500 Ω
c 2 kΩ
d 100 Ω

18 Using **Figure 3.112**, determine the total power dissipated by the circuit when the supply voltage is 56 V.

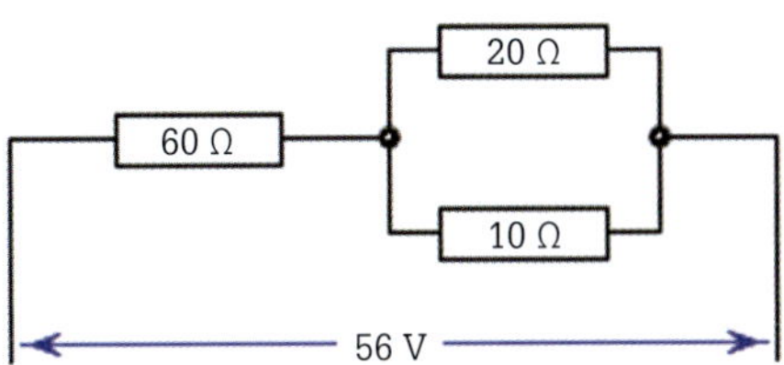

FIGURE 3.112 Series–parallel problem

a 0.84 W
b 47.04 W
c 3733.52 W
d 34.84 W

19 If an open circuit occurs in a branch of the parallel network of resistors in a series–parallel circuit, the current through the series network will:

a stay at the same value
b be zero as no current flows
c increase
d decrease

20 With electrical measuring instruments, what does the term 'analogue' mean?

a on or off
b changing continuously
c meter
d a full scale deflection

21 What type of meter measures current?

a voltmeter
b ammeter
c ohmmeter
d insulation tester

22 Name the meter that can measure current without breaking the circuit.
 a ammeter
 b multimeter
 c clamp meter
 d ohmmeter

23 Which of the following statements applies to a moving-iron instrument?
 a non-linear scale, measuring d.c.
 b a linear scale, measuring a.c.
 c a linear scale, measuring d.c.
 d a non-linear scale, measuring a.c. and d.c.

24 Which resistance extends the range of a millivolt meter to read voltages of the order of 1000 V?
 a a parallel high-value resistor
 b a parallel low-value resistor
 c a series high-value resistor
 d a series low-value resistor

25 The sensitivity of a voltmeter is expressed as:
 a amperes per ohm
 b ohms per volt
 c amperes per volt
 d volts per ampere

26 Which of the following is used to extend the measuring range of an ammeter?
 a multiplier
 b analogue to digital convertor
 c shunt resistor
 d dynamometer

27 An instrument suitable for measuring the resistance of a conductor is:
 a continuity tester
 b clamp meter
 c ohmmeter
 d insulation tester

28 An instrument suitable for measuring the effectiveness of insulation is:
 a multimeter
 b ohmmeter
 c volt/ammeter measurement
 d insulation resistance tester

29 What is the voltage reading (see **Figure 3.113**) recorded on the following instrument display?

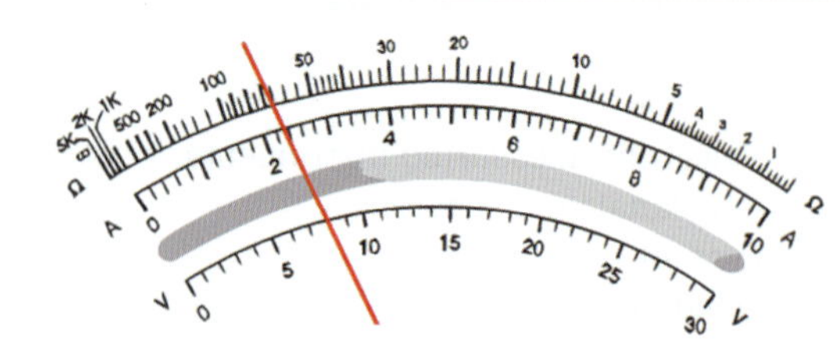

FIGURE 3.113 Voltage reading

30 What is the resistance reading (see **Figure 3.114**) recorded on the following instrument display?

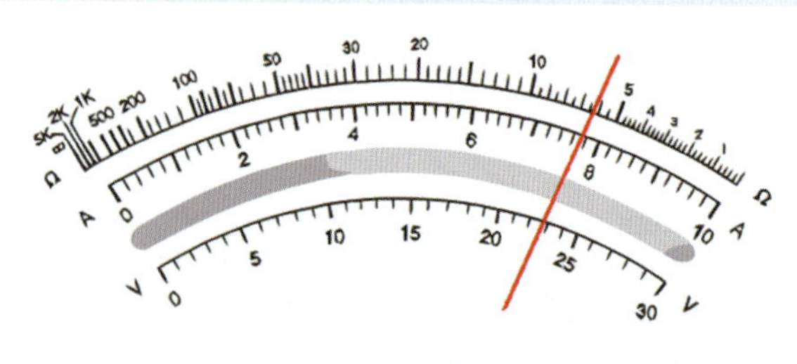

FIGURE 3.114 Resistance reading

31 What is the current reading (see **Figure 3.115**) recorded on the following instrument display?

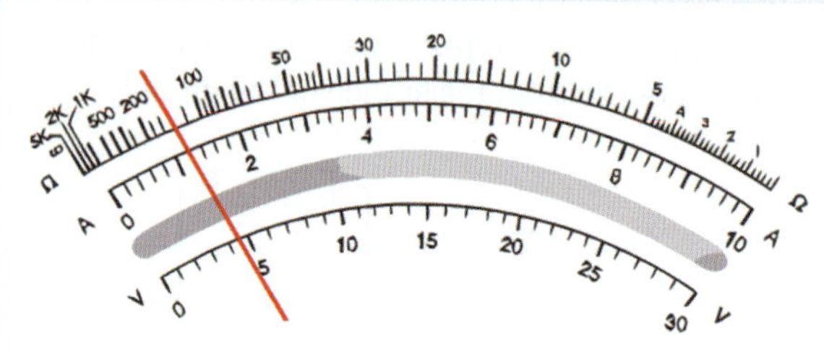

FIGURE 3.115 Current reading

32 Calculate the value of the unknown resistor (R_x) in the balanced bridge as shown in **Figure 3.116**.

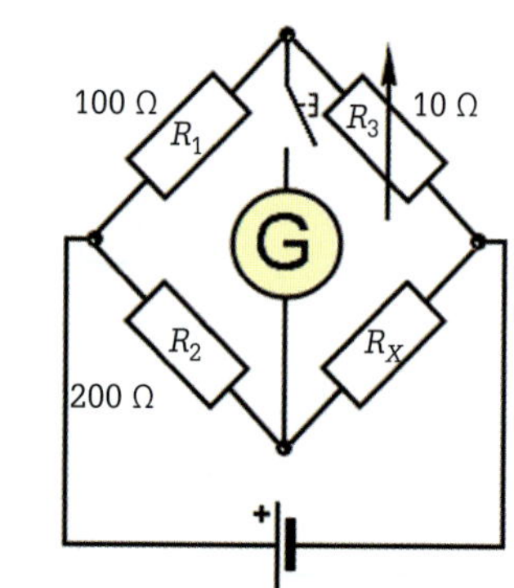

FIGURE 3.116 Bridge problem

 a 2000 Ω
 b 1000 Ω
 c 20 Ω
 d 5 Ω

33 A device designed to be capable of storing electric charge is a:
 a resistor
 b inductor
 c battery
 d capacitor

34 The percentage change in current after one time constant when charging a fully charged capacitor is:
 a 14%
 b 36.8%
 c 63.2%
 d 86%

35 The unit of measurement for capacitance is the:
a ohms/volt
b coulomb
c tau
d farad

36 A bleed resistor:
a provides the rate of charge
b provides a voltage drop
c discharges a capacitor
d limits the charging current

37 The relationship between the capacitance and potential difference across a capacitor is the:
a working voltage of the capacitor
b charge of the capacitor
c time constant of the capacitor
d effective capacitance

38 What factor does not determine the capacitance of a capacitor?
a area of the plates
b distance between the plates
c working voltage
d type of dielectric

39 Which of the following capacitors is polarised?
a polyester
b polystyrene
c ceramic
d electrolytic

40 Determine the capacitance of four 33 µF capacitors connected in parallel across 20 V.
a 660 µF
b 8.2 µF
c 5 µF
d 132 µF

41 Determine the capacitance of four 10 µF capacitors connected in series across 16 V.
a 40 µF
b 1.6 µF
c 2.5 µF
d 0.6 µF

42 Determine the charge stored on a 10 µF capacitor connected across a 65 V supply.
a 47 mC
b 670 µC
c 364 mC
d 86 µC

43 Using the values in question 42, determine the charging current of the capacitor after 10 time constants?
a zero
b 12 mA
c 37.92 mA
d 7.584 mA

44 The most common use of capacitors in low-voltage systems is:
a the provision of lagging currents
b power factor correction
c rectifying a.c.
d voltage stabilisation

4 Magnetic and electromagnetic devices

This chapter provides electrotechnology workers with knowledge of magnetic and electromagnetic principles and the laws of electromagnetism and inductance. In addition, they learn how these laws apply to the operation of electromagnetic devices. Electrotechnology workers also gain knowledge of the undesirable effects and hazards of induced voltages. This chapter provides underpinning knowledge for the unit UEEEL0021 from the UEE training package.

LEARNING OBJECTIVES

Magnetism

- Describe how materials are magnetised
- State the characteristics of magnetic fields
- Explain the purpose of magnetic screening

Electromagnetism

- State the characteristics of magnetic fields around a conductor and a coil
- Apply the right-hand thumb rule
- Calculate the force between current-carrying conductors

Magnetic circuit and nomenclature

- Describe magnetic quantities and their units
- Calculate various magnetic quantities in a magnetic circuit

Losses in magnetic circuits

- Describe a magnetisation curve
- Explain the meaning of a permeability curve and hysteresis loop

Electromagnetic induction

- State Faraday's law
- Apply Fleming's right-hand rule
- Apply Lenz's law

Inductance

- Explain the time delay caused by the inductive effect
- Calculate the energy stored within the electromagnetic field

Magnetic devices

- Describe an inductor
- Explain the meaning of inductance, self-inductance and mutual inductance
- Explain the meaning of time constant in a series LR circuit

4.1 Magnetism

Magnetic materials

Magnetite

Magnetism has a history dating back over 3000 years to the discovery of the peculiar property of a natural ore called magnetite. In addition, magnetite as illustrated in **Figure 4.1** is a black oxide of iron expressed chemically as Fe_3O_4.

FIGURE 4.1 Magnetite

The familiar name for magnetite is lodestone or leading stone, which resulted from its use as an aid for navigation. The Chinese found that a piece of this ore would tend to orient itself in relation to a star referred to as the leading star. The ore, when allowed to rotate freely on a string or float by assistance on water, would point in the direction of the star. The Greeks also knew of magnetite but only as a matter of curiosity.

The magnetic phenomenon of magnetite was regarded for centuries as supernatural, but in modern times we understand that the properties of magnetite are due to the force of magnetism. Natural magnets like magnetite no longer have any practical use because artificial magnets possess stronger magnetic abilities together with a variety of shapes. These artificial magnets can keep their magnetic properties for a long period and are permanent magnets.

Magnets in some way influence all substances but the extent of magnetisation varies greatly with the kind of substance involved. For most substances, the response is feeble, and they lose their magnetic properties once the removal of the magnetisation force occurs. These substances are temporary magnets.

Classifications of magnets

Magnetic substances have three classifications: diamagnetic, paramagnetic and ferromagnetic. However, the greater majority of substances are diamagnetic which become only feebly magnetised when subjected to the power of magnetism and in the opposite direction to that force. Examples of diamagnetic substances include glass and sulphur, and the metals silver, gold and copper.

Paramagnetic and ferromagnetic

The second type of magnetic substances is paramagnetic. Paramagnetic substances become magnetised only slightly more strongly than diamagnetic substances; however, they are magnetised in the same direction as the applied magnetic force. Examples of paramagnetic substances include the metals platinum and aluminium. The third type of magnetic substances is ferromagnetic. Ferromagnetic substances become strongly magnetised in the same direction as the applied magnetic force. Examples of ferromagnetic substances include the metals iron, cobalt and nickel, together with alnico alloys of these metals and aluminium.

Ferromagnetic substances retain a majority of the magnetic properties on removal from the magnetising effect of the magnet. However, different ferromagnetic substances are magnetised to varying extents by the same magnetising force.

When a magnetic substance encounters iron filings, most of the filings attach themselves to the ends of the magnetic substance with few in the middle. Refer to **Figure 4.2**.

FIGURE 4.2 Bar and horseshoe magnet with iron filings

Magnets

Poles

Every substance that exhibits magnetic properties has two poles where most of its magnetic strength concentrates. In addition, a magnetic substance containing two poles is a magnetic dipole. The Earth also behaves as a magnetic dipole. There is a magnetic pole close to the Earth's geographic North Pole, which essentially forms one end of an extremely large but imaginary bar magnet that passes through a north–south axis of the Earth.

A bar magnet freely suspended by a thread as shown in **Figure 4.3** always swings around until its magnetic axis (a straight line passing through the ends of the magnet) lies north–south. The north pole (or north-seeking pole) of the suspended bar magnet points roughly (but not exactly) towards the geographic North Pole of the Earth. If we accept that the end of the bar magnet that points north as the north pole of the magnet, then it must be attracted to a south magnetic pole. It follows then that the magnetic pole in the northern hemisphere of the Earth is actually a magnetic south pole.

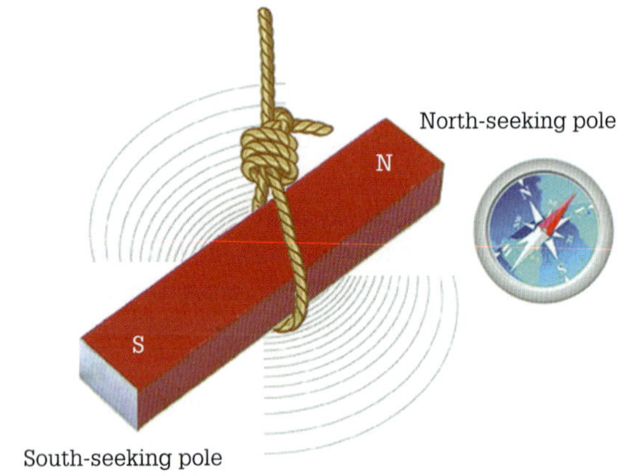

FIGURE 4.3 South- and north-seeking poles

The relationship between the poles of a magnet can be summed up as follows:

SWITCH ON

A fundamental law of magnetism states that like poles repel each other while unlike poles attract each other. Refer to **Figure 4.4**.

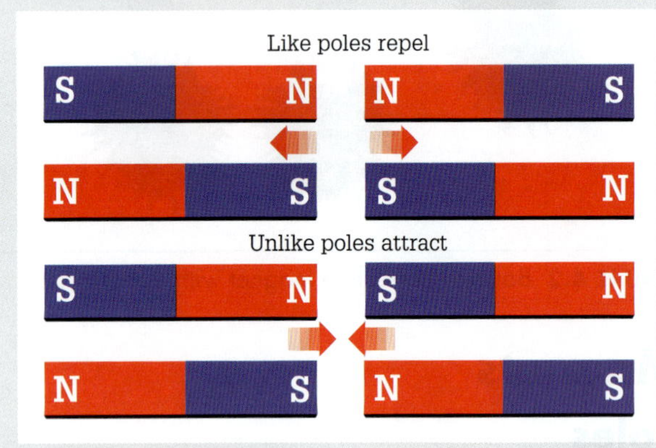

FIGURE 4.4 Relationship between poles

Inducing magnetic properties

If a ferrous material such as iron or steel (see **Figure 4.5a**) is exposed to a pole of a permanent magnet (see **Figure 4.5b**) it becomes magnetised. The iron and steel in **Figure 4.5b** are induced magnets and their 'magnetism' is only temporary; as such they are not permanent magnets. A polarity test on these induced magnets shows that the induced pole adjacent to the pole of the permanent magnet has an opposite polarity to the magnet pole. Refer to **Figure 4.5**.

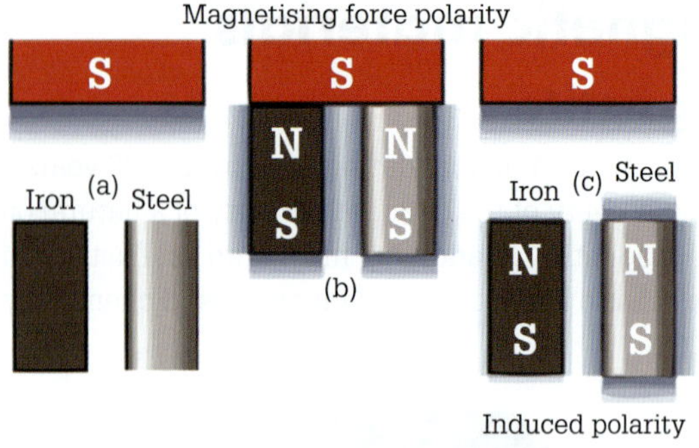

FIGURE 4.5 Iron and steel material being induced

When the two induced magnets move away from the permanent magnet (**Figure 4.5c**) the steel substance retains some of its magnetism and becomes a weak permanent magnet. The iron substance, however, is a temporary induced magnet possessing only residual magnetism, called remanence, and its magnetism gradually weakens and disappears.

Making permanent magnets

It is possible to make a permanent magnet based on the principle of induced magnetism by the application of either of two methods. The first method involves a stroking technique while the other relies upon the magnetic properties of a solenoid.

Stroking technique

The stroking technique as shown in **Figure 4.6** works for any ferromagnetic substance and has several steps that occur.

- The permanent magnet requires stroking repeatedly in the same direction along the ferromagnetic substance.
- At the end of each stroke, there must be broad sweeps away from the substance to be magnetised.
- The same pole on the permanent magnet must be the one used when stroking the ferromagnetic substance.

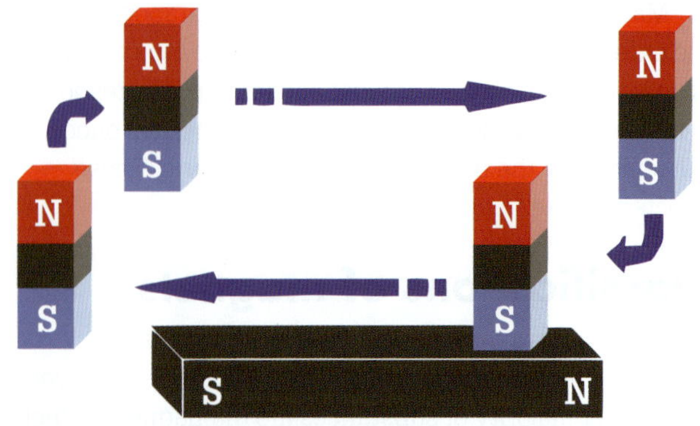

FIGURE 4.6 Inducing magnetic poles by the stroking method

Solenoid method

The most effective method to create a magnet occurs with the use of a solenoid. With this method, the ferromagnetic substance requiring magnetisation enters into a coil. In addition, the coil has many turns of conducting wire so that it produces a very strong magnetic field. Consequently, when the solenoid energises, the magnetic field created induces magnetic properties in the ferromagnetic substance. Refer to **Figure 4.7**.

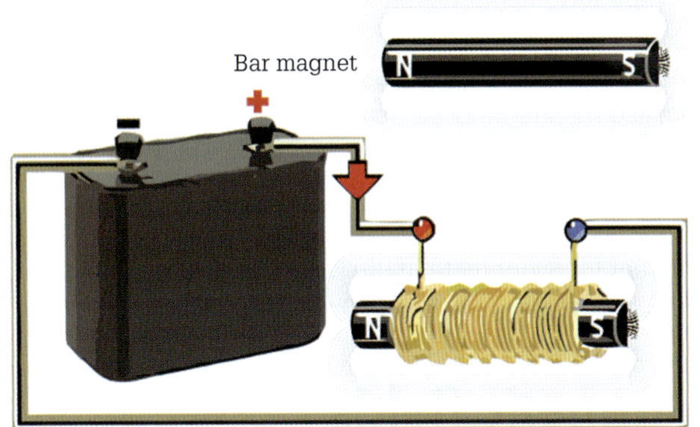

FIGURE 4.7 Solenoid method of making a magnet

Permanent magnets have good retentivity whereby they are able to retain their magnetic properties better than temporary magnets. Furthermore, if a permanent magnet is broken into pieces, as in **Figure 4.8**, each piece becomes a complete smaller and weaker permanent magnet.

FIGURE 4.8 Magnet broken into pieces

Magnetisation process

Classical theory

The classical theory of magnetism, the Weber-Ewing theory, describes the magnetisation process as the aligning of the molecules of a substance in the direction of the magnetisation force. Their theory assumes that every molecule of a magnetic substance is a small magnet possessing a north and south pole. This implies that the magnetisation of a substance consists of changing the direction of these molecular magnets so that they work in unison and in the same direction. **Figure 4.9** illustrates the classical theory of magnetism.

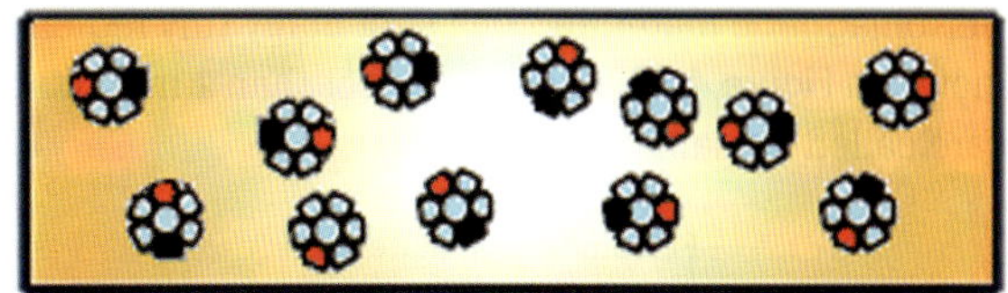

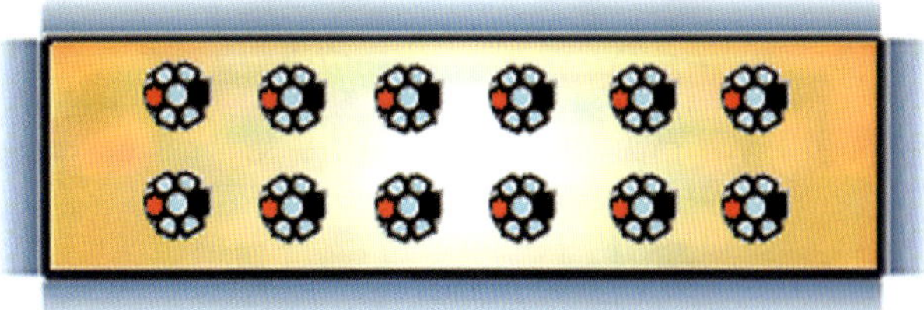

FIGURE 4.9 Unmagnetised and magnetised molecules

Support for this theory occurs by the fact that when a magnet breaks into pieces the pieces become smaller but weaker permanent magnets. If this process continued downwards the last molecule would exhibit its own magnetic properties. Additional support transpires in that when a magnet has heat applied or is repeatedly struck, the molecular alignment scatters and the substance becomes demagnetised.

Domain theory

The second theory of magnetism, called the domain theory, describes magnetic properties as being created by the spinning movement of electrons in the atoms of the substance. Electrons orbit within the electron cloud and around the nucleus of each atom and revolve or rotate on their axes. This spin, as shown in **Figure 4.10**, of the electron enables the electron to possess a magnetic field in addition to the electric field gained in its orbit around the nucleus.

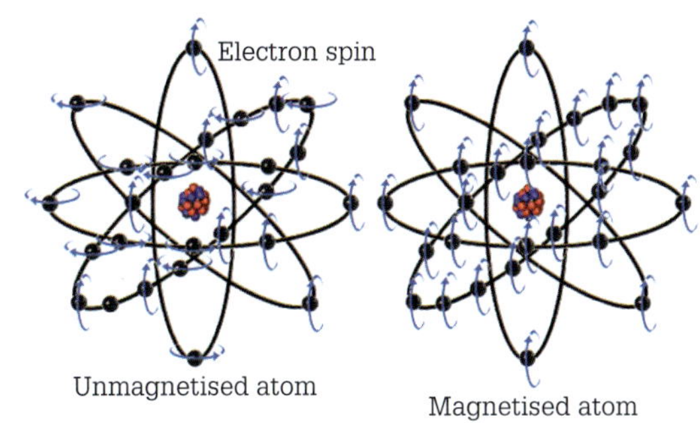

FIGURE 4.10 Electron spin

It is the number of electrons spinning in the same direction that determines the effectiveness of the magnetic field of an atom. If the majority of electrons spin in the same direction then the atom of the substance becomes magnetised. When many atoms of the same substance are grouped together, there is an interaction between them. It follows that this interaction produces groups of atoms with parallel magnetic fields. Finally, such a grouping of magnetically parallel atoms is a domain.

Ferromagnetic materials consist of many tiny regions, as shown in **Figure 4.11**, usually smaller than 1 mm in area, and are called domains.

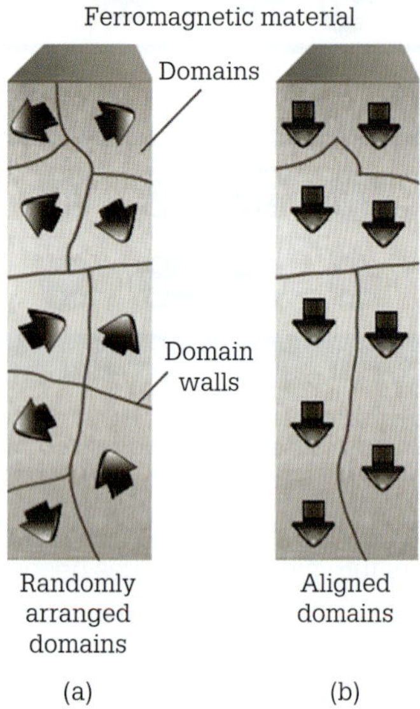

FIGURE 4.11 Domains

When randomly arranged the domains are not magnetised (**Figure 4.11a**). However, when an external magnetic field applies to a ferromagnetic material, two events occur. First, the size of the domain boundaries aligned with the external magnetic field direction grows. Second, the domain boundaries of the domains not aligned reduce.

The alignment (**Figure 4.11b**) gives magnetism to the ferromagnetic material. In addition, the aligned domains form closed loops of magnetic flux as shown in **Figure 4.12**.

FIGURE 4.12 Magnetic field around a bar magnet

The magnetic properties of a ferromagnetic material depend on how easy it is for the domain boundaries to move. Hard ferromagnetic material such as alnico and ferrite contain carbon and other impurities, and these make it difficult for the domain boundaries to move. However, once the domains align, the impurities prevent the domains from moving out of alignment, and the hard ferromagnetic material remains permanently magnetised. **Figure 4.5** illustrated this effect with iron as a soft ferromagnetic material and steel as a hard ferromagnetic material.

When an external magnetic force applies to a substance, the domains work in unison and in the same direction as the external magnetic force. Moreover, the number of domains acting in unison determines the strength of the magnetised substance. If this theory is correct, then domains are a magnetic dipole. In contrast, the existence of a magnetic monopole has been theorised in which the magnetic field is radially outwards from the domain; however, there is as yet no evidence for its existence.

Hard ferromagnetic materials like steel lose their magnetic properties if hit continually or heated to a critical temperature. The loss of magnetic properties occurs because the magnetised domains will organise themselves randomly after being struck or heated. In addition, the temperature at which a ferromagnetic material loses its magnetic properties is the Curie temperature and it is different for every metal (500–600 °C for steel alloys and sintered samarium–cobalt 750 °C).

Magnetic fields

From the study of particle physics, that is, the study of the basic building blocks of all the substances or matter we know of, physicists understand that the magnetic field has an enormous number of magnetic lines of force. We can see the existence of a magnetic field by sprinkling iron filings over a magnet as shown in **Figure 4.12**.

The space around the magnet in which it exerts a force is its magnetic field. The iron filings map out this magnetic field in the form of lines of force. So, the lines of force are a vector quantity. The accepted view is that the lines of force emanate from the North Pole and enter the South Pole. **Figure 4.13** shows a three-dimensional magnetic field permeating all the space around the magnet.

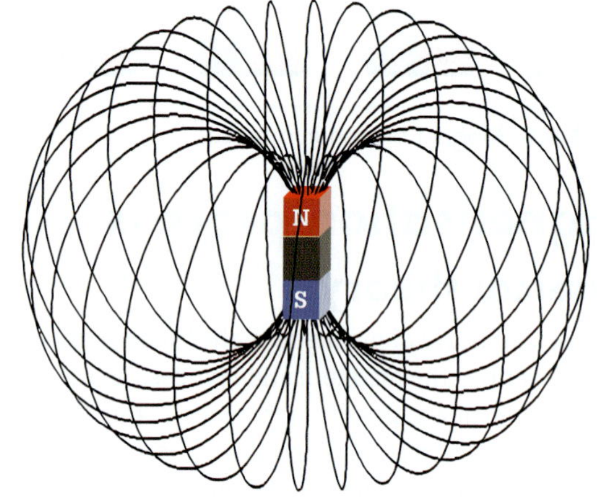

FIGURE 4.13 Three-dimensional nature of a magnetic field

Figures 4.14 and **4.15** illustrate the magnetic fields of various magnets and pole positions. The point marked X is the neutral point where the two magnetic forces are equal.

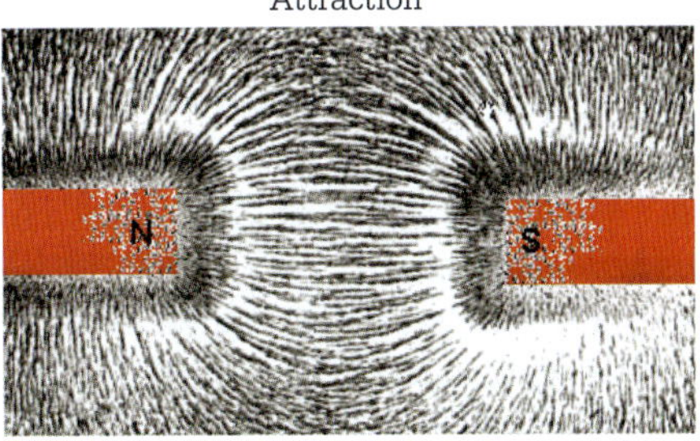

FIGURE 4.14 Magnetic fields of bar magnets

FIGURE 4.15 Horseshoe magnet and its magnetic field

Detecting magnetic fields

A magnaprobe moves gyroscopically to accurately map out the three-dimensional nature of a magnetic field as shown in **Figure 4.16**.

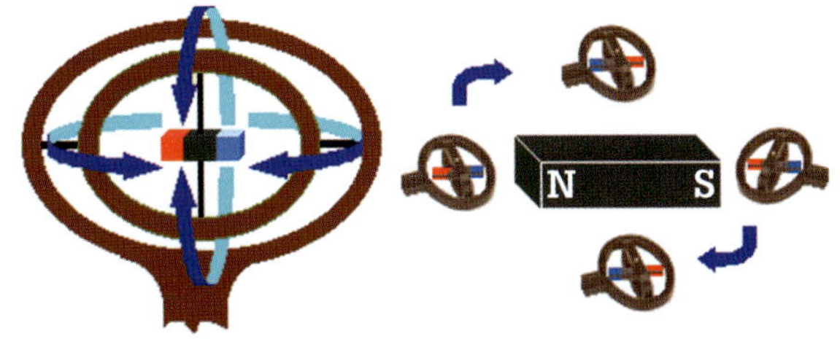

FIGURE 4.16 Magnaprobe

The magnaprobe consists of a small alnico bar magnet that is gimbal mounted (behaves like a gyroscope) so that it freely rotates as you move it around a magnetic substance. It is very sensitive, and its function is the detecting of stray fields or magnetic leakage.

Magnetic keeper

In order to assist permanent magnets retain their magnetism when stored, soft iron keepers attach to the poles of the magnets as shown in **Figure 4.17**. They help maintain the strength of the magnet by providing a complete magnetic circuit.

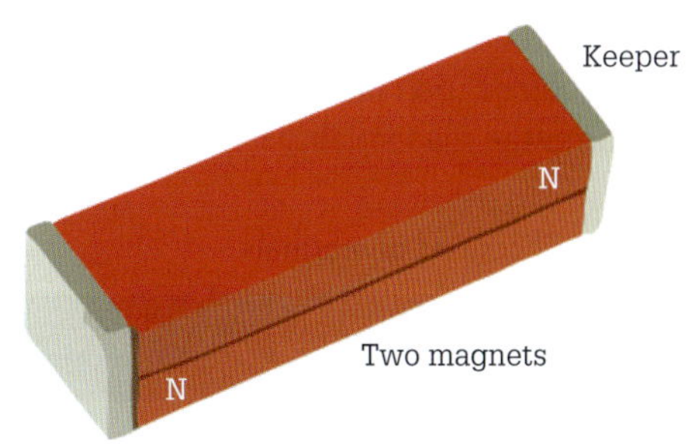

FIGURE 4.17 Role of the keeper

Magnetic circuit characteristics

The path taken by the lines of force, or magnetic flux, is a magnetic circuit. Within this magnetic circuit as shown in **Figure 4.18**, lines of force exist that possess six characteristics:

- Magnetic lines of force originate from the North Pole through the surrounding space and enter the South Pole.
- Each line of force forms an independent closed loop.
- Lines of force do not merge with or cross other lines of force.
- Lines of force that are parallel and work in the same direction repel one another.
- Lines of force concentrate where the field is the strongest.
- Lines of force try to become as short as possible without breaking.

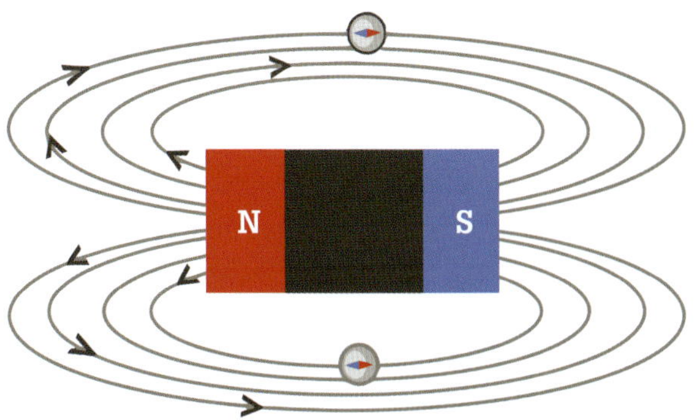

FIGURE 4.18 Magnetic lines of force

Classification of magnetic materials

When magnetic substances develop in the presence of a magnetic field, the magnet is *anisotropic*. This magnetic treatment orients the magnetic substance to take on maximum magnetisation and enables a higher magnetic density. By contrast, a magnetic substance not created in a magnetic field is *isotropic*.

Substances have a classification according to their ability to retain their magnetism. Substances that are easy to magnetise but lose their induced magnetism within a short period after the magnetising force is removed are magnetically soft substances, hence the term 'soft iron' in this context.

Those substances that offer some reluctance initially to the magnetising force but retain nearly all of their induced magnetism are magnetically hard substances. These hard substances have two classes.

The two classes of permanent magnets are non-rare-earth and rare-earth magnets. The family of non-rare-earth magnets includes ceramic or ferrite magnets and alnico magnets. Rare-earth magnets on the other hand, include neodymium–iron–boron and samarium–cobalt types. In addition, grades of their compositions are another classification. **Table 4.1** summarises the classifications of magnets.

TABLE 4.1 Classifications of magnets

Family of permanent magnets	Subtype	Characteristics
Non-rare-earth magnets	**Ceramic (or ferrite)**	Made from iron oxide together with barium and strontium and manganese–zinc or nickel–zinc elements. They have the highest flux density and coercive force (the force required to demagnetise a material) as well as the best opposition to demagnetisation and oxidation compared to other magnets. Also, they have a crystalline structure and are very hard and brittle.
	Alnico	Alnico magnets consist of an alloy of aluminium, nickel and cobalt. Their features include high remanence and temperature stability, and low coercive force ability.
Rare-earth magnets	**Neodymium–iron–boron**	These are the strongest of the permanent magnets possessing the highest coercive magnetic field strength, the lowest temperature stability characteristic and a relatively low corrosion and oxidation resistance.
	Samarium–cobalt	Advanced technology allows construction of these magnets from a magnetic powder form that bonds together with polymers such as epoxy. This process allows the magnetic powder to mould into different complex shapes.

Applications of magnets

Permanent magnets find use in the following applications in the Switch on box below.

SWITCH ON

Applications of permanent magnets

Ceramic magnets:

- speaker magnets
- magnetos used on lawnmowers and outboard motors
- d.c. brushless motors and permanent magnet motors as used in motor vehicles
- separators (to remove ferrous material from nonferrous material)
- stepper motors

Alnico magnets:

- telephones
- microphones
- generators
- loudspeakers
- clutches
- electron tubes (magnets shape and control the electron beam)
- reed switches

Neodymium–iron–boron magnets:

- d.c. starter motors
- servomotors
- dot-type printers
- speakers
- separators
- linear actuators

Samarium–cobalt magnets:

- sensors
- computer disk drives
- satellite systems
- linear actuators

»

Magnetic screening and shielding

Without magnetic fields, technology that is electrically related could not exist. Magnetic fields of one form or another permeate all areas of the world and generally do not pose a threat to humans. However, there are effects produced at low frequencies of 50–60 hertz where the fields generated can interfere with the operation of magnetic tape, computer monitors, power supplies and transformers, together with transmission and distribution lines. In spite of this the magnetic interference from both a.c. and d.c. sources can be effectively trapped at the source of the generated field or shielded from where the affected device is located. Highly permeable materials such as Mumetal® (a magnetic alloy with high nickel content) conduct the flux lines around the shielded space.

Prevention of magnetic fields cannot occur. That is, there is no such thing as a magnetic insulator. As such, there is no way to stop them, as all magnetic field lines must terminate on the opposite pole. However, magnetic fields can be re-routed around objects. This is a form of magnetic shielding or screening. Highly permeable material conducts the magnetic flux around the shielded space within which an object of lower permeability can be placed. Refer to **Figure 4.19**.

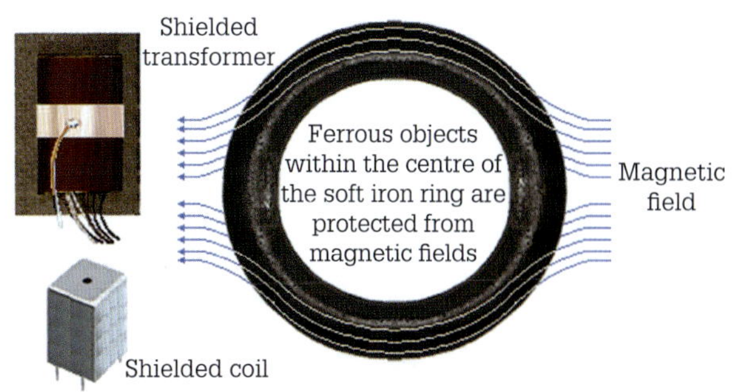

FIGURE 4.19 Shielded transformer and coil

REVIEW QUESTIONS

1. What is magnetite?
2. What are two examples of diamagnetic substances?
3. What classification of magnetic substances retain most of their magnetism when the magnetising force is removed?
4. Which end of a suspended bar magnet aligns itself with the Earth's geographic North Pole?
5. What is a magnetic substance containing two poles called?
6. What happens magnetically when a piece of soft iron is brought close to a permanent magnet?
7. Name two methods used to produce permanent magnets.
8. What is stated by a fundamental law of magnetism?
9. What is the purpose of a magnetic 'keeper'?
10. Name two applications of Alnico permanent magnets.

4.2 Electromagnetism

Magnetic field around current carrying conductor

The convention used to show the direction of conventional current flow, and the resulting magnetic field is shown in **Figure 4.20**.

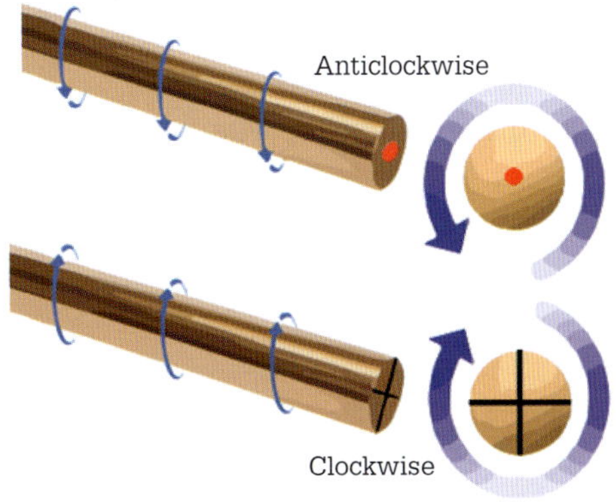

FIGURE 4.20 Symbols showing the direction of current

Figure 4.21 shows the magnetic field or flux surrounding the current flowing through a straight conductor. The field occurs at right angles to the direction of current.

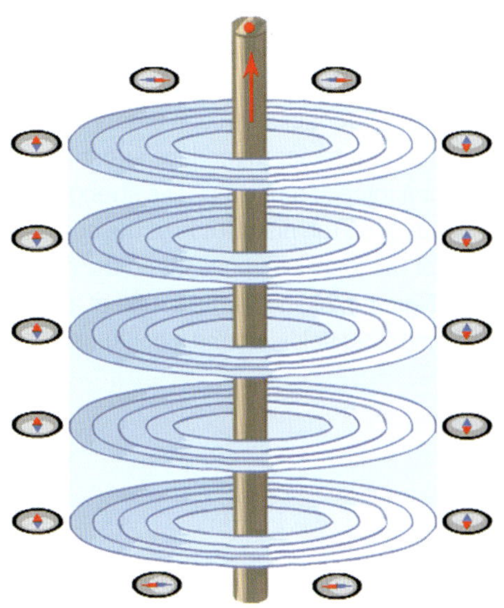

FIGURE 4.21 Magnetic field around a straight conductor

The direction of the magnetic field around the conductor may be determined by using the right-hand grip rule:

SWITCH ON

With knowledge of which direction the current is flowing in a conductor hold the conductor with the right hand so that the thumb points in the direction of conventional current flow; the curled fingers then point in the direction of the magnetic field around the conductor. Refer to **Figure 4.22**.

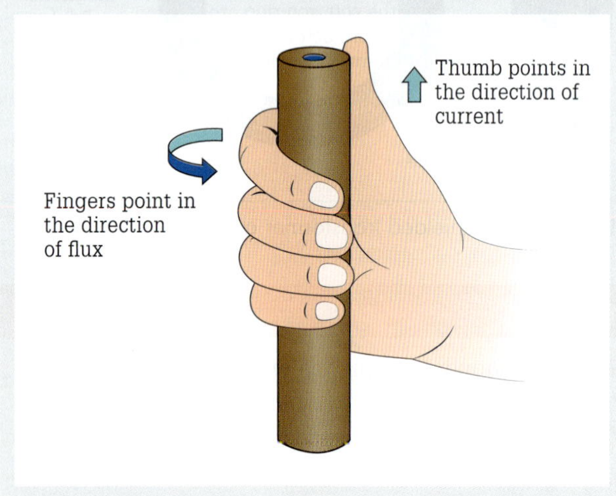

FIGURE 4.22 Right-hand grip rule

The main characteristics of the nature of a magnetic field around a conductor are:

- The flux is circular and concentric around the conductor.
- The strength of the field decreases away from the conductor.
- The direction of the magnetic field reverses when the direction of the current reverses.
- The strength of the magnetic field is proportional to the size of the current.

Parallel current carrying conductors

The magnetic field produced by each parallel current carrying conductor interact to produce either a force of attraction or a force of repulsion. The strength of the force is dependent on the magnitude of the current and the direction of the force is dependent on the direction of current flow. **Figure 4.23** shows the action of attraction and repulsion of magnetic fields around a conductor.

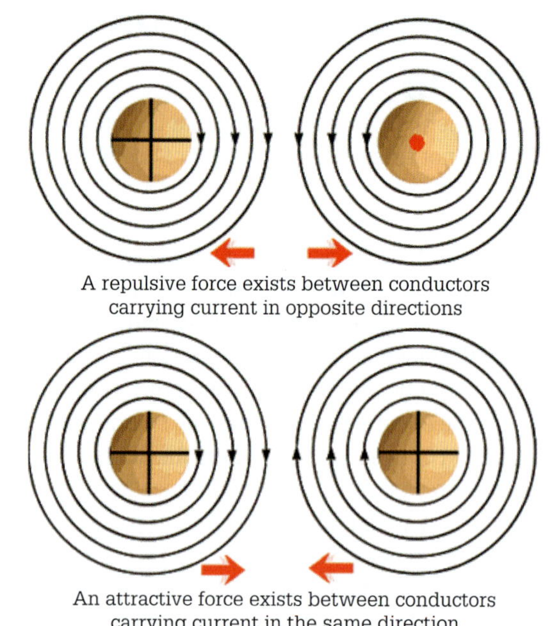

FIGURE 4.23 Action of attraction and repulsion of magnetic fields around a conductor

Forces between conductors

If two parallel conductors 1 m apart carry the same current and the force per unit length on each conductor is 2×10^{-7} N/m, then the current is 1 A.

Under short-circuit conditions the force exerted by conductors can cause mechanical damage and stress on connections.

Consider two long, straight, parallel wires separated by a distance r and carrying currents I_1 and I_2 in opposite directions. We can easily determine the force on one conductor due to a magnetic field set up by the other conductor. Conductor 1, which carries a current I_1, creates a magnetic field B at the position of conductor 2. The direction of B is perpendicular to conductor 2.

From **Figure 4.24** the magnetic field at conductor 2 can be determined from the current in conductor 1 using:

$$B = \frac{\mu_0 I_1}{2\pi r}$$

where B = magnetic field strength, measured in tesla
μ_0 = permeability of free space = $4\pi \times 10^{-7}$ measured in henrys per metre (Hm^{-1})
I_1 = current flowing through the wire, measured in amperes
r = distance from the wire, measured in metres

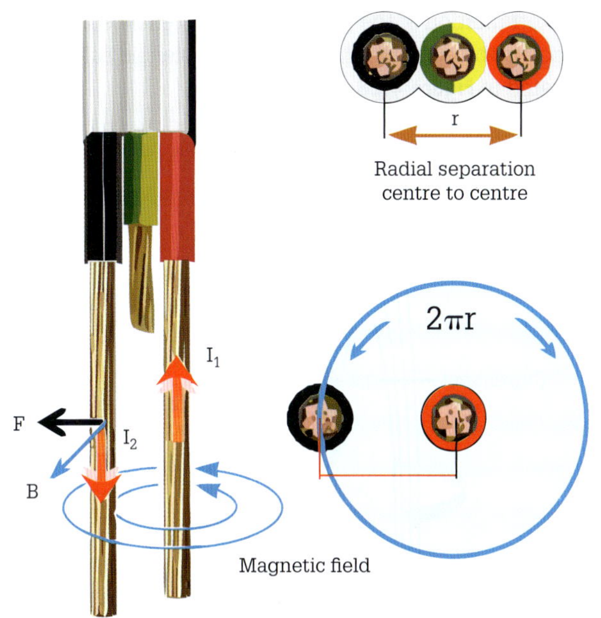

FIGURE 4.24 Force between conductors

Since each conductor lies in the magnetic field set up by the other, each conductor experiences force which depends on the current flowing through both conductors.

The magnitude of the force on a length ΔL of conductor 2:

$$F = I_2 \Delta L B$$

The force per unit length in terms of the currents:

$$\frac{F}{\Delta L} = \frac{\mu_0 I_1 I_2}{2\pi r}$$

The numerical value of 2×10^{-7} N/m in the former definition of the ampere is obtained from the equation above with

$I_1 = I_2 = 1$ A and $r = 1$ m.

Where $\mu_0 = 4\pi \times 10^{-7}$ and the magnetic field of I_1 circumference is $2\pi r$:

$$\frac{4\pi \times 10^{-7}}{2\pi \times r} = \frac{2 \times 10^{-7}}{r}$$

where r is the radial distance separation between current-carrying conductors I_1 and I_2.

Once the magnetic field has a value, the magnetic force equation applies. (**Note:** this equation only applies to long straight conductors.) The right-hand rule provides the direction of the force.

EXAMPLE 4.1

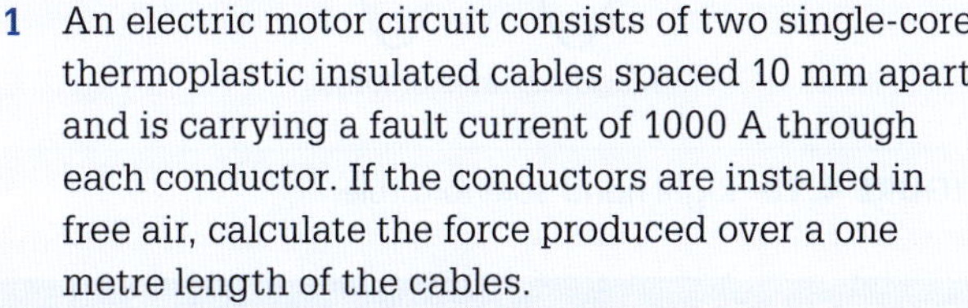

1 An electric motor circuit consists of two single-core thermoplastic insulated cables spaced 10 mm apart and is carrying a fault current of 1000 A through each conductor. If the conductors are installed in free air, calculate the force produced over a one metre length of the cables.

$$B = \frac{\mu_0 I_1}{2\pi r}$$

$$= \frac{4\pi \times 10^{-7} \times 1000}{2\pi \times 25 \times 10^{-3}}$$

$$= \mathbf{8\ mT\ (0.008\ T)}$$

As the current I_2 is 1000 A then:

$$F = I_2 \Delta L B$$

$$= 1000 \times 1 \times 8 \times 10^{-3}$$

$$= \mathbf{8\ newtons\ per\ metre}$$

Since the currents flow in opposite directions, the conductors repel one another.

2 Two parallel conductors separated by 5.0 cm repel each other with a force per unit length of 6.0×10^{-4} N. The current in one conductor is 15.0 A.

a Find the current in the other conductor.

$$\frac{F}{\Delta L} = \frac{\mu_0 I_1 I_2}{2\pi r}$$

$$\frac{6 \times 10^{-4}}{1} = \frac{4\pi \times 10^{-7} \times 15 \times I_2}{2\pi \times 0.05}$$

$$I_2 = 10\ \text{A}$$

b Are the currents in the same direction or in the opposite direction?

Opposite currents repel each other.

c What would happen if one current reversed in direction and doubled in size?

They would attract each other with twice the force per unit length.

3 A switchboard is fed by two 16 mm² SDI cables separated from each other by a distance of 25 mm. In addition, the earth fault loop impedance measured at the 230 V switchboard is 0.2 Ω. If a short circuit occurs at the switchboard, what force per metre occurs?

$$\frac{F}{\Delta L} = \frac{\mu_0 I_1 I_2}{2\pi r}$$

Fault current I in each conductor $= \frac{V}{Z} = \frac{230}{0.2} = 1150$ A

$$F = \frac{4\pi \times 10^{-7} \times 1150 \times 1150}{2\pi \times 0.025}$$

$$= \mathbf{10.58\ N}$$

Electromagnet

Figure 4.25 shows a conductor formed into a single circular turn. The turn is made up of many links of a chain, each link adding its individual magnetic field together at the centre of the turn where the magnetic field is the strongest. The effect is a combined field similar to the field of a bar magnet. For this reason, the solenoid behaves as if it has a north pole at one end and south pole at the other. An application of the right-hand screw rule reveals the direction in which the magnetic field acts.

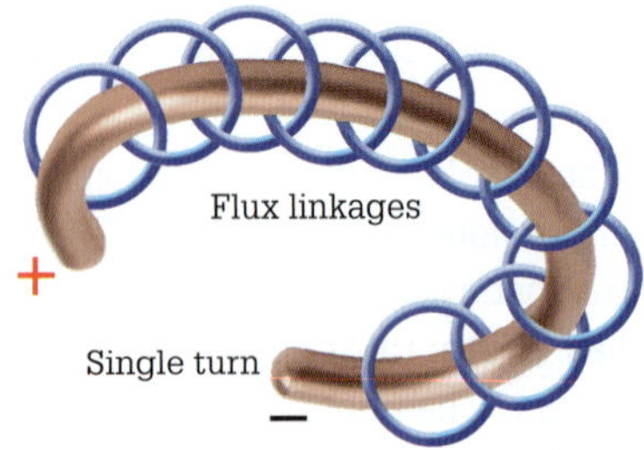

FIGURE 4.25 Flux linkages with a single turn

A solenoid forms when turns of wire occur around a cylindrical former. When a current passes through the solenoid conductors, each turn produces its own magnetic field. In addition, each field links with each other to concentrate their combined field strength within the hollow space of the solenoid as shown in **Figure 4.26**.

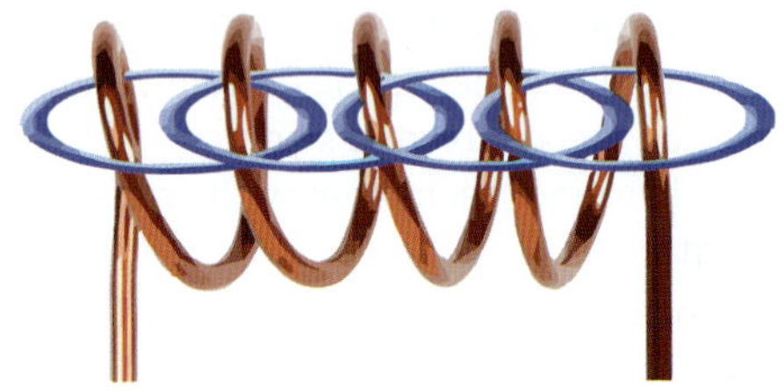

FIGURE 4.26 Flux linking of individual fields

The magnetic field concentrated within the hollow space has direction and strength dependent on the current value flowing, and the number of turns per unit length of the solenoid as shown in **Figure 4.27**.

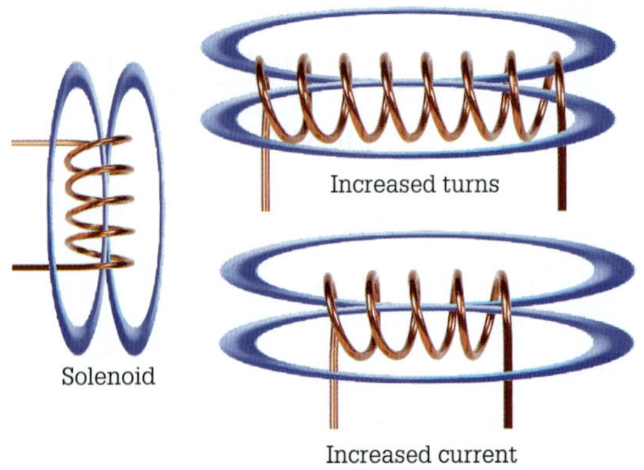

FIGURE 4.27 Solenoid – increased turns and increased current effect

Determination of the polarity of the solenoid's magnetic field occurs by applying the right-hand solenoid rule as illustrated in **Figure 4.28**.

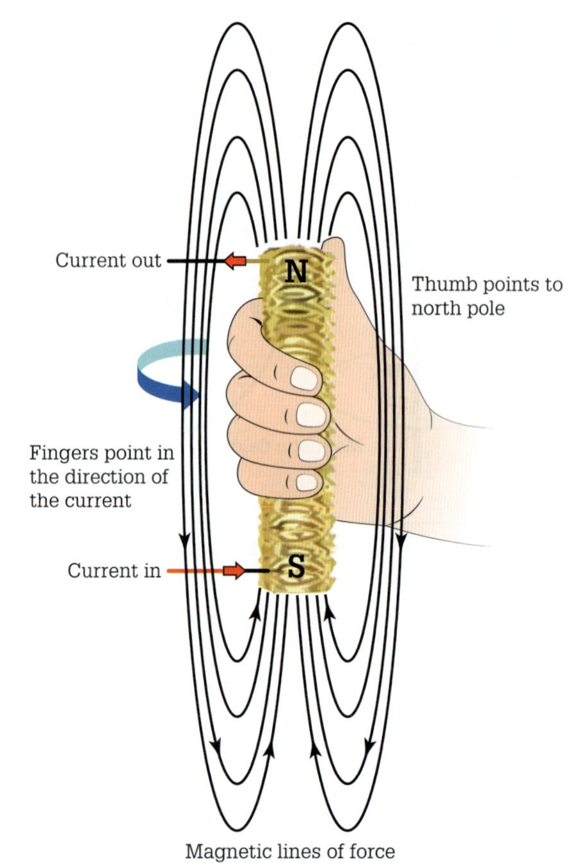

FIGURE 4.28 Right-hand solenoid rule

SWITCH ON

The right-hand solenoid rule states:

If the right hand holds the solenoid, with the fingers pointing in the direction of the current, then the outstretched thumb parallel with the axis of the solenoid points in the direction of the North Pole.

The solenoid turns into an electromagnet if a ferromagnetic soft substance inserts into the hollow core of the solenoid as in **Figure 4.29**.

FIGURE 4.29 Electromagnet

Furthermore, the core concentrates the magnetic flux and increases the magnetic strength of the solenoid. It is evident that in relation to air, the magnetic flux has an easier reluctance path because of the soft iron core. The strength of the field builds up to a level determined by the magnetic characteristics of the soft ferromagnetic substance.

Magnetomotive force

The magnetomotive force (symbol F_m) is the force that establishes and maintains the magnetic field in the magnetic circuit. The magnetomotive force can be created by a permanent magnet or by an electric current flowing in the electric circuit of a coil. Refer to **Figure 4.30**.

FIGURE 4.30 Magnetomotive force

The magnetomotive force is the force developed by one ampere flowing through one turn of the coil. The SI unit of measurement used for magnetomotive force is the ampere-turn. The rule that applies to the magnetomotive force is:

SWITCH ON

The magnetomotive force (F_m) of a coil is proportional to the product of the number of turns (N) and the current (I).

This statement expressed as an equation is:

$$F_m = IN$$

where F_m = magnetomotive force in ampere-turns (At)
I = current in amperes
N = number of turns

EXAMPLE 4.2 

Determine the magnetomotive force developed by a coil comprising 2000 turns of a copper conductor carrying a current of 500 mA.

$$F_m = IN$$
$$= (500 \times 10^{-3}) \times 2000$$
$$= 1000 \text{ At}$$

EXERCISE 4.1 

a Referring to Example 4.2, what is the magnetomotive force when:
  i the number of turns on the coil is reduced to 1000?
  ii the current is increased to 800 mA?
b What is the magnetomotive force in the magnetic circuit of a contactor coil that has 8000 turns of a copper conductor carrying 30 mA?
c What current does a relay coil of 3000 turns draw to produce a magnetomotive force of 500 At?
d A current of 20 mA flows through the coil of a flow switch. How many turns does the coil have in order to produce a magnetomotive force of 500 At?
e Assuming that all other factors are constant, describe how the magnetomotive force in a magnetic circuit is affected when:
  i the number of turns of the coil is halved
  ii the current through the coil is doubled
  iii the potential difference across the coil is doubled
  iv the ohmic resistance of the conductor is halved
  v the number of turns and the resistance of the coil are both halved.

Applications of electromagnets

Applications for electromagnets include:

- electric motors
- transformers
- solenoids for relays and contactors
- circuit breakers
- electromagnetic brakes and clutches.

REVIEW QUESTIONS

1 Draw the symbol to show the direction of current flow as it flows toward the voltage source.
2 What rule enables the determination of the magnetic field around a conductor?
3 Does a force of attraction or repulsion exist between two parallel conductors carrying current in the same direction?
4 A switchboard is supplied by two 16 mm^2 SDI cables separated from each other by a distance of 50 mm. If a fault current of 2500 A flows in each conductor, what force per metre occurs between the conductors?
5 What is the name given to the device whereby turns of wire are wound around a cylindrical former?
6 Name the two factors that determine the concentration of the magnetic field within the hollow space of a solenoid.
7 What effect does a soft ferromagnetic substance have when inserted into the hollow space of a solenoid?
8 In the right-hand solenoid rule, what does the thumb represent?
9 Calculate the magnetomotive force produced by a 1500 turn coil carrying 650 mA.
10 Name two applications for electromagnets.

4.3 Magnetic circuit and nomenclature

A magnetic circuit comprises at least one closed loop path that contains a magnetic flux. Permanent magnets or electromagnets provide the magnetomotive force to establish the magnetic flux, which is confined to the magnetic circuit. The magnetic circuit is usually made up of a magnetically conductive material like iron. The magnetic circuit may also include air gaps or other materials in the path. **Figure 4.31** shows a simple magnetic circuit comprising a source of magnetomotive force (electromagnet) and a path for the magnetomotive flux, which consists of an iron path and an air gap.

Some practical examples of common magnetic circuits include:

- horseshoe magnet with iron keeper (low-reluctance circuit)
- horseshoe magnet with no keeper (high-reluctance circuit)
- electric motor (variable-reluctance circuit).

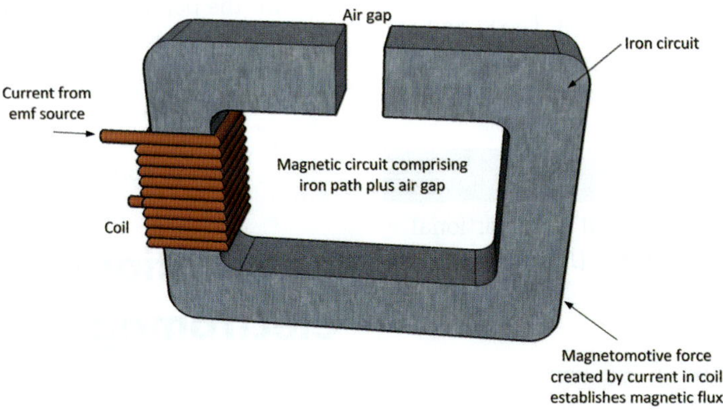

FIGURE 4.31 Magnetic circuit

With reference to **Figure 4.31**, current flowing in the coil produces a magnetomotive force (F_m) that establishes a magnetic flux (Φ) against the reluctance (R_m) of the core and air gap. This is analogous with the electrical circuit where a battery provides an electromotive force (E) that establishes a current (I) against the resistance (R) of the circuit. **Table 4.2** summarises this analogy.

TABLE 4.2 Comparing magnetic and electric circuits

Magnetic circuit			Electric circuit		
Term	Symbol	Unit	Term	Symbol	Unit
Magnetomotive force	F_m	Ampere-turn	Electromotive force	E	Volt
Reluctance	R_m	Ampere-turn/weber	Resistance	R	Ohm
Magnetic flux	W_b	Weber	Current	I	Ampere

Terminology

Magnetic flux

Magnetic flux (symbol Φ, which is the Greek symbol for phi) is the total number of lines of force in a magnetic circuit. The SI unit of measurement used for magnetic flux is the weber.

As an equation:

$$\Phi = \frac{F_m}{R_m}$$

where Φ = the magnetic flux in webers

F_m = the magnetomotive force in ampere-turns

R_m = the reluctance of magnetic circuit in ampere-turns per weber

As $F_m = IN$ then the equation becomes:

$$\Phi = \frac{IN}{R_m}$$

EXAMPLE 4.3

A magnetic circuit is 400 mm in length and has a CSA of 5 mm². Calculate the flux in the magnetic circuit if a current of 100 mA is flowing in the 8000 turns copper coil and the reluctance of the magnetic circuit is 1273 × 10⁶ At/Wb.

$$\Phi = \frac{IN}{R_m}$$

$$= \frac{(100 \times 10^{-3}) \times 8000}{1273 \times 10^{6}}$$

$$= \mathbf{628 \times 10^{-7}\ Wb}$$

EXERCISE 4.2

a Referring to Example 4.3, what is the value of the magnetic flux in the circuit when:
  i the number of turns is increased to 10 000?
  ii the current increases to 200 mA?

b A relay coil has 4000 turns and carries a current of 50 mA. What is the magnitude of the magnetic flux when the total reluctance of the magnetic circuit is 1.5 At/Wb?

c What current must flow through 2000 turns to produce a flux of 500 mWb in a magnetic circuit having a reluctance of 2.5 At/Wb?

d The coil of a contactor has 1000 turns with a resistance of 300 Ω and operates from a 120 V d.c. supply. Calculate the flux in the magnetic circuit if the coil has a reluctance of:
  i 120 At/Wb
  ii 30 At/Wb

Flux density

Flux density (symbol *B*) is the number of lines of force per square metre of cross-sectional area of the magnetic circuit. **Figure 4.32** illustrates how total flux can remain unchanged but flux density changes due to narrowing of the magnetic circuit causing the lines of magnetic force to be more concentrated.

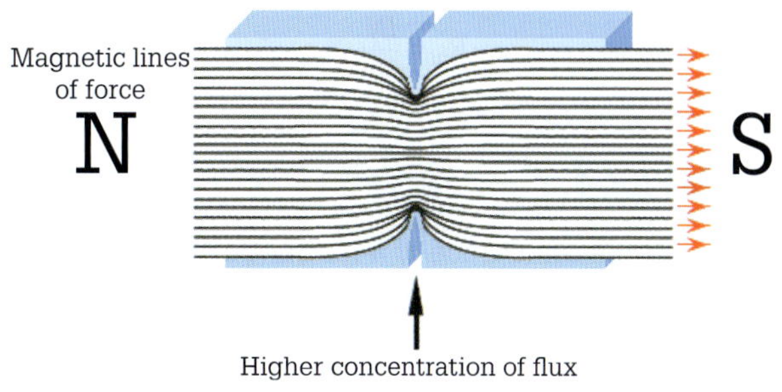

FIGURE 4.32 Flux density

The SI unit of measurement for flux density is the tesla (Wb/m²). One tesla is equal to 10⁸ lines of force or one weber per square metre. The equation for determining flux density is:

$$B = \frac{\Phi}{A}$$

where B = flux density in teslas (T)

Φ = the magnetic flux in webers (Wb)

A = the cross-sectional area in metres squared (m²)

EXAMPLE 4.4

Find the flux density of an air-cored former that has a cross-sectional area of 200 mm² and a total magnetic flux of 0.0000025 webers. (**Note:** 1 mm² = 1 × 10⁻⁶ m²)

$$B = \frac{\Phi}{A}$$

$$= \frac{0.0000025}{200 \times 10^{-6}}$$

$$= \mathbf{0.0125\ T}$$

EXERCISE 4.3

a Find the flux density of an air-cored former that has a cross-sectional area of 100 mm² and a total magnetic flux of 150 µWb.

b Determine the cross-sectional area of an air-cored coil having a flux density of 250 mT and a total magnetic flux of 50 µWb.

Reluctance

Reluctance (symbol R_m) is the opposition a substance offers to the passage of flux in a magnetic circuit.

The SI unit of measurement used for reluctance is the ampere-turn/weber. The rule that applies to magnetic circuits concerning reluctance follows:

$$R_m = \frac{F_m}{\Phi}$$

where R_m = reluctance in ampere-turns per weber (At/Wb)

F_m = magnetomotive force in ampere-turns (At)

Φ = magnetic flux in weber (Wb)

Substituting IN for F_m, the equation becomes:

$$R_m = \frac{IN}{\Phi}$$

EXAMPLE 4.5

Determine the reluctance of a magnetic circuit that results in a total magnetic flux of 400 mWb being established in the core from a 1500 turn coil carrying 800 mA.

$$R_m = \frac{IN}{\Phi}$$

$$= \frac{(800 \times 10^{-3}) \times 1500}{400 \times 10^{-3}}$$

$$= \mathbf{3000\ At/Wb}$$

SWITCH ON

The reluctance of a magnetic circuit is directly proportional to the circuit's length and inversely proportional to the cross-sectional area and the permeability of the substance.

Note the similarity of this rule and that for the resistance of a conductor in an electric circuit. Based on the properties of the particular magnetic material, the equation for reluctance becomes:

$$R_m = \frac{l}{\mu_0\ \mu_r\ a}$$

where R_m = reluctance in ampere-turns per weber (At/Wb)
l = length of the magnetic circuit in metres (m)
μ_0 = permeability of free space in henrys/m (Hm^{-1})
μ_r = relative permeability (for air $\mu_r = 1$)
a = cross-sectional area in square metres (m^2)

Note that μ_a represents the absolute permeability of the magnetic circuit in henrys per metre (H/m), which is the product of μ_0 and μ_r.

EXAMPLE 4.6

What is the reluctance of a non-magnetic coil former 100 mm long and 100 cm^2 in CSA?
Note: 1 cm^2 = 1 × 10^4 m^2 and $\mu_0 = 4\pi \times 10^{-7}\ Hm^{-1}$.

$$R_m = \frac{1}{\mu_0\ \mu_r\ a} = \frac{100 \times 10^{-3}}{(4\pi \times 10^{-7}) \times 1 \times (100 \times 10^{-4})}$$

$$= \mathbf{7.96 \times 10^6\ At/Wb}$$

EXERCISE 4.4

a What is the reluctance of a non-magnetic coil former 100 mm long and 50 cm^2 in CSA?

b What is the reluctance of a magnetic core measuring 70 mm in length with a CSA of 15 cm^2 if the relative permeability of the core is 150?

c What is the length of an air gap in a control relay when the magnetic circuit has a CSA of 14 cm^2 and a reluctance of 284 × 10^3 At/Wb?

d What is the CSA of a 0.15 mm long air gap in a contactor having a reluctance of 250 × 10^3 At/Wb?

e Assuming all other factors are constant, how is the reluctance of an air gap of circular CSA in a magnetic circuit affected when:

i the CSA doubles

ii the length doubles.

Magnetising force

The magnetising force (symbol H), also known as field strength or field intensity, is concentration of magnetomotive force per unit length. For an air-cored coil the length corresponds to the length of coil and for a ferromagnetic core the length is the average length of the magnetic circuit.

In comparison to an electric circuit, the magnetomotive force is like the electromotive force, while the magnetising force compares to the potential difference across one section of an electric circuit. Furthermore, the SI unit of measurement for magnetising force is the ampere-turns per metre (At/m). The equation for determining magnetising force is as follows:

$$H = \frac{F_m}{l}$$

where H = the magnetising force in ampere-turns per metre (At/m)
F_m = the magnetomotive force in ampere-turns (At)
l = the length of the magnetic circuit in metres (m)
As $F_m = IN$, an alternative method for determining magnetising force is as follows:

$$H = \frac{IN}{l}$$

EXAMPLE 4.7

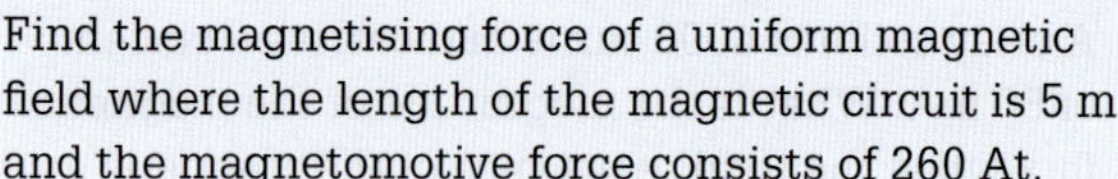

Find the magnetising force of a uniform magnetic field where the length of the magnetic circuit is 5 m and the magnetomotive force consists of 260 At.

$$H = \frac{F_m}{l}$$

$$= \frac{260}{5}$$

$$= \mathbf{52\ At/m}$$

EXERCISE 4.5

a Calculate the magnetising force of a uniform magnetic field where the length of the magnetic circuit is 150 mm and a current of 250 mA is flowing in the coil of 1800 turns.

b Calculate the length of a magnetic circuit when a current of 500 mA in a 1200 turn coil produces a magnetising force of 4500 At/m.

»

c Assuming all other factors are constant, explain how the magnetising force in a magnetic circuit changes when:
- i the current through the coil halves
- ii the voltage across the coil doubles
- iii the length of the magnetic circuit decreases.

Permeability

The permeability (symbol μ, which is the Greek symbol for mu) of a substance is the measure of the ease with which magnetic lines of force pass through a substance. The permeability of open space or vacuum (symbol μ_0) is the standard against which the permeability of other substances is measured. This standard is:

$$\mu_0 = 4\pi \times 10^{-7} \text{ in henrys per metre (H m}^{-1})$$

Note: One henry is equal to one weber per ampere.

Permeability is the ratio of the number of lines of force through a unit area of 1 m^2 of the substance to the number of lines of force through the same space when air replaces the substance.

The permeability ratio is the factor by which permeability increases above the standard (μ_0) and is the relative permeability (μ_r) of the substance concerned. For air, the relative permeability is given the value of unity ($\mu_r = 1$). However, relative permeability is not a constant and changes with flux density. Note that relative permeability is a ratio and has no unit of measurement.

Diamagnetic substances have a relative permeability slightly below unity, while paramagnetic substances have a relative permeability slightly above unity. Ferromagnetic substances, however, such as cast iron, can have a relative permeability of up to 900, while other ferromagnetic substances can have a relative permeability value expressed in the thousands. Approximate maximum relative permeability of different materials is shown in Table 4.3.

TABLE 4.3 Approximate maximum relative permeabilities of various materials

Material	μ_r
air	1
ferrite U60	8
ferrite M33	750
cast iron	900
nickel	2000
ferrite N41	3000
ferrite T38	10000
silicon GO steel	40000
supermalloy	100000

Note: Unlike μ_0, μ_r is not constant and changes with flux density.

The absolute or actual permeability (μ_a) of a magnetic substance is the product of the permeability of free space and the relative permeability of the magnetic substance. The following equation provides the value of actual permeability of a substance.

$$\mu_a = \mu_r \times \mu_0 \text{ in henrys per metre (H m}^{-1})$$

where μ_a = actual permeability in henrys/m (H m^{-1})
μ_r = relative permeability
μ_0 = permeability of free space in henrys/m (H m^{-1})

EXAMPLE 4.8

Calculate the actual permeability of silicon grain oriented (GO) steel used for the laminations in the core of a solenoid if the relative permeability of the substance is 40000 at the required flux density ($\mu_0 = 4\pi \times 10^{-7}$ H m^{-1})

$$\mu = \mu_r \times \mu_0$$
$$= 40000 \times 4 \times 3.142 \times 10^{-7}$$
$$= 0.050\,272\,0 \text{ H m}^{-1}$$

Note: This value of permeability applies only for a predetermined magnetising force.

The higher the permeability value that a magnetic substance possesses, the higher is that substance's magnetic flux density at that specific magnetising force. Figure 4.33 shows the relative permeability curve.

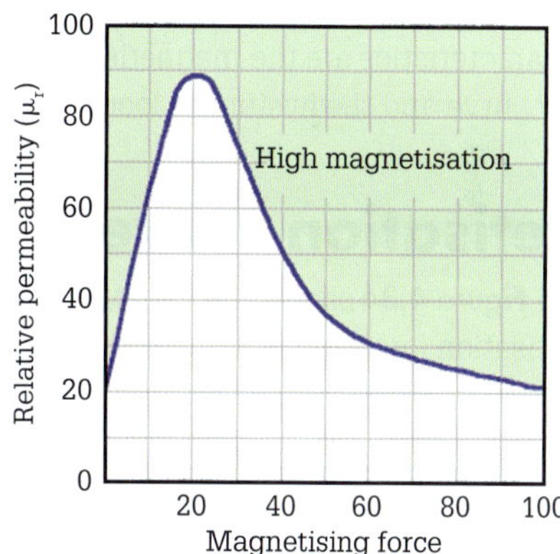

FIGURE 4.33 Permeability curve

Actual permeability is the ratio of flux density to field strength. Calculation of the actual permeability of a substance uses the following method.

$$\mu_a = \frac{B}{H}$$

where μ_a = actual permeability in henry per metre (H m^{-1})
B = the flux density in tesla (T)
H = the magnetising force in ampere-turns per metre (At/m)

EXAMPLE 4.9

Calculate the actual permeability of a ferromagnetic core of a solenoid that has a magnetising force of 4800 At/m producing a flux density of 30 mT.

$$\mu_a = \frac{B}{H} = \frac{(30 \times 10^{-3})}{4800} = \mathbf{6.25\ \mu H\ m^{-1}}$$

If a ferromagnetic core with a high value of permeability is the core of a solenoid, then a coil for a given value of inductance requires fewer turns.

REVIEW QUESTIONS

1. Define the following magnetic terms and state their SI units of measurement:
   a. magnetising force
   b. permeability
   c. magnetic flux
   d. flux density
   e. reluctance
2. A relay coil has 3000 turns and carries a current of 10 mA. What is the magnitude of the magnetic flux when the total reluctance of the magnetic circuit is 1.2 At/Wb?
3. Find the flux density of an air-cored former that has a cross-sectional area of 150 mm² and a total magnetic flux of 150 µWb.
4. What is the reluctance of a magnetic core measuring 100 mm in length with a CSA of 12 cm² if the relative permeability of the core is 150?
5. Calculate the magnetising force of a uniform magnetic field where the length of the magnetic circuit is 250 mm and a current of 150 mA is flowing in the coil of 2800 turns.

4.4 Losses in magnetic circuits

Due to the non-linear characteristics of many ferromagnetic substances, graphical techniques describe their magnetic characteristics. Consequently, the three graphical characteristics are the magnetisation curve, the permeability curve and the hysteresis loop.

Magnetisation curve

The graph in **Figure 4.34** shows the variation of flux density with magnetisation force and is a 'magnetisation' or *B/H* curve.

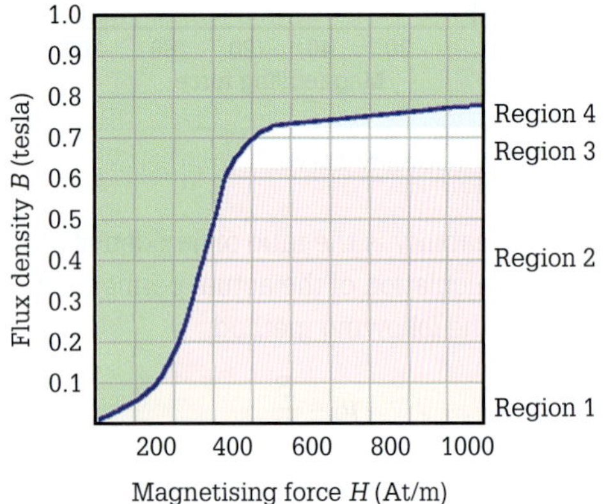

FIGURE 4.34 Initial magnetisation curve or *B/H* curve

Manufacturers of ferromagnetic materials usually provide this curve because the shape of the curve shows how the ferromagnetic material in any component made from it responds to changes in the magnetisation force.

When a magnetising force (*H*) is first applied to a ferromagnetic material the flux density (*B*) created is at its weakest level. As the applied magnetising force gradually increases, the flux density also increases, swiftly at first, and then more slowly until magnetic saturation happens. At this point, further increases in the magnetising force produce no significant change in flux density. The graph of **Figure 4.34** is an initial magnetisation curve because it begins from an unmagnetised ferromagnetic material and reveals how the flux density increases as the magnetising force increases. Four distinct regions occur on most magnetisation curves; a result of changes to the domains of the ferromagnetic materials. The four regions are:

- **Region 1** Near to the point of origin a slow increase due to initial absorption of energy.
- **Region 2** A longer consistent, rapid growth occurring with little increase in magnetising force. Region two is where most magnetic materials operate.

- **Region 3** The slowing of the growth of flux, representing the 'knee of change'.
- **Region 4** The levelling of growth as the ferromagnetic material saturates and cannot absorb any more increase in flux.

The original magnetisation curve (**Figure 4.34**) tells us that the flux density must be below 0.67 T to avoid saturation. In order to make best use of the magnetic properties of ferromagnetic material, magnetic devices operate in region 2, where consistent magnetic growth transpires with little magnetising effort. This type of curve may be prepared by making a sample of the ferromagnetic material into a closed loop called a ring core or toroid, as shown in **Figure 4.35**. In addition, the coil has a set number of turns and a predetermined current passed through the coil. These two factors create the magnetic flux in the sample that is measured. Measurement takes place a number of times at different excitation levels with a digital tesla meter called a magnetometer.

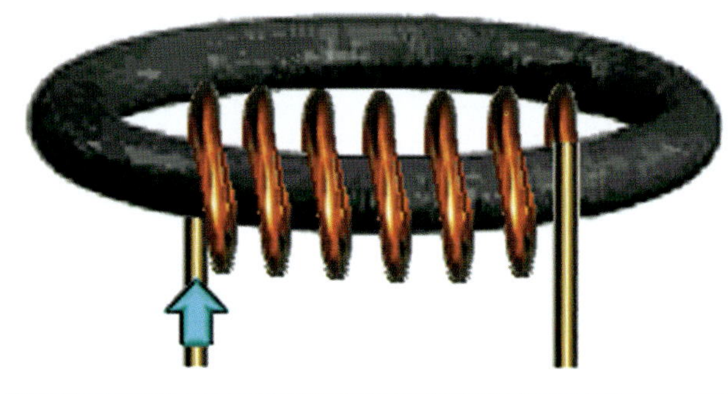

FIGURE 4.35 Toroid

EXERCISE 4.6

a **Table 4.4** shows measured values of *B* and *H* for a ferromagnetic material. Using these values, plot a magnetisation curve showing how the flux density varies for different values of magnetising force.

TABLE 4.4

B	0.1	1.0	1.8	2.4	2.6	2.7	2.8	2.9	2.94	2.95	2.97	2.98	3.0	tesla
H	1	4	6	8	10	12	14	16	18	20	22	24	30	At/m

Your graph should resemble **Figure 4.36**. From your magnetisation curve find:

b The magnetising force produced by a flux density of:
  i 1.8 T
  ii 2.8 T
  iii 3.0 T

c The flux density needed to produce a magnetising force of:
  i 4.0 At/m
  ii 10.0 At/m
  iii 22.0 At/m

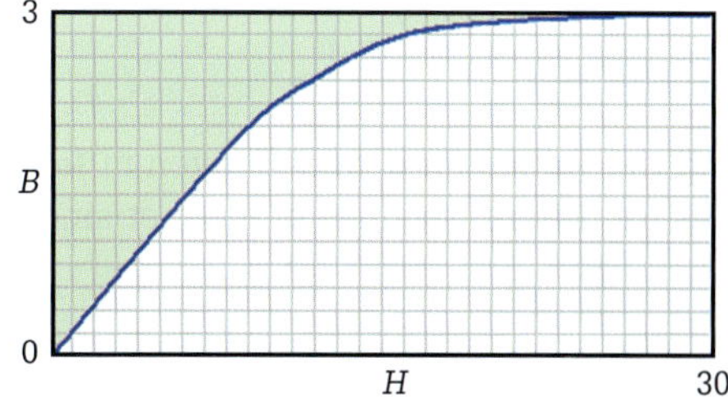

FIGURE 4.36 *B*/*H* curve

The choice of a ferromagnetic material is of prime importance if the expected magnetic results happen from any design using ferromagnetic materials. Therefore, the magnetisation curve is one element in the choice of ferromagnetic materials. **Figure 4.37** shows the magnetisation curves for various ferromagnetic materials.

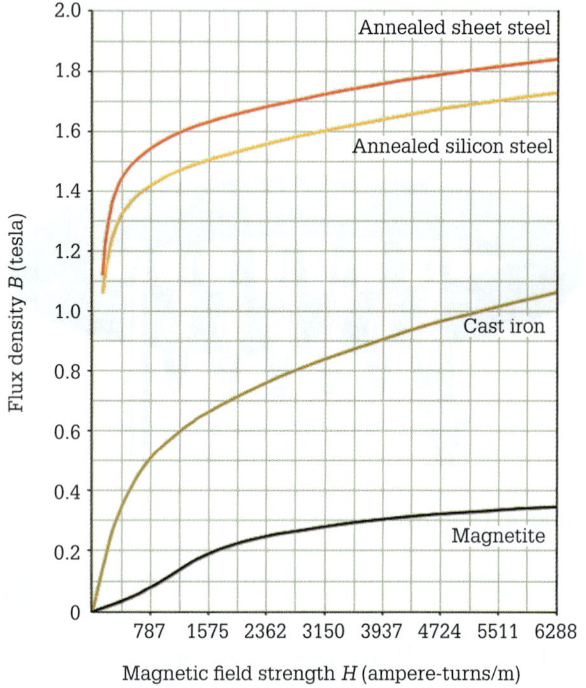

FIGURE 4.37 Magnetisation curves of different materials

Permeability curve

The second graphical characteristic of interest is the magnetic conductivity or permeability of a ferromagnetic material. Permeability varies with different values of magnetising force and the level of magnetisation of the material. Therefore, for a defined set of operating conditions, if the required value of flux density in the ferromagnetic material or the magnetising force transpires, locate the other value from the magnetisation curve. The result is that calculation of the absolute permeability (μ_a) occurs from the B and H values.

Furthermore, a relative permeability curve (the red curve) as shown in **Figure 4.38** shows how relative permeability varies for different values of magnetic flux density and the magnetising force.

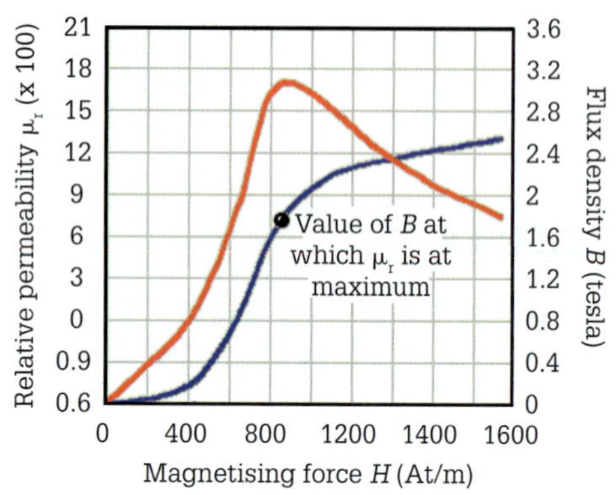

FIGURE 4.38 Relative permeability curves (μ_r)

EXERCISE 4.7

a Using the values for μ_r and H given in **Table 4.5**, plot a μ_r/H curve showing how the relative permeability varies for different values of magnetising force.

TABLE 4.5

μ_r	1.0	1.5	2.75	3.8	4.25	4.4	4.2	3.8	3.45	3.15	2.85	
H	100	200	300	250	400	420	500	600	700	800	900	At/m

Your graph should resemble **Figure 4.39**. From your B/H curve determine:

b The relative permeability when the magnetising force is:
  i 100 At/m
  ii 250 At/m
  iii 800 At/m

c The magnetising force when the relative permeability is:
  i 1.5
  ii 4.25
  iii 2.85

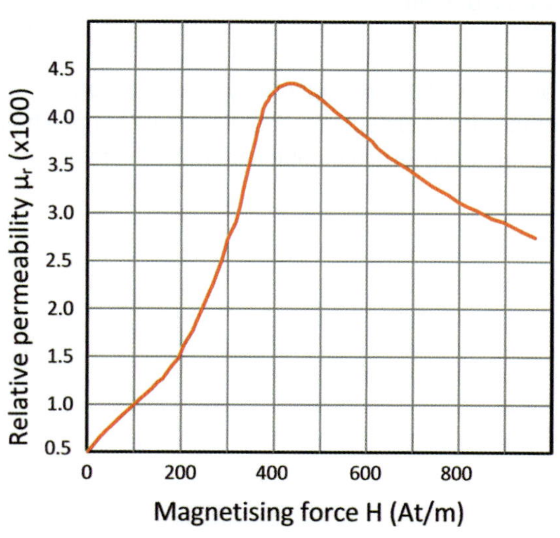

FIGURE 4.39 μ_r/H curve

Hysteresis loop

The third graphical characteristic of importance concerning ferromagnetic materials is the hysteresis curve or loop. When a ferromagnetic material is subject to a complete magnetisation cycle, such as the core of a solenoid through which an alternating current is passing, the flux produced lags behind the magnetising force that created it. This effect is termed hysteresis and is shown graphically in **Figure 4.40**.

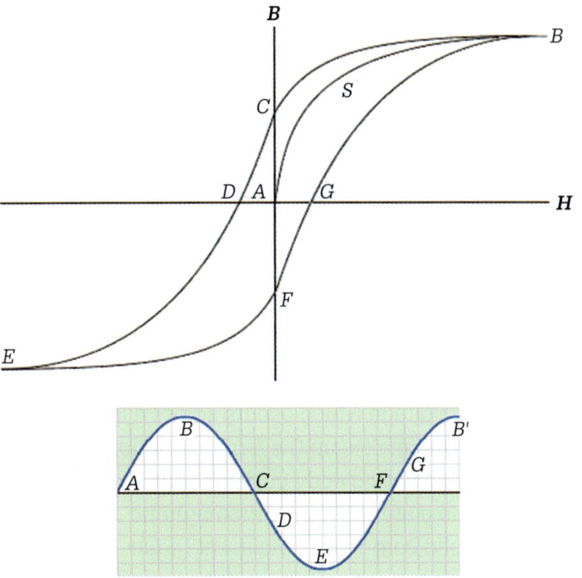

FIGURE 4.40 Hysteresis loop

A hysteresis loop defines the flux density of a ferromagnetic material, the coercive force, the remanence and the required level of magnetising force needed to saturate the material. In addition, explanation of the factors that comprise a hysteresis curve can be had by referring to the curve in **Figure 4.40**.

Saturation

A characteristic of the ferromagnetic material related to this curve is saturation. Saturation is the maximum value of magnetic induction at specified field strength. It is the point where very little gain in flux density occurs in relation to increasing magnetising force.

When a ferromagnetic material saturates, it loses its linearity, and this point occurs near the centre of the 'knee' indicated as point *S*. If the magnetising force reduces to zero, the flux density does not return to zero along the same curve but reduces along a new path as shown by the curve *BC*.

Further observations show that the flux density has not fallen to zero, and the ferromagnetic material has retained its magnetic properties.

Remnant flux

This remnant flux, or residual flux density, also called remanence, remaining in the ferromagnetic material occurs between points *A* and *C*. In order to reduce this remanence to zero, the magnetising force *A* to *D* is increased in a negative direction.

Consequently, the negative force is the coercive force and is a measure of magnetising force required to de-magnetise the material. High coercivity is essential for permanent magnetic material in order to stay magnetised in the presence of the opposing magnetic field. However, for transformers, high coercivity is not desirable because it increases the amount of lost energy.

When the magnetising force continues its path in the negative direction a corresponding magnetic flux density curve develops, as represented by *DE*. As the magnetising force continues its progress the alternating current waveform completes one cycle, and another curve establishes between points *E, F, G* and *B* that is a similar curve to *BCDE*.

The points *B* to *C, D, E, F, G* and back to *B* again form a closed figure called a hysteresis loop.

Energy stored in the magnetic material

The closed figure of the hysteresis loop shows the energy required to drive a sample of a ferromagnetic material through a complete cycle of magnetisation. The area of the hysteresis loop gives a measure of the energy stored in the magnetic material. This energy appears as heat within the ferromagnetic material and is hysteresis loss measured in watts per kilogram.

Hysteresis loops for various ferromagnetic materials

All ferromagnetic materials possess hysteresis. Hysteresis is a term given to the effect which causes a ferromagnetic material to lag behind in its magnetism when exposed to excitation forces. The lag results from energy losses due to internal friction within the material. The area of the loop and the frequency of reversal of the magnetic field per second determine the magnitude of the loss.

Hysteresis loops for various ferromagnetic materials are shown in **Figure 4.41**.

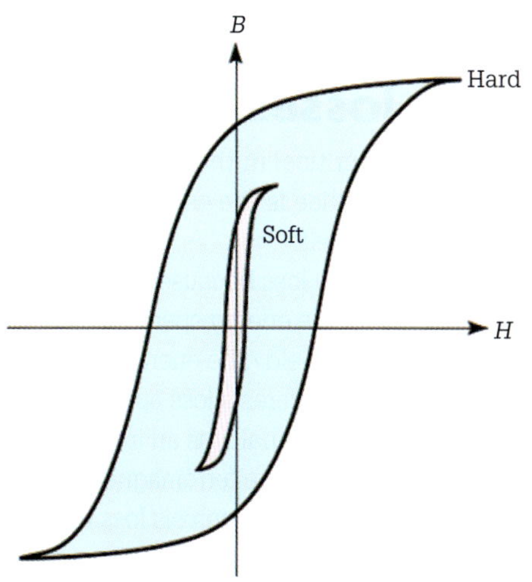

FIGURE 4.41 Hysteresis loops of soft and hard materials

A small loop area indicates a low hysteresis loss, which means that the ferromagnetic material is a 'soft substance' magnetically and would be ideal for use as a transformer core, for example, silicon steel. By contrast, a large hysteresis loop indicates a 'hard' ferromagnetic material indicating that the material is suitable for use as a permanent magnet, for example, alnico.

Furthermore, square *B/H* curves indicate that the ferromagnetic material is useful for permanent magnets, as the remanent magnetism approximates that of saturation as seen in **Figure 4.42**. This loop also indicates that the material is suitable in electromagnetic switching circuits – low current when the ferromagnetic material is unsaturated and very high current when the material saturates.

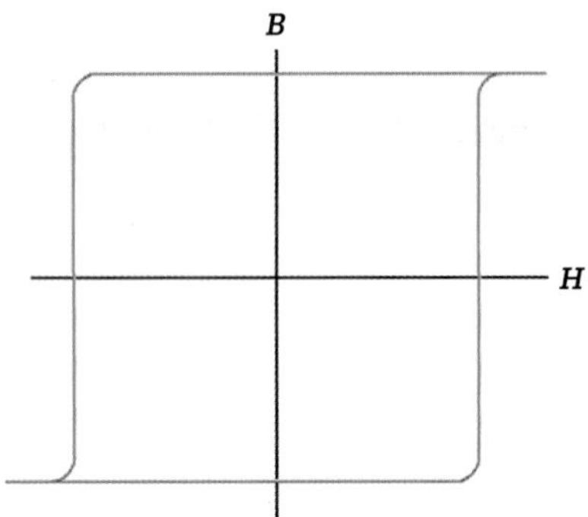

FIGURE 4.42 Square *B/H* curve – ferrite

Because each family of ferromagnetic materials has several grades with a range of magnetic properties, de-magnetisation curves play an important role in the selection of the most suitable ferromagnetic material. The de-magnetisation curve is located in the second quadrant of the hysteresis curve. The intersection of the curve with the magnetising force (horizontal line) in the second quadrant is known as the coercive force and is a measure of a ferromagnetic material's ability to withstand de-magnetisation from external magnetic fields.

Magnetic losses

A characteristic that is critical in the energy efficiency of an electromagnetic device is the energy within the ferromagnetic material itself.

Hysteresis loss is energy loss because there is no useful return of energy for the energy spent on each reversal of direction of the magnetic field. The energy spent occurs in the form of heat. Because hysteresis loss appears in the form of heat in a ferromagnetic material, it is an iron loss. Iron loss refers to the energy wasted by the ferromagnetic material undergoing magnetisation. The hysteresis loss happens each time the ferromagnetic material cycles from a positive value to a negative value. The energy loss is directly proportional to the frequency (*f*) of the excitation force. Consequently, hysteresis loss increases with amplified frequency.

The other iron loss that occurs in a ferromagnetic material is the eddy current loss. This energy loss happens because of induced currents called eddies since they flow in closed paths within the ferromagnetic material itself. In addition, eddy current losses occur because the ferromagnetic material is electrically conductive. The material is conductive because the magnetic field is contained within the outside edge of the material in the same manner as it is contained within a turn of the coil. Around the outside edge of the ferromagnetic material an induced electromotive force (emf) happens as shown in **Figure 4.43**.

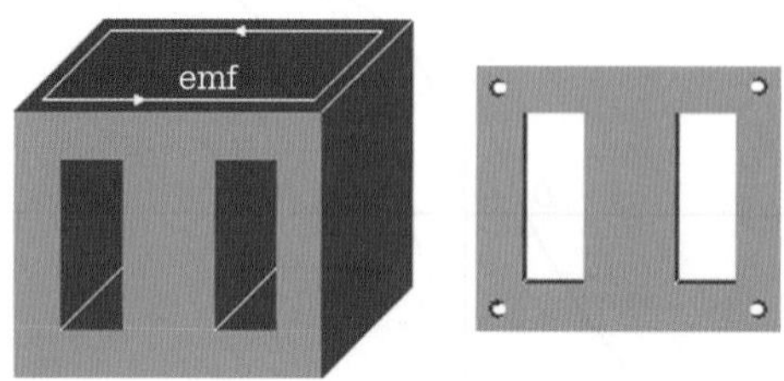

FIGURE 4.43 Solid core and a lamination

This induced voltage which occurs across a low-resistance path that makes up the ferromagnetic material, results in large circulating eddy currents that heat up the ferromagnetic material. If the ferromagnetic material remained as a solid core the eddy current induced would be very large. However, the effect of eddy currents can be minimised by laminating the ferromagnetic material as shown in **Figure 4.43**.

The eddy currents can still circulate within the laminations, but because each lamination has an insulating oxide coating, the eddy current cannot flow from one lamination to another. If the ferromagnetic material was laminated into 10 sheets each lamination would carry one-tenth (1/10) of the total flux. So, the voltage induced in each lamination is one-tenth (1/10) of what it would have been in the solid ferromagnetic material.

The power loss in each lamination is one-hundredth (1/100) of what would have been the energy loss with a solid material. Thus 1/100 × ten laminations = 1/10 of the original loss in total. Laminating reduces eddy current energy loss because this energy loss is inversely proportional to the square of the number of laminations. However, there is a method to reduce this eddy current loss even further. Because resistance is inversely proportional to the cross-sectional area of material, then if the laminations reduced in width, say by half, the resistance of the lamination doubles and, therefore, the induced eddy current halves.

It follows that eddy current energy loss is the result of the flux density, the ferromagnetic quality, its thickness and the effectiveness of the insulation between the sheets of the ferromagnetic material. Finally, both hysteresis and eddy currents loss are constant and independent of load conditions. A typical loss value for a 0.3 mm thick silicon steel lamination at a maximum flux density of 1.7 T at 50 cycles per second is 1.32 watts per kilogram.

Magnetic leakage

Where a portion of a generated magnetic flux passes through regions of space, such as air, insulation or the structural metal members of a magnetic device – instead of along its intended path of the magnetic circuit, it is deemed to have 'leaked' away as magnetic leakage. Leakage is characteristic of all magnetic circuits. Magnetic leakage means that wastage of a number of ampere-turns transpires because the total flux produced is never 100 per cent efficient. Refer to **Figure 4.44**.

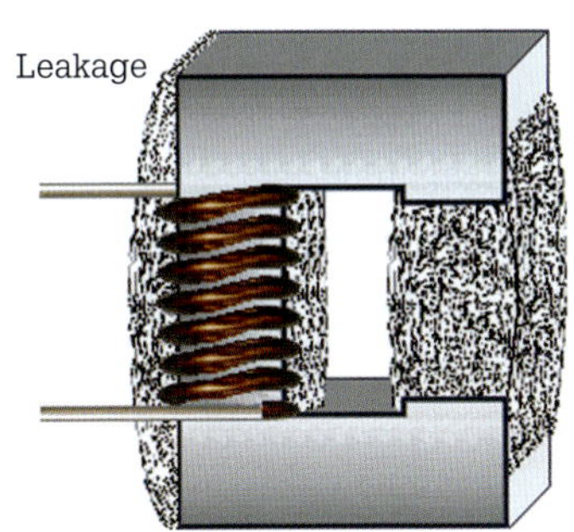

FIGURE 4.44 Leakage and fringing

Magnetic fringing

Magnetic fringing is similar to leakage; however, it represents the spreading of the flux lines through the air gap and along the sides of the magnetic circuit. The spreading of the flux occurs because of the repulsion between lines of flux that act in the same direction causing the flux to bulge out at the edges of the air gap as shown in **Figure 4.44**. Many magnetic devices, both alternating current and direct current, contain a magnetic flux that varies cyclically in both magnitude and direction. Because of its cyclic nature, the devices have a constant, in some cases substantial, and continuous energy loss throughout the whole of the time the magnetic device energises regardless of the load. Energy losses reduce the efficiency of the magnetic device.

Air gap

Many electromagnetic devices such as relays and contactors have air gaps, and they are of primary importance in these magnetic circuits. A magnetic field has to cross the air gap in these devices in order that work occurs. An air gap is nothing else but the volume of air between two surfaces of a ferromagnetic material. An air gap has a length that is the distance between the magnetic surfaces and an area pertaining to the surface area of one of the magnetic surfaces. Flux can traverse the air gap and create a flux density within the air gap as a direct result of the magnetomotive force (F_m) produced by a defined number of ampere-turns. Refer to **Figure 4.45**.

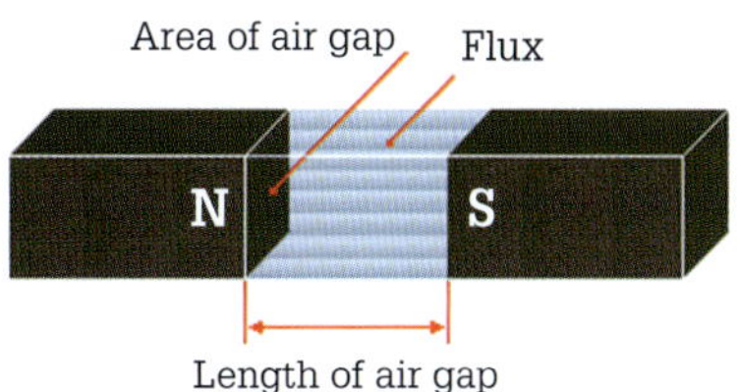

FIGURE 4.45 Air gap relationship

Usually the mating surfaces of an inductor magnetic core are flat for minimum reluctance but some magnetic cores have shorter limbs in order to provide an air gap. In a typical inductor core, the highly permeable ferromagnetic core guides the magnetic flux around the windings. However, in a gapped core as shown in **Figure 4.46** a small section of the ferromagnetic substance is replaced by another permeable material, such as air.

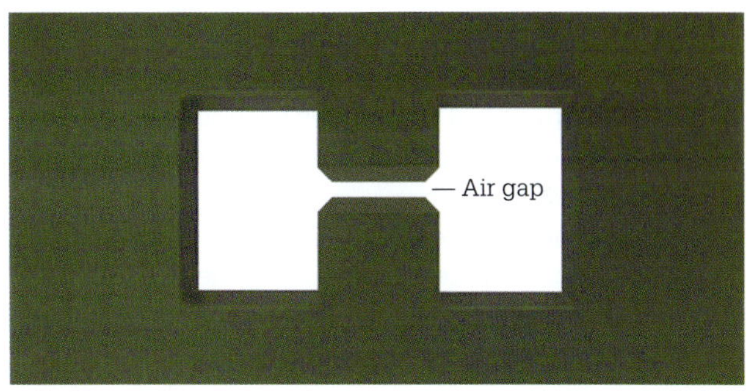

FIGURE 4.46 Air gap in a ferrite core

An air gap minimises any excessive flux (or core saturation) produced by a high level of current in the windings. The air gap's task is to increase the reluctance of the core so that less flux flows for any given level of magnetomotive force. The air gap's reluctance depends only on its length and cross-sectional area. Consequently, when you add an air gap to increase the reluctance of the core it is as if you have decreased the core's permeability, thereby reducing the inductance of the winding wound on the core.

EXAMPLE 4.10

An annealed silicon steel ring of CSA 10 cm² has a radial air gap of 2 mm. If the mean length of the silicon steel ring is 30 cm and is wound with 1000 turns, calculate the magnetomotive force to produce a flux of 16 mWb and the current produced. Refer to **Figure 4.37** for the magnetisation curve for annealed silicon steel. **Note:** 1 cm² = 1 x 10^{-4} m².

For the annealed silicon steel

$$B = \frac{\Phi}{A}$$

$$= \frac{16 \times 10^{-3}}{10 \times 10^{-4}}$$

$$= \mathbf{16\ T}$$

»

From **Figure 4.37**, when $B = 1.6$ T, $H = 2992$ At/m. Therefore, the magnetomotive force for the annealed silicon steel path:

$$F_m = H \times l$$
$$= 2992 \times 0.3$$
$$= \mathbf{897.6\ At}$$

For the air gap

The flux density is the same in the air gap as in the annealed silicon steel. For air:

$$\mu = \mu_r \times \mu_0 = 1 \times 4\pi \times 10^{-7} = 4\pi \times 10^{-7}\ \text{H m}^{-1}$$

$$H = \frac{B}{\mu}$$
$$= \frac{1.6}{4\pi \times 10^{-7}}$$
$$= \mathbf{1\,273\,075\ At/m}$$

Therefore, the magnetomotive force for the air gap is:

$$F_m = H \times l$$
$$= 1\,273\,075 \times 2 \times 10^{-3}$$
$$= \mathbf{2546.15\ At}$$

The total magnetomotive force to produce a flux of 16 mWb:

$$897.6 + 2546.15 = 3443.75\ \text{At}$$

A magnetomotive force of 3443.75 At with 1000 turns requires a current of:

$$I = \frac{F_m}{N}$$
$$= \frac{3443.75}{1000}$$
$$= \mathbf{3.44375\ A}$$

REVIEW QUESTIONS

1. Why are graphical techniques used to describe the magnetic characteristics of many ferromagnetic substances?
2. What is the main reason for manufacturers of ferromagnetic materials providing the *B/H* curve for the material?
3. Why is a graph of the magnetic conductivity or permeability of a ferromagnetic material of use?
4. What does a hysteresis loop define?
5. What is magnetic saturation?
6. Which term describes the flux remaining in the ferromagnetic material?
7. What is meant by the term 'coercive force'?
8. The hysteresis curve for a particular magnetic material is said to have a small loop area. To what application is the material suited?
9. What is meant by the term 'eddy current loss'?
10. Explain what is meant by the term 'magnetic leakage'.
11. Explain what is meant by the term 'magnetic fringing'.
12. Why would a magnetic circuit include an air gap?

4.5 Electromagnetic induction

Electromotive force

When a conductor moves through a magnetic field or the magnetic field is moved through the conductor as in **Figure 4.47** at right angles a small electromotive force (emf) is induced across the ends of the conductor. The polarity of the induced emf is found using Fleming's right-hand generator rule (see **Figure 4.48**).

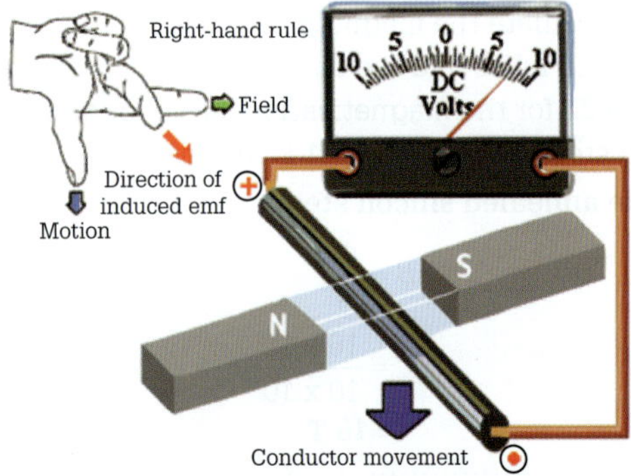

FIGURE 4.47 Electromagnetic induction

Faraday's Law

This electromagnetic induction is known as Faraday's law. If the conductor is part of an electric circuit, the induced electromotive force value is determined using a sensitive centre-zero micro-voltmeter connected across the ends of the conductor. In addition, the induced electromotive force exists only while the conductor is moving and cutting through the magnetic field.

Fleming's right-hand generator rule

Fleming's right-hand generator rule enables the determination of the direction of induced current. Refer to **Figure 4.48**.

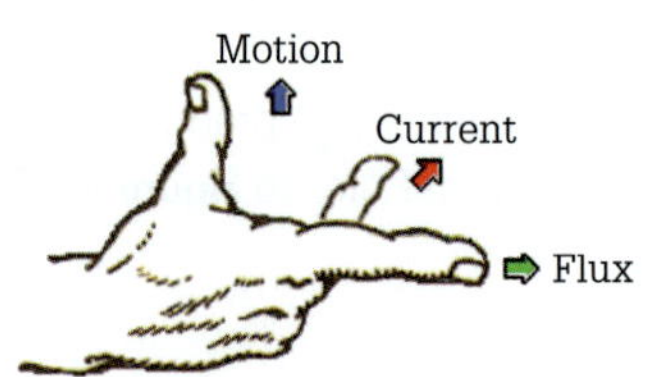

FIGURE 4.48 Fleming's right-hand generator rule

Note that there is no emf induced if the conductor is stationary within the field or if the conductor moves parallel to the magnetic field as shown in **Figure 4.49**.

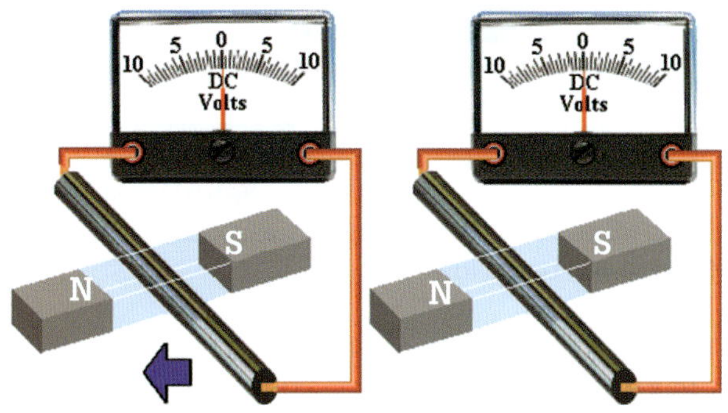

FIGURE 4.49 No emf induced as conductor moves sideways or is stationary

Hence, an induced voltage only occurs when there is relative motion of either the conductor or the magnetic field. The direction of the induced emf is also determined by applying Fleming's right-hand rule.

With a moving field, the thumb points in the relative direction of motion of the conductor. For example, if the field moves downwards, then the relative direction of the conductor is upwards, and the thumb will be pointing upwards. The right-hand thumb rule has practical application with generators. A generator is an electromechanical machine where conductors rotate in a magnetic field.

If the polarity of the magnetic field north in relation to south reverses (in relation to **Figure 4.47**), as in **Figure 4.50**, the polarity of the induced emf reverses. In addition, by reversing the direction of motion of the conductor through a magnetic field (in relation to **Figure 4.47**), as illustrated in **Figure 4.51**, the polarity of the induced emf can also be changed.

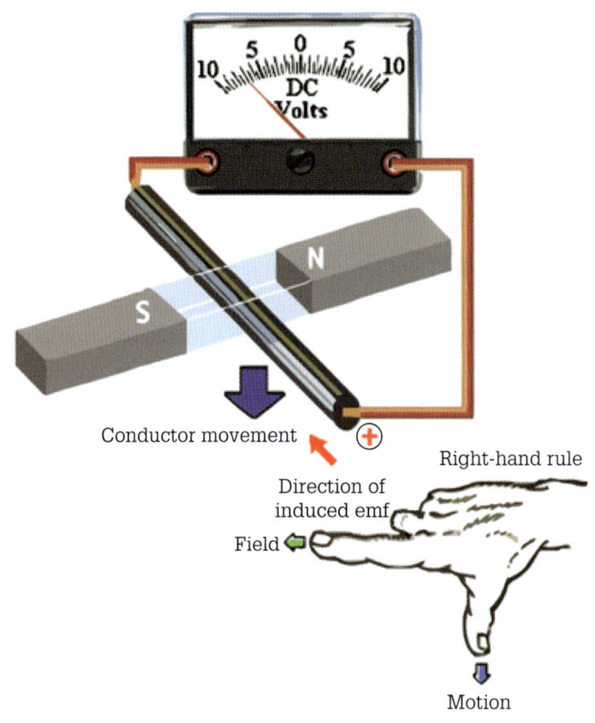

FIGURE 4.50 Reversal of induced emf through polarity change

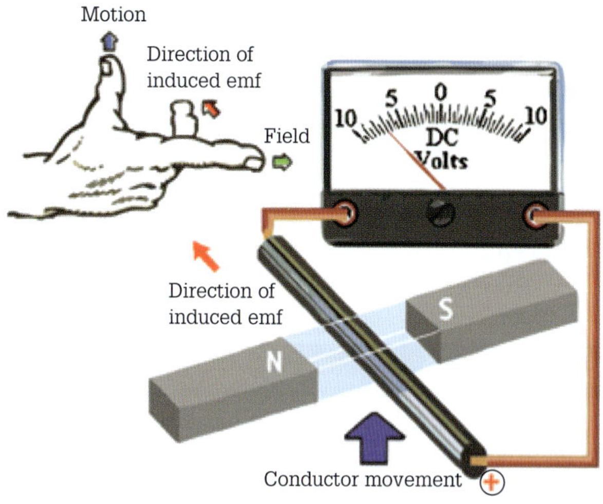

FIGURE 4.51 Reversal of induced emf through motion change

The direction of the induced emf depends on either the polarity of the magnetic field or the direction of motion of the conductor as it cuts through the magnetic lines of force. For this reason, the magnitude of the induced emf depends on four factors:

1. the flux density (B)
2. the length (l) of the conductor actually cutting through the magnetic field
3. the rate of motion or velocity (v) of the conductor through the magnetic field
4. the angle at which the conductor cuts through the magnetic field.

Referring to **Figure 4.52**, a single conductor is cutting through a magnetic field and the maximum induced emf is produced when the conductor moves through path A or B at right angles to the magnetic lines of force.

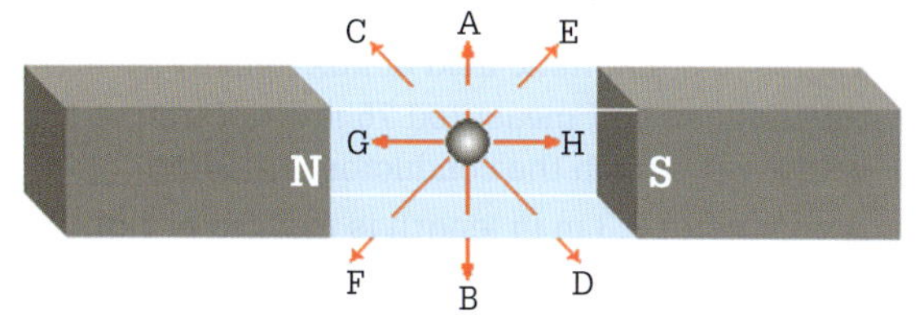

FIGURE 4.52 Dependence of induced emf on an angle of cutting

No induced electromotive force happens when a conductor moves through path G or H parallel to the lines of force, as no cutting of these lines occurs. However, for conductor motion at angles between 0° and 90°, such as paths C, D, E and F, the induced electromotive force provides values between zero and maximum because some cutting of the lines of force does occur. Finally, when the cutting of the lines of force occurs at right angles, calculation of the magnitude of the induced electromotive force occurs by combining the length, flux density and velocity into a simple equation:

$$e = Blv \sin\ \theta$$

where e = the instantaneous value of induced electromotive force in volts
B = the flux density of the magnetic field in tesla
l = the length of the conductor at right angles to the magnetic field in metres
v = velocity of the conductor in metres per second (ms^{-1})
$\sin\theta$ = sin of the angle conductor is moving relative to magnetic field

SWITCH ON

If the magnetic field of one tesla has a conductor 1 m in length moving through it for a distance of 1 m at right angles to the magnetic lines of force at a velocity of 1 m per second, then 1 volt will be induced in the conductor.

EXAMPLE 4.11

A conductor 0.2 m in length is moving at right angles to a magnetic field of flux density 50 mT at a velocity of 10 ms^{-1}. Determine the magnitude of the induced voltage.

$$e = Blv \sin\theta$$
$$= (50 \times 10^{-3}) \times 0.2 \times 10 \times \sin 90°$$
$$= \mathbf{100\ mV}$$

EXERCISE 4.8

a A conductor 0.15 m in length is moving at right angles to a magnetic field of flux density 5 mT at a velocity of 25 ms^{-1}. Determine the magnitude of the induced voltage.

b A conductor 0.25 m in length is moving at right angles to a magnetic field of flux density 15 mT. If the induced emf measured 75 mV, determine the velocity at which the conductor is moving.

c A conductor 0.3 m in length is moving at right angles to a magnetic field at a velocity of 25 ms^{-1}. If the induced emf measured 90 mV, determine the strength of the magnetic field.

Lenz's law

If a bar magnet is moved into a solenoid, as illustrated in **Figure 4.53**, an emf is induced in the conductors. In this example, the conductor is stationary and the magnetic field is doing the moving. The current can only flow when the solenoid connects in an external circuit as shown in **Figure 4.53**. There are three methods whereby the magnitude of the induced voltage and the magnitude of the induced current can be increased:

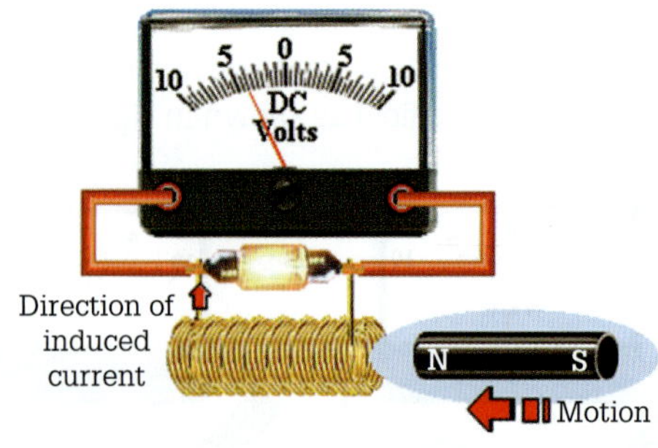

FIGURE 4.53 Lenz's law

1 using a stronger magnet which delivers an increased flux density
2 increasing the velocity of the magnet into the solenoid
3 increasing the number of turns of the solenoid, as this increases the length of the conductor being cut by the magnetic field.

Figure 4.54 shows the induced current direction change when the magnet withdraws from the solenoid.

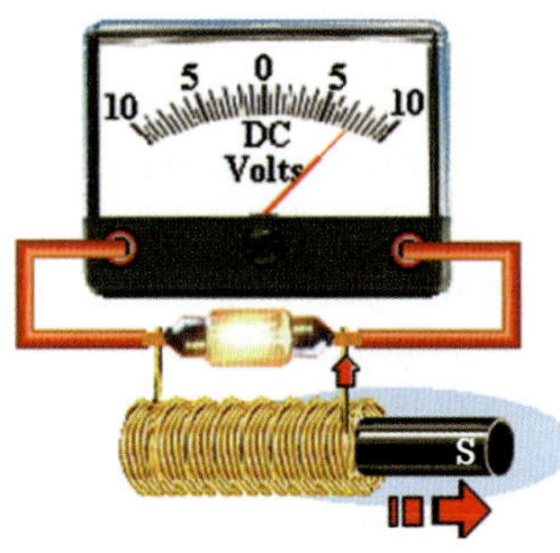

FIGURE 4.54 Direction of induced current is reversed

The direction of the induced emf and the induced current can be found by applying Lenz's law.

SWITCH ON

The direction of the induced emf is always such that it establishes an induced current opposing the motion or change of magnetic flux responsible for inducing the emf.

In **Figure 4.55** the motion of the magnet into the solenoid has opposition because the induced current has the effect of turning the solenoid into a feeble electromagnet with its north pole repelling the approaching north pole of the magnet.

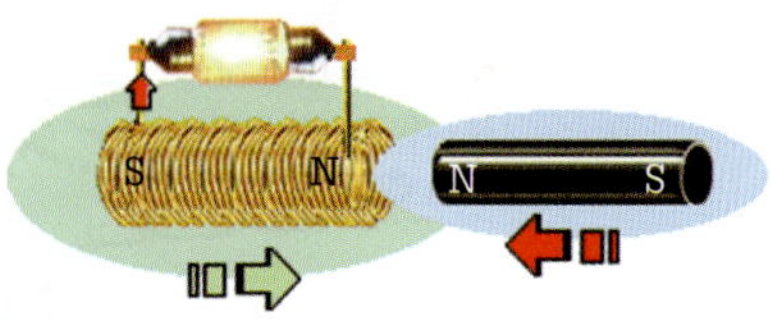

FIGURE 4.55 Motion of the magnet is opposed by the solenoid's feeble magnetic field

In **Figure 4.56** the motion of the magnet has opposition again. However, this time the solenoid attracts the north pole of the magnet as the magnet moves out of the solenoid. Consequently, because we know which ends of the solenoid have become north and south poles, the direction of the induced current can be determined by applying the right-hand grip rule for solenoids.

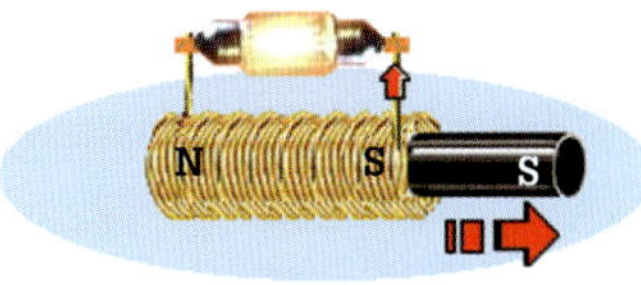

FIGURE 4.56 Motion of the magnet and the direction of the induced current

Figure 4.57 shows the induced emf in a conductor of a generator and the direction of the induced current. This induced current produces a magnetic field around the conductor that has a direction that strengthens the magnetic field of the magnets below the conductor and this strengthened field opposes the motion of the driving force of the conductor in that it tries to repel the conductor.

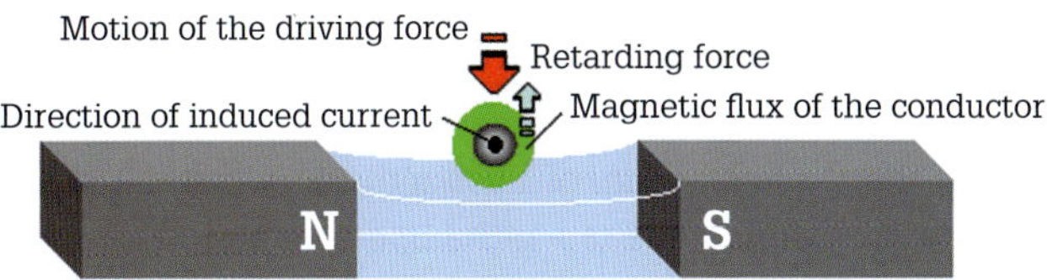

FIGURE 4.57 Generator action

Figure 4.58 shows the induced emf in the secondary winding of a transformer and the direction of the induced current.

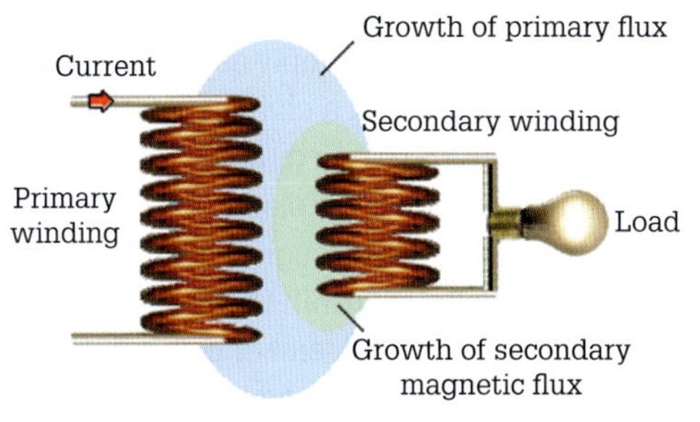

FIGURE 4.58 Induced emf in the secondary winding of a transformer

Bifilar winding

Bifilar winding (meaning two filaments) is a method of manufacturing a coil, as illustrated in **Figure 4.59**, in which two magnet-wires wind simultaneously onto a core instead of a single, large winding wire; afterwards, the wire ends are separated to form two independent windings of an equal inductance. Bifilar windings increase the magnetic coupling. Magnetic coupling arises due to the presence of mutual inductance between the two windings (a weak transformer effect). The wires go in the same direction creating a tightly coupled effect. They produce less leakage flux and hence lower leakage inductance.

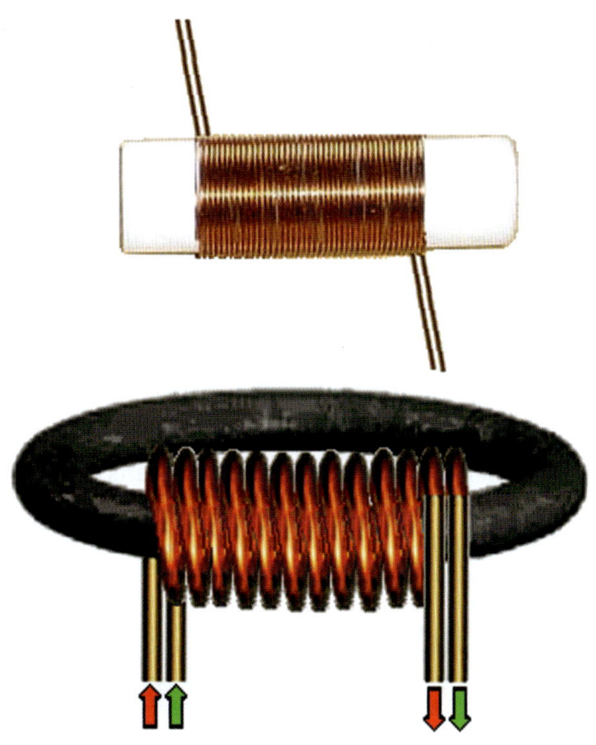

FIGURE 4.59 Bifilar winding inductors

Some bifilar windings are wound so that the current flows in opposite directions. The electromagnetic field created by one winding is equal and opposite to that created by the other, resulting in no electromagnetic field. This means that the windings have no self-inductance.

Bifilar-wound coils have applications for some relays, transformers for switched-mode power supplies and each stator pole of a step motor.

REVIEW QUESTIONS

1 Which rule allows us to determine the polarity of emf induced in a conductor as it moves through a magnetic field?

2 What factors determine the magnitude of emf induced in a conductor as it moves through a magnetic field?

»

3 A conductor 0.45 m in length is moving at right angles to a magnetic field of flux density 45 mT at a velocity of 25 ms^{-1}. Determine the magnitude of the induced voltage.

4 The direction of the induced emf is always such that it establishes an induced current opposing the motion or change of magnetic flux responsible for inducing the emf is a summary of which law?

5 By what methods is it possible to increase the magnitude of emf induced in a conductor as it moves through a magnetic field?

6 What winding method increases the magnetic coupling between turns on a coil?

7 A conductor moving through a magnetic field has an emf induced of a specific polarity. How is it possible to reverse this polarity?

8 At what angle must a conductor move relative to the magnetic field in order to provide maximum induced emf?

4.6 Inductance

Inductance is defined as that property of an electrical conductor that opposes a change in the electric current flowing through it.

A straight conductor or a conductor manufactured in the form of a coil is an inductor. If current passes through the straight conductor or coil, energy stores in the form of an electromagnetic field established around the straight conductor or coil. When induction occurs in an electric circuit and affects the flow of current, inductance has occurred.

Whenever current in an electric circuit starts or stops or increases or decreases in magnitude, an electromagnetic field that surrounds the conductor increases or decreases as seen in **Figure 4.60**.

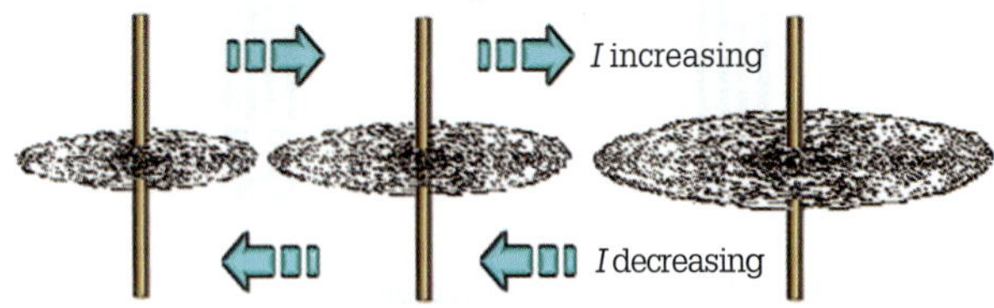

FIGURE 4.60 Electromagnetic field around a conductor

The changing field causes an inductive effect that delays the change in current.

SWITCH ON

Circuit symbols

The principal method of classifying inductors is according to the type of material used in their core. **Figure 4.61** illustrates different inductor symbols.

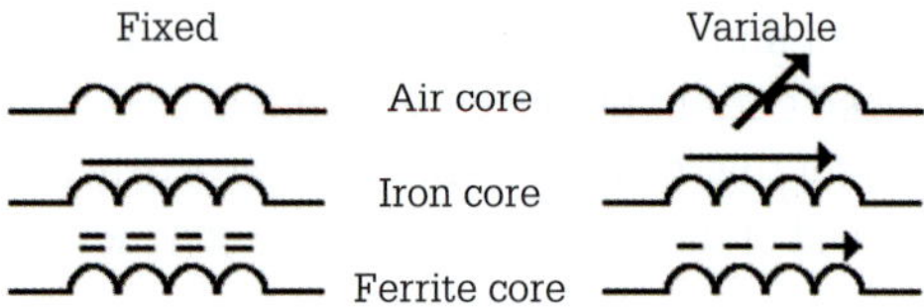

FIGURE 4.61 Circuit symbols for inductors

Unit of inductance

A conductor or a component has an inductance (symbol L) of one henry when an emf of 1 volt induces, when the current is changing at a rate of 1 ampere per second.

SWITCH ON

The henry is a large unit, and in practical inductors the inductance is in the order of millihenry (1 mH = 0.001 H) or microhenry (1 µH = 0.000 001 H).

Self-inductance

Self-inductance (also known simply as inductance) is the property of a circuit by which a change in current causes a change in voltage in the same circuit. If the current through the inductor varies, the electromagnetic field also varies. When this field cuts through the straight conductor or coil, it induces a voltage within them. This voltage is a back electromotive force (back emf or self-induced emf).

Thus, the emf generated around the circuit, due to its own current, is directly proportional to the rate at which the current changes. Lenz's law tells us that if the current is increasing then the emf should always act to reduce the current, and vice versa. Otherwise, if the emf acted to increase the current when the current was increasing, then an unreal feedback effect would occur, whereby the current continues to increase without limit.

With d.c., self-inductance is present only at switching on and switching off. However, with a.c., self-inductance is present all the time because the current has a sinusoidal waveform.

Induced emf

Calculation of the induced emf uses the following equation:

$$e = -L\frac{\Delta I}{\Delta t}$$

Where e = average induced emf in volts (V)
$-L$ = inductance in henrys (H) (the – preceding indicates the induced emf is opposing)
ΔI = the change in current in amperes (A)
Δt = the change in time in seconds (s)

EXAMPLE 4.12

If the current through a conductor of inductance of 0.3 H increases from 0 A to 5 A in 50 ms, calculate the average induced emf in the conductor.

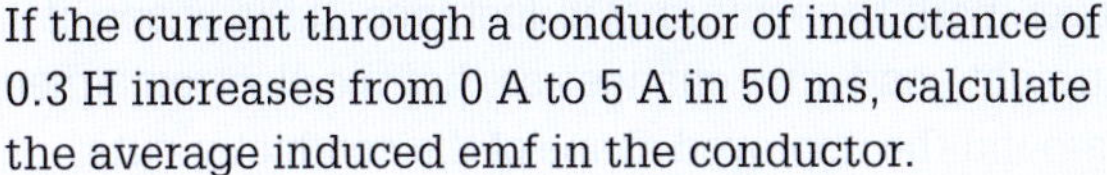

$$e = -L\frac{\Delta I}{\Delta t}$$

$$= -0.3 \times \frac{(5-0)}{(50 \times 10^{-3})}$$

$$= \mathbf{-30\ V}$$

EXERCISE 4.9

a If the current through a conductor of inductance of 0.15 H increases from 0 A to 2 A in 5 ms, calculate the average induced emf in the conductor.

b If the current through a conductor increases from 0 A to 3 A in 2 ms and creates a back emf of 90 V, calculate the inductance of the conductor.

Note: The minus sign is a reminder that the emf always acts to oppose the change in magnetic flux that generates the emf and is known as a counter emf (cemf).

Mutual inductance

When one energised straight conductor or a coil induces an emf in a second straight conductor or coil, mutual inductance (symbol M) has occurred. Mutual inductance occurs because the electromagnetic flux from the energised straight conductor or coil cuts through the other straight conductor or turns of the coil. As shown in **Figure 4.62** the electromagnetic field produced by current flow (I_1) in the 'circuit A' cuts the conductor turns of the inductor in circuit B and induces an emf. In addition, the induced emf creates a current flow (I_2) through the load resistor (R).

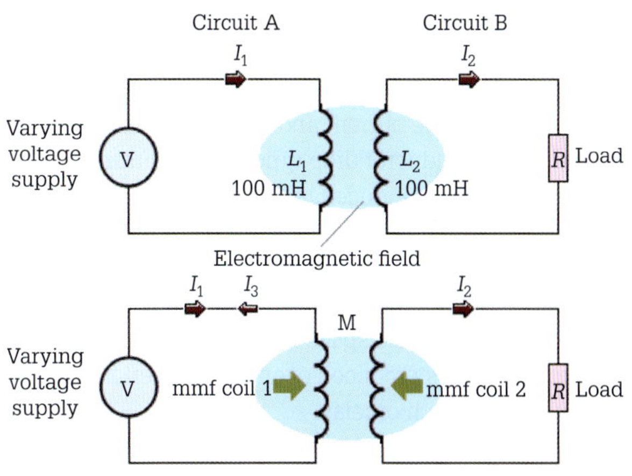

FIGURE 4.62 Mutual inductance

If the coil of circuit B experiences the same electromagnetic flux change as the coil of circuit A and it has the same number of conductor turns, the voltage of equal magnitude and phase to the applied voltage will be induced. This effect is mutual inductance: the induction of a voltage in one coil responds to a change in current in the other coil.

As the load draws current from circuit B it may seem that additional magnetic flux exists, but this does not occur. If additional flux existed, more voltage would be induced in the coil of 'circuit A'. However, this does not occur, because the induced voltage in the coil of the 'circuit A' must remain at the same level and phase in order to balance the varying voltage supply, in accordance with Kirchhoff's voltage law. Consequently, the magnetic flux is unaffected by the load current of 'circuit B'. However, what does change is the quantity of magnetomotive force (mmf) in the magnetic circuit.

As additional flux cannot exist, the only way circuit B's coil mmf may exist is if a counteracting mmf is generated by the coil of 'circuit A' of equal magnitude and opposite phase. This is what happens. The current (I_3) forms in the coil of 'circuit A', 180° out of phase with coil B's current, generating counteracting mmf and preventing additional electromagnetic flux.

The term 'mutual induction' refers to the electromagnetic condition in which two circuits are sharing the energy of one of the circuits. Sharing means that the energy transfers from one circuit to the other. A device that uses the principle of mutual induction is the transformer.

Undesirable effects of self-induction and mutual induction

Effect of self-induction

If the current has an interruption as, for example, by opening a switch, the current and the electromagnetic flux around the conductor drop quickly. A resulting electromotive force is induced in the conductor, and a large potential difference develops across the two contacts of the switch. The energy stored in the electromagnetic field of the conductor dissipates as heat and radiation in an electric arc across the space between the contacts of the switch. The arcs produced eventually deteriorate the contacts of the switch or relay.

Effect of mutual induction

Conductors, where they pass through metal surrounds such as switchboards, can induce circulatory currents called eddy currents in the surrounding metal, as a result of mutual induction. The surrounding metal acts as a shorted conductor producing an undesirable by-product – heat – in the surrounding metal. However, this effect is overcome by using a slotted metal plate which effectively open-circuits the metal surrounding the conductors.

The mutual inductance between two conductors placed close enough that the electromagnetic field induced by a current flowing into one can inject a voltage spike onto the other conductor is proportional to the rate of change of the current in the other conductor. For example, lightning strikes near transmission lines can induce a voltage spike through mutual induction across the line, resulting in damage or destruction of switches, cables and transformers.

Factors affecting inductance

There are a number of interrelated factors determining the inductance of an electromagnetic circuit. The four main interrelated factors are:

1 the length of the coil (l)
2 in the case of a coil, the number of turns (N)
3 the CSA of the core (A)
4 the permeability of the core material (μ_a).

Calculation of the inductance in henrys uses the following equation.

$$L = \frac{N^2 \mu_a A}{l}$$

where L = inductance in henrys (H)
N = number of turns on the coil
μ_a = absolute permeability of the core material in henrys per metre (H m^{-1})
A = cross-sectional area of the coil in square metres (m^2)
l = length of the coil in metres (m)

EXAMPLE 4.13

A 50 mm long coil has 100 turns, a CSA of 200 mm^2 and the iron core has an absolute permeability of 6×10^{-3} H m^{-1}. Calculate the inductance of the coil.

$$L = \frac{N^2 \mu_a A}{l}$$

$$= \frac{100^2 \times (6 \times 10^{-3}) \times (200 \times 10^{-6})}{50 \times 10^{-3}}$$

$$= \mathbf{240\ mH}$$

EXERCISE 4.10

a A 150 mm long coil has 500 turns, a CSA of 50 mm^2 and the iron core has an absolute permeability of 6×10^{-3} H m^{-1}. Calculate the inductance of the coil.

b Calculate the inductance of the coil in (a) if the iron core is removed.

A straight conductor of x length has half the inductance of conductor $2x$ in length, all other factors being equal. However, both have minimal inductance. If the shape of the straight conductor changes so that the electromagnetic field around the improved shape of the conductor cuts across some other portion of the same conductor, the inductance increases. In **Figure 4.63** a straight conductor is now a coiled system so that the turns lie adjacent and parallel to one another.

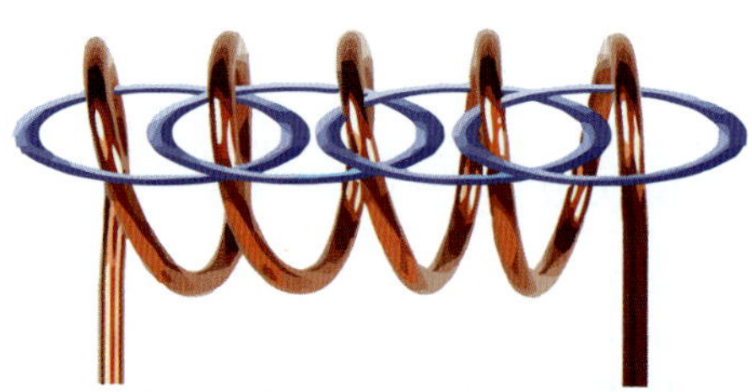

FIGURE 4.63 Electromagnetic field cutting through adjacent turns

Energy capacity of a coiled system

Consequently, when current flows, an electromagnetic field establishes around each of the turns. If we consider that the electromagnetic field starts in turn one and expands outwards we can see that its electromagnetic field also cuts across turn two, inducing another emf that adds to the existing induced emf which turn two has created by itself. In addition, the electromagnetic field produced by turn two also expands across turn one inducing another emf in that turn.

Furthermore, the effect on inductance of creating a coiled system is that it increases the induced emf that in turn enables the storing of higher energy within the final electromagnetic field. It follows that, if the energy capacity of the coiled system increases, then the inductance also increases.

When a coiled shape is so constructed, we understand the electromagnetic fields of each turn as 'linking in' with each other like the links of a chain, creating greater inductive strength. If all other factors remain constant, then by doubling the number of turns in a coil system, as shown in **Figure 4.64**, we find that the inductance has increased fourfold. This is because inductance is proportional to the square of the number of turns.

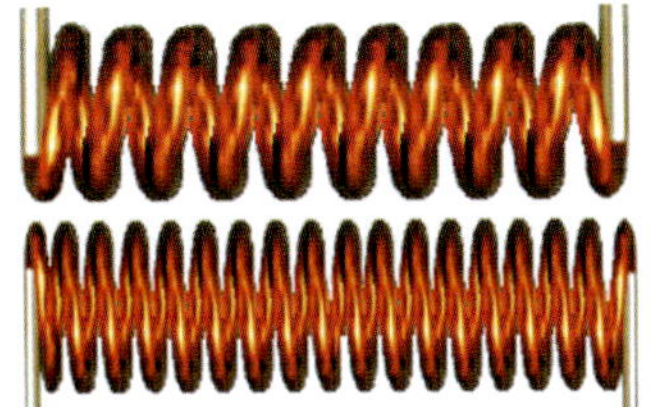

FIGURE 4.64 Effect of turns on inductance

The closer the turns, the more effective are the flux linkages and the stronger the self-inductance. **Figure 4.65** demonstrates the effectiveness where two coils with the same number of turns have only one variance and that is the length of the electromagnetic circuit. Consequently, doubling the length of the coil while maintaining the same number of turns, halves the inductance.

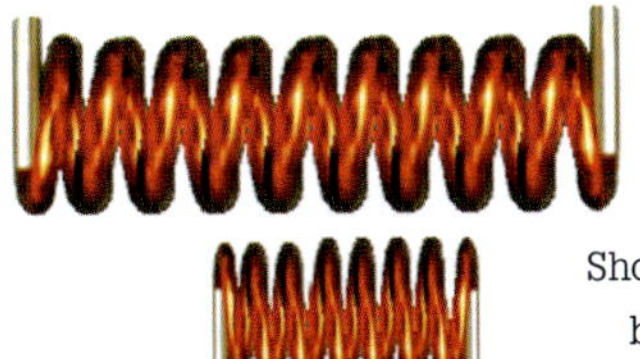

FIGURE 4.65 Effect of length on inductance

Coil diameter

Another factor affecting inductance is the coil diameter. Large diameter coils require more magnet wire compared to a smaller coil consisting of the same number of turns. As a result, a higher cemf in the coil with the larger diameter occurs because more lines of force exist. Therefore, CSA of a coil affects inductance.

$$\text{area of a circle: } A = \pi r^2$$

Hence, doubling the radius of a coil increases the inductance by a factor of four.

The term 'air-core coil' describes an inductor without a ferromagnetic core. Air-cored coils have several advantages when compared with ferromagnetic-core coils. First, like all air-core coils, it is free from iron losses. Second, single-layer coils have the additional advantage of low self-capacitance and thus high self-resonant frequency. If the coil system has turns wound in layers to increase the cross-sectional area of the coil, as in **Figure 4.66**, the flux linkages are also increased, which in turn increases the counter-electromotive force. For this reason, the increase of coil cross-sectional area allows the property of inductance to rise.

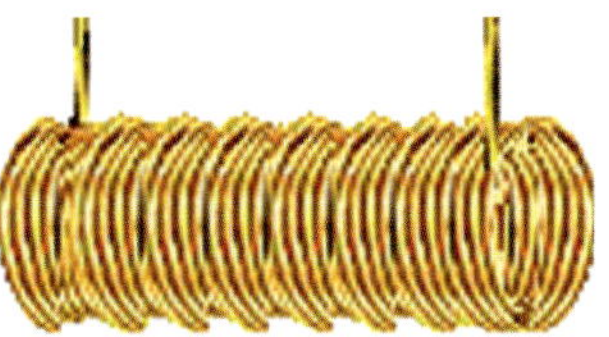

FIGURE 4.66 Coil system wound in layers

This coil system becomes even more self-inductive by providing a highly permeable (quickly entered) core that can be easily magnetised. The coil is then constructed around the core as shown in **Figure 4.67**.

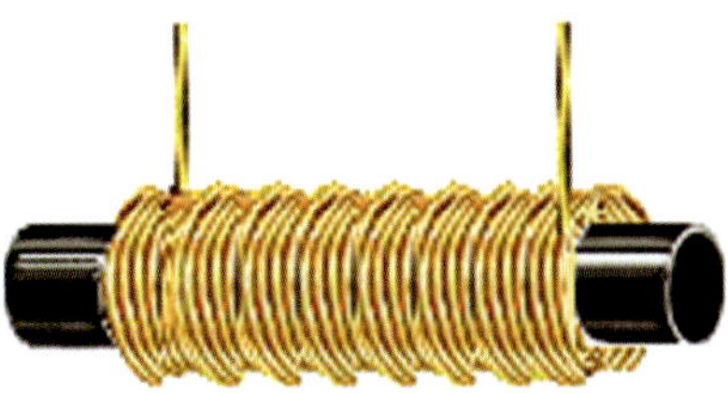

FIGURE 4.67 Coil with a permeable core

Types of core material

The main types of core materials used with inductors are air, iron, molybdenum permalloy powder cores, silicon steel and ferrite. An air core has a reducing effect on the electromagnetic strength of the coil as it is reluctant to allow the movement of the electromagnetic field through it, and the field in doing work to overcome this reluctance loses energy. A core of a ferromagnetic substance that can be magnetised (e.g. iron) allows the electromagnetic flux to concentrate through it. In addition, because of their high permeability, the total electromagnetic flux strength is many times stronger than that of an air core. Cores of these materials offer various excellent electromagnetic properties, such as high saturation flux density, high resistivity and a stable permeability. Refer to **Figure 4.68**.

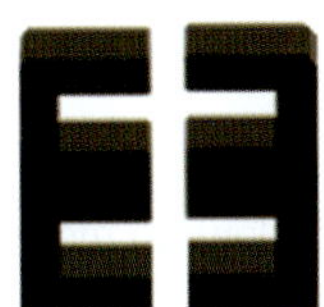  

FIGURE 4.68 Powdered iron material

Molybdenum permalloy powder (MPP) cores are 81% nickel, 2% molybdenum and 17% iron. MPP has high resistivity and minimal iron losses. These materials find use as inductors for high-quality, low-loss filter circuits, transformers, chokes and inductors. Refer to **Figure 4.69**.

FIGURE 4.69 Molybdenum cores

When low carbon steel alloys with small amounts of silicon, resistivity increases, which helps to reduce eddy current losses in the core. Furthermore, the hysteresis loss reduces because the addition of silicon affects the grain structure of the steel and thereby improves the material's permeability. With its high permeability and increased resistivity, silicon steel as a soft-magnetic material minimises the loss of electric energy.

This material is capable of being magnetised to a high value of flux density when exposed to relatively small fields but loses this magnetism on removal of the field. Transformers, motors and generators are some applications that utilise the ferromagnetic properties of silicon steel. Refer to **Figure 4.70**.

FIGURE 4.70 Silicon steel laminations

Ferrites are ceramic, ferromagnetic substances that are dark grey or black in appearance and are very hard and brittle. They are crystalline oxides containing oxygen, iron and other metallic elements such as nickel, zinc or magnesium. They may also contain non-metallic additives such as silicon and cobalt. However, these materials exhibit magnetic properties only when subject to a magnetising force. Refer to **Figure 4.71**.

FIGURE 4.71 Ferrite substances

These materials have high permeability allowing more lines of force to develop. However, the effect is an increase in inductance.

Types of inductors

Inductors in the low μH range etch onto printed circuit boards while field windings for d.c. motors and generators are a significant use of inductors. In addition, the electrical property of inductance occurs in the stator, rotor and armature design. The deeper the windings are in the laminations, the more inductive are their characteristics.

The transformer, toroidal coils and the relay are other electrical devices that utilise the electrical property of inductance. Toroidal coils and transformers confine the electromagnetic field to the inner regions of the coils, while an inductor called 'ballast' is used to ionise the argon gas within a fluorescent lamp. The highly inductive ballast injects a high voltage across the electrodes.

There are many more applications for inductors in the electronics industries. Some of these inductive devices prevent EMI (electromagnetic impulse) from entering into telephone circuits. Other inductors reduce noise propagated from the direct current output of a.c. adaptors. A typical application is radio frequency interference (RFI) suppression on light dimmers.

SWITCH ON

Hazards and safety precautions

Inductors have the ability to release stored energy at a substantially higher voltage than that used to create it. This energy may release accidentally if the inductors have no proof of their de-energised state. Electric motors contain inductors and must come to a full stop after switching off before isolation procedures occur. A full stop allows time for the residual voltage across the motor windings to fall to zero. In addition, inductors can produce large eddy currents in adjacent conductive material if suddenly de-energised, causing excessive heating.

With high-energy magnetic fields, cardiac pacemakers will be affected and metal body parts will be heated. All types of magnetic strip plastic cards degrade when exposed to high magnetic fields.

For circuit protection, surge-suppressing diodes, varistors or other automatic shorting devices occur with many types of inductors to provide a path for the current when an interruption of excitation happens.

Inductors in a d.c. circuit

Figure 4.72 shows several turns of wire and the ohmic resistance of the turns connected in series with a voltage source, a switch and a load resistor.

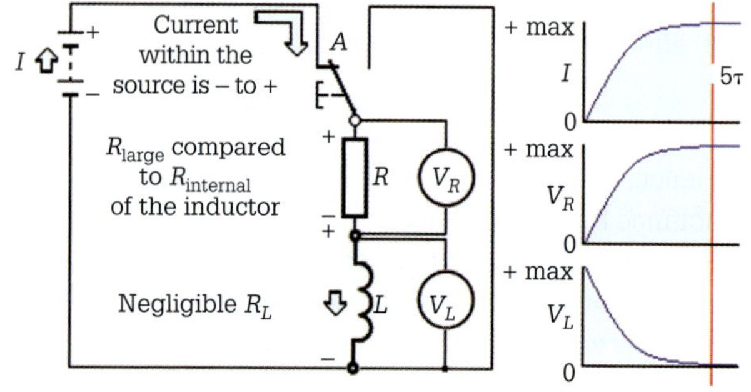

FIGURE 4.72 Energy stored in an inductor

When the switch closes to position A, the inductance of the coil delays the build-up of the final value of current in the circuit. The delay demonstrates the effect of Lenz's law: the induced emf opposes the flow of current now when switching on. The final value of current depends nearly entirely on the magnitude of the driving voltage and the ohmic resistance of the resistor.

Curves showing the build-up of current with time and the decay of the induced emf with time for the circuit at the instant of switching on are shown in **Figure 4.73**.

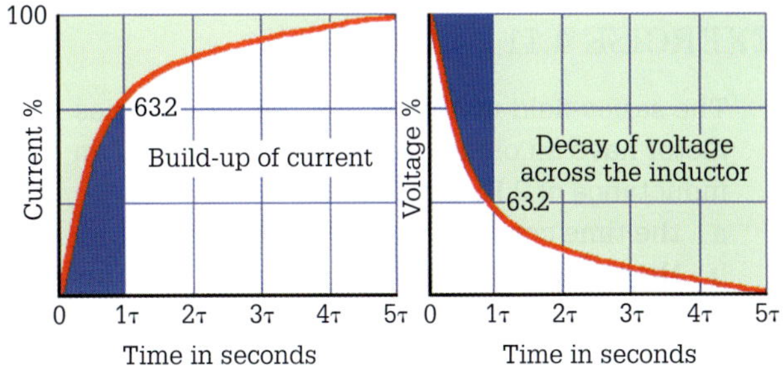

FIGURE 4.73 Build-up of current and decay of induced emf in an inductor

The part of the curves extending over a period beginning with the closing or opening of the switch and ending when the change is almost complete (5τ) is described as a transient.

The time delay experienced in the build-up of current directly relates to the combination of the ohmic resistance and the inductance in the circuit and may be in the order of microseconds, seconds or even minutes. The following equation determines the time delay caused by the inductive effect.

$$\tau = \frac{L}{R}$$

where τ = time constant in seconds
L = inductance in henrys (H)
R = resistance in ohms (Ω)

The equation gives the time constant (symbol τ, which is the Greek symbol for tau) for the circuit. The τ of an inductive circuit is the time it takes for the transient current to make a 63.2% change in value. After five time constants (5τ), the change is 99.3% complete. As shown in **Figure 4.73**, the inductance does not prevent the current from changing, it delays the change. After a time interval equal to five time constants, the current flow equals its ohmic value.

Returning energy to the circuit

When the supply current stops, the stored energy in the electromagnetic field starts to decay. **Figure 4.74** illustrates this decay using the same circuit used to store energy within the electromagnetic field.

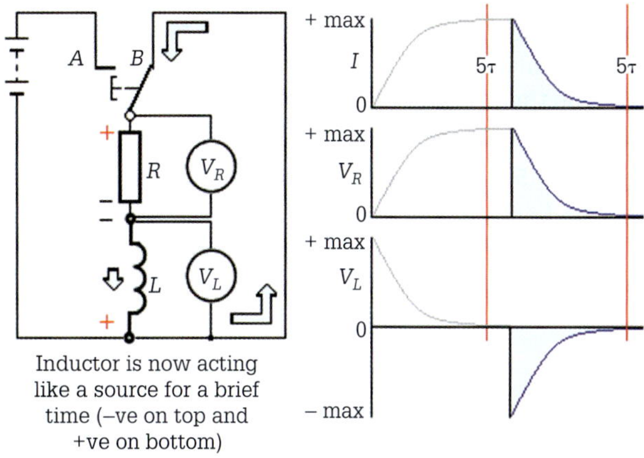

FIGURE 4.74 Energy released from the inductor

When switch *A* closes to position *B*, the circuit switches to 'field discharge' path, and the electromagnetic field begins to decay. Because the inductor is now a current source, the induced emf has a polarity opposite to the polarity it had when first created. However, the current still varies, in the same way, as when the electromagnetic field was building.

At the instant the switch closes to position *B*, both the current and the induced emf are at their maximum value. As the electromagnetic field discharges its stored energy into the circuit both the induced emf and the current fall at the same rate, finally reaching zero current and zero induced voltage. **Figure 4.75** shows the decay of induced emf and circuit current with time for the circuit.

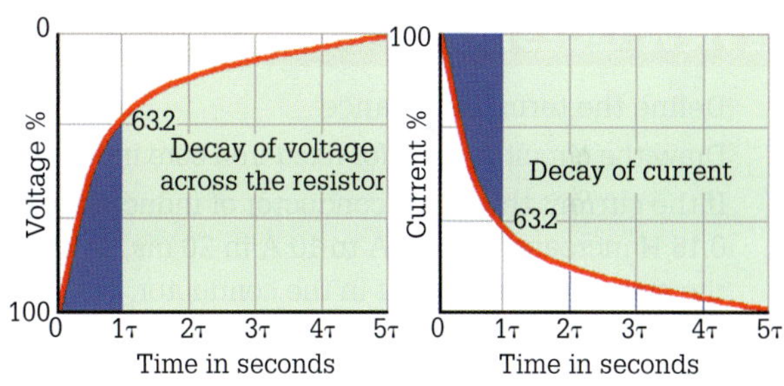

FIGURE 4.75 Decay of voltage and current with time

Note that the induced emf graph shows a falling and not a rising voltage. The graph is shown this way because the induced emf across the inductor has reversed its polarity. The current graph occurs in this manner because the current's direction has not changed.

EXAMPLE 4.14

The series-field windings of a 120 V d.c. series generator have an ohmic resistance of 50 Ω and an inductance of 5 H. Determine:

a the time constant

$$\tau = \frac{L}{R} = \frac{5}{50} = \mathbf{100\,ms}$$

»

b the time it will take for the current to attain its full value

The time it will take to reach its final value is five time constants (5τ):

$$5 \times 0.100 = 0.500 \text{ seconds (500 ms)}$$

c the final value of current

Calculation of the approximate final magnitude of current after 500 ms uses Ohm's law.

$$I = \frac{V}{R}$$

$$= \frac{120}{50}$$

$$= \mathbf{2.4\,A}$$

Energy stored in an electromagnetic field

Energy is stored in the electromagnetic field during the build-up of current and releases when the current decays. The magnitude of the energy stored depends on the inductance of the circuit and the square of the circuit current multiplied by a factor of 0.5.

$$W = \tfrac{1}{2}LI^2$$

where W = energy stored in joules (J)
L = inductance in henrys (H)
I = the value of current in amperes (A)

EXAMPLE 4.15

Determine the energy stored within the electromagnetic field when a current of 5 A flows through a coil having an inductance of 1.5 H.

$$W = \tfrac{1}{2}LI^2$$

$$= \tfrac{1}{2} \times 1.5 \times 5^2$$

$$= \mathbf{18.75\,J}$$

EXERCISE 4.11

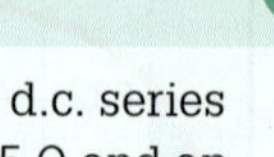

1 The series-field windings of a 200 V d.c. series motor have an ohmic resistance of 25 Ω and an inductance of 2 H. Determine:
   a the time constant
   b the time it will take for the current to attain its full value
   c the final value of current
2 Determine the energy stored within the electromagnetic field of a coil when a current of 2 A flows through the coil having an inductance of 2.5 H.

REVIEW QUESTIONS

1 Define the term 'inductance'.
2 Draw the circuit symbol for fixed iron core inductor.
3 If the current through a conductor of inductance of 0.15 H increases from 0 A to 10 A in 20 ms, calculate the average induced emf in the conductor.
4 Define the term 'mutual induction'.
5 What is an undesirable effect of self-induction?
6 A 100 mm long coil has 1000 turns, a CSA of 150 mm^2 and the iron core has an absolute permeability of 6×10^{-3} H m^{-1}. Calculate the inductance of the coil.
7 The series-field windings of a 120 V d.c. series generator have an ohmic resistance of 150 Ω and an inductance of 0.5 H. Determine the time constant.
8 Determine the energy stored within the electromagnetic field when a current of 25 A flows through a coil having an inductance of 0.5 H.

4.7 Magnetic devices

Magnetostriction devices

Magnetostriction is a magnetic property found only in ferromagnetic materials such as iron, nickel, cobalt and their alloys. These ferromagnetic materials when exposed to a magnetic field undergo a reorientation of their molecular structure, which causes a change in their physical dimensions. Magnetostriction materials exhibit small reversible strains and changes in their physical properties when exposed to a magnetic field. Refer to **Figure 4.76**.

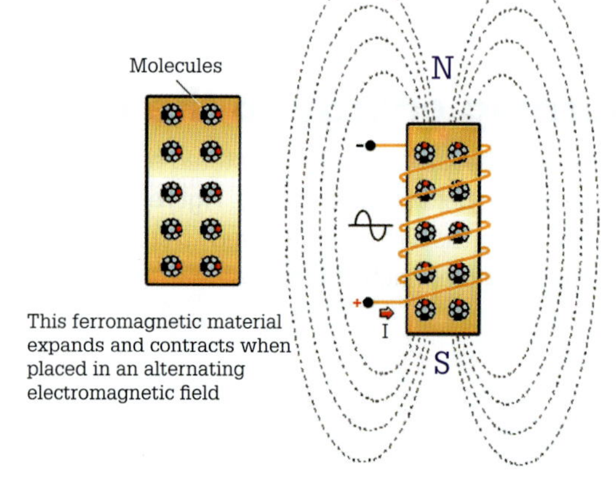

FIGURE 4.76 Magnetostriction

In ferromagnetic materials, an applied magnetic field causes rotation of molecules towards the magnetic field direction. This reorientation causes a small change in the length of the magnetostriction material. Consequently, when the molecules are completely aligned, saturation occurs and the applied magnetic field can produce no further magnetostriction.

Magnetostriction materials make devices such as actuators (stepper motor), transducers (ultrasonic cleaning) and sensors (stress and pressure sensors). **Figure 4.77** shows a block diagram of the energy transformation in magnetostriction materials. When the magnetostriction material is used as a sensor it is stressed by a force and produces a magnetic field. When the material is used as an actuator and is stressed by a magnetic field it produces a force.

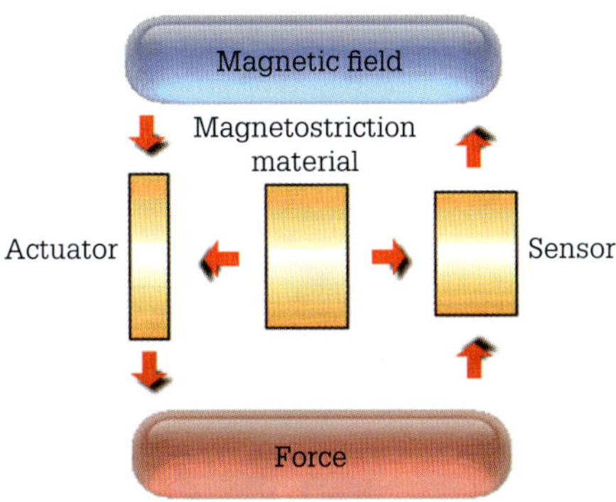

FIGURE 4.77 Block diagram of the energy transformation in magnetostriction

Magnetostriction sensors are used for currency validation, in vending machines, navigation, object detection, speed of moving parts and in micro-electromechanical machines.

Magnetic sensing devices

Most conventional sensors directly measure a physical property such as temperature, flow rate, strain, light or pressure. These types of sensors provide an output signal that directly indicates the state of the physical property.

Magnetic sensors such as those that use the Hall Effect (proximity and antilock brake sensors), do not directly measure a physical property. They detect changes or instability in magnetic fields and from these changes obtain information on properties such as rotation, direction, angle, presence or electric current. The output signal from magnetic sensors requires signal processing to indicate the state of the property detected. Refer to **Figure 4.78**.

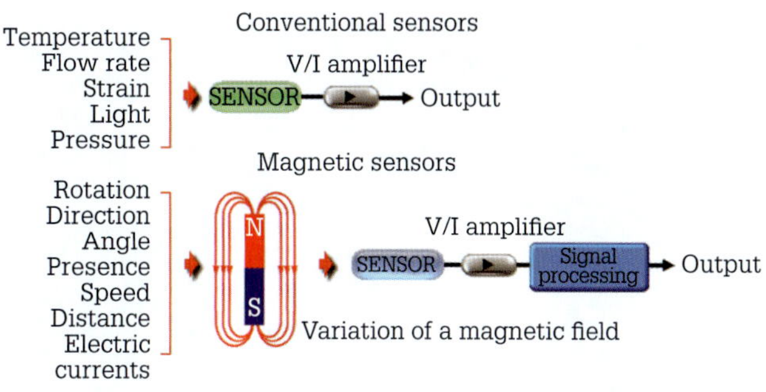

FIGURE 4.78 Conventional and magnetic sensors

Magnetic methods employed to extinguish switching arc

An arc is formed between the opening contacts associated with circuit breakers, contactors and other similar high-current switching devices when carrying current. As the arc is the flow of electrons, it has magnetic properties. It is possible therefore to use a device known as a magnetic blowout coil to push the arc into another device known as an arc chute.

Arc chute

The electric arc (ionised air) is an effect of current. In switching devices the arc struck between contacts needs extinguishing. A standard method used to extinguish an arc is by the use of an arc-extinguishing device. An arc-extinguishing device consists of a number of metal plates called an arc grid. When the arc, under the influence of its own magnetic field, enters the grid it rapidly stretches and is then cooled by the metal plates and extinguished.

Blowout coil

In some circuit breakers or contactors, the rapid extinction of the arc occurring between the separated contacts increases by means of magnetic blowout. The purpose of blowout coils, as illustrated in **Figure 4.79**, is to create a magnetic field across the contacts in a contactor or circuit breaker with the intention of lengthening and extinguishing the electromagnetic arc formed as the contacts open to interrupt the current. The interaction of the electromagnetic fluxes causes the arc to be repelled up and away from the electromagnetic field of the blowout coil. The expanded arc draws into an arc chute where it extinguishes.

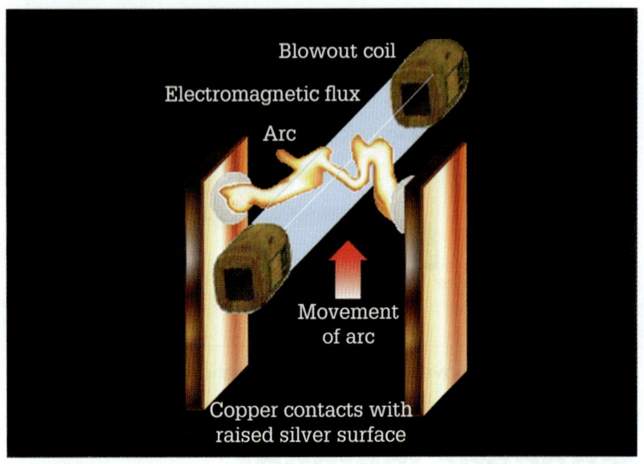

FIGURE 4.79 Movement of arc in a blowout coil

Applications of electromagnetism and inductance

Many devices make use of the effects of electromagnetism and inductance. These include generators, motors, solenoids, relay coils, transformers, circuit breakers, moving-coil and moving-iron ammeters and voltmeters, the write and read heads on video recorders and hard disk drives, loudspeakers and dot-matrix printers.

Permanent magnet generator

Because a permanent magnet generator has a stationary field and a revolving armature, the permanent magnet generators are commonly used in wind energy technologies. The development of better permanent magnetic substances has enabled the field for their application to increase. For small wind and hydropower generators, the use of permanent magnet generators is essential. These generators use a toroidal stator that incorporates a brushless neodymium nickel-plated permanent magnet producing a radial flux together with an airfoil, thus capturing the most energy available.

Tachogenerators

Tachogenerators are other devices that use the effects of electromagnetism and inductance. These devices enable the control of machinery where precise rotation speeds occur. The tachogenerator takes the form of a small electric motor. In addition, the device can be directly coupled, in-line, via a flexible coupling to a driven spindle, or it can be belt driven by means of a timing belt and pulley arrangement.

A tachogenerator operates using the process of inducing an electromotive force by a permanent magnetic circuit into the windings of an iron-cored rotor while this is revolving. The result is that the voltage output at the terminals of the tachogenerator is a d.c. voltage that is a precise function of rotation speed.

Consequently, this voltage provides a speed feedback signal for the d.c. power drive or a.c. inverter powering the main motor. In this way, accurate maintaining of the motor speed occurs.

Squirrel cage motor

A squirrel cage motor consists of a rotor and stator winding. The stator winding consists of a number of poles through which a.c. flows, thus causing an induced electromotive force to rotate at synchronous speed. In addition, the number of poles depends upon the electrical build – two-, four-, six- or eight-pole motors are available; however, the most common machines are four poles, with synchronous speed of 1500 rpm when supplied at 50 Hz.

Brakes and clutches

Disc friction brakes, as shown in **Figure 4.80**, apply the principles of friction. These brakes stop and hold loads. Some of the most common applications include cranes, conveyors and hoists.

FIGURE 4.80 Disc friction brake

There are three types of non-friction electric clutches and brakes – eddy current, hysteresis and magnetic particle. Applications requiring variable slip use these types of electromechanical devices. The devices employ electromagnetic attraction, rather than friction as in the shoe brake, to carry out their function.

Eddy current brake

An eddy current is an opposing circulating current set up in a conductor in response to a changing magnetic field. By Lenz's law, the current circulates in such a way as to create a magnetic field opposing the change. Because of this effect, eddy currents cause energy loss. More accurately, eddy currents transform kinetic energy into heat. In many applications, the loss of useful energy is not particularly desirable, but there are some practical applications: one is in the brakes for both crane and hoist machinery.

During braking, a metal rotor has eddy current generated on it, the result of a magnetic field. The magnetic interaction between the applied field and the eddy current fields act to slow the disc down. Consequently, the faster the disc is turning, the stronger the effect. In other words, as the crane or hoist slows; the braking force is reduced, producing a smooth stopping motion. Refer to **Figure 4.81**.

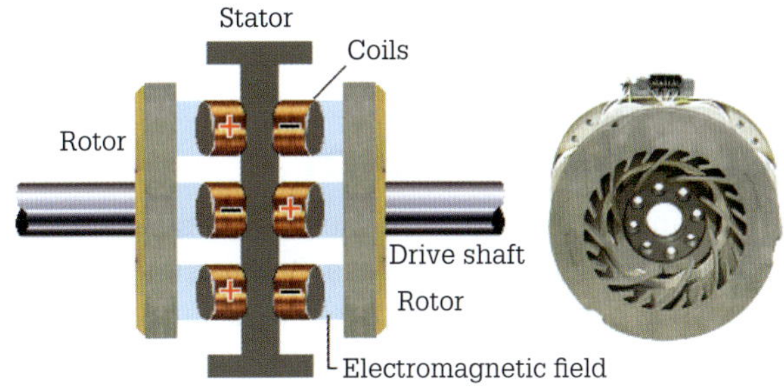

FIGURE 4.81 Eddy current brake

The eddy current clutch consists of two rotating devices employing the basic operation of a simple generator: a cylinder driven at constant speed by the a.c. motor and, within the cylinder, a rotor that connects to the load. In addition, torque is transmitted from the cylinder to the rotor through an adjustable magnetic field that attracts the two devices. As the rotor revolves, eddy currents generate. So, the faster the rotor turns, the more eddy currents develop, and increased opposition occurs. For this reason, a linear correlation exists between speed and torque.

Hysteresis brake

The hysteresis brake, as shown in **Figure 4.82**, provides torque through the use of two basic components – a pole structure divided into mesh-like compartments and a specially designed steel rotor assembly – fitted together, but not in physical contact with each other. Until the pole structure energises, the rotor can spin freely on its shaft bearings. However, when a magnetising force from the field coil is applied to the pole structure, the air gap becomes saturated magnetically, which in turn restrains the rotor, providing a braking action between the pole structure and rotor.

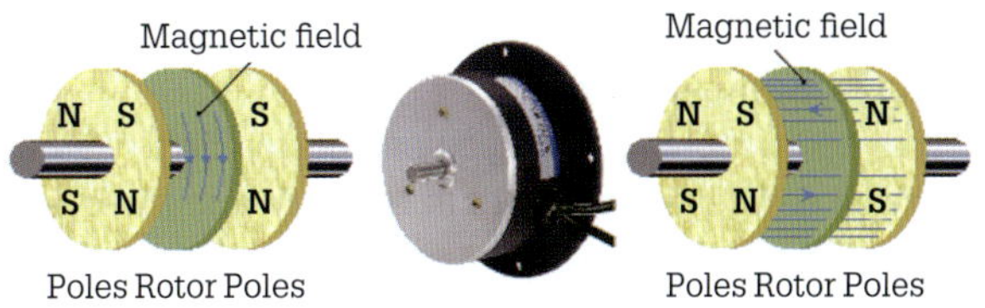

FIGURE 4.82 Hysteresis brake

When poles of opposite polarity face each other minimum saturation of the rotor occurs, as the magnetic lines of the force permeate straight through the rotor, and the shaft is free to turn. However, when similar poles are opposite each other it provides maximum magnetic saturation of the hysteresis rotor. Magnetic lines of force travel circumferentially through the rotor, thus producing maximum braking.

Magnetic particle brake

The magnetic particle brake, as shown in **Figure 4.83**, consists of a rotor disc contained within the cavity, an excitation coil and magnetic powder. In addition, the cavity in between the excitation coils contains a fine, dry stainless steel powder. When the supply voltage is off, the stainless steel powder sits unbound in the cavity; however, when an exciting voltage is applied to the coil, magnetic flux develops. Because of this, current flow tries to bind the stainless steel powder together.

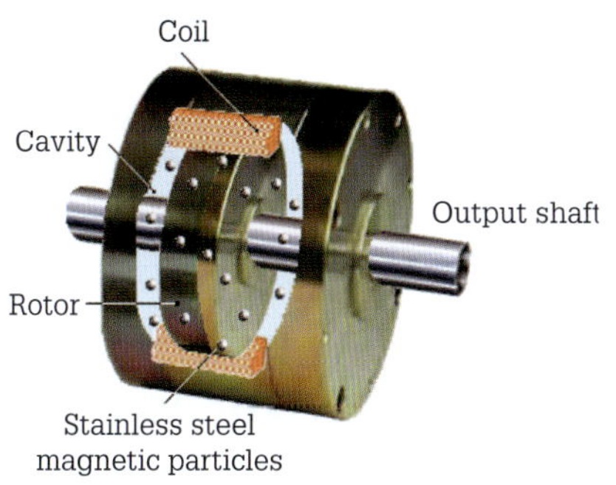

FIGURE 4.83 Magnetic particle brake

As the current increases, the binding of the stainless steel powder becomes stronger. At the same time, the brake rotor passes through these bound powder particles. As the individual powder particles start to bind together, they form chains along the magnetic field lines, linking the rotor disc to the housing, enabling a resistant force to develop on the rotor, slowing and eventually stopping the output shaft. When the exciting voltage turns off, the shaft is free to turn.

Measuring instruments

The moving-coil and moving-iron meter movements are covered in **Section 3.4 Meters in a circuit**. Both of these meter movements utilise magnetism in their operation.

Wattmeter

The wattmeter, as illustrated in **Figure 4.84**, is a dynamometer. It measures a.c. power (symbol *P*, unit watt, symbol W) but is itself rated in volts and amperes. Power is the product of voltage and current and power factor. A wattmeter measures both voltage and current, but not power factor. In addition, a wattmeter is similar to the moving-coil meter, having a moving coil attached to a pointer, but has two fixed electromagnets instead of permanent magnets to create a magnetic field.

The two fixed electromagnets connect in series with each other (called the current coils) and with a load so they are energised by the load current in the circuit. Regarding the moving coil (voltage coil), it connects in parallel with the load and is therefore load voltage operated.

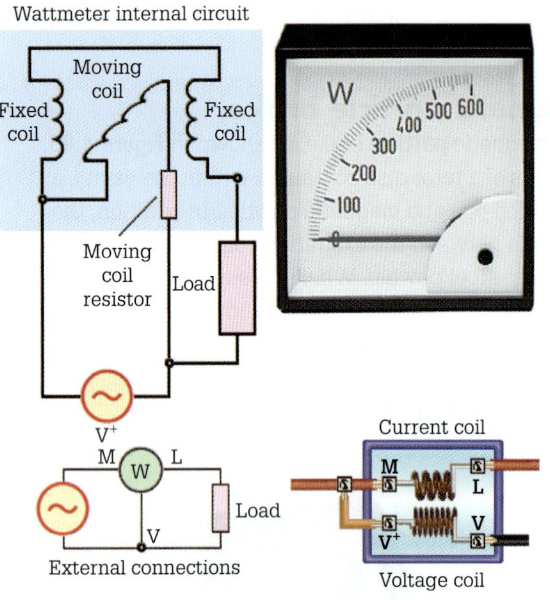

FIGURE 4.84 Wattmeter

Induction kilowatt-hour meter

Kilowatt-hour meters measure the power expended over a period. The kilowatt-hour meter, as illustrated in **Figure 4.85**, consists of a motor (split-phase induction motor) where the speed of the rotating part (eddy current disc) is proportional to the measured power. There is also a counting mechanism connected to the motor via gearing. The eddy current disc provides a load for the motor. When the disc rotates in a magnetic field, induced eddy currents occur in the disc. Consequently, the eddy currents produce fluxes of their own that oppose the motion that produced the eddy currents.

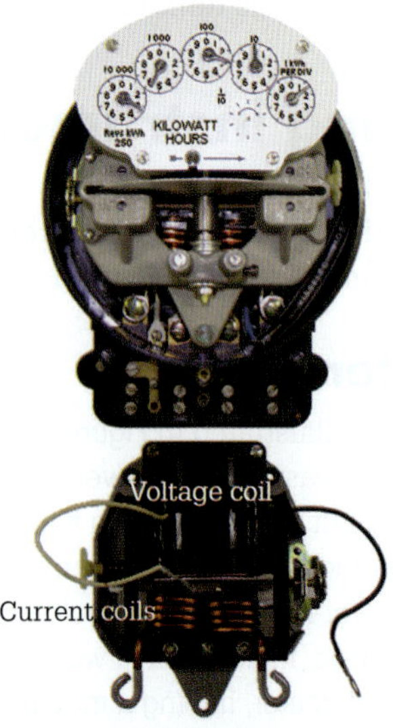

FIGURE 4.85 Kilowatt-hour meter

The motor element has two windings: a current coil (small number of turns, virtually non-inductive) connected in series with the load, and a high-impedance voltage coil (large number of turns) connected across the supply. This feature allows the flux from the voltage coil to lag 90° behind the flux from the current coil when the power factor of the load is unity. The driving torque resulting from the interaction of all the fluxes is directly proportional to the product of volts times amperes times power factor, which is the true power delivered to the load.

Unwanted effects of inductance and electromagnetism

A significant amount of the electrical power generated supplies alternating current motors. Because the windings of alternating current motors possess inductance, the magnetic flux produced within the windings increases and decreases. As the magnetic flux decreases, the energy stored in the electromagnetic field releases and an induced current surges along the supply cables.

In order to create a magnetic field, the motor requires inductive power and this power does no useful work. However, the current required to produce this inductive power can be relatively high, and surges backwards and forwards along the supply cable, creating unwanted effects.

SWITCH ON

Hazards and safety precautions

Inductors have the capacity to release stored energy at a much higher voltage than that used to produce the stored energy. Consequently, release of this energy could occur if the inductors have no proof that they are de-energised.

Electric motors must to come to a full stop after switching off before isolation procedures occur. A full stop is important because of the residual voltage that exists across the motor windings.

Inductors can produce large eddy currents in adjacent conductive material if suddenly de-energised, causing excessive heating.

In industry and commercial workplaces, inductive equipment and machines require additional current from the supply lines for their operation. This current causes a higher voltage drop across the supply cables and extra power loss in the supply lines themselves due to the I^2R heating of the supply cables.

This additional inductive power calls for higher-rated protection devices such as circuit breakers and fuses. In addition, control devices such as switches would need replacing, as their current rating may no longer be of a suitable value. The circuits feeding the inductive loads would need larger cross-sectional cables installed.

As a result, the circuit cables carry both the inductive and working current with ease.

Where workplaces have their own transformer, the additional inductive power load could overload the transformer. In order to eliminate the effects of overloading a larger transformer, higher-rated protection and control devices would be required.

Fluorescent lighting systems with ballasts, induction motors on low load and welding machines all require more inductive power than their working power output.

Manufacturer specifications

Two common electromagnetic devices are the relay and contactor, which are available in a vast variety of ratings and configurations. It is important to be able to competently read the manufacturer's data to ensure selection of an appropriate device.

Relay selection

Selection of an appropriate relay for a particular application requires evaluation of many different factors:

- Number and type of contacts – normally open, normally closed, double-throw.
- There are two types – this style of relay can be manufactured two different ways: 'make before break' and 'break before make'. The old-style telephone switch required make-before-break so that the connection didn't get dropped while dialling the number. The railroad still uses them to control railroad crossings.
- Rating of contacts – small relays switch a few amperes, large contactors are rated for up to 3000 amperes, a.c. or d.c.
- Voltage rating of contacts – typical control relays rated 300 VAC or 600 VAC, automotive types to 50 VDC, special high-voltage relays to about 15000 V.
- Coil voltage – machine-tool relays usually 24 VAC or 120 VAC, relays for switchgear may have 125 V or 250 VDC coils; 'sensitive' relays operate on a few milliamperes.
- Package/enclosure – open, touch-safe, double-voltage for isolation between circuits, explosion proof, outdoor, oil-splash resistant.
- Mounting – sockets, plug board, rail mount, panel mount, through-panel mount, enclosure for mounting on walls or equipment.
- Switching time – where high speed is required.
- 'Dry' contacts – when switching very low level signals, special contact materials may be needed such as gold-plated contacts.
- Contact protection – suppress arcing in very inductive circuits.
- Coil protection – suppress the surge voltage produced when switching the coil current.
- Isolation between coil circuit and contacts.
- Aerospace or radiation-resistant testing, special quality assurance.
- Expected mechanical loads due to acceleration – some relays used in aerospace applications are designed to function in shock loads of 50 *g* or more.
- Accessories such as timers, auxiliary contacts, pilot lamps, test buttons.
- Regulatory approvals.
- Stray magnetic linkage between coils of adjacent relays on a printed circuit board.

Table 4.6 illustrates some of the information you can expect to see in the manufacturer's data for a relay.

TABLE 4.6 Sample manufacturer's data – Relay

General purpose relay GPR-001			
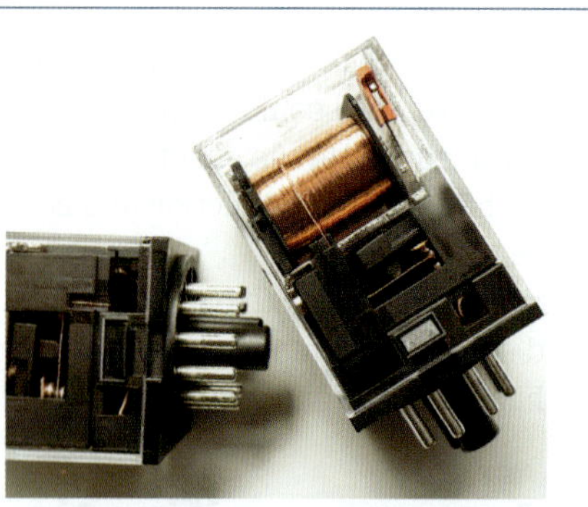			
Description			
Reliable general purpose relay featuring mechanical indicator / pushbutton.			
Capable of breaking fairly large current despite small size.			
Long-life silver contacts ensuring reliable operation (minimum 100 000 electrical operations)			
In-built operation indicator, pushbutton and diode surge protection.			
Conforms to CENELEC standards.			
Terminal	Plug-in	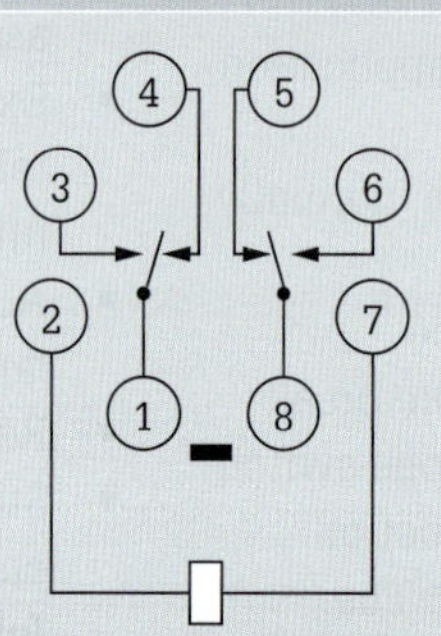	
Contact form	DPDT		
Coil rating			
Rated voltage	240 V a.c. 50 Hz	**Rated current**	10.5 mA
Must operate voltage	80% max of rated voltage	**Must release voltage**	30% min of rated voltage
Maximum voltage	90–110% of rated voltage	**Power consumption**	Approx. 3 VA
Contact rating			
Load	Resistive (Cos Φ = 1)	Inductive (Cos Φ = 0.3)	
Contact material	Ag		
Rated Load	10 A	7 A	
Maximum switching voltage	250V		
Maximum switching power	2500 VA	1750 VA	
Characteristics			
Contact resistance	50 mΩ max		
Operate time	18 ms max		
Release time	20 ms max		
Max operating frequency	Mechanical: 15 000 operations/hour Electrical: 1500 operations/hour (under rated load)		
Insulation resistance	100 MΩ min (at 50–0 V d.c.)		
Dielectric strength	2500 V a.c., 50 Hz for 1 min between coil and contacts 1000 V a.c. for 1 min between contacts and terminals of same polarity		
Endurance	10 000 000 operations minimum (at operating frequency of 18 000 operations per hour)		
Ambient temperature	Operating −10 °C to 40 °C		
Ambient humidity	Operating 5% to 85%		
Weight	Approx. 90 g		

Source: Shutterstock.com/

Contactor selection

The utilisation categories allow for initial selection of a device that can meet the demands of the purpose the motor is designed for. However, there are certain other constraints to consider. These are all the factors which have nothing to do with the purpose itself, such as climatic conditions (temperature, humidity), geographical setting (altitude, salt mist) and so on.

In certain situations, the reliability of the equipment can also be a critical factor, especially if maintenance is difficult. The electrical life (durability of contacts) of the device (contactor) therefore becomes an important feature. It is therefore necessary to have detailed and accurate catalogues to ensure the product chosen complies with all these requirements.

Contactors are characterised by one or more operating conditions such as:

- currents
- voltages
- power factor or time constant
- circuit making and breaking conditions
- type of load (squirrel cage motor, brush motor, resistor)
- conditions in which making and breaking take place (motor running, motor stalled, starting process, counter-current breaking and the like)

Table 4.7 illustrates some of the information you can expect see in the manufacturer's data for a contactor.

TABLE 4.7 Sample manufacturer's data – Contactor

Motor Contactor MC-001	Source: Shutterstock.com/Wongsakorn Napaeng
Rating	AC-3
Coil voltage	230 V ac
Number of poles	3
Contact current rating	9 A
Power rating	4 kW
Range	MotStart
Series	DILM
Normal state configuration	3NO
Contact voltage rating	400 V ac
Auxiliary contacts	2 [1 × NO, 1 × NC]
Terminal type	Screw
Max operating temperature	+40 °C
Min operating temperature	−25 °C
Width	45 mm
Length	68 mm
Depth	117 mm
Pick-up voltage	0.8–1.1 × nominal
Drop-out voltage	0.3–0.6 × nominal
Power consumption pick-up	24 VA
Power consumption sealing	3.4 VA
Closing delay	15–20 ms
Opening delay	10–20 ms

REVIEW QUESTIONS

1 What is magnetostriction?
2 Name an application for magnetostriction materials.
3 Name three properties measured by magnetic sensors.
4 What is the purpose of a magnetic blowout coil?
5 Name four devices that make use of effects of electromagnetism and inductance.
6 List the three types of non-friction electric clutches and brakes.
7 What type of measuring is an analogue wattmeter?
8 What is a major hazard associated with switching inductive loads?

CHAPTER REVIEW

4.1 Magnetism

- Magnets in some way influence all substances but the extent of magnetisation varies greatly with the kind of substance involved.
- Magnetic substances have three classifications: diamagnetic, paramagnetic and ferromagnetic.
- Like poles repel each other, while unlike poles attract each other.
- Magnetic lines of force possess six characteristics.
- To prevent the loss of magnetism in a bar magnet, use a piece of soft iron called a keeper across opposite magnetic poles.
- Highly permeable materials called shields conduct flux lines around a shielded object.

4.2 Electromagnetism

- A magnetic field exists whenever electric charges are empowered to move.
- The direction of the magnetic field around the conductor may be determined by using the right-hand grip rule.
- There are four main characteristics of a magnetic field around a conductor.
- The right-hand solenoid rule is used to determine the polarity of the solenoid's magnetic field.
- A force exists between parallel current-carrying conductors.
- The magnetomotive force is the force that establishes and maintains the magnetic field in the magnetic circuit.

4.3 Magnetic circuit and nomenclature

- The magnetising force, or the field strength or field intensity, is the magnetic force drop per metre length of a consistent magnetic circuit.
- The permeability of a substance is the measure of the ease with which magnetic lines of force pass through a substance.
- Magnetic flux is the total number of lines of force in a magnetic circuit.
- Flux density is the number of lines of force per square metre of cross-sectional area of the magnetic circuit.
- Reluctance is the opposition a substance offers to the passage of flux in a magnetic circuit.

4.4 Losses in magnetic circuits

- Three characteristics of ferromagnetic materials can be shown graphically and are the magnetisation curve, the permeability curve and the hysteresis loop.
- Permeability of a ferromagnetic material is its magnetic conductivity.
- In the *B/H* curve, a small loop area indicates a low hysteresis loss, while a large hysteresis loop indicates a 'hard' ferromagnetic material.
- Hysteresis loss is an iron loss. Magnetic leakage means that the effect of the ampere-turns is limited because the total flux generated is never 100 per cent effective.
- Magnetic fringing describes the spreading of the flux lines through the air gap and along the sides of a magnetic circuit.

4.5 Electromagnetic induction

- When a conductor moves through a magnetic field and cuts the magnetic field at right angles, a small electromotive force is induced between the ends of the conductor.
- The direction of the induced emf depends on either the polarity of the magnetic field or the direction of motion of the conductor as it cuts through the magnetic lines of force.
- An induced voltage occurs while there is relative motion of either the conductor or the magnetic field.
- If a bar magnet moves into a solenoid, an electromotive force is induced in the conductors that make up the solenoid.
- The direction of the induced emf is always such that it sets up an induced current opposing the motion or change of magnetic flux responsible for inducing the emf.
- Bifilar winding (meaning two filaments) is a method of manufacturing a coil.

4.6 Inductance

- Inductance is a feature that makes itself evident by opposing the starting, stopping or changing of current in a circuit.
- Self-induction is the property of a circuit whereby a change in current causes a change in voltage in the same circuit.
- When one energised straight conductor or coil induces an emf in a second straight conductor or coil, it is mutual inductance.
- The principal method of classifying inductors is according to the type of core material.
- The time delay experienced in the build-up of current directly relates to the combination of the ohmic resistance and the inductance in the circuit and may be in the order of microseconds, seconds or even minutes.
- Energy is stored in the electromagnetic field during the build-up of current and releases when the current decays.
- The magnitude of the energy stored depends on the inductance of the circuit and the square of the circuit current multiplied by a factor of 0.5.

4.7 Magnetic devices

- A common method used to extinguish an arc is by the use of an arc chute.
- Blowout coils develop a magnetic field across the contacts in a contactor or circuit breaker.

- An eddy current is an opposing circulating current set up in a conductor in response to a changing magnetic field.
- The hysteresis brake provides torque by the use of two basic components.
- The magnetic particle brake consists of a rotor disc contained within a cavity, an excitation coil and stainless steel powder.
- The current required to produce inductive power can be relatively high, and surges backwards and forwards along the supply cable, creating unwanted effects.
- Kilowatt-hour meters measure alternating current expended over a period.

TRIAL EXAM

For Chapter 4 knowledge assessment, please complete the following trial exam.

1 The popular name for magnetite is:
   a magnetic substance
   b ferromagnetic
   c alnico
   d lodestone

2 An example of a ferromagnetic substance is:
   a copper
   b silver
   c iron
   d platinum

3 The induced pole adjacent to the pole of a permanent magnet has:
   a the same polarity as the magnet pole
   b an opposite polarity to the magnet pole
   c a stronger magnetic flux than the magnet pole
   d few domains

4 Ferromagnetic materials consist of many tiny regions called:
   a keepers
   b boundaries
   c alignments
   d domains

5 A magnet may be created in soft iron by
   a stroking or placing in an energised coil
   b tapping firmly with a magnetic material
   c placing a magnetic keeper on it
   d heating to 400 °C

6 Magnetic lines of force:
   a are assumed to emanate from the north pole and enter the south pole
   b merge with or cross other lines of force
   c are a scalar quantity
   d that are parallel and act in the same direction attract one another

7 To prevent the loss of magnetism, when a bar magnet is not in use:
   a a magnaprobe is used
   b a keeper is used
   c a magnetic shield is used
   d a reluctance material is used

8 Remanence is:
   a lines of force not used for a magnetic application
   b the magnetism left in soft magnetic material
   c the heat remaining in a material when subjected to a magnetic field
   d a reluctance material

9 Magnetic screening uses:
   a keepers
   b highly permeable material
   c magnetically hard substances
   d magnetic insulators

10 A magnaprobe:
   a prevents the loss of magnetism
   b redirects magnetic fields
   c measures the lines of force
   d detects stray fields or magnetic leakage

11 Magnetically soft materials:
   a are easily magnetised and retain their magnetism for periods
   b are easily magnetised but lose their magnetism within a short period
   c are hard to magnetise and lose their magnetism over time
   d are hard to magnetise and retain their magnetism

12 The convention used to show the direction of electric current in a conductor:
   a a + and a – sign
   b the letters N and S
   c a cross and a dot
   d the letters A and N

13 When using the right-hand grip rule, the thumb points in the direction of the:
   a magnetomotive force
   b south pole
   c magnetic field
   d conventional current flow

14 In the right-hand solenoid rule, the fingers point in the direction of:
   a magnetising force
   b south pole
   c magnetic field
   d conventional current flow

15 A solenoid and an iron core creates:
- a a coercive force
- b an electromagnet
- c high magnetic leakage
- d a magnetomotive force

16 The strength of the magnetic field that surrounds a current carrying conductor is proportional to:
- a length of the conductor
- b the direction of current flow
- c the magnitude of the current
- d diameter of the conductor

17 What is the magnetomotive force developed by 5000 turns of copper conductor carrying a current of 100 mA?
- a 0.005 At
- b 16.67 At
- c 500 At
- d 50 000 At

18 The SI unit of measurement used for magnetising force is:
- a henrys per metre
- b ampere-turns
- c ampere-turns per metre
- d tesla

19 The SI unit of measurement for magnetic flux is:
- a henrys per metre
- b tesla
- c weber
- d henry

20 Reluctance is:
- a the opposition a substance offers to the establishment of magnetic flux
- b the total number of lines of force in a magnetic circuit
- c directly proportional to the density of the magnetic substance
- d the ease by which magnetic flux is established in a material

21 The reluctance of a magnetic circuit is inversely proportional to:
- a the length of the magnetic circuit
- b magnetic flux
- c the permeability of the magnetic substance
- d flux density

22 A *B/H* curve indicates:
- a the variation of magnetic flux with magnetic circuit length
- b the variation of inductance with electromotive force
- c the variation of magnetomotive force with current
- d the variation of flux density with magnetisation force

23 In a *B/H* curve, a ferromagnetic soft substance has:
- a high magnetic leakage
- b a small loop area
- c a high coercive force characteristic
- d an application as a permanent magnet

24 Laminations reduce the effect of:
- a magnetic reluctance
- b magnetic fringing
- c magnetic leakage
- d eddy currents

25 Magnetic leakage:
- a increases the effectiveness of the total flux generated
- b causes the attraction of lines of flux acting in the same direction
- c increases the efficiency of the magnetic device
- d causes wastage of ampere-turns

26 The spreading of the flux lines through the air gap and along the sides of a magnetic circuit is:
- a coercive force
- b magnetic fringing
- c magnetic leakage
- d magnetic conductivity

27 An air gap is included in a magnetic circuit to:
- a increase the flux density
- b increase the number of lines of force
- c increase the reluctance
- d increase the effect of the ampere-turns

28 Electromagnetic induction is summarised in:
- a flux density
- b generator action
- c Lenz's law
- d Faraday's law

29 What type of winding is used to increase magnetic coupling?
- a series
- b bifilar
- c field
- d compensating

30 In a coil the effect of inductance at the instant of energisation is to:
- a reduce the magnitude of the induced emf
- b delay the establishment of current
- c enhance the effect of the driving voltage
- d increase the ohmic resistance of the coil conductors

31 The series-field windings of a 100 V d.c. series generator have an ohmic resistance of 150 Ω and an inductance of 10 H. Determine the time constant.
- a 15 s
- b 6.67 s
- c 66.7 ms
- d 0.667 s

32 The SI unit of inductance is:
- a henry
- b ampere-turns
- c tesla
- d joule

33 If the current through a conductor having an inductance of 0.2 H increases from 0 A to 5 A in 40 ms, calculate the average emf induced in the conductor.

a −25 V
b 250 V
c 1 V
d 100 V

34 If all other factors remain constant, the inductance of a coil can increase fourfold by:

a increasing the length of the coil
b increasing the CSA of the coil conductor
c doubling the number of turns
d using an air core

35 A hazard that can exist with inductors is their:

a capacity to release stored energy at a much higher voltage than that used to create it
b bleed resistor
c magnetic property
d surge-suppressing diodes

36 Magnetostriction is a:

a transducer
b magnetic device
c sensor
d magnetic property

37 Magnetic sensors:

a measure light
b directly measure a physical property
c detect changes or instability in magnetic fields
d determine pressure

38 A common magnetic sensor is:

a Hall Effect device
b transformer
c solenoid
d generator

39 A blowout coil is used to

a protect a motor from excessive load
b provide protection for the supply
c extinguish an arc
d cool the motor windings

40 An eddy current is an opposing circulating current set up in a conductor in response to a:

a counter-electromotive force
b kinetic energy
c transmitted torque
d changing magnetic field

5 Direct current (d.c.) rotating machines

This chapter provides electrotechnology workers with knowledge of the correct operation of direct current (d.c.) machines. This chapter provides underpinning knowledge for the unit UEEEL0019 from the UEE training package.

LEARNING OBJECTIVES

Construction of direct current machines

- Explain the function of various components of direct current machines

Direct current generator

- State the principle of operation of direct current generators
- Explain the meaning of and calculate the voltage regulation for a generator

Types of direct current generators

- State the various methods of excitation used in generators
- Explain the voltage and current relationships for various generators
- Draw equivalent electrical circuit for various generators
- Calculate generated voltage

Direct current motor

- Explain how torque is produced in a motor
- Apply Fleming's motor rule to a simple motor

Types of direct current motors

- Outline uses for various d.c. motors
- Recognise various circuit diagrams and connections for d.c. motors

Specialty direct current machines

- Outline uses for various speciality d.c. motors and explain the operation

Machine efficiency

- Determine machine efficiency
- Explain the meaning of maximum efficiency

Machine maintenance and testing

- Outline common machine testing procedures
- Describe applicable safety measures when maintaining d.c. machines.

5.1 Construction of direct current machines

Direct current (d.c.) machines comprise generators and motors. Direct current generators and motors are similar in structure. Consequently, with few adjustments, they can interchange. A d.c. generator converts mechanical energy into electrical energy when driven by a prime mover and may be made to function as a motor by applying a voltage across the output terminals.

Solid-state devices that convert alternating current to direct current for application in d.c. drive systems are replacing the d.c. generator.

A d.c. motor converts electrical energy back into mechanical energy and is the most comprehensive of all types of electric motors. Its speed easily adjusts in small steps, ranging from standstill to rated full-load speed and above.

The major components of the d.c. machine are:

- frame
- armature (comprising shaft, laminated core and commutator)
- brushes and brushgear
- poles (comprising pole core and windings).

Frame construction

A rolled-steel frame is standard with d.c. motors and has the rolled frame butt welded with the feet located where needed. Cast frames use grey iron or customised alloys of nickel, molybdenum and vanadium. However, larger d.c. motors may have a laminated steel frame or consist of a split-frame construction.

A laminated frame structure allows the electromagnetic flux to flow more smoothly when changes occur in the load current. This results in better commutation (less sparking at the brushes). With split-frame construction, the top half of the machine removes, allowing for inspection of all the components and their cleaning without interference with the alignment of the motor itself.

The design of all d.c. motor frames is for optimal electromagnetic-flux-carrying capability, together with the mechanical strength to withstand the distortion due to flux and the transfer of the torque forces to the motor foundations.

Armature

Shaft

An armature steel shaft is thicker in the middle than at the ends. Its construction enables the shaft to withstand the bending effects that the weight of the armature, the electromagnetic side-pull and the sideways drag of the pulley exert on the shaft. Refer to **Figure 5.1**.

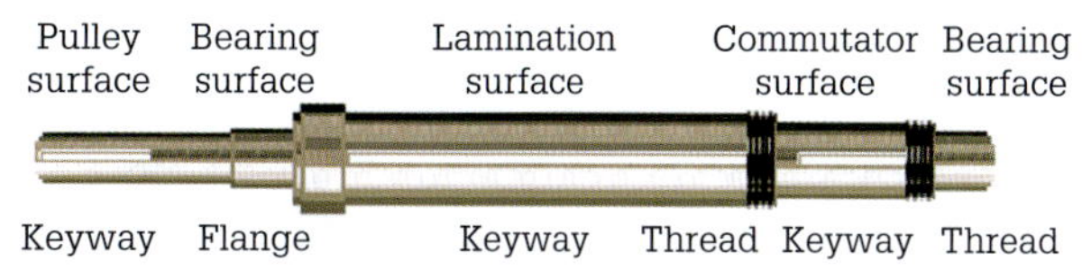

FIGURE 5.1 Armature shaft

In addition, the shaft has to withstand the twisting strain that it is subject to and the out-of-balance stresses that occur on the shaft if an unbalanced armature core attaches.

Core

The armature core consists of thin (0.27, 0.3 or 0.5 mm thick) individually insulated sheets of high-permeability, grain-oriented, non-ageing silicon steel called laminations. The silicon additive (approximately 4%) produces soft, low-remanence magnetic steel. This softening reduces the hysteresis loss and increases the resistances of the lamination thereby limiting the effects of eddy currents.

Steel that is subject to ageing will experience an iron loss increase from 10% to 30% after about a year's operation.

Figure 5.2 illustrates two styles of die cut lamination stampings. The first shows a solid stamping and the second a ventilated stamping which, in addition to reducing weight, provides passages for the circulation of air to carry away some of the heat generated by the armature windings.

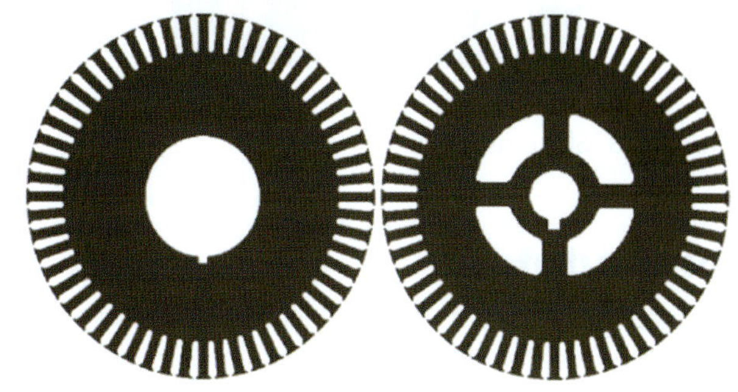

FIGURE 5.2 Lamination stamping

The laminations are stamped (or laser-cut) from electrical-grade sheet steel, without burrs, and hand stacked into a cylindrical arrangement. The ferromagnetic circuit occurs layer by layer to the width required by the armature design. The existence of burrs creates air gaps in the core that in turn have the effect of increasing the magnetising current because air has a much lower permeability than steel. In addition, burrs also prevent the achievement of the optimum stacking factor.

All laminations have a coating of insulating material such as varnish or oxide. The purpose of the surface insulation is to reduce the magnitude of eddy currents within the core. The laminations installed on the shaft are secured by either bolting or clamping between end plates. Refer to **Figure 5.3**.

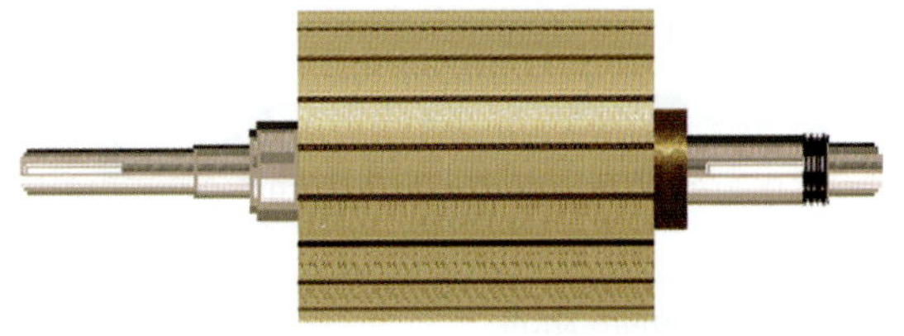

FIGURE 5.3 Laminations clamped together

The teeth formed in the lamination have different shapes depending upon the method chosen to secure the armature coils in the slots. **Figure 5.4** shows various forms of armature teeth.

FIGURE 5.4 Forms of armature teeth

Slot wedges

Timber, fibre or metal slot wedges hold the armature coils in place. In addition, with open-slot-type armatures, steel wire or glass tape contain the coils in the armature. **Figure 5.5** illustrates typical slot cross-sections showing the coil conductors and retaining wedges in place.

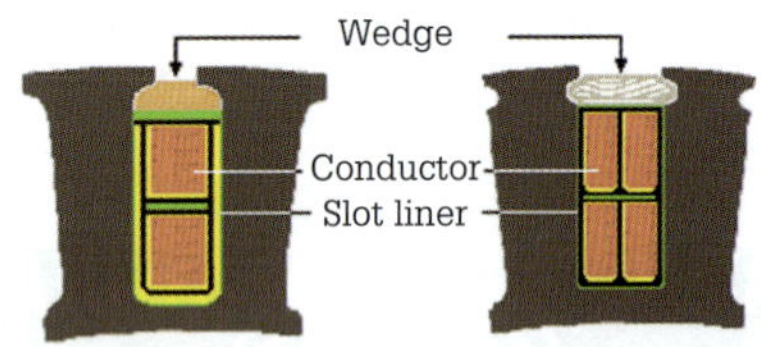

FIGURE 5.5 Slot cross-section

Commutator

The commutator consists of a number of hard-drawn L-shaped copper segments interleaved with high heat-resistant mica separators assembled around a tubular steel hub. The purpose of the mica separators is to insulate each segment from each other. Refer to **Figure 5.6**.

FIGURE 5.6 L-shaped copper segment

Commutators can consist of copper segments held together with V-rings or as a moulded composite construction. An assembled armature showing the commutator's position on the shaft is shown in **Figure 5.7**.

FIGURE 5.7 Assembled armature

The commutator segments have sufficient radial depth so that the commutator can be turned down to maintain its circular form. The mica separators are commonly 'undercut' by about 1 mm to prevent the wearing away of the carbon brushes which ride on the commutator. In addition, each commutator segment has a chamfer along its sides to improve brush and commutator life. Commutator colour uniformity on its brush-riding surface indicates its serviceable condition. Good film is burnished bronze to dark brown or black and is uniform in colour depending upon the grade of brushes used. **Figure 5.8** shows examples of assembled commutators and mica insulation.

FIGURE 5.8 Commutators and mica insulation

Armature coils

Armature coils can consist of rectangular copper conductors individually insulated and bonded together with separators. Formed coils are either lap or wave formed. The formed full coils have a wrapping of insulation to protect the coil from the steel core. After winding, multiple coats of insulating varnish create a smooth surface that resists moisture, chemical contamination and heat. Finally, the conductors of the armature in a d.c. machine connect to the commutator in a definite sequence.

Lap windings

The lap-wound armature winding forms a circle similar to a runner doing a lap of an oval. Refer to **Figure 5.9**.

FIGURE 5.9 Lap-type armature coil

In lap windings, the leads of each coil connect to adjacent commutator segments as illustrated in **Figure 5.10**.

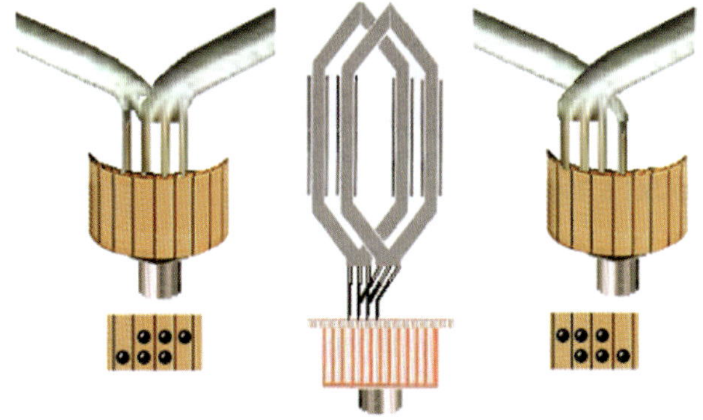

FIGURE 5.10 Progressive and retrogressive lap windings

If the leads connected to the commutator progress around the armature in the same direction as the coils, then the winding is a progressive winding. However, if the leads cross the winding, it is a retrogressive winding. Direct current motors with lap-wound armatures frequently have as many brush sets as there are main poles. In addition, the armature circuit usually has the same number of parallel paths, as there are main poles. For example, a four-pole d.c. motor has four brush sets on its armature and four parallel paths for armature current to flow.

Wave winding

The wave-type winding forms a wave around the core of the armature. Refer to **Figure 5.11**.

FIGURE 5.11 Wave-type armature coil

In a wave-wound armature, the wave circuit returns to a commutator segment adjacent to the start segment after traversing the number of poles. Refer to **Figure 5.12**.

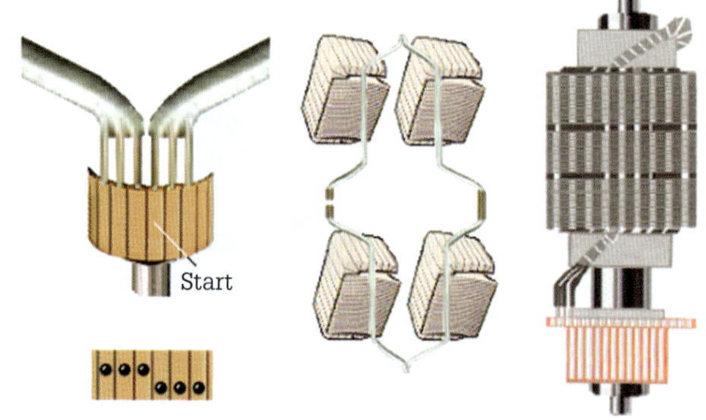

FIGURE 5.12 Wave commutator connections

Direct current motors with wave-wound armatures require only two brush sets. However, there are some exceptions to this rule. In addition, regardless of the number of main poles, there are only two parallel paths for armature current to flow.

Brushes

Brushes provide the connection between the rotating commutator and the external voltage supply. They occur in a number of varieties that differ in resistance and abrasive quality. There are four grades of brushes used with d.c. machines: carbon graphite, resin-bonded graphite, metal graphite and electrographitics.

Surface speeds of the commutator and loads placed on the armature affect the service life of brushes. In addition, sparking, a result of high current loads, not only wears brushes but damages the commutator as well. Brushes have a high coefficient of friction and in service produce carbon and metallic dust. However, once a commutator film is established the effect of friction minimises.

Brushes are placed 'end on' the commutator and feed forward as they wear away by means of a spring. In addition, they connect to a brush-holder by means of a copper braid conductor called a pigtail.

Brush-holders

Brush-holders (also called brush boxes) are devices clamped to a rocker arm, designed to maintain brush pressure against the commutator at the desired angle. Refer to **Figure 5.13**.

FIGURE 5.13 Brush-holders

The brush-holders in any d.c. machine are a set. The number of brush sets depends upon the number of field poles while the physical arrangement of the brush-holders around the commutator depends on the armature coil placement. In the case of a four-pole d.c. machine, all positive brush-holders connect together, and all negative brush-holders connect together, as in **Figure 5.14**, resulting in four parallel paths through the armature.

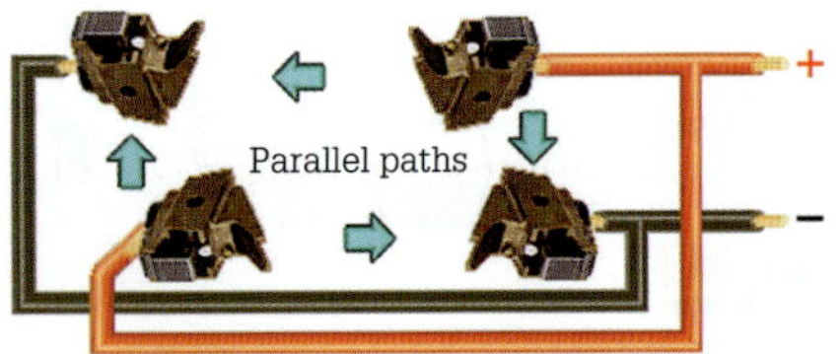

FIGURE 5.14 Brush-holder arrangement for a four-pole d.c. machine

Field poles

Main field

The main magnetic field of a d.c. machine is produced by a permanent magnet or by an electromagnet as shown in **Figure 5.15**.

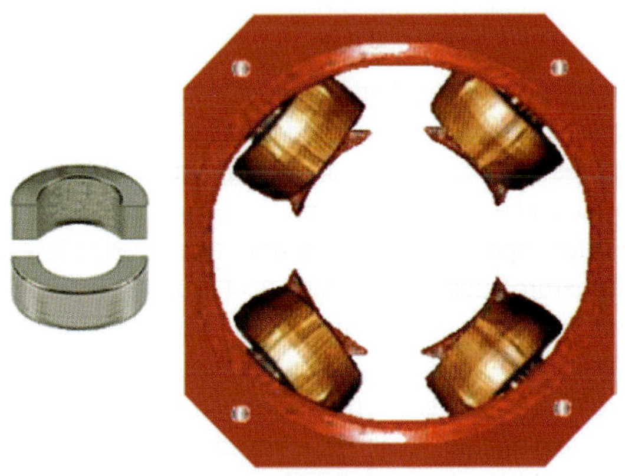

FIGURE 5.15 Permanent magnet and electromagnetic main fields

The field assembly of an electromagnet consists of pole pieces and field coils. The pole pieces that fit the curvature of the armature are bolted to the inside of the yoke or frame. Refer to **Figure 5.16**.

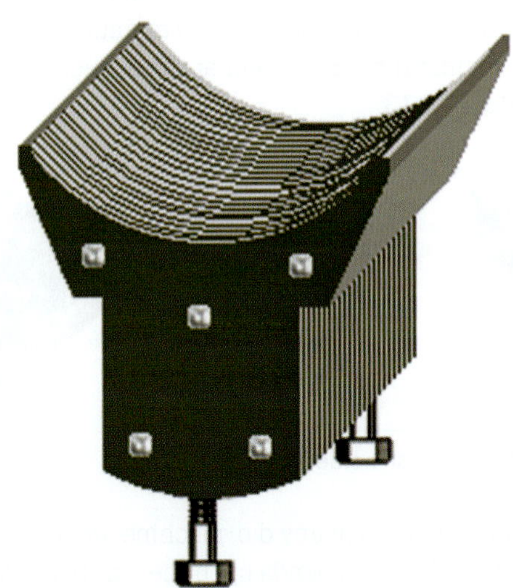

FIGURE 5.16 Laminated field pole

The purpose of the main field is to develop the magnetic flux. The pole face fits the curve of the armature and provides a uniform flux distribution from the main field. The main fields, when assembled on their pole piece, occur in a definite sequence of magnetic polarity around the yoke of a d.c. machine. Refer to **Figure 5.17**.

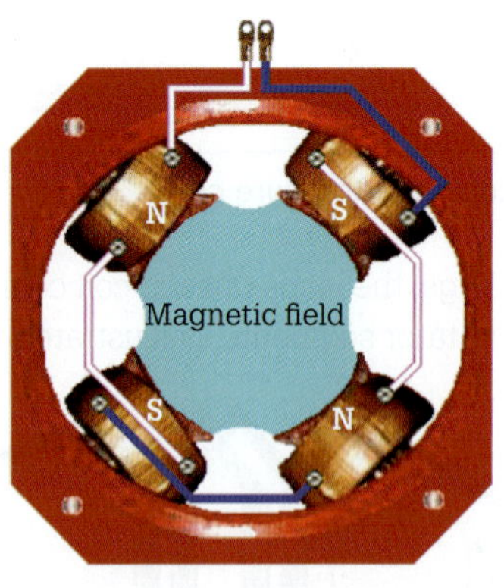

FIGURE 5.17 Magnetic sequence

Field windings occur as shunt, series or compound. Shunt windings have many turns of fine-diameter magnet wire, while series windings have fewer turns of larger-diameter magnet wire. Compound windings consist of both shunt and series-type windings. Refer to **Figure 5.18**.

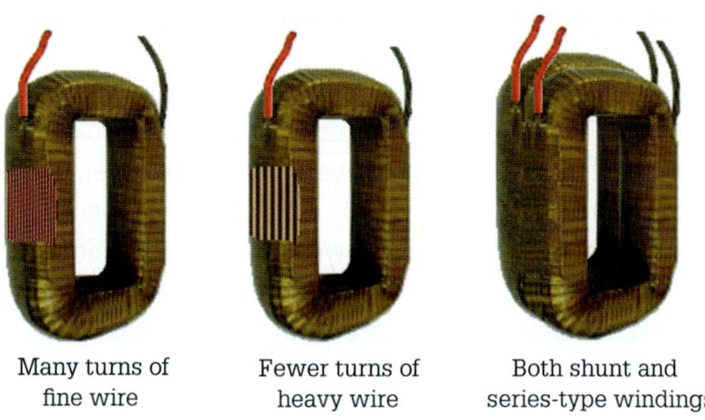

FIGURE 5.18 Field coils – shunt, series and compound

Compensating winding

Compensating windings as illustrated in **Figure 5.19** are a pole-face winding used on large d.c. machines.

FIGURE 5.19 Compensating winding on top of the field winding

These windings compensate for and terminate the adverse effect of the armature reaction on the main magnetic field by producing a magnetomotive flux that is always equal to the armature magnetomotive flux.

They are used to prevent flashovers between the commutator bars or brush to brushes (high-current arcs) when d.c. machines experience vast and rapid load swings. Rapid load swings cause the armature flux to induce high transient voltages in all the armature coils by transformer action creating the arcing problem.

Armature reaction

When a generator supplies a load, the induced current in the armature winding develops an electromagnetic field of its own at right angles to the main field of the generator. This cross-magnetisation effect of the armature tends to distort the uniform magnetic field produced by the main field coils. The result is armature reaction. **Figure 5.20** shows the distortion of the magnetic field from cross-magnetisation by the armature.

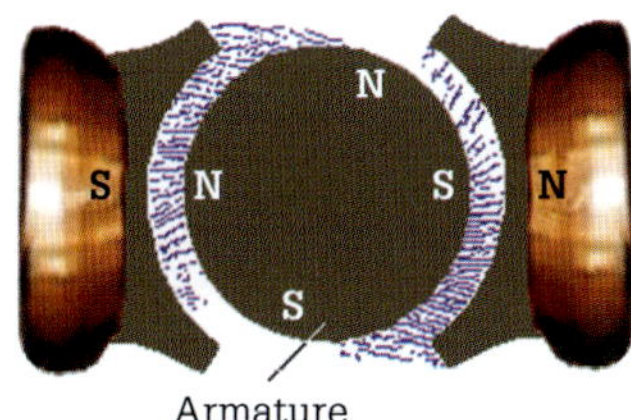

FIGURE 5.20 Armature reaction

The induced north poles created in the armature attract the south poles of the main field whereby the south-seeking flux enters the main fields. This effect also occurs with the south poles induced in the armature. According to Lenz's law, the direction of the current in the armature of a d.c. generator is such as to oppose the motion producing it. So, when the armature rotates, the main field flux distorts and creates a dragging effect in relation to the armature rotation. This effect is armature reaction which increases as armature increases with the result that the generator requires more power from its prime mover to drive it.

In a d.c. motor, the direction of the actuating current is the reverse of the armature current of a d.c. generator. So, the armature reaction assists the direction of rotation of the armature and adds to the torque used by the armature shaft to do mechanical work. As the motor load increases, the current taken from the supply also increases, creating additional armature reaction with its additional torque.

Interpoles

In order to rectify the effects of armature reaction, the main poles have interpoles placed between them.

The interpoles are wound and connect so that their resultant magnetic field opposes that of the cross-magnetic field. However, the main purpose of interpoles is to counteract the electromotive force of self-induction in the armature coils undergoing commutation. There are frequently the same numbers of interpoles as main poles. Refer to **Figure 5.21**.

FIGURE 5.21 Interpole

Interpoles connect in series with the armature winding. For generators, the polarity of an interpole is the same as that of the main pole, following the interpole in the direction of rotation. Conversely, the polarity of the interpole for a motor is the same as that of the main pole in front of the interpole in the direction of rotation. Refer to **Figure 5.22**.

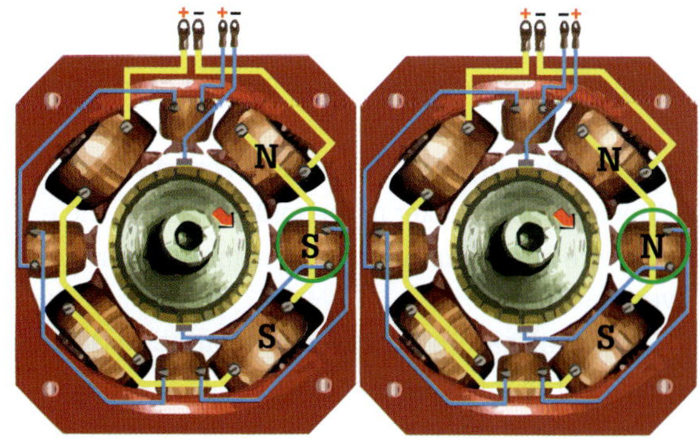

FIGURE 5.22 Interpole connections for a shunt generator and motor

Interpoles eliminate sparking caused by the self-induced electromotive force, created by the armature coil undergoing commutation due to armature reaction. The interpole generates a neutralising voltage in the armature coil that opposes the self-induced electromotive force.

Nameplate

A nameplate (see **Figure 5.23**) provides information about the machine, which includes:

- manufacturer's name
- output power (Watts or HP)
- speed at rated output (RPM)
- voltage rating (Volts)
- armature current
- winding connection
- field current
- field resistance.

MYMOTORCO		
kW 10	RPM 1200	VOLTS 400
ARM AMPS 15	WOUND SHUNT	
FLD AMPS 1.5/3.0	FLD OHMS @25°C 133	
INSUL CLASS F	DUTY CONT	MAX AMBIENT 40°C
SUPPLY DC		FLD VOLTS 400/200
TYPE E	ENCL DP	INSTR
MOD DCM002	SER 1234XYZ	

FIGURE 5.23 Sample machine nameplate

Machine safety

SWITCH ON

When working on or near rotating machines, observe the following:

- Not placing any part of your body into moving machinery.
- Not wearing jewellery, neckties or loose-fitting clothing.
- Wearing proper protective clothing and equipment suitable for the operation being performed.
- Before attempting to perform repairs or maintenance on any machine, make sure that it is de-energised.
- Compressed air may be used to clean machinery parts that have been properly disassembled provided that the supply air pressure does not exceed 30 psi and a safety shield tip is used.
- Not tampering with or permanently disabling safety devices such as sensors, cut-out switches or guards.
- Being aware that under non-compensated fluorescent lighting, machines rotating at certain frequencies may not be detected as moving.

»

REVIEW QUESTIONS

1. What is the energy conversion process in a motor?
2. Name the type of frame construction used for large d.c. machines.
3. What is the purpose of the 'frame' with d.c. machines?
4. Why is a shaft thicker in the middle than at the ends?
5. Why are some laminations for armature cores ventilated?
6. How are armature laminations secured to the shaft?
7. What is the purpose of armature slot wedges?
8. What type of material separates the commutator segments from each other?
9. What do brushes provide?
10. What are two methods of establishing the main magnetic field in a d.c. machine?
11. Name the three types of field windings used in d.c. machines.
12. What type of winding is a compensating winding?
13. How do interpoles connect?
14. What is the function of interpoles?
15. Name the two methods used to connect the conductors around the armature in a d.c. machine.

5.2 Direct current generator

The generator or dynamo transforms mechanical energy into electrical energy. **Figure 5.24** shows alternating current generators, known also as synchronous generators or alternators.

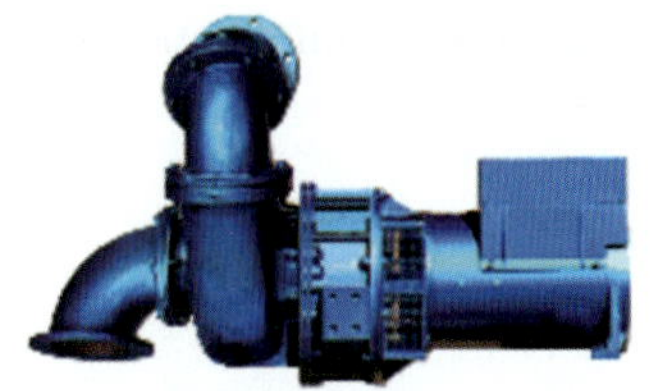

FIGURE 5.24 Alternators

The basic a.c. alternator consists of a coil of wire that has its ends connected to slip rings. Both rotate by a prime mover in a magnetic field produced by electromagnetism or permanent magnets. As the coil rotates, it cuts the magnetic lines of force produced by the field poles, enabling an electromotive force to occur across the ends of the coil. Consequently, the induced a.c. voltage causes a current to flow when the alternator connects to an external circuit.

Figure 5.25 shows the principle of generating alternating current via the rotation of a single turn of a copper conductor, called the armature, in a magnetic field.

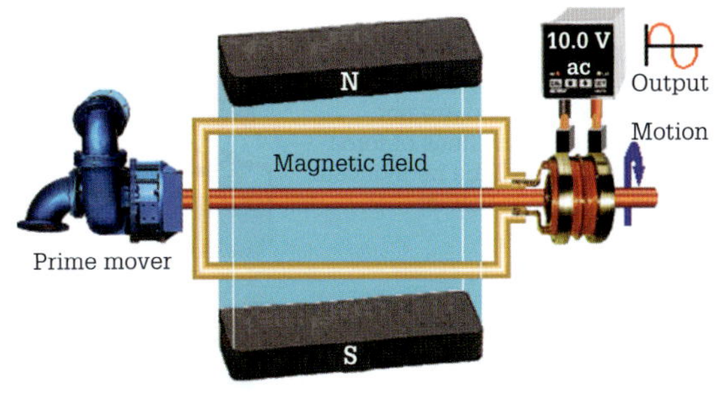

FIGURE 5.25 Permanent magnet alternator

Simple alternator

Referring to Figure 5.26, assume that the armature conductor turn rotates in a clockwise direction through one complete revolution, which is also called 360 electrical degrees. (One revolution with four poles = 720 electrical degrees.) In addition, the external load (the lamp) connects to the armature by graphite brushes and slip rings.

In examining position A the sides of the turn are moving parallel with the magnetic field and are not cutting the magnetic lines of force. Therefore there will be no induced emf and no current flow at this instant in time.

As the turn travels along its circular path it will have an increasingly larger emf induced in it. At position B maximum induced emf and current flow is reached.

From position B to position C the rate of cutting perpendicularly to the magnetic lines of force becomes diminished until at position C no cutting occurs. Therefore, once again there will be no induced emf and no current flow at this instant in time.

From position C to D the turn has an increasingly larger emf induced but one which is acting in the opposite direction. This has occurred because the motion of the turn has reversed in relation to the magnetic field. At position D maximum induced emf is reached.

From position D to A the induced emf gradually diminishes until at position A no induced emf and no current flow exists at this instant in time.

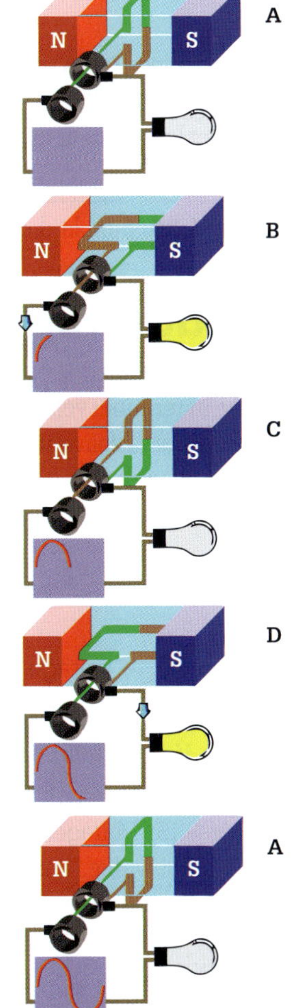

FIGURE 5.26 Simple alternator

The induced electromotive force and current alternate twice during a complete revolution, reaching two maximum values that are in opposite directions and two zero values.

When the electromotive force for position B is positive, then it is negative for position D. So, the induced electromotive force alternates from positive to negative and the current that this electromotive force produces is alternating current.

A graphical curve called a sinusoidal waveform represents one revolution of the basic alternator. The direction of current flow in the direction of the generated current of Figure 5.26 can be verified by applying Fleming's right-hand generator rule as shown in Figure 5.27.

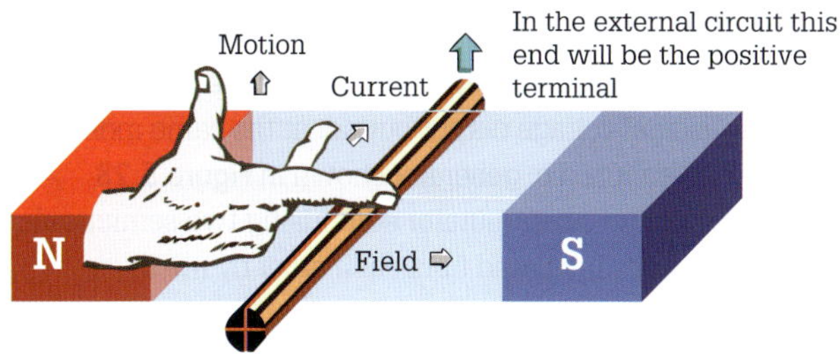

FIGURE 5.27 Right-hand generator rule

Sinusoidal current and voltage

Sinusoidal current and voltage are important for many electrical devices; practical alternators are designed so that their generated output does have a sinusoidal waveform. The current generated by the simple a.c. alternator is single-phase alternating current. When the armature windings compose of three groups of coil windings aligned at 120 electrical degrees to each other, the alternator produces three-phase alternating current. The three-phase alternators provide power generation and distribution. A typical alternator output used in power stations would be 500 MW at 24 kV, three-phase 50 Hz output.

Power or high-voltage alternating current alternators are electrically excited. Electrical excitation means that electromagnets replace permanent magnets to create a stronger magnetic field. In addition, a direct current connects to these coils through the graphite brushes on the slip rings and the total arrangement is rotated by means of the prime mover. Consequently, the induced electromotive force develops in a stationary armature. Note that the device delivering the electromotive force is an armature if it is the rotating element, while the fields are stationary. By contrast, if the device delivering the electromotive force is stationary, then the device is a stator, in which case the rotating field is now called the rotor. The use of electromagnets enables adjustment of the generated voltage. The current in the electromagnet coils can be increased or decreased, in turn increasing or decreasing the electromagnetic flux that is produced. Having control over the electromagnetic flux means that the electromotive force produced at the output terminals of the armature is variable.

Direct current generator

The principle of generating direct current is the same principle that applies to generating alternating current. However, as there is a requirement for a d.c. output instead of an a.c. output there must be some means to change the polarity of the output each time the a.c.-generated emf reaches zero. If this does not occur then the emf output remains an alternating sinusoidal current.

The means to change the polarity of the output is through the use of a commutator and brushes. Therefore, a d.c. generator is an a.c. alternator with a commutator. The commutator allows the generated alternating voltage of the armature to become a direct current.

The commutator consists of individual hard-drawn copper segments insulated from one another with mica insulation. This arrangement enables the commutator to act as a switch, switching the negative half of the sinusoidal electromotive force output to the same output terminal as the positive half of the sinusoidal voltage output but not at the same moment in time. The simple d.c. generator shown in **Figure 5.28** consists of a simple commutator comprising two semicircular copper segments insulated from each other by mica.

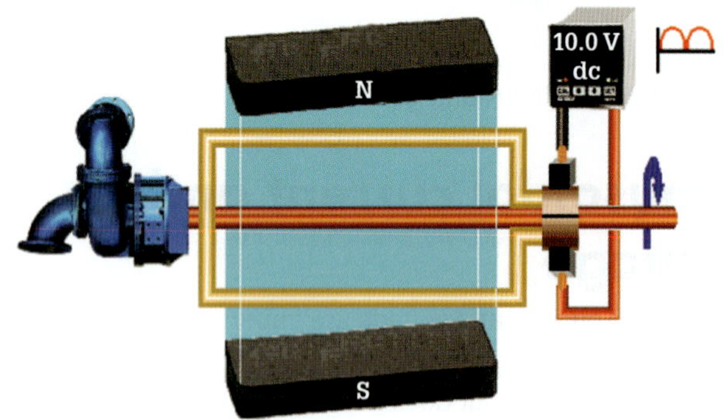

FIGURE 5.28 Simple d.c. generator

Operation of the commutator

1. One end of the rotating armature conductor (coil) connects to one of these copper segments and the other end of the conductor to the other segment.
2. Graphite brushes make contact with the commutator collecting the induced current and direct it via an external circuit to the load. The graphite brushes are located in a position that enables them to break contact with one segment and make contact with the other segment at the instant at which the induced current changes direction in the armature conductor.
3. The commutator reverses the current flow in the armature conductor as it passes from one pole to the next so that the current in either side of the conductor is always the same value as it moves from a given pole.
4. Finally, the induced current passes to an external circuit through the same graphite brush and is called a unidirectional current.

When a d.c. generator contains only a single coil, it provides a pulsating output with a broad ripple. The pulsating effect and the ripple as shown in **Figure 5.29** are reduced in practical generators by using a commutator with a large number of segments and an armature with a system of coils employed to give overlapping outputs. These coils are at angles to each other on the armature to deliver a combined output.

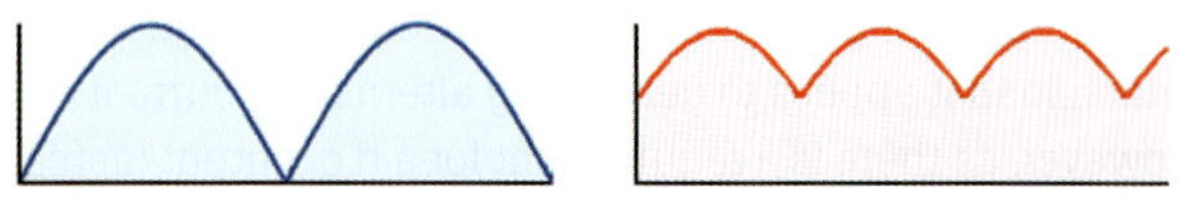

FIGURE 5.29 Large and reduced ripple d.c. waveforms

Figure 5.30 shows the action of a simple d.c. generator.

In examining position A the sides of the turn are moving parallel with the magnetic field and are not cutting the magnetic lines of force. Therefore, there will be no induced emf and no current flow at this instant in time.

As the turn travels along its circular path it will have an increasingly larger emf induced in it. At position B maximum induced emf and current flow is reached.

From position B to position C the rate of cutting perpendicularly to the magnetic lines of force becomes diminished until at position C no cutting occurs. Therefore, once again there will be no induced emf and no current flow at this instant in time.

From position C to D the turn has an increasingly larger emf induced but one which is acting in the opposite direction. This has occurred because the motion of the turn has reversed in relation to the magnetic field. At position D maximum induced emf is reached.

From position D to A the induced emf gradually diminishes until at position A no induced emf and no current flow exists at this instant in time.

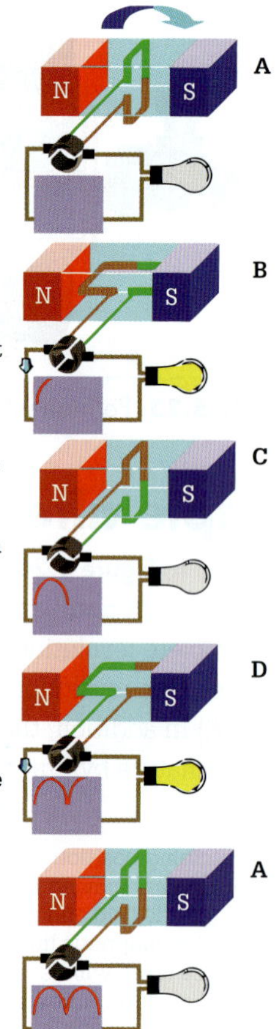

FIGURE 5.30 Simple d.c. generator

Voltage regulation

The voltage regulation of a d.c. generator is the percentage change in terminal voltage, from no load to rated load when the generator is delivering its terminal voltage at rated speed.

The following equation determines the percentage voltage regulation:

$$V_{Reg} = \frac{V_{NL} - V_{Rated}}{V_{Rated}}$$

where V_{Reg} = voltage regulation

V_{NL} = no-load or open-circuit voltage (V)

V_{Rated} = rated nameplate or full-load voltage (V)

Note that the result of this equation is a decimal value (such as 0.15). Voltage regulation is often expressed as a percentage value. This is achieved by multiplying the decimal value (the result of the above equation) by 100.

EXAMPLE 5.1

A 75 kW 1500-rpm generator operating at rated load has a terminal voltage of 455 V. If the no-load voltage is 485 V, determine the voltage regulation.

$$V_{Reg} = \frac{V_{NL} - V_{Rated}}{V_{Rated}}$$

$$= \frac{485 - 455}{455}$$

$$= \mathbf{0.0659 \text{ or } 6.59\%}$$

EXERCISE 5.1

a A 55 kW 1500-rpm generator operating at rated load has a terminal voltage of 452 V. If the no-load voltage is 475 V, determine the voltage regulation.

b A 50 kW 1500-rpm generator operating at rated load has a terminal voltage of 448 V. If the machine has a voltage regulation of 4.5%, determine the no-load terminal voltage.

REVIEW QUESTIONS

1 What is another name for an alternating current generator?

2 Describe how a simple single-turn coil generator produces an emf.

3 What device on a direct generator converts the generated alternating current to direct current?

4 In Fleming's right-hand generator rule, what does the thumb indicate?

5 A 35 kW 1000-rpm generator operating at rated load has a terminal voltage of 232 V. If the no-load voltage is 255 V, determine the voltage regulation.

5.3 Types of direct current generators

Generator classification

Direct current generators convert mechanical energy into electric energy. They are d.c. motors used as generators, the difference being the direction of power through them. There are four main types of d.c. generators, classified according to how their field flux develops.

1 shunt
2 separately excited
3 series
4 compound
   - cumulatively compound
   - differentially compound.

These types of d.c. generators differ in their voltage–current terminal characteristics, power ratings, efficiency, excitation method used and voltage regulation, and in their applications.

Shunt generator

A shunt generator has its field windings connected in a parallel configuration (across the armature) and develops its initial electromotive force from the residual magnetism (self-excitation) in the ferromagnetic pole cores. The ferromagnetic cores retain enough residual magnetism from each generator shutdown to develop about 5% of the generator rated voltage when the generator is brought up to its rated operating speed.

As the generator comes up to speed the armature conductors cut the residual flux present, and a low-value electromotive force develops. The generated voltage so produced occurs across the parallel-connected field winding causing a rheostat-controlled current to flow through the field circuit. This current causes the voltage to increase again and that in turn pushes more current through the field winding causing a further rise in voltage and so forth.

However, the voltage of any self-excited generator may fail to build up for several reasons. For example, the resistance of the shunt field windings and the permeability of the ferromagnetic pole cores limit the build-up of the no-load voltage of the generator. In addition, if the ferromagnetic field poles have lost their residual magnetism, the generator cannot establish an electromotive force. If residual loss occurs the field coils need to be 'flashed' (connect a low-voltage source across the field coils) in order to restore the magnetic presence.

Other factors causing the failure of voltage build-up include low armature speed, reversed field connections (causes an opposing flux field to the residual flux field) and reversed armature rotation by the prime mover. In addition low brush pressure on the commutator, dirty brushes or commutator, and open- or short-circuited field winding circuit also prevent the voltage build-up. Refer to **Figure 5.31**.

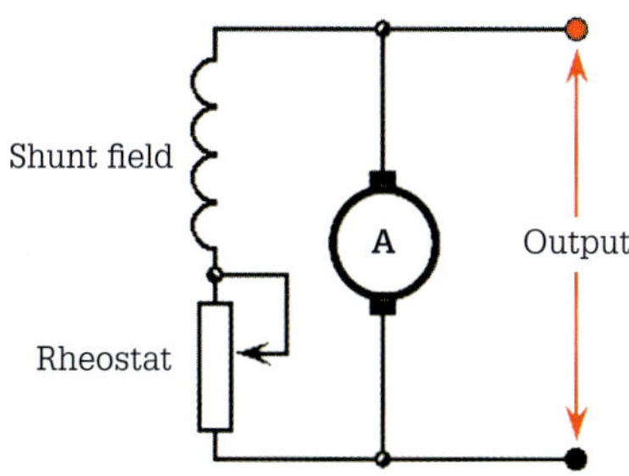

FIGURE 5.31 Circuit diagram of a self-excited shunt generator

Figure 5.32 shows the wiring diagram of a self-excited shunt-connected generator.

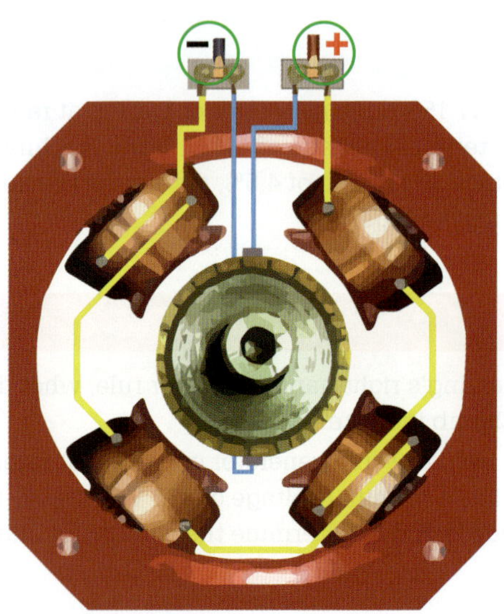

FIGURE 5.32 Shunt generator connections

Separately excited generator

A separately excited d.c. generator has its field windings supplied by a separate external d.c. voltage source. These generators are usually small and driven by a separate motor or from the main generator shaft. **Figure 5.33** illustrates the circuit diagram of a separately excited generator.

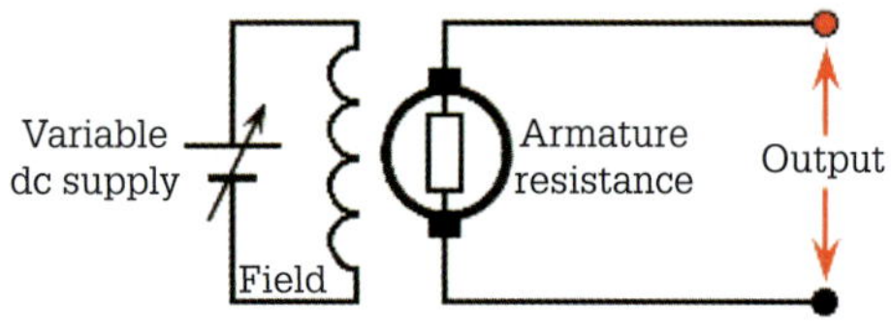

FIGURE 5.33 Circuit diagram of a separately excited generator

Figure 5.34 shows a separately excited shunt generator, equipped with interpoles and a compensating winding.

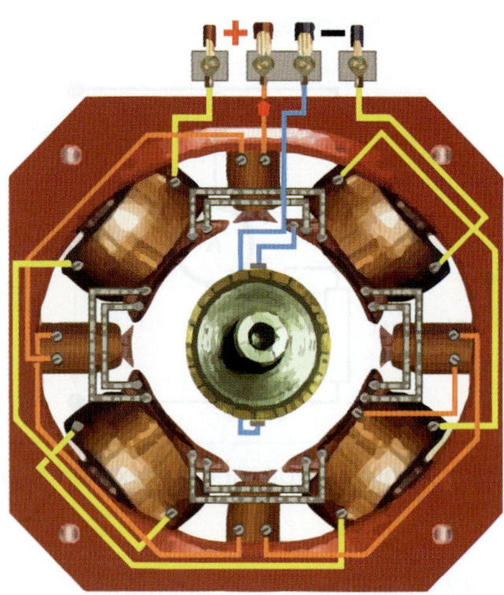

FIGURE 5.34 Separately excited shunt generator with interpoles and a compensating winding

The compensating winding connects in series with the armature. Consequently, the magnetic field it produces does not affect the magnetic flux of the main pole. The external excitation of the field means that the generated electromotive force is dependent only on the field-winding electromagnetic flux and the armature shaft speed. A prime mover develops the armature shaft speed, and this rate is usually constant. Because the supply voltage to the field windings is constant, control of the generated voltage occurs by connecting a rheostat in series with the field winding.

The rheostat provides a means of varying the field excitation through current control. Consequently, the separately excited generator operates in a stable condition with any field excitation to provide a wide range of output voltages. Since the generator in this example has compensating windings, armature reaction does not occur. No armature reaction means that the electromotive force induced in the armature is equal to the no-load or open-circuit voltage of the generator. Whereas, if compensating windings are omitted, then the electromotive force induced in the armature reduces due to the de-magnetising effects of armature reaction.

Series generator

The series generator is also a self-excited generator with its field windings, armature circuit and load connected in series. As all the connections are series configured, the field current is also the load current. Therefore, the field winding must have minimum ohmic resistance to prevent a large voltage drop across the field windings.

The series field windings consist of a few turns of large cross-sectional area wire to prevent a significant voltage drop. **Figure 5.35** illustrates the circuit diagram of a series-connected generator. **Figure 5.36** shows the series-connected generator connections.

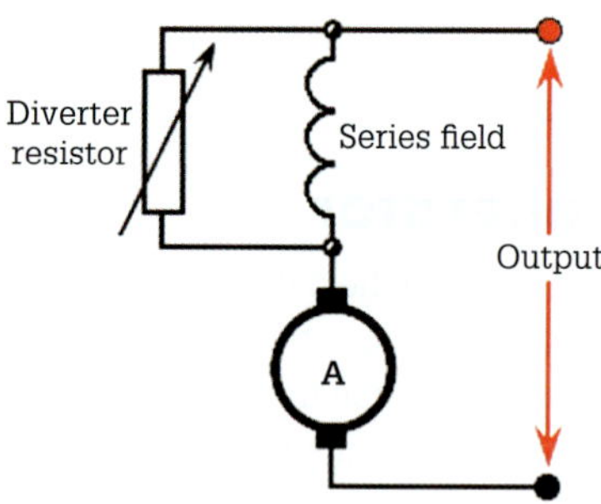

FIGURE 5.35 Series-connected generator – circuit diagram

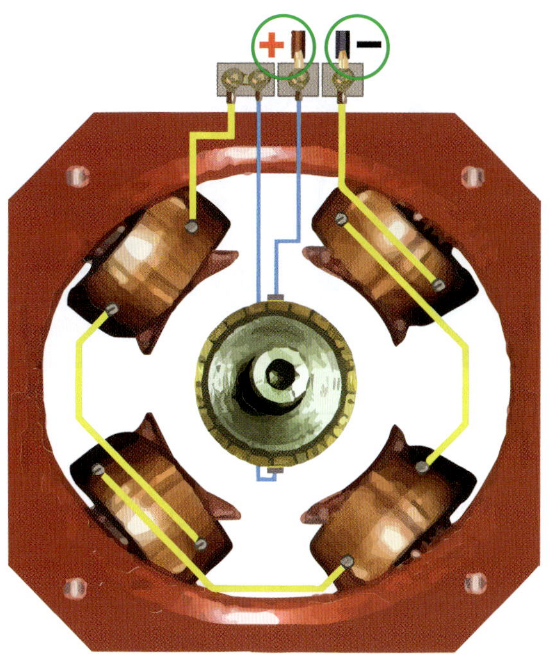

FIGURE 5.36 Series generator

Series-connected generators can experience considerable variations of voltage across their output terminals. Voltage control of these generators occurs by connecting a diverter resistor across the series field, thereby diverting load current around the field. Control is difficult because voltage change is the result of load current variations that in turn were the result of load conditions. Consequently, the series generator is not a constant voltage source and so finds little use in industry. However, one use is in arc welding.

Compound generators

The compound generator has a magnetic field established by residual magnetism and strengthened by a shunt winding. In addition, the series winding connects to the armature circuit and carries load current.

The shunt field of compound generators can be connected 'short shunt' across the armature or 'long shunt' across the output terminals of the generator. The difference between the two connections is the amount of current flowing through the series field winding.

Generators are usually connected short shunt because this arrangement tends to maintain the shunt-field current constant on variable loads as a voltage drop across the series winding does not directly affect the voltage on the shunt field.

Figure 5.37 shows compound generator long-shunt connections while the circuit diagram of a short-shunt compound generator occurs in **Figure 5.38**, and the long-shunt compound generator in **Figure 5.39**.

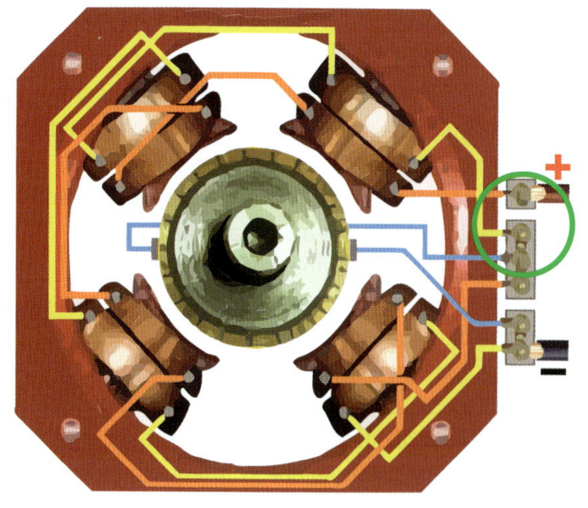

FIGURE 5.37 Short-shunt compound generator

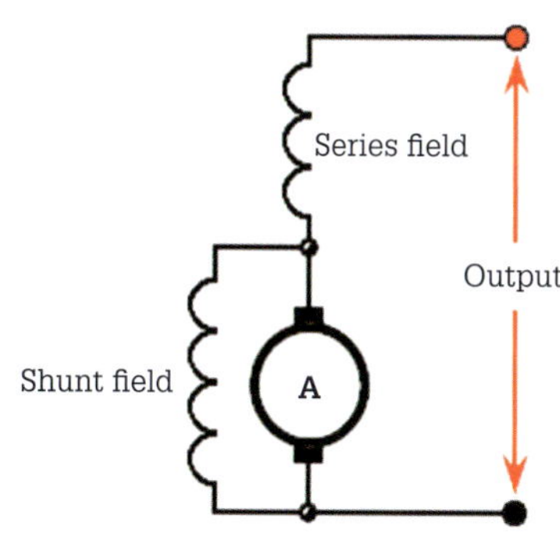

FIGURE 5.38 Circuit diagram of a short-shunt compound connected generator

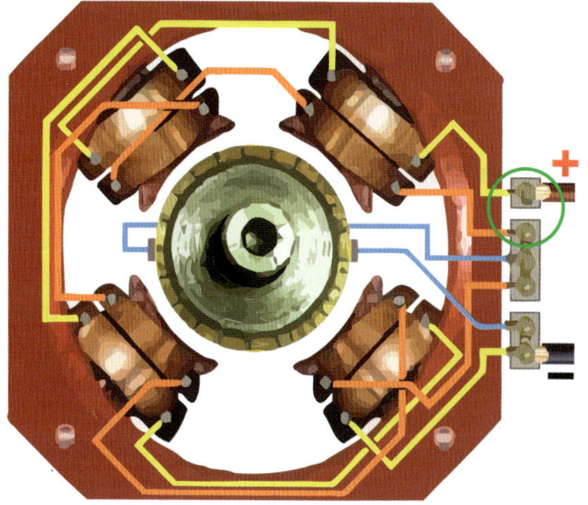

FIGURE 5.39 Long-shunt compound generator

Generator characteristics

The usefulness of any d.c. machine is evaluated according to its performance when a load is applied. One factor that governs machine performance is the method of field excitation employed.

The field's excitation, such as separate, permanent magnet, series or shunt has an impact on the characteristics of the machine, with each machine demonstrating different characteristics. For this reason, performance characteristics comprise a set of graphs showing the relevant dependent variables of operation as a function of load. The performance characteristic graphs include excitation of current–voltage, speed–voltage, load voltage–load current, torque–load current, and torque–speed and speed–load current.

Separately excited generator characteristics

Figure 5.40 shows a typical magnetisation graph of a separately excited generator. The values of armature voltage versus field current at a constant rate enable the plotting of the magnetisation characteristic of a d.c. generator at no load and constant speed.

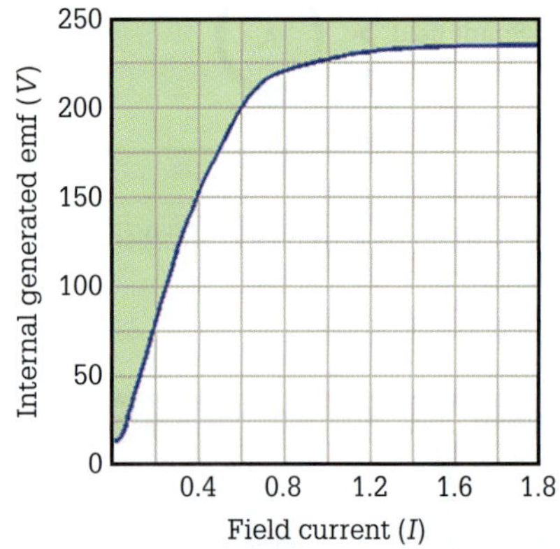

FIGURE 5.40 Magnetisation characteristic

Referring to **Figure 5.40**, with no field current applied the internal generated electromotive force is about 12 V. This is because of the low flux density produced by the residual magnetism in the pole cores. As the field current increases, the generated electromotive force increases in a near-linear progression until the ferromagnetic core saturates at about 0.6 A. After this point, greater increases in field current are required to produce gains in generated voltage. For example, between 0 and 0.4 A, the generated voltage rose from 12 V to 150 V, a rise of 138 V. By contrast, the same increase in current from 1.2 A to 1.6 A resulted in an 8 V gain. The magnetisation characteristic curve is also the no-load characteristic. **Figure 5.41** illustrates the connection diagram for the determination of the load characteristic.

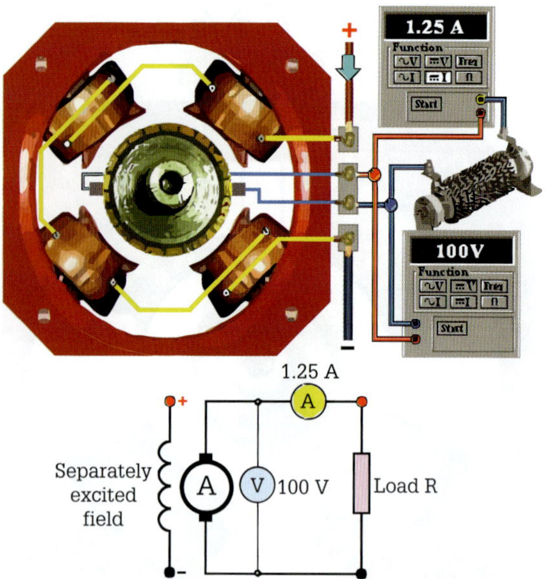

FIGURE 5.41 Load characteristic connections – separately excited generator

The load characteristic of a separately excited generator is determined by holding the field current at a constant value while the armature rotates at a steady speed. Under these conditions, measurement of the terminal voltage and load current occurs at various loading conditions and their progression is plotted on a graph. A typical load characteristic or regulation curve of a separately excited generator is shown in **Figure 5.42**.

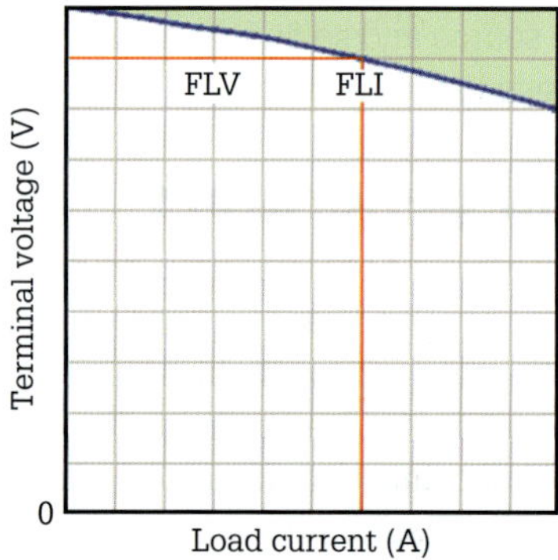

FIGURE 5.42 Load characteristic – separately excited generator

The load characteristic curve shows the terminal voltage dropping slowly as the load increases. In addition, the output voltage drop is due to the effect of armature resistance, brush resistance and armature reaction. The voltage–speed characteristic of a separately excited generator shows that these generators produce a linear terminal voltage. The linear voltage remains while the field flux is constant because the generated terminal voltage is directly proportional to speed. **Figure 5.43** shows a voltage–speed characteristic of a separately excited generator.

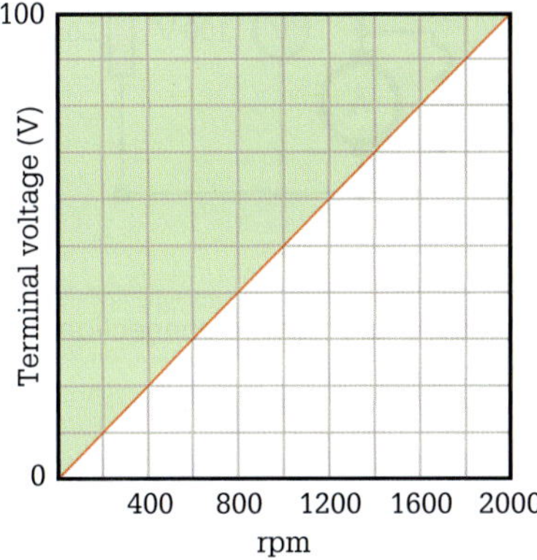

FIGURE 5.43 Voltage–speed characteristic – separately excited generator

Shunt generator characteristics

The magnetisation curve of a shunt-connected generator is the same as for the separately excited generator as shown in **Figure 5.40**. **Figure 5.44** shows the circuit diagram for the determination of the load characteristic of a shunt-connected generator.

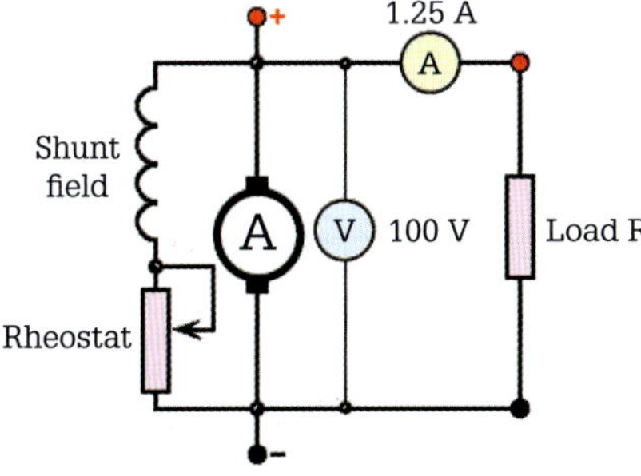

FIGURE 5.44 Load characteristic connections – shunt generator

The voltage–speed characteristic of a shunt-connected generator shows that these generators do not produce a linear terminal voltage in relation to speed. This load characteristic shows that, similar to the separately excited generator, the terminal voltage drops with increasing load because of the generator's internal voltage drop. In addition, because the generator provides its own field excitation a drop in terminal voltage reduces the voltage applied across the shunt field. Consequently, a decrease in field current results in a decrease in field flux and a further decrease in terminal voltage. Refer to **Figure 5.45** for the loads characteristic of a shunt-wound generator.

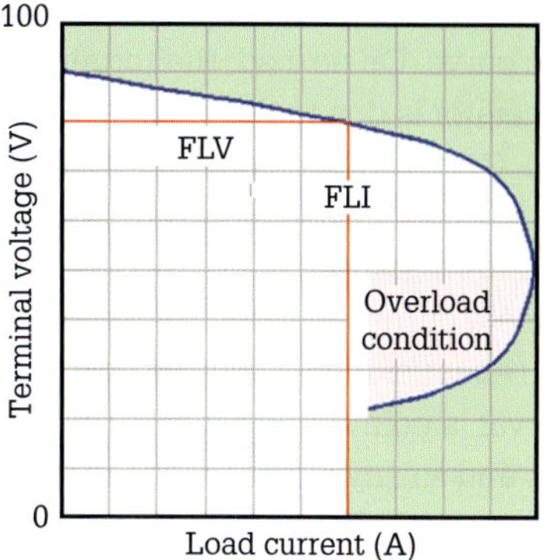

FIGURE 5.45 Load characteristic – shunt generator

The load characteristic of the terminal voltage of the shunt-wound generator falls off more rapidly under load than the separately excited generator. The rapid falling of voltage occurs because the low terminal voltage is unable to drive the necessary current through the shunt field to maintain an active field flux for the armature conductors. A reduced field flux means that the armature conductors have a lower current induced and, therefore, are unable to sustain the required load current. Control of the terminal voltage of a shunt-wound generator transpires by a rheostat inserted in series with the field windings. As the rheostat resistance increases, the field current reduces and the generated voltage falls. For a given setting of the field rheostat, the terminal voltage at the armature brushes is approximately equal to the generated voltage minus the *IR* drop produced by the load current in the armature. The voltage at the terminals of the generator lowers as the load increases. Only the shunt generator operates without damage under short-circuit conditions. Where the load demands are relatively constant, the shunt-wound generator is the generator of choice.

Series generator characteristics

Refer to **Figure 5.46** for the circuit diagram regarding the determination of the load characteristic of a series generator.

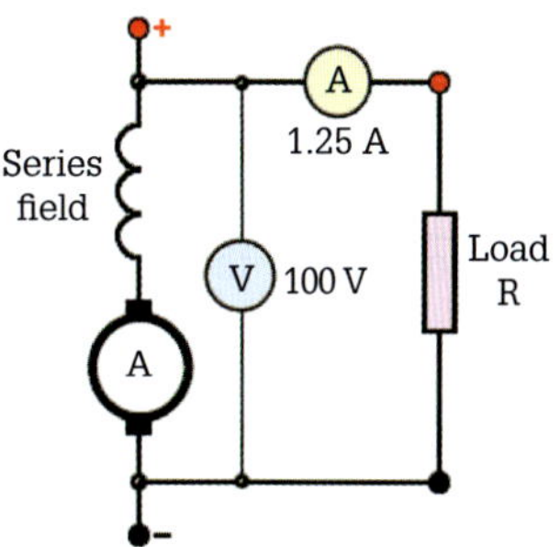

FIGURE 5.46 Load characteristic connections – series generator

The load characteristic curve is also the magnetisation characteristic curve. On no load, the terminal voltage of the series generator is due to the residual magnetism in the field cores. Once the load connects, the flux developed in the cores increases. Consequently, this increase in flux causes the terminal voltage to increase and this in turn pushes more current through the series windings, which has the effect of increasing the field flux even further. However, a point arrives where the field cores are saturated, and so any increase in load causes the terminal voltage to fall. Refer to **Figure 5.47** for the load characteristic of a series generator.

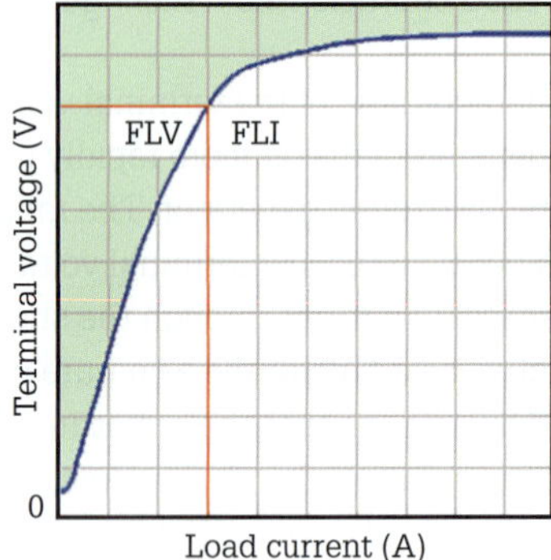

FIGURE 5.47 Load characteristic – series generator

This generator is unsuited for conditions where the load current varies because the terminal voltage also changes. However, terminal voltage control occurs by changing the armature speed or by connecting a diverter resistor in parallel with the series windings.

The voltage–speed characteristic shows that these generators also produce a non-linear terminal voltage. Refer to **Figure 5.48** for the voltage–speed characteristic of a series generator.

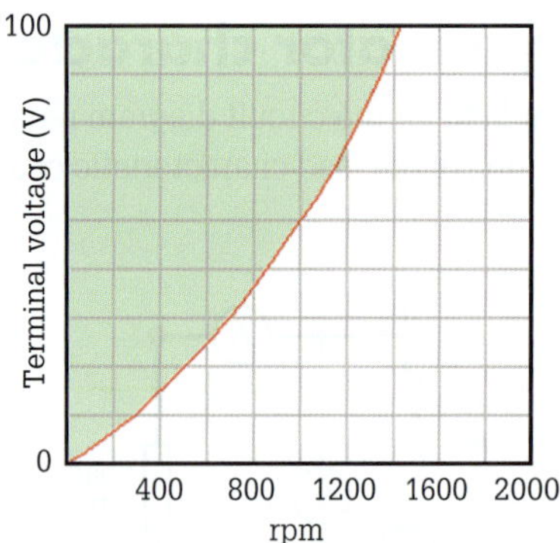

FIGURE 5.48 Voltage–speed characteristic – series generator

Cumulatively compound generator characteristics

Figure 5.49 shows a schematic connection diagram for the determination of the load characteristic of a cumulatively compound generator.

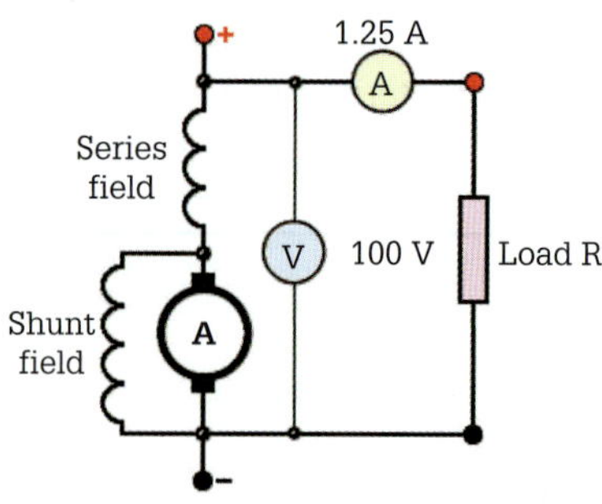

FIGURE 5.49 Load characteristic connections – cumulatively compound generator

In the cumulatively compound generator, the series-field and the shunt-field windings assist each other. The larger the load, the stronger is the flux and the higher the generated electromotive force. When the number of turns in the series winding can change, the generator can be over-, level- or under-compounded.

When the generator is level-compounded, the terminal voltage is practically constant between no load and full load. However, if the number of turns reduces, there is a definite fall of terminal voltage between no load and full load, and the generator is under-compounded. When the number of series turns increases above its level-compound number, there is a distinct rise of terminal voltage between no load and full load.

Figures 5.50(A)–(C) show the diverter variable resistor connected in parallel with the series field for the various degrees of compounding.

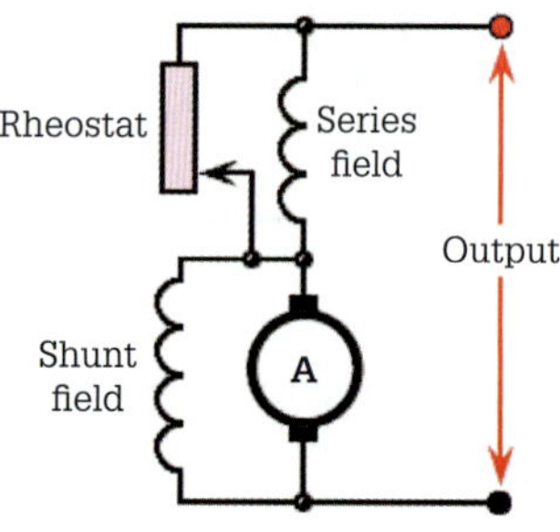

FIGURE 5.50(A) Over-compound

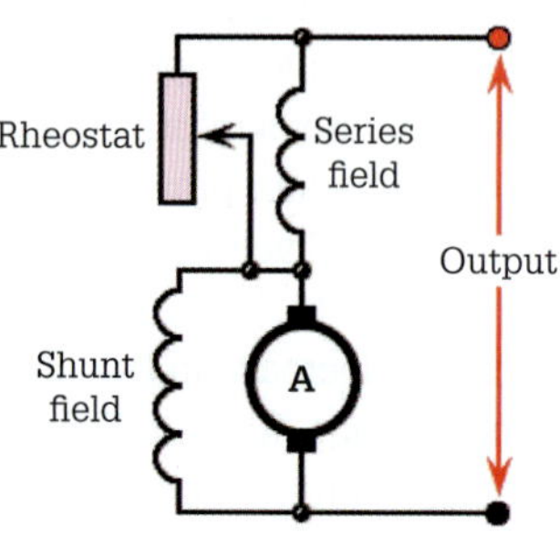

FIGURE 5.50(B) Level-compound

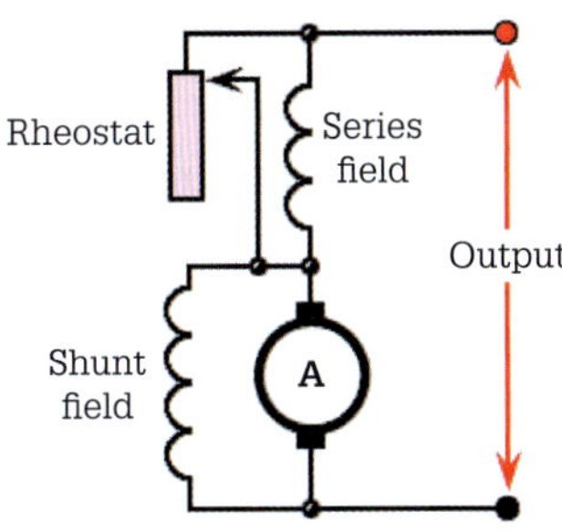

FIGURE 5.50(C) Under-compound

Figure 5.50(A) shows the series field operating at maximum current because the diverter is at full resistance. Full resistance means that minimum current goes through the diverter, and maximum current goes through the series field, creating an over-compound condition. In **Figure 5.50(B)**, the design of the generator is for higher voltage at full load than at no load. The maximum resistance position compensates for extreme changes in current demands and prevents a drop in terminal voltage. **Figure 5.50(C)** shows the diverter adjusted for the under-compound condition. The diverter has minimum resistance, so most of the current goes around the series field, and the generator operates with the characteristics of a shunt generator. Refer to **Figure 5.51** for the loads characteristic of a cumulatively compound generator.

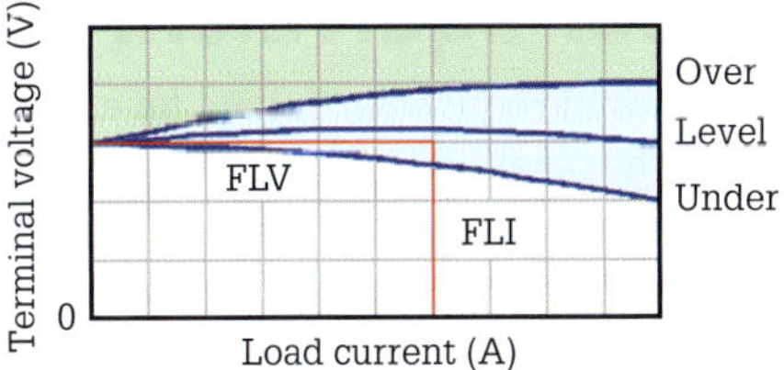

FIGURE 5.51 Load characteristic – cumulatively compound generator

The voltage–speed characteristic of a cumulatively compound generator shows a non-linear terminal voltage in relation to speed as shown in **Figure 5.52**.

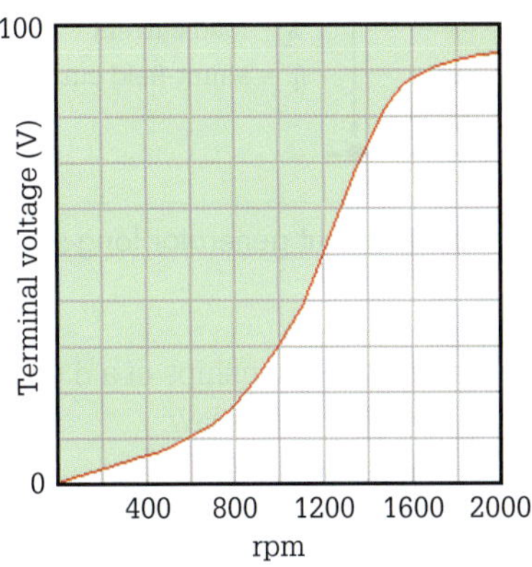

FIGURE 5.52 Voltage–speed characteristic – cumulatively compound generator

In the differentially compound generator of **Figure 5.53**, the series field and the shunt-field winding magnetomotive forces oppose each other. So, the larger the load, the weaker is the flux and the lower the generated electromotive force.

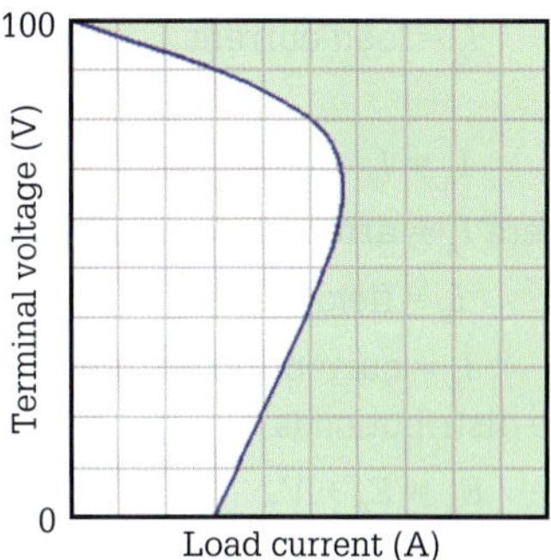

FIGURE 5.53 Load characteristic – differentially compound generator

Differential compound generators have the same characteristics as series generators in that they are fundamentally constant current generators. However, they generate rated voltage at no load with the voltage dropping rapidly as the load current increases.

V and *I* relationships of d.c. generators

Series generators have current and voltage relationships similar to those of series-connected d.c. circuits. The current flow through a series generator is the same as that through each component in the circuit as shown in **Figure 5.54**.

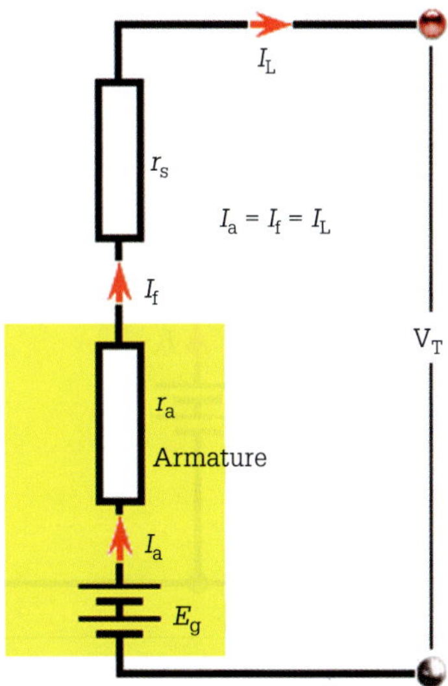

FIGURE 5.54 Series-connected generator components

where E_G = generated voltage (open-circuit voltage or armature voltage)

V_T = terminal voltage

I_A = armature current

I_F = field current

I_L = load current

Therefore:

$$I_A = I_F = I_L$$

where I_A = armature current

I_F = field current

I_L = current flow through the load

The voltage distribution is:

$$E_A = E_F + V_T$$

where E_A = voltage across the armature

E_F = voltage across the field windings

V_T = terminal voltage

Also note that:

$$E_G = V_T + I_A R_A$$

where E_G = voltage generated across the armature

V_T = terminal voltage

$I_A R_A$ = voltage drop across the armature

And:

$$E_A = E_G - I_A R_A$$

Shunt generators have current and voltage relationships similar to those of parallel-connected d.c. circuits as shown in **Figure 5.55**. The current through a shunt generator is the sum of that through each component in the circuit.

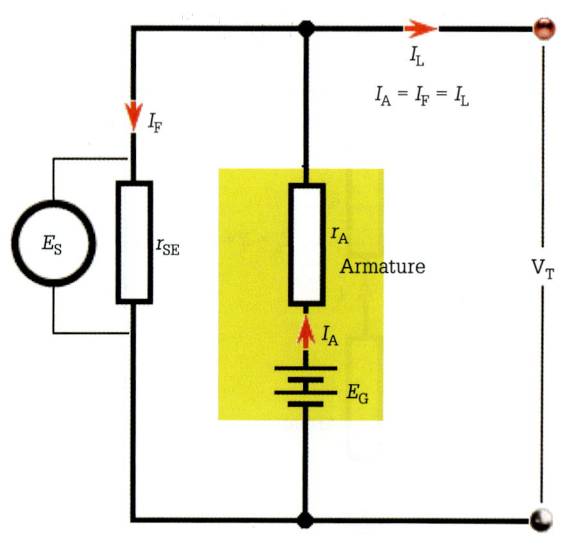

FIGURE 5.55 Shunt-connected generator components

where E_G = generated voltage (open-circuit voltage or armature voltage)

E_S = shunt field voltage

V_T = terminal voltage

I_A = armature current

I_F = field (shunt) current

I_L = load current

r_A = armature circuit resistance

Therefore:

$$I_A = I_F + I_L$$

The voltage in a parallel circuit is the same across each path, therefore:

$$E_A = E_S = V_T$$

Also note that the voltage relationship is:

$$E_G = V_T + I_A R_A$$

Compound generators have two methods of connection: short and long shunt. These generators have current and voltage relationships similar to those of series–parallel-connected d.c. circuits.

In the short shunt:

$$I_A = I_F + I_S \text{ and } I_S = I_L$$

and the voltage relationship is:

$$E_G = E_F = E_S + V$$

In the long shunt:

$$I_A = I_F + I_L = I_S$$

and the voltage relationship is:

$$E_G = E_S + E_F \text{ and } E_F = E$$

Generator equivalent circuit

Figure 5.56 shows the equivalent circuit for a long-shunt d.c. generator. The notation E_G represents the armature conductors in which an induced voltage occurs. This voltage is proportional to the strength of the field flux and the rotation speed of the armature.

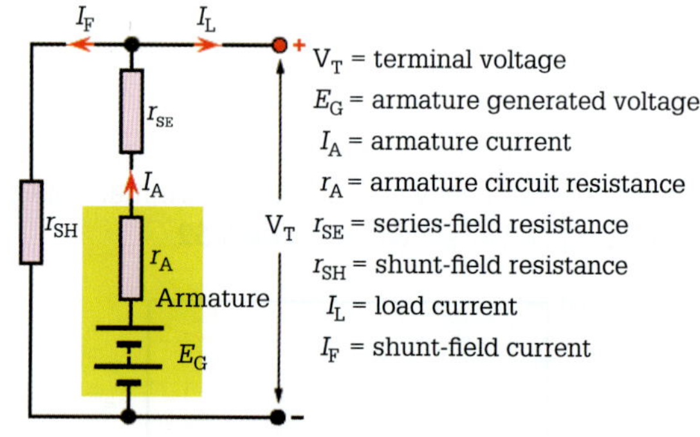

FIGURE 5.56 Direct current generator long-shunt equivalent circuit

Voltage and current relationships of a d.c. generator equivalent circuit refer to Ohm's law:

$$E_G = V_T + I_A (r_A + r_{SE})$$

$$V_T = E_G - I_A (r_A + r_{SE})$$

$$I_A = I_L + I_F$$

where R_A = armature resistance

R_{SE} = series-field resistance

EXAMPLE 5.2

Determine the generated voltage of a 220 V series generator that has an armature current of 250 A and an armature resistance, including brushes, of 0.025 ohms and a series-field resistance of 0.004 ohms.

$$E_G = V_T + I_A(r_A + r_{SE})$$
$$= 220 + 250(0.025 + 0.004)$$
$$= \mathbf{227.25\ V}$$

EXERCISE 5.2

a Determine the generated voltage of a 250 V series generator that has an armature current of 150 A and an armature resistance, including brushes, of 0.045 ohms and a series-field resistance of 0.04 ohms.

b Determine the terminal voltage of a series generator that has an armature current of 300 A, an armature resistance, including brushes, of 0.045 ohms and a series-field resistance of 0.055 ohms when the no-load terminal voltage is 230 V.

Generated voltage (compound generator)

In a compound generator, the shunt field connects across the armature circuit in either the short-shunt or the long-shunt method while the load connects in parallel across the generator terminals. If the load increases, the generated current also increases, which causes a fall in the terminal voltage of the generator. This fall in terminal voltage happens because of three factors:

1 armature reaction
2 reduction in shunt-field current
3 armature resistance (R_A).

When an armature flux develops, its direction reduces the shunt field poles flux. The outcome is a reduced generated voltage and terminal voltage.

When the terminal voltage of the generator reduces because of armature reaction, the voltage across the shunt field and its current falls. The reduced current lessens the strength of the shunt-windings flux with the result that the generated and terminal voltages fall.

The armature circuit (R_A) of a generator contains the following resistances connected in series:

- armature winding resistance of the conductors
- brush resistance
- contact resistance between the brushes and the commutator.

Commutator resistance

When the generator has no load, no current exists in the armature circuit. Furthermore, the terminal voltage is the same value as the generated voltage which in this case is the open-circuit voltage (no $IR\ V_D$ exists). Once a load is placed on the generator, a voltage drop exists across the armature circuit due to the series resistances. The terminal voltage becomes less than the generated voltage. Calculation of the generated voltage uses the following equation.

$$E_G = V_T + I_A R_A$$

where E_G = generated voltage (open-circuit voltage or armature voltage)
V_T = terminal voltage
I_A = armature current
R_A = armature circuit resistance

If there is a small change from no load to full load, the generator has good regulation.

EXAMPLE 5.3

1. A short-shunt compound generator, as illustrated in **Figure 5.57**, has armature and series-field and shunt-field resistances of 0.5 Ω, 0.025 Ω and 125 Ω respectively. The generator supplies a 5 kW load at 220 V (terminal voltage). Calculate the generated electromotive force (E_G).

Armature resistance (R_A) = 0.05 Ω
Series-field resistance (R_{SE}) = 0.025 Ω
Shunt-field resistance (R_{SH}) = 125 Ω
Load = 5 kW
Terminal voltage = 220 V

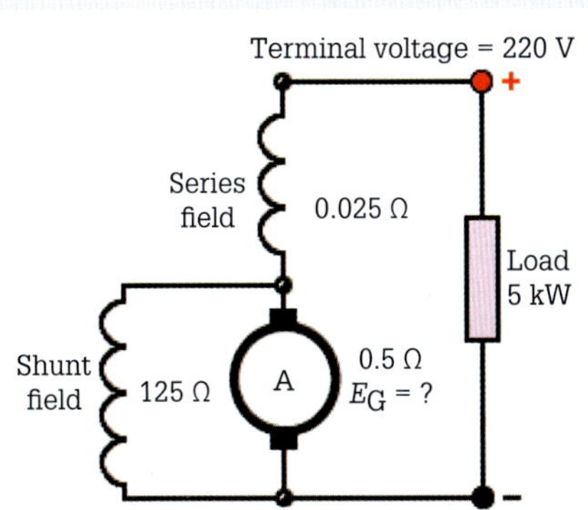

FIGURE 5.57 Short-shunt compound generator

»

Generated emf (E_G)
Load current:

$$I = \frac{P}{V} = \frac{5000}{220} = \mathbf{22.73\ A}$$

Voltage drop in the series winding:

$$V_{D_{SE}} = IR_{SE} = 22.73 \times 0.025 = 0.57\ V$$

Voltage across shunt-field winding:

$$V_{d_{sh}} = V + IR_{se} = 220 + 0.57 = \mathbf{220.57\ V}$$

Shunt-field current:

$$I_{SH} = \frac{V_{d_{SH}}}{R_{SH}} = \frac{220.57}{125} = 1.76\ A$$

Armature current:
Generated emf:

$$I_A = I + I_{SH} = 22.73 + 1.76 = 24.5\ A$$

$$E_G = V + IR_{SE} + IR_A = 220 + 0.57 + (24.5 \times 0.05) = \mathbf{221.8\ V}$$

Note: There is an additional voltage drop due to brush resistance and its contact resistance. If we assume a 0.9 V drop for each brush, then the actual generated emf is:

$$221.8 + (0.9 \times 2) = \mathbf{223.6\ V}$$

2. A long-shunt compound generator, as illustrated in **Figure 5.58**, has an armature and series-field and shunt-field resistances of 0.06 Ω, 0.04 Ω and 200 Ω respectively. The generator supplies a 200 A load at 400 V (terminal voltage). Calculate the generated electromotive force (E_G).

Armature resistance (R_A) = 0.06 Ω
Series-field resistance (R_{SE}) = 0.04 Ω
Shunt-field resistance (R_{SH}) = 200 Ω
Load current = 200 A
Terminal voltage = 400 V

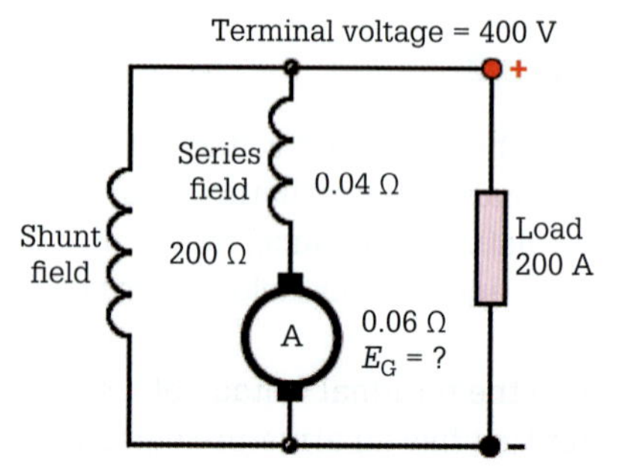

FIGURE 5.58 Long-shunt compound generator

Generated emf (E_G)
Shunt current:

$$V_T = \frac{V}{R_{SH}} = \frac{400}{200} = 2.00\ A$$

Armature current:

$$I_A = I + I_{SH} = 200 + 2 = \mathbf{202.0\ A}$$

Generated emf:

$$E_G = V_T + I_A R_A + I_A R_{SE} = 400 + (202.0 \times 0.06) + (202.0 \times 0.04) = \mathbf{420.2\ V}$$

Note: There is an additional voltage drop due to brush resistance and its contact resistance. If we assume a 0.9 V drop for each brush, then the actual generated emf is:

$$420.2 + (0.9 \times 2) = \mathbf{422.0\ V}$$

EXERCISE 5.3

a A short-shunt compound generator, similar to that shown in **Figure 5.57**, has armature and series-field and shunt-field resistances of 0.15 Ω, 0.035 Ω and 175 Ω respectively. The generator supplies a 2.5 kW load at 230 V (terminal voltage). Calculate the generated electromotive force (E_G).

b A long-shunt compound generator, similar to that shown in **Figure 5.58**, has an armature and series-field and shunt-field resistances of 0.20 Ω, 0.15 Ω and 250 Ω respectively. The generator supplies a 150 A load at 250 V (terminal voltage). Calculate the generated electromotive force (E_G).

Generator ratings and applications

A generator has its rating in power output. Since the generator is at a specified voltage, the rating is the current the generator can safely supply at its rated voltage. Generator rating and performance data are on the nameplate attached to the generator.

The rotation of generators is termed clockwise or anticlockwise as viewed from the driven end. Usually the direction of rotation is on the data plate. If no direction stamp occurs on a plate, an arrow on the cover plate of the brush housing may mark the rotation.

Although the use of solid-state converters and control systems has changed the demand for d.c. generators, they still have application in various industries. Series-wound generators are of little use for general power work. However, they do operate as voltage boosters in long supply lines because of large current loads and for experimental work in laboratories.

Shunt generators have applications only where the load is entirely predictable, and the generator selected to carry that load is without serious voltage drop. Some aeroplanes have shunt generators to provide d.c. power needs.

Compound generators have application for shipboard d.c. power because they are versatile and respond to a wide variety of loads. Compound generators also find use as electric traction generators for electric and diesel-electric traction locomotives and are used as the power supply of the d.c. motors and the auxiliary services. In addition, some aeroplanes use them for d.c. power.

Use of a level-compound generator happens when a constant voltage is required, and the distance to the load is minimal. In comparison, over-compound generators are used when d.c. power needs transmitting over long distances. The rise in the terminal voltage compensates for the voltage drop in the service line.

The differentially compound generator occurs where overload protection is more necessary than constant voltage. Examples are electric winches and dredges where an overload or short circuit causes a voltage drop to limit the output current to a safe value. In addition, they have an application as welding generators where the terminal voltage must fall once the arc strikes.

REVIEW QUESTIONS

1. What are the four main classifications for d.c. generators?
2. How does a shunt-connected generator produce the initial generated emf?
3. How does a separately excited generator differ from a shunt generator?
4. Why does the terminal voltage of a separately excited generator drop slightly as the load increases?
5. How does a compensating winding connect?
6. How is the generated voltage in a series generator controlled?
7. Draw the circuit diagram of a shunt-connected generator.
8. Draw the load characteristic curve for a shunt-connected generator.
9. Draw a typical magnetisation curve for a separately excited generator.
10. What are two possible shunt field connections for a compound-connected generator?
11. How is the rating of a generator specified?
12. A long-shunt compound generator, similar to that shown in **Figure 5.58**, has an armature and series-field and shunt-field resistances of 0.10 Ω, 0.08 Ω and 180 Ω respectively. The generator supplies a 200 A load at 220 V (terminal voltage). Calculate the generated electromotive force (E_G).

5.4 Direct current motor

The d.c. motor's operating principle is such that, when a current passes through a conductor in a magnetic field, it experiences a mechanical force. This is termed motor action.

Motor action

Motor action as illustrated in **Figure 5.59** is the result of the interaction between magnetic fields.

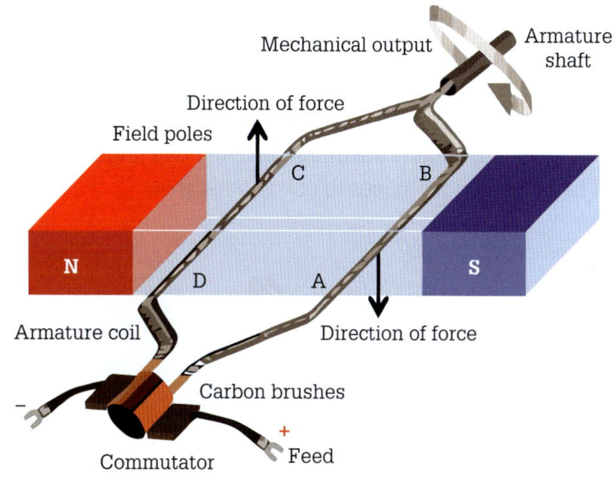

FIGURE 5.59 Motor action

The two sides of the armature coil carry current in opposite directions. The electromagnetic field around the armature conductor AB has a different direction from that of armature conductor CD. The electromagnetic field around the armature conductors interacts with the electromagnetic field created by current in the field coil windings (poles) or the magnetic field of permanent magnet poles. The armature revolves because of the repulsion and attraction interaction between the fields. This interaction produces a turning moment or torque that causes the armature to rotate in a clockwise direction towards the next main pole. Because the commutator is also revolving, the commutator, in reversing the electromagnetic fields of the armature conductors, switches the polarity of the armature current. This action ensures that the torque always acts in the same direction.

Figure 5.60 illustrates that when the current through the conductor flows in one direction there is an upward force action on the conductor. If either the direction of the current or the magnetic field reverses, the force on the conductor acts downwards. In addition, the force acts at right angles to both the current and field directions, causing the conductor to move.

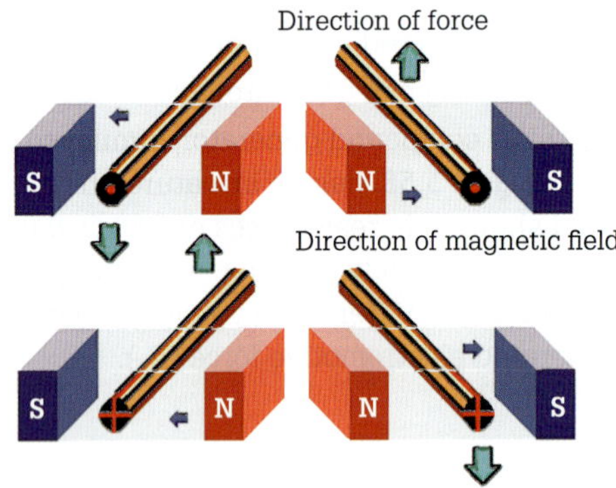

FIGURE 5.60 Motor action

Fleming's left-hand motor rule

Knowing the direction of current flow and the magnetic field allows application of Fleming's left-hand motor rule as shown in **Figure 5.61**. This rule indicates the relationship between the conventional current flow, the magnetic field and the direction of the force. If the thumb and first two fingers of the left hand are at right angles to one another then:

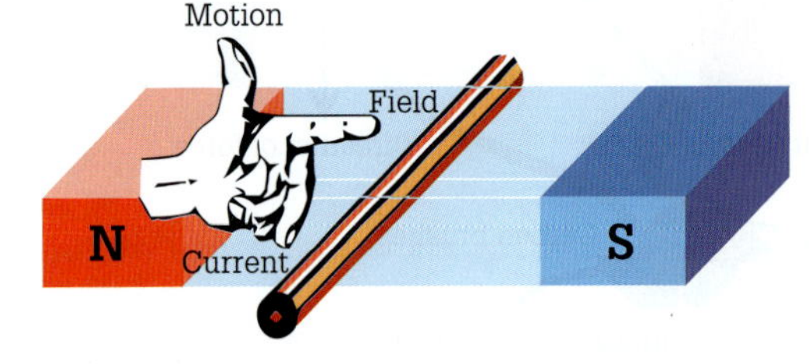

FIGURE 5.61 Fleming's left-hand motor rule

This rule is only valid where the field and current directions are at right angles to each other. A force still acts on the conductor if the field and current directions are at some other angle, but the direction in which the force acts is more difficult to predict. We can understand why this rule works by considering the magnetic forces that the current in the conductor exerts on the magnetic field of the magnets. Refer to **Figure 5.62**.

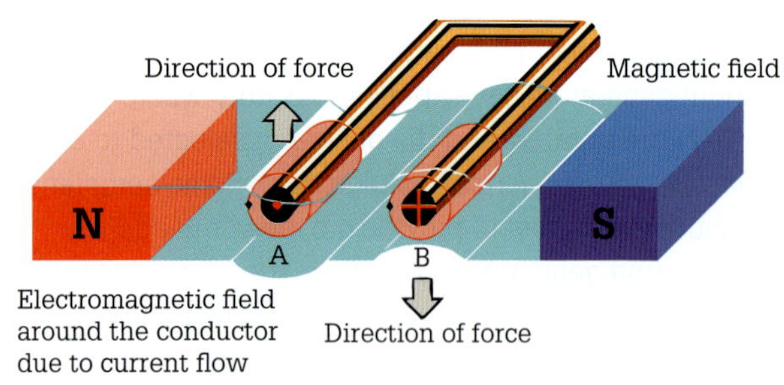

FIGURE 5.62 Forces acting on a conductor

Forces acting on a conductor

Figure 5.62 shows an end view of the electromagnetic field around a loop of copper wire carrying current as it exists within the magnetic field of two permanent magnets. The red dot represents current flowing towards the observer, while the cross represents current flowing away from the observer. The red circular pattern around the conductor represents the electromagnetic field created by the current in these conductors.

Below 'conductor A' the magnetic field of the permanent magnets combines with the electromagnetic field of the conductor creating a stronger field below the conductor. Above this same conductor, the direction of the electromagnetic field of the copper conductor is such that it opposes the magnetic field of the permanent magnets, thereby making the field weak above the conductor. Conductor A is, therefore, lifted upwards by the stronger magnetic field under the conductor. A similar effect exists with conductor B; however, a stronger field exists above the conductor and, therefore, pushes conductor B downwards.

The direction of the force on the conductors in a magnetic field can be determined by applying the left-hand motor rule. The lifting of one side of the loop and the pushing down of the other side creates a turning effect, or torque, on the copper loop. This combination of magnetic forces is the torque that turns the armature or rotor of electric motors.

Figure 5.63 illustrates the direction of the force experienced by the conductors.

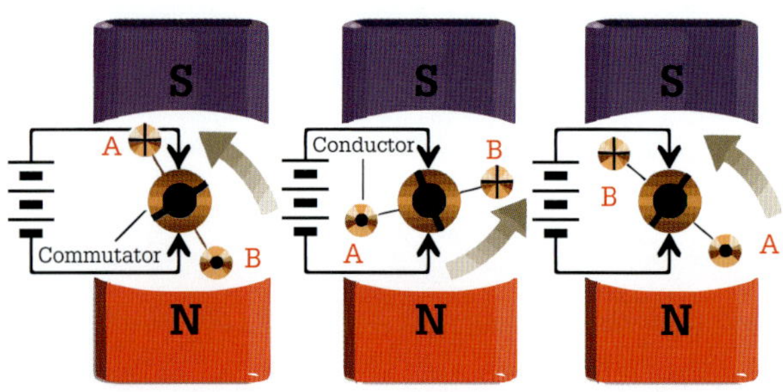

FIGURE 5.63 The direction of force on the conductors

The two-segment commutator reverses the current flow in the conductors (A and B) as they pass from one pole to the next so that the current in either conductor is always the same as it passes from a given pole. Consequently, the direction of the force is the same and the armature loop rotates continuously in a given direction. At the instant when the brushes are passing over the insulated joints between the segments in the commutator, the conductors are moving parallel to the lines of magnetic flux of the field poles; the conductors, therefore, experience no force. However, the speed of rotation of the armature loop keeps it moving until the 'point of no force' occurs. In practice, the armature has many coils and a multi-segment commutator. As a result, the force on the armature is nearly constant, and the 'point of no force' does not occur.

Generated emf

When the armature starts rotating in a d.c. motor, its conductors cut the field flux and hence there is an emf induced in the armature called the back emf or counter-electromotive force. It is called back emf because it moves in the opposite direction to the applied voltage according to Lenz's law.

The back emf (E_G) is always less than the applied voltage and its magnitude, and can be determined if the values of the terminal voltage (V_T), armature current (I_A) and resistance (R_A) are available.

$$E_G = V_T - I_A R_A$$

The back emf provides the d.c. motor to be a self-regulating machine by regulating the flow of armature current. The back emf allows the armature to draw just enough armature current to develop the torque required by the load. There must be slightly more supply voltage than back emf to allow sufficient armature current to flow. Once the load engages, the armature slows down, reducing the back emf, which then allows increased armature current to flow, with the result of improved power developed to drive the load.

Calculation of the generated emf produced by the armature of a d.c. generator or motor uses the following equation.

$$E_G = \frac{\Phi ZNP}{60\ a}$$

where E_G = generated emf in volts (V)
Φ = flux per pole in webers (Wb)
Z = total number of active armature conductors
N = speed of the armature in rpm
P = number of poles
a = number of parallel paths through the armature (two parallel paths for wave and equal to the number of poles for a lap winding)

EXAMPLE 5.4

A four-pole wave-wound d.c. motor armature has 750 active conductors in a field pole flux of 50 mWb. Calculate the generated emf when the motor rotates at 1500 rpm.

$$E_G = \frac{\Phi ZNP}{60\ a} = \frac{(50 \times 10^{-3}) \times 750 \times 1500 \times 4}{60 \times 2}$$

$$= \mathbf{1.875\ kV}$$

EXERCISE 5.4

a A four-pole lap-wound d.c. motor armature has 650 active conductors in a field pole flux of 50 mWb. Calculate the generated emf when the motor rotates at 1000 rpm.

b A wave-wound d.c. motor armature has 600 active conductors in a field pole flux of 25 mWb. Calculate the number of poles if the motor has a generated emf of 1200 V when rotating at 1200 rpm.

c A four-pole lap-wound d.c. motor armature rotates in a field pole flux of 50 mWb. Calculate the number of active armature conductors if the generated emf is 800 V when the motor rotates at 1000 rpm.

A d.c. motor increases its speed, increasing its back emf and reducing its armature current until a balance transpires. A balance means that the armature current generates just enough power to overcome friction and load requirements. At that point, the armature no longer accelerates, and its speed remains constant.

Torque production

All d.c. machines when in service produce torque and generate voltage at the same moment in time. If operating as a motor the d.c. machine develops torque and a counter-electromotive force (cemf). If the d.c. machine is a generator, it generates an electromotive force and if servicing a load it develops a counter-torque. Torque is the turning effort of a motor expressed in newton-metres (Nm). It is a measure of how much force acting on an object causes that object to rotate. The force on the current-carrying conductor in a magnetic field depends upon:

1 the flux density of the field, B, in teslas
2 the strength of the current, I, in amperes
3 the length of the conductor perpendicular to the magnetic field, l, in metres.

When the magnetic field, the current and the conductor are mutually at right angles then:

$$\text{force, } F = BIl \text{ in newtons}$$

When the conductor and the field are at an angle θ° to each other then:

force, $F = BIl \sin \theta$ in newtons

EXAMPLE 5.5

A conductor as shown in **Figure 5.64** carries a current of 15 A and is at right angles to a magnetic field having a flux density of 0.5 T. If the length of the conductor in the field is 45 cm, calculate the force acting on the conductor.

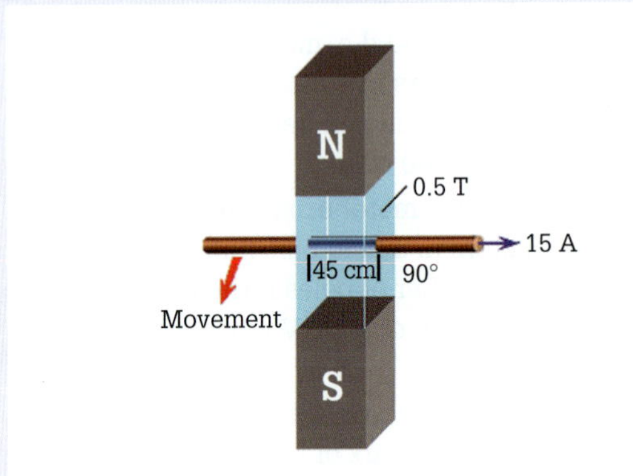

FIGURE 5.64 Force example

$$F = B \times I \times l$$
$$= 0.5 \times 15 \times 0.45$$
$$= \mathbf{3.375\ N}$$

EXERCISE 5.5

A conductor as illustrated in **Figure 5.65** carries a current of 25 A and is inclined at an angle of 30° to the direction of the magnetic field. If the field has a flux density of 0.75 T and the length of the conductor in the field is 35 cm, calculate the force acting on the conductor.

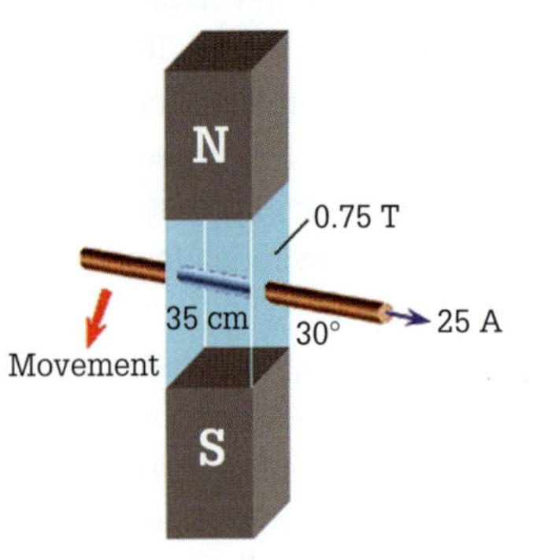

FIGURE 5.65 Force exercise

When the effective radius between the conductor and the pivot point has a value, determination of the magnitude of the torque produced by each conductor in a d.c. motor can be determined using the equation:

$T = BIlr$ in newton-metres (Nm)

where T = torque in newton-metres (Nm)

B = flux density in teslas (T)

l = conductor length in metres (m)

I = current in amperes (A)

r = effective radius between the conductor and the pivot point in metres (m)

EXAMPLE 5.6

The two-pole armature in **Figure 5.66** has an effective radius of 0.2 m and a single armature turn with sufficient length in each armature slot of 0.25 m. If the armature conductor carries a current of 5 A when exposed to a flux density of 0.5 T, calculate the torque developed by one side of a single conductor loop.

$$T = BIlr$$
$$= 0.5 \times 0.25 \times 5 \times 0.2$$
$$= \mathbf{0.125\ Nm}$$

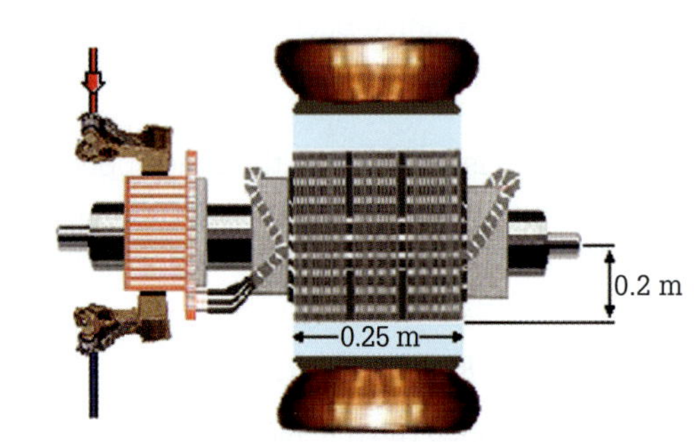

FIGURE 5.66 Two-pole armature

EXERCISE 5.6

a A simple single-turn coil forms the armature in a two-pole motor and has an effective radius of 0.15 m and a length influenced by the magnetic field of 0.25 m. If the armature conductor carries a current of 2 A when exposed to a flux density of 0.5 T, calculate the torque developed by the conductor loop.

b A simple single-turn coil forms the armature in a two-pole motor and has an effective radius of 0.2 m and a length influenced by the magnetic field of 0.15 m. If the armature coil produces 0.050 Nm of torque, calculate the conductor current when exposed to a flux density of 0.5 T.

This equation for torque applies only to a d.c. machine with a single conductor armature turn. In practice, armatures have multiple coils, each with many turns and each operating to produce a portion of the total torque. The equation to determine the torque must consider the number of poles, the number of active armature conductors, the number of parallel paths that the armature winding has and the flux per pole. The torque developed by a d.c. motor also has a relationship with the power output and motor speed:

$$T = \frac{60P}{2\pi N}$$

where T = torque in newton-metres (Nm)
P = power output in watts (W)
N = the speed of rotation in revolutions per minute (rpm)

EXAMPLE 5.7

The power available at the shaft of a d.c. motor is 15 kW at a rotational speed of 1500 rpm. Calculate the available shaft torque.

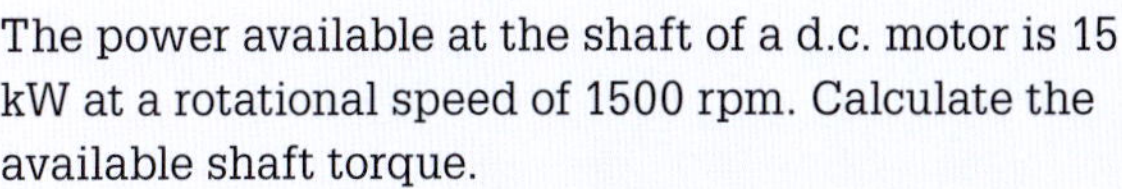

$$\begin{aligned} T &= \frac{60P}{2\pi N} \\ &= \frac{60 \times 15\ 000}{2\pi \times 1500} \\ &= 95.5\ \text{Nm} \end{aligned}$$

EXERCISE 5.7

a The power available at the shaft of a d.c. motor is 10 kW at a rotational speed of 1200 rpm. Calculate the available shaft torque.

b A d.c. motor is producing 65 Nm of torque at the shaft when rotating at 1200 rpm. Calculate the power rating of the motor.

The graph in **Figure 5.67** shows the ideal torque speed-curve for a d.c. motor. Note that the torque is inversely proportional to the speed of the armature shaft. The starting torque point on the graph represents where the motor torque is at maximum. At this point, the armature shaft is not rotating. The second point indicated on the graph is where the speed of the motor is at maximum. At this point, there is no torque applied to the armature shaft.

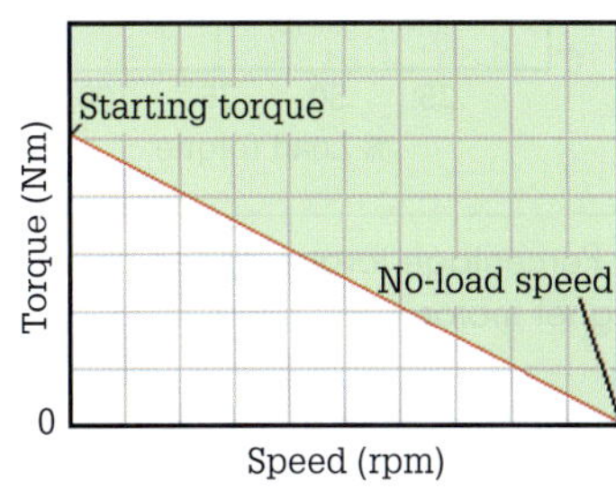

FIGURE 5.67 Torque–speed characteristic curve

Commutation

The commutator and the graphite brushes constitute a rotary switch that provides the switching action for the process of commutation, which is the reversing of armature coil current.

Successful operation of a d.c. machine requires that during the brief interval in which any copper segments of the commutator are short circuited by the brushes, the current in the shorted armature coils ceases and starts again in the reverse direction through the armature coils. This is commutation and occurs in an area of the d.c. machine called the neutral plane as shown in **Figure 5.68**.

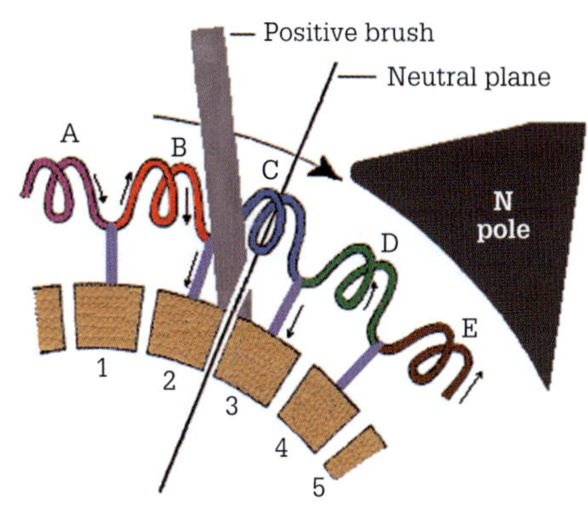

FIGURE 5.68 Generator commutation

The armature coils A, B, C, D and E connect commutator segments 1, 2, 3, 4 and 5, with the positive commutator brush in the neutral position. In addition, the brush is in contact with segments 2 and 3. Consequently, armature currents in the coils on either side of the neutral position flow into the brush through segments 2 and 3 while coil C is short circuited. As the armature turns, the commutator segments come successively into contact with the brush and upon leaving experience a reversal of armature current.

SWITCH ON

In practical applications, the brushes are usually set at an angle of lead or lag to the neutral plane. For this reason, there is minimal sparking at the point where the brushes and commutator surface meet. Moreover, the sparking is due to the self-induction in the coil undergoing commutation. In addition to the resetting of the brushes, interpoles in large d.c. machines reduce or eliminate the effects of self-induction.

REVIEW QUESTIONS

1. What is the operating principle of a d.c. motor?
2. What term describes the interaction detailed in question 1?
3. State Fleming's left-hand motor rule.
4. A four-pole wave-wound d.c. motor armature has 650 active conductors in a field pole flux of 20 mWb. Calculate the generated emf when the motor rotates at 1200 rpm.
5. What factors determine the magnitude of the force developed by a current carrying conductor?
6. What is the term that describes the turning effort of a motor?
7. A two-pole armature with an effective radius of 0.3 m has a single armature turn with an effective length in each armature slot of 0.45 m. If the armature conductor carries a current of 20 A when exposed to a magnetic field of 10 T, calculate the torque developed by the coil.
8. The power available at the shaft of a d.c. motor is 20 kW at a rotational speed of 3000 rpm. Calculate the available shaft torque.
9. Referring to the 'ideal' torque–speed characteristic curve of **Figure 5.67**, at what point is there no torque applied to the shaft?
10. To what does commutation refer?

5.5 Types of direct current motors

Permanent-magnet machines

Magnetic substances have applications that make use of the magnetic field either to convert electric energy to mechanical energy or to convert mechanical energy to electric energy. A permanent-magnet d.c. machine can have a conventional armature together with field poles made of a ferromagnetic material with no conductor windings. Direct current is applied to the armature only. Refer to **Figure 5.69**.

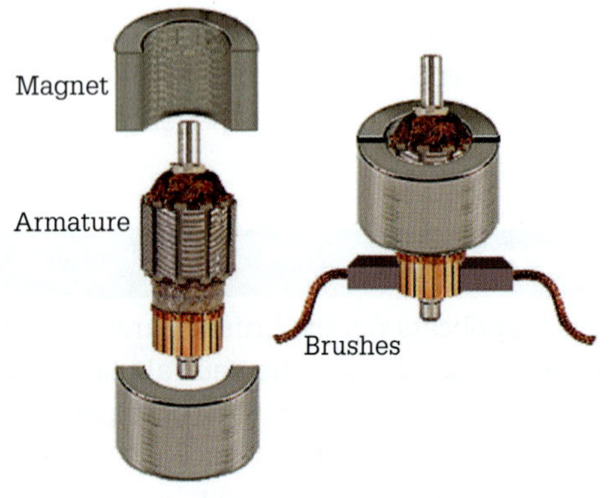

FIGURE 5.69 Permanent-magnet motor

The primary permanent-magnet motor has a stationary magnetic field assembly, and rotating assembly supported by bearings. The field assembly is the stator consisting of the yoke and permanent magnets. The rotating assembly is an armature and consists of the wound core, shaft and armature. The brushes, brush-holders and lead connections are in the commutator end shield. Interchanging the supply leads can reverse the direction of rotation of permanent-magnet d.c. motors.

Permanent-magnet d.c. motors use conventional armatures and permanent magnets rather than windings in the field section. Like other motor elements, the choice of a magnet is critical, and magnets with different properties used for permanent-magnet d.c. motors include the non-rare-earth and rare earth magnets.

These motors have substantially lower noise than other types of d.c. motors; high torque to volume ratio; enclosed construction; and high reliability. Direct current permanent-magnet motors produce high torque at low speed (up to 150% of full-load torque at starting) and are self-braking upon disconnection of electrical power. They have power ratings ranging from 15 watts to 400 kilowatts and with rotational speeds of up to 40 000 rpm. Refer to **Figure 5.70** for the percentage speed–load torque characteristic of a permanent-magnet d.c. motor.

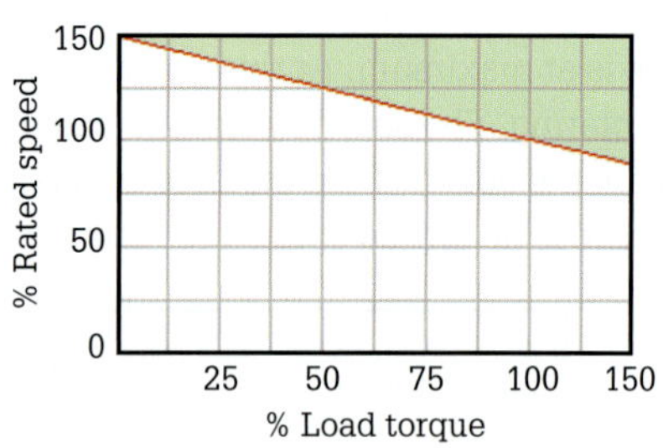

FIGURE 5.70 Percentage torque–speed characteristic – permanent-magnet motor

As the field is constant at all times, the performance curve is linear, and the armature current drawn varies linearly with torque. The only methods for speed control with these motors are armature voltage variation with a variable d.c. supply and armature resistance control with a rheostat as illustrated in **Figure 5.71**.

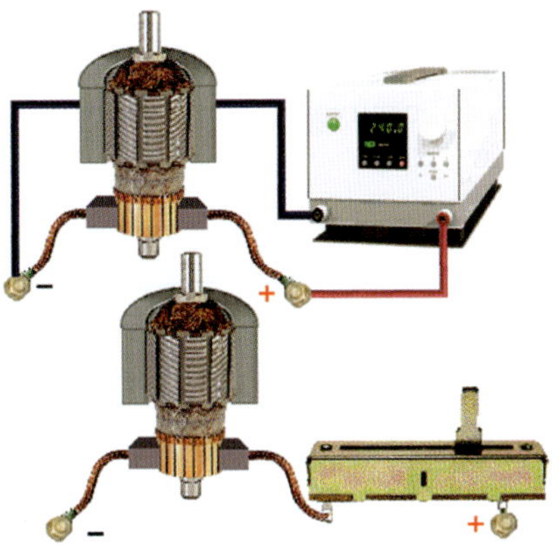

FIGURE 5.71 Speed control with armature voltage variation

The magnetic flux of a permanent-magnet d.c. motor is fixed; it is not possible to vary the field and control its speed. The only method of speed control is armature voltage or armature resistance control that produces motor speed characteristics similar to those of shunt-wound motors.

Some permanent-magnet motors cannot endure continuous operation because they overheat rapidly, which destroys the permanent magnets. Permanent-magnet motors find use in many battery-operated devices and in applications such as printing, packaging, pharmaceutical, plastics, textiles, food-processing machinery, domestic appliances, conveyors, elevators and as main drives or control motors. Motor vehicle manufacturers install d.c. permanent-magnet motors in their vehicles to operate power seats, windows and windscreen wipers.

Shunt-wound motors

Shunt-wound d.c. motors are more common than any other d.c. motor. In this motor, the shunt-field winding connects across the d.c. supply and the armature. Consequently, the shunt winding provides the electromagnetic flux that interacts with the armature current to produce torque resulting in motor action. **Figure 5.72** shows a four-pole shunt-wound motor connection using a lap-wound armature.

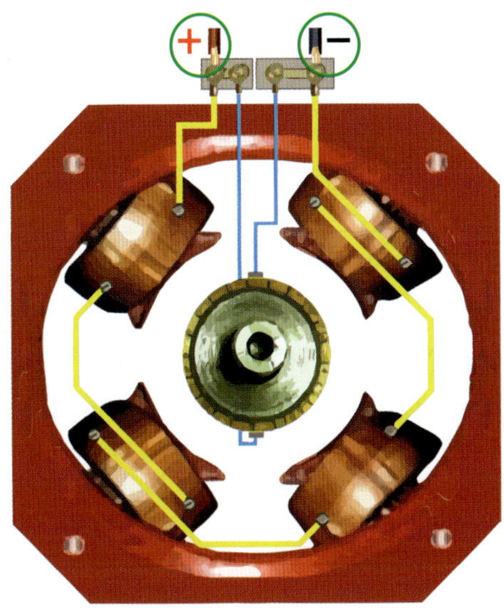

FIGURE 5.72 Four-pole shunt motor

Figure 5.73 shows the circuit diagram of a shunt-connected motor incorporating a shunt-field rheostat to provide speed control.

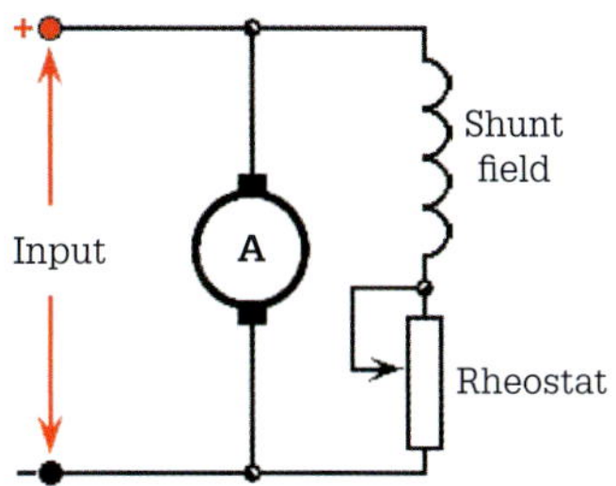

FIGURE 5.73 Circuit diagram of a shunt-connected motor

An illustration of a four-pole shunt-wound motor connection with interpoles and a wave-wound armature is shown in **Figure 5.74**.

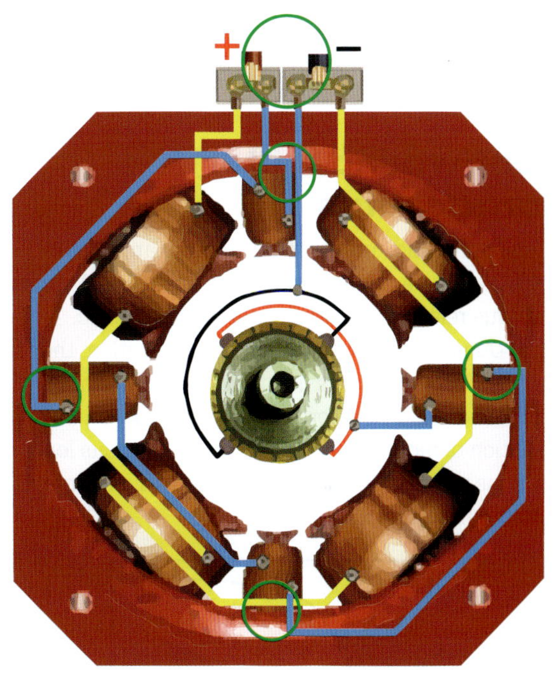

FIGURE 5.74 Four-pole shunt-wound motor with interpoles

The shunt field coils are wound with many turns of a small-cross-sectional-area copper conductor to produce a coil with a relatively high resistance. The large number of turns in the shunt-field coils produces a strong electromagnetic field.

This electromagnetic field consumes energy only as a copper or I^2R loss; it does no useful work of itself. However, it does provide the necessary environment for the armature conductor's electromagnetic field to thrust against when producing torque. If wattmeters connect in both the armature circuit and the shunt-winding circuit, the shunt-winding wattmeter indicates no additional energy drawn by it as the motor load increases. However, the armature circuit shows a definite increase in energy demanded from the supply as the load increases. The result demonstrates that the armature conductor electromagnetic field engages in useful work in developing torque to cause rotation. A shunt-wound motor is a constant-speed motor.

When a load applies to the armature shaft, the armature tends to slow. This minimal loss in speed reduces the counter-electromotive force and results in an increase in armature current drawn from the supply. This action continues until the increased armature current produces a larger electromagnetic field to create a torque capable of meeting the demand of the increased load. Because of this action, the shunt-wound motor can compensate for any changes in load conditions and still maintain a constant speed. Shunt-wound d.c. motors provide medium starting torque, 125% to 200% full load, and are capable of delivering 300% of full-load torque for short periods. The direction of armature rotation changes in the shunt motor by reversing the direction of supply current in either the shunt-field-winding circuit or the armature circuit. If the motor has an interpole winding or a stabilised shunt winding (refer to compound-wound d.c. motors), the same relative direction of current needs to be maintained in these windings. For this reason, it is always simpler to reverse the direction of the armature current at the brush-holders.

Shunt-wound motor characteristics

The characteristic curves for a d.c. motor are a set of graphs that relate the magnetisation, speed, load current, and torque relationships at a constant voltage d.c. supply. Due to the effects of magnetic saturation, the flux established in the poles is not directly proportional to the applied magnetomotive force. For this reason, an accurate calculation of motor speed and torque for various load conditions requires the use of a magnetisation curve as illustrated in **Figure 5.75**.

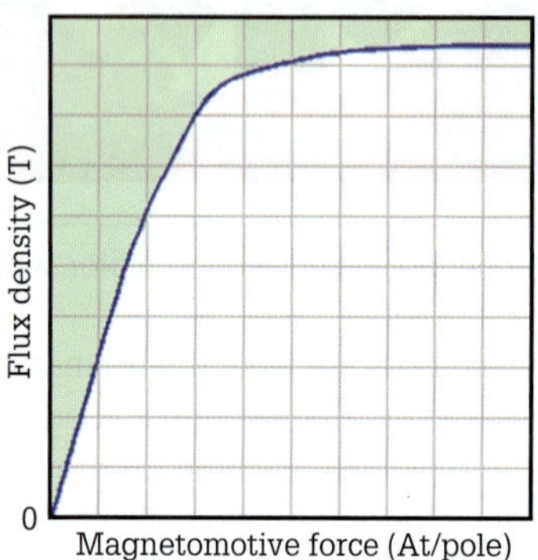

FIGURE 5.75 Magnetisation curve – d.c. motor

Speed–load characteristic

A separately excited d.c. motor has its field circuit energised by a separate constant voltage supply, while a shunt motor has its field circuit directly energised by the voltage supply to the armature circuit. When the voltage supply to these motors is constant, their speed characteristics are the same. Shunt-wound motors have their field coils connected in parallel (shunt) with the armature. The voltage across the field coil is independent of the armature. By connecting a rheostat in series with the field winding, excellent speed control occurs by varying the current flow through the field. **Figure 5.76** shows the speed characteristic curve of a shunt motor demonstrating good regulation.

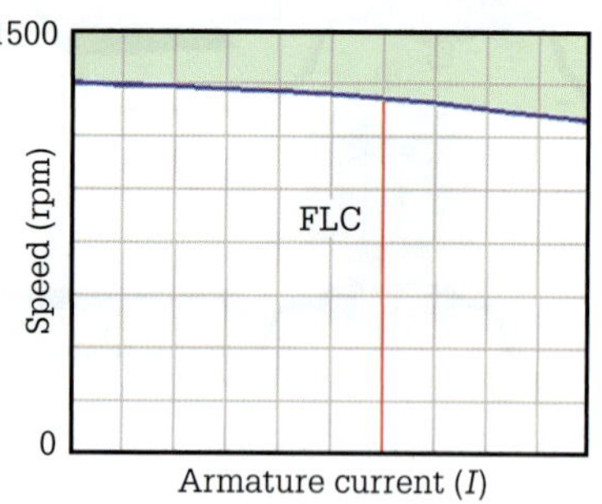

FIGURE 5.76 Speed–load characteristic – shunt motor

Although the speed characteristic falls slightly, the shunt motor is a constant-speed machine. The speed drops slightly with an increase in armature current due to armature reaction weakening the shunt field flux. The weakened flux reduces the induced voltage due to increasing armature voltage drop, and this tends to reduce speed. Under normal operating conditions, the speed of the shunt motor is set with the field rheostat. A shunt-wound motor can have its direction changed by reversing either the armature or field current. Applications include fans, blowers, centrifugal pumps, conveyors, elevators, printing presses, woodworking machines and metalworking machines.

Torque characteristic

The torque characteristic indicates that the torque increases in a linear relationship with an increase in armature current. Shunt motors are capable of delivering 300% of full-load torque for short periods. **Figure 5.77** shows the torque characteristic for a shunt motor.

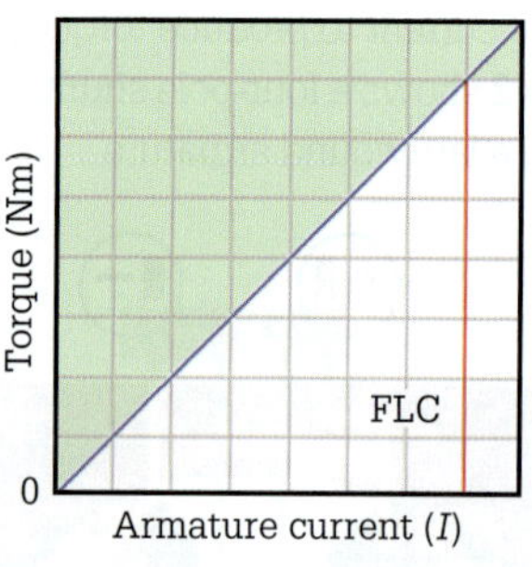

FIGURE 5.77 Torque characteristic – shunt motor

Series-wound motors

A series-wound d.c. motor has a field winding of relatively few conductor turns, but with a large-cross-sectional area giving the field a low resistance. In addition, the armature and the field circuit connect in series, providing only one path for current to flow. There is no shunt-field winding to provide the base electromagnetic flux against which the armature field reacts. **Figure 5.78** shows the circuit diagram of a series-connected motor and **Figure 5.79** the series d.c. motor connections.

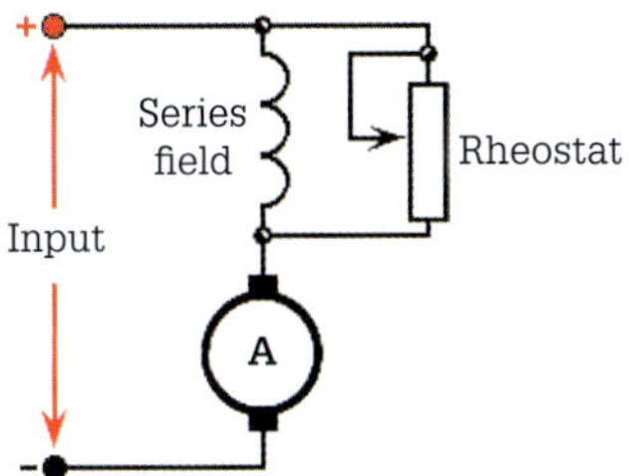

FIGURE 5.78 Circuit diagram of a series-connected motor

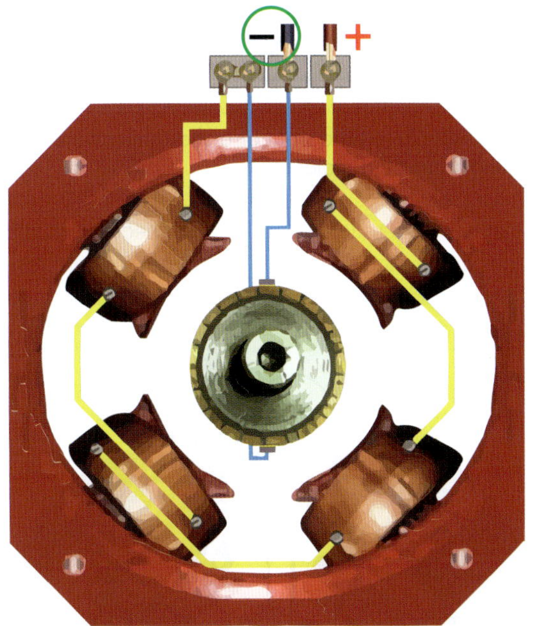

FIGURE 5.79 Four-pole series-connected motor

With the shunt motor, the speed of the shaft is constant because any increase in torque causes an increase in armature current. By contrast, in a series motor, any load increase causes an increase in armature and field current and their fluxes remain in harmony with each other. When the load increases the armature slows and the counter-electromotive force reduces. At the same time the armature current increases, and the field strength increases, reducing the motor speed. With a decrease in speed an increase in torque occurs. Since the torque depends on the interaction of armature and field flux, the torque produced increases as the square of the increased current drawn from the supply. Consequently, an increasing load is limited because of the decrease in armature speed. In comparison to a shunt motor, a series motor develops a greater starting torque (when the current is doubled the torque increases four times) than will a shunt motor for the same increase in current.

With no shunt winding to provide the controlled electromagnetic flux environment, a series motor accelerates to destructive speeds if there is no attached load. Removing the load causes the developed torque of the armature to be greater than the load torque and that results in an increase in speed. As the counter-electromotive force begins to increase, the current drawn from the supply decreases, resulting in a reduced series-field flux. So, the weakened field causes the armature to turn faster. These synchronised opposing actions of decreasing flux and increasing speed effectively prevent the counter-electromotive force from checking the increasing speed by rising in proportion to it. The direction of rotation of a series motor occurs by reversing the current through the series winding or the direction of armature rotation. **Figure 5.80** shows a four-pole series-connected motor with interpoles.

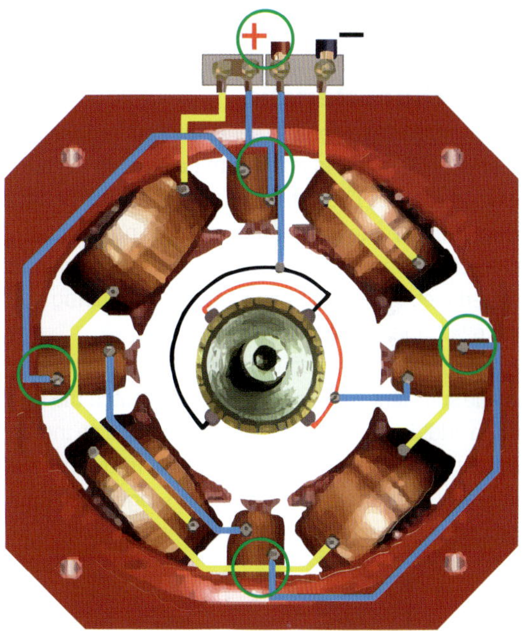

FIGURE 5.80 Four-pole series-connected motor with interpoles

The series motor's high starting torque and its high ratio of light-load to full-load speed characteristic makes this motor suitable for traction motors on locomotives, hoists and cranes.

Series-wound motor characteristics

Speed characteristic

Speed regulation in series motors is less precise than in shunt-wound motors. If the motor load decreases, current flowing in both the armature and series-field circuits reduces, resulting in a greater increase in speed. However, if the load and supply voltage is constant, speed regulation occurs by connecting a diverter rheostat in parallel with the field winding to bypass some of the armature current around the field. **Figure 5.81** shows the speed characteristic curve of a series motor.

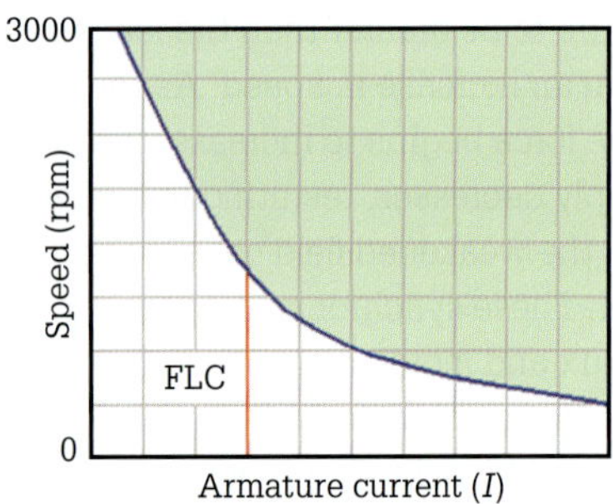

FIGURE 5.81 Speed–load characteristic – series motor

Removal of the mechanical load from series motors results in an indefinite speed increase that can destroy the motor or its taper roller bearings or sleeve bearings.

Torque characteristic

Starting torque developed in series motors ranges between 300% and 375% of full load but can be as high as 800% of full-load torque. These high values of torque occur at starting because the armature generates no counter-electromotive force. As the armature develops speed the generated electromotive force increases, reducing the effective voltage, current and torque.

When the armature shaft load increases, the armature slows to provide sufficient voltage and current to match the load torque. In addition, an increase in load results in an increase in both armature and field current. As a result, torque increases by the square of the current increase ($T \propto I^2$). **Figure 5.82** shows a typical torque characteristic curve of a series motor.

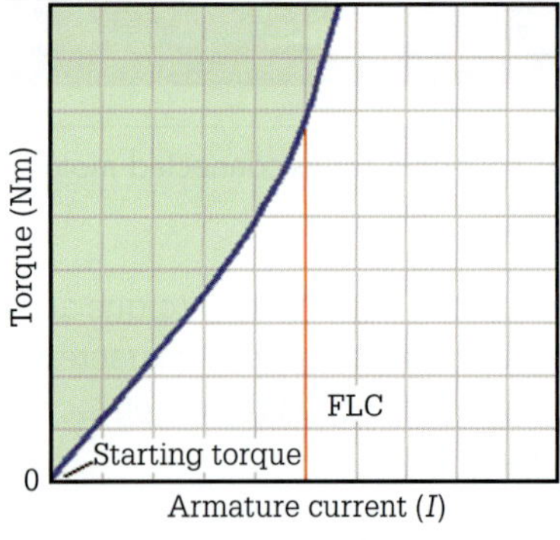

FIGURE 5.82 Torque characteristic – series motor

The armatures of large kW series motors are usually lap wound. These windings are the preferred connection type for high current, low-voltage applications because of the number of parallel paths provided for current flow. Series d.c. motors are ideal for traction work where the load requires a high breakaway torque to get it moving. Applications are traction motors in diesel-electric locomotives, hoists, ships, cranes and automobile starters, and in oil drilling rig applications.

Compound-wound motors

The compound motor has two sets of field windings on each field pole, one in parallel with the armature and one in series with the armature. This conjunction combines the desired characteristics of the shunt and series motors into one motor. Refer to **Figure 5.83** for the circuit diagram of a short-shunt cumulative compound motor.

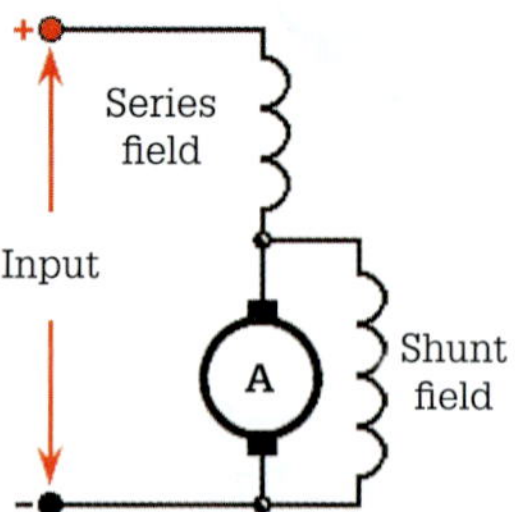

FIGURE 5.83 Circuit diagram of a short-shunt cumulative compound-connected motor

When the series-field winding is connected so that its electromagnetic field assists the shunt-field-winding's electromagnetic field, the motor is a cumulative compound motor. Refer to **Figure 5.84**.

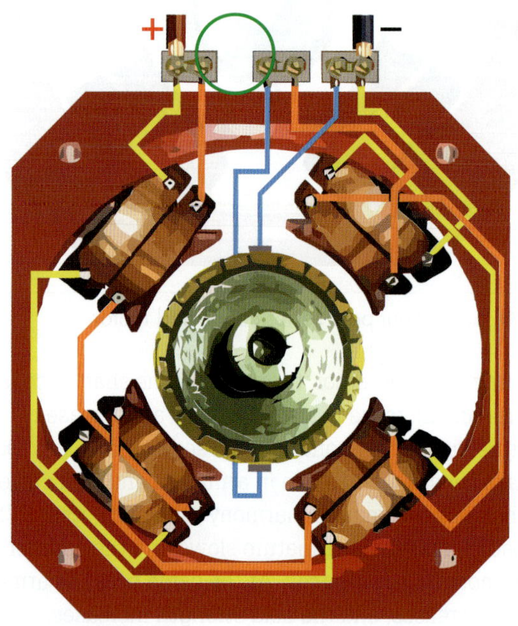

FIGURE 5.84 Connections for a four-pole cumulative compound motor

In the cumulatively compound motor, there are two field fluxes: a constant shunt-field flux and the series-field flux, which is proportional to the armature current, and therefore to the load. As a result, this motor has a lower starting torque than the series motor but a higher starting torque than the shunt motor and does not over-speed at no-load.

When the series-field winding is connected so that its flux field opposes the shunt-field-windings' flux field, the motor is a differential compound motor. In other words, as the load on the armature shaft increases, the armature current increases and the total electromagnetic motor flux decreases.

However, as the flux decreases the armature speed increases and this causes another increase in load that draws more armature current from the supply, leading to a further decrease in the electromagnetic flux and another increase in speed. The result is that the differentially compounded motor is unstable, especially at starting, and is almost never used.

Compound-wound motor characteristics

Speed–load characteristic

Compound motors connect either cumulatively or differentially. When connected cumulatively, the series field connects to assist the electromagnetic flux produced by the shunt winding, providing added torque and speed. By contrast, the speed of the compound motor falls more sharply than the shunt-wound motor. **Figure 5.85** shows the speed characteristic curve of a cumulatively compounded-connected motor.

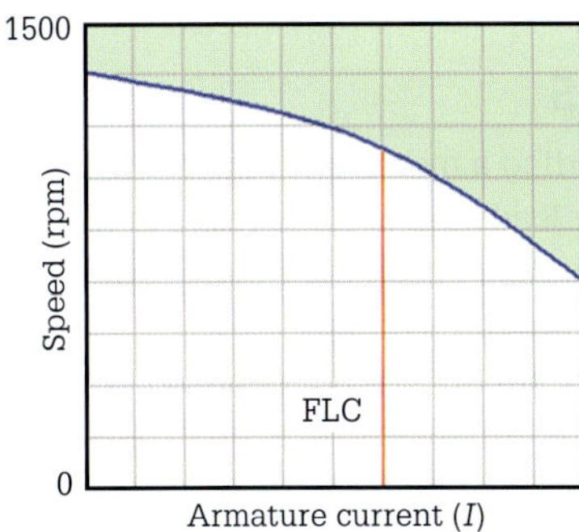

FIGURE 5.85 Speed–load characteristic – cumulatively compound motor

The compound-wound d.c. motor employs both a series field and a shunt field and combines the operating characteristics of both the series- and shunt-wound motors. By varying the number of turns of the two windings the operating characteristics may be varied to suit particular load requirements.

The motor's performance is roughly between that of the series-wound and the shunt-wound motor, with a high starting torque and excellent speed control. It has an inherently controlled no-load speed, making it safer than a series motor for devices such as cranes, which may lose their loads. Common uses of the cumulatively compound motor include elevators, air compressors, conveyors, cranes and presses.

Torque characteristic

Figure 5.86 shows a typical torque characteristic curve of a cumulatively compound motor.

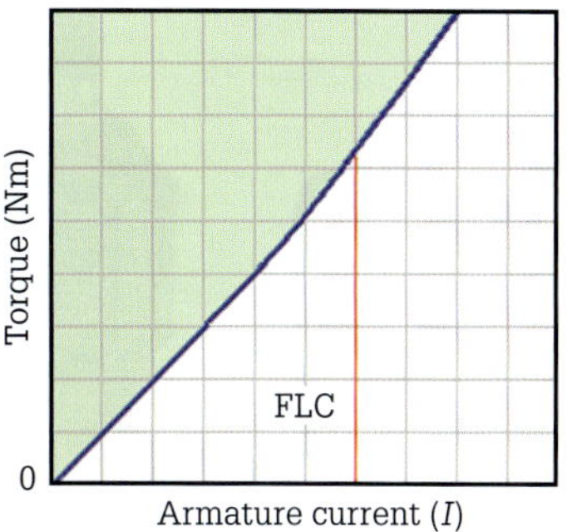

FIGURE 5.86 Torque characteristic – cumulatively compound motor

Other d.c. motors

Separately excited motors

A separately excited d.c. motor has its field-winding circuit energised from a voltage input independent from the armature circuit. If this separate voltage is constant, the motor exhibits the same characteristics as the shunt-connected d.c. motor.

Stabilised shunt motors

Stabilised shunt motors are similar to a cumulative compound motor and connect in the same manner. However, the series fields of these motors create an electromagnetic field that opposes the de-magnetising effect of armature reaction. Furthermore, the series-field winding has a small number of turns, up to one and a half turns per pole. These motors exhibit a slight drop in speed with increasing load and have applications where a moderate starting torque and a constant speed is required.

Series universal motors

A series universal motor is a small motor specially designed to operate at nearly the same power output and speed on either d.c. or single-phase a.c. This motor is the only type of d.c. motor that can achieve a unidirectional torque from an a.c. supply.

The series connection results in the same current flowing through the armature and field windings. When the current reverses, it reverses in both the field and armature windings, producing torque in the same direction. No-load speeds in excess of 12 000 revolutions per minute are achievable without damage. Reversing the direction of rotation of a universal motor occurs by reversing the direction of current in the series field or the armature, but not in both. Universal motor connections have the armature connected in series with the series-field winding as illustrated in **Figure 5.87**.

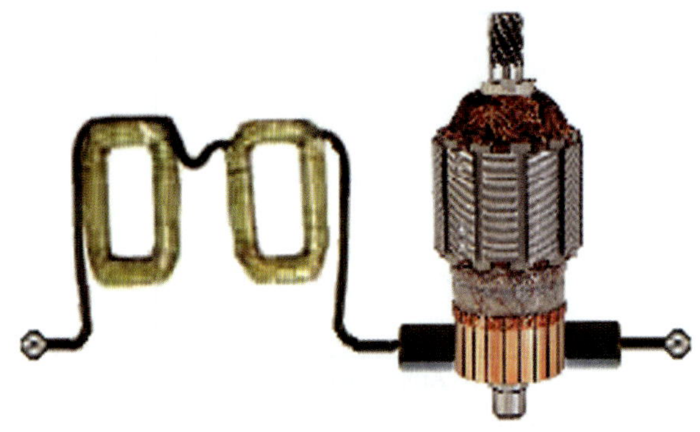

FIGURE 5.87 Series universal motor connections

In order for the universal motor to operate on a.c., its field poles, stator frame and armature core require laminating. If the field poles are solid, the energy losses occurring in the core are excessive due to the effect of eddy currents. These motors have applications in portable power machines.

Torque-speed characteristic

Figure 5.88 shows a typical torque–speed characteristic of a universal motor.

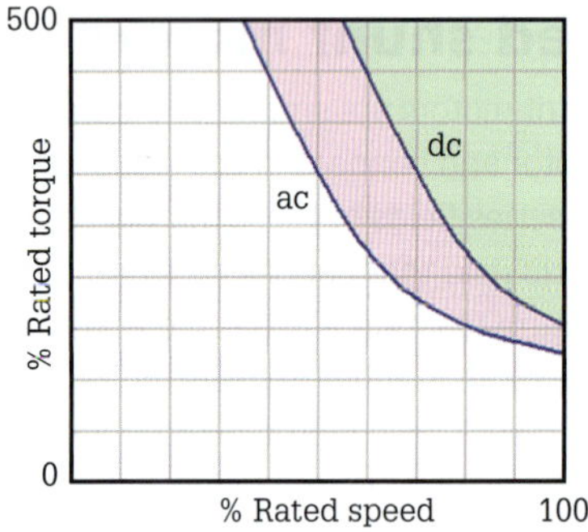

FIGURE 5.88 Torque–speed characteristic – universal motor

The graph shows two curves, one for direct current application, and the other for alternating current use. On alternating current, the motor produces a more drooping characteristic than on direct current. This characteristic occurs because when operating on 50 hertz both the series-field windings and the armature windings produce significant impedance. Consequently, the impedance causes a large voltage drop across these windings. The result of this is that the armature voltage is smaller for an a.c. supply voltage than for a d.c. supply voltage.

With a.c., the reduced armature voltage means that the speed is slower for a given armature current and torque than it would be on direct current. Also, with a.c. the field and armature current are always in phase, so the flux reverses with armature current. Consequently, the torque is always in the same direction. Varying the supply voltage can control the speed. For its physical size, the universal motor can develop higher torques and shaft speed than any other single-phase motor of the same power rating.

Mechanical loads for d.c. motors

In understanding a motor, it is important to understand what a motor load means. Load refers to the torque output and corresponding speed required. Loads can generally be categorised into three groups:

1 **Constant speed** The motor provides constant speed from no load to full load over an extensive range of torque loading. (Example: machine tools.)
2 **Constant torque** The motor works against a constant force. (Example: conveyors.)
3 **Constant power** Mechanical characteristics of the load change (size and weight) and the speed and torque change. (Example: machine tools.)

REVIEW QUESTIONS

1 What method is used for speed control of permanent-magnet motors?
2 How does the shunt-field winding connect in a shunt-wound motor?
3 How can the direction of armature rotation change in a shunt motor?
4 What percentage of full-load torque is a shunt-wound motor able to provide at start?
5 How does a compound motor differ in construction from other d.c. motors?
6 State the two ways the shunt field can connect in compound motors.
7 Describe a separately excited motor.
8 State the purpose for a magnetisation curve of a d.c. motor.
9 How does speed control occur with series-connected motors?
10 With shunt-wound motors, why does the speed drop slightly with an increase in load current?
11 Describe the torque characteristic curve of a series motor.
12 Draw the speed characteristic of a shunt motor.
13 What is the effect on a series motor if removal of the mechanical load occurs?
14 What are two applications for series motors?
15 Why is the shaft speed of a universal motor different when the motor is operating on alternating current rather than direct current?

5.6 Specialty direct current machines

Industry has need for machines that are suitable for specific purposes. The following examines briefly some of these 'specialty' machines.

Electromagnetic pulse generator

The electromagnetic pulse generator generates an intense momentary pulsed d.c. magnetic field. However, to release a pulse, the voltage needs charging on a capacitor first. This voltage comes from a full-wave rectifier circuit. Once charged the energy stored on the capacitor discharges into the pulsing coil.

One purpose of the pulse generator is to locate faults within cables. A series of pulsed electromagnetic waves are injected into the cable. Any faults in the cable change its electric properties, which creates reflections for the pulsed waves. The delay times between the pulses sent on the faulted cable and the reflected pulse signals are used to calculate the distance to the cable fault. The amplitude of the reflections contains information about the type of fault.

Stepper motor

A stepper motor has applications for accurately repeatable tasks such as positioning. Stepper motors convert electrical pulses into mechanical movements or 'steps' (one digital pulse is equivalent to one step) in fixed angular increments. The sequence of the electrical pulses directly relates to the rotation of the rotor shaft while the speed at which the rotation occurs depends upon the frequency of the applied pulses. Stepper motors can operate continuously at rated full-load current; however, steppers represent the least efficient among the specialist motor types because of continuous current draw. Their operating life expectancy is in excess of 10 000 hours.

There are three types of stepper motors: variable reluctance, permanent magnet and standard or hybrid. Refer to **Figure 5.89**.

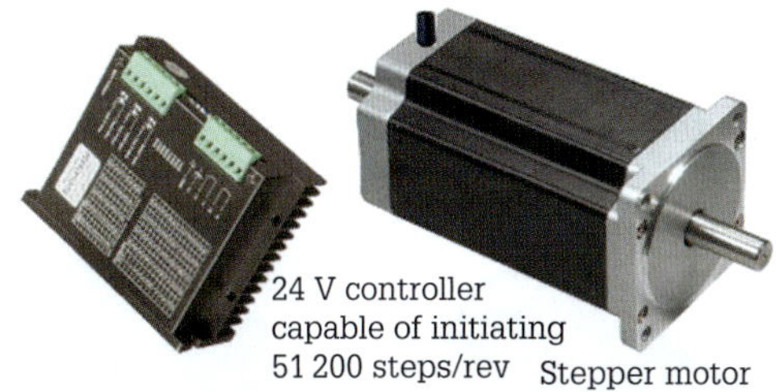

FIGURE 5.89 Hybrid stepper motor and controller

The hybrid stepper motor consists of multi-toothed stator poles and a three-part rotor. The term 'hybrid' results from the fact that this motor operates under the combined principles of the permanent magnet and the variable reluctance stepper motors. The rotor contains two toothed parts (a front and back set consisting of up to 200 teeth) separated by an axially magnetised permanent magnet, with the opposing teeth offset as shown in **Figure 5.90** by a half of one tooth pitch.

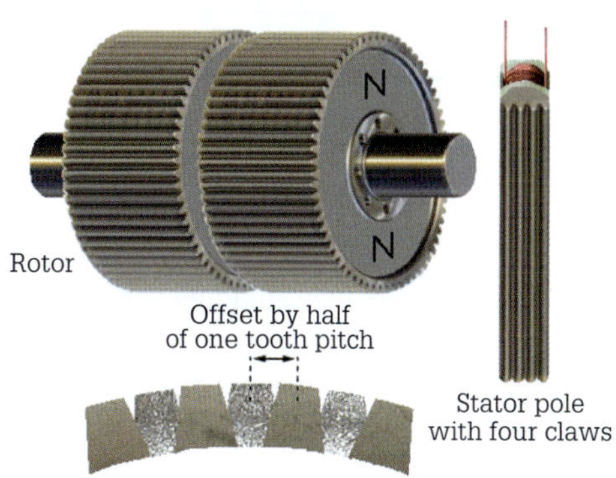

FIGURE 5.90 Rotor, stator pole and tooth-pitch offset of a hybrid stepper motor

The stator generally has the same number of teeth as the rotor, but may have two fewer, depending upon the motor's design. The stator has coils of two different phases wound on the same main pole using a bifilar configuration. The permanent magnet means that the two-toothed sections as shown in **Figure 5.90** have different magnetic poles. This configuration enables the motor to have 200 full steps (resulting from 50 pole pairs generated by 50 teeth on each rotor lamination) per revolution of the motor shaft.

Dividing the 200 steps into a single rotation of 360° equals a 1.8° full step angle (0.9° in half-stepping mode – 400 steps/revolution). The stepper motor is also capable of micro-stepping (0.72° per step – 500 steps/revolution). In addition, determination of the step size, or step angle, depends on the number of teeth, the motor construction and the type of drive system used for control. The stator also has teeth to build up an equivalent number of poles in relation to the main pole. Usually eight main poles occur for 1.8° hybrids.

Stepper motor applications are computer peripherals, medical equipment, machine tools and stage lighting devices; and to move paper in printers as well as various industrial control applications.

Servomotor

There are three basic types of servomotors: a.c. servomotors, d.c. servomotors and a.c. brushless servomotors. **Figure 5.91** shows the d.c. brushed permanent-magnet servomotor.

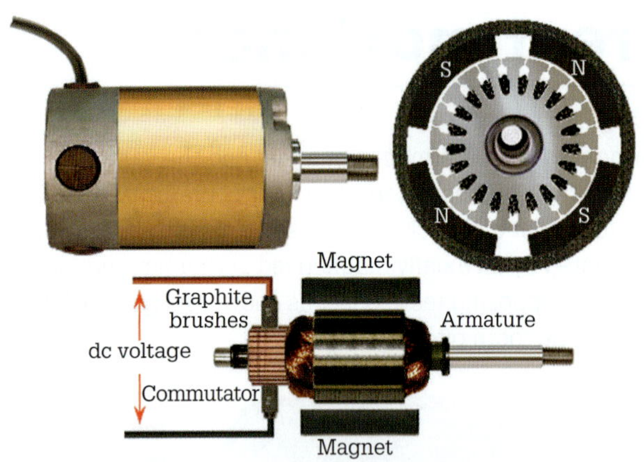

FIGURE 5.91 Direct current brushed permanent-magnet servomotor

The d.c. brushed permanent-magnet servomotor has two main parts:

1 A housing containing the field magnets (stator).
2 A rotor made up of coils of magnet wire wound in the rotor slots and connected to a commutator. Graphite brushes are in contact with the commutator and carry d.c. current to the rotor coils.

A d.c. servomotor system includes a motor, a feedback device and a d.c. drive. The motor operates on direct current pulses received from the drive. These motors have two interacting magnetic fields. One field electrically operates while the other is the result of permanent magnets. The interaction between the two magnetic fields produces a torque that turns the rotor. Direct current servos can operate at speeds greater than 10 000 rpm.

The feedback device sends performance data such as rotor position and rotor speed back to the drive. The drive generates an output positioning signal consisting of a series of pulses (pulse train) of variable width, as illustrated in **Figure 5.92**, which influences the angular position of the rotor shaft.

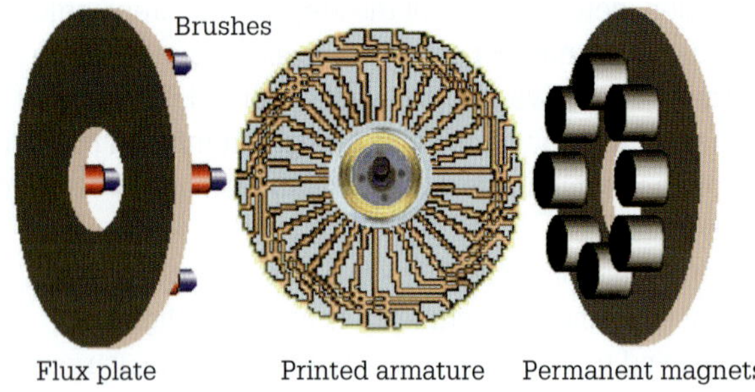

FIGURE 5.92 Printed armature servomotor

When the servo has a command to move, it goes to the position and holds that position. However, servos do not hold their position forever; the position pulse must repeat every 20 milliseconds to instruct the servo to stay in place. Direct current servomotors find application in ticket vending machines, in pumps in medical facilities, in plain paper copiers and machine operations.

Printed armature servomotor

The armature of a printed circuit servomotor, also called a torque motor, is disc shaped and non-ferrous in construction. The armature conductors occur on both sides of the laminated non-conductive disc by printed circuit techniques. Precious metal composition brushes bear directly on the armature conductors that also serve as the commutator. A single revolution controls shaft movement. The complete armature assembly uses a 'pancake'-type motor frame. Refer to **Figure 5.92**.

Some printed armature servomotors are capable of providing over 13 500 Nm of peak torque. The printed armature servomotor finds application as a positioning servomotor where intermittent movement without torque fluctuations is required. The precise speed and lack of torque disturbances and cogging make these motors suitable for application in video, analogue and digital recording equipment. Refer to **Figure 5.93**.

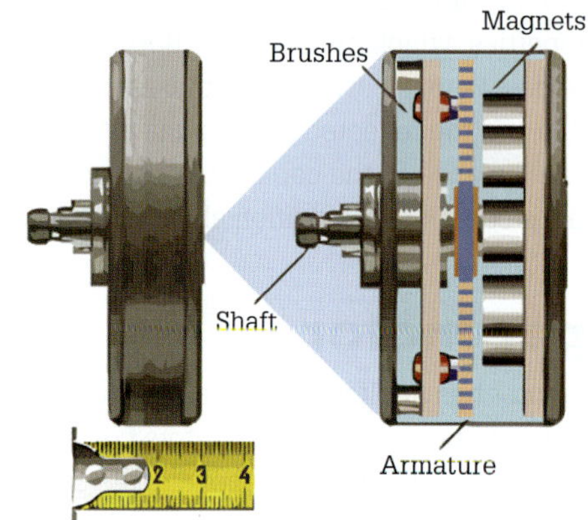

FIGURE 5.93 Pancake printed armature servomotor

Tachogenerator

Devices that convert angular speed to some other form of reading or usable signal are tachometers. Some tachometer devices are direct contact units that press against a rotating shaft to measure the rpm on a dial. Non-contact devices on the other hand, use strobe lights or infrared technology to sense the speed of the rotating shafts. However, when the signal produced is an emf the device is a tachogenerator. The tachogenerator is a precision d.c. generator, and the output voltage signal is a precise function of the rotational speed. Refer to **Figure 5.94**.

FIGURE 5.94 Tachogenerator armature and permanent magnet field

Tachogenerator applications include the measurement and control of the rotational speeds of variable-speed drives used in paper and steel mills, machine tools and process plants.

A tachogenerator gives a polarised output voltage, positive voltage for one direction and negative for the reverse direction, so a feedback control system has to be able to cope with either polarity if the controller is to reverse the motor. Information gained by a tachogenerator travels as a signal to the operator, for use in a signal feedback control system such as a d.c. power drive or an a.c. inverter powering the drive motor.

Electronic commutation

Traditional d.c. motors are inherently power efficient but have high maintenance due to the carbon brush method of commutation. The term EC means a d.c. motor having electronic commutation achieved with a microprocessor and integrated speed control. EC motor technology, which is also known as brushless d.c., provides high efficiency, long service life and speed control options.

Brushless d.c. motor

Brushless d.c. motors consist of a stator (armature), a controller and a rotor that has two or more permanent magnets that produce a magnetic field. However, for the rotor to turn continuously, the currents flowing through the coils must switch, depending on the rotor position. In addition, this switching relies on an electronic-controlled commutation system for its operation. The system regulates and controls the speed and directions of the rotor by precisely delivering a rapidly changing current to the stator coils and at the same time determining the rotor's orientation and position. Feedback devices such as Hall sensors provide data to the controller in relation to the rotor's positioning and speed. Whenever the Hall magnets pass near the sensor, they give a high or low signal, indicating that the north or south pole is passing. Based on the combination of the Hall sensor signals, the exact sequence of commutation can be determined. Refer to **Figure 5.95**.

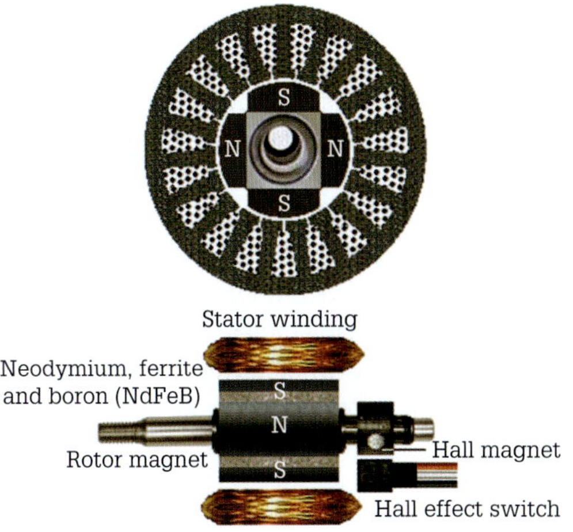

FIGURE 5.95 Brushless d.c. motor

Applications for brushless motors are: paper feed drives, mixing and stirring equipment, medical pumps and general industrial automation.

Linear motor

Linear machines have a thrusting or levitating action instead of a rotary motion while the two forms of linear machines are the flat and the tubular. In addition, concerning the stator and rotor, either can move. **Figure 5.96** shows a flat linear motor creating a thrusting movement.

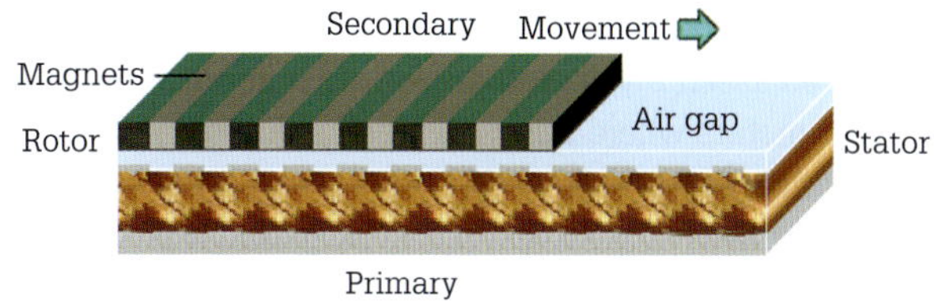

FIGURE 5.96 Linear flat motor

A linear motor consists of a slider and a stator. The stator element contains the primary windings that conduct current within a core material and sensors for position detection and temperature monitoring. A slider consists of a steel element in which the rare-earth magnets (the secondary) mount in alternating polarity on a steel plate to generate magnetic flux density. In addition, the plate has provision for loads to attach.

The electromagnetic interaction between the current in the primary and the magnetic field of the secondary causes the development of an electromagnetic force. This electromagnetic force may be an attraction or a repulsion force between the primary and the secondary.

For the establishment of electromagnetic interaction, it is necessary for the direction of the magnetic fields of the primary and secondary elements to be at right angles to each other. The right angle produces direct linear force instead of rotating movement. If the interaction between the two flux fields occurs for thrust, then the machine is a linear motor. Varying the stator current controls the force or torque of the linear motor. When the current and the flux density interact (B), force (F) develops.

To prevent the slider and the stator adhering to each other an air cushion exists between them by a pneumatic system similar to an air hockey table. A linear machine in which the electromagnetic force is such that the secondary hovers above the primary is a linear levitation machine as shown in **Figure 5.97**. An application for linear levitation is in hovercraft-type high-speed trains.

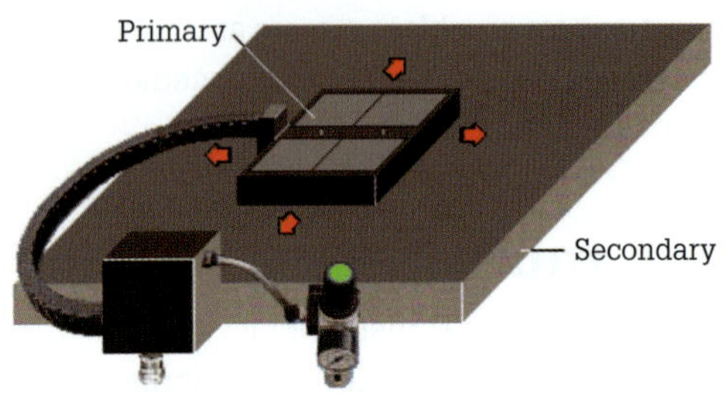

FIGURE 5.97 Slider movement of a flat linear motor

Linear motors are ideal for short-stroke linear motions. For example, in label printing applications where the label prints on the product as the product moves on a conveyor. Automated assembly operations are other applications for linear motors where they provide positioning and force control of products. Additional sensors increase the effectiveness of the linear motor by giving it probing, compressing, sensing and measuring capabilities.

A tubular linear machine illustrated in **Figure 5.98**, shows how the rare-earth magnets orientate to the stator winding with an air gap between the two.

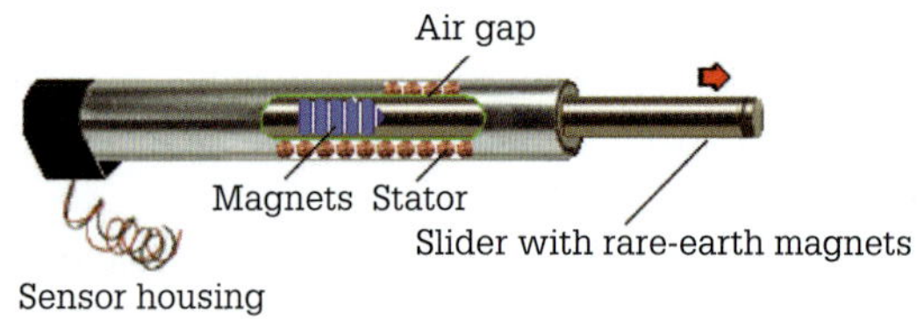

FIGURE 5.98 Tubular linear motor

The circumferential current in the stator windings cuts orthogonally the air gap radial flux produced at each pole, thereby producing an axial force as shown by the red arrow. The information obtained by the sensors within the tubular linear motor feeds a programmable logic controller (PLC) for analysis. Finally, tubular motors can have a stroke movement of up to 1.45 m with a speed of up to 4.2 m/s. Applications for tubular motors include the detection of the depth of machined holes in components and placement of stoppers in bottles in assembly operations.

REVIEW QUESTIONS

1 What is a typical application for the pulse generator?
2 What type of motor converts electrical pulses into mechanical movements?
3 What type of motor operates under the combined principles of permanent magnet and variable reluctance?
4 Name the three types of servomotors.
5 A printed armature servomotor is also known as ______________.
6 What device is a precision d.c. generator?
7 Name two forms of linear machine.
8 List two applications for tubular linear motors.

5.7 Machine efficiency

The power consumption of a d.c. motor (input power) is equal to its terminal voltage times the current. However, every motor has losses, which means that the motor consumes more power than it delivers at its shaft.

Improvement of motor efficiency can occur at the design stage by adding more active material (lamination stack length, better lamination coating to reduce iron loss and more copper turns with a larger CSA) than necessary to achieve a particular kilowatt rating. In addition, an anti-condensation heater, fans and better bearings reduce the mechanical losses within a d.c. motor. Apart from these, better efficiency also depends on the following factors:

- **Capacity –** larger motors have a higher efficiency
- **Number of pairs of poles –** the more pairs of poles, the lower the efficiency
- **Load –** the smaller the load, the lower the efficiency.

Machine losses

Direct current motor losses separate into three main categories:

1 **Iron losses** These are magnetic losses in laminations (hysteresis, eddy current), inductance and eddy currents. Also included are stray magnetic losses that occur in the air gap between armature and field as a result of armature reaction. Hysteresis and eddy current losses vary with flux density and speed.
2 **Copper losses** This is the power loss that occurs in the armature or field coils when the motor is on load. This power wastes in the form of heat due to the resistance of the coils. The copper loss varies with the load in proportion to the current squared (I^2R loss). Implied are a light load, low current, heavy load, and high current and greatly increased copper loss. Copper loss is a variable loss whereas the iron loss is a steady loss because the supply voltage and speed are usually constants.

3 **Mechanical losses** Mechanical losses develop as a result of friction in the motor bearings, the fan for air-cooling and the brushes riding on the commutator. Power losses due to friction increase as the square of the speed, and those due to windage (fan) increase as the cube of the speed. Consequently, the total of these losses affects motor efficiency.

Figure 5.99 shows the losses that occur in a d.c. machine.

FIGURE 5.99 Machine losses

Determining machine losses

It is possible to determine machine losses from the following:

1 The no load power represents the constant losses of the machine. This includes iron and mechanical losses.

2 The copper losses at full load can be determined experimentally or by calculation.

Figure 5.100 shows power flow and the losses that occur in a d.c. generator and **Figure 5.101** shows power flow and the losses that occur in a d.c. motor.

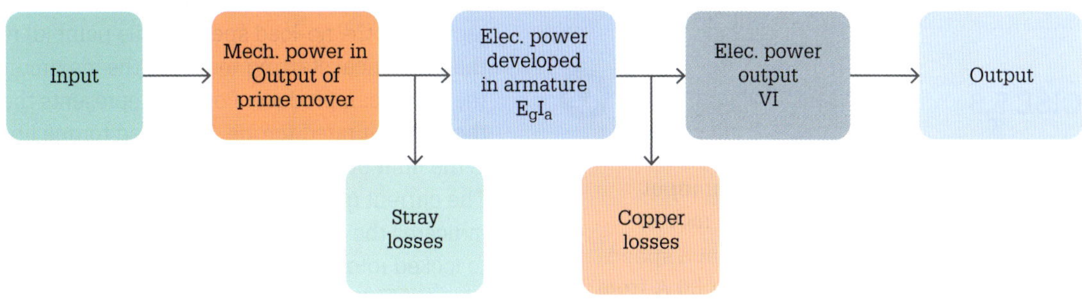

FIGURE 5.100 Power flow and energy transformation stage for a generator

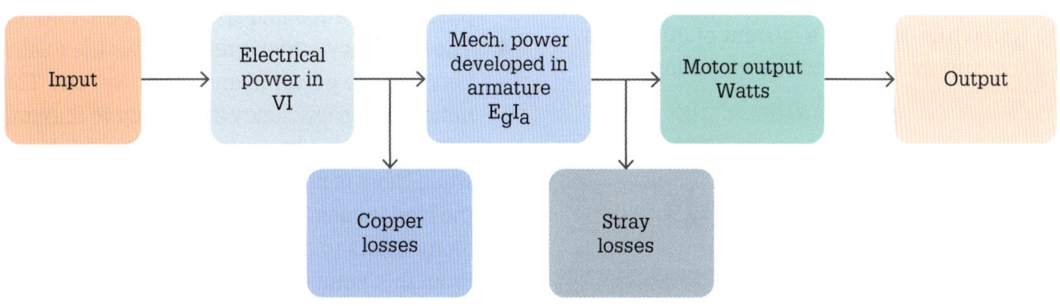

FIGURE 5.101 Power flow and energy transformation stage for a motor

Certain modifications on site, such as high-altitude, abnormal ambient temperature and lower than standard temperature rise, causes motor efficiency to vary from stated values. In addition, dust-laden, abrasive and corrosive environments can prevent the free exchange of air, thereby preventing heat from the motor dissipating. High-humidity operating environments can also affect the efficiency. The heating (run-time) and cooling (downtime) cycles tend to draw moisture in the form of condensation into the motor as it cools down and remains idle. The result is that moisture causes gradual deterioration of insulation.

Determining machine efficiency

For a d.c. machine, efficiency is the ratio of the output power to the input power:

$$\eta = \frac{P_{Out}}{P_{In}}$$

where η = efficiency (normally expressed as a percentage)

P_{Out} = power output in watts (W)

P_{In} = power input in watts (W)

Furthermore,

$$\text{Power in} = \text{power out} + \text{total losses}$$
$$= \text{power out} + \text{iron loss} + \text{copper loss}$$

Noting also that,

$$P_{out} = P_{in} - \text{total losses}$$

EXAMPLE 5.8

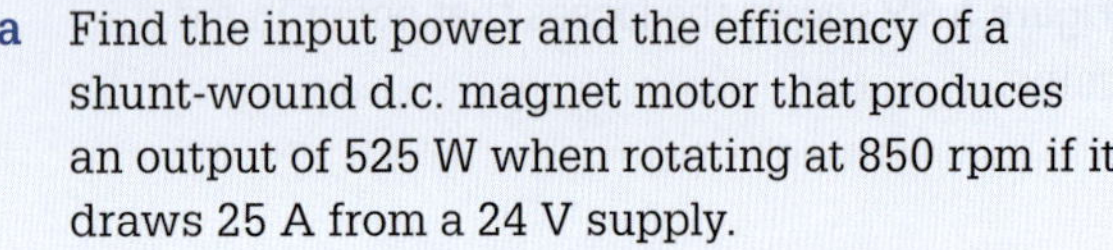

a Find the input power and the efficiency of a shunt-wound d.c. magnet motor that produces an output of 525 W when rotating at 850 rpm if it draws 25 A from a 24 V supply.

$$P_{In} = VI$$
$$= 24 \times 25$$
$$= \mathbf{600\ watts}$$

$$\eta = \frac{P_{Out}}{P_{In}} = \frac{525}{600}\ \mathbf{0.875\ or\ 87.5\%}$$

b Find the input power and the efficiency of a 120 V d.c. brush permanent-magnet motor supplying a 4.5 kW load at a speed of 2750 rpm when the total power losses are 0.35 kW.

$$P_{In} = P_{out} + \text{losses}$$
$$= 4.5 + 0.35$$
$$= \mathbf{4.85\ kW}$$

EXERCISE 5.8

a Calculate the efficiency of a generator that delivers 40 kW of output power from an input power of 48 kW provided by the prime mover.

b A motor is producing 70 Nm of torque at a speed of 1400 rpm while drawing a current of 30 A from a 400 V d.c. supply. Calculate the motor input power.

c A motor is producing 70 Nm of torque at a speed of 1400 rpm while drawing a current of 30 A from a 400 V d.c. supply. Calculate the motor losses.

Minimum energy performance standards

SWITCH ON

Minimum energy performance standards (MEPS) are part of the Australian reduction in greenhouse gases commitment and are a regulatory tool used to ensure that Australians have more efficient appliances and equipment. Certain products containing magnetic and electromagnetic devices and machines must comply with specific standards for energy efficiency. MEPS programs are mandatory in Australia by state government legislation and regulations that give force to the relevant Australian standards.

Characteristic curves

The graph in **Figure 5.102** represents the characteristics of a typical d.c. specialist motor. As long as these motors operate with high efficiency (shaded area), long service life occurs. However, motor operation below the shaded area, results in reduced service life and motor failure.

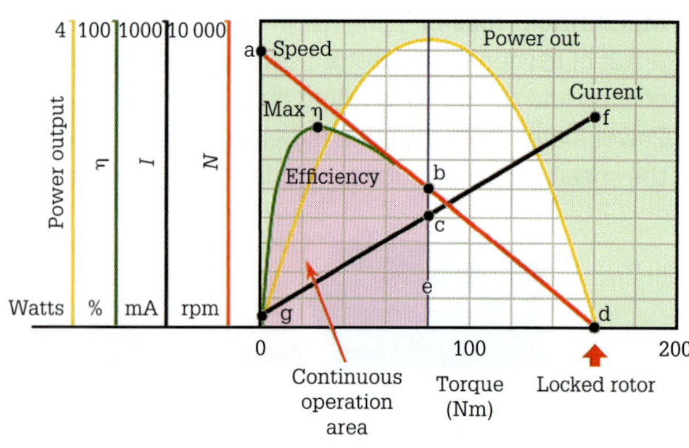

FIGURE 5.102 d.c. specialist motor characteristic curves

Line (e) in **Figure 5.102** represents the torque developed by the motor operating continuously in ambient temperature of 20 °C.

There are four curves shown in **Figure 5.102**:

1. The speed graph (red) is a straight line. Point (a) represents the no-load speed, while point (d) represents the theoretical starting torque or the maximum torque the motor can produce. Point (b) represents the speed of the motor at rated torque. The rated torque line (green) is the limit of continuous, operation (shaded pink).
2. The current graph (black) is a straight line and indicates the current values changing from no load (g) to locked rotor or starting current (f). In addition, point (c) represents the current drawn from the supply at rated torque.
3. The output power curve (yellow) shows the power (the product of torque and speed) over the torque range.
4. The efficiency curve (green) shows the relationship between the power in and the power out. The curve indicates the efficiency at start-up to maximum efficiency. In addition, the efficiency of the motor is zero when there is no load (it is doing no useful work). As the load increases, the efficiency improves because the motor starts to do work. Finally, the efficiency levels. From this point onwards, addition loading only causes the motor to draw a larger current while the efficiency remains the same.

Maximum efficiency

In general, for the efficiency to be at maximum for any motor, the losses must be at minimum. Because the iron loss is constant, and the copper loss changeable, maximum efficiency transpires when the value of the copper loss equals the iron loss. All motors have a voltage, current and rpm at which the motor's maximum efficiency happens. These values are available in the manufacturer's data sheets.

REVIEW QUESTIONS

1 Identify three factors that contribute to the efficiency of a d.c. machine.
2 What are three sub-categories of machine losses?
3 Which machine loss is determined by a no-load test?
4 Calculate the efficiency of a generator that delivers 30 kW of output power from an input power of 38 kW provided by the prime mover.
5 A motor is producing 45 Nm of torque when operating at a speed of 1200 rpm while drawing a current of 28 A from a 220 V d.c. supply. Calculate the motor losses.
6 What is the purpose of minimum energy performance standards (MEPS)?

5.8 Machine maintenance and testing

Terminal identification

The leads of a compound motor are always marked before delivery. **Figure 5.103** shows typical markings, which are:

- F1–F2 Shunt field circuit
- S1–S2 Series field circuit
- A1–A2 Armature circuit.

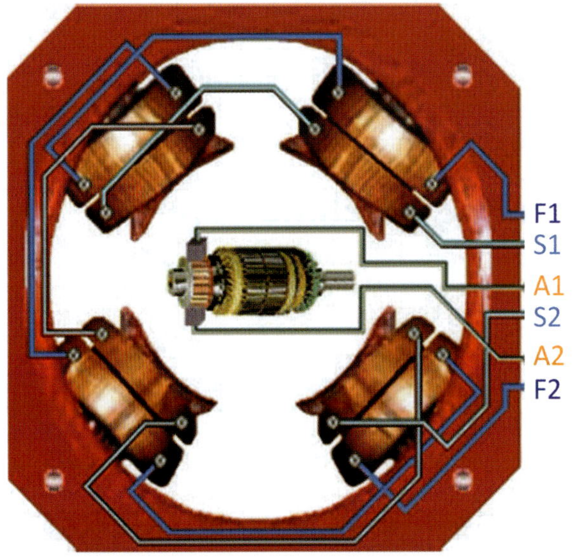

FIGURE 5.103 Typical winding markings for a d.c. motor

If the lead markings are no longer visible, it is necessary to test the six leads for re-marking before connecting the motor. Identify the leads in the following manner:

- Use an ohmmeter to determine the three circuits of the armature, the series field and the shunt field. One pair of leads will have a higher resistance compared to the other two pairs. These connect to the shunt field. Both of the remaining pairs have a lower resistance.
- Remove the carbon brushes and the ohmmeter will indicate an open loop (OL) when applied to one pair. These leads connect to the armature. The remaining pair are the series-field leads.

Testing

To detect defects in a d.c. motor you need to test the field and armature windings for the following:

- earth fault
- open circuit windings
- short circuit windings
- reversed coils.

Earth fault

An earthed field coil will either cause the motor over-current protection to operate or produce a weak magnetic field that will not turn the armature. A visual inspection usually reveals a completely burned field, but testing is necessary to find an earthed field. An earthed field may cause the motor to run faster than normal and spark badly under no-load.

A Hi-Pot tester, such as that shown in **Figure 5.104**, is used to test for an earthed field and the principle of operation is similar to that of the insulation resistance tester.

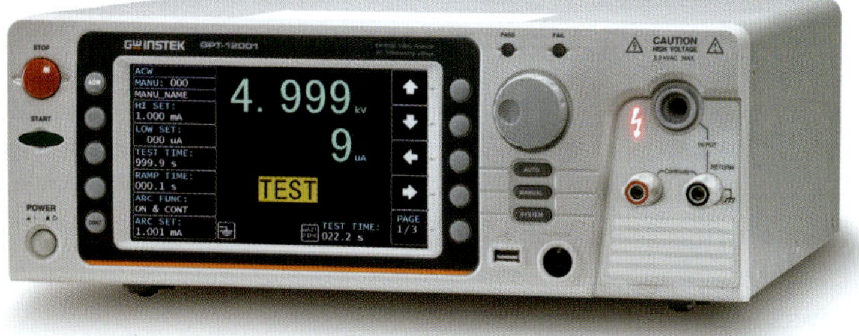

FIGURE 5.104 Hi-Pot tester for testing for earth faults

Source: GOOD WILL INSTRUMENT CO., LTD.

Open circuit winding

Open circuits in either series or shunt field will prevent the motor from starting. If a shunt field coil is open circuit while the motor is running, it may cause the motor to accelerate and 'run away' if the motor is not fully loaded. The source of the open circuit may be the field connecting lead. These leads break easily, especially when not tied securely to the coil. The open circuit may also occur in the lead extending out of the motor or be due to a poor connection at one of the field poles. Locate an open circuit either by inspection or by testing using a continuity test.

Open circuit armature coils

An open circuit in an armature coil will cause severe sparking at the commutator and will prevent the motor running at rated speed. An inspection will reveal burn marks on the commutator bars to which the open coil connects. On a lap winding, one open coil will cause one burned bar. On a four-pole wave winding, one open coil will cause two burned bars. The cause of the open circuit may be loose leads in the commutator bars or poorly soldered leads. Remove the leads from the bar, clean them, then replace and resolder them. If a broken wire in the coil causes the open, jump the two bars on either side of the burned bar. When more than one burned bar appears on the commutator, jump the bars in only one place and run the motor. If this eliminates the sparking, do not jump any more bars.

Short circuit winding

A shorted field coil may cause the motor over-current protection to operate or produce a weak magnetic field that will not turn the armature. A visual inspection usually reveals a completely burned field, but testing is necessary to find a shorted field. Often a shorted field may cause the motor to run faster than normal and spark badly under no-load.

Identify shorted field coils by undertaking a resistance measurement test with an ohmmeter. All field coils in a given motor are alike, so the resistance should be the same for each. Test the resistance of each coil with an ohmmeter. A lower reading on one field than the others indicates a shorted coil. It is necessary to rewind or replace the shorted coil.

Short circuit armature or commutator

If there are many shorted coils in an armature the armature may not rotate. In some motors, the armature will move a half turn or turn over very slowly. A special test tool known as a 'growler' is used to test for shorted coils. Before performing this test it is necessary to clean the mica between commutator bars to eliminate this as the possible cause of the short circuit.

A shorted armature coil may show the visible signs of heat and smoke. Smoke emanating from a motor is nearly always a sign of shorted or burned coil. Sometimes the smoke is obvious and at other times, it is hardly noticeable. The odour of burning coils is, however, quite noticeable. If this condition exists for a short time, adjacent coils will suffer damage. Prompt corrective measures may save the winding. Whenever smoke comes from a motor, turn off the supply and locate the faulty coil by feeling the armature for the hot spot.

Reversed armature leads

This defect can occur only in a rewound armature. Sparking at the brushes is an indication of reversed armature leads. After ruling out other possibilities, the only way definitely to determine reversed leads is to test the armature.

Machine maintenance

To keep a d.c. machine running at optimum efficiency it is important to conduct routine inspection and servicing. This involves:

- checking for dust and corrosion
- applying appropriate lubrication
- checking for excessive heat, noise or vibration
- checking the winding insulation for fraying, breakdown and burning
- checking the brushes and commutator for excessive wear, pitting, or damage – replace brushes if necessary, or skim and undercut the commutator.

Rotating machine safety

While carrying out any maintenance on a d.c. machine it is important to consider safety risks associated with rotating machinery, including:

- rotating parts
- lethal voltages
- high inductance
- high impact kinetic energy.

SWITCH ON

Hazards and safety precautions

Inductors have the capacity to release stored energy at a much higher voltage than that used to produce the stored energy. Consequently, release of this energy could occur if the inductors have no proof that they are de-energised.

Electric motors must come to a full stop after switching off before isolation procedures occur. A full stop is important because of the residual voltage that exists across the motor windings.

Inductors can produce large eddy currents in adjacent conductive material if suddenly de-energised, causing excessive heating.

Manufacturer specifications

A wealth of information is available from machine manufacturers in the form of specifications, drawings and nameplates. **Figure 5.105** shows a sample specification sheet for a shunt-wound d.c. motor and **Figure 5.106** shows the associated nameplate detail.

GPM15	
1.1 kW, 1800 RPM, DC, 184C, TEFC, F1	
Armature Voltage	200 V
Base Speed	1800 rpm
Enclosure	TEFC
Field Voltage	100 V 200 V
Frame	184C
Frame Material	Steel
Output Power	1.1 kW
XP Class and Group	None
Agency Approvals	CSA UR
Ambient Temperature	40 °C
Armature Current	8.0 A
Armature Inertia	1.27 $Kg.m^2$
Base Indicator	Rigid
Bearing Grease Type	Polyrex EM (-20F +300F)
Drip Cover	NONE
Duty Rating	CONT
Feedback	NONE
Field Winding Type	SHUNT
Heater	None

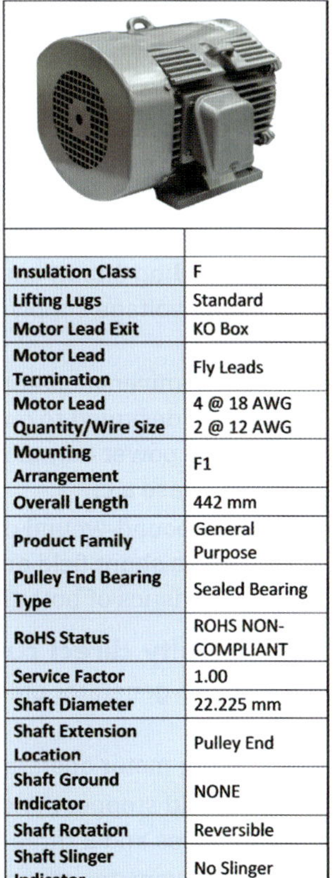

Insulation Class	F
Lifting Lugs	Standard
Motor Lead Exit	KO Box
Motor Lead Termination	Fly Leads
Motor Lead Quantity/Wire Size	4 @ 18 AWG 2 @ 12 AWG
Mounting Arrangement	F1
Overall Length	442 mm
Product Family	General Purpose
Pulley End Bearing Type	Sealed Bearing
RoHS Status	ROHS NON-COMPLIANT
Service Factor	1.00
Shaft Diameter	22.225 mm
Shaft Extension Location	Pulley End
Shaft Ground Indicator	NONE
Shaft Rotation	Reversible
Shaft Slinger Indicator	No Slinger
Motor Standards	NEMA
XP Division	Not Applicable

FIGURE 5.105 Shunt-wound d.c. motor specification

Source: Shutterstock.com/Surasak_Photo

CAT NO.	GPM15			
SPEC	GPM-123			
kW	1.1		**ENCL**	TEFC
RPM	1800			
FRAME	184C		**TYPE**	3636D
ARM V	200		**ARM A**	8.0
FLD V	200/100		**FLD A**	.25/.5
INSUL	F	**AMB**	40	
DUTY	CONT		**SUPPLY**	———
BRG/DE	6206	**BRG/ODE**	6205	
BRUSHES	2/BP5125			
SERIAL	123456			
APRV-CSA		**APRV-UL**		

FIGURE 5.106 Shunt-wound d.c. motor nameplate

REVIEW QUESTIONS

1 What markings would appear on the leads of a compound motor and what do they identify?

2 What are four defects that should be tested for in a d.c. machine?

3 What test instrument is suitable for testing for earth faults in a d.c. machine?

4 How may a d.c. motor exhibit an open circuit shunt winding when operating under light load?

5 Name four risks associated with working on rotating d.c. machines.

CHAPTER REVIEW

5.1 Construction of direct current machines

- Direct current generators and motors are similar in construction and can be interchanged.
- The armature core consists of thin individually insulated sheets of high-permeability, grain-oriented, non-ageing silicon steel.
- The commutator consists of a number of hard-drawn L-shaped copper segments interleaved with mica separators assembled around a tubular steel hub.
- Brushes provide the connection between the rotating commutator and the external voltage supply.
- Field windings are shunt, series or compound.
- Compensating windings are pole-face windings used on large d.c. machines.
- Armature reaction is the distortion of the magnetic field by cross-magnetisation by the armature.
- Interpoles eliminate sparking caused by the self-induced electromotive force. The conductors around the armature in a d.c. machine connect together and to the commutator in a definite sequence by one of two methods called lap and wave.

5.2 Direct current generator

- The generator or dynamo transforms mechanical energy into electrical energy.
- The different types of d.c. generators vary in their voltage–current terminal characteristics, power ratings, efficiency, excitation method used and voltage regulation, and in their applications.
- The voltage regulation of a d.c. generator is the percentage change in terminal voltage from no load to rated load in relation to rated voltage and speed.

5.3 Types of direct current generators

- A separately excited d.c. generator has its field windings supplied by a separate external d.c. voltage source.
- The series generator is also a self-excited generator with its field windings, armature circuit and load connected in series.
- The shunt field of compound generators can be connected 'short shunt' across the armature or 'long shunt' across the output terminals of the generator.
- Compound generators have two methods of connection: short and long shunt.
- In a compound generator, a fall in terminal voltage results by armature reaction, reduction in shunt-field current or armature resistance.

5.4 Direct current motors

- A motor converts electrical energy into mechanical energy.
- Knowledge of the directions of current flow and the magnetic field enables Fleming's left-hand motor rule.
- The developed torque by a direct current motor has three dimensions.

5.5 Types of direct current motors

- The primary permanent-magnet motor has a stationary magnetic field assembly and a rotating assembly supported by bearings.
- In a shunt motor, the shunt-field winding connects across the d.c. supply and the armature.
- A series d.c. motor has the armature and the field circuit connected in series. The compound motor has two sets of field windings on each field pole, one in parallel with the armature and one in series with the armature.
- A separately excited d.c. motor is a motor whose field-winding circuit energises from a separate constant voltage supply from that of the armature circuit.
- A series universal motor is a small series motor specially designed to operate at approximately the same power output and speed on either d.c. or single-phase a.c.
- The compound-wound d.c. motor has a series field and a shunt field and combines the operating characteristics of both the series and shunt motors.

5.6 Specialty direct current machines

- The pulse generator produces electromagnetic waves.
- A stepper motor operation is for positioning.
- The hybrid stepper motor consists of multi-toothed stator poles and a three-part rotor.
- A d.c. servomotor system includes the motor, a feedback device and a d.c. drive.
- Devices that convert angular speed to some other form of reading or usable signal are tachometers.

5.7 Machine efficiency

- Machine losses are categorised as iron losses, copper losses or mechanical losses.
- Iron and mechanical losses are considered constant losses.
- Efficiency is the ratio of useful output power to input power.
- Maximum efficiency in a motor occurs when iron loss equals copper loss.

5.8 Machine maintenance and testing

- The leads of a d.c. machine are identified as A1 and A2 for armature, F1 and F2 for shunt field, and S1 and S2 for series field.
- Testing involves checking for earth faults, open circuit windings, short circuit windings and reversed coils.

TRIAL EXAM

For Chapter 5 knowledge assessment, please complete the following trial exam.

1 The current generated at the output terminals of a simple a.c. alternator is:
  a direct current
  b three-phase alternating current
  c displacement current
  d single-phase alternating current

2 The means to change the polarity of the output of a d.c. generator is by the use of:
  a brush-holders
  b a commutator
  c graphite brushes
  d slip rings

3 The percentage change in terminal voltage from no load to rated load when the generator is delivering its terminal voltage at rated speed is:
  a voltage regulation
  b open-circuit voltage
  c no-load voltage
  d nameplate voltage

4 Motor action is the result of the interaction between:
  a the armature and the prime mover
  b field current and applied voltage
  c armature conductors
  d magnetic fields

5 When applying the left-hand motor rule, the first finger points to:
  a the direction of the induced emf
  b the direction of the magnetic field
  c the direction of the force
  d the direction of the current

6 If the supply voltage, armature current and resistance have values then:
  a the speed of the armature can be calculated
  b the flux per pole can be calculated
  c the generated emf can be calculated
  d the supply current can be calculated

7 With d.c. motors, why is there slightly more supply voltage than back emf?
  a to limit the no-load speed
  b to improve commutation
  c to maintain flux density
  d to allow armature current to flow

8 Sparking is an effect of:
  a an excessive load
  b self-induction in a coil undergoing commutation
  c the reduction in the speed of a series motor
  d an increase in field flux

9 A machine that delivers a d.c. voltage that is a precise function of rotation speed is:
  a a d.c. motor
  b a d.c. generator
  c a tachogenerator
  d a squirrel cage motor

10 Why is an armature shaft thicker in the middle than at the ends?
  a to improve permeability
  b to withstand bending effects
  c to reduce eddy current losses
  d to allow magnetic saturation

11 How are the effects of eddy currents limited in an armature core?
  a by using a silicon additive in the iron
  b by using different forms of armature teeth
  c by using a solid core
  d by using a laminated core

12 How is brush and commutator life of a d.c. machine improved?
  a by using mica separators between commutator segments
  b by tightening the 'V-ring'
  c by chamfering each commutator segment
  d by increasing self-induction in the coil undergoing communication

13 Few turns of heavy-gauge wire indicates that a winding is:
  a armature field
  b series field
  c commutator field
  d shunt field

14 Interpoles are connected:
  a in series with the armature winding
  b in parallel with the field windings
  c across the armature brush-holders
  d in series–parallel with the compensating winding

15 How are d.c. generators classified?
  a according to the number of brush sets
  b by the number of field poles
  c according to how their field flux is created
  d by the magnitude of the armature current

16 The circuit diagram of Figure 5.107 represents a:

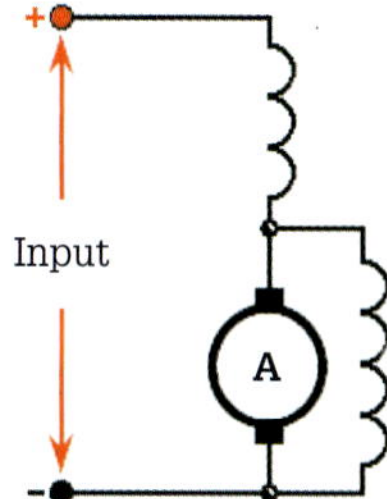

FIGURE 5.107

  a short-shunt compound generator
  b long-shunt compound generator
  c short-shunt compound motor
  d long-shunt compound motor

17 What type of device varies the field excitation of a self-excited shunt generator?
a interpoles
b field diverter
c field rheostat
d load

18 What allows a self-excited generator to 'build-up' emf?
a generated electromotive force
b field coil inductance
c armature resistance
d residual field magnetism

19 In which generator is the armature current the sum of the field current and the load current?
a separately excited generator
b series generator
c permanent magnet generator
d shunt generator

20 Varying load conditions cause considerable variations in the output voltage of a:
a series generator
b compound generator
c shunt generator
d separately excited generator

21 What type of generator has the labels 'over-', 'level' or 'under'?
a separately excited generator
b series generator
c compound generator
d shunt generator

22 A plot of terminal voltage against load current for a separately excited generator is known as:
a magnetisation curve
b load characteristic curve
c voltage–speed characteristic
d open circuit characteristic

23 The product of terminal voltage and supply current yields:
a motor input power
b motor output power
c generator input power
d generator output power

24 Which d.c. motor has its input power supplied to the armature only?
a series motor
b permanent-magnet d.c. motor
c shunt motor
d compound motor

25 Which d.c. motor has application for accurately repeatable tasks, such as positioning?
a pulse generator
b stepper motor
c servo motor
d linear actuator

26 In which d.c. motor will a small current flow through the field coils while the majority of the motor current flows through the armature circuit?
a series motor
b permanent-magnet d.c. motor
c shunt motor
d compound motor

27 A constant speed motor is:
a a universal motor
b series wound
c compound wound
d shunt wound

28 Which d.c. motor has two sets of field windings on each field pole, one in parallel with the armature and one in series with the armature?
a series motor
b permanent-magnet d.c. motor
c shunt motor
d compound motor

29 Which type of specialist machine locates faults within cables?
a pulse generator
b stepper motor
c servomotors
d printed armature motor

30 A tachogenerator:
a converts electrical energy into torque
b converts an electrical pulse into shaft movement
c converts angular speed to some other form of reading or usable signal
d converts field flux into armature control

31 Machine losses have three main categories:
a iron, hysteresis and eddy current
b friction, copper and hysteresis
c windage, eddy and mechanical
d iron, copper and mechanical

32 A 55 kW 1500-rpm generator operating at rated load has a terminal voltage of 448 V. If the no-load voltage is 470 V, determine the voltage regulation.

33 Determine the generated voltage of a 220 V series generator that has an armature current of 100 A and an armature resistance, including brushes, of 0.035 ohms and a series-field resistance of 0.06 ohms.

34 A four-pole lap-wound d.c. motor armature has 520 active conductors in a field pole flux of 20 mWb. Calculate the generated emf when the motor rotates at 1000 rpm.

35 A simple single-turn coil forms the armature in a two-pole motor and has an effective radius of 0.35 m and a length influenced by the magnetic field of 0.35 m. If the armature conductor carries a current of 5 A when exposed to a flux density of 0.5 T, calculate the torque developed by the conductor loop.

36 The power available at the shaft of a d.c. motor is 7.5 kW at a rotational speed of 1450 rpm. Calculate the available shaft torque.

37 Calculate the efficiency of a generator that delivers 850 kW of output power from an input power of 920 kW provided by the prime mover.

Single-phase low voltage a.c. circuits

This chapter provides electrotechnology workers with an introduction to basic concepts of alternating current and the terms used to describe waveforms such as sine waves. Electrotechnology workers will develop an understanding of how resistive, inductive and capacitive components perform in an a.c. circuit. This chapter provides underpinning knowledge for the unit UEEEL0020 from the UEE training package.

LEARNING OBJECTIVES

Alternating current quantities

- Apply Pythagoras' theorem and solve trigonometric ratios
- Identify sine waves and explain associated terms
- Explain the basic operating principles of an oscilloscope
- Outline the procedure of using an oscilloscope to take waveform measurements

Phasor diagrams

- Show how phasors are used to represent alternating quantities
- Use phasor diagrams to add two alternating quantities

Single element a.c. circuits

- Calculate voltage, current and power in single element a.c. circuits
- Define the terms 'inductive reactance' and 'capacitive reactance'

RL and RC series a.c. circuits

- Calculate voltage, current and power in RL and RC series a.c. circuits

Resistance, inductance and capacitance in combination

- Calculate voltage, current and power in RLC series and parallel a.c. circuits

Power in an a.c. circuit

- Explain the effect of low power factor in a.c. circuits
- Outline the effect of low or under voltage in a.c. circuits

Power factor improvement

- Outline methods employed to compensate for low power factor in a.c. circuits

Harmonics and resonance in a.c. systems

- Outline the effect of harmonic currents in a.c. circuits
- Describe resonance and calculate circuit parameters at resonance

6.1 Alternating current quantities

In order to develop an understanding of alternating current quantities it is necessary to have a good grounding in some mathematical theorems and trigonometry. One useful mathematical theorem is Pythagoras' theorem, which equates the three sides of a right-angled triangle.

Pythagoras' theorem

Pythagoras' theory evaluates the relationship between the lengths of the sides of a right-angled triangle. Triangles are plane shapes having three straight sides. A right-angled triangle contains an angle of 90° as shown in **Figure 6.1**. In a right-angled triangle, the longest side is called the hypotenuse and is located opposite the right angle.

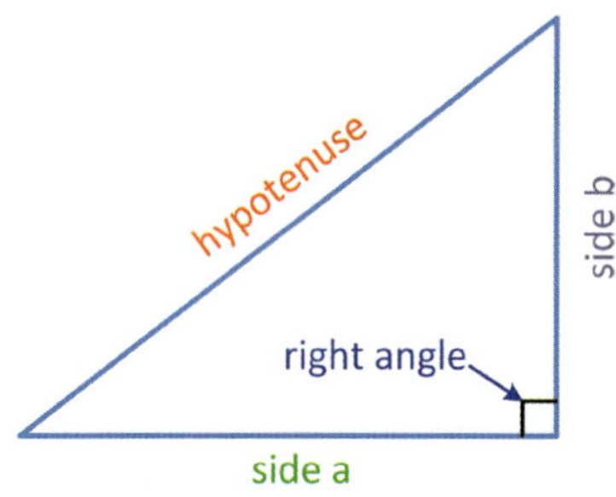

FIGURE 6.1 Right-angled triangle

Pythagoras' theorem states that, in a right-angled triangle, 'the square of the length of the hypotenuse is equal to the sum of the squares of the other two sides'. As an equation:

$$hypotenuse^2 = (side\ a)^2 + (side\ b)^2$$

If we use h to represent the hypotenuse, a to represent side a, and b to represent side b, then solving for h yields:

$$h = \sqrt{a^2 + b^2}$$

EXAMPLE 6.1

Calculate the length of the hypotenuse (h) for the triangle of **Figure 6.2**.

FIGURE 6.2 Example 6.1

$$h = \sqrt{a^2 + b^2}$$
$$= \sqrt{8^2 + 6^2}$$
$$= 10$$

EXERCISE 6.1

a A right-angled triangle has side a of length 160 mm and side b of length 120 mm. Calculate the length of the hypotenuse (h).

b The length of the hypotenuse in a right-angled triangle is 35 mm. If the length of side a is 20 mm, what is the length of side b?

c The length of the hypotenuse in a right-angled triangle is 15.26 cm. If the length of side b is 8 cm, what is the length of side a?

Trigonometric functions

A function is a quantity in mathematics so related with another quantity that if any alteration should be made to the latter there will be a consequent change in the former. The dependent quantity is said to be a function of the other (e.g. the circumference of a circle is a function of the diameter). The six trigonometric functions are the six different ratios that you can set up from a Pythagorean right-angled triangle ($a^2 + b^2 = c^2$) as shown in **Figure 6.3**.

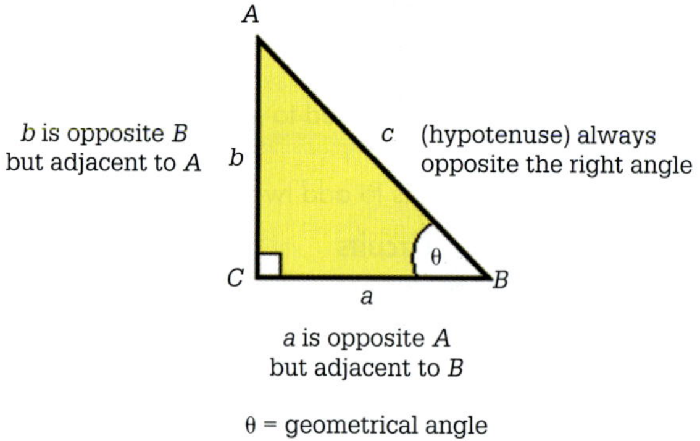

FIGURE 6.3 Relationship of sides in a right-angled triangle

Let the right angle in **Figure 6.3** be labelled C and the hypotenuse c. Let A and B denote the other two angles, and a and b the sides opposite them respectively. In this triangle the trigonometric functions, expressed as ratios are summarised in **Table 6.1**.

TABLE 6.1 Trigonometric ratios

Trigonometric Ratio	For angle A	For angle B
$Sine = \frac{Opposite}{Hypotenuse}$	$Sin\ A = \frac{a}{c}$	$Sin\ B = \frac{b}{c}$
$Cosine = \frac{Adjacent}{Hypotenuse}$	$Cos\ A = \frac{b}{c}$	$Cos\ B = \frac{a}{c}$

»

Trigonometric Ratio	For angle A	For angle B
$Tangent = \frac{Opposite}{Adjacent}$	$Tan\ A = \frac{a}{b}$	$Tan\ B = \frac{b}{a}$
$Secant = \frac{Hypotenuse}{Adjacent}$	$Sec\ A = \frac{c}{b}$	$Sec\ B = \frac{c}{a}$
$Cosecant = \frac{Hypotenuse}{Opposite}$	$Cosec\ A = \frac{c}{a}$	$Cosec\ B = \frac{c}{b}$
$Cotangent = \frac{Adjacent}{Opposite}$	$Cotan\ A = \frac{b}{a}$	$Cotan\ B = \frac{a}{b}$

EXAMPLE 6.2

a Calculate the length of side b for the triangle of **Figure 6.4**.

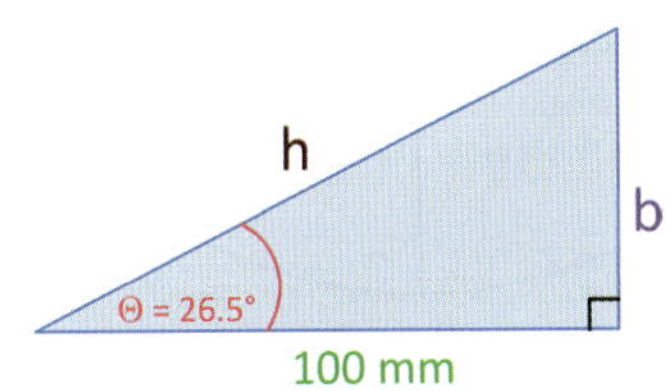

FIGURE 6.4 Example 6.2

$$Tan\ \theta = \frac{Opposite}{Adjacent}$$

$$\therefore \text{Opposite} = \text{Adjacent} \times Tan\ \theta$$

$$= 100 \times Tan\ 26.5°$$

$$= \mathbf{49.9\ mm}$$

b Calculate the length of the hypotenuse (h) for the triangle of **Figure 6.4**.

$$Cos\ \theta = \frac{Adjacent}{Hypotenuse}$$

$$\therefore Hypotenuse = \frac{Adjacent}{Cos\ \theta}$$

$$= \frac{100}{Cos\ 26.5°}$$

$$= \mathbf{111.7\ mm}$$

EXERCISE 6.2

a Calculate the length of side b for the triangle of **Figure 6.5**.

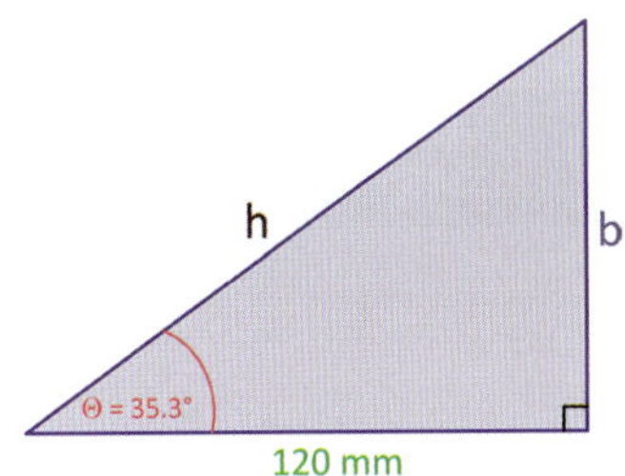

FIGURE 6.5 Exercise 6.2

b Calculate the length of the hypotenuse (h) for the triangle of **Figure 6.5**.

The cosine, cotangent and cosecant of a geometrical angle, represented by the symbol θ (theta) are the sine, tangent and secant of the complement of that angle. For most purposes the three trigonometric functions, sine, cosine and tangent, are enough. All equations for the solution of triangles are based on the law of sines and the law of cosines, together with the fact that the sum of the three angles in a triangle equals 180°. Also, a tangent is an angle's sine divided by its cosine.

Inverse trigonometry functions

When analysing alternating current circuits, it is often the case that the length of the sides of a right-angled triangle are known and we need to determine the angle, say θ. For all the trigonometric functions there is a function that allows us to determine the angle from the trigonometric ratio. These prepend the word 'arc', which is short for arcus, to the aforementioned trigonometric function. These are generally referred to as being the inverse, so the inverse of sin is arcsin. When written as 'arcsin A', we understand it means 'the angle whose sin is A'.

Consider the sine of the angle 60°, which resolves to 0.866. If we take the ratio, in this case 0.866, and want to find the angle we solve for arcsin 0.866, which yields 60°. This function is used when the trigonometric ratio of the angle is known and we want to know the angle.

On a calculator

The scientific calculator is indispensable for finding the values of trigonometric functions and their inverses. On a calculator the inverse buttons may be marked; for example, as [arcsin], [asin] or [sin^{-1}]. The latter form, sin^{-1} is misleading as raising a value to the power negative one [$^{-1}$] implies the reciprocal, which is not the same thing as the inverse function. **Figure 6.6** illustrates the steps involved when using a calculator to solve for an angle knowing the trigonometric ratio.

FIGURE 6.6 Using a calculator to solve for an angle knowing the trigonometric ratio

EXAMPLE 6.3

a Calculate the angle θ for the triangle of **Figure 6.7** if the length of side a is 130 mm and the length of side b is 65 mm.

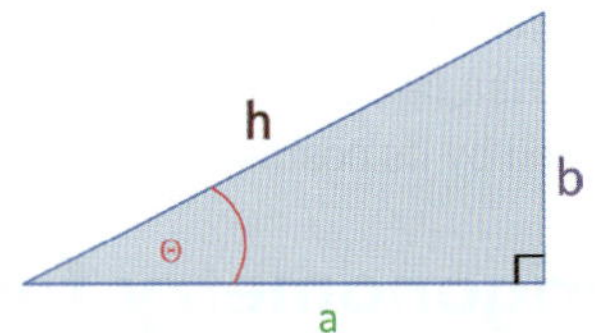

FIGURE 6.7 Example 6.3

$$Tan\ \theta = \frac{Opposite}{Adjacent}$$

$$\therefore \theta = [tan^{-1}]\left(\frac{Opposite}{Adjacent}\right)$$

$$= [tan^{-1}]\left(\frac{65}{130}\right)$$

$$= 26.6°$$

b Calculate the angle θ for the triangle of **Figure 6.7** if the length of side a is 150 mm and the length of the hypotenuse (h) is 180 mm.

$$Cos\ \theta = \frac{Adjacent}{Hypotenuse}$$

$$\therefore \theta = [cos^{-1}]\frac{Adjacent}{Hypotenuse}$$

$$= [cos^{-1}]\frac{150}{180}$$

$$= 33.6°$$

EXERCISE 6.3

a Calculate the angle θ for the triangle of **Figure 6.7** if the length of side a is 180 mm and the length of side b is 105 mm.

b Calculate the angle θ for the triangle of **Figure 6.7** if the length of side a is 110 mm and the length of the hypotenuse (h) is 150 mm.

c Calculate the angle θ for the triangle of **Figure 6.7** if the length of side b is 105 mm and the length of the hypotenuse (h) is 150 mm.

Rectangular coordinates

Trigonometric ratios or functions apply to angles within a circle. **Figure 6.8** shows angles of 30°, 150° and 210°. If we use a scientific calculator to find the sin, cos, and tan of these angles, we get the results summarised in **Table 6.2**.

TABLE 6.2 Summary of trigonometric values

Angle Θ	Sin Θ	Cos Θ	Tan Θ
30°	0.5	0.866	0.577
150°	0.5	−0.866	−0.577
210°	−0.5	−0.866	0.577

You will notice that the magnitude of each of the values tabulated for each trigonometric function is the same. The value is either positive or negative. This is because each angle can be equated to 30°; 150° is found from 180° – 30° and 210° equals 180° + 30°. **Figure 6.8** shows how to determine whether the trigonometric function yields a positive or negative result. In quadrant 1, all functions are positive.

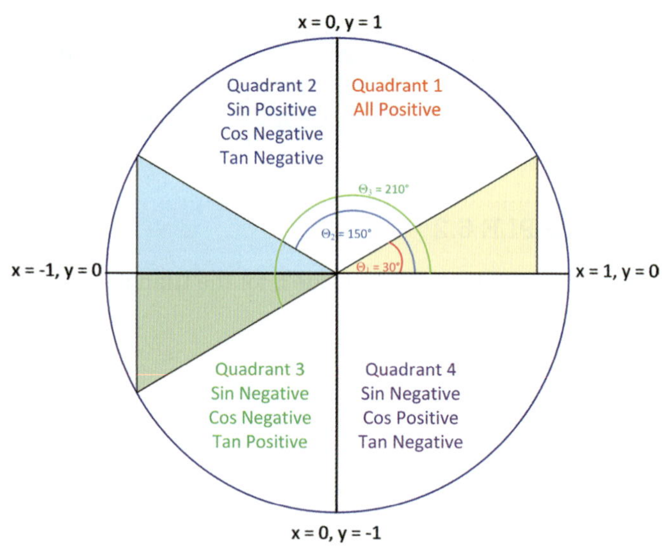

FIGURE 6.8 Rectangular coordinates

EXAMPLE 6.4

a Use a scientific calculator to determine the sin of 250° and in which quadrant does this lie?

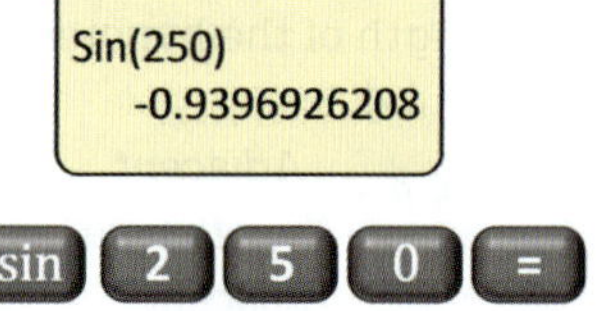

FIGURE 6.9 Using scientific calculator

Answer: Quadrant 3

EXERCISE 6.4

- Use a scientific calculator to determine the trigonometric functions listed in **Table 6.3**.

TABLE 6.3 Result of trigonometric functions

Sin 330°	Cos 230°	Tan 345°	Sin 135°	Cos 120°	Tan 210°

Sine wave

The term 'sine wave' is used to express the mathematical relationship of the shape of the complete wave produced by one revolution of the conductor in a uniform magnetic field

(Φ) in a defined amount of time (t). A sine wave is created when the instantaneous values of voltage generated (e) by the conductor at various angular positions are plotted against time as shown in **Figure 6.10**.

$$e = \frac{\Delta\theta}{\Delta t}$$

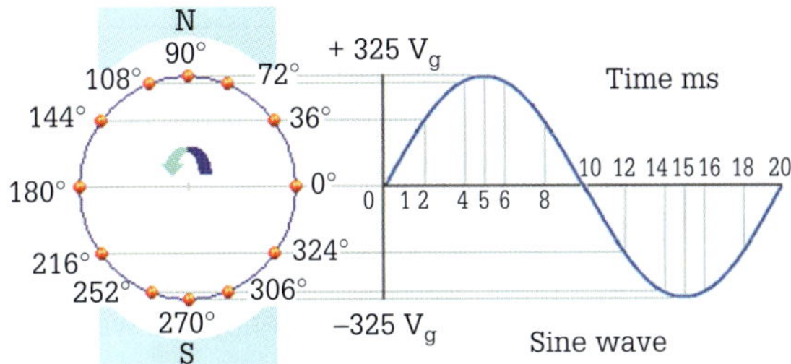

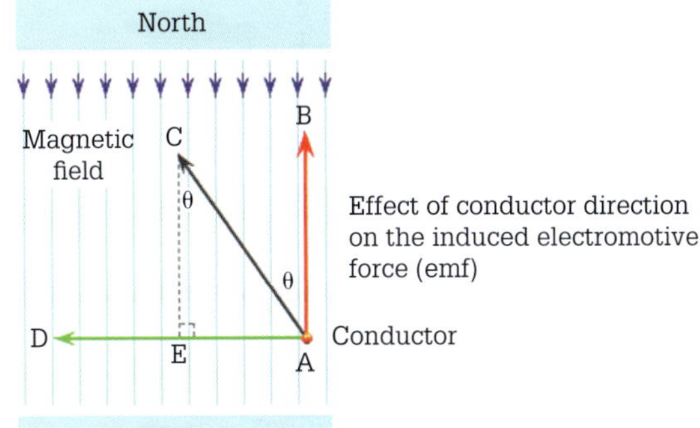

FIGURE 6.10 Sine wave and effect of conductor direction

Using trigonometry, the instantaneous value of emf (e) induced in the conductor moving from A to C can be determined by:

$$e = E \sin\theta$$

where e = instantaneous value of emf in volts
E = maximum value of emf in volts
θ = angle of rotation in degrees

The wave shape of **Figure 6.10** is therefore a graph of $E \sin\theta$ and is referred to as a 'sine wave', and is said to be 'sinusoidal' in shape.

EXAMPLE 6.5

Calculate the instantaneous voltage generated from a maximum emf of 340 V when the angle of rotation is 120°.

$$e = E\sin\theta$$
$$= 340 \sin 120°$$
$$= \mathbf{294.45\ V}$$

EXERCISE 6.5

Calculate the instantaneous voltage generated from a maximum emf of 120 V when the angle of rotation is:

a 60°
b 160°
c 240°
d 330°

An emf and current that are alternating change polarity regularly. To indicate these reversals graphically, it is usual to regard the emf and current in one direction as having a positive value and the voltage and current in the opposite direction as having a negative value. If we were to tabulate the sine of all angles between 0° and 360° in 10° increments and then plot these sine values as points on a graph, then joining the points to form a continuous curved line would look like **Figure 6.11**. On the vertical or y-axis of the graph positive positions are displayed above and negative positions are displayed below the horizontal x-axis as illustrated.

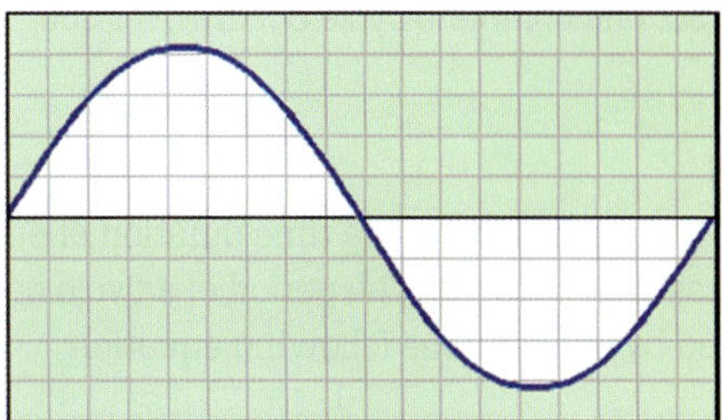

FIGURE 6.11 Graph of sine function

Waveforms

The sine wave of the a.c. voltage is not the only kind of the wave that exists and is commonly produced by a.c. circuits. The waveform produced by an a.c. circuit can be a square wave, triangle wave or sawtooth wave as illustrated in **Figure 6.12**.

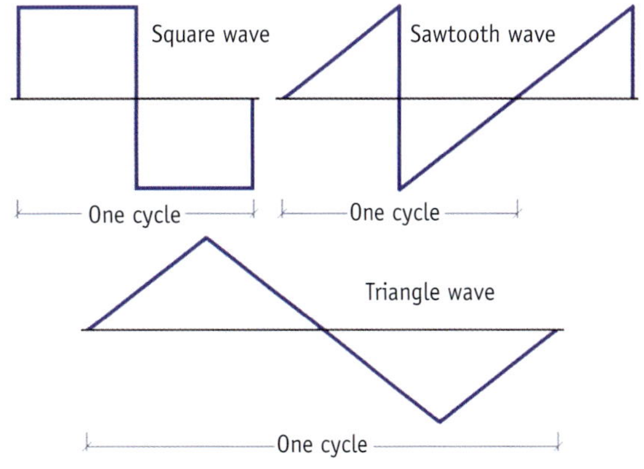

FIGURE 6.12 a.c. waveform

A distinguishing feature of alternating waves is that equal areas are enclosed above and below the x-axis. The square, sawtooth and triangle waves play an important role in digital signal processing applications such as clock signals, trigger pulses and timing signals.

Waveform terminology

Cycle

A cycle is any repetition of a variable quantity recurring at equal intervals of time. A cycle with a sine wave is the set of positive and negative values of an alternating voltage or current occurring in one wave shape. A cycle is illustrated in **Figure 6.13**.

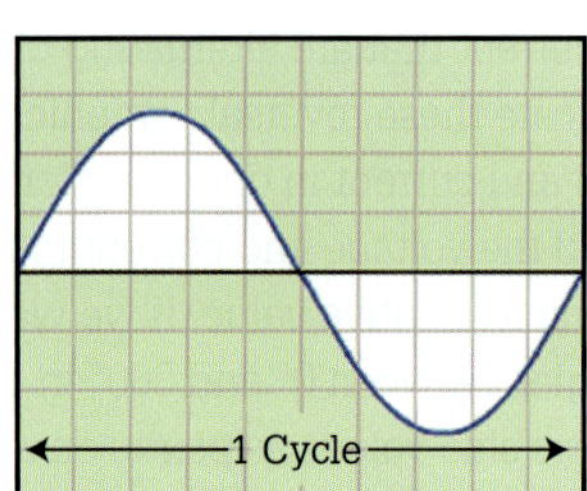

FIGURE 6.13 One cycle

A cycle may also be specified in terms of angular measure. So, one complete cycle occurs over 360° or 2π radians.

Period

A period or periodic time is the time duration of one cycle. It is expressed in seconds. The period of a cycle can be determined by applying the following equation:

$$t = \frac{1}{f}$$

where t = period (time) in second (s)
f = frequency in hertz (Hz)

EXAMPLE 6.6

What is the period of a sine wave that has a frequency of 50 Hz?

$$t = \frac{1}{f}$$
$$= \frac{1}{50}$$
$$= \mathbf{20\ ms}$$

EXERCISE 6.6

Calculate the period of sine waves having the following frequencies:

- a 25 Hz
- b 100 Hz
- c 1 kHz
- d 20 kHz

Frequency

The frequency (symbol f) is the number of whole cycles completed in one second. It is expressed in units of hertz (Hz) and is the number of complete cycles per second.

In Australia, the power line frequency is 50 Hz, meaning that the alternating current voltage of 230/400 V, +10%, −6% oscillates at a rate of 50 complete cycles every second.

A wavelength (symbol λ, which is the Greek symbol for lambda) is the distance between corresponding points on two successive cycles and is the distance travelled in the period of one cycle. An illustration of wavelength is shown in **Figure 6.14**.

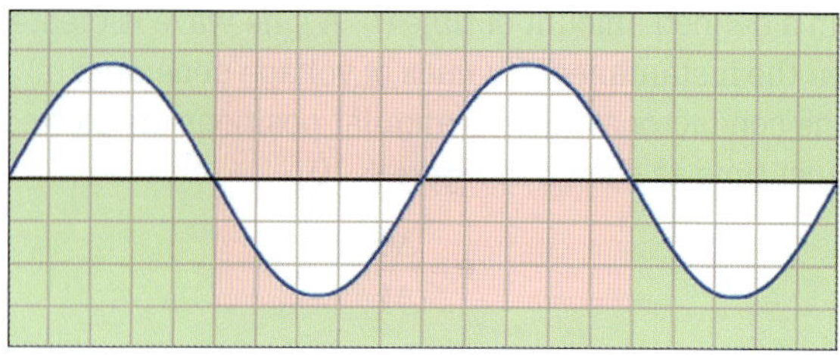

FIGURE 6.14 Wavelength

Frequency of a wave

The period and frequency are mathematical reciprocals of one another. For example, if a sine wave has a period of 100 s, its frequency will be 0.01 Hz. The frequency of a wave can be determined by:

$$f = \frac{1}{t}$$

where t = period (time) in second (s)
f = frequency in hertz (Hz)

EXAMPLE 6.7

What is the frequency of a sine wave that has a period of 50 ms?

$$f = \frac{1}{t}$$
$$= \frac{1}{50 \times 10^{-3}}$$
$$= \mathbf{20\ Hz}$$

EXERCISE 6.7

Calculate the frequency of sine waves having the following periods:

- a 4 ms
- b 100 ms
- c 20 µs
- d 500 µs

Peak value

The maximum or peak value (symbol V_{max}, I_{max} or V_p, I_p) is also known. The amplitude of the wave is the highest value that the voltage or current reaches in one direction. It is necessary to know this value in a circuit to ensure that insulation does not deteriorate when stressed by the peak values of voltage and current.

A conductor has the maximum electromotive force induced in it when it cuts directly across the magnetic lines of force. The peak-to-peak value ($V_{p\text{-}p}$ or $I_{p\text{-}p}$) is the highest value that the voltage or current reaches in both directions. The peak value is shown in **Figure 6.15**.

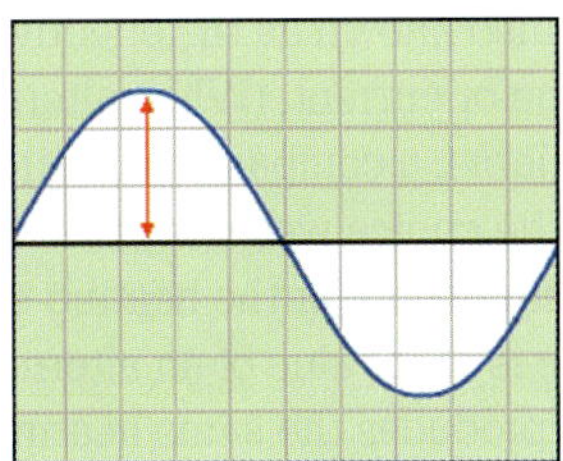

FIGURE 6.15 Peak value

An oscilloscope is used to measure the peak and peak-to-peak values. This is because the oscilloscope displays the crests of the waveform with a high degree of accuracy. The peak-to-peak value measurement is shown in **Figure 6.16**.

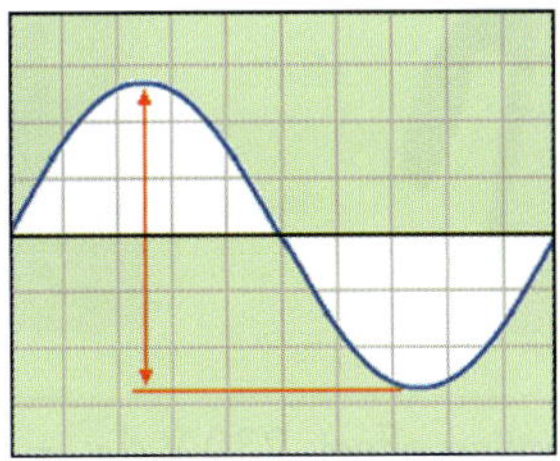

FIGURE 6.16 Peak-to-peak value measurement

Instantaneous values

The instantaneous values as shown in **Figure 6.17** of voltage or current (*c* or *i*) may vary from zero to a peak value many times per second. Each instantaneous value is proportional to the rate at which the conductor cuts the magnetic lines of force at that instant. The instantaneous emf generated in a conductor rotating at a uniform speed in a constant magnetic field is proportional to the sine of the angle (θ) through which the conductor has rotated from the zero value. The instantaneous values of voltage or current for that conductor when plotted on a graph form a curve that coincides with the trigonometry sine table.

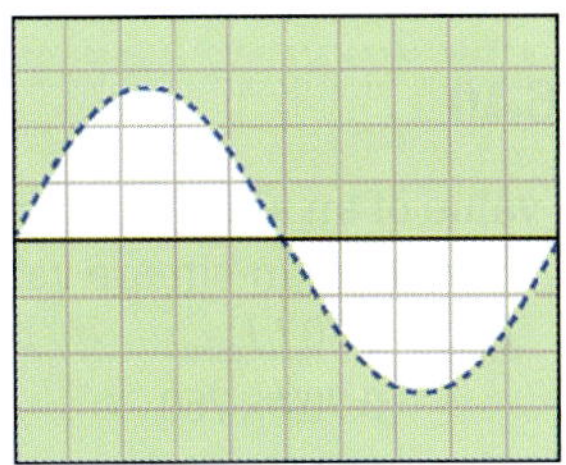

FIGURE 6.17 Instantaneous values

The instantaneous values of voltage and current can be determined by applying the equations:

$$e = V_{max} \times \sin\theta \text{ and } i = I_{max} \times \sin\theta$$

where e = instantaneous value of voltage (volts)

V_{max} = maximum or peak value of voltage (V)

i = instantaneous value of current (A)

I_{max} = maximum or peak value of current (A)

$\sin\theta$ = trigonometric function for the angle at which the flux is being cut

EXAMPLE 6.8

a An alternating voltage has a maximum value of 140 V. Find the instantaneous value of voltage at 60° after the commencement of the cycle.

$$\begin{aligned} e &= V_{max} \times \sin\theta \ (\sin 60° = 0.87) \\ &= 140 \times 0.87 \\ &= \mathbf{121.8\ volts} \end{aligned}$$

b Calculate the instantaneous value of current at 223° after the cycle has commenced for a generated output of 40 A maximum.

$$\begin{aligned} i &= I_{max} \sin\theta \\ &= 40 \times \sin 223° \\ &= 40 \times -0.68199836 \\ &= \mathbf{-27.28\ A} \end{aligned}$$

The minus sign indicates that this current is produced on the negative portion of the sine waveform.

EXERCISE 6.8

A sine wave with a maximum value of 200 V is applied to a 100 Ω resistive load. Show graphically how the instantaneous values of current vary as the angle varies from 0° to 360° in 15° intervals. Complete your table using the following equations:

$$i = I_{max} \times \sin\theta \text{ and } i = \frac{e}{R}$$

After tabulating your answers, your plot should resemble **Figure 6.18**.

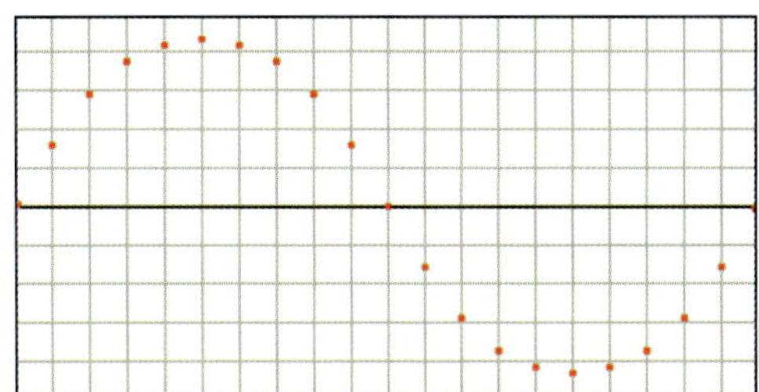

FIGURE 6.18 Instantaneous values of current

Average value

The average or mean value (symbol V_{av} or I_{av}) is the average of all the instantaneous values for one-half of a cycle. The mean is used to compare alternating current with direct current and it is the quantity of alternating current that gives or supplies the same electromagnetic field as an equal amount of direct current. Its importance is in its relation with the root-mean-square value as this ratio gives the form factor.

SWITCH ON

For a sine wave the mean is equal to 0.637 times the maximum value. The mean of a complete cycle is taken over half a cycle only because the mean of a complete cycle of a sine wave is zero. The mean of a sine wave is shown in **Figure 6.19**.

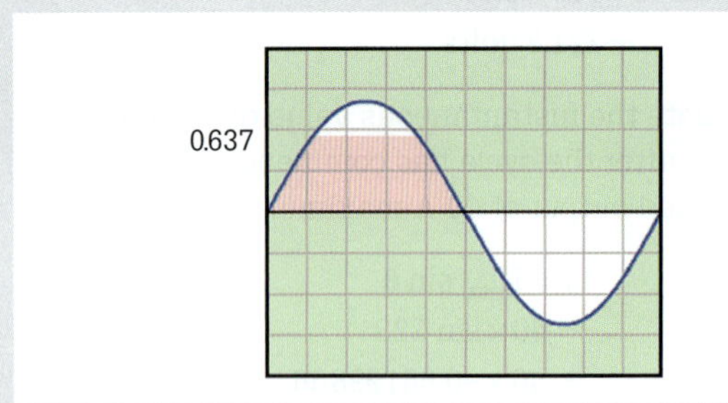

FIGURE 6.19 Average value – 0.637 of the peak

The average values of voltage and current can be calculated by applying the following equations:

$$V_{av} = 0.637\ V_{max} \text{ and } I_{av} = 0.637\ I_{max}$$

where V_{av} = average value of voltage (V)

V_{max} = maximum or peak value of voltage (V)

I_{av} = average value of current (A)

I_{max} = maximum or peak value of current (A)

EXAMPLE 6.9

A sinusoidal voltage wave has a maximum value of 100 V. What is the average value of voltage for a half cycle?

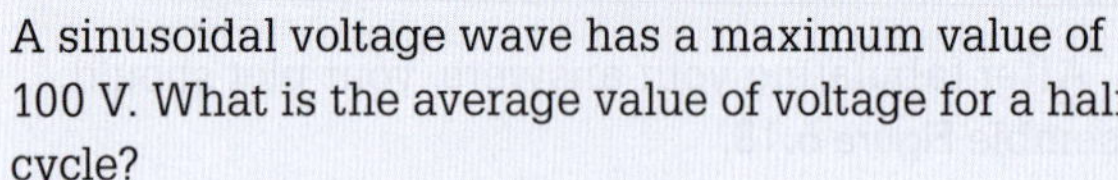

$$V_{av} = 0.637\ V_{max}$$
$$= 0.637 \times 100$$
$$= \mathbf{63.7\ volts}$$

EXERCISE 6.9

a A sinusoidal voltage wave has a maximum value of 340 V. What is the average value of voltage for a half cycle?

b A sinusoidal voltage wave has an average value of 160 V for a half cycle. What is the maximum value of voltage?

RMS value

The root-mean-square value (abbreviation rms) is the most useful and practical value of alternating voltage or current for a sine wave. It is determined by taking the square root of the mean (average) of the sum of the squares of all the instantaneous values and is equal to 0.707 $\left(\frac{1}{\sqrt{2}}\right)$ times the maximum value.

This value is also known as the effective value of an alternating voltage or current. (When alternating current first became available, a comparison to direct current was necessary. The choice was which value to use – the magnetic effect equivalent or the heating effect?)

The rms value is that value of an alternating voltage or current which produces the same heating effect as the same value of direct voltage or current. Consider the heating elements of the jugs in **Figure 6.20**. If a direct voltage of 230 V energises the element, a certain heating effect will be produced. If the element is connected to an alternating 230 V rms supply, the same heating effect will be produced.

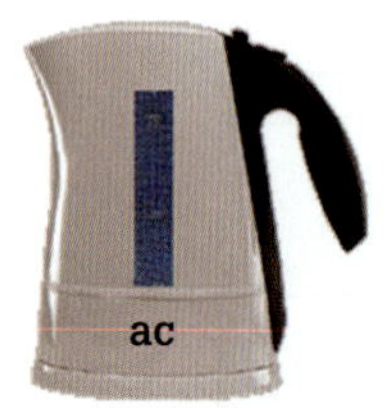

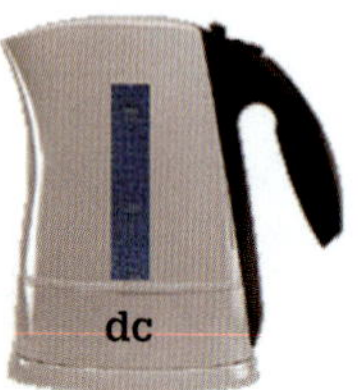

FIGURE 6.20 Heating effect: 230 V d.c. = 230 V rms

Unless otherwise specifically stated, values of alternating voltage and current are given in their rms values. In Australia, the mains voltage is 230 V and the maximum voltage of the mains is approximately 325 V. Meters used for measuring alternating voltage or current indicate the rms values of alternating voltage or current unless otherwise specified. The rms value of a sine wave can be calculated by applying the following equation:

$$V_{rms} = 0.707\ V_{max} \text{ and } I_{rms} = 0.707\ I_{max}$$

where V_{rms} = rms value of voltage (V)

V_{max} = maximum or peak value of voltage (V)

I_{rms} = rms value of current (A)

I_{max} = maximum or peak value of current (A)

EXAMPLE 6.10

The maximum value of an alternating current is 750 A. What is its root-mean-square value?

$$I_{rms} = 0.707\ I_{max}$$
$$= 0.707 \times 750$$
$$= \mathbf{530.25\ amperes}$$

EXERCISE 6.10

a The maximum value of an alternating voltage is 340 V. What is its rms value?

b The rms value of an alternating current is 100 A. What is its maximum value?

A voltage conversion summary is illustrated in **Figure 6.21**.

Convert from	Convert to				
	Peak	Peak to peak	rms	Instantaneous	Average
Peak		$V_{pp} = 2V_p$	$V_{rms} = 0.707\ V_p$	$V = V_p \sin\theta$	$V_{avg} = 0.637\ V_p$
Peak to peak	$V_p = \frac{V_{pp}}{2}$		$V_{rms} = \frac{V_{pp}}{2.828}$	$V = 0.5\ V_{pp} \sin\theta$	$V_{avg} = 0.318\ V_{pp}$
rms	$V_p = 1.414\ V_{rms}$	$V_{pp} = 2.828\ V_{rms}$		$v = \frac{V_{rms} \sin\theta}{0.707}$	$V_{avg} = 0.9\ V_{rms}$
Instantaneous	$V_p = \frac{v}{\sin\theta}$	$V_{pp} = \frac{2\ v}{\sin\theta}$	$V_{rms} = \frac{0.707\ v}{\sin\theta}$		$V_{avg} = \frac{0.637\ v}{\sin\theta}$
Average	$V_p = \frac{V_{avg}}{0.637}$	$V_{pp} = 1.274\ V_{avg}$	$V_{rms} = 1.1\ V_{avg}$	$v = \frac{V_{avg} \sin\theta}{0.637}$	

FIGURE 6.21 Voltage conversion summary

Angular measurement of a sine wave

A sine wave can also be expressed in terms of an angular measurement. This angular measurement is expressed in degrees and radians as shown in **Figure 6.22**.

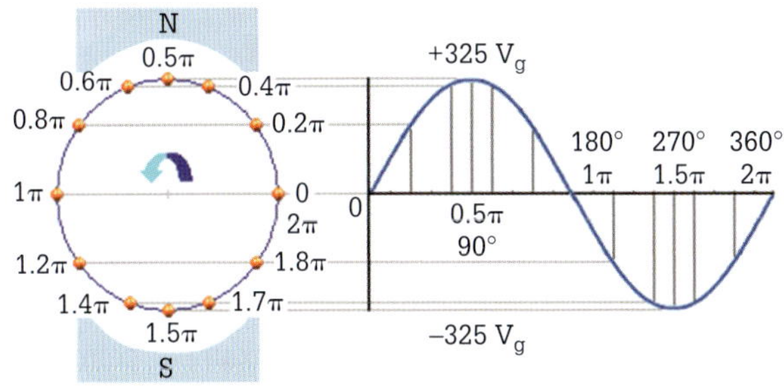

FIGURE 6.22 Angular measurement of a sine wave

A radian is a unit of measurement defined as 360°/2π. The radian is the angle formed by two radii, lines from the centre to the outside circumference of a circle, where the arc formed is equal to the radius. An angle in radians can be calculated by dividing the length of the arc the angle cuts out by the radius of the circle. Since the circumference of a circle is 2π × radius, there are 2π radians in 360°, so:

$$1 \text{ radian} = \frac{360}{2\pi} = 57.3° \text{ approx}$$

As a full revolution of a circle is expressed as 2π radians, so:

$$90° = 0.5\pi \text{ radians}$$
$$180° = 1\pi \text{ radians}$$
$$270° = 1.5\pi \text{ radians}$$
$$360° = 2\pi \text{ radians}$$

Generally, if we have an angle measured in degrees, it can be converted to radians using the following equation:

$$\text{radians} = \frac{\text{angle in degrees}}{57.3°}$$

This can also be expressed as:

$$\text{radians} = \text{angle in degrees}\ \frac{\pi}{180}$$

EXAMPLE 6.11

Express the 35° point of a sine wave as an equivalent number of radians.

$$\text{radians} = \frac{\text{angle in degrees}}{57.3°}$$
$$= \frac{35°}{57.3°}$$
$$= \mathbf{0.6108\ rads}$$

EXERCISE 6.11

Express the following angular quantities in their radian form:

a 50°
b 200°
c 320°
d 90°
e 150°

The oscilloscope

An oscilloscope, such as that shown in **Figure 6.23**, is a test and measurement that graphically displays electrical signals in the time domain. This means it quickly measures the amplitude of a signal at a particular instant in time and displays this as a dot on the screen. It takes a number of samples in quick succession and displays each of these so quickly that the signal appears as a line on the screen such as that shown in **Figure 6.23**.

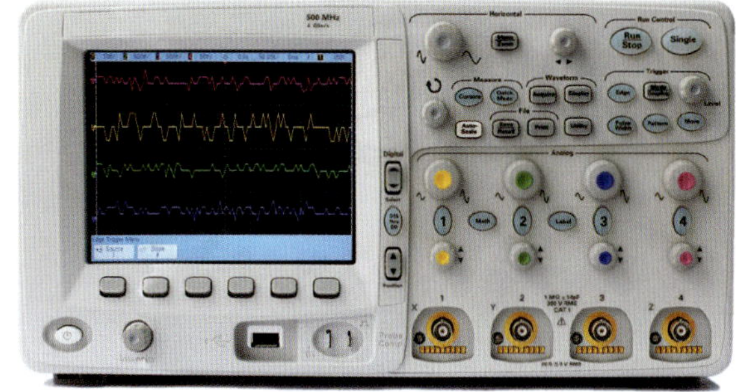

Source: Shutterstock.com/Dmitry Strizhakov

FIGURE 6.23 Digital oscilloscope

Oscilloscopes are an invaluable tool for designing, manufacturing or repairing electronic equipment. They also find use in power quality analysis when used to measure electrical phenomena.

Oscilloscope operation

The main operational blocks of an oscilloscope are the vertical, horizontal and trigger systems. These individual systems work harmoniously to provide information about the electrical signal, which allows the oscilloscope to accurately reconstruct the signal. **Figure 6.24** shows the architecture of a digital oscilloscope.

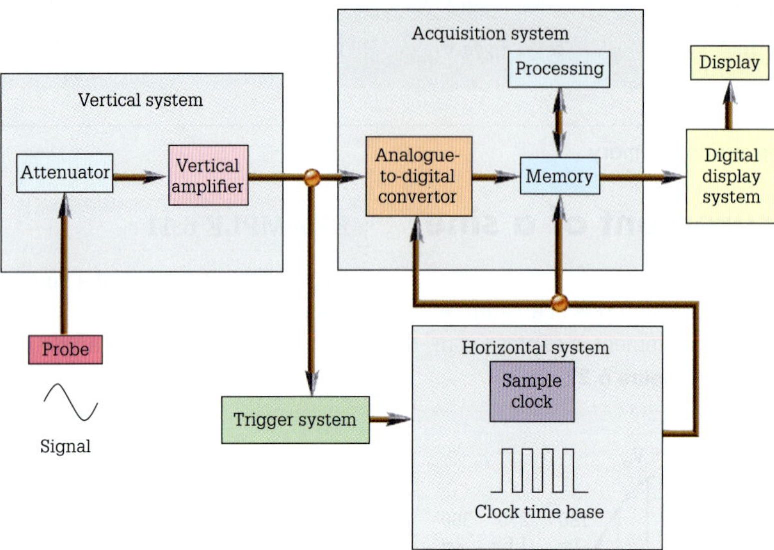

FIGURE 6.24 Architecture of a digital oscilloscope

A probe (**Figure 6.25**) connects to the point under test and conveys the signal to the first stage of the oscilloscope, which either attenuates or amplifies the signal to optimise the amplitude of the signal. This system is called the vertical system as it applies vertical scale control. The signal then travels to the acquisition block, where the analogue-to-digital converter (ADC) samples the signal voltage [analogue] and converts it to a digital value [digital]. The horizontal system contains a sampling clock and takes a snapshot of the signal sample at a precise time, which provides the horizontal coordinate. The sampling clock triggers the ADC, which stores its digital output in the acquisition memory as a record point. The trigger system detects a specific signal condition as specified by the user in the incoming signal and applies it as a time reference in the waveform record. The display shows the event that met the trigger criteria as well as the waveform data preceding or following the event.

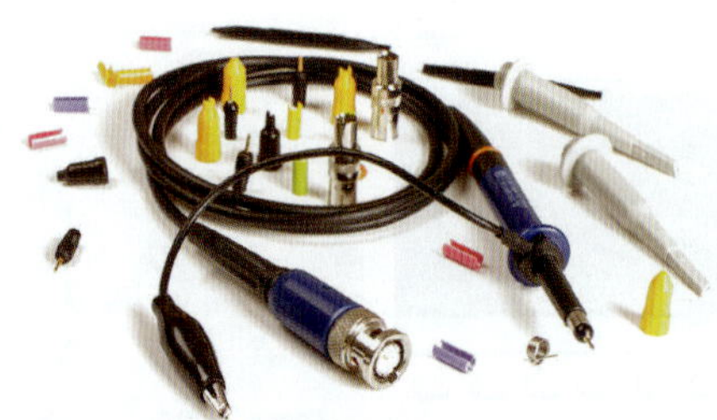

Source: Shutterstock.com/Audrius Merfeldas

FIGURE 6.25 Oscilloscope probe and lead

Oscilloscopes are capable of measuring voltages over a fairly narrow range. In order to read higher magnitude it is necessary to use attenuator probes. For example, commonly available 10:1 probes only allow one-tenth of the original voltage through to the oscilloscope. Consequently, you must multiply the resulting measurement by a factor of 10 giving rise to the name of ×10 probes.

Measurements with an oscilloscope

A voltmeter measures the potential difference between two points in an electrical circuit. A multimeter is able to measure potential difference (voltmeter), current (ammeter), resistance (ohmmeter), and generally supports other functions such as frequency and capacitance measurements. An oscilloscope displays graphically how the potential difference changes over time. Essentially an oscilloscope measures voltage waveforms. On an oscilloscope screen, voltage is displayed vertically on the *y*-axis and time is represented horizontally on the *x*-axis. The intensity or brightness of the display is sometimes referred to as the *z*-axis. The displayed graph, similar to that shown in **Figure 6.26**, can provide information pertaining to:

- time and voltage values of a signal
- frequency of an oscillating signal
- the 'moving parts' of a circuit represented by the signal
- frequency with which a particular portion of the signal is occurring relative to other portions
- whether or not a malfunctioning component is distorting the signal
- how much of a signal is direct current (d.c.) or alternating current (a.c.)

- the portion of the signal that is noise
- whether noise is changing over time.

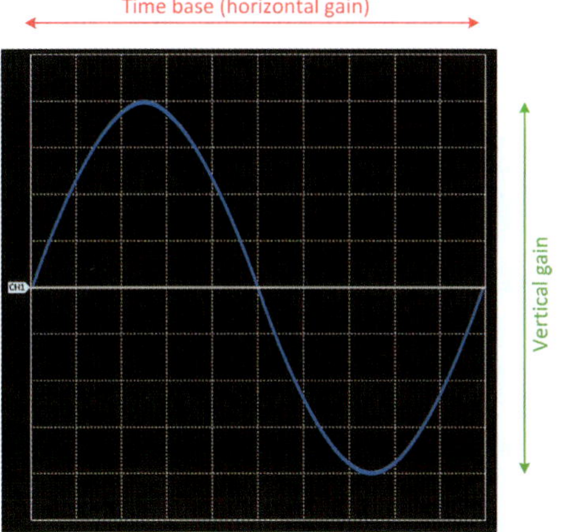

FIGURE 6.26 Oscilloscope display

Oscilloscope measurement controls

The basic measurement controls for oscilloscopes are:

- horizontal scale adjustment
- vertical scale adjustment
- trigger.

Vertical scale

A rotary knob, located in the vertical portion of the oscilloscope controls, provides adjustment to the number of volts per division. **Figure 6.27** shows this control along with a 1 kHz square wave with the vertical scale adjusted to 1V/DIV. This vertical setting means that every division on the vertical scale (up and down) represents 1 volt. The channel 1 marker to the left of the display graticule shows the location of the 0 volt or ground level. **Figure 6.27** also shows that the waveform has a height of 5 divisions (2½ divisions above zero and 2½ divisions below zero). Therefore, the waveform has an amplitude of 5 volts (5 divisions × 1 volt/DIV).

If the vertical control is adjusted to 2V/DIV (turned clockwise), then for the same waveform depicted in **Figure 6.27**, the display shows a shorter waveform as **Figure 6.28** indicates. Note that the signal input has not changed. As the vertical scale is now 2V/DIV, the waveform has a height of 2½ divisions (1¼ divisions above zero and

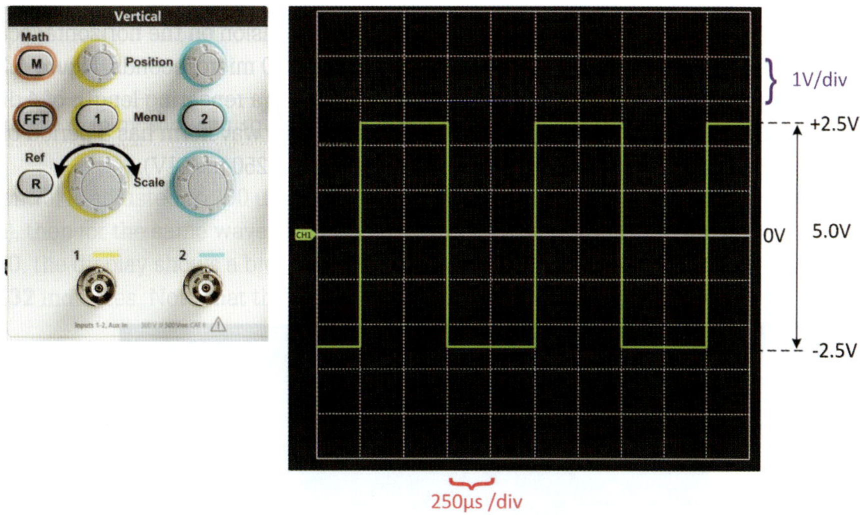

FIGURE 6.27 Oscilloscope display showing square wave at 1 volt per division

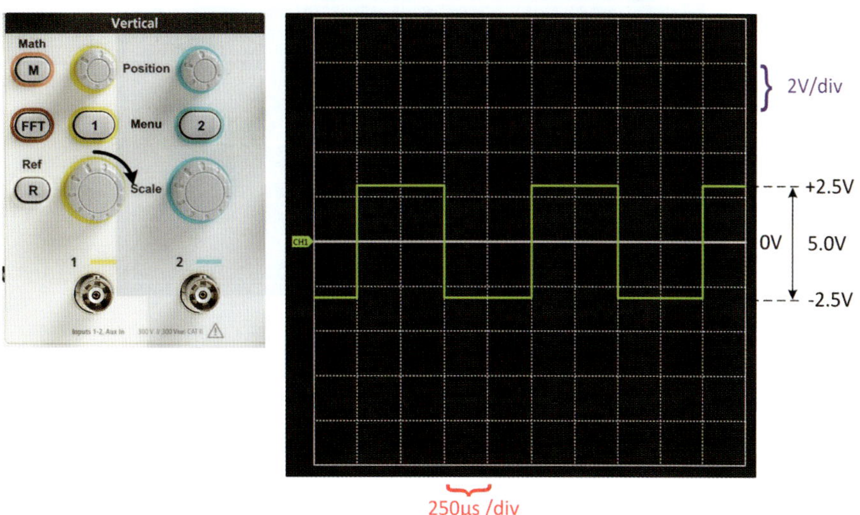

FIGURE 6.28 Oscilloscope display showing square wave at 2 volts per division

Trigger control

The triggering system essentially tells the oscilloscope when to start drawing the waveform on the display. **Figure 6.33** shows the effect of an incorrectly adjusted trigger level. Fortunately, oscilloscopes have an automatic trigger function, which makes it easy to display a signal.

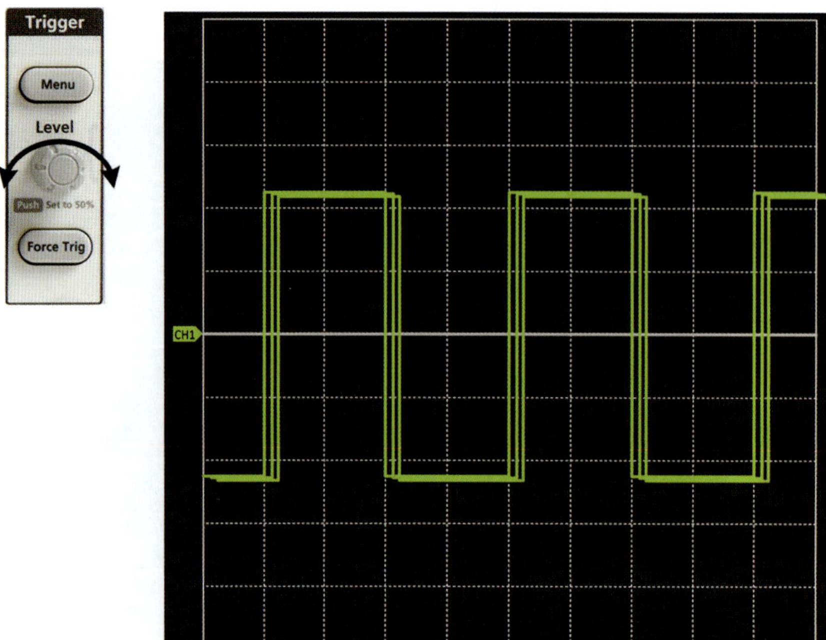

FIGURE 6.33 Oscilloscope display showing incorrect trigger setting

Portable oscilloscope

Portable or hand-held digital oscilloscopes, such as that shown in **Figure 6.34**, have multimeter functions such as:

- max display: 4000 counts
- d.c. voltage: 400.0 mV, 4.000 V, 40.00 V, 400.0 V, 1000 V
- a.c. voltage: 4.000 V, 40.00 V, 400.0 V, 1000 V
- d.c. current: 40.00 mA, 4000.0 mA
- a.c. current: 40.00 mA, 4000.0 mA
- resistance: 400.0 MΩ – 40.00 MΩ
- capacitance: 5.12 nF, 51.2 nF, 512 nF, 5.12 μF, 51.2 μF, 100 μF
- diode: 0 V–1.5 V
- open-short test: beeping.

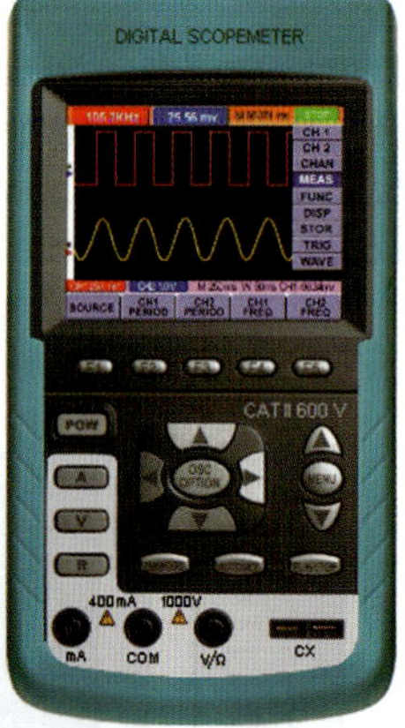

FIGURE 6.34 Portable digital oscilloscope

REVIEW QUESTIONS

1. A right-angled triangle has side a of length 120 mm and side b of length 80 mm. Calculate the length of the hypotenuse (h).
2. The length of the hypotenuse (h) in a right-angled triangle is 45 mm. If the length of side a is 25 mm, what is the length of side b?
3. Calculate the angle θ for a right-angled triangle if the length of the adjacent side is 150 mm and the length of the opposite side is 95 mm.
4. Calculate the angle θ for a right-angled triangle if the length of the adjacent side is 135 mm and the length of the hypotenuse is 150 mm.
5. Calculate the instantaneous voltage generated from a maximum emf of 90 V when the angle of rotation is 60°.
6. Calculate the period of a sine wave having a frequency of 40 Hz.
7. Calculate the frequency of a sine wave having a period of 6 ms.
8. A sinusoidal voltage wave has a maximum value of 280 V. What is the average value of voltage for a half cycle?

»

9 A sinusoidal voltage wave has an average value 150 V for a half cycle. What is the maximum value of voltage?
10 The maximum value of an alternating voltage is 300 V. What is its root-mean-square value?
11 The rms value of an alternating current is 70.7 A. What is its maximum value?
12 Express 450° in its radian form.
13 Complete the following for the data given in each diagram of **Figure 6.35**.
14 When using a 10X probe what factor must the resulting measurement be multiplied by?
15 List three waveforms that can be produced by an a.c. circuit.
16 Over how many radians will one sine wave cycle occur?
17 What is the number of whole cycles completed in one second known as?
18 What is the name given to the highest value that the voltage or current in sine waveform reaches in either direction from zero?
19 State the average value of a complete cycle of a sine wave.
20 A period or periodic time is the time duration of ________________.

a.

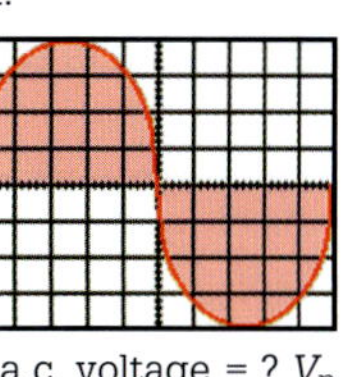

a.c. voltage = ? V_{p-p}
Frequency = ? Hz
Gain control = 1 V/div
Time base = 10 μs/div

b.

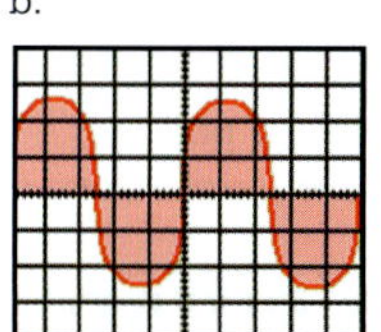

a.c. voltage = ? V_{p-p}
Frequency = ? Hz
Gain control = 5 V/div
Time base = 50 μs/div

c.

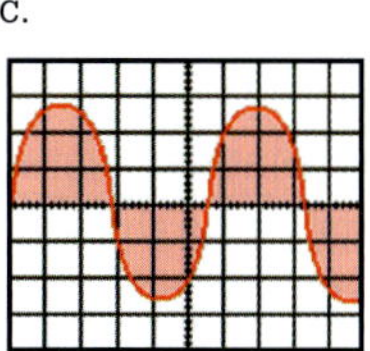

a.c. voltage = ? V_{p-p}
Frequency = ? Hz
Gain control = 0.5 V/div
Time base = 20 μs/div

d.

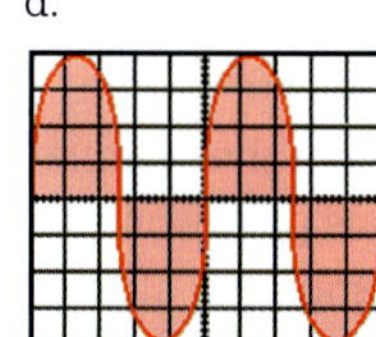

a.c. voltage = ? V_{p-p}
Frequency = ? Hz
Gain control = 1 V/div
Time base = 0.5 μs/div

e.

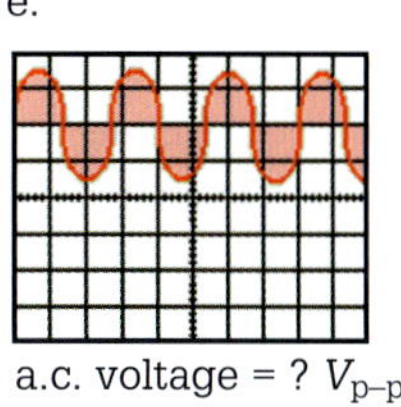

a.c. voltage = ? V_{p-p}
Frequency = ? Hz
Gain control = 20 V/div
Time base = 2 μs/div

f.

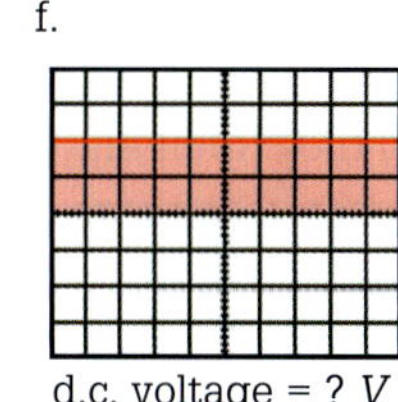

d.c. voltage = ? V
Frequency = ? Hz
Gain control = 1 V/div
Time base = 20 μs/div

FIGURE 6.35 Waveform diagrams

6.2 Phasor diagrams

Representing a.c. quantities

As soon as a load draws a current from an alternating supply it has the same frequency as the supply voltage driving it. When the voltage and current waveform pass through their zero values and increase to their peak values in the same direction and at the same instant in time, the current is regarded as being in phase with the voltage. If this does not happen the current is regarded as being out of phase with the driving voltage. We saw earlier that there are three possible phase relationships between the sine wave current and a sine wave voltage in an electrical circuit. These were in phase, lag and lead. When specifying phase relationships it is the time difference of the waveforms expressed in electrical degrees that is important. In order to convert time to electrical degrees, the time taken for one complete cycle is divided into 360 equal divisions. One division therefore equals one degree. The relationship of time to electrical degrees is illustrated in **Figure 6.36**.

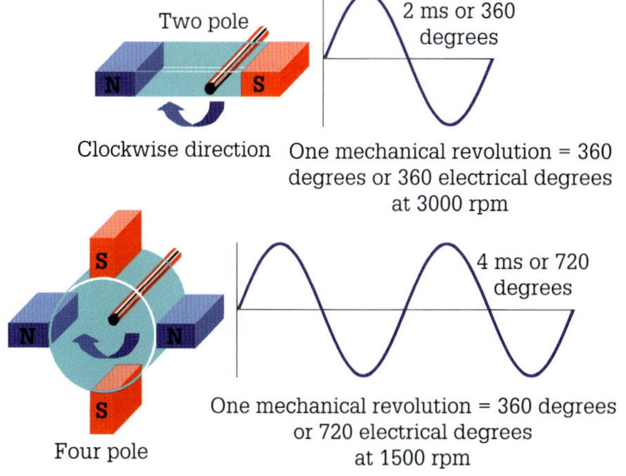

FIGURE 6.36 Electrical degrees

The amount of time that a circuit current lags or leads the applied voltage varies in different circuits according to the components or devices requiring the current. As time can also be measured in electrical degrees this difference in time is expressed in electrical degrees and is called the phase difference or phase angle (symbol ϕ which is the Greek symbol for phi).

Phase difference

Phase denotes the particular point in the cycle of a waveform, usually measured as an angle in degrees. The phase difference is also known as phase shift or phase angle. It consists of two or more waveforms of the same frequency and is determined from the horizontal difference along the x-axis in degrees or radians that one waveform has moved from their similar reference point (two similar instantaneous points on each waveform) along the x-axis.

When comparing alternating voltages and currents of the same frequency they are either in phase or out of phase with one another, depending on the type of electrical components in the circuit.

For expediency, waveforms of voltage and current are drawn on the same graph by plotting all the instantaneous values over a period of one cycle. It is then easier to see how the voltage and current values vary from instant to instant in the circuit and to determine the phase relationship between these values.

Figure 6.37 illustrates graphically the voltage and current phase relationships.

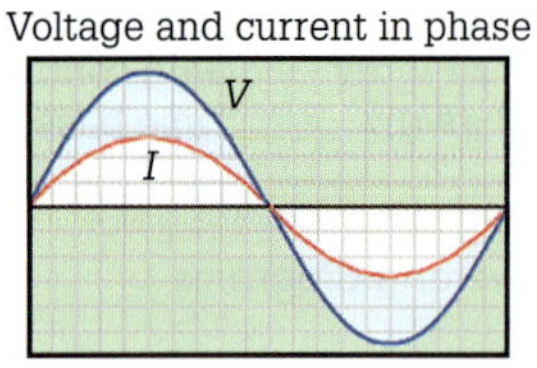

Voltage and current in phase

V

I

FIGURE 6.37 Phase relationships between voltage and current

As seen in **Figure 6.37**, when the voltage and current waves reach their maximum positive, maximum negative and zero values at the same instant in time they are 'in phase'.

When these two waves do not reach these maximum values at the same instant in time they are 'out of phase' as illustrated in **Figure 6.38**.

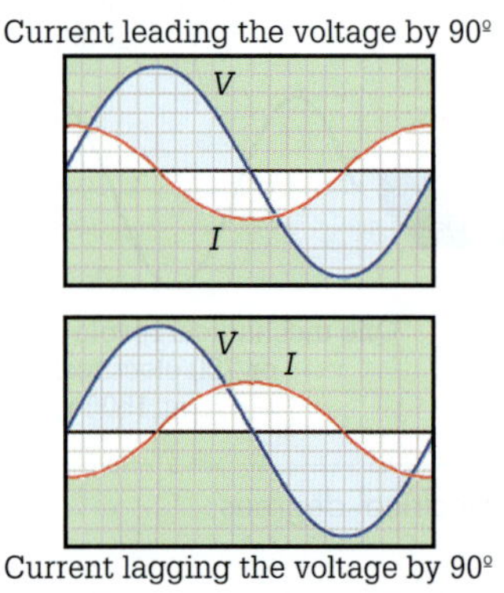

FIGURE 6.38 Out-of-phase relationships between voltage and current

When the current maximum instantaneous value occurs before the maximum voltage value, the current 'leads' the voltage. When the current maximum instantaneous value occurs after the maximum voltage value, the current 'lags' the voltage.

Phase difference between two alternating quantities

Phase difference is the difference, expressed in electrical degrees, time or radians, between two waves having the same frequency and referenced to the same point in time. In **Figure 6.39** there are two waveforms with a phase difference of 45°: the red waveform leading the blue waveform. This difference is measured between two similar instantaneous points on each waveform. They are not independent waveforms and they constitute one phase.

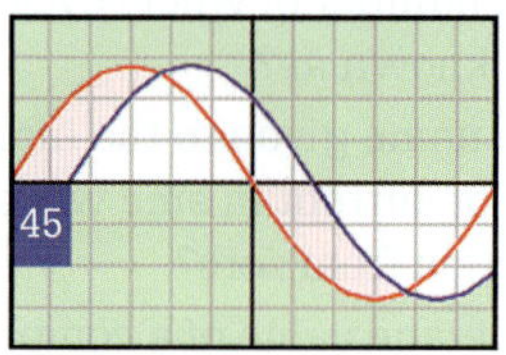

FIGURE 6.39 Phase difference of 45°

The phase difference of two waveforms displayed on the grid of an oscilloscope can be determined by applying the following equation:

$$\phi = \frac{d \times 360}{L} \text{ in degrees}$$

where ϕ = phase difference in degrees

d = distance between two equal points of the waveform in graticule divisions

L = length of one cycle in graticule divisions

EXAMPLE 6.12

Find the phase difference between the two waveforms displayed in **Figure 6.40**.

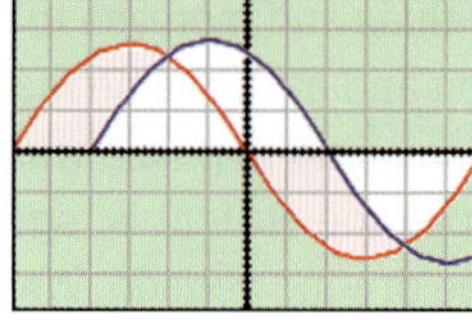

FIGURE 6.40 Phase difference between two waveforms

$$\text{phase difference} = \frac{d \times 360}{L} = \frac{2 \times 360}{12} = \mathbf{60\ degrees}$$

EXERCISE 6.12

Find the phase difference between the two sine waveforms displayed in **Figure 6.41**.

FIGURE 6.41 Phase difference between two sine waveforms

Representing phasor quantities

Graphical solutions of waveforms are lengthy processes. Problems involving the addition of in-phase and out-of-phase waveforms can be simplified by using phasor diagrams. A phasor is a straight line used to represent an electrical quantity such as voltage or current that has a magnitude and direction. Each phasor must have a length that represents the magnitude of the electrical quantity drawn to a suitable scale. It must also have the same direction from a reference point as that of the quantity it represents. An arrowhead indicates the direction. A voltage phasor is drawn with an open arrowhead and the current phasor has a solid arrowhead. Phasor diagrams usually represent the rms values and are marked *V* or *I* as shown in **Figure 6.42**.

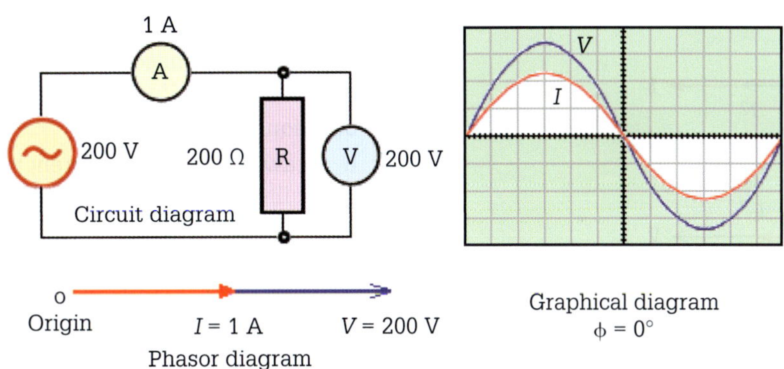

FIGURE 6.42 Phasor diagram of a.c. quantities in phase

Figure 6.43 illustrates graphically and with phasor diagrams a current of 1 A and a voltage of 200 V for out-of-phase conditions.

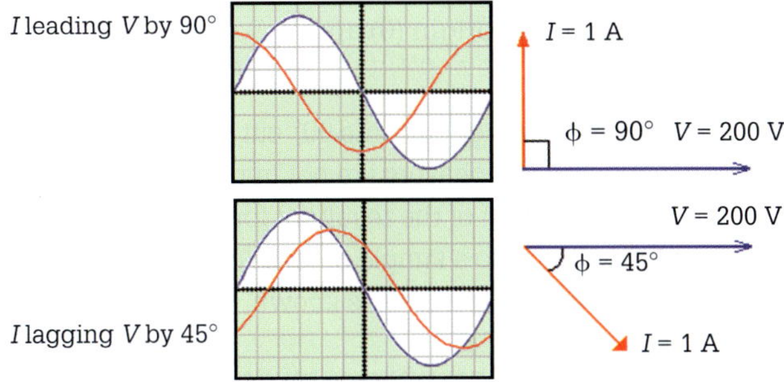

FIGURE 6.43 Out-of-phase conditions

Phasor diagrams, as well as indicating the phase difference and magnitude of the quantity, also indicate whether the current is in phase, leading or lagging the voltage.

Conventionally, phasors rotate anticlockwise. When representing a number of voltages or currents in a circuit with phasors, a reference phasor is drawn. This reference phasor is drawn horizontally to the right and must represent a common element in the circuit.

Therefore, in **Figure 6.43** the voltage phasor is drawn horizontally to the right. The current phasor when leading is drawn upwards and when lagging is drawn downwards. The reference phasor can be drawn to any length. Examples of common elements are:

- voltages in series circuits are referred to a common reference phasor (current).
- currents in a parallel circuit are referred to a common reference phasor (voltage).

EXAMPLE 6.13

Use phasors to represent two voltages, $V_1 = 5$ V and $V_2 = 2$ V, showing V_1 in phase with the current and V_2 lagging the current by 90° as shown in **Figure 6.44**.

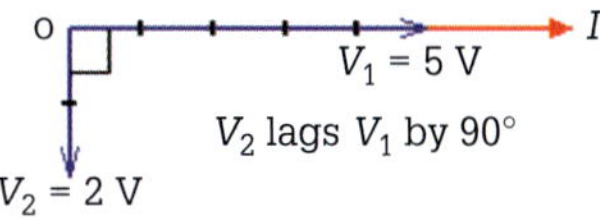

FIGURE 6.44 Phasor representation of two voltages

EXERCISE 6.13

a Use a phasor diagram to represent the following circuit quantity conditions:

i two sinusoidal voltages of 10 V and 15 V both in phase with the current

ii two currents branching into different circuits, I_1 = 15 A and I_2 = 10 A, with I_1 in phase with the applied voltage and I_2 lagging the applied voltage by 90°

iii two voltages 180° out of phase with each other, V_1 = 30 V and V_2 = 20 V; V_1 leads the circuit and V_2 lags the circuit current by 90°

iv a current of 15 A in phase with the circuit voltage and a current of 10 A leading the circuit voltage by 45°

v three sinusoidal voltages of 10 V, 15 V and 20 V all in phase with the current.

b Determine the phase relationships of current with respect to voltage in the following phasor diagrams in **Figure 6.45**.

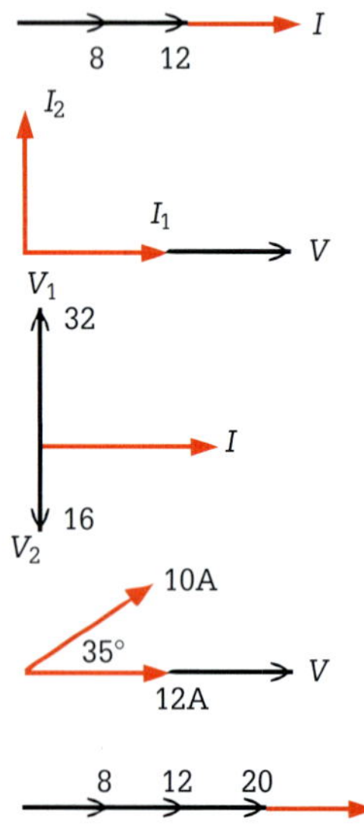

FIGURE 6.45 Phase relationships of current and voltage

Addition of phasor quantities

The result of adding sinusoidal voltages or currents can be easily determined using phasor diagrams. This result provides not only the magnitude but also the phase difference and whether the voltage or current is in phase, lagging or leading. The method of determining the resultant of two phasors is simple when the phasors represent sinusoidal voltages or currents in phase or 180° out of phase.

EXAMPLE 6.14

Find the resultant magnitude of two voltages V_1 = 5 V and V_2 = 3 V when:

a both voltages are in phase with each other.

b V_1 is in phase and V_2 is 180° out of phase with the current.

Solution

a Draw the reference current phasor I to any length as illustrated in **Figure 6.46**.

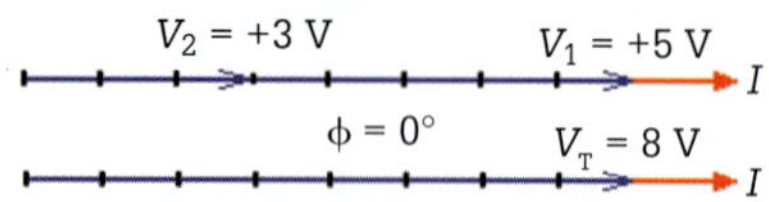

FIGURE 6.46 Addition of phasors

- Draw V_2 to scale and in phase with the current.
- Draw V_1 to scale on the end of V_2.
- Obtain the resultant voltage (V_T) of these two voltage phasors by measuring the total length of V_1 and V_2 .

Solution

b Draw the reference current phasor I to any length as illustrated in **Figure 6.47**.

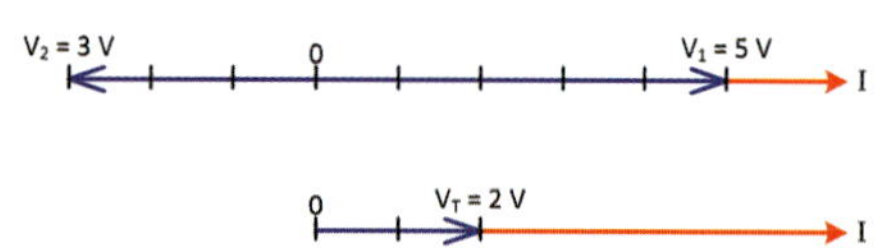

FIGURE 6.47 Addition of phasors

- Draw V_2 to scale in phase with the current.
- Draw V_1 to scale in the opposite direction to V_2.
- Obtain the resultant voltage (V_T) of these two voltage phasors by measuring the length between the origin (0) and V_1.

When two voltages or currents 90° out of phase are being investigated the phasors are not in the same straight line. The determination of the resultant for these out-of-phase conditions is also a simple process once the phase difference between the phasors is known.

EXAMPLE 6.15

Resolve the resultant magnitude of two voltages, V_1 = 4 V in phase with the circuit current and V_2 = 4 V leading the circuit current by 90°.

Instructions

Draw the reference current phasor I to any length as illustrated in **Figure 6.48**.

»

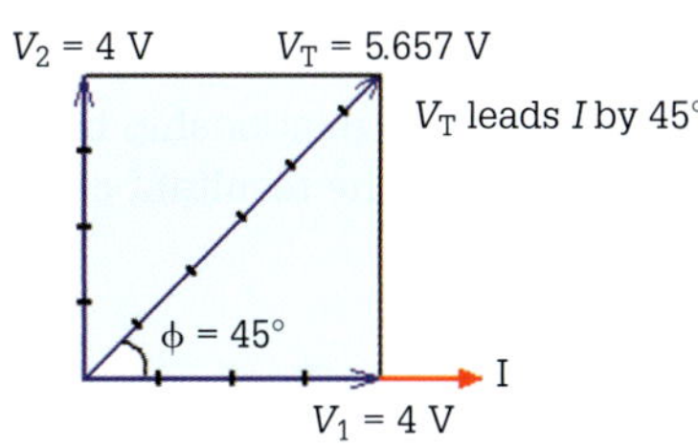

FIGURE 6.48 Addition of phasors – parallelogram method

- Draw V_1 to scale in phase with the current.
- Draw V_2 to scale at 90° out of phase with the current.
- Obtain the resultant voltage phasor by completing the parallelogram and drawing the diagonal.
- Measure the resultant to determine the resultant V_T.

The resultant voltage phasor (V_T) also has a phasor relationship with the current. This relationship is expressed by the phase difference (ϕ) the resultant voltage (V_T) makes with the current (I). This angle can be measured using a protractor or calculated by applying trigonometry.

Applying the cosine equation to the example in **Figure 6.48**:

$$\text{cosine } \phi = \frac{\text{adjacent}}{\text{hypotenuse}} = \frac{4}{5.657} = \mathbf{0.707}$$

The resultant voltage (V_T) leads the current by 45°.

EXERCISE 6.14

a Using phasors drawn to scale, represent the following voltages and currents together with their resultant magnitude and direction:
   i V_1 = 30 V, V_2 = 26 V, both being in phase with the current (I).
   ii I_1 = 18 A in phase with the voltage (V) and I_2 = 8 A also in phase with the voltage (V).
   iii V_1 = 18 V in phase with the current (I) and V_2 = 16 V, 180° out of phase with the current (I).
   iv I_1 = 4 A in phase with the voltage (V) and I_2 = 6 A, 180° out of phase with the voltage (V).

b Using phasors drawn to scale, calculate the resultant and resolve the phase difference in the following circuit conditions:
   i I_1 = 30 mA in phase with the voltage (V) and I_2 = 18 mA, leading the voltage (V) by 90°.
   ii I_1 = 18 A in phase with the voltage and I_2 = 24 A, lagging the voltage (V) by 90°.

Resolution of three phasor quantities

Depending upon the components connected within a circuit it is necessary with alternating current theory to resolve three different phasor voltages or currents in the same circuit.

EXAMPLE 6.16

Find the resultant voltage and its phase difference with the current of the following three circuit voltages:

a V_1 = 6 V in phase with the current (I)
b V_2 = 3 V leading the current (I) by 90°
c V_3 = 5 V lagging the current (I) by 90°.

Instructions

- Draw the reference current phasor I to any length as illustrated in **Figure 6.49**.

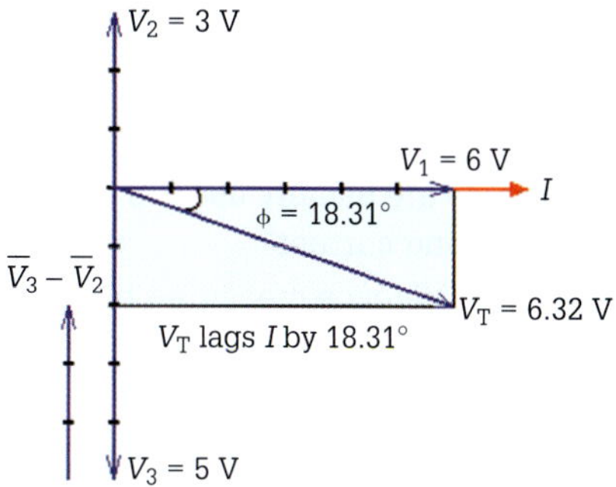

FIGURE 6.49 Resolution of three phasors

- Draw V_1 to scale and in phase with the current.
- Draw a leading V_2 to scale at 90° out of phase with the current.
- Draw a lagging V_3 to scale at 90° out of phase with the current.
- Resolve V_2 and V_3 (indicated as $\bar{V}_3 - \bar{V}_2$) to indicate phasor quantities.
- Obtain the resultant voltage phasor by completing the parallelogram and drawing the diagonal.
- Measure the diagonal to determine the resultant voltage (V_T). Apply trigonometry to calculate the phase difference (ϕ). V_T = 6.32 V Lagging I by 18.3°.

EXERCISE 6.15

a i State the phasor relationship between the current and each of the voltages represented by the phasor diagram of **Figure 6.50**.

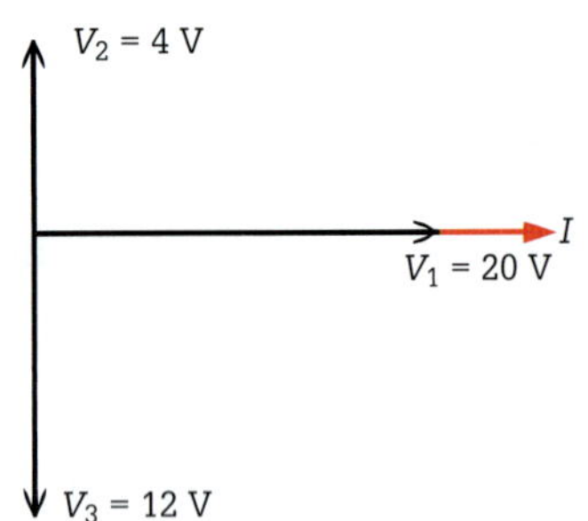

FIGURE 6.50 Resolution of three voltage phasors

ii Draw in the parallelogram and resolve the resultant voltage and the phase difference.

b i From the following phasor diagram of **Figure 6.51** determine the phase relationship between the driving voltage and the resultant current.

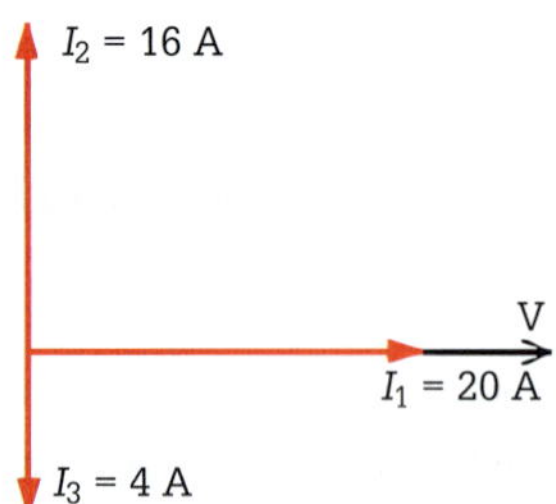

FIGURE 6.51 Resolution of three current phasors

ii By what amount would the current I_3 have to be increased to make the phase difference zero?

REVIEW QUESTIONS

1 Phasor diagrams are usually used to represent what value of alternating current?
2 What is phase difference measured between?
3 When representing phasor quantities, how is current represented?
4 Phasors conventionally rotate in which direction?
5 The result of adding sinusoidal voltages or currents can be easily determined using what kind of diagrams?

6.3 Single element a.c. circuits

Resistive a.c. circuits

Some alternating current circuits consist only of resistance, and for these circuits the same rules and laws apply as for direct current circuits. In most problems root-mean-square (rms) values are used. The following equations apply to resistive components such as resistors, stove elements and lamp filaments when connected in various combinations.

$$I = \frac{V}{R} \quad V = IR \quad R = \frac{V}{R} \qquad \text{Ohm's law}$$

$$I_T = I_1 = I_2 = I_3 \text{ etc.}$$
$$V_T = V_1 + V_2 + V_3 \text{ etc.} \qquad \text{Series circuits}$$
$$R_T = R_1 + R_2 + R_3 \text{ etc.}$$

$$I_T = I_1 + I_2 + I_3 \text{ etc.}$$
$$V_T = V_1 + V_2 + V_3 \text{ etc.} \qquad \text{Parallel circuits}$$
$$\frac{1}{R_T} = \frac{1}{R_1} + \frac{1}{R_2} + \frac{1}{R_3} \text{ etc.}$$

$$P = VI \quad P = \frac{V^2}{R} \quad P = I^2R \qquad \text{Power}$$

EXERCISE 6.16

a Two resistors are connected in series to a 36 V sinusoidal supply. Given that $R_1 = 320\ \Omega$ and the circuit current is 50 mA, calculate the resistance of the other resistor (R_2) and the voltage across R_1.

b Determine the value of resistance that must be connected in series with a 24 V lamp to limit the current to 40 mA when connected to the 48 V output terminals of a transformer.

c Three resistors drawing 50 mA are connected in series to a 40 V output transformer. If $R_1 = 200\ \Omega$ and $R_2 = 100\ \Omega$ calculate:
i the resistance of R_3
ii the voltage across R_2
iii the power dissipated by R_1.

d A circuit consists of two resistors connected in series across a 120 V sinusoidal supply. If $R_1 = 40$ kΩ and R_2 is a rheostat, calculate:

»

i the change in the voltage drop across R_1 when the value of the rheostat is changed from 40 kΩ to 80 kΩ
ii the power dissipated by the rheostat when it is set to a value of 8 kΩ
iii the total resistance of the combination when the rheostat is set to 16 kΩ.

e A circuit contains three resistors consisting of 270 Ω, 120 Ω and 100 Ω connected in parallel to a 10 V sinusoidal supply. Calculate:
i the total resistance
ii the power dissipated by the 120 Ω resistor
iii the current flowing through the 270 Ω resistor
iv the total power dissipated by the circuit.

f Three resistors are connected in parallel to the 6 V tapping of a transformer. Given that the current through R_1 is 4 mA, the power dissipated by R_2 is 36 mW and the resistance of R_3 is 6 kΩ, determine:
i the total current through the circuit
ii the total resistance of the circuit
iii the peak value of the voltage
iv the total power dissipated by the resistors.

g Determine the supply current drawn in Figure 6.52 from a 230 V mixed circuit when three 2 kW elements, three 200 W soldering irons and three 200 W lamps are connected in parallel.

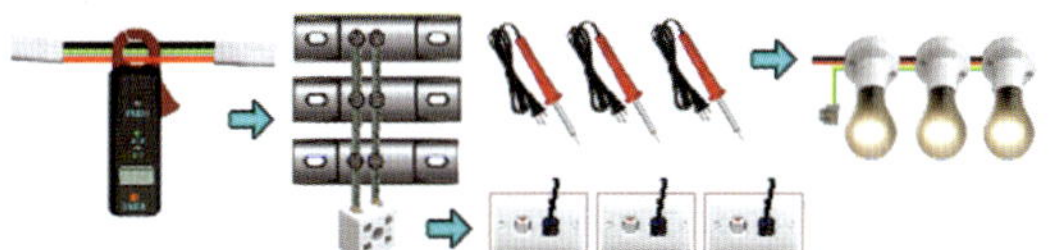

FIGURE 6.52 Electrical equipment in parallel

h Two resistors, R_1 and R_2, are connected in parallel with each other and this combination is connected in series with a 1120 Ω resistor, R_3, across a sinusoidal voltage of 60 V. When the current flow through R_1 is 8 mA and the current through R_3 is 20 mA calculate:
i the voltage drop across R_3
ii the resistance of R_2.

i Three resistors, R_1 = 200 Ω, R_2 = 300 Ω and R_3 = 1k2 Ω, are connected in series to the 15 V output terminals of a transformer. If a 1 kΩ loading circuit is connected across R_3 as shown in Figure 6.53, calculate:

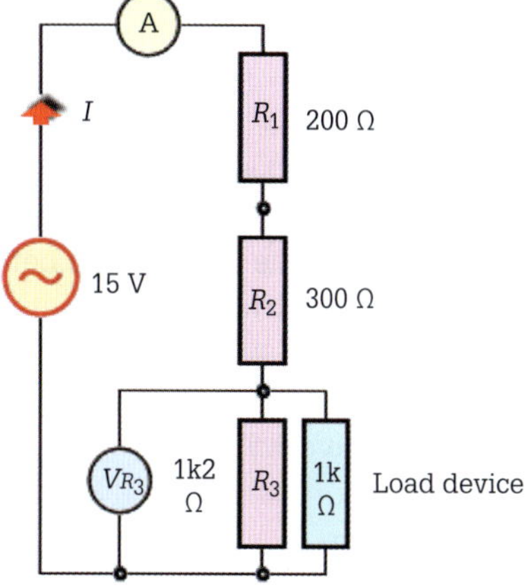

FIGURE 6.53 Loading device in a series–parallel circuit

i the current through R_2
ii the voltage drop across R_3.
When the loading circuit is removed calculate:
iii the new current through R_2
iv the new voltage drop across R_3.

Voltage – current phase relationship of resistors

In an alternating current circuit consisting of resistance only, the voltage and current have no phase difference between them and are in phase. Consequently, when the current is at its peak value the voltage across the resistor is at the same value. The result is that when the current falls to zero the voltage across the resistor is zero.

Figure 6.54 shows a circuit, phasor and a graphical representation of the electrical quantities of voltage and current in a simple resistive circuit.

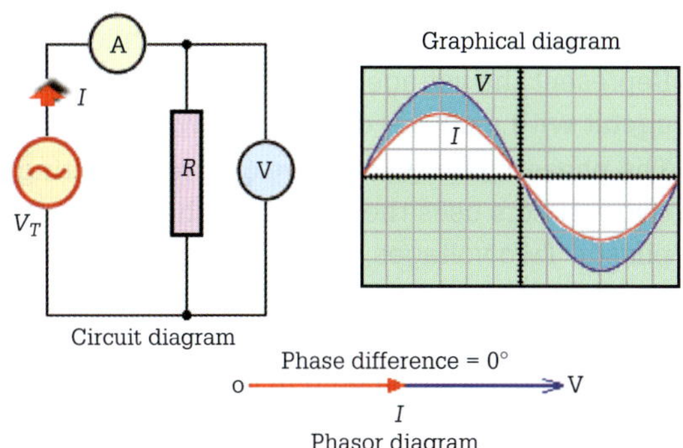

FIGURE 6.54 Phase relationship of resistors

When a circuit contains two or more resistors in series, the phase relationship of the current with the three voltages established in the circuit is difficult to illustrate graphically, but is easily illustrated with a phasor diagram. The phase relationship of two resistors connected in series is shown in Figure 6.55.

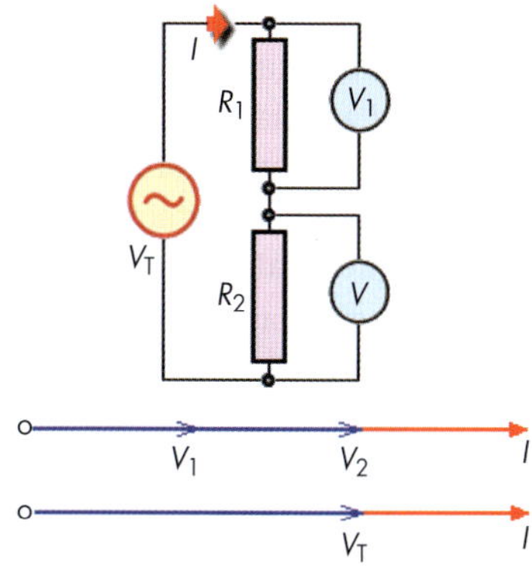

FIGURE 6.55 Phase relationship of two resistors in series

When a circuit contains two or more resistors in parallel the voltage across each resistor in the combination is the same as the applied voltage. This voltage becomes the reference phasor and the various currents are drawn as shown in Figure 6.56 in relation to the supply voltage.

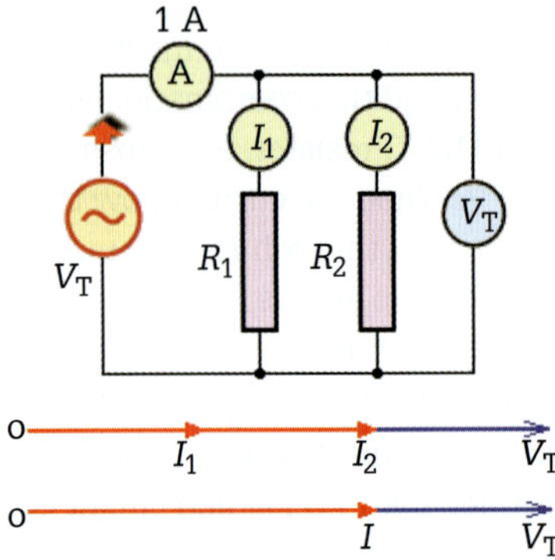

FIGURE 6.56 Phase relationship of two resistors in parallel

EXAMPLE 6.17

a Two resistors, R_1 = 39 Ω and R_2 = 59 Ω, are connected in series to a 9.8 V supply. Represent via a phasor diagram the relationship between the voltages and the current.

Instructions

- Draw the circuit diagram from the description as shown in Figure 6.57.

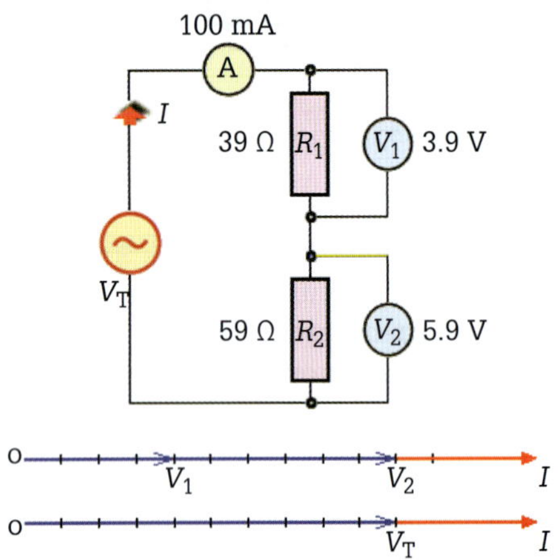

FIGURE 6.57 Phasor diagram representation of a resistive series circuit

- Calculate V_1 and V_2 using Ohm's law.
- Draw V_1 and V_2 in phase with each other to scale along the current reference phasor.
- Mark on the combined voltage phasor V_T, representing the total applied voltage.

b The following phasor diagram represents the voltage and current in a circuit with two lamps connected to a transformer. Calculate the power dissipated by each lamp as shown in Figure 6.58.

$$P_1 = V \times I_1$$
$$= 12 \times 0.2$$
$$= 2.4 \text{ W}$$
$$P_2 = V \times I_2$$
$$= 12 \times 0.1$$
$$= 1.2 \text{ W}$$

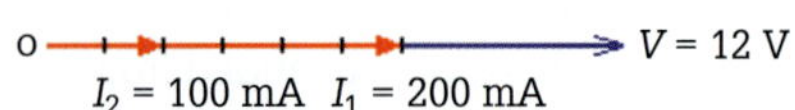

FIGURE 6.58 Phasor diagram representation of a resistive parallel circuit

EXERCISE 6.17

a A circuit containing two series resistors of 250 Ω and 400 Ω respectively draws a current of 25 mA from an alternating current supply. Calculate the voltage across each resistor and represent them drawn to scale as a phasor diagram.

b A circuit has two resistors in parallel drawing 2 A from the 24 V supply. If one resistor has a value of 48 Ω:

  i calculate the current through each component in the circuit

  ii draw to scale a phasor diagram representing the voltage and currents of the circuit.

c A circuit contains a 400 Ω resistor and a 49 Ω lamp in parallel. Complete the circuit diagram of Figure 6.59 and calculate:

  i the total resistance of the circuit

  ii the supply voltage

  iii the peak-to-peak value of the total current.

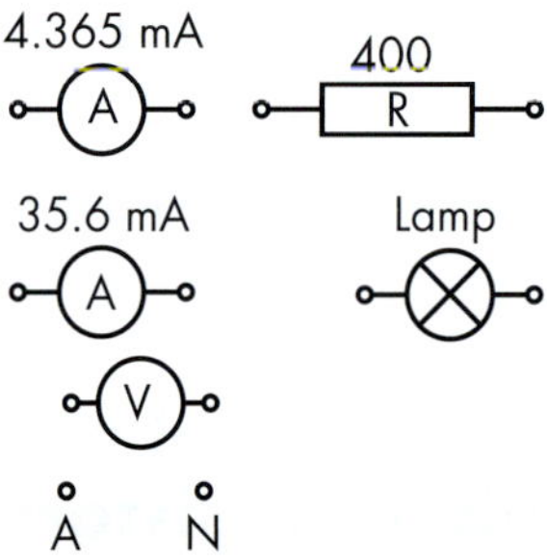

FIGURE 6.59 Incomplete circuit diagram

d Three resistors consisting of 22 Ω, 33 Ω and 56 Ω are to be connected to a 440 V generator. Draw two phasor diagrams to scale to represent the voltage and current when the resistors are connected in series and parallel.

e Several electrical devices such as a 150 W soldering iron, a 1200 W lamp load and a 1200 W load resistor are being operated in parallel. Represent the voltage and currents in the 460 V circuit with a phasor diagram drawn to scale.

f The phasor diagram of Figure 6.60 representing two resistors uses the current as the reference phasor. Two voltages, V_1 = 32 V and V_2 = 48 V, are shown in phase with the 2 A current. Complete the circuit diagram and calculate R_1 and R_2 and the total power dissipated in the circuit.

»

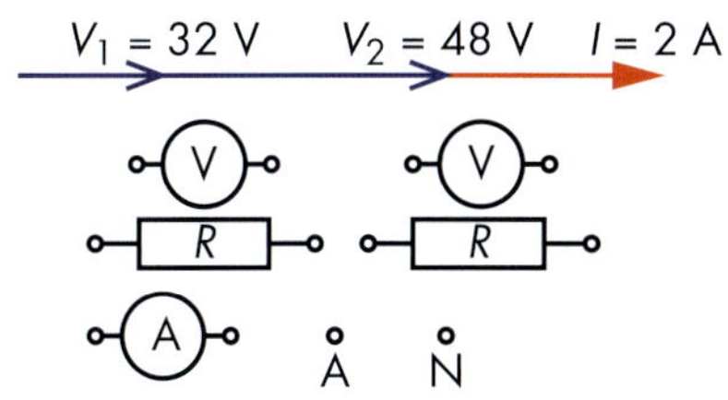

FIGURE 6.60 Incomplete circuit diagram

Power in resistive circuits

Alternating current when doing work through a resistor produces heat irrespective of the direction of the current. If rms values of current are used then the heating effect is the same as the equivalent direct current values. When we examine a graph of power we see that the curve displayed remains above the zero axis of the graph.

EXERCISE 6.18

a Complete a table (at 20° increments) using the headings below. From the results obtained plot a graph of instantaneous power ($P = ei$) given that e = 10 V peak and i = 4 A peak.

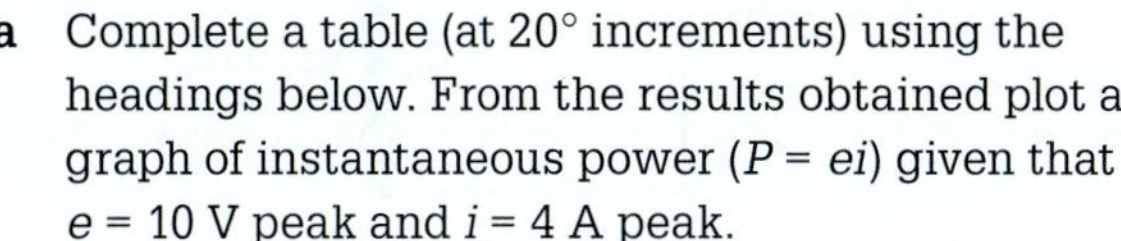

$$e = V_{max} \times \sin\phi \quad i = I_{max} \times \sin\phi$$

Your graph should resemble the graph in Figure 6.61. If we examine the graph of power we see that the curve displayed remains above the zero axis of the graph. This is because the product of two negative values ($-e$ times $-i$) produces a positive result.

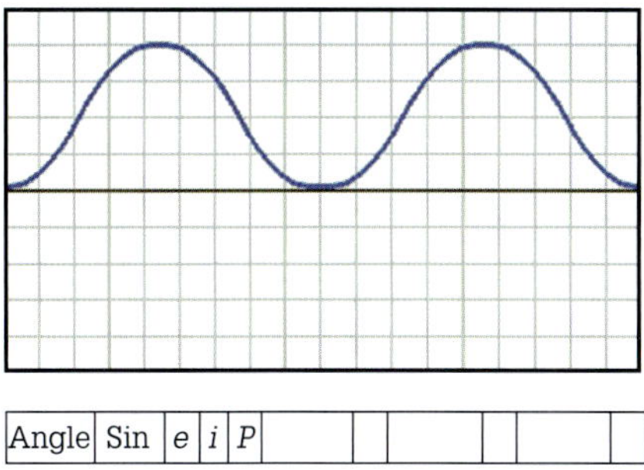

Angle	Sin	*e*	*i*	*P*						

FIGURE 6.61 Instantaneous power in a resistive circuit

b A 1 kΩ resistive load draws a current of 100 mA peak when connected to a transformer. Complete a table using the headings below and draw a graph of the instantaneous power dissipated using values calculated from these equations:

$$P = i^2R$$
$$i = I_{max} \times \sin\phi$$

Your graph should resemble the graph in Figure 6.62. Notice that the power curves of the graph have twice the frequency of the voltage and current waveforms, and that the power waveform is located in the positive half of the graph.

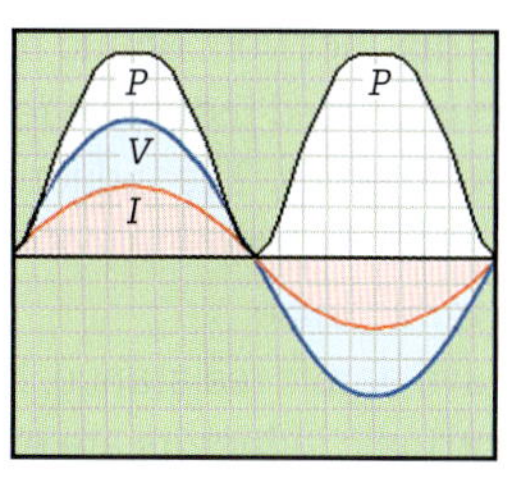

Angle	Sin	*i*	*P*	Angle	*P*	Angle	*P*	Angle	*P*

FIGURE 6.62 Power, voltage and current in a resistive circuit

Alternating current resistance

The assumption of intrinsic and constant resistance is not justified in practice. In addition, a definition of resistance by Ohm's law says that all the electrical power delivered to a circuit is directly converted to heat. While this is the case for direct current (only d.c. resistive circuits), it is not true for alternating current.

Eddy currents and self-induction cause energy loss in an alternating current conductor. The eddy currents set up voltages within the conductor and cause electrons to be repelled towards the surface of the conductor, causing more current to flow near the surface of the conductor than at the centre of the conductor.

This condition has the effect of decreasing the effective cross-sectional area of the conductor. As you have learnt, a decrease in cross-sectional area causes an increase in resistance. The development of high current flow near the surface of a conductor is called the skin effect.

The energy losses in alternating current conductors due to eddy currents and the skin effect are directly proportional to the frequency of the current flowing within the conductor. The higher the frequency, the greater is the energy loss. Power cables are usually constructed of stranded conductors to reduce the skin effect.

This reduced energy loss occurs because the surface area of the entire individual conductor strands added together is greater than the surface area of a solid conductor. The concept of resistance as being only the 'ohmic' or the direct current equivalent resistance in respect to alternating current flow in a conductor is not true. Resistance in alternating current, as we have seen, no longer obeys Ohm's law because it varies with current and frequency. Resistance, when referring to both alternating current and direct current, is directly associated with the power consumed by the equation $P = I^2R$. It is necessary to define the resistance of a conductor when referring to alternating current as:

$$R = \frac{P}{I^2}$$

Inductive a.c. circuits

The electrical property of a circuit to oppose any change in current is referred to as inductance. When a sinusoidal waveform energises an inductive circuit, the continually

»

changing current creates a continuous opposition to current flow that is distinct from ohmic resistance.

This opposition is termed inductive reactance (symbol X_L) and is measured in ohms. The inductive reactance is directly proportional to the frequency of the sinusoidal waveform and the inductance.

All inductors have ohmic resistance; however, in this section only intrinsic inductors are examined. In an intrinsic inductor the relationship between the inductive reactance and the inductance can be determined by applying the following equation:

$$X_L = 2\pi f L$$

where X_L = inductive reactance in ohms (Ω)
f = frequency in hertz (Hz)
L = inductance in henry (H)

EXAMPLE 6.18

Calculate the inductive reactance of a coil that has an inductance of 200 mH when connected to a 50 Hz supply as shown in **Figure 6.63**.

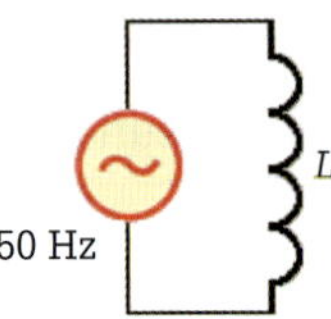

FIGURE 6.63 Inductive reactance

$$\begin{aligned} X_L &= 2\pi fL \\ &= 2 \times 3.142 \times 50 \times 0.2 \\ &= \mathbf{62.84\ \Omega} \end{aligned}$$

To determine the current through an inductor we apply Ohm's law. When inductors are connected in series or parallel the equations that are used are the same as those for similarly connected resistors as long as no magnetic flux is shared with an adjacent inductor. When connecting inductors in various combinations the equations listed in **Table 6.4** apply.

TABLE 6.4 Equations for connecting inductors

Equation type	Equation
Ohm's law	$I = \frac{V}{X_L}$ $V = IX_L$ $X_L = \frac{V}{I}$
Series circuits	$I_T = I_1 = I_2 = I_3$ $V_T = V_1 + V_2 + V_3$ $X_{L_{Total}} = X_{L_1} + X_{L_2} + X_{L_3}$
Parallel circuits	$\frac{1}{X_{L_T}} = \frac{1}{X_{L_1}} + \frac{1}{X_{L_2}} + \frac{1}{X_{L_3}}$ etc.

EXAMPLE 6.19

Calculate the current in a circuit when two inductors, L_1 = 200 mH and L_2 = 600 mH, are connected to a 32 V 50 Hz supply in the following configurations of **Figures 6.64(A)** and **(B)**.

a Series

$$\begin{aligned} X_{L_1} &= 2\pi fL \\ &= 2 \times 3.142 \times 50 \times 0.2 \\ &= \mathbf{62.84\ \Omega} \\ X_{L_2} &= 2\pi fL \\ &= 2 \times 3.142 \times 50 \times 0.6 \\ &= \mathbf{188.52\ \Omega} \end{aligned}$$

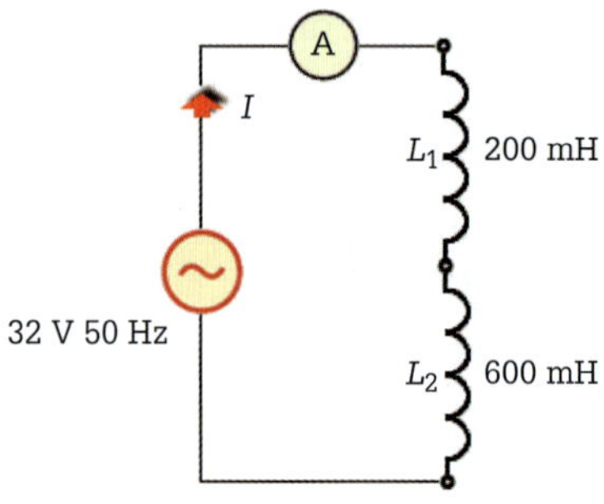

FIGURE 6.64(A) Inductors in series

$$\begin{aligned} \text{Total } X_{L_1} &= 62.84 + 188.52 = 251.36 \\ I &= \frac{V}{X_L} \\ &= \frac{32}{251.36} \\ &= \mathbf{0.127\ A\ or\ 127\ mA} \end{aligned}$$

b Parallel

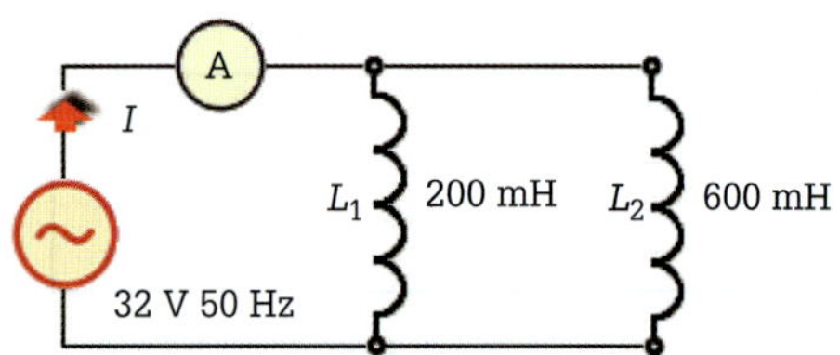

FIGURE 6.64(B) Inductors in parallel

$$\begin{aligned} X_{L_1} &= 2\pi fL \\ &= 2 \times 3.142 \times 50 \times 0.2 \\ &= \mathbf{62.84\ \Omega} \\ X_{L_2} &= 2\pi fL \\ &= 2 \times 3.142 \times 50 \times 0.6 \\ &= \mathbf{188.52\ \Omega} \end{aligned}$$

$$\begin{aligned} \frac{1}{X_{L_T}} &= \frac{1}{X_{L_1}} + \frac{1}{X_{L_2}} \\ &= \frac{1}{62.84} + \frac{1}{188.52} \\ &= 47.13\ \Omega \end{aligned}$$

$$\begin{aligned} I &= \frac{V}{X_L} \\ &= \frac{32}{47.13} \\ &= 0.679\ \text{A } \textit{or } 679\ \text{mA} \end{aligned}$$

EXERCISE 6.19

a Calculate the inductive reactance of each of the following inductors, L_1 = 150 mH and L_2 = 250 mH, connected in series if the supply is 32 V 50 Hz.

b Find the inductive reactance of a 100 mH contactor coil at a frequency of 50 Hz.

c What is the inductance of a universal motor-pole winding which has a reactance of 2 kΩ at a frequency of 50 Hz?

d At what supply frequency does a 100 mH coil have a reactance of 100 Ω?

e A 150 mH coil experiences a fluctuating frequency of 2 kHz and 3 kHz. Determine the difference in reactance between the low frequency and the higher frequency.

f Calculate the current drawn by a 24 V, 250 mH relay coil when connected to a 50 Hz supply.

g Find the value of inductance that would limit the current to 30 mA when connected to a 32 V 50 Hz supply.

h A current of 12 µA flows through a 60 mH inductor at a frequency of 20 kHz. At what frequency would the current be limited to 4 µA?

i A 4 H inductor is connected in series to a 10 H inductor across the output terminals of a 120 V, 50 Hz transformer. Calculate the current drawn by the inductors.

j Two inductors consisting of 120 mH and 480 mH are connected in parallel across a transformer supplying a circuit current of 1.2 A at a frequency of 50 Hz. Determine the supply voltage.

Frequency and inductors

For a given value of inductance the frequency increases and the inductive reactance will increase, and as frequency decreases the smaller will be the inductive reactance. An increase in frequency causes a decrease in current flow and a decrease in frequency causes an increase in current flow. The quantity $2\pi f$ in the equation represents the rate of change of the current.

EXERCISE 6.20

a Complete a table and plot graphs of inductive reactance against frequency for a 100 mH and a 200 mH inductor over a 0 Hz to 1 kHz frequency range. Your completed graph should look like **Figure 6.65**. From the graph find the frequencies at which the inductive reactance is 280 Ω.

b Complete a table and plot graphs of current against frequency for a 100 mH and then a 200 mH inductor over a 40 Hz to 1 kHz frequency range when the supply voltage is 40 V. Your completed graph should look like **Figure 6.66** when using suitable scales. From the graph find the current through each inductor at 600 Hz.

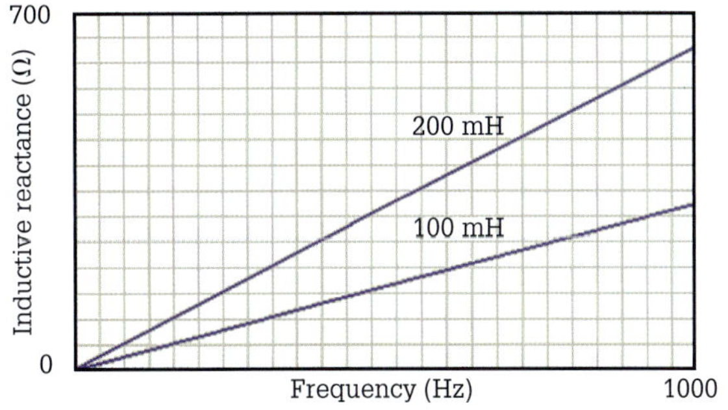

FIGURE 6.65 Graph of inductive reactance against frequency

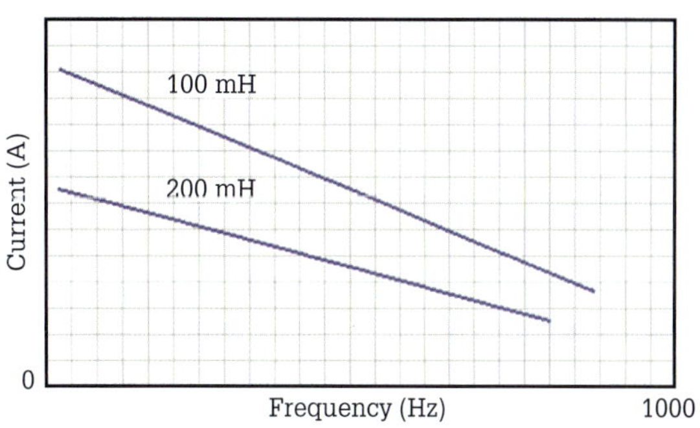

FIGURE 6.66 Graph of current against frequency

Volt–current phase relationship of inductors

The property of inductance has another characteristic and that is its effect on the phase relationship between voltage and current. When a sinusoidal voltage is applied to an inductor a circuit current flows. When this voltage varies from one instantaneous value to another, the current also varies but at a constant interval of time after the voltage variation. This lagging effect of the current does not exceed 90°. This only occurs when the inductor is assumed to be pure inductance. A circuit containing pure inductance is illustrated graphically and with a phasor diagram in **Figure 6.67**.

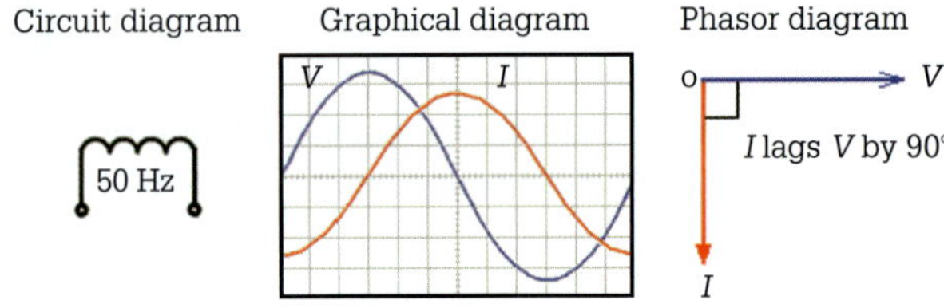

FIGURE 6.67 Pure inductance

When a circuit contains two or more inductors it is difficult to show the phase relationships between voltage and current graphically. However, they are easily represented using phasor diagrams.

When the inductors are connected in series the supply current is the same throughout the circuit and it becomes the reference phasor. In parallel circuits the voltage is the reference phasor as the voltage drop across each inductor is the same as the supply voltage.

EXAMPLE 6.20

Represent the voltages in the following circuit of **Figure 6.68** with phasors drawn to scale illustrating their relationship to the supply current.

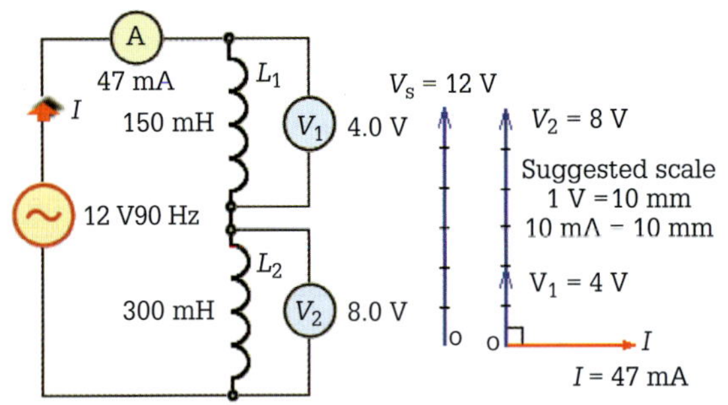

FIGURE 6.68 Phasor and circuit diagram of inductors in series

EXERCISE 6.21

When two inductors of 25 mH and 45 mH are connected in series across a 100 Hz supply the current drawn is 25 mA. Calculate the voltage across each inductor and draw the phasor diagram to represent the relationship of the voltages to the current.

EXAMPLE 6.21

Draw the circuit diagram of two inductors, 300 mH and 150 mH, connected across a 30 V 100 Hz supply and represent the voltage and currents in a phasor diagram (**Figure 6.69**).

»

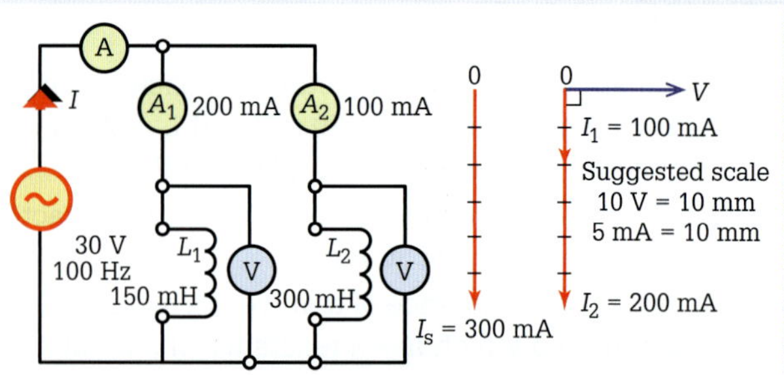

FIGURE 6.69 Phasor and circuit diagram of inductors in parallel

EXERCISE 6.22

a Draw a phasor diagram to represent the currents and voltage in a circuit when two inductors, one 100 mH and the other 250 mH, are connected in parallel to a 24 V 100 Hz supply.

b A circuit consists of a 450 mH inductor connected in parallel with a 120 mH inductor. If the supply current through the 450 mH inductor is 10 mA at 1.6 kHz:
   i determine the total current
   ii draw a phasor diagram to represent the currents and voltage in the circuit.

c An inductor of 1.5 H and an inductor of 500 mH are connected across a 10 V 50 Hz transformer. Draw phasor diagrams to represent the voltage and current when the inductors are connected in parallel.

Power in inductive circuits

When a sinusoidal waveform flows through an inductor, a magnetic field builds up and collapses every half cycle. Over the course of each waveform cycle, energy is taken from the electrical state of the circuit during the first 90° growth of the waveform to create the electromagnetic field.

This stored energy is then restored to the electrical state of the circuit when the electromagnetic field collapses during the next 90° decline of the waveform. This means that in an intrinsic inductor for one complete cycle of the sinusoidal waveform the total energy taken from the electrical circuit equals the amount returned. Therefore, an intrinsic inductor does not absorb or dissipate power when connected across a sinusoidal voltage for a given period of time.

The rate at which an inductive device stores or returns energy is called its reactive power (symbol Q_L). The unit of measurement of reactive power at any instant of time is volt-amperes reactive (symbol VA_R). The state of power in an inductive circuit energised by a sinusoidal voltage waveform is illustrated by completing the following exercise.

EXERCISE 6.23

The sinusoidal curves in Figure 6.70 represent an intrinsic inductive circuit in which the peak current = 1 A and the voltage = 2 V. Using trigonometry for the sine of an angle, determine the instantaneous values of voltage and current, then calculate the instantaneous values of power using the following equations:

$$P = ei$$
$$e = V_{max} \sin \theta$$
$$i = I_{max} \sin \theta$$

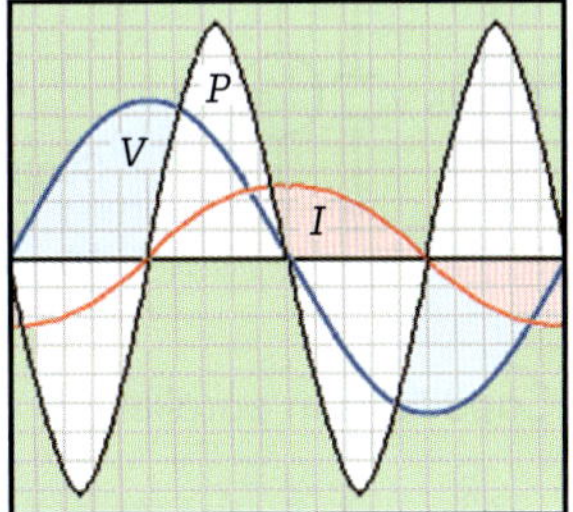

FIGURE 6.70 Sinusoidal curves

The graph illustrates that no energy is dissipated or consumed in an intrinsic inductive circuit. Also, in examining the power graph with respect to the voltage (blue) and current (red) waveforms, it can be seen that the power fluctuates at a frequency twice that of the voltage or current waveform.

Capacitive a.c. circuits

When a sinusoidal voltage is applied to a capacitor it charges via the flow of electrons to and from the plates, which gives rise to the flow of current through the circuit and the development of a voltage across the plates. The effect of the capacitor is to oppose the change in the applied sinusoidal voltage and limit the current flow in the circuit. This opposition to the sinusoidal current is called capacitive reactance (symbol X_C) and is measured in ohms. The capacitive reactance is inversely proportional to the capacitance and the frequency of the supply voltage and can be determined by applying the following equation:

$$X_C = \frac{1}{2\pi fC}$$

where X_C = capacitive reactance in ohms (Ω)
f = frequency in hertz (Hz)
C = capacitance in farads (F)

EXAMPLE 6.22

Calculate the capacitive reactance of a 20 µF capacitor connected to a sinusoidal waveform with a frequency of 50 Hz.

$$X_C = \frac{1}{2\pi fC} = \frac{1}{2\pi \times 50 \times (20 \times 10^{-6})} = \mathbf{159\ \Omega}$$

EXERCISE 6.24

a Calculate the capacitive reactance of a 100 µF capacitor connected to a sinusoidal waveform with a frequency of 50 Hz.

b Calculate the frequency to which a 50 µF capacitor must connect to have a capacitive reactance of 12.73 Ω.

c Calculate the capacitance of a capacitor that has a capacitive reactance of 25 Ω when connected to a 150 Hz supply.

Capacitors find application in series and parallel circuits and in series–parallel combination circuits. When connected in these various combinations the following equations apply:

$X_C = \frac{V}{I}$ $\quad I = \frac{V}{X_C}$ $\quad V = IX_C$	Ohm's law
$X_{C_{Total}} = X_{C_1} + X_{C_2} + X_{C_3}$ etc.	Series
$\frac{1}{X_{C_T}} = \frac{1}{X_{C_1}} + \frac{1}{X_{C_2}} + \frac{1}{X_{C_3}}$ etc.	Parallel

EXAMPLE 6.23

Find the total current drawn by the following capacitor combination in Figure 6.71.

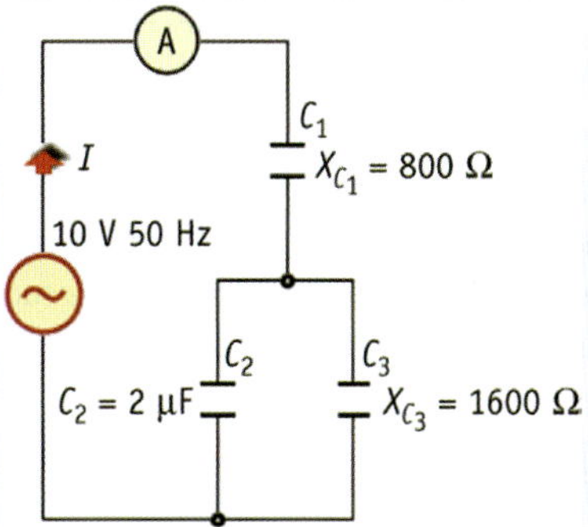

FIGURE 6.71 Capacitor combination

$$X_{C_2} = \frac{1}{2\pi fC}$$
$$= \frac{1}{2 \times 3.142 \times 50 \times 2 \times 10^{-6}}$$
$$= \mathbf{1591\ \Omega}$$

$$\frac{1}{X_{C_{Par}}} = \frac{1}{X_{C_2}} + \frac{1}{X_{C_3}}$$
$$= \frac{1}{1591} + \frac{1}{1600}$$
$$= \mathbf{798\ \Omega}$$

$$X_{C_T} = X_{C_1} + X_{C_{Par}}$$
$$= 800 + 798$$
$$= 1598\ \Omega$$

$$I = \frac{V}{X_{C_T}}$$
$$= \frac{10}{1598}$$
$$= \mathbf{6.26\ mA}$$

EXERCISE 6.25

a A capacitor has a capacitive reactance of 2 kΩ at a frequency of 200 Hz. Calculate the new reactance when the frequency is adjusted to 150 Hz.

b Calculate the current that a circuit containing two capacitors, C_1 = 40 µF and C_2 = 80 µF, in series will draw from a 10 V 30 Hz supply.

c Two capacitors, C_1 = 15 µF and C_2 = 20 µF, connect in parallel across a 15 V 50 Hz supply. Calculate the total current drawn by the circuit.

Volt–current phase relationship of capacitors

The capacitance in a sinusoidal circuit not only opposes the change in voltage and limits the current flow but also causes the voltage and current to be out of phase. In an intrinsic capacitive circuit the current is said to lead the voltage by 90°. This is illustrated graphically and by a phasor diagram in **Figure 6.72**.

Circuit diagram

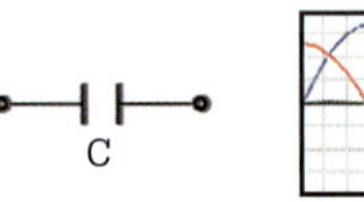

Graphical diagram

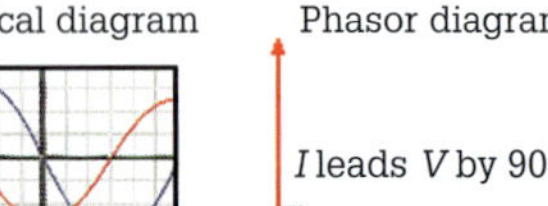

Phasor diagram

I leads *V* by 90°

FIGURE 6.72 Pure capacitance

In circuits with two or more capacitors, the representation of voltage and current is shown more readily with phasor diagrams than with graphical diagrams. When capacitors are connected in series the current becomes the reference phasor because it is the same in all parts of the circuit. In parallel circuits the voltage is the reference phasor because the voltage across each capacitor is the same as the supply voltage.

EXAMPLE 6.24

a Represent the voltages in **Figure 6.73** with a phasor diagram drawn to scale and show their relationship to the current.

Instructions

- Calculate by Ohm's law I = 100 mA, V_{C1} = 20 V and V_{C2} = 80 V.
- Draw the reference phasor I to a suitable length as shown in **Figure 6.73**.

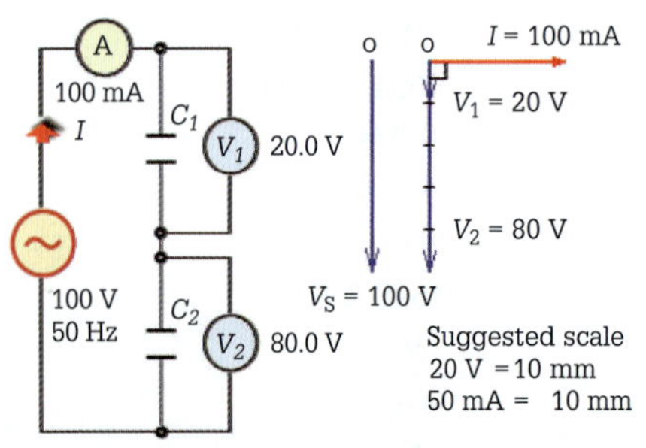

FIGURE 6.73 Phasor diagram for a series capacitance circuit

- Draw V_{C1} to scale lagging the current by 90°.
- Draw V_{C2} to scale on the end of V_{C1} as shown.
- The total length of V_{C1} and V_{C2} represents the supply voltage.

b Draw the circuit diagram represented by the phasor diagram in **Figure 6.74** and calculate the value of each capacitor in microfarads.

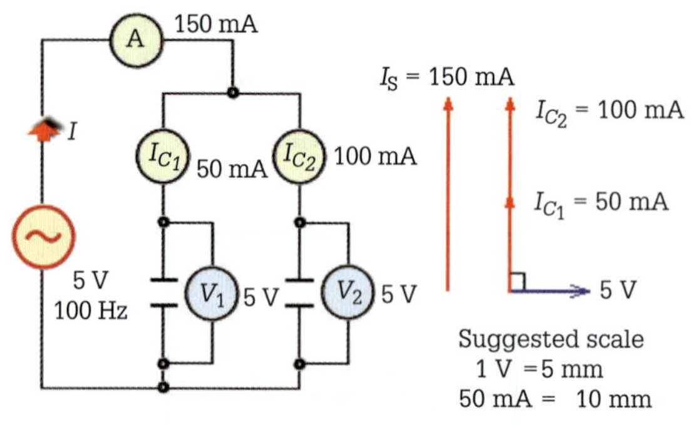

FIGURE 6.74 Capacitors in parallel

As the voltage is the reference phasor the capacitors are connected in parallel.

$$X_{C_1} = \frac{V}{I_{C_1}}$$
$$= \frac{5}{0.05}$$
$$= \mathbf{100\ \Omega}$$
$$C_1 = \frac{1}{2\pi f X_{C_1}}$$
$$= \frac{1}{2 \times 3.142 \times 100 \times 100}$$
$$= \mathbf{15.9\ \mu F}$$
$$X_{C_2} = \frac{V}{I_{C_2}}$$
$$= \frac{5}{0.1}$$
$$= \mathbf{50\ \Omega}$$
$$C_2 = \frac{1}{2\pi f X_{C_2}}$$
$$= \frac{1}{2 \times 3.142 \times 100 \times 50}$$
$$= \mathbf{31.8\ \mu F}$$

EXERCISE 6.26

a Two capacitors, C_1 = 10 µF and C_2 = 22 µF, are connected in series to a 24 V 50 Hz sinusoidal supply.

i Calculate the voltage across each capacitor.

ii Complete a phasor diagram representing the voltages and current in the circuit.

b Complete a phasor diagram representing the voltage and currents when a 5 µF and a 10 µF capacitor are connected in parallel to a 24 V 2 kHz sinusoidal supply.

»

c Two capacitors, 6 μF and 4 μF, are connected in series across a 500 Hz sinusoidal supply drawing 100 mA.
  i Calculate the voltage across each capacitor.
  ii Complete a phasor diagram representing voltage and current in the circuit.

d Three capacitors are connected in series to a 20 V 50 Hz supply that delivers 20 mA. If the capacitance of two capacitors is 15 μF and 22 μF:
  i calculate the capacitance of the unknown capacitor
  ii complete a phasor diagram.

Power in capacitive circuits

When a sinusoidal circuit waveform energises a capacitor an electrostatic field builds up and collapses every half cycle. Over the course of each waveform cycle, energy is taken from the electrical state of the circuit during the first 90° growth of the waveform to create the electrostatic field. This stored energy is then restored to the electrical state of the circuit when the electrostatic field collapses during the next 90° decline of the waveform. This means that for one complete cycle of the sinusoidal waveform in an intrinsic capacitor the total energy taken from the electrical circuit equals the amount returned. Therefore, an intrinsic capacitor does not absorb or dissipate power when connected across a sinusoidal voltage for a given period of time. The rate at which a capacitor stores or returns energy is called its reactive power (symbol *Q*) and is measured in volt-amperes reactive (abbreviation VA_R).

EXERCISE 6.27

The sinusoidal curves of a pure capacitive circuit have a peak current of 1 A at a supply voltage of 2 V. Using trigonometry for the sine of an angle, determine the instantaneous values of voltage and current at 10° intervals and calculate the instantaneous values of power using these equations:

$$P = ei$$

$$e = V_{max} \sin \theta$$

$$i = I_{max} \sin \theta$$

Plot the power waveform on your graph from the instantaneous values of power calculated in your table. The completed graph should resemble Figure 6.75.

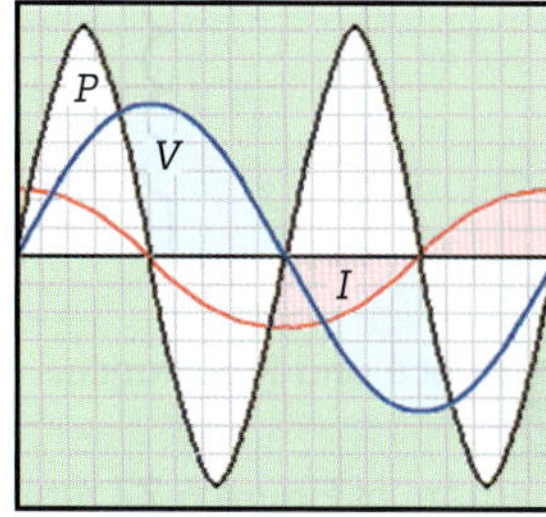

FIGURE 6.75 Sinusoidal curves for a pure capacitive circuit

The graph illustrates that no energy is dissipated or consumed in an intrinsic capacitive circuit and that the power waveform has a frequency twice that of the voltage or current.

REVIEW QUESTIONS

1 In an alternating current circuit consisting of resistance only, the voltage and current are said to be ______________.

2 Calculate the current a 120 Ω resistor will draw from a 230 V, 50 Hz supply.

3 Determine the power dissipated by a 60 Ω resistor when connected to a 230 V, 50 Hz supply.

4 A circuit contains three resistors consisting of 180 Ω, 120 Ω and 100 Ω connected in parallel to a 100 V sinusoidal supply. Calculate:
  i the total resistance
  ii the power dissipated by the 120 Ω resistor
  iii the current flowing through the 180 Ω resistor
  iv the total power dissipated by the circuit.

5 Describe the term 'inductance'.

6 Find the inductive reactance of a 250 mH contactor coil at a frequency of 50 Hz.

7 A current of 120 μA flows through a 60 mH inductor at a frequency of 20 kHz. At what frequency would the current be limited to 40 μA?

8 What is the phase relationship between voltage and current in a purely inductive circuit?

9 What is meant by the term 'reactive power'?

10 Calculate the power dissipated by a 100 mH inductor when connected to a 230 V, 50 Hz supply.

11 What is the opposition to the sinusoidal current in a capacitive circuit called?

12 In a capacitive circuit, what is the current directly proportional to?

13 When examining phase difference in an intrinsic capacitive circuit, the current is said to lead the voltage by how many degrees?

14 Calculate the power dissipated by a 100 μF capacitor when connected to a 230 V, 50 Hz supply.

6.4 RL and RC series a.c. circuits

Resistance and inductance in series

In the series circuit of **Figure 6.76** containing resistance and inductance, the phasor sum of the voltages across the resistive and the inductive components equals the applied voltage.

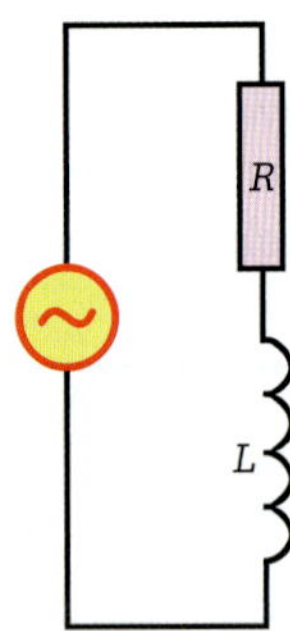

FIGURE 6.76 Resistance and inductance

The phasor diagram of **Figure 6.77** represents the voltages in the pictorial diagram in relation to the current (*I*), which is the reference phasor because it is a series circuit.

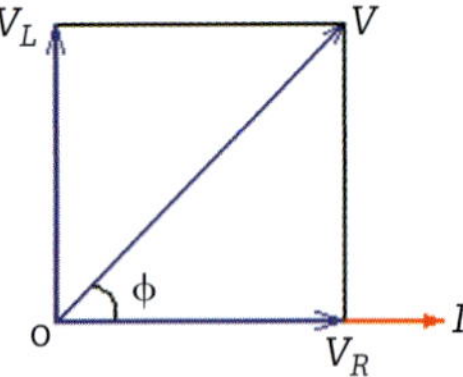

FIGURE 6.77 Phasor diagram of *R* and *L* in series

The phasor diagram shows that the current (*I*) is in phase with the resistive voltage (V_R) but lags the inductive voltage (V_L) by 90°. The applied voltage (*V*) is the resultant phasor sum of the two voltages V_R and V_L and is determined by completing the parallelogram, drawing the diagonal and measuring its length.

The same result can be achieved by applying trigonometry. The voltages that occur in a series RL circuit are calculated by applying the following equations:

$$V_R = IR \quad V_L = IX_L \quad V = \sqrt{V_R^2 + V_L^2}$$

where V_R = the voltage across the resistance in volts (V)
I = the circuit current in amperes (A)
R = resistance in ohms (W)
V_L = the voltage across the inductance in volts (V)
X_L = inductive reactance in ohms (Ω)
V = the applied voltage in volts (V)

EXAMPLE 6.25

Determine the voltages in the circuit of **Figure 6.78** and represent them as a phasor diagram.

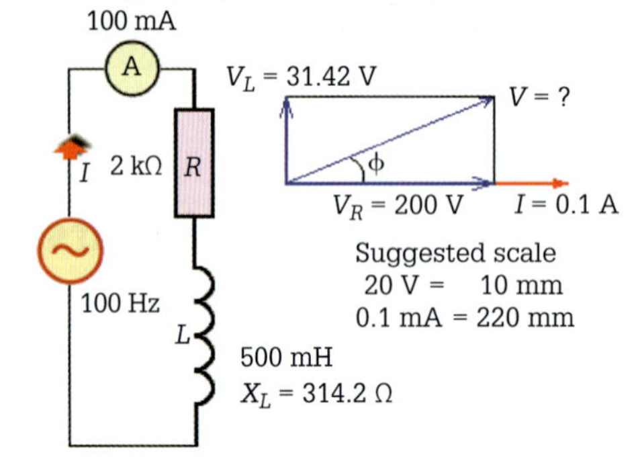

FIGURE 6.78 Resistance and inductance in series

$$\begin{aligned} V_R &= IR \\ &= 0.1 \times 2000 \\ &= \mathbf{200\ V} \end{aligned}$$

$$\begin{aligned} X_L &= 2\pi fL \\ &= 2 \times 3.142 \times 100 \times 0.5 \\ &= \mathbf{314.2\ \Omega} \end{aligned}$$

$$\begin{aligned} V_L &= IX_L \\ &= 0.1 \times 314.2 \\ &= \mathbf{31.42\ V} \end{aligned}$$

$$\begin{aligned} V &= \sqrt{V_R^2 + V_L^2} \\ &= \sqrt{200^2 + 31.42^2} \\ &= \mathbf{202.45\ V} \end{aligned}$$

Impedance in RL series circuits

In intrinsic resistive circuits resistance provides the only opposition to current drawn from the sinusoidal supply. With intrinsic inductive circuits all the opposition to the current is in the form of inductive reactance, whereas resistance is an unchanging quantity and is unaffected by the circuit voltage or current. Inductive reactance is affected because its value depends upon the frequency of the applied voltage.

Although a voltage drop occurs when current flows through either a resistance or an inductance, the phase relationship between the current and the voltage drop is different for a resistance than it is for an inductance. Because the voltage effects of resistance and inductance in a sinusoidal circuit have a phase relationship, we regard the resistance and inductive reactance themselves as having the same phase relationship.

For this reason their combined effect in opposing current flow in RL circuits is called impedance (symbol *Z*) and is measured in ohms. The impedance of a series RL circuit is the opposition to the supply current by the circuit resistance and its inductive reactance.

Impedance triangle

The phasor sum of the resistance and inductive reactance, the impedance, can be calculated by applying Pythagoras' theorem:

$$Z = \sqrt{R^2 + X_L^2}$$

where Z = impedance in ohms (Ω)

R = resistance in ohms (Ω)

X_L = inductive reactance in ohms (Ω)

This equation is derived from the phasor diagram of the current and voltage conditions shown in **Figure 6.78**. The impedance in series RL circuits varies with frequency due to the inductive component and increases as the frequency increases.

The values of resistance, inductive reactance and impedance can be represented in an impedance triangle as shown in **Figure 6.79**.

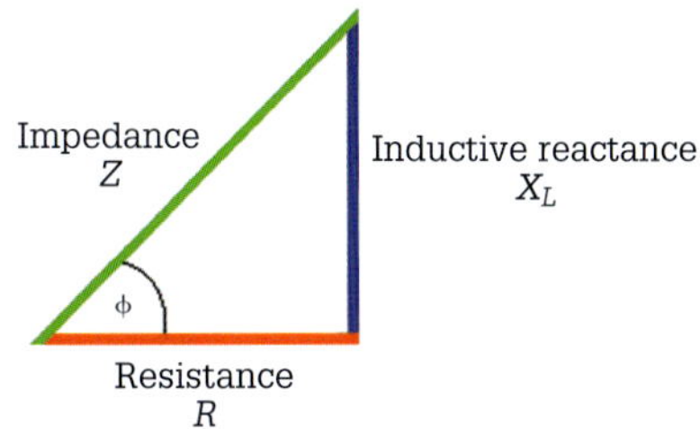

FIGURE 6.79 Impedance triangle — inductive

EXAMPLE 6.26

Calculate:

a the impedance of the circuit in **Figure 6.80** at 100 Hz

$$Z = \sqrt{R^2 + X_L^2}$$
$$= \sqrt{2000^2 + 314.2^2}$$
$$= \mathbf{2024.5\ \Omega}$$

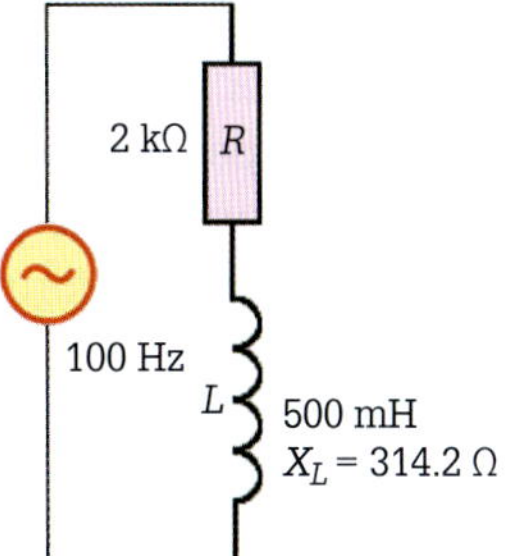

FIGURE 6.80 Increasing frequency in an inductive circuit

b the new impedance when the frequency is increased to 200 Hz

$$X_L = 2\pi f L$$
$$= 2\pi \times 200 \times (500 \times 10^{-3})$$
$$= \mathbf{628.3\ \Omega}$$
$$Z = \sqrt{R^2 + X_L^2}$$
$$= \sqrt{(2 \times 10^3)^2 + 628.3^2}$$
$$= \mathbf{2096\ \Omega}$$

Note: X_L has to be determined before Z can be resolved.

EXERCISE 6.28

a Calculate the applied voltage in a series RL circuit when the voltage across R = 25 V and the voltage across L = 35 V.

b A circuit that has a 600 Ω resistor and a purely inductive coil having a reactance of 300 Ω connected in series to a sinusoidal supply draws a current of 60 mA. Represent the voltages with a phasor diagram and calculate the applied circuit voltage.

c By means of a phasor diagram, represent the voltages in a series circuit containing a 160 mH inductor and a 100 Ω resistor drawing a current of 200 mA at a frequency of 200 Hz.

d In a series RL circuit X_L = 150 Ω and R = 50 Ω. Calculate the impedance of the circuit.

e If the applied voltage to a series RL circuit is 25 V, calculate the voltage across the resistor connected in series with a purely inductive 300 mH coil when the voltage across the coil is 10 V.

f In a series circuit containing a 40 Ω resistor and a purely inductive 100 mH coil, the voltage across the resistor is 100 V at a frequency of 50 Hz. Determine the applied voltage.

g Calculate the impedance of a series circuit that has a 500 mH coil and a 2.5 kΩ resistor connected to a sinusoidal supply with a frequency of 100 Hz.

h Calculate the impedance of a coil with a resistance of 100 Ω and an inductance of 2 mH at a frequency of:

  i 2 kHz

  ii 10 kHz.

i In a phasor diagram representing the voltages in a series 10 kHz RL circuit, the applied voltage is V = 20 V, V_R = 10 V and the current drawn is 10 mA. Calculate the values of resistance and inductance in the circuit.

j The resistance of a coil is 100 Ω. If the impedance is 200 Ω at 5 kHz, what is the inductance of the coil?

Volt–current relationship in series RL circuits

The angle by which the current lags the applied voltage in an RL circuit is called the phase difference (symbol ϕ). This angle is calculated by applying the trigonometry ratios of sine, cosine or tangent to the phasor diagram or impedance triangle.

To determine the phase difference from the values of resistance and inductive reactance the following tangent equation is used:

$$\tan \phi = \frac{X_L}{R}$$

where ϕ = phase difference in degrees (°)

X_L = inductive reactance in ohms (Ω)

R = resistance in ohms (Ω)

After the value of the angle, the word 'lagging' is added to indicate that the circuit is inductive.

EXAMPLE 6.27

Using the impedance triangle of **Figure 6.81**, calculate the phase difference of the inductive circuit.

$$\tan \phi = \frac{X_L}{R}$$
$$= \frac{1600}{800}$$
$$= 2.0$$
$$\phi = 63.4° \text{ lagging}$$

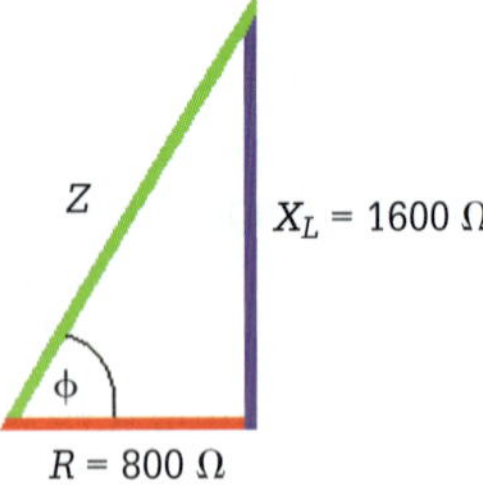

FIGURE 6.81 Impedance triangle of an inductive circuit

EXERCISE 6.29

a Calculate the phase difference in a series RL circuit that has an inductive reactance of 150 Ω and a resistance of 50 Ω.

b A 200 mH coil has a resistance of 200 Ω. What is the phase difference when the coil is connected across a sinusoidal waveform with a frequency of 1 kHz?

c A coil with a resistance of 200 Ω has an impedance of 450 Ω when connected across a sinusoidal supply of 1 kHz. Determine the phase difference.

Power factor

An armature coil has an instantaneous voltage induced in it of a value that is dependent on the angle theta (θ) at which the coil cuts the field flux. This is shown in **Figure 6.82**.

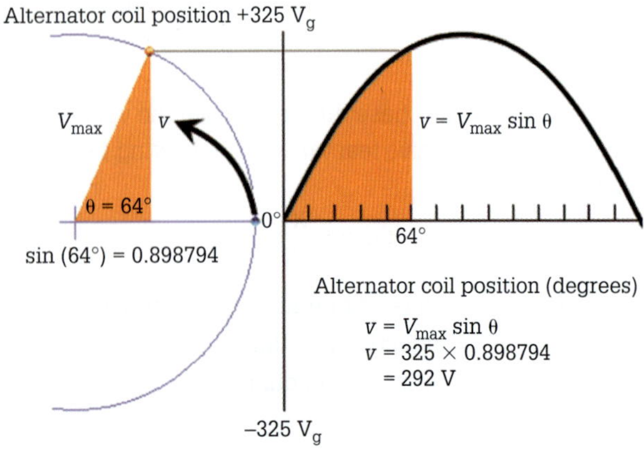

FIGURE 6.82 Instantaneous induced voltage at 64°

The value of the instantaneous voltage at an angle theta of 64° can be determined by using trigonometry. The waveform produced at the alternator terminals is called the fundamental frequency waveform.

In order to distinguish between geometrical degrees (θ) and electrical degrees we assign the Greek symbol phi (ϕ) to indicate electrical degrees.

The instantaneous fundamental waveform is generated at an angle θ of 64°. When current is drawn by the load, the instantaneous fundamental frequency waveform splits into two instantaneous frequency components of voltage and current as shown in **Figure 6.83**.

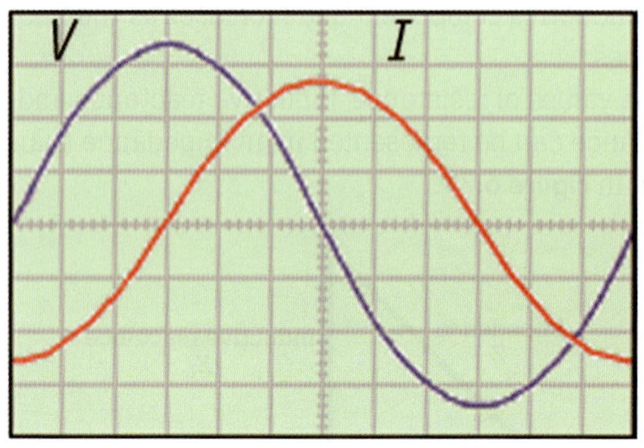

FIGURE 6.83 Two frequency components of the fundamental waveform

Assuming that the current waveform has shifted to the right of the voltage waveform we can say that the phase difference, shift or angle (ϕ) between voltage and current is 90 electrical degrees. The electrical angle ϕ is called the power factor angle. The cosine of this electrical angle defines the power factor. The cosine of 90 electrical degrees is zero.

The power factor's range is from one to zero because these are the containing values for the range of the cosine of an angle. The higher the value of the cosine (closer to 1), the more efficient is the electrical system. A 'low' power factor (closer to zero) requires a larger delivery current and therefore causes increased demand on electrical infrastructure.

Another definition of power factor concerns the ratio between active power and apparent power. *P*, *Q* and *S* form a right-angled triangle as shown in **Figure 6.84**. Because reactive power and apparent power form the adjacent and hypotenuse sides of the right-angled triangle, the power factor ratio is also equal to the cosine of that angle (ϕ).

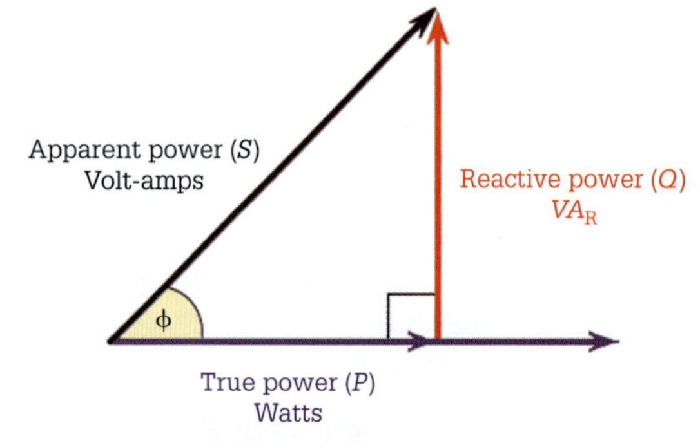

FIGURE 6.84 Power triangle

Supply authorities require that the power factor of an installation measured at the authority's metering position at full load be kept above 0.8 to 0.9. If the installation is less than this, corrective measures must be undertaken. The reason is as follows.

A lightly loaded motor as illustrated in Figure 6.85 draws a current of 5 A 230 V with a power factor of 0.5 (meaning that I lags V by 60°).

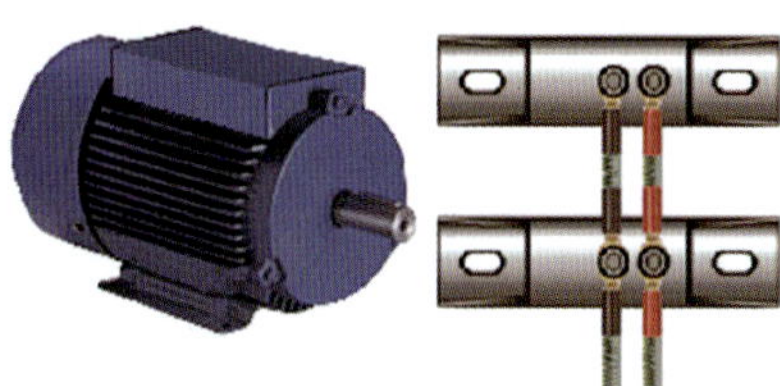

FIGURE 6.85 Comparison of power dissipated by a single-phase motor and a heating element

The power dissipated by this motor is 575 W (P) not the 5 A × 230 V = 1150 VA (S) you would expect. If we compare this power dissipated with a 575 W heating element we will find that the current drawn from the supply will be 2.5 A (pf = unity) because V_R and I_R are in phase.

The current drawn is half that of the motor yet the power dissipated is the same. A power factor of 0.5 lagging means that only half the current is available to do actual work. The other half is providing energy for the electromagnetic field and does not contribute to actual work done. Higher currents drawn from the supply mean that there will be a greater voltage drop occurring across the supply and increased power (I^2R) losses from the supply conductors.

Power in RL circuits

- **True power (*P*)** is measured in watts (W) and is dissipated by the resistive component in a series RL circuit. It is equal to the product of the circuit current, the applied voltage and the power factor of the circuit.
- **Reactive power (*Q*)** is measured in volt-amperes reactive (VA_R) and is the product of the circuit current and the voltage across the inductive component.
- **Apparent power (*S*)** is measured in volt-amperes (VA) and is the product of the circuit current and the applied voltage across the RL circuit.

  Equations for power factor and true power are:

$$\text{power factor (pf)} = \frac{\text{true power}}{\text{apparent power}}$$

$$= \text{cos of the angle } \theta \text{ (theta)}$$

$$P = I^2R = VI\cos\phi$$

where P = true power in watts (W)

V = applied voltage in volts (V)

I = current in amperes (A)

ϕ = phase angle in degrees (°)

R = resistance in ohms (Ω)

Note: In series circuits, pf may be found using:

$$\frac{VR}{V} = \frac{R}{Z} = \cos\phi$$

EXAMPLE 6.28

In the circuit shown in Figure 6.86 calculate:

a the power factor

b the true power dissipated

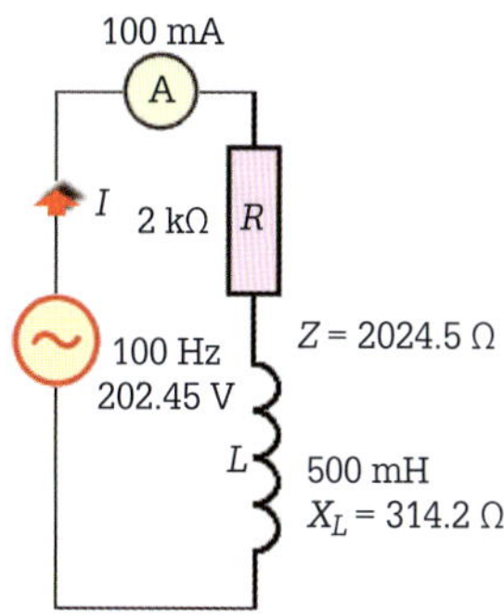

FIGURE 6.86 Resistance and inductance in a series circuit

$$\cos\phi = \frac{R}{Z} = \frac{2000}{2024.5} = \mathbf{0.9879\ lagging}$$

$$P = VI\cos\phi = 202.45 \times 0.1 \times 0.9879 = \mathbf{20\ W}$$

The power triangle as shown in Figure 6.87 can be used to express power relationships in an alternating-current RL series circuit.

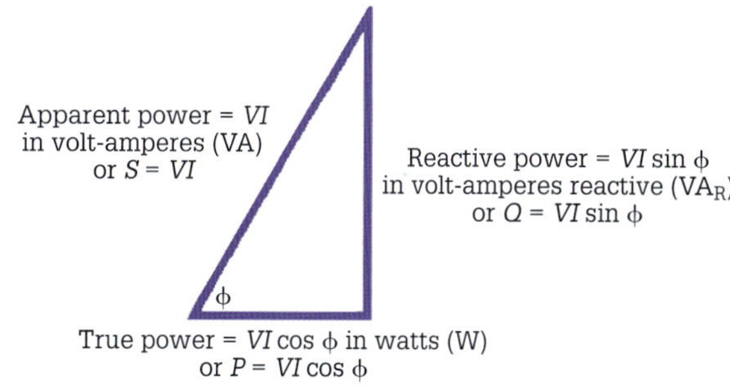

FIGURE 6.87 Power triangle of an RL series circuit

EXERCISE 6.30

a Find the power factor when a coil with a resistance of 100 Ω and an inductive reactance of 200 Ω is connected to a 100 V 50 Hz sinusoidal supply.

b A circuit consists of a 1 kΩ resistor and an inductor connected in series to a 60 V 50 Hz sinusoidal supply. Calculate the power factor when the voltage across the resistor is 20 V.

c In a circuit containing resistance and inductance connected in series to a 200 V sinusoidal supply the current is 100 mA and the power factor is 0.9 lagging. Determine the true power dissipated by the circuit.

d A coil with a resistance of 250 Ω has a power factor of 0.85 lagging when connected across an 8 kHz supply. Calculate the new power factor when the coil is connected across a 6 kHz supply.

Resistance and capacitance in series

In the series circuit of **Figure 6.88** containing resistance and capacitance, the phasor sum of the voltages across the resistive and the capacitive components equals the applied voltage.

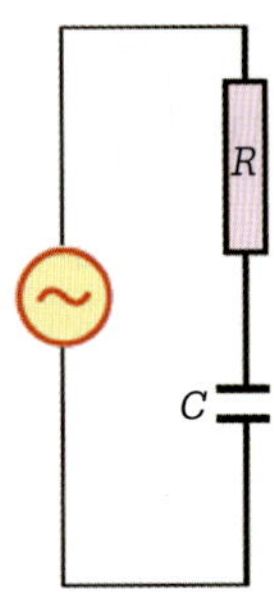

FIGURE 6.88 Resistance and capacitance in series

The phasor diagram of **Figure 6.89** represents the voltages in the above pictorial diagram in relation to the current (I) which is the reference phasor.

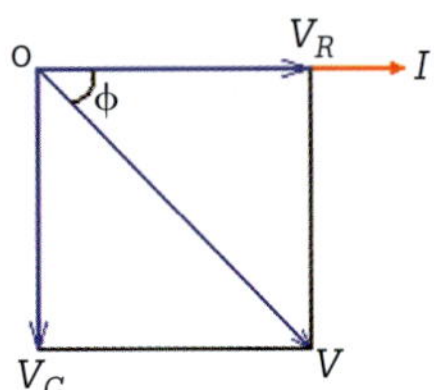

FIGURE 6.89 Phasor diagram for R and C in series

The phasor diagram shows that the current (I) is in phase with the resistive voltage (V_R) but leads the capacitive voltage (V_C) by 90°.

The applied voltage (V) is the resultant phasor sum of the two voltages, V_R and V_C, and is determined by completing the parallelogram, drawing the diagonal and measuring its length. The same result can be achieved by applying trigonometry.

The voltages that occur in a series RC circuit are calculated by applying the following equations:

$$V_R = IR \quad V_C = IX_C \quad V = \sqrt{V_R^2 + V_C^2}$$

where V_R = the voltage across the resistance in volts (V)
I = the circuit current in amperes (A)
R = resistance in ohms (Ω)
V_C = the voltage across the capacitor in volts (V)
X_C = capacitive reactance in ohms (Ω)
V = the applied voltage in volts (V)

EXAMPLE 6.29

Determine the voltages in the circuit in **Figure 6.90** and represent them as a phasor diagram.

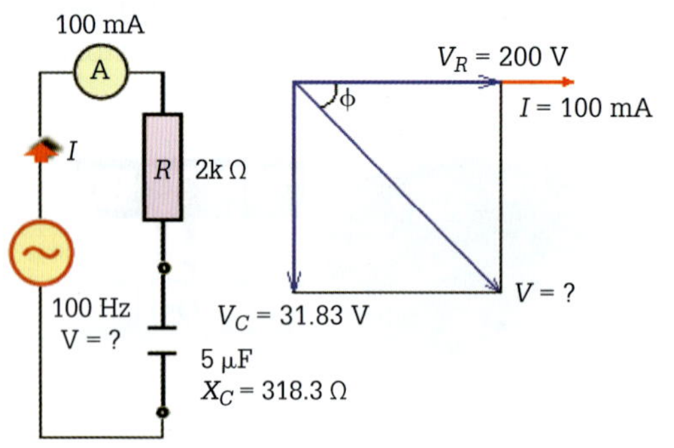

FIGURE 6.90 Phasor diagram

$$\begin{aligned} V_R &= IR \\ &= 0.1 \times 2000 \\ &= \mathbf{200\ V} \\ X_C &= \frac{1}{2\pi fC} \\ &= \frac{1}{2 \times 3.142 \times 100 \times 0.000005} \\ &= \mathbf{318.3\ \Omega} \\ V_C &= IX_C \\ &= 0.1 \times 318.3 \\ &= \mathbf{31.83\ V} \\ V &= \sqrt{V_R^2 + V_C^2} \\ &= \sqrt{200^2 + 31.83^2} \\ &= \mathbf{202.5\ V} \end{aligned}$$

Impedance in RC series circuits

In pure resistive circuits resistance provides the only opposition to current drawn from the sinusoidal supply. With intrinsic capacitive circuits all of the opposition to the current is in the form of capacitive reactance while resistance is an unchanging quantity and is unaffected by the circuit voltage or current. However, capacitive reactance is affected because its value depends upon the frequency of the applied voltage.

Although a voltage drop occurs when current flows through either a resistance or a capacitance the phase relationship between the current and the voltage drop is different for a resistance than it is for a capacitance.

For this reason their combined effect in opposing current flow in RC circuits is called impedance (symbol Z) and is measured in ohms. The impedance of a series RC circuit is the total opposition to current drawn from the supply by the circuit resistance and capacitive reactance.

Impedance triangle

The impedance can be calculated by applying Pythagoras' theorem:

$$Z = \sqrt{R^2 + X_C{}^2}$$

where Z = impedance in ohms (Ω)
R = resistance in ohms (Ω)
X_C = capacitive reactance in ohms (Ω)

The impedance in series RC circuits varies with frequency due to the capacitive component and decreases as the frequency increases. The values of resistance (R), capacitive reactance (X_C) and impedance (Z) can be represented in an impedance triangle. An impedance triangle of an RC series circuit is shown in **Figure 6.91**.

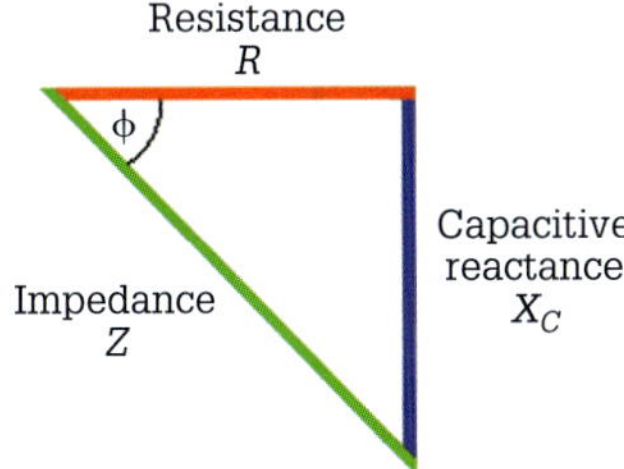

FIGURE 6.91 Impedance triangle for an RC series circuit

EXAMPLE 6.30

Calculate:

a the impedance of the circuit in **Figure 6.92** at 100 Hz, when the capacitive reactance is 318.3 Ω

$$Z = \sqrt{R^2 + X_C^2}$$
$$= \sqrt{2000^2 + 318.3^2}$$
$$= \mathbf{2025.2\ \Omega}$$

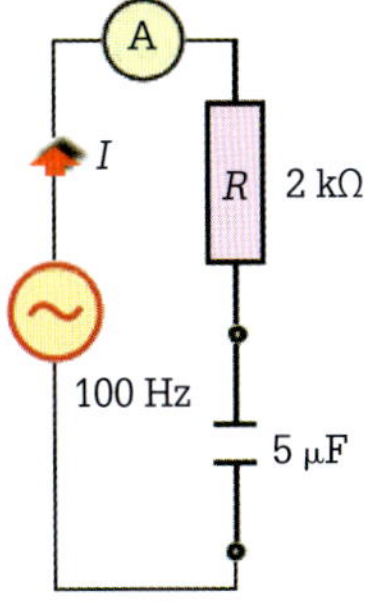

FIGURE 6.92 *R* and *C* in series

b the new impedance when the frequency is increased to 200 Hz

$$X_C = \frac{1}{2\pi fC}$$
$$= \frac{1}{2\pi \times 200 \times (5 \times 10^{-6})}$$
$$= \mathbf{159\ \Omega}$$
$$Z = \sqrt{R^2 + X_C^2}$$
$$= \sqrt{(2 \times 10^3)^2 + 159^2}$$
$$= \mathbf{2006\ \Omega}$$

EXERCISE 6.31

a The voltages across a resistor and a capacitor connected in series are 50 V and 35 V respectively. Calculate the supply voltage.

b Represent the voltages across a capacitor (V_C = 15 V) and a resistor (V_R = 25 V) on a phasor diagram and determine the applied circuit voltage.

c A series RC circuit is connected across a supply voltage of 250 V. If the voltage across the capacitor is 140 V, what is the voltage across the resistor?

d A series RC circuit has a 120 Ω resistor connected in series with a capacitive reactance of 70 Ω drawing a current of 1 A. Complete a phasor diagram and determine the applied circuit voltage.

e A resistor and a capacitor are connected in series to a 100 V sinusoidal supply. If the voltage across the capacitor is 55 V and the supply current is 200 mA determine the value of the resistor.

f Calculate the circuit impedance when a 200 Ω resistor and a 10 μF capacitor are connected in series to a 50 Hz supply.

g Calculate the circuit impedance when a 22 Ω resistor and a 120 μF capacitor are connected in series to 200 Hz and then 300 Hz.

h A series RC circuit has an impedance of 40 Ω at 2 kHz. If the reactance of the capacitor is 10 Ω calculate the value of the resistance.

i What value of capacitor is required to be connected in series with a 1 kΩ resistor in order to provide an impedance of 2 kΩ at a frequency of 50 Hz?

Volt–current relationship in series RC circuits

The angle (symbol ϕ) by which the current leads the applied voltage in an RC circuit is called the phase difference or phase angle. This angle is calculated by applying the trigonometry ratios of sine, cosine or tangent to the phasor diagram or impedance triangle. To determine the phase difference from the values of resistance and capacitive reactance the following tangent equation is used:

$$\tan\phi = \frac{X_C}{R} \text{ in degrees leading}$$

where ϕ = phase difference in degrees (°)
X_C = capacitive reactance in ohms (Ω)
R = resistance in ohms (Ω)

After the value of the angle the word 'leading' is added to indicate that the circuit is capacitive.

EXAMPLE 6.31

1 Calculate the phase difference of the RC series circuit in **Figure 6.93**.

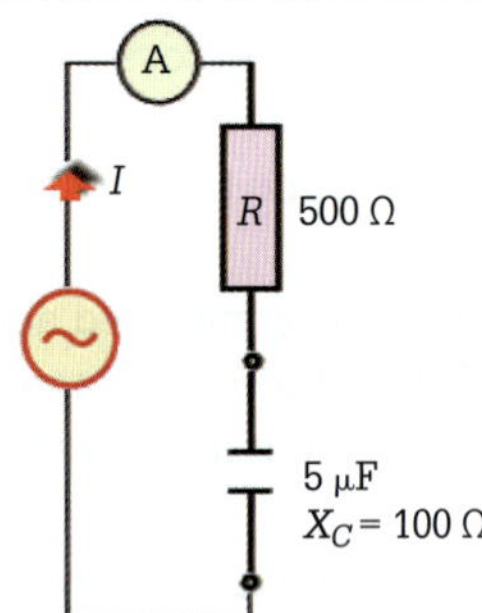

FIGURE 6.93 Resistance and capacitance in series

$$\tan\phi = \frac{X_C}{R}$$
$$= \frac{100}{500}$$
$$= 0.2$$
$$\phi = \mathbf{11°}$$

2 Calculate:

a the power factor in **Figure 6.94**

$$\cos\phi = \frac{R}{Z}$$
$$= \frac{2000}{2025}$$
$$= \mathbf{0.9877\ leading}$$

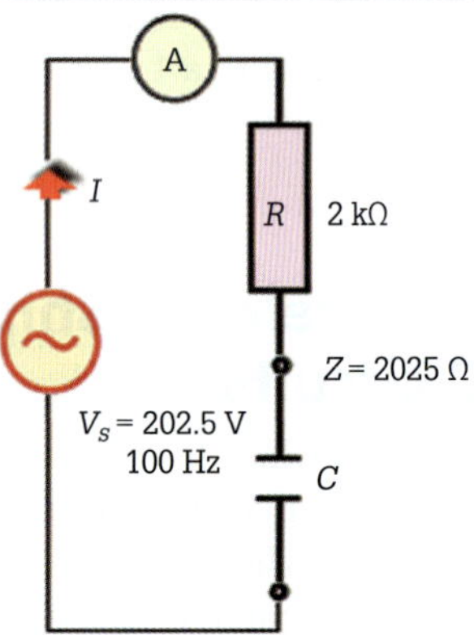

FIGURE 6.94 RC circuit

b the power dissipated

$$I = V_s / Z$$
$$= 202.5 / 2025$$
$$= \mathbf{0.1\ A}$$
$$P = VI\cos\phi$$
$$= 202.5 \times 0.1 \times 0.9877$$
$$= \mathbf{20\ W}$$

Power can be represented in a right-angled triangle called the power triangle. The power triangle as shown in **Figure 6.95** can be used to express power relationships in an alternating current RC series circuit.

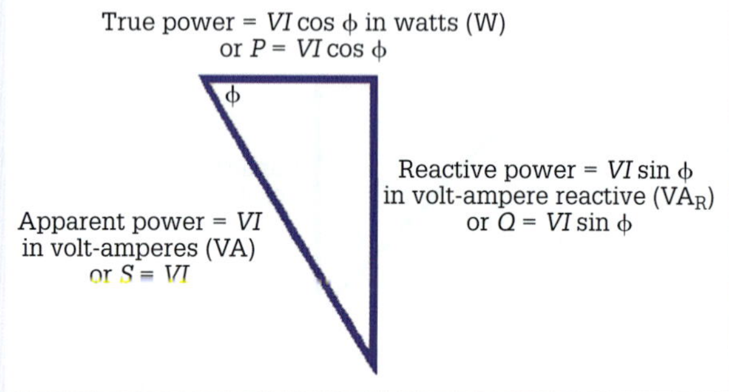

FIGURE 6.95 Power triangle for an RC series circuit

EXERCISE 6.32

a A 100 Ω resistor and a 100 μF capacitor are connected in series to a sinusoidal voltage with a frequency of 50 Hz. Calculate the phase angle.

b What is the phase angle in a series RC circuit when the capacitive reactance has a value of 500 Ω and the resistance has a value of 1 kΩ?

c What is the phase angle when a 200 Ω resistor and a 1.8 μF capacitor are connected across a 2 kHz supply?

REVIEW QUESTIONS

1 In a series circuit containing resistance and inductance, the supply voltage is equal to the phasor sum of what two quantities?

2 What effect does increasing the frequency applied to a series RL circuit have on circuit impedance?

3 Between what values will the phase angle of a practical series RL circuit range?

4 In a practical RL circuit, the current ________ the applied emf.

5 In order to distinguish between geometrical degrees (θ) and electrical degrees we assign what Greek symbol?

6 What is true power (P) measured in?

7 What is the combined effect of resistance and capacitance in opposing current flow in RC circuits called?

8 Name the word that is added to indicate that a circuit is capacitive when measuring phase difference.

9 The voltages across a resistor and a capacitor connected in series are 20 V and 30 V respectively. Calculate the supply voltage.

10 Calculate the circuit impedance when a 100 Ω resistor and a 150 µF capacitor are connected in series to a 50 Hz supply.

11 What is the phase angle when a 200 Ω resistor and an 18 µF capacitor are connected across a 50 Hz supply?

12 Find the power factor when a coil with a resistance of 100 Ω and an inductive reactance of 200 Ω is connected to a 100 V 50 Hz sinusoidal supply.

6.5 Resistance, inductance and capacitance in combination

Resistance, capacitance and inductance can be connected in series to form an RLC series circuit or parallel to form a parallel RLC circuit. Different methods are used to analyse each of these circuit combinations.

RLC series circuits

Figure 6.96 shows a circuit containing resistance, inductance and capacitance in series in which the phasor sum of the voltages across the resistance (V_R), the inductance (V_L) and the capacitance (V_C) equals the applied circuit voltage (V).

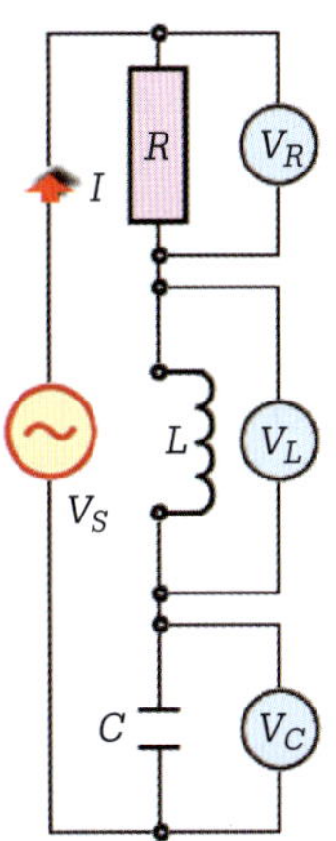

FIGURE 6.96 Resistance, inductance and capacitance in series

The phasor diagram of **Figure 6.97** represents the voltages in the circuit in relation to the current (I) that is the reference phasor. The current (I) is in phase with V_R, leads V_C by 90° and lags V_L by 90°. The applied voltage (V) is determined by resolving V_L and V_C and then ($V_L - V_C$) or ($V_C - V_L$) and V_R.

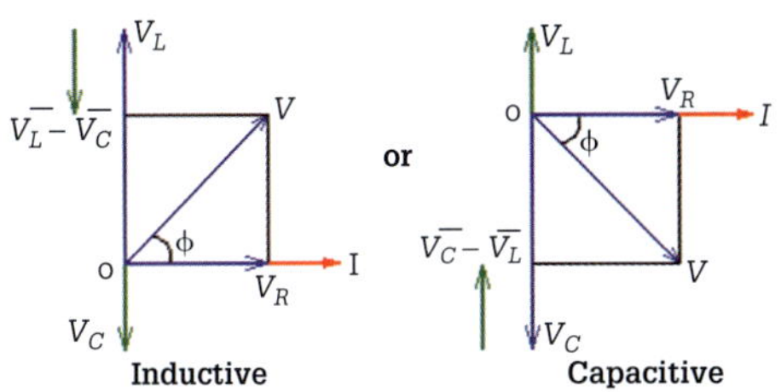

FIGURE 6.97 Phasor diagrams for an RLC series circuit

The phasor diagram of **Figure 6.97** illustrates two possibilities of the inductive and capacitive components in the circuit. If the inductive reactance is greater than the capacitive reactance then the circuit is seen as an inductive circuit.

If the capacitive reactance is greater than the inductive reactance then the circuit is capacitive. The voltages in a series RLC circuit are expressed as follows:

$$V_R = I_R \quad V_L = IX_L \quad V_C = IX_C$$

$$V = \sqrt{V_R^2 + (V_L - V_C)^2} \text{ or } V = \sqrt{V_R^2 + (V_C - V_L)^2}$$

where V_R = voltage across the resistive component in volts (V)
I = circuit current in amperes (A)
R = resistance in ohms (Ω)
V_L = voltage across the inductive component in volts (V)
X_L = inductive reactance in ohms (Ω)
V_C = voltage across the capacitive component in volts (V)
X_C = capacitive reactance in ohms (Ω)
V = applied voltage in volts (V)

EXAMPLE 6.32

Determine the voltages in the circuit shown in **Figure 6.98** and complete the phasor diagram.

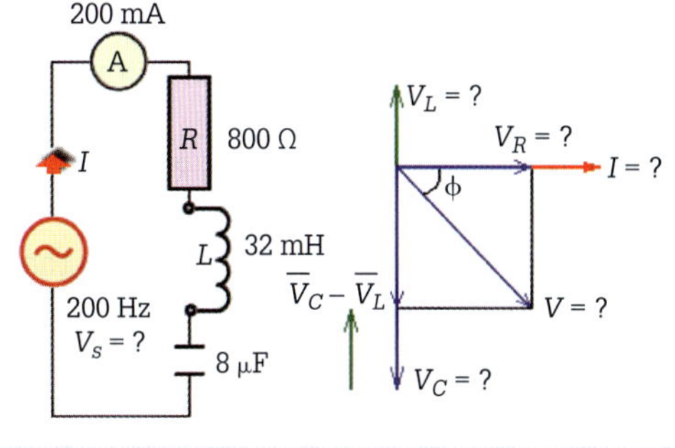

FIGURE 6.98 Circuit and phasor diagram of a resistive, inductive and capacitive circuit

»

$$V_R = IR$$
$$= 0.2 \times 800$$
$$= \mathbf{160\ V}$$
$$X_L = 2\pi fL$$
$$= 2 \times 3.142 \times 200 \times 0.032$$
$$= \mathbf{40.2\ \Omega}$$
$$V_L = IX_L$$
$$= 0.2 \times 40.2$$
$$= \mathbf{8.04\ V}$$
$$X_C = \frac{1}{2\pi fC}$$
$$= \frac{1}{2 \times 3.142 \times 200 \times 0.000008}$$
$$= \mathbf{99.5\ \Omega}$$
$$V_C = IX_C$$
$$= 0.2 \times 99.5$$
$$= \mathbf{19.9\ V}$$
$$V = \sqrt{V_R^2 + (V_C - V_L)^2}$$
$$= \mathbf{160.44\ V}$$

EXERCISE 6.33

a A series RLC circuit has the voltages $V_R = 40$ V, $V_L = 20$ V and $V_C = 40$ V across the circuit components. Calculate the applied voltage.

b A series RLC circuit has the voltages $V_R = 80$ V and $V_L = 60$ V. If the supply voltage measures 200 V, what is the potential across the capacitor?

Impedance in RLC series circuits

The combined opposition to the flow of current in a series RLC circuit is called impedance (symbol Z). The combined values of resistance, inductive reactance and capacitive reactance can be represented in the form of an impedance triangle as in **Figure 6.99**.

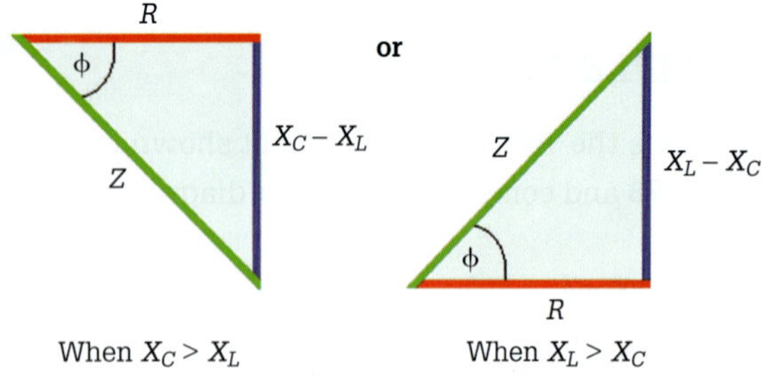

FIGURE 6.99 Impedance triangle for an RLC circuit

From these impedance triangles the value of the combined opposition to the flow of current called the impedance can be calculated:

$$Z = \sqrt{R^2 + (X_L - X_C)^2} \text{ or } Z = \sqrt{R^2 + (X_C - X_L)^2}$$

As both X_L and X_C vary with frequency the impedance (Z) will be dependent upon frequency.

EXAMPLE 6.33

Calculate the impedance of the 200 Hz circuit in **Figure 6.100**.

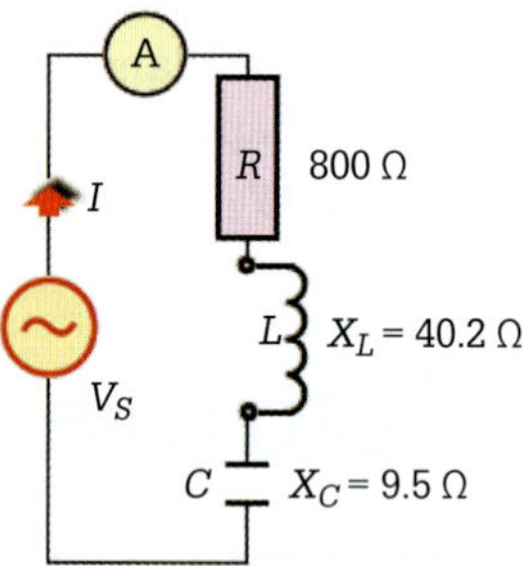

FIGURE 6.100 RLC in series

$$Z = \sqrt{R^2 + (X_L - X_C)^2}$$
$$= \sqrt{800^2 + (40.2 - 9.5)^2}$$
$$= \mathbf{800.59\ \Omega}$$

EXERCISE 6.34

a In a series RLC circuit, determine the impedance when the resistance is 50 Ω, the inductive reactance is 40 Ω and the capacitive reactance is 20 Ω.

b Find the impedance of a circuit when a 100 Ω resistor, a 150 mH inductor and a 100 µF capacitor are connected in series across a sinusoidal supply of 50 Hz.

c A series circuit has an impedance of 250 Ω at 50 Hz. Calculate the resistance when the inductive reactance is 150 Ω and the capacitive reactance is 100 Ω.

Phase difference in RLC series circuits

The phase difference (symbol φ) is the angle by which the current drawn from the supply lags or leads the applied voltage. The supply current is the reference quantity for series circuits. In a series RLC circuit, when the inductive reactance is greater than the capacitive reactance the inductor has a higher potential across it meaning the circuit is predominantly inductive and the current lags the inductor voltage.

When the capacitive reactance is greater than the inductive reactance the circuit is capacitive and the current leads the voltage. The phase difference is calculated by applying the trigonometry ratios of sine, cosine or tangent to the voltage phasor diagram or the impedance triangle. In many problems the tangent ratio is used:

$$\tan \phi = \frac{X_L - X_C}{R} \text{ or } \tan \phi = \frac{X_C - X_L}{R}$$

EXAMPLE 6.34

Calculate the phase difference of an RLC circuit illustrated by the impedance triangle in **Figure 6.101**.

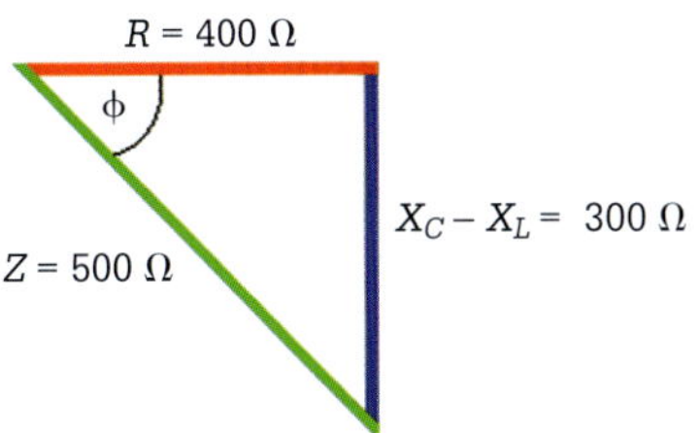

FIGURE 6.101 Impedance triangle when $X_C > X_L$

$$\tan \phi = \frac{X_C - X_L}{R}$$
$$= \frac{300}{400}$$
$$= 0.75$$
$$\tan^{-1} 0.75 = 37°$$

EXERCISE 6.35

a Calculate the phase difference in a sinusoidal circuit where the resistance has a value of 100 Ω, the inductive reactance a value of 150 Ω and the capacitive reactance a value of 200 Ω.

b A circuit contains a 2 kΩ resistor, a 400 mH inductor and an 8 µF capacitor connected in series. Calculate the phase difference when the frequency of the supply is 50 Hz.

c The phase difference in a series RLC circuit is 30° lagging. Calculate the value of the resistance when the inductive reactance equals 2 kΩ and the capacitive reactance equals 800 Ω.

d A circuit consists of a 1 kΩ resistor, a 180 mH inductor and a 15 µF capacitor connected in series across a 50 Hz sinusoidal supply. Determine the power dissipated in the circuit if the circuit draws 300 mA from the supply.

e In a series RLC circuit the impedance is 1.5 kΩ and the current leads the applied voltage of 50 V by 45°. Calculate the power factor and the true power dissipated.

Resistance and inductance in parallel

When an alternating current circuit contains resistance and inductance connected in parallel the total supply current is equal to the phasor sum of the currents in the resistive branch and the inductive branch of the phasor diagram. Phasor diagrams for parallel circuits are constructed by using the supply voltage as the reference phasor.

The resistive current (I_R) is in phase with the supply voltage (V) while the inductive current (I_L) lags the supply voltage by 90°. The total supply current (I_T) is the result of the phasor relationship between these two currents. **Figure 6.102** illustrates an RL parallel circuit.

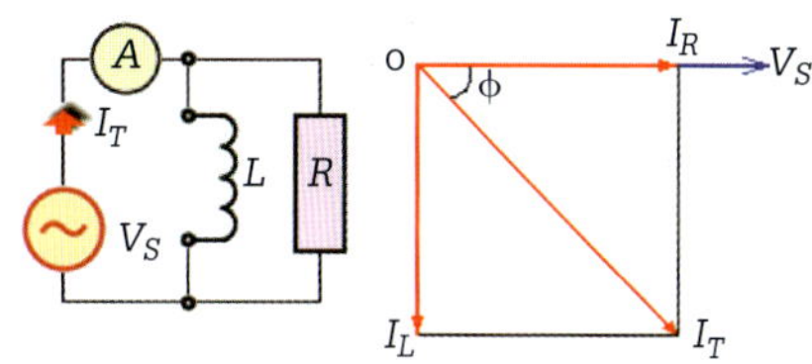

FIGURE 6.102 Resistance and inductance in parallel

SWITCH ON

Voltage and impedance phasors are not valid in parallel circuits.

Currents in RL circuits

Currents in RL circuits are calculated by applying the following equations:

$$I_R = \frac{V}{R} \qquad I_L = \frac{V}{X_L} \qquad I_T = \sqrt{I_R^2 + I_L^2}$$

where I_R = current through R in amperes (A)
V = supply voltage in volts (V)
R = resistance in ohms (Ω)
I_L = current through L in amperes (A)
X_L = inductive reactance in ohms (Ω)
I_T = circuit current in amperes (A)

EXAMPLE 6.35 

Represent the currents in the circuit in **Figure 6.103** with phasors and show their relationship to the supply voltage.

$$I_R = \frac{V}{R} = 20 \text{ mA} \qquad X_L = 2\pi fL = 1005.4\ \Omega$$

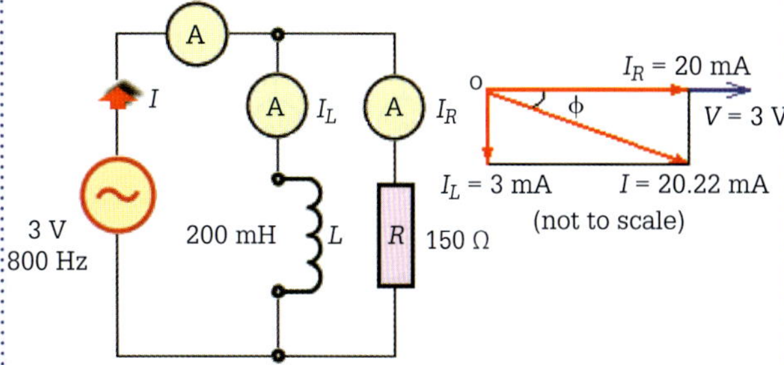

FIGURE 6.103 Phasor and circuit diagram of resistance and inductance in parallel

»

Instructions

Draw the reference phasor V to any length as shown in **Figure 6.104**.

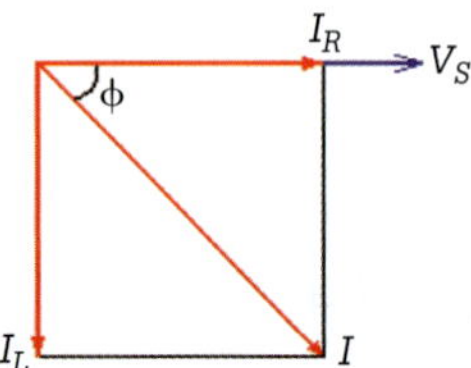

FIGURE 6.104 Construction of a parallel RL phasor diagram

- Draw I_R to scale and in phase with V.
- Draw I_L to scale lagging the V by 90°.

Note: To complete the phasor diagram and so find the circuit current I, a parallelogram is drawn using the phasors I_R and I_L.

The length of the diagonal of the parallelogram represents the supply current (I).

Impedance in parallel RL circuits

The impedance of a parallel RL circuit is determined by using the supply current obtained by the phasor addition of the branch currents and the supply voltage, according to the following equation:

$$Z = \frac{V}{I}$$

where Z = impedance in ohms (Ω)
V = supply voltage in volts (V)
I = supply current in amperes (A)

EXAMPLE 6.36

Calculate the impedance of the circuit in **Figure 6.105** when the frequency is 800 Hz.

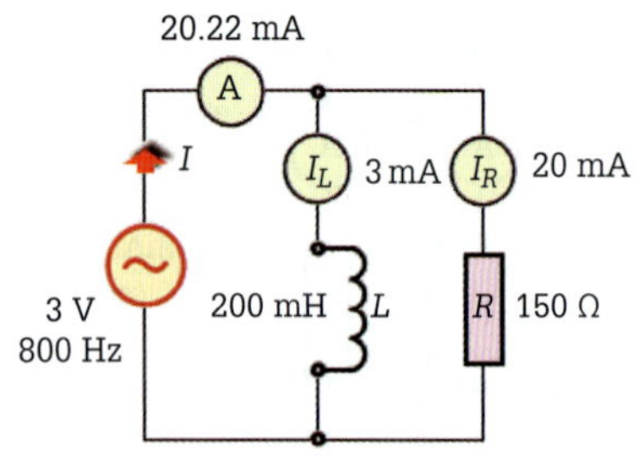

FIGURE 6.105 R and L in parallel

$$I = \sqrt{I_R^2 + I_L^2} = 20.22 \text{ mA}$$

$$Z = \frac{V}{I} = 148.37\ \Omega$$

EXERCISE 6.36

a A circuit contains a resistor and an inductor connected in parallel to the output terminals of an alternator. If the current in the resistive branch is 20 A and 45 A in the inductive branch, calculate the total current in the circuit.

b Determine the total current when a 2.4 H lamp ballast and a 1.2 kΩ resistor are connected in parallel to a 230 V 50 Hz sinusoidal supply.

c A resistor is connected in parallel with an inductor and draws 200 mA from a 50 V 50 Hz supply. If the total current is 500 mA calculate the value of the inductance.

Phase difference in parallel RL circuits

The angle by which the current lags the applied voltage in an RL circuit is called the phase difference (symbol ϕ). This angle is calculated by applying the trigonometry ratios of sine, cosine or tangent to the phasor diagram or impedance triangle. To determine the phase difference from the values of resistance and inductive reactance, the following tangent equation is used:

$$\tan \phi = \frac{I_L}{I_R}$$

where ϕ = phase difference in degrees (°)
I_L = inductive current in amperes (I)
I_R = resistive current in amperes (I)

EXAMPLE 6.37

Using the phasor diagram given in **Figure 6.106** calculate the phase difference of the RL parallel circuit.

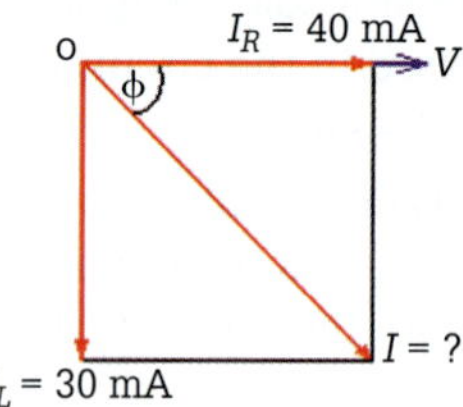

FIGURE 6.106 Phase angle of an RL parallel circuit

$$\tan \phi = \frac{I_L}{I_R} = \frac{30}{40} = 0.75$$

$$\phi = 37°$$

EXERCISE 6.37

a Calculate the phase difference in a parallel RL circuit in which the current in the inductive branch is 150 mA and the current in the resistive branch is 250 mA.

b A resistor connected in parallel with an inductor draws 190 mA from the supply. Calculate the phase difference if the total current drawn from the supply is 245 mA.

c What is the phase difference in a parallel RL circuit when the current in each branch equals 100 mA?

d An inductor with an inductance of 500 mH is connected in parallel with a resistance of 500 Ω across a 150 V 50 Hz supply. Calculate:

 i the power factor

 ii the true power dissipated.

Resistance and capacitance in parallel

In the parallel circuit of **Figure 6.107** containing resistance and capacitance, the phasor sum of the currents through the resistive and the capacitive components equals the total current drawn from the supply.

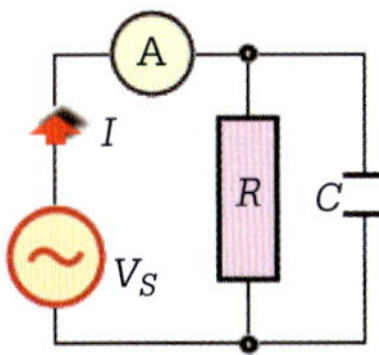

FIGURE 6.107 Resistance and capacitance in parallel

The phasor diagram of **Figure 6.108** represents the currents in the above diagram in relation to the applied voltage (V) that is the reference phasor.

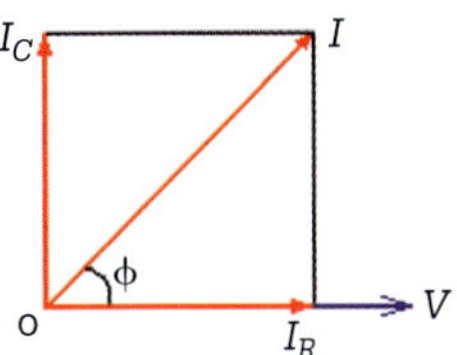

FIGURE 6.108 Phasor diagram for an RC parallel circuit

The phasor diagram shows that the applied voltage (V) is in phase with the resistive current (I_R) but that the capacitive current (I_C) leads it by 90°. The total supply current (I) is the resultant phasor sum of the two currents, I_R and I_C, and is determined by completing the parallelogram, drawing the diagonal and measuring its length. The same result can be achieved by applying trigonometry. The currents drawn from the supply in a parallel RC circuit are calculated by applying the following equations:

$$I_R = \frac{V}{R} \quad I_C = \frac{V}{X_C} \quad I = \sqrt{I_R^2 + I_C^2}$$

where I_R = the current drawn by the resistance in amperes (A)
V = the applied voltage in volts (V)
R = resistance in ohms (Ω)
I_C = the current drawn by the capacitance in amperes (A)
X_C = capacitive reactance in ohms (Ω)
I = the circuit current in amperes (A)

EXAMPLE 6.38

Determine the currents in the circuit in **Figure 6.109** and represent them as a phasor diagram.

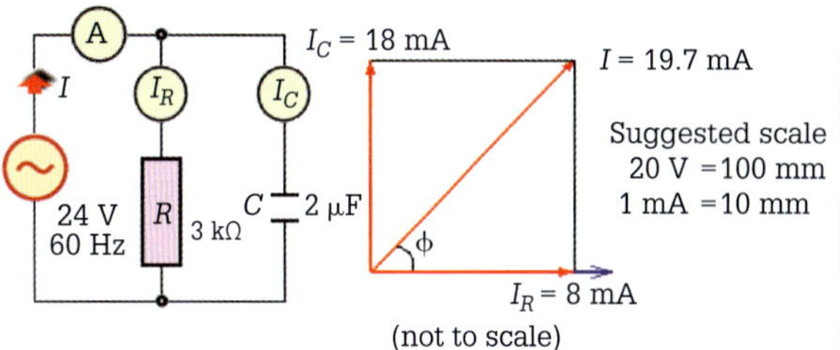

FIGURE 6.109 R and C in a parallel circuit

$$I_R = \frac{V}{R}$$
$$= \frac{24}{3000}$$
$$= \mathbf{8\ mA}$$
$$X_C = \frac{1}{2\pi fC}$$
$$= \frac{1}{2 \times 3.142 \times 60 \times 2 \times 10^{-6}}$$
$$= \mathbf{1326\ \Omega}$$
$$I_C = \frac{V}{X_C}$$
$$= \frac{24}{1326}$$
$$= \mathbf{18\ mA}$$
$$I = \sqrt{I_R^2 + I_C^2}$$
$$= \sqrt{8^2 + 18^2}$$
$$= \mathbf{19.7\ mA\ leading}$$

Impedance in parallel RC circuits

The impedance of a parallel RC circuit is determined by using the supply current that is obtained by the phasor addition of the branch currents and the supply voltage in the following equation:

$$Z = \frac{V}{I}$$

where Z = impedance in ohms (Ω)
V = supply voltage in volts (V)
I = supply current in amperes (A)

EXAMPLE 6.39

Calculate the impedance of the circuit in **Figure 6.110** when the frequency is 400 Hz.

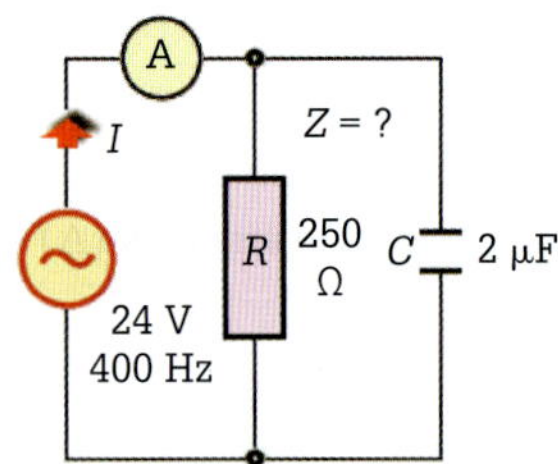

FIGURE 6.110 *R* and *C* in parallel

$$X_C = \frac{1}{2\pi fC}$$
$$= \frac{1}{2 \times 3.142 \times 4 \times 10^2 \times 2 \times 10^{-6}}$$
$$= \frac{1}{50.272 \times 10^{-4}}$$
$$= \mathbf{198.92\ \Omega}$$
$$I_C = \frac{V}{X_C}$$
$$= \frac{24}{198.92}$$
$$= \mathbf{120.65\ mA}$$
$$I_R = \frac{V}{R}$$
$$= \frac{24}{250}$$
$$= \mathbf{96\ mA}$$
$$I = \sqrt{I_R^2 + I_C^2}$$
$$= \sqrt{96^2 + 120.65^2}$$
$$= \mathbf{154.183\ mA\ leading}$$
$$Z = \frac{V}{I}$$
$$= \frac{24}{154.183 \times 10^{-3}}$$
$$= \mathbf{155.66\ \Omega}$$

EXERCISE 6.38

a Calculate the total current in a parallel RC circuit when the current in the resistive branch is 4.8 A and the current drawn by the capacitive branch is 6.4 A.

b A 16 µF capacitor and a 100 Ω resistor are connected in parallel to a 230 V 50 Hz supply. Determine the current drawn by the capacitor, resistor and total current drawn from the supply.

c When a resistor is connected in parallel with a 100 µF capacitor to a 50 Hz 120 V supply the combination has an impedance of 90 Ω. Determine the value of the resistance.

Phase difference in parallel RC circuits

The angle by which the current leads the applied voltage in an RC circuit is called the phase difference (symbol ϕ). This angle is calculated by applying the trigonometry ratios of sine, cosine or tangent to the phasor diagram.

To determine the phase difference from the values of resistance and capacitive reactance the following tangent equation is used:

$$\tan\phi = \frac{I_C}{I_R}$$

where ϕ = phase difference in degrees (°)
I_C = capacitive reactance in ohms (Ω)
I_R = resistance in ohms (Ω)

EXAMPLE 6.40

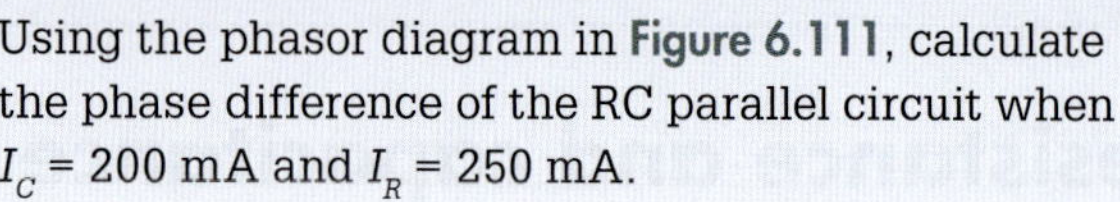

Using the phasor diagram in **Figure 6.111**, calculate the phase difference of the RC parallel circuit when I_C = 200 mA and I_R = 250 mA.

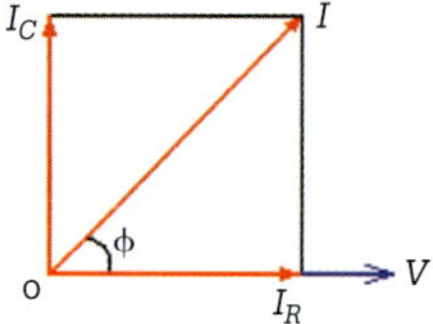

FIGURE 6.111 Parallel RC circuit

$$\tan\phi = \frac{I_C}{I_R}$$
$$= \frac{200}{250}$$
$$= 0.80 \qquad \phi = \mathbf{38.6^\circ\ leading}$$

EXERCISE 6.39

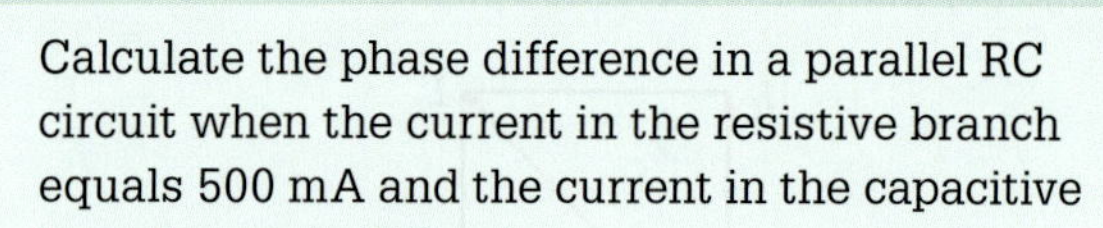

a Calculate the phase difference in a parallel RC circuit when the current in the resistive branch equals 500 mA and the current in the capacitive branch equals 200 mA.

b What is the phase difference in a parallel RC circuit when the current in the resistive branch equals 500 mA and the current in the capacitive branch equals 500 mA?

Power in RC circuits

See true power, reactive power and apparent power in the 'Power in RL circuits' section earlier in the chapter.

Resistance, inductance and capacitance in parallel

Figure 6.112 shows a circuit containing resistance, inductance and capacitance in parallel in which the phasor sum of the currents drawn by the resistance (I_R), the inductance (I_L) and the capacitance (I_C) equals the total circuit current (I).

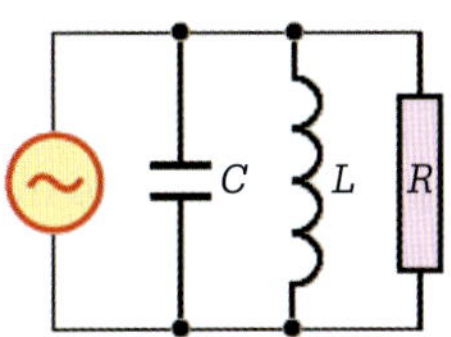

FIGURE 6.112 Resistance, inductance and capacitance in parallel

The phasor diagram of **Figure 6.113** represents the currents in the circuit in relation to the applied voltage (V) which is the reference phasor. The voltage (V) is in phase with I_R, but I_C leads it by 90° and I_L lags it by 90°. The total current (I) is determined by resolving I_L and I_C and then ($I_L - I_C$) or ($I_C - I_L$) and I_R.

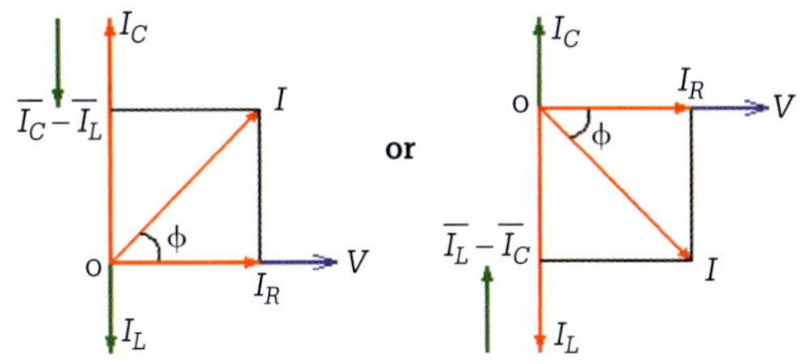

FIGURE 6.113 Current phasors for a parallel RLC circuit

The currents in a parallel RLC circuit are expressed as follows:

$$I_R = \frac{V}{R} \qquad I_L = \frac{V}{X_L} \qquad I_C = \frac{V}{X_C}$$

$$I = \sqrt{I_R^2 + (I_L - I_C)^2} \quad \text{or} \quad I = \sqrt{I_R^2 + (I_C - I_L)^2}$$

where I_R = current through the resistive component in amperes (A)
I = circuit current in amperes (A)
R = resistance in ohms (Ω)
I_L = current through the inductive component in amperes (A)
X_L = inductive reactance in ohms (Ω)
I_C = current through the capacitive component in amperes (A)
X_C = capacitive reactance in ohms (Ω)
V = applied voltage in volts (V)

EXAMPLE 6.41

A parallel RLC circuit with an applied voltage of 16 V 50 Hz has an inductive reactance of 1600 Ω and a capacitance of 10 μF. If the current through the resistor is 80 mA, calculate the currents drawn by the circuit and complete the phasor diagram (see **Figure 6.114**).

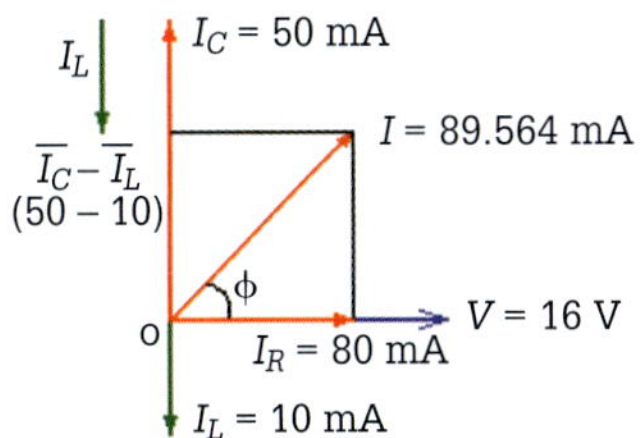

FIGURE 6.114 Current phasors for a parallel RLC circuit

$$I_L = \frac{V}{X_L} = \frac{16}{1600} = \mathbf{10\ mA}$$

$$X_C = \frac{1}{2\pi fC} = \frac{1}{2 \times 3.142 \times 50 \times 1.0 \times 10^{-5}} = \mathbf{318.27\ \Omega}$$

$$I_C = \frac{V}{X_C} = \frac{16}{318.27} = \mathbf{50.27\ mA}$$

$$I = \sqrt{I_R^2 + (I_C - I_L)^2} = \sqrt{80^2 + (50.27 - 10)^2} = \mathbf{89.564\ mA}$$

EXERCISE 6.40

a A parallel RLC circuit has the currents I_R = 3 A, I_L = 4 A and I_C = 6 A. Calculate the total current.

b The total current drawn by a parallel RLC circuit is 5 A. Calculate the current in the resistive branch given that the current drawn by the inductive branch is 2 A and the capacitive branch is 1 A.

c In a parallel RLC circuit the resistive current is 160 mA, the inductive current is 100 mA and the capacitive current is 400 mA. Complete a phasor diagram and find the total circuit current.

Impedance – parallel circuits

The impedance triangle that was used to evaluate impedance in a series circuit cannot be used with a parallel circuit because the phasors that represent current are inversely proportional to resistance, reactance and impedance.

Phase difference in a parallel RLC circuit

The phase difference (symbol ϕ) is the angle by which the current drawn from the supply lags or leads the applied voltage in a parallel RLC circuit. The phase difference is calculated by applying the trigonometry ratios of sine, cosine or tangent to the voltage phasor diagram or the impedance triangle. In many problems the tangent ratio is used, for example:

$$\tan\phi = \frac{I_L - I_C}{I_R} \quad \text{or} \quad \tan\phi = \frac{I_C - I_L}{I_R} \quad \text{or} \quad \cos\phi = \frac{I_R}{I_T}$$

»

EXAMPLE 6.42

Calculate the phase difference illustrated by the RLC impedance triangle in **Figure 6.115**.

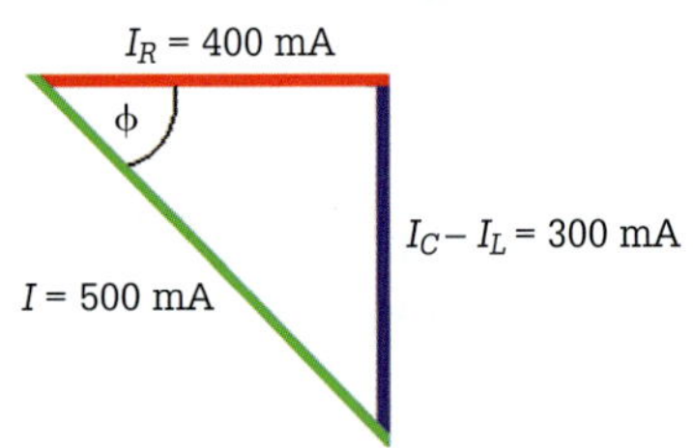

FIGURE 6.115 RLC impedance triangle

$$\tan\phi = \frac{I_C - I_L}{I_R} = \frac{300}{400} = 0.75 \qquad \phi = \mathbf{37^\circ}\text{ lagging}$$

EXERCISE 6.41

a Calculate the phase difference in a parallel circuit when the current in the resistive branch is 10 A, the current in the inductive branch is 20 A and the current in the capacitive branch is 15 A.

b A 200 Ω resistor, an inductor with a reactance of 180 Ω and a capacitor with a reactance of 100 Ω are connected in parallel to a 120 V supply. Calculate the phase difference in the circuit.

c Calculate the phase difference in a parallel circuit when the current in the resistive branch is 150 mA, the current in the inductive branch 300 mA and the current in the capacitive branch is 200 mA.

Power in RLC parallel circuits

REVIEW QUESTIONS

1 If the inductive reactance is greater than the capacitive reactance in an RLC circuit, then what is the circuit seen as?

2 Phasor diagrams for parallel circuits are constructed by using the supply voltage as ____________________.

3 How is it that the impedance triangle that was used to evaluate impedance in a series circuit cannot be used with a parallel circuit?

4 A series RLC circuit has the voltages V_R = 50 V, V_L = 30 V and V_C = 40 V across the circuit components. Calculate the applied voltage.

5 In a series RLC circuit, determine the impedance when the resistance is 100 Ω, the inductive reactance is 50 Ω and the capacitive reactance is 80 Ω.

6 Calculate the phase difference in a series circuit where the resistance has a value of 50 Ω, the inductive reactance a value of 100 Ω and the capacitive reactance a value of 150 Ω.

7 A circuit contains a resistor and an inductor connected in parallel to the output terminals of an alternator. If the current in the resistive branch is 20 A and 35 A in the inductive branch, calculate the total current in the circuit.

8 What is the phase difference in a parallel RL circuit when the current in each branch equals 5 A?

9 Calculate the total current in a parallel RC circuit when the current in the resistive branch is 4 A and the current drawn by the capacitive branch is 6 A.

10 A parallel RLC circuit has the currents I_R = 5 A, I_L = 3 A and I_C = 6 A. Calculate the total current.

6.6 Power in an a.c. circuit

Reactive circuit loads such as inductors and capacitors acquire some of the energy delivered to the circuit. This energy is then returned by the inductor and capacitor, giving the impression that they consume energy. This impression results from the fact that reactive loads do draw current from the supply and experience a voltage drop.

The power description for reactive loads is reactive power. The unit of measurement is volt-amperes reactive (VA_R). Reactive power is the product of the reactive current and the voltage across the reactive loads and is a function of the load's reactance (X_L or X_C).

Some of the energy delivered to an alternating current circuit dissipates heat due to the circuit's ohmic resistance. The energy consumed is the actual amount of power dissipated by the circuit. For this reason, the energy consumed is 'true power' and is measured in watts (W). In addition, the true power is equal to the product of the circuit current and the voltage across the circuit.

In comparison, apparent power is the power that appears supplied to the circuit and is the phasor sum of both the reactive and true power. The unit of measurement is volt-amperes (VA). Apparent power is the product of supply current and the voltage across the circuit and is a function of the circuit's impedance (Z). **Table 6.5** summarises the power equations.

TABLE 6.5 Calculating power

Power type	Equation
True power	$P = I^2R \quad P = \dfrac{V^2}{R} \quad P = VI\cos\phi$
Reactive power	$Q = I^2X_L$ or I^2X_C $Q = \dfrac{V^2}{X_L}$ or $\dfrac{V^2}{X_c}$ $Q = VI\sin\phi$
Apparent power	$S = I^2Z \quad S = \dfrac{V^2}{Z} \quad S = VI$

where P = true power in watts (W)

Q = reactive power in volt-amperes reactive (VA_R)

S = apparent power in volt-amperes (VA)

The three types of power – true, reactive and apparent – demonstrate a relationship to one another in a right-angled triangle called the power triangle. The power triangle shows the power relationships in any alternating current circuit. Refer to **Figure 6.116**.

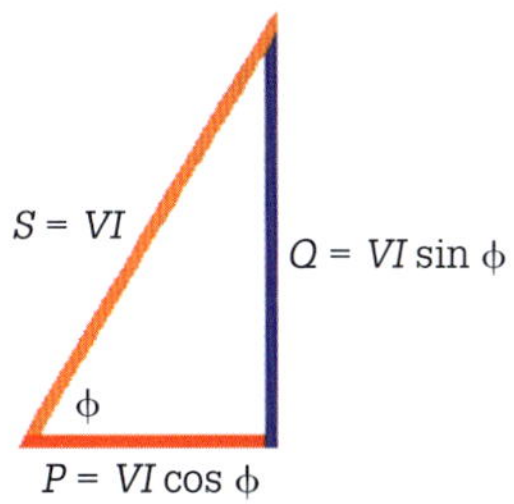

FIGURE 6.116 Power triangle for an inductive load

By applying the laws of trigonometry we can solve the value of any side, given the value of the other two sides or the value of one side and the angle ϕ.

$$P = \sqrt{S^2 - Q^2} \quad S = \sqrt{P^2 + Q^2} \quad Q = \sqrt{S^2 - P^2}$$

EXAMPLE 6.43

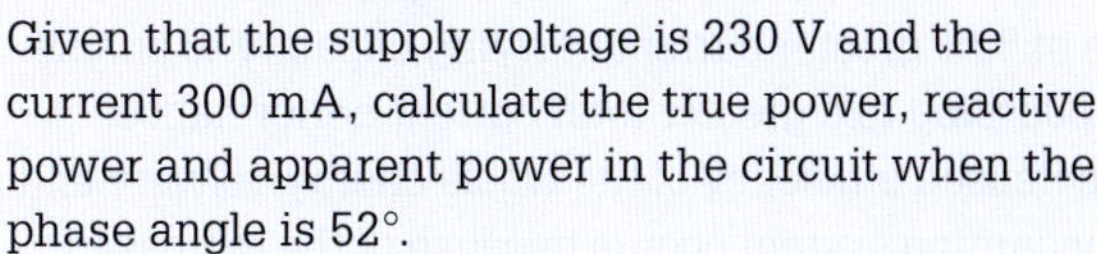

Given that the supply voltage is 230 V and the current 300 mA, calculate the true power, reactive power and apparent power in the circuit when the phase angle is 52°.

cos 52° = 0.616 and sin 52° = 0.788

$P = VI\cos\phi$

$= 230 \times 0.3 \times 0.616$

$= \mathbf{42.5\ W}$

$Q = VI\sin\phi$

$= 230 \times 0.3 \times 0.788$

$= \mathbf{54.37\ VA_R}$

$S = VI$

$= 230 \times 0.3$

$= \mathbf{69\ VA}$

EXERCISE 6.42

a A single-phase 230 V system delivers 15 A to an inductive load at a phase angle of 65°. Calculate:
   i the true power dissipated by the inductive load
   ii the reactive power used to establish the electromagnetic field
   iii the apparent power drawn by the single-phase system.

b A 230 V single-phase circuit delivers 2 A at a phase angle of 40° to a fluorescent lamp load. Calculate:
   i the true power dissipated by the inductive load
   ii the reactive power used to establish the electromagnetic field
   iii the apparent power drawn by the single-phase system.

Power factor

The determination of the power in a direct current circuit is a straightforward matter since it is only necessary to multiply the current and voltage to obtain true power measured in watts. With alternating circuits, this holds true only when the current is in phase with the applied voltage. When the current is not in phase with the voltage, multiply the product of current and voltage by a coefficient called the power factor (symbol pf) to obtain the true power. Power factor can be leading or lagging depending upon the type of load (inductive or capacitive) connected.

Power factor is a number quantifying the relationship between apparent power and real power as shown for the inductive load in **Figure 6.117** and is a measure of the power loss in a circuit.

$$\text{pf} = \frac{\text{true power } (P)}{\text{apparent power } (S)} = \frac{\text{kW}}{\text{kVA}} = \text{cos of the angle}(\phi)$$

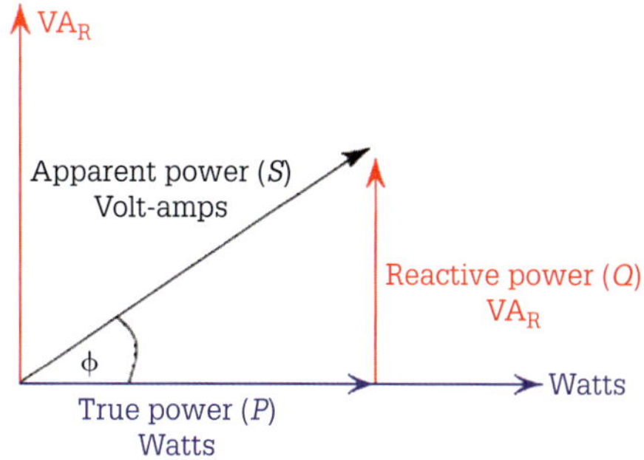

FIGURE 6.117 Power factor for an inductive load

In trigonometrical terms, the power factor is equal to the cosine of the angle (ϕ) established between the true power and the apparent power. This suggests that the power factor scope occurs as decimal values between zero to one inclusive. **Figure 6.117** shows how the calculation of reactive power (Q) due to the inductive reactance X_L of the inductive load occurs.

$$Q = I^2X_L \text{ or } \frac{V^2}{X_L} \text{ or } VI\sin\phi$$

Requirements of supply authorities

The major application of power factor correction is in reducing the maximum demand of apparent power (kVA) delivered to an installation as measured by the supply authority. There is also a sustainability benefit, as a result of more efficient electricity use.

Supply authorities require that the power factor of an installation measured at the authority's metering position at full load remain above 0.8 in some states and 0.9 in others. If the power factor is less than this, corrective measures must be undertaken. The reason for this follows. A lightly loaded motor, as illustrated in **Figure 6.118**, draws a current of 5 A at a supply voltage of 230 V at a power factor of 0.5.

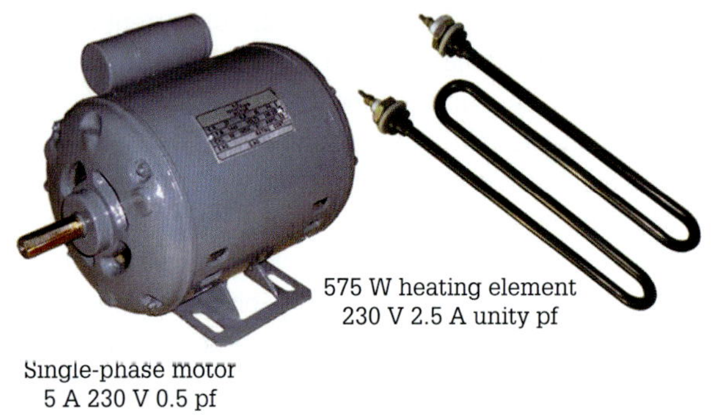

FIGURE 6.118 Comparison of motor and heater power and pf

The power dissipated by this motor is 575 W not the 5 A × 230 V = 1150 W you would expect. Furthermore, if we compare the power dissipated with a 575 W heating element we will find that the current drawn from the supply will be 2.5 A (pf = unity) because V_R and I_R are in phase. The current drawn is half that of the motor yet the power dissipated is the same. A power factor of 0.5 lagging means that only half the current is available to do the actual work. The other half provides energy for the electromagnetic field and does not contribute to real work done.

Higher currents drawn from the supply mean that there is a greater voltage drop occurring across the supply and consequently, power (I^2R) losses from the supply conductors. It follows that this is an investment loss for supply authorities that would prefer to have a system power factor of unity.

Effects of low power factor

A low power factor degrades the level of power quality and efficiency. **Figure 6.119** illustrates the effect on power utilised by a load system as the power factor becomes lower.

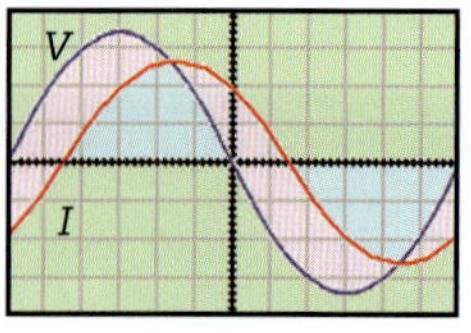

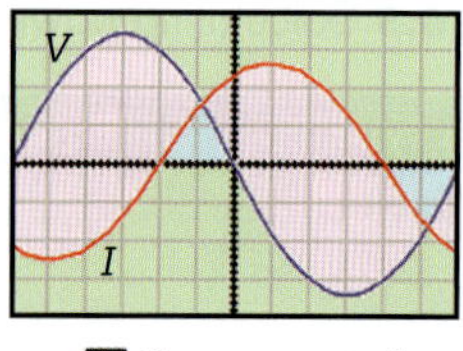

FIGURE 6.119 Effects of low power factor

In **Figure 6.119** the cosine of a 45° angle between V and I is 0.7071. Therefore, the power factor is 0.7071. This means the circuit has used approximately 71% of the energy supplied by the source and wasted approximately 29%.

The cosine of a 120° angle between V and I is −0.8142 (second quadrant). Consequently, the power factor is −0.8142. This means the circuit has used approximately 18.5% of the energy supplied by the source and wasted approximately 81.5%.

Low power factor loads cause a larger current to flow indicating poor utilisation of electric power. The result is additional heating of the load. Consequently, the load and its conductors incur damage. In addition, a low power factor causes low-voltage conditions, resulting in 'brown outs' (dimming of lights) and sluggish motor operation. By contrast, a high power factor states that the electrical power is being utilised efficiently. It follows that a low power factor causes a larger apparent power drawn from the distribution system. This implies that the lower the power factor, the higher are the power losses in the system.

A low power factor in an installation is due to the magnetising currents (I_{reactive}) taken by consuming devices, such as induction motors, discharge lighting and transformers. If these consuming devices are under-loaded, an immense wasteful lagging reactive power develops. In the example of **Figure 6.119**, the effects of low power factor result in many unnecessary energy and system inefficiencies of which the following are some of the primary examples:

- Relatively large and costly electrical equipment including alternators, switchgear, transformers and cables, which in the example in **Figure 6.119** have to be of 50% greater current carrying capacity to enable processing to a greater total current performance.
- Decreased efficiency of the whole energy delivery and consuming system occurs because of the increased copper losses. Copper losses for a given load are inversely proportional to the square of the power factor.

$$\text{efficiency} = \frac{\text{power in} - \text{losses}}{\text{power in}} \times 100$$

- Accordingly, any power factor improvement results in increased efficiency of the system.

Finally, a low power factor causes a significant voltage drop across the consuming devices as well as a reduction in the line voltage. The reduction in voltage results in the overheating of consuming devices, reduction in the kilowatt rating of induction motors, together with a decline in their starting torque and unsteady speed characteristics. Moreover, a 0.5 power factor results in 20% energy loss in the system when the measured voltage drop is 5%.

$$\text{For a 0.5 pf at a 5\% voltage drop} = \frac{5}{0.5^2} = 20\% \text{ energy loss}$$

$$\text{For a 0.95 pf at a 1\% voltage drop} = \frac{1}{0.95^2} = 1.11\% \text{ energy loss}$$

The power factor of an induction motor is at maximum when the motor is running at full load.

Low or under-voltages

Low or under-voltages in the short term prevent the electrical equipment or machines of an installation from operating, or from operating correctly.

Typical symptoms include:

- the 'hold-in' ability of contactors and relays is affected causing equipment to shut down
- a drop in speed occurs with electric motors
- protection devices are operating because of the increased load current required by equipment and machines trying to meet their rated power at a lower voltage.

REVIEW QUESTIONS

1 Why do reactive loads give the impression that they consume energy?

2 The term 'VA' refers to what kind of power?

3 An industrial circuit draws 80 A at a phase angle of 65° from a 230 V supply. Calculate:
   a the true power dissipated by the inductive load
   b the reactive power used to establish the electromagnetic field
   c the apparent power drawn by the industrial circuit.

4 What is the quantity of measurement for true power?

5 What value of power factor is the minimum required by supply authorities?

6 Name some effects of low power factor on an electrical system.

7 What is the percentage energy loss in a circuit with a power factor of 0.65 at a 5% voltage drop?

8 Under what condition is the power factor of an induction motor at maximum?

6.7 Power factor improvement

The reactive component ($I_{reactive}$) dissipates no power; however, this current transmits along the distribution lines and wastes energy because of the resistive components in the distribution system. These resistive components include the cables themselves, the contacts of the relays, and inductive loads such as induction motors, a transformer within an installation and lighting circuits without pf correction capacitors. When using power factor correction, the distribution system provides less power, resulting in lower energy losses and improved transmission on the distribution lines.

Power factor correction occurs by the addition of capacitors in parallel with the associated motor circuits and they can connect at the starter or the switchboard. Static pf correction uses capacitors connected and controlled by a motor starter. Bulk correction uses capacitors connected at a switchboard but controlled by individual motor starters. Switched capacitor banks provide bulk correction to the entire installation with control equipment, switching the amount of capacitance to improve the power factor.

Figures 6.120 and **6.121** illustrate the effect of power factor correction for an inductive load, such as a motor.

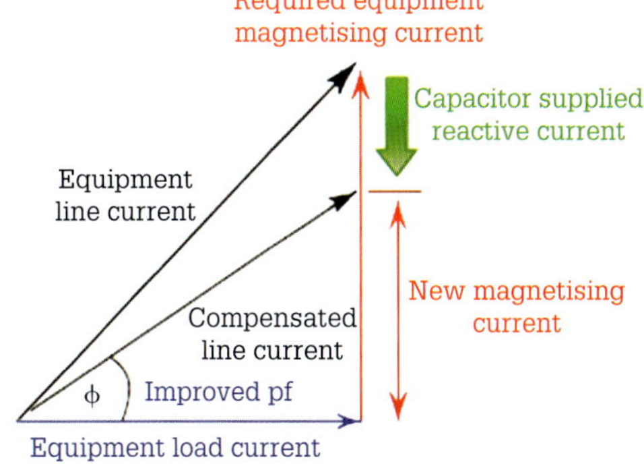

FIGURE 6.120 Power factor correction

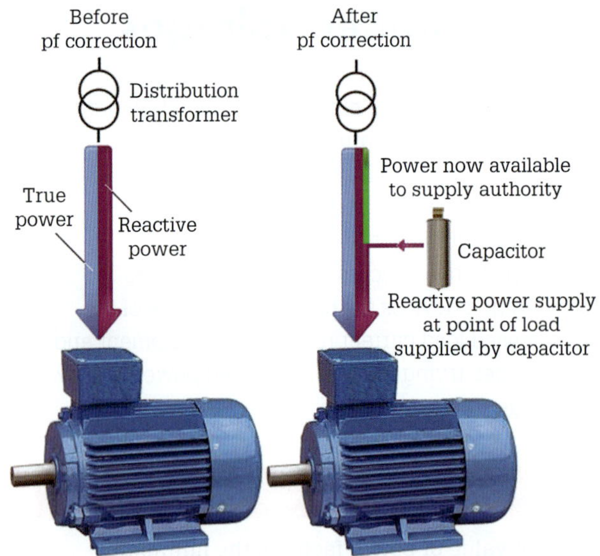

FIGURE 6.121 More power available to the distribution network

Figure 6.120 shows the effectiveness of power factor correction in reducing the equipment line current and the decreases in linked energy losses. The capacitance provided by the correction equipment provides reactive power at the equipment or machine load terminals reducing the power angle (ϕ). The result is a decline in the line current between the power factor equipment and the distribution system. The overall effect is a decreased electrical load as seen by the distribution system illustrated in **Figure 6.121**, with more power now available to the supply authority.

Note that while the line current between the distribution network and the correction device reduces, the load current between the correction device and the load remains unaffected.

Compensating for low power factor

Compensating for low power factor correction involves the installation of capacitor banks, fixed capacitors, switched capacitors, static VA_R compensators and synchronous capacitors. All these devices supply reactive power. Consequently, they have the effect of reducing the magnitude of the line current as shown in **Figure 6.120**.

Fixed-value capacitors

In order to correct the effect of low power factor an equal amount of leading reactive power should be connected across the circuit. The simplest method is to hard wire a fixed-value power factor correction capacitor (PFC) across the offending consuming device or circuit. Note these capacitors are rated in kilovars (symbol kVA_R) instead of farads.

Figure 6.122 shows how power correction capacitors with a fixed kVA_R rating improve low power factor.

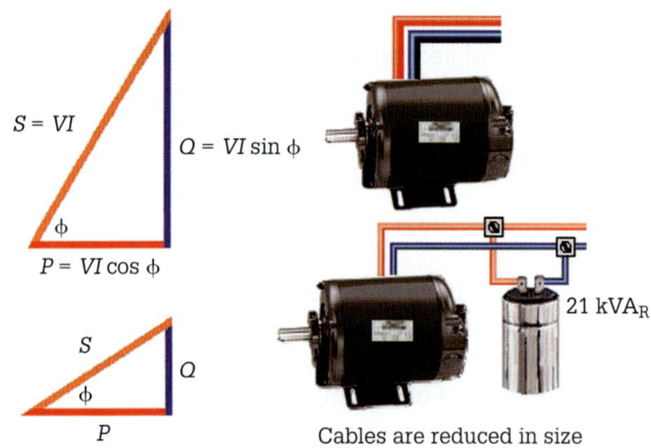

FIGURE 6.122 Power factor improvement

Fixed capacitors deliver a fixed value of reactive power to the electrical system. Fixed capacitors occur with different fluorescent lighting and motor loads. However, they can connect to the main power bus. The characteristics of single-phase pf correction capacitors are:

- self-healing metallised polypropylene dielectric
- biodegradable non-toxic filling
- overpressure safety device.

Self-healing metallised polypropylene dielectric

A self-healing dielectric consists of sheets of metallised polypropylene film. The sheets are formed by a very thin layer of aluminium deposited directly onto plastic film by evaporation under vacuum. In the event of degradation of the dielectric under working conditions, the consequent electric arc produces evaporation of a small area of the aluminium and thus restores the insulation.

This property, known as self-healing, allows the capacitor to continue operation without interruption of the service while retaining the capacitor's properties. The process of self-healing does not damage the polypropylene dielectric, enabling these capacitors to have a very long life with high reliability.

The polypropylene dielectric has very low dielectric losses (typically 0.5% or less of the capacitor's kVA_R rating). Furthermore, this characteristic allows a much lower capacitor operating temperature, with a consequent reduction in ageing through heat.

Biodegradable non-toxic filling

Single-phase power factor correction capacitors impregnated with a non-polychlorinated biphenyl (PCB) oil dielectric solution do not cause pollution of the environment.

Other capacitors using 'dry technology' have an inert gas to eliminate all traces of air and moisture. In addition, the use of an inert gas, as opposed to oil, reduces fire risks caused by leaking oil and avoids damage to sensitive equipment.

Overpressure safety device

The capacitors include an internal pressure disconnection system that adequately protects the capacitors against bursting in case of fault. In the event of an internal failure, caused by a permanent overload or a non-healing dielectric breakdown, the internal rise of pressure would trigger an overpressure break-action mechanism of a safety device.

Protection from stored energy

Power factor correction capacitors fitted with discharge resistors reduce voltage to less than 50 V 60 seconds after disconnection from the power supply.

Switched capacitors

Switched capacitors are ideally suited for correction applications where the load changes in an installation. A typical correction system includes a number of capacitor connection stages. In addition, the number of capacitors required depends on the characteristics and the reactive power requirements of the installation.

To correct the power factor of a load automatically, a number of the capacitors connect between the electrical bus and neutral to give the 'reactive' current required. A microprocessor-based controller continuously monitors the reactive power demand on the supply and controls the capacitors. The controller then offsets the reactive power of the load and power factor correction reduces the overall demand on the supply by maintaining the best power factor possible.

For this reason, switching stages of set amounts of kVA_R occurs. In addition, large set values transpire by cascading a number of smaller kVA_R capacitors. Smaller capacitors have the beneficial effect of reducing the in-rush current to the capacitors thereby minimising supply disturbances.

Nevertheless, a system of power factor correction that adds the capacitors in steps can cause the capacitors to develop over-voltages and deliver high-voltage transients on the electrical system. Consequently, when transients occur, the capacitors are subject to deterioration from overheating. In that case, to overcome heating problems, inductors are included with the switched capacitors to form a filter that reduces voltage transient disturbances. Refer to **Figure 6.123**.

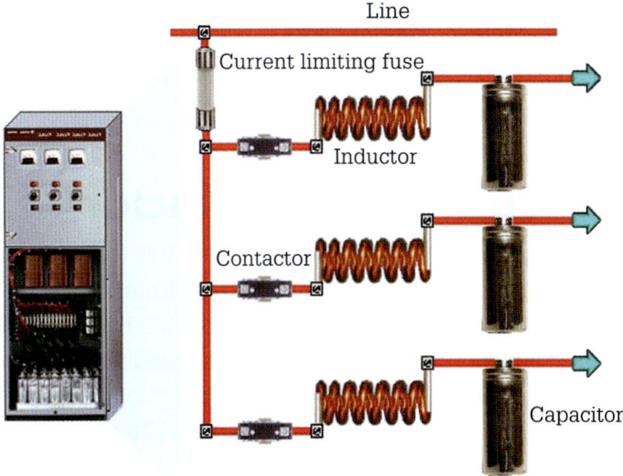

FIGURE 6.123 Switched capacitors

Static VA_R compensator

The static VA_R compensator is a highly developed correction technology, which provides fast response and continuous infinite variability of capacitive and inductive reactive power to the installation's power system.

Static VA_R compensators consist of a combination of fixed shunt-connected capacitor banks and variable thyristor-controlled shunt inductor banks. This system is a silicon-controlled rectifier bridge that either directs the continuously variable VA_Rs from the capacitor bank onto the electrical system or dissipates the VA_Rs across the inductor banks as required to maintain a predetermined system power factor.

This infinitely variable VA_R compensator maintains a constant power factor or maintains a constant busbar voltage. Unlike fixed and electro-mechanically switched capacitor banks, they automatically adjust to the dynamic behaviour of loads to provide the correct amount of VA_R support required each instant, effectively eliminating the reactive power.

In Australia, installation of large static VA_R compensators occurs at Bairnsdale, Victoria, where a 55 MVA_R compensator was required to maintain the supply voltage range of Eastern Energy's 66 kV distribution systems. Prior to installation, the distribution had experienced excessive voltage swings due to seasonal loads.

Installed at Armidale, New South Wales, is another static VA_R compensator. The installation required a 400 MVA_R compensator to stabilise the interconnection of the Queensland and New South Wales power systems at Transgrid's 330 kV substation. Although very expensive, static VA_R compensators find application in electrical systems using large motors and drives. **Figure 6.124** shows an example of a static VA_R compensator.

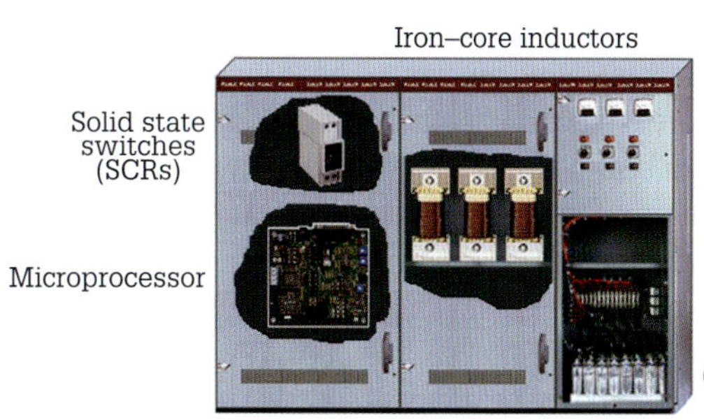

FIGURE 6.124 Static VA_R compensator

Typical applications of static VA_R compensators are:

- **mining:** elevators, crushers, conveyors, shovels, etc.
- **commercial:** elevators in office buildings, etc.
- **steel industries:** arc furnaces, rolling mills, conveyors, cranes, etc.
- **pulp and paper industries:** sawmills, wood chippers, pumps, etc.
- **recreational:** linear induction, motor-controlled amusement rides.

Synchronous capacitors

Instead of using capacitors, the leading reactive power required to compensate for the lagging reactive power system occurs by using synchronous motors. The result is an improvement in the total power factor of the system. Because these motors are rated up to 0.8 pf they can be used to supply leading kVA_Rs to counteract lagging power caused by inductive loads. When used for power factor correction, these motors are synchronous capacitors.

Synchronous capacitors provide stepless automatic power factor correction thereby producing no switching voltage transients as occurs in the switched capacitor system. The synchronous capacitor can match any supply voltage by means of transformers. If there are large, rapid and random swings in load kVA_R, continuous adjustment of the capacitor to compensate can be actioned by changing its field excitation. **Figure 6.125** illustrates a synchronous capacitor.

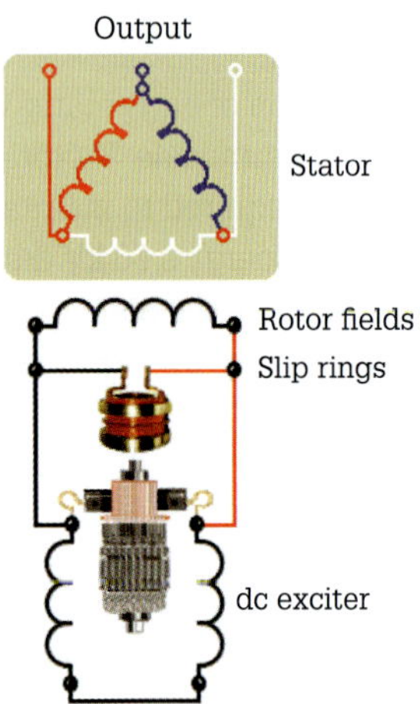

FIGURE 6.125 Synchronous capacitor

When quick response to reactive loads is necessary, synchronous capacitors are used. These motors run on a very light load or on an intermittent load such as a compressor that charges a receiver only two or three times per day and runs on no load for the remainder. Consequently, the light load ensures that the torque angle is at a minimum, and the machine has a pf of unity or very slightly lagging. For a leading pf, the field excitation of the motor need only be increased a few per cent above normal (100%). The machine is then well within its stable operating range, and the magnetic link between rotor and stator is not so stiff as to endanger the shaft and bearings if there is a sudden load change.

Indirectly, power factor correction saves fuel for the prime mover to turn a generator, reduces transformer and transmission losses and improves voltage regulation. An increase for power capacity throughout the distributor's system transpires without additional investment in generators, transformers or distribution network.

EXAMPLE 6.44

A small industrial workshop has a load of 350 kW at a power factor of 0.8. How much capacitor kVA_R is required to improve the power factor to 0.95?

In this example $\tan\phi = \dfrac{kVA_R}{kW}$ and

$kVA_R = kW \times \tan\phi$

The difference between these kVA_R values will be the capacitor kVA_R needed to improve the power factor from 0.8 to 0.95 as shown in **Figure 6.126**.

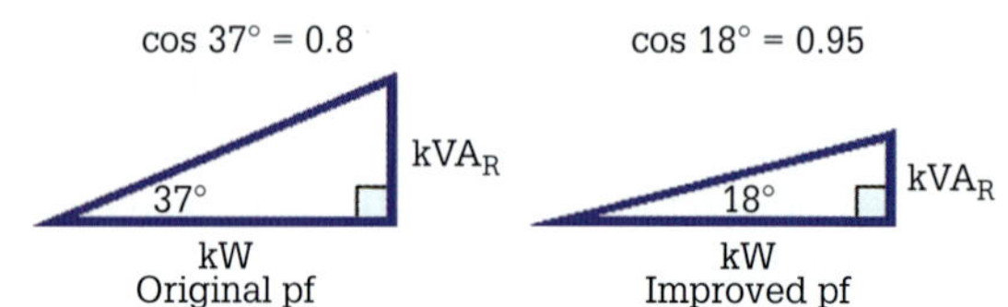

FIGURE 6.126 Improving the power factor

$$\begin{aligned} \text{capacitor } kVA_R \text{ required} &= kW(\tan\phi_1 - \tan\phi_2) \\ &= 350(\tan 37° - \tan 18°) \\ &= 350(0.7536 - 0.3249) \\ &= \mathbf{150.045\ kVA_R} \end{aligned}$$

$\cos 37° = 0.8 \qquad \cos 18° = 0.95$

EXERCISE 6.43

A small industrial workshop has a load of 250 kW at a power factor of 0.7. How much capacitor kVA_R is required to improve the power factor to 0.9?

Electronic lighting ballasts

Soft-start electronic control equipment, such as low-loss electronic ballasts, offer close to unity power factor. The use of these ballasts instead of low-loss iron-core ballasts, particularly when used in domestic and commercial installations, will increase the efficiency of the distribution network.

The installation of power factor correction is a widely recognised way to reduce energy consumption, thereby reducing electricity distribution costs and benefiting the environment.

Measuring power factor

The best method to determine power factor is to measure power, voltage and current, using meters. Enter the results into a calculator so that the power factor calculated can be determined by dividing the power by the product of current and voltage. **Figure 6.127** shows the connections for the meters for a single-phase system. **Figure 6.128** shows a pf meter.

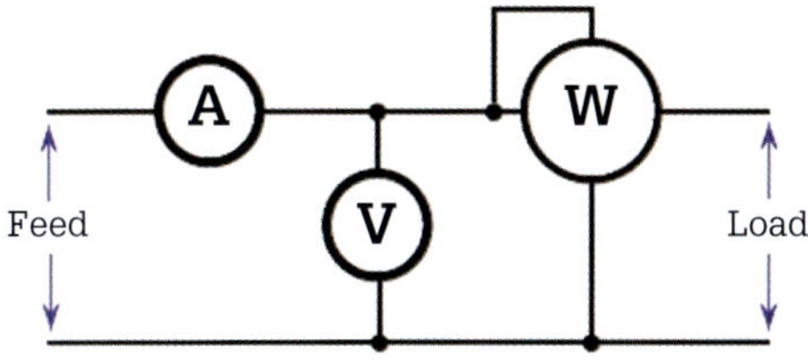

FIGURE 6.127 Connection of instruments to enable calculation of power

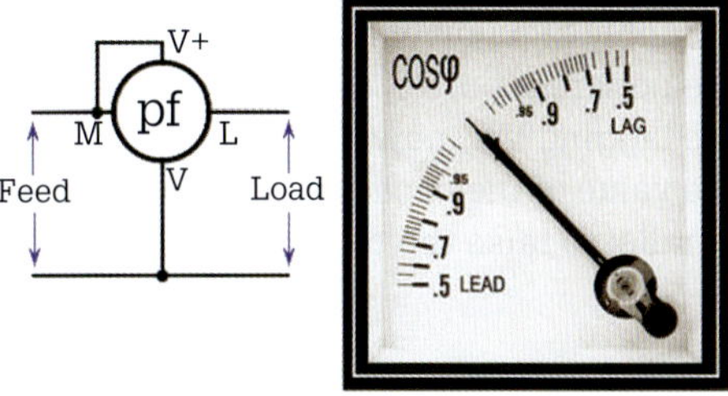

FIGURE 6.128 Connection for single-phase power factor meter

The digital power factor meter shown in **Figure 6.129** uses a system that measures the phase angle between voltage and current at the point (reference point) where each wave (*I* and *V*) passes through zero and calculates the cosine of the angle electronically. However, better digital systems include a detector that determines if the pf is leading or lagging.

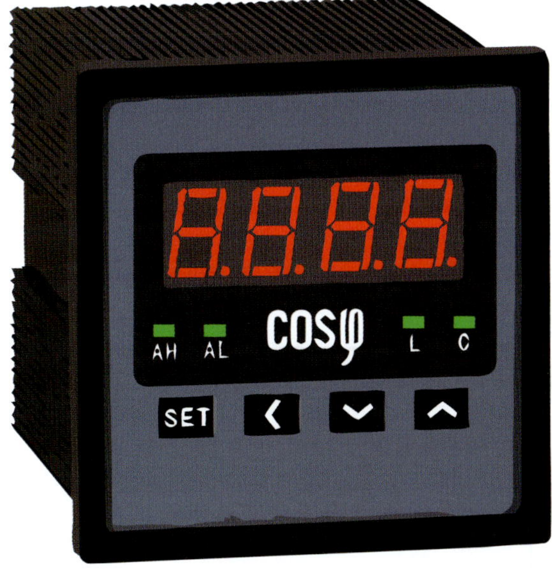

FIGURE 6.129 Digital power factor meter

EXAMPLE 6.45

A 200 V a.c. circuit contains a 40 Ω resistor connected in series with a capacitor of reactance 30 Ω. Calculate the power factor.

$$Z = \sqrt{(R^2 + X_C^2)} = \sqrt{40^2 + 30^2} = 50\ \Omega$$

$$I = \frac{V}{Z} = \frac{200}{50} = 4\ \text{A}$$

$$\text{pf} = \frac{P}{V \times I} = \frac{640}{200 \times 4} = 0.8\ \text{leading}$$

Alternatively, calculate pf from values of true and apparent power.

$$VI = 200 \times 4 = 800\ \text{VA}$$

$$P = VI \cos\phi = VI\frac{R}{Z} = 800 \times \frac{40}{50} = 640\ \text{W}$$

$$\text{pf} = \frac{P}{VI} = \frac{640}{200 \times 4} = 0.8\ \text{leading}$$

EXERCISE 6.44

Instruments connected to a single-phase motor provide the following readings: wattmeter 2000 W, voltmeter 230 V and ammeter 12 A. Using these values calculate the pf of the motor.

AS/NZS 3000: 2018 *Wiring Rules* requirements relating to power factor

Read the following in conjunction with AS/NZS 3000: 2018 *Wiring Rules*.

Clause 2.3.7.2 Functional switching devices. This clause requires these devices to be de-rated when used to control loads with substantial low power factor.

For example, under normal operating conditions certain electrical loads (e.g. induction motors, welding equipment, arc furnaces and fluorescent lighting) take active power and inductive reactive power (kVA_R) from the supply. If a 3 kW load connects to a 230 V 50 Hz, supply, the current drawn for cos ϕ = 1 is 13 A but if a load having the same power rating but operating at cos ϕ = 0.6 would draw a current of 21.7 A. This means that the switch must be 'de-rated' due to the higher switching current.

Clause 4.1.4 Adverse effects and interference. This clause refers to equipment. No adverse effects should occur to other installed equipment when the proposed new equipment is installed.

An element that has an adverse effect, as a result of installing new equipment that needs consideration, is power factor. For example, if proposed equipment is not compatible to the pf of the supply then decreased service life of all equipment occurs.

Table C2 Maximum demand – non-domestic electrical installations. When determining maximum demand in a non-domestic installation that has welders, then column 1H of Table C2, Maximum demand – non-domestic electrical installations, informs that welding machines in accordance with paragraph C2.5.2 must take into account power factor correction. In addition, when using this table refer to C2.5.2.1.

REVIEW QUESTIONS

1. Name two methods available for power factor correction.
2. Name typical industry sectors that employ static VA_R compensators.
3. What do synchronous capacitors provide?
4. What is identified as the 'best method' to determine power factor?
5. An industrial workshop has a load of 500 kW at a power factor of 0.65. What kVA_R rating of capacitor bank is needed to improve the power factor to 0.85?
6. An induction motor draws 1.5 kW from a 230 V 50 Hz supply at a power factor of 0.55. Determine:
   a. the apparent power
   b. the lagging VA_R power
   c. the value of the capacitor kVA_R needed in parallel to raise the total pf to 0.95.

6.8 Harmonics and resonance in a.c. systems

Two conditions that can have an adverse effect on a.c. circuits are system harmonics and resonance, both series and parallel.

System harmonics

Harmonics is the name given to contaminating waveforms that are sinusoidal in shape but are multiples of the fundamental waveform. Harmonics are the result of electronic devices and equipment. Furthermore, these devices and equipment draw current from the supply in terms of short pulses rather than as a smooth sinusoidal growth and decay. The voltage supplied to customers by energy distribution companies is as close as possible to a pure sinusoidal waveform called the fundamental and is described in terms of frequency and amplitude.

Fundamental waveforms

With a fundamental sinusoidal waveform, the frequency of the current is identical to the frequency of the voltage as long as the load impedance does not vary. **Figure 6.130** illustrates a fundamental voltage and current waveform.

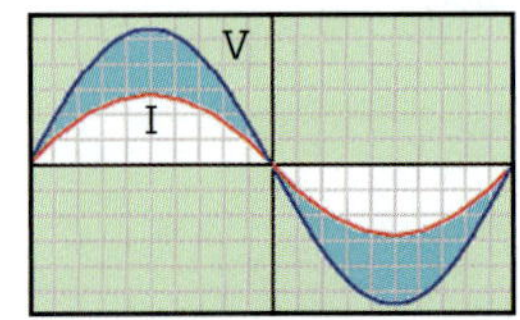

FIGURE 6.130 Fundamental waveform

Linear and non-linear loads

When load impedance does not vary, the load is linear. With linear loads that are resistive, capacitive or inductive the voltage across these loads, and the current drawn by them, will have the same fundamental frequency of 50 Hz. The non-linear load is the cause of harmonics. With non-linear loads, the frequency of the current drawn changes as the load varies from near zero to a maximum value, causing the amplitude of the voltage across the load to change as well. Consequently, a flow of current occurs at a frequency that deviates from the fundamental waveform (50 Hz). **Figure 6.131** illustrates how the current waveform may look in a non-linear load.

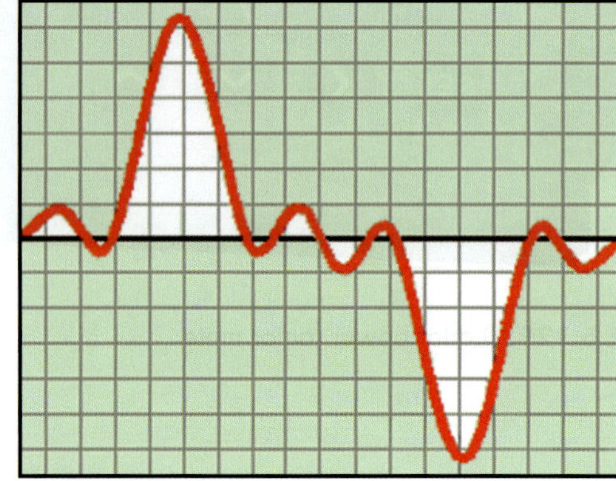

FIGURE 6.131 Current waveform in a non-linear load

This current drawn by the load has a waveform that is a composite of multiple frequencies and is a complex waveform. The deviation of this current waveform from the fundamental waveform occurs multiple times during each cycle of 50 Hz supply, causing harmonic distortion. When working with linear loads we are concerned only with the fundamental waveform of 50 Hz.

The graphic representation of the fundamental waveform is a simple sine *X–Y* graph indicating magnitude and direction with time. Because non-linear loads create from the input sine wave distorted waveforms, containing multiple frequencies, Fourier analysis is used to develop a graphic representation of the resulting waveform.

Fourier analysis

Fourier analysis is a procedure developed by the French mathematician and physicist Jean Baptiste Fourier (1768–1830), and is a means of separating a periodic function or waveform into its essential components, called fundamental and harmonics, by resolving the distorted waveform into a range of sine waves of varying frequencies and amplitudes. It is possible to determine mathematically from a Fourier analysis how much of each harmonic exists in a distorted waveform, and consequently the degree of distortion produced. All distorted waveforms repeat themselves in relationship to some basic frequency such as 50 Hz and the sine wave related to that frequency is the fundamental or 1st harmonic, as shown in **Figures 6.132(a)–(e)**.

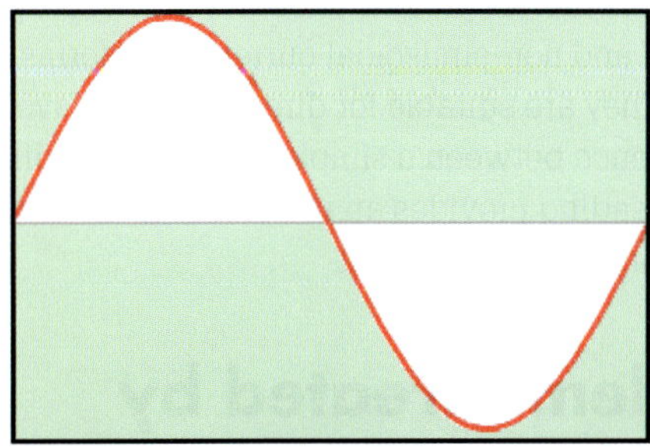

FIGURE 6.132(a) Fundamental, or 1st harmonic, 50 Hz

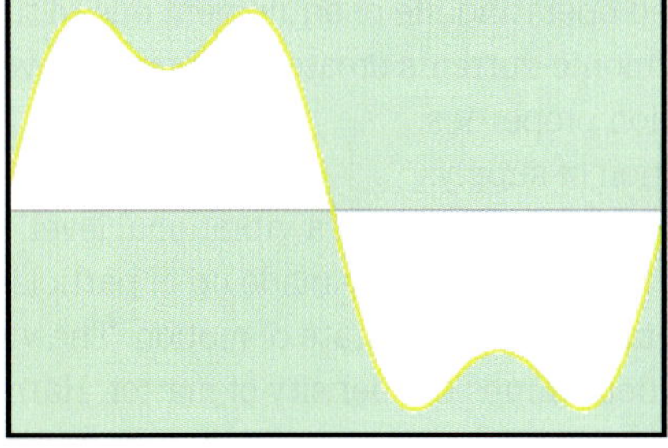

FIGURE 6.132(b) 3rd harmonic, 150 Hz

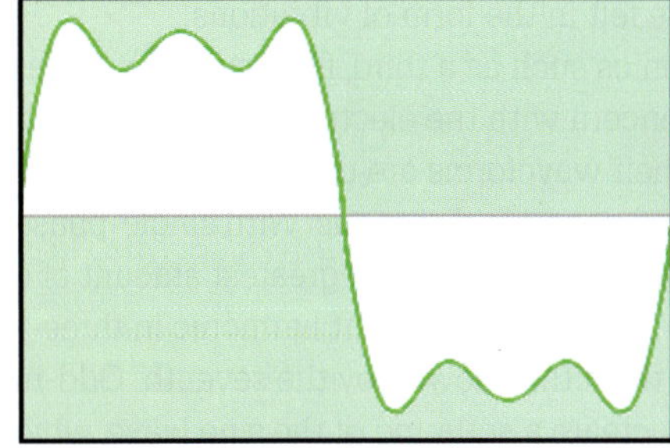

FIGURE 6.132(c) 5th harmonic, 250 Hz

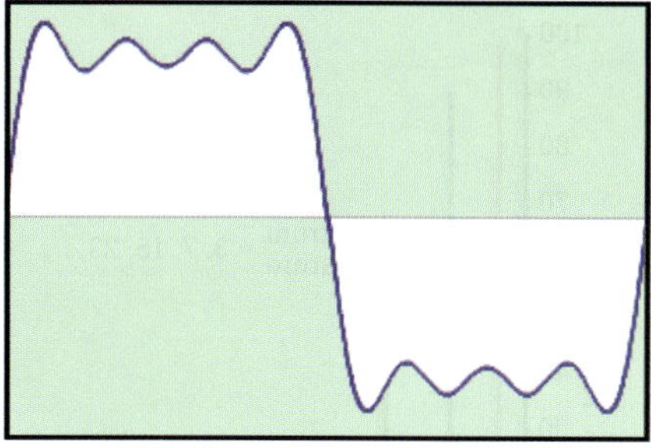

FIGURE 6.132(d) 7th harmonic, 350 Hz

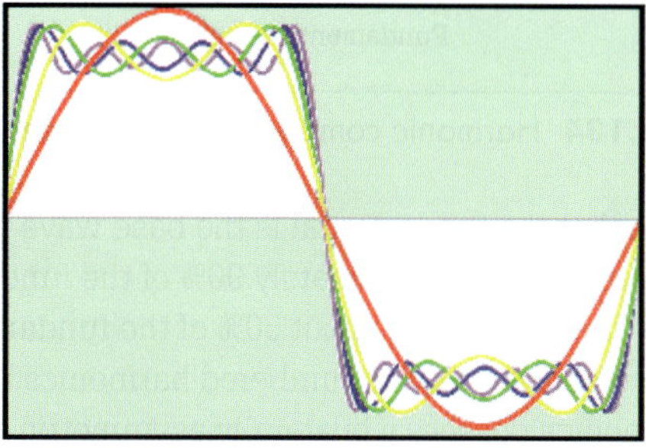

FIGURE 6.132(e) Fundamental plus odd harmonics

Each consecutive sine wave harmonic of a particular distorted waveform has a frequency that is an integer multiple of the fundamental. So, the 2nd harmonic has a frequency of 100 Hz (two times the fundamental), the 3rd is at 150 Hz, the 4th at 200 Hz and so on. Finally, these add up to produce the final waveform as shown in **Figure 6.133**.

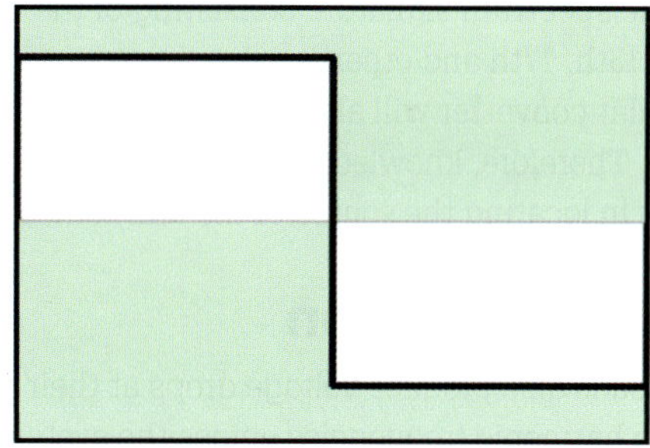

FIGURE 6.133 Final distorted waveform

The primary feature of distortion waveforms is frequency, as shown in **Figures 6.132(a)–(e)** and **Figure 6.133**. Another feature of a distorted waveform is magnitude, which is the harmonic distortion factor (HDF). Each sine wave of the set making up the final distorted waveform may have a different significance from the others. In addition, the size of each harmonic is a proportion of the fundamental.

Consequently, each harmonic component appears as a percentage of the first harmonic or fundamental harmonic. Refer to **Figure 6.134**.

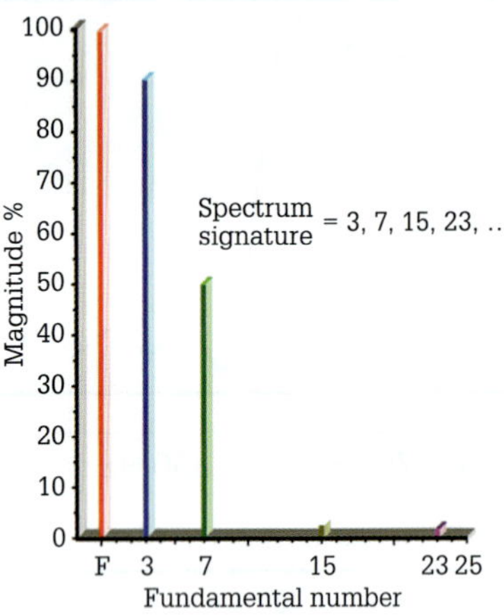

FIGURE 6.134 Harmonic components of a current

Notice that the fundamental is the base wave (F) and the 3rd harmonic is approximately 90% of the fundamental while the 7th harmonic is about 50% of the fundamental. The table displays no even-numbered harmonics because even harmonics possess a bilaterally asymmetric waveform (even-numbered harmonics are mitigated).

The aggregate effect of all harmonics is total harmonic distortion (THD). This result equals the root-mean-square value of the fundamental expressed as a percentage.

All loads generate their own harmonic spectrum signature and similar load types generate compatible spectrum signatures. For example, a three-phase, full-wave, six-pulse (number of d.c. current pulses produced each cycle) silicon-controlled rectifier (SCR) converter has a harmonic spectrum signature consisting of the 1st, 5th, 7th, 11th, 13th, 17th and others.

A similar converter will also carry the same spectrum signature. Therefore, knowledge of spectrum signatures can assist in locating the source of the harmonics.

Voltage distortion

Variable loads also produce voltage drops at their respective harmonic frequencies across the system impedance, distorting the voltage waveform. This voltage drop is a characteristic of the fifth harmonic and can be as much as 1% of average supply voltage. An illustration of a distorted voltage waveform is in **Figure 6.135**.

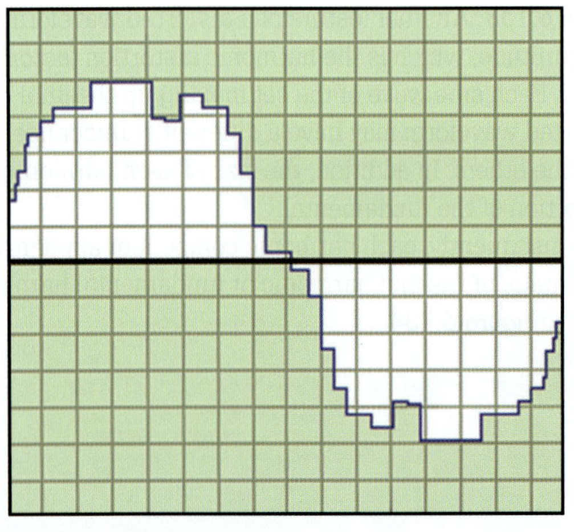

FIGURE 6.135 Voltage waveform across a non-linear load

Electronic loads are susceptible to distortion of their supply voltage waveform and may break down when the distortion is excessive.

The effects of current and voltage distortion differ. Current distortion mainly affects equipment that transmits or conducts energy. On the other hand, voltage distortion is generally an effect of current distortion. It is the result of semiconductors turning on suddenly, allowing the current to rise quickly causing the supply voltage to drop because of changed circuit impedance.

Once the voltage is distorted, all of the current driven by it will be distorted, amplifying the current effects. Voltage distortion affects equipment that consumes energy such as electric motors and electronic-controlled equipment such as computers.

Measuring current in harmonic systems

Standard ampere clamp-on ammeters lack the ability to perform basic electrical troubleshooting for today's electrical systems. The fact is these meters are only sensitive to 50 Hz current. Consequently, a suitable ammeter is one that provides true rms and instantaneous peak current readings.

True rms meters provide accuracy when measuring sinusoidal and non-sinusoidal current waveforms. In addition, they are suitable for direct current waveforms. The difference between a simple sinusoidal reading and a complex reading provides an indication of the magnitude of the harmonic current present.

Problems created by harmonic currents

Harmonic distortion creates three major problems for the whole of the electrical system.

1 Reduced operating life of equipment due to the heat the harmonic currents create that breaks down their insulation properties.
2 Disruption of supply.
3 An exchange of energy on a vibrational level.

Matter is a form of energy made up of particles in the form of atoms that are in a state of motion. The vibrational frequency determines the density of matter. Harmonics modify the matter they are passing through by their vibrational interactions, which lead to higher than normal wear and tear on equipment and machines. This effect can be heard as the affected material emits a buzzing sound and can be felt in the form of vibrations.

Harmonics such as a third, fifth and seventh cause the highest concern with the electrical distribution system because their waveforms are usually uniform in both positive and negative directions. With single-phase circuits, the third harmonic causes the greatest amount of current distortion. The most significant harmonic in three-phase circuits is the fifth, followed by the seventh. Odd-number harmonics create a squaring of the sine wave, which increases power losses due to the d.c. nature of the wave.

Specific electrical equipment or loads that produce harmonics

Non-linear loads in the form of electronic-based equipment have proliferated since the late 1980s and now generate significant harmonics in the power system. In addition, many of these loads rely on an internal d.c. power source to drive their operation.

Discharge lighting

As ballasts are non-linear inductors – they can be the source of harmonics. Furthermore, electronic ballasts can have very significant harmonic outputs in the order of 30% of the fundamental. The assessment is that these ballasts operate like switching power supplies. Single-phase lighting circuits that use magnetic or electronic controls are a primary contributor to harmonic currents flowing in a distribution system.

Arc furnaces

Electric arc furnaces draw a harmonic influenced current during the melting process. Consequently, the voltage waveform is affected.

Power factor correction capacitors

Power factor correction capacitors do not generate harmonics themselves but provide circuit connections for possible resonant conditions. If these conditions occur at a frequency close to a harmonic frequency already present in the load or system voltage or current, large voltages or currents occur at that frequency. These large harmonic currents circulate between the supply network and the capacitor equipment. In addition, these currents create further harmonic voltage disturbances. The effect is a higher voltage across the capacitor exceeding the maximum voltage rating of the capacitor. The result is dielectric failure. As capacitors offer less impedance to these higher frequency harmonic voltages, there are resultant harmonic currents in the power factor correction capacitors. These capacitors have to be able to absorb the additional harmonic energy generated and may be forced to conduct up to twice as much current as they normally would. Consequently, the wiring and any other components connected in relation to the capacitor need to be rated correctly in order to handle the currents involved.

Electronic appliances and equipment

The timing circuits of electronic appliances and equipment vary because of multiple zero crossings (turning on and off) of the voltage waveform caused by voltage distortion. Personal computer switched mode power supplies are the largest contributor of harmonics within a commercial environment, creating high variations in power quality.

Telephones

Harmonics can create interference called noise on telephone circuits due to electromagnetic induction coupling between the power circuits of an installation carrying harmonics and the telephone circuit.

Motors

These are linear loads but are affected when their voltage supply creates harmonics causing the motor to draw harmonic current. Harmonic currents increase I^2Z heat losses in electric motors and distribution systems. Induction motor impedance is frequency dependent. Therefore, harmonics affect impedance. The main harmonic frequency that causes harm to induction motors is the fifth harmonic. Consequently, the fifth harmonic results in an impedance that is five times the value caused by the fundamental frequency. More to the point, every ampere of fifth harmonic current develops five times as much heating as an ampere of fundamental current.

In addition, the fifth harmonic can cause significant voltage problems for three-phase motors. The fifth is a negative sequence harmonic that produces a reverse-torque in induction motors slowing down their rotation. Consequently, the motor tries to overcome the reverse torque by drawing more current. The additional current can create two effects. Either protection devices operate or the motor develops a higher temperature resulting in reduced motor life.

Skin effect

Because harmonic frequencies are higher than the fundamental frequency, 'skin effect' becomes a factor. The skin effect is a conductor condition where the high frequency causes the electrons to flow in the outer surface of the conductor, effectively reducing its cross-sectional area. A reduction in cross-sectional area (CSA) alters the ohmic resistance of the conductor, thereby reducing the current-carrying capacity of the conductor.

Skin effect increases as the frequency and magnitude of the harmonic increases, resulting in a thermally stressed conductor (e.g. copper). In addition, the determination of the depth of skin effect in copper uses the following formula:

$$\text{skin depth of copper} = \frac{66.04}{\sqrt{f}}$$

where skin depth = depth in millimetres (mm)

f = frequency in hertz (Hz)

With circular conductors, the depth of the skin effect can vary from 9.34 mm for a circular copper conductor at 50 Hz to 2.95 mm at 500 Hz. A graph applying the above formula showing the depth of skin effect from the fundamental (50 Hz) to the 10th harmonic (500 Hz) is illustrated in **Figures 6.136(a)** and **(b)**.

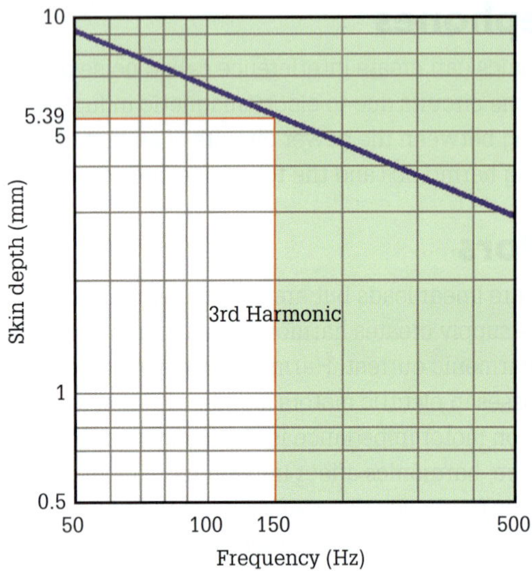

FIGURE 6.136(A) Graph of skin depth versus frequency

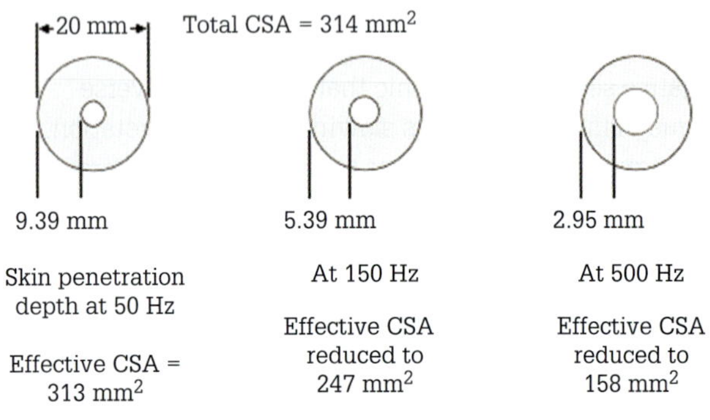

FIGURE 6.136(B) Conductor cross-sectional area (CSA) and skin depth

Figure 6.136(b) informs us that conductors of up to 20 mm in diameter or less are unaffected at 50 Hz as the conducting skin penetrates to the centre. However, if the 3rd harmonic is present (150 Hz), the skin penetration is only 5.39 mm and provides only 78% of the original CSA for conduction. An 11 mm diameter conductor, however, would also be suitable as the conducting skin penetrates almost to its centre. So, a conductor having a diameter of up to three times the skin depth at the harmonic frequency still possesses adequate current-carrying capacity.

SWITCH ON

Techniques used to reduce the effects of harmonics:

1. By using oversize neutral conductors rated at 173% of the current of the phase conductors.
2. Isolation transformers that feed harmonic-producing loads and switch mode power supplies should only be loaded to 50% of their kVA rating. The consequence is an installation not used to its full capacity.
3. Replace transformers with K-rated transformers. K transformers have unique magnetic cores and coil windings designed to accommodate harmonics. The K factor is an index of the transformer's ability to handle non-linear load current without abnormal heating occurring: the higher the K factor, the better the transformer's ability to handle harmonic currents.
4. Replace power factor capacitors with detuned filter banks. These banks consist of reactors connected in series with capacitors of the correct capacitance (kVA_R) to reduce the effect of resonance below the critical order (5th, 7th, 11th and 13th) harmonics.
5. Where power factor is not a problem, harmonic filters are used. The filter is a capacitor in series with an inductor tuned to a particular harmonic frequency. Consequently, the impedance of the filter is zero at that frequency, enabling the filter to absorb the harmonic current.
6. Harmonic compensators reduce harmonic voltages. These devices are a large inverter that cancels out the harmonics. VA_R compensators use power-electronic switching to provide variable amounts of reactive compensation to absorb the harmonic current on a cycle-by-cycle basis. In addition, they continuously monitor load conditions and are able to match their output to the requirement of the load.

Resonance in a.c. circuits

The quantity 'resonance' is also termed 'natural frequency'. It is an electrical property possessed by some series and parallel connected a.c. circuits. Whenever the characteristics of inductance, capacitance and resistance are found in a circuit, they may create the phenomenon resonance: a resonant circuit that resonates at one frequency. The resonance of a circuit represents the frequency at which electrical energy is oscillating back and forth between a capacitor and an inductor.

Within the parallel resonant circuit, the energy of the circuit is stored as electromagnetic energy in the inductor and then a quarter of a cycle later as electrostatic energy in the capacitor. Because of the 'ohmic' resistance of the circuit this energy that is oscillating back and forth between the inductor and the capacitor gradually declines as it is wasted as heat. This effect is illustrated in **Figure 6.137**.

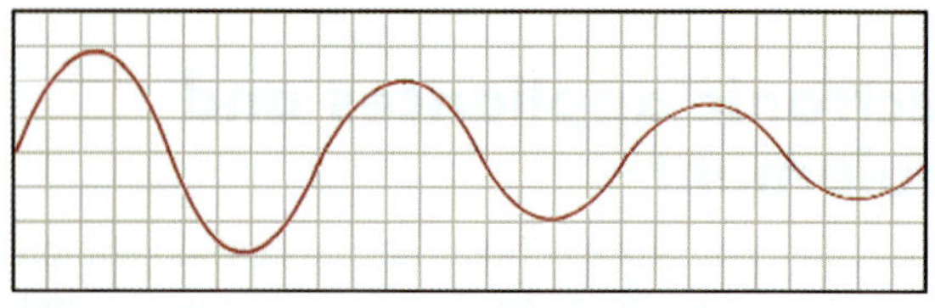

FIGURE 6.137 Damped oscillations in an RLC circuit

Circuits that are resonant at a given frequency are called tuned circuits. They may also be called filters, because they are used to 'filter' one set of frequencies out from all the others within a given band of frequencies.

Resonance in series circuits

A series circuit that contains resistance, inductance and capacitance is at resonance when the inductive reactance equals the capacitive reactance. At the resonant frequency the three identifying characteristics of a series circuit are that the impedance (Z) is equal to the circuit resistance, the current flow is at maximum and the phase angle is zero. The resonant frequency for any combination of inductance and capacitance connected in series is determined by using the following equation:

$$f_R = \frac{1}{2\pi\sqrt{LC}}$$

where f_R = resonant frequency in hertz (Hz)
L = inductance in henrys (H)
C = capacitance in farads (F)

EXAMPLE 6.46

Calculate the resonant frequency of the circuit in **Figure 6.138**.

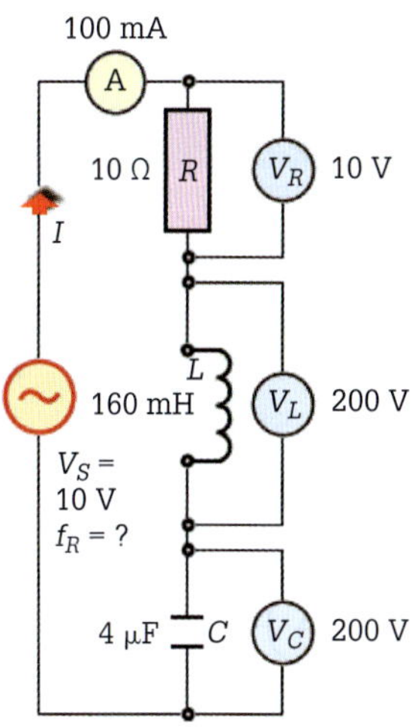

FIGURE 6.138 Series resonant circuit

$$f_R = \frac{1}{2\pi\sqrt{LC}}$$

First resolve

$$\begin{aligned}\sqrt{LC} &= \sqrt{160 \times 10^{-3} \times 4 \times 10^{-6}}\\ &= \sqrt{160 \times 4 \times 10^{-9}}\\ &= \sqrt{640 \times 10^{-9}}\\ &= 8 \times 10^{-4}\end{aligned}$$

Now

$$f_R = \frac{1}{2 \times 3.142 \times 8 \times 10^{-4}} = \mathbf{198.92\ Hz}$$

Frequency and reactance in RLC series circuits

In an RLC series circuit an increase in the frequency of the input waveform causes the inductive reactance (X_L) to increase and the capacitive reactance (X_C) to decrease:

$$X_L \propto f \quad \text{and} \quad X_C \propto \frac{1}{f}$$

At resonant frequency the inductive reactance equals the capacitive reactance:

$$X_L = X_C$$

Above resonance the inductive reactance is of a higher value than the capacitive reactance and the circuit is classified as inductive. Below resonance the capacitive reactance is greater than the inductive reactance and the circuit is capacitive. If the frequency of an RLC circuit is varied and the values of current at these frequencies are plotted graphically the result is a resonance curve of the circuit. This curve is illustrated in **Figure 6.139**.

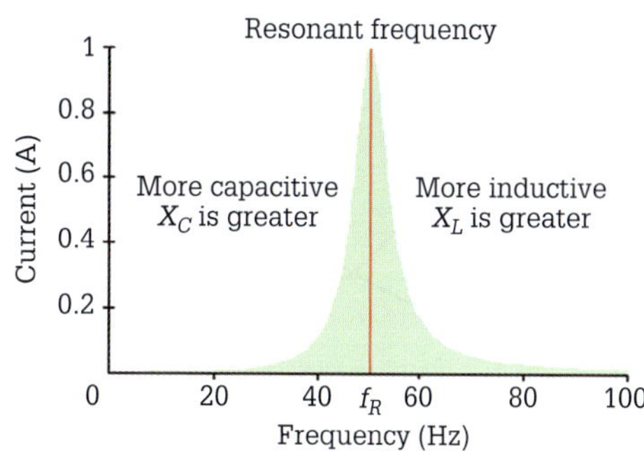

FIGURE 6.139 Resonant frequency in a series circuit

EXERCISE 6.45

1. From the circuit in **Figure 6.140(A)**, complete a table using the following headings:

TABLE 6.6 Exercise 6.45

	Reactance (Ω)	
f (Hz)	X_L	X_C

»

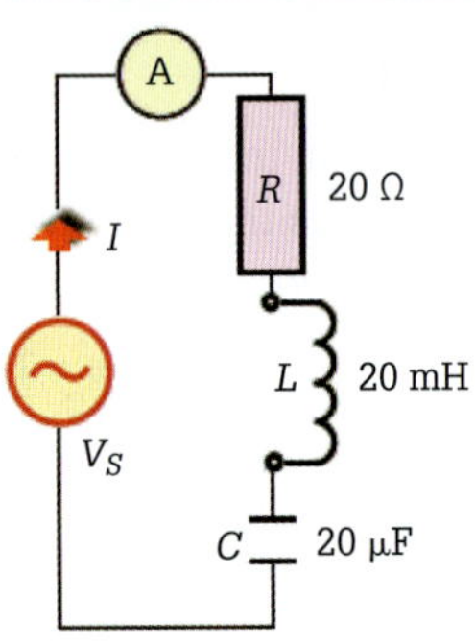

FIGURE 6.140(A) RLC series circuit

and plot the graph of:

- X_L versus frequency on the reference axis
- X_C versus frequency on the reference axis in six steps of 100 Hz to 600 Hz.

On the graphs of **Figure 6.140(B)** locate:

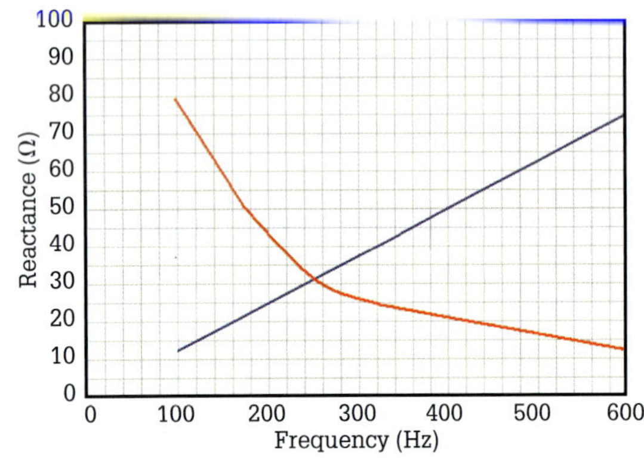

FIGURE 6.140(B) Series resonance graph

a the point of resonant frequency for the circuit
b the frequency range over which the circuit is inductive
c the frequency range over which the circuit is capacitive.

2.

a Calculate the resonant frequency of a circuit consisting of a 140 mH inductor and a 2.9 µF capacitor connected in series with a 5 Ω resistor.
b A 100 Ω resistor, a 250 mH inductor and a 100 µF capacitor are connected in series to a 100 V sinusoidal supply. At what frequency is maximum current drawn from the supply and what is the current at this frequency?
c A circuit containing a 4 µF capacitor is resonant at 200 Hz. Calculate:
  i the reactance of the inductance
  ii the value of the inductance in mH.
d What value of capacitance is needed in series with a 50 mH inductor to make the circuit resonant at 250 Hz?
e A capacitor of 10 µF and a 100 mH inductor are connected in series to form a resonant circuit. Determine the value of the inductance needed to reduce the resonant frequency by 100 Hz.

Frequency and current in a series RLC circuit

The current in a series RLC circuit is at maximum value when the circuit is at resonance since the only opposition to current flow at this frequency is resistance. When the frequency moves from the resonant wave, the current decreases.

EXAMPLE 6.47

From the circuit in **Figure 6.141(A)**, complete **Table 6.7** using the headings below and plot a graph of current versus frequency (in six steps of 100 Hz, from 100 to 600 Hz).

Step 1 Calculate value of X_L for each frequency using the equation $X_L = 2\pi fL$, which yields the results 12.5664 Ω, 25.1327 Ω, 37.6991 Ω, 50.2655 Ω, 62.8319 Ω, 75.3982 Ω

Step 2 Calculate value of X_C for each frequency using the equation $X_C = \frac{1}{2\pi fC}$, which yields the results 79.5775 Ω, 39.7887 Ω, 26.5258 Ω, 19.8944 Ω, 15.9155 Ω, 13.2629 Ω

Step 3 Calculate value of Z for each frequency using the equation $Z = \sqrt{R^2 + X^2}$ where X is the difference between X_L and X_C, which yields the results shown for Z in **Table 6.7**

TABLE 6.7 Example 6.47

f (Hz)	Z	I
100	69.93202 Ω	1.43 A
200	24.79512 Ω	4.03 A
300	22.90944 Ω	4.37 A
400	36.36488 Ω	2.75 A
500	51.00142 Ω	1.96 A
600	65.27478 Ω	1.53 A

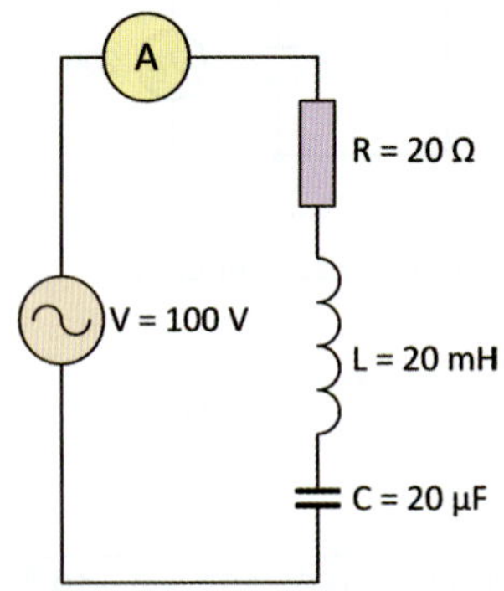

FIGURE 6.141(A) RLC in series

When graphed the current varies as shown in **Figure 6.141(B)**.

»

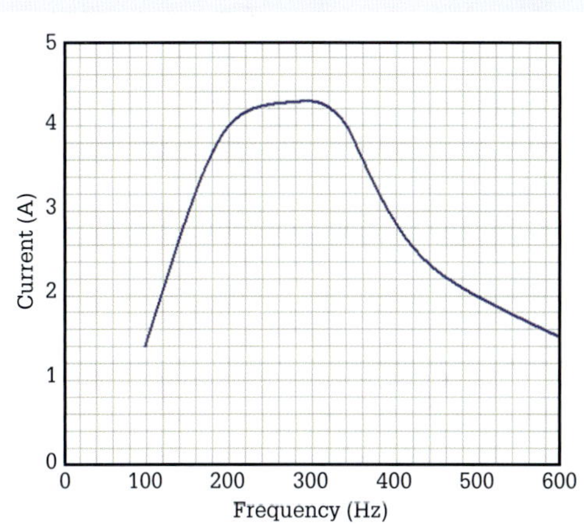

FIGURE 6.141(B) Completed graph of current versus frequency

The result is a curve known as the resonance curve of the circuit, which shows a resonant frequency of about 250 Hz in this case.

Resonance in a parallel LC circuit

For the same values of inductance and capacitance the frequency at which resonance takes place in a parallel LC circuit is identical to the frequency at which series resonance occurs using the same value components. Therefore, the resonant frequency for any combination of inductance and capacitance connected in parallel is determined by:

$$f_R = \frac{1}{2\pi\sqrt{LC}}$$

where f_R = resonant frequency in hertz (Hz)
L = inductance in henrys (H)
C = capacitance in farads (F)

EXAMPLE 6.48

Calculate the resonant frequency of the parallel circuit in **Figure 6.142** consisting of a 200 mH inductor and an 8 µF capacitor.

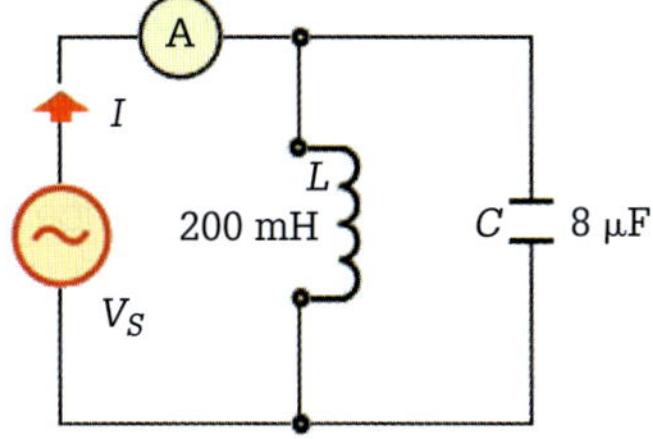

FIGURE 6.142 Resonance in a parallel LC circuit

$$f_R = \frac{1}{2\pi\sqrt{LC}}$$

First resolve

$$\sqrt{LC} = \sqrt{200 \times 10^{-3} \times 8 \times 10^{-6}}$$
$$= \sqrt{200 \times 8 \times 10^{-9}}$$
$$= \sqrt{1600 \times 10^{-9}}$$
$$= \mathbf{0.001265}$$

Now

$$f_R = \frac{1}{2 \times 3.142 \times 1.265 \times 10^{-3}} = \mathbf{125.81\ Hz}$$

EXERCISE 6.46

a What is the resonant frequency of a parallel circuit consisting of an 18 µF capacitor and a 15 mH inductor?

b At what frequency will a 4 µF capacitor resonate when connected in parallel to an inductance of 20 mH?

Frequency and current in a parallel circuit

When the frequency of a sinusoidal voltage connected across a parallel combination of pure inductance and capacitance is decreased, the current drawn by the inductive branch increases and the current in the capacitive branch decreases. An increase in frequency causes a decrease of current drawn by the inductive branch and the current in the capacitive branch increases.

$$I_L \propto \frac{1}{f} \quad \text{and} \quad I_C \propto f$$

At resonant frequency the current drawn by the inductive branch equals the current drawn by the capacitive branch and the circuit current is zero.

$$I_L = I_C$$

Resonance in a parallel RLC circuit occurs when the reactive current in the inductive branches is equal to the reactive current in the capacitive branches (when $X_L = X_C$). Thus, in these circuits the current through the circuit is at minimum because of the high impedance presented by parallel X_L and X_C. Only the resistive current path is effectively in circuit. Because the X_L and X_C currents are equal and opposite in phase, they cancel one another at resonance. But these currents are real and circulate back and forth between the inductor and capacitor, and in power circuits (230 V/400 V 50 Hz) they may cause damage to components.

SWITCH ON

Hazards and safety precaution

In power circuits parallel resonance is nowhere near as dangerous as series resonances, but caution must be used due to possible heavy circulating currents between *L* and *C*.

Above resonance, the capacitive current is of a higher value than the inductive current and the circuit is classified as capacitive. Below resonance, the inductive current is greater than the capacitive current and the circuit is inductive. These relationships are illustrated in **Figure 6.143**.

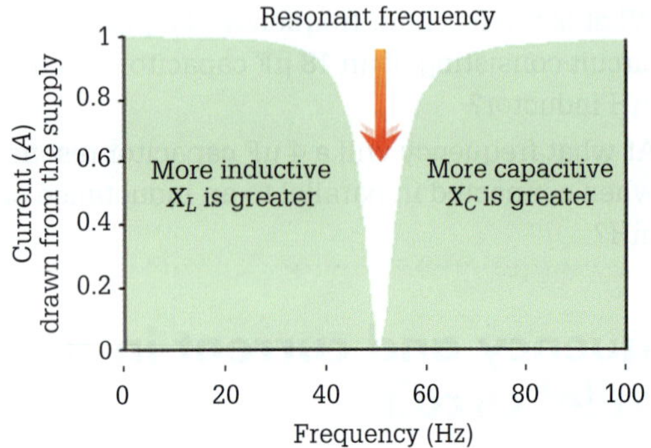

FIGURE 6.143 Resonance in a parallel circuit

These relationships will now be examined by completing the following exercise.

EXERCISE 6.47

From the circuit of **Figure 6.144(A)**, complete a table using the headings shown below and plot the graph using six steps of 100 Hz (i.e. 100 to 600 Hz).

- X_L versus frequency on the reference axis
- X_C versus frequency on the reference axis

TABLE 6.8 Exercise 6.47

f (Hz)	X_L	I_L	X_C	I_C	$I_L - I_C$
100					
200					
300					
400					
500					
600					

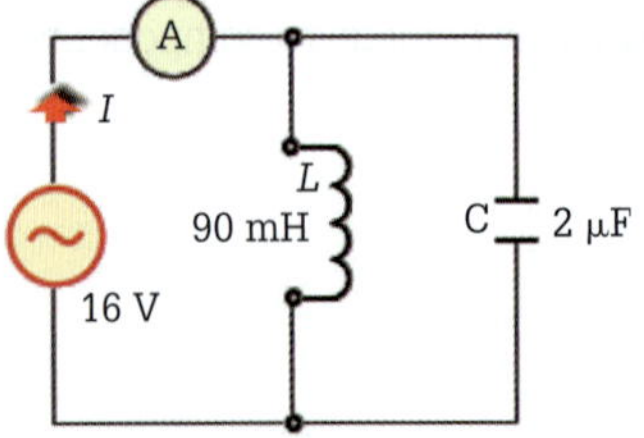

FIGURE 6.144(A) Parallel LC circuit

Your graph should resemble **Figure 6.144(B)**. On the graphs locate:

a the point of resonant frequency for the circuit
b the frequency ranges over which the current lags the voltage
c the frequency ranges over which the current leads the voltage.

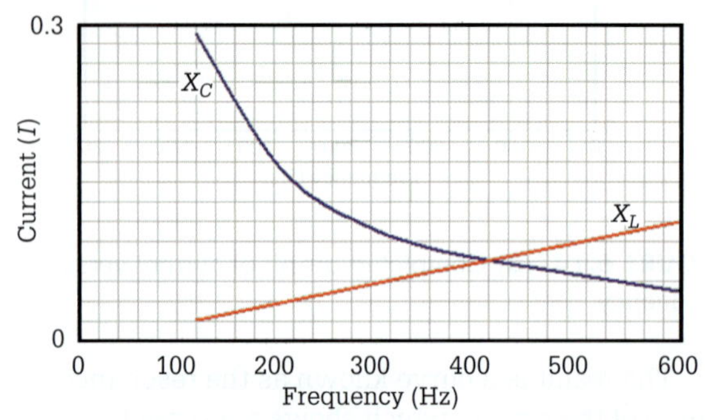

FIGURE 6.144(B) Parallel LC circuit

Note: When the frequency departs from resonance the current drawn by the circuit increases.

Difference between series and parallel resonance circuits

The important difference between series and parallel resonant circuits is that a series circuit offers low impedance to currents at resonant frequency and high impedance for currents at all other frequencies. The parallel circuit is the opposite of the series. It provides a high impedance path for current at resonant frequency and lower impedance at all other frequencies. Resonance is what makes the radio and television tuner work.

The antenna of the radio and television receives transmitted signals from every station in the area, but only the station whose frequency matches the resonant frequency of the tuning circuit will cause currents to flow in the circuit. These currents, when amplified, are the ones that produce the sound you hear and the picture you see.

Magnetic resonance imaging, used for example in spinal injuries, allowing medical persons to establish whether a signal commanding a finger to move originates in the correct region of the brain, and microwave cooking, are other practical applications of resonance phenomena.

»

REVIEW QUESTIONS

1 What specific devices produce harmonics?
2 What is the 'skin effect'?
3 What is total harmonic distortion (THD)?
4 What is Fourier analysis?
5 What is the alternative name given to the 1st harmonic?
6 Name three major problems created by harmonic distortion.
7 What type of ammeter is necessary in trouble-shooting today's electrical systems?
8 What electrical quantity of a circuit represents the frequency at which electrical energy is oscillating back and forth between a capacitor and an inductor?
9 In a series a.c. circuit, what two quantities are equal at the resonant frequency?
10 Calculate the resonant frequency of a circuit consisting of a 120 mH inductor and a 5 μF capacitor connected in series with a 5 Ω resistor.
11 What is the resonant frequency of a parallel circuit consisting of a 22 μF capacitor and a 25 mH inductor?
12 What is the important difference between series and parallel resonant circuits?

CHAPTER REVIEW

6.1 Alternating current quantities

- Pythagoras' theorem allows for calculating an unknown length of a side of a right-angled triangle if lengths of other two sides are known.
- Trigonometric functions relate an angle of a right-angled triangle to ratios of two side lengths.
- The waveform produced by an a.c. circuit can be a square wave, triangle wave or sawtooth wave.
- A cycle is any repetition of a variable quantity recurring at equal intervals of time.
- A period or periodic time is the time duration of one cycle.
- A wavelength is the distance between corresponding points on two successive cycles and is the distance travelled in the period of one cycle.
- The frequency is the number of cycles completed in one second.
- The maximum or peak value is the highest value that the voltage or current reaches in one direction.
- The peak-to-peak value is the highest value that the voltage or current reaches in both directions.
- The instantaneous value may vary from zero to a peak value many times per second.
- The average value of a complete cycle is taken over half a cycle only because the average value of a complete cycle of a sine wave is zero.
- The root-mean-square (rms) value is the most useful and practical value of alternating voltage or current for a sine wave.
- An oscilloscope can display the wave shape of a varying voltage, allowing both voltage and time measurements.
- A digital oscilloscope samples the waveform and then reconstructs it.

6.2 Phasor diagrams

- A phasor is a straight line used to represent an electrical quantity such as voltage or current that has a magnitude and direction.
- Conventionally, phasors rotate in an anticlockwise direction.
- When representing a number of voltages or currents in a circuit with phasors a reference phasor is drawn.

6.3 Single element a.c. circuits

- Some alternating current circuits consist of resistance only and for these circuits the same rules and laws apply as for direct current circuits.
- In an alternating current circuit consisting of resistance only, the voltage and current have no phase difference between them and are in phase.
- Alternating current when doing work through a resistor produces heat irrespective of the direction of the current.
- The electrical property of a circuit to oppose any change in current is referred to as inductance.
- The inductive reactance is directly proportional to the frequency of the sinusoidal waveform and the inductance.
- The rate at which an inductive device stores or returns energy is called its reactive power (Q_L).
- The effect of the capacitor is to oppose the change in the applied sinusoidal voltage and limit the current flow in the circuit.
- In an intrinsic capacitive circuit the current is said to lead the voltage by 90°.
- The rate at which capacitors store or return energy is called their reactive power (Q_C).

6.4 RL and RC series a.c. circuits

- In a series circuit containing resistance and inductance, the phasor sum of the voltages across the resistive and the inductive components equals the applied voltage.
- The impedance of a series RL circuit is the total opposition to current drawn from the supply by the circuit resistance and the inductive reactance.
- In a series circuit containing resistance and capacitance, the phasor sum of the voltages across the resistive and the capacitive components equals the applied voltage.
- The impedance of a series RC circuit is the opposition to the supply current by the circuit resistance and the capacitive reactance.

6.5 Resistance, inductance and capacitance in combination

- The angle (ϕ) by which the current lags the applied voltage in a circuit is called the phase difference.
- The power triangle can be used to express power relationships in an alternating current series circuit.
- The impedance triangle allows the determination of total opposition to current in series a.c. circuits.
- Phasor diagrams are used to solve problems in parallel a.c. circuits.

6.6 Power in an a.c. circuit

- The power description attributed to reactive loads is reactive power measured in volt-amperes reactive (VA_R).
- The energy consumed is the actual amount of power being dissipated by a circuit and is true power measured in watts (W).
- Apparent power is the power that appears supplied to the circuit and is the phasor sum of both the reactive and the true power. Apparent power has volt-amperes (VA) as its unit of measurement.
- Power factor is the ratio of true power and apparent power and is a measure of the power loss in a circuit.
- A high power factor means that electrical power is being utilised efficiently, while a low power factor indicates poor utilisation of electrical power.

6.7 Power factor improvement

- There are several methods available for power factor improvement. These include fixed capacitors, switched capacitors, static VA_R compensators and synchronous capacitors.
- The static VA_R compensator is a power factor correction technology.
- Synchronous capacitors provide step-less automatic power factor correction.
- Soft-start electronic control equipment, such as low-loss electronic ballasts, offer close to unity power factor.

6.8 Harmonics and resonance effect in a.c. systems

- Harmonics is the name given to contaminating waveforms that are sinusoidal in shape but are multiples of the fundamental waveform.
- The non-linear load is the root cause of harmonics.
- A characteristic of a distorted waveform is magnitude, which is the harmonic distortion factor (HDF).
- Knowledge of spectrum signatures assists in locating the source of the harmonics.
- Current distortion mainly affects equipment that transmits or conducts energy, while voltage distortion is generally an effect of current distortion.
- Skin effect is a conductor condition where the high frequency causes the electrons to flow in the outer surface of the conductor, effectively reducing its cross-sectional area.
- Standard ampere clamp-on ammeters lack the ability to perform basic electrical troubleshooting on electrical systems.
- Resonance is an electrical property possessed by some series- and parallel-connected alternating current circuits.
- The resonance of a circuit represents the frequency with which electrical energy is oscillating back and forth between a capacitor and an inductor.
- A circuit that contains resistance, inductance and capacitance is at resonance when the inductive reactance equals the capacitive reactance.
- Above resonance, the capacitive current is of a higher value than the inductive current and the circuit is classified as capacitive. Below resonance, the inductive current is greater than the capacitive current and the circuit is inductive in parallel circuits.
- The important difference between series and parallel resonant circuits is that a series circuit enables an easy or accepting path for currents at resonant frequency and a rejecting path for all other frequencies.

TRIAL EXAM

For Chapter 6 knowledge assessment, please complete the following trial exam.

1 What is the period of the sine wave shown in **Figure 6.145**?

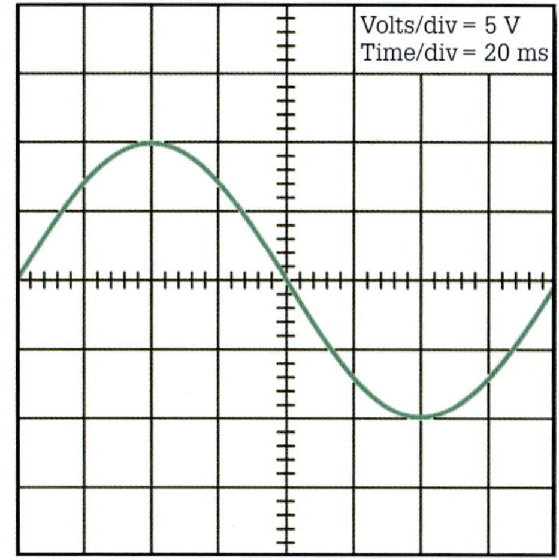

FIGURE 6.145 Measuring the period and voltage of a sine wave

a 80 ms
b 4 ms
c 100 ms
d 160 ms

2 Determine the peak (maximum) voltage of the sine wave in **Figure 6.145**.

a 10 V
b 7.07 V
c 5 V
d 6.637

3 What is the frequency of the sine wave shown in **Figure 6.146**?

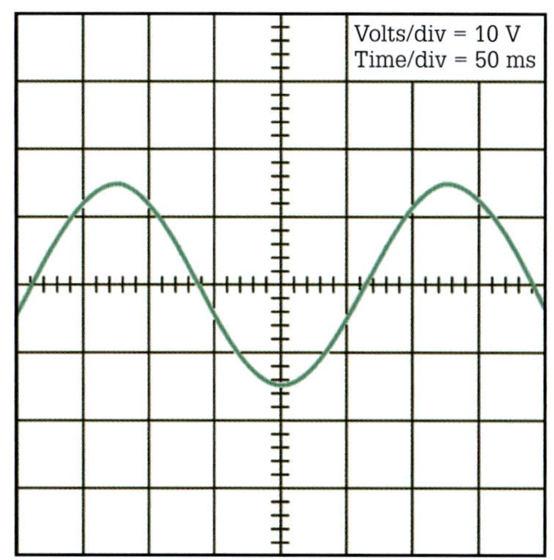

FIGURE 6.146 Measuring the frequency and voltage of a sine wave

a 4 Hz
b 2.5 Hz
c 200 Hz
d 500 Hz

4 Determine the rms voltage of the sine wave shown in **Figure 6.146**.

a 80 V
b 15 V
c 500 V
d 10.6 V

5 Determine the peak-to-peak voltage of the sine wave shown in **Figure 6.147**.

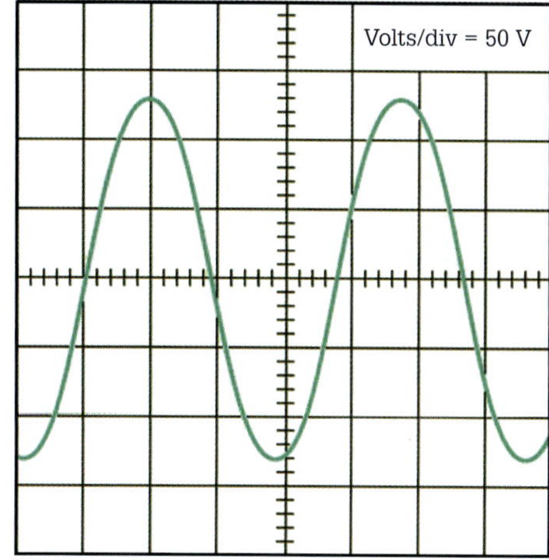

FIGURE 6.147 Measuring the peak-to-peak voltage of a sine wave

a frequency
b sinusoidal waveform
c 260 V
d 150 V

6 Determine the average voltage for a half cycle of the sine wave shown in **Figure 6.148**.

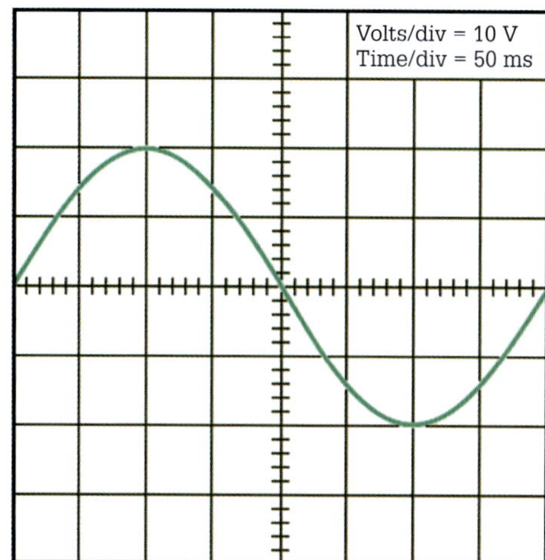

FIGURE 6.148 Measuring the average voltage of a sine wave

a 20 V
b 12.74 V
c 31.85 V
d 25.48 V

7 Determine the frequency of the sine wave in **Figure 6.149**.

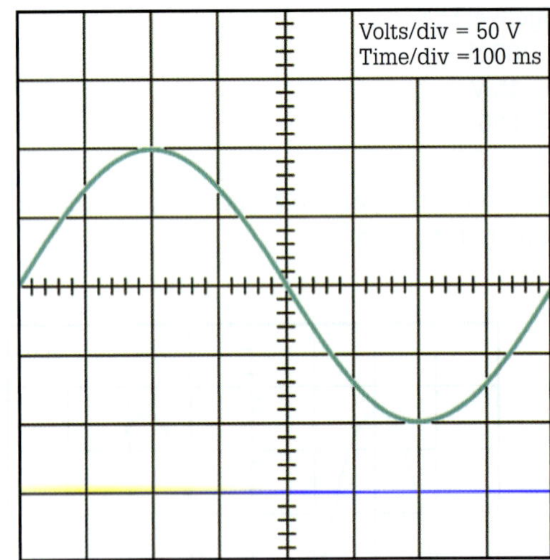

FIGURE 6.149 Measuring the frequency of a sine wave

a 2.5 Hz
b 10 Hz
c 1.25 Hz
d 5 Hz

8 The repetition of a variable quantity recurring at equal intervals of time is called:
a frequency
b sinusoidal waveform
c period
d cycle

9 The time duration of one cycle is called:
a frequency
b sinusoidal waveform
c period
d cycle

10 The frequency of a period of 100 ms is:
a 1 kHz
b 100 Hz
c 0.1 Hz
d 10 Hz

11 The number of cycles completed in one second is called:
a frequency
b sinusoidal waveform
c period
d cycle

12 The period of a 400 Hz waveform is:
a 10 s
b 1 s
c 2.5 ms
d 5 ms

13 What electrical value is necessary to know to ensure that insulation does not deteriorate when stressed by voltage and current?
a rms
b instantaneous
c peak voltage
d average

14 The distance between corresponding points on two successive cycles is called:
a peak to peak
b wavelength
c angular measurement
d rms

15 An a.c. voltage has a peak value of 160 V. Determine the instantaneous value of voltage at 39.57° after the commencement of the cycle.
a 113.12 V
b 101.92 V
c 226.3 V
d 251.2 V

16 Determine the rms value of a sinusoidal voltage wave that has an average value of 76.44 V for a half sine wave.
a 188.38 V
b 84.84 V
c 76.44 V
d 169.73 V

17 The a.c. value of current or voltage which produces the same heating effect as the same value of d.c. voltage is called:
a peak
b average
c instantaneous
d rms

18 Determine the rms value of a waveform that has a peak-to-peak value of 270 V.
a 212.1 V
b 95.46 V
c 191.1 V
d 106.05 V

19 The number of radians in 180° is:
a 0.5 π radians
b 1 π radians
c 1.5 π radians
d 2 π radians

20 The equation to determine true power (P) in an a.c. circuit is:
a $P = V^2I$
b $P = VI\cos\theta$
c $P = VI\tan\theta$
d $P = VI\sin\theta$

21 The power factor is a numerical value between:
a 0 and 1
b 0° and 90°
c volt-amperes and reactive power sides of a power triangle
d the sine and cosine of an angle

22 The symbol for impedance is:
a R
b Z
c X_C
d X_L

23 The SI unit of impedance is:
a watt
b ampere
c volt
d ohm

24 Apparent power is measured in:
a watts
b lambdas
c volt-amperes
d vars

25 Capacitive reactance produces a current:
a proportional to the resistance
b in phase with the voltage
c lagging the voltage by 90°
d leading the voltage by 90°

26 The characteristics of a resonant series RLC circuit are:
a high current with high impedance
b low current at resonant frequency
c twice the value of X_L compared to X_C
d minimum impedance and unity pf

27 The characteristics of a resonant parallel RLC circuit are:
a high current with high resistance
b low impedance at resonant frequency
c minimum current at maximum Z
d high voltage and low pf

28 Reactive power is measured in:
a watts
b VA_R
c VA
d degrees

29 The equation used to determine apparent power in an a.c. circuit is:
a $S = VI$
b $S = VI \cos \phi$
c $S = VZ$
d $S = VI \sin \phi$

30 Given that the supply voltage is 230 V and the current drawn from the supply is 1.63 A at a phase angle of 72°, calculate the true power in the circuit.
a 141.1 W
b 115.85 W
c 374.9 W
d 356.55 W

31 The minimum power factor at full load allowed by a supply authority for an installation is:
a 0.5
b 0.6
c 0.7
d 0.8

32 What is the electrical parameter that is a measure of the power loss in a circuit?
a efficiency
b reactive current
c power factor
d line voltage

33 When an installation has its power factor improved, the line current of the installation:
a increases in proportion with the increase in pf
b remains the same
c decreases
d varies with the apparent power

34 A low power factor in an installation is generally due to:
a resistive currents
b capacitive currents
c magnetising currents
d displacement currents

35 The connection of a capacitor in parallel with an inductive load decreases the:
a power factor
b supply current
c true power
d load voltage

36 The hold-in ability of contactors and relays is affected by:
a reduced load current
b full-load voltage
c low or under-voltages
d a reduction of power losses

37 A series RC circuit is connected across a supply voltage of 220 V. If the voltage across the capacitor is 120 V, what is the voltage across the resistor?

38 A series RLC circuit has the voltages $V_R = 45$ V, $V_L = 25$ V and $V_C = 40$ V across the circuit components. Calculate the applied voltage.

39 An industrial workshop has a load of 400 kW at a power factor of 0.7. What kVA_R rating of capacitor bank is needed to improve the power factor to 0.9?

40 What is the resonant frequency of a parallel circuit consisting of a 25 μF capacitor and a 15 mH inductor?

7 Three-phase low voltage a.c. circuits

This chapter provides electrotechnology workers with knowledge of three-phase star and delta alternating current (a.c.) systems and circuits, why three-phase is used, and the relationships between line and phase values. Electrotechnology workers will gain knowledge of calculating fault-loop impedance and explain its effect in determination of fault-current levels. In addition, electrotechnology workers will develop problem-solving skills that relate to balanced and unbalanced a.c. systems.

LEARNING OBJECTIVES

Three-phase systems

- Explain the principles of multiphase systems
- Describe how three-phase is generated
- Define phase sequence

Three-phase star connections

- Explain the fundamentals of three-phase star connections

Three-phase four-wire systems

- Explain the fundamentals of three-phase four-wire systems

Three-phase delta and interconnected systems

- Explain the fundamentals of three-phase delta connections
- Explain how star and delta devices can be interconnected
- Explain the fundamentals of interconnected star and delta devices

Energy and power requirements of a.c. systems

- Describe how a.c. power systems and loads respond to energy and power needs
- Explain how power factor can be improved

Fault-loop impedance

- Explain why knowledge of short-circuit capability is essential
- Determine the earth-loop impedance

Industry standards

- Outline common *Wiring Rules* requirements applicable to alternating current supplies

7.1 Three-phase systems

Multiphase systems

Electrical circuits containing only one sinusoidal voltage source are single-phase a.c. With a single-phase system, the electrical energy produced occurs from a voltage source such as an a.c. generator by the load via two conductors, active and neutral. By contrast, a multiphase or polyphase system contains more than one alternating current voltage source delivering electrical energy to the load via three or more conductors. These multiphase systems have source voltages that are equal in magnitude at the same frequency, but separate from each other in time.

Multiphase systems find application in the generation and transmission of energy where the demand is for high power. The conventional multiphase system has three voltage sources and is a three-phase system.

Single voltage generation

The primary a.c. generator consists of a coil of wire that has its ends connected to slip rings and a prime mover rotates the assembly in a magnetic field produced by permanent magnets. As the coil rotates, it cuts the magnetic lines of force enabling an induced emf to develop across the ends of the coil. The induced emf causes a current to flow when the generator connects to an external circuit. This type of generator is a permanent-magnet generator.

Sinusoidal current and voltage are necessary for many electrical devices. Therefore, practical generators are designed so that their generated output has a sinusoidal waveform. **Figure 7.1** shows the current produced by the simple generator.

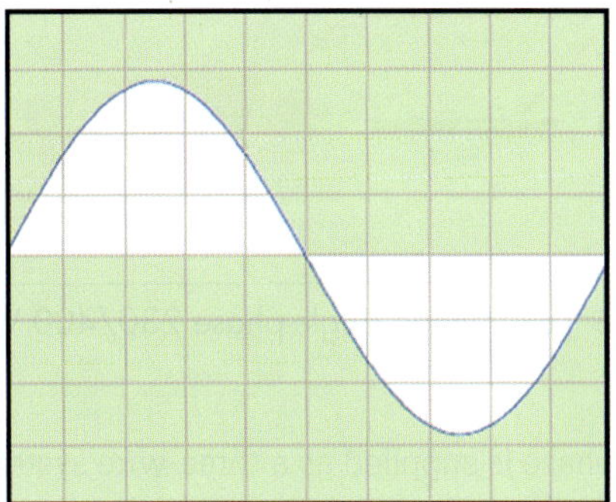

FIGURE 7.1 Sinusoidal waveform

Multiphase voltage generation

A single-phase generator can become a multiphase or polyphase generator by increasing the number of armature windings. In fact, multiphase generators produce as many autonomous voltage waveforms as there are armature windings or phases.

A three-phase generator has three independent armature windings displaced from one another by 120 electrical degrees. Consequently, the voltages induced in the three separate windings are 120° apart in time-phase. Apart from two-phase (90°) windings, the electrical displacement between the armature windings can be determined by:

$$\frac{360^\circ}{\text{number of armature windings}}$$

Power or high-voltage generators are electrically excited. Electrically excited means that electromagnets (coil windings) create the necessary magnetic field.

A direct current feeds these coils through the graphite brushes on the slip rings and the total arrangement is rotated by means of the prime mover, creating a rotating magnetic field. Consequently, the induced electromotive force develops in a stationary armature.

Note that the device that makes available the electromotive force is an armature if it is the rotating element, while the fields are stationary fields. In contrast, if it is stationary, the device is a stationary armature, or a stator, in which case the rotating field is the rotor.

The use of electromagnets enables adjustment of the generated emf. In other words, the current in the electromagnet coils can be increased or decreased, thereby controlling the electromagnetic flux produced. Having control over the electromagnetic flux means control over the emf produced.

When the stationary armature or stator windings consisting of three groups of coil windings align at 120 electrical degrees to each other, the generator produces three single-phase alternating current outputs.

An illustration of a three-phase generator with a stator and a rotor occurs in **Figure 7.2**.

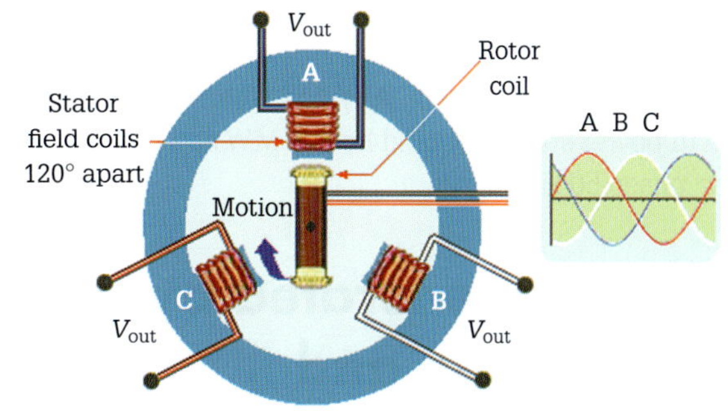

FIGURE 7.2 Three-phase generator with stator and rotor

An illustration of a three-phase generator with an armature and stationary fields occurs in **Figure 7.3**.

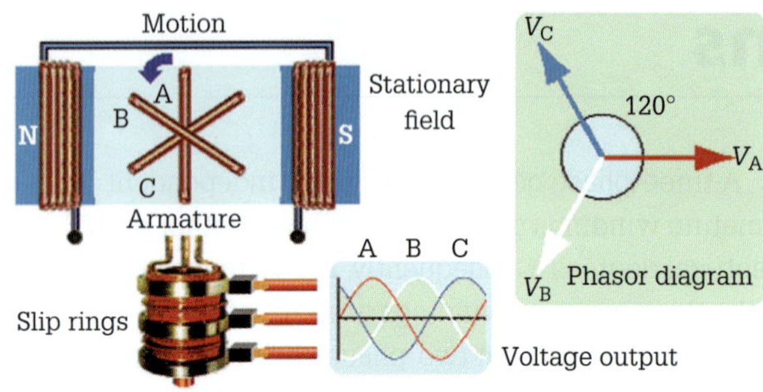

FIGURE 7.3 Three-phase generator with armature and stationary field

The generation of multiphase voltages transpires in the same manner as that for single-phase voltages. When a prime mover turns at a constant speed, the rotor has a sinusoidal voltage induced across each of the stator field coils. These induced voltages have the same magnitudes if the stator coils have the same number of conductor turns. Because the stator coils are mechanically fixed in position the induced sinusoidal voltages will have the same frequency. Consequently, the only factor that is different between the three induced voltages is their phase relationship. One hundred and twenty electrical degrees separate each voltage wave. As shown in **Figure 7.4**, each induced voltage reaches its peak value at a different instant in time.

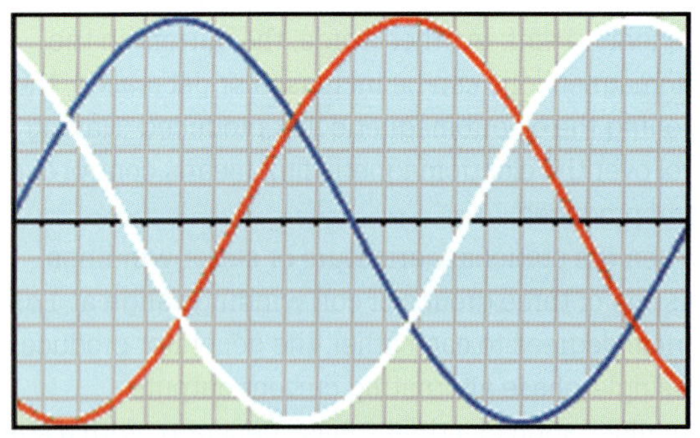

FIGURE 7.4 Three-phase waveform

The three-phase generator is for electrical power generation and distribution. A typical alternator output used in power stations would be 500 MW at 24 kV three-phase 50 Hz output.

Voltages generated by single-phase and polyphase alternators

Electrical power generation usually occurs as three-phase alternating current voltage. The power transmits to distribution substations at high voltages. For three-phase transformers to meet the necessary load requirements they must step down the voltage. In fact, the load conditions demand a 230 V single-phase supply or a three-phase plus neutral 400 V supply.

Single-phase a.c. is a two-wire system (**Figure 7.5**) or a three-wire system (**Figure 7.6**). The two-wire system consists of two conductors of 230 V potential between the active conductor and:

- the neutral conductor
- the earth conductor
- the ground.

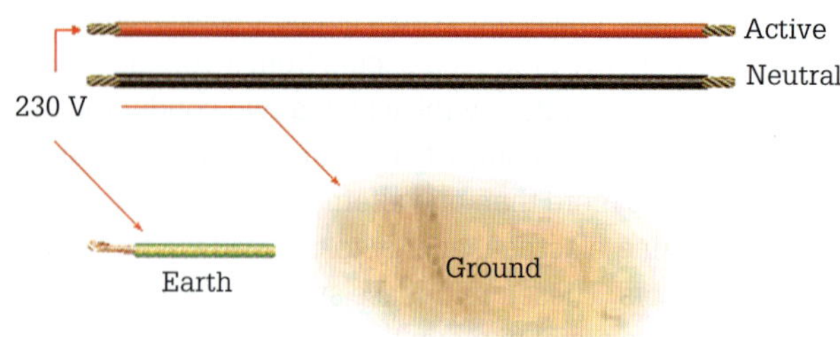

FIGURE 7.5 Low-voltage single-phase 230 V two-wire system

The three-wire system consists of two active conductors and a neutral conductor. This system provides the following potentials:

- 230 V potential between either of the active conductors and:
  - the neutral conductor
  - the earth conductor
  - the ground.
- 400 V potential between the active conductors.

Single-phase voltage distribution systems are used to connect residential properties and small commercial establishments. **Figure 7.6** illustrates the three-wire system.

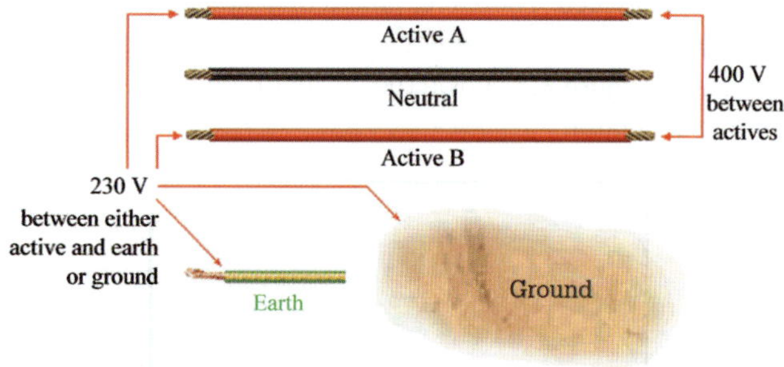

FIGURE 7.6 Low-voltage single-phase 230/400 V three-wire system

Three-phase is supplied as a three-wire system or as a four-wire system. The three-wire system is typically used for high-voltage delivery using bare aerial cables. In Australia, the most common voltages for the distribution system measured between active conductors are 11 kV, 22 kV and 33 kV. There is also an active-to-earth or ground voltage, which equals the voltage measured between active conductors divided by $\sqrt{3}$ (for an 11 kV distribution system, this equates to a potential of 6351 V). **Figure 7.7** illustrates a high-voltage three-wire distribution system using bare aerial cables.

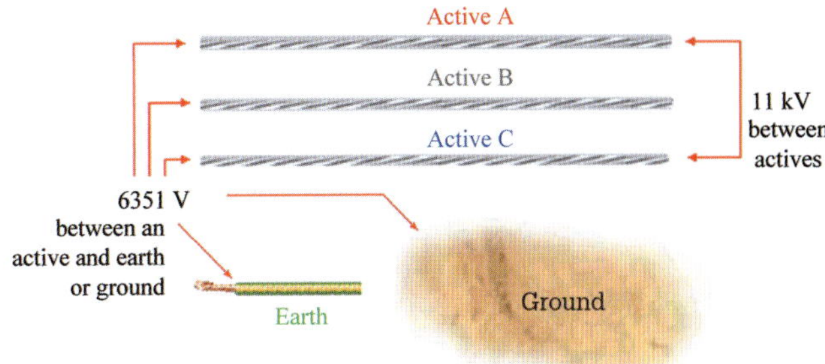

FIGURE 7.7 High-voltage three-phase 11 kV three-wire system

The four-wire system consists of three active conductors of 230 V between each active conductor and the neutral conductor, and 400 V potential between each of the active conductors themselves. There is also 230 V potential between each active conductor and the earth or ground. **Figure 7.8** illustrates the three-phase four-wire system.

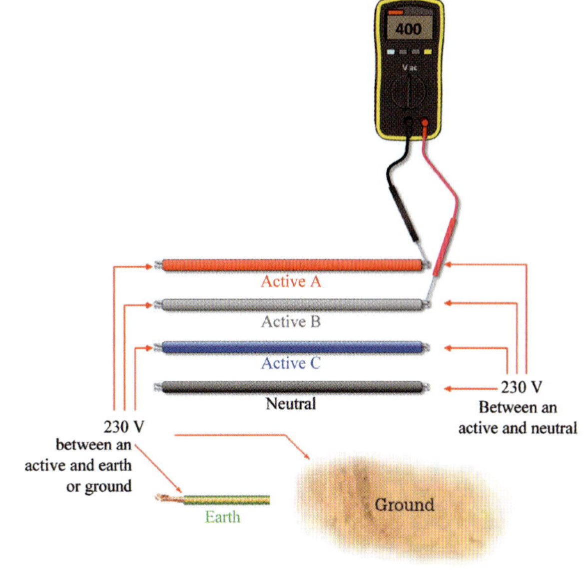

FIGURE 7.8 Three-phase 400/230 V four-wire system

Figure 7.9 illustrates the feeds obtainable from a three-phase four-wire system.

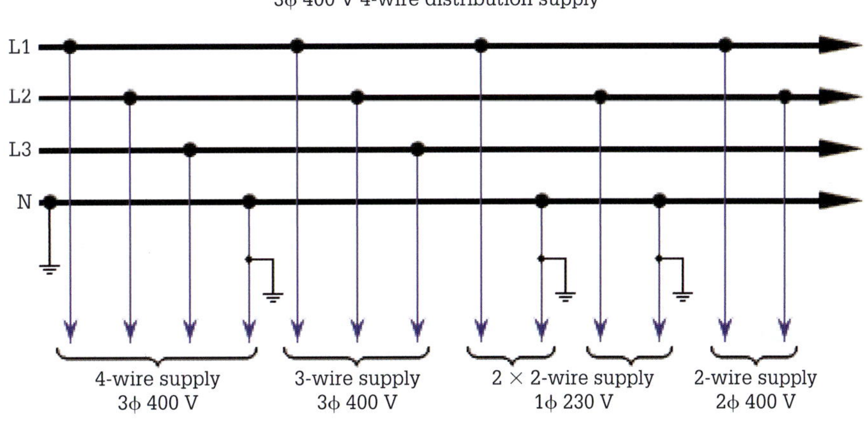

FIGURE 7.9 Feeds obtainable from a three-phase four-wire system

A four-wire feed is distributed to industries and commercial establishments and, recently, domestic installations due to an increase in the need for air-conditioners.

Two-phase is a two-wire system for high-voltage delivery. The two-wire high-voltage system consists of two active conductors of 11 kV, 22 kV or 33 kV between each active conductor. With the 11 kV systems, each active conductor has a voltage to earth and ground of 6351 V. The 22 kV system has 12 702 V between each active conductor and the earth and ground while the 33 kV system has 19 053 V between each active conductor and the earth and ground. **Figure 7.10** illustrates the two-phase, two-wire system.

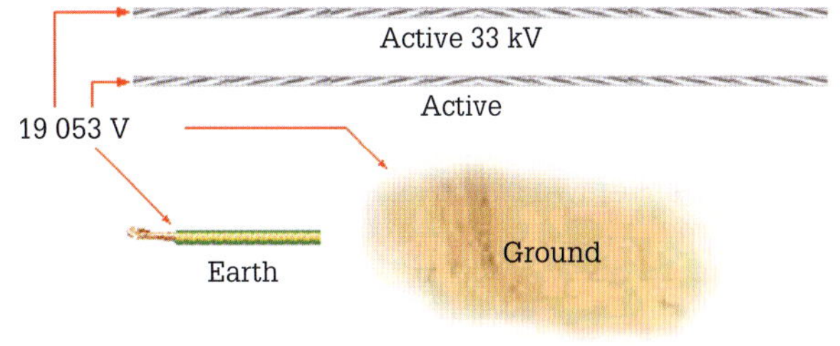

FIGURE 7.10 Two-phase 33 kV two-wire system

The single-wire earth return (SWER) is a high-voltage single-phase system. In addition, this system utilises an isolation transformer fed from a three-phase 33 kV transformer. The isolation transformer feeds a 19.1 kV single-phase voltage to the consumer's SWER distribution transformer.

This transformer converts the high-voltage single-phase input into a three-phase 230/460 V output. Furthermore, the return path for current with the SWER line is via the general mass of ground.

This system occurs in rural areas where the cost of installing other systems would be prohibitive. However, the main difficulty with this system is in maintaining the effectiveness of the earth return circuit. Refer to **Figure 7.11**.

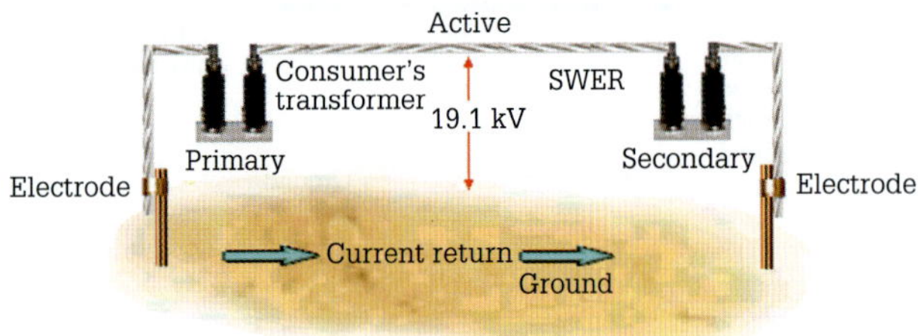

FIGURE 7.11 Single-wire earth return system

Advantages of a three-phase system

In order to understand the advantages of three phases for power distribution, comparison between a single-phase power distribution system and a three-phase system must occur. The starting point is to assume that both systems are delivering the same power to their respective loads at unity power factor.

Single-phase power distribution

Figure 7.12 illustrates a single-phase 230 V rms supply across a two ohm resistive-type load.

The power given by the single-phase system is:

$$P = \frac{V^2}{R}$$
$$= \frac{230^2}{2}$$
$$= \mathbf{26\,450\ W}$$

The current drawn by the load is:

$$I = \frac{V}{R}$$
$$= \frac{230}{2}$$
$$= \mathbf{115\ A}$$

FIGURE 7.12 Single-phase power distribution

Figure 7.13 illustrates a three-phase balanced supply across three 6 Ω resistive-type loads.

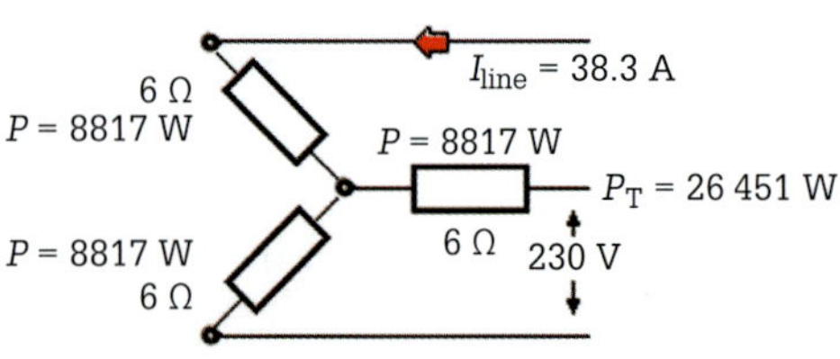

FIGURE 7.13 Three-phase power distribution

Three-phase power distribution

The power given by each phase of the three-phase system is:

$$P = \frac{V^2}{R}$$
$$= \frac{230^2}{6}$$
$$= \mathbf{8817\ W}$$

The total power delivered by the three-phase system is 26 451 watts, which is the same as for the single-phase system.

The current drawn by the load in each phase is:

$$I = \frac{V}{R}$$
$$= \frac{230}{6}$$
$$= \mathbf{38.3\ A}$$

From these calculations, it is evident that the line conductors of the three-phase system have one-third the current-carrying capacity of the line conductors in the single-phase system. Consequently, the conductor size required for single-phase bare copper aerial conductors carrying a current of 120 A would be 42 mm^2 in still air.

The conductor size required for the three-phase bare copper aerial conductor carrying a current of 38.3 A in still air is 8.6 mm^2. If copper conductors carry the current then that means a three-phase system has 31% less copper than a single-phase system. For example, 8.6 × 3 conductors divided by 42 × 2 conductors × 100% = 31%.

Less copper translates into a material saving of 70% in conductor costs of the three-phase system when compared with the single-phase system. This significant difference of material costs between the two systems is an important consideration when transmitting power over a distance. Another factor considered when choosing a system is the power loss in the form of heat from the copper aerials. For example, the single-phase power loss for the two copper aerials of **Figure 7.12** would be (assuming a 1000 m run):

$$\rho \text{ of copper} = 1.72 \times 10^{-8} \text{ and } 42\ \text{mm}^2 = 42 \times 10^{-6}\text{m}^2$$

$$R = \frac{\rho l}{a}$$
$$= \frac{(1.72 \times 10^{-8})}{(42 \times 10^{-6})} \times 1000$$
$$= \mathbf{0.409\ \Omega}$$

$$P = I^2R \times 2 \text{ conductors}$$
$$= 115^2 \times 0.409 \times 2$$
$$= \mathbf{10.8\ kW}$$

The three-phase power loss for three copper aerials is:

ρ of copper $= 1.72 \times 10^{-8}$ and 8.6 mm^2 $= 8.6 \times 10^{-6}$m^2

$$R = \frac{\rho l}{a}$$
$$= \frac{(1.72 \times 10^{-8})}{(8.6 \times 10^{-6})} \times 1000$$
$$= \mathbf{2.0\ \Omega}$$

$$P = I^2R \times 3 \text{ conductors}$$
$$= 38.3^2 \times 2 \times 3$$
$$= \mathbf{8.8\ kW}$$

From these calculations, the power loss via heat in the conductors of the three-phase system is 81.4% of that of a single-phase system. This percentage translates into a major saving with generation costs.

Pulsating nature of single-phase

In addition, the instantaneous power drawn by a load connected to a single-phase supply pulsates. At unity power factor the power reduces to zero twice during each cycle. **Figure 7.14** shows the pulsating nature of single-phase.

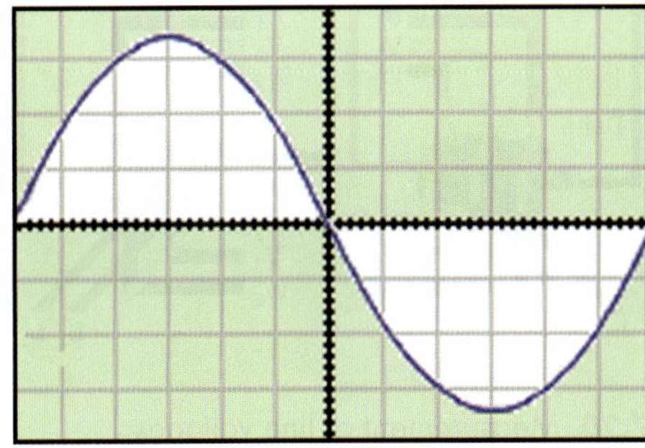

FIGURE 7.14 Pulsating nature of single-phase power

By contrast, the instantaneous power drawn by a balanced load in a three-phase system is a constant value of 0.707 peak and is not pulsating, in the same way, as single-phase supply. Refer to **Figure 7.15**.

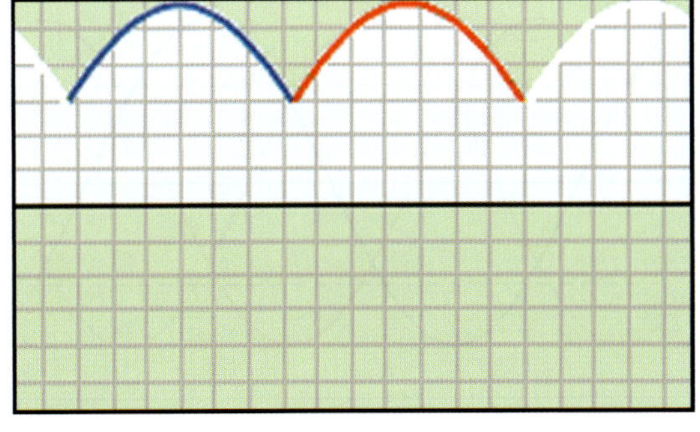

FIGURE 7.15 Constant nature of three-phase power

Constant nature of three-phase power

The steady nature of three-phase power allows for a constant and efficient energy conversion of electrical power to mechanical power when applied to electric motors. Three-phase power creates a smooth, uniform rotating magnetic field of constant density. In fact, such fields are necessary for the efficient operation of electric motors. With single-phase motors, the power drawn by the run winding creates a variable rotating magnetic field.

Single-phase motors require a starting winding, switching mechanisms and, in some applications, capacitors. The pulsating rotating magnetic field makes the single-phase motor less efficient than the three-phase motor.

Concerning the stator or armature winding connections of three-phase alternators, one of the two types available permits the selection of two output voltages, 400/230 volts. The single-phase generator does not have this capability. In general, the operating characteristics of the three-phase apparatus are superior to those of single-phase machinery. The physical size and weight of a three-phase motor is typically about 30% less than a similar power rated single-phase motor.

Three-phase connections

In **Figure 7.16**, three coils, A, B and C, are mechanically fixed 120° apart on an armature that is free to turn. The three coils connect to form a three-phase voltage source. One method of doing this is by joining the three finishes of each coil together. At the instant in time shown in **Figure 7.16**, coil A is moving parallel to the lines of force and the induced electromotive force across its ends is at zero potential. Coils B and C are cutting lines of force, and they have an equal in magnitude but opposite in polarity electromotive force induced across their conductor ends. When the rotor shown revolves in an anticlockwise direction, the voltages in each winding reach their positive peak values in the order of A, B, C.

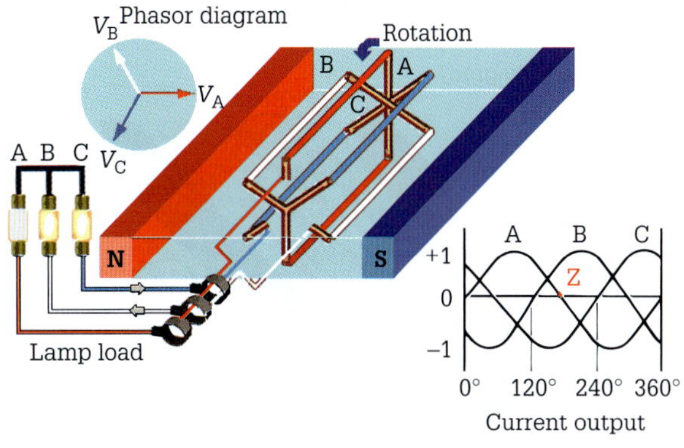

FIGURE 7.16 Three-phase voltage generation

Figure 7.16 shows the waveform diagram for a balanced load. Under this condition, the line currents are also 120° out of phase with one another. If we assume a maximum current of 1 A when V_A is zero at instantaneous point Z then the current I_A is also zero. However, current I_B is +0.866 A and current I_C is −0.866 A. Although the three currents are changing in value and direction as their conductors are rotated through the magnetic field, the phasor sum of the instantaneous currents is zero ($\bar{i}_A + \bar{i}_B + \bar{i}_C = 0$). Therefore, there is no out-of-balance current flowing in the circuit. If the load was not balanced (load varying on one phase) then the system would be out of balance and a neutral conductor necessary.

With a three-phase alternator, the neutral conductor taken from the star point of the armature connects to another slip ring. This feature enables the neutral to carry any out-of-balance currents that unbalanced loads produce. For a three-phase alternator not to have a neutral available, the load must be balanced by having the same current loading on each phase with the same power factor.

After a rotation of 120°, the induced electromotive force across coil B is at zero potential at the time shown. Coils A and C are cutting lines of force, and they have an equal in magnitude but opposite in polarity electromotive force across their conductor ends. Refer to **Figure 7.17**.

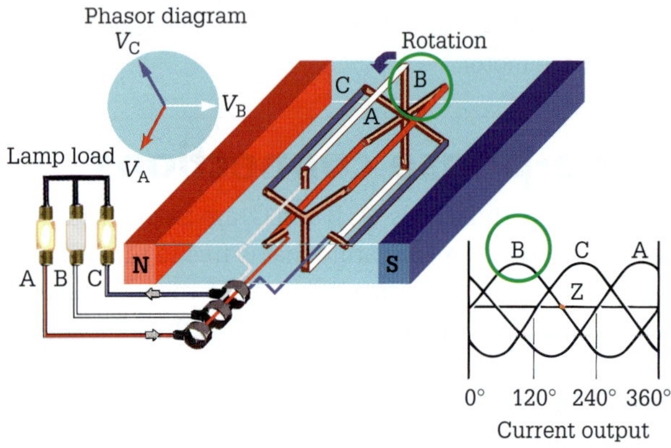

FIGURE 7.17 After a rotation of 120°

After a rotation of 240°, the induced electromotive force across coil C is at zero potential at the time shown. Coils A and B are cutting lines of force, and they have an equal in magnitude but opposite in polarity electromotive force induced across their conductor ends. Refer to **Figure 7.18**.

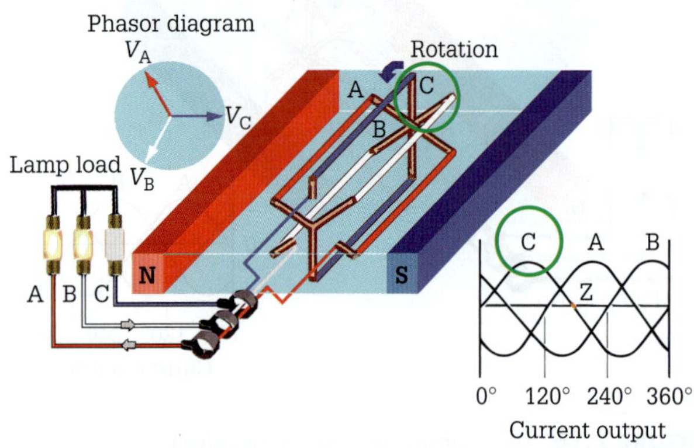

FIGURE 7.18 After a rotation of 240°

There are six significant voltages developed by three-phase a.c. generators. Three of these voltages are phase voltages (V_{Ph}) and the other three are line voltages (V_L). The manner in which these six voltages develop depends upon the two primary three-phase coil connections. These connections are delta and star. Refer to **Figure 7.19**.

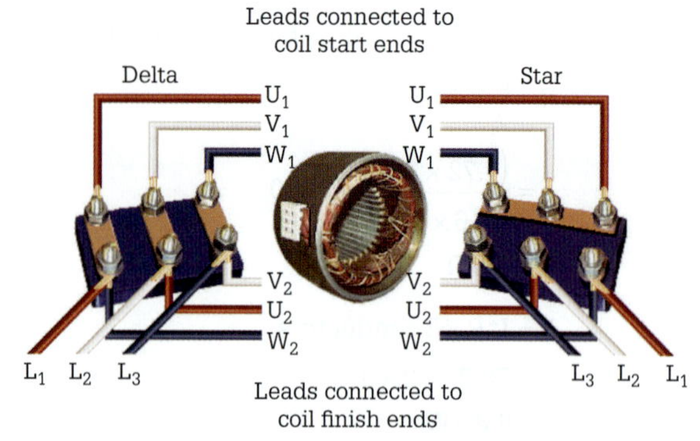

FIGURE 7.19 Three-phase coil lead connection

Line and phase voltages

Voltage measurement concerning three phase relate to line and phase voltages.

Measure line voltages between lines as illustrated in **Figure 7.20**.

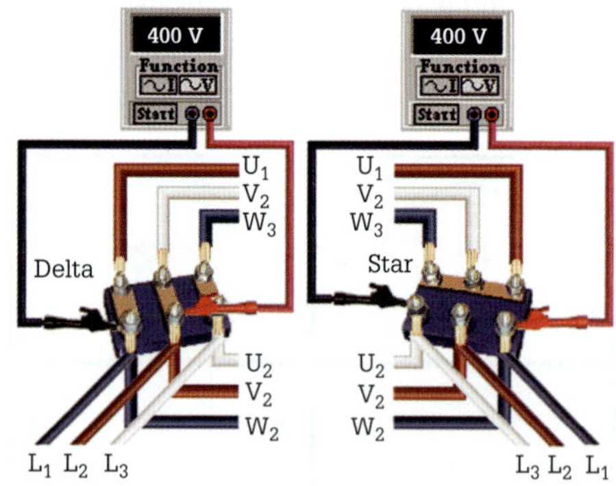

FIGURE 7.20 Measurement of line voltages

In this diagram, voltage measurements across U_1 and W_1, U_1 and V_1, V_1 and W_1 are line measurements in a delta-connected system. Voltage measurements across V_2 and W_2, V_2 and U_2 and U_2 and W_2 are line measurements in a star-connected system. **Figure 7.21** shows the line voltage of a three-phase waveform.

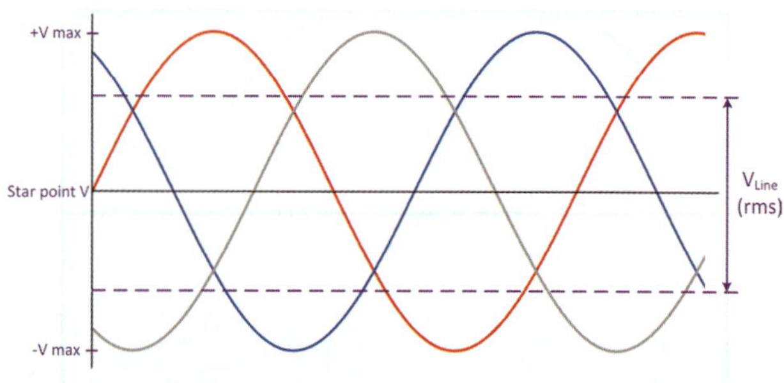

FIGURE 7.21 Line voltage of a three-phase waveform

Measure phase voltages across a single coil group or phase as illustrated in **Figure 7.22**.

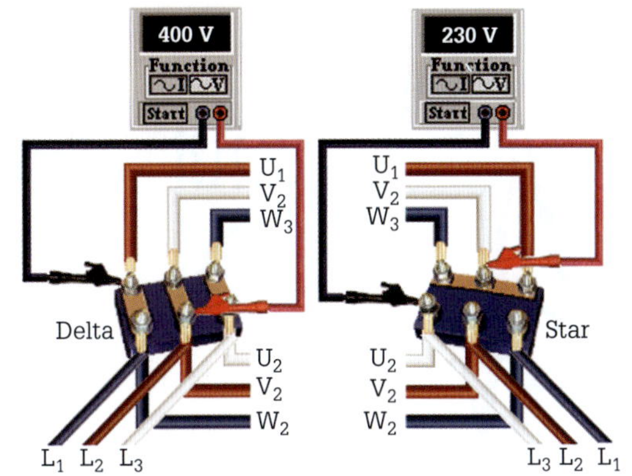

FIGURE 7.22 Measurement of phase voltages

In this diagram, voltage measurements across U_1 and U_2, V_1 and V_2 and W_1 and W_2 are phase measurements in a delta-connected system. Voltage measurements across U_2, V_2 and W_2 and the star point are phase measurements in a star-connected system. In addition, the voltage measurements are root-mean-square (rms) values. In Australia, line voltages transmit at 33 kV, 22 kV, 400 V and 230 V. This means that at an instant in time the peak voltage of the sinusoidal waveform generated is 46.662 kV, 31.108 kV, 15.554 kV, 586.81 V and 339.36 V. **Figure 7.23** shows the phase voltage of a three-phase waveform.

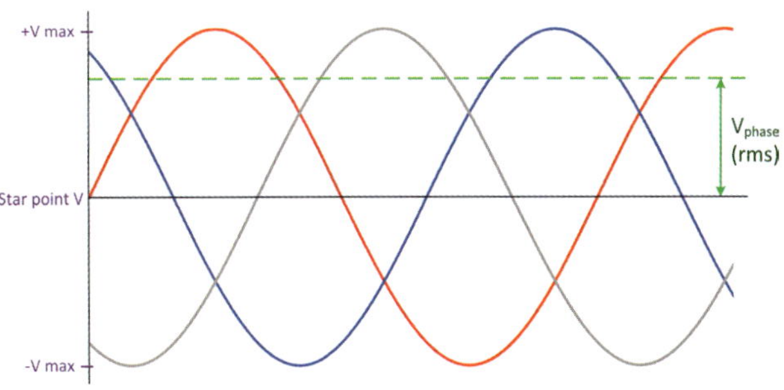

FIGURE 7.23 Phase voltage of a three-phase waveform

Figure 7.24 illustrates the peak and rms voltage positions on a three-phase waveform.

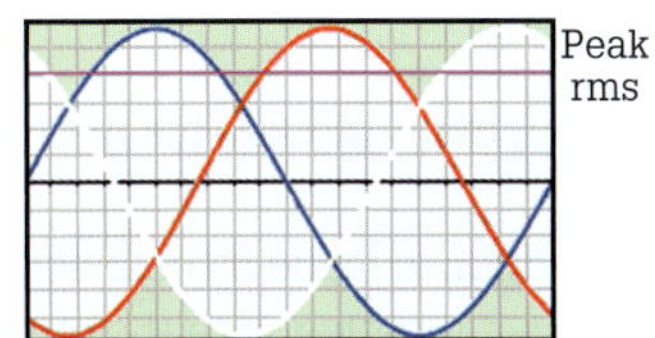

FIGURE 7.24 Three-phase waveform

Phase sequence

A three-phase system of voltages or currents has a cyclic direction of phase known as phase sequence in which the phase voltages and currents reach their peak value 120° after each other. For this reason, identification of the three phases is required. Identification allows a stated sequence and enables the balancing of loads across the phases. Some identification methods use the letters A, B and C while others use U, V and W for phase identification, and L1, L2 and L3 for line identification. A colour coding applies in Australia and the insulation colours used for 400 V three-phase supply and distribution are brown, black and grey (or red, white and blue respectively for older installations).

When the rotating field of **Figure 7.25** revolves in an anticlockwise direction, the voltages across the ends of each stator coil group or phase reach their peak value in the positive phase sequence of A, B and C or when using insulation colours brown, black and grey or red, white and blue. By contrast, if the phase order is ACB, a negative phase sequence results.

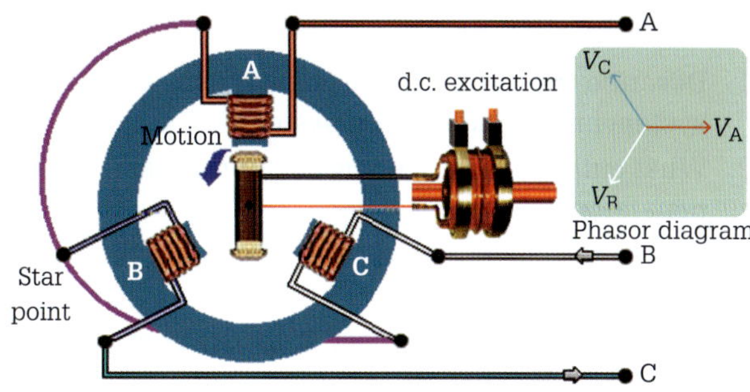

FIGURE 7.25 Natural phase sequence

The phase sequence is the order in which the three induced voltages follow one another. The phasor diagram of **Figure 7.25** shows this. The three sinusoidal voltages generate at the same peak values but displaced in phase by 120°. However, interchanging any two of the three line conductors connected to a star or delta terminal block also reverses the phase rotation. **Figure 7.26** shows line interchanging for a delta-connected terminal block.

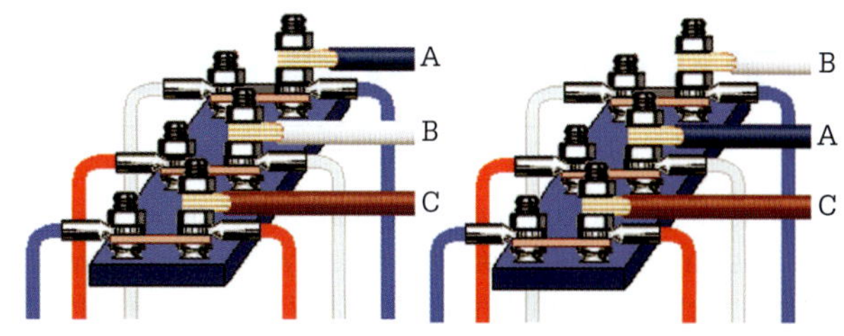

FIGURE 7.26 Changing phase sequence at the terminal block

Phase sequence indicators

The order of the cyclic direction is established by using a three-phase sequence indicator. A three-phase sequence indicator shows the order of the phase sequence. Knowledge of the phase sequence prior to starting electrical motors and other equipment is important. The consequence is an incorrect connection that could cause damage to the equipment. **Figure 7.27** shows an illustration of a three-phase sequence indicator.

Australia follows an anticlockwise phase sequence and the convention for this sequence is ABC or L1, L2, L3 connected left to right on the terminal block.

A phase indicator shows the sequence of phase rotation on its digital display. The presence of all three phases is the illumination of three letters and numbers with each phase clearly identified as L1, L2 and L3 in accordance with current practice. An open phase illuminates two phases only. Proper phase sequence displays L1, L2 and L3 while a reverse sequence illuminates L1, L3 and L2.

FIGURE 7.27 Phase sequence indicator

REVIEW QUESTIONS

1 Describe the basic difference between a polyphase system and a single-phase system.
2 What output is produced by most practical generators?
3 How is the magnitude of the output a.c. voltage varied in a generator?
4 At what angle do stationary armature or stator windings align to each other in a three-phase generator?
5 State the potential differences measured in a two-wire, 230 V, single-phase system.
6 What would be the purpose of a three-phase system that comprises three conductors?
7 State the most common phase-to-phase voltages used for distribution in Australia.
8 What is an advantage of a three-phase system?
9 What is a SWER system?
10 State the conventional phase sequence used in Australia.
11 A generator develops six significant voltages. Three of these are line voltages. What are the other three?
12 What are the two main methods for connecting three-phase coils and loads?
13 State the colour coding phase identification method used in Australia.
14 Determine the peak value of voltage for an 11 kV rms generated voltage.
15 If the peak value of voltage generated by a three-phase machine is 46.662 kV, calculate the root-mean-square value.
16 Name the instrument used to determine the phase sequence in a three-phase system.

7.2 Three-phase star connections

The generation and distribution of energy for general purposes happens using either a three-wire or a four-wire star-connected system. There are many advantages of using a star-connected system, one of which is that the phase voltages are lower in this system when compared with a delta system, requiring less insulation than the stator.

The star connection is made by connecting one end of each phase coil group together with the other end brought out as the active line terminal end as shown in **Figure 7.28**.

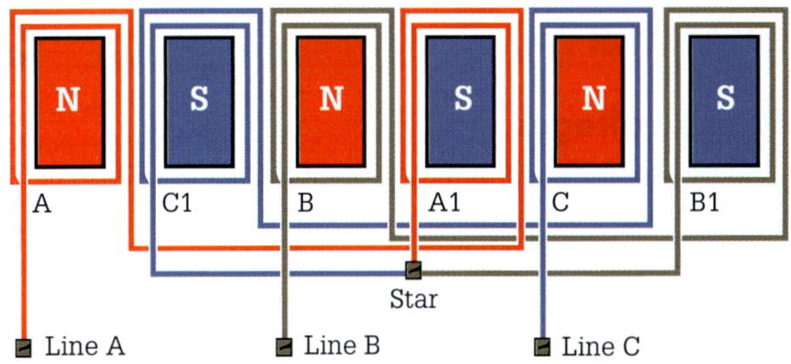

FIGURE 7.28 Three-phase two-pole generator stator star connections

If the star connection occurs inside the generator, it is usual for an earth to be connected to the star-point as shown in **Figure 7.29**. This ensures that the star-point is at zero volts to earth and that the three line voltages are at a constant value to earth.

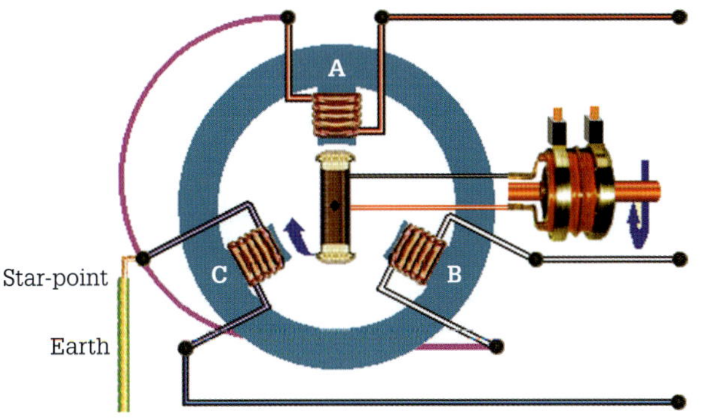

FIGURE 7.29 Star connection

The voltage measured across a single coil group in a star connection is the phase voltage as illustrated in **Figure 7.30**. Notice that the phase voltage measurement is across one phase and the star-point. Similarly, the voltage measured between the lines is known as the line-to-line voltage or line voltage

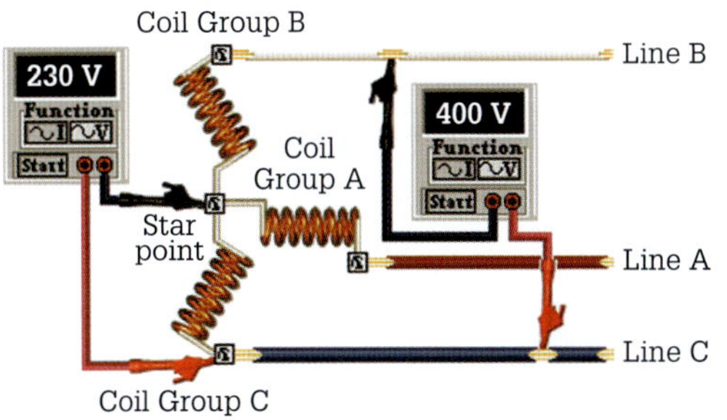

FIGURE 7.30 Phase and line voltages in a star connection

Voltage relationships in a star-connected system

Figure 7.31 shows a star-connected alternator supplying a balanced star-connected load through a three-phase line to a star-connected three-phase motor. In addition, voltmeters connect across a distribution line and phase of the generator and motor. Consequently, the voltage across the line indicates the value of 400 V but the voltmeter connected across the generator and motor phase indicates a value of 230 V.

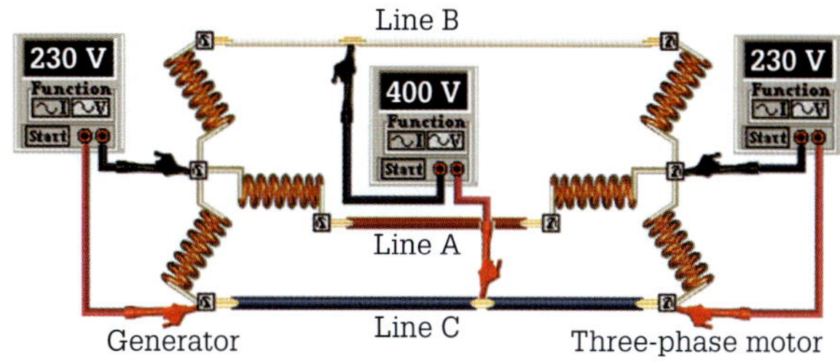

FIGURE 7.31 Voltage relationships

In a star-connected generator and system, the line voltage is of greater potential than the phase voltage by a factor of $\sqrt{3}$ (1.732).

Square root of 3 ($\sqrt{3}$)

EXAMPLE 7.1

Take the triangle of **Figure 7.32**.

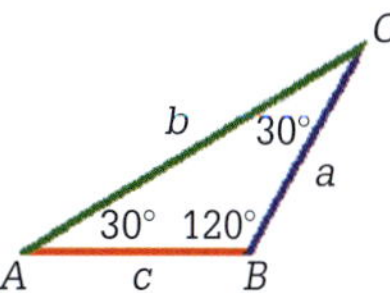

FIGURE 7.32 Square root of 3 ($\sqrt{3}$)

Use the sine rule for non-right-angled triangles:

$$\frac{a}{\sin A} = \frac{b}{\sin B}$$

$$\textit{therefore } b = \frac{a \sin B}{\sin A}$$

Assigning a value of one unit to side a we get:

$$\text{side } b = \frac{1 \times \sin 120^\circ}{\sin 30^\circ} = \frac{1 \times 0.866025...}{0.5}$$

$$= 1.7320508...$$

$$= 1.7320508 \text{ is } \sqrt{3}$$

Thus the ratio between sides a and b is **1 : 1.732**.

In the triangle of **Figure 7.32** the ratio becomes 230:400 (rounded off).

The $\sqrt{3}$ voltage difference occurs because measured line voltage occurs across two phase coil groups. As three-phase voltages are 120° apart, and the line voltage occurs across two phase coil groups, the line voltage is the phasor addition of the voltages generated by each phase coil group. **Figure 7.33** shows a phasor diagram of the phasor addition of the voltages.

EXERCISE 7.1

If the triangle of **Figure 7.32** has the length of sides *a* and *c* of 1270, determine the length of side *b*.

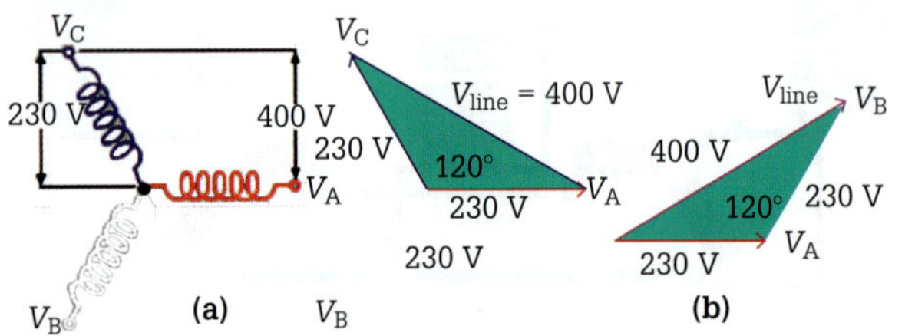

FIGURE 7.33 Phasor addition

In **Figure 7.33(a)**, there is the star-connected coil group indicating the voltage measurement. Next to this diagram is the phasor drawing demonstrating that the phase voltages are 120° out of phase with each other. The resultant line voltage phasor starts from the end of one phasor to the end of the other phasor. With **Figure 7.33(b)**, the parallelogram method of phasor addition drawn to a suitable scale is used.

When the resultant of each pair of phase voltages occurs in a phasor diagram as illustrated in **Figure 7.34**, the line voltage leads the phase voltage by 30° and the phasor addition of the phase voltages is zero.

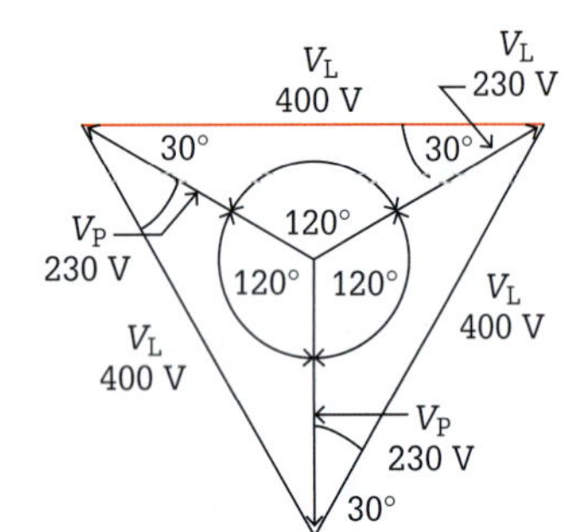

(a) Phasor sum of voltages in a three-phase star

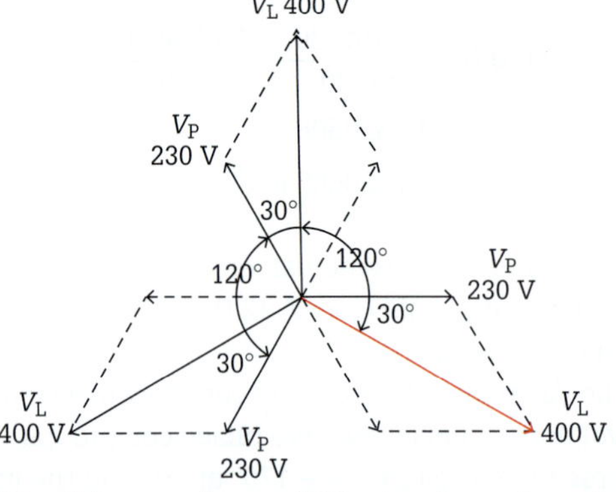

(b) Parallelogram phasor addition in a three-phase star

FIGURE 7.34 Relationship between star phase and line voltages

The equations used to determine voltages in a star-connected system are:

$$V_{line} = \sqrt{3}\,V_{phase} \qquad V_{phase} = \frac{V_{line}}{\sqrt{3}}$$

EXAMPLE 7.2

Phase voltages

The phase voltages in a star-connected three-phase generator are as follows:

A phase = 6351 V; B phase = 6351 V; C phase = 6351 V

Find the line voltage.

$$V_{line} = \sqrt{3}\,V_{phase}$$
$$= \sqrt{3} \times 6351$$
$$= \mathbf{11\ kV}$$

Line voltages

The line voltages in a star-connected three-phase generator are as follows:

Line A–B = 33 kV; Line C–A = 33 kV; Line B–C = 33 kV

Find the phase voltage.

$$V_{phase} = \frac{V_{line}}{\sqrt{3}} = \frac{33\,000}{\sqrt{3}} = 19\,052.6\text{V}$$

EXERCISE 7.2

a The phase voltages in a star-connected three-phase generator are A phase = 1270 V; B phase = 1270 V; C phase = 1270 V. Determine the line voltage.

b The line voltages in a star-connected three-phase generator are Line A–B = 22 kV; Line C–A = 22 kV; Line B–C = 22 kV. Determine the phase voltage.

Current relationships in a star system

In **Figure 7.35** ammeters have been inserted in the phase winding of a star-connected load and in the line supplying energy to the load.

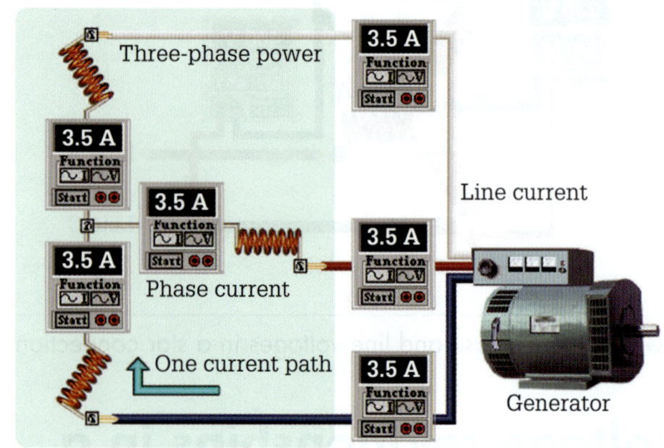

FIGURE 7.35 Relationship between star phase and line currents

From **Figure 7.35** it is apparent that the line current equals the phase current in the phase coil group to which the line is connected. This is because there is one path for current to flow. In a star-connected system:

$$I_{line} = I_{phase}$$

Each phase current is 30° out of phase with the line voltage at unity power factor but in phase with its phase voltage as illustrated in **Figure 7.36**.

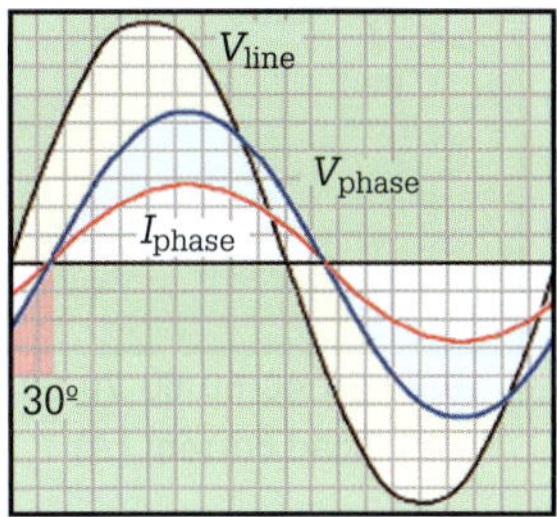

FIGURE 7.36 Phase current at unity pf

Negative phase sequence

Three conditions confirm a balanced three-phase system.

- The sequential order of a three-phase voltage is correct.
- The phase voltages are comparable
- The phase displacement is 120 electrical degrees

Consequently, when those conditions occur, the supply contains no negative phase sequence (or phase imbalance) currents. It follows that phase imbalance currents result from unbalanced loads and line faults. In relation to motors, phase imbalance currents cause the development of a second but reversed magnetic field by the stator windings. The interaction of the two magnetic fields results in reduced motor torque. The implication is that the motor requires more current to achieve the same work. Another effect resulting from phase imbalance currents is the heating of the rotor surface. In short, both effects can damage the motor; a negative phase sequence relay is necessary for the protection of generators and motors.

Reversed phase

To protect the system from reverse phase connections and phase failure, use a phase sequence relay. These relays ensure that the sequence is right when loads connect to three-phase.

Balanced load

A distribution power system has star-connected generators and usually includes star- and delta-connected loads. Moreover, their load may be balanced or unbalanced. **Figure 7.37** shows a star-connected generator driving star-connected loads of equal magnitude.

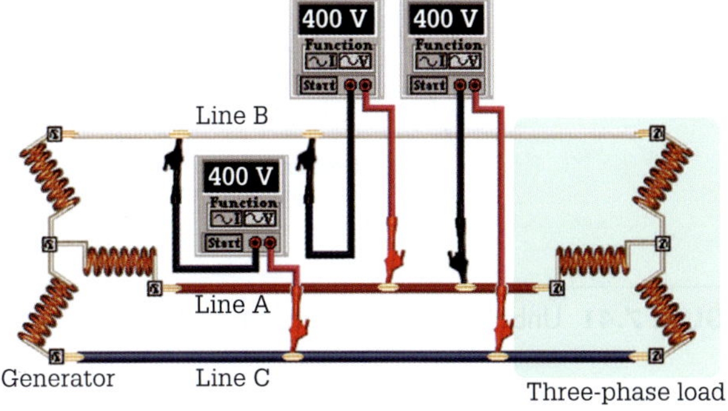

FIGURE 7.37 Balanced load

In examining **Figure 7.37** it is apparent that the phase voltage across each load equals the phase voltage of the generator. Therefore, each phase current in the load equals the line current feeding the load. **Figure 7.38** shows the three-phase voltages of the load together with the current phasor.

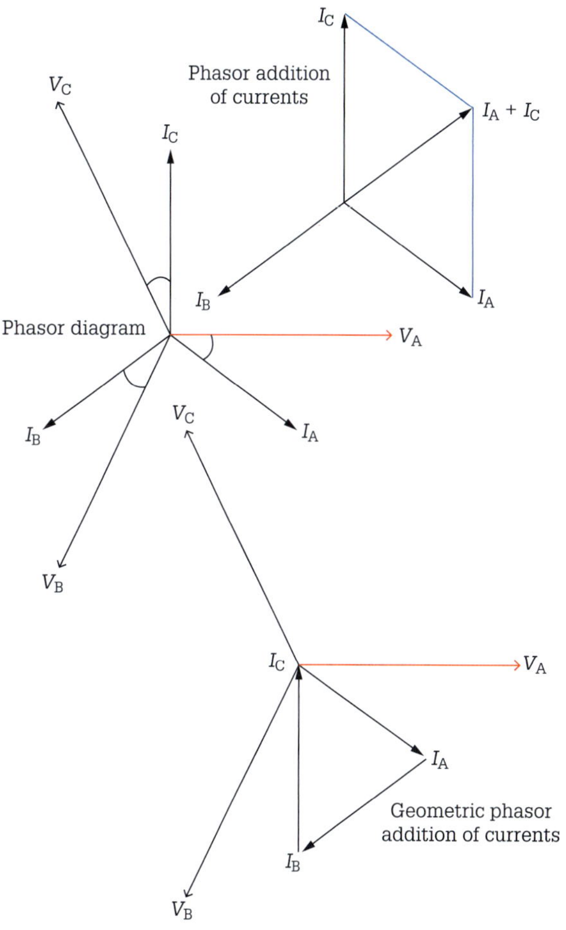

FIGURE 7.38 Phasor addition of currents

The currents are equal in magnitude and lag the phase voltages by angle ϕ. Using the phasor addition method of resolving phasors, if $\bar{I}_C$ is added to $\bar{I}_A$ we find the resultant $I_C + I_B$ to be equal in magnitude and the opposite direction to I_B. Adding the resultant $\bar{I}_C + \bar{I}_A$ to $\bar{I}_B$ gives zero, proving that there are no out-of-balance currents and the system balances.

SWITCH ON

For a balanced star system, the load must be drawing the same current at the same angle ϕ on each phase. Consequently, when these two conditions happen there is no neutral current.

Effect of unbalanced loads on a three-wire star system

Many three-phase systems represent an unbalanced condition due to uneven loading on the different phases. With such a system, knowledge of the magnitude of voltage across and current in one phase does not notify us of the voltages across and currents in the other phases. Consequently, each voltage and current requires independent determination. When a star-connected load is unbalanced, and there is no neutral, then the load's star-point is isolated from the star-point of the alternator. Because of the unbalanced load, the phase voltages (these are not $1/\sqrt{3}$ of the line voltage) and currents are different, with the result that the star-points (neutral point) are not at the same potential. The potential of the star-point of a load is subject to variation, 'a floating star-point', according to the unbalance of the load. **Figure 7.39** shows a star-connected generator driving star-connected single-phase loads of different magnitudes.

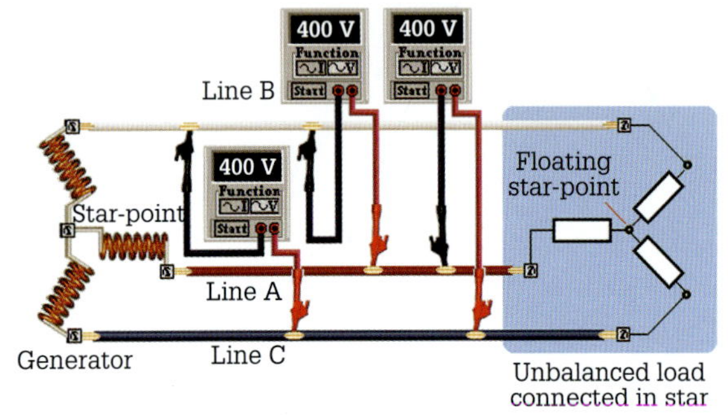

FIGURE 7.39 Unbalanced load

The standard phase to star-point voltage in Australia is 230 V. The resulting voltages between phases are ($\sqrt{3} \times 230$) which is equivalent to 400 V nominal. Provided the loads remain balanced, the voltage between each phase and the star-point remains unaffected. Most residential consumers in Australia connect to a single-phase system derived from a distributed three-phase system.

If the three-phase load currents deviate from the ideal balanced condition, the voltage between each phase and the star-point will experience wide variations. Depending upon the degree of unbalance between the three single-phase loads the three voltages across the loads vary from zero to any voltage up to phase-to-phase voltage.

Uneven loading on the three phases causes problems for single-phase circuits. From **Figure 7.39** if we assume that the load impedance connected to the white phase is 15 Ω, red phase is 10 Ω and blue phase is 6 Ω at various power factors, then, the currents drawn by the loads also vary according to the changing voltage across them.

A graphic representation in **Figure 7.40** shows the voltage variations between each phase and the star-point and their effect on each load current.

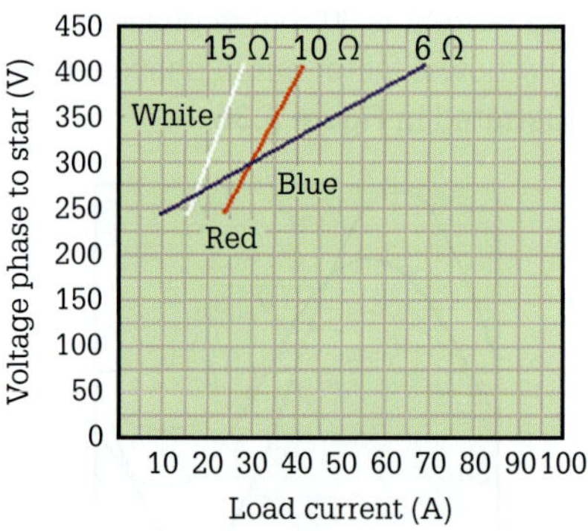

FIGURE 7.40 Voltage and current variations

The star or neutral point of the convergence of the three phases causes a shift, depending on the ratio of the load currents drawn by the three phases in an unbalanced condition. This change results from the variations of phase to star-point voltage. Such a consequence causes damage to connected loads due to over-voltage on the circuit.

Most connected equipment and appliances are only designed to withstand a 10% over-voltage (230 V × 110% = 253 V), plus some additional margin, in order to cater for the probability of short-duration temporary over-voltages. **Figure 7.41** shows a phasor diagram illustrating an unbalanced load in a star system in which the loads have different power factors and current magnitudes. Under these circumstances, the neutral conductor must return the imbalance in the phase currents to the supply.

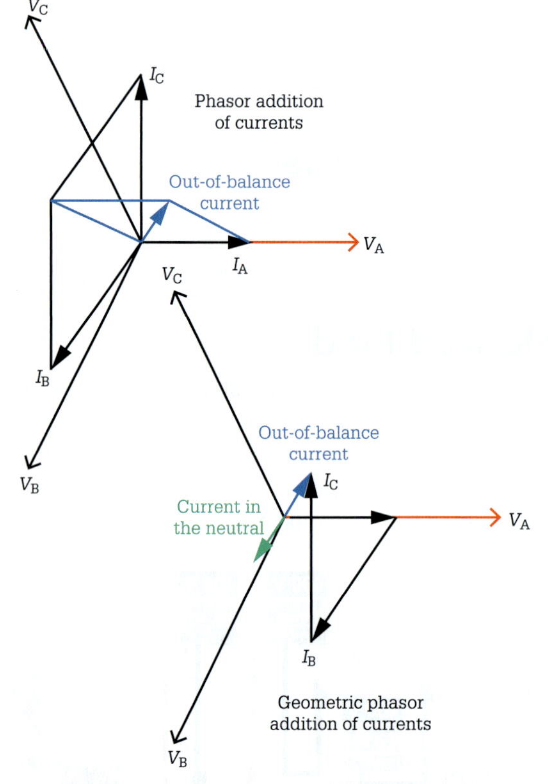

FIGURE 7.41 Unbalanced load in a star system

REVIEW QUESTIONS

1 Why is the generation and distribution of energy carried out using a three-phase star connection?
2 How is a star connection effected in a three-phase generator?
3 What is the voltage measurement between the star point and each line terminal called?
4 State the angle of relationship between the line voltage and the phase voltage in a star-connected system.
5 If the phase voltage in a star generator is 6351 V, determine the line voltage.
6 If the line voltage in a star generator is 22 kV, determine the phase voltage.
7 State the current relationship between the line current and phase current in a star-connected system.
8 What are the two electrical conditions that must exist for a balanced three-phase system supplying a load?
9 What electrical condition causes problems for single-phase circuits connected across a three-phase, three-wire star-connected system?
10 When the three-phase currents deviate from the ideal balanced conditions, what effect does this have on the phase voltage across the loads?

7.3 Three-phase four-wire systems

The generation and distribution of energy to consumers occurs using the star connection because the star-point provides a terminal for a neutral line. The establishment of a neutral allows another voltage to exist between it and any active line. In addition, the neutral at the star-point connects to the general mass of earth. This connection ensures that the neutral in relation to the earth is at zero potential. Consequently, if an earth fault occurs, the system has protection. **Figure 7.42** shows a four-wire three-phase generator.

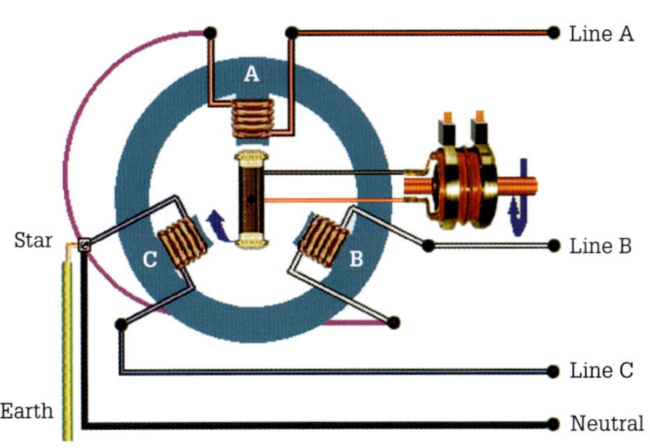

FIGURE 7.42 Star connection with neutral

MEN system of earthing

With a multiple earth neutral (MEN) system of earthing, the distribution transformer neutral point connects to the soil. In addition, approximately every fourth transmission pole has an earth conductor. These earthing conductors attach to earth electrodes embedded into the soil. A timber batten in some states in Australia provides mechanical protection for the grounding conductor and protects persons and animals from possible 'touch potentials'.

Touch potential is a term that is used to describe the voltage difference that occurs across a person or animal by touching an exposed grounding conductor and the soil (or ground to ground, a differential potential) when fault current is flowing. **Figure 7.43** illustrates the grounding conductor.

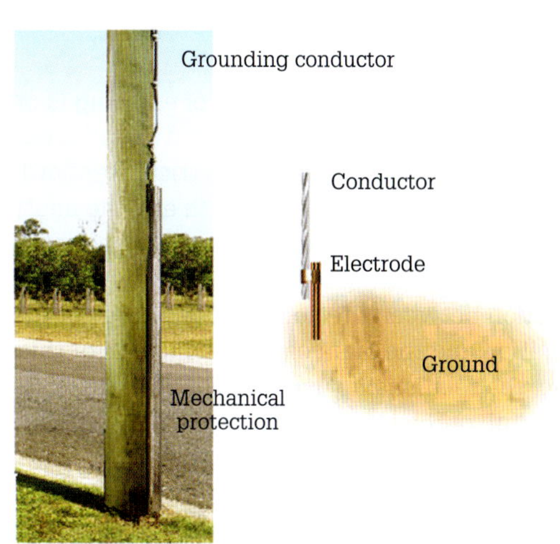

FIGURE 7.43 Grounding conductor

At the consumer's switchboard, the main neutral connects to the main earth of the installation to create another MEN link. Refer to **Figure 7.44**.

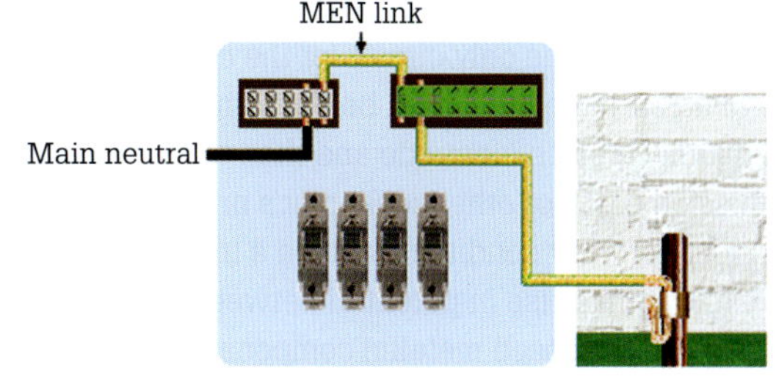

FIGURE 7.44 MEN connection at the switchboard

The MEN system of earthing uses the distribution neutral conductor for any fault currents that may flow because of the active conductor passing current to earth. Moreover, the number of parallel paths that the fault current can access when an earth fault occurs assist this flow of fault current. **Figure 7.45** illustrates the many parallel paths created between the distribution neutral and the ground.

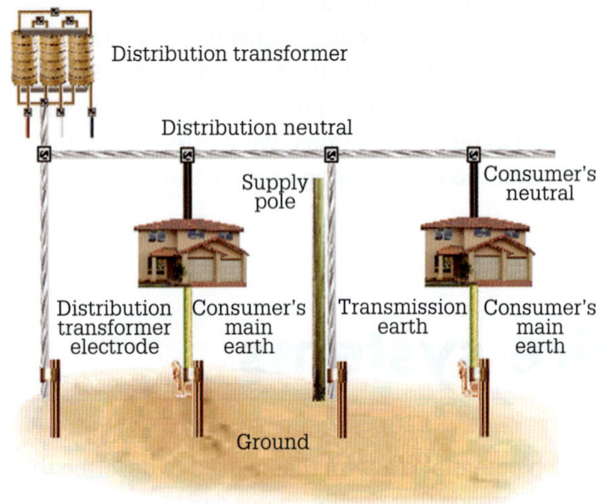

FIGURE 7.45 Parallel paths in a MEN system

SWITCH ON

MEN defined

The multiple earth neutral system of earthing is one in which defined parts of an electrical installation that require earthing under an electrical standard connect to the general mass of earth and the neutral conductor of the distribution system. Consequently, this system of earthing requires an electrical connection between the neutral connections of the installation and the main earth conductor. For this reason, the relationship usually occurs at the main neutral link of the installation.

The effectiveness of an MEN earthing system

When the active conductor passes current to earth because of a fault, the protection device operates. If the neutral becomes open circuited at any point, the fault current conducts by the parallel earth paths. However, reliance on the MEN system alone is insufficient for personal safety. In the event that a neutral was open circuited or the MEN link was missing, the only circuit for the current to return to the distribution point would be through the ground at the consumer's earth electrode and then to the supply pole transmission earth or other consumer's main earths. This path under very dry conditions can be a high-impedance path resulting in a rise in potential between the earthing system and all exposed metallic components that are in contact with it. Such a situation creates hazardous touch voltages on exposed metallic conductive parts of the earth circuit as they become 'live' in such circumstances if they are not bonded to the main earth conductor.

The effectiveness of the MEN system is reliant on the continuity of the main neutral conductor, the MEN link and the connection of the main earthing conductor to the earth electrode. In addition, use of core balance earth leakage switches minimises the effects of touch voltages and their potential lethal effects. **Figure 7.46** illustrates a residential MEN earthing system.

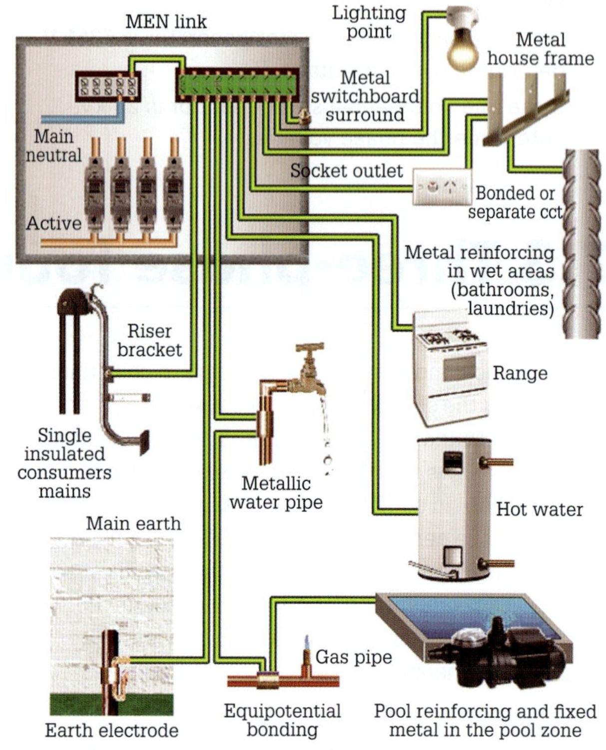

FIGURE 7.46 Residential MEN earthing system (cct = circuit)

The best way to connect the earths would be to run all the earths separately back to a single earth bar. However, the practical way is to arrange the earths as a strict 'tree' structure, with the circuit and bonded earths connected only to the main earth conductor.

High impedance in the neutral conductor

As the electrical environment is very dynamic, it is not realistic to balance any three-phase four-wire system loads correctly. Balancing is difficult because the loads are being turned on and off. These changing electrical conditions create imbalances in the system. Consequently, when currents become unbalanced as shown in **Figure 7.47** phase currents no longer cancel each other, and a neutral current begins to flow. This current in the neutral conductor is equal to the phasor sum of the instantaneous values of current in the three distribution lines. Because this current flows through the neutral conductor in the reverse direction as the line currents, it requires a negative sign.

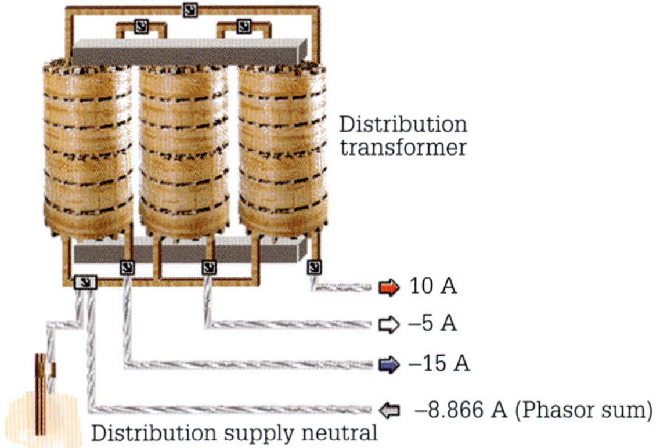

FIGURE 7.47 Unbalanced currents in a four-wire system

Consider an unbalanced three-phase four-wire system that has three single-phase loads of 30 A, 60 A and 40 A connected as shown in **Figure 7.48**.

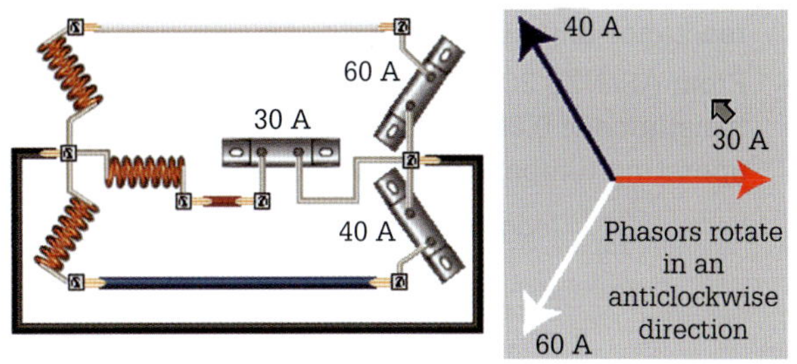

FIGURE 7.48 Unbalanced single-phase loads and phasor diagrams

The unbalanced current returns to the power distribution point via the neutral conductor. **Figure 7.49** shows geometric phasor addition.

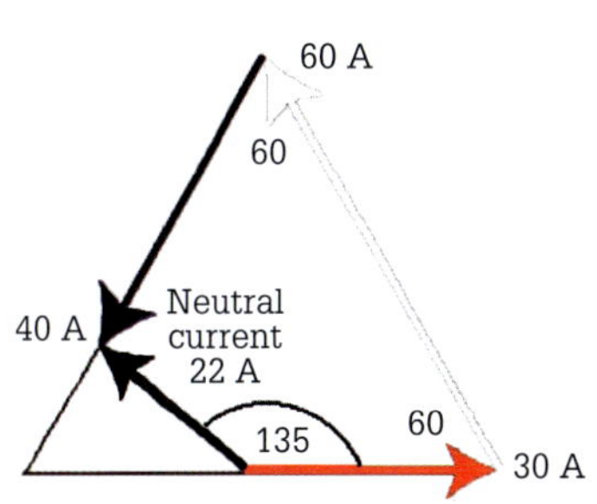

FIGURE 7.49 Geometric phasor addition

The neutral conductor carries a 22-ampere unbalanced current at an angle of 135° back to the distribution supply star-point.

With no current occurring in the neutral wire there is no voltage drop and the neutral is at ground potential. Conversely, in unbalanced systems the neutral carries current causing a voltage drop to transpire between it and the general mass of earth.

In the event of a high impedance or open-circuited neutral, the current returns to the supply source via the earth connections. Therefore, the MEN system must have a low impedance path to the general mass of earth. A low-impedance path enhances the ability of the MEN circuit to reduce neutral currents. Once a fault current flows into the general mass of earth the existence of various MEN earths ought to hold the neutral at very low voltages. **Figure 7.50** illustrates the conveying of the neutral current to the supply star-point through multiple MEN connections.

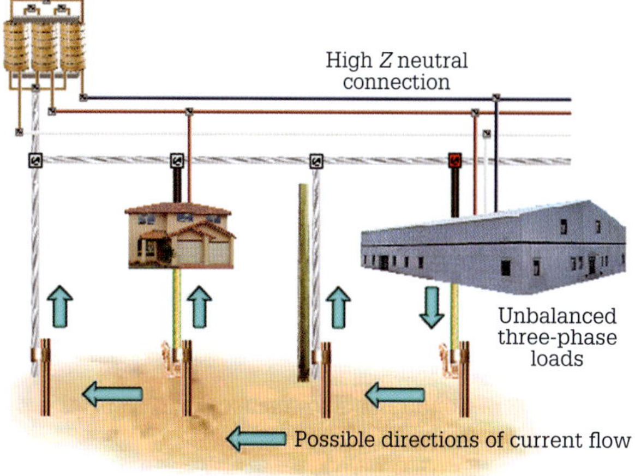

FIGURE 7.50 Shunting of neutral current due to high impedance connection

The voltage across each active to the neutral depends on the loads on each phase and the ground resistance. The worse the balancing of the load over the three phases, the worse the voltage disturbance. The result is that mains voltage may occur across earthed terminations and the neutral.

The need for a neutral conductor

In practice, three-phase three-wire systems occur rarely for end users of electrical energy because of the lack of a neutral conductor that compensates for the out-of-balance conditions. The neutral is included to enable another voltage to exist between it and any active conductor. The neutral can also carry any out-of-balance currents.

Effects of a break in the neutral conductor

In a single-phase system derived from a single-phase source, a break in the neutral conductor directly results in a loss of the energy supply. For a three-phase system the effects of a broken neutral are:

1 The lines to star-point voltages (phase voltages) are no longer 230 V – this could be higher or lower by quite a large amount. The voltages between phases (phase-to-phase) remain unaffected by a break in the neutral conductor.
2 A high potential difference develops across the breach in the neutral.

The effects of an open circuit neutral vary, depending mainly on the load balance of a three-phase system, but also on the category of earthing system used and the location of the open circuit relative to the load. The effects could include damage to loads due to over-voltages on single-phase circuits or the establishment of dangerous touch voltages on exposed conductive parts.

Neutral conductor and wiring standards

The Australian and New Zealand standard AS/NZS 3000: 2018 *Wiring Rules* sets out requirements for all installations together with the requirements of any regulatory authority. Some of the mandatory requirements for the neutral follow.

- Section 1.4 'Definitions', sub-clause 1.4.84 'Neutral (neutral conductor or mid-wire)'
- Section 2.5 'Protection against overcurrent', sub-clause 2.5.1
- Clause 2.3.2.1.2 'Alternating current systems'
- Section 1.5 'Fundamental principles', clause 1.5.2 'Control and isolation'
- Section 3.5 'Conductor size', clause 3.5.2 'Neutral conductor'.

The following clauses refer to identification of the neutrals on a switchboard. Please refer to the standard for their descriptions:

- Clause 2.10.5 'Equipment identification', sub-clause 2.10.5.4 'Terminals of switchboard equipment'
- Sub-clause 2.10.5.3 'Bars and links'
- Sub-clause 2.10.5.5 'Common neutral'.

Voltage drop requirements

AS/NZS 3008.1.1: 2017 'Electrical installations, Selection of cables', subsection 4.6, has the voltage drop requirements for an unbalanced three-phase circuit.

In these circumstances of unbalance, a rule of thumb is to assume balanced three-phase load conditions and calculate the current flowing in the heaviest loaded phase.

Alternatively, determination of the voltage drop value can occur on a single-phase basis by geometrically summing the voltage drop in the heaviest loaded phase and the voltage drop in the neutral.

V_d = voltage drop in heaviest loaded active + voltage drop in neutral

$$V_d = I_A L_A Z_{cA} + I_N L_N Z_{cN}$$

where I_A = active current, in amperes
L_A = length of the active in kilometres
Z_{cA} = impedance of active cable, in ohms/km
I_N = neutral current in amperes
L_N = length of the neutral in kilometres
Z_{cN} = impedance of neutral cable, in ohms/km

The voltage drop in each conductor requires assessing, with knowledge of a particular conductor material, size, temperature and length, the magnitude and phase angle of the current in each conductor, and the phase angle of the load.

The reactance X_c and resistance R_c of cables are specified in ohms per kilometre, which allows calculation of the total impedance Z_c for any given cable route length L.

$$V_d = I Z_C$$

where V_d = voltage drop in cable, in volts
I = current flowing in cable, in amperes
Z_c = impedance of cable, in ohms = $\sqrt{(R_c^2 + X_c^2)}$
R_c = cable resistance, in ohms; a function of the material, size and temperature of the conductors
X_c = cable reactance, in ohms; a function of the conductor shape and cable spacing

EXAMPLE 7.3

A 65 A resistive load is supplied through a 25 mm², 4-core, V90 cable having circular conductors. Determine the voltage drop in one line conductor if it has a route length of 50 m.

- From AS/NZS 3008.1.1 Table 30, 25 mm² has a reactance of 0.0853 Ω/km
- From AS/NZS 3008.1.1 Table 35, 25 mm² has a resistance of 0.884 Ω/km at 75 °C, which is the operating temperature of V90 cables.

$$R_C = \frac{length\text{ (in m)} \times Resistance/km}{1000} = \frac{50 \times 0.884}{1000} = 0.0442\ \Omega$$

$$X_C = \frac{length\text{ (in m)} \times Reactance/km}{1000} = \frac{50 \times 0.0853}{1000} = 0.004265\ \Omega$$

$$Z_C = \sqrt{R_C{}^2 + X_C{}^2} = \sqrt{0.0442^2 + 0.004265^2} = 0.0444\ \Omega$$

$$V_d = I Z_C = 65 \times 0.0444 = \mathbf{2.88\ V}$$

EXERCISE 7.3

A 55 A resistive load is supplied through a 16 mm², 4-core, V90 cable having circular conductors. Determine the voltage drop in one line conductor if it has a route length of 45 m.

From AS/NZS 3008.1.1 Table 30, 16 mm² has a reactance of 0.0861 Ω/km

From AS/NZS 3008.1.1 Table 35, 16 mm² has a resistance of 1.40 Ω/km at 75 °C, which is the operating temperature of V90 cables.

REVIEW QUESTIONS

1 In a four-wire system, what connection ensures that the neutral in relation to the earth is at zero potential?
2 What is the electrical relationship between the star-point and the general mass of earth in an MEN system?
3 What is meant by the term 'touch potential'?
4 Name an advantage of a four-wire star-connected three-phase system.
5 On what does the effectiveness of the MEN system of earthing rely?
6 What additional device minimises the effects of touch voltage in an electrical installation?
7 What functions does the neutral conductor serve in a four-wire star-connected three-phase system connected to unbalanced single-phase loads?
8 In the event of a high impedance or open-circuited neutral, how does the neutral current return to the supply source?
9 A 50 A resistive load is supplied through a 16 mm², 4-core, V90 cable having circular conductors. Determine the voltage drop in one line conductor if it has a route length of 65 m.
10 Where would an electrician locate regulatory requirements relating to the wiring of a neutral in a consumer's installation?

7.4 Three-phase delta and interconnected systems

Three-phase power occurs with delta-connected generators or transformers. The delta connection derives its name from the resemblance of its schematic symbol to the Greek letter delta (Δ). Connecting one end of each coil phase group to the start of another coil phase group to which the active-line conductors connect, as shown in **Figure 7.51**, makes the delta connection.

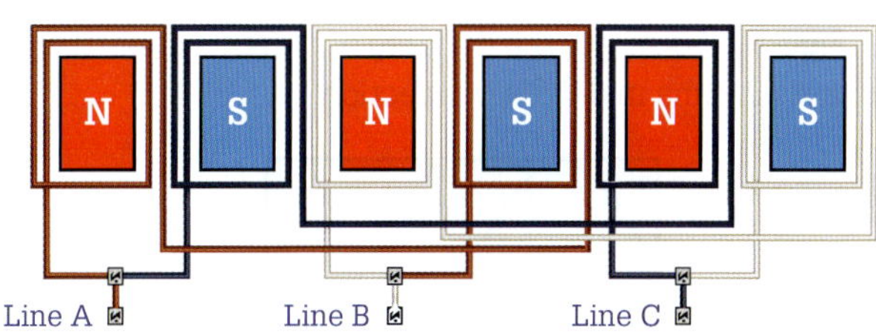

FIGURE 7.51 Three-phase two-pole generator stator delta connection

A two-pole delta-connected generator is shown in **Figure 7.52**.

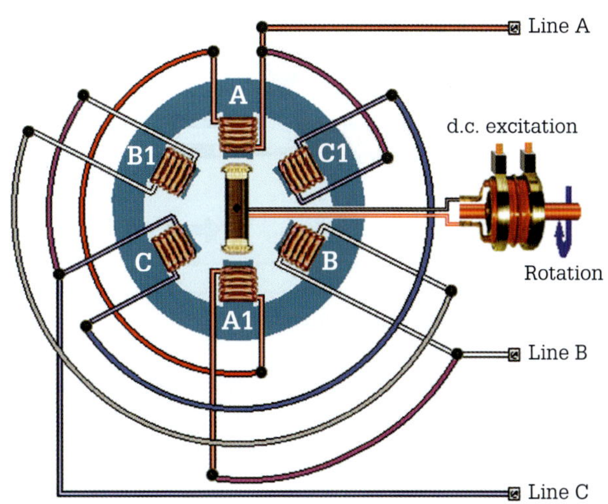

FIGURE 7.52 Two-pole delta connection

An advantage of the delta-connected generator is that if one phase open circuits, the other two phases continue to supply three-phase. However, the capacity of the generator reduces to 57.7%.

The primary winding of the delta-star-connected distribution transformer in **Figure 7.53** is a load on the input line ABC. Moreover, the star-connected secondary simulates a generator supplying the load lines XYZ and N.

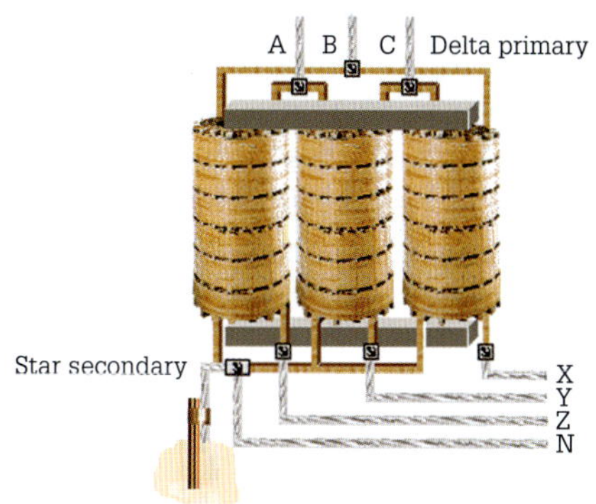

FIGURE 7.53 Three-phase delta-star distribution transformer

The voltage measured across a single coil group is the phase voltage as illustrated in **Figure 7.54**. Notice that the phase voltage and the line-to-line voltage have the same value.

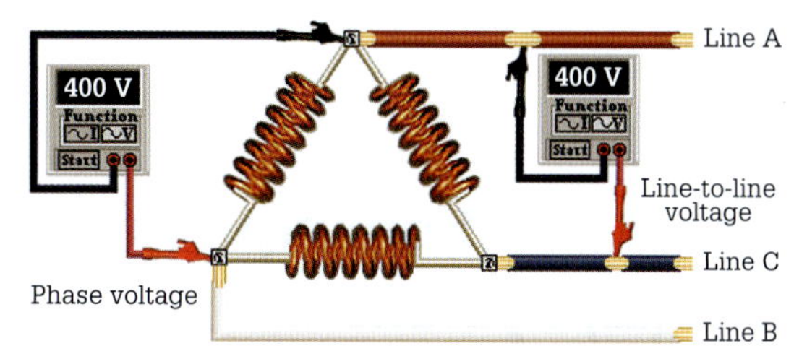

FIGURE 7.54 Phase and line voltages in a delta connection

Voltage relationships in a delta-connected system

Because the three-phase delta voltages are 120° apart, and unlike ends of the coil phase groups connect, the resultant voltage is equal to the phasor addition of the individual voltages. A phasor diagram illustrating the phasor addition of voltages is shown in **Figure 7.55**.

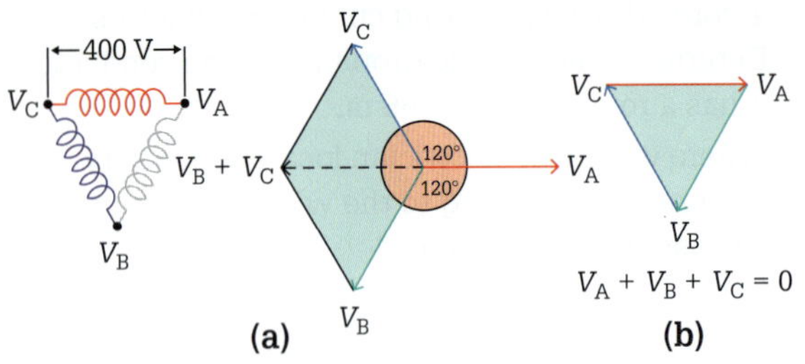

FIGURE 7.55 Phasor addition

Figure 7.55(a) shows the parallelogram method of phasor addition and that the phasor addition of the voltages $\overline{V}_B$ and $\overline{V}_C$ is equal to, and acts in an opposite direction to, voltage V_A. The phasor addition of the three voltages is $\overline{V}_A + \overline{V}_B + \overline{V}_C = 0$.

In **Figure 7.55(b)** it can be shown by geometric phasor addition that the three voltages around the delta system are equal to zero volts.

As the voltages add up to zero at every instant in time, there is no voltage potential available to drive current around through the coil phase groups. As a result, there is no circulating current. In contrast, an unbalanced system generates circulating current.

To verify that three voltages can connect together in a delta configuration without causing circulating current, open one coil phase group connection and measure the voltage across the break. **Figure 7.56** shows that there are zero volts across the coil phase groups indicating that no current circulates within the delta connection when the connection restores.

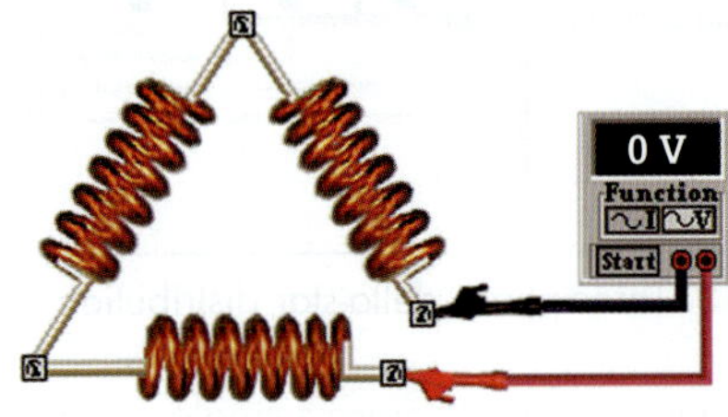

FIGURE 7.56 Zero voltage across coil phase groups

In a delta connection, line voltage and phase voltage are the same. In a delta-connected system:

$$V_{line} = V_{phase}$$

Current relationships in a delta (Δ) system

Figure 7.57 shows a delta-connected generator supplying a balanced delta-connected motor load through a three-phase line. Ammeters connect in the distribution line and within a phase of the motor.

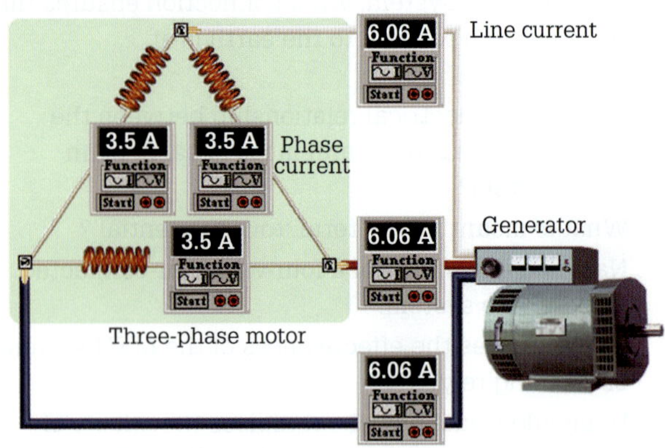

FIGURE 7.57 Relationship between delta phase and line currents

The ammeters in the distribution line indicate a value of 6.06 A, while the ammeters connected in the coil phase groups indicate 3.5 A.

In a delta-connected generator and system the line current is of greater capacity than the phase current by a factor of $\sqrt{3}$ (1.732). This current difference occurs because the current flows through the different coil phase groups at dissimilar instants in time because they are 120° out of phase with each other. During one moment in time, current flows between two lines only. At other moments in time, current flows from two distribution lines to the third.

The delta connection is a parallel circuit because there is more than one path for current.

Figure 7.58 shows the dividing of currents at particular instants in time in a delta connection.

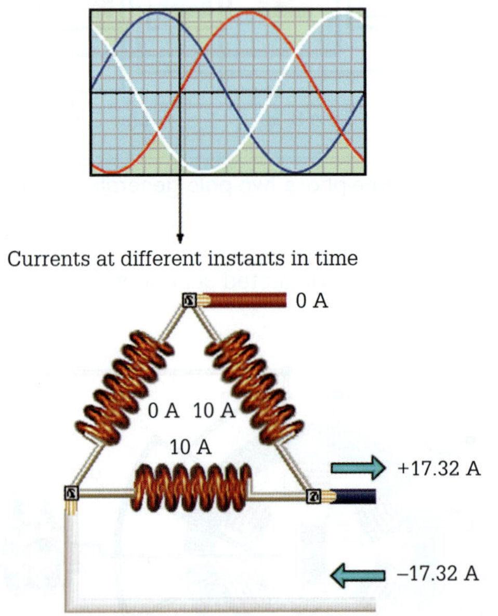

FIGURE 7.58 Three-phase waveform

Because the currents in a delta system are 120° out of phase with each other, use phasor addition when determining the line current flowing in the system. In

Figure 7.59, the drawn-to-scale phase currents are 120° out of phase with each other. Furthermore, representation of the resultant line current phasor occurs as a line drawn from the end of one phasor to the end of the other phasor.

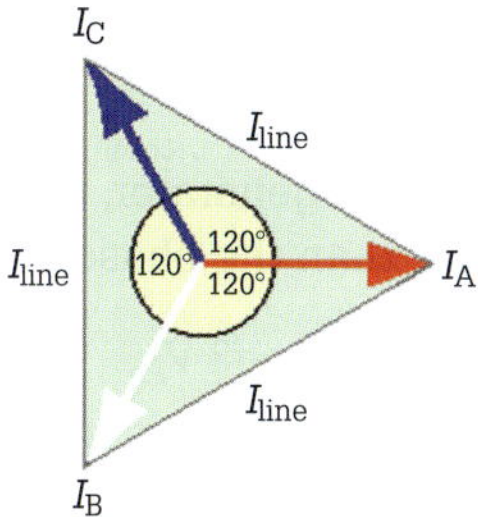

FIGURE 7.59 Phasor addition

When the resultant of each pair of phase currents is represented in a phasor diagram, as illustrated in Figure 7.60, it can be seen that the line current leads the phase current by 30° and the phasor addition of the phase currents is zero, and the phasor addition of the line currents is also zero.

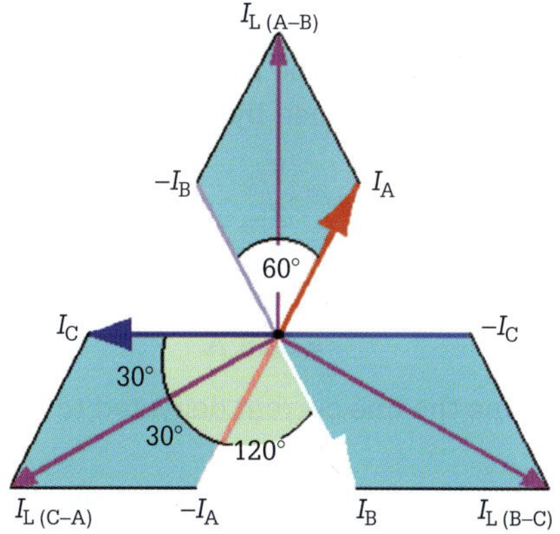

FIGURE 7.60 Relationship between delta phase and line currents

The phase currents in Figure 7.60 are I_A, I_B and I_C. To determine the line current, first, determine the phasor addition of the currents in the two coil phase groups to which that line connects. For example, the current drawn from line $I_{L(A-B)}$ must be:

$$\bar{I}_L = \bar{I}_A + (-\bar{I}_B)$$

Since $\bar{I}_A$ and $-\bar{I}_B$ are two equal current phasors 60° apart in a balanced system, their phasor sum is $\sqrt{3}$ or 1.732 times the value of either $\bar{I}_A$ or $\bar{I}_B$. Therefore, $\bar{I}_{L(A-B)} = \sqrt{3}\,\bar{I}_A$ as shown in Figure 7.60. The equations used to determine currents in a delta-connected system are:

$$I_{Line} = \sqrt{3}\,I_{Phase} \text{ and } I_{Phase} = \frac{I_{Line}}{\sqrt{3}}$$

Reversed phase in delta-connected generator

When connected correctly, the sum of the phase voltages equals zero and hence there is no circulating current in the phase winding as shown in Figure 7.61 (a) and (b). If, however, one coil connects in reverse, such that the connections are A1 to B2, A2 to C2, and C1 to B1, then the sum of the phase voltages is much greater than zero as shown in Figure 7.61 (c) and (d). In this example, C phase coil is connected in reverse, resulting in 800 V being generated within the closed circuit of the windings of a 400 V, three-phase generator. This increased voltage will cause high circulating currents to flow in the windings, which will cause them to overheat and rapidly burn out.

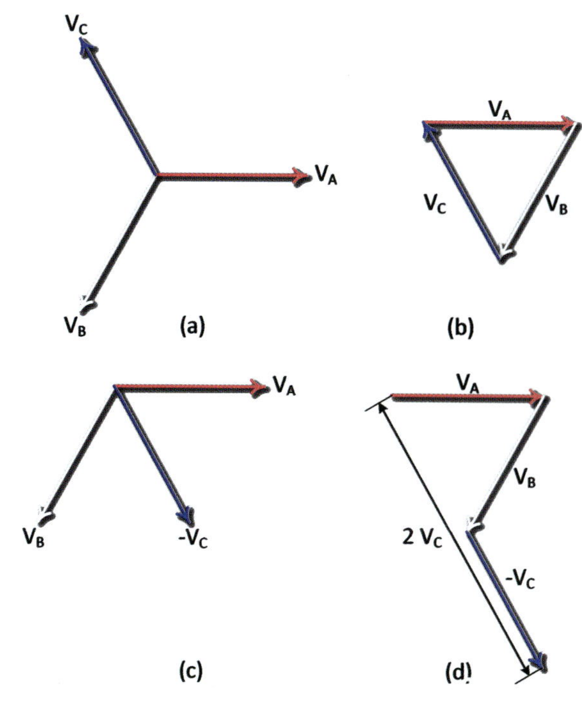

FIGURE 7.61 Effect of reversed coil in delta connection

One method of testing for this condition before connecting the coils is to connect two leg pairs of the coils and measure the potential between the open legs as shown in Figure 7.62, where C-phase coil is reversed. A reversed coil will indicate twice phase voltage.

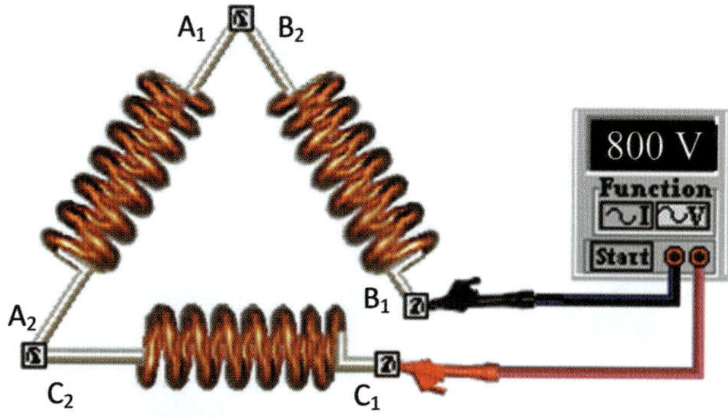

FIGURE 7.62 Testing for reversed coil in delta connection

Delta-connected loads

Equipment designed to operate from delta-connected power, such as motors, can also run from star-connected systems without a problem, since the phase-to-phase voltages are available in both systems. However, equipment that requires star-connected power cannot operate from a delta-connected system. The phase-to-neutral voltages are not available. Besides electric motors, delta-connected transformers as used in high-voltage transmission systems have their primary winding connected in delta configuration.

SWITCH ON

Balanced loads such as three-phase-connected motors do not need a neutral conductor and can be connected across the three lines regardless of whether it is a delta system or not.

Interconnected star and delta devices

In the following examples, values of phase and line voltages and phase and line currents are determined for different three-phase configurations.

Star–delta (Y–Δ)

EXAMPLE 7.4

Figure 7.63 shows a star-connected three-phase generator with a rated voltage of 230 V per coil phase group driving a balanced load consisting of three 10 Ω impedance windings connected in delta. The following values are determined by applying various equations.

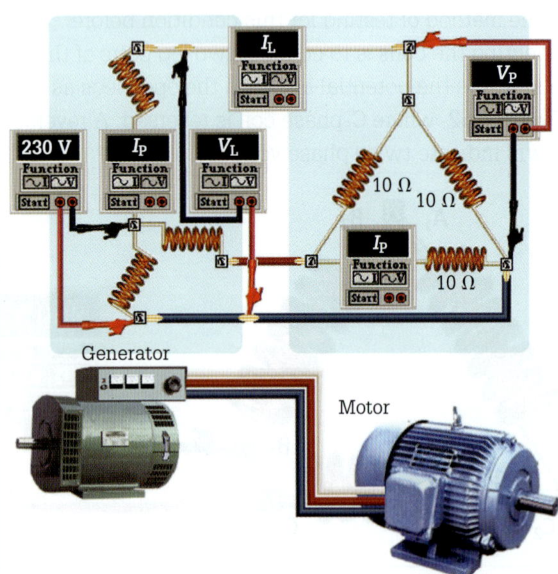

FIGURE 7.63 Interconnected star–delta devices

a Line voltage of the star-connected generator: $V_{L\,(Load)}$
b Phase voltage of the load: $V_{Ph\,(Load)}$
c Phase current drawn by the load: $I_{Ph\,(Load)}$
d Line current delivered to the load: $I_{L\,(Load)}$
e Phase current in the generator: $I_{Ph\,(Gen)}$

Solutions

a Determine the line voltage of the star-connected three-phase generator: $V_{L\,(Load)}$.
In a star-connected generator, the line voltage is 1.732 times greater than the phase voltage:

$$V_L = \sqrt{3}\,V_{Ph} = \sqrt{3} \times 230 = \mathbf{398\ V}$$

b Determine the phase voltage of the load: $V_{Ph\,(Load)}$.
The three-phase delta-connected load connects directly to the generator. Therefore, the phase voltage across the load is equal to the line voltage supplied by the generator.

$$V_{Ph} = V_L = \mathbf{398\ V}$$

c Determine the phase current drawn by the load: $I_{Ph\,(Load)}$.
Now that the phase voltage of the load has a value, the magnitude of the phase current drawn by the load can be determined by using Ohm's law:

$$I_{Ph} = \frac{V_{Ph}}{Z_{Ph}} = \frac{398}{10} = \mathbf{39.8\ A}$$

d Determine the line current delivered to the load: $I_{L\,(Gen)}$.
The three 10 Ω impedance phase-group windings connected in delta experience 39.8 A flowing in each phase. As the load connects in delta, the line current drawn by the load must be $\sqrt{3}$ times greater than the phase current:

$$I_L = \sqrt{3}\,I_{Ph} = \sqrt{3} \times 39.8 = \mathbf{68.9\ A}$$

e Determine the phase current in the generator: $I_{Ph\,(Gen)}$.
The coil phase groups of the alternator connect in star. In a star connection, the phase current and the line current are equal to each other:

$$I_{Ph} = I_L = \mathbf{68.9\ A}$$

EXERCISE 7.4

A star-connected three-phase generator has a rated voltage of 220 V per coil phase group and supplies a balanced load consisting of three 15 Ω impedance windings in a delta motor. Calculate the following:

a Line voltage of the star-connected generator: $V_{L\,(Load)}$
b Phase voltage of the load: $V_{Ph\,(Load)}$
c Phase current drawn by the load: $I_{Ph\,(Load)}$
d Line current delivered to the load: $I_{L\,(Load)}$
e Phase current in the generator: $I_{Ph\,(Gen)}$

Delta–star (Δ–Y)

EXAMPLE 7.5

Figure 7.64 shows a delta-connected three-phase generator with a rated voltage of 210 V per coil phase group driving a balanced load consisting of three 20 Ω impedance windings connected in star. The following values are determined by applying various equations.

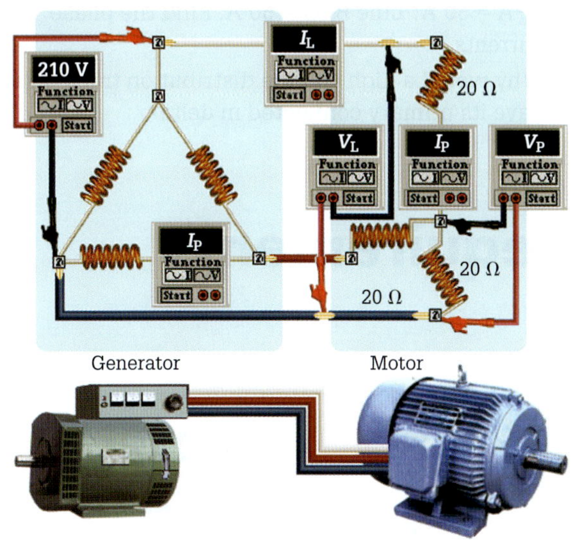

FIGURE 7.64 Interconnected delta–star devices

a Line voltage of the delta-connected generator: $V_{L\,(Load)}$

b Phase voltage of the load: $V_{Ph\,(Load)}$

c Phase current drawn by the load: $I_{Ph\,(Load)}$

d Line current delivered to the load: $I_{L\,(Load)}$

e Phase current in the generator: $I_{Ph\,(Gen)}$

Solutions

a Determine the line voltage of the delta-connected three-phase generator: $V_{L\,(Load)}$.

In a delta-connected system, the line voltage is the same magnitude as the line voltage:

$$V_L = V_{Ph}$$
$$= \mathbf{210V}$$

b Determine the phase voltage of the load: $V_{Ph\,(Load)}$.

$$V_{Ph} = \frac{V_L}{\sqrt{3}}$$
$$= \frac{210}{\sqrt{3}}$$
$$= \mathbf{121.2\ V}$$

The star-connected three-phase load connects directly to the generator. Therefore, the phase voltage across the load is less than the line voltage supplied by the generator by a factor of $\sqrt{3}$

c Determine the phase current drawn by the load: $I_{Ph\,(Load)}$.

Now that the phase voltage of the load has a value, the magnitude of the phase current drawn by the load can be determined by using Ohm's law:

$$I_{Ph} = \frac{V_{Ph}}{Z_{Ph}}$$
$$= \frac{121.2}{20}$$
$$= \mathbf{6.06\ A}$$

d Determine the line current delivered to the load: $I_{L\,(Load)}$.

The three 20 Ω impedance coil phase group windings connected in star experience 6.06 A in each phase. As the load connects in star the line current drawn by the load is of the same magnitude as the phase current:

$$I_L = I_{Ph}$$
$$= \mathbf{6.06\ A}$$

e Determine the phase current in the generator: $I_{Ph\,(Gen)}$.

The coil phase groups of the generator connect in delta. In a delta connection the phase current is less than the line current by a factor of $\sqrt{3}$:

$$I_{Ph} = \frac{I_L}{\sqrt{3}}$$
$$= \frac{6.06}{\sqrt{3}}$$
$$= \mathbf{3.5\ A}$$

EXERCISE 7.5

A delta-connected three-phase generator has a rated voltage of 250 V per coil phase group and supplies a balanced load consisting of three 10 Ω impedance windings in a star-connected motor. Calculate the following:

a Line voltage of the delta-connected generator: $V_{L\,(Load)}$

b Phase voltage of the load: $V_{Ph\,(Load)}$

c Phase current drawn by the load: $I_{Ph\,(Load)}$

d Line current delivered to the load: $I_{L\,(Load)}$

e Phase current in the generator: $I_{Ph\,(Gen)}$

REVIEW QUESTIONS

1. How is a delta connection effected in a three-phase generator?
2. What voltage is measured across a single coil group in a three-phase generator?
3. What is the displacement between the three-phase delta voltages?
4. Draw and label a phasor diagram illustrating the line currents and phase currents in a delta system.
5. What is the displacement between the line current and phase current in a three-phase delta system?
6. Determine the phasor sum of the phase voltages in a delta-connected system.
7. Give a practical example of a three-phase balanced load.
8. The phase currents in a delta-connected three-phase generator are as follows: A phase = 25 A; B phase = 25 A; C phase = 25 A. Find the line current.
9. The line currents in a delta-connected three-phase generator are as follows: Line A–B = 50 A; Line C–A = 50 A; Line B–C = 50 A. Find the phase currents.
10. Why would a high voltage distribution transformer have its primary connected in delta?

7.5 Energy and power requirements of a.c. systems

The energy created in each phase of a star-connected circuit is the same as the energy developed in a single-phase circuit. The loads connected to the circuit consume this energy. Furthermore, the power consumed is determined by applying the equation:

$$P = VI\cos\phi \text{ per phase}$$

Cos ϕ is the angle that indicates displacement between phase voltage and phase current. The total power delivered to the load by the three-phase supply is:

$$P_1 + P_2 + P_3 \text{ (unbalanced loads)}$$

or simply 3 × *P* if it is a balanced load

$$P = 3VI\cos\phi$$

Phase values of voltage and current are difficult to obtain in an operating a.c. system. It is simpler to measure line values of voltage and current. In a star system:

$$I_{Ph} = I_L \text{ and } V_{Ph} = \frac{I_L}{\sqrt{3}}$$

Substituting line values for phase values in a star system gives:

$$P = 3\frac{V_L}{\sqrt{3}} I_L \cos\phi$$

Hence:

$$P = \sqrt{3}\, V_L I_L \cos\phi$$

EXAMPLE 7.6

The three-phase star-connected 400 V motor in Figure 7.65 draws a current of 20 A at a power factor of 0.86 from the supply. Determine how much power the motor draws from the supply.

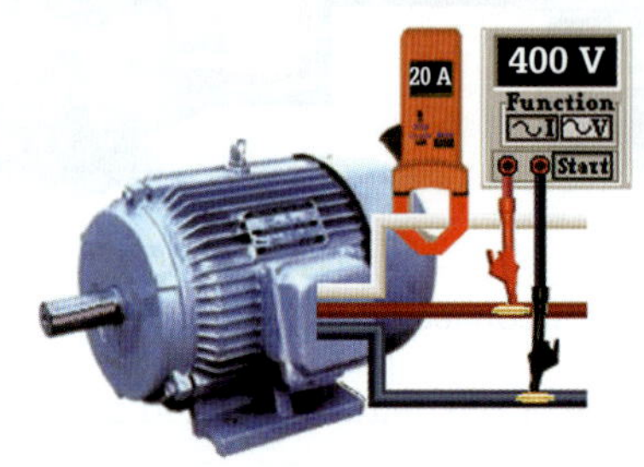

FIGURE 7.65 Three-phase star-connected motor

$$P = \sqrt{3}\, V_L I_L \cos\phi$$
$$= 1.732 \times 400 \times 20 \times 0.86$$
$$= \mathbf{11\,916.2\ W\ or\ 11.916\ kW}$$

EXERCISE 7.6

The three-phase star-connected 400 V motor draws a current of 25 A at a power factor of 0.78 from the supply. Determine how much power the motor draws from the supply.

Calculating power

When a three-phase circuit contains resistance, inductance and capacitance, some of the power is stored and then returned by the inductor or capacitor (reactive power), and some of the power becomes dissipated by the resistance (true power). As to apparent power, it is the phasor sum of both the reactive and the true power as shown in **Figure 7.66**. It follows that apparent power can be determined by applying the following equation:

$$S = \sqrt{P^2 + Q^2}$$

where S = apparent power in volt-amperes (VA)

P = true power in watts (W)

Q = reactive power in volt-amperes reactive (VA_R)

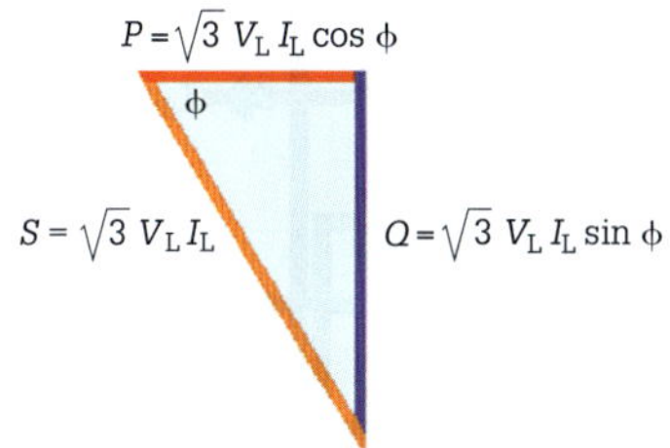

FIGURE 7.66 Power triangle SPQ

For a balanced three-phase system, apparent power can restate as:

$$S = \sqrt{3}\, V_L I_L \text{ in volt-amperes (VA)}$$

Or apparent power can be expressed in kVA:

$$S = \frac{\sqrt{3}\, V_L I_L}{1000} \text{ in kVA}$$

Reactive power can be restated in terms of a three-phase balanced system:

$$Q = \sqrt{3}\, V_L I_L \sin\phi \text{ in volt-amperes reactive } (VA_R)$$

The power factor of a balanced three-phase system is the ratio of true power to apparent power and is a measure of the power loss in a system.

EXAMPLE 7.7

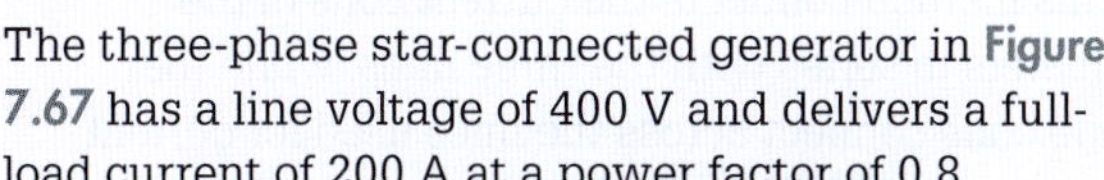

The three-phase star-connected generator in **Figure 7.67** has a line voltage of 400 V and delivers a full-load current of 200 A at a power factor of 0.8.

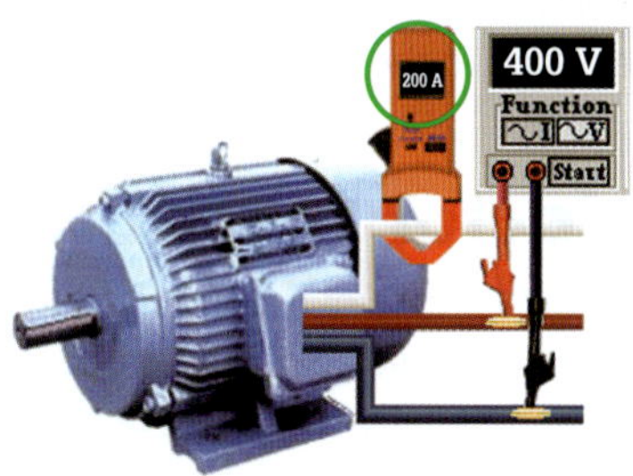

FIGURE 7.67 Three-phase star-connected generator

a Find the value of the kVA rating of the generator at the load power factor

$$S = \frac{\sqrt{3}\, V_L I_L}{1000}$$
$$= \frac{1.732 \times 400 \times 200}{1000}$$
$$= \mathbf{138.56\ kVA}$$

b Find the value of the full-load power drawn by the load in kilowatts

$$P = \sqrt{3}\, V_L I_L \cos\phi$$
$$= 1.732 \times 400 \times 200 \times 0.8$$
$$= \mathbf{110.848\ kW}$$

EXERCISE 7.7

A three-phase star-connected generator has a line voltage of 400 V and delivers a full-load current of 180 A at a power factor of 0.85.

a Find the value of the kVA rating of the generator at the load power factor.

b Find the value of the full-load power drawn by the load in kilowatts.

Methods used to measure power

Measurement of power in a three-phase system occurs by measuring the voltage and current, as is the case with single-phase systems. In fact, three-phase systems are more complicated since total power is the sum of the power in all three phases.

The instrument used to read the power in three-phase circuits is a wattmeter. Wattmeters find the average of the product of voltage and current and this is done either with the interaction of magnetic fields, as happens in meters with potential and current coils, or electronically.

One-wattmeter method

If it is required to measure power in either a star or delta three-phase system where the phases are balanced in relation to load and power factor, a single wattmeter measures the power in one phase, and the result is multiplied by three.

If no neutral is available, as is the case with three-wire systems, power happens using a 'Y box'. A Y box has two branches created from its artificial star-point which have the same impedance and power factor as the wattmeter's voltage circuit that is also the third branch of the Y box. Refer to **Figure 7.68**.

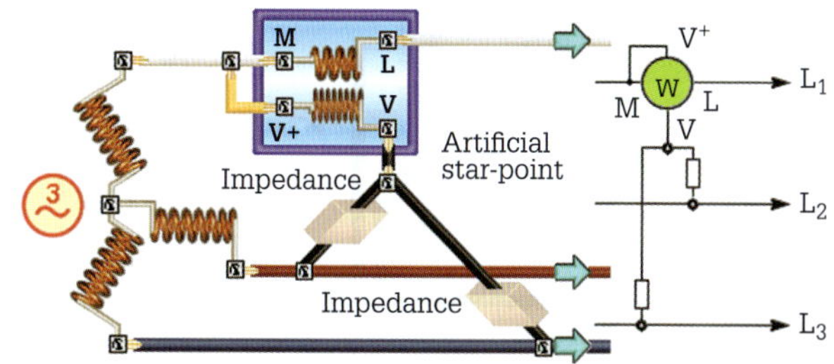

FIGURE 7.68 One-wattmeter method

In the one-wattmeter method of three-phase power measurement, the current coil of the meter connects in series with one of the lines and the voltage coil between one line and the artificial star-point. In fact, the total power is three times the reading of the wattmeter. Because most three-phase balanced systems have slight voltage variations between their phases this method of power measurement is inaccurate. Consequently, to measure power accurately, the power measurement must occur in all three phases and be added together.

Three-wattmeter method: three-wire system

Figure 7.69 illustrates the three-wattmeter method of measuring power in three-phase, three-wire systems. Observe that the current coils connect in series with the lines, and that one terminal of each voltage coil connects to a line; while the other terminals connect together in a star configuration.

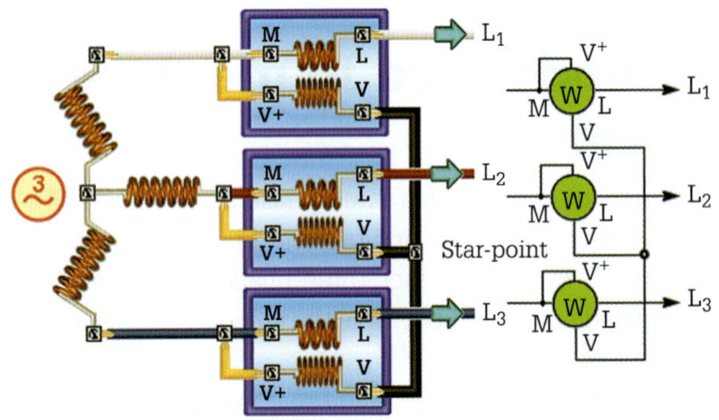

FIGURE 7.69 Three-wattmeter method: three-wire system

The total power is the sum of the wattmeter readings. This method is **valid for both balanced and unbalanced loads**.

Two-wattmeter method: three-wire system

Figure 7.70 illustrates how two wattmeters connect into a three-phase three-wire system to measure the total power delivered to the load, irrespective of whether the load is delta or star connected, **balanced or unbalanced**. With this method $P_T = P_1 + P_2$.

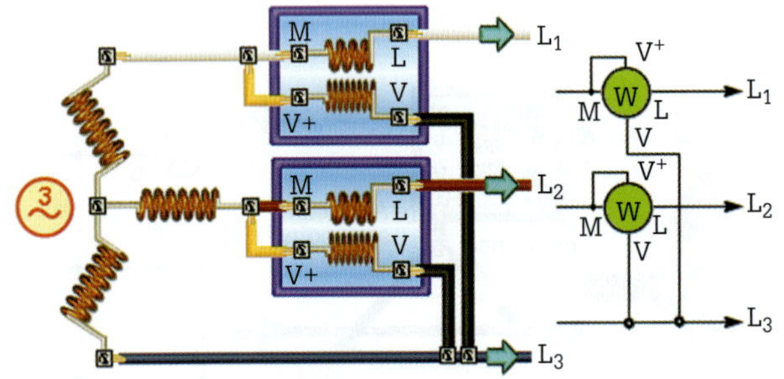

FIGURE 7.70 Two-wattmeter method: three-wire system

This method connects the current coils of the wattmeters in series with any two lines, while the voltage coils connect across the lines used for current measurement and the third line.

Three-wattmeter method: four-wire system

Figure 7.71 illustrates how three wattmeters connect to measure the total power delivered to the load, irrespective of whether the load is delta or star, balanced or unbalanced.

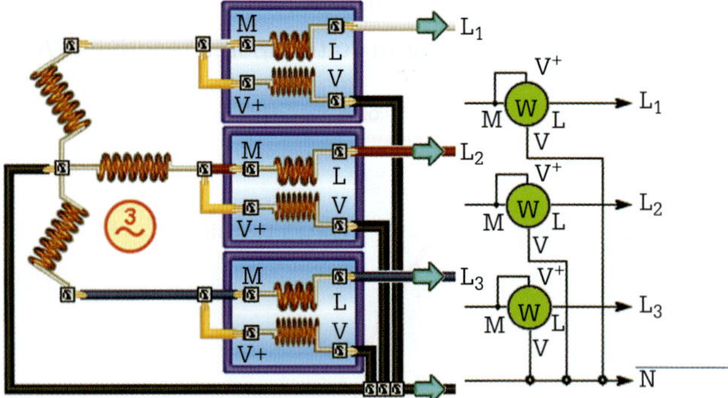

FIGURE 7.71 Three-wattmeter method: four-wire system

SWITCH ON

All electrical machines have a power rating. The power rating of the machine is an excellent quality indicator of machine performance, especially when compared to an identical model. For example, consider a motor generating excessive internal heat. A slightly binding rotor bearing or a loose terminal connection may still allow the motor to perform regularly. Just because the motor appears normal does not mean thermal deterioration and insulation failure are not occurring. In fact, a motor may be meeting its basic specifications for operation yet consuming 10% more power than normal. It follows that power analysis detects otherwise invisible problems.

This method connects the current coils of the wattmeters in series with the three lines, while the voltage coils connect across the lines used for current measurement and the neutral. Because of this, the algebraic sum of the three wattmeter readings equals the total power delivered to the load.

Power factor measurement

The two-wattmeter method of power measurement enables the calculation of the phase power factor. In addition, the phase angle and thereby the power factor can be determined, when the wattmeters are connected as shown in **Figure 7.72** and the installation has a phase sequence of L_1–L_2–L_3, by applying the following equation.

Note: This applies only to balanced loads.

$$\tan\phi = \sqrt{3}\left(\frac{W_2 - W_1}{W_2 + W_1}\right)$$

Therefore, to calculate power factor:

$$Power\ factor = \cos\left(tan^{-1}\left(\sqrt{3}\left(\frac{W_2 - W_1}{W_2 + W_1}\right)\right)\right)$$

where W_1 and W_2 are the wattmeter readings.

Note that the placement of the wattmeters and the phase sequence has an impact on the equation.

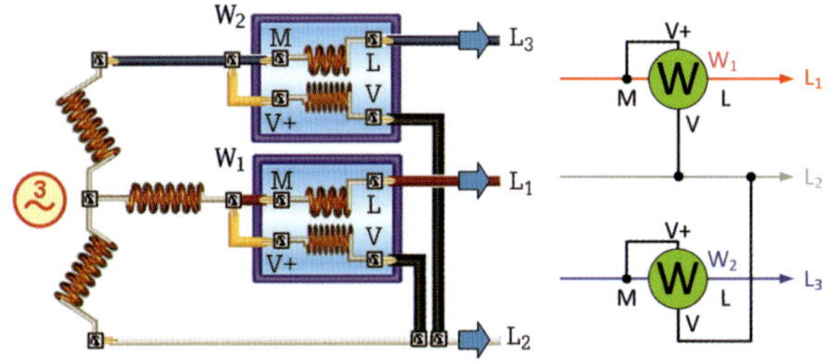

FIGURE 7.72 Two-wattmeter method to determine power factor: three-wire system

EXAMPLE 7.8

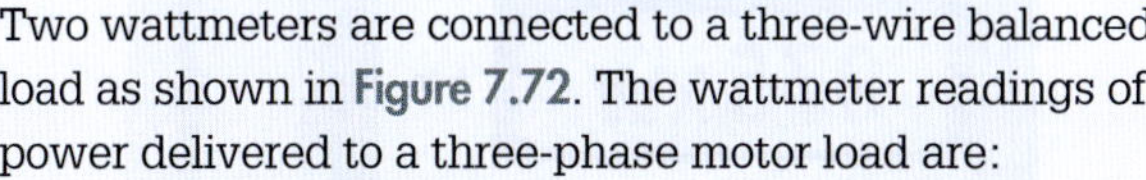

Two wattmeters are connected to a three-wire balanced load as shown in **Figure 7.72**. The wattmeter readings of power delivered to a three-phase motor load are:

$$W_1 = 16 \text{ kW and } W_2 = 20 \text{ kW}$$

Calculate the phase power factor of the load.

$$\text{Power factor} = \cos\left(\tan^{-1}\left(\sqrt{3}\left(\frac{W_2 - W_1}{W_2 + W_1}\right)\right)\right)$$

$$= \cos\left(\tan^{-1}\left(\sqrt{3}\left(\frac{20 - 16}{20 + 16}\right)\right)\right)$$

$$= \mathbf{0.982}$$

EXERCISE 7.8

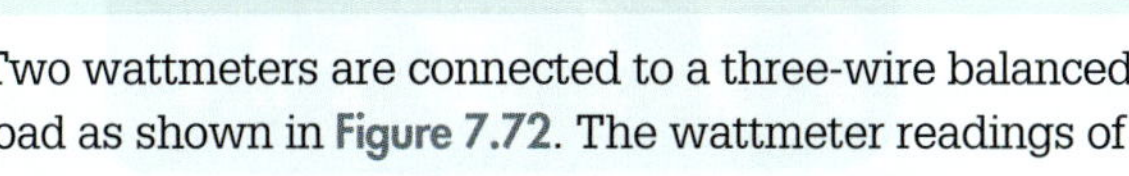

Two wattmeters are connected to a three-wire balanced load as shown in **Figure 7.72**. The wattmeter readings of power delivered to a three-phase motor load are:

$$W_1 = 5 \text{ kW and } W_2 = 10 \text{ kW}$$

Calculate the phase power factor of the load.

Power factor meter

The power factor of a three-phase system reads directly from a power factor meter connected into the circuit. An illustration of a power factor meter occurs in **Figure 7.73**.

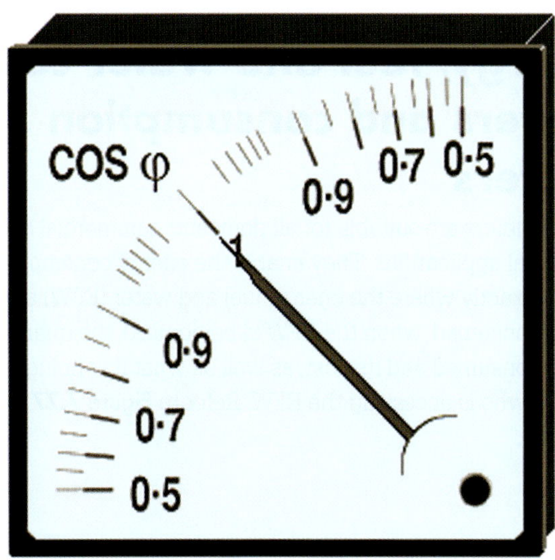

FIGURE 7.73 Power factor meter

Watt-hour meter

The watt-hour meter or consumption meter is an instrument for measuring the actual amount of electrical energy used. Since energy is equal to the integral of average power and time, the watt-hour meter must utilise both these factors. The unit of measurement for energy is the joule (J) which is equal to the energy supplied by a one-watt power source in one second.

The watt-hour (Wh) is the energy transmitted when one watt is delivered over one hour. However, this unit is too small when measuring actual energy transmission so the kilowatt-hour (kWh) is used. In fact, one kilowatt-hour equals 1000 watt-hours.

Tariff watt-hour meters are, by far, the most common instrument for measuring electricity demand and consumption.

Electronic kilowatt-hour meters

Electric energy retailers are very aware of the limitations of electromechanical meters, as well as the rapidly increasing presence of harmonic generation inside their customers' facilities. Consequently, they are replacing electromechanical meters with the latest analogue-to-digital microprocessor-based electronic kilowatt-hour meters. Refer to **Figure 7.74**.

FIGURE 7.74 Electronic polyphase watt-hour meter

High-tech electronic instruments see all current passing through them: fundamental, reactive and harmonic. In fact, these meters can accurately measure and record reactive and distortion current content for specific kW demand, kVA demand and power factor. More to the point, lack of knowledge of distortion current by a consumer may result in possible harmonic penalty billing.

The flow of reactive power (VA_{Rs}), through both the distribution network and the customer's facility results in energy losses. Reactive power means that the conductors and transformer capacity require upgrading in order to carry the reactive power. This lost energy can now be measured by the electronic meters and a charge in the form of a penalty is applied to the consumer to recover costs. The meters provide an automated meter reading that minimises operating costs for billing requirements.

Power factor correction

In addition to energy losses, reactive power drawn from a three-phase supply can cause excessive voltage drop. Consequently, the reactive power requires correction by either the application of shunt capacitors or load-tap changing transformers and synchronous capacitors.

Maximum-demand indicators

Demand reflects the maximum electrical power required to operate a customer's facility. Demand occurs in either kilowatts (kW) or kilovolt amperes (kVA).

Installing maximum-demand indicators enables measurement of demand and demand growth. The demand meter effectively monitors electricity supply and consumption by means of current transformers, which occur around the feeder conductors of a load and require no live contact. It is evident that no live contact provides all-round safety for the user of the instrument. An illustration of a maximum-demand current meter with adjustable slave pointer occurs in **Figure 7.75**. (This pointer moves upscale by the main needle and stays at its highest reading. Consequently, resetting to zero is required.)

FIGURE 7.75 Maximum-demand current meter

These meters measure the maximum demand on current in relation to loads. The maximum demand is the highest rate of current recorded over any 15-minute interval. This knowledge is useful when expanding existing electrical installations to determine if the existing feeder cables, such as consumer mains and sub-mains and final sub-circuits, can carry the increased current. Rather than having an analogue reading, an electronic demand meter indicates energy (kWh) and maximum demand (kW max or kVA max) on a liquid crystal display (LCD). In addition, the display can be actioned to scroll through these and a series of other programmed values. **Figure 7.76** shows an example of an electronic maximum-demand meter.

FIGURE 7.76 Electronic maximum-demand meter

These meters can measure and monitor many energy parameters. For example, they can measure current kWh used, as well as current kW demand, maximum kW demand and date and time. A communication port allows downloading of the measurements to a remote personal computer.

Energy, fuel and water cost meters and consumption meters

These meters are suitable for all domestic, commercial and industrial applications. They enable the energy consumer to locate exactly where the energy, fuel and water (EFW) are being consumed, when the EFW is being used, the quantity being consumed and its cost, as well as what it is being used for and who is accessing the EFW. Refer to **Figure 7.77**.

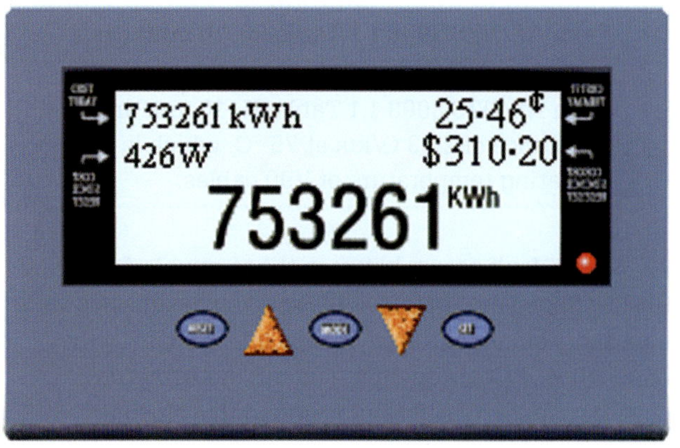

FIGURE 7.77 Energy, fuel and water cost meter

At present EFW meters occur in the United Kingdom and find application in many diverse installations. These meters display the running costs over any prescribed time interval, together with the EFW consumption and access times. EFW meters enable the energy consumer to obtain and interpret information, thereby allowing them to control their costs and to improve their efficiency with the use of electrical resources.

REVIEW QUESTIONS

1 A star-connected three-phase generator with a rated voltage of 250 V per coil phase group supplies a balanced load consisting of three 50 Ω impedance coil phase group windings connected in delta. Find the following values:
   a line voltage of the generator: $V_{L\ (Gen)}$
   b phase voltage of the load: $V_{Ph\ (Load)}$
   c phase current drawn by the load: $I_{Ph\ (Load)}$
   d line current delivered to the load: $I_{L\ (Load)}$
   e phase current in the generator: $I_{Ph\ (Gen)}$

2 A delta-connected three-phase generator with a rated voltage of 440 V per coil phase group supplies a balanced load consisting of three 120 Ω impedance coil phase group windings connected in star. Find the following values:
   a line voltage of the generator: $V_{L\ (Gen)}$
   b phase voltage of the load: $V_{Ph\ (Load)}$
   c phase current drawn by the load: $I_{Ph\ (Load)}$
   d line current delivered to the load: $I_{L\ (Load)}$
   e phase current in the generator: $I_{Ph\ (Gen)}$

3 A three-phase star-connected 400 V motor draws 36 A at a power factor of 0.78. Determine how much power the motor draws from the supply.

4 A three-phase star-connected generator has a line voltage of 660 V and delivers a full-load current of 200 A at a power factor of 0.85. Find the following values:
   a the kVA rating of the generator at the load power factor
   b the full-load power drawn by the load in kilowatts.

5 Name the instrument used to measure power in a three-phase circuit.

6 A single wattmeter is used to measure the power in a balanced star connected three-phase system. If the meter reads 2.5 kW, determine the total power drawn from the supply.

7 Describe how the wattmeters connect when measuring power using the three-wattmeter method.

8 How is the total power determined using the two-wattmeter method of measuring power?

9 In a three-wattmeter four-wire system, to what lines are the voltage coils of the wattmeters connected?

10 Two wattmeters are connected to a three-wire balanced load as shown in Figure 7.72. The wattmeter readings of power delivered to a three-phase load are: W_1 = 9 kW and W_2 = 15 kW. Calculate the phase power factor of the load.

11 What type of meter sees fundamental, reactive and harmonic currents?

12 State an advantage of electronic meters for supply authorities.

13 What is the function of a maximum demand indicator?

14 What units of measurement express the maximum demand of a system?

15 Over what time interval does the maximum demand indicate the highest rate of current recorded?

7.6 Fault-loop impedance

Short-circuit capability applies to various applications such as the sizing of transformers and the selection of interruption ratings for circuit breakers and fuses. The value of current caused by a short circuit depends on the system voltage and the total impedance of the fault circuit.

EXAMPLE 7.9

A 16 A single-phase circuit as shown in **Figure 7.78** supplying a load is 15 m long. This circuit uses 4 mm² twin + E TPS cable with V90 insulation. Determine the value of the short-circuit current if a fault of 1 Ω impedance (Z_{Fault}) occurred at the load terminals.

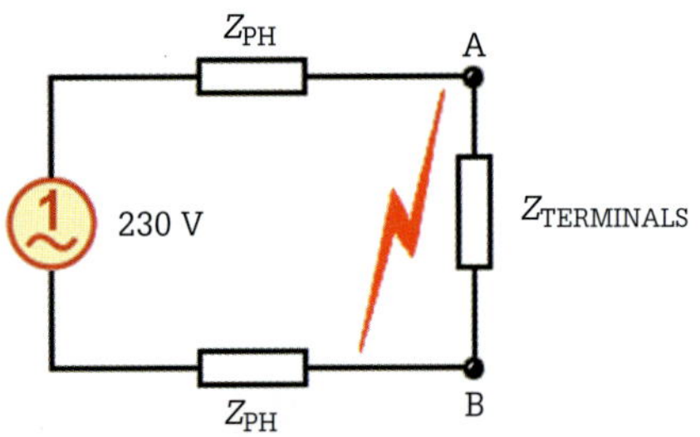

FIGURE 7.78 Short-circuit example

- From AS/NZS 3008.1.1 Table 30, 4 mm² has a reactance of 0.102 Ω/km
- From AS/NZS 3008.1.1 Table 35, 4 mm² has a resistance of 5.61 Ω/km at 75 °C, which is the operating temperature of V90 cables.

$$R_{PH} = \frac{length \text{ (in m)} \times Resistance / km}{1000}$$
$$= \frac{15 \times 5.61}{1000}$$
$$= 0.08415\ \Omega$$

$$X_{PH} = \frac{length \text{(in m)} \times Reactance / km}{1000}$$
$$= \frac{15 \times 0.102}{1000}$$
$$= 0.00153\ \Omega$$

$$Z_{PH} = \sqrt{R_{PH}^2 + X_{PH}^2}$$
$$= \sqrt{0.08415^2 + 0.00153^2}$$
$$= 0.084164\ \Omega$$

Impedance of the circuit comprises (2 supply conductors and the fault impedance) $Z_{PH} + Z_{PH} + Z_{Fault}$

$$Z_{Circuit} = Z_{PH} + Z_{PH} + Z_{Fault}$$
$$= 0.084164 + 0.08464 + 1$$
$$= 1.168\ \Omega$$

The short-circuit current, I_{sc}, can be determined as follows:

$$I_{SC} = \frac{V}{Z_{Circuit}}$$
$$= \frac{230}{1.168}$$
$$= \mathbf{197\ A}$$

EXERCISE 7.9

A 35 A single-phase circuit as shown in **Figure 7.78** supplying a load is 30 m long. This circuit uses 10 mm² twin + E TPS cable with V90 insulation. Determine the value of the short-circuit current if a fault of 0.5 Ω impedance (Z_{Fault}) occurred at the load terminals.

- From AS/NZS 3008.1.1 Table 30, 10 mm² has a reactance of 0.0906 Ω/km
- From AS/NZS 3008.1.1 Table 35, 10 mm² has a resistance of 2.23 Ω/km at 75 °C, which is the operating temperature of V90 cables.

When a fault occurs between the terminals A and B, the low impedance between these points results in a very high short-circuit current I_{sc} that is limited only by the impedance Z. Consequently, a circuit protective device must be capable of safely interrupting this amount of current.

Earth-loop impedance

To make certain that a fault current is low enough to operate the circuit protective device (fuse or miniature circuit breaker) within a safe time limit, knowledge of the fault-loop impedance is necessary.

Note: A loop is not a circuit. In fact, a circuit carries current on a planned path. This implies that a loop is not a planned path, and fault current may take many different routes.

Clearing a short circuit to earth requires sufficient fault current to enable the protective device to operate quickly. Quick operation provides sufficient protection for persons when a protective earthing conductor or exposed conductive parts become live under fault conditions (indirect contact).

Earth-fault-loop impedance measurement test

Clause 8.3.9 of AS/NZS 3000:2018 requires an earth-fault-loop impedance measurement test and Section B4.6 'Earth fault-loop impedance measurement' provides guidance on performing this test.

Circuit not connected to supply

This method uses an ohmmeter that can measure low values of resistance. Certain procedures must happen:

- Isolate the circuit under test.
- Connect the circuit active conductor to its protective earthing conductor at the switchboard.
- At the furthest point on the circuit, measure the resistance between the active and earth conductor.
- The resistance value obtained (R_{phe}) should not exceed the value given in Table 8.2 for the conductor size and associated protection device.

Circuit connected to supply

Where supply is available, the fault-loop impedance may be evaluated with a microprocessor-controlled loop impedance test instrument and the result verified by Table 8.1 of the *Wiring Rules*. Microprocessor-controlled loop impedance testers as illustrated in **Figure 7.79** are the most accurate means of measuring impedance.

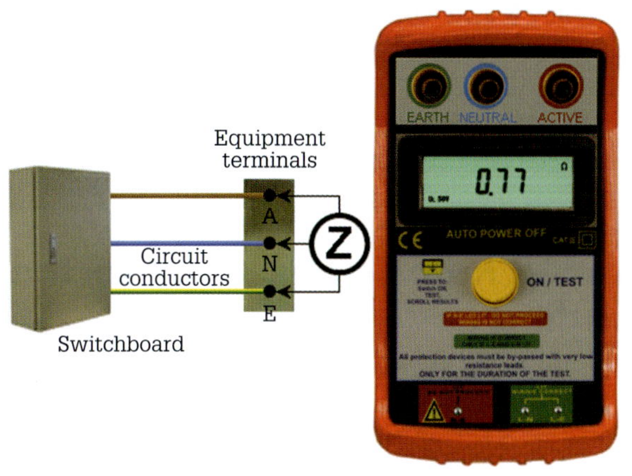

FIGURE 7.79 Fault-loop tester

The electronic tester examines actual circuit conditions and influences that contribute to the performance of the protective device under fault conditions such as temperature. The tester simulates a fault from active to earth or from active to neutral.

Because loop impedance measurement occurs with the supply on, safety measures apply. Consequently, prior to this test, protective earths must be tested to ensure their continuity and resistance level. Otherwise, a faulty earth may create a hazardous situation for persons undertaking the loop impedance test and for other persons within the installation.

When in operation, the tester in the first instance measures the unloaded voltage, then connects a known resistance in series with the circuit being tested, thereby simulating a fault. It follows that the tester measures the voltage drop across the resistor. The result is that the tester completes the analysis and disconnects the simulated fault before the protective device has time to react. However, it may be necessary to bypass RCDs during the loop test, as the type of tester used may not prevent tripping.

Undertaking the test itself is very simple and consists of three steps.

1. Ensure all main equipotential bonding is in place.
2. Connect the instrument to the point to be tested, active, neutral and earth.
3. Press the test button on the instrument and record the reading.

The tester provides a direct readout of the prospective short-circuit current. In addition, the earth-loop impedance of each circuit at the furthest point from the protection device requires measurement. The measured value must not exceed the value shown in Table 8.1, when tested in accordance with Clause 8.3.9.3 (b).

The measurement of the external loop impedance of the installation occurs at the incoming supply point or main switchboard.

REVIEW QUESTIONS

1. What factors determine the magnitude of current caused by a short circuit?
2. A 38 A single-phase circuit as shown in Figure 7.78 supplying a load is 45 m long. This circuit uses 10 mm^2 twin + E TPS cable with V90 insulation. Determine the value of the short-circuit current if a fault of 0.5 Ω impedance (Z_{Fault}) occurred at the load terminals.
3. Why is it necessary to have knowledge of the fault-loop impedance?
4. What is the most accurate means of measuring fault-loop impedance?
5. With what is it necessary to compare the result obtained from a fault-loop impedance test?

7.7 Industry standards

The main industry standard of concern for electrotechnology workers is AS/NZS 3000: 2018 *Wiring Rules*. Within this standard is information pertaining to earth fault-loop impedance.

Wiring Rules requirements for alternating current principles

To make certain that the circuit protection device operates within the specified disconnection time, it is essential that the earth fault-loop impedance is below permissible levels (see Table 8.1, AS/NZS 3000: 2018 *Wiring Rules*). The smaller the value of impedance, the higher is the value of fault-current and consequently the faster the disconnection time.

Section B4 provides guidance and conditions for the fault-loop impedance circuit.

The earth fault-loop as illustrated in Figure 7.80 consists of the following elements.

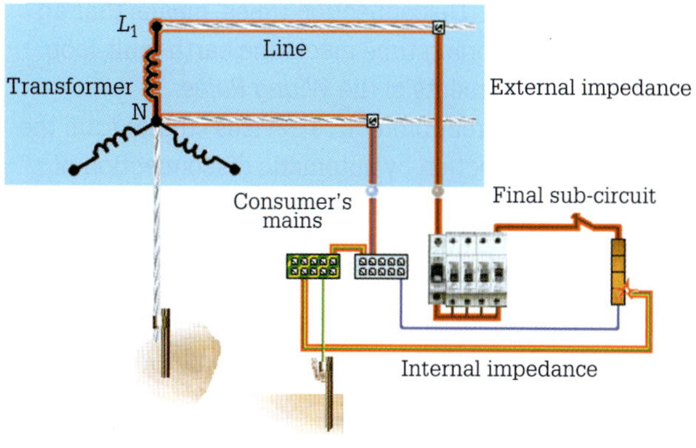

FIGURE 7.80 Earth fault-loop

The selection of a suitable overload protection device depends on the value of the prospective short-circuit current. Therefore, measuring or calculating the loop impedance is necessary. In addition, the protective device is selected so that the fault current,

$$I_a = \frac{U_o}{Z_s}$$

ensures its operation within the required time. The efficient operation of the protective device requires the calculation of the earth-fault-loop impedance, Z_s. Transposing the equation gives the maximum value of impedance that ensures operation of the protection device in the event of an active to earth fault.

$$Z_s = \frac{U_o}{I_a}$$

where Z_s = earth-fault-loop impedance

U_o = nominal phase voltage (230 V)

I_a = current causing automatic operation of the protective device

I_a for circuit breakers is the mean tripping current as follows:

- Type B = 4 × rated current
- Type C = 7.5 × rated current
- Type D = 12.5 × rated current.

The values of Z_s in *Wiring Rules*, Table 8.1, 'Maximum values of earth-fault-loop impedance (Z_s at 230 V)' transpired using the above equation.

Table 8.1 provides the maximum values of circuit fault-loop impedance provided for circuit breakers (use 0.4 s disconnection times) and fuses. Circuit-breaker time–current characteristics indicate that the 0.4 s disconnection time is in the instantaneous (short circuit or magnetic) tripping zone. This zone is for fault currents of five to 10 times the rated current of circuit breakers that have a rating at 63 A or less (nominally taken as 7.5 times).

Note that Table 8.1 does not provide maximum circuit impedance values for circuit breakers operating under the 5 s situation. The time–current characteristics of circuit breakers indicate that a 5 s time is in the thermal (overload) tripping zone. For circuit breakers of 63A or less, the tripping current is two to five times the rated current (nominally taken as four times).

Note that in Table B4.1 the values provided in column two apply for five seconds disconnection time for Type C circuit breakers.

In Section B4.3, 'Disconnection times' means that an RCD of suitable tripping time meets the earth-fault-loop impedance requirements of the *Wiring Rules*.

Where over-current protective devices cannot fulfil the conditions for protection by automatic disconnection of supply, such protection occurs by RCDs having a proper tripping time. An example where the conditions may occur is with circuits supplying socket-outlets, the length of which is unknown, or circuits of great length and small cross-sectional area thus having high impedance.

- **Clause 8.3.10 'Operation of RCDs'** With the supply connected, test each final sub-circuit to make certain that the RCD operates. This suggests that it is best practice to have the impedance lower than the acceptable value before testing to confirm that the RCD protects the circuit.

Wiring Rules clauses for further reading

- Clause 1.5.5.1 Fault protection – General
- Clause 1.5.5.2a Methods of protection – Automatic
- Clause 1.5.5.3.b Protection by automatic disconnection of supply – Touch voltage
- Clause 1.5.5.3c Protection by automatic disconnection of supply – Earthing system impedance
- Clause 1.5.5.3d Protection by automatic disconnection of supply – Disconnection times

Measuring fault-loop impedance examples

EXAMPLE 7.10

a A single-phase circuit supplying 15-ampere socket outlets installed in a commercial installation has no RCD protection. This circuit uses 30 m of 4 mm² twin + E TPS cable with V90 insulation and a 25 A Type C circuit breaker. Determine the maximum fault-loop impedance allowed.

First, we must determine if the circuit meets the length requirements for the size conductor and circuit breaker type used. Table B1, AS/NZS 3000: 2018 shows the maximum circuit lengths for various combinations of conductor and earth sizes and circuit protection devices. If circuits are under these lengths, the required impedance transpires.

The 30-metre length meets the requirements of Table B1 for a Type C circuit breaker. In addition, Table 8.1, *Wiring Rules* indicates that a Type C circuit breaker operates when a loop has a maximum earth-fault-loop impedance of 1.23 Ω. Confirmation of this value occurs by using the equation:

$$Z_s = \frac{U_o}{I_a} = \frac{230}{7.5 \times 25} = 1.226\ \Omega$$

b If the circuit of (a) required an addition of another point to be added 10 m further on, will the circuit still meet fault-loop requirements?

Yes, because the total length is now 40 m which is still under the maximum allowed by Table B1 for a Type C 25 A circuit breaker protecting a 4 mm² cable.

EXAMPLE 7.11

A domestic installation has a socket outlet circuit not protected by an RCD of 16 A wired with 2.5 mm² twin + E TPS cable with V90 insulation and copper conductors. Determine the maximum internal fault-loop resistance provided when the protection device is a Type C circuit breaker rated at 16 A. Also, determine the maximum external and internal impedance allowed.

Table 8.2, *Wiring Rules* provides resistance values that enable a measurement with a low-reading ohmmeter.

The maximum internal fault-loop resistance of the 2.5 mm² active and protective earthing conductors allowed for a **Type C circuit breaker is 1.2 Ω** (R_{phe}).

If this circuit was connected to the supply, the maximum value of the earth-fault-loop impedance, both internal and external allowed using a Type C circuit breaker, would be as follows.

From Table 8.1 the maximum value of Z_s allowable at 230 V for both **internal and external impedance is 1.9 Ω.**

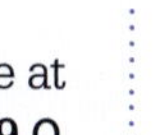

EXAMPLE 7.12

An installation has the following data as shown in Figure 7.81. Calculate the total earth-fault-loop impedance when an earth fault occurs on a connected appliance.

Total earth-fault-loop impedance

$$= Z_T + Z_L + Z_{CM} + Z_{FSC}$$

$$= 0.042 + 0.002 + 0.25 + 0.138$$

$$= \mathbf{0.432\ \Omega}$$

If the final sub-circuit had a Type C circuit breaker then the calculated loop impedance is below the maximum 1.9 Ω allowed if an earth fault occurs.

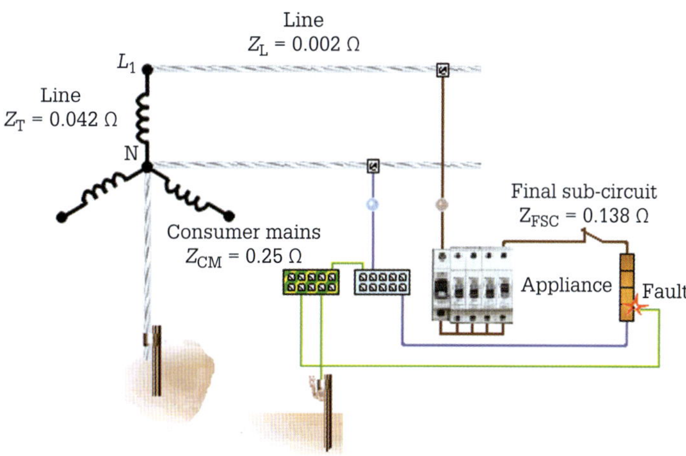

FIGURE 7.81 Fault-loop impedance, Example 7.12

EXAMPLE 7.13

A domestic installation has a 15 A socket outlet circuit not protected by an RCD and wired with 4 mm² twin + E TPS cable with V90 insulation and copper conductors.

a Determine the maximum internal fault-loop resistance allowed with the supply connected when the protection is a single-pole 32 A 4.5 kA Type C circuit breaker. Use Table 8.1, *Wiring Rules*. From Table 8.1 the maximum value of Z_s at 230 V is **1.0** Ω. Confirm this using the equation below:

$$Z_s = \frac{U_o}{I_a} = \frac{230}{7.5 \times 32} = 0.958\ \Omega$$

Note: Type *C* = 7.5 × rated current.

b Determine the maximum internal fault-loop resistance (R_{phe}) allowed. Use Table 8.2, *Wiring Rules*.

The maximum allowable internal fault-loop impedance measured with a low-reading ohmmeter for a Type C breaker is 0.6 Ω (R_{phe}). Confirm this in the notes to Table 8.1.

The values of Table 8.2 have been rounded to one decimal place, and are approximately 64% of the values given in Table 8.1. Therefore:

$$R_{int} = Z_{int} \times 0.64$$

R_{int} = 1.0 (from Table 8.1) × 0.64 = **0.64 Ω**

The measured value of resistance (R_{phe}) must not exceed the value in Table 8.2 for the appropriate conductor size and type of protective device.

EXAMPLE 7.14

An installation has the following data as shown in **Figure 7.82**. Calculate the total earth-fault-loop impedance.

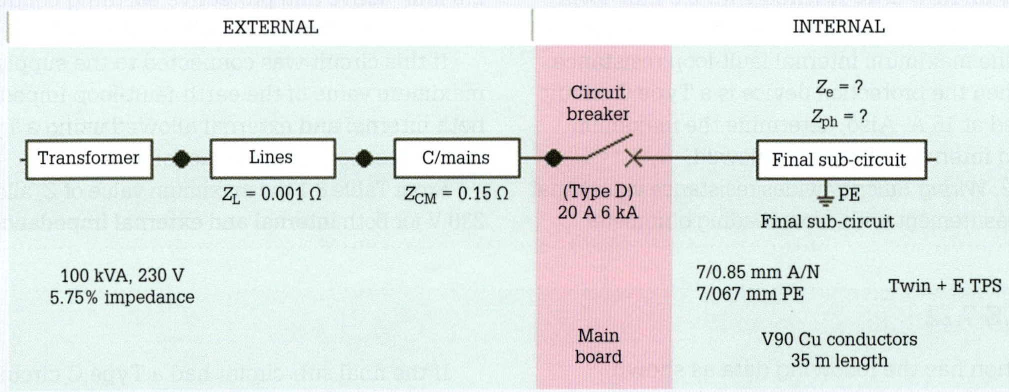

FIGURE 7.82 Fault-loop impedance, Example 7.14

The 5.75% impedance rating indicates that 25 101.5229 amperes of current will flow in the secondary if the line to line secondary is short-circuited when the primary voltage is raised from zero volts to a point at which 5.75% of 230 V, or 13.225 V, appears at the secondary terminals. The impedance (Z) of the transformer secondary may now be calculated.

As the three-phase VA (S) = $\sqrt{3}\ V_L\ I_L$ then:

$$I_L = \frac{S}{\sqrt{3}V_L}$$

$$= \frac{100\ 000}{\sqrt{3} \times 400}$$

$$= 144 \text{ A}$$

Calculate the available short-circuit current

$$I_{S/C} = \frac{I_L \times 100}{Z\%}$$

$$= \frac{144 \times 100}{5.75}$$

$$= 2504 \text{ A}$$

Impedance of transformer (Z_{TXF})

$$Z_{TXF} = \frac{V_{PH}}{I_{S/C}}$$

$$= \frac{230}{2504}$$

$$= 0.0918\ \Omega$$

Note: Some tables may quote other values for reactance (Ω/m) @ 20 °C for 1 mm². The reactance varies because of the type of cable – SDI, Flat TPS, Circular, XPLE and SWG – that the copper conductor is manufactured for. The table following is for annealed copper.

From the table in **Figure 7.83** the resistance of 4 mm² which is 7/0.850 (Z of active) equals 0.0043 Ω/m. Therefore, 35 m equals 0.1505 Ω.

Nominal CSA mm²	Number of strands	ohms/m @ 20° C
1.0	3	0.0172
1.5	3	0.0115
2.5	3	0.00688
4	7	0.0043
6	7	0.002866
10	7	0.00172
16	7	0.001075
25	3	0.000688

FIGURE 7.83 Cable impedance table

From the table the resistance of 2.5 mm² which is 7/0.67 (Z earth) equals 0.00688 Ω/m. Therefore, 35 m equals 0.2408 Ω.

Total earth-fault-loop impedance

$$= Z_{TXF} + Z_L + Z_{CM} + Z_{PH} + Z_E$$

$$= 0.0918 + 0.001 + 0.15 + (0.1505 + 0.2408)$$

$$= \mathbf{0.6341}\ \Omega$$

This value is well below the maximum earth-fault-loop impedance permitted as shown by:

$$Z_S = \frac{U_o}{I_a}$$

$$= \frac{230}{12.5 \times 20}$$

$$= 0.92\ \Omega$$

Note: Type D = 12.5 × rated current.

Confirm from Table 8.1 of the *Wiring Rules* that the maximum earth-fault-loop impedance allowed is 0.9 Ω.

EXAMPLE 7.15

When an electrical installation is being planned, Z_{ext} may not be available (it depends on the transformer and supply cables). If not available, Z_{int} is determined by applying:

$$Z_{int} = 0.8 \times \frac{U_o}{I_a}$$

where Z_{int} = the internal earth-fault-loop impedance
U_o = nominal phase voltage (230 V)
I_a = current causing automatic operation of the protective device

When the length and CSA of the supply mains are unknown, the assumption is that there is 80% or more of the nominal phase voltage available at the point of the circuit protective device. Therefore, Z_{int} should not have a value greater than 0.8 Z_s.

Determine the maximum internal fault-loop resistance allowed when the circuit protection is a single-pole 63 A 4.5 kA Type D circuit breaker.

$$Z_{Int} = 0.8 \frac{U_o}{I_a}$$

$$= 0.8 \times \frac{230}{12.5 \times 63}$$

$$= 0.2337\ \Omega$$

Note: Type D = 12.5 × rated current.

REVIEW QUESTIONS

1 Why is it important to know the short-circuit capability?

2 A single-phase circuit supplying a fixed appliance not protected by an RCD is installed in a manufacturing facility. This circuit uses 40 m of 2.5 mm² twin + E TPS cable with V90 insulation and a 25 A Type 'C' circuit breaker. Determine the maximum fault-loop impedance allowed.

3 A single-phase circuit supplying 10 A socket-outlets is installed in a domestic installation. The circuit uses 2.5 mm² twin + E TPS cable with V90 insulation and a 20 A Type 'C' circuit breaker. Determine the maximum permitted value of the internal fault-loop impedance and the maximum permitted value of the fault-loop impedance.

4 A domestic installation has a 15 A socket-outlet circuit not protected by an RCD and wired with 4 mm² twin + E TPS cable with V90 insulation and copper conductors.

a Determine the maximum internal fault-loop resistance allowed with the supply connected when the protection is a single-pole 25 A 4.5 kA Type 'C' circuit breaker.

b Determine the maximum internal fault-loop resistance (R_{phe}) allowed.

5 An industrial installation as shown in Figure 7.84 consists of the following:

- a transformer rated at 500 kVA with 4% impedance
- three-phase consumer's mains consisting of 20 m of 70 mm² PVC 600/1000 V SDI cables (0.000 245 Ω/m)
- three-phase sub-mains consisting of 30 m of 35 mm² PVC 600/1000 V SDI cables (0.000 491 Ω/m) and 10 mm² earth (0.00172 Ω/m)
- a final single-phase sub-circuit twin + E TPS supplying a fixed appliance consisting of 20 m of 2.5 mm² (0.00688 Ω/m) and a 2.5 mm² protective earth (0.00688 Ω/m) TPS twin + E protected by a 25 A type C circuit breaker.

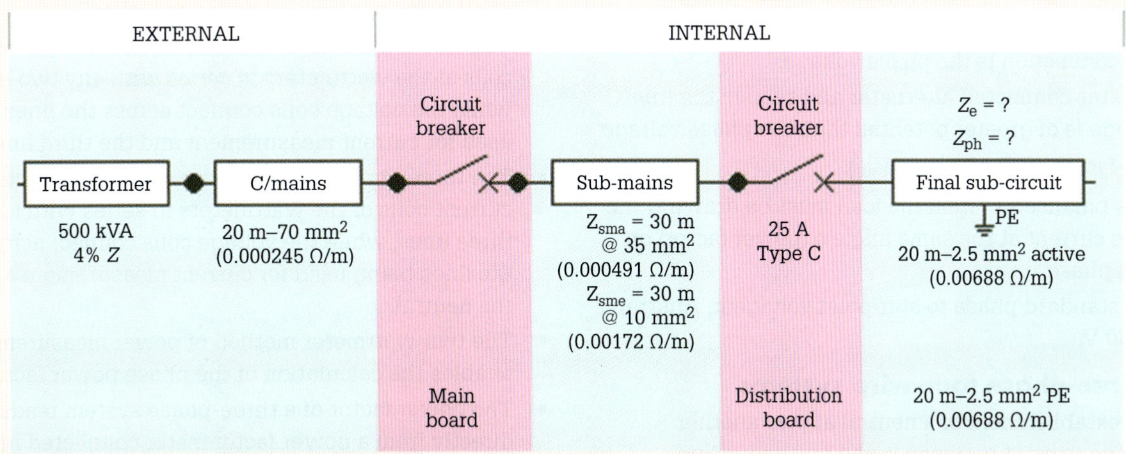

FIGURE 7.84 Review question 5

Determine:

a the prospective fault current at the transformer terminals

b the prospective fault current at the main board

c the prospective fault current at the distribution board

d the earth-fault-loop impedance

e the earth-fault-current due to a short to earth on the appliance

CHAPTER REVIEW

7.1 Three-phase systems

- A multiphase or polyphase system contains more than one alternating current voltage source delivering electrical energy to the load via three or more conductors.
- The current generated by the simple a.c. generator is a single-phase alternating current.
- A three-phase generating system has three sinusoidal waves, each wave separated from each other by 120 electrical degrees.
- Single-phase a.c. is available as a two-wire system or a three-wire system.
- Three-phase a.c. is available as a three-wire system or as a four-wire system.
- The single-wire earth return (SWER) is a high-voltage single-phase system.
- The power drawn from the three-phase supply is constant and not falling to zero as with the power drawn from a single-phase supply.
- There are six significant voltages developed by generators; three of these voltages are phase voltages (V_{Ph}) and three are line voltages (V_L).
- A three-phase system of voltages or currents has a cyclic direction of phase known as phase sequence in which the phase voltages and currents reach their peak value 120° after each other.
- Australia follows an anticlockwise phase sequence and the convention for this sequence is ABC or L1, L2, and L3 connected left to right on the terminal block.

7.2 Three-phase star connections

- There are many advantages of using a star-connected system, one of which is that the phase voltages are lower in this system when compared with a delta system, requiring less insulation in the stator.
- The voltage measured across a single coil group in a star connection is the phase voltage.
- In a star-connected alternator and system the line voltage is of greater potential than the phase voltage by a factor of $\sqrt{3}$ $\left(V_L = \sqrt{3}V_{Ph}\right)$.
- For a balanced system the load must be drawing the same current at the same angle ϕ (power factor) on each phase.
- The standard phase to star-point voltage in Australia is 230 V.

7.3 Three-phase four-wire systems

- The establishment of a neutral allows another voltage to exist between it and any active line.
- Use of a neutral enables two voltages in the distribution system – phase to phase (400 V) and phase to neutral (230 V).
- The multiple earth neutral (MEN) system of earthing is one where all parts of an installation that need to be earthed are connected to the general mass of the earth and connected to the main neutral conductor.
- The MEN system of earthing uses the distribution neutral conductor for any fault currents that may flow because of the active conductor passing current to earth.
- The effectiveness of the MEN system is reliant on the continuity of the main neutral conductor, the MEN link and the connection of the main earthing conductor to the earth electrode.
- Changing electrical conditions can create imbalances in the system.

7.4 Three-phase delta and interconnected systems

- Delta-connected generators or transformers supply three-phase power.
- An advantage of the delta-connected alternating current generator is that if one phase open circuits, three-phase is still obtainable from the other two phases.
- The voltage measured across a coil phase group is the phase voltage.
- In a delta-connected system $V_{\text{line}} = V_{\text{phase}}$ and $I_{\text{line}} = \sqrt{3}I_{\text{phase}}$
- The energy created in each phase of a star-connected circuit is the same as the energy developed in a single-phase circuit.
- $P = V_L I_L \cos \phi$
- The instrument used to read the power in three-phase circuits is a wattmeter.
- Where the phases are balanced as to load and power factor, a single wattmeter could be used to measure the power in one phase and the result multiplied by three.

7.5 Energy and power requirements of a.c. systems

- To measure power accurately, the power is measured in all three phases and added together.
- The two-wattmeter method connects the current coils of the wattmeters in series with any two lines, while the voltage coils connect across the lines being used for current measurement and the third line.
- The three-wattmeter four-wire system connects the current coils of the wattmeters in series with the three lines, while the voltage coils connect across the lines being used for current measurement and the neutral.
- The two-wattmeter method of power measurement enables the calculation of the phase power factor.
- The power factor of a three-phase system reads directly from a power factor meter connected into the circuit.
- The watt-hour meter or consumption meter is an instrument for measuring the actual amount of electrical energy used.
- Demand reflects the maximum electrical power required to operate a customer's facility.
- Maximum-demand indicators enable measurement of demand and demand growth.

- The purpose of determining maximum demand in consumer's mains is for the selection of the conductor's CSA in relation to the current it will carry.

7.6 Fault-loop impedance

- It is necessary to know the impedance of the path that any fault current would take to ensure that it is low enough to operate the circuit protective device (fuse or miniature circuit breaker) within a safe time limit.
- Clearing a short circuit to earth requires a fault current high enough to cause the protective device to operate quickly.
- Measuring or calculating loop impedance is essential when determining the prospective short-circuit current in a circuit so that the selection of the correct overload protection occurs.
- Microprocessor-controlled loop impedance testers are the most accurate means of measuring impedance.

7.7 Industry standards

- To make certain that the circuit protection device operates within the specified disconnection time, it is essential that the earth fault-loop impedance is below permissible levels.
- The selection of a suitable overload protection device depends on the value of the prospective short-circuit current.
- Where over-current protective devices cannot fulfil the conditions for protection by automatic disconnection of supply, such protection occurs by RCDs having a proper tripping time.

TRIAL EXAM

For Chapter 7 knowledge assessment, please complete the following trial exam.

1 A three-phase generator has three independent armature windings displaced from one another by:
   a 120°
   b 90°
   c 180°
   d 360°

2 How can the electromotive force produced at the output terminals of a generator be increased or decreased?
   a by control of the number of field windings
   b by control of the electromagnetic flux
   c by control of the load current
   d by control of the number of armature windings

3 How many voltage levels are available in a three-phase four-wire a.c. system?
   a 1
   b 2
   c 3
   d 4

4 A three-phase supply has the advantage of:
   a lower power loss than single phase
   b larger motors than equivalent single phase
   c higher power loss than single phase
   d in star-connected systems, the line current is smaller than the phase current

5 The transmission of a set amount of power by a three-phase system compared to a single-phase system to their respective loads at unity power factor requires:
   a a variable rotating magnetic field
   b smaller CSA conductors
   c increased prime mover speed
   d a starting winding

6 In a four-wire star-connected system, the line current equals?
   a phase current
   b 0.707 × phase current
   c 1.414 × phase current
   d 0.637 × phase current

7 The phase voltage in a delta-connected system is equal to:
   a line voltage
   b $\frac{1}{\sqrt{3}}$ line voltage
   c 0.637 × line voltage
   d 0.707 × line voltage

8 Phase sequence is:
   a the order in which the line voltages of a three-phase system achieve their minimum voltage
   b the order in which the phase voltages of a three-phase system achieve 0.707 of line voltage
   c the order in which the load voltages of a three-phase system achieve their 1.414 times value of line voltage
   d the order in which the separate voltages of a three-phase system achieve their maximum voltage

9 Determine the rms value of a 325.3 V peak voltage three-phase waveform.
   a 720 V
   b 400 V
   c 360.4 V
   d 230 V

10 Determine the average value of a 565.8 V peak voltage three-phase waveform for one complete cycle.
   a 0 V
   b 400 V
   c 360.4 V
   d 230 V

11 The line voltage in a three-phase star-connected system:
a lags the phase voltage by 30°
b leads the phase voltage by 30°
c is in phase with the line voltage
d is 120° out of phase with the line voltage

12 The phase voltage in a star-connected three-phase system is equal to:
a IZ
b $\sqrt{3}V_L$
c $\frac{V_L}{\sqrt{3}}$
d IR

13 Determine the phase voltage in a star-connected generator that has a line voltage of 2449 V.
a 2121 V
b 3000 V
c 5196 V
d 1732 V

14 The line current in a star-connected three-phase system is equal to:
a I_{PH}
b $\sqrt{3}I_{PH}$ I_{PH}
c 0.707 I_{PH}
d 0.637 I_{PH}

15 In the event of a high impedance or open-circuit neutral conductor, the neutral current will return to the source via:
a earthing system and general mass of earth
b the other active conductors
c negative phase excursions
d no current flows in the neutral conductor

16 If a star system is balanced with the line currents drawing 80 A at the same phase angle, then the current in the neutral is:
a 0 A
b 70.7 A
c 63.7 A
d 173.2 A

17 Three-phase four-wire distribution systems are used to:
a provide power factor correction
b provide varying phase currents
c provide power to an unbalanced system
d provide different voltages

18 Which conductor in a three-phase four-wire supply carries out-of-balance currents?
a earth
b neutral
c for clockwise rotation line L1
d for anticlockwise rotation L3

19 What type of generator still produces three-phase power if one phase opens?
a four-wire star connected
b star connected
c delta connected
d single-phase

20 The phase voltage in a delta-connected three-phase system is equal to:
a V_L
b $\sqrt{3}V_L$
c 0.707 V_L
d 0.637 V_L

21 Determine how much power a three-phase 400 V connected motor draws from the supply when drawing 10.2 A at a pf of 0.9 from the supply.
a 9 kW
b 15 558 W
c 6363 W
d 5733 W

22 The selection of a suitable overload protection device depends on the value of the:
a load current
b indirect current
c prospective short circuit current
d mean tripping current

23 The purpose of determining maximum demand in consumer's mains is:
a determining the number of final sub-circuits
b selecting correct protective devices
c resistance measurement
d CSA determination

24 Why is it necessary to know the impedance of the path (loop) that any fault current would take?
a for operation of the circuit protective device
b to determine the size of the main earth
c to size the main switch
d to balance the circuit load

25 A 70 A resistive load is supplied through a 25 mm^2, 4-core, V90 cable having circular conductors. Determine the voltage drop in one line conductor if it has a route length of 65 m.

26 A star-connected three-phase generator has a rated voltage of 230 V per coil phase group and supplies a balanced load consisting of three 10 Ω impedance windings in a delta motor. Calculate the following:
a Line voltage of the star-connected generator: $V_{L\,(Load)}$
b Phase voltage of the load: $V_{Ph\,(Load)}$
c Phase current drawn by the load: $I_{Ph\,(Load)}$
d Line current delivered to the load: $I_{L\,(Load)}$
e Phase current in the generator: $I_{Ph\,(Gen)}$

27 A domestic installation has a 15 A socket outlet circuit not protected by an RCD and wired with 4 mm^2 twin + E TPS cable with V90 insulation and copper conductors. Calculate the maximum internal fault-loop resistance allowed with the supply connected when the protection is a single-pole 32 A 4.5 kA Type D circuit breaker.

Alternating current (a.c.) rotating machines

This chapter provides electrotechnology workers with knowledge of the operating principles, characteristics and application of three-phase and single-phase induction motors as well as synchronous motors and generators. In addition, electrotechnology workers will develop an understanding of electric motor selection, starting requirements and motor starters, types of motor connections and motor protection, causes of motor malfunction, fault diagnosis and testing. This chapter provides underpinning knowledge for the unit UEEEL0024 from the UEE training package.

LEARNING OBJECTIVES

Three-phase induction motor operating principles

- Describe the principles of operation of three-phase induction motors

Three-phase induction motor construction

- Describe the construction of a three-phase induction motor

Three-phase induction motor characteristics

- Explain how torque, speed and power relate as applied to three-phase induction motors

Single-phase split-phase motor

- State the operating principles and characteristics of single-phase split-phase motors
- List applications for single-phase split-phase motors

Single-phase shaded-pole and capacitor motors

- State the operating principles and characteristics of single-phase shaded pole and capacitor motors
- List applications for single-phase shaded pole and capacitor motors

Series universal single-phase motor

- State the operating principles and characteristics of single-phase series universal motors
- List applications for single-phase series universal motors

Motor protection

- Describe how to protect three-phase and single-phase motors against various adverse conditions
- Outline *Wiring Rules* requirements for a.c. motors and controls

Synchronous motors

- Describe the construction details of synchronous motors
- Explain the principles of operation of synchronous motors

Alternators and generators

- Explain the construction details of synchronous alternators
- Explain the principles of operation of synchronous alternators

Testing of low voltage a.c. machines

- Outline various a.c. machine tests

Mechanical faults in LV a.c. rotating machines

- List the various mechanical faults that occur in LV a.c. rotating machines

Driven load and coupling faults

- List the various driven load and coupling faults that occur in LV a.c. rotating machines

Electrical faults and symptoms in LV a.c. rotating machines

- Outline the various electrical faults and symptoms that occur in LV a.c. rotating machines

8.1 Three-phase induction motor operating principles

Three-phase induction motors are the most commonly used electric motors. They are reliable, frequency-dependent constant-speed machines, simple in construction and easy to maintain, with high efficiency.

Induction motors

The name 'induction motor' comes from the electromotive force (emf) induced via transformer action (there is no electrical connection between the stator and the rotor) into the rotor by means of the rotating magnetic flux produced in the stator. The induced emf causes a current in the rotor that develops an electromagnetic field. Consequently, the rotor field with the rotating electromagnetic field of the stator causes rotor rotation. Refer to **Figure 8.1**.

FIGURE 8.1 Three-phase induction motor

Principles of induction-motor action

The turning action of an induction motor is the result of Faraday's law and Lenz's force law applied to a conductor. **Figure 8.2** shows a simple three-phase, two-pole motor containing three stator coils spaced 120° apart around the surface of the stator. The three coils connected in a star configuration have their free ends connected to a three-phase source.

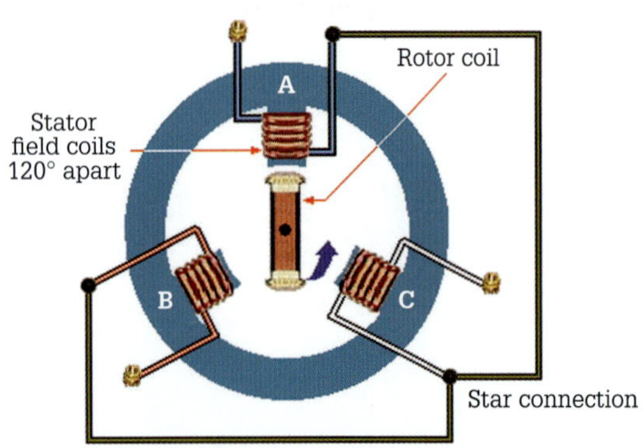

FIGURE 8.2 Three-phase motor connections

When energised, the currents drawn by each stator phase group reach their maximum value at different instants in time and magnetise the stator. The three currents are displaced in time from each other by 120 electrical degrees, which means the electromagnetic flux created by each coil is displaced by 120 electrical degrees. **Figure 8.3** illustrates the instantaneous values of current and direction (anticlockwise) of the stator flux as it relates to time.

At zero degrees, phase-A draws maximum current and establishes a very strong electromagnetic north pole. The currents drawn by the other two phases B and C are both only at 50% of phase-A and flow in the opposite direction to phase-A. Consequently, their electromagnetic poles are weak and are the opposite polarity (south poles) to the pole of phase-A. At 60 degrees in time, phase-C is at maximum value and produces a very strong south pole while phases A and B establish weak north poles. At 120 degrees in time, phase-B establishes a strong north pole while phases A and C are weak south poles.

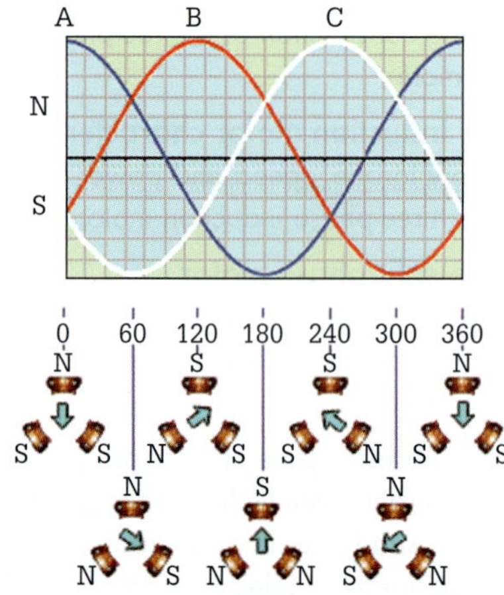

FIGURE 8.3 Three-phase flux waveform

The electromagnetic flux established by each coil is an alternating flux (rising to a maximum value then falling to zero then rising to a peak with an opposite polarity and falling to zero again). It is the combined flux of the three coils at different instants in time that produce a two-pole rotating flux around the stator at a constant speed called synchronous speed. When a rotor is located within this rotating field it develops rotational motion. When the electromagnetic lines of force of the stator cut the rotor conductors, an emf, which is induced by transformer action, drives current through the rotor conductors to energise them magnetically. Refer to **Figure 8.4**.

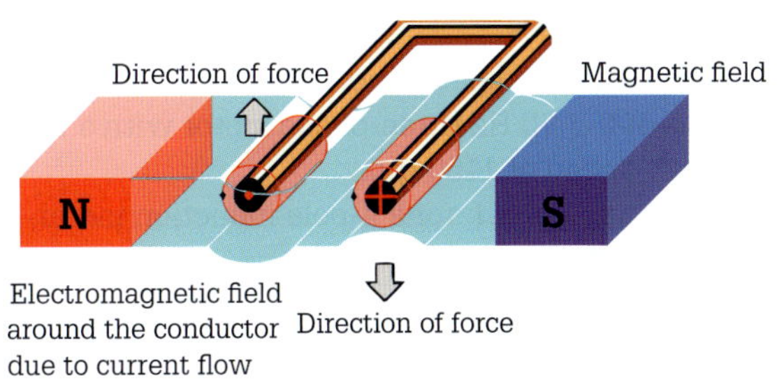

FIGURE 8.4 Forces acting on a rotor conductor

Fleming's left-hand motor rule as illustrated in **Figure 8.5** shows the turning direction of the rotor.

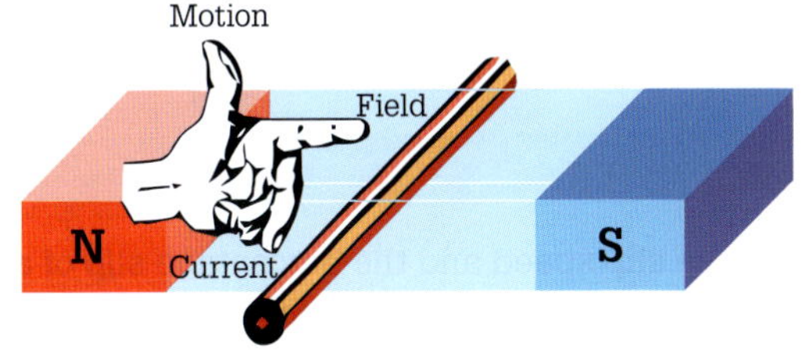

FIGURE 8.5 Left-hand motor rule

SWITCH ON

Fleming's left-hand motor rule

- The thumb gives the direction of the force.
- The first finger points in the same direction as the field.
- The second finger points in the direction of the current.

The three-phase supply creates a rotating electromagnetic field in the stator winding. Consequently, the induced electromagnetic field in the rotor follows it, thereby producing torque. According to Lenz's law, the rotor conductors must develop a turning effort in the same direction as the rotating stator flux. If the rotor did catch up to the stator field, the rotor conductors would not experience the cutting effect of a varying rotational flux. In other words, the induced emf producing current would not exist and thus no torque. Induction motors operate at a speed somewhat less than the synchronous speed of the stator flux, usually at about 95%. This difference between the rotor speed and the speed of the stator flux defines a motor parameter called the 'motor slip'.

At standstill an induction motor behaves as a transformer with a shorted secondary winding. As such it draws current of a significant magnitude compared to its operating current. As the rotor rotates, it causes a counter emf to be induced in the stator windings, which has the effect of reducing the current drawn from the supply. This counter emf behaves similarly to that examined in Chapter 5 for d.c. rotating machines. A motor produces torque and also generates an emf.

Air gap

The link between the electromotive fields of the stator and rotor is influenced by the air gap between the rotor and the stator. This air gap depends upon motor design, is a uniform width, and can be from 0.4 mm to 2.5 mm. However, the best designs keep the air gap to a minimum because the wider the gap, the greater is the magnetising force required from the stator to establish in it an electromagnetic flux.

If the air gap is not uniform around the 360 degrees of the rotor then uneven electromagnetic fields develop. These electromagnetic imbalances can cause movement of the stator windings resulting in winding failure. An electrically induced vibration is produced which results in bearing failure.

Speed

The speed of the rotating flux of an induction motor is the synchronous speed (N_s). The synchronous speed (rev/min) is directly proportional to the frequency (Hz) of the supply voltage and inversely proportional to the number of poles. If a motor has p-pairs of poles and if the speed (N_s) of the magnetic flux is in revolutions per second, at a definite frequency (f) in hertz then:

$$N_s - \frac{f}{p}$$

This equation provides the synchronous speed of the rotor in revolutions per second based on the number of pole pairs (p). So, to provide revolutions per minute based on number of poles the equation becomes:

$$N_s = \frac{120f}{P}$$

where N_s = synchronous speed in rev/minute (rpm)

f = frequency in hertz (Hz)

P = number of poles

120 = converts cycles per second to minutes (the multiplier of 120 results from changing the cycles/second to cycles/minute (× 60) and changing the pairs of poles (a four-pole motor has two pairs of poles) to number of poles. In the previous equation the number of poles ÷ 2 provides the pairs of poles while in this equation the ÷ 2 becomes a × 2 to the frequency (60 × 2 = 120)

EXAMPLE 8.1

Determine the synchronous speed in rpm of a four-pole 400 V 50 Hz three-phase induction motor.

$$N_s = \frac{120f}{P}$$

$$= \frac{120 \times 50}{4}$$

$$= \mathbf{1500\ rpm}$$

EXERCISE 8.1

Determine the synchronous speed in rpm of the following three-phase induction motors:

a six-pole 400 V 50 Hz
b twelve-pole 400 V 50 Hz
c two-pole 400 V 50 Hz.

The synchronous speed of a three-phase motor varies by altering the number of poles or by changing the frequency of the supply voltage. An increase in supply frequency causes the electromagnetic flux to rotate at a higher speed. Most standard squirrel-cage motors can withstand an over-speed occurrence of 1.2 times their rated speed.

Slip

The difference between the synchronous speed and the actual speed of the rotor is the slip speed. Without the slip, there would be no relative motion between the rotating stator electromagnetic flux and the rotor conductors. As a result, no current induces into the rotor conductors, and no torque develops.

$$\text{slip} = N_S - N$$

Slip speed is usually expressed as a percentage of synchronous speed.

$$\%\text{slip} = \frac{N_S - N}{N_S} \times 100$$

N_s = synchronous speed in rev/minute (rpm)
N = rotor speed in rev/minute (rpm)

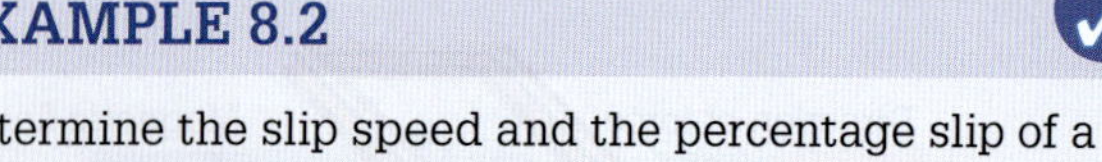

EXAMPLE 8.2

Determine the slip speed and the percentage slip of a four-pole 400 V 50 Hz three-phase motor with a full-load shaft speed of 1450 rpm.

From Example 8.1, for a 4-pole induction motor, N_s = 1500 rpm

$$\begin{aligned}\text{slip} &= N_S - N \\ &= 1500 - 1450 \\ &= \mathbf{50\ rpm} \\ \%\text{slip} &= \frac{N_S - N}{N_S} \times 100 \\ &= \frac{1500 - 1450}{1500} \times 100 \\ &= \mathbf{3.33\%}\end{aligned}$$

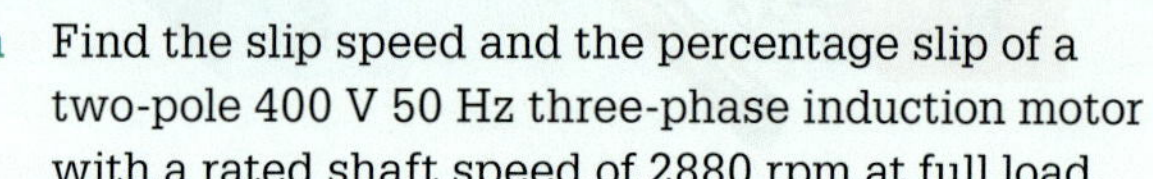

EXERCISE 8.2

a Find the slip speed and the percentage slip of a two-pole 400 V 50 Hz three-phase induction motor with a rated shaft speed of 2880 rpm at full load.
b Find the slip speed and the percentage slip of a six-pole 400 V 50 Hz three-phase induction motor with a rated shaft speed of 950 rpm at full load.

The slip of three-phase induction motors depends on the mechanical load connected to the rotor shaft. Furthermore, an increase in the load decreases the shaft speed thereby increasing the slip. Slip is near zero with an unloaded motor and is in the range of 3% to 13%, depending upon motor design, at full load.

REVIEW QUESTIONS

1 On what principle does an induction motor operate?
2 What is produced by the stator of a three-phase induction motor?
3 Which two laws concern the turning action of a motor?
4 What design element of a motor affects the relationship between the stator and rotor's electromagnetic fields?
5 Determine the synchronous speed in rpm of an 8-pole 400 V 50 Hz induction motor.
6 What is meant by the term 'slip speed'?
7 Calculate the slip speed in rpm of a ten-pole 400 V 50 Hz three-phase induction motor with a full-load shaft speed of 560 rpm.
8 Calculate the slip speed percentage of a ten-pole 400 V 50 Hz three-phase induction motor with a full-load shaft speed of 576 rpm.

8.2 Three-phase induction motor construction

Regardless of the power rating of the motor, the components used in the construction of three-phase induction motors are the same.

Components of a three-phase induction motor

The three-phase induction motor consists of a ribbed frame that contains the stator and the rotor. The stator is made up of thin sheet metal called laminations which are pressed together to form the stator. Around the internal periphery of the stator are slots into which a winding is placed. The rotor also consists of laminations and cooling fins attached to a shaft which delivers the mechanical power to the load. An end shield closes one end of the frame at the drive end, and a second end shield closes the second end, or non-drive end. The shaft is journeyed in at least one of the end shields. Furthermore, it can extend through at least one of the end shields. A fan cover (also called a cowl) encloses the cooling fan, while the fan attaches to the shaft. Leads extend through the motor frame from the stator assembly to the terminal block housing. **Figure 8.6** shows an exploded view of a three-phase induction motor.

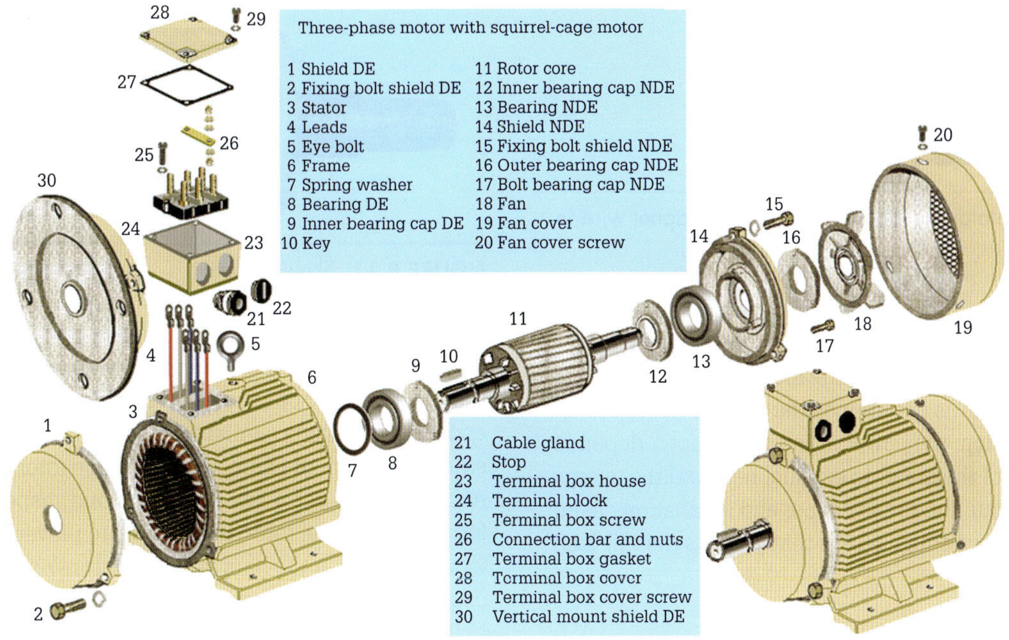

FIGURE 8.6 Exploded view of a three-phase induction motor (DE = drive end; NDE = non-drive end)

Stator

The stationary member (the stator) of the induction motor consists of stacked laminated 1.0–1.5 mm thick silicon-alloy sheet steel stampings mounted in a light alloy, grey cast iron or steel housing called the frame.

The use of silicon steel reduces the energy loss incurred by hysteresis. Each lamination as shown in **Figure 8.7** has a coat of oxide or varnish to minimise eddy current energy loss. Refer to **Figure 8.8** for stator lamination assembly.

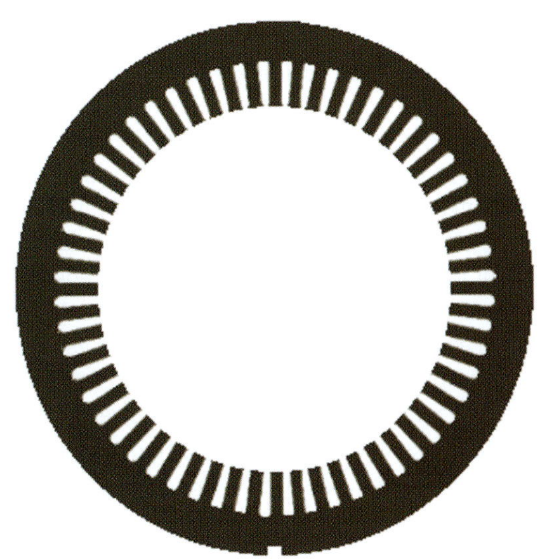

FIGURE 8.7 Lamination stamping

The stator lamination assembly is illustrated in **Figure 8.8**.

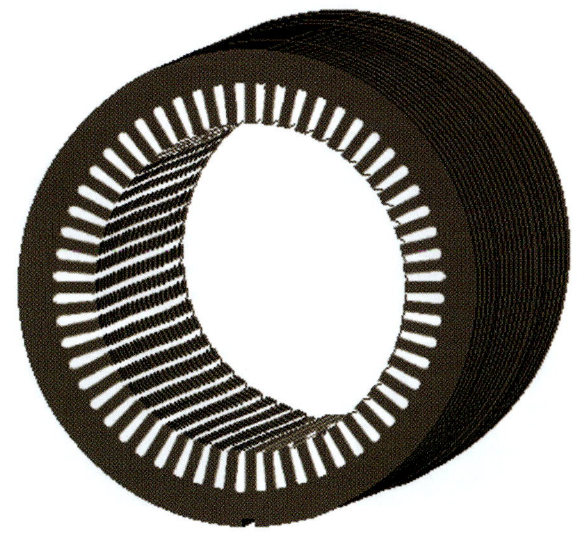

FIGURE 8.8 Induction motor stator assembly

Wound within the evenly spaced lamination slots are three-phase sets of coils spaced 120 electrical degrees apart. These phase sets of coils can connect in series or parallel, forming phase groups. The separate phase groups of coils can then connect in the fundamental star or delta configuration and are energised from a three-phase supply. The purpose of the stator coils, as shown in **Figure 8.9**, is to produce a rotating electromagnetic field when energised from a three-phase supply.

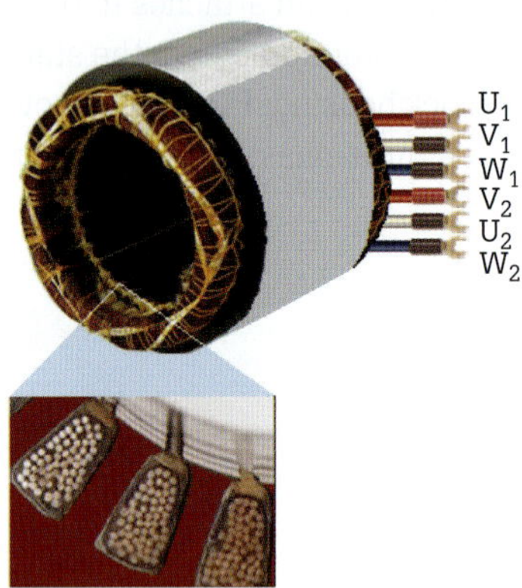

FIGURE 8.9 Stator winding (inset showing magnet wire in stator slots)

The design of the coils and their span across the stator slots as well as the type of lamination all have an effect on the effectiveness of the motor. For example, the number of coil turns determines the voltage rating of the motor. In addition, the power rating of the motor depends on the ability of the motor to exhaust heat and the energy losses that occur due to the copper and lamination design. Finally, it is the winding design that governs the full-load speed of the motor.

Rotor

The rotor also consists of stacked laminated iron stampings but mount on a steel shaft in a similar fashion to armature laminations in the d.c. machine discussed in Chapter 5. In addition, these stampings have slots to take the rotor conductors. **Figure 8.10** shows different types of rotor stampings.

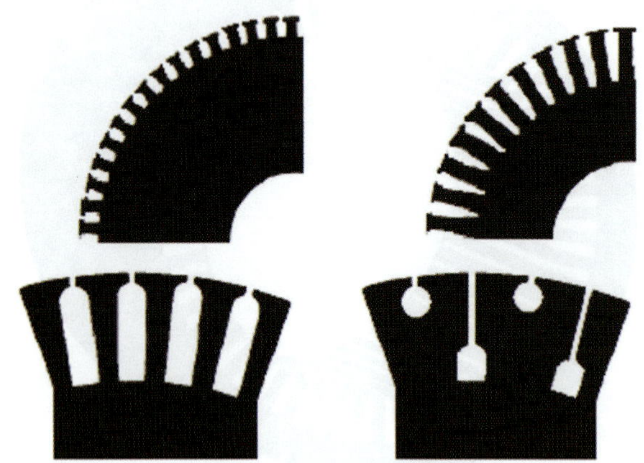

FIGURE 8.10 Rotor stampings

The rotor assembly is normally skewed as shown in **Figure 8.11** to increase the starting torque and to prevent electromagnetic cogging (rotation occurring in jerks rather than a smooth motion), electrical noise and rotor vibration due to the interaction of the two electromagnetic fields. Remember that there is no electrical connection to the rotor conductors in an induction motor; when the rotor is at standstill the relative speed of the rotating magnetic field is high, which induces current in the rotor conductors that in turn produces a magnetic field. It is this rotor field that needs to follow the stator field, so skewing the rotor conductors effectively broadens the conductor influenced by the stator field.

There are two types of induction motor rotors, a squirrel-cage rotor and a wound rotor.

FIGURE 8.11 Skewed slots

Squirrel-cage rotor

In small and medium-sized induction motor rotors, the uninsulated rotor conductors join at the rotor assembly ends to form a cage similar to that used for exercising squirrels, hence the name squirrel-cage rotor. Standard rotor windings use a few bars of various shapes inserted or formed within the rotor. In addition, to prevent magnetic locking between the bars and the stator winding the rotor design enables the bars to skew to the longitudinal axis of the shaft. A simplified squirrel-cage rotor winding, illustrated in **Figure 8.12**, consists of several bars connected at both ends by end rings to form a closed cage.

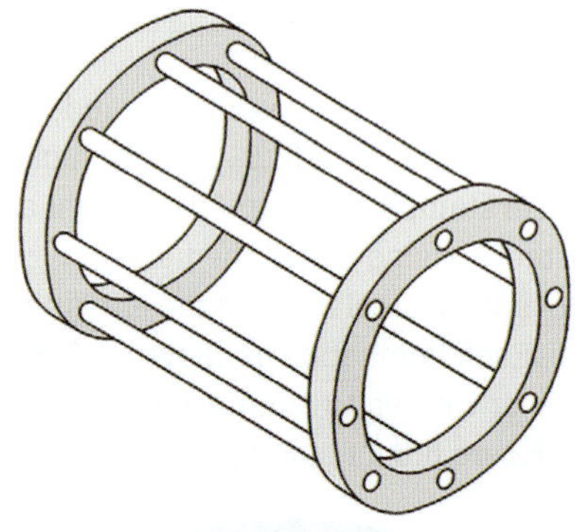

FIGURE 8.12 Simplified squirrel-cage rotor winding

The squirrel cage rotor design allows the rotor to operate at high speeds and withstand significant electrical and mechanical overloads. Another type of rotor is wire wound; however, wound-rotor induction motors are being phased out by controlled drives with squirrel-cage motors. Most rotors have molten aluminium poured into the lamination assembly to form the rotor conductors, shorting rings and simple fan blades as one piece. By contrast, large squirrel-cage rotors use copper alloy bars inserted into the rotor slots that are brazed onto end rings to form the squirrel cage. **Figure 8.13** shows a cast conductor squirrel cage rotor.

FIGURE 8.13 Squirrel-cage rotor

Squirrel-cage rotors may be manufactured with deep-bar and double-cage rotors. Refer to **Figure 8.14**.

The material of the bars, their shape, cross-sectional area and location from the surface of the rotor determine the rotor performance characteristics. In addition, the rotor circuit contains the electrical characteristics of resistance, inductance and impedance. However, while the resistance of the rotor circuit is dependent on the physical and material make-up of the bars (large CSA = low resistance), the other electrical characteristics are dependent on the size of the current induced into the rotor circuit. If the rotor bars are positioned deep into the rotor, the effect on the rotor circuit is of increased inductance due to the amount of the rotor iron that surrounds the bars. By contrast, rotors whose bars are close to the surface demonstrate a high resistive effect on the rotor circuit.

FIGURE 8.14 Bar-type squirrel-cage rotor

Moreover, the impedance of both bars will effectively change as the frequency of the induced current changes.

For example, the flux developed in the top of the bar links firmly with the stator. However, at the bottom of the bar, the flux loosely links to the stator.

Wound rotor

A wound rotor has a three-phase winding installed within its slots that mirrors the windings of the stator in that it has as many poles as the stator. Wound-rotor windings connect in a star configuration internally with the other three lead ends connected to slip rings. Refer to **Figure 8.15**.

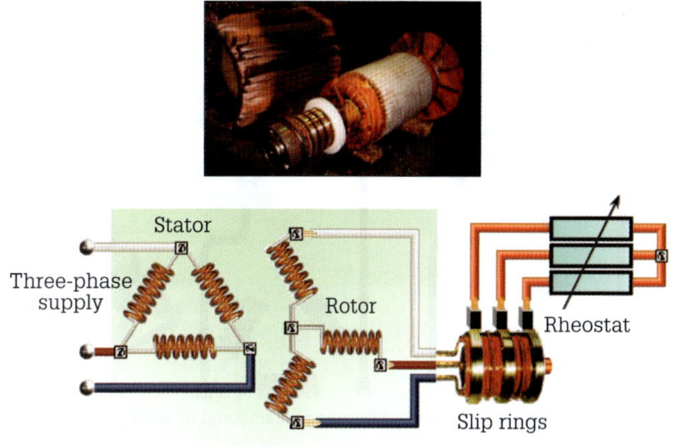

FIGURE 8.15 Wound rotor

The stator windings of a wound-rotor motor usually connect in delta with some stators connected in star. In addition, the terminal block has six terminals – three for the stator leads and three for the star-connected rotor windings via slip rings. These three rotor terminals connect to the variable star-connected rheostat while the three-stator terminals connect to a three-phase supply. Rotor windings are star-connected due to less insulation needed to insulate the coils.

The central feature of wound rotors is the modification of the torque–speed characteristic of the motor to suit the load. Change happens by using external rheostats in the form of wire, cast-iron grids or electrolyte liquid. Consequently, the rheostat resistance increases the resistance of the rotor circuit, lowering the rotor current and the stator current.

Inserting high resistance into the rotor circuit results in a low value of starting current together with the development of a high torque. As the motor accelerates, the value of the resistance reduces, enabling maximum torque to move towards the motor's synchronous speed.

This type of starting is ideal for very high inertia loads allowing the motor to accelerate the load slowly and smoothly. Typical applications include elevators, conveyor belts and hoists.

Terminal block connections

Most three-phase stators have six leads brought out to a six-terminal block. However, there exist some small special-purpose motors that are designed to operate only in star or in delta configuration and only have three leads brought out from the stator as shown in **Figure 8.16**.

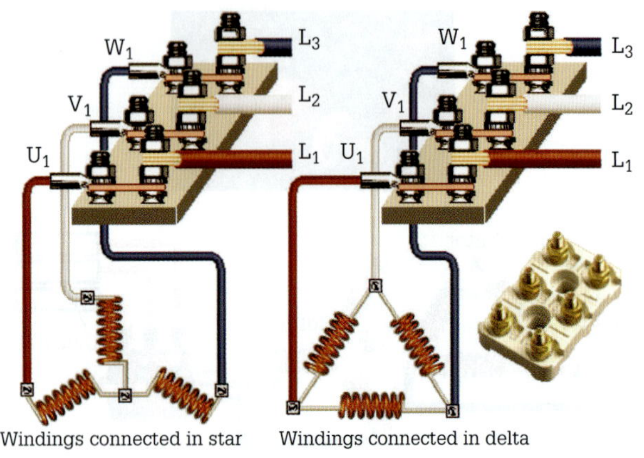

FIGURE 8.16 Terminal block connections for star- and delta-connected stators

Motors connected in either configuration of star or delta with three leads from the stator connect directly on line.

Stators with six leads also connect to a six-lead terminal block as illustrated in **Figure 8.17**. These leads can connect in star or delta configuration at the terminal block by using brass links to effect the termination. Motors connected in this fashion also connect directly on line. With larger motors, no links occur because six-line leads connect to the terminal block so that the motor can connect to various types of motor starters.

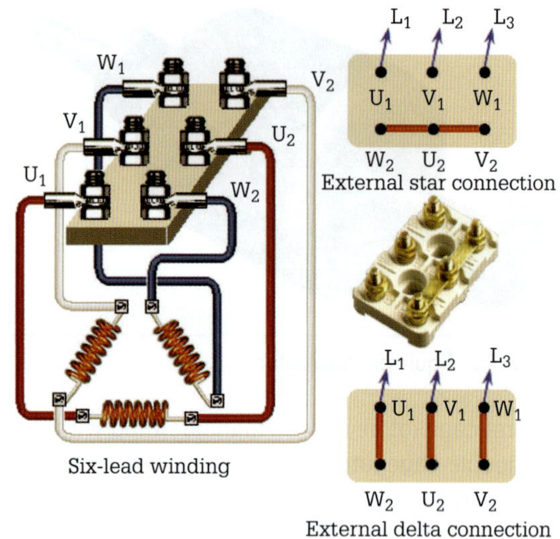

FIGURE 8.17 Terminal block connections for six-lead stator connections

When viewed from the drive shaft end, the motor rotor should rotate in a clockwise direction when the R-W-B supply leads connect to the U-V-W motor terminals.

REVIEW QUESTIONS

1. Why are stator laminations made of silicon steel?
2. What is the purpose of the stator coils?
3. Why would a squirrel cage rotor have skewed rotor bars?
4. What elements in relation to the stator determine motor performance?
5. What type of induction motor rotor mirrors the windings of the stator in that it has as many poles as the stator?
6. What is the central feature of wound rotor induction motors?
7. How many leads do most three-phase induction motors have brought out to the terminal block?
8. What is the reason for induction motors having the number of wires of question 7 brought out?

8.3 Three-phase induction motor characteristics

The main characteristics when comparing or selecting three-phase induction motors are torque, output power, and operating power factor.

Torque

The induction motor operates due to the torque produced by the interaction of the stator electromagnetic field and the rotor's electromagnetic field. In addition, both of these fields are due to currents that have resistive and reactive components.

Consequently, the torque produced is the result of the interaction of these current components and is directly related to the I^2R of the rotor. **Figure 8.18** represents a typical torque–speed characteristic graph of a standard three-phase induction motor.

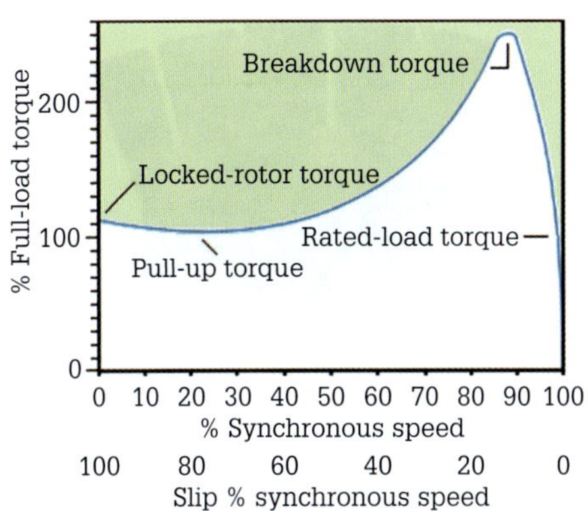

FIGURE 8.18 Starting characteristics – standard three-phase induction motor

These graphs allow a comparison of different induction motors and their characteristics in order to determine the motor most suitable for the load.

Induction motors provide extra torque at starting in order to overcome the inertia of starting from rest. The starting characteristic curve of **Figure 8.18** shows the starting or locked-rotor torque as 115% of the rated-load torque. The locked-rotor torque is the smallest measured torque established when the rotor is locked when the rated voltage and frequency is applied. Induction motors at rest behave like a short-circuited transformer and they draw a very high current identified as the 'locked-rotor current'.

Both these parameters are a function of the terminal voltage and the motor design. As the rotor accelerates, both the torque and the current alter with the rotor speed as long as the voltage remains constant. In addition, the current drawn by the stator decreases slowly as the rotor accelerates and only begins to fall significantly when the rotor has reached at least 80% full speed. As illustrated by the graph in **Figure 8.18**, the torque drops a little to the smallest torque produced by the rotor between zero speed and the speed that corresponds to the breakdown torque. In addition, the pull-up torque is the point where the rotor begins to accelerate rapidly.

The breakdown torque is the stalling point of the rotor and is the maximum torque the rotor can generate at rated voltage and frequency. This parameter also indicates the maximum overload that the motor can withstand without excessive heat building up in the stator windings causing a burnout. Finally, the rated-load torque is the designed value for the motor. This value is the torque necessary for the motor to produce its rated kilowatt output at rated-load speed (typically about 96% of synchronous speed).

Without a load, the rotor almost reaches the synchronous speed of the rotary field, since only a small counter-torque (no-load losses) is present. If it were to turn exactly synchronously, induced voltage would not occur, current would cease to flow, and there would no longer be any torque.

Calculating torque from output power

It is possible to calculate the load torque an induction motor develops based on the output power from the equation:

$$T = \frac{9.55P}{N}$$

where T = torque in newton metres (Nm)

P = motor output power in watts (W)

N = motor speed in rpm

Note that 9.55 is derived from $\frac{60}{2\pi}$

EXAMPLE 8.3

Determine the full-load torque developed by a three-phase induction motor having a rated power output of 3.6 kW when operating at 1450 rpm.

$$T = \frac{9.55P}{N}$$

$$= \frac{9.55 \times (3.6 \times 10^3)}{1450}$$

$$= \mathbf{23.7\ Nm}$$

EXERCISE 8.3

a Determine the full-load torque developed by a three-phase induction motor having a rated power output of 3.6 kW when operating at a shaft speed of 2880 rpm at full load.

b Determine the full-load torque developed by a three-phase induction motor having a rated power output of 4.5 kW when operating at a shaft speed of 950 rpm at full load.

Rotor impedance and torque

Induction motors produce maximum torque when rotor resistance equals rotor reactance (X_L). Established values of rotor resistance and inductance values occur in the design stage by the physical and electrical properties of the rotor and its winding.

At the instant of start, the rotor is stationary (slip = 100%) and the stator flux cuts it at synchronous speed and induces a rotor voltage and current at a frequency (rotor frequency) equal to the supply frequency. As the rotor accelerates, the percentage slip decreases (to, say, 90%). Consequently, the rotor current frequency is also 90% of supply frequency. This decrease in rotor frequency as the rotor continues to accelerate to full-load speed when slip is typically about 4% is expressed by the following:

$$\text{rotor frequency} = \text{supply frequency} \times \%\ \text{slip}$$

Therefore, rotor frequency = 50 Hz × 100% = 50 Hz at start, decreasing to rotor frequency = 50 Hz × 4% = 2 Hz at full-load speed.

Now, rotor resistance is unaffected by rotor frequency, but rotor reactance is directly proportional to the frequency ($X_L = 2\pi fL$). Rotor reactance (X_L) is at maximum at start (when rotor frequency is at maximum) and decreases as the rotor accelerates. Consequently, at some speed (and associated rotor frequency), rotor reactance equals rotor resistance, and it is at this point that maximum or breakdown torque occurs. Refer to **Figure 8.19**.

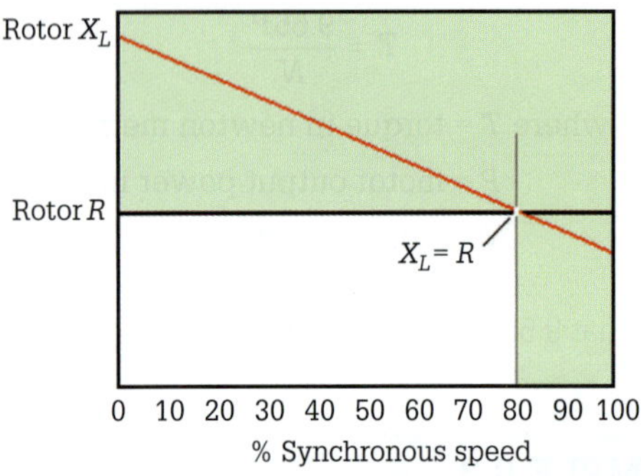

FIGURE 8.19 Maximum torque when $X_L = R$

The type of a rotor cage and the shape of the rotor slot determine the torque characteristic and the magnitude of the starting current. For example, squirrel-cage rotors with a typical cage of single slot and round, rectangular or trapezoidal conductors produce a relatively high starting torque and a high starting current.

Deep-bar rotor

The rotor of the deep-bar squirrel-cage induction motors uses rectangular cross-sectional copper bars designed so that the current flow directs towards the top of the bar when the motor starts to increase the active resistance. Refer to **Figure 8.20**.

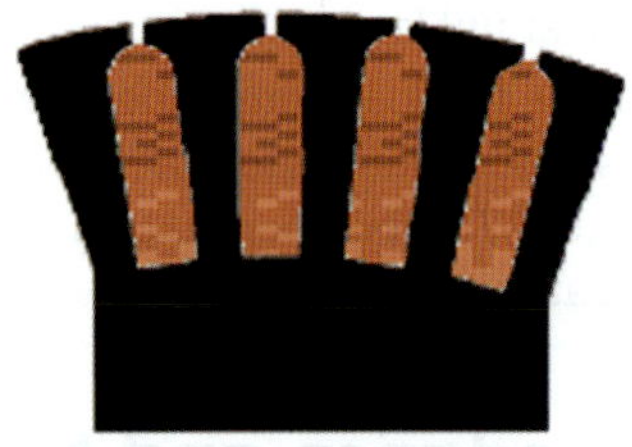

FIGURE 8.20 Deep-bar rotor

At starting, the rotor current frequency is high and, because of leakage flux and the skin effect, the rotor has induced currents moving towards the top of the bars close to the rotor's surface.

The current flows through a smaller cross-sectional area of high resistance, resulting in exceptional starting torque. As the rotor increases in speed, the skin effect diminishes (the change in resistance is a resultant based upon the change of frequency of the rotor currents with speed – rate of stator flux cutting rotor conductors has fallen) and rotor frequency falls to about 3 Hz, allowing the rotor currents to penetrate more deeply into the rotor bars. Consequently, a deeper penetration means that at rated speed the rotor conductors offer a low-resistance path for current, allowing the rotor to produce a low full-load slip and good operating efficiency.

Double-cage rotor

Figure 8.21 shows a cross-section of a double-cage rotor. These rotors consist of high-resistance bars near the rotor surface with low-resistance bars set deep within the rotor laminations. It operates like the deep-bar rotor.

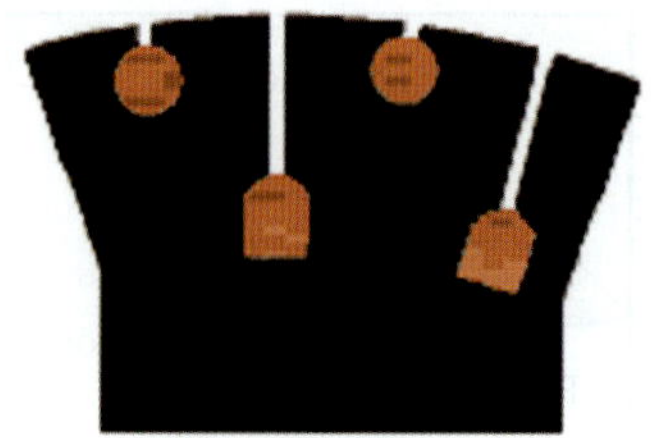

FIGURE 8.21 Double-cage rotor

At starting, the high-resistance bars are sufficient because the deeper low-resistance bars have a high reactance ($X_L = 2\pi fL$) thereby ensuring that the rotor currents flow in the outer bar. At rated speed, both bars exhibit low reactance, and the rotor current is diversified between them resulting in a high-efficiency motor with low slip per cent. Double-cage induction motors – because they offer a higher starting torque with lower starting current and real efficiency at normal operating conditions than standard squirrel-cage motors – find use as the prime mover for pumps and compressors that require starting under load. However, these motors are not suitable for some loads due to the rotor I^2R energy losses that occur during start-up. These energy losses concentrate in the outer bar, which may overheat and melt.

Motor output power

The power output for any type of induction motor is stated on the motor nameplate and is its full-load output power. Output power can be determined by applying the following equation:

$$P = \frac{2\pi NT}{60}$$

where P = rated power in watts or kilowatts
N = rotor speed in revolutions per minute (rpm)
T = load torque in newton-metres (Nm)

EXAMPLE 8.4

Calculate the power output of a three-phase induction motor that develops a torque of 26 Nm at 1440 rpm.

$$P = \frac{2\pi NT}{60}$$
$$= \frac{2\pi \times 1440 \times 26}{60}$$
$$= \mathbf{3.92\ kW}$$

EXERCISE 8.4

a Calculate the power output of a three-phase induction motor that develops a torque of 52 Nm at 1450 rpm.

b Calculate the power output of a three-phase induction motor that develops a torque of 36 Nm at 950 rpm.

Efficiency

The efficiency rating of an induction motor takes into account the energy losses that dissipate in both the stator and the rotor. These energy losses can be categorised primarily as stator and rotor copper losses (I^2R loss), iron losses, mechanical losses and stray flux losses as illustrated in **Figure 8.22**. High-efficiency motors should have totals for these losses from 20% to 30% less than equivalent standard motors.

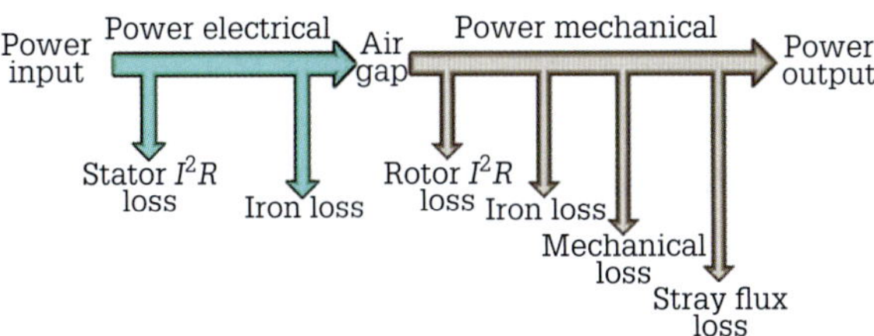

FIGURE 8.22 Power transfer diagram

The rotor output, measured in watts, is equal to the total mechanical power developed less the losses experienced by the rotor. The per-unit efficiency (symbol η) of an induction motor is the ratio of the usable rotor output to the input power.

$$\eta = \frac{P_{out}}{P_{in}}$$

The input power or true power (P_T) to the stator is determined by connecting a wattmeter in line with the motor or it can be determined by applying the following equation:

$$P_{in} = \sqrt{3}\,VI\cos\phi$$

where P_{in} = input power in watts or kilowatts
V = supply voltage in volts (V)
I = stator current in amperes (A)
ϕ = power factor

EXAMPLE 8.5

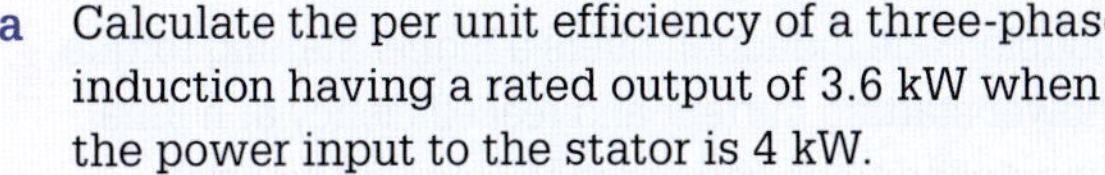

a Calculate the per unit efficiency of a three-phase induction having a rated output of 3.6 kW when the power input to the stator is 4 kW.

$$\eta = \frac{P_{out}}{P_{in}}$$
$$= \frac{3.6}{4}$$
$$= 0.9 \textbf{ or } 90\%$$

b Calculate the per unit efficiency of a 400 V, 50 Hz, three-phase induction having a rated output of 3.6 kW when the line current input to the stator is 7 A at a lagging power factor of 0.8.

$$P_{in} = \sqrt{3}\ V\ I\cos\phi$$
$$= \sqrt{3} \times 400 \times 7 \times 0.8$$
$$= 3.88\ \text{kW}$$

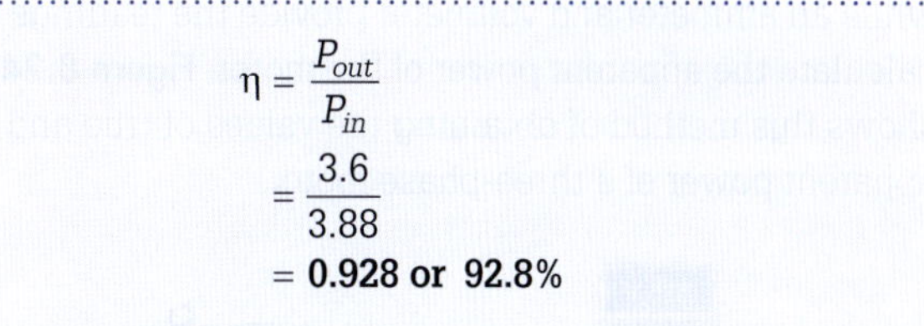

$$\eta = \frac{P_{out}}{P_{in}}$$
$$= \frac{3.6}{3.88}$$
$$= \mathbf{0.928 \text{ or } 92.8\%}$$

EXERCISE 8.5

a Calculate the per unit efficiency of a three-phase induction having a rated output of 5.6 kW when the power input to the stator is 6.2 kW.

b Calculate the per unit efficiency of a 400 V, 50 Hz, three-phase induction having a rated output of 5.6 kW when the line current input to the stator is 10.5 A at a lagging power factor of 0.85.

c From the nameplate details in **Figure 8.23**, determine the efficiency of the motor.

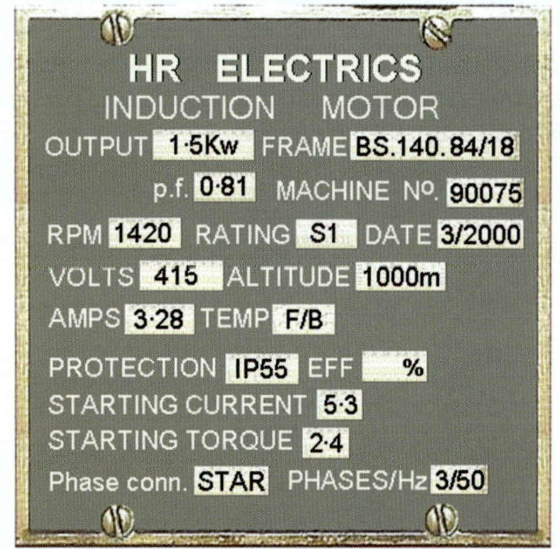

FIGURE 8.23 Three-phase nameplate

Power factor

The power factor describes a percentage of current drawn by the motor that does some form of work. By contrast, the remaining percentage is the reactive or wattless current. Power factor is the ratio of true power to apparent power.

$$pf = \frac{P}{S}$$

where S = apparent power in volt-ampere (VA)
P = true power in watts

Apparent power (S) for a three-phase circuit is calculated by using the equation:

$$S = \sqrt{3}\ V\ I$$

where S = apparent power in volt-ampere (VA)
V = line voltage in volts (V)
I = line current in amperes (A)

Using the values noted in **Figure 8.24** the power factor of the motor is 0.84 lagging. A wattmeter connected in the motor circuit reads the true power drawn by the motor

»

while an ammeter and voltmeter provide the readings to calculate the apparent power of the motor. **Figure 8.24** shows this method of obtaining the values of true and apparent power of a three-phase motor.

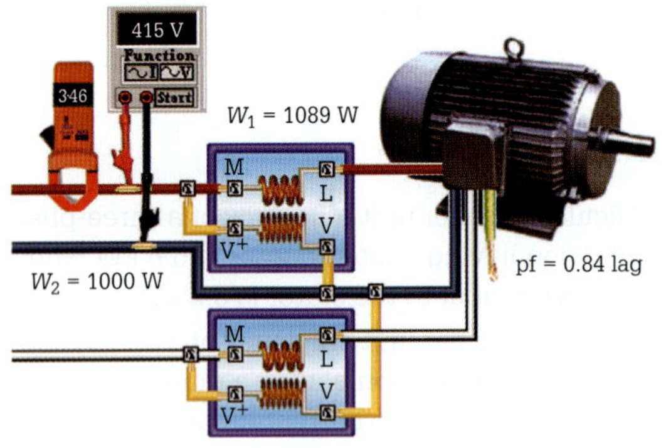

FIGURE 8.24 Test circuit to determine true and apparent power

EXAMPLE 8.6

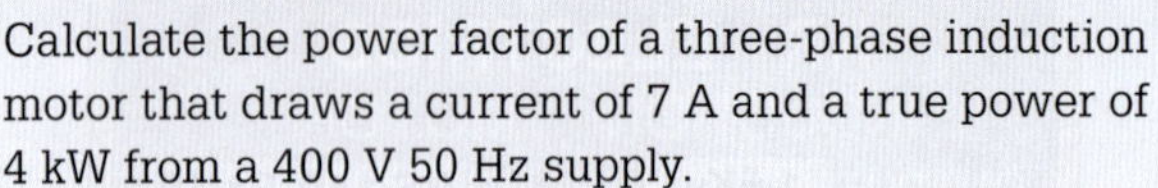

Calculate the power factor of a three-phase induction motor that draws a current of 7 A and a true power of 4 kW from a 400 V 50 Hz supply.

$$S = \sqrt{3}\ VI$$
$$= \sqrt{3} \times 400 \times 7$$
$$= 4.85 \text{ kVA}$$
$$pf = \frac{P}{S}$$
$$= \frac{4}{4.85}$$
$$= 0.825$$

EXERCISE 8.6

a Calculate the power factor of a three-phase induction motor that draws a current of 10 A and a true power of 6.5 kW from a 400 V 50 Hz supply.

b A three-phase induction motor draws a current of 15 A from a 400 V 50 Hz supply when operating at a full-load power factor of 0.87. Calculate the true power drawn from the supply.

High-efficiency motors

For many standard induction motors, maximum motor efficiency occurs around 75% of rated load. However, the range of the motor's efficiency varies with individual motors and can extend over a broader range. Induction motors are under-loaded when operating at a level where efficiency lowers significantly with decreasing load demands. Running a motor at values higher than rated load causes heating problems with the motor's organic insulation material that in turn affects efficiency and motor life.

Since 2001, three-phase induction motors ranging from 0.73 kW to 185 kW manufactured in or imported into Australia must comply with Minimum Energy Performance (MEPS) requirements that are set out in Australia and New Zealand Standard AS/NZS 1359.5: 2004. Motors classified as high efficiency have design features such as reduced resistance windings and low-loss ferromagnetic materials that improve their performance over standard motors.

Improvements ensure lower magnetisation levels resulting in reduced energy losses and lower operating temperature together with a high coefficient of heat dissipation. High-efficiency motors also run more quietly when compared to a standard motor due to better and smaller cooling fan design. All the motor improvements enhance electrical performance and improve efficiency. The establishment of these motors is part of a worldwide attempt to improve the efficiency of processes of energy conversion obtained from non-renewable energy sources. A direct outcome of their design is the reduction in harmful environmental greenhouse gas emissions by the energy supply source. These motors operate at efficiencies that are 10% better than equivalent standard motors.

Some difficulties happen with high-efficiency motors besides their establishment and implementation cost. Some of these motors operate at a slightly higher speed (when compared to a standard motor) which can affect the coupling ratio when connected to existing machines and loads. Their torque and starting current are also different. Some manufacturers consider the energy-efficient focus on induction motors as ineffective because the energy savings relate to energy losses within the motor. When compared to a standard motor the overall improvement is tiny. Inefficiencies in the driven machinery waste the power. The redesigning of machines allows a lower kilowatt motor to deliver real energy gains.

Double-cage induction offers a higher starting torque with lower starting current and real efficiency at normal operating conditions compared to standard squirrel-cage motors.

REVIEW QUESTIONS

1. State the purpose of torque-speed characteristic graphs.
2. What is breakdown torque?
3. What is locked-rotor torque?
4. What advantages are afforded by a double-cage rotor?
5. Determine the full-load torque developed by a three-phase induction motor having a rated power output of 2.8 kW when operating at 2940 rpm.
6. Calculate the power output of a three-phase induction motor that develops a torque of 4 Nm at 2960 rpm.
7. Calculate the per unit efficiency of a three-phase induction having a rated output of 6.5 kW when the power input to the stator is 7.2 kW.
8. Calculate the power factor of a three-phase induction motor that draws a current of 8.5 A and a true power of 5.2 kW from a 400 V 50 Hz supply.

8.4 Single-phase split-phase motor

One of the most common single-phase motors is the split-phase induction motor, which has constructional similarities with the three-phase induction motor.

Principles of operation

Unlike a three-phase supply, which develops a rotating magnetic field, a single-phase supply only produces a pulsating magnetic field.

Rotating magnetic field

The stator windings of single-phase motors are unable to produce a rotating magnetic field in order to create starting torque. As such, additional methods are employed to start the rotor turning. The single-phase motor with its rotor stationary produces only a pulsating electromagnetic field with no tendency of the rotor to start in either direction. **Figure 8.25** shows the pulsating electromagnetic field of a single-phase induction motor.

FIGURE 8.25 Pulsating electromagnetic field

The diagram shows that the single-phase motor is unable to self-start with a pulsating electromagnetic field. This is because the axis of the electromagnetic field of the induced rotor current is always in phase with the axis of the stator's pulsating field flux and the torque is therefore zero. However, if the rotor turns manually or mechanically, a magnetic field generates in the rotor by induction, and motor action occurs causing the rotor to accelerate to near synchronous speed. Various auxiliary methods are employed as a means of starting single-phase induction motors. A common method uses some form of 'phase splitting'. Phase splitting occurs when the motor 'sees' two voltage sources with one voltage displaced from the other by 90 electrical degrees. The provision of an auxiliary winding or shading coils, in addition to the main stator winding, enables phase splitting to occur.

Phase splitting

Phase splitting means that the main winding and an auxiliary winding have different electrical properties so that their respective currents are out of phase with each other combined with being positioned such as to create a rotating magnetic field. **Figure 8.26** shows two methods of achieving this; leftmost is a split-phase induction and rightmost is a shaded pole motor. In the split-phase induction motor, to enable self-starting, the auxiliary or start winding connects in series with a starting switch and this combination is in parallel with the main or run winding as illustrated in **Figure 8.27**.

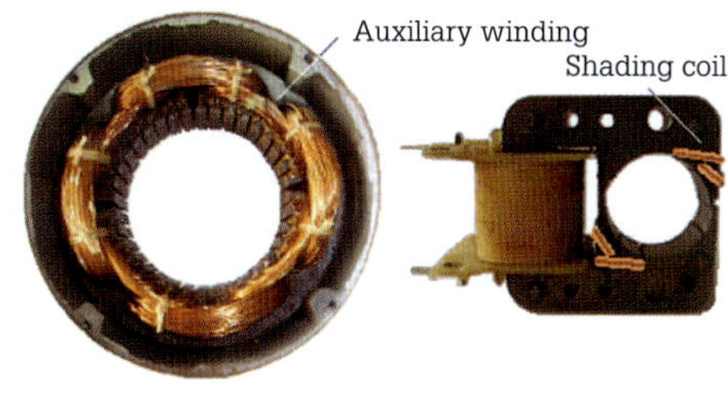

FIGURE 8.26 Phase splitting

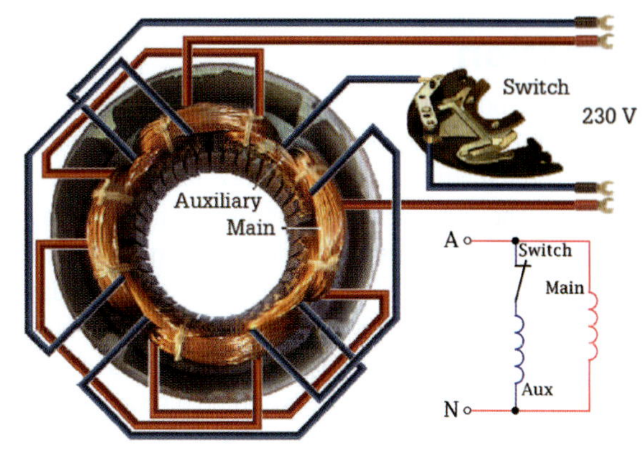

FIGURE 8.27 Connections for a split-phase motor

A failed auxiliary winding circuit results in a motor that makes a low humming sound. In addition, the motor starts in either direction when spun by hand.

The object of phase splitting is the creation of an elliptical electromagnetic field similar to that of a polyphase flux field. **Figure 8.28** shows the current relationship between the two windings of the split-phase motor at starting.

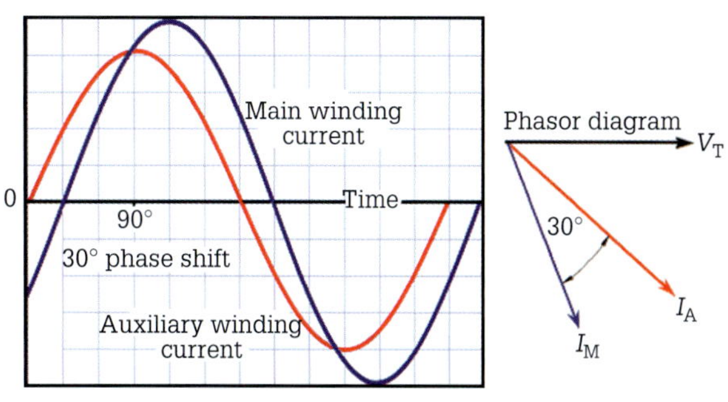

FIGURE 8.28 Graphical and phasor diagram for a split-phase motor

Phase displacement

The phase displacement illustrated shows the highly inductive main winding current lagging the auxiliary winding current by 30 electrical degrees. Due to energy losses in the form of copper loss, the main winding always has a high number of large-CSA conductors. This winding lies deep within the stator core slot, almost filling it. The result is a winding with high inductance and low ohmic resistance drawing a current (I_M) that lags the supply voltage. By contrast, the auxiliary winding consists of a much smaller CSA conductor with relatively few turns. A winding with high resistance, low reactance and low impedance occurs. In addition the winding draws a current (I_A) more in phase with the supply voltage (V_T).

The phase difference allows the current to develop a magnetic field in the auxiliary winding before the run winding. As the auxiliary winding current starts to diminish and its magnetic field decreases, the current and the magnetic field in the run winding is increasing. This phase difference in currents drawn by the two windings from the supply creates an elliptical electromagnetic field to which the rotor wants to align its poles.

Operation

The magnetic polarities of the rotor winding first developed under the start-winding move towards the pull of the run winding as the magnetic field of the auxiliary winding diminishes. The movement towards alignment with the run winding is sufficient to develop a starting torque causing the motor to turn in one direction when started. The turning in one direction implies that the direction of rotation is from the auxiliary winding to the adjacent run winding of the same polarity. Once the auxiliary winding disconnects from the circuit, the momentum of the rotor and the oscillating field of the run winding continue rotor rotation.

If the motor has a design for low starting torque and pull-up torque, the auxiliary winding can withstand a continuous supply voltage. However, if the motor has a design for high-starting torque (usually twice the full-load torque) and pull-up torque, then the auxiliary winding requires winding for high current values. The auxiliary winding draws several times (up to eight times) the full-load rated current starting from the supply. Consequently, there must be a means of removing the auxiliary winding when the rotor reaches a speed at which the torque, due only to the main winding, is capable of driving the load to its rated speed. Moreover, without disconnection, the auxiliary winding would burn out as a result of its low impedance. Since the rotor is required to reach a particular speed (usually 75% of rated speed) before the running winding develops sufficient torque on its own (see **Figure 8.29**), a centrifugal switch, as shown in **Figure 8.33**, is used to disconnect the auxiliary winding from the motor circuit.

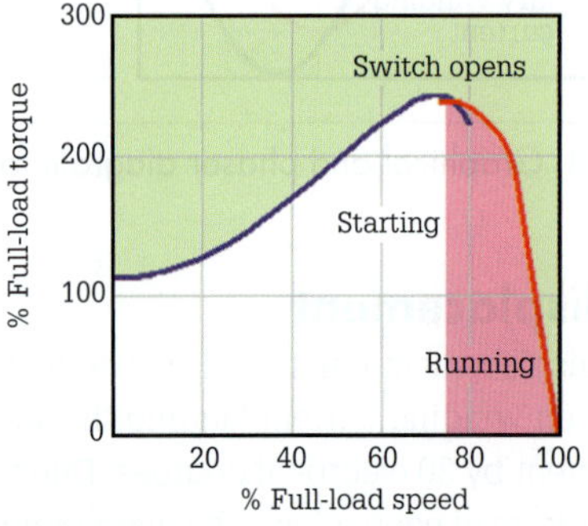

FIGURE 8.29 Torque–speed characteristic

As can be seen from the graph of **Figure 8.29**, the torque increases with speed until reaching maximum torque, thereby producing rapid acceleration, to allow the motor to bring up to speed any load it can start.

The rotor as illustrated in **Figure 8.32** of a split-phase motor is a simple squirrel-cage arrangement made up of aluminium bars cast in one piece with end rings into the core lamination slots as is the case of the three-phase squirrel cage rotor. When the rotor is at rest, a split-phase motor is similar to a transformer with its secondary (rotor) short circuited. This accounts for the high starting current for this type of motor. As the rotor begins to turn, a counter-electromotive force develops that gradually limits the current to its full-load rated value at full-rated speed.

Reversing direction of rotation

A split-phase motor reverses by interchanging the leads to either the auxiliary or main winding but not both.

Speed control

When connected to a fixed-frequency supply, the motor has no speed control capability other than that obtainable by reconnecting for a different pole make-up. Typical motor sizes range up to 0.5 kW.

Components of a split-phase induction motor

The single-phase induction motor consists of a frame housing a stator assembly and a rotor assembly. The rotor assembly includes an axially extending rotating shaft. An end shield closes one end of the frame at the drive end (DE) and a second end shield closes the second end, or non-drive end (NDE). The shaft is journeyed in the non-drive-end shield, and it extends through the drive-end shield. A governor for the centrifugal switch attaches to the shaft to rotate with said shaft. Leads extend within the motor frame from the stator assembly to the cut-out switch assembly.
Figure 8.30 shows an illustration of a split-phase motor and **Figure 8.31** shows an exploded view of the motor.

FIGURE 8.30 Split-phase motor

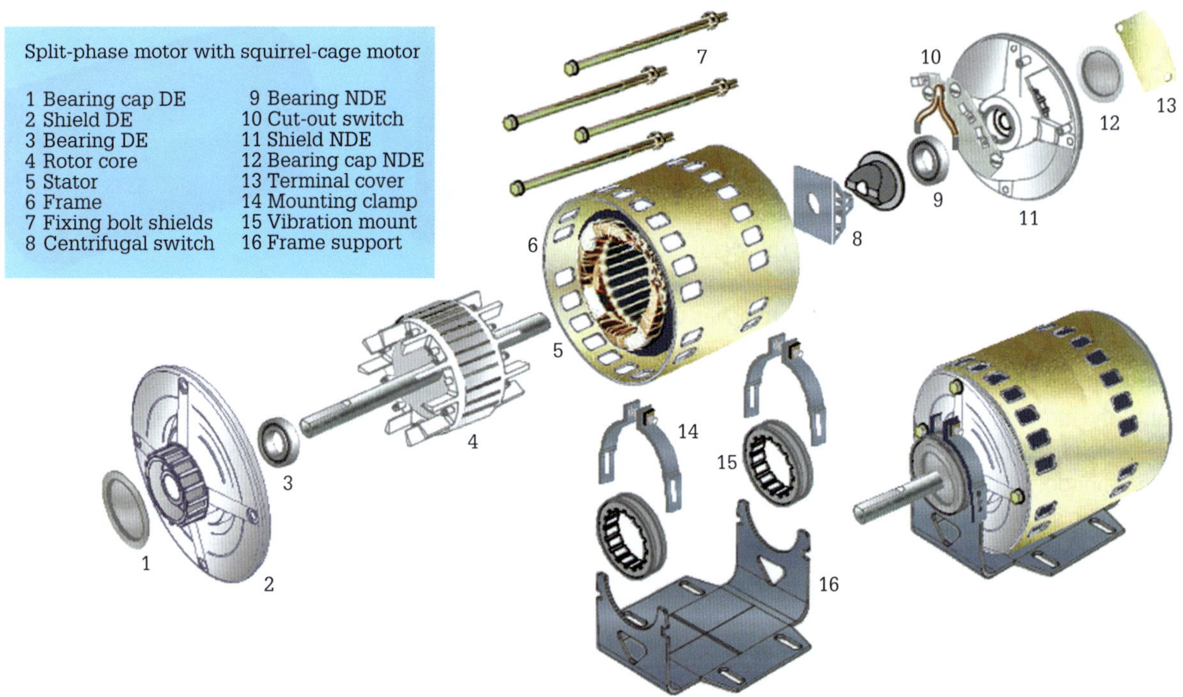

FIGURE 8.31 Exploded view of a split-phase induction motor

Stator

As for the three-phase induction motor, the stator assembly of a single-phase motor consists of a number of silicon-iron lamination stampings with slots available for the stator coils. Refer back to **Figure 8.7** for an induction motor lamination stamping.

The stator assembly consists of a number of these lamination stampings that are stacked end to end and clamped. Refer back to **Figure 8.8** for an induction motor stator.

Rotor

The rotor of a single-phase motor consists of a simple squirrel cage whereby molten aluminium pours directly into the lamination slots to cast a complete cage. The rotors are dynamically balanced to provide exceptionally smooth velocity. **Figure 8.32** shows a rotor with cast integral fins for cooling.

FIGURE 8.32 Rotor

The design of the rotor determines the starting and running characteristics of a single-phase motor. Some rotors have skewed slots to prevent power distortions in the windings due to transformer action.

A skewed rotor reduces cogging whereby the rotor tends to lock onto a position where it is aligned with the stator poles. Cogging is undesirable because it introduces vibration and noise. However, the result of skewing is an electrically quiet motor with very smooth shaft rotation.

Starting switch

A number of different types of switches are used to open circuit the auxiliary or start winding when the rotor attains about 70% of rated speed. These are:

1. mechanical centrifugal switch
2. solid-state centrifugal switch
3. current relay
4. potential relay – see capacitor-start and capacitor-start capacitor-run split-phase motors.

Mechanical centrifugal switch

The centrifugal switch consists of a rotating governor and stationary switch assembly. This construction enables the starting circuit to energise for a brief period to get the motor up to running speed quickly thereby limiting the starting current to a short time interval. **Figure 8.33** shows an illustration of a governor and the switch assembly.

FIGURE 8.33 Governor and switch

The governor mounts directly on the end of the rotor core or the rotor shaft. The arrangement of the weighed down governor means that the switch contacts hold in the normally closed position by spring tension. The stationary switch assembly attaches to the motor's end shield and has a particular arrangement and connections depending upon the motor's design application. As the rotor gains velocity after starting, centrifugal force throws out the weighted governor arms, which overcomes the spring tension and opens the switch contacts. This action removes the auxiliary winding from the motor circuit. Once the motor operates solely on the main winding the switch remains open until the motor stops.

The switch needs to make-and-break cleanly. If a clean break does not occur, the result may be contact bounce that causes the switch contacts to cycle on and off several times before making or breaking the starting circuit. The bouncing could cause the contacts to arc unnecessarily, and eventually the switch would fail. If the centrifugal switch jams in the open position the motor hums but does not start turning; but it runs if the shaft is turned by hand. If the switch remains in the closed position, the motor starts and runs but becomes very noisy and hot.

Solid-state centrifugal switch

Solid-state centrifugal switches replace the mechanical switch and actuator mechanism in some single-phase motors. This device duplicates the function of connecting and disconnecting the auxiliary winding from the motor circuit at particular speeds. It can do this by sensing the voltages present in the main and auxiliary windings rather than the centrifugal force of rotation.

When the rotor of a single-phase motor is at standstill, there is no magnetic coupling between the windings and no voltage induced across the rotor bars. However, when the rotor starts to rotate the main winding induces current in the rotor bars that creates an electromagnetic flux. This flux via transformer action induces a voltage that is directly proportional to the rotor speed across the start winding. The voltage across both the main and the auxiliary windings feeds a comparator to sample the voltages.

The solid-state circuitry interrupts the start circuit after the rotor has accelerated to 75% of synchronous speed. At this speed, a cut-out voltage develops causing the logic circuit to shut down the power stage feeding the starting circuit. Once the starting circuit disconnects, the main winding drives the rotor to its operating speed. These switches have detection circuitry that constantly monitors the auxiliary winding voltage.

If the motor experiences an overload or a reduction in rotor velocity, a change in voltage occurs across the auxiliary winding. When this voltage falls to a predetermined value, the solid-state switch automatically reconnects the starting circuit, and the motor goes through its normal start-up procedure. **Figure 8.34** shows an illustration of solid-state centrifugal switches.

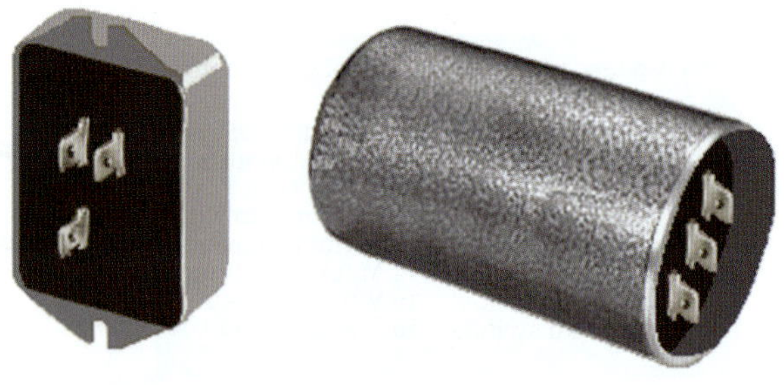

FIGURE 8.34 Solid-state centrifugal switches

Current-operated relays

Current-operated relays as shown in **Figure 8.35** open the auxiliary winding of some split-phase motors. These motors operate in hazardous areas where an arc across switch contacts creates difficulties. Other areas include most refrigeration compressors in which the stator windings immerse in oil within a hermetically sealed unit where contamination of the refrigerant is unacceptable.

FIGURE 8.35 Current-operated relays

Figure 8.36 shows a current-operated starting relay circuit for a split-phase motor.

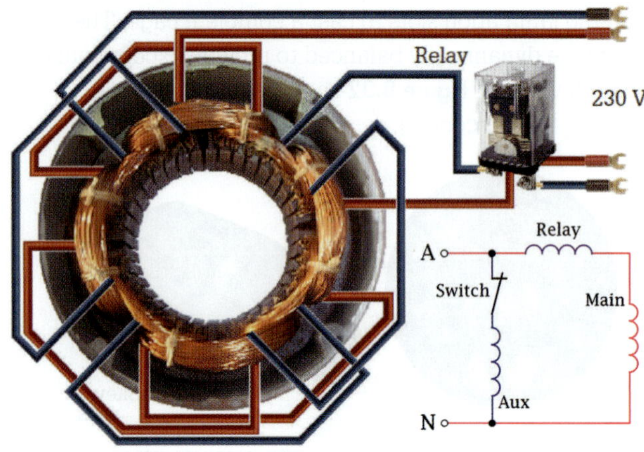

FIGURE 8.36 Current-operated starting circuit

The magnetic core of a.c. relays use laminated steel to reduce eddy currents, and a shading coil is employed to prevent the electromagnetic flux from falling to zero each time the current through the coil goes through zero. When the single-phase motor energises, the initial current drawn by the main winding has a sufficient magnitude to establish an electromagnetic force within the relay coil. This electromagnetic force closes the relay contacts allowing the auxiliary winding to connect.

Once the motor starts, the back electromotive force generated in the main winding limits the current through the series-connected relay coil. At approximately 75% of full-load speed, the relay has insufficient current passing through it to maintain its contacts in the closed position.

The contacts open, disconnecting the auxiliary winding from the motor circuit and the motor operates on the main winding only. Current-type motor starting relays assist in the starting function of permanent split-phase, capacitor-start split-phase and single-phase a.c. motors. However, the size of the motor is limited to approximately 0.5 kW, 230 V. The most common applications are household refrigerator and freezer compressors, automatic dishwasher motors and automatic washing machine motors.

Solid-state current relays (SCR) are also available as external starting relays. These SCRs, or triac-based and snubber circuit relays, can be used to start all types of split-phase motors. Refer to **Figure 8.37**.

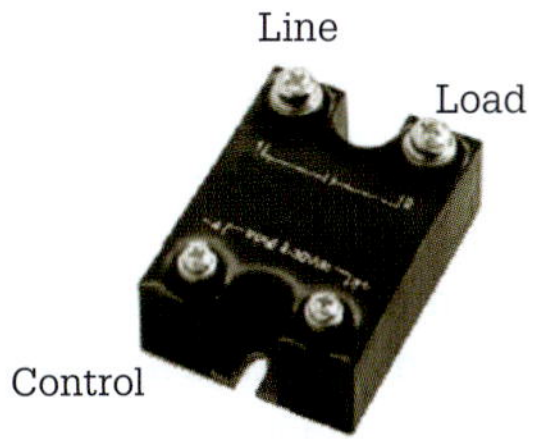

FIGURE 8.37 Solid-state relay

Identification of stator windings

Identification of the auxiliary and main windings of a split-phase motor happens at the motor terminals by measuring winding resistance. **Figure 8.38** shows this process for a 0.25 kW split-phase motor. The auxiliary or start winding has higher resistance than the main or run winding.

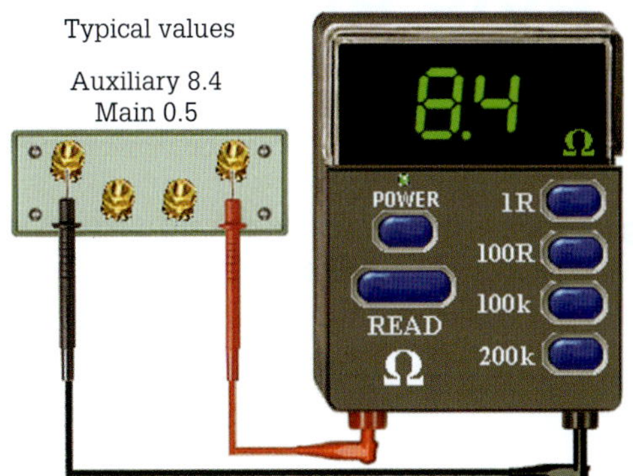

FIGURE 8.38 Identification of stator windings

Single-phase motor starters

Single-phase motors use either motor starting switches or a.c. magnetic starters. A starter must incorporate under-voltage release and overload protection. Magnetic contactors must be of the correct voltage and be non-reversing unless required, together with resettable integral motor overload protection. An enclosure appropriate for the working area must contain the starter. The front cover of the housing should display the start, stop and reverse buttons (the start/stop button control provides the under-voltage protection), a green indicator light and the overload-reset button.

Thermal overload relays connected to magnetic contactors are versatile, reliable and accurate bimetal-type protection relays for protection of single-phase motors. Every relay has a mode selector where the resetting of the relay happens manually or automatically. Refer to **Figure 8.39**.

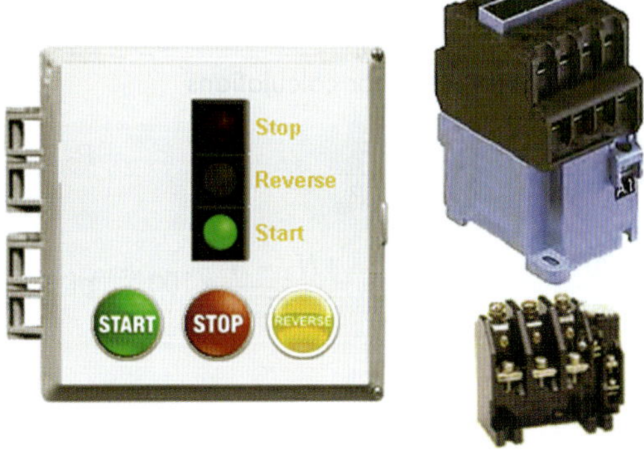

FIGURE 8.39 Single-phase motor starter with magnetic contactor and overload relay

The high rush of current associated with every start or attempted start causes thermal stress on the insulation of the stator winding. The effects of thermal stress are cumulative and have a limiting effect on the life of a single-phase motor. For this reason, the number of starts for split-phase motors is usually limited to eight per hour. Local supply authorities determine the allowable starting current for all single-phase 230 V motors. Except for single wire earth return (SWER) areas of supply, the starting current allowable by many authorities in Australia and New Zealand is 45 A.

Split-phase motor applications

The split-phase motor's uncomplicated design makes it less expensive than other single-phase motor types manufactured for industrial and commercial use. However, the design limits motor performance.

Starting torques for split-phase motors are low, typically 75% to 200% of rated load, while the high starting current of the motor, approximately six to eight times its full-load rating, is a definite disadvantage when compared to other single-phase motors. If the starting times are delayed, the start winding overheats and burns out. In addition, because of the high starting current, thermal protection is difficult due to the fast thermal trip time required to protect the start winding. On the positive side, these motors can be speed controlled by varying their supply voltage. However, a 3% decrease in voltage results in a 10% de-rating of their kVA output. If speed lowers too much, the centrifugal switch closes and the motor stops.

These motors are suitable for loads such as bench grinders and range hood fan motors that require moderate torques and constant speed.

Comparison of split-phase and three-phase motors

In comparison with three-phase induction motors, single-phase motors have higher maintenance costs and relatively shorter service life. Single-phase motors produce less starting torque than the equivalent three-phase motor, draw a higher line current and are less efficient.

The voltage drop on starting is higher with single-phase motors than it is with a three-phase motor. The higher voltage drop causes power supply authorities to restrict the starting current of all single-phase motors to 45 A.

SWITCH ON

Always refer to the local supply authority's electricity connection and metering manual for the value of allowable starting current.

Split-phase motor calculations

The calculations for split-phase induction motors are the same as for three-phase induction motors. **Table 8.1** summarises these equations, but for a full treatment see the section on three-phase motors at the beginning of this chapter.

TABLE 8.1 Split-phase motor calculations

$P_{Out} = \frac{2\pi NT}{60}$	$\eta = \frac{P_{Out}}{P_{In}}$	$P_{Out} = P_{In} - P_{Loss}$
$N_s = \frac{120f}{p}$	$\text{slip } \% = \frac{N_S - N}{N_S} \times 100$	$\text{pf} = \frac{\text{true power}}{\text{apparent power}}$

REVIEW QUESTIONS

1. What type of electromagnetic field does a single-phase supply produce?
2. Outline the object of 'phase-splitting'.
3. What is the phase difference between currents in the main and auxiliary windings in a single-phase, split-phase motor?
4. Describe the effect of a failed auxiliary winding circuit in a single-phase, split-phase motor.
5. Describe how to reverse the direction of rotation of a single-phase, split-phase motor.
6. Name typical applications for single-phase, split-phase motors.
7. Name the mechanical device that open circuits the auxiliary winding at the appropriate time for single-phase, split-phase motors.
8. At approximately what percentage of full-load speed is the auxiliary disconnected in a single-phase, split-phase motor?
9. Outline the function of current-operated relays.
10. Which winding of a single-phase, split-phase motor has the lower resistance?

8.5 Single-phase shaded-pole and capacitor motors

Two other types of single-phase motor are the shaded pole and capacitor motor. Some capacitor motors are similar in construction to the single-phase induction motor.

Shaded-pole motor

Another method of starting single-phase motors that results in a low starting torque is by means of a shading coil. **Figure 8.40** shows the action of a shading coil in establishing a rotating magnetic field.

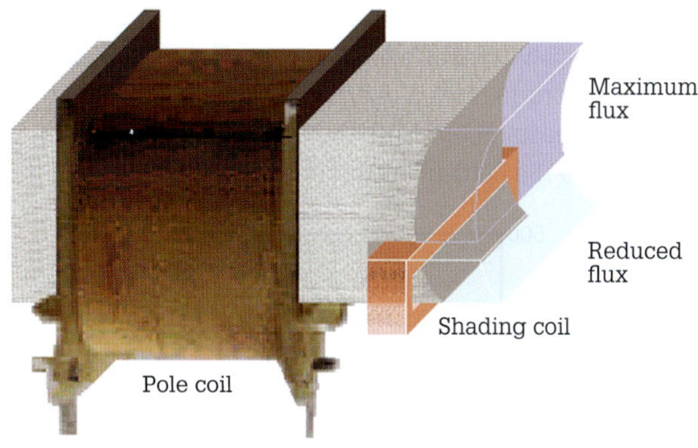

FIGURE 8.40 Action of the shading coil

The extremity of the pole core divides into two sections (two-thirds and one-third) in one of which a copper coil called a shading coil occurs. When the pole coil energises, electromagnetic lines of force pass through the shading coil causing a voltage to develop. This voltage causes a current to flow in the shading coil in a direction that always opposes the electromagnetic flux that produced it (Lenz's law). Consequently, the pole core's flux is stronger in relation to the shading coil's flux. The result is that the flux in the unshaded portion of the pole reaches its maximum strength sooner than does the flux in the shaded portion. This difference in time causes the pole flux to move from the unshaded portion of the pole towards the shaded portion, creating the split-phase action of the motor. **Figure 8.41** shows the sliding flux nature of a shaded-pole motor.

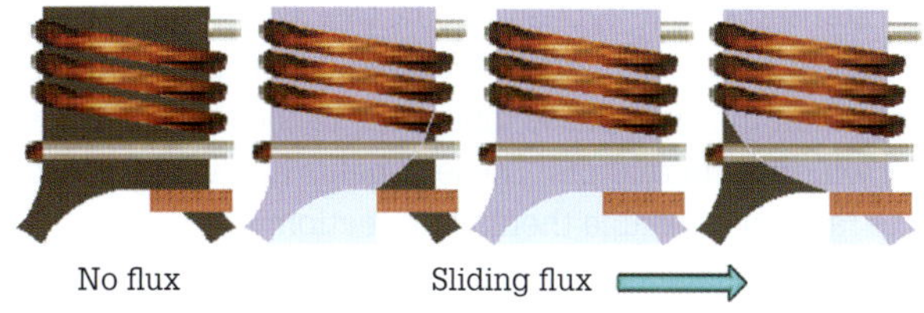

FIGURE 8.41 Sliding flux nature of a shaded-pole motor

Reversing direction of rotation

The rotation of a shaded-pole motor is always towards the shading coil. Interchanging the leads cannot reverse the direction of rotation of a shaded-pole motor. Reversing the pole core mechanically, as shown in **Figure 8.42**, can do this. There are shaded-pole motors that have two sets of shading coils together with two switches, one set for each direction of rotation.

FIGURE 8.42 Reversing a shaded-pole motor

Torque

The starting and running torque of a shaded-pole motor is low but is sufficient for most applications since the starting torque required only has to overcome static friction (50% to 75% of full-load torque). Shaded-pole motors have low efficiency, about 20% to 33%, with a relatively constant speed characteristic. They are available in power ratings from 3 watts to 1500 watts. **Figure 8.43** shows the torque–speed characteristic of a shaded-pole motor.

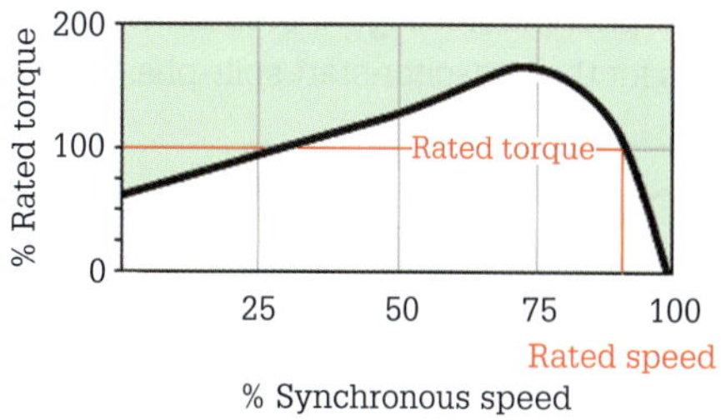

FIGURE 8.43 Torque–speed characteristic

Applications

Mechanically, the shaded-pole motor construction suits high-volume production. These motors are 'throwaway' motors, meaning that they are easier to replace than to repair. The disadvantages of shaded-pole motors include their low starting torque, typically 50% to 100% of full-load torque. They are high-slip motors with their operating speed 7% to 10% below synchronous speed, with efficiency usually below 30%. The speed is easily controlled by varying the supply voltage, or through a multi-tap stator winding.

Due to their low manufacturing cost, shaded-pole motors are ideal as the driving motor in light-duty applications such as multi-speed fans. **Figure 8.44** shows a shaded-pole motor.

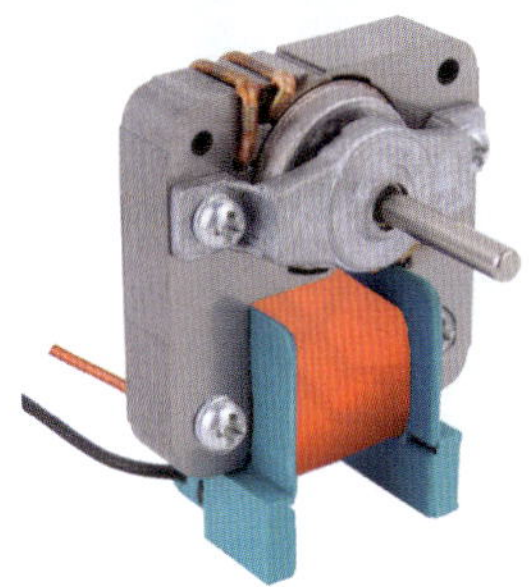

FIGURE 8.44 Shaded-pole motor

Capacitor-start split-phase induction motor

Capacitor-start split-phase induction motors are similar to split-phase induction motors, except that an a.c. electrolytic capacitor connects in series with the auxiliary or start winding to increase the phase shift between the main and auxiliary windings to allow the motor to start with higher torque. The capacitor causes the current to lead the supply voltage at an angle of between 76° and 89° compared to 30° phase displacement for the split-phase motor. The time phase difference in currents drawn by the two windings from the supply creates a near-circular rotating electromagnetic field similar to a polyphase induction motor that develops a strong starting torque.

This phase displacement together with extra turns in the auxiliary winding allows the auxiliary winding to develop a higher starting current and hence larger starting torque than the split-phase motor. **Figure 8.45** shows the phase relationships for the capacitor-start split-phase motor.

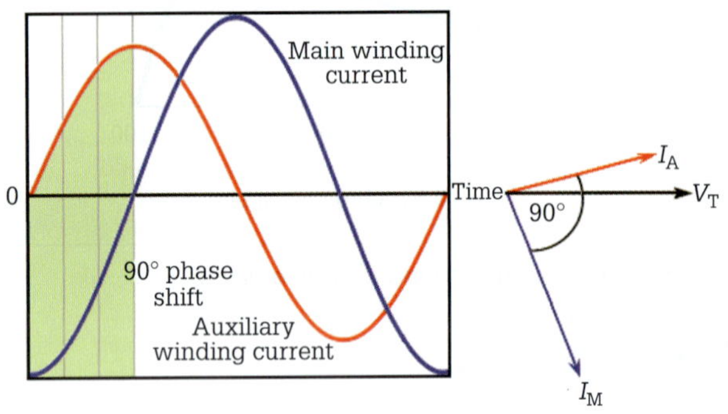

FIGURE 8.45 Phasor diagram for a capacitor-start split-phase motor

The short-duty-rated capacitor connects in series with the starting switch and the auxiliary winding as illustrated in the diagram of **Figure 8.46**.

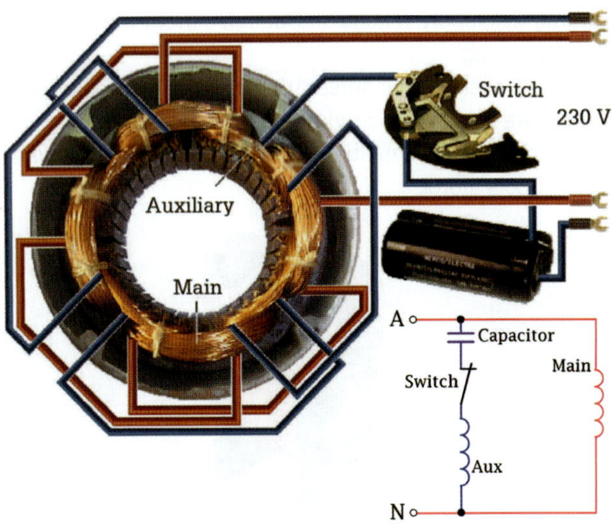

FIGURE 8.46 Connections for a capacitor-start split-phase motor

Reversing direction of rotation

As with the split-phase induction motor, the capacitor-start split-phase induction motor reverses by interchanging the leads to either the auxiliary or main winding but not both.

Torque

Start capacitors increase motor starting torque and allow the motor to be cycled on and off rapidly. These capacitors are rated for momentary use: they stay energised long enough to bring the motor quickly to 75% of full speed and are then switched from the circuit. **Figure 8.47** shows the typical torque–speed characteristic for a capacitor-start split-phase motor. Note the higher starting torque compared to the split-phase induction motor.

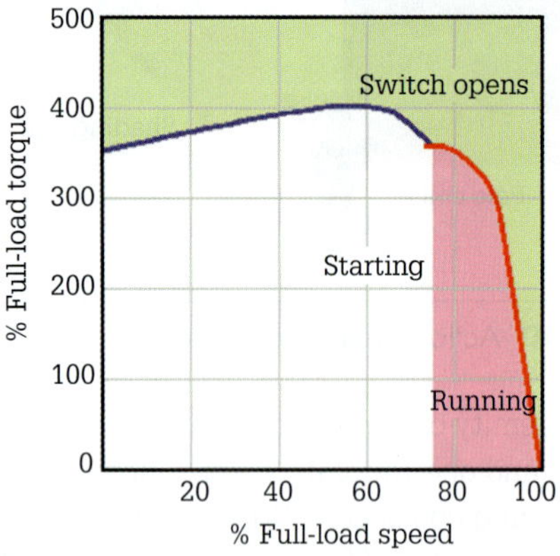

FIGURE 8.47 Torque–speed characteristic for a capacitor-start split-phase motor

The starting curve shows the motor characteristics with both the auxiliary and main windings energised. The running curve illustrates run features after the starting winding switches out of the circuit.

Applications

Split-phase capacitor-start motors have some benefits when compared with split-phase motors. Because the capacitor connects in series with the start circuit, the motor develops high starting torque, typically 250% to 350% of rated load. The starting current, usually 450% to 600% of rated current, is less than the split-phase motor due to the stator's larger CSA winding conductors. These benefits allow for increased cycle rates and reliable thermal protection.

Their applications range is much wider than the split-phase motor because of their higher starting torque and lower starting current. Split-phase capacitor-start motors are the general-purpose motor for many industrial and commercial applications, and are suitable for most applications requiring constant speed under steady load. They have good efficiency and excellent power factor (almost equal to unity), which makes them suitable for shaft-mounted fans, compressors, pumps, machine tools, pedestal drilling machines, air-conditioners, conveyors, blowers, fans and other hard-to-start applications. **Figure 8.48** shows the capacitor mounted externally to the frame of the motor.

FIGURE 8.48 Capacitor-start split-phase motors

Although the capacitor-start split-phase induction motor has a high starting torque, motor starts are limited to about eight times per hour because of overheating.

Potential-operated relays

Potential-operated relays find application in capacitor-start and capacitor-start capacitor-run split-phase motors. The coil of this relay has many turns of small-diameter wire and connects in parallel with the auxiliary winding. However, in order to use a potential relay, a single-phase motor must have a starting capacitor. Refer to **Figure 8.49**.

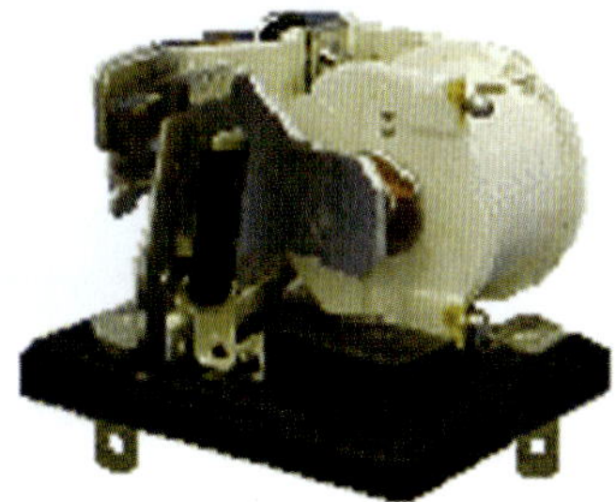

FIGURE 8.49 Potential-operated relay

The potential relay operates by sensing an increase in the voltage developed across the auxiliary winding when the motor is running. When power connects to the motor, the rotor begins to turn, creating its own magnetic field, which induces a voltage across the ends of the auxiliary winding. The induced voltage produces a higher voltage across the auxiliary winding than just the relevant supply voltage. When the motor accelerates to about 75% of its rated speed, the pick-up voltage across the auxiliary winding is high enough to energise the coil of the potential relay. The energised coil operates the relay contacts, disconnecting the starting capacitor from the motor-starting circuit.

The coil of the potential relay remains energised by transformer action (rotor and auxiliary winding) keeping the relay contacts open as long as the motor is running.

Figure 8.50 shows a potential-operated relay circuit for a capacitor-start split-phase induction motor.

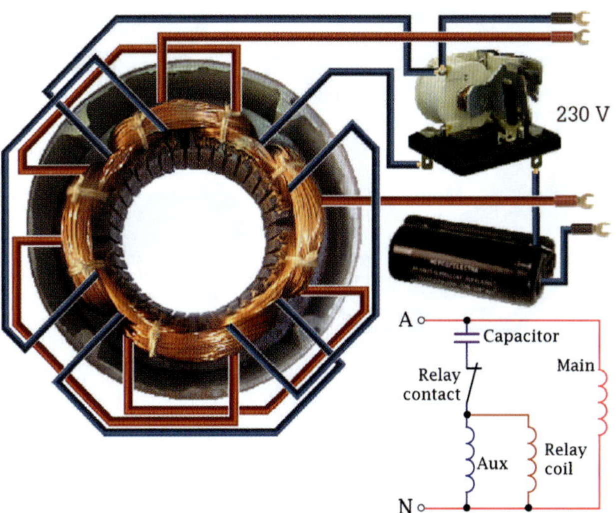

FIGURE 8.50 Connections for a potential-operated relay circuit for a capacitor-start split-phase induction motor

Potential relays have four ratings: continuous coil voltage, minimum pick-up voltage, maximum pick-up voltage and drop-out voltage. The exact rating designed for that particular motor must occur if the relay fails. The relays must be replaced if contacts remain open and not in their normally closed state. The relay is difficult to fault-find and replace when replacing the start capacitor.

The purpose of current and potential relays is to replace the centrifugal switch. Their advantage is that they can be installed remotely from the motor and outside any hazardous areas if required.

Permanent split-phase capacitor induction motor

The auxiliary and the main windings of a permanent split-phase capacitor motor are identical, having the same cross-sectional area, number of turns and resistance. However, a continuously rated metallised polypropylene film or oil capacitor permanently connects across the auxiliary winding. Refer to **Figure 8.51** for the connection diagram of a permanent split-phase capacitor motor.

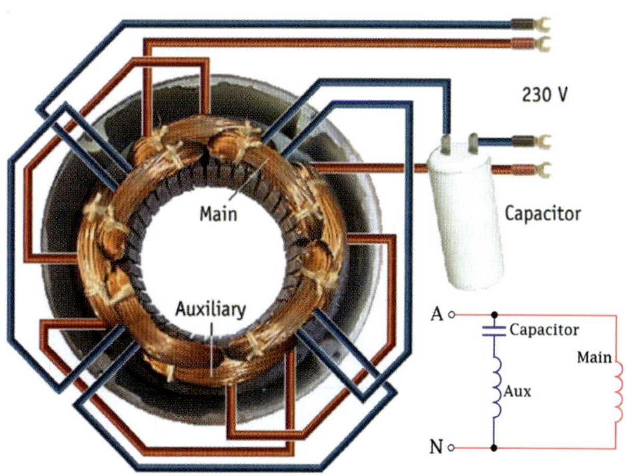

FIGURE 8.51 Connections for a permanent split-phase capacitor motor

The value of the capacitance of the capacitor is a compromise between the best starting and best running performances. This results in a smoother running motor. In addition, because the two windings connect to the supply, the electromagnetic field produced is similar to that of a rotating polyphase induction motor. This effect occurs because the capacitance allows the current to lead the voltage in the auxiliary winding. Consequently, the electromagnetic field develops in that winding first.

The permanent-capacitor motor speed is voltage dependent. Furthermore, because both windings remain energised, there is no need for a centrifugal switch. **Figure 8.52** shows the phase relationship for the permanent split-phase capacitor motor.

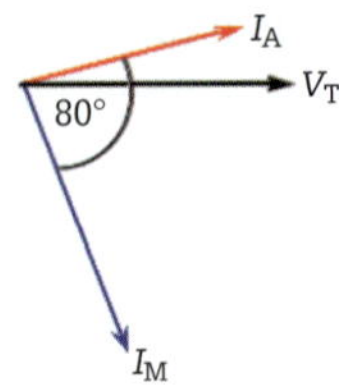

FIGURE 8.52 Phasor diagram for a permanent split-phase capacitor motor

Reversing direction of rotation

A reversing switch can connect to allow the capacitor to interchange from one winding to the other. The direction of rotation of the electromagnetic field results in a direction change for the rotor. Figure 8.53 shows the reversing-switch action.

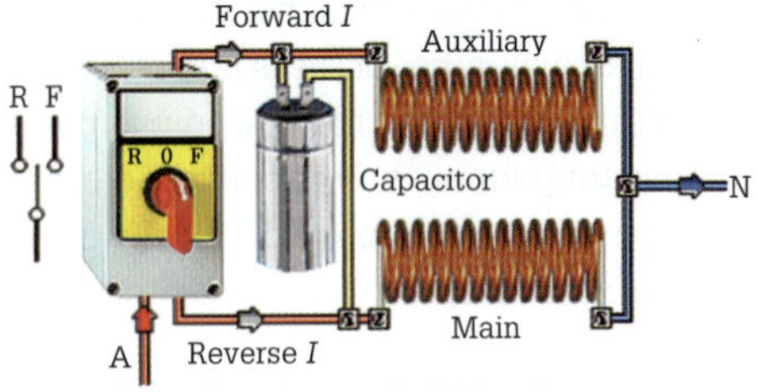

FIGURE 8.53 Reversing-switch action

Torque

The starting torque of a permanent split-phase capacitor motor is quite low, ranging from 30% to 150% of rated load, which makes the motor unsuitable for hard-to-start loads. Figure 8.54 shows a typical torque–speed characteristic for a permanent split-phase capacitor motor.

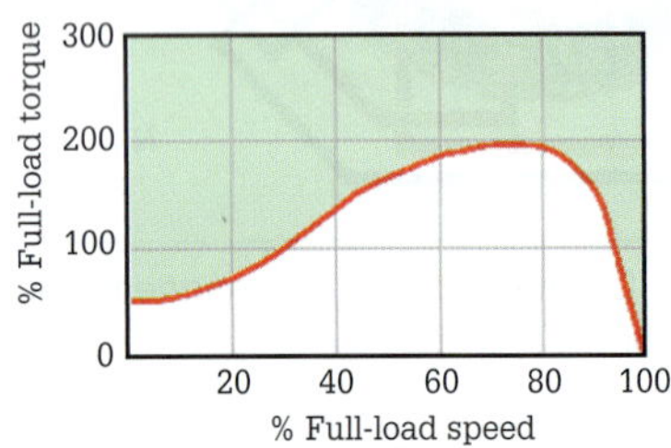

FIGURE 8.54 Torque–speed characteristics for a permanent split-phase capacitor motor

How much current is delivered to the capacitor-connected winding depends on the capacity of the capacitor and the system voltage. Any fluctuation in line voltage affects the speed of the motor. Speed changes from no load to full load can be severe.

Applications

Permanent split-phase capacitor motors have several advantages. They need no starting centrifugal switch and so can be reversed easily without physically changing connections. The low starting currents of permanent split-phase capacitor motors (50% to 100% of FLC) make them suitable for applications requiring high cycle rates (up to ten times per minute).

They are very reliable, with low starting current and torque together with smooth acceleration. These motors are the most reliable of the range of single-phase motors. Permanent split-phase capacitor motors have a wide variety of applications depending on the design. These include garage door motors, fans and blowers, and roof ventilators. A permanent split-phase capacitor motor is shown in Figure 8.55.

FIGURE 8.55 Permanent split-phase capacitor motor

Capacitor-start capacitor-run motor

A capacitor-start capacitor-run induction motor combines the best of the capacitor-start characteristics with the best of the permanent split-phase capacitor motor features. Like a capacitor-start motor, it has an electrolytic capacitor in series with the auxiliary winding. The action of the capacitor enables the motor to produce enough torque to start the motor.

However, it also has a run-type capacitor that connects in series with the auxiliary winding after the start capacitor switches off. With the run capacitor connected, the motor always has the auxiliary winding operating, and increased torque is available. Refer to Figure 8.56 for the electrical connections in a capacitor-start capacitor-run induction motor.

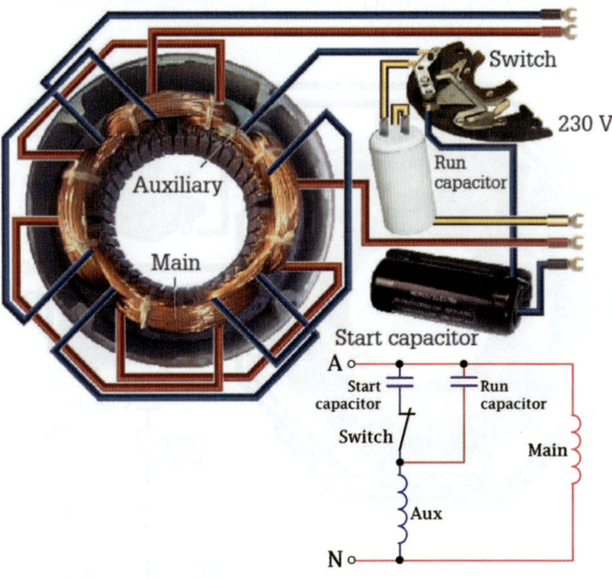

FIGURE 8.56 Connections for a capacitor-start capacitor-run motor

Reversing direction of rotation

In order to reverse a capacitor-start capacitor-run motor interchange the connections to either the auxiliary winding or the main winding but not both.

Torque

As each capacitor has a particular purpose, motor performance optimises with high breakdown torque, lower full-load current and higher efficiency. **Figure 8.57** shows a typical torque–speed characteristic for a capacitor-start capacitor-run motor.

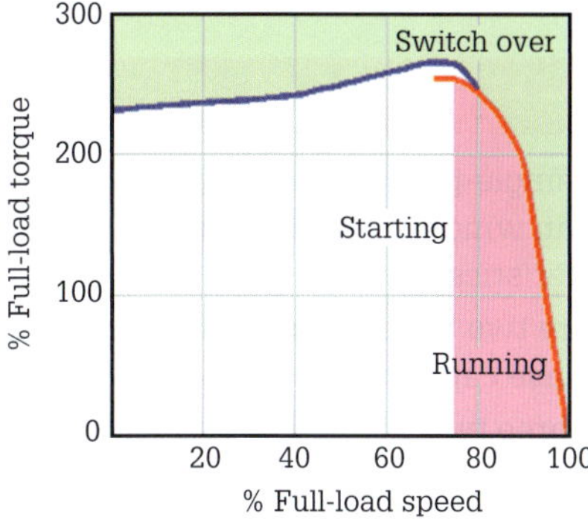

FIGURE 8.57 Torque–speed characteristic for a capacitor-start capacitor-run motor

Applications

Capacitor-start capacitor-run split-phase motors are more efficient than split-phase capacitor-start motors and require less running current. Among other things, this means that the motor operates at a temperature lower than other single-phase motor types of comparable kilowatt power output. However, these motors are 58% to 77% efficient. These motors are suitable for most applications requiring constant speed under varying loads. Their applications include drive motors for various types of milling machines, lathes, cranes, hoists, concrete mixers, air-compressors and other high-torque applications requiring 1 kW to 7.5 kW. Note the two capacitors on the motor shown in **Figure 8.58** that typifies the capacitor-start capacitor-run induction motor.

FIGURE 8.58 Capacitor-start capacitor-run motor

Motor capacitors

Starting capacitors

Capacitors used for starting single-phase motors use a.c. aluminium electrolytics. They are two capacitors connected back to back so that the capacitor is not polarised and can conduct in two directions. These capacitors have a high value of capacitance, ranging from 20 µF to 700 µF at a frequency of 50 Hz, and have a rating that allows them to operate in ambient temperatures from –40 °C up to 65 °C. However, above this temperature, their practical life reduces. Unlike d.c. electrolytics, these motor-starting capacitors have a rating for 'duty cycle'. A typical duty-cycle rating is 20 three-second starts per hour or their equivalent (60 × 1 second starts per hour). Exceeding the duty cycle causes a reduction in the active life of the capacitor.

If these capacitors fail, a capacitor with the same capacitance value must replace them. An incorrect capacitor causes the starting torque of the motor to alter. In addition, the auxiliary winding and the capacitor may resonate, causing other adverse effects. **Figure 8.59** shows an illustration of motor-starting capacitors.

FIGURE 8.59 Motor-starting capacitors

Run capacitors

Run capacitors are paper-spaced, oil-filled or metallised polypropylene film. These capacitors have only a few microfarad capacitances from 0.5 µF to 50 µF in order to produce a small out-of-phase current. However, they have a rating for continuous duty (up to 85 °C) with a voltage rating two to three times the voltage rating of the motor. The metallised polypropylene capacitor is smaller than the paper-spaced oil-filled type and is less hazardous because of its low leakage current. These capacitors have a very high current load capacity coupled with a very low loss factor. **Figure 8.60** shows a paper-spaced oil-filled capacitor and a metallised polypropylene capacitor.

FIGURE 8.60 A paper-spaced, oil-filled capacitor and a metallised polypropylene run capacitor

Comparison of split-phase motor characteristics

Comparison graphs act as a guide for the selection of single-phase motors for various applications. The most useful comparison graph is that for the starting torques of the various split-phase motors as shown in **Figure 8.61**.

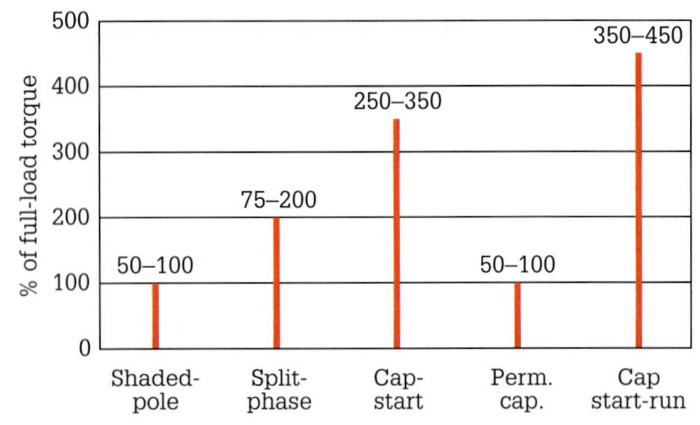

FIGURE 8.61 Starting torque

REVIEW QUESTIONS

1. How is the direction of rotation in a shaded-pole motor reversed?
2. What is the typical efficiency of a shaded-pole motor?
3. State the range of output power of a shaded-pole motor.
4. What type of capacitor is used in the starting circuit of a single-phase capacitor-start induction motor?
5. State the typical phase displacement between current in the auxiliary and main winding of a single-phase capacitor-start induction motor.
6. Name three applications for single-phase capacitor-start induction motors.
7. Which single-phase motor employs auxiliary and main windings having the same electrical characteristics?
8. What are two typical applications for permanent split-phase capacitor motors?
9. Name three typical applications for capacitor-start capacitor-run motors.
10. Where is the coil of a potential relay connected in the motor circuit of a capacitor-start split-phase induction motor?

8.6 Series universal single-phase motor

A universal motor is a small series motor that is not an induction motor, but specially designed to operate at approximately the same power output and speed on either d.c. or single-phase a.c.. This motor is the only type of d.c. motor that can achieve a unidirectional torque from an a.c. supply. Refer to **Figure 8.62**.

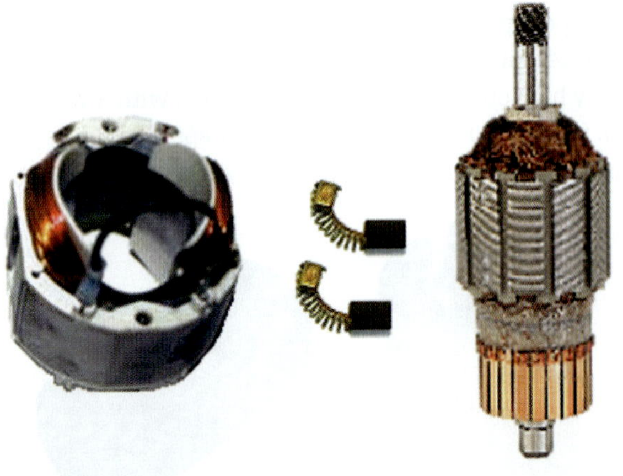

FIGURE 8.62 Universal motor

As the armature current and the field current are in series, the alternating current reverses at the same instant in time resulting in an induced torque in the same direction as before. The torque developed by the universal motor is proportional to the square of the armature current. Universal motor connections have the armature connected in series with the 'series field' winding (see **Figure 8.63**). No-load speeds in excess of 12 000 revolutions per minute are achievable without damage. These motors can have their speed varied by different analogue methods:

- The tapped field method varies the field electromagnetic flux; the stronger the flux the slower the speed and vice versa.
- The series variable resistor method controls speed by varying the circuit current.
- The armature diverter variable resistor method controls the speed by varying the current through the armature.

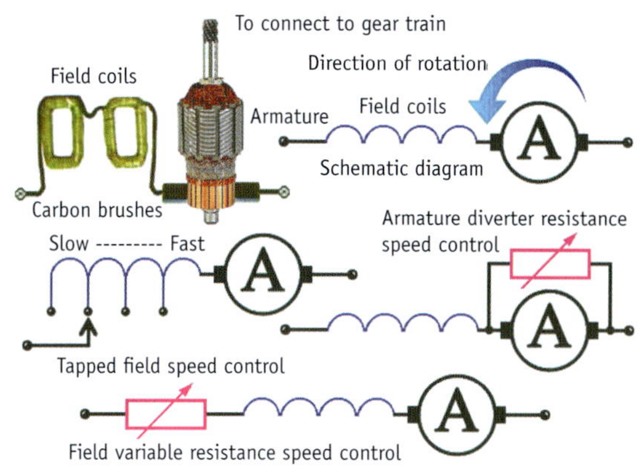

FIGURE 8.63 Universal motor and speed control

Reversing direction of rotation

Reversing the direction of rotation of a universal motor happens by reversing the direction of current in either the series field or the armature but not both.

Torque

Figure 8.64 shows a typical torque–speed characteristic of a universal motor.

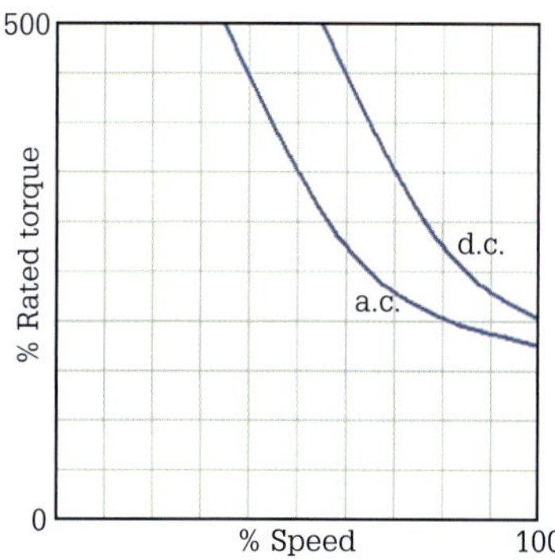

FIGURE 8.64 Torque–speed characteristic of a universal motor

The graph shows two curves, one for direct current application, and the other for alternating current use. On alternating current, the motor produces a more drooping characteristic than on direct current. The a.c. characteristic occurs because both the series field and armature windings develop significant impedance due to the inductive effect of the windings. A large voltage drop occurs across these windings.

The result of this is that the armature voltage is smaller for an a.c. supply voltage than for a d.c. supply voltage. For a.c. current, the reduced armature voltage means that the armature shaft speed is slower for a given armature current and torque than it would be on direct current.

In the case of a.c., the field and armature current are always in phase; therefore, the flux reverses with armature current ensuring that the torque is always in the same direction.

For its physical size, the universal motor can develop higher torques and shaft speed than any other single-phase motor of the same power rating. Varying the supply voltage can control the speed. Refer to **Figure 8.65**.

FIGURE 8.65 Universal motor

Electronic speed control of universal motors

Microcontrollers are replacing analogue speed controls for a.c. universal motors. The microcontroller controls a triac to vary armature current as shown in **Figure 8.66**.

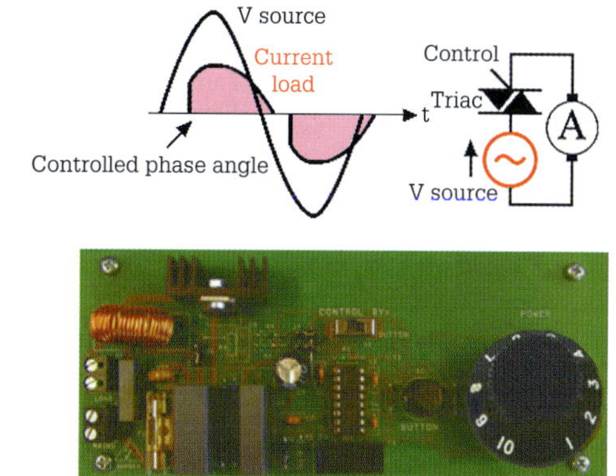

FIGURE 8.66 Electronic speed control of universal motors

Applications

Electronically controlled universal motors occur in home appliances or industrial applications for various purposes, such as motor regulation in power tools, vacuum cleaner control, sewing machines and in some washing machines.

Brushless permanent-magnet synchronous motors

Split-phase motors and universal-type motors occurring in many appliances are under threat from electronically controlled brushless permanent-magnet synchronous motors (BLDC) and permanent-magnet alternating current motors (PMAC). Six-step inverters drive these motors with commutation controlled by either position sensors or a logic integrated circuit.

Comparison of single-phase motors

Figure 8.67 shows a summary of the characteristics of single-phase motors.

Single-phase motors standard range						
	Split-phase	Capacitor start	Capacitor start/run	Permanent split-phase	Shaded pole	Universal
Peak starting torque	1 to 1.75 Tn	2 to 3 Tn	1.8 to 2.5 Tn	0.25 to 3 Tn	0.25 to 1 Tn	3 to 5 Tn
Peak starting current	4 to 6 In	3 to 5 In	3 to 5 In	1.25 to 4 In	2 In	3 to 5 In
Control	On–Off	On–Off	On–Off	On–Off	On–Off	On–Off
Speed range rpm	1000 to 3000	1000 to 3000	1330 to 2830	750 to 3000	1500 to 3000	3000 to 20 000
Advantages	Good starting torque	High power High starting torque All-purpose motor	High power High starting torque All-purpose motor	No starting switch Reversed easily Can be used with speed controllers Quiet running	Inexpensive Long life	Can be driven by a.c. or d.c. Suitable to drive many types of hand-held appliances
Disadvantages	High starting current	More expensive than split-phase	Expensive due to additional start capacitor and full-time capacitor	Low starting torque Speed varies under load	Low starting torque Small ratings	High construction complexity Low reliability Brushes create sparks and ozone
Applications	Pedestal drills Washing machines Pumps Small grinders	Compressors Fans Blowers Small conveyors Gear applications	Air-compressors High pressure water pumps Cleaning machines Air-conditioners	Adjusting mechanisms Gate operators Garage door openers	Multi-speed fans Small appliances	Vacuum cleaners Portable power tools Blenders Hair dryers

FIGURE 8.67 Summary of the characteristics of single-phase motors

REVIEW QUESTIONS

1 What type of motor develops a high starting current and a large starting torque?
2 State the shaft speed a series universal motor is able to achieve.
3 How is the direction of rotation reversed in a series universal motor?
4 Why does a series universal motor operate at lower speed on a.c. than d.c.?
5 State an application for an electronically controlled series universal motor.

8.7 Motor protection

Motor protective devices safeguard the motor's control equipment and the supply circuit conductors against excessive heating due to various starting conditions, load conditions, internal motor failures and supply variations.

The following AS/NZS 3000: 2018 *Wiring Rules* clauses provide essential information in relation to motor protection devices.

- Clause 2.5
- Clause 2.5.2

Furthermore, automatic protection devices used to open-circuit conductors carrying overload current, short-circuit current or the prospective fault current must not be able to close on the fault spontaneously. Types of protective devices meeting the requirements include:

- sealed fuse links
- miniature circuit breakers (MCBs)
- moulded case circuit breakers which provide circuit over-current protection with inverse time and instantaneous tripping characteristics and cannot close on a fault.

Refer to Clause 2.4.3 of AS/NZS 3000: 2018.

Protection devices

Over-current protection devices interrupt the supply to the induction motor when the motor draws excessive current. These devices include HRC fuses and circuit breakers. Overcurrent protection devices operate when a short circuit exists within the motor circuit or an overload causes the current drawn to exceed the normal parameters of operation.

Fuses

In principle, the operation of a fuse-type protection device is deliberately connecting a weak link in series with an electrical circuit. In the event of an over-current, this link activates first and thus protects the circuit conductors and the load.

Standard HRC fuses carry the average load current of the circuit without nuisance interruptions. However, when an over-current occurs the fuse must interrupt the over-current, limit the energy surge and resist the voltage across the fuse during arcing. Standard HRC fuses ought to remove a fault current in less than half a cycle, thus limiting the magnitude of the current flow. Selection occurs according to the current-carrying capacity of the conductors and the requirements of the load.

Standard HRC fuses are not suitable for over-current protection of motors. If the fuse is rated according to the full-load current rating of the motor then it interrupts the circuit during the in-rush starting current cycle, which could be up to eight times the full-load current. If the fuse only has a rating to contain the starting current, then the fuse would offer no protection under overload conditions.

A motor-start or time-delay fuse (aM type, where a = partial range and M = protection of electric motors, which can withstand motor starts), as shown in **Figure 8.68**, has a thermally controlled element that delays the opening of the element in the case of overload. The overloads element provides protection against low-level over-current and holds the in-rush current that is five times greater than the ampere rating of the fuse for a minimum of 10 seconds. Therefore, two-element-type motor-start fuses have a thermally controlled element for overload conditions connected in series with a fusible element that operates under short-circuit conditions. Moreover, the size of the over-current determines which element functions.

FIGURE 8.68 Motor-start fuse

Motor-start HRC fuses are refined and reliable protection devices with very high interrupting and current-limiting capability. In addition, they have excellent cycling ability for frequent motor starts and stops.

Selecting HRC motor fuse links

Table 8.2 shows the full-load currents (FLCs) of typical three-phase induction motors at 400 V up to 18.5 kW. This table also provides HRC motor fuse link (type aM) data for direct on line (DOL) starting and assisted starting conditions. The table uses various factors to determine the suitable fuse link for a motor rating under DOL and assisted starting conditions as **Figure 8.69** shows.

TABLE 8.2 Fuse link data for DOL and assisted starting of induction motors

Motor rating	Full-load current	DOL starting fuse link	Assisted starting fuse link
kW	A	A	A
0.37	1.08	4	2
0.55	1.395	4	2
0.75	1.875	6	4
1.1	2.75	10	6
1.5	3.75	16	10
2.2	5.5	16	10
3	7.5	20	16
4	10.0	25	16
5.5	13.75	32	20
7.5	18.75	50	32
11	27.5	63	35
15	37.5	80	50
18.5	46.25	100	63

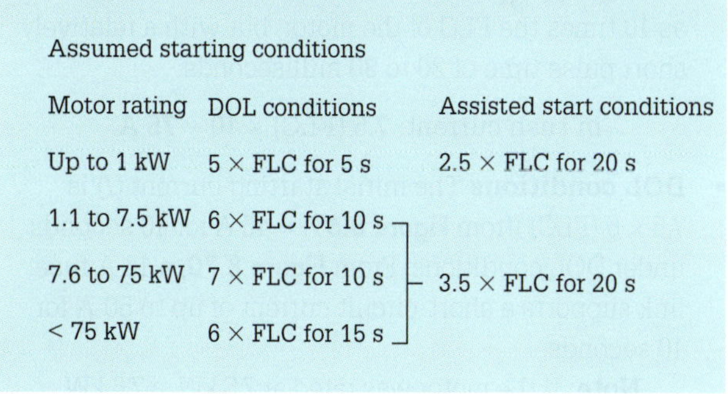

Assumed starting conditions

Motor rating	DOL conditions	Assisted start conditions
Up to 1 kW	5 × FLC for 5 s	2.5 × FLC for 20 s
1.1 to 7.5 kW	6 × FLC for 10 s	3.5 × FLC for 20 s
7.6 to 75 kW	7 × FLC for 10 s	
< 75 kW	6 × FLC for 15 s	

FIGURE 8.69 HRC fuse link 'aM' type assumed starting conditions

Assisted start conditions includes star–delta, autotransformer, soft start.

Note: Some motor manufacturers may have 3.5 × FLC for assisted starting conditions for standard duty motors and 5 × FLC for heavy-duty assisted starting applications.

To choose the recommended fuse link select the motor full-load current and starting condition from **Table 8.2**. For example, a 3 kW motor has a full-load current of 7.5 A.

Consequently, for DOL conditions, the feed to the motor has 20 A motor-start HRC fuse protection. The feed (at a reduced CSA) to the motor has protection when a 16 A motor-start HRC fuse for assisted starting conditions is installed. Fuse-link manufacturers provide data on the selection of a fuse link for various kW-rated induction motors and their starting methods. However, it is important to use links of the same manufacturer. The same manufacturer is necessary because time–current curves as shown in **Figure 8.70** refer to testing by the fuse-link manufacturer for their fuse links. There is no available data for mixing fuse links of different manufacturers.

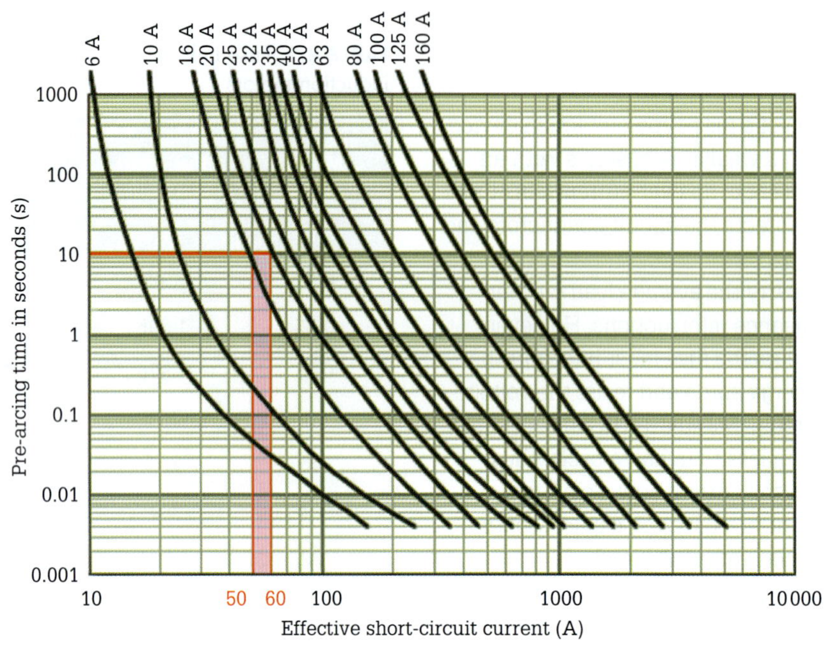

FIGURE 8.70 Data sheet for 'aM' HRC fuses

If we use the same data for the 3 kW motor and the data sheet in **Figure 8.70** we find the following:

- **In-rush current** The in-rush current (not starting current) appears at switch-on and could be as great as 10 times the FLC of the motor, but with a relatively short pulse time of 20 to 30 milliseconds.

  in-rush current: 7.5 (FLC) × 10 = **75 A**

- **DOL conditions** The initial starting current (*I*) is: 7.5 × 6 (FLC) (from **Figure 8.69**) = 45 A for 10 seconds under DOL conditions. From **Figure 8.70**, a 16 A fuse link supports a short-circuit current of up to 50 A for 10 seconds.

  Note: If the motor was rated at 7.5 kW, a 7.6 kW rating in **Figure 8.69** is sometimes acceptable, therefore we obtain 7.6 × 7 (FLC) = 53.2 A for 10 seconds under DOL conditions.

  From **Figure 8.70** a 20 A fuse link supports an effective short-circuit current of up to 60 A for 10 seconds.

- **Assisted-start conditions** Under assisted-start conditions (from **Figure 8.69**) the initial starting current (*I*) is:

  7.5 × 3.5 (from **Figure 8.69**) = 26.25 A for 20 seconds

From **Figure 8.70**, a 16 A fuse link can support 46 A for approximately 30 seconds.

This illustrates the importance of using the same manufacturer of fuse links when using these protective devices for motor protection.

EXAMPLE 8.7

Using **Figure 8.71**, determine the required fuse links and effective short-circuit current for a 30 kW 380 V 1470 rpm motor with an FLC of 57.4 A at a pf of 0.82 with an efficiency of 92% started DOL over five seconds.

in-rush current: 57.4 × 10 = 574 A

»

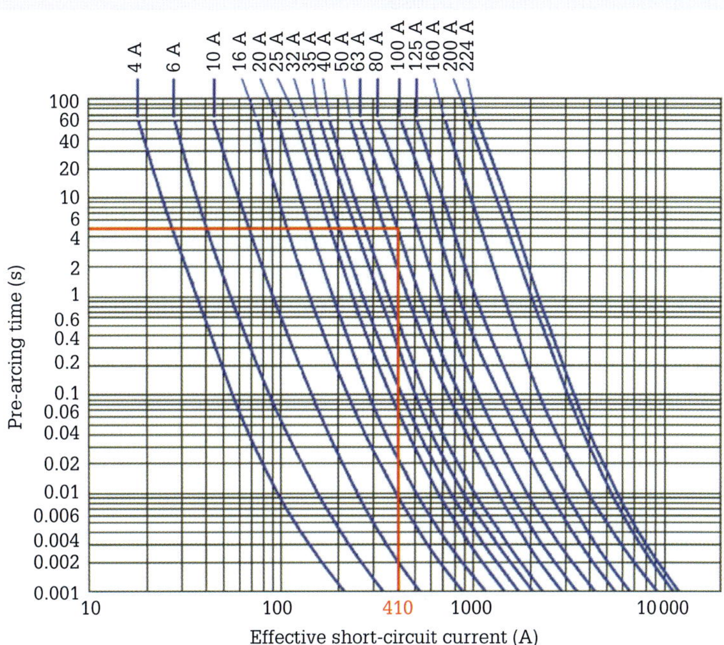

FIGURE 8.71 Data sheet No. 2 for 'aM' HRC fuses

The initial starting current (I) is:

57.4 × 7 (from **Figure 8.69**) = 401.8 A for 10 seconds

Using **Figure 8.71** a 63 A fuse link can support 410 A for 5 seconds without opening.

EXERCISE 8.7

a Using the data sheet in **Figure 8.71** determine the required aM fuse links for a 55 kW 380 V 740 rpm motor with an FLC of 117 A at a pf of 0.81 with an efficiency of 93% started by assisted starting over six seconds.

b Using the data sheet in **Figure 8.71** determine the required aM fuse links for a 22 kW 400 V 1470 rpm motor with an FLC of 39 A at a pf of 0.84 with an efficiency of 92.4% started by DOL starting over five seconds.

Circuit breakers

The operating principle of a circuit breaker differs entirely from that of HRC fuses. Circuit breakers carry out double duty as on-and-off switches as well as providing over-current protection. Furthermore, with motor-control circuit breakers, the means of operation is twofold, there being an adjustable thermal element for overload currents and an electromagnetic unit for short circuit or rapid high-current faults.

The adjustable thermal element usually consists of a bimetallic strip that bends as a result of the heating effect of an overload current tripping a release mechanism. However, in the event of short circuit the coil of the electromagnetic element energises, which causes the metal armature of the solenoid to release the same tripping device.

Fuse-type protection devices assure current interruption without fail while the operation of a motor-control circuit breaker is dependent on the mechanical inertia of the tripping mechanism. Although fuses respond well, they can be subject to thermal ageing, which may change the melting characteristics of the fuse element.

Circuit breakers interrupt all phases while fuse over-current protection devices protect only the connected phase. This fact implies limitations with fuses in their overall protection of three-phase motors. Single phasing of the three-phase motors is an outcome of one phase experiencing an interruption. However, if the protection system also contains a phase relay trip activated by either voltage or current, then the fuse has support in protecting the system. Circuit breakers also have this capacity for additional system protection.

The apparent difference between a circuit breaker and a fuse is that the circuit breaker can reactivate after an interruption while a fuse requires replacing. However, this feature of circuit breakers can lead to false security concerning their protective ability. If the breaker experiences many interruptions, the circuit and the breaker require testing. Refer to **Figure 8.72**.

FIGURE 8.72 Thermal-magnetic circuit breaker

Thermal–magnetic circuit breakers are general-purpose protection devices suitable for the majority of industrial applications.

EXAMPLE 8.8

1. A 400 V 5.5 kW three-phase motor with a full-load speed of 2930 rpm has a full-load current of 13.75 A at an efficiency of 89%. The load connected to this motor requires a run-up time of 10 s with the motor started DOL. Determine the minimum size circuit breaker rating suitable for this particular motor. In addition, determine the circuit breaker rating for assisted starting conditions.

Using Table 8.3 the initial start-up under DOL conditions is six times full-load current for 10 seconds.

The initial starting current (I) is:

13.75 × 6 (from Table 8.3) = 82.5 A for 10 seconds

Therefore, the chosen circuit breaker must control this current for 10 seconds without unlatching.

TABLE 8.3 Starting condition values for various induction motor ratings

Assumed starting conditions		
Motor rating	**DOL conditions**	**Assisted start conditions**
Up to 0.75 kW	5 × FLC for 6 s	2.5 × FLC for 15 s
1.1 to 7.5 kW	6 × FLC for 10 s	2.5 × FLC for 15 s
11 to 75 kW	7 × FLC for 10 s	2.5 × FLC for 15 s
90 to 160 kW	6 × FLC for 15 s	2.5 × FLC for 20 s

Figure 8.73 shows the time–current characteristics of Type D and C miniature circuit breaker.

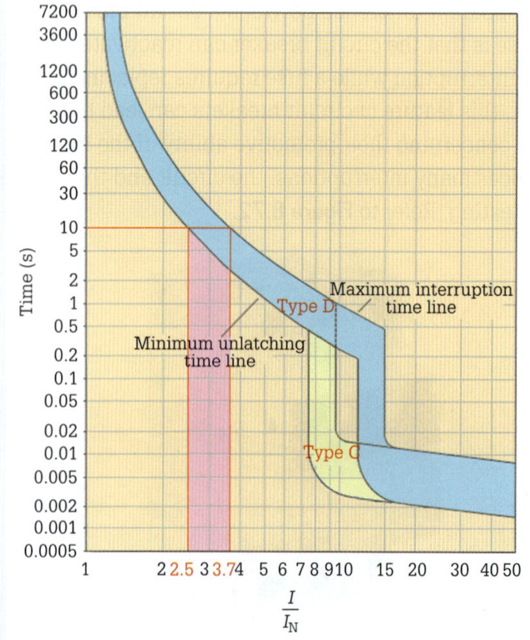

FIGURE 8.73 Type D and C miniature circuit breaker time–current characteristics

From the characteristic curve, the circuit breaker unlatches (clears the fault) automatically at some point in time within the minimum and maximum timeline shown on the curve. The higher current portion of the curves (minimum unlatching time) depicts the time to switch due to thermal action and indicates overload protection of the feed and connected load. By contrast, the lower current portion of the curves (maximum interruption time) depicts switching due to the magnetic action and indicates a short circuit.

Comparing the data for the motor condition to the characteristic curve we see that at 10 s the ratio of I to I_N is 3.7. The 3.7 gives us the DOL unlatching time line ratio.

The minimum breaker current (I_N) for DOL starting over 10 s for 82.5 A (I) is:

$$I_N = \frac{I}{ratio} = 82.5 / 2.5 = \mathbf{33A}$$

Hence, the circuit breaker selected must be a size closer to and above this calculated value. Circuit breakers are available in the following sizes: 6, 10, 16, 20, 25, 32, 40, 50, 63, 80, 100, 125, 160 and 200 A.

A 40 A continuous current rating Type D circuit breaker could be used as it can support an initial starting current of up to 100 A (2.5 × 40) for 10 seconds. Note that the feed to the motor requires a CSA capable of carrying 82.5 A for 10 seconds.

For assisted starting, the initial starting current (I) is 13.75 × 2.5 = 34.375 A for up to 15 seconds (see Table 8.3). The minimum breaker current (I_N) for assisted starting is:

$$\frac{34.375}{2.5} = \mathbf{13.75A}$$

A 16 A Type D circuit breaker supports an initial starting current of up to 40 A (2.5 × 16) for 10 seconds.

Note: Assisted starting requires smaller CSA cables for the motor feed than a DOL feed.

2. Select an appropriate feeder circuit breaker (Type C) to protect a 1.5 kW three-phase motor DOL start. The motor has the following data: FLC is 3.75 A, starting current 6 × FLC with a run-up time of five seconds.

Starting current: 3.75 × 6 = 22.5 A for 5 s

In-rush current: 3.75 × 10 = 37.5 A

The minimum breaker current (I_N) for DOL starting over 10 s for 22.5 A (I) is:

$$I_N = \frac{I}{ratio} = \frac{22.5}{2.5} = \mathbf{9A}$$

Comparing the data against the time–current characteristics of a Type C miniature circuit breaker (Table 8.3) we see that at 5 s the breaker will carry 3 × I_N without tripping. Therefore, a 7.5 A miniature

»

circuit breaker would carry 22.5 A for five seconds. However, the smallest available miniature circuit breaker is 10 A and as this value is greater and nearest to the minimum I_N allowed, a 10 A breaker is suitable. The types of breakers that could be suitable using the manufacturer's data are:

- a Type B circuit breaker which has an instantaneous trip (in-rush current) of 4 × I_N
- a Type C circuit breaker which has an instantaneous trip of 5 × I_N
- a Type D circuit breaker which has an instantaneous trip of 10 × I_N.

Any of the three circuit breakers are suitable as long as the instantaneous trip (in-rush) current was not a problem; and with all three types, it is not a problem.

If B3.2.2.2, 'Protection by circuit-breakers', AS/NZS 3000: 2018 *Wiring Rules* was consulted then the instantaneous tripping current is as follows:

- Type B = 4 × rated current
- Type C = 7.5 × rated current
- Type D = 12.5 × rated current.

EXERCISE 8.8

a Select an appropriate feeder circuit breaker to supply a 4 kW three-phase motor with DOL start. The motor has the following data: FLC 12 A, starting current 6 × FLC with a run-up time of three seconds.

b Select an appropriate feeder circuit breaker (Type C) to supply a submersible pump motor 11 kW three-phase motor soft start. The motor has the following data: FLC 27 A, starting current 2.5 × FLC with a run-up time of seven seconds.

Type D circuit breakers

These circuit breakers occur in applications where high in-rush currents (highly inductive circuits) are likely, such as transformers, welding machines and induction motors. The starting or run-up time of induction motors can vary between 1 and 20 seconds depending on starting current and the type of load.

Type D circuit breakers have magnetic trip settings which are 10 to 20 times the continuous current ratings (I_N). Note that Type C circuit breakers also protect inductive circuits (inductive lighting and control circuits incorporating coils) and these breakers have magnetic trip settings which are five to 10 times the continuous current rating (I_N). In some circumstances, a Type B could be used as long as its instantaneous trip is not a problem.

Circuit breakers do not provide current protection for motors. In practice, they combine with a motor over-current protective device. Under these conditions, the circuit breaker protects the cable feed to the motor while the motor's own protective device protects the motor itself.

To accurately select the correct circuit breaker for protecting a motor it is essential to know the correct full-load current, the starting current, the run-up time and the type of starting conditions (DOL or assisted). This information is plotted against the time–current characteristic curve of the type of circuit breaker selected.

Induction motor protection

In the 2018 edition of AS/NZS:3000 *Wiring Rules* there are four clauses with particular importance to induction motor protection.

- Clause 2.3.4.5
- Clause 4.13.2
- Clause 4.13.3
- Clause 4.13.3.3.

Motor protection relays

Motor protection relays include the following:

- magnetic overload relays
- under- and over-voltage relays
- current controlled relays
- integrated circuit and digital relays.

Magnetic overload relays

Magnetic overload relays are over-current or differential relays. They provide over-current protection for motors and activate the power contactor, thereby de-energising the motor power circuit. Most over-current relays energise from the output of current transformers and have operating times dependent on the value of the transformer signal. In addition, they have both an instantaneous short-circuit protection capability and an inverse time characteristic, which means that the time delay is long for low values of current and rapid for larger values of current. The magnetic overload relay consists of three electromagnetic coils with pistons, dashpots and normally closed control circuit contacts. Refer to **Figure 8.74**.

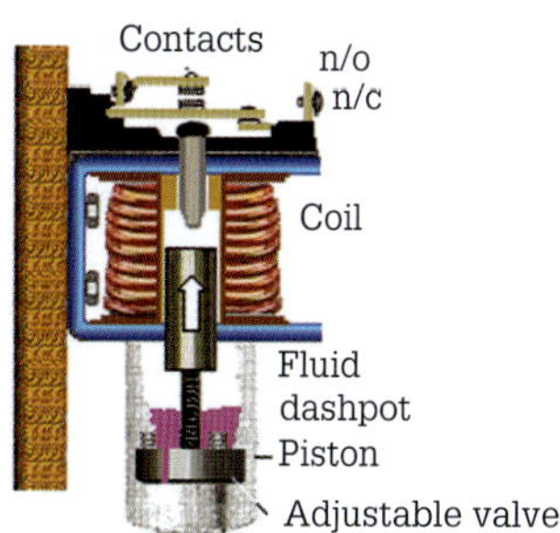

FIGURE 8.74 Electromechanical overload relay

The current transformer energises the solenoid and the electromagnetic field created exerts a force upon the piston. When the solenoid current rises to the trip-point value, the piston moves into the solenoid, and its movement operates the contact mechanism. The threaded dashpots can move up and down to vary the piston position thereby providing an adjustable current trip point. However, in order to prevent nuisance tripping, all three dashpots have to be set at the same level. Time delay occurs because of an adjustable valve in the bottom of the piston. These relays reset automatically once the current has fallen to 20% of the trip-point value. Magnetic overload relays detect over-current conditions caused by seized bearings, a jammed fan, locked rotor, as a result of a blocked pump or compressor or simply the load trying to exceed the capabilities of the motor.

Under- and over-voltage relays

Polyphase motors burn out due to over-voltage, and the same motors overheat or produce low torque due to low voltage. Over-voltage breaks down motor insulation due to excessive heat during start-up, causing the winding to fail prematurely. High-surge voltages can cause arcing between the first conductor's turns on the stator, causing short circuits. However, three-phase voltage relays such as that shown in **Figure 8.75** provide protection against under-voltage, over-voltage and an unbalanced voltage by continuously monitoring the three-phase voltages. The relay operates when an externally adjustable trip-point voltage happens. The trip-point voltage setting of some voltage relays is user adjustable from 85% to 115% of the nominal voltage rating. The differential voltage setting occurs as a fixed value between 1% and 15% below the trip-point voltage setting (allows for a time delay). Consequently, the relay de-energises when the input voltage signal is above the trip-point setting and the relay de-energises when the input voltage signal is below the dropout setting. With some relays the under-voltage drop-out can be delayed for up to 300 seconds and will de-energise after that period or when the voltage drops below the low-voltage trip-point setting. Finally, the relay remains in the de-energised state until the voltage rises above the low-voltage trip point.

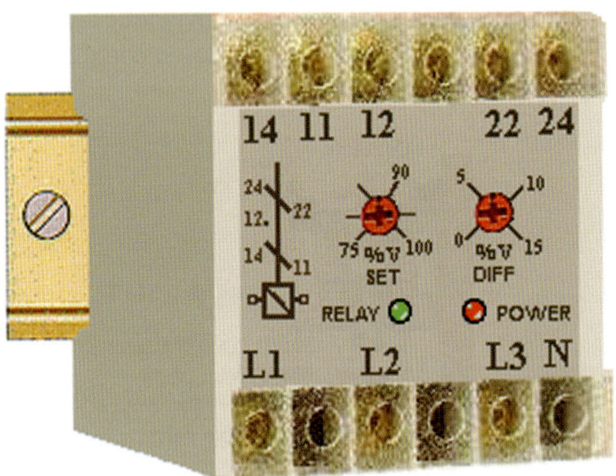

FIGURE 8.75 Under-voltage relay

Under-voltage protection allows the supply contactor to operate in response to dips in three-phase voltage, phase imbalance, loss of phase or a loss of all three phases. The over-voltage relay has no delay and energises when the voltage exceeds the trip-point high-voltage setting. The relay remains in the energised state until the voltage drops below the high-voltage trip-point. LED indicators show when conditions are normal, and the supply to the motor energises. **Figure 8.76** shows an over- and under-voltage monitor.

The yellow LED indicates the state of the relay while a green LED indicates the condition of the power supply.

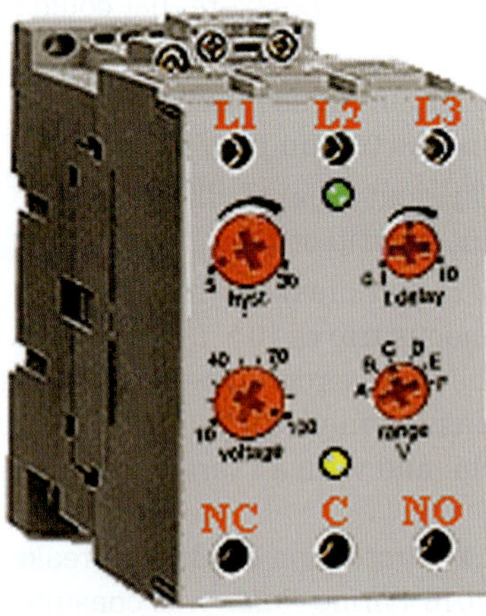

FIGURE 8.76 Over- and under-voltage relay

Current-controlled relays

Over-current relays initiate a circuit breaker tripping sequence to isolate the motor from the rest of the system. Some over-current relays have motor voltage input parameters to distinguish between short circuit and overload conditions. A short circuit always causes a significant voltage decline while an overload creates a moderate voltage decline. Some relays operate from the output of a standard ratio-type current transformer (CT), with 5 A secondary current at rated primary current.

Integrated circuit and digital relays

Integrated circuit (IC) and digital relays are smaller than their corresponding electromechanical relays. For example, you can use a three-phase IC over-voltage relay in place of three single-phase mechanical over-voltage relays.

The accuracy of electronic relays is greater than that of electromechanical relays, which enables a motor protection system to have a more sensitive and defined discrimination (difference between opening and closing currents). In an effective protection system, the closest protection device protecting the motor must operate first.

Discrimination transpires by using different current ratings and different time-delay curves of the same or different circuit breaker types. In addition, electronic relays draw less operating power from the supply than electromechanical relays. IC relays have excellent accuracy. They are dependable and consistent and are able to provide many different relay functions in one complete package. Digital (microprocessor- and software-based) motor protection relays provide overload, locked-rotor, unbalanced current, short circuit, earth fault, thermal, number of starts and correct phase-sequence protection.

They achieve this degree of protection by calculating a thermal model of the heat generated by the current within the motor. Some relays have memory (i.e. store the last trip event) and diagnostic capability. Thermal overload relays simulate the heating effect in the motor windings by sending the motor current through resistive elements in the relay. Heat generation acts upon a bimetallic element to open a control-circuit contact. Once the relay trips, it can be manually or automatically reset to allow the motor circuit to energise.

Resistive elements called heaters have calibration for specific current ranges to correspond to the actual motor nameplate current. An overload provides single-phasing protection in a delta-connected motor, in which case the overload is set to trip at 58% of full-load line current.

Protection from environmental effects

Environmental factors that require consideration for motor protection include:

- over-temperature
- humidity
- enclosures
- ingress protection (IP).

Over-temperature protection

When a motor experiences an overload, it runs hotter than its design operating temperature. The increase in temperature stresses the stator insulation and shortens motor life. There are several factors contributing to an increase in motor operating temperatures and the heat developed, aside from overloading and under-voltage operation. Some of these factors include ambient temperature, elevation above sea level (reduced atmosphere density), sunlight (direct radiation), blocked ventilation ducts and too frequent starting and stopping. A device used to monitor the effects of these temperature agents is a PTC thermistor.

PTC thermistors are thermally sensitive resistors made of poly-crystalline ceramic material and are manufactured using a composition of barium and strontium titanate doped with additives such as manganese or silica. This type of thermistor has a base resistance value at 25 °C and a resistance–temperature characteristic that displays a small negative temperature coefficient until the thermistor reaches the 'curie point' or changeover temperature (the temperature at which the resistance equals twice the base resistance). At this critical point, the thermistor exhibits a substantial increase in resistance; that is, positive temperature coefficient of resistance. PTC thermistors are for over-current and over-temperature protection and provide low resistance to current under normal conditions, increasing their resistance in relation to a temperature rise.

Three series-connected thermistors (one per phase) are embedded in the three-phase stator winding and form one part of the motor's control circuitry. When an excessive current flows through the stator windings, the resistance of the thermistor will be increased by the heat generated, causing the thermistor to limit the current flow through its circuit. Reduced motor control current flow causes a relay to de-energise, allowing normally closed contacts to open. This causes the power contactor to release and disconnect the motor. **Figure 8.77** shows the three series-connected PTC thermistors and their characteristic curve

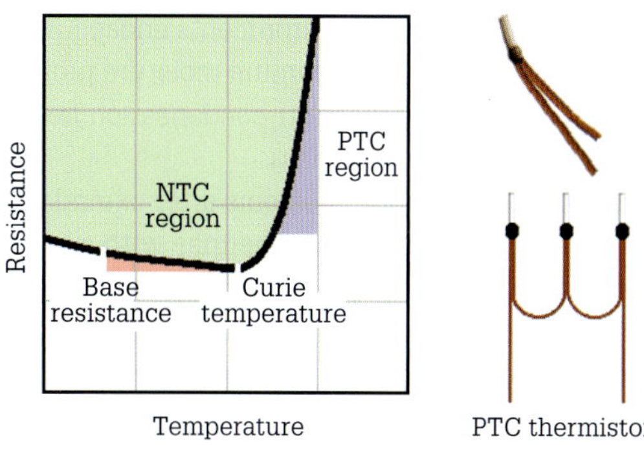

FIGURE 8.77 PTC thermistors and characteristics

All motors with ratings of 25 kW and higher, as well as all motors subjected to run-up times in excess of 15 seconds, should have thermistors for over-temperature protection installed in the stator windings. In addition, three thermistors, one per phase, are required in single-wound motors and six thermistors required for pole-changing (double-winding) motors. Thermistor protection is used to protect the motor from:

- heavy-duty starting (overloading)
- restricted cooling
- frequent switching
- blocked rotor
- phase failure, with a consequent increase in the temperature of the windings.

A glass-fibre sleeve fits over the thermistor leads, for mechanical protection. The sleeve must not cover the thermistor bead. The thermistor inserts in the windings to ensure the best thermal contact with adjacent conductors of the winding. The bonding of the thermistor to the winding must be sufficient, so that thermal cycling does not cause the bonding to fail. If thermal cycling occurs, an air gap occurs between the thermistor and the winding, resulting in reduced thermistor sensitivity. All leads from thermistors to their protection control relays should be twisted pairs to minimise stray voltage pick-up. Where the control relays are far from the motor, use screened cables.

When a fault temperature occurs, the signal from the thermistors feeds a PTC thermistor-tripping unit that isolates the power circuit from the motor.

Humidity

A cause of some motor failures is moisture, which can enter a motor in a number of ways; for example, the humidity in the atmosphere or moisture from high-pressure cleaning and sanitising sprays. Moisture enters the motor via the shaft, end shields and terminal box.

The humidity in the atmosphere causes condensation to form inside the motor when the motor cools and the environment surrounding the motor frame becomes warmer.

To overcome moisture problems non-friction seals fit on the shaft, and the bottom of the motor frame has special drain fittings. The terminal boxes have gaskets fitted and the power lead entry sealed with silicone. Bearings are double sealed and have moisture-resistant grease. However, for high-humidity environments encapsulation of the stator winding offers maximum moisture protection. With some motors an internal anti-condensation heater prevents moisture from developing.

In order to prevent motor problems after extended shutdowns, some motors' stators energise with a low d.c. voltage to heat and dry out the windings. Finally, motors in humid conditions require regular testing by a high-voltage insulation resistance tester to check on stator insulation condition.

Enclosures

The motor enclosure refers to the casing of the motor and matches the motor to its operating environment. The two most common types of enclosures for electric motors are open drip proof (ODP) and totally enclosed fan cooled (TEFC).

An ODP motor consists of a sheet steel stator enclosure with vent openings stamped into the end shields to allow enough airflow to the inside of the motor. The vents prevent water dripping on the motor from flowing into the motor. In addition, a fan mounts on the motor's rear shaft to pull cool air through the motor.

TEFC motors are probably the most commonly used motors in ordinary industrial environments. These motors have enclosed casings to prevent moisture and particles entering. They have an integral cooling fan mounted on the non-drive end of the shaft enclosed in its own housing.

Totally enclosed non-ventilated (TENV) motors are built with a thick frame body with extra fins (cooling fins increase surface area for maximum heat dissipation) to allow active radiation of heat. They are not equipped with a cooling fan and depend upon the convection currents of the surrounding air to cool the motor. Consequently, usage of these motors occurs in environments where dust and fibrous material would clog a fan.

A driven device (fan or blower) cools totally enclosed air over (TEAO) motors or another device provides a cooling air stream over the motor.

Explosion-proof (XPRF) motors prevent the ignition of any materials, liquids or fumes that surround the motor. The motor design contains any explosion resulting from a spark inside the motor. The heavy-duty frame, end shields and associated fixing bolts, terminal box housing and cover, terminal plate and cable entry form an explosion-proof enclosure of the motor. There are several external surface temperature classes of explosion-proof motors (T1 – 450 °C to T6 – 85 °C) that relate to the ignition temperature of any contacted material. Motors selected for hazardous or explosion-type environments must be suitable for their operating zone classification. For example, Zone 1 is an area where an explosive condition exists permanently while Zone 2 covers an area where an explosive situation rarely occurs. The level of explosion risk determines zone classification.

Ingress protection

The degree of protection provided by enclosures for rotating machines is the IP (ingress protection) code, for example IP55 – the numerals indicate degrees of protection against the ingress of solids and liquids. The first numeral indicates the degree of protection against contact and ingress of foreign bodies and protection against hazardous live, and moving, parts. The second figure indicates the degree of protection against water entry.

The designation IP55 means that the enclosure is dust protected and protected against low-pressure spraying water from any direction. Other IP ratings include IP56, IP65 and IP66. Most standard TEFC three-phase motors are IP55 rated while ODP motors are IP22 or IP23 rated. For further information on ingress protection consult the Australian standard AS 1939–1990, 'Degrees of Protection Provided by Enclosures for Electrical Equipment' (IP Code).

Starting and reversing

Interchanging the connections of any two of the three power conductors on a three-phase motor reverses the direction of the electromagnetic stator field thus reversing the direction of shaft rotation. However, the regular number of starts per day over a period of months or years influences motor life.

Excessive cycling affects the life of control components such as starters, sensors and relays. The number of starts and reverses that a motor sees (on and off) can also cause motor shaft damage (twisting stress), bearing damage, stressed insulation and motor overheating (for every 10 °C rise the motor insulation life is reduced by half).

Intermittent operation of a reversible motor for a short period results in a significant flow of current when the motor starts or reverses, creating increased heat generation. Therefore, the motor should run for a short time in order to dissipate heat build-up from the starting current. Additionally, the current surge in the windings at start-up causes the stator windings to experience mechanical stress. The current surge causes the windings to bend slightly at the point where the winding conductors leave the stator slot, thereby placing mechanical pressure on the insulation at those points. Consequently, the effects of heat and mechanical stresses mean that there are a limited number of starts per hour.

Standard three-phase motors below 1 kW can deliver up to 12 starts while motors above 30 kW would be limited to four starts or less per hour. However, these values depend upon the design and duty application of the motor.

AS/NZS 3000: 2018 *Wiring Rules* requirements

Reference to the *Wiring Rules* is necessary. In Section 4.13, 'Motors', requirements for protection against injury from mechanical movement, protection against overload and protection against over-temperature are explained. Refer to the following Rules:

- Clause 4.13.1.1
- Clause 4.13.1.2
- Clause 4.13.1.3
- Clause 4.13.1.4
- Clause 4.13.2
- Clause 4.13.3
- Clause 4.13.3.2
- Clause 4.13.3.3.

Other relevant clauses are Clause 7.2.9 and Clause 6.6.4.5.

Figure 8.78 shows switching devices capable of starting and stopping the motor together with provision for an emergency stopping switching device and an isolating switching device.

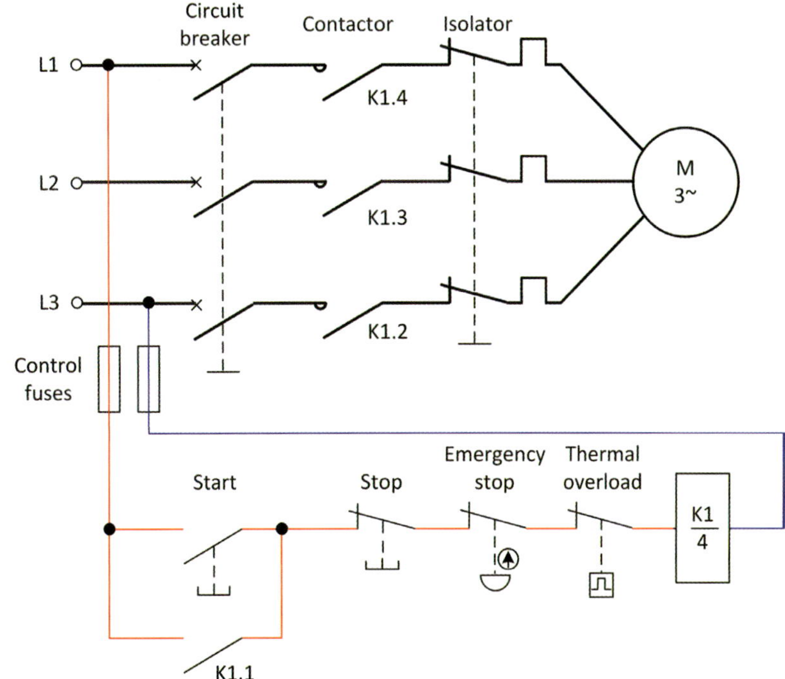

FIGURE 8.78 Power and control circuit

Single-phase motor protection

In many single-phase motor applications, protection occurs to prevent thermal damage to the motor because of high currents caused by mechanical overloads. Thermal overload is a major cause of single-phase motor malfunction. All unattended single-phase motors greater than 230 VA, together with 480 VA and above for shaded-pole motors, must have over-temperature protection devices. These devices must comply with the Australian Standard AS 60947.8: 2005, and for single-phase motors must be able to disconnect the active supply conductor.

These protection devices have an internal thermal switch that utilises a snap action bimetal trip embedded in the main winding. The thermal switch opens the current circuit when the temperature or current exceeds an individual level. These motors are referred to as being internally protected, and do not require separate protection in the supply contactor. The motor manufacturer sets the trip or release temperature of the sensor to match the insulation class of the stator winding. However, the sensor cannot change the trip temperature. Once activated in a fault condition these devices maintain their open states until the motor circuit is de-energised, whereby the device cools and resets. Examples of thermal switches capable of switching 12 A at 230 V at their switching temperature (from 55 °C up to 145 °C) are shown in **Figure 8.79**.

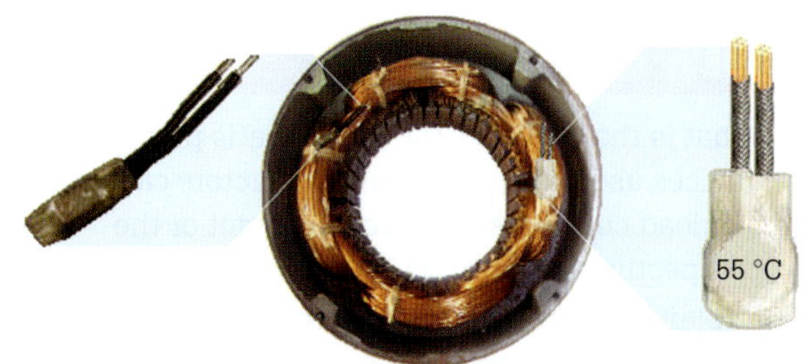

FIGURE 8.79 Thermal switches

Thermal cut-out switches connect in series with the main winding and monitor the actual winding temperature. Other types of thermal switches mount on the end shields and can be self-resetting or manually reset. Refer to **Figure 8.80**.

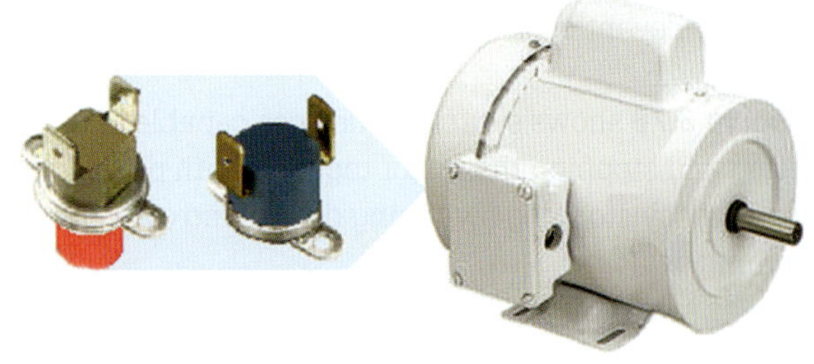

FIGURE 8.80 Manual and self-resetting thermal switches on end shields

The end-shield thermal switches sense the temperature within the motor and are not as fast in action as the winding-embedded-type sensor. However, both types of sensors detect changes in ambient temperature, as a result of changed cooling conditions such as a broken fan or blocked ventilation. Rotors have protection, but indirectly, via the stator temperature.

Single-phase motors should be capable of withstanding an overload current equal to 1.5 times the full-load rated current for two minutes after attaining normal operating temperature. Therefore, in order to respond quickly to temperature variations, sensors require immediate contact with the stator windings. PTC thermistors, as shown in **Figure 8.81**, are over-temperature protection devices for various types of motors.

FIGURE 8.81 PTC thermistor-wire ended 130 °C

Placement of the thermistor must occur in the connection end-winding overhang, and at least 8 mm below the surface of the winding. **Figure 8.82** shows a control circuit using a thermistor for over-temperature protection for a single-phase motor.

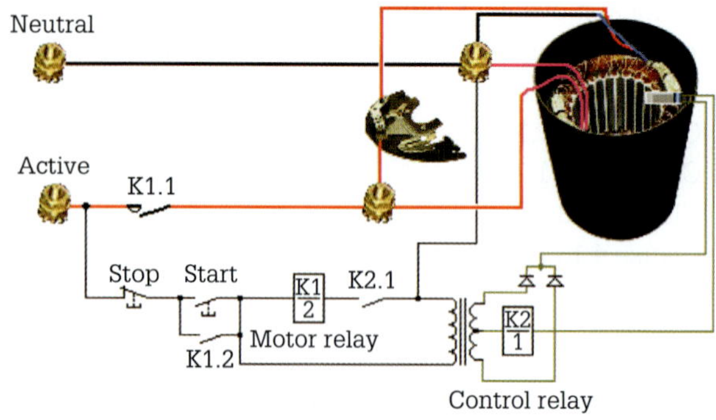

FIGURE 8.82 Single-phase motor protected by thermistors

Thermistors are very low signal devices and their output needs amplification via solid-state circuitry in order to energise a thermal protection control relay to disconnect the motor from the supply. Finally, consult the Australian and New Zealand standard AS/NZS 3000: 2018 *Wiring Rules* about protection against over-temperature of motors.

Overload-protection devices

Overload-protection devices interrupt the supply upon excessive demand by single-phase motors on the energy supply system. These devices are thermal overload protection devices. All motors exceeding 0.37 kW, if not protected by other means, should have protection against overload. In addition, each motor requires an isolating switch within sight of the motor.

Other protection devices associated with single-phase motors include bearing temperature monitors and vibration monitors.

REVIEW QUESTIONS

1 What is the requirement of automatic protection devices used to open-circuit conductors carrying overload current, short-circuit current or the prospective fault current?
2 Explain why a standard HRC fuse is not suitable for over-current protection of motors.
3 What is the purpose of the thermally controlled element in a motor-start or time-delay fuse?
4 What is the expected full-load current of a 4 kW, three-phase induction motor?
5 What are the two means of protection afforded by a circuit breaker?
6 How does single-phasing occur with three-phase motors?
7 State the main advantage of a circuit breaker over a fuse.

»

8 What circuit breaker is best suited to highly inductive loads?
9 Which motor protection device employs three electromagnetic coils?
10 What do over-current relays initiate when excessive current flows in the motor circuit?
11 Name the main advantage of a digital relay over an electromechanical relay.
12 An effective protection system must have good protection discrimination. What does this mean?
13 What type of device affords over-temperature protection for three-phase motors?
14 Name the best protection method for motors operating in high-humidity environments.
15 What IP rating do most TEFC motors afford?
16 What effect does interchanging the connections of any two of the three power conductors on a three-phase motor have?
17 What is a major cause of single-phase motor malfunction?
18 To where do thermal cut-out switches connect in single-phase motors?

8.8 Synchronous motors

Another type of alternating current motor is a synchronous motor. These motors are not induction motors. Synchronous motors have high operating efficiency with smooth constant starting, accelerating torque, constant speed and versatile power factor control. Large-kilowatt synchronous motors are the drive motors for ship propulsion, refinery pumps and other substantial drive applications.

The stator winding of a three-phase synchronous motor is identical to that of a three-phase polyphase motor utilising a distributed winding. The stator, as shown in **Figure 8.83** is referred to as the armature. The armature when energised by a three-phase supply creates a rotating electromagnetic field in exactly the same manner as that of a polyphase motor.

The synchronous speed developed by three-phase synchronous machines occurs because of the voltage frequency applied across the armature winding and the number of poles. Furthermore, the synchronous speed is equal to the speed of the rotating armature flux. It is, therefore, constant for a given frequency.

$$N_s = \frac{120f}{P}$$

where N_s = synchronous speed in rev/minute (rpm)
f = frequency in hertz (Hz)
P = number of poles

FIGURE 8.83 Synchronous motor armature

Types of synchronous motor

Synchronous motors have variations in their rotor design such as reluctance rotors (salient and smooth), hysteresis rotors and permanent magnet rotors.

Reluctance synchronous motor

The reluctance (magnetic resistance) synchronous motor is like a polyphase motor in that the stators are the same. However, the rotor is different in that it has electromagnets or salient magnetic poles. Its salient construction creates a preferred magnetic path that allows the rotor to align itself with the rotating electromagnetic field of the armature. These rotors have the same number of poles as there are stator poles, thereby allowing them to be suitable for a wide range of speed applications.

Synchronous motors designed for high-inertia, low-speed loads have many salient poles projecting radially outwards from the rotor shaft. Low-speed synchronous motors have eight or more salient poles. Refer to **Figure 8.84**.

FIGURE 8.84 Salient pole construction showing the damping and pole winding

Rotor field coils and spider

The rotor field coils are coil formed or edge-bent strap-wound on laminated pole cores that dovetail to a cast-steel spider mounted on the rotor shaft. The rotor spider consists of steel laminations machined to shrink-mate with the shaft to form the rotor core assembly as shown in **Figure 8.85**.

FIGURE 8.85 Rotor core assembly

Dovetail slots milled into the metal spider accept the dovetailed pole pieces. However, for low-speed synchronous motors, a fabricated spider with bolted-on poles is used.

The solidly braced field coils are staggered around the spider to allow for efficient heat transfer. These coils connect to an external direct current supply by slip rings and carbon brushes and when energised form the electromagnetic poles of the rotor. Salient pole motors are the large drive motors for ship propulsion, cement mills and oilrig pumping.

Damper winding

The damper winding, as shown in **Figure 8.86** has a construction of round copper or copper alloy bars set into slots in the salient pole cores above the salient field pole coils. A shaped copper connection bar between adjacent poles short circuits all the damper winding bars. The connection bar alleviates thermal stresses that occur while starting.

FIGURE 8.86 Damper winding showing the connection bar

The damper winding helps to repress hunting (mechanical oscillation – where the motor oscillates about the power angle) caused by torque oscillations when the motor is driving certain types of loads. Current produced in these windings during the oscillation period provides a counter-torque that dampens the effect of hunting. No voltage generates in the damper winding during synchronous operation.

On starting reluctance synchronous motors, a three-phase supply connects to the armature to produce the necessary rotating electromagnetic field. When a d.c. excitation voltage is supplied across the rotor salient poles, these poles develop their own electromagnetic flux. These poles are attracted by the armature's rotating electromagnetic field poles and lock in synchronism with this rotating electromagnetic field once they are driven up to near synchronous speed. Other forms of rotor excitation include brushless excitation and static frequency changers called cyclo-converters.

As the salient poles of the rotor have a fixed polarity, and the armature polarity alternates sinusoidally as it rotates, the net effect is the maintaining of the rotor at standstill. Unlike a three-phase induction motor, a three-phase reluctance synchronous motor develops no torque and is not self-starting. Therefore, the rotor must match the same speed as the armatures rotating electromagnetic field.

Self-starting

For some synchronous motors, self-starting is achieved by using the inductive effect of damper windings (starting windings, pole-face windings, amortisseur windings) that enables the rotor to emulate the function of a squirrel-cage rotor (obtain accelerating torque from eddy currents) and accelerate to near synchronous speed. When this method of self-starting occurs, application of the d.c. excitation voltage takes place when the rotor reaches about 90% of its synchronous speed. Both the low- and high-speed reluctance synchronous motors deliver a constant speed from no load to full load with no slip.

If the salient poles energise before the rotor reaches its maximum speed, the rotor may not synchronise, and excessive vibration causing pole slipping occurs. Large motors start using a pony motor (auxiliary motor) or a cyclo-converter in order to overcome the rotor's inertia and to bring the rotor up to required speed. **Figure 8.87** shows a salient pole rotor with integral exciter rotor.

FIGURE 8.87 Salient pole rotor with exciter rotor

With the cyclo-converter method of starting, the armature winding energises with a variable frequency supply. Variable frequency allows the armature field to rotate slowly so that the rotor poles can follow. As the supply frequency increases to the desired load frequency, the rotor increases in speed.

The variable frequency exciter consists of a three-phase alternator and rectifier unit mounted on one end of the rotor shaft. The direct current output from the rectifier feeds directly into the salient pole windings without going through brushes and slip rings as **Figure 8.87** shows.

Synchronous speed

All synchronous motors have a constant speed (synchronous) that is independent of the load they are driving. However, they do not tolerate an increasing load that becomes greater than the starting power required between the rotor and the rotating electromagnetic field of the stator. When the load exceeds this starting power, the synchronism between the fields established by the rotor and armature fails, and the rotor stops turning.

The rotor of a reluctance synchronous motor can simply be a squirrel-cage rotor with portions milled or ground out. The portions milled out are asymmetrical in order to provide better locked-rotor torque. The sections on the rotor not milled out become the poles. Because the direct current rotor flux does not vary as it does with alternating current flux, the rotor can be of a substantial construction or made using 2 mm thick non-insulated, high-grade silicon steel lamination stampings.

If the rotor has lamination stampings, magnetic flux barrier slots that help to align the flux are provided. Barrier slots, as shown in **Figure 8.88**, enable the rotor to establish increased synchronising torque.

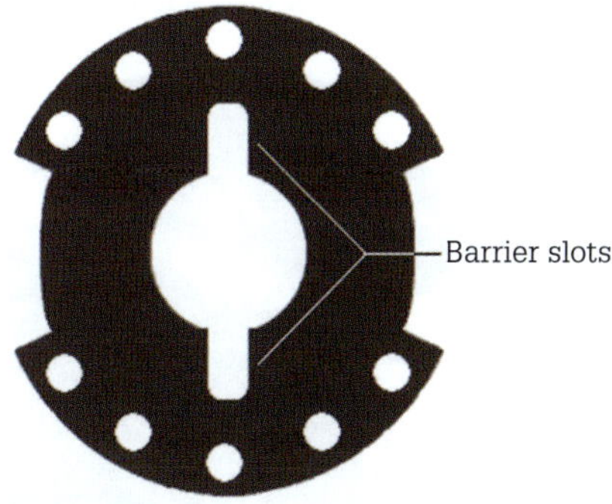

FIGURE 8.88 Two-pole reluctance rotor lamination with barrier slots

Application

Reluctance synchronous motors have applications where several motors must operate in synchronisation for coordinating-type machines and where a variable-frequency drive requires a full range of speed control. The types of applications include synchronised conveyors, wrapping and folding machines, and as proportioning drive motors on pumps.

High-speed (two- to eight-pole) round or cylindrical rotors have a high-strength alloy steel forging to withstand the centrifugal stresses created. These types of rotors have low starting torque characteristics, suitable for pump and fan applications.

Hysteresis synchronous motor

The hysteresis synchronous motor consists of an induction-type stator (armature) and a smooth rotor without windings or slots as shown in **Figure 8.89**. The rotor is a very hard, highly permeable permanent-magnet alloy material carried on a supporting non-magnetic (brass) arbor.

The rotor design eliminates the pole saliencies required by the reluctance motors. Due to the uniform rotor structure, synchronism can occur at any random angular position for the supply voltage (any point of the 360° voltage waveform and any random angular position of the rotor).

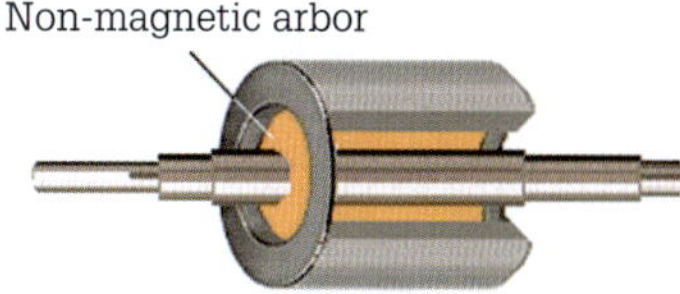

FIGURE 8.89 Cutaway view of a hysteresis rotor showing a non-magnetic arbor

As the armature electromagnetic poles rotate, the induced magnetic poles in the rotor material continually recreate in new positions following the rotating electromagnetic field of the armature.

However, because of the effect of hysteresis, the induced rotor poles always lag behind the armature poles by a slight angle. This unvarying lag angle results in a fixed force of attraction and a constant accelerating torque brings the rotor to synchronous speed.

Torque production

At synchronous speed, the hysteresis torque is zero, and the rotor becomes magnetised along some random axis. The motor runs as a permanent-magnet synchronous motor. As the rotor does not have pole saliencies, the same rotor type occurs in single-speed or multispeed motors.

The smooth rotor construction provides a constant magnetic path for the electromagnetic flux and communicates very low electromagnetic noise and vibration. The torque produced by hysteresis synchronous motors is uniform throughout every revolution, thereby eliminating the effects of cogging.

Figure 8.90 shows the torque–speed characteristic of a hysteresis motor. The blue line represents the hysteresis operation of the motor while the red line represents the eddy current or magnetising torque range of the motor while running at synchronous speed. Magnetising torque only occurs at synchronous speed while hysteresis torque only develops during the period when the shaft speed is increasing to synchronous speed.

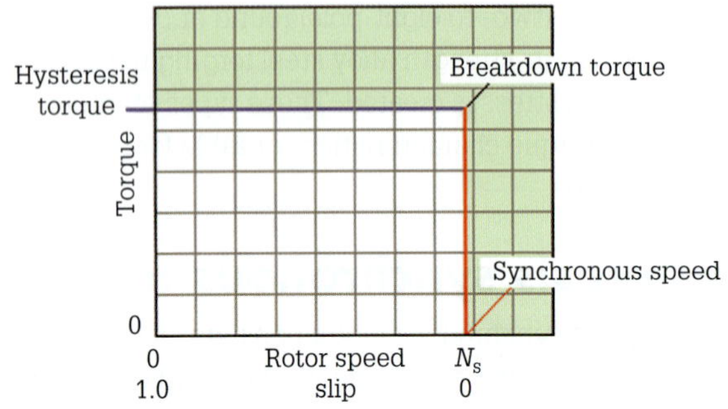

FIGURE 8.90 Torque–speed characteristic of a hysteresis motor

Application

These motors demonstrate great strength by pulling into synchronism high-inertia loads that would require a reluctance-type synchronous motor several times its kilowatt rating. However, their starting current is limited to approximately 150% of rated full-load current because of the high resistance and reactance of the rotor. Hysteresis synchronous motors have an output rating of less than 1 kW.

FIGURE 8.91 Hysteresis synchronous motor

Hysteresis synchronous motors like that shown in **Figure 8.91** find application in appliances such as rotating grills and electric ovens, as timing devices and valve control devices, and in instruments such as scanners and chart recorders.

Permanent-magnet synchronous motor

The rotor in permanent-magnet synchronous motors (PMSM) is a permanent magnet as **Figure 8.92** shows, requiring no external direct current excitation source. The magnetic circuit determines the rotor excitation. The armatures of these motors are a conventional three-phase distributed stator winding.

Torque production

The total torque developed by these motors varies according to the machine parameters such as the saliency ratio (defines reluctance torque) and the magnet thickness, which represents the magnetic flux and, therefore, determines the magnetic torque.

Permanent-magnet synchronous motors use permanent magnets bonded on the periphery of the rotor core or buried inside the rotor core together with flux barriers.

A special bonding technique ensures secure fixing of the magnets during operation with high centrifugal and accelerating forces.

FIGURE 8.92 Bonded PMSM rotor

Setting rare-earth magnets inside the rotor core, rather than bonded on the rotor surface, improves mechanical strength and magnetic protection. By positioning the magnets and the flux barriers appropriately, this motor develops both magnetic and reluctance torque. A higher generated torque occurs than is possible with a bonded PMSM. The rotor uses a geometrical arrangement of flux barriers to prevent de-magnetisation of the permanent magnets. De-magnetisation is possible during synchronisation or at pull-out torque. The flux barriers also prevent loss of magnetic strength when the rotor exits from the armature assembly.

As there is no field or rotor current, the rotor is not subject to energy losses in principle. No energy loss significantly improves efficiency given the ratio of input power to output power.

These motors as illustrated in **Figure 8.93** operate at approximate unity power factor at full load.

FIGURE 8.93 Permanent-magnet synchronous motor

A prototype permanent-magnet synchronous motor has an additional circumferential field winding in the middle of the armature winding to provide for a variable d.c. excitation. This will enable variation of the flux per pole by inducing another iron magnetic pole on the rotor without de-magnetising the permanent magnets.

Application

A type of synchronous motor is the consequent pole permanent-magnet motor (CPPM) shown in **Figure 8.94**. Applications for these motors are in hoisting systems, conveyor systems and printing machines and as the drive motor for pumps. Some electric vehicles use the rare-earth (neodymium iron boron) permanent-magnet synchronous motor (50–70 kW) as their traction motor.

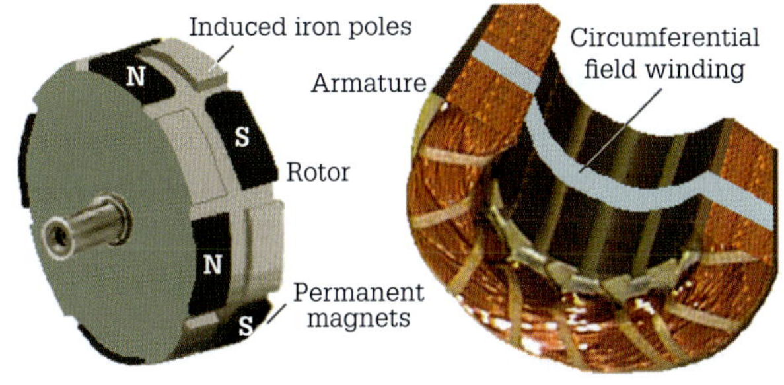

FIGURE 8.94 Consequent pole permanent-magnet motor

Operating characteristics of a synchronous motor

A synchronous motor operates at the same average speed for all values of load from no load to its pull-out load value. When a synchronous motor runs at no load, the rotor poles are directly opposite the armature poles and their axes coincide. However, when the load is increased the rotor magnetic field changes its angular displacement, with the armature's rotating magnetic field lagging behind by several electrical degrees. This displacement angle (symbol δ) is the power angle, load angle or torque angle as **Figure 8.95** shows.

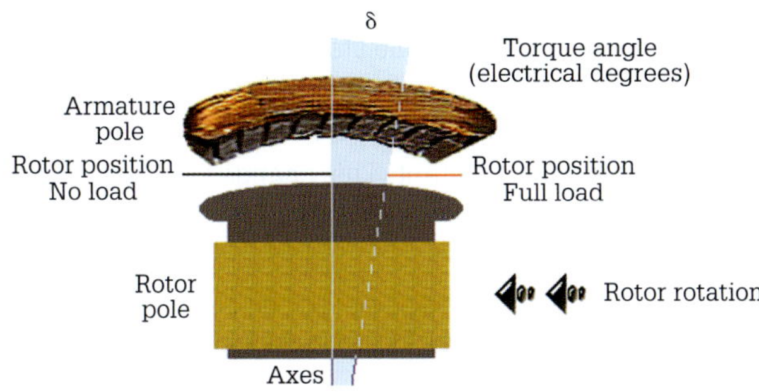

FIGURE 8.95 Power angle

The alternative name for δ, torque angle, is because the torque developed is proportional to the sine of this angle.

V-curves

A set of curves (constant-load curves) for a synchronous motor generated by varying the armature current with the excitation voltage, or armature current with the field current at different load conditions, are V-curves. The V-curves of **Figure 8.96** represent the effect of different values of field excitation on armature current and power factor for various loads.

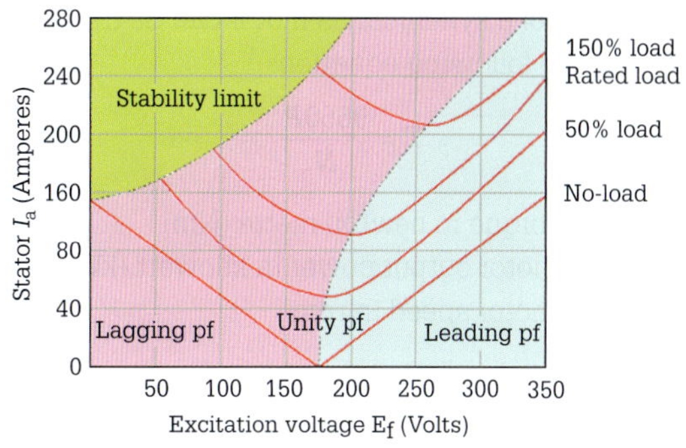

FIGURE 8.96 V-curves for a synchronous motor

The no-load curve illustrates the large reactive power that can be absorbed or delivered by simply changing the excitation. The minimum point of the V-curves represents typical excitation or unity power factor. The left-hand side from this point represents the under-excited condition producing a lagging power factor. The right-hand side represents the over-excited condition, which produces a leading power factor.

The line joining the focal points of the V-curves represents the stability limit of the motor. Any reduction in the field excitation below this limit for a particular load prevents the rotor from maintaining its steady-state power angle. Hunting can also be present for short periods after sudden variations in load. Moreover, if the reduction in field excitation is too severe, or the load is excessive, the rotor will pull out of synchronisation. As the load increases, the V-curve shifts upward and to the right. Note that in order to maintain unity power factor with increasing shaft load, the field excitation voltage must increase.

Torque

Starting torque is the turning force delivered by a rotor at the instant it energises. The starting torque is often higher than rated running or full-load torque. Pull-in torque is the maximum constant torque that enables the rotor to lock into synchronism at rated voltage and frequency. The pull-out torque is the critical value of torque whereby the rotor pulls out of synchronisation (breakdown) with the armature's rotating electromagnetic field. Synchronous motors should be able to contain the pull-out torque for at least 60 seconds before breakdown. The kilowatt rating and speed establish the breakdown torque of a synchronous motor. **Figure 8.97** shows a torque–speed graph of the driving squirrel-cage motor and a synchronous motor.

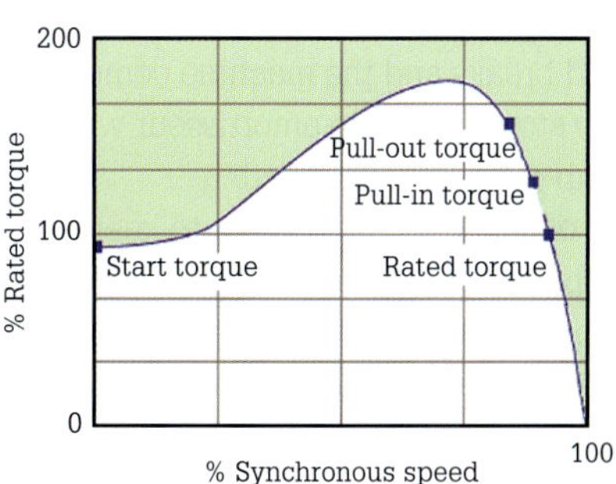

FIGURE 8.97 Torque–speed graph

The torque of a synchronous motor can be determined by applying the following equation:

$$T = \frac{9550P}{N}$$

where T = torque in newton metre (Nm)
P = motor output power in kilowatt (kW)
N = motor speed in rpm

Note that 9550 is derived from $\frac{60}{2\pi} \times 1000$

Since the kilowatt output of a synchronous motor is related to torque and speed, all synchronous motors must provide rated torque at full load. However, an individual motor may draw more current than its nameplate rating when it is producing rated torque, but it should not exceed its rated temperature rise.

EXAMPLE 8.9

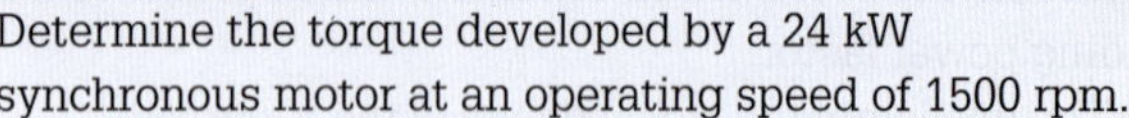

Determine the torque developed by a 24 kW synchronous motor at an operating speed of 1500 rpm.

$$T = \frac{9550P}{N}$$
$$= \frac{9550 \times 24}{1500}$$
$$= 152.8 \textbf{ Nm}$$

EXERCISE 8.9

Determine the torque developed by a 16 kW synchronous motor at an operating speed of 1000 rpm.

At no load, the torque angle (δ) is zero degrees (electrical) in theory. However, in practice, there exists some small angle, as torque is required to overcome friction and windage. The torque angle increases as the physical load on the shaft increases, and more torque is demanded to drive it. Consequently, torque is proportional to the sine of the angle. However, note that the angle changes in response to changes in load.

The theoretical limit for torque increases occurs when δ = 90°(electrical). Any further increase in load increases δ beyond 90°, which means torque decreases (sin 91° = 0.9998). The consequence of a further increase in load is that the rotor slows down and drops out of synchronisation. Consequently, the magnetic linkage between stator field and rotor field breaks and the machine comes to a sudden and dramatic stop unless the amortisseur winding can keep it going at sub-synchronous speed.

The goal of field excitation is not to keep the two fields at 90° to each other, as shown in **Figure 8.98** but rather to provide enough magnetic linkage so that the expected maximum torque develops at a torque angle of somewhat less than 90°. At the same time, it must not make the system so stiff as to endanger the shaft and the drive train. (As previously mentioned, 90° is the theoretical limit; in practice, you keep well below that.)

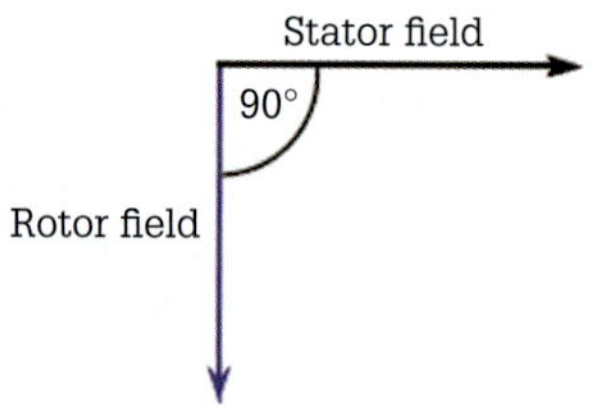

FIGURE 8.98 Maximum torque angle between stator and rotor fields

Power factor correction

Figure 8.99 shows the consequence of modifications in field excitation on power angle, power factor and armature current of a synchronous motor operating with a stable load with a consistent supply voltage and frequency.

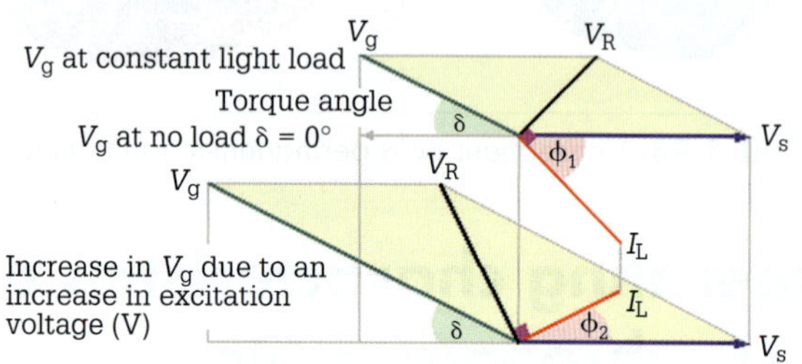

FIGURE 8.99 Phasor diagram of unvarying field excitation

The rotor field flux cuts the stator windings and induces a voltage (V_g) in them. Because the speed is constant, the value V_g is governed by the rotor field strength, rotor field current and ultimately rotor excitation voltage (V). The angle that V_g creates with the supply voltage (V_S) is a function of the torque angle (δ). If the angle δ is 0° then the angle of V_g is 180° – 0° when referred to V_S.

The phasor sum of V_g and V_S is V_R, and it is this voltage that controls the current in the stator windings (I_L lags VR by 90°). By altering rotor field excitation (V), control of the amplitude of V_g changes angle δ minimally, if at all, provided the load is constant. This changes the amplitude of V_R and its angle to V_S and hence both the line current and the phase angle ϕ.

Reversing synchronous motor direction of rotation

The direction of rotation of a synchronous motor occurs by its starting direction as established by the armature's rotating electromagnetic field. To reverse the direction of rotation, only the phase sequence of the three-phase supply connected to the armature terminals needs changing. Note that depending on the starting method; for example pony motor, which may also need reversing.

Synchronous capacitor

When synchronous motors become synchronous capacitors for power factor control, they lack an external rotor shaft.

These motors operate at a leading power factor in order to compensate for the lagging power factor of commercial or industrial power systems. By placing VAR onto the power systems, as required, synchronous capacitors support a system's voltage or sustain the system power factor at a definite value. Refer to **Figure 8.100**.

A power system develops voltage variability when a disruption causes an emerging and uncontrollable decline in voltage. The main factor that causes this form of unsteadiness is the lack of ability of the supply system to meet the demand for reactive power. Synchronous capacitors rated in MVAR connect in parallel with the supply system. The capacitor produces no switching transients and is unaffected by electrical harmonics.

FIGURE 8.100 Synchronous capacitor

Synchronous capacitors do not produce excessive voltage levels and are not subject to electrical resonances. A synchronous capacitor provides step-less (no voltage transients occur) automatic power factor correction. Because of its rotating inertia, the synchronous capacitor can provide voltage support even during a short power outage.

To enable the motor to act as a capacitor, adjustment of the d.c. current supplied to the field winding is needed. Consequently, the motor overexcites and the armature draws a leading current from the three-phase supply to restore the power factor. Over-excitation enables the synchronous capacitor to adjust to slight changes in the load, and it can overload for short periods. However, because running the motor in an overexcited state requires a high field current and flux, the motor may be lightly loaded to take care of the temperature rise caused by rotor heating.

When reactive demands require an immediate response, a synchronous capacitor is used. A synchronous capacitor is capable of supplying VAR equal to its rating to the system instantly as well as absorbing up to 50% of its rating. It also can address supply quality issues such as voltage regulation and flicker.

Nameplate details

Figure 8.101 shows a typical synchronous motor nameplate.

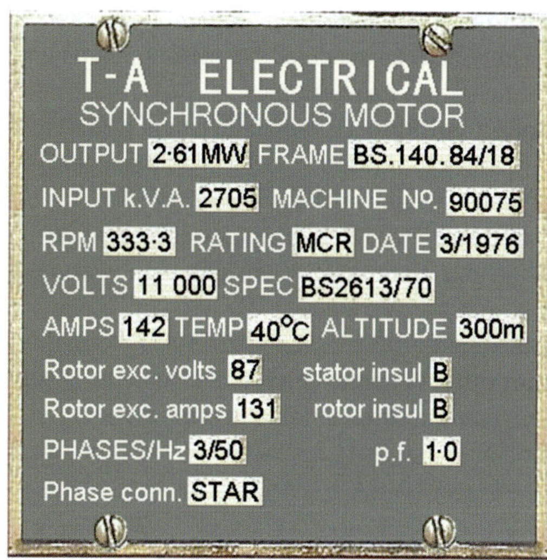

FIGURE 8.101 Synchronous motor nameplate

The nameplate in **Figure 8.101** has the following designations.

- **Output:** Shaft power (2.61 MW) is a measure of the synchronous motor's mechanical output rating and is its ability to deliver the torque required for the load at rated speed. Its designation is watts (W), kW or MW on the nameplate.
- **Frame size:** Generally, the frame size gets larger with increasing output power rating or decreasing speeds. World standards associations prescribe standard frame sizes for particular power output, speed and enclosure combinations. The frame size provides mounting dimensions such as the shaft diameter, its height above mounting level and the hole outline.
- **Full-load speed:** Full-load torque occurs at the rated power output at this speed. It appears as 'RPM' on the nameplate and is the speed appropriate to the number of poles.
- **Rating:** This defines the length of time during which the motor can carry its power output rating safely. Frequently, this is MCR (motor continuously rated).
- **Insulation class:** This indicates the resistance of the insulating components from the effects of heat. The four major classifications, in order of increasing thermal capability, are A, B, F and H (Class A, 105 °C; B, 130 °C; F, 155 °C; H, 180 °C).
- **Temperature:** This indicates the maximum continuous operating temperature of the motor at full load.
- **Full-load amps:** This is the current the motor draws under full load (142 A).
- **Frequency:** Input frequency is usually 50 Hz. Changes in frequency affect motor speed.
- **Phase:** This represents the number of alternating current power lines supplying the motor.
- **Power factor:** This is the power factor for the motor at full load.

- **Altitude:** This shows the maximum height above sea level at which the synchronous motor remains within its temperature rise design.
- **Applied voltage:** The applied voltage (11 kV) allows knowledge of torque ($T \propto V^2$). The torque changes with variations in supply voltage.
- **Rotor data:** This information states rotor applied d.c. voltage and d.c. excitation current. Input watts = 87 V × 131 A = 11 397 W.
- **Input:** Input = $\sqrt{3} \times V \times I \times$ pf.
- **Other data:** A nameplate also includes the motor's brand name, a serial number and the type of connection unique to the motor.

Efficiency

The efficiency of a synchronous motor is determined at rated output voltage, frequency and power factor. The following energy losses are included in the efficiency determination:

- I^2R loss of the armature
- I^2R loss of the field
- core loss
- stray-load loss
- friction and windage loss
- exciter loss.

Figure 8.102 summarises these losses and shows the power transfer in a synchronous motor.

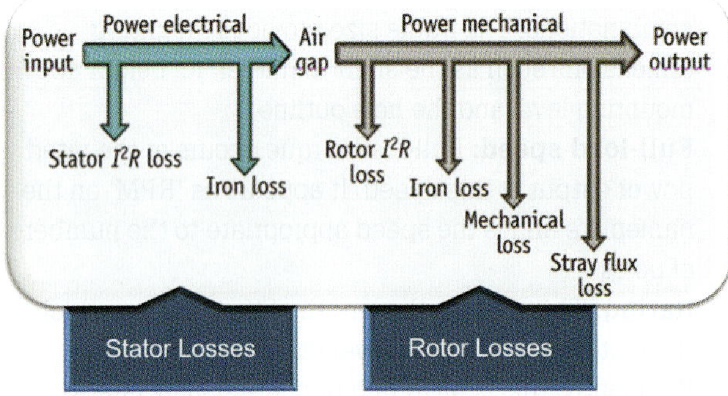

FIGURE 8.102 Synchronous motor power transfer diagram

Standard production tests for synchronous motors determine the efficiency values at 50%, 75% and 100% of full load.

If the nameplate provides only the line voltage, current and power factor, calculation of the input power is determined using the following equation.

$$P_{in} = \sqrt{3}\ V_L I_L pf$$

where P_{in} = input power in watt (W)
V_L = line voltage in volt (V)
I_L = line current in ampere (A)
pf = operating power factor

EXAMPLE 8.10

Using the voltage, current and power factor values of **Figure 8.101**, calculate the input power if the motor is operating at unity power factor.

$$P_{in} = \sqrt{3}\ V_L I_L\ pf$$
$$= \sqrt{3} \times 11\,000 \times 142 \times 1$$
$$= 2705\ \textbf{kW}$$

EXERCISE 8.10

A synchronous motor has the following details:

- Line voltage = 400 V
- Line current = 120 A
- pf = 1.0
- Power output = 87.5 kW

Calculate the input power if the motor is operating at unity power factor.

EXAMPLE 8.11

Using the information from the nameplate of **Figure 8.101**, the efficiency of the synchronous motor can be determined by:

$$\eta = \frac{P_{out}}{P_{in}}$$
$$= \frac{2610}{2705}$$
$$= 0.965 \textbf{ or } 96.5\%$$

Note: 2705 kVA × pf is power; since pf = 1 then this number is the true power input.

EXERCISE 8.11

A synchronous motor has the following details:

- Line voltage = 400 V
- Line current = 120 A
- pf = 1.0
- Power output = 87.5 kW

Calculate the efficiency of the synchronous motor.

Calculation of the line current drawn per phase by a synchronous machine uses the following equation.

$$I_L = \frac{S}{\sqrt{3}V_L}$$

where I_L = line current in ampere (A)
S = input apparent power in volt-ampere (VA)
V_L = line voltage in volt (V)

EXAMPLE 8.12

Determine the line current per phase of a three-phase 10 kVA 400 V four-pole 50 Hz star-connected synchronous motor operating at unity power factor.

$$I_L = \frac{S}{\sqrt{3}V_L}$$

$$= \frac{10\ 000}{\sqrt{3} \times 400}$$

$$= 14.4 \text{ A}$$

EXERCISE 8.12

Determine the line current per phase of a three-phase 12 kVA 400 V four-pole 50 Hz star-connected synchronous motor operating at unity power factor.

Braking

Due to the inertia of the rotor and its connected load, large synchronous motors may take several minutes to come to a complete standstill. Various braking methods reduce this coasting time.

- Maintain full d.c. excitation, with the armature leads short circuited.
- Maintain full d.c. excitation, with the armature connected to three load resistors.
- Apply mechanical braking.
- With the first two methods, the synchronous motor slows down because it functions as an alternator, dissipating its energy in the ohmic elements of the armature winding and the braking resistors.

REVIEW QUESTIONS

1. Outline the characteristics of synchronous motors.
2. What is another name for the stator in synchronous motors?
3. Name three styles of synchronous motor rotors.
4. What type of pole is associated with high-inertia, low-speed synchronous motors?
5. State the purpose of a damper winding in a synchronous motor.
6. What is a consequence of the salient poles in a synchronous motor energising before the rotor reaches its maximum speed?
7. Name two methods employed to start large synchronous motors.
8. What does the term 'hunting' mean?
9. How can a rotor of a reluctance synchronous motor be designed to establish increased synchronising torque?
10. Name two applications for hysteresis motors.
11. Describe the construction of the rotor of a permanent-magnet synchronous motor.
12. What happens to the rotor of a synchronous motor if the load increases beyond the motor's capability?
13. What information does a set of V-curves provide?
14. Name the critical value of torque whereby the rotor pulls out of synchronisation (breakdown) with the armature's rotating electromagnetic field.
15. Determine the torque developed by a 20 kW synchronous motor at an operating speed of 1500 rpm.
16. How can synchronous motors reverse?
17. Do synchronous capacitors operate with leading or lagging power factor?
18. Why is braking necessary with some synchronous motors?
19. A synchronous motor has the following details: Line voltage = 400 V; Line current = 80 A; pf = 1.0; Power output = 57.5 kW. Calculate the input power if the motor is operating at unity power factor.
20. A synchronous motor has the following details: Line voltage = 400 V; Line current = 80 A; pf = 1.0; Power output = 57.5 kW. Calculate the efficiency if the motor is operating at unity power factor.

8.9 Alternators and generators

The synchronous alternator is the most important means of producing electric energy. Another term for these machines is synchronous generators. They range in size from a fraction of a kVA to substantial MVA. The nominal line output voltage of an alternator depends upon its kVA rating (the alternator windings are limited by the current through them) – the larger the rating, the higher the voltage. **Figure 8.103** shows a Caterpillar 200 kW 250 kVA alternator.

FIGURE 8.103 Caterpillar 200 kW 250 kVA alternator

Most of the electrical energy in the world is produced by high-speed prime movers such as steam turbines driving cylindrical rotors and low-speed hydraulic turbines driving salient pole rotors. Internal combustion engines, natural or LP gas, electric motors and wind or waves can also drive alternators. Various load types require driven alternators. They find applications in ships, locomotives, as standby alternators for main alternators and portable power applications. The construction of synchronous alternators is fundamentally the same as for the synchronous motors described in the previous section (8.8). However, the speed at which the alternator must operate determines their construction – salient pole rotor or cylindrical rotor. Cylindrical rotors, for example are small in diameter but long in length and are usually made of solid steel forging with the rotor coil slots milled into the surface of the cylindrical mass as shown in **Figure 8.104**.

FIGURE 8.104 Milled cylindrical rotor

Furthermore, concentric field coils are wedged into the slots and retained in place by high-strength end rings. The high speed of these rotors produces very strong centrifugal forces which restrict the diameter of the rotor to between 1.2 m to 1.8 m. These rotors can be up to 7.5 m in length and can have a total mass of up to 204 tonnes.

Alternator principles

Alternators may have a stationary field and a rotating armature (stator). However, it is more practical to build them with a rotating field and a stationary armature. The rotating field occurs by exciting windings on the rotor with d.c. power supplied in the same manner as with synchronous motors.

Power or high-voltage alternators are electrically excited. A direct current feeds these coils through graphite brushes on the slip rings and the total arrangement is rotated by means of the prime mover creating a rotating magnetic field. The induced electromotive force develops in the stationary armature.

The advantage of having a stationary armature is ease of design – the rotor has a low-voltage d.c. winding which is easier to manage both electrically (insulation) and mechanically (centrifugal forces). Only two slip rings are required for the d.c. feed because the a.c. output is taken from the stationary armature.

The use of electromagnets enables the adjustment of the generated output. The current in the electromagnetic coils can be increased or decreased, increasing or decreasing the electromagnetic flux. Having control over the electromagnetic flux means that a variable electromotive force is available at the output terminals.

When the stationary armature or stator windings contain three groups of coil windings aligned at 120 electrical degrees to each other, the generator produces three single-phase alternating current outputs as shown in **Figure 8.105**.

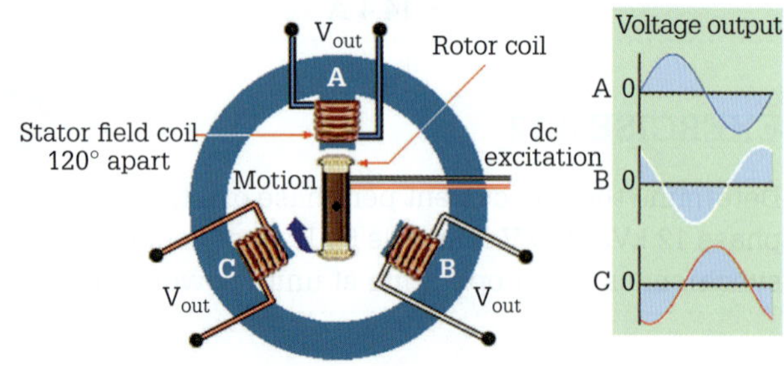

FIGURE 8.105 Three-phase alternator

Note: In this figure the mechanical degrees and the 120 electrical degrees coincide.

Figure 8.106 shows a representation of a three-phase alternator with a rotating armature and stationary fields.

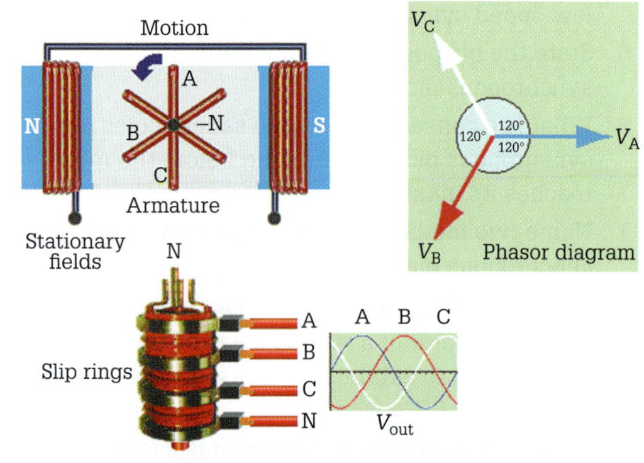

FIGURE 8.106 Three-phase generation

The generation of multiphase voltages occurs in the same manner as that for single-phase voltages. When the rotor turns by a prime mover at a constant speed, there will be a sinusoidal voltage induced across each of the stator field coils. These induced voltages have the same magnitudes if the stator coils have the same number of conductor turns. In addition, because the stator coils mechanically fix at the same angle to each other, the induced sinusoidal voltages have the same frequency. The only factor that is different between the three induced voltages is their phase relationship. They locate from each other by 120°. As shown in **Figure 8.107**, each induced voltage reaches its peak value at a different instant in time.

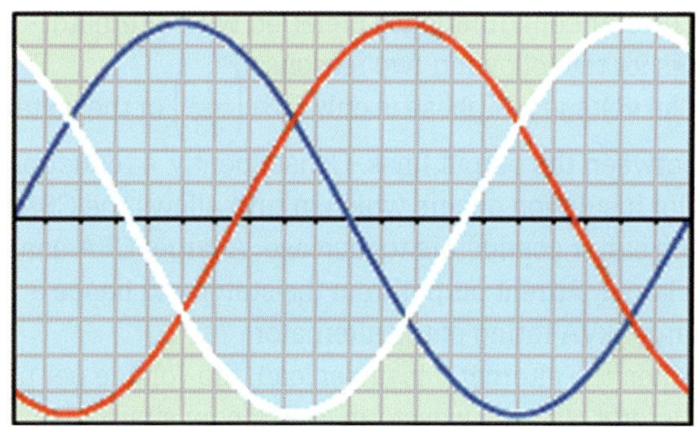

FIGURE 8.107 Three-phase waveform

Electrical power generation and distribution use three-phase synchronous generators. A typical alternator output for power stations would be 500 MW at 24 kV, three-phase 50 Hz.

The alternator stator

The alternators of both the salient pole and the cylindrical type have the armature windings in the stator and the field windings on the rotor. The armature build contains high quality, low-loss silicon steel laminations insulated on both sides and stacked under hydraulic pressure. The stator slots skew to reduce slot harmonics in the induced voltage, and ventilating ducts subdivide the lamination stack into discrete packets, thereby ensuring efficient cooling. **Figure 8.108** shows the stator of a synchronous alternator.

FIGURE 8.108 Synchronous alternator stator

The armature windings in some alternators are of a distributed double layer construction and are chorded (two-third pitch windings) to reduce or eliminate third-order harmonic content of the a.c. output waveform.

Moreover, the stator conductors are made of high-hydrolysis-resisting enamelled wire or flat copper straps insulated with mica paper. The insulating system usually satisfies Class F or Class H insulation requirements. The classification means that the insulation materials are non-hygroscopic and non-tracking and can withstand severe thermal stressing.

A two-component resin impregnates the stator, resulting in a winding assembly that has high mechanical strength and vibration resistance together with excellent dielectric strength. However, some stators have an elastic coat that provides the stator winding with resistance against salt, fungus and abrasive contamination.

Stator connections

There are two essential means available by which to connect the three separate phase group windings. These are delta or star.

Delta

The delta connection derives its name from the resemblance of its schematic symbol to the Greek letter delta (symbol Δ). Connecting one end of each phase coil group to the start of another phase coil group, to which the active line conductors connect as shown in **Figure 8.109**, makes the delta connection.

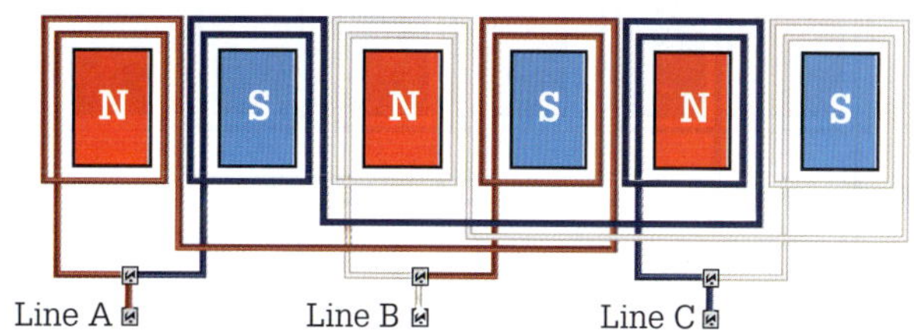

FIGURE 8.109 Three-phase two-pole stator delta connection

Figure 8.110 shows a two-pole delta-connected alternator using a single coil to represent a pole-winding group.

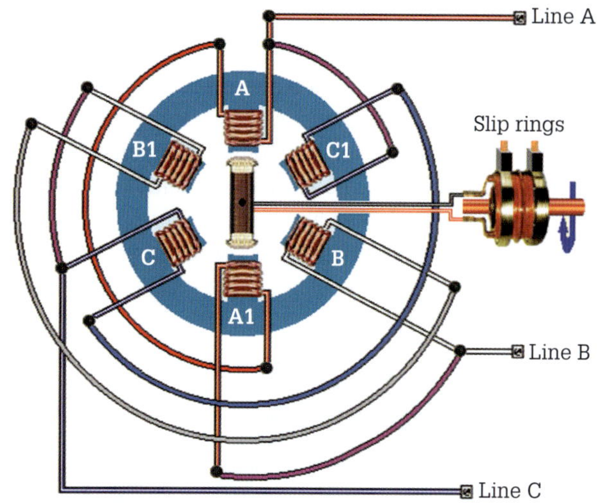

FIGURE 8.110 Two-pole delta connection

In the delta-connected alternator of **Figure 8.110** the voltage across any two of the phases (called line voltage) is the same as the voltage generated in any one phase. However, the current provided in any line is $\sqrt{3}$ the phase current. Consequently, a delta-connected alternator will require insulation capable of withstanding both the generated voltage and the load current.

There are two major disadvantages for an alternator connected in delta configuration. An unbalanced load will make all three line voltages unbalanced, and secondly, only the voltages across each phase (no neutral) are available.

An advantage of the delta-connected alternator compared to the star-connected alternator is that if one phase becomes open circuited the remaining two phases can still deliver three-phase power. However, the capacity of the alternator requires de-rating to 57.7% of its full-load capacity.

Star

The generation and distribution of energy for general purposes occurs using either a three-wire or a four-wire star-connected system. The star connection occurs by connecting one end of each phase coil group together and bringing the other end out as the active line terminal end as shown in **Figure 8.111**.

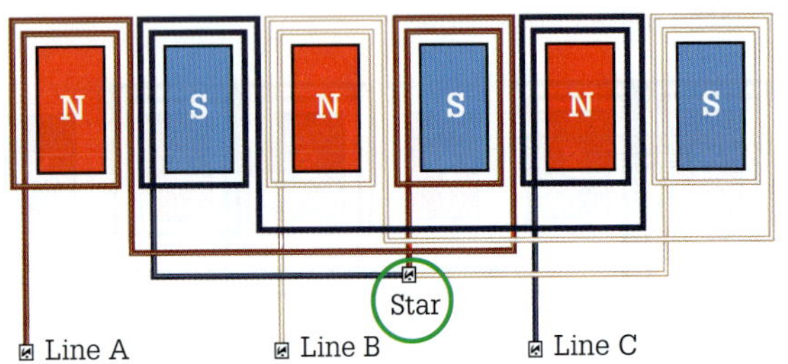

FIGURE 8.111 Three-phase two-pole stator star connections

All induction motors may operate very effectively as alternators by driving them at a speed greater than their motor's synchronous speed. Alternator stators are identical to that of a three-phase induction motor, and most are a three-phase lap winding connected in a star configuration with the star-point or neutral point connected to earth. However, the alternator is not a motor – it does not have to develop any starting torque. Alternators are usually designed with windings of lower ohmic resistance values and higher efficiency at rated load when compared to an induction motor used as an alternator.

If the star connection occurs inside the generator, it is usual for an earth to be connected to the star-point as shown in **Figure 8.112**.

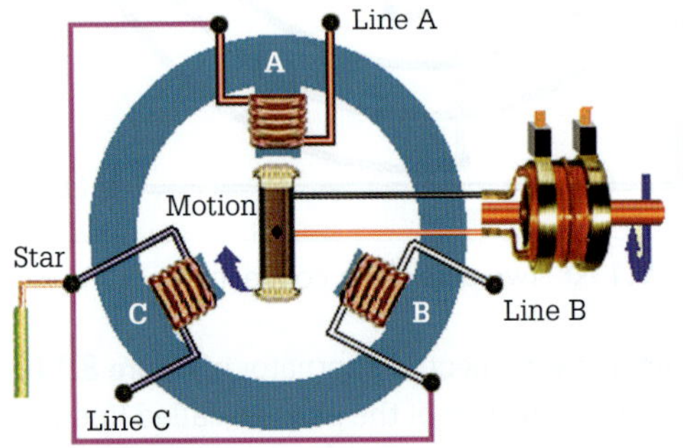

FIGURE 8.112 Star connection

The star configuration is preferred with alternator stators for the following two reasons:

1. The voltage per phase is only 58% $\left(\frac{1}{\sqrt{3}}\right)$ of the voltage between the output lines. Consequently, a reduction in slot insulation occurs which in turn allows the CSA of the stator conductors to increase. A larger CSA enables a larger current output from the stator and hence a larger kVA rating of the alternator.
2. When the alternator is under load, the voltage per phase distorts, and the waveform is no longer sinusoidal. The waveform distortion happens because of the third harmonic voltage. However, with a star connection, the distorting line-to-neutral harmonics do not occur between the lines because they effectively cancel each other out.

Nameplate

The nameplate gives the manufacturer's data, the alternator electrical information, brush reference and serial number. The alternator frame has an arrow marking to indicate the rotation direction, which is usually clockwise when looking at the shaft end of the alternator.

An alternator has a rating according to its frequency, terminal voltage and full-load current output. Some alternators also have a volt-ampere rating (VA, kVA or MVA).

Figure 8.113 shows a 3.58 kVA alternator nameplate.

FIGURE 8.113 Alternator nameplate

Cooling

Most alternators have a centrifugal fan at the driven end, which axially draws cold air in from the rear side and radially throws hot air out at the front side. This ensures optimum ventilation and keeps the temperature of the windings within limits.

However, on some alternators an electric-driven, speed-controlled fan is used to minimise the alternator's fan losses during low-speed operation. Other alternators have a water–heat exchanger with closed-circuit cooling fitted or have their armature and field windings immersed in hydrogen because of their heat-absorption properties.

Alternator frequency

Electric equipment uses electrical energy with a fixed frequency such as 50 hertz (Hz) in Australia. The frequency output of an alternator depends on the fixed speed of a prime mover. To produce a 50 Hz output, most prime movers operate at 1500 or 3000 revolutions per minute. For example, a two-pole alternator has two field poles on the rotor (rotating field construction). Therefore, each time the rotor makes one revolution, one complete voltage cycle develops at the alternator terminals. In order to produce 50 Hz power, the rotor revolves by a prime mover 50 revolutions every second, or 3000 rpm. In order to maintain the frequency at its required value under load conditions, a governor on the prime mover detects any drop in speed and responds by increasing the speed of the prime mover. The speed of the alternator is an important factor affecting the frequency and the voltage output from the alternator. A 10% change in prime-mover speed causes a corresponding 10% change in frequency.

A four-pole alternator has four field poles, so each time the rotor makes one revolution, two complete cycles are produced at the alternator terminals. The rotor would have to turn 1500 rpm to produce 50 Hz power. Each of these alternators has its advantages and disadvantages. For example, 1500 rpm, four-pole sets are the most common and least expensive. They offer the best balance concerning noise, efficiency, and cost and engine life. However, the 3000 rpm, two-pole sets are smaller and lightweight, best suited for portable, light-duty applications, but the higher rpm means that the prime mover may not last.

Excitation

There are various methods of alternator excitation classified as brushed or static, brushless, permanent magnet or self-excited system depending on the method used to transfer the excitation energy to the field winding. The most common type is the brushless exciter.

Self-excitation

A self-excited alternator depends upon the residual magnetism contained within the rotor. When the prime mover drives the alternator, the residual magnetism induces a speed-dependent voltage across the rotor winding thereby building up the rotor's magnetic field. Consequently, when the prime mover accelerates following the start, the alternator rotor must reach a certain speed in order to ensure that self-excitation occurs.

Permanent-magnet excitation

Permanent-magnet alternators have their excitation provided by rare-earth, high-field neodymium iron boron (NdFeB) magnets contained in the rotor.

Brushless excitation

Brushless exciters use a small three-phase alternator mounted on the same shaft as the main exciter rotor. This small alternator has shaft-driven permanent magnets that provide an efficient source of exciter field energy.

A voltage regulator assembly mounted on the exciter's rotor rectifies the three-phase, high-frequency alternating current output. The controlled d.c. voltage output from the regulator energises the stationary field of the main rotating exciter. Refer to **Figure 8.114**.

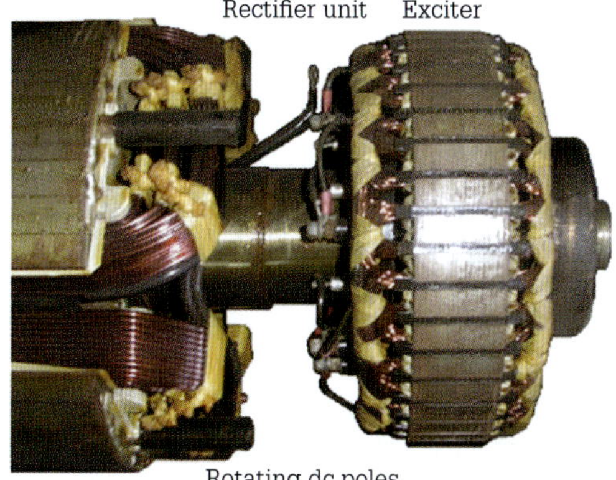

FIGURE 8.114 Brushless exciter

Figure 8.115 shows the block diagram of a voltage regulator on a three-phase alternator.

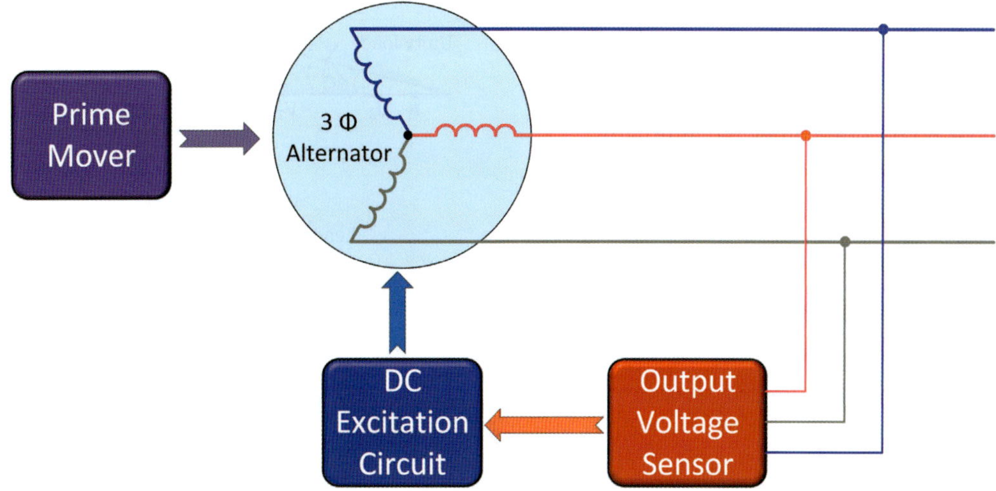

FIGURE 8.115 Voltage regulator block diagram

Static excitation

A static exciter system feeds the alternator's output terminals via an excitation transformer and capacitors connected across each phase. The rotor field windings receive the rectified current through slip rings. Some static exciters interface with a microcomputer controller that monitors the excitation system.

Generated voltage

The electromotive force developed by an alternator is directly proportional to speed, field flux density and the number of stator conductors. From Faraday's law (which defines the relationship of voltage and flux) an equation for the generated voltage across each phase of the armature (stator winding) occurs:

$$V_g = 4.44 K_W \Phi f n$$

where V_g = generated rms voltage per phase
Φ = flux per pole in webers (Wb)
f = frequency in hertz (Hz)
n = number of conductor turns per phase
K_w = winding constant

Since the speed must be stable to maintain a constant frequency and because of the fixed number of conductors, only the field flux density is adjusted to control the output voltage. The field flux density adjusts by varying the d.c. current supplied to the rotating field. Adjustment occurs manually or automatically.

The manual excitation control is operator dependent. In other words, the operator adjusts the voltage output of the alternator based on load demands. However, the automatic excitation control uses electronics to monitor the desired alternator output voltage against an internal reference setting and adjusts the d.c. excitation current required to maintain constant alternator terminal voltage.

Adjustment of the alternator output voltage is necessary because the constant frequency output voltage changes as the energy demand on the alternator varies from no load to full load. The output voltage varies if the frequency (speed) changes because of varying connected loads.

EXAMPLE 8.13

Calculate the rms line voltage produced by a three-phase, four-pole, 50 Hz, star-connected alternator having 220 turns per phase, a winding constant of 0.85 and 150 mWb flux per pole.

$$V_g = 4.44 K_W \Phi f N$$

$$= 4.44 \times 0.85 \times \left(150 \times 10^{-3}\right) \times 50 \times 220$$

$$= \mathbf{6.23\,kV}$$

EXERCISE 8.13

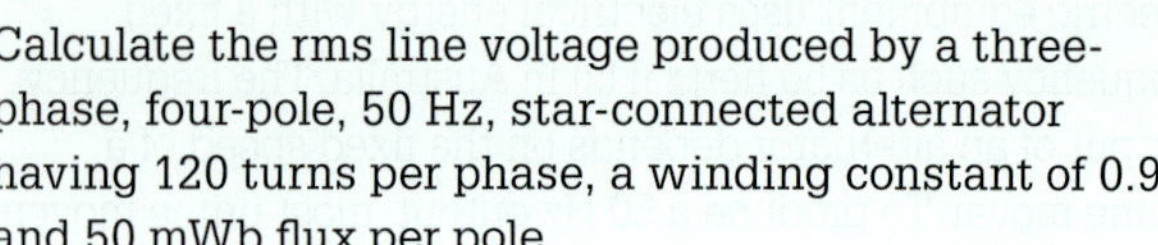

Calculate the rms line voltage produced by a three-phase, four-pole, 50 Hz, star-connected alternator having 120 turns per phase, a winding constant of 0.9 and 50 mWb flux per pole.

Synchronous alternator under load

The behaviour of a synchronous alternator under load varies according to the load's power factor and if the alternator acts alone or in parallel with other alternators.

Figure 8.116 shows the equivalent circuit diagram of one phase of a single alternator driven at a constant speed with the d.c. field current at a constant value under load.

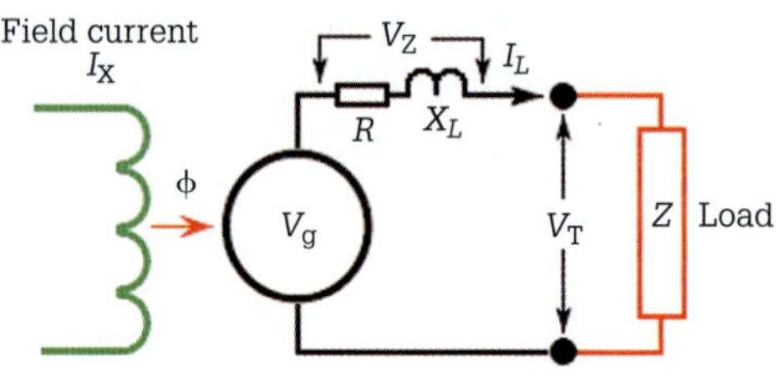

FIGURE 8.116 Equivalent circuits – alternator under load

An alternator is a voltage source (V_g) in series with internal resistance and inductive reactance. If, for simplicity, we assume that R is small compared to X_L then the internal voltage drop V_Z can be considered as being produced at 90° to I load. Further to this, we can say that the terminal voltage (V_T) across the load voltage is the phasor sum of V_g and V_Z.

The generated voltage (V_g) is affected by speed and field strength and because load current affects the field flux distribution (due to armature reaction) it too will affect the voltage (V_g).

However, the effect is different for resistive, inductive and capacitive loads. **Figure 8.117** shows unity, lagging and leading operating characteristics in phasor diagrams of various load conditions.

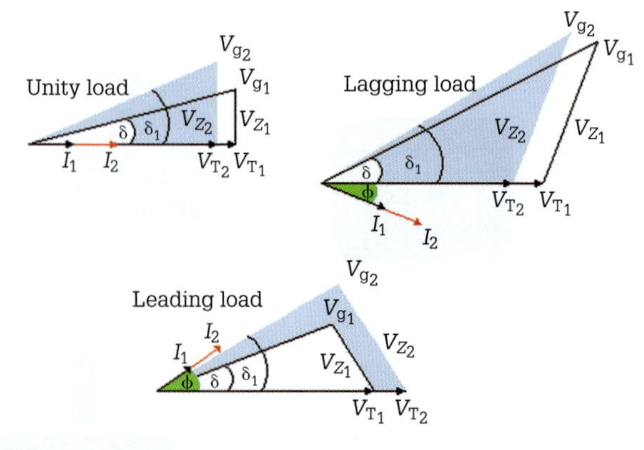

FIGURE 8.117 Effects of alternator loads

Resistive (unity) load

Armature reaction has only a minimal effect.

Now V_T is the phasor sum of V_Z and V_{g_1} (I_{Load} will be in phase with V_T), thus initially I_1 produces V_{Z_1} which added to V_1 produces V_{T_1}. Increasing the load to I_2 produces V_{Z_2} which added to V_{g_2} gives us V_{T_2} (note: $V_{g_1} \approx V_{g_2}$). That is, the terminal voltage decreases as the load current increases.

Inductive (lagging) load

With inductive loads, armature reaction distorts the field flux in such a way that it is effectively reduced so that V_g, at the same load current as before, is also reduced. In an inductive circuit, load current will lag V_T, thus voltages V_{g_1} and V_{g_2} are a little smaller than in the resistive load and $V_{g_2} < V_{g_1}$. That is, the terminal voltage drops dramatically with increases in load current.

Capacitive (leading) load

With capacitive loads, armature reaction distorts the field in such a way that it increases. This results in a larger generated voltage (V_g) for the same load current as compared with the resistive and inductive loads.

In a capacitive circuit the load current will lead the terminal voltage (V_T); thus voltages V_{g_1} and V_{g_2} are larger than their resistive and inductive counterparts and $V_{g_2} > V_{g_1}$.

These voltages, coupled with voltages V_{Z_1} and V_{Z_2} at 90° to the load current, result in a terminal voltage that increases with increased current (however, it does fall as rated full load is approached).

Voltage regulation curves

Voltage regulation curves illustrate how the terminal voltage of an alternator changes as a function of the load current. Plotting of the curves occurs with the d.c. field excitation fixed at a steady value for the three types of power factor loads. **Figure 8.118** shows regulation curves.

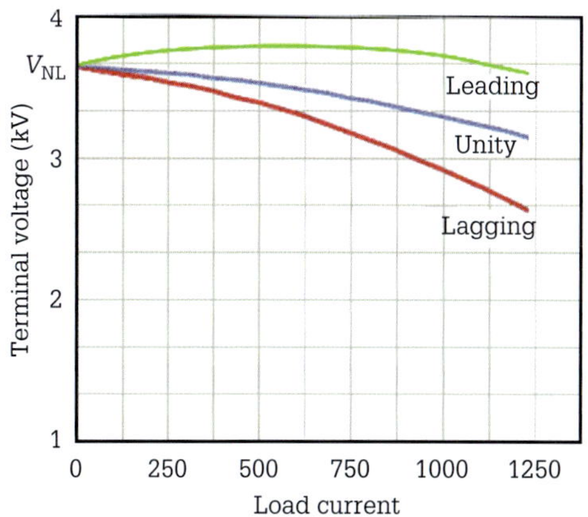

FIGURE 8.118 Voltage regulation curves

The voltage regulation of a synchronous alternator is the percentage change in terminal voltage, from no load to rated load when the generator is delivering its terminal voltage at rated speed.

The following equation determines the percentage voltage regulation:

$$V_{Reg} = \frac{V_{NL} - V_{Rated}}{V_{Rated}}$$

where V_{Reg} = voltage regulation

V_{NL} = no-load or open-circuit voltage (V)

V_{Rated} = rated nameplate or full-load voltage (V)

Note that the result of this equation is a decimal value (such as 0.15). Voltage regulation is often expressed as a percentage value. This is achieved by multiplying the decimal value (the result of the above equation) by 100.

An alternator operating at a lagging power factor has a large positive voltage regulation. However, an alternator operating at unity power factor has a small positive voltage regulation, while an alternator operating at a leading power factor normally has a negative voltage regulation.

EXAMPLE 8.14

A synchronous alternator operating at rated load and unity power factor has a terminal voltage of 458 V. If the no-load voltage is 480 V, determine the voltage regulation.

$$V_{Reg} = \frac{V_{NL} - V_{Rated}}{V_{Rated}}$$

$$= \frac{480 - 458}{458}$$

$$= 0.048 \textbf{ or } 4.8\%$$

EXERCISE 8.14

a A synchronous alternator operating at rated load and unity power factor has a terminal voltage of 448 V. If the no-load voltage is 465 V, determine the voltage regulation.

b A synchronous alternator operating at rated load and leading power factor has a terminal voltage of 478 V. If the no-load voltage is 465 V, determine the voltage regulation.

Single-phase alternators

These alternators are for portable applications using a small petrol or diesel engine as the prime mover. The prime mover drives the alternator at 1500 or 3000 rpm depending upon the alternator design. A governor is also included to maintain correct frequency output. Single-phase alternators use a brushless exciter system.

Most single-phase alternators are the rotating field type. These units are self-exciting and need residual magnetism to get terminal voltage initiated. The voltage developed by the residual magnetism is fed into an exciter winding and then through a diode bridge converting it to direct current. The d.c. voltage supplies the field that rotates in the centre of the armature and induces voltage into the stator windings. The induced 230 V establishes at a frequency of 50 cycles per second if the engine speed is correct.

Many portable single-phase alternators use voltage regulators or individual capacitors to maintain a stable terminal voltage under load conditions. The voltage regulator samples the output voltage and if required adjusts the d.c. excitation current automatically to improve the output voltage. Note that because these alternator sets are self-exciting they may not produce a voltage output if a load connects prior to starting. **Figure 8.119** shows an illustration of a portable single-phase alternator without its roll cage.

FIGURE 8.119 Single-phase alternator

Single-phase alternators may have a stationary field and a rotating armature (stator). However, it is more practical to build them with a rotating field and a stationary armature.

The rotating field is obtained by exciting windings on the rotor with direct current power supplied by residual magnetism or imbedded permanent magnets. **Figure 8.120** shows an illustration of the stator and rotor of a single-phase alternator circuit.

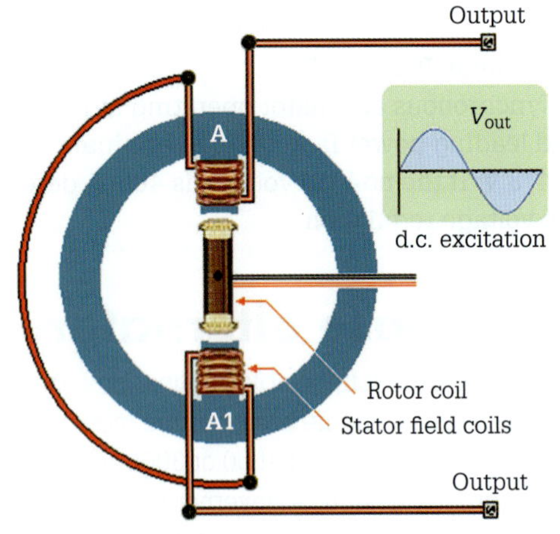

FIGURE 8.120 Single-phase alternator circuit

Flashing the d.c. field

Some self-excited single-phase alternators may need to have the fields flashed to establish the residual magnetism necessary to start the exciter induction process. However, if this process transpires, follow the manufacturer's recommendations – otherwise there is damage to the alternator and the voltage regulator. Flashing the field happens by connecting the battery to the exciter field terminals. By allowing direct current to flow through the exciter field windings, residual magnetism establishes in the iron of the field poles. The direct current source must connect correctly – positive-to-positive and negative-to-negative. Failure to do so can damage the voltage regulator semiconductors.

Floating earth

Single-phase alternators configure differently to the mains supply. The single-phase alternator has a 'floating earth' (some voltage value other than zero and completely isolated from the general mass of earth), while the mains has an earthed neutral (zero voltage). When a short to earth (metal) occurs, electrical safety is assured because there is no earth conductor to return the fault current back to the source. However, contact with both active and neutral at the same time results in electric shock.

Choosing an alternator

Alternator manufacturers have standard designs based on a number of physical and operating conditions. Some of the physical features that control the voltage and kilowatt rating of an alternator design are the size and material used for the laminations, the length of the lamination stack, and the quantity of copper wire used to create the windings. The alternator frequency is controlled by the speed of the prime mover rotating the alternator shaft. Direct current is supplied to create a magnetic field in the rotor through various methods called brush-type, brushless or permanent-magnet excitation. All these items together with the amount of ventilation air driven through the alternator, affect alternator output and temperature performance.

The two main criteria used in selecting an alternator are temperature and rating. There are two temperatures to consider when selecting an alternator:

1. Ambient temperature. This refers to the temperature of the air entering the alternator air inlet and determines which continuous or standby rating applies.
2. Temperature rises. The temperature rise results from the heat generated by the resistance of the conductors in the windings and other losses in the alternator.

Ratings of alternators

Continuous rating

Continuous rating or prime rating means an unlimited number of annual operating hours (24 hours a day, 7 days a week) in full-load conditions. A 10% overload facility is available for 1 hour in every 12.

Standby rating

Each alternator set has a standby rating. Standby alternators provide an emergency supply to critical loads when a supply failure occurs. Critical loads may include elevators, hospital facilities, communication systems, fire services and emergency lighting. The standby rating is suitable for supplying continuous supply, at variable load, for the duration of any utility power failure. However, no overload facility occurs on standby ratings. While most standby alternators do not have any designated duty limits (some manufacturers may have duty limits for their alternators), the accepted standard of operation for a continuously rated alternator is much less per year.

The difference between the two ratings is the maximum allowable temperature for alternator stator insulation. **Table 8.4** shows the maximum operating temperatures for the various classes of insulation for a continuously rated alternator.

TABLE 8.4 Temperature rise

	Class A	Class B	Class F	Class H
Max. temp °C	100	120	145	165

Table 8.4 gives us the temperature rise allowed for both the heat generated by the windings and other losses in the alternator and the ambient temperature. Assuming an ambient temperature of 27 °C for a Class A insulated alternator, the alternator can develop a heat rise of 98 °C (125 – 27).

Since standby duty requires fewer operating hours than continuous duty uses, the standby alternator operates at a temperature of up to 25 °C warmer as shown in **Table 8.5**.

TABLE 8.5 Standby alternator temperature

	Class A	Class B	Class F	Class H
Max. temp °C	125	145	170	190

The additional operating temperature allows alternators with a standby rating to generate more kilowatts. For example, a continuous rated 50 Hz 380–415 V three-phase alternator of 50 kW (62 kVA) would have a standby rating of 55 kW (68 kVA) with a standby rating current of 96 A at a power factor of 0.8 lag. This alternator would require a water-cooled 55 kW diesel-fuelled 1500 rpm prime mover to provide turning power for the alternator.

All the above ratings occur from the following reference conditions as determined by the manufacturer of the alternator:

- ambient temperature; for example, 27 °C
- altitude above sea level; for example, 150 m
- relative humidity; for example, 60%.

There are, however, a number of other factors to be considered when choosing an alternator. These concern the kW rating of the alternator. For example, factors such as load imbalance on three-phase, high starting current loads. When a.c. appliances such as refrigerators, washing machines, pumps and power tools start, they can create a short-term power surge demand far in excess of their power rating. In addition, single-phase loading on three-phase alternators, non-linear loading, load sensitivity to voltage variation and high neutral currents are just a few of the reasons for looking at the kilowatt rating of the alternator. An exploded view of a single-phase alternator is shown in **Figure 8.121**.

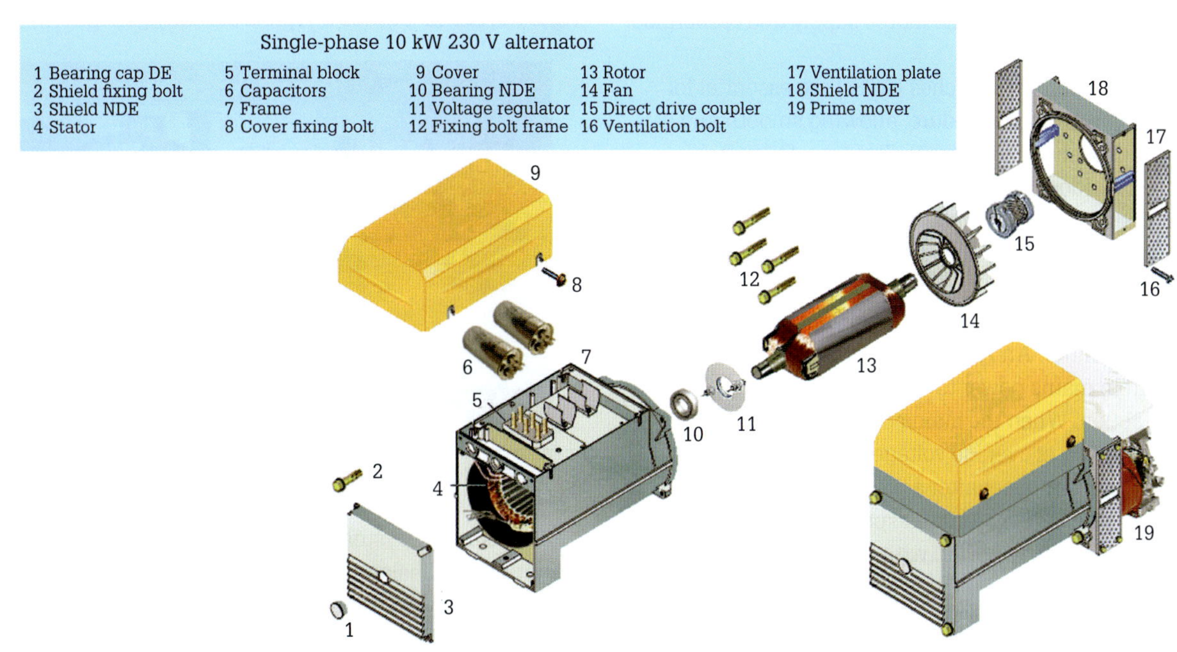

FIGURE 8.121 Exploded view of a standby single-phase alternator

The vast majority of single-phase loads have power factors approaching one. Therefore, single-phase alternator power ratings are taken at power factor of unity or one. By contrast, three-phase loads tend to have lower power factors, approaching 0.8 lag; therefore, three-phase alternator power ratings occur at a power factor of 0.8 lag and are in VA or kVA.

If using a single-phase alternator to start an electric motor, as a rough guide only allow for an alternator that has a continuous rating of 2.5 to 3 times the motor rating.

Standby operation

An important consideration when designing an installation incorporating a standby alternator is to prevent back-feeding when the main supply comes back on-line.

Back-feeding

Back-feeding can happen when an alternator connects to the installation wiring system without disconnecting the system from the mains. The most common way this could occur with single-phase alternators is if the alternator is connected directly to a sub-circuit or at the switchboard.

If you feed power back into the distribution system during an outage, you will energise the transformer serving the installation. Feedback poses an electric-shock hazard for line crews and network assets, and for other consumers connected to the electricity distribution system. If power restores while the alternator is back-feeding, the alternator may be severely damaged. Additional protection is required with alternators to prevent unsafe conditions when connected to the installation during an outage.

Auto-transfer switch

An auto-transfer switch and automatic alternator controls are usually included to automatically energise a standby alternator set, disconnect the mains supply and reenergise the load.

Automatic transfer switches (ATSs) are essential for the power generation procedure, allowing smooth and speedy handover of load current. The transfer switch prevents hazardous feedback of current to the mains system. It ensures that different power sources completely synchronise before their power combines, or loads are transferred.

For emergency power in buildings, an auto-transfer switch is essential. An ATS can automatically transfer the load from mains power to emergency power without operator action. When the mains power fails, or voltage drops below 80% of average voltage, the ATS starts the emergency generator set after a pre-set time of zero to 10 seconds (adjustable). At rated speed, the ATS transfers the load to emergency power, with the available size ranging from 40 A to 4000 A.

Break-before-make switch

A break-before-make switch prevents two independent voltage sources from short circuiting while switching. On actuation, the break-before-make switch movable contact breaks contact with a fixed contact before making contact (non-shorting) with another fixed contact. **Figure 8.122** shows the circuit symbol of a break-before-make switch showing one normally closed and one normally open contact.

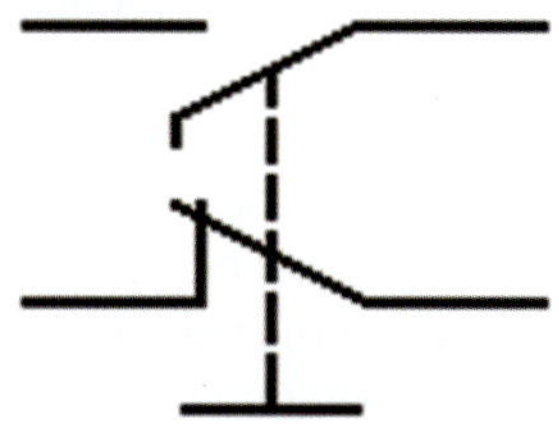

FIGURE 8.122 Break-before-make symbol

A break-before-make switch is used as a transfer switch to 'break' the electrical connection with utility mains power before it 'makes' the connection between the load and the emergency alternator power supply. The transfer switch prevents back-feeding on the alternator into the mains. A break-before-make switch prevents restored mains power from damaging the alternator.

Standby alternators

A standby generator ensures a continuous, uninterrupted energy supply even if the regular supply fails.

The recommended noise limit for standby generators during the daytime is background noise level + 10 dBA and it should not exceed 55 dBA. At night-time the requirement is background noise level + 5 dBA and it should not exceed 45 dBA. A standby electricity generator normally operates only during grid power failures and test runs or for special purposes.

A standby generator system has two primary subsystems: the generator, consisting of the prime mover and the governor; and the distribution system, containing the automatic transfer switch and associated switchgear. **Figure 8.123** illustrates a typical standby generator.

FIGURE 8.123 Standby alternator

The standby system should support the alternator operating as an island in the event of a credible contingency event. Usually the electrical installation is supplied from the mains supply as shown in wiring system-A and wiring system-B of **Figure 8.124**. Both of these systems use a manual transfer switch standby configuration.

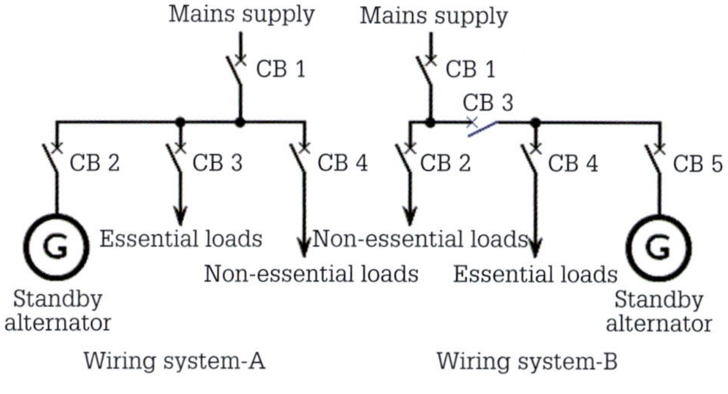

FIGURE 8.124 Standby island alternator wiring systems

The transfer switch allows the generator to operate as an island separate to the mains supply. In examining wiring system-A, we can see that upon the detection of loss of mains supply:

1 the mains circuit breaker CB 1 is manually tripped
2 non-essential loads are disconnected by opening CB 4
3 the alternator set prime mover is started and brought up to speed
4 load is transferred from the mains to the standby alternator set by closing CB 2.

Wiring system-B illustrates a system where essential and non-essential loads divide into separate sub-mains. Consequently, with wiring system-B, upon detection of loss of mains supply:

1 mains circuit breaker CB 1 remains closed
2 sub-mains circuit breaker CB 2 is tripped
3 remote start of alternator set is initiated manually and brought up to speed
4 load is transferred to the standby alternator set by closing CB 5
5 non-essential loads are disconnected by the opening of CB 3.

To shut down the alternator set, a typical operating sequence is:

1 initiate remote shutdown of alternator set (manually)
2 synchronise alternator to mains supply
3 close mains circuit breaker CB 1 (circuit A)
4 transfer load from alternator back to mains
5 trip alternator circuit breaker CB 2 (circuit A) or CB 5 (circuit B) to island alternator
6 initiate alternator cool-down sequence
7 shut down prime mover.

Distribution system containing a standby alternator

Figure 8.125 shows a factory with one 310 kVA generator set that receives 433 V from one LV switchboard connected to the mains supply kiosk sub-station by one incoming cable (4 × 1C × 400 mm^2) through a 1000 A circuit breaker (located at the 400 V switchboard). The cable is fed by a 750 kVA, 11/0.433 kV transformer, which connects to the 11 kV network by ring main units (RMUs). In addition, this RMU connects to the 132/11 kV zone sub-station by an 11 kV radial feeder.

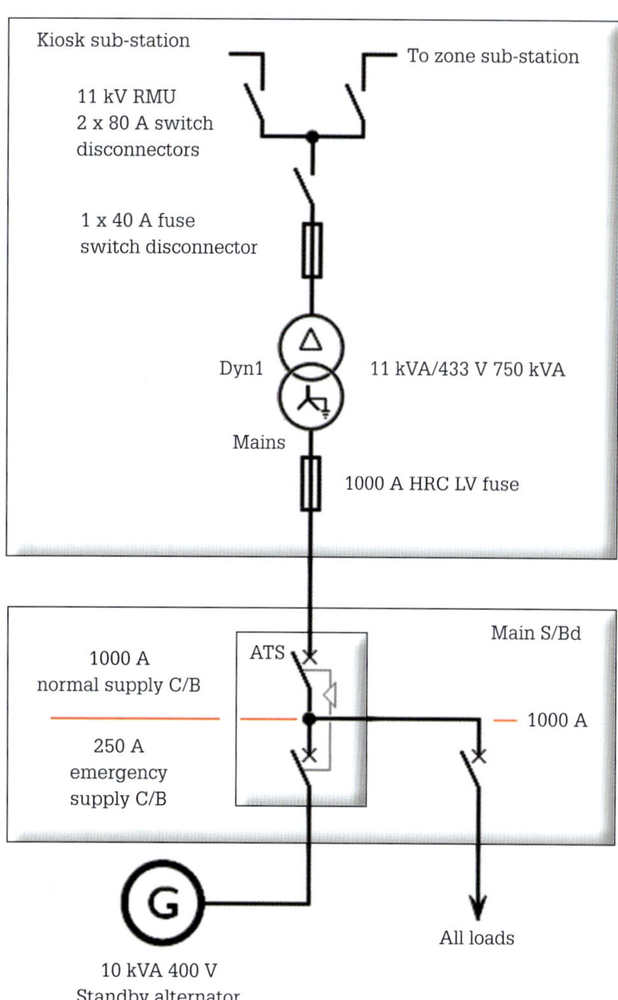

FIGURE 8.125 Distribution system of an industrial factory

The standby generator set (10 kVA) provides an emergency supply to the factory's essential load as the non-essential load (non-essential lighting and air-conditioning) is load-shed when emergency interruption of the mains supply occurs. The generator connects to the main switchboard by means of an incoming cable connected to a 250 A circuit breaker. The automatic transfer switch (ATS) automatically switches between the mains supply and the emergency standby supply. A programmable logic controller (PLC) manages the connected load. The PLC senses when mains power interrupts and starts up the generator if the mains power remains absent. For example, after about five to ten seconds, when the generator is producing full power, the transfer switch disconnects the load from the mains and connects it to the generator, restoring power to the load. The PLC continues to monitor mains power and, when restored, switches the load from the generator back to the mains. For dependable starting of the generator in an emergency, six-monthly maintenance and running of the generator set must occur. Test running of the generator requires the connection of a load bank unless momentary interruption of supply to essential loads can happen.

Parallel operation

Parallel operation involves the synchronisation and connection of the generator to the mains supply for the period, time, or term required for network restoration. Parallel operation requires that the output voltage and the frequency from each generator must be the same as the mains voltage. The phase rotation of each generator must reflect the mains rotation, and the voltage output from each generator must be in phase with the mains voltage. **Figure 8.126** illustrates the distribution system of an industrial factory with two standby generator sets.

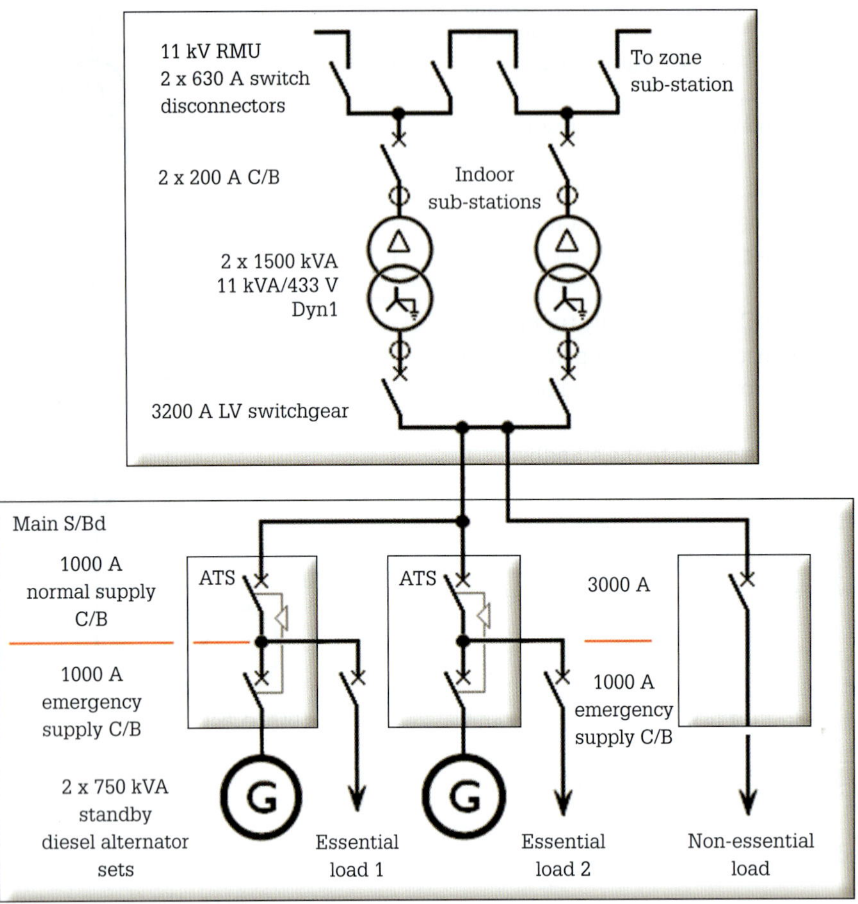

FIGURE 8.126 Parallel operation

REVIEW QUESTIONS

1. Describe the armature construction of a synchronous alternator.
2. What is the more practical method of synchronous alternator construction?
3. Why are electromagnets used to create the magnetic field in synchronous alternators?
4. Upon what parameter does the frequency of the output voltage of an alternator depend?
5. Name the two essential means available to connect the three separate phase group windings of a three-phase alternator.
6. What are two major disadvantages for an alternator connected in delta configuration?
7. Most alternators have a centrifugal fan at the driven end. What is the purpose of this fan?
8. There are various methods of alternator excitation. Name four classifications of excitation.
9. Calculate the rms line voltage produced by a three-phase, four-pole, 50 Hz, star-connected alternator having 120 turns per phase, a winding constant of 0.92 and 80 mWb flux per pole.
10. What do voltage regulation curves illustrate?
11. A synchronous alternator operating at rated load and leading power factor has a terminal voltage of 462 V. If the no-load voltage is 440 V, determine the voltage regulation.
12. Some self-excited single-phase alternators may need to have the fields flashed. Why would this be necessary?
13. What is meant by the continuous rating or prime rating for an alternator?
14. What does a break-before-make switch prevent?
15. A standby generator system has two primary subsystems. Name these subsystems.
16. What are the requirements for parallel operation of alternators?

8.10 Testing of low voltage a.c. machines

Before thoroughly testing a low voltage a.c. machine an important first step is to complete a repair specification. This specification covers routine repair and rewind of low-voltage single-phase and three-phase squirrel-cage induction motors and lists and describes the minimum requirements for repair and overhaul of such machines.

Repairing low voltage a.c. machines

The repair process usually involves:

- lifting
- dismantling
- stator winding removal
- stator rewinds
- reassembly.

Lifting

For lifting and carrying induction motors, one or more lifting eyebolts are required. Eyebolts support only the weight of the motor, not the weight of the machine that incorporates the motor. Induction motors require lifting as illustrated in **Figure 8.127**.

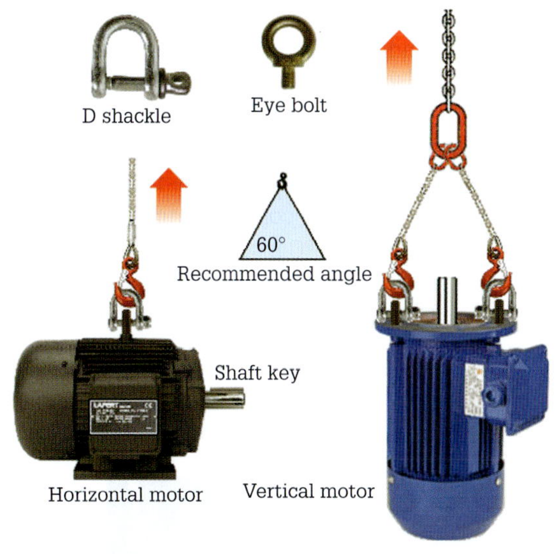

FIGURE 8.127 Lifting of induction motors

When lowering induction motors always make sure that they will rest on safe, stable supports. Note that horizontal sling angles of less than 30° are not recommended.

Dismantling

Before dismantling, an induction motor must be:

- disconnected from the mains
- de-coupled from driven machine
- free of all non-motor attachments such as coupling halves, pulleys, gear cogs, etc.

Before removing the fan cowl and the end shields from the motor frame, they should be clearly alignment marked by a centre punch as shown in **Figure 8.128**. One centre punch mark for the non-drive end and two centre punch marks for the drive end. Alignment marking allows for correct assembly of the serviced induction motor.

FIGURE 8.128 Alignment marking of motor end shield and fan cowl

Rotor removal

For rotors too heavy for removal by hand, a crane utilising a special lifting device intended for rotor removal is used. In **Figure 8.129**, a steel pipe is placed over the non-shaft end to allow easier removal of the rotor from the motor frame.

Attention is required to the following:

- Take care not to damage the journals (where the end shield fits the motor frame) or the stator windings and laminations.
- Use a stitched sling as it does not damage the bearing surfaces of the rotor.

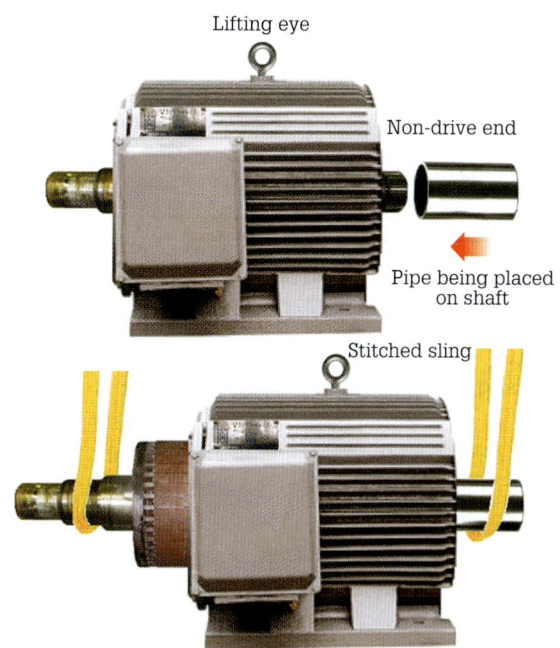

FIGURE 8.129 Removal of large rotor from the motor frame

Vertical motors require dismantling according to the manufacturer's instructions. Pay particular attention to:

- the amount of rotor lift (end play)
- the number of thrust bearings and bearing spacers, including the orientation of thrust and mating surfaces.

Stator winding removal

The majority of induction motor failures occur in the stator windings of the motor, the result of a combination of thermal, chemical and mechanical interactions.

Prior to stripping a stator of its windings, winding data needs recording to permit replicating the initial configuration. The data collected includes the class of insulation (slot and phase), the CSA of the magnet wire, the number of turns per coil, the number of coils per pole phase group, the coil end extension, the coil span and the type of coil connection. A motor repair identification card as shown in **Figure 8.130** is suitable for recording this information.

Furthermore, with a new winding, the length of the coil overhang should not increase and the CSA of the conductors should be the same (or slightly increased, if possible). By doing this, similar winding resistance and energy losses occur, thereby preserving or increasing stator service life and energy efficiency.

Application ________ Data taker ________
Volts ____ Amps ____ ϕ ____ Hz ____
RPM ____ kW ____ Temp rise ____
Type ____ Frame ____ Class ____ Duty ____
MFG ________ Serial No. ________
Bearing front ________ Bearing back ________
Coil shape ________
No. poles ________
No. slots ________
No. coils ________
Coil side/slot ________
Connection ________
No. groups ________
Coils/group ________
Coil span ________
Wire size ________
No. of turns ________
End projection front ________ Back ________
Comments ________
____
____

FIGURE 8.130 Motor repair identification card

Burn-out

The old windings from the stator core require removing without damaging the laminations. One workshop practice is first to degrade the winding insulation thermally in a temperature-controlled oven while monitoring the temperature in order to prevent damage to the stator core. The thermal effect causes the softening of the hardened varnish and the loosening of the insulating materials within a stator to allow for the winding removal. Temperatures of up to 400 °C are used for some stators. At this temperature, the stator laminations have protection from excessive heat. By contrast, strong chemicals strip windings when burn-out facilities are not available.

Core loss test

A core loss test occurs on stators before and after stripping to check for damaged laminations or hot spots. Core loss testing determines the core's condition and is a tool used to avoid rewinding unacceptably high core loss stators. Core loss is the primary cause of wasted electrical energy and can result from overheating during operation, motor burnout and physical damage such as rotor drag on the stator core. It is impossible to know if the motor is capable of providing service at rated values without core testing.

In addition, all obvious lamination damage and significant frame damage, plus any problems indicated by core loss tests, require repairing. Methods of lamination repair include:

- selective grinding of the lamination stack with a small air or power grinder followed by inspection to verify no surface shorting of the laminations
- inserting split mica between the laminations
- restacking with new de-burred laminations.

Stator rewinds

Original stator windings have a particular thermal class. Therefore, the windings require the same thermal class system, including leads, sleeving and insulation materials.

Magnet wire techniques

Replacement of the magnet wire, as illustrated in **Figure 8.131**, must have the correct coating of insulation such as Plain Enamel (105 °C insulation, Class A), Polyurethane Nylon 155 (155 °C, Class F) and Polyester 200 (200 °C, Class H). Magnet wire can be round, square or rectangular.

FIGURE 8.131 Magnet wire

The techniques and methods used to insert or wind the magnet wire into the stator slots are more important than the coating of insulation used on the wire. The coating cannot be useful when nicked or scratched by the laminations. Magnet wire connections need sleeving with a material (fibreglass sleeving) providing a high level of electrical and mechanical strength according to the insulation class of the motor. Uninsulated connections poking through insulation can be a failure point for the motor winding. Sleeving is necessary for the magnet wire because coil leads have to pass over other coils from other phases where voltage differences exist. In addition, coil leads require sleeving to extend from inside the stator slot to where it joins another lead.

Insulation materials

Insulating materials such as slot liners, tapes and phase insulation must meet or exceed the temperature class of the motor and must be compatible with the impregnation material used to insulate the stator windings. Nomex®, Mylar®, Kapton®, mica, fibreglass, silicone, Kevlar® and varnished glass cloth insulating materials are used in many stators. All insulation materials should be stored in a clean, dry location, with materials that degrade at room temperature kept in a refrigerator.

Insulation requirements for a stator

An induction motor stator has the following insulation requirements:

- A slot liner extending at least 5 mm past each end of the slot together with an insulated separator between the top and bottom coil sides in each slot.
- A stator slot wedge to retain the coil sides securely in the stator slots. Slot wedges are made from a variety of insulating materials (Mylar®, fibre and wood, Nomex® and fibreglass) to meet the various temperature classifications required. They are available in two configurations: curve formed or square formed.
- Phase insulation must be placed between the coil end turn positions for the different phases because high-voltage differences exist between the phases. The phase insulation acts as a physical and dielectric barrier to prevent short circuits between the phases. It must be trimmed to permit effective air flow over the stator windings.
- Lacing and tying cords and tapes (twisted glass cords, cotton cords or braided flat tapes using cotton or glass yarns). End turns must be fully compacted so that there are no loose wires. End turns, leads and jumpers must be laced or tied tightly together so that each coil is tied securely to adjacent coils.
- Materials that do not corrode in the motor's operating environment. For example, coil group connections must be insulated, have no sharp edges and be fused or brazed with this type of material. Refer to **Figure 8.132**.

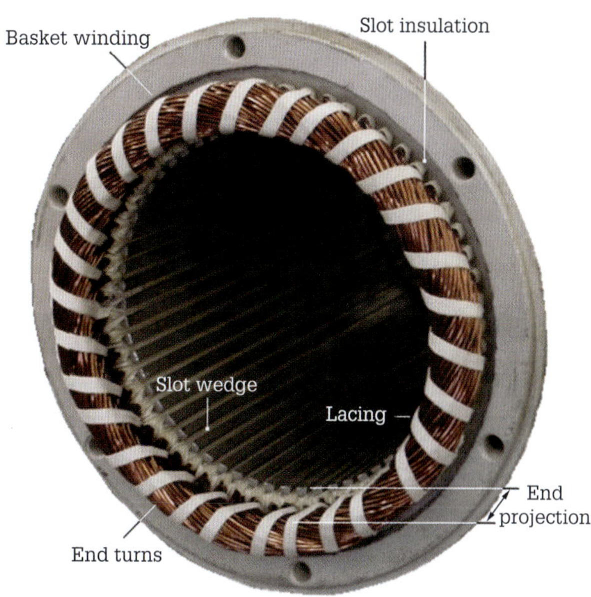

FIGURE 8.132 Stator insulation requirements

Method of winding

The stator core slots have a sheet of insulation material inserted, designated as a slot liner. As the applied voltage divides between the turns within a coil, crossover conditions can cause the stator to fail prematurely (voltage difference between turns). Consequently, coil winding must prevent crossover or twisting of the magnet wires when forming the coils and keep the beginning and end turns away from each other. Refer to **Figure 8.133**.

FIGURE 8.133 Coil group

The first coil side inserts into the stator core slots to sit at the bottom of the slot. A strip of insulation called the separator inserts on top of this bottom coil side to provide phase insulation. A second coil follows in the next adjacent slot until all the coils are in place. Finally, the slot liner folds over the top coil and slot wedges insert into the slot to hold the windings in place. Leads connected to the pole-phase groups should be flexible and multi-stranded, and have the same CSA as the original leads. The finished winding is called a basket winding.

Note: There is no 'best way' to wind stators. The technique is to select an established method, design the tooling for it and perfect it.

Winding tests

Before impregnation, the winding requires testing to verify that there are no wrong connections or shorted turns. Most stator failures develop through a combination of voltage surges that occur at start-up and normal insulation deterioration. Stator failures often begin as a turn-to-turn short that eventually goes to earth.

Winding testing uses a voltage surge test, a d.c. high-potential test, and insulation resistance test. The surge test reproduces the voltage surge produced when the motor switches on. An insulation failure means the rewinding of the stator.

The digital surge comparison tester (capacitor) uses a series of surge voltage pulses simultaneously to test turn-to-turn, coil-to-coil and phase-to-phase for insulation defects. The test surge voltage is twice the circuit rating and 1000 V for several microseconds to ensure no damage occurs to the windings. Each phase test occurs as A–B, B–C and A–C.

A high-potential test injects twice the working voltage of the motor plus 1000 V d.c. between each phase winding and earth for one minute. If insulation fails rewinding of the stator occurs.

The insulation resistance (IR) test determines the quality of insulation to earth. In this test, the motor frame containing the stator is the earth, and the test instrument imposes a d.c. voltage on the motor windings with the resulting measurement occurring in megohms.

Impregnation

Impregnation protects the stator winding by minimising the amount of air able to ionise or become ozone. Ozone chemically attacks some types of insulation materials leading to partial current discharges between coil turns. Consequently, a rewound stator impregnated with varnish, epoxy or polyester resin must be chemically compatible with the insulation materials. Sealing of the stator occurs using one of the following methods:

- dip-and-bake
- trickle
- vacuum pressure impregnation (VPI).

A dip-and-bake treatment requires the stator to dip in a clear solvent baking-type varnish until no air bubbles permeate through the varnish. The dipping cycle repeats twice, to replace the air surrounding and between the wires of the stator winding and then the stator bakes in a temperature-controlled oven. However, because solvent varnishes lose 50% to 70% of their volume during baking voids, air pockets may occur in the windings during the baking process.

The trickle method of stator impregnation uses a clear air-drying varnish, epoxy or polyester resin, and the impregnation material is poured into the end turns and slots of the (vertically inclined) stator, which has been heated via its winding to assist in curing the impregnation material.

The use of 'solventless' epoxy varnish and vacuum pressure impregnation treatment entirely seals the stator windings against moisture, oil and chemicals and vibration. The method gives the stator windings excellent mechanical strength and effective heat dissipation.

Reassembly

The assembly of the motor is the reverse of the disassembling process.

- Alignment marks must line up.
- The reinsertion of the rotor must not damage the journals or the stator windings and core laminations.
- Fixing bolts should go back into the same holes that they came from.
- On motors with insulated bearings, the insulation requires checking.
- On vertical motors, the end play must be the same as the original manufacturer setting.
- Motors for use in hazardous environments must have all explosion-proof features maintained.

Locating faults in low voltage a.c. machines

Faults that may occur include:

- shorted or partially shorted stator windings
- earth faults
- insulation faults
- rotor faults.

Testing for short-circuited stator windings

Figure 8.134 and **Figure 8.135** show an illustration of a short-circuit tester or internal growler used to detect short-circuit faults.

FIGURE 8.134 Short-circuit tester

FIGURE 8.135 Testing for short circuits

A phase-to-phase fault occurs when a short circuit rises between two different stator phase windings. Consequently, a high current result allows the motor protection circuit to de-energise the motor.

Earth faults

Earth faults can easily be determined by using a high-voltage insulation resistance tester ('Hi-Pot' test). This test is performed by applying twice the working voltage in d.c. volts between each phase conductor of a three-phase motor, the starting and running windings of a single-phase motor and the earthing point on the frame of the motor. A reading of less than one megohm (1 MΩ) indicates an earth fault. The earthing point requires checking for continuity with the installation earthing circuit using a low value reading ohmmeter. The resistance should be low enough to allow the passage of current to operate the protective device as **Figure 8.136** shows.

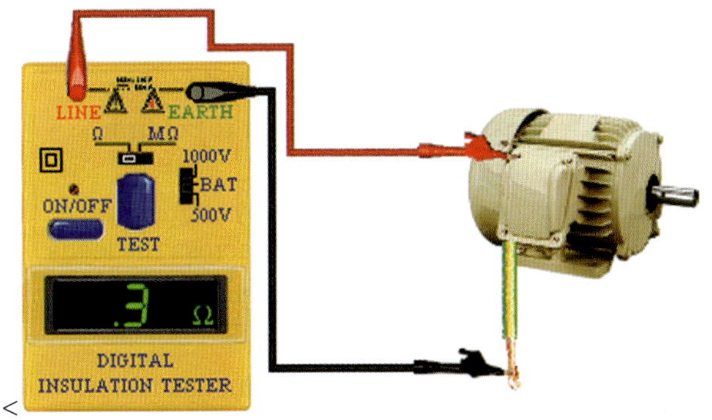

FIGURE 8.136 Testing for earth continuity

Insulation faults

Insulation faults such as short circuits between pole-phase groups of a three-phase winding and between the auxiliary winding and the main winding of single-phase stators can be determined by applying a voltage pressure of twice the working voltage in d.c. volts between these winding elements. An insulation reading of less than one megohm (1 MΩ) indicates insulation failure. Due to safety factors, the minimum insulation resistance for 230 V motors should not be less than 25 MΩ and 100 MΩ for 400 V motors (drying out the motor may improve these values). However, an insulation test cannot find short circuits between conductor turns of a coil-pole group. **Figure 8.137** shows an insulation test with the meter set to the MΩ scale.

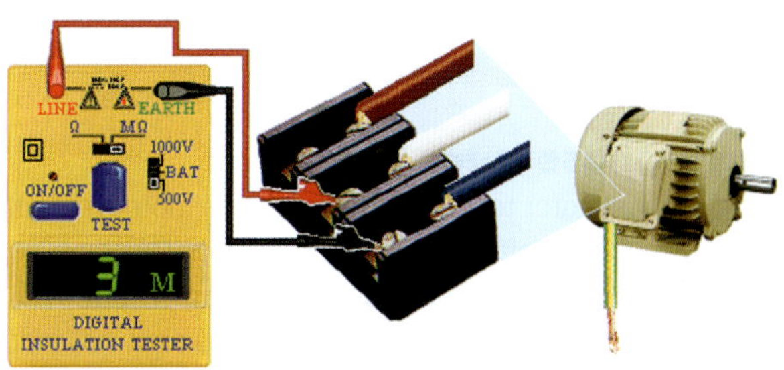

FIGURE 8.137 Insulation resistance test (must be more than 1 MΩ)

SWITCH ON

Any winding internal connections require disconnection for the insulation resistance test.

With three-phase motors, low voltage connects across each phase group one at a time, and the current through the group is measured. If the three readings are the same, the motor is 'balanced', and the windings are in good order. However, if there is a difference between the phase current readings, the motor needs dismantling, and the windings subjected to short-circuit testing. The problem may not be in the stator windings but the rotor. Location of rotor faults requires the current balance method of testing for shorted turns as shown in **Figure 8.138**.

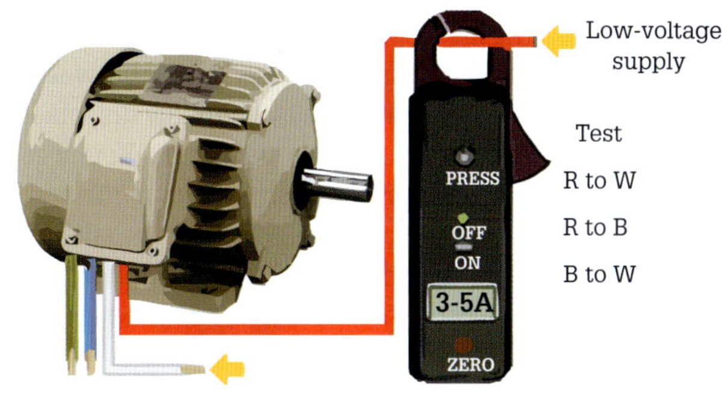

FIGURE 8.138 Current balance method

SWITCH ON

No internal winding disconnection occurs for the current balance test.

Rotor faults

All bar-type rotors require testing for damaged bars. This test applies a single-phase voltage to the stator of the assembled motor while the shaft slowly turns through at least one revolution. Variation of the stator current in excess of 3% is an indication of a rotor defect. Additional tests that can be performed to locate bar-type rotor problems include short circuit and insulation resistance tests. Finally, the rotor requires dynamically balancing before assembly of the motor.

Single-phase rotor test

A single-phase rotor test happens while the motor remains assembled. A low-voltage single-phase supply with an ammeter connected is applied across the run winding so that the winding draws a current slightly less but not more than its rated full-load current. The motor shaft is then rotated slowly by hand and observations made as to the deflection of the ammeter and if the tendency of the rotor is to 'stay' or 'cog'. Consequently, if the observed meter deflections exceed 5% of the test current, then the motor needs to be disassembled, and the rotor tested for short circuits, voids and cracks. Refer to **Figure 8.139**.

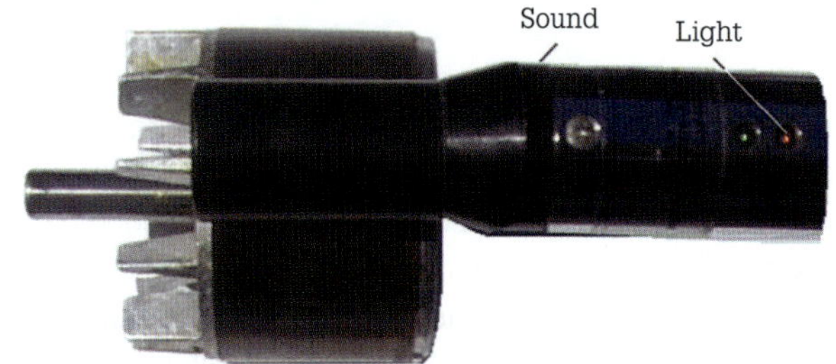

FIGURE 8.139 Testing for rotor bar faults – voids

If a three-phase motor with a bar-constructed rotor produces a noise of erratic strength and diminishing frequency while the load increases, the reason, in most cases, is a lop-sided rotor winding. If the air gap between the rotor and the stator laminations is unequally dispersed, uneven magnetic fields occur. These magnetic differences cause the stator windings to move, resulting in insulation abrasions leading to winding failure. Magnetic imbalances can create an electrically induced vibration that directly affects the integrity of the shaft bearings, resulting in microscopic cracks.

Rotors require placement between centres and the total indicated run-out of the rotor assembly being measured. Ensure that the balancing correction weights have an appropriate fixing method. Metal cooling fans should be 'dye' tested in order to emphasise any structural defects such as stress and vibrational cracks. Finally, frequent cleaning of cooling fans, especially in dirty environments, is essential to maintain the airflow over the stator and rotor.

Coil resistance

With the motor isolated from the supply, use an ohmmeter to measure and compare the resistance of the windings.

- For three-phase machines the resistance values should be identical.
- For single-phase motors the resistance should be in keeping with the type of winding. That is, the start or auxiliary winding has higher resistance than the run or main winding.

Centrifugal switch

The centrifugal switch is probably the worst offender in producing open circuits. The switch becomes worn, defective, and dirty. Insufficient pressure of the rotating part of the switch against the stationary part will prevent the contacts from closing, producing an open circuit.

Possible problems with the centrifugal switch include:

- welded or stuck contacts
- other faulty parts may be causing the contacts to remain closed
- rotating part of the switch may not release the contacts on the stationary part due to improper placement of fibre washers on the rotor shaft.

With the motor disassembled and the start winding disconnected to the centrifugal switch, test the centrifugal switch as follows:

1. Connect an ohmmeter across the contacts of the centrifugal switch. The resistance should be open loop (∞ Ω).
2. Press the two contacts on the centrifugal switch together. The resistance should be zero ohms. If the resistance is still open loop (∞ Ω), examine the centrifugal switch carefully. Clean all contacts and adjust the pressure of the rotating part.

Testing start and run capacitors

SWITCH ON

A good capacitor stores an electrical charge and may remain energised after removing power. Before making measurements and before touching the leads of a capacitor:

- isolate the capacitor from the supply
- use a voltage presence indicator to confirm isolation
- safely discharge the capacitor by connecting a suitable load such as a resistor across the leads
- wear appropriate personal protective equipment.

A capacitor-start, single-phase induction motor that fails to start may indicate a faulty capacitor. The motor will continue to run once operating, which makes troubleshooting problematic. The motor may present with noisy operation. Motor start capacitors will have the capacitance value marked on the capacitor.

Before discharging a capacitor it is necessary to isolate it from the supply and confirm isolation. Once proved to be isolated, connect a 20 kΩ, 5-watt resistor across the capacitor terminals for five seconds. Use a voltmeter to confirm the capacitor is fully discharged.

To test the capacitor for functionality:

1 Visually inspect the capacitor for leaks, cracks, bulges or other signs of deterioration and if evident, replace the capacitor.
2 Select Capacitance Measurement mode on the DMM.
3 Connect the test leads to the capacitor terminals. Keep test leads connected for a few seconds to allow the multimeter to automatically select the proper range.
4 Compare the measured value with that specified on the capacitor.

Checking motor bearings for wear

Checking bearings basically involves:

1 Rotating the shaft by hand to ensure smooth and quiet operation.
2 Attempting to move the shaft up and down and side to side to check for lateral movement, which is indicative of a worn bearing.
3 Checking for end play by attempting to push and pull the shaft backwards and forwards.

If sleeve bearings are resisting rotation or produce rasping or scraping sounds whenever the shaft is rotated, then they may be seized. It may be possible to remedy this by lubricating the bearings with the appropriate lubricant.

Worn ball bearings may possess a distinctive dry rolling sound when rotating or moving the shaft. They can also manifest a raspy feel and resist rotation of the shaft. In this case, lubrication cannot fix the problem. The only solution to these faulty bearings is to replace them fully.

Locked rotor

Badly worn bearings may result in the rotor being unable to turn. This is a fairly obvious condition and may manifest by the operation of circuit protective devices such as over-current protection devices or thermal protection devices. It is necessary to disassemble the motor to ascertain the root cause.

Load test

Excessive load on the motor can be determined by measuring the line currents and comparing to nameplate details. **Figure 8.138** shows how a clamp ammeter is used to measure line currents for a three-phase motor.

Thermal overload

Testing the operation of thermal overloads is not usually necessary. If, however, it is suspected that the thermal overloads are nuisance tripping, follow this procedure:

1 Measure the normal motor running current (I_{motor}).
2 Stop the motor and allow it cool for at least 10 minutes.
3 Calculate the following ratio: $\frac{I_{motor}}{I}$ where I is the overload minimum full-load current.
   For example, if I_{motor} = 4.0 A, and the starter incorporates a 1 – 5 A overload, I (minimum full-load current) is 1 and the ratio is therefore 4. If the ratio is less than 1.2, this test cannot be run. For best results, the shortest test times are obtained when the ratio is 2.0 or greater.
4 Set the overload to its minimum full-load current and turn on the motor.
5 Wait for the overload to trip. The trip time should be within the range defined by the trip curves on the overload instruction sheet for the ratio calculated. The trip time should be close to the cold trip time defined on the curves as shown on the instruction sheet.

REVIEW QUESTIONS

1 When using a sling to lift a motor or rotor, what horizontal sling angle is recommended?
2 When disassembling a motor, what is the purpose of alignment markings?
3 When lifting a rotor it is recommended to use a stitched sling. Why is this?
4 What is the purpose of core loss testing?
5 What condition does the surge test reproduce?
6 What tool is used to test for shorted stator coils?
7 How are rotor faults determined?
8 What is the likely result of insufficient pressure of the rotating part of the centrifugal switch against the stationary part?
9 How might worn ball bearings manifest?
10 How might excessive load on the motor be determined?

8.11 Mechanical faults in LV a.c. rotating machines

Mechanical faults that occur in low voltage a.c. rotating machines involve:

- bearings
- fans
- bent shaft
- locked rotor
- blocked air vents
- centrifugal switch failure
- environmental factors.

Bearing faults

There are several reasons why a bearing may fail prematurely. Improper lubrication accounts for around 40% of all premature bearing failures. Both too much lubrication and insufficient lubrication are damaging to bearings. Another important consideration is using the correct lubricant for the particular application. It is important to carefully follow the bearing manufacturer's guidelines when lubricating bearings. The correct amount of lubricant should be applied at the appropriate intervals.

Bearing fatigue is responsible for about 35% of all bearing failures. Bearing fatigue usually occurs due to the bearing being overloaded, which happens when the bearing exceeds its load, speed or temperature requirements. This may cause the bearing components to fracture and subsequently fail. The greater the overload on the bearing, the greater the effect on bearing life.

Bearings should be replaced at the first sign of fatigue, to avoid damage to associated equipment.

A bearing puller (hydraulic if available) removes bearings to avoid damage to the motor shaft. Once removed, match the new bearing to the old and always replace a sealed or shielded bearing with one of similar protection.

Install bearings correctly by using a bearing press or by tapping into place by means of a metal pipe that fits against the inner race. Forcing the bearings onto the shaft via the outer race should not occur. In addition, with thrust bearings always ensure that they act in the correct direction.

Bearings require heating, without the use of a direct flame, to approximately 100 °C to permit them to slide easily onto the shaft up to the shaft shoulder. An induction heater or machine oil baths are suitable methods to heat the bearings. Furthermore, with large bearings, pack the shaft in dry ice to allow easy fitting of the heated bearings. However, bearings with bores of less than 45 mm are press fitted.

A deep-groove ball bearing or roller bearing axially locates the rotor in horizontally mounted induction motors. By contrast, in vertically mounted induction motors, a radial ball bearing or an angular contact ball bearing axially locates the rotor. See **Figure 8.140** for some common bearing types.

With new induction motors, the nameplate should indicate the type of bearings and the re-lubrication data.

FIGURE 8.140 Bearings

When a current passes through a bearing, a welding arc could establish, causing pitting of the balls and races. Therefore, some induction motors may have insulated bearings (provided with an aluminium-oxide-coated outer race). The bearing resistance should be at least 1 MΩ at 500 V d.c.

O-rings, radial oil seals, lock nuts, preload washers, locking clips and circlips associated with the rotor or end shields require replacing when reassembling the induction motor. **Figure 8.141** shows some common bearing fittings.

FIGURE 8.141 Fittings associated with bearings

Figure 8.142 illustrates the effect that a failed bearing has on the stator laminations of a single-phase motor. The failed bearing has caused the rotor to engage with the laminations, resulting in significant damage, which in turn has resulted in burnout of the windings.

FIGURE 8.142 Stator lamination damage and stator winding burn-out

End shields

Another associated bearing problem is where the bearings do not fit securely in the end-shield bearing house. This can occur if the bearing seizes on the motor shaft and slips in the end-shield bearing housing, resulting in wear. This is particularly the case where bearings are secured through an interference fit in the end-shield bearing housing. Worn end-shield bearing housings require building up by welding or metal spraying, and then should be machined.

Fan faults

Problems associated with fans include a build-up of dust and foreign matter that greatly restrict air movement through the blades of the fan and can lead to over-heating of the motor. Fans should be checked for cracks and other defects before fitting to the motor shaft or rotor. Fans should also fit firmly on the shaft and not free-wheel on the shaft. This is particularly important where the fan is secured through an interference fit to the shaft.

Bent shaft

In a perfectly straight shaft, the centres of each shaft cross-section from end-to-end of the shaft lie in a straight line. If this is not the case, then the shaft is bent.

In a bent shaft, the axis of the shaft is different than its axis of rotation. The range of gyration that the bent shaft produces is known as shaft runout, which is typically measured in terms of 'Total Indicator Reading'. Refer to **Figure 8.143**.

FIGURE 8.143 Measuring shaft runout

The consequences of operating a motor with a bent shaft include:

- equipment vibration due to unbalance
- damage to bearings, seals and couplings
- contact, and possible seizure with, close-clearance surfaces such as stator and windings
- material fatigue.

Causes of motor shaft bending include mechanical overload, impact during operation, misalignment of motor and driven load, and elevated operating temperature causing thermal expansion.

Locked rotor

When the rotor of a motor is turning, it generates a 'back EMF' that opposes the current drawn from the supply. It is this emf, not the impedance of the coils, that limits the magnitude of current the motor draws from the supply. When the rotor is locked, there is no back emf to limit the flow of current through the motor. All the power drawn from the supply is converted to heat by the resistance of the windings. Bear in mind that an induction motor, when not rotating, behaves as a transformer with a shorted secondary output. This quickly overheats the coils, melting insulation, creating short-circuit windings, reducing the winding resistance and further increasing the current. Without suitable over-current and thermal protection a fire may ensue.

Generally, a locked rotor is attributable to failed bearings or seized driven load.

Blocked air vents

Blocked or restricted air vents will restrict air flow over the motor windings, resulting in over-heating. If the motor has an open design, it is possible to look into the motor housing and ascertain if there is a build-up of dust and oil that may be blocking the air vents and causing the motor to overheat. Over-heating can cause the motor's thermal overloads to operate and disconnect the motor from the supply. Prolonged operation in over-temperature conditions can cause the winding insulation to fail.

Centrifugal switch failure

If the centrifugal switch contacts fail to close when the motor stops, then the auxiliary or start winding circuit will remain open. When the motor is re-energised it will not start as a split-phase motor requires both main and auxiliary windings to be energised at start in order to produce starting torque. If the motor emits a low humming sound and fails to start, it is quite possible that the auxiliary winding circuit is open circuit. Either the centrifugal switch contacts are not closed, or there is a break in the coils of the starting windings. The main or run winding will draw excessive current from the supply.

If the centrifugal switch contacts fail to open when the motor is operating then the auxiliary winding will remain in circuit. The auxiliary is not rated for full-time operation and will quickly over-heat and open circuit.

Environmental factors

Electric motors are prone to failure when operating in extreme temperature, either hot or cold. Alternating current motors have optimal temperature ranges for peak operational performance that are determined by the manufacturer. It is important therefore to operate the motor within these specifications. In general, industrial a.c. motors operate best at ambient temperatures between +40° to −20 °C.

In operating environments where the air temperature will fall below –20 °C, it is essential for the motor to be correctly lubricated. Grease-lubricated bearings will need lubricant that is recommended by the manufacturer as suitable for low-temperature conditions. Oil-lubricated bearings will need a thermostat-regulated oil-sump heater to warm the oil.

Where the air temperature exceeds +40 °C and there are no other ambient cooling/ventilation measures, the motor may need to be rewound to a higher class of insulation (or an oversized motor selected at the start).

The altitude at which an a.c. motor must operate is also a major concern. Industrial a.c. motors are typically designed for operation up to an altitude of 1000 metres. Above this altitude the ambient air has less cooling capacity. A rule of thumb states that a motor needs to be de-rated 1% for every 100 metres above 1000 metres. This means that a 10 kW rated motor would have an operational rating of 9 kW if required to operate at 2000 metres (10 × 1% × 10 kW equals a 1 kW reduction).

Poor ventilation impacts negatively on the operation of an a.c. motor. Dirty or salt-laden air can block air vents and, in the case of salt-laden air, lead to corrosion. An enclosed motor with good ventilation is preferred if it is necessary to operate an a.c. motor in an extremely dirty or salt-laden environment. A totally enclosed fan cooled (TEFC) motor is probably the best choice in these conditions. TEFC motors use a shaft-mounted fan to blow cooling air over a ribbed frame to dispel heat. Other viable options are totally enclosed air-to-air cooled (which has an air-to-air heat exchanger) or totally enclosed water-to-air cooled (which has a water-to-air heat exchanger) motor type.

Fault summary

In general, motor failures can be categorised into several types: stator failures due to arc tracking and thermal ageing, rotor failures, contaminants and mechanical failures. However, the significant reasons why three-phase and single-phase motors fail in service are ineffective preventive maintenance schedules, lack of protection devices and the reaction time of the protective devices fitted.

Winding failures that occur are due to the age of the insulation, contamination by the operating environment, such as moisture or chemical agents, thermal overload, damaged coil conductors and incorrect class of insulation after rewind, power surges from the supply, and weak motor fixing causing vibration. All these factors increase the insulation conductivity. Energy contained within the conductor is always trying to find release from its confinement in order to seek an easier conducting path. The elements of motor failure all work together to stress the winding conductors until one conductor responds by failing. Consequently, this event leads to higher stress on the conductor creating a hot spot that eventually affects adjoining conductors resulting in arc-type conditions. The result is the complete failure of the motor windings. **Figure 8.144** shows an arc-type condition.

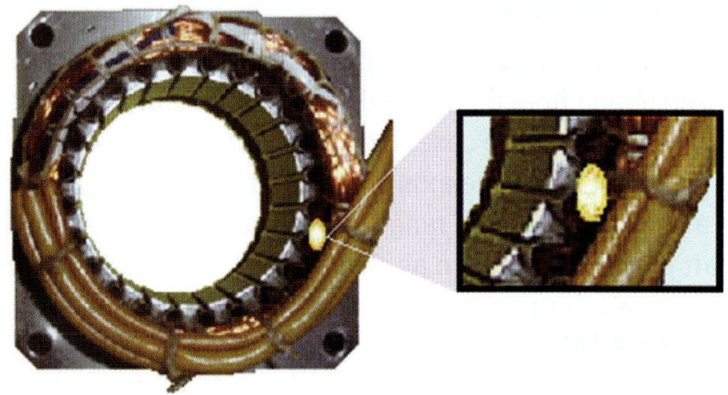

FIGURE 8.144 Arc-type conditions

REVIEW QUESTIONS

1. What is the main cause of premature bearing failure?
2. Name the tool used to remove bearings without damaging the motor shaft.
3. With new induction motors, where is it possible to get information as to the type of bearings and the re-lubrication data?
4. What might electric current passing through a bearing cause?
5. Identify checks that should be performed on fans before fitting to the motor shaft or rotor.
6. What is meant by the term 'shaft runout'?
7. Name four causes of motor shaft bending.
8. What are possible causes of locked rotor?
9. What is the consequence of having blocked motor air vents?
10. State the likely consequence of the centrifugal switch contacts failing to open when the motor is operating.
11. Name two environmental factors that affect motor operation.

8.12 Driven load and coupling faults

Driven load faults include:
- slipping belts
- coupling alignment
- vibration
- bearing failure
- load stalling.

Slipping belts

Worn or damaged drive pulleys can cause spin burn, which aside from making annoying noise can result in premature belt failure. Incorrect belt tension can also cause drive belts to slip. Other less common causes are incorrect belt cross-section or type, excessive oil, excessive grease or moisture and insufficient wrap on a small pulley.

If this is occurring then it is time to check pulleys and belt tension as under-tensioned belts and smooth pulleys result in unnecessary wear. The solution may be to replace worn pulleys with new pulleys or to re-tension the v-belt.

V-belt drives do not require much in-service attention to be efficient. Getting the basics right however can avoid premature failure and improve the life span of v-belts and pulleys, ensuring optimum efficiency from these components.

Some simple steps for maintaining drive belt health are:
- Do not place new belts on to worn pulleys as it can reduce the life of a belt by up to 50%. If the internal wall of the pulley is bright and shiny then it is time to replace it.
- If the pulley is only worn on one side, then it is prudent to check the alignment of the pulleys (see next point).
- Check pulley alignment; laser alignment tools are the best method for checking alignment, which can greatly increase belt life.
- Most v-belt drives require post-installation tensioning and subsequent re-tensioning within 24–48 hours. Over- and under-tensioning can lead to belt failure.
- Do not mix old and new belts on the same drive.
- If the drive is located in a dusty environment or where there is oil and grease present, it is likely that these contaminants can combine to create a paste that eats away at the most robust of components. A simple guard can prevent foreign matter from contaminating the drive.
- Extreme temperatures, when there is either a high or very low ambient temperature, can reduce belt life. In these environments consider using an EPDM rubber belt. EPDM stands for ethylene propylene diene monomer, which is a synthetic rubber used in a range of applications.

Coupling alignment

A coupling is used to connect two rotating shafts for the transfer of rotary motion and torque. For a coupling to work at its optimum efficiency, it must match all required conditions, including performance, environmental, use and service factors. The coupling should not prematurely fail if it is correctly selected and installed. If, however, one of these factors is not met, a coupling can prematurely fail, resulting in anything from a small inconvenience, to a significant financial loss, and even the potential of personal injury.

The axis of the shaft of the induction motor must align with that of the shaft of the driven equipment. Accurate shaft alignment and coupling connection guarantees faultless operation of both the motor and the driven equipment. **Figure 8.145** illustrates a flexible half-coupling arrangement.

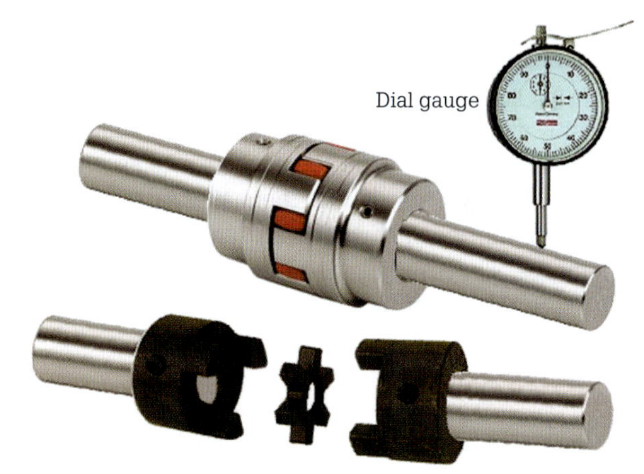

FIGURE 8.145 Flexible half-couplings

While couplings are typically flexible to accommodate misalignment, do not rely just on their flexibility. For ideal shaft alignment, you should first install the coupling on the load shaft. Then move the motor into proper alignment and secure it. When using half-couplings, the maximum permissible alignment error is 0.1 mm. The clearance between the coupling halves must be a minimum of 1 mm.

To check if the external surface of the half-couplings is coaxial use a steel rule or a dial concentric gauge. The test occurs in four diametrically opposite points. Finally, correct any alignment errors by using shims (thin sheet metal) placed between the feet of the motor and its mounting surface.

Misalignment between the motor shaft and the shaft of the load is a common cause of bearing failure. Misalignment introduces excessive vibration and internal bearing loads, which shorten the working life of an electric motor.

Vibration

Vibration is simply a back and forth movement or oscillation of machines and components in driven equipment. Vibration in industrial motors can be a symptom, or cause, of a problem, or it can be associated with normal operation. As an example, oscillating sanders and vibratory tumblers vibrate as part of normal operation.

Vibration can be indicative of problems and if not investigated can lead to damage or expedited deterioration. There are several causes of vibration in driven equipment, but common causes are imbalance, misalignment, wear and looseness.

Imbalance in a rotating component will cause vibration when the unbalanced weight rotates around the machine's axis, which creates a centrifugal force. This imbalance may be the result of manufacturing defects (such as machining errors or casting flaws) or maintenance issues (such as deformed or dirty fan blades, or missing balance weights). The effects of imbalance become more noticeable as machine speed increases. Imbalance can severely reduce bearing life as well as cause undue machine vibration.

Misalignment or shaft runout can cause vibration as a result of machine shafts being out of alignment. Angular misalignment occurs when the axes of a motor and driven load are not parallel. When the axes are parallel but not exactly aligned, the condition is known as parallel misalignment. Misalignment can occur during assembly or may develop over time, due to thermal expansion, components shifting or improper reassembly after maintenance. The resulting vibration can be radial or axial (in line with the axis of the machine) or both.

As components such as ball or roller bearings, drive belts or gears become worn, they may cause machine vibration. When a roller bearing race becomes pitted, the bearing rollers will cause a vibration every time they traverse the damaged area. Other sources of vibration due to wearing components include a heavily chipped or worn gear tooth, or a drive belt that is degrading.

Vibration can become noticeable and destructive if the component that is vibrating has loose bearings or is loosely attached to its mounts. The looseness may be attributable to the underlying vibration. Looseness can permit any vibration that may be present to cause further damage, such as further bearing wear, wear and fatigue in equipment mounts and other components.

Effects of vibration

Vibration can accelerate machine wear, result in excess power draw, and result in equipment being prematurely removed from service, which results in unplanned downtime. Other effects of vibration include safety issues and reduced working conditions. When measured and correctly analysed, vibration can play an important role in preventive maintenance programs. It can serve as an indicator of machine condition and allow preventive maintenance before major damage occurs.

REVIEW QUESTIONS

1 What may be caused by worn or damaged drive pulleys?
2 How may misaligned pulleys be detected?
3 Outline the purpose of a drive coupling.
4 What is guaranteed by having accurate shaft alignment and coupling connection?
5 Where are shims (thin sheet metal) placed to correct any alignment errors?
6 What is machine vibration?
7 How does imbalance in a rotating component cause vibration?
8 Name two detrimental effects of vibration.

8.13 Electrical faults and symptoms in LV a.c. rotating machines

A number of electrical faults can occur in a.c. machines, including:

- single phasing
- phase reversal
- short circuits
- core and rotor lamination faults
- voltage faults
- earth faults – polarisation index.

Single phasing

Single phasing occurs when one phase of a three-phase supply is open circuited and that causes a primary voltage unbalance in the windings of the stator. If a three-phase motor is operating with a single-phase fault, the motor attempts to deliver its full output capacity to the load. Consequently, the motor drives the load, until it burns out, or the overload protection devices take the motor offline.

Phase reversal

When two out of three motor feed lines interchange, phase reversal of the motor occurs. When connected online the motor turns in the wrong direction, causing significant damage.

A balanced three-phase motor means that its electrical characteristics, including passive properties, such as ohmic resistance, and active properties, such as inductive reactance and impedance, are also balanced. However, as faults develop within the stator, one or more of these electrical properties changes.

A short circuit between turns or phase to earth due to insulation failure does occur. Depending on the magnitude of the short circuit, a magnetic hum radiating from the motor becomes audible. If short-circuit turns occur then the single-phase motor must be dismantled, and short-circuit testing devices applied to the stator windings. At the same time, the rotor requires examining for any faults such as casting voids or cracks on the end cage. However, these faults with poured rotors are very rare.

Short circuits

There are several types of short circuits that relate to motor windings. These are:

1 between adjacent conductor turns
2 between coils in a pole-phase group
3 between coils in different phase groups
4 between a coil conductor and earth.

In the early stages, short-circuit faults consist of low intensity. However, the motor may display signs by way of increased operating temperature (insufficient to trip a thermal overload), intermittent tripping of circuit breakers and blown motor fuses.

A shorted turn develops by insulation breakdown between two turns in the same phase. This failure causes a very high current (several times higher than average current) to arise in this coil. The magnetic field produced by this shorted turn affects the total magnetic flux in the phase. If a motor with short circuits between conductor windings stays in service, circulation currents may produce hot spots, causing the stator insulation to deteriorate, resulting in a complete burn-out of the motor. Short circuits between conductor turns that have not fused together are the cause of nuisance tripping, especially during the period of motor starting. Finally, as the short-circuit faults begin to increase the motor may display signs of reduced turning effort, power output and speed.

Core and rotor lamination faults

During a motor failure, high temperatures can occur. These temperatures affect the electrical characteristics of the stator and rotor core steel and result in increased iron losses and lower motor efficiency. Moreover, high temperatures cause hot spots to develop in the stator's iron core leading to premature winding failures.

Bearing failure causing 'poling' is another cause of motor failure with the rotor abrading against the stator iron. Rewinding a stator with iron loss affects the motor performance characteristics. However, if the stator teeth in the affected area have light damage, separate the teeth using a wedge-shaped tool. The objective is to break the weld at the top of the tooth created by the abrasion of the rotor against the stator core.

Core loss testing occurs to verify the rebuild ability of the core iron and the removal of hot spots. These testers have a sensing coil head to detect any circulating fault currents caused by imperfections in the core when the core is excited to a set level of tesla (flux density). Examination of these fault currents occurs by digital means, and the result is displayed on the meter. Rewinding a damaged stator core results in reduced motor efficiency and higher running temperature. Consequently, these factors increase the likelihood of another failure when back in service.

Voltage faults

Unequal voltage between phases caused by unbalanced loads on the power source, weak connections at the motor terminals or a high-resistance switch contact results in thermal degradation of insulation in one phase of a three-phase stator winding. Some of the fault mechanisms that cause high-resistance connections include loose or corroded motor terminals and incorrect sizing of motor supply conductors. For instance, a 2.5 Ω loose terminal connection on a 5 kW 415 V motor with a pf of 0.8 and efficiency of 80% drawing 11 A at full load would have a 27 V voltage drop across the connection. Furthermore, a resistive fault gives off heat as an energy loss in the motor circuit.

One method of detecting high-resistance connections is by performing phase-to-phase resistance testing at the load side of the motor protection device. On a three-phase motor, the three resistance measurements should be nearly identical. If all three readings are the same, no imbalance exists. However, if one phase provides a high-resistance reading, a resistive imbalance exists indicating a fault.

Three-phase motors require de-rating when the voltage imbalance exceeds 1% and motor service should stop where the voltage imbalance exceeds 5%. A small voltage imbalance significantly increases the motor energy losses and at the same time decreases motor efficiency.

An unbalanced voltage supply affects the motor's current, speed, torque, temperature rise and efficiency. A small imbalance in the supply voltage can result in an extensive current disturbance between the three phases. As a rule, 1% change in voltage results in approximately 6% to 10% variation in current. In addition, reduced voltage means that the motor does not develop its rated torque and the motor cannot reach its rated speed. The increased slip results in higher ohmic losses within the motor windings arising from the I^2R effect. The effect leads to an increase in motor temperature and the resulting carbonisation of the stator insulation.

Voltage supply problems are usually the result of incorrect balancing of energy loads across the installation system used to provide the three-phase motor. Other voltage problems causing insulation failure are the result of lightning strikes, capacitor discharges and solid-state switching devices.

Earth faults – polarisation index

Knowledge of the polarisation index of a motor is another measure of the condition of the insulation of a motor or generator in service. The index transpires from measurements of the stator's insulation resistance. The polarisation analysis is a relatively new test that looks at the motor insulation acting as the dielectric of a capacitor. All external connections to the motor must disconnect before insulation measurement occurs, and the stator winding grounded to the motor frame in order to remove any capacitive charge within the stator windings.

Applying either 500 or 1000 V d.c. (twice the working voltage of the motor) between the winding and earth using a high-voltage insulation resistance tester or a high-potential test set carries out the test. The voltage happens over a 10-minute duration. However, after one minute, a value is measured and after another nine minutes, the insulation resistance measured again.

The purpose of the applied voltage is to attract any free electrons within the insulation. Higher resistance after ten minutes, as compared to the one-minute reading, indicates good insulation. Calculation of the polarisation index can be made by applying the following equation.

$$\text{polarisation index} = \frac{\text{resistance after 10 minutes}}{\text{resistance after 1 minute}}$$

The recommended minimum value of the polarisation index for single and polyphase motors and generators is 2.0. An index of less than this number indicates that the motor or generator is not suited for service. **Table 8.6** demonstrates how the polarisation test is evaluated.

TABLE 8.6 Polarisation index and insulation evaluation

Polarisation index	Insulation evaluation
< 1	extremely poor
1–1.5	poor
1.5–2.0	reasonable
3–4	good
> 4	excellent

However, if the index is five or greater, then the insulation may be dry and fragile.

Motor condition analysis instruments

Solid-state motor condition analysis instruments evaluate the service condition of large, expensive motors in critical applications without shutting down essential equipment.

The instruments available from the various manufacturers have different combinations of analytic ability. In general, they are able by power analysis to glean relevant information from the incoming power quality, the motor circuit condition, stator and rotor health, and motor efficiency.

The information gathered can be stored for diagnosis and analysis to provide historical comparisons for the motor under examination over time. Examination of the power quality received by the motor occurs by using total harmonic distortion and crest factor measurements to determine distortions in the alternating current input signals to the motor. Evaluation of the input power circuit and the rotor and stator condition occurs by examination of negative sequence currents. This form of current indicates high resistive connections and high impedance imbalance within the motor.

Continuous signature analysis of motor voltage, current, speed and vibration, with time or frequency domain analysis, are an active fault-detection method for electric motors. Solid-state instruments using signature analysis (thermal and vibrational signatures) use the parameters taken from the motor when first placed in service. Further readings are obtained until the instrument has mean and standard deviation values saved. Consequently, with future measurements, any motor parameters that are then outside the saved thresholds are faulty.

Motor condition analysis instruments can provide informative graphs, through which detection of rotor defects and eccentricity (air gap) can be detected. The instruments can also graph the start-up characteristic of the motor for current and time. These motor condition analysis instruments engage reporting abilities immediately after testing allowing for quick and efficient assessment of the motor condition.

Thermal 'hot spot' identification

A technique for the detection and location of thermal hot spots in electrical machines and hazardous atmospheres now exists. Fluorescence spectroscopy detects sub-micron particles generated by hot surfaces. Microencapsulated fluorophores that have an individual and distinctive fluorescent 'fingerprint' applies to any surface exposed to the atmosphere. Consequently, if the surface develops hot spots, the fluorophores release and a simple computerised luminescent spectrometer detects and identifies the liberated particles.

Future trends

The development of an insulation sniffer using gas analysis and chromatographs for early detection of internal hot spots, or arcing and sparking, will provide further test equipment needed for maintaining the effective operation of alternating current machines.

Minimum energy performance standards

Electric motors consume the greater bulk of electrical energy in the industrial division. Electric motors account for almost 30% of overall electricity consumption in Australia, or some 11% of total greenhouse gas emissions.

Since 2001, three-phase electric motors from 0.73 kW to 185 kW manufactured in or imported into Australia must comply with Minimum Energy Performance (MEP) requirements of the Australian and New Zealand standard AS/NZS 1359.5: 2004. This standard sets out methods for determining the rated output, thermal performance, pull-up torque, motor efficiency and other tests in relation to three-phase motors. The minimum energy performance standard enables the electrical industry to make advances in reducing the levels of greenhouse gas emissions.

Machine maintenance

To keep an a.c. machine running at optimum efficiency it is important to conduct routine inspection and servicing. This involves:

- checking for dust and corrosion
- applying appropriate lubrication
- checking for excessive heat, noise or vibration
- checking the winding insulation for fraying, breakdown and burning.

Rotating machine safety

While carrying any maintenance on an a.c. machine, it is important to consider safety risks associated with rotating machinery, including:

- rotating parts
- lethal voltages
- high inductance
- high impact kinetic energy.

SWITCH ON

Safety behaviour around rotating machines involves:

- Not placing any part of your body into moving machinery.
- Not wearing jewellery, neckties or loose-fitting clothing.
- Wearing proper protective clothing and equipment suitable for the operation being performed.
- Before attempting to perform repairs or maintenance on any machine, make sure that it is de-energised.
- Compressed air may be used to clean machinery parts that have been properly disassembled provided that the supply air pressure does not exceed 30 psi and a safety shield tip is used.
- Not tampering with or permanently disabling safety devices such as sensors, cut-out switches or guards.
- Being aware that under non-compensated fluorescent lighting, machines rotating at certain frequencies may not be detected as moving.

Manufacturer specifications

A wealth of information is available from machine manufacturers in the form of specifications, drawings and nameplates. **Figure 8.146** and **Figure 8.148** show a sample specification sheet for a TEFC three-phase motor and OPEN single-phase motor respectively and **Figure 8.147** and **Figure 8.149** show the associated nameplate details.

GPACM10	
7.5 kW, 1900 RPM, AC, 215T, TEFC, 50 Hz	
Catalogue Number	GPACM10
Enclosure	TEFC
Frame	215T
Frame Material	Steel
Frequency	50 Hz
Output / Frequency	7.5 kW / 50 Hz
Phase	3
Synchronous Speed / Frequency	3000 RPM / 50 Hz
Voltage / Frequency	400 V / 50 HZ
XP Class and Group	None
XP Division	Not Applicable
Agency Approvals	UR CSA
Ambient Temp	40 °C
Auxiliary Box	None
Auxiliary Box Lead Termination	None
Base Indicator	Rigid
Bearing Grease Type	Polyrex EM (-20F +300F)
Blower	None
Current / Voltage	13.1 A / 400 V
Design Code	B
Drip Cover	None
Duty Rating	CONT
Efficiency @ 100% Load	90.2 %
Electrically Isolated Bearing	Not Electrically Isolated
Feedback Device	None
Front Face Code	Standard
Front Shaft Indicator	None
Heater Indicator	No Heater
Insulation Class	F
Inverter Code	Inverter Ready

Lifting Lugs	Standard
Locked Bearing Indicator	None
Motor Lead Exit	KO Box
Motor Lead Termination	Fly Leads
Motor Lead Quantity/Wire Size	6 @1 mm^2
Mounting Arrangement	F1
Number of Poles	2
Overall Length	485 mm
Power Factor	0.89
Product Family	General Purpose
Pulley End Bearing Type	Ball
Pulley Face Code	Standard
Pulley Shaft Indicator	Standard
Rodent Screen	None
RoHS Status	RoHS COMPLIANT
Service Factor	1.15
Shaft Diameter	35 mm
Shaft Extension Location	Pulley End
Shaft Ground Indicator	No Shaft Grounding
Shaft Rotation	Reversible
Shaft Slinger Indicator	No Slinger
Speed	2900 rpm
Speed Code	Single Speed
Motor Standards	NEMA
Starting Method	Star Start - Delta Run
Thermal Device - Bearing	None
Thermal Device - Winding	None
Vibration Sensor Indicator	No Vibration Sensor
Winding Thermal 1	None
Winding Thermal 2	None

Source: Shutterstock.com/Surasak_Photo

FIGURE 8.146 TEFC three-phase motor specification

CAT NO.	GPACM10				
SPEC	GPACM3PH40050				
kW	7.5			**PH**	3
Volts	400				
Amp	13.1				
RPM	2900				
FRAME	215T	**Hz**	50	**IP**	54
Serv Factor	1.15	**DES**	B	**CL**	F
Nom Eff	90.2				
PF	0.89		**Usable at 208 V**		
Rating	40C AMB-S1 CONT		**CC**		
DE Brg	6307		**NDE Brg**	6206	
Encl	TEFC	**Serial**	GPACM1234xy		
APRV-CSA		**APRV-UR**			

FIGURE 8.147 TEFC three-phase motor nameplate

Current / Voltage	5.3 A / 220 V
Design Code	N
Drip Cover	None
Duty Rating	CONT
Efficiency @ 100% Load	64.0%
Electrically Isolated Bearing	Not Electrically Isolated
Feedback Device	None
Front Shaft Indicator	None
Heater Indicator	No Heater
Insulation Class	B
Inverter Code	No Inverter

Shaft Rotation	Reversible
Shaft Slinger Indicator	No Slinger
Speed	2850 rpm
Speed Code	Single Speed
Motor Standards	NEMA
Starting Method	Direct on line
Thermal Device - Bearing	None
Thermal Device - Winding	None
Vibration Sensor Indicator	No Vibration Sensor
Winding Thermal 1	None
Winding Thermal 2	None

FIGURE 8.148 OPEN single-phase motor specification

SPACM075

0.56 kW, 2850 RPM, AC, 56, OPEN, 50 Hz

Catalogue Number	SPACM075
Enclosure	OPEN
Frame	56
Frame Material	Steel
Frequency	50 Hz
Output @ Frequency	0.56 kW / 50 Hz
Phase	1
Synch Speed / Frequency	3000 RPM / 50 HZ
Voltage / Frequency	220 V / 50 HZ
XP Class and Group	None
XP Division	Not Applicable
Agency Approvals	CSA UR
Ambient Temp	40 °C
Auxiliary Box	None
Auxiliary Box Lead Termination	None
Base Indicator	Rigid
Bearing Grease Type	Polyrex EM (-20F +300F)
Blower	None

Lifting Lugs	None
Locked Bearing Indicator	None
Motor Lead Termination	Fly Leads
Motor Lead Quantity/Wire Size	6 @0.75 mm^2
Motor Type	3424 L
Mounting Arrangement	F1
Number of Poles	2
Overall Length	280 mm
Power Factor	0.70
Product Family	General Purpose
Pulley End Bearing Type	Ball
Pulley Face Code	Standard
Pulley Shaft Indicator	Standard
Rodent Screen	None
Service Factor	1.25
Shaft Diameter	16 mm
Shaft Ground Indicator	No Shaft Grounding

CAT NO.	SPACM075				
SPEC	GPACM3PH40050				
kW	0.75			**PH**	1
Volts	220				
Amp	5.3				
RPM	2850				
FRAME	56	**Hz**	50	**IP**	
Serv Factor	1.25	**DES**	N	**CL**	B
Nom Eff	64%				
PF	0.70		**Usable at 208 V**		
Rating	40C AMB-CONT		**CC**		
DE Brg	6203		**NDE Brg**	6203	
Encl	OPEN	**Serial**	GPACM1234sp		
APRV-CSA		**APRV-UR**			

FIGURE 8.149 OPEN single-phase motor nameplate

EXERCISE 8.15

Use the information presented in the motor specification of **Figure 8.146** to answer the following questions.

a What is the output power of the motor?

b How many poles does the motor have?

c What is the full-load current for this motor?

d Does this motor have anti-condensation heaters?

e How many operating speeds are provided for this motor?

Use the information presented in the motor nameplate of **Figure 8.147** to answer the following questions.

f Calculate the motor input power when operating on full load.

g Use the value calculated in Question f and the specified motor output power to calculate the motor full-load efficiency.

h Compare the calculated efficiency from Question g with the specified efficiency.

Use the information presented in the motor specification of **Figure 8.148** to answer the following questions.

i What bearings are specified for DE and NDE?

j Is it possible to operate this motor in the reverse direction?

REVIEW QUESTIONS

1 What is single-phasing of a three-phase motor?

2 List the four types of short circuits that relate to motor windings.

3 How may the effect of a short-circuit coil fault manifest?

4 What type of test checks for damaged laminations or hot spots?

5 Name two causes of unequal operating voltages with respect to three-phase motors.

6 What is the polarisation index a measure of?

7 What is the function of solid-state motor condition analysis instruments?

8 Name a technique that detects thermal hot spots with respect to electric motors.

9 What does the MEP standard set out?

CHAPTER REVIEW

8.1 Three-phase induction motor operating principles

- The turning action of an induction motor is the result of Faraday's law and Lenz's force law applied to a conductor.
- Fleming's left-hand motor rule shows the turning direction of the rotor.
- The three-phase supply creates a rotating electromagnetic field in the stator winding, which the induced electromagnetic field in the rotor follows thereby producing torque.
- At standstill an induction motor behaves as a transformer with a shorted secondary winding.
- The speed of the rotating flux of an induction motor is the synchronous speed (Ns). The synchronous speed (rev/min) is directly proportional to the frequency (Hz) of the supply voltage and inversely proportional to the number of poles.
- The difference between the synchronous speed and the actual speed of the rotor is the slip speed.

8.2 Three-phase induction motor construction

- The three-phase induction motor consists of a ribbed frame that contains the stator and the rotor. The stator is made up of thin sheet metal called laminations which are pressed together to form the stator.
- The rotor assembly is normally skewed to increase the starting torque and to prevent electromagnetic cogging.
- There are two types of induction motor rotors: a squirrel-cage rotor and a wound rotor.
- Most three-phase stators have six leads brought out to a six-terminal block.

8.3 Three-phase induction motor characteristics

- The induction motor operates due to the torque produced by the interaction of the stator electromagnetic field and the rotor's electromagnetic field.
- Induction motors provide extra torque at starting in order to overcome the inertia of starting from rest.
- The breakdown torque is the stalling point of the rotor and is the maximum torque the rotor can generate at rated voltage and frequency.
- Induction motors produce maximum torque when rotor resistance equals rotor reactance (X_L).

- The type of rotor cage and the shape of the rotor slot determine the torque characteristic and the magnitude of the starting current.
- The power output for any type of induction motor is stated on the motor nameplate and is its full-load output power.
- The efficiency rating of an induction motor takes into account the energy losses that dissipate in both the stator and the rotor.
- The power factor describes a percentage of current drawn by the motor that does some form of work.

8.4 Single-phase split-phase motor

- The stator windings of single-phase induction motors are unable to develop a rotating magnetic field in order to create starting torque unless additional methods start the rotor turning.
- Phase splitting occurs when the motor 'sees' two voltage sources with one voltage displaced from the other by 90 electrical degrees.
- The synchronous speed of a single-phase motor changes when the number of poles, or the frequency of the supply voltage, varies.
- Single-phase motors produce less starting torque than the equivalent three-phase motor, draw a higher line current and are less efficient.

8.5 Single-phase shaded-pole and capacitor motors

- Capacitor-start split-phase motors are similar to split-phase motors, except that an a.c. electrolytic capacitor causes the current shift to start the motor.
- Capacitor-start split-phase motors develop a higher starting current and hence larger starting torque.
- Start capacitors increase motor starting torque and allow a motor to be cycled on and off rapidly.
- Shaded-pole motors are ideal as the driving motors in light-duty applications such as multispeed fans.
- Split-phase capacitor-start motor applications include belt-drive prime movers for small conveyors, large blowers and pumps.
- Capacitor-start capacitor-run motors are ideal for most applications requiring constant speed under varying loads.
- Capacitors used for starting single-phase motors are a.c. aluminium electrolytics.
- Run capacitors have a very high current load capacity coupled with a very low loss factor.

8.6 Series universal single-phase motor

- A universal motor is a small series motor that is not an induction motor.

8.7 Motor protection

- Motor protective devices are required in order to safeguard the motor, motor control equipment and the supply circuit conductors against excessive heating.
- The operation of a fuse-type protection device is based on deliberately connecting a weak link in series with an electric circuit.
- A motor-start or time-delay fuse has a thermally controlled element that delays the opening of the element in the case of overload.
- Circuit breakers will interrupt all phases while fuse over-current protection devices protect only the phase that they connect in series with.
- Under-voltage protection allows the supply contactor to operate in response to dips in three-phase voltage, a phase imbalance, loss of phase or a loss of all three phases.
- Over-current relays initiate a circuit breaker tripping sequence to isolate the motor from the rest of the system.
- Thermal overload relays simulate the heating effect in the motor windings by passing the motor current through resistive elements in the relay.
- Interchanging the connections of any two of the three power conductors reverses the motor direction.

8.8 Synchronous motors

- Synchronous motors have high operating efficiency with smooth, constant starting, accelerating torque and versatile power factor control. The damper winding helps to repress hunting caused by torque oscillations.
- All synchronous motors have a constant speed (synchronous) that is independent of the load they are driving.
- The torque produced by hysteresis synchronous motors is uniform throughout every revolution thereby eliminating the effects of cogging.
- V-curves represent the effect of different values of field excitation on armature current and the power factor for various loads.
- The maximum torque a synchronous motor can develop is the pull-out torque and it occurs when the power angle is at 90°.
- Synchronous capacitors operate at a leading power factor in order to compensate for the lagging power factor of a commercial or industrial power system.
- The efficiency of a synchronous motor is determined at rated output voltage, frequency and power factor.

8.9 Alternators and generators

- The alternating current synchronous alternator is the most important means of producing electric energy.
- There are two basic means available by which to connect the three separate pole-phase group windings: these are delta or star.
- Energy from an exciter is necessary to establish the magnetic field in a revolving field alternator.
- The electromotive force developed by an alternator is directly proportional to speed, field flux density and the number of stator conductors.
- Regulation curves illustrate how the terminal voltage of an alternator changes as a function of the load current.
- Some self-excited single-phase alternators may need to have the fields flashed to establish the residual magnetism.

8.10 Testing of low voltage a.c. machines

- Before thoroughly testing a low voltage a.c. machine an important first step is to complete a repair specification.
- Faults that may occur in low voltage a.c. machines include: shorted or partially shorted stator windings, earth faults, insulation faults and rotor faults.

8.11 Mechanical faults in LV a.c. rotating machines

- Mechanical faults that occur in low voltage a.c. rotating machines involve: bearings, fans, bent shafts, locked rotors, blocked air vents, centrifugal switch failure and environmental factors.

8.12 Driven load and coupling faults

- Driven load faults in low voltage a.c. machines include: slipping belts, coupling alignment, vibration, bearing failure and load stalling.

8.13 Electrical faults and symptoms in LV a.c. rotating machines

- A number of electrical faults can occur in a.c. machines, including: single phasing, phase reversal, short circuits, core and rotor lamination faults, voltage faults and earth fault.
- The polarisation index of a motor is another measure of the condition of the insulation of a motor or generator in service.

TRIAL EXAM

For Chapter 8 knowledge assessment, please complete the following trial exam.

1 The directions of the stator field, rotor rotation and rotor currents are determined by:
  a Fleming's right-hand rule
  b Faraday's right-hand rule
  c Fleming's left-hand rule
  d Faraday's left-hand rule

2 Factors that determine the speed of the rotating magnetic field in an induction motor are:
  a the supply frequency and the number of current paths
  b the number of poles and the supply frequency
  c the rotor frequency and the slip of the motor
  d the supply frequency and the number of turns in each coil

3 As the speed of an induction motor increases, the slip of the rotor:
  a decreases
  b remains the same
  c increases
  d surges

4 The stator core of an induction motor is laminated to:
  a improve starting torque
  b provide silent running
  c reduce hysteresis loss
  d reduce eddy current loss

5 How many slip rings would you expect to find on the shaft of a squirrel cage rotor?
  a none
  b one
  c two
  d three

6 The wound rotor windings of a three-phase induction motor are usually connected in:
  a star configuration
  b delta configuration
  c split-phase configuration
  d long shunt configuration

7 Torque developed by a squirrel cage induction motor is at a maximum:
  a when the rotor is at standstill
  b at synchronous speed
  c when rotor resistance equals rotor reactance
  d at rated speed

8 If the supply voltage applied to a three-phase induction motor is doubled, the torque will:
  a halve
  b double
  c triple
  d quadruple

9 The results obtained from a locked rotor test would be:
  a starting torque and current
  b maximum torque and current
  c full load torque and current
  d no-load torque and current

10 The centrifugal switch of a split-phase motor disconnects the:
  a run winding at rated full speed
  b start winding at rated full speed
  c start winding at 75% of rated full speed
  d run winding at 75% of rated full speed

11 When the split-phase motor is running at rated full speed, the stator produces a:
  a rotating magnetic field
  b square magnetic field
  c triangular magnetic field
  d pulsating magnetic field

12 If the centrifugal switch on a split-phase motor becomes open circuit:
  a the motor will burn out
  b the motor will draw more current
  c the motor will fail to restart
  d the starting torque will improve

13 The single-phase motor which is most likely to be fitted with a reversing switch is the:
  a capacitor-start capacitor-run motor
  b permanently split capacitor motor
  c split-phase motor
  d capacitor-start motor

14 The single-phase split-phase motor capable of developing the most torque is the:
- a permanent split capacitor motor
- b capacitor-start motor
- c shaded-pole motor
- d capacitor-start capacitor-run motor

15 For a shaded-pole motor, the direction of flux movement and rotor rotation is always:
- a based on the active and neutral connection
- b from shaded pole to unshaded pole
- c from unshaded pole to shaded pole
- d determined by the number of poles

16 The single-phase motor which develops the most torque for its size is the:
- a capacitor-start capacitor-run motor
- b series universal motor
- c split-phase motor
- d capacitor-start motor

17 If the brushes on a universal motor became open circuit it will:
- a run at reduced torque
- b run at reduced speed
- c continue to run but will not re-start
- d cease to run

18 For the universal motor to run efficiently in a.c. mode the field poles, stator frame and armature core must be:
- a made from paramagnetic material
- b made from diamagnetic material
- c laminated
- d integrated

19 If the load on a universal motor is doubled, the torque is:
- a quartered
- b halved
- c doubled
- d quadrupled

20 The armature and field circuits of a universal motor are connected in:
- a series mode
- b short shunt mode
- c long shunt mode
- d compound mode

21 The inverse current-time characteristic of thermal overloads means that the overload will react:
- a faster to a larger overload current than to a small overload current
- b to any overload current regardless of the current value
- c slower to a larger overload current than to a small overload current
- d faster to a smaller overload current than to a larger overload current

22 Electronic type thermal protection units operate in conjunction with a/an:
- a NTC thermistor
- b ZTC thermistor
- c PTC thermistor
- d bi-metal strip

23 In-line thermal overload **heaters** are connected:
- a in two of the motor supply lines
- b in series with the motor windings
- c in series with the motor control circuit
- d in the control circuit

24 A three-phase induction motor is started DOL and locks up at start. The expected response of the protection device would be:
- a rapid tripping of the motor overload protection
- b instantaneous tripping of the short circuit protection
- c instantaneous tripping of the motor overload protection
- d slow tripping of the motor overload protection

25 Typically, motors above 30 kW would be limited to:
- a 2 starts per hour
- b 4 starts per hour
- c 6 starts per hour
- d 8 starts per hour

26 The number of electromagnetic poles on the rotor of a synchronous motor is always:
- a two more than the number of stator poles
- b the same as the number of stator poles
- c two less than the number of stator poles
- d the same as the number of pony motor poles

27 For a four-pole synchronous alternator to produce a 50 Hz output, the alternator would need to run at:
- a 375 rpm
- b 1000 rpm
- c 1500 rpm
- d 3000 rpm

28 Thermal overload **contacts** are connected:
- in the motor supply lines
- in series with the motor windings
- in parallel with the motor windings
- in the control circuit

29 The synchronous motor that has smooth poles which slide onto a rotor is a/an:
- a reluctance type
- b hysteresis type
- c permanent magnet type
- d induction type

30 The rotor speed of a synchronous machine:
- a is the same as the supply frequency
- b is the same as the synchronous speed
- c is determined by the per cent slip
- d is determined by the load

31 Determine the number of poles required by a 50 Hz alternator that is driven by a prime mover with a speed of 125 rpm.
- a 50
- b 48
- c 40
- d 24

32 Determine the input power of a three phase 415 V, four-pole synchronous motor drawing 14.23 A at a power factor of 0.88.
- a 10 kW
- b 9 kW
- c 8 kW
- d 6 kW

33 Determine the armature current per phase of a three-phase 10 kVA 415 V four-pole 50 Hz star-connected synchronous motor operating at unity power factor.
a 24 A
b 19.6 A
c 13.9 A
d 8 A

34 When an inductive load on a synchronous generator increases, terminal voltage:
a increases
b decreases
c remains the same
d doubles

35 For a self-excited generator to develop output voltage:
a it must be run at synchronous speed
b it must be started via a variable speed drive
c there must be residual magnetism in the rotor
d the rotor must have salient poles

36 If the d.c. excitation voltage applied to an alternator is increased, the result will:
a increase the power factor of the load
b increase the terminal voltage
c increase the efficiency of the machine
d increase the minimum energy performance standard

37 A prime mover must:
a operate at constant speed
b adjust its speed for load variations
c never reach synchronous speed
d be an induction motor

38 A stand-by alternator is switched on-line via a:
a make-before-break switch
b break-before-make switch
c an electromechanical overload relay
d a TOL equipped circuit breaker

39 To prevent magnetic locking the rotor bars of squirrel cage rotors are:
a skewed
b parallel to the stator slots
c star connected
d open circuited

40 To produce a rotating field in a three-phase induction motor, the currents drawn by each stator phase group must be displaced by:
a 0 electrical degrees
b 60 electrical degrees
c 90 electrical degrees
d 120 electrical degrees

41 A double cage rotor enables an induction motor to:
a have a higher starting torque than standard motors
b have higher starting current than standard motors
c deliver pulsating torque
d have variable speed

42 Calculate the power output of a three-phase induction motor that develops a torque of 20 Nm at 1440 rpm.
a 9500 W
b 3 kW
c 59717 W
d 855 W

43 Most standard TEFC motors have an ingress protection classification of:
a IP22
b IP55
c IP65
d IP23

44 The difference between the rotor speed and the synchronous speed is used to define a motor parameter called the:
a slip speed
b speed frequency
c average speed
d nominal speed

45 Determine the synchronous speed of a six-pole 400 V 50 Hz three-phase motor.
a 1000 rpm
b 3000 rpm
c 1500 rpm
d 750 rpm

46 Determine the percentage slip of a four-pole 400 V, 50 Hz three-phase motor with a full-load shaft speed of 1440 rpm.
a 92%
b 4%
c 8%
d 104%

47 At the instant of start the rotor slip is:
a 92%
b 40%
c 10%
d 100%

48 Misalignment between an electric motor shaft and the shaft of the driven equipment represents a common cause of:
a rotor bar failure
b stator failure
c bearing failure
d line current failure

49 What type of winding makes a synchronous motor self-starting?
a an auxiliary winding
b salient winding
c stator windings
d damper winding in the pole face

50 How can the rotation of a three-phase motor reverse?
a by interchanging the connections of any two of the three power conductors
b by reversing the rotor
c by turning the motor frame around
d by connecting in delta rather than in star

9 Transformers

This chapter provides electrotechnology workers with knowledge of the basic construction, principle of operation and characteristics, applications, connection methods and testing requirements of various transformers and their ancillary equipment. In addition, electrotechnology workers will gain knowledge of the safety issues relating to transformer usage and high voltage. This chapter provides underpinning knowledge for the unit UEEEL0025 from the UEE training package.

LEARNING OBJECTIVES

Transformer operation

- Outline the principles of mutual induction as applied to transformer operation
- Determine the value of secondary voltage and current in a transformer
- Explain the principles of power transference from primary to secondary

Transformer core construction, laminations and insulation

- Describe common lamination types and applications
- Explain the construction of common transformer windings

Application of transformers

- Explain common uses of transformers

Voltage transformers

- Explain the main classifications of voltage transformers

Transmission and distribution transformers

- Name types of auxiliary equipment used with transformers
- Explain the operation of tap changers
- Outline the function of various transformer ancillary equipment

Autotransformers and instrument transformers

- Explain the principle of operation of autotransformers
- Explain the principle of operation of potential and current instrument transformers
- State the safety requirements with respect to instrument transformers

Transformer losses and efficiency

- Calculate efficiency, losses and all-day efficiency
- Draw the equivalent circuit of a transformer

Transformer cooling

- State the methods used to cool a transformer

Voltage regulation and percentage impedance

- Calculate the percentage impedance of a transformer
- Explain the meaning of voltage regulation
- Explain how harmonics are generated in transformers

Parallel operation of transformers

- Explain the requirements for connecting single- and three-phase transformers in parallel
- Explain how transformers are connected
- Describe the term 'vector grouping'
- Describe a tertiary winding
- State the effects of incorrect transformer connections

Industry standards and *Wiring Rules* requirements

- Define high voltage
- Understand the requirements of an access permit and switching sheet
- Explain the difference between working earths and operational earths
- Describe the purpose of signage in relation to high voltage
- State the requirements of safe approach distances for high-voltage work
- Define step and touch potential

Transformer nameplates and specifications

- Outline the purpose of transformer nameplates
- Explain the procedure for performing basic transformer tests of proving insulation, testing continuity, and identifying windings

9.1 Transformer operation

The transformer is one of the essential pieces of apparatus effecting the transmission of electrical energy. The use of transformers enables electric power to be generated at any convenient voltage or current level required. A transformer consists primarily of:

- a soft-iron core
- a primary winding
- a secondary winding.

The basic transformer as shown in **Figure 9.1** consists of two coupled coils in close proximity to each other, wound on a ferromagnetic substance and sharing a common electromagnetic circuit. A transformer is a device for transferring electrical energy at the same frequency from one sinusoidal circuit to another by means of the common electromagnetic circuit. The transformer's purpose is to modify the value of voltage. Actually, electrical energy at a high voltage may be transformed to energy with a lower voltage, and energy with a lower voltage may be transformed to energy at a higher voltage. Similarly, the current value in the energising circuit may change to a different value in the secondary circuit. A transformer can also transform impedances. Finally, the transformer can provide isolation between the supply and the load while maintaining the electrical supply to the load.

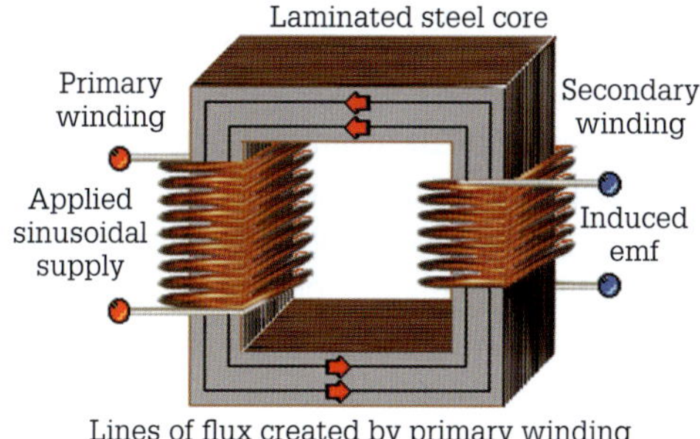

FIGURE 9.1 Basic transformer

The basic transformer consists of a rectangular steel core built up from thin individual steel sheets called laminations. Two conductor windings called coils are located in close proximity on opposite limbs of the core. The coil connected to the energising circuit is the 'primary winding' and that connected to the load is the 'secondary winding'.

Transformer operation

Consider the diagram in **Figure 9.1**. The primary coil obtains energy from the applied sinusoidal supply. When the primary coil is energised current starts to flow, and an alternating electromagnetic field expands out of the coil. The primary coil then changes the electrical energy of the sinusoidal supply into electromagnetic energy of a magnetic field. The alternating electromagnetic field expands outwards from the centre of the primary coil and collapses into the coil as the sinusoidal supply through the coil varies from zero to a maximum and back to zero again. While the electromagnetic field of the primary coil is expanding and collapsing, it cuts through the turns of the secondary coil, inducing a voltage across it. The changing emf from the primary coil links with the secondary coil and mutual inductance causes an induced emf across the linked coils. Since the same electromagnetic flux links both coils, the same voltage is induced in each turn of both coils.

The induced voltage across the secondary coil is the voltage of mutual induction, and the action of inducing this voltage is transformer action. In transformer action, electrical energy transfers from the primary coil to the secondary coil by means of a varying electromagnetic field.

Whenever the secondary coil of a transformer disconnects from its load, there is no current drawn by the secondary coil. The primary coil, when the secondary coil is open circuited, draws only a small amount of current. This current supplies the magnetomotive force, which produces the transformer electromagnetic flux. The current is the excitation or magnetising current.

When a load connects to the secondary coil of a transformer, the secondary current flowing through the secondary turns produces a counter-magnetomotive force (cmf). According to Lenz's law, the cmf acts in a direction that opposes the flux from the primary coil that produced it. This opposition of the cmf tends to reduce the transformer flux and results in a reduction in the cmf in the primary coil. Since both the impedance of the primary coil and the cmf in the core restrict the primary current, whenever the cmf reduces, the primary current continues to increase until the original transformer flux reaches a state of equilibrium.

Coupling

Coupling is a quantity of measurement used in transformer design to measure the effect of the position of the individual windings and the mutual induction in the transformer circuit.

For example, transformers work because of induction. This implies that induction will be greater if the two coils are closer together, as the electromagnetic flux produced by the energised coil affects the maximum capability of the secondary coil. If the coils are further apart or the secondary coil aligns at an angle to the primary coil, less electromagnetic flux is available to do work in the secondary coil.

If all the electromagnetic flux of the primary coil cuts across the secondary, the coupling quantity has a value of one, known as 'unity coupling'. **Figure 9.2** shows the effect of coupling.

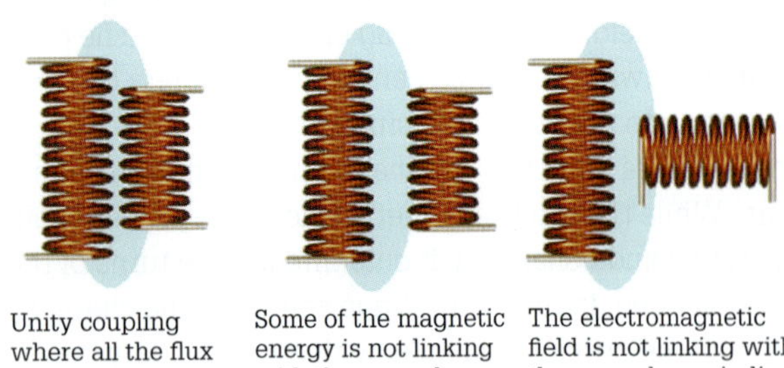

FIGURE 9.2 Effect of coupling

If the two coils separate so that only half the electromagnetic energy transfers from the primary to the secondary coil, the coupling quantity is 0.5. This value indicates the percentage of electromagnetic flux available for mutual induction and is the 'coefficient of coupling' (symbol k). For maximum mutual induction to transpire, both coils must be parallel to each other. If there is an angular difference between the primary coil and the secondary coil, the mutual induction decreases. This relationship is shown in **Figure 9.2**.

Idealising the transformer can bring out the most important aspects of transformer action. **Figure 9.3** illustrates an ideal transformer connected to a load that has the following electrical and electromagnetic properties:

- the ohmic resistance of both coils is zero
- the electromagnetic flux confines itself to the laminated core and links completely with both coils
- the laminated core offers no opposition to the magnetic flux.

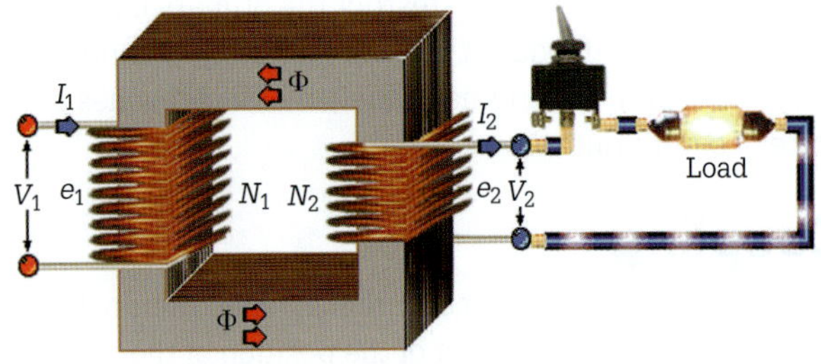

FIGURE 9.3 Ideal transformer

Circuit symbols

Transformer symbols are located in AS/NZS1102.106: 1997 'Graphical symbols for electrotechnical documentation', Part 106, Production and conversion of electrical energy. Two graphical symbols usually occur for each device in the standard as shown in **Figure 9.4**.

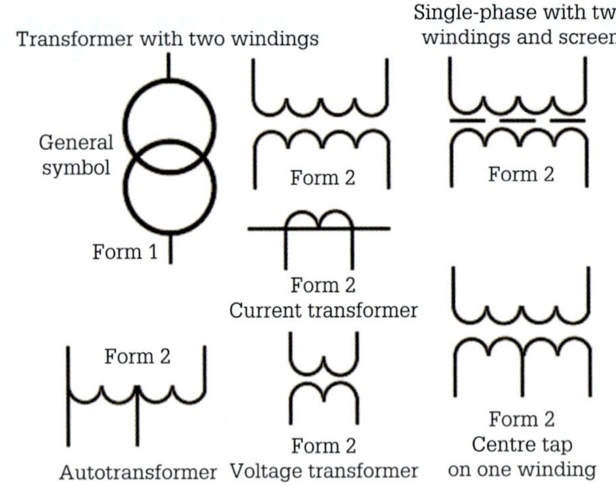

FIGURE 9.4 Transformer symbols

Relationship between voltage, current and turns

Referring to **Figure 9.3**, when the input voltage (V_1) connects across the primary coil, a magnetic flux (Φ) develops in the core. Consequently, the voltage (e_1) induced by the coil's self-induction is approximately equal to the input voltage (V_1).

The induced voltage (e_1) is the result of the number of conductor turns (N_1) and the strength of the magnetic flux.

$$V_1 = e_1 = N_1 \frac{\Delta\Phi}{\Delta t} \tag{1}$$

The magnetic flux developed by the primary coil links with the secondary coil by means of the ferromagnetic core. The effect of this is the induction of a voltage (V_2) across the ends of the secondary coil. Consequently, voltage V_2 by the self-induction property of the coil induces an electromotive force (e_2) that has the same value as the secondary terminal voltage (V_2).

The induced voltage (e_2) is the result of the number of secondary turns (N_2) and the electromagnetic flux in relation to time.

$$V_2 = e_2 = N_2 \frac{\Delta\Phi}{\Delta t} \tag{2}$$

From equations 1 and 2 we obtain:

$$\frac{V_1}{V_2} = \frac{N_1}{N_2} \tag{3}$$

Note that an unloaded transformer acts as an inductor. The primary coil only draws a tiny magnetising current because of the impedance of the primary circuit. No current occurs in the secondary coil.

When a load connects across the secondary winding, current (I_2) flows establishing a magnetic flux.

Consequently, a magnetomotive force (mmf) develops. The value of the magnetomotive force is dependent upon the number of conductor turns (N_2) and the load current (I_2).

The development of the mmf causes the primary current (I_1) to increase, creating a counter-magnetomotive force (N_1I_1) that opposes N_2I_2. As the transformer (**Figure 9.3**) is ideal, there is no magnetomotive force required to establish an electromagnetic flux.

$$N_1I_1 = N_2I_2 = \text{net mmf} = 0 \text{ and } \frac{N_1}{N_2} = \frac{I_2}{I_1} \qquad (4)$$

SWITCH ON

Note that the currents in the primary and secondary windings are inversely proportional to the turns of the coils and that if higher current is demanded by the load, more current flows from the supply into the primary winding. It is this magnetomotive-balancing characteristic ($N_1I_1 = N_2I_2$) that enables the primary coil to know of the presence of current in the secondary winding and to respond to the loading needs of the secondary coil.

As equations 3 and 4 relate to the number of turns, these two equations can be expressed as:

$$V_1I_1 = V_2I_2 \text{ and } \frac{V_1}{V_2} = \frac{I_2}{I_1} \qquad (5)$$

The equation $V_1I_1 = V_2I_2$ is another way of expressing power ($P = VI$) in volt-amperes. In an ideal transformer, the instantaneous power input to the transformer equals the instantaneous power out ($P_{in} = P_{out}$) from the transformer because there are no power losses experienced.

When the load dissipates power, an equivalent amount of power is provided by the supply. The interconnection of these transformer equations is the transformation ratio and is expressed as follows:

$$\frac{V_1}{V_2} = \frac{N_1}{N_2} = \frac{I_2}{I_1}$$

This equation states that the ratio of the primary current in relation to the secondary current of an ideal transformer equals the inverse ratio of the number of turns. As current and voltage can vary (but not their ratios) depending upon load and input conditions, the constant relationship is the turns ratio.

EXAMPLE 9.1

The transformer in **Figure 9.5** has 1000 turns on the primary winding and 500 turns on the secondary winding.

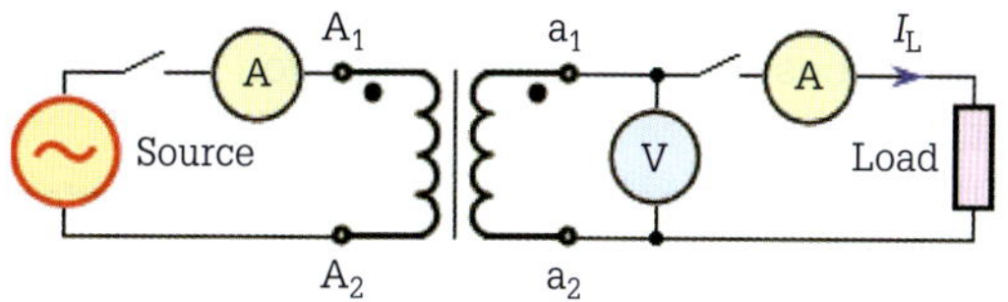

FIGURE 9.5 Transformer for Example 9.1

If a 230 V 50 Hz supply is connected to the primary winding and a resistive load of 200 Ω connects across the terminals of the secondary, calculate:

a secondary voltage

$$\frac{V_1}{V_2} = \frac{N_1}{N_2}$$

Transposing the equation:

$$\text{Secondary voltage } V_2 = \frac{V_1N_2}{N_1} = \frac{230 \times 500}{1000} = \mathbf{115\ V}$$

$$\text{Secondary current } I_2 = \frac{V_2}{R} = \frac{115}{200} = \mathbf{575\ mA}$$

b primary current

$$\frac{N_1}{N_2} = \frac{I_2}{I_1}$$

Transposing the equation:

$$\text{Primary current } I_1 = \frac{N_2 I_2}{N_1} = \frac{500 \times (575 \times 10^{-3})}{1000} = \mathbf{287.5\ mA}$$

Note: The constant ratio (turns ratio) for this transformer is 2:1. The voltage relationship is 230:115 which is 2:1. The current relationship is 0.575 amperes (I_2) to 0.2875 amperes (I_1), which is the converse of the turns ratio.

EXERCISE 9.1

A transformer has 920 turns on the primary winding and 96 turns on the secondary winding.

If the primary connects to a 230 V 50 Hz supply and the secondary connects to a resistive load of 16 Ω, calculate:

a secondary voltage

b secondary current

c primary current.

REVIEW QUESTIONS

1 What are the three main components of a transformer?
2 What is the purpose of a transformer?
3 How is a voltage produced in the secondary winding of a transformer?
4 How does coupling affect induction in a transformer?
5 Draw the general circuit symbol for a transformer.
6 A transformer has 920 turns on the primary winding and 48 turns on the secondary winding. If the primary connects to a 230 V 50 Hz supply and the secondary connects to a resistive load of 24 Ω, calculate:
 a secondary voltage
 b secondary current
 c primary current.

9.2 Transformer core construction, laminations and insulation

The major components of a transformer are:

- the core
- the windings
- insulation.

Core construction

The core assembly of a transformer may be:

- toroidal-type
- shell-type
- core-type

Toroidal-type construction

The toroidal-type transformer is illustrated in **Figure 9.6**.

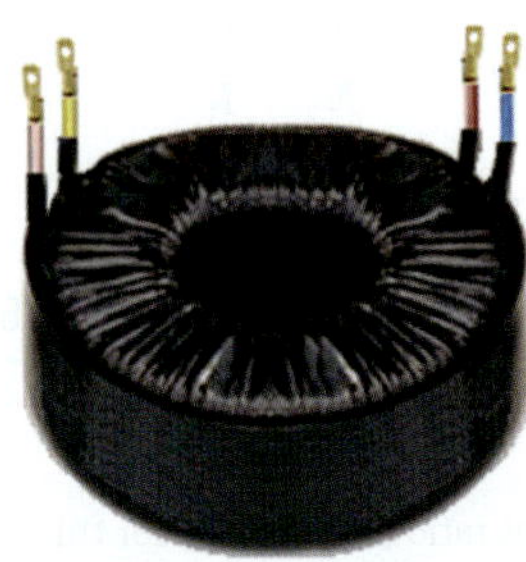

FIGURE 9.6 Toroidal power transformer

Toroidal transformers as shown in **Figure 9.6** use tape-wound circular ferromagnetic cores. This class of transformer offers many advantages over a conventional laminated transformer. For example, they have less weight and are smaller.

The type of core provides an almost perfect magnetic circuit having its primary and secondary windings uniformly distributed around the core. The ring core minimises losses, fringing, leakage and distortion, and provides excellent magnetic shielding.

These transformers also decrease the magnetising force required to produce a given flux density, and they are much more efficient than E-type lamination cores (refer to page 406).

Shell-type construction

The name 'shell' results from the fact that a shell of steel surrounds the coil windings. The primary and secondary windings occur on the centre leg of the transformer core.

In general, the shell-type transformer design is more economical for low-voltage transformers while the core-type design is more economical for high-voltage transformers. Laminations for high-voltage transformers are commonly mitre-corner-cut (45° angle) and V-notched. **Figure 9.7** shows the shell-type construction using E & I laminations.

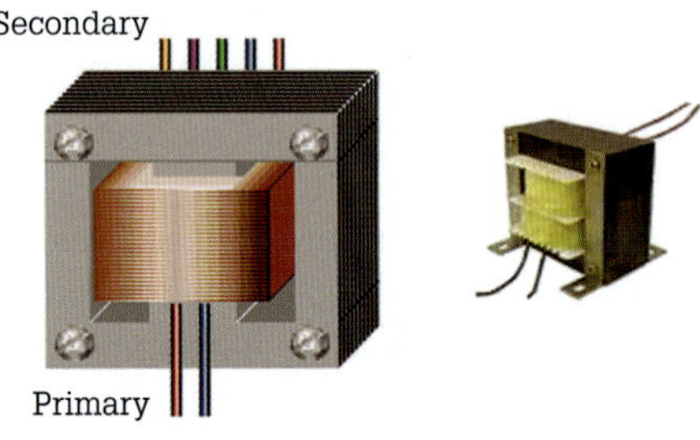

FIGURE 9.7 Shell-type construction

The windings of the shell-type transformer are rectangular, and the ferromagnetic core occurs through the opening and around the outside of the windings to form a shell. Larger transformers utilise pancake-type coils, while smaller VA types use a bobbin coil method as **Figure 9.8** shows.

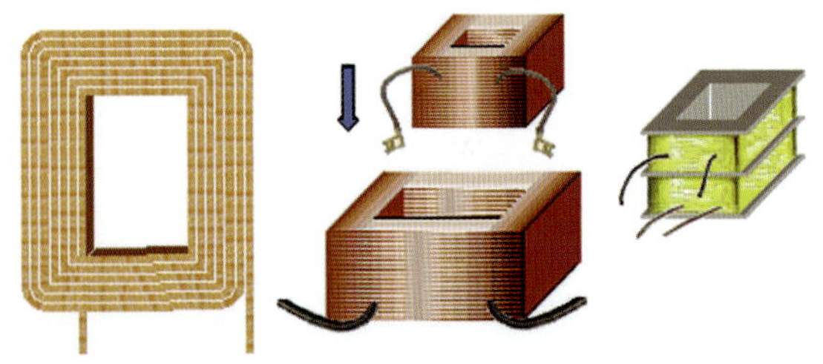

FIGURE 9.8 Shell coils – pancake, layer and bobbin

A characteristic of the shell-type transformer is a short electromagnetic circuit consisting of two parallel paths as illustrated in **Figure 9.9**.

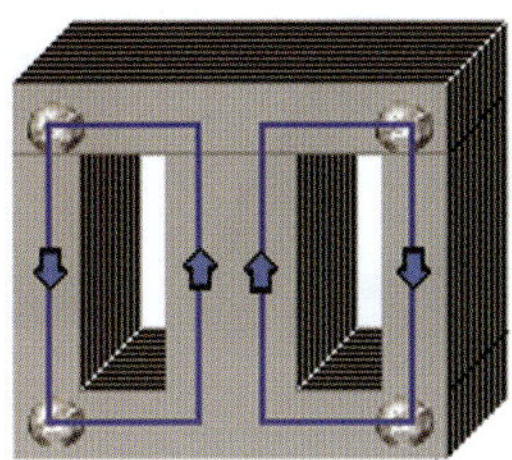

FIGURE 9.9 Electromagnetic circuit showing the two parallel paths

The width of the core's centre means that the mean length of the turn is longer than with core-type construction. In contrast, the shell-type transformer has a high voltage per turn together with fewer turns due to the primary and secondary windings occupying the same leg. The shell transformer has less magnetic fringing and fewer losses when compared with the core-type transformer.

Core-type construction

Core-type transformers usually have the low-voltage winding on the inside next to the ferromagnetic core. Around the outside of the low-voltage coil there is an insulating barrier and the high-voltage winding is outside this. The characteristics of the core-type transformer are a long electromagnetic circuit and a shorter length mean turn for the winding than for the shell-type transformer.

The ferromagnetic core is in the shape of 'legs' surrounded by the windings and joined at the ends by 'yokes' as shown in **Figure 9.10**.

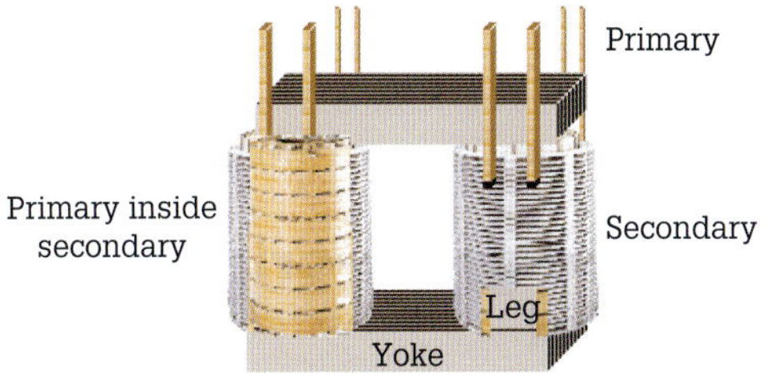

FIGURE 9.10 Core-type showing the yoke and legs

Transformer winding types

Efficient transformer winding happens when the required number of wire turns uses a minimum of space. At the same time, the cross-section of the conductor must be able to carry the rated current without overheating. There are four winding types used depending on the transformer design. These are:

- layer
- pancake
- bobbin
- toroidal.

Layer-wound coil

A layer-wound coil consists of single layers of conductors separated by layers of sheet insulation such as Kraft paper, Mylar or Nomex. Here, the insulation serves a twofold purpose: it is a support platform for the conductor layer and provides electrical isolation from the layer of turns above and below. Layer-wound coils require accurate placement of conductors and insulation material, while the winding process itself is labour intensive.

Pancake coils

Pancake coils are used with large shell-type transformers. They are edge wound with rectangular copper strap. The winding uses one turn per layer, similar to a clock spring.

Bobbin-wound coil

Bobbin-wound coils are usually for transformers with power ratings of less than 1 kVA. The bobbin is the platform support for the windings and provides electrical isolation between the primary and secondary coils. The conductor turns are precision wound using the layer method of winding but with thin film-type insulation installed where required.

Toroidal windings

Toroidal windings have a complicated winding technique because of the ring nature of the core. Due to the complexity, toroidal-type constructions are expensive and are reserved for situations where the size and weight benefits are superior to that for a transformer built using the E & I lamination method.

Lamination design

The most significant single factor in the design of the shell-type and core-type transformers is the lamination design. Lamination design affects transformer size and volume. Once a design happens, the lamination itself cannot alter paying a hefty penalty in modifications to tooling for an entirely new die.

Joints in the lamination circuit must be as tight as possible, with the joints in adjacent laminations overlapped by alternating layer (a) in **Figure 9.12** with layer (b). Overlapping provides a small air gap for the electromagnetic flux to cross and then develop a comparatively small magnetising current. Such attention to air gaps also reduces the amount of electromagnetic hum (noise) that transformers generate.

In small volt-ampere (VA) transformers, the mean length of the magnetic circuit is small and so the air gap is of more relative importance than in larger transformers. To minimise the magnetising current in these small VA transformers, E & I laminations are used. Another method of minimising the magnetising current is to increase the width of the steel legs of the laminations that do not pass through the coil windings. Widening does not affect the size of the coil windings, but it does decrease the magnetic flux density in the widened legs of the laminations. The decrease in magnetic flux density decreases the iron loss and magnetising current. For larger power transformers, the laminations are straight pieces and mitre-corner-cut with no widening used because the air gap is of less relative importance. However, smaller distribution transformers benefit from the lamination widening process. To limit the effects of eddy currents, laminations have an oxide or varnish coating to keep the laminations separate.

Shell-type transformer core

The shell-type transformer core consists of thin (0.27, 0.3 or 0.5 mm thick), individually insulated sheets of high-permeability, grain-oriented, non-ageing silicon steel. The silicon additive (approximately 4%) creates soft, low-remanence magnetic steel. This softening reduces the hysteresis loss and increases the resistance of the lamination, thereby limiting the effects of eddy currents. Steel that is subject to ageing will experience an iron loss increase from 10% to 30% after about a year's operation.

The laminations are punched, or laser-cut, from electrical grade sheet steel, without burrs, and hand stacked into a rectangular arrangement of different width laminations. The ferromagnetic circuit develops layer by layer to the width required by the transformer design. The existence of burrs creates air gaps in the core that in turn have the effect of increasing the magnetising current because air has a much lower permeability than steel. Burrs also prevent the optimum stacking factor to be met (the number of laminations required). **Figure 9.11** shows an example of a transformer demonstrating the width.

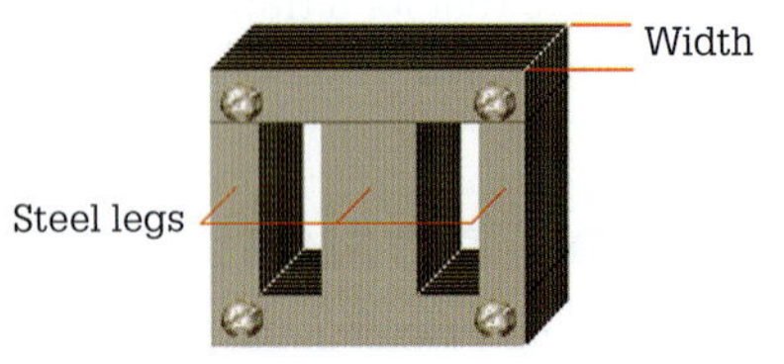

FIGURE 9.11 Transformer width

Examples of the shapes of various stampings of laminations are shown in **Figure 9.12**.

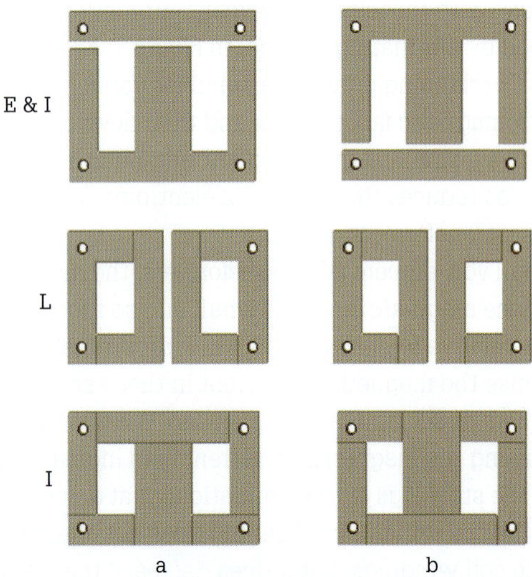

FIGURE 9.12 Lamination stampings – E & I, L and I

Other core designs include the U & I and C-type cores as illustrated in **Figure 9.13**.

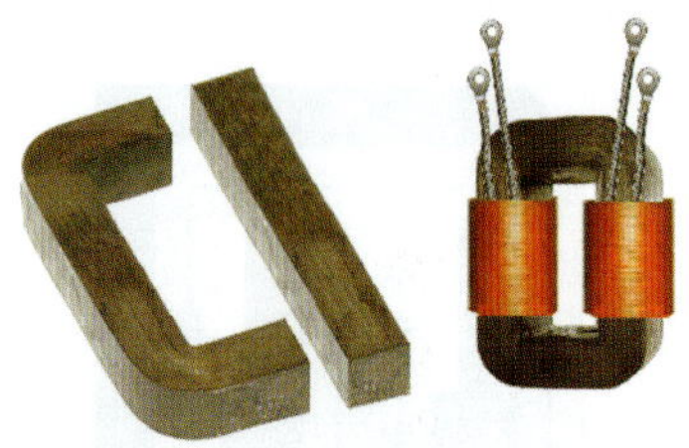

FIGURE 9.13 U & I and C-type cores

Lamination materials

Materials used for transformer laminations include:

- laser-scribed cold-rolled grain-oriented silicon steel
- amorphous magnetic metal
- cold-rolled grain-oriented silicon steel
- cold-rolled non-grain-oriented silicon steel
- amorphous steel.

Laser-scribed cold-rolled grain-oriented silicon steel

Laser-scribed cold-rolled grain-oriented silicon steel provides low core losses and is suitable for meeting sustainability requirements. Even a 0.1% efficiency improvement in iron losses for a single transformer provides a substantial carbon reduction when considering the number of power transformers in Australia.

Amorphous magnetic metal

Another lamination material is amorphous magnetic metal. This metal has a much higher electrical resistance than standard transformer steel and it reduces eddy current loss. However, amorphous metal is very difficult to cut down into sections for stackable laminations, such as those used traditionally in cores. Instead, amorphous metal cores are wound cores.

Cold-rolled grain-oriented silicon steel

Cold-rolled grain-oriented silicon steel (CRGO) has a cold process to establish thickness and flatness necessary for laminations. The process allows the magnetic 'grain' of the silicon steel to be aligned in one common direction, and a high permeability results. This steel is ideal for toroids and C cores, since the grain and the magnetic flux align with each other (i.e. in a circular pattern around the core). However, permeability is reduced for E & I laminations because the flux must flow across the 'grain' at the ends of the laminations.

Cold-rolled non-grain-oriented silicon steel

More suited to E & I laminations, cold-rolled non-grain-oriented silicon steel mirrors the CRGO process.

However, the magnetic grain is left random, resulting in no alignment of the magnetic domains – this reduces permeability. However, the permeability is more effective with stamped laminations as opposed to rolled (toroids and C cores).

Amorphous steel

Amorphous metals or metallic glasses form from a liquid state by a rapid quenching technique. The atomic arrangement of amorphous metals is out of order; it is not arranged on a crystalline lattice as occurs in ordinary metals. **Figure 9.14** shows the atomic arrangement of silicon and amorphous iron.

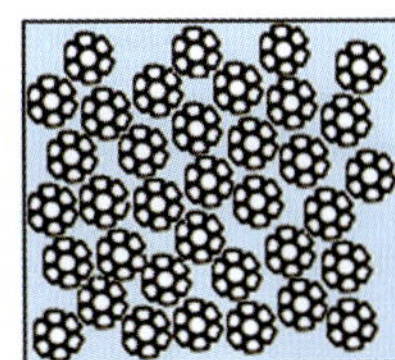

FIGURE 9.14 Crystal arrangement – silicon and amorphous iron

Amorphous alloys of iron (Fe), manganese (Mn), zirconium (Zr) and bismuth (Bi), are iron based and contain manganese, refractory metals (Zr and molybdenum (Mo)) and metalloids (boron (B)). Laminations made from an amorphous alloy are characterised by high saturation induction density, low coercive force, low energy consumption, low exciting current, anticorrosion characteristics, excellent temperature stability and long service life. Being a unique, soft magnetic material, it can greatly enhance a transformer's efficiency; make its physical size more compact; reduce its weight; and lower its energy consumption.

Design problems

When current flows in a transformer, heat generates in the windings and the ferromagnetic core. Heat can destroy the insulation, and one of the design engineer's primary problems is to prevent a transformer from attaining too high a temperature.

Another major difficulty is to provide sufficient insulation and to position it so that the transformer will withstand any voltage condition that it may experience in service. Other major problems are to build a transformer that will operate with reasonably low losses and to keep the construction material cost effective.

To obtain the best solution, various methods of winding and ferromagnetic cores have developed. These arrangements fall within one or other of three general types: the toroidal type, shell type and core type.

REVIEW QUESTIONS

1. List the three essential items of a transformer.
2. Name the three general classes of the arrangements of transformer cores.
3. What are two of the main advantages a toroidal transformer has over a conventional laminated transformer?
4. Where are the primary and secondary coils of a shell-type transformer located?
5. Name two shapes of lamination stampings used in the construction of shell-type transformers.
6. Name the core construction that is more economical for high-voltage transformers.
7. Describe the construction of the core-type transformer.
8. Name the four types of transformer winding.
9. Typically, what is the maximum rating for a bobbin-wound transformer?
10. Which transformer parameters are affected by lamination design?
11. What do laser-scribed cold-rolled grain-oriented silicon steel laminations provide?
12. What type of silicon steel is suited for toroids and C cores?

9.3 Application of transformers

Transformers find use in a variety of applications ranging from impedance matching to various forms of potential (voltage transformers).

Impedance matching

Transformers are able to change voltages and currents to other values (step up or step down) and in doing so provide power transfer between the primary and the secondary. These characteristics of a transformer allow it to serve as an impedance-matching device utilising the principle of maximum power transfer. This principle states that the maximum amount of power dissipated by the load impedance occurs when it equals the impedance of the circuit supplying the power. To help explain this, consider the case of sinusoidal voltage and the secondary impedance (load) as shown in **Figure 9.15**.

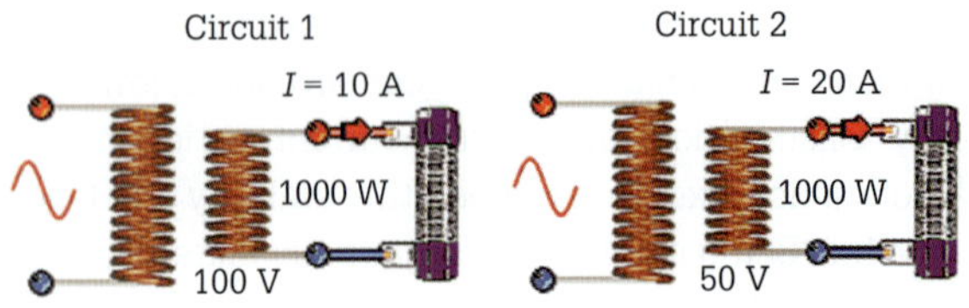

FIGURE 9.15 Sinusoidal voltage and the secondary impedance

Both impedances (the resistors) dissipate the same level of power, but this occurs at different secondary voltages and current outputs. When using Ohm's law to determine the impedances of the loads, the following results occur:

$$Z = \frac{V}{I} = \frac{100}{10} = \mathbf{10\ \Omega} \qquad Z = \frac{V}{I} = \frac{50}{20} = \mathbf{2.5\ \Omega}$$

Circuit one has a higher impedance load connected than circuit two. However, if the load of circuit one were transferred to circuit two, the current drawn would be 5 A and the power dissipation would be only 250 W or one-quarter of its rated power. The heavier load could work efficiently by using a step-up transformer as shown in **Figure 9.16**.

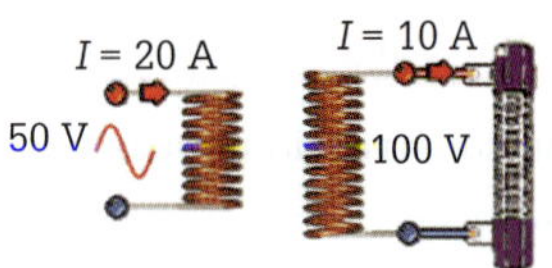

FIGURE 9.16 Step-up transformer

Examination of **Figure 9.16** and its primary circuit tells us that the voltage and current values are indicative of 2.5 Ω load impedance not the actual 10 Ω impedance of the load itself. Not only has the step-up transformer provided the correct secondary voltage and current, it has transformed impedance. This transformer energises the load to just the right level of drawing power at the correct voltage and current values to satisfy the maximum power transfer considerations and enable the most efficient power delivery to the load. The function of an impedance-matching transformer is to transform the high impedance of the primary winding to equal the impedance of the load.

Magnethermic effect of transformer action

The magnethermic effect of transformer action has heating applications as illustrated in the following types.

In-line induction water heaters

Alternating current from the primary winding of an in-line induction power unit, as shown in **Figure 9.17**, creates an electromagnetic field that couples with a coil made of hollow copper tubes. It follows that the coupling effect induces a current to flow through the metal of the electrically shorted copper tubes, producing heat. If water flows through these tubes the heat produced by the copper tubes raises its temperature. Water convection or a pump carries the heated water to a heat exchanger for removal. Induction heating provides fast, consistent heat for manufacturing applications requiring the heating of vessels, tanks, fluid or gas. Typical heating applications include air and gas heating for chemical and food processes, hot oil heating for process and edible oils, and vaporising and superheating of water for instant steam production.

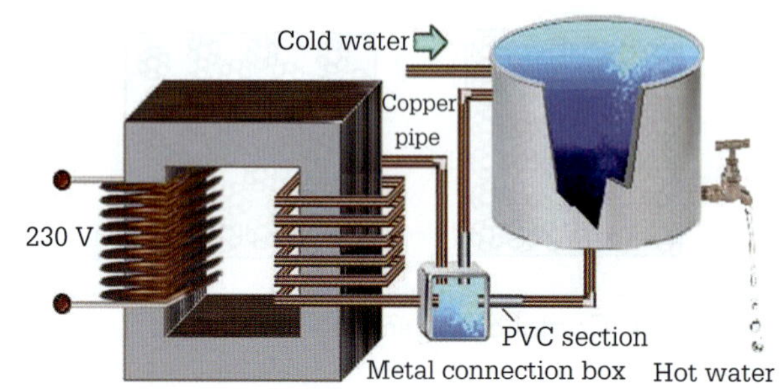

FIGURE 9.17 Induction water heating

Induction furnaces

Induction furnaces apply the same principle as the induction heater. Copper coils acting as the primary of a transformer encircle a layer of refractory material (material not susceptible to electromagnetic influences) surrounding the entire length of the furnace interior as **Figure 9.18** shows. Cooling water flowing through the copper coils prevents them from melting. Energising the coils creates a magnetic field that penetrates the refractory material. The lines of flux surrounding the work coil cut into the surface of the mass of conductive material requiring heat and induce circulating currents. These circulating currents quickly melt the mass of metal material inside the furnace.

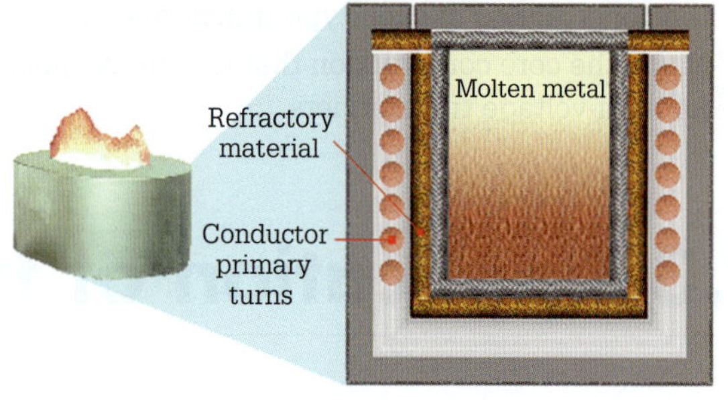

FIGURE 9.18 Induction furnace

Induction cooking

With induction cooking, as shown in **Figure 9.19**, the electromagnetic field becomes the invisible cooking zone projected above the induction plate surface. Heat is the result of eddy currents induced into the base of a metal pan. The eddy currents make the pan heat up quickly, as well as heating the contents inside it.

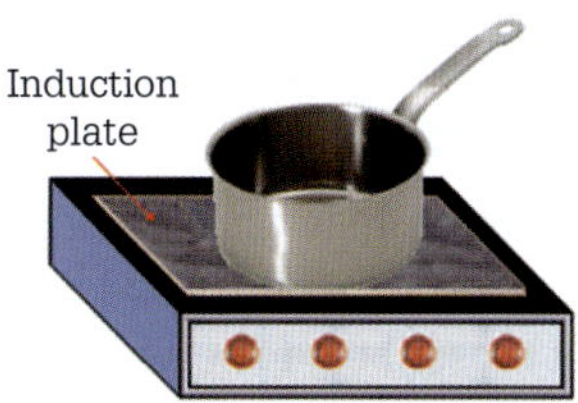

FIGURE 9.19 Induction cooking

When a small metal object, such as a key, is placed on the induction plate, the induction heater will not heat it because the induction appliance has a small-article-detection sensor.

Induction-bearing heaters

The induction-bearing heater as shown in **Figure 9.20** is an excellent and efficient way to expand the inner race of bearings and other devices for placement on round shafts. The bearing acts as a shorted secondary turn which causes a significant circulating current to flow, resulting in substantial heat being generated.

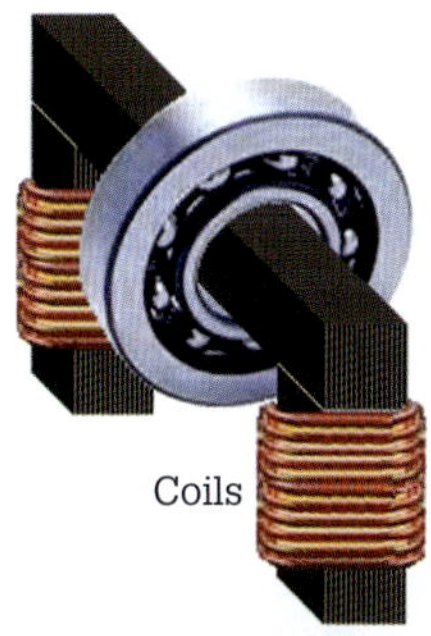

FIGURE 9.20 Induction-bearing heater

Saturable core reactors

A saturable core reactor, as shown in **Figure 9.21**, consists of a control winding connected to a direct current circuit which is used to vary the magnetic flux of the reactor iron core around which another alternating current coil, the power winding, exists. For this reason, varying the d.c. current through the control winding drives the iron core in and out of magnetic saturation. Control of the magnetic flux enables the flux transfer ratio between the two coils to change, and this varies the inductance of the power winding. An increase in current drives the iron core near to saturation, thereby decreasing the power winding's inductance and lowering its impedance. Consequently, the output power of the saturable reactor is controlled.

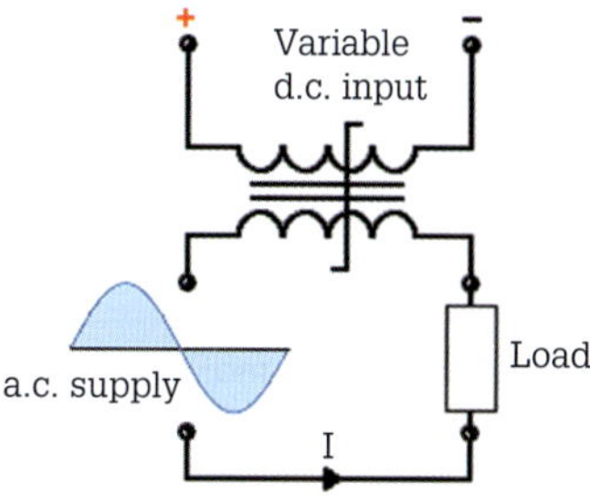

FIGURE 9.21 Circuit diagram – saturable core reactor

Saturable core reactors control the voltage and current for high-power loads such as industrial furnaces and ovens, welders and high-power voltage regulators.

REVIEW QUESTIONS

1 When a transformer is operating as an impedance-matching device what electrical principle is being utilised?
2 What does the term 'magnethermic effect' mean in relation to transformer action?
3 How is heat produced by an induction cooktop?
4 State the purpose of a saturable core reactor.
5 Name three applications for saturable core reactors.

9.4 Voltage transformers

The great variety of transformers makes it necessary to classify them under four general classes, called power, volt-ampere, custom and instrument.

Power transformers

Power transformers find application in the power transmission and distribution industries. They have the highest power rating and continuous voltage performance rating of all classes of transformers. Power transformers have a tank of welded steel plate construction suitably strengthened where necessary for the protection of, and as a container for, the cooling medium. Lifting lugs are welded to the sides of the tank to facilitate the lifting of the transformer bodily by means of a hoist or crane. **Figure 9.22** shows two styles of power transformer.

FIGURE 9.22 Power transformers

Power transformers feed the generated voltage from the alternator through a distribution system to the final consumer. Such a system showing the various stages of transformation is shown in **Figure 9.23**.

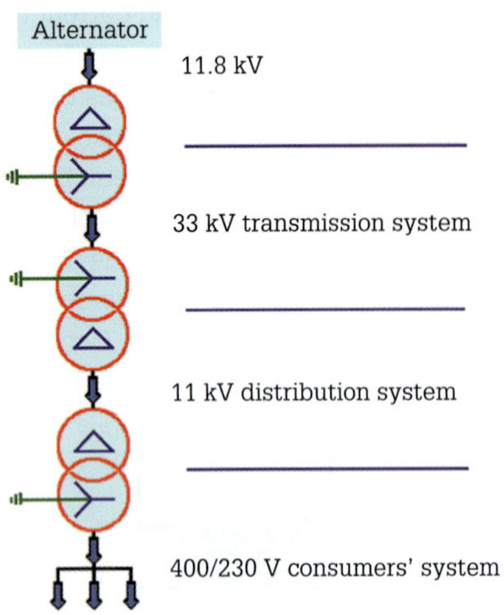

FIGURE 9.23 Stages of transformation

- Stage 1: Power transformers at generating stations step up the generated voltage to high levels (11.8 kV to 33 kV or even to 400–500 kV) for transmission.
- Stage 2: At district sub-stations, the transmission voltages are stepped down (33 kV or 66 kV) for distribution to the various zones.
- Stage 3: At the zone sub-station, distribution voltages step down (from 33 kV or 66 kV to 11 kV or 22 kV).
- This power is then distributed using pole or pad-based distribution transformers (11 kV to 400/230 V) for delivery to the customer.

Note that power or distribution transformers can create noise pollution. The maximum noise level should be below 61 dB (A).

Volt-ampere transformers

Volt-ampere transformers have ratings below 300 VA. Transformers in this class consist of many different types, with many different applications. For example, volt-ampere transformers, as shown in **Figure 9.24**, provide power to other electrical and electronic systems.

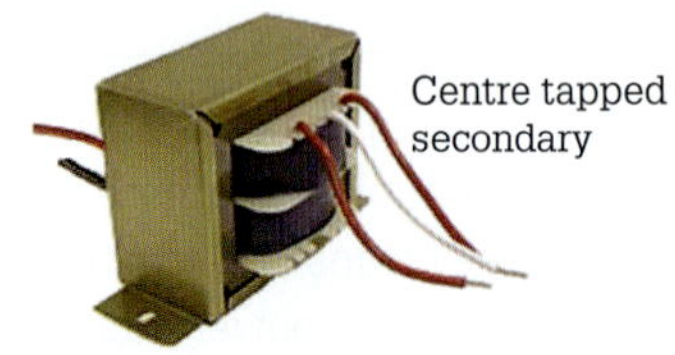

FIGURE 9.24 Volt-ampere transformer

Custom transformers

There are different applications for the electromagnetic property of induction. This is evident with custom transformers that include high-reactance (leakage), constant-voltage/ferroresonant, autotransformers and isolation transformers.

High-reactance (leakage) transformers

The transformer leakage reactance associated with that part of the magnetic flux developed by the primary winding does not cross the gap between the primary and secondary windings and so does not reach the secondary winding. Magnetic shunts incorporated into the construction of the transformer divert some of the primary flux from coupling with the secondary winding. Reactance transformers have a moderate degree of leakage reactance on no load that results in a high secondary no-load voltage and a low secondary current. As the transformer loads, the secondary flux opposes the primary flux, and some of the primary flux is either unable to cross the gap between the windings or diverts through the magnetic shunts. The effect of this is the reduction of the secondary voltage, together with increasing load current output.

The high-reactance (leakage) transformer, as shown in **Figure 9.25**, energises neon signs. Other lighting applications include starting sodium vapour and metal halide lamps, which also require a high voltage to start the flow of current and then maintain output load current within a defined range. Additional applications include sintering furnaces and some welding machines (these have a movable secondary winding). In power transmission and distribution systems, some transformers have a higher leakage reactance than normal to keep prospective fault currents of the secondary within limits that switching devices can handle.

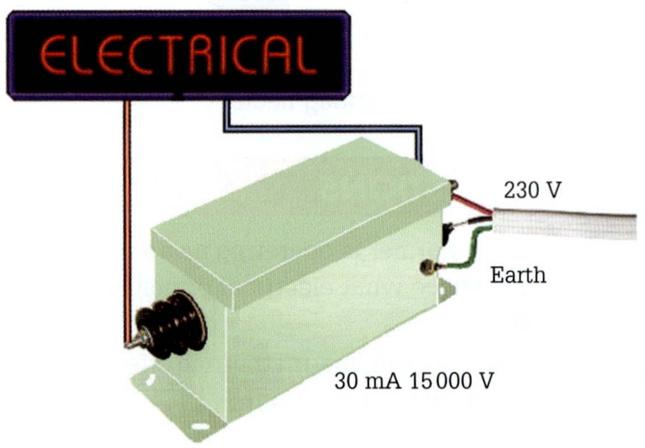

FIGURE 9.25 High-reactance neon sign transformer

Constant-voltage/ferroresonant (CVF) transformers

CVF transformers allow the primary voltage to vary while the secondary winding provides a relatively constant output voltage within 1%. CVF transformers use a principle called saturation to regulate voltage. Saturation prevents severe increases in current from passing to the secondary winding and then on to sensitive equipment.

CVF transformers differ from conventional transformers in that they have a large leakage inductance, a saturating shunt inductance, and a capacitor winding that resonates with the transformer secondary winding. However, some CVF transformers may utilise capacitors in parallel with the load rather than a capacitor winding. These transformers do not produce a true sine wave but a square wave. The waves have their top clipped and their sides flatter than a true sine wave.

CVF transformers are excellent at controlling voltage sags – the most common power quality problem encountered. All equipment is prone to stop if the supply voltage reduces or falls to zero volts. CVF transformers control surges, brownouts, noise and distortion. They tend to be big and heavy because of the size of the transformer and the amount of grain-oriented silicon steel needed for the core to reach saturation. These transformers also make audible noise and produce heat. **Figure 9.26** shows an example of a constant-voltage/ferroresonant transformer.

FIGURE 9.26 Constant-voltage/ferroresonant transformer

Isolation transformers

An isolation or linear transformer, as shown in **Figure 9.27**, is a specialty transformer because it has a 1:1 primary to secondary winding turns ratio. These transformers do not step voltage up or down but act as a protection device. In this capacity, they isolate the earth conductor of the supply input from any metal or any portion of the load circuit. For example, no current flows to the general mass of earth if a short exists between the secondary output of the transformer and any conductive material.

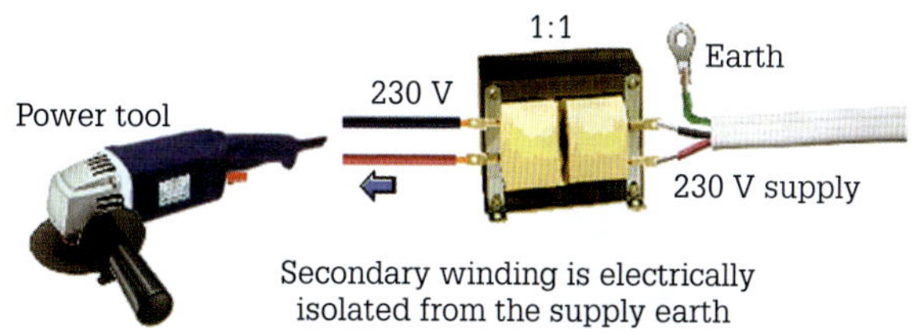

FIGURE 9.27 Isolation transformer

SWITCH ON

An isolation transformer does not limit the danger of electric shock if contact occurs across the transformer's secondary winding.

REVIEW QUESTIONS

1. Name the four general classes of voltage transformer.
2. Which class of transformer has the highest power rating and continuous voltage performance rating of all classes of transformers?
3. What is the function of power transformers?
4. Name two typical applications of the high-reactance (leakage) transformer.
5. What type of transformer has a 1:1 primary-to-secondary winding turns ratio?

9.5 Transmission and distribution transformers

Transmission and distribution transformers operate in much the same manner as volt-ampere transformers. The main differences are in the construction of transmission and distribution transformers. These constructional differences can be categorised as:

- ancillary equipment
- cooling
- tapped windings.

Ancillary equipment

Transmission and distribution transformers have ancillary equipment installed to ensure safe operation at the higher operating voltages. This ancillary equipment includes:

- bushings
- explosion vents
- rapid pressure rise relays
- surge diverters.

Bushings

Transformers have their high- and low-voltage leads brought out through bushings manufactured in accordance with standards and rated to suit the voltage of the load with a high factor of safety. Alternatively, transformers are fitted with a cable box with glands suitable for paper-insulated cable, or compression glands suitable for PVC/XLPE cables.

Effective transformer bushings are essential for the reliability and continuity of service of transformers. Bushings require designing with adequate safety factors to protect against normal and abnormal voltages at the operating frequency under all climatic conditions. The bushing design must offer a degree of protection against lightning and other high-voltage disturbances.

Some bushings redirect abnormal voltage stresses by flashing over to the ground via their arcing horns. The features occurring in transformer bushings are:

- durability
- high dielectric strength
- mechanical strength.

Bushings differ in shape and size for the particular requirements of the transformer application as shown in **Figure 9.28**.

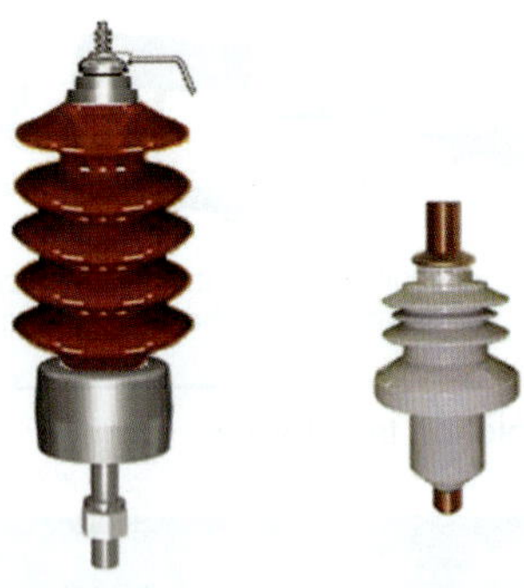

FIGURE 9.28 High-voltage (with arcing horn) and low-voltage transformer bushings

Bushings consist of two types of material: porcelain and polymer. Silicious porcelain based on alkaline alumina silicates is used in standard bushing construction while aluminous porcelain is for high-strength bushings. Porcelain bushings have high electrical and mechanical strength with dimension stability at high temperatures.

Polymer bushings are a very tough epoxy cast resin product. For example, bisphenol-a-epoxy bushings have application indoors, while cycloaliphatic epoxy has excellent ultraviolet resistance for outdoors use. **Figure 9.29** shows high- and low-voltage polymer bushings.

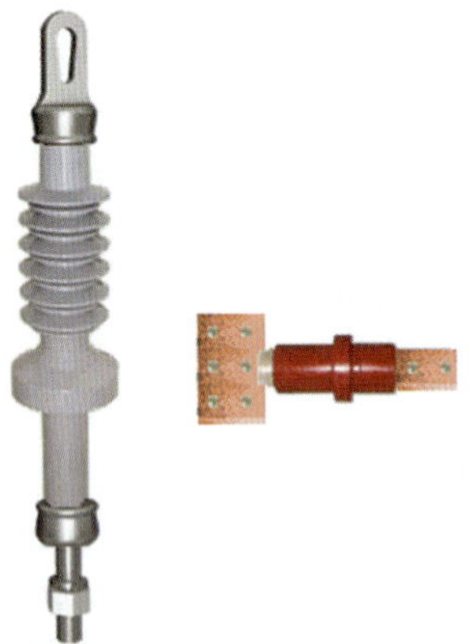

FIGURE 9.29 High- and low-voltage polymer bushings

Epoxy bushings have outstanding chemical resistance, excellent electrical insulating properties, and continuous service temperature ratings. Transformer bushings with chips or broken skirts show mechanical stress and if not removed from service may fail in future service.

Arc burns on bushings indicate that they have been subject to electrical and thermal stress, which shows that complete failure of the bushing could occur later.

Explosion vents

Sometimes transformers experience a short circuit in the windings that vaporise a portion of the cooling fluid. This coolant gas when mixed with the air above the coolant forms an explosive mixture that can ignite from the arc struck by the short circuit.

To guard against the possibility of a sudden high pressure caused by the ignition of an explosive gas, a resealable diaphragm explosion vent, also known as a pressure relief device, fits on the tank.

A pressure relief device, such as that shown in **Figure 9.30**, consists of an opening to the atmosphere covered by a resetting diaphragm which keeps out moisture and contaminants but which can be easily lifted by a predetermined pressure (preferred operating pressure is 56 kPa) to vent the tank.

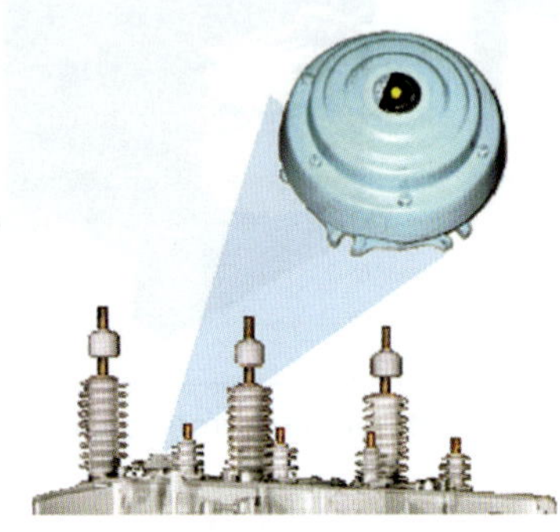

FIGURE 9.30 Explosion relief vent

The explosion relief device mounts on top of the transformer on either the main cover or the manhole. It is located above the coolant to prevent overflow of coolant in case the relief device operates. Some transformers have a sidewall-mounted pressure relief valve as shown in **Figure 9.31**.

FIGURE 9.31 Sidewall pressure relief valve

Rapid pressure rise relays

Some transformers use a rapid pressure rise relay device to detect abnormal internal tank pressure if the conductors of the windings arc. The rapid pressure rise relay function is to deliver transformer protection signals to the protective relay system during a potentially destructive over-pressure event. The protective relay system connects in a circuit that sends a signal to a control device which trips a circuit breaker to de-energise the transformer.

The rapid pressure device is of robust construction to prevent disturbance by vibration, pump surges, mechanical knocks or pressure variations due to coolant temperature changes within the tank. The device mounts near the bottom of the tank as **Figure 9.32** shows.

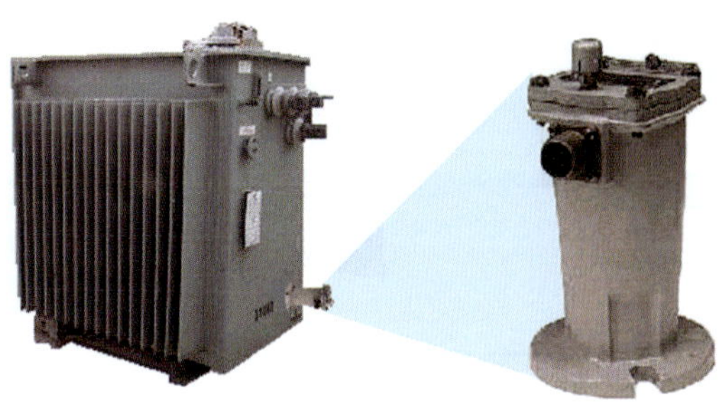

FIGURE 9.32 Rapid pressure rise relay

Surge diverters

A power surge is an increase of voltage lasting longer than a millisecond that, when considering the damage that can occur to equipment, is a long time interval. In addition, power surges can damage and even destroy transformers.

There are many causes of power surges: manmade occurrences, such as power grid network switching, and the starting and then stopping of large electrical motors, alternators, transformers, reactors and capacitor banks, as well as atmospheric disturbances.

When a power surge occurs in the neighbourhood of a transmission or distribution line a high-voltage impulse waveform appears on the line. If it reaches a transformer protected by surge diverters and is large enough, the impulse will spill over the diverter to ground, impressing on the transformer only the part of the voltage waveform that occurred before the spill-over. The transformer insulation must be able to withstand a voltage at least equal to the actuation voltage of the diverter. A surge diverter redirects any increase of voltage above nominal value to ground by acting as a short circuit.

Valve-type surge diverter

The valve-type surge diverter consists of two types of elements connected in series as shown in **Figure 9.33**. The first element is a series of 'spark gaps' followed by the valve element comprising voltage-dependent resistors (VDRs). When a high-voltage surge impulse waveform appears on the line to the transformer, the impulse flashes over the series-connected air gaps and then discharges to ground through the voltage-dependent resistor.

After the impulse has spilled over to the ground, the standard supply potential tends to support the flashover across the elements of the diverter. The consequence is that the follow-on current at normal supply voltage reduces due to the current-limiting nature of the diverter's elements. As a result, the flashover clears. The design of the series gaps and voltage-grading resistors is such that the gap setting is unaffected by vibration, mechanical shock or changes in temperature.

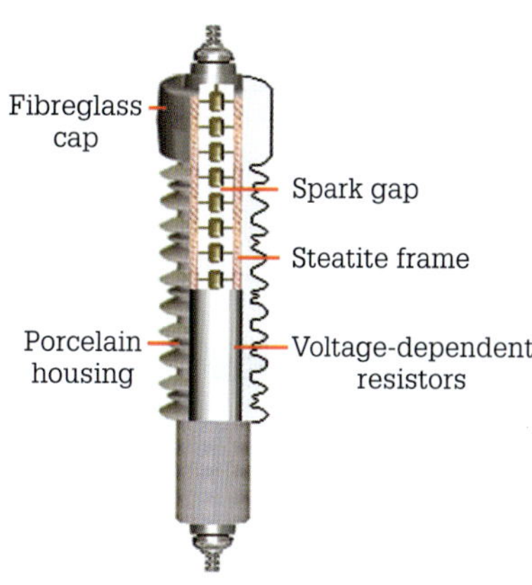

FIGURE 9.33 Valve-type surge diverter

Metal oxide varistor (MOV)

A metal oxide varistor (MOV) surge-protection device is a non-linear gapless device using MOV discs in a column as **Figure 9.34** shows. During normal operating conditions, the nominal line-to-ground voltage applies continuously across the diverter terminals.

When over-voltages occur, the arrestor limits the over-voltage by diverting the surge current to ground. Once the over-voltage condition passes, the diverter returns to its non-linear state.

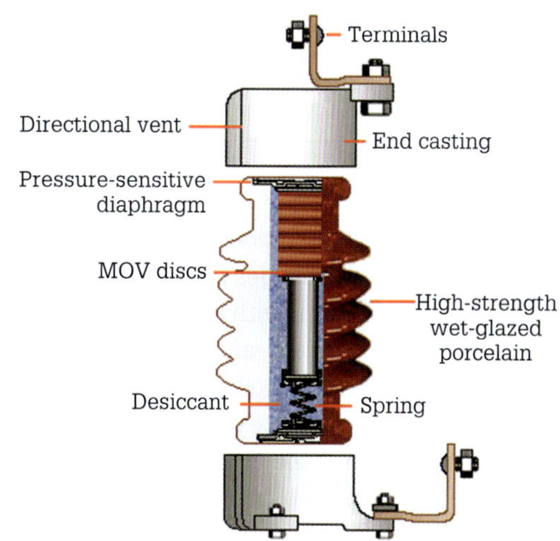

FIGURE 9.34 Metal oxide varistor

The spring protects the MOV discs from mechanical shocks and vibration, while the wet-glazed porcelain enables protection against thermal shocks. In addition, the role of the desiccant is to ensure a dry internal atmosphere by absorbing any ambient moisture trapped in the device. A pressure-sensitive diaphragm can vent internal pressures before they reach dangerous levels. Finally, directional vents in the end castings ensure safe transfer of any fault arc outside any damaged diverter.

Cooling equipment

The majority of transmission and distribution transformers employ some method of removing heat from the windings and distributing the heat. Some form of coolant is used in these transformers, and requires additional ancillary equipment, including:

- conservator tanks
- gas relays
- temperature indicators.

Conservator tanks

Some coolant-filled transformers have a conservator tank or expansion tank, which is a small reservoir tank mounted on top of the main transformer tank. The conservator allows the dielectric coolant to flow into, and out of the reservoir as the coolant in the main tank expands and contracts. The conservator tank, as shown in **Figure 9.35**, is a coolant preservation system designed to isolate the transformer windings from the degrading influence of oxygen and moisture. Conservators also act as sumps to collect impurities, preventing their entry into the main tank. They can be free breathing, nitrogen-gas-sealed or hermetically sealed.

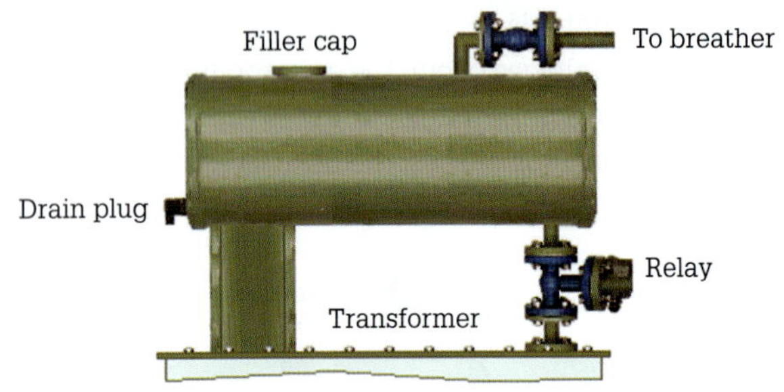

FIGURE 9.35 Conservator tank

This reduction in contamination occurs because only a small amount of coolant exchanges between the main tank and the conservator. However, with bladder- or diaphragm-designed conservators a coolant–air interface still exists in the conservator, exposing the coolant to the atmosphere.

With these conservators, the expanding and contracting process enables the coolant and the air to maintain their balance at atmospheric pressure.

In due course, coolant in the conservator exchanges with coolant in the main tank, and oxygen and other contaminants gain access to the insulation. However, nitrogen-gas-sealed or hermetically sealed conservators do not experience this problem.

Normally, the minimum capacity of the conservator is 7.5% of the coolant capacity of the transformer tank. For this reason, the conservator is of sufficient volume to take care of expansion of coolant in the main tank.

Breathers and desiccants

The interior of the conservator above the coolant connects to the atmosphere through a dehydrating breather. These devices by visual assessment provide an efficient means of controlling the level of moisture entering the conservator during the change of volume of the coolant or airspace caused by temperature changes. One of its purposes is to dry the air that passes into the transformer tank when the air above the oil contracts on cooling. A breather also vents the internal air space within the conservator to the atmosphere to equalise pressure. **Figure 9.36** shows a beaded silica gel (desiccant – sustains a state of dryness) breather.

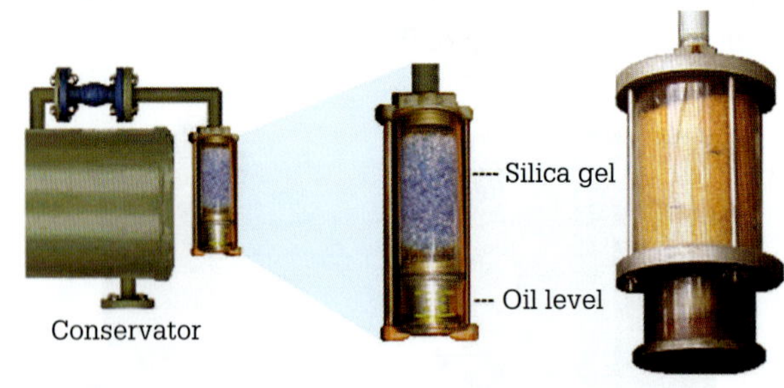

FIGURE 9.36 Beaded silica gel breather

The bottom of the silica gel breather requires filling to the correct level to ensure efficient operation. The silica gel desiccant changes from blue to pink when charged with moisture. Other brands of desiccants are orange or green in colour and become translucent when charged. Oil-contaminated desiccant crystals require discarding. However, heating in an oven for 12 hours at 100 °C can regenerate the pink desiccant as shown in **Figure 9.37**.

FIGURE 9.37 Desiccant recharging

Gas relays

Arcing occurs in coolant-filled tap changers whenever they operate. The result of arcing in coolant-filled equipment is the production of acetylene and hydrogen fault gases. Acetylene, the main gas associated with arcing, is found in on-load tap-changer equipment tanks at varying quantities depending upon the operation cycle of the tap changer. Ethylene, methane and ethane, referred to as the 'hot metal gases', are associated with pyrolysis or overheating.

The presence of hot metal gases indicates the overheating of the arcing contacts of the tap changer. A gas analyser determines the presence and volume of gases within a coolant-type transformer. A device that monitors the presence of gases is a gas and oil actuated relay also known as a Buchholz relay as shown in **Figure 9.38**.

FIGURE 9.38 Buchholz relay

Double-element relays act as a warning system for major faults. The alarm element will operate after a specified volume of gas collects and also operates in the event of oil leakage or if air enters the cooling system. Examples of possible faults are:

- degraded core lamination bolt insulation
- shorted laminations
- degraded arcing contacts
- heat-stressed coil windings.

The trip element operates by a coolant surge in the event of more serious faults such as:

- earth faults
- short circuits within the coil windings
- bushing puncture
- short circuit between phases.

The trip element will also operate in the event of rapid coolant loss.

Temperature indicators

The service life of a transformer relies upon the effectiveness of its insulation and this effectiveness depends on the maximum temperature the insulation experiences and the time interval in which the maximum temperature is maintained. It is essential to keep the temperature of the hottest spot in the coil windings below insulation degradation level. So, it is important to know the temperature of the coil windings.

Winding-temperature sensors in dry-type transformers sense the temperature of the coil lead. Switches in these devices engage a cooling fan or trigger an alarm. All temperature indicators should incorporate a dial with a pointer indicator and a separate resettable pointer to register the maximum temperature reached since the last reset. An obvious indicator of the coolant-type transformer winding temperature is the temperature of the transformer coolant. Several instruments indicate this temperature. **Figure 9.39** shows a fixed-stem dial-type indicator.

FIGURE 9.39 Dial-type indicator

The dial temperature indicator is excellent for convenience but is not a reliable measure of coil-winding temperature. A sudden surge of power causes the winding-conductor temperature to rise more rapidly than the coolant temperature and it is possible for the conductor to reach an excessive temperature before the coolant indicates any warning.

It is necessary to have a temperature indicator that shows the actual temperature of the hot spot in the coil windings. Accurate winding-temperature indicators have capillary extension and thermal imaging ability. The instruments measure the load current through the windings and the temperature of the cooling medium. The coolant temperature is measured with a capillary-connected sensing bulb that is positioned in the oil-tight pocket provided in the tank.

The measuring system also has a specially designed heating element to measure the transformer load. This heating element is a thermal model of the transformer windings and connects to a current transformer via a matching resistance or matching unit, to allow the setting of the correct winding-temperature gradient. An illustration of an accurate winding-temperature indicator is shown in **Figure 9.40**.

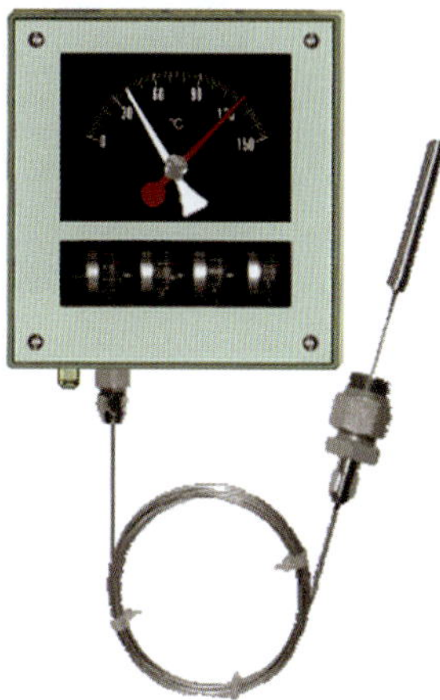

FIGURE 9.40 True winding-temperature indicator

Tapped windings

The voltage ratio of a transformer depends on the ratio of the number of turns in the primary winding to the number of turns in the secondary winding. The secondary can have a number of separate windings. All of the secondary windings occur on the same magnetic core as shown in **Figure 9.41**.

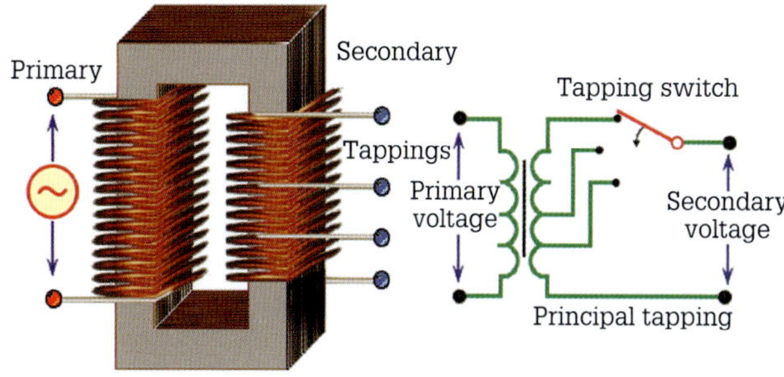

FIGURE 9.41 Transformer tappings

Tappings on a transformer winding allow different voltages to occur. In addition, one of the tappings is a principal tapping. That implies that the principal tapping is the tapping to which the others refer.

EXAMPLE 9.2

An 11 kV distribution transformer as illustrated in Figure 9.42 has a turns ratio of 26.5:1. Confirm the primary voltage at the principal tap, the +2.5% tap, the −2.5% tap, the −5% tap and the −7.5% tap.

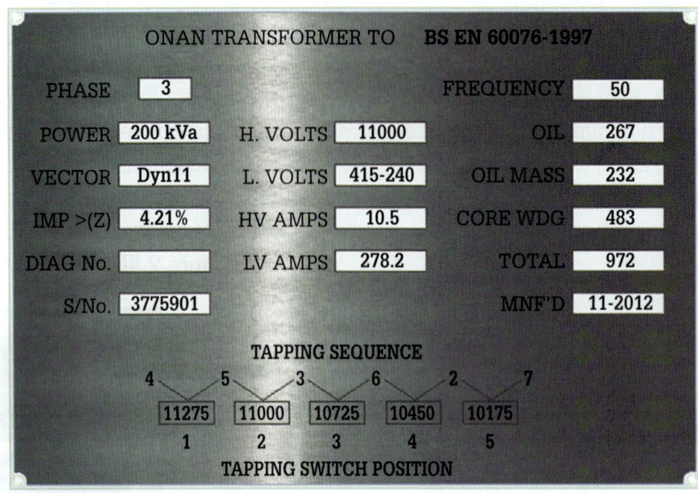

FIGURE 9.42 11 kV distribution transformer nameplate showing primary winding tappings

V_1 at the principal tap = 11 000 V as this is the nominal primary voltage

$$V_1 \text{ at the } +2.5\% \text{ tap} = \frac{102.5}{100} \times 11\,000 = \mathbf{11\,275\ V}$$

$$V_1 \text{ at the } -2.5\% \text{ tap} = \frac{97.5}{100} \times 11\,000 = \mathbf{10\,725\ V}$$

$$V_1 \text{ at the } -5\% \text{ tap} = \frac{95}{100} \times 11\,000 = \mathbf{10\,450\ V}$$

$$V_1 \text{ at the } -7.5\% \text{ tap} = \frac{92.5}{100} \times 11\,000 = \mathbf{10\,175\ V}$$

EXERCISE 9.2

A 22 kV distribution transformer has a turns ratio of 55:1. Confirm the primary voltage at the principal tap, the +2.5% tap, the −2.5% tap, the −5% tap and the −7.5% tap.

Taps affect the leakage flux in the windings, which can lead to higher eddy losses and circulating currents. Hence, these losses in turn produce higher localised heating and hot spots. In addition, there is higher transient voltage stress around the tap point, which affects the thermal performance of the transformer insulation, resulting in dielectric failure. It follows that taps and their connecting leads result in more points of potential failure.

Tap changers

It is sometimes desirable to vary the transformation ratio of a transformer. Variation is done to adjust for voltage sag (drop) in a feeder or main, or to adapt the secondary voltage to the varying requirements of a particular load. A tap changer is a mechanical switching device (manual or motor controlled) whose function is to change the turns ratio of a transformer. There are two types of tap changers, the off-load and the on-load.

Taps on primary windings compensate for variations in the supply voltage. Because of this, the transformer maintains a more even secondary output to the load. The taps have a rating at a fixed percentage of the nominal voltage. For example, a fixed percentage of 2.5% or 5% above or below the average primary or secondary voltage is usual. Taps occur in the electrical centre of the windings, to permit variation of the number of winding turns without any variation in the kVA rating. The rating of all the transformer taps is the full kVA capacity of the transformer.

The high-tension (HT) side of a distribution transformer is preferable as the tap changing winding. Reducing the primary turns has the same effect on the secondary voltage as increasing the secondary turns. This is because the current-carrying components are smaller, leading to less complicated and expensive support equipment. If the low-tension (LT) winding is the tapped winding, then this winding must be placed on the outside of the HT winding. However, this arrangement causes problems around protective insulation and its dielectric strength because the HT winding is next to the laminated core.

Tap changing happens with either disconnection of the transformer from the network (off-circuit tap changer) or on-load (OLTC). In the latter case, the power supply has no interruption. The switching of the tap can occur in mediums such as air, oil and vacuum.

No-load or off-circuit tap changers

These tap changers consist of links or a selector switch immersed in oil, operated by an insulated handle reaching outside the transformer case. However, the link-type tap changer requires the removal of the door or hatch to enable access to the transformer windings. The links on the phase or phase windings connect to the appropriate tap connection.

The tap positions for the windings have identification on the outside of the tank. **Figure 9.43** shows a diagram of a six-position tap changer.

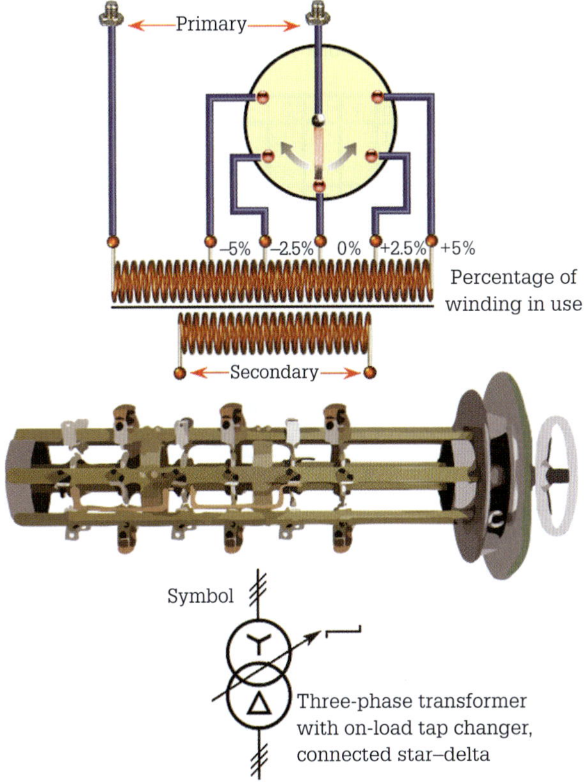

FIGURE 9.43 No-load tap changer

The tap selector consists of fixed contacts on an insulated structure together with a movable contact in the shape of a circular arc. The structure forms an enclosure around the central shaft that carries the moving contact. The movable contact mounts on the shaft with springs. This type of tap changer is for use under the coolant.

No-load tap changers are not for breaking the circuit and should never operate until after the transformer de-energises and is disconnected from the line.

The method used to determine the correct tapping for a no-load tap changer is to measure the voltages at the primary and secondary terminals of the transformer under full-load conditions. If the secondary voltage is more than 3% higher or lower than required, then the primary or secondary connection transfers to the appropriate terminal on the selector plate.

Tap changers can be manufactured for various voltage and current ratings and can have many tap positions and decks as shown in **Table 9.1**.

TABLE 9.1 Variations in tap changer specifications

Characteristic	Specification
Voltages	1.1, 2.2, 11, 22, 33, 66, 132, 220 kV
Current	50, 100, 200, 400, 600, 800, 1000, 1600, 3200 A
Positions	5, 7, 9, 11, 17 and 21
No. of decks	1, 2, 3, 6 and 9
Type of connections	Linear selector, single bridging, double-bridging star, delta and series–parallel.

Figure 9.44 illustrates the positioning of a no-load tap changer within the transformer tank.

FIGURE 9.44 Location of no-load tap changer

As can be seen in **Figure 9.44**, the thermal operating environment for no-load tap changers is in or near the top part of the transformer coolant. Therefore, the tap changer must operate agreeably in this environment.

No-load tap changers can fail in service for several reasons:

- The subjection of the tap changer to over-voltages due to switching surges or lightning transmission line strikes produce a voltage stress across the tap connections, leading to insulation breakdown.
- Incorrect contact pressure at the tap contacts, can cause hot spots, which result in the thermal degradation of the contact surfaces.
- Temperature cycling due to load and ambient temperature changes leads to expansion and contraction of the tap changer's solid insulation barriers. Mechanical stress and cracking of the insulation may occur.

On-load tap changers

On-load tap changers (oil natural tap changers (ONTC)) possess two fundamental features:

1 a built-in impedance load to prevent short circuiting of the tapped section

2 an alternative circuit so that one circuit can carry the load current while switching is being carried out on the other.

The impedance that exists in the configuration of a tap changer can be either resistive or reactive in nature. A tap changer with a resistive type of impedance has high-speed switching as an important characteristic, whereas the reactive tap changer uses slow-moving switching. However, high-speed resistor switching is the preferred method of tap changing worldwide.

All on-load tap changers enable the continued operation of a transformer without disconnecting the transformer from the supply and the load. On-load tap changers have a rating in terms of specific criteria regardless of their features. These criteria are as follows:

- maximum continuous rated current
- maximum double current for 30 minutes per day
- maximum symmetric through-fault current
- maximum crest fault current.

When a tap change initiates, the selector switch preselects the desired tap on the regulating winding of the transformer. The diverter switch then transfers the current from the in-service tap to the preselected tap.

While this quick changeover is occurring, a resistance or reactance cuts into the circuit to ensure a no-break transfer of current. The cutting in limits the circulating current between the two taps of the regulating winding switching.

Reactive-type tap changer

One principle on which some on-load tap changers operate is the use of a preventive autotransformer as the reactive impedance in stepping from tap to tap on the transformer winding as shown in **Figure 9.45**. The preventive autotransformer connects in series with the load to limit the circulating current while the tap changer is in the bridging tap position.

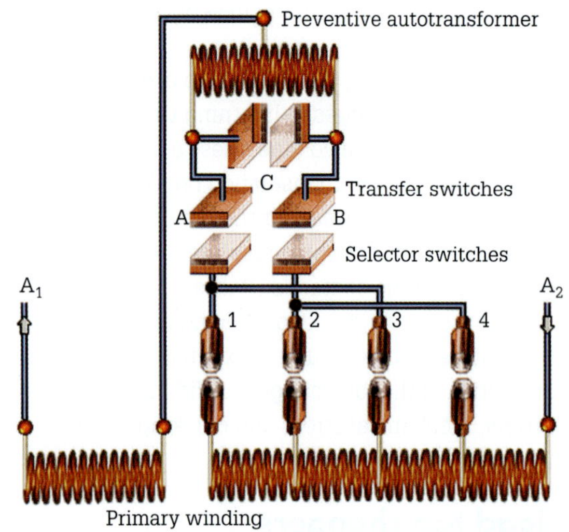

FIGURE 9.45 On-load reactive tap changer

The switches 1, 2, 3 and 4 are selector switches and they are never used to open or close the load circuit. Their location can be within or outside the transformer tank. In addition, the switches A, B and C are transfer switches. Transfer switches serve to open or close the load circuit and are contained within a separate compartment from the selector switches. Separation is required because the effects of arcing can contaminate the immersing oil for these switches. When selector switch 1 and the transfer switches A and C close, the voltage at the centre of the preventive autotransformer is at the same potential as the voltage at the end of tap 1. To move to the next tap, the selector switches operate as shown in the sequence-switching diagram in **Figure 9.46**.

Position	1		2		3		4
Selector switch no.							
1	•	•					
2		•	•	•			
3				•	•	•	
4						•	•
Transfer switch							
A	•	•		•	•	•	
B		•	•	•		•	•
C	•		•		•		•

• = Switch closed

FIGURE 9.46 Sequence of switching

The order of switching from position no. 1 to position no. 2 is coloured pink in **Figure 9.46** and is as follows:

- open transfer switch C and close selector switch 2
- close transfer switch B and open transfer switch A
- close transfer switch C and open selector switch 1.

The middle tap of the autotransformer is now at the same potential as selector tap 2 and the section of the winding from 1 to 2 is at rest. The selector switches always close before the corresponding transfer switches and always open after the transfer switch has opened. This sequence of switching is controlled automatically and takes four transfer-switch operations and two selector-switch operations per transformer tap change.

Resistive-type tap changer

The resistive-type tap changer uses resistors to bridge the tappings during load tap changes. These resistors have a value between 0.2 Ω and 100 Ω depending upon the type of duty required. When a tap change initiates, the selector switch preselects the desired tap on the transformer winding.

The diverter switch then transfers the current from the in-service tap to the preselected tap. While this quick changeover is occurring, a resistance placed into the circuit ensures a no-break transfer of current.

The resistor limits the circulating current between the two taps of the transformer winding switching. **Figure 9.47** shows a three-phase resistive tap changer showing the connections for one phase.

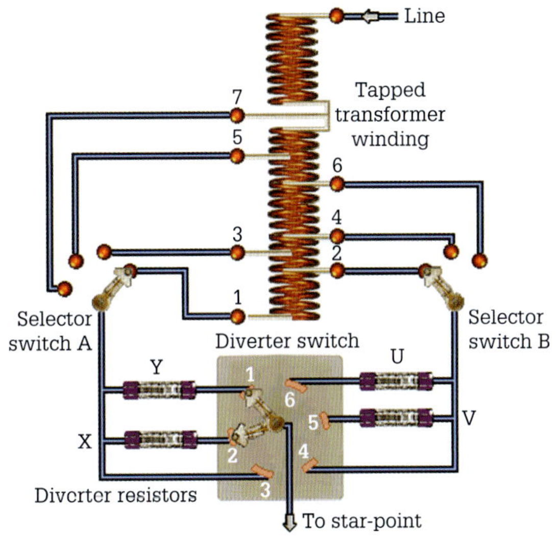

FIGURE 9.47 Resistive-type tap changer

The resistive-type tap changer consists of selector switches, a diverter switch and diverter resistors. The winding that is tapped in this case is the secondary. The diverter switches are quick-acting changeover switches, using resistances for the bridging between running positions during a tap change. The following switching sequence outlines the procedure for switching in the preselected tap 2 by engaging the diverter switch in a clockwise direction from its original position whereby contact 3 is engaged to the star-point.

- **Step 1:** With the selector switches A and B positioned as indicated, contacts 1 and 2 of the diverter switch are bridged, causing resistors X and Y to carry full-load current and limit any circulating current.
- **Step 2:** Contacts 1 and 6 of the diverter are now bridged, causing resistors U and Y to carry full-load current plus the circulating current due to the inter-tap voltage between tap 1 and tap 2. Circulating current at this step in the sequence is at its highest intensity because the current has two paths in which to flow.
- **Step 3:** Contacts 6 and 5 of the diverter are now bridged, causing resistors U and V to carry full-load current. The number of turns in the winding at this point in the sequence has now been decreased, and there is only one path for the current to flow.
- **Step 4:** The diverter switch rotates in a clockwise direction; therefore, full-load current flows through diverter contact 4.
- **Step 5:** Selector switch A moves from tap position 1 to tap position 3 so that a further tap change in the same direction occurs if required. The diverter switch would then rotate in an anticlockwise direction from its existing position on contact 4 to sequence a change to tap 3.

The resistive-type tap changer can be located in the main tank of the transformer or on its side. All the parts mount separately in oil-filled and isolated compartments. If the tap changer has a location within the tank, the selector switch compartment requires separation from the main transformer tank by oil-tight insulating panels. The diverter switch operation happens so that the oil of this section is separate from the oil in the main tank of the transformer. The on-load tap changer device, as illustrated in **Figure 9.48**, actuates by a motor drive mechanism that can have local, automatic, remote and manual control. The speed of this drive mechanism allows the resistors to temporarily carry the load without overheating.

FIGURE 9.48 Resistive on-load tap changer and three-phase transformer coils

Types of transformer faults

The following is a short rundown of the types of faults that can occur in a power transformer:

- high-voltage and low-voltage phase-to-phase fault (external) from bushing flashovers
- high-voltage winding earth fault
- low-voltage winding earth fault
- inter-turn shorting fault – insulation breakdown
- core-damage fault – laminations have moved, creating additional core losses
- tank fault – loss of oil, sludge preventing cooling.

REVIEW QUESTIONS

1. Name the three properties of transformer bushings.
2. List the two types of materials used for transformer bushings.
3. What may happen when a transformer experiences a short circuit in the windings?
4. Why would a transformer have an explosion relief device located above the coolant?
5. What is the function of a rapid pressure rise relay?
6. What defines a power surge?

»

7 What type of device enables excess voltages to be transferred to ground?
8 What are two types of surge diverter?
9 Describe the purpose of a conservator.
10 What gas is produced from arcing in on-load tap-changer equipment tanks?
11 What are the 'hot metal gases' produced in on-load tap-changer equipment tanks?
12 Name the device that monitors the presence of 'hot metal gases'.
13 Why is it necessary to monitor the temperature of transformer coil windings?
14 State the purpose of a tap changer.

9.6 Autotransformers and instrument transformers

Two special purpose transformers are autotransformers and instrument transformers. Instrument transformers may be either potential transformers or current transformers.

Autotransformers

An autotransformer enables voltage transformation with only one winding. **Figure 9.49** shows a single-phase autotransformer which consists of a single continuous winding with one end regarded as a common point. As this transformer does not have the advantage of double-wound transformers (dividing circuits into electrically isolated sections), a person encountering either of the secondary leads may receive an electric shock equal to the primary voltage if the lead is common to both open circuits. To give electrical protection to people using these transformers, autotransformers in Australia connect only to an installation supplied from a multiple earth neutral (MEN) system of earthing.

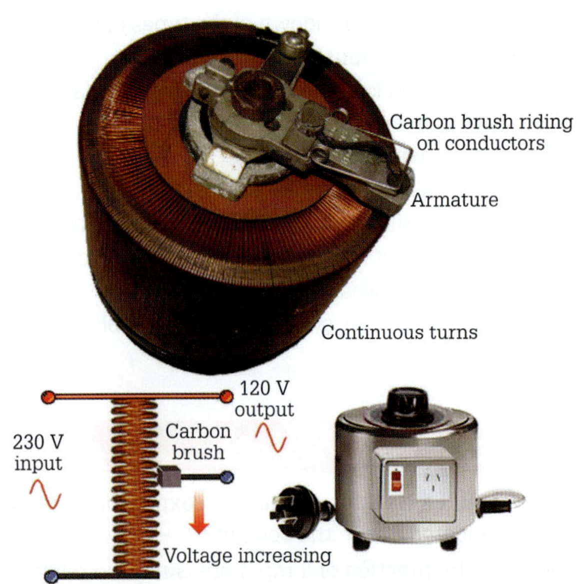

FIGURE 9.49 Autotransformer

Autotransformer windings use magnet wire wound on the outside of a substantial toroidal ferrite core. One terminal of the autotransformer is common to both the primary and the secondary, while the other connects to a moving carbon contact that rotates to produce a variable voltage. Autotransformers find use as reduced-voltage fan control and motor speed control. High-reactance autotransformers are also available.

Step-down autotransformers

In **Figure 9.50**, the autotransformer winding (terminals A and C) has a tapping at terminal B, with a load connected across terminals B and C.

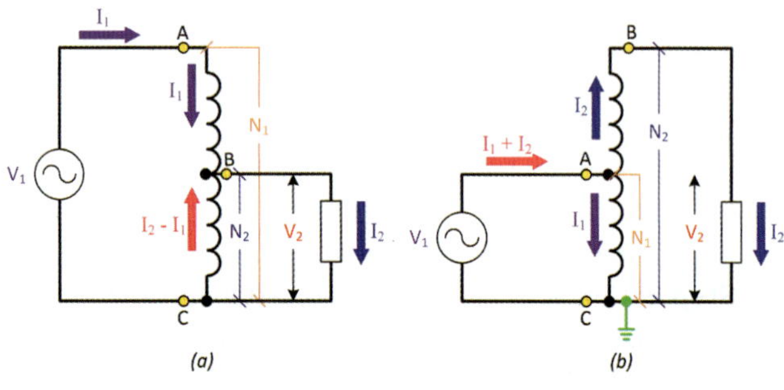

FIGURE 9.50 Step-down (a) and step-up (b) autotransformers

With the step-down autotransformer, the current (I_1) flowing into the autotransformer from the supply enters terminal A and flows into terminal B. For this reason, resultant current ($I_2 - I_1$) flows through the common winding (terminals B and C) from terminals C to B. Furthermore, the load current I_2 is the sum of currents I_1 and ($I_2 - I_1$).

The induced emf across any given number of turns depends on the turns/volt ratio of the winding between terminals A and C as it relates to the winding between terminals B and C. Therefore, if the winding between terminals A and C has a tapping at terminal B, a nominal voltage (V_2) is available across terminals B and C.

Neglecting energy losses, the following voltage ratio applies:

$$\frac{V_2}{V_1} = \frac{N_2}{N_1} = \frac{I_1}{I_2}$$

where V_1 = primary potential in volts
V_2 = secondary potential in volts
N_1 = the number of turns between A and C
N_2 = the number of turns between B and C
I_1 = primary current in amperes
I_2 = secondary current in amperes

EXAMPLE 9.3

Figure 9.50(a) represents a step-down autotransformer with 460 turns. If the applied voltage (V_1) is 230 V, the turns/volt ratio is 2:1 and a tapping occurs at terminal B at 200 turns, determine:

a the output voltage (V_2) of the transformer.

$$\text{As } \frac{V_2}{V_1} = \frac{N_2}{N_1} \text{ therefore,}$$

$$V_2 = V_1 \times \frac{N_2}{N_1}$$
$$= 230 \times \frac{200}{460}$$
$$= \mathbf{100\ V}$$

As the turns/volt ratio is 2:1, then 200 turns divided by 2 equals 100 V.

b the output current (I_2) of the transformer if the load resistance equals 50 Ω.

$$I_2 = \frac{V_2}{R_L}$$
$$= \frac{100}{50}$$
$$= \mathbf{2\ A}$$

c the input current (I_1) of the transformer if the load resistance equals 50 Ω.

$$I_1 = I_2 \times \frac{N_2}{N_1}$$
$$= 2 \times \frac{200}{460}$$
$$= \mathbf{870\ mA}$$

d the current in the common portion of the winding.

$$I_{Common} = I_2 - I_1$$
$$= 2 - \left(870 \times 10^{-3}\right)$$
$$= \mathbf{1.13\ A}$$

EXERCISE 9.3

A step-down autotransformer comprises 920 turns. If the applied voltage (V_1) is 230 V, calculate:

a the turns/volt ratio

b if a tapping occurs at 300 turns, determine the output voltage (V_2) of the transformer

c the output current (I_2) of the transformer if the load resistance equals 15 Ω

»

d the input current (I_1) of the transformer if the load resistance equals 15 Ω

e the current in the common portion of the winding.

Step-up autotransformers

With the step-up autotransformer as shown in **Figure 9.51(b)**, the current drawn from the supply is the sum of the primary and secondary currents ($I_1 + I_2$). The current in the common portion of the winding between terminals A and C is I_1. The secondary current (I_2) flows between terminals B and C.

SWITCH ON

Safety hazard

Due to the risk of serious electric shock, it is essential to earth the common winding of a step-up transformer.

EXAMPLE 9.4

Figure 9.50(b) represents a step-up autotransformer with 460 turns. If the applied voltage (V_1) is 230 V and a tapping occurs at terminal A at 200 turns, determine the output voltage (V_2) and supply current (I_1) of the transformer with a load demand of 20 A.

$$\text{As } \frac{V_2}{V_1} = \frac{N_2}{N_1} \text{ therefore,}$$

$$V_2 = V_1 \times \frac{N_2}{N_1}$$
$$= 230 \times \frac{460}{200}$$
$$= \mathbf{529\ V}$$

$$I_1 = \frac{I_2 V_2}{V_1}$$
$$= \frac{20 \times 529}{230}$$
$$= \mathbf{46\ A}$$

EXERCISE 9.4

A step-up autotransformer has 920 turns. If the applied voltage (V_1) is 230 V and a tapping occurs at 460 turns, determine:

a the output voltage (V_2)

b the supply current (I_1) of the transformer with a load demand of 15 A.

Pros and cons of autotransformers

Table 9.2 summarises the relative advantages and disadvantages of autotransformers.

TABLE 9.2 The advantages and disadvantages of the autotransformer compared to the double-wound transformer

Disadvantages	Advantages
The autotransformer is electrically connected between its input and the output. By contrast the double-wound transformer is connected via electromagnetism between its input and output.	The regulation, leakage inductance and physical size of an autotransformer for a given VA rating are all less than a double-wound transformer delivering the same power.
If an open-circuit occurs in the winding both the input and output incur the same voltage value.	The copper loss (I^2R) of an autotransformer is lower than for a comparable double-wound transformer.
The autotransformer has no application where isolation between the primary and secondary circuits is required.	The efficiency of an autotransformer is higher than a comparable double-wound transformer.

Applications of autotransformers

Applications of autotransformers include:

- speed control of induction motors
- lighting control in theatres, hotels and photographic studios
- starting polyphase induction motors
- power supplies
- industrial process and heating control
- control of rectifiers in electroplating
- supply voltage adjustment.

Instrument transformers

There are limits to the magnitude of voltage and current values that can be measured directly by electrotechnology workers using hand-held measuring instruments. Transformers provide the medium to change high values of voltage and current to a level that is easily monitored.

Known as instrument transformers these devices act as isolating transformers enabling the person doing the electrical monitoring to be isolated from contact with the high-voltage line pressure and current. The two types of instrument transformers are:

- **Current:** Used with an ammeter to measure current in a.c. energised lines.
- **Potential:** Used with a voltmeter to measure potential difference across a.c. lines.

Current transformer

A current transformer (CT) has a primary coil of one or more turns of heavy conductor, or an actual busbar as the primary. As for the secondary coil, it has many turns of fine magnet wire. The secondary steps down the current to a maximum of five, one or 0.1 amperes. Actually, the primary current contains two components:

1 an exciting current to magnetise the transformer core
2 a larger primary current component that is transformed to a secondary current.

The load connected to a CT is a 'burden', expressed in VA. The burden is the sum of the impedance of the connecting conductors, terminations and the impedance of the measuring instrument.

$$\text{burden} = I^2Z$$

EXAMPLE 9.5

A CT has a full-load secondary current of 5 A. If the CT's circuit including the ammeter's impedance has a total impedance of 0.4 Ω, calculate the CT's burden rating.

$$\begin{aligned}\text{burden} &= I^2Z \\ &= 5^2 \times 0.4 \\ &= \mathbf{10\ VA}\end{aligned}$$

EXERCISE 9.5

A CT has a full-load secondary current of 1 A. If the CT's circuit including the ammeter's impedance has a total impedance of 0.15 Ω, calculate the CT's burden rating.

For the ammeter to provide accurate readings, the burden must not exceed the CT's rating. For this reason, a standard burden rating for a CT is 15 VA.

There are two basic types of low-current, iron-core transformers: the 'through-type' primary and the 'busbar' primary. **Figure 9.51** illustrates a through-type current transformer.

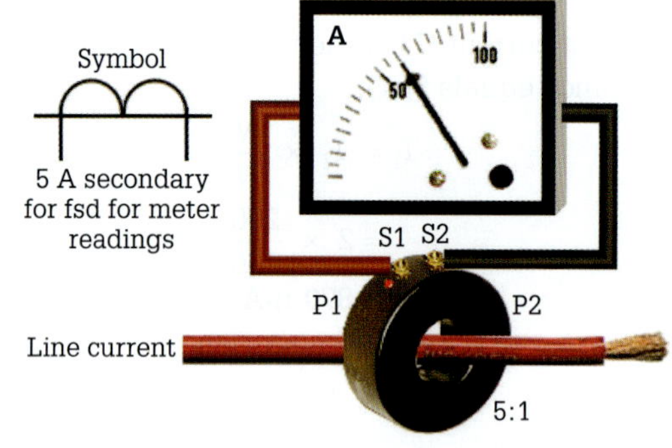

FIGURE 9.51 Through-type current transformer

The polarity marks of a current transformer indicate that when the primary current enters at the polarity mark (P1) of the primary, a current in phase with the primary current and proportional to it in magnitude leaves the polarity terminal of the secondary (S1). The red dot indicates the S1 (phase-side) connection of the current transformer. **Figure 9.52** includes a busbar-type current transformer showing the polarity notation.

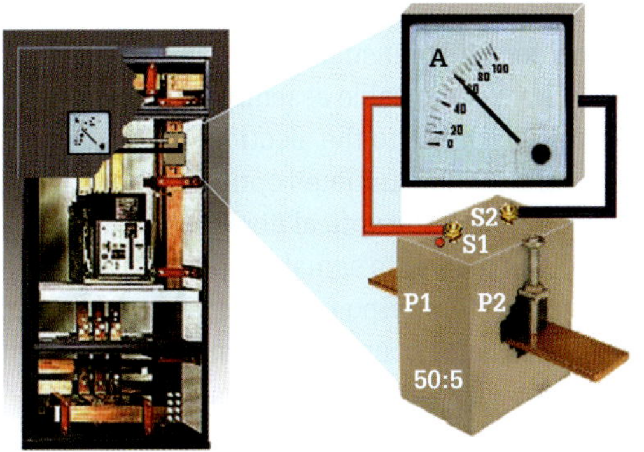

FIGURE 9.52 Busbar current transformer

SWITCH ON

Hazards and safety precautions

The secondary circuit of a current transformer must remain closed, to prevent a very high electromotive force being induced in the secondary. Not only will this induced emf break down the insulation of the transformer, but it creates a hazardous situation for the person monitoring the circuit.

However, many current transformers have a 'make-before-break' contact switch to ensure that the transformer is not open circuited when equipment is being connected or removed. Fuses should not occur on the secondary side of a CT because an open circuit could result in a very high voltage induced in the windings.

Current transformers mediate at an accurate ratio to allow an attached instrument to gauge the current without actually drawing full power through the instrument. They are required to transform relatively small amounts of power because their only load, called a burden, is the moving elements of an analogue ammeter or the electronics of a digital ammeter.

Current transformers are available in several ratios such as 50:5, 100:5, 300:5. For example, if a conductor carrying 100 A passes through a 100:5 ratio CT, a current of 5 A develops in the CT. Current transformers are capable of measuring 50 Hz to 400 Hz currents of 5 A to 15 000 A.

Current transformers are used to:

- transform the line current to a value that is suitable for measuring instruments
- isolate the measuring instrument's meters from the supply.

Broadband-terminated current transformer

With a broadband-terminated current transformer (sometimes termed a current monitor or current probe) you can accurately measure pulse current waveforms of the circuit under examination. This transformer is essentially a transducer that converts the current under test to a voltage signal for an instrument to decode. **Figure 9.53** shows a broadband-terminated current transformer.

Broadband-terminated current transformers are appropriate for use in measuring very high capacitor discharge currents. They enable the accurate measurement of turn-on and turn-off values in power semiconductor switches. Broadband-terminated current transformers also find application in current measurement with the lightning simulation testing of surge arrestors.

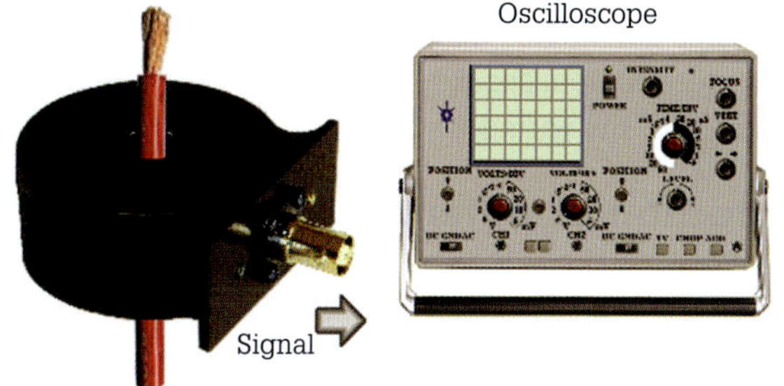

FIGURE 9.53 Broadband-terminated current transformer

The conductor carrying the monitored current passes through the aperture in the transformer. Its voltage output is directly proportional to current and acts as a voltage generator in series with a 50 Ω resistor over its bandwidth. The output is taken via a matching cable and fed into the oscilloscope.

Potential transformer

A potential transformer is a step-down transformer (step down to 110 V). It connects line-to-line or line-to-neutral, in the same way as a voltmeter. Moreover, the secondary voltage bears a fixed relationship with a primary voltage so that any change in potential in the primary circuit accurately reflects in the meter or other device connected across the secondary terminals. The convention is to match the colour of instrument wire to the connected phase.

Potential transformers connect to either an analogue or a digital voltmeter. By multiplying the reading on the voltmeter by the ratio of transformation, the person monitoring the circuit can determine the voltage across the supply or primary side.

Common transformation ratios are 10:1, 20:1, 40:1, 80:1, 100:1 and even higher. **Figure 9.54** illustrates a potential transformer connected to a high-voltage line.

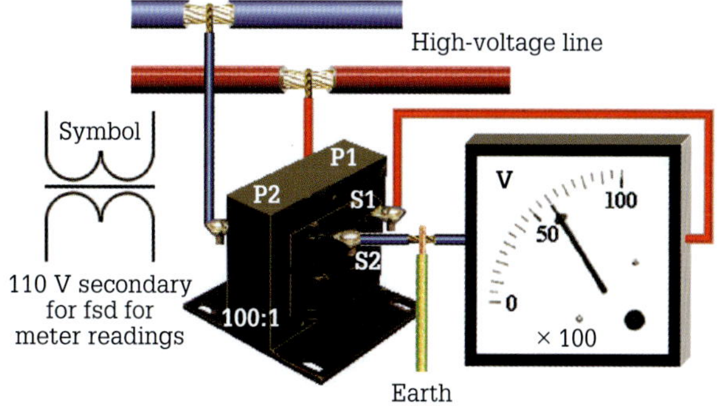

FIGURE 9.54 Potential transformer

In general, a potential transformer is like a standard two-winding step-down transformer, except that it transforms a tiny amount of power. Transformers for this application are always of the shell type because this construction provides better accuracy with voltage measurement. For protection, the secondary circuit is insulated from the high-voltage primary. In addition, an earth prevents the secondary circuit from reaching dangerous potentials. The person monitoring the circuit has protection from shock hazard in case of accidental contact with the wiring.

Protection of potential transformers

A potential transformer can be protected from a secondary short circuit by incorporating fuses into the secondary circuits. A short circuit on the secondary winding creates only a few amperes in the primary winding and is not sufficient to rupture a high-voltage fuse on the primary side. However, the voltage drop in the secondary circuit is of importance because voltage drops across the secondary fuses (a 6 A fuse can have a resistance of 0.048 Ω) and long circuit conductors can change the accuracy of the measurement.

Potential transformers such as those shown in **Figure 9.55** can be used with voltmeters for voltage measurements or they can be used in combination with current transformers for wattmeter or watt-hour meter measurements. They are also used to operate protective relays and devices, and for many other applications. Outdoor oil-cooled, high-voltage, single-phase potential transformers occur on systems ranging from 11 kV to 33 kV.

FIGURE 9.55 High-voltage potential transformers

Optical fibre current and potential transformers

Optical fibre current and voltage sensors are an alternative to conventional current and potential transformers for high-voltage measurements. Their significant advantages include higher accuracy over a broader load range, broad bandwidth coverage, and minimal risk of failure.

Additionally, ferroresonance (ferroresonance appears as persistent oscillations in the secondary output and can adversely affect system protection, resulting in false trips) and current transformer saturation, as well as the danger of an open secondary with conventional, potential and current transformers, are eliminated.

Fibre-optic transformers have a digital output signal and are interfaced with data acquisition systems to provide low- and high-energy analogue signals into instruments for measurement.

The transformer consists of optical sensors and a hollow-core composite insulator column filled with dry nitrogen. A fibre-optic cable assembly connects the passive optical sensors to the electronic control module.

The control module initiates optical signals that are transmitted through the optical fibres to the column where the electric and magnetic signals of the line current or voltage they are monitoring influence the polarity of the light. The electronics control deciphers the returning signals and creates digital outputs that represent the primary current or voltage.

Accuracy is one of the main advantages of optical sensors. All indications suggest that the optical sensor provides a more accurate representation of actual line conditions. Units in operation overseas have proven to be reliable and robust. Optical current sensing uses two linearly polarised light waves sent from the control module within optic-fibre cable to the sensing head.

The two light waves sent from the control module experience a phase shift in proportion to the magnetic field produced by the current being measured. This information redirects back to the control module where it produces an output signal that represents the current under measurement.

Optical voltage sensing uses several miniature optical electric field sensors positioned within the hollow nitrogen-filled insulator. A voltage always creates an electromagnetic field between a conductor and ground. When a circular light signal enters the column, it experiences a change in its form, as a result of the action of the electromagnetic field. Three light sensors interpret this change in form and this information enters the control module where it is combined to give an accurate voltage measurement. An illustration of a fibre-optic transformer is shown in **Figure 9.56**.

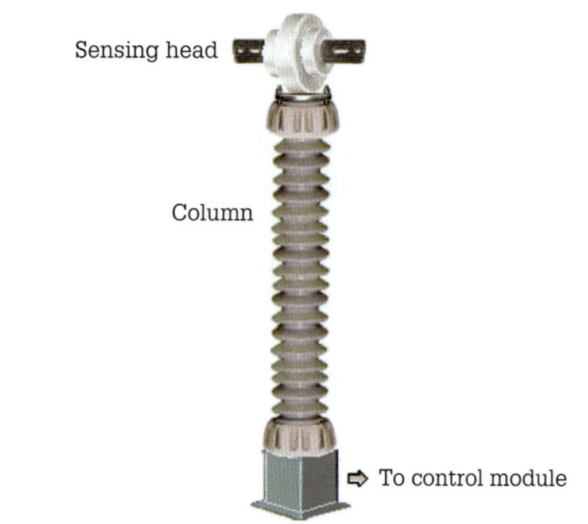

FIGURE 9.56 Fibre-optic transformer

Selection of current and potential transformers

These devices are selected according to the load rating of the device to be evaluated and its line-to-line voltage.

EXAMPLE 9.6

1 A load estimated to be 3000 kVA requires measurement. If the load connects to a three-phase, three-wire 11 000/415 V line-to-line potential, select a suitable current and potential transformer. (Select from 50:5, 75:5, 100:5, 200:5, 300:5 CT; step down to 240 V with a 10:1, 20:1, 40:1, 50:1, 80:1, 100:1 PT.)

$$I_{Line} = \frac{kVA}{\sqrt{3} \times V_{Line}} = \frac{3000 \times 10^3}{\sqrt{3} \times 11\,000} = \mathbf{157.5\ A}$$

Current transformers with 200:5 ratios would be suitable. The line-to-line voltage of 11 000 V requires that an 11 000:240 or a 50:1 ratio would be suitable for the two potential transformers.

2 A three-phase four-wire 11 000 V line supplies a 1000 kVA load. Select suitable current and potential instrument transformers. (Select from 50:5, 75:5, 100:5, 200:5, 300:5 CT; step down to 240 V with a 10:1, 20:1, 50:1, 80:1, 100:1 PT.)

$$I_{Line} = \frac{kVA}{\sqrt{3} \times V_{Line}} = \frac{1000 \times 10^3}{\sqrt{3} \times 11\,000} = \mathbf{52.5\ A}$$

A current transformer with a 75:5 ratio would be suitable.

The line-to-neutral voltage is:

$$\frac{11\,000}{\sqrt{3}} = \mathbf{6351\ V}$$

Therefore, a 50:1 ratio would be suitable for the two potential transformers.

EXERCISE 9.6

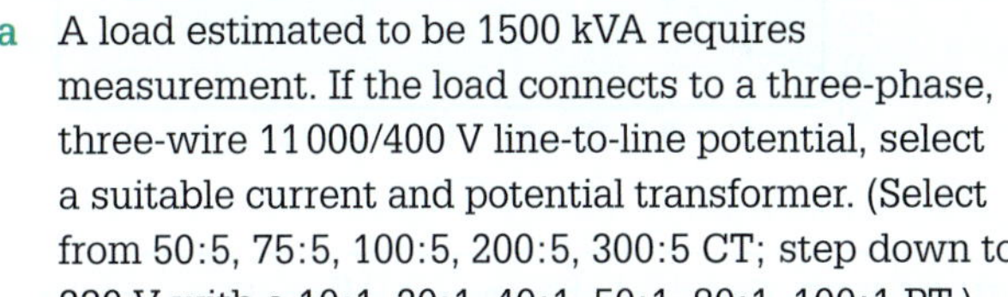

a A load estimated to be 1500 kVA requires measurement. If the load connects to a three-phase, three-wire 11 000/400 V line-to-line potential, select a suitable current and potential transformer. (Select from 50:5, 75:5, 100:5, 200:5, 300:5 CT; step down to 230 V with a 10:1, 20:1, 40:1, 50:1, 80:1, 100:1 PT.)

b A three-phase four-wire 11 000 V line supplies a 2500 kVA load. Select suitable current and potential instrument transformers. (Select from 50:5, 75:5, 100:5, 200:5, 300:5 CT; step down to 240 V with a 10:1, 20:1, 50:1, 80:1, 100:1 PT.)

REVIEW QUESTIONS

1 Which transformer enables voltage transformation with only one winding?

2 Name two applications for autotransformers.

3 A step-down autotransformer comprises 920 turns. If the applied voltage (V_1) is 230 V, calculate:
   a the turns/volt ratio
   b if a tapping occurs at 160 turns, determine the output voltage (V_2) of the transformer
   c the output current (I_2) of the transformer if the load resistance equals 8 Ω
   d the input current (I_1) of the transformer if the load resistance equals 8 Ω
   e the current in the common portion of the winding.

4 Why would the common portion of a step-up autotransformer be earthed?

5 In what way do instrument transformers provide protection for the person reading the meter connected to the transformer?

6 The red dot indicates which connection terminal of the current transformer?

7 Calculate the current value provided by a current transformer that has a ratio of 100:5, if the measuring ammeter is recording 4 A.

8 What type of device ensures that the transformer is not open circuited when equipment is being connected or disconnected?

9 What is a 'burden' with current transformers?

10 What type of transformer is a potential transformer?

11 Why is the shell-type transformer construction suitable for iron-core potential transformers?

12 State two advantages fibre-optic transformers have when compared with iron-core instrument transformers.

9.7 Transformer losses and efficiency

In contrast to the ideal transformer, practical transformers have losses. These are open circuit or iron losses and short circuit or copper losses.

The iron losses are independent of the power used by the transformer. Better design and improved construction material reduce these losses. For example, the cross-sectional area of the core can be increased, or a superior ferromagnetic material used. Moreover, the determination of iron losses occurs by applying an excitation or open-circuit test.

The copper losses (I^2R) increase with the square of the current (doubling the current gives four times the I^2R losses). Consequently, increasing the cross-sectional area of the conductors of the windings reduces both conductor resistance and copper loss. Copper loss can be determined by applying an impedance or short-circuit test. Furthermore, the load losses of a transformer can also be reduced by improving the power factor of loads connected to the transformer.

Efficiency

The efficiency (symbol η) of an ideal transformer is 100% – the result of power in equalling power out. With practical transformers, this level of efficiency is not achieved. However, efficiencies of greater than 97% do occur.

Working out efficiency is part of the manufacturer's transformer test. In order to work out efficiency, quantities from measured values are included in the calculation. However, quantities such as unity power factor and rated voltage and frequency are assumed, and the power loss is taken at a particular temperature (e.g. 75 °C).

Two power losses are of concern – they are the iron and the copper loss. The iron loss in watts is constant for all loads on the transformer, being dependent on the voltage and frequency only; the copper loss varies as the square of the load current. This means that if the load drops by half, the copper loss decreases to a quarter. Efficiency may be calculated as follows:

$$\eta = \frac{P_{out}}{P_{in}}$$

The value obtained from this equation is a decimal value. Multiply this value by 100 to express the efficiency as a percentage.

Transformers are rated in terms of output kVA (or VA or MVA), as such efficiency can be obtained from the equation:

$$\eta = \frac{P_{out}}{P_{out} + P_{loss}}$$

EXAMPLE 9.7

What is the full-load efficiency of a 600 kVA transformer having an iron loss of 1.8 kW and a full-load copper loss of 5 kW?

$$\eta = \frac{P_{out}}{P_{out} + P_{loss}}$$

$$= \frac{600}{600 + (1.8 + 5)}$$

$$= \mathbf{0.989 \text{ or } 98.9\%}$$

EXERCISE 9.7

What is the full-load efficiency of a 1000 kVA transformer having an iron loss of 3.5 kW and a full-load copper loss of 9.5 kW?

Maximum operating efficiency of a transformer occurs when the iron losses are equal to the copper losses. Since no-load losses are constant and load losses are variable, maximum efficiency will only occur at one particular load. In practice, high power rated transformers are more efficient because of the quality and type of materials used in their construction.

All-day efficiency

All-day efficiency is the ratio of the total energy output of the transformer during a 24-hour period to the total energy input for the same time interval. This efficiency parameter is necessary when the transformer carries little or no load during large time intervals of the 24-hour period.

Figure 9.57 displays a graph showing the efficiency of a transformer at different times of the day over a 10-hour period.

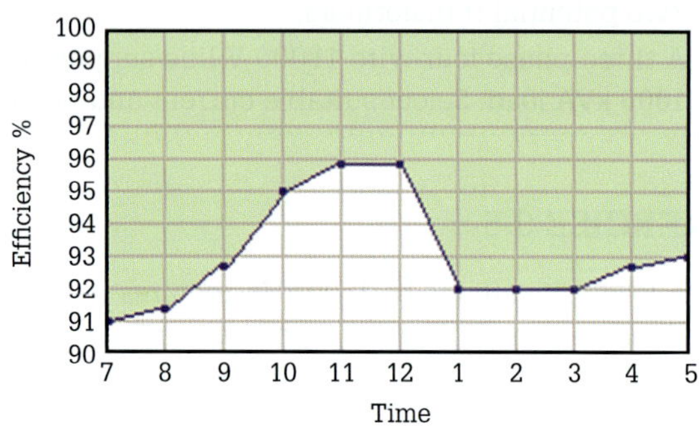

FIGURE 9.57 Graph of efficiency at different times of the day

The load curve of the graph indicates the energy wasted in copper loss during the 10-hour period. The all-day efficiency is plotted daily over a number of days to give an indication of transformer usage as illustrated in **Figure 9.58**.

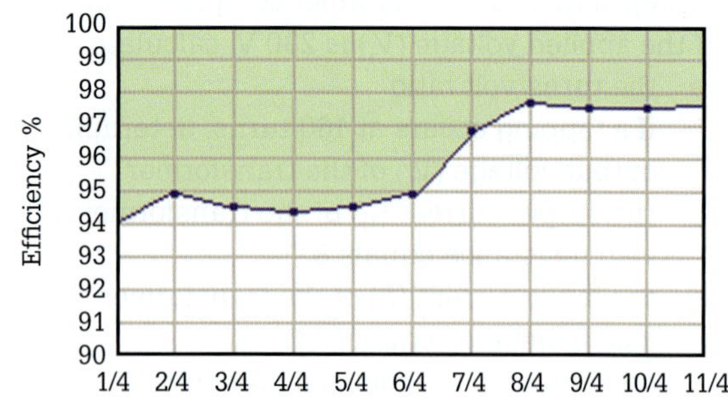

FIGURE 9.58 Transformer usage graph

The efficiency of a transformer is at maximum at a specific power factor when the copper losses are equal to the iron losses. All-day efficiency can be determined by applying the following equation:

$$\text{all-day efficiency} = \frac{\text{daily output energy}}{\text{daily output energy} + \text{losses}}$$

Daily energy is expressed in kilowatt hours (kWh).

As the load on a transformer varies so does the copper loss (P_{Cu}). In order to determine the copper loss at other than full-load:

$$P_{Cu} @ \ part\ load = P_{Cu} \times \left(\frac{part\ load\ kVA}{full\ load\ kVA}\right)^2$$

All-day efficiency calculations are essential for effective load management of transformers. Such load management of transformers will lead to substantial energy and cost savings. To highlight the importance of effective load management, consider the following example.

EXAMPLE 9.8

Find the all-day efficiency of a 1000 kVA, 22 000/400 V distribution transformer having full load copper and iron losses of 9 kW and 7 kW respectively. During the 24-hour day the transformer has the following duty cycles:

No. of hours	Loading in kW	Power factor
10	800	0.85
5	600	0.75
9	200	0.9

A load of 800 kW at a power factor of 0.85 is equal to:

$$\text{kVA} = \frac{\text{loading}}{\text{pf}} = \frac{800}{0.85} = \mathbf{941.2\ kVA}$$

A load of 600 kW at a power factor of 0.75 is equal to:

$$\text{kVA} = \frac{\text{loading}}{\text{pf}} = \frac{600}{0.75} = \mathbf{800\ kVA}$$

A load of 200 kW at a power factor of 0.9 is equal to:

$$\text{kVA} = \frac{\text{loading}}{\text{pf}} = \frac{200}{0.9} = \mathbf{222.2\ kVA}$$

Copper loss (Cu loss) at full load of 1000 kVA = 9 kW

Copper loss at 941.2 kVA:

$$P_{Cu} @ \ 941.2\ kVA = P_{Cu} \times \left(\frac{part\ load\ kVA}{full\ load\ kVA}\right)^2 = 9 \times \left(\frac{941.2}{1000}\right)^2 = \mathbf{7.97\ kW}$$

Copper loss at 800 kVA:

$$P_{Cu} @ \ 800\ kVA = P_{Cu} \times \left(\frac{part\ load\ kVA}{full\ load\ kVA}\right)^2 = 9 \times \left(\frac{800}{1000}\right)^2 = \mathbf{5.76\ kW}$$

Copper loss at 222.2 kVA:

$$P_{Cu} @ \ 222.2\ kVA = P_{Cu} \times \left(\frac{part\ load\ kVA}{full\ load\ kVA}\right)^2 = 9 \times \left(\frac{222.2}{1000}\right)^2 = \mathbf{0.44\ kW}$$

Total copper loss in 24 hours = (10 × 7.97) + (5 × 5.76) + (9 × 0.44) = 112.46 kWh.

The iron loss occurs throughout the day irrespective of the load on the transformer because the primary remains energised over the 24 hours.

Total iron loss over 24 hours = 24 × 7 = 168 kWh.

Transformer output over the 24 hours = (10 × 800) + (5 × 600) + (9 × 200) = 12 800 kWh

$$\eta \text{ all-day} = \frac{\text{output}}{\text{output} + \text{losses}} = \frac{12\ 800}{12\ 800 + 112.46 + 168} = \mathbf{0.9786 \text{ or } 97.86\%}$$

Note: Input = output plus the losses.

The all-day efficiency is always less than the nameplate efficiency of a transformer.

EXERCISE 9.8

Find the all-day efficiency of a 1000 kVA, 22 000/400 V distribution transformer having full load copper and iron losses of 10 kW and 8 kW respectively. During the 24-hour day the transformer has the following duty cycles:

No. of hours	Loading in kW	Power factor
12	850	0.85
7	650	0.75
5	200	0.9

EXAMPLE 9.9

An engineering workshop contains two transformers each having iron and copper losses at full load of 3.6 W and 15 W respectively. The transformers have 40% and 15% loading and both loads must connect to one transformer. Compare the savings in energy if only one transformer is used.

The losses before this occurred were:

- full-load copper loss was 15 W per transformer at full load
- for the 40% loaded unit, copper loss = 0.4^2 × 15.0 = 2.4 W
- for the 15% loaded unit, copper loss = 0.15^2 × 15.0 = 0.3375 W
- full-load iron loss was 3.6 W per transformer at full load
- total losses = 2.4 + 0.3375 + 3.6 + 3.6 = **9.9375 W**.

»

With one transformer only on load:

- iron loss = 3.6 W
- copper loss = $(0.4 + 0.15)^2 \times 15 = 4.5375$ W
- total loss = 3.6 + 4.5375 = **8.1375 W**.

Over a 12-month period of 260 working days, 24 hours per day:

Original loss:

$$\text{loss}_{\text{orig}} = \frac{260 \times 24 \times 9.9375}{1000}$$

$$= \mathbf{62.01\ kWh}$$

New loss:

$$\text{loss}_{\text{new}} = \frac{260 \times 24 \times 8.1375}{1000}$$

$$= \mathbf{50.78\ kWh}$$

Saving = 62.01 – 50.78 = **11.23 kWh**, which is an 18.1% saving in energy loss.

EXERCISE 9.9

An engineering workshop contains two transformers each having iron and copper losses at full load of 5 W and 18 W respectively. The transformers have 45% and 25% loading. Determine the saving over a 12-month period of 260 working days, 24 hours per day if both loads connect to one transformer.

Equivalent circuit of a practical transformer

Practical transformer designs approximately meet the conditions that we assumed in our study of the ideal transformer. A better model is required, however, for calculations that are more accurate.

Calculations concerning voltages and currents in a practical transformer use a phasor diagram.

Transformer operation under no load

When an a.c. voltage occurs across the primary, a primary magnetising current (I_m) and a primary iron loss current (I_e) are established resulting in a primary no-load current (I_0). The current I_m is a purely reactive current and lags the primary terminal voltage (V_1). The mmf due to current I_m sets up an alternating magnetic flux (Φ_m) in the transformer core. The alternating flux links both the primary and secondary windings and induces an emf (E_1 and E_2) in each winding. In relation to the primary winding, the emf (E_1) is the counter or back emf. This emf (E_1) sets up a volts/turn value. If the transformer were ideal then the back emf would oppose the rate of change of current in the primary winding and no I_m would flow. However, in a practical transformer I_m does flow.

The induced emf across the secondary winding is the secondary open-circuit voltage. The value of the secondary voltage is proportional to the number of secondary turns and the rate of change of the magnetic flux.

The relationship between the magnetising flux (Φ_m) and the induced emfs is found in Faraday's law, which states that the value of the induced emf is proportional to the rate of change of magnetic flux linkage. Mathematically this is expressed as:

$$e = N\frac{\Delta\Phi}{\Delta t}$$

where e = the induced emf in volts

N = the number of turns in the coil

$\Delta\Phi$ = the change in magnetic flux in webers

Δt = the change in time in seconds

The equation states that the induced emf is equal to the number of turns multiplied by the rate of change of flux (the frequency of an a.c. supply).

In addition, Lenz's law affects the relationship. This law states that the polarity of the induced emf creates a magnetic field that opposes the flux linkage change if current is allowed to flow (the direction of E_1 is such as to limit the current that produces it). **Figure 9.59** illustrates a phasor diagram for a single-phase transformer on no load.

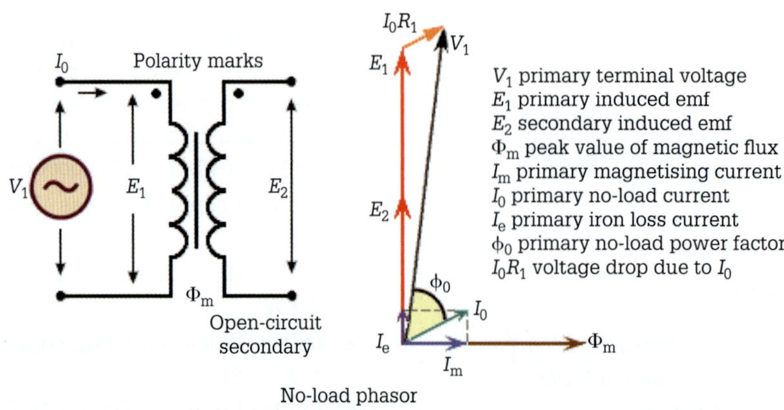

FIGURE 9.59 No-load phasor diagram

The magnetic flux (Φ_m) is the reference phasor because it is common to both windings. Note also that this flux links both the primary and secondary windings and therefore it is a mutual flux. The flux Φ_m induces emfs E_1 and E_2 across the primary and secondary windings. These two emfs are in phase with each other because of the mutual flux linkage.

The iron loss current (I_e) is in phase with the primary induced emf (E_1). I_e is tiny in comparison with I_m and supplies the hysteresis and eddy-current losses in the core. The phasor addition of the magnetising current (I_m) and the core loss current (I_e) provides the magnitude of the no-load current (I_0). The magnetising current and the core loss current are magnified in **Figure 9.60** so their effects can easily be seen. The voltage drop I_0R_1 is also greatly magnified in **Figure 9.60** which indicates that the magnitude of the primary induced emf (E_1) is approximately equal to the magnitude of the primary terminal voltage (V_1). The no-load current (I_0) is also referred to as the excitation current.

Transformer operation under load

The phasor diagram for a transformer at full load is shown in **Figure 9.60** with the flux (Φ_m) as the reference phasor. This complex diagram can be understood by examining how it was constructed. We start by assuming that the secondary is supplying a lagging current I_2 at a terminal voltage V_2 with phase angle Φ_2. These are the first two phasors drawn. Now, to V_2 we add the active component (I_2R_2, voltage drop due to secondary resistance) in phase with I_2, and the reactive component (I_2X_2, voltage drop due to secondary leakage reactance) to find the induced voltage E_2 in the secondary. Induced in the primary is the voltage E_1.

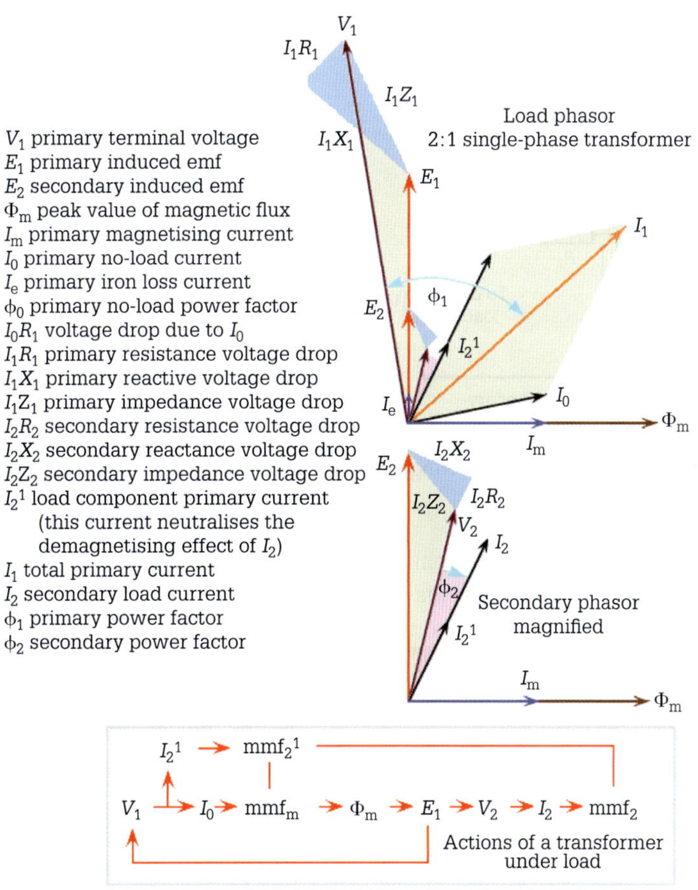

FIGURE 9.60 Phasor diagram of a transformer under load

The flux (Φ_m) is at right angles to E_1, as shown. The current I_0 is necessary to create the flux, and is drawn with its proper relation to Φ_m. The current I_2 reflects to the primary as I_2^1. Consequently, I_2^1 adds to I_0 to find the total primary current I_1.

Now that we know I_1, we can add the phasor addition of the active component (I_1R_1, voltage drop of primary resistance) and the reactive component (I_1X_1, voltage drop due to primary leakage reactance) to E_1 to find the primary terminal voltage, V_1. In **Figure 9.60**, the magnetising current and the voltages caused by leakage flux and winding resistance are greatly magnified to show their effects. The difference between the terminal voltage V_2 and the induced voltage E_2 becomes less as they approach one another.

Note that the secondary current only produces mmf and not flux. This secondary mmf (mmf_2) opposes the original mmf (mmf_m) thereby reducing it, the flux (Φ_m) and the voltage (E_1) it was producing. Because the voltage has been temporarily reduced, more primary current is allowed to flow which produces additional primary mmf (mmf_2^1). This additional mmf exactly equals and opposes the secondary mmf (mmf_2) and so their resultant mmf is zero. Consequently, the original primary mmf is the only mmf (mmf_m) producing flux (Φ_m). Therefore, there is no tendency for the iron to saturate (you cannot saturate a transformer by overload – the overload limit is heat).

The phasor diagram method is complicated to employ. A better method is the substitution of a simplified equivalent circuit in place of the practical transformer.

Equivalent circuit of a transformer

The operation of a transformer, as shown in **Figure 9.61**, can be considered to be an ideal transformer without any energy losses, magnetic leakage or magnetising current. Then, the variations within a practical transformer by means of additional circuits are inserted between the applied voltage and the primary winding and between the secondary winding and the load.

Figure 9.61 shows a circuit diagram of the equivalent circuit of a transformer.

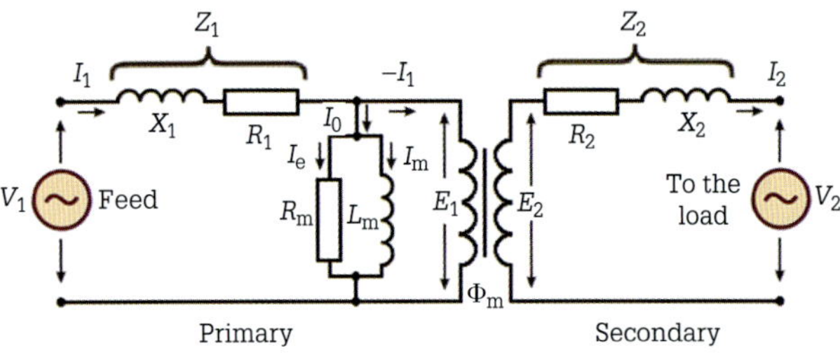

FIGURE 9.61 Equivalent circuit of a transformer showing the losses that occur

In the equivalent circuit, R_1 and R_2 represent the ohmic resistances of the primary and the secondary windings of a practical transformer. X_1 and X_2 represent the reactance of transformer windings due to leakage and fringing flux. The transformer core losses due to eddy currents and hysteresis are indicated by the resistor R_m. This resistor draws an energising current I_e equal to the iron core loss component of the primary current of a practical transformer. In addition, L_m represents the inductive reactor effect which draws a magnetising current I_m (produces the flux) in a practical transformer. Finally, the resultant current of I_e and I_m is I_0 and the final current is the no-load current.

Open-circuit test

The open-circuit test measures the iron loss of a transformer. To measure iron loss the high-voltage winding of the transformer is open circuited and measuring instruments connect as illustrated in **Figure 9.62**.

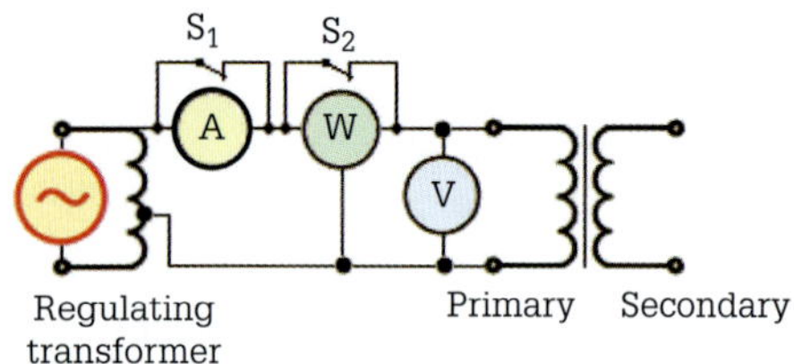

FIGURE 9.62 Connections for open-circuit test

With the connections shown and the ammeter and wattmeter shunted out (S_1, S_2) of the circuit, adjust the regulating transformer to the rated voltage of the transformer using the voltmeter. After switching off the regulating transformer, disconnect the voltmeter. Energise again and open switch (S_1) and record the current flowing. De-energise again. Shunt out (S_1) the ammeter and the open switch (S_2) across the wattmeter and record the wattmeter reading. The ammeter measures the no-load current while the wattmeter measures the iron loss within the transformer.

REVIEW QUESTIONS

1 Name the main two energy losses that a transformer experiences.
2 Which of the two energy losses is constant for all loads?
3 What is the full-load efficiency of a 200 kVA transformer with an iron loss of 0.5 kW and a full-load copper loss of 2 kW?
4 When does maximum operating efficiency of a transformer occur?
5 What is all-day efficiency?
6 Why are all-day efficiency calculations useful?
7 Draw a schematic diagram of the equivalent circuit for a practical transformer.
8 State the purpose of an open-circuit test.

9.8 Transformer cooling

The design of power transformers is such that they operate at, or close to, full-load. As such, they attain high operating temperatures. This heat needs to be dissipated in the surrounding environment. Various methods are employed to assist in dissipating the heat developed.

The energy losses (iron and copper) in transformers appear as heat, and one of the difficulties for the transformer designer is how to provide the most effective means for getting rid of this heat. Heat accelerates the chemical action naturally occurring within insulating materials, deteriorating the insulation. Over time, the transformer windings develop loose turns. Because transformer turns are subject to electromagnetic repulsion and attraction forces, the loose turns start to wear mechanically, and earth faults or short circuits occur. To maximise the service life of a transformer various cooling methods are utilised. The method chosen depends upon the physical size of the transformer and operating conditions.

Self-cooled (AN)

Small, dry, air-insulated, air natural (AN) VA-type, as shown in **Figure 9.63**, and larger kVA-type (to 10 000 kVA) transformers operate without oil, and the coils are wound without cooling ducts. Cooling ducts are not required because this class of transformer has a surface area that is large in comparison with the volume, and the energy loss dissipated per square centimetre is small.

The heat generated has only a short distance to travel in order to reach the surface of the transformer. The reduced distance means that the interior of the transformer is not much hotter than its surface.

With these transformers, the heat caused by the transformer losses dissipates into the surrounding atmosphere. This dissipation occurs by means of radiation and natural convection currents of air circulating around the transformer windings or the transformer tank or radiators.

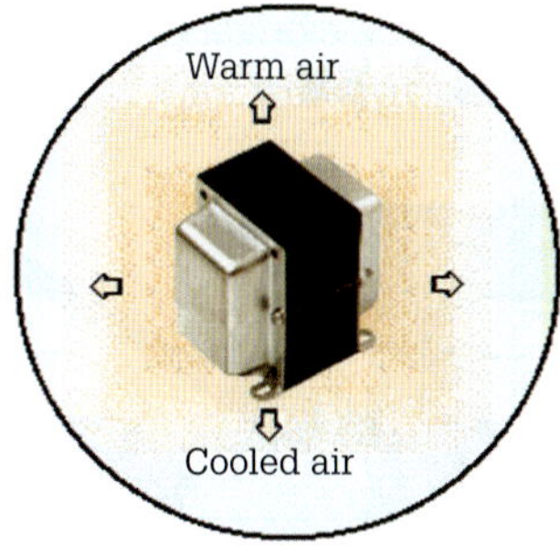

FIGURE 9.63 Dissipation of heat from dry-type transformers

Dry-type transformers include most instrument types as well as transformers of the distribution class for low voltages and ratings. Large, air-cooled dry-type transformers, as shown in **Figure 9.64,** are used within buildings where oil-type transformers could present a fire or possible leakage hazard.

FIGURE 9.64 Dry self-cooled transformers

When AN transformers are of more than 10 kVA capacity, it becomes progressively more difficult to keep them cool. In order to increase the amount of transformer winding surface exposed to the cooling air, the transformer windings have air ducts between them as shown in **Figure 9.65**. The heat generated from the coils warms the air within the ducts.

As cooler air enters the ducts from below, convection currents are set up which carry the heat away from the inside of the coils and thereby supplement the cooling provided by the exterior cooling surface of the transformer.

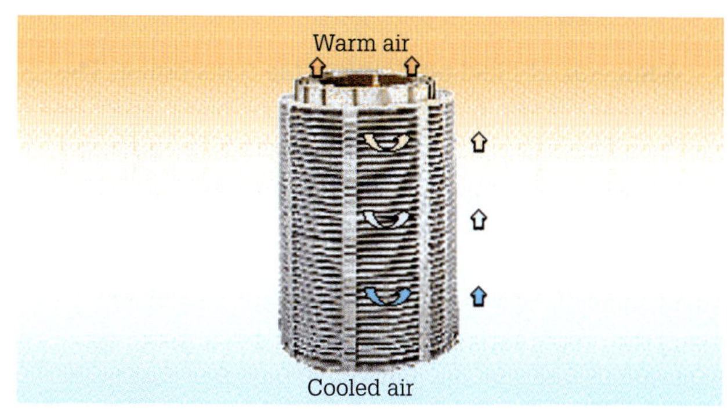

FIGURE 9.65 Air-duct cooling

An extension of the convection principle of the self-cooled dry transformer is an air-blast cooling process (AF). This cooling method uses a metal cabinet to contain a transformer that has many ducts built into the coil windings. The cabinet connects to a heat exchanger or to an air blower.

Air-blast transformers, like that shown in **Figure 9.66**, exist where fire hazards prevent the use of oil-immersed transformers.

FIGURE 9.66 Air-blast cooling

Oil cooled

In general, transformers too large for cooling by the self-cooled method use a process of immersion in oil or other synthetic fluids called the oil natural method (ON). **Figure 9.67** shows the transformer tank with flat sides for a small distribution transformer.

FIGURE 9.67 Transformer tank and oil

A mineral oil-filled tank houses the transformer. The oil not only helps to cool the transformer, but it enriches the cellulose-based insulation of the windings through impregnation. The mineral oil has a low viscosity that enables it to act as a cooling agent by means of convection currents that circulate the mineral oil up along the energised coils where it absorbs heat and then down along the cooler surface of the tank where it gives up heat. In other words, the mineral oil transfers heat from the transformer windings to the flat tank sides where it dissipates into the surrounding air.

Another method (ONAN – oil natural air natural) as shown in **Figure 9.68** increases the exposed surface of the transformer tanks by the use of tubes. These tubes are 'C' shaped and welded at the top and bottom into holes in the tank wall. The number of tubes welded to the tank depends upon the amount of dissipated heat.

FIGURE 9.68 Transformer tank with tubes showing convection current movement of mineral oil

As transformer sizes increased, the number of coil turns and core material increased and as a result heat increased. To compensate for this, the transformer had an increase in width and height to provide a larger cooling surface. However, to keep the physical size of modern transformers within manageable limits the surface of the tank has external radiators connected as shown in **Figure 9.69**.

FIGURE 9.69 Transformer tank with external radiators

The radiators connect to the tank at the top and bottom so that the heated mineral oil flows from the top of the tank into the radiators. Here the mineral oil cools and returns to the bottom of the tank, ready to rise through the coil ducts of the windings and repeat the process. If the convection current of air along a heated surface is disturbed, heated air moves away. Consequently, cooler air reaches the transformer tank surface. The use of fan-cooled transformers (ONAF) demonstrates the cooling effect. Above or beside the radiators are motor-driven fans as shown in **Figure 9.70**.

FIGURE 9.70 Fan-cooled transformer

Direct water-cooled transformers (ONWF), like that shown in **Figure 9.71** are used for high-current applications where direct air cooling is not practical and/or where space is limited. Transformers using this type of cooling find application as rectifier transformers for process industries such as electroplating and welding.

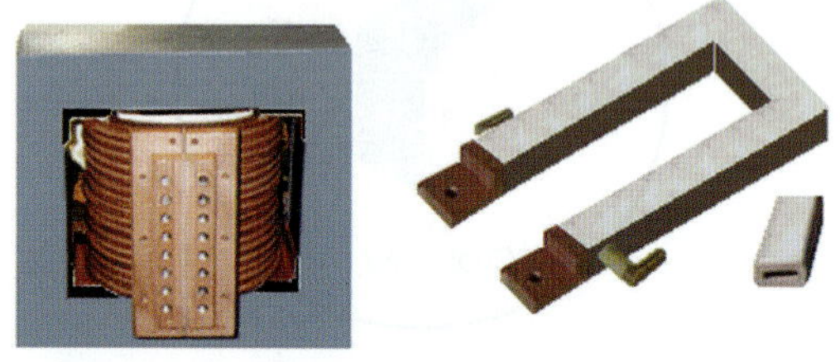

FIGURE 9.71 Water-cooled transformer using hollow copper or aluminium tube

The cooling water feeds into the high-current windings by individual manifolds to provide protection against corrosion. These hollow copper or aluminium tube conductors insulated with heat-shrinking polyester tape become the primary or secondary transformer windings or both, with the coolant flowing through their hollow centre. The cooling system is of the closed type, using distilled water, and includes a storage tank and circulating pumps.

Another type of water-cooled transformer has the oil and water heat exchanger attached to the outside of the tank as shown in **Figure 9.72**. The cool water is pumped through a heat exchanger to cool the heated oil.

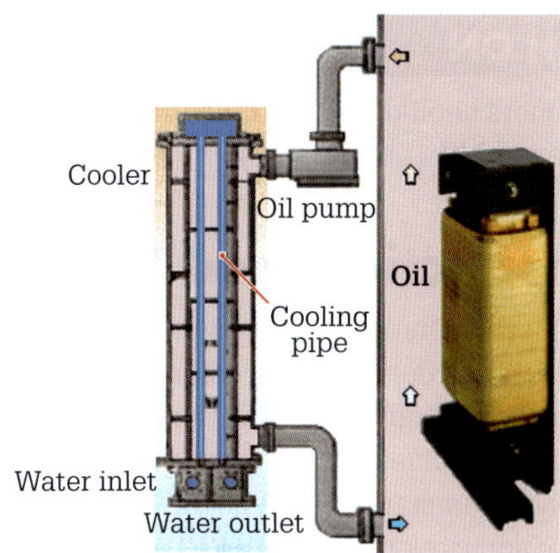

FIGURE 9.72 External water-cooled transformer

As the windings heat the oil, the oil rises to the top and exits through piping to the radiator. As it cools, the oil descends through the radiator and re-enters the transformer tank at the bottom. Many water-cooled transformers have cooling tubes of double-wall construction. The intent of this construction is to bleed off any leaking water or oil into the space between the two walls of the tube to avoid contamination.

Figure 9.73 shows another method of transformer cooling is forced-oil circulation. In this system, hot oil moves from the top of the transformer tank by a circulating pump, forced through a radiator attached to the outside of the tank and returned to the bottom of the tank. At the same time, fans force air over the cooling surface of the radiator.

FIGURE 9.73 Forced-oil cooling

SWITCH ON

Transformer paint

In transformers that use the conduction method of cooling, tank colour has a negligible effect on heat loss by conduction. However, colour does affect transformers using a radiation method of cooling.

Cooling mediums

The most important factor that determines the life and practical operation of a transformer is the immersing coolant. Some of the characteristics desirable in transformer coolants are:

- high dielectric strength to reduce the insulation distances from the walls of the tank
- low viscosity to enable rapid circulation of the fluid and dissipation of heat
- high resistance to emulsion in order to prevent the holding of moisture in suspension in the fluid
- freedom from the formation of sludge under normal operating conditions
- freedom from acids, alkalis and sulphur compounds to prevent corrosion of the conductors or damage to the insulation material
- low freezing point to keep fluid in cold weather.

Mineral oil

Mineral oils, such as Shell Diala B or Shell Diala BX, are transformer oils that contain inhibitors. Inhibitors slow down the oxidation of oil by delaying the formation of acids that attack the insulation and increase the ability of the oil to absorb water. They impede the formation of sludge due to insulation breakdown.

Mineral oil is about as environmentally safe a choice of dielectric fluid as can be found. Transformer mineral oil performs at least four functions for the transformer. Oil provides additional insulation for the transformer windings.

Mineral oil also dissolves gases produced by oil degradation, moisture, gas from cellulose insulation deterioration, gases and moisture. Gases generated depend on the amount of dissolved oxygen in the oil and the temperature, and how close bare copper conductors are to the heating. Moisture contamination is a common cause of deterioration in the insulating quality of mineral oil (insulation quickly degrades by excess moisture and the presence of oxygen). As oil temperature increases, it is easier for moisture to break down in the oil.

Another mineral oil-attacking agent is oxidation. Oxidation causes the formation of acids and sludge. The exclusion of oxygen is of prime importance. However, in self-cooled transformers, the oxygen supply is virtually unlimited, and oxidative deterioration occurs rapidly. The rate of oxidation with oil-cooled transformers depends on the temperature of the oil – the higher the temperature, the faster the oxidative breakdown of the mineral oil. More to the point do not overload transformers.

In transformers, sludge sticks to all surfaces. This sludge forms an insulating barrier to the flow of heat from the windings. Due to sludge, the transformer insulation is overheated, and the insulation develops reduced dielectric strength. This effect eventually leads to short circuiting.

Any maintenance program for mineral-oil transformers must include observation of the dissolved gases in the oil and other oil properties. It is evident that the three principal causes of gas generation are ageing, thermal faults and electrical faults. Observation of the dissolved gases is an important transformer diagnostic tool.

Synthetic liquids

Silicone oil

Silicone oils used in transformers are different from mineral oils. They meet the demand of a dielectric coolant for transformers with high-temperature properties, low toxicity and low flammability. Silicone liquids have a higher flash point than mineral oils and have excellent electrical and thermal stability. These liquids keep their insulating and cooling properties over time, even at very high temperatures.

Silicone oils have excellent oxidation resistance and fewer degradation products or sludge when compared with mineral oil. They possess high chemical stability and have clarity for easy contamination checks. In the event of a spill or leakage, silicone liquids have very low water solubility, are degradable in soils and do not deplete oxygen in surface water. Some silicone-cooled transformers are hermetically sealed and pressurised with nitrogen. An advantage of the hermetically sealed transformer is that the coolant is never in contact with the atmosphere thus avoiding periodic oil analysis.

Midel

Midel is a premium green-coloured insulating liquid, based on synthetic ester (formed from an organic acid and an alcohol) and is a pentaerythritol ester liquid. It is a safer alternative to mineral oils in transformers where fire safety and protection of the environment are primary considerations. It is fire resistant up to 300 °C. Midel has a slow heating rate because of its high specific heat capacity and thermal conductivity. It is also non-toxic and non-corrosive, as well as being biodegradable and environmentally friendly. Midel cooling is especially suited for situations where transformers operate in water treatment plants. Synthetic organic esters occur in both sealed and breathing transformers.

Other coolants

R-Temp

R-Temp is a non-toxic fire-resistant hydrocarbon-based liquid. This liquid is biodegradable and possesses excellent dielectric and arc-quenching properties. In addition, R-Temp demonstrates excellent stability, with no sludge by-products.

Envirotemp RF3

The RF3 fluid is non-toxic and biodegradable and possesses excellent fire-resistant properties and a high flash point.

SWITCH ON

Transformers should operate only with the coolant recommended by the transformer manufacturer.

Oil inspection and testing

A transformer's operation is dependent on the condition of its oil so, given the dangers of oil contamination, periodic inspection and testing of the oil in transformers in service is necessary. The time interval between inspections depends on circumstances such as the importance of the transformer application and the load it carries. Primary inspection of transformer oil in service should include the following three checks:

1. **Oil level check**. The correct oil level has a marking on either the oil gauge or the tank wall.
2. **Sludge check**. A visual examination indicates whether sludge is present.
3. **Viscosity check** (resistance to flow). Good oil should have low viscosity.

Oil testing

Transformer oil testing is a proven prevention technique that should be a part of any maintenance program. There are five basic tests on transformer oil which, when considered together, give the tester a reasonably accurate idea of the serviceability of the oil.

1. dielectric test
2. acidity test
3. power factor test
4. IFT (interfacial tension) test
5. dissolved-gas-in-oil test.

Other tests such as moisture content and oxidation levels may be required, depending on the environment in which the transformer operates and its age.

Dielectric strength test

The dielectric strength test measures the insulating strength of transformer coolants. The test subjects the coolant to electric stress. The value of the breakdown voltage is indicative of the amount of contaminant in the oil. Because of this, low dielectric values indicate oil contaminants such as water, dirt or conducting particles in the oil.

Acidity test

The acidity test (pH test) measures the content of organic acids resulting from oxidation. For this reason, the test reveals the ageing status of the coolant as well as the coolant's tendency to form sludge.

Power factor test

The power factor test is an accepted preventative maintenance test for insulating oil that shows the dielectric loss in the oil. A low power factor reveals deterioration or contamination, or both.

IFT test

The IFT (interfacial tension) test is an indication of the sludging characteristics. New oil exhibits high interfacial tension while oil in service that has oil-oxidation contaminants has low IFT.

Dissolved-gas-in-oil test

This test provides an early indication of transformer problems by analysing the type and quantity of gases dissolved in the transformer oil. Certain quantities and combinations of gases are indicative of insulation overheating, coolant overheating, partial discharge (corona) or arcing taking place in the transformer.

Transformer rating and cooling

Auxiliary equipment, such as pumps or fans, or both, improves the cooling efficiency of the transformer. In addition, the kVA rating could increase by 33% at 150 °C.

However, if the auxiliary equipment fails and the transformer operating temperature rises by 10 °C above the stated transformer temperature rating, the increased heat can cut transformer life by up to 50%. Where additional cooling methods are used, means for remote control and temperature monitoring should be provided.

REVIEW QUESTIONS

1. State the two transformer conditions that influence a designer's choice in transformer cooling method.
2. Why can small VA-type transformers operate without oil and cooling ducts?
3. How does the heat generated by dry-type transformers dissipate?
4. Where are large, air-cooled dry-type transformers generally located and why?
5. Mineral oil not only cools the transformer when in service, it also provides another service to the coil windings. What is this service?
6. In what applications are water-cooled transformers utilised?
7. Describe the 'forced-oil circulation' cooling method.
8. Why do transformer cooling oils contain inhibitors?
9. What is a common cause of deterioration in the insulating quality of mineral oil?
10. There are five basic transformer oil tests. What are they?
11. What is the effect of sludge on transformer coolant operation?
12. State the three principal causes of gas generation in a transformer.
13. Name two properties of silicone oils.
14. Name the auxiliary equipment that improves cooling efficiency of the transformer.

9.9 Voltage regulation and percentage impedance

Two important transformer operating parameters are voltage regulation, which defines how well a transformer can maintain output voltage as load changes, and percentage impedance, which is used to determine fault current levels.

Another important operating consideration is the effect harmonic currents have on a transformer.

Voltage regulation

If a transformer delivers at a specified power factor its rated VA, kVA or MVA output at its rated secondary terminal voltage, this voltage changes when the load disconnects. Moreover, the voltage difference together with a particular power factor provides the voltage regulation of the transformer. Calculating voltage regulation for transformers differs from the equation used for generators and alternators and is determined using the following equation:

$$V_{Reg} = \frac{V_{no\text{-}load} - V_{full\text{-}load}}{V_{no\text{-}load}}$$

This equation results in a decimal value, which is multiplied by 100 to yield an answer expressed as a percentage.

Voltage regulation is a term used to give a measure of the performance of the transformer when it is operating between no-load and full-load conditions. The power factor of the load requires specification when providing the voltage regulation because it differs widely with the type of connected load and its requirements.

EXAMPLE 9.10

A 2000:400 V transformer (**Figure 9.74**) operating at 0.8 lagging power factor has a no-load secondary terminal voltage of 415 V. Calculate the voltage regulation.

»

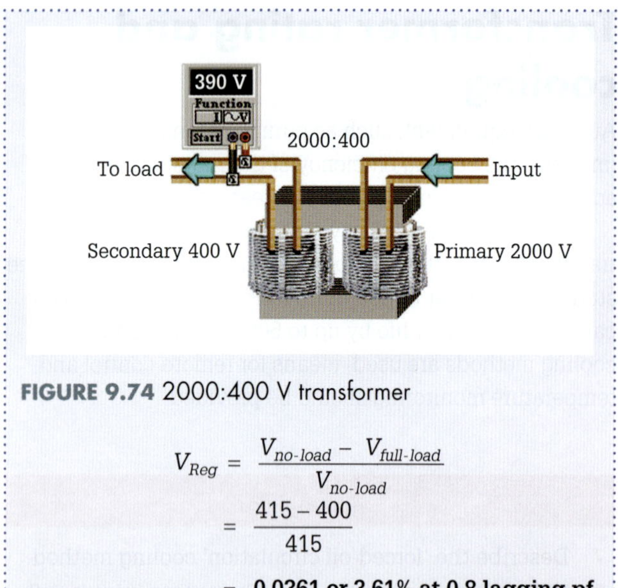

FIGURE 9.74 2000:400 V transformer

$$V_{Reg} = \frac{V_{no\text{-}load} - V_{full\text{-}load}}{V_{no\text{-}load}}$$

$$= \frac{415 - 400}{415}$$

$$= \mathbf{0.0361 \text{ or } 3.61\% \text{ at } 0.8 \text{ lagging pf}}$$

EXERCISE 9.10

a A 50 kVA 2300:230 V distribution transformer has a no-load voltage of 238 V at a power factor of 0.95 lagging. Calculate the voltage regulation.

b A distribution transformer has the following data: 200 kVA, 2400:240 V at a power factor of 0.89 lagging. If the no-load secondary terminal voltage rose by 12 V, calculate the voltage regulation.

Transformer output as shown in **Figure 9.75** has a continuous full load (10 V @ 2.5 A). Lighter loads will result in higher output voltages (11 V @ 1.0 A), while heavier or intermittent loads result in proportionately lower voltage sags (9 V @ 3.6 A).

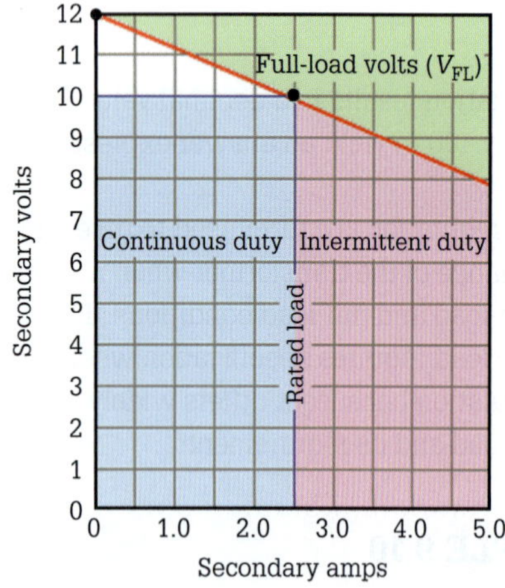

FIGURE 9.75 Transformer load graph

Larger transformers have better load regulation than smaller transformers: typically, 500 VA = 5%; 100 VA = 10%; 50 VA = 15%; 10 VA = 20%; 5 VA = 30%. If improved regulation is required for lower VA ratings, the transformer requires redesigning. Accordingly, the regulation improves by decreasing leakage flux (by increasing core size) and winding resistance (decreasing the copper loss).

Transformers transmit variations in the primary or line voltage directly to the secondary. That is, a 230 V fluctuating from –5% to +10% (218.5 V to 253 V) would make a 10 V rated secondary to fluctuate from 9.5 V to 11 V. The no-load voltage would respond similarly, with the 12 V becoming 11.4 V to 13.2 V.

Voltage regulation in transformers is a critical consideration when supplying voltage-sensitive loads, such as 12 V ELV lighting.

Percentage impedance

The impedance percentage of a transformer is marked on most transformer nameplates and is defined as:

SWITCH ON

The percentage impedance (Z%) of a transformer is the voltage drop on full load due to the winding resistance and leakage reactance expressed as a percentage of the rated voltage.

The resistance and leakage reactance are the load energy loss of a transformer. In addition, energy losses include the copper loss (I^2R) in the windings and the iron losses caused by eddy currents and stray fields that cause energy losses in the transformer case and winding clamping frame. The impedance voltage drop is the voltage required to circulate rated-load current through the resistance and reactance of the windings. It is also the percentage of the standard terminal voltage required to circulate full-load current under short-circuit conditions of the secondary.

Measuring impedance

Measurement of the impedance is done by means of a short-circuit test. To measure impedance the secondary winding of the transformer is short circuited and measuring instruments connected as in **Figure 9.76**.

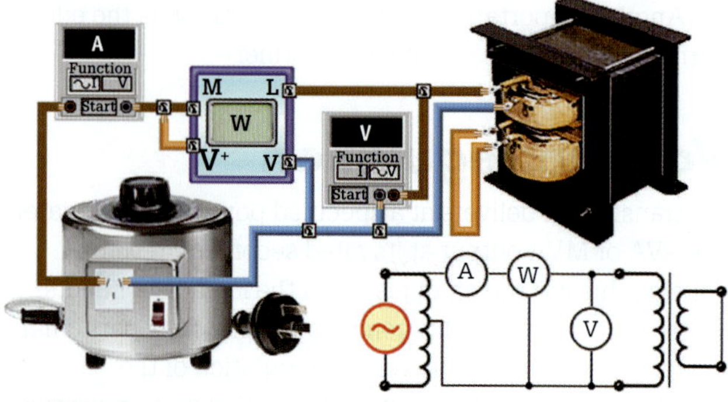

FIGURE 9.76 Impedance test connections

The regulating transformer voltage adjusts until the transformer under test is carrying its rated full-load current. Readings on the voltmeter and wattmeter are noted. The power recorded by the wattmeter is the I^2R or copper loss in the windings. Consequently, the percentage impedance can be calculated.

$$Z\% = \frac{\text{impedance voltage}}{\text{rated voltage}} \times 100$$

Effect of higher and lower transformer impedances

The impedance of the transformer has a significant impact on system fault levels. It determines the maximum amount of current occurring under fault conditions. For a single-phase transformer, the rated (full-load) current is calculated from:

$$I_{Full\text{-}load} = \frac{VA}{V_{Line}}$$

For a three-phase transformer, the rated (full-load) current is calculated from:

$$I_{Full\text{-}load} = \frac{VA}{\sqrt{3} \times V_{Line}}$$

The fault current is calculated taking into account the percentage impedance:

$$I_{Fault} = I_{Full\text{-}load} \times \frac{100}{Z\%}$$

The following example outlines the significance of percentage impedance.

EXAMPLE 9.11

a Determine a minimum circuit breaker fault current interrupting capacity for a 10 kVA single-phase transformer having 4% impedance, to be operated from a 230 V 50 Hz supply.

$$I_{Full\text{-}load} = \frac{VA}{V_{Line}}$$

$$= \frac{10\,000}{230}$$

$$= \mathbf{43.5\ A}$$

The fault current is calculated taking into account the percentage impedance:

$$I_{Fault} = I_{Full\text{-}load} \times \frac{100}{Z\%}$$

$$= 43.5 \times \frac{100}{4}$$

$$= \mathbf{1087.5\ A}$$

The circuit breaker or fuse would have a minimum fault current capacity of 1087.5 A at 230 V.

b Determine a minimum circuit breaker fault current interrupting capacity for a 100 kVA three-phase transformer having 5% impedance, to be operated from a 400 V 50 Hz supply.

$$I_{Full\text{-}load} = \frac{VA}{\sqrt{3} \times V_{Line}}$$

$$= \frac{100\,000}{\sqrt{3} \times 400}$$

$$= \mathbf{144.3\ A}$$

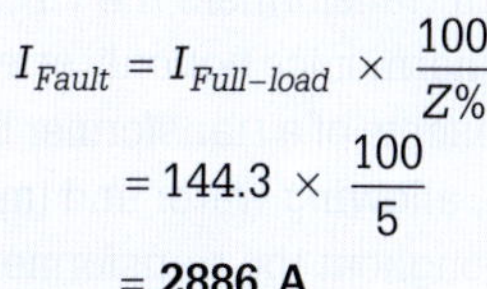

$$I_{Fault} = I_{Full\text{-}load} \times \frac{100}{Z\%}$$

$$= 144.3 \times \frac{100}{5}$$

$$= \mathbf{2886\ A}$$

The circuit breaker or fuse would require a minimum fault current capacity of 2886 A at 400 V.

EXERCISE 9.11

a Determine a minimum circuit breaker fault current interrupting capacity for a 50 kVA single-phase transformer having 5% impedance, to be operated from a 230 V 50 Hz supply.

b Calculate the available short-circuit current for a three-phase, 400 V, 1000 kVA transformer with an impedance of 4.5%, that is adjacent to and supplying the 400 V busbars on a factory switchboard.

A transformer with lower impedance leads to a higher fault level and a transformer with higher impedance leads to a lower fault level. The figures calculated in **Example 9.11** are maximum. In actual operation there are many impedances between the transformer and the fault that reduce the fault level.

The primary factors that determine the possible fault current are transformer rating, the impedance of the cable, and that of the connections. These factors, in addition to the fault resistance, determine the actual fault current.

Although a transformer can provide a severe limiting effect on fault current, the resistance of the conductors and terminations becomes very significant as distance increases. The resistance of a few metres of cable can reduce the fault current considerably. The maximum short-circuit current obtained from the output of a transformer can be determined by multiplying the reciprocal of the impedance by the full-load current. Thus, if a transformer has 5% impedance, the reciprocal of 0.05 is 20 and maximum short-circuit current is 20 times the full-load current. As well as fault-level considerations, the impedance value also determines the voltage drop that occurs under load (referred to as voltage regulation) and impacts on load sharing when two or more transformers operate in parallel.

Harmonic currents in transformers

When transformers are energising, the magnetising current drawn is different from the operating current. The harmonics created during this time interval vary over the life of the transformer. The primary effect is the heating of the transformer above its nameplate rating caused by the additional iron losses (higher eddy-current losses in the core) and copper losses (higher coil winding resistance caused by skin effect) generated by the harmonics.

In addition, with transformers the vibrational interactions of the harmonics not only stress the mechanical components of a transformer but also stress the transformer oil, allowing water and impurities to circulate freely throughout the transformer tank. The result of this can be insulation failure and possible burnout of the transformer windings.

Signs of current harmonics include higher than average temperatures in the transformer, voltage distortion and high crest factor. Crest factor is a waveform measure that indicates the ratio of peak values to the rms value. Crest factor is used as an indication of how extreme the peaks are in a waveform. In an ideal sinusoidal waveform, the crest factor is 1.414 ($\sqrt{2}$). In practical transformers, crest factors different from this figure indicate distortion in the waveform. Current distortion has a high crest factor while voltage distortion has a lower crest factor. Current harmonics that increase energy loss are only harmful to transformers when the transformers are not correctly de-rated.

Some transformers supplying mainly single-phase loads are de-rated based on crest factor information. This information comes from the secondary of the transformer. The transformer harmonic de-rating factor is the ratio of 1.414 to the transformer crest factor. Alternatively, 1.414 times the rms load current divided by peak load current also provides the value for the harmonic de-rating factor. The de-rating kVA of the transformer is the nominal kVA of the transformer multiplied by the harmonic de-rating factor.

Total harmonic distortion

As harmonics repeat every cycle and cause distortions from the ideal sinusoidal waveform of current and voltage, harmonic content, or total harmonic distortion (THD), causes poor power quality and can be a hazard to a transformer. Harmonic distortion adds excessive heat and stress, shortening the life of the transformer with the possibility of an electrical fire. In reality, the heating effect of harmonics is proportional to the square of the current and the square of the harmonic. The level or amplitude of the harmonic waveform is the factor determining if it is a hazard. THD expresses the number of harmonics as a percentage of the fundamental. This number for an ideal sine wave is zero. **Figure 9.77** illustrates a distorted waveform composed of the fundamental and the fifth harmonic.

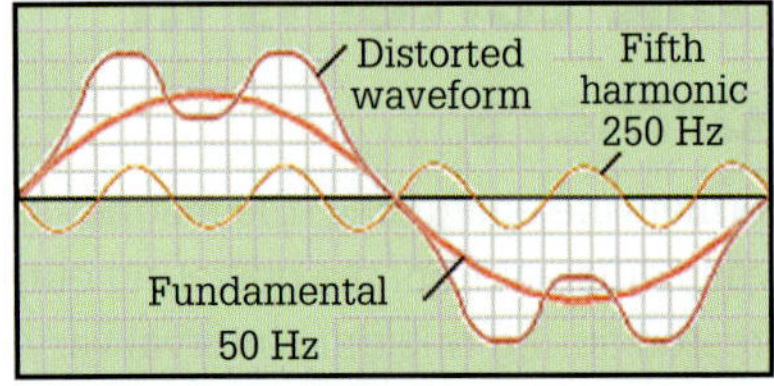

FIGURE 9.77 Harmonic distorted waveform

EXAMPLE 9.12

If the current delivered by a transformer is 350 A at the supply frequency of 50 Hz and the measured current at the fifth and seventh harmonics is 45 A and 25 A respectively, calculate the total harmonic distortion.

$$I_{THD} = \frac{\sqrt{I_5^2 + I_7^2}}{I_1}$$

$$= \frac{\sqrt{45^2 + 25^2}}{350} = \mathbf{14.7\%\ THD}$$

EXERCISE 9.12

If the current delivered by a transformer is 250 A at the supply frequency of 50 Hz and the measured current at the fifth and seventh harmonics is 35 A and 18 A respectively, calculate the total harmonic distortion.

THD is a measure of the amount of distortion produced as current flow from the service line. This line current can flow at the fundamental frequency, or combine with odd harmonic currents (multiples of the fundamental) such as 150 Hz (third harmonic), 250 Hz (the fifth harmonic) and 350 Hz (seventh harmonic).

The THD value is the effective value of all the harmonic currents added together, compared with the value of the fundamental current. For example, 10% THD means that the total harmonic current is equal to 10% of the total 50 Hz current.

If the THD exceeds 15%, then the transformer's capacity to supply the connected load requires re-evaluation.

From a power supply viewpoint, high harmonic distortion increases the volt-ampere (VA) demand on transformers, transmission lines and generators, which decreases the energy reserve on the entire power distribution system and the ability to add more customers on existing distribution transformers.

Triplen harmonics

Triplen harmonics have a frequency that is a multiple of three times the fundamental such as the 3rd, 9th and 15th harmonics. These harmonics affect delta–star-connected three-phase transformers. The triplen current occurs only in the delta winding of the transformer due to mutual induction, resulting in a circulating current within the delta winding.

Triplen currents increase the resistive losses within the delta winding, causing a higher operating temperature and a reduction in effective load capacity. An illustration of triplen harmonic current components carried by the neutral is shown in **Figure 9.78**.

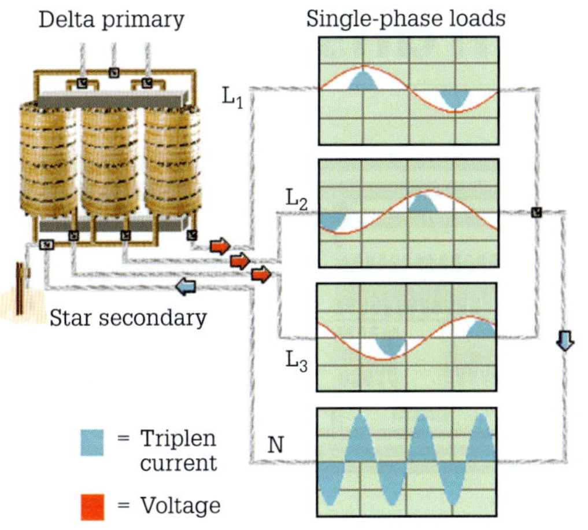

FIGURE 9.78 Triplen harmonics

Triplen harmonic currents are distinctive in that they are additive in the neutral conductor of three-phase four-wire star connected circuits. The fact is that the triplen current pulse is not in phase with the line current and so cannot return via one of the lines. The only return path for the current is through the neutral. The individual harmonic currents feed into the neutral from each load at a frequency that enables the harmonics to be in phase with each other. As a result, there are three harmonic current pulses accumulated in the neutral for every single line current pulse.

The Australian and New Zealand standard AS/NZ 3000: 2018 *Wiring Rules* requires the over-sizing of neutral conductors serving loads because harmonics distort the voltage and thereby increase neutral return currents. Neutral busbars and connecting lugs need resizing to carry the additional current. As the neutral conductor could carry up to 1.73 times the line current, under-sized neutral conductors can lead to a serious fire hazard.

Voltage harmonics

Current harmonics can also distort the voltage waveform and cause voltage harmonics. For example, a three-phase transformer that has one phase feeding a non-linear load causes voltage harmonics to be developed in the other two phases by mutual induction, thereby feeding the harmonics to other circuits connected to the transformer, such as electric motor circuits.

With electric motors the negative sequence harmonics (e.g. 5th, 11th, 17th) have a sequence of rotation (ACB) that is opposite to the fundamental sequence (ABC), thereby producing a rotating electromagnetic field in the opposite direction to the fundamental. This conflict causes heating within the stator windings and contributes to mechanical fluctuations within a motor-connected load system. In Australia the recommended maximum voltage harmonic levels are:

- total harmonic distortion 3%
- 3rd, 5th and 7th harmonics 2%
- 2nd, 11th and 13th harmonics 1.5%
- 23rd and 25th harmonics 0.7%.

In order to develop a better solution to power quality problems, it is necessary to measure the harmonics using a harmonics analyser. These measurements provide detailed information on the full spectrum of harmonic currents and voltages. High-speed spectrum analysers determine the waveforms and help isolate the cause of a harmonic problem. Harmonic distortion caused by the installation or appliances connected to a supply authority's service should not be in excess of the limits prescribed in the Australian and New Zealand standard AS/NZS 61000.4.7: 2007, 'Electromagnetic Compatibility (EMC)–Testing and measurement techniques'. This standard is a guide on harmonics and inter-harmonics, and measurement and instrumentation for power supply systems and equipment related thereto.

Harmonic currents have a significant impact on transformer load losses. About 5% of load loss is due to eddy currents in the windings, and these losses are proportional to the square of the frequency. As a result, the losses taking place from the current at the third harmonic are nine times that due to a fundamental of the same size. The load losses in a transformer supplying non-linear loads can easily be twice the rated (fundamental frequency) losses. Some of the harmonics cancel as the effects of the independent loads add up, thereby reducing the effects of harmonics upstream in the system.

In reality, there are two strategies for dealing with harmonics. Use a larger transformer than required for the load conditions to allow for the extra losses, or use a specially designed transformer to minimise losses with non-linear loads. Any reduction in transformer loss means a lower generation requirement. Another result is reduced greenhouse gas emissions.

REVIEW QUESTIONS

1. Define the term 'voltage regulation'.
2. An 11 000:400 V transformer operating at 0.85 lagging power factor has a no-load terminal voltage of 420 V. Calculate the voltage regulation.
3. Define the term 'percentage impedance'.
4. Name the test used to measure transformer impedance.
5. State the impact that the impedance levels of a transformer have on system fault levels.
6. Calculate the available short-circuit current for a three-phase, 400 V, 1500 kVA transformer with an impedance of 2.5%, that is adjacent to and supplying the 400 V busbars on a factory switchboard.
7. What are signs of current harmonics?
8. What is a consequence of high 'total harmonic distortion'?
9. At what percentage of THD should a transformer be re-evaluated?
10. What are triplen harmonics?

9.10 Parallel operation of transformers

An important consideration before paralleling transformers is to have an understanding of possible connection combinations.

Transformer connections

Transformer windings may connect to either a single-phase, two-phase or three-phase power supply. Combinations such as three-to-two phase, three-to-single phase, six-phase and twelve-phases occur.

Single-phase

Figure 9.79 shows two single-phase transformers connected across a low-voltage input and supplying an extra-low-voltage output. The two secondary windings on the transformer shown to the right connect in parallel.

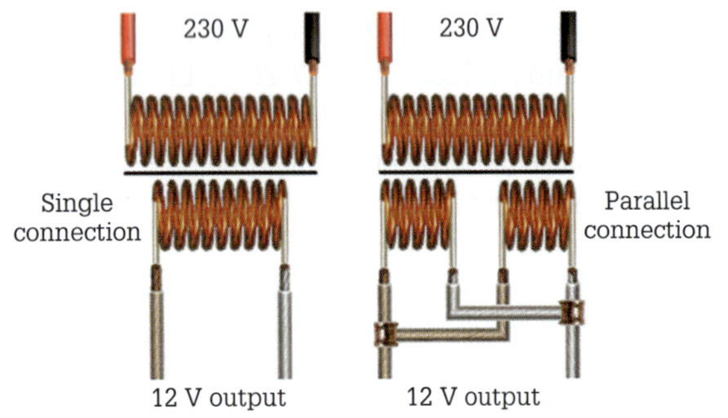

FIGURE 9.79 Single-phase connections

Two-phase

Figure 9.80 shows a two-phase conversion of a three-phase supply using two single-phase transformers supplying a three-wire output.

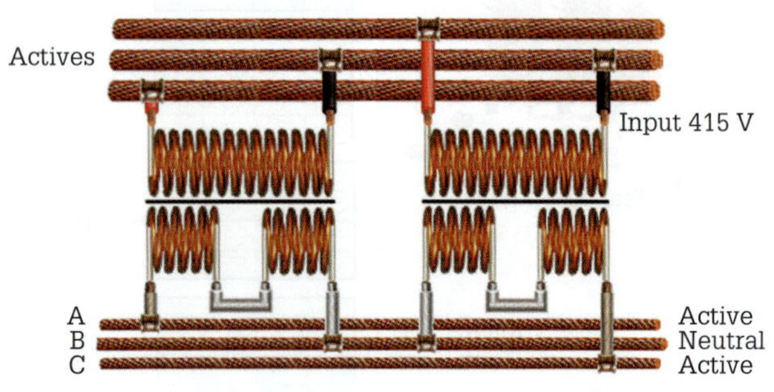

FIGURE 9.80 Two-phase connections

Two-phase transformer windings consist of two single-phase circuits, the voltages of which are 90 electrical degrees apart. This means that the voltage across one winding passes through its maximum value at the instant when the voltage across the other winding is passing through zero.

The common-output conductor B in **Figure 9.80** must carry $\sqrt{2}$ times the current in output conductors A and C.

Three-phase

There are different types of inter-phase connections available for transformer three-phase voltages and currents to increase or lower output voltages or current. The most common connections are star–star (Y–Y), delta–delta (Δ–Δ), star–delta (Y–Δ), delta–star (Δ–Y), open delta (V–V) or Scott (T–T) connections.

Star–star (Y–Y)

The star–star connection is very economical for small, high-voltage output core-type transformers because the coil conductor turns per phase and the total quantity of coil insulation per phase is at a minimum when compared with other connection types. **Figure 9.81** shows an example of the star–star connection.

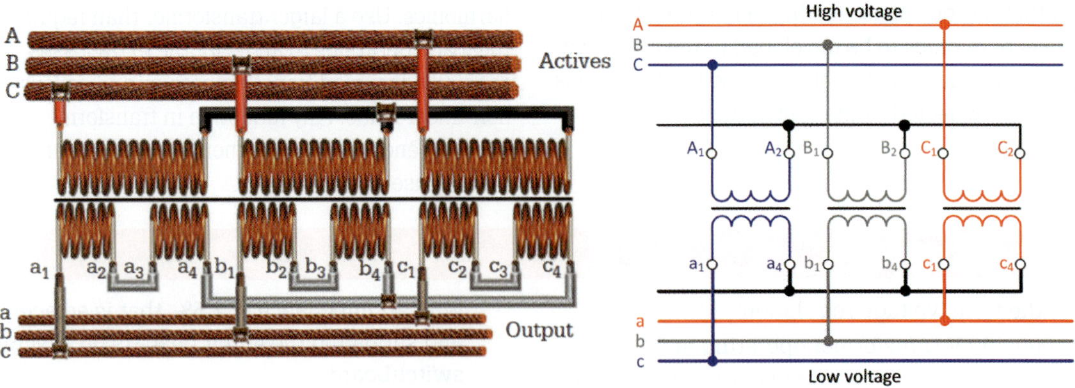

FIGURE 9.81 Three-phase transformer connected star–star

Delta–delta (Δ–Δ)

The delta–delta form of connection is economical for high current, low-voltage transformers. In design, it has smaller cross-sectional-area conductors with a large number of conductor turns, together with more coil insulation per phase compared with the star–star connection. **Figure 9.82** illustrates an example of the delta–delta form of connection.

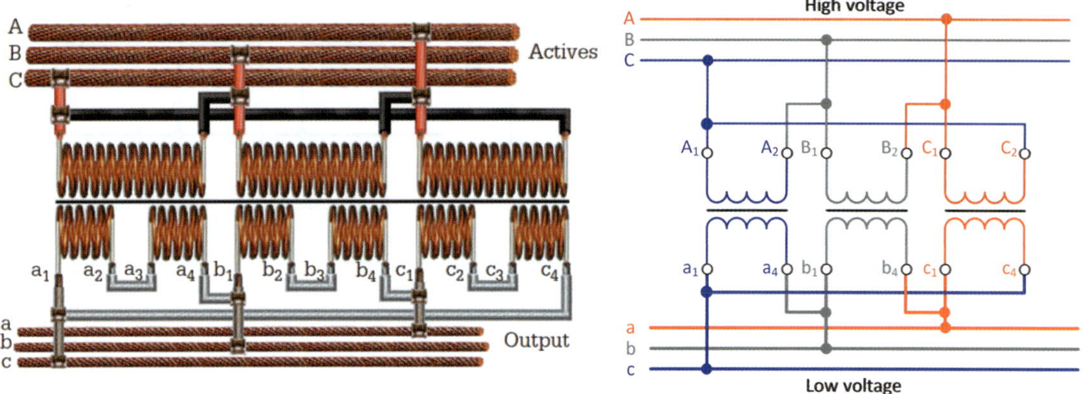

FIGURE 9.82 Three-phase transformer connected delta–delta

Star–delta (Y–Δ)

In the star–delta connection, the primary side connects in star, which means fewer numbers of turns are required due to the reduced phase voltage. This makes for an economical connection method for large high voltage step down power transformers. The neutral conductor allows for any current imbalance. **Figure 9.83** shows a three-phase star–delta transformer.

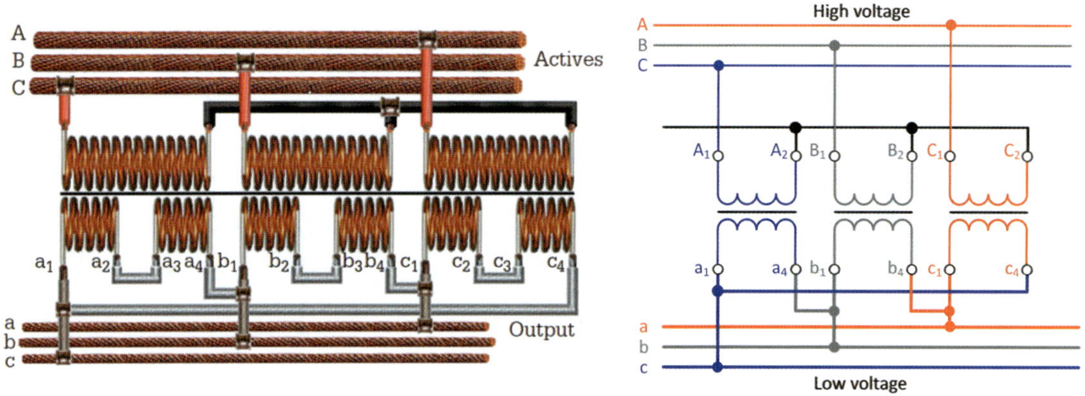

FIGURE 9.83 Three-phase transformer connected star–delta

The main application of the connection illustrated in **Figure 9.83** is at the sub-station end of a high-voltage transmission line where the voltage steps down. The high voltage is impressed on the star-connected primary windings and the delta connections are for the low voltage secondary connection.

high-voltage circuit and a lower voltage per transformer than required for delta connection. It follows that a smaller number of high-voltage turns with a large cross-sectional area conductor and less coil insulation occurs with a star winding.

Delta–star (Δ–Y)

Step-up transformers use the delta-star connection as illustrated in **Figure 9.84**. The star connection on the high-voltage side gives a neutral point for earthing the

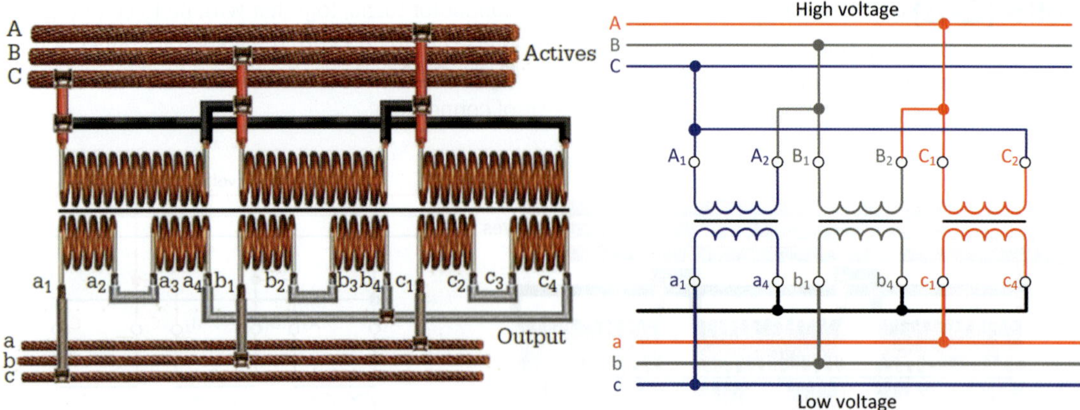

FIGURE 9.84 Three-phase transformer connected delta–star

Open delta (V–V)

The advantage of the open delta connection is that it takes two single-phase transformer units to produce a three-phase output. However, a disadvantage lies in the fact that the currents lead the voltage by 30° in one phase and lag by 30° in the other. As a result, the power factor of the different phases becomes unbalanced and results in high transformer power losses and unbalanced secondary load voltages. Another disadvantage is that its kVA output requires 15% higher rating than an equivalent three-phase transformer. **Figure 9.85** illustrates the open delta connection.

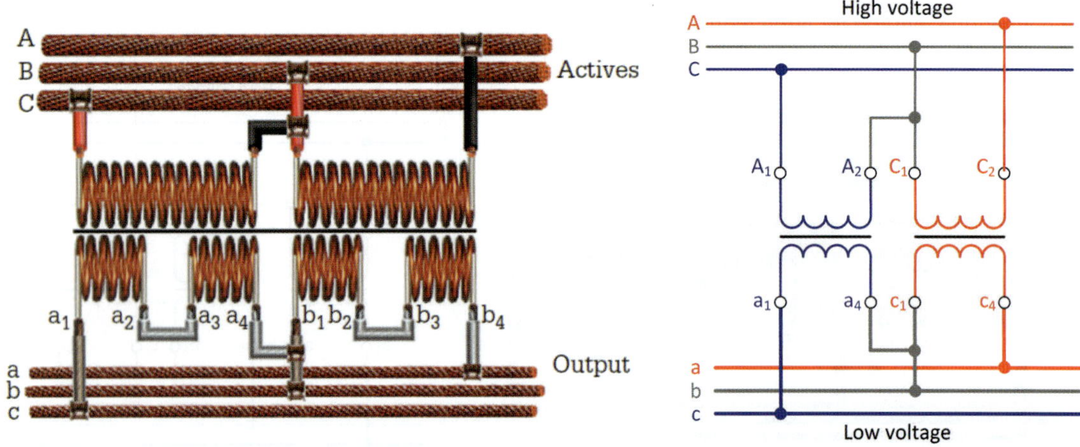

FIGURE 9.85 Three-phase transformer connected open delta

Scott (T–T)

A Scott connection as shown in **Figure 9.86** is used with autotransformers to achieve a two-phase output from a three-phase input.

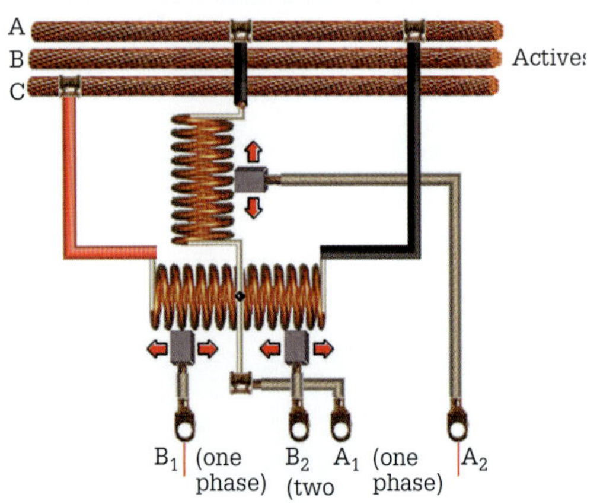

FIGURE 9.86 Scott or T–T connection

In such connections, the autotransformer connected between two of the three-phase lines is the main winding and the one connected from the middle of its winding to the other three-phase line is called a teaser.

A three-phase-to-three-phase transformation happens with two single-phase transformers. However, Scott connections in double-wound transformers are for three-phase to two-phase and two-phase to six-phase transformations.

Zigzag connection

A zigzag winding consists of two winding sections: the first section connected in star, the second connected in series between the first section and the line terminals.

The arrangement of the two parts is that each phase of the second section winds on a different limb of the transformer to the part of the first section it connects.

Vector groups

The vector group notation on a transformer nameplate consists of a capital and a lower-case letter in addition to a code number or alpha setting (α) as shown in **Figure 9.87**.

FIGURE 9.87 Vector group notation

With the vector group the capital letter refers to the primary winding (Y = star connected and D = delta connected) and the lower-case letter (y = star connected and d = delta connected) to the secondary winding. The neutral point (star-point) is always marked N and its position if included in the vector group is shown in **Figure 9.88**.

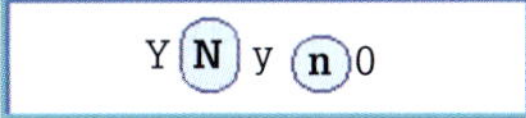

FIGURE 9.88 Vector group notation neutral point

With vector diagrams, the higher voltage in either primary or secondary winding is designated by the number 1 in front of a capital letter and lower voltage by the number 2. The code number communicates the phase angle or phase displacement in multiples of 30° by which side one (1) leads side two (2), as shown in **Table 9.3**.

TABLE 9.3 Code number and phase shift

Code	Phase shift	Code	Phase shift
0	0°	6	180°
1	30°	7	210°
2	60°	8	240°
3	90°	9	270°
4	120°	10	300°
5	150°	11	330°

Using **Table 9.3**, the vector group Yy6 is a star–star transformer with a 180° (6 × 30°) phase shift or phase displacement. A vector diagram of a transformer shows the winding configuration and the lead marking.

The lead marking 2a designates the lead nearest in phase with lead 1A so that the phase sequence (ABC, abc) is in the same order as on the high-voltage side. **Figure 9.89** shows a vector diagram of a high-voltage primary and a low-voltage secondary.

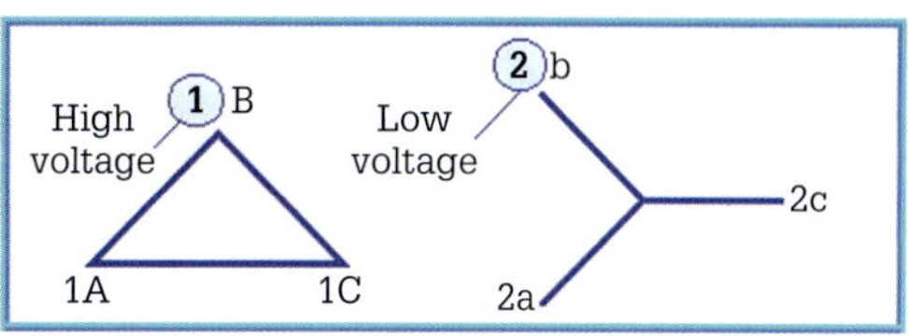

FIGURE 9.89 High-voltage primary and low-voltage secondary delta–star

A number 1 after the letter marks the beginning of a winding; a number 2 after the letter signifies the end of a winding, as shown in **Figure 9.90**.

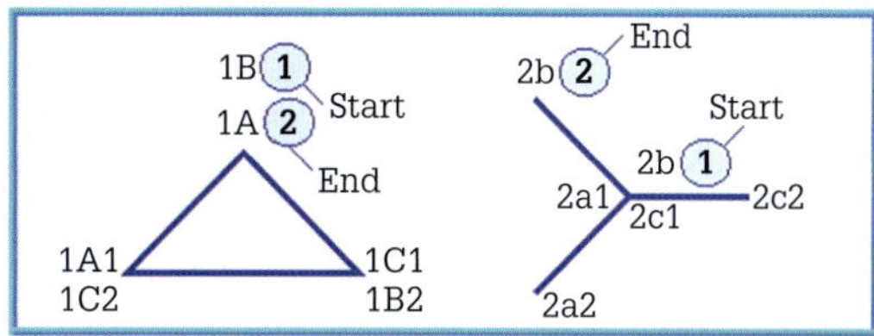

FIGURE 9.90 Numbers indicate start and end

When taps are provided in a three-phase transformer winding, the leads require recording on the vector diagram as shown in **Figure 9.91**.

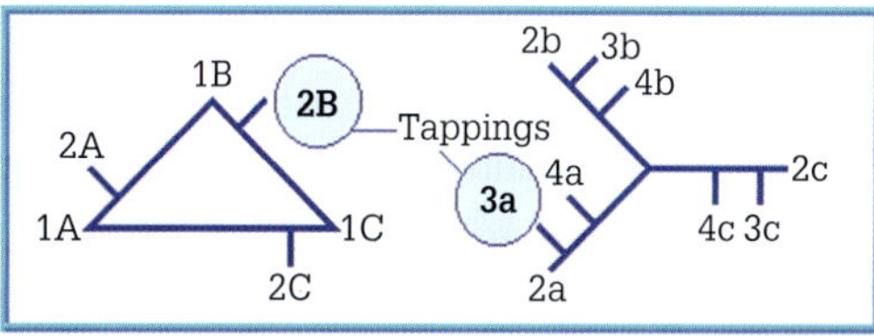

FIGURE 9.91 Vector diagram showing tapping marking

Tappings are marked 2A, 2B, 2C, 3a, 3b, 3c, 4a, 4b and 4c.

Consider a delta–star-connected power transformer, as shown in **Figure 9.91**. If the high-voltage side leads or lags the low-voltage side by either 30° or 330°, then the corresponding code number or alpha settings would be Dy1 or Dy11.

The equivalent phase shift or displacement between high and low voltage can be based on a clock dial: 1 = one o'clock, or 30° lagging; 11 = eleven o'clock (330°), or 30° leading. Phase shift depends on phase rotation.

A Dy11 vector group for a phase rotation of A–B–C becomes Dy1 vector group for a phase rotation of A–C–B.

Some of the most common vector groups are summarised in **Figure 9.92** with the high-voltage winding on the left and the low-voltage winding on the right.

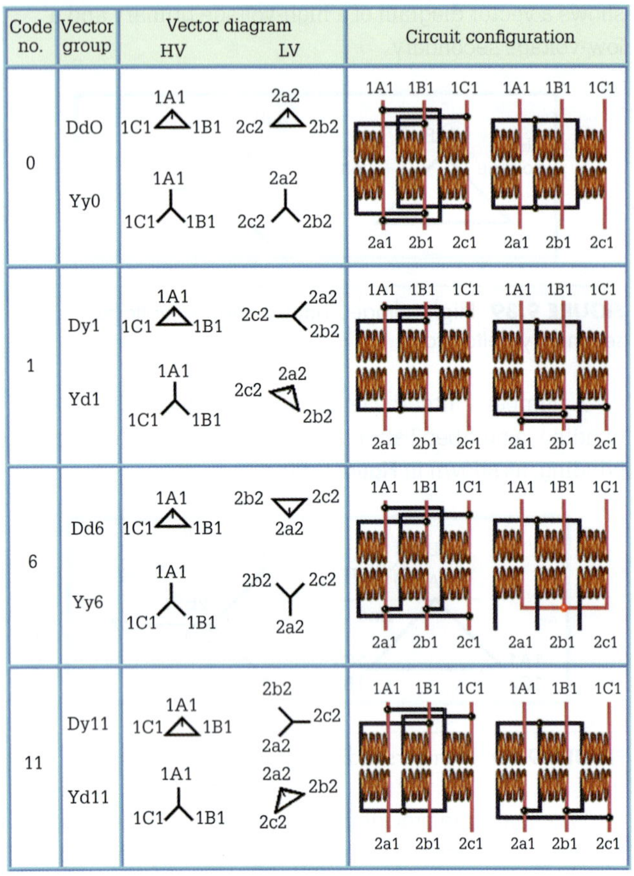

Code no.	Vector group	Vector diagram HV	Vector diagram LV	Circuit configuration
0	Dd0	1A1, 1C1, 1B1 (delta)	2a2, 2c2, 2b2 (delta)	1A1 1B1 1C1 / 2a1 2b1 2c1
	Yy0	1A1, 1C1, 1B1 (star)	2a2, 2c2, 2b2 (star)	1A1 1B1 1C1 / 2a1 2b1 2c1
1	Dy1	1A1, 1C1, 1B1 (delta)	2a2, 2c2, 2b2 (star)	1A1 1B1 1C1 / 2a1 2b1 2c1
	Yd1	1A1, 1C1, 1B1 (star)	2a2, 2c2, 2b2 (delta)	1A1 1B1 1C1 / 2a1 2b1 2c1
6	Dd6	1A1, 1C1, 1B1 (delta)	2b2, 2c2, 2a2 (delta)	1A1 1B1 1C1 / 2a1 2b1 2c1
	Yy6	1A1, 1C1, 1B1 (star)	2b2, 2c2, 2a2 (star)	1A1 1B1 1C1 / 2a1 2b1 2c1
11	Dy11	1A1, 1C1, 1B1 (delta)	2b2, 2c2, 2a2 (star)	1A1 1B1 1C1 / 2a1 2b1 2c1
	Yd11	1A1, 1C1, 1B1 (star)	2a2, 2b2, 2c2 (delta)	1A1 1B1 1C1 / 2a1 2b1 2c1

FIGURE 9.92 Common vector groups

Interconnected star or zigzag standard vector groups occur in industry. These supply the neutral conductor for a three-wire operation when a standard neutral is not available.

Figure 9.93 shows the vector group Dz0 with the high-voltage winding on the left and the low-voltage winding on the right.

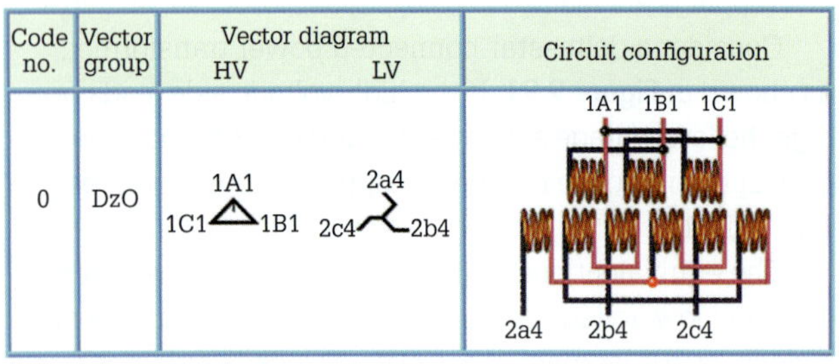

Code no.	Vector group	Vector diagram HV	Vector diagram LV	Circuit configuration
0	Dz0	1A1, 1C1, 1B1 (delta)	2a4, 2c4, 2b4 (zigzag)	1A1 1B1 1C1 / 2a4 2b4 2c4

FIGURE 9.93 Vector group Dz0

Other common vector groups for the interconnected star are Zd0, Yz1, Zy1, Dz6, Zd6, Yz11 and Zy11. **Figure 9.94** shows the marking of line terminals, the vector diagram and the circuit configuration of a single-phase transformer with the high-voltage winding to the left, and on the right the low-voltage winding.

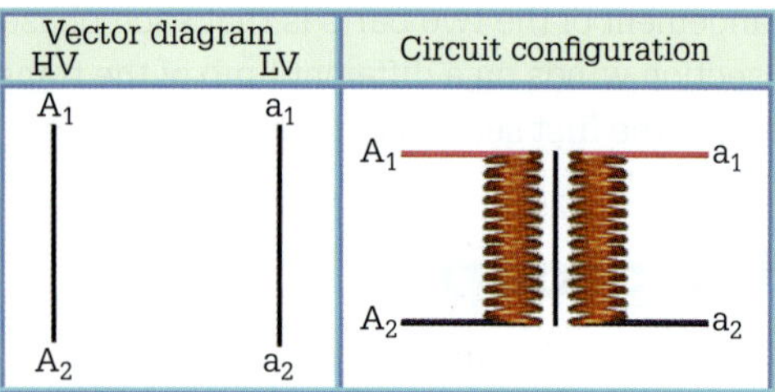

FIGURE 9.94 Line terminal marking

In order to put transformers in parallel, it is necessary to know their respective vector groupings, as well as the polarity of the leads. **Figures 9.87–9.94** show inter-phase connections standards. With three-phase transformers, there are various other combinations of inter-phase connections resulting in other phasor relationships. These include delta–interconnected star, star–interconnected star, interconnected star–star, star–double star, star–double delta, delta–double star, Le Blanc and double Scott.

Tertiary windings

The tertiary connection consists of a single winding per phase; the three connect to form a closed delta circuit with terminals that make it capable of carrying loads, mostly dimensioned for one-third of the transformer-rated power. A transformer with a star–star, tertiary delta winding is illustrated in **Figure 9.95**.

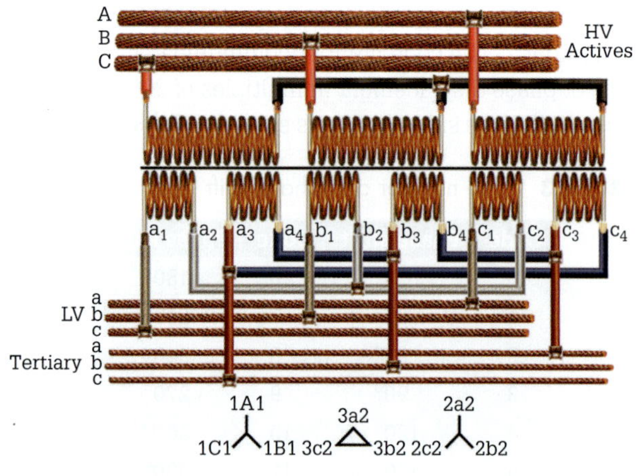

FIGURE 9.95 Three-phase transformer connected star–star with tertiary delta

In some transmission sub-stations, power transformers are typically equipped with tertiary windings that supply distribution transformers with primary voltages to provide power for the sub-station. The uses of tertiary windings depend upon load conditions and other design criteria that require the use of tertiary windings. The design criteria include:

- Stabilising the neutral point during a line-to-earth fault, and improvement of voltage regulation with unbalanced line-to-earth loads.
- Protecting the system and the transformer from excessive third-harmonic voltages when a synchronous condenser connects to the tertiary winding.
- Reducing telephone interference due to third-harmonic currents in the lines and earth.
- Supplying a local load at a third voltage.
- Maintaining an earth system.
- Facilitating the relaying of system faults (mirrors what the primary winding is doing).
- Preventing a series resonant circuit developing.

The benefits of a tertiary winding are offset by:

- higher transformer losses
- increased cost of bus work and protection when terminals are terminated externally
- possibility of tertiary failure causing the loss of the transformer to service.

SWITCH ON

Hazards and safety precautions

Incorrect connections

An incorrect transformer connection can result in short circuits between winding turns and the breakdown of the winding to earth and between the high- and low-voltage windings. The surge of current through the transformer may produce high-voltage rises within the transformer windings and stress the end coils to voltages that cause insulation failure. Incorrect connections produce high temperatures through the transformer, resulting in brittle, contaminated insulation. They may also cause the puncture of insulators and high-voltage and low-voltage windings. The power frequency arc produced by incorrect connections can explode the transformer, causing secondary damage to surrounding equipment and buildings.

Transformer polarity

It is necessary to determine the polarity of transformer windings before effecting parallel connections. Failure to parallel transformers with correct polarity can result in very high circulating currents.

Polarity of single-phase transformers

Polarity is a term used only with single-phase transformers. Polarity refers to the instantaneous voltage direction obtained from the primary winding in relation to the secondary winding. Knowing the polarity is essential when connecting single-phase transformers in parallel.

Consider the transformers as shown in **Figure 9.96**. If the voltage is applied to the primary windings, a magnetic flux is produced in the silicon steel core that links both windings. As a result, a voltage is induced across both secondary windings.

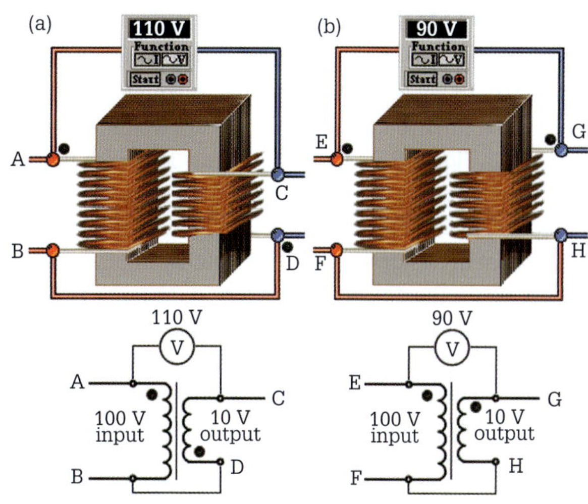

FIGURE 9.96 Additive and subtractive polarity

Since the secondary winding in **Figure 9.96(a)** is wound in the same direction as the primary winding, the voltage across both windings has the same instantaneous polarity. That is, point D is positive to point C at the same instant in time as point A is positive to point B. However, if the direction of one of the windings reverses as shown in **Figure 9.96(b)**, the electromagnetic flux induces voltages in opposite directions in the two windings. That is, point E is positive to point F when point G is positive to point H.

If the transformers illustrated in **Figure 9.96** connect to a sinusoidal input, and adjacent leads from the primary and secondary windings connect, measurement of the voltage between the two remaining leads can occur. When the voltmeter reading is greater than the voltage across the primary windings, the transformer is classified as having additive polarity.

Similarly, if the measured voltage is less than the supply voltage the polarity is subtractive. In addition, because of the relative direction of the two windings, additive polarity has the disadvantage of developing high-voltage stresses in the insulation between the windings. Accordingly, the high voltage may be the sum of the primary and secondary voltages and is most evident where the end turns of the primary winding are adjacent to the end turns of the secondary winding.

For this reason, most transformer terminations have an arrangement for subtractive polarity. Subtractive polarity is the Australian standard with single-phase power transformers. In Australia, the polarity indication happens via a black dot beside the transformer leads indicating the instantaneous positive of both primary and secondary windings.

If a transformer operates in isolation, polarity is unimportant. However, when two or more transformers require operation in parallel, polarity is crucial.

Phase relationship of three-phase transformers

In order to place three-phase transformers in parallel it is necessary to know the relative phase sequence of the three high-voltage leads and the three low-voltage leads. As shown in **Figure 9.97** the high-voltage leads are marked A_2, B_2 and C_2 respectively, and the phase sequence is in that order. The letters a_2, b_2 and c_2 apply to the low-voltage leads so that a_2 designates the lead nearest in phase with A_2. As a result, the phase sequence is in the same order as on the high-voltage side. When three-phase transformers connect in parallel all terminals with the same letter and subscript must be joined together (A_2 to a_2, etc.).

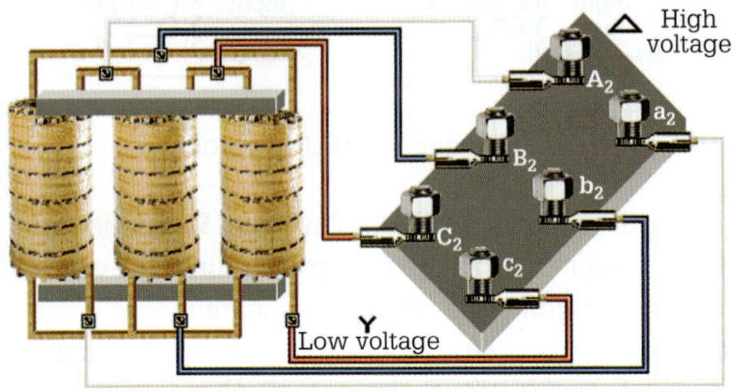

FIGURE 9.97 Phase relationship

The neutral connection when connected to an external terminal is marked Y_N for a high-voltage neutral and y_n for a low-voltage neutral. The relative position of terminal connections for three-phase transformers has been standardised.

The terminal positions when viewed from the high-voltage side of the transformer are located left to right and are N A B C. If facing the transformer from the low-voltage side they are c b a n. Terminal marking is illustrated in **Figure 9.98**.

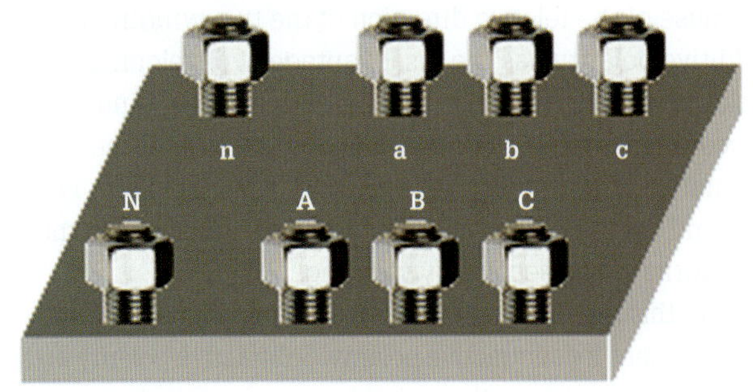

FIGURE 9.98 Terminal marking

Parallel operation of transformers

Efficient parallel operation of two or more items of electrical equipment that deliver an output means that their load shares in proportion to the transformer's rated capacity. Moreover, the sum of the load supply currents from the individual items equals the line current required by the load. With three-phase transformers, there are four conditions – phase sequence, vector grouping, turns ratio and percentage impedance that must be satisfied for efficient parallel operation.

Three-phase transformers require testing for reverse coils before they connect to a load. To test a three-phase transformer for a reversed coil, refer to the following steps.

Star connection

Primary coils

- Connect the primary coils in *star* and to the supply, but leave the secondary coils unconnected.
- Switch on and measure the *individual secondary coil* voltages; if one is higher than the other two, its associated primary coil is *reversed*.

Secondary coils

- Having tested (and corrected) the primary side, connect the secondary coils in *star*, switch on and measure the *line-to-line* voltages.
- If one is higher than the other two, determine which coil gives a *low* reading with respect to the other two – this one is the *reversed* coil.

Delta connection

Both the primary and secondary checks are treated in the same way as the star. Note that the primary side must be checked first.

1. Connect the coils in *delta* but leave the *final connection point open*.
2. Connect to the supply and switch on, then using a voltmeter set to *twice* the expected line voltage, measure the voltage appearing across this open point.
3. If the reading obtained is *zero volts* all is well.
4. If the reading obtained is approximately *twice* the expected line voltage, a *reversed* coil exists.
5. To correct a reversed coil, simply reverse any *one* of the coils and *test again*. Repeat this until *zero volts* is found across the connection point. This implies that the open point closes to complete the delta connection. This test is the 'closing delta connection test'.

Phase sequence

Consider the phase sequence of the two transformers in **Figure 9.99**. If the incoming transformer has a different phase sequence (ACB) to the transformer already online (ABC), a circulating current will flow between the two transformers.

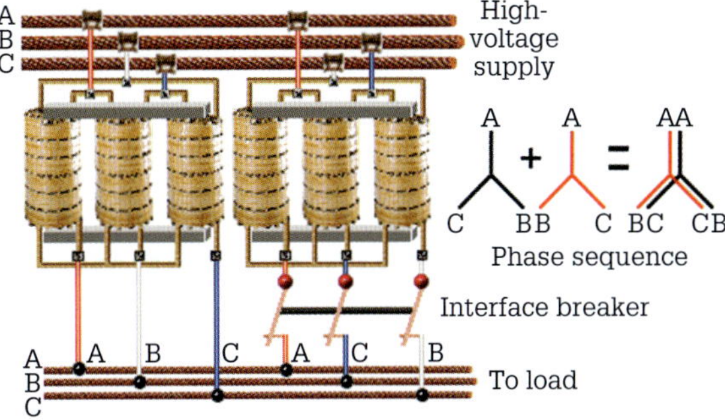

FIGURE 9.99 Incorrect phase sequence

The incoming transformer has the potential of the terminal C, reaching a particular value in the cycle every 120° later than A, and is followed 120° after by B. Therefore, A could be connected to the line that A is feeding because their phasors coincide. The result is both will have the same potential at all times.

However, the voltages B–B and C–C do not coincide and their difference in potential would cause a circulating current to flow. Therefore, for effective paralleling, phase sequence requires attention.

Advantages of paralleling

Separate energy banks of transformers provide continuity of service in the event of preventative maintenance or emergency repair that requires removal of one transformer bank from a service. Indeed, separate energy banks allow the old single transformer (the cost of a replacement manufactured to the new load may be uneconomical) to remain in service when its existing load increases.

Vector grouping

Figure 9.100 shows two transformers of which one has zero phase displacement (Yy0) and the other a phase displacement of 30° lagging (Dy1).

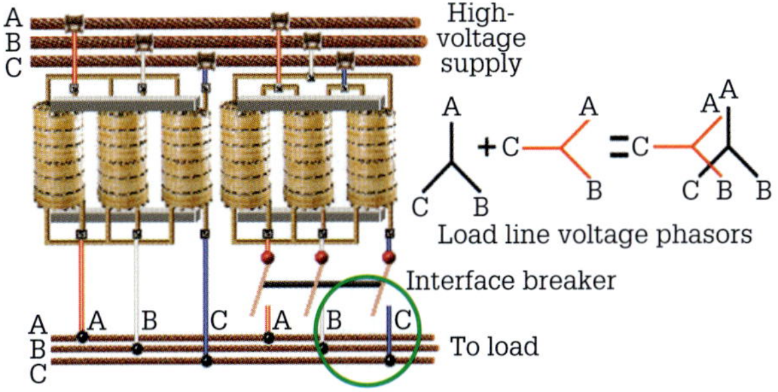

FIGURE 9.100 Incorrect vector grouping

Consider A–A without any pathway being created for circulating currents. A voltage corresponding to the two phasors B–B and C–C equal to the full-load voltage exists across the interface breaker. If the breaker closes, the difference between the voltages results in a net voltage which would cause current greater than the load current to circulate on the closed network between the two transformers. It is essential to ensure that the phase displacement (vector group) of transformers connecting in parallel is the same.

All transformers in the same vector group can be parallel, and those with a 30° lead and a 30° lag can be parallel by reversing the high-voltage and low-voltage phase sequence of either transformer. Another condition for paralleling concerns the turns or voltage ratio. When two or more transformers have to operate in parallel, the ratio of high-voltage turns to low-voltage turns must be the same in all transformers under all conditions of operation. The voltage drop from no load to full load must be the same in all transformers both in magnitude and phase sequence. The consequence is that a difference in ratio causes a variance in terminal voltage at all loads. Therefore, the transformer with the higher voltage output will be doing more work and at no load a circulating current will flow between the transformers.

EXAMPLE 9.13

Concerning percentage impedance, suppose that two transformers have the same no-load or open-circuit low-voltage value of 1000 V but that transformer 1 has a full-load voltage of 950 V and transformer 2 a full-load voltage of 980 V.

When these two transformers connect across the load, the voltage at the terminals of both must be the same and be between 950 V and 980 V. The voltage drop in transformer 1 will be less than 50 V; therefore it has a higher voltage output across its terminals. Thus, its current output (current is inversely proportional to the voltage at a given load) must be less than its rated full-load current. In transformer 2, the voltage drop will be more than 20 V and the current drawn by the load will be greater than its rated current. Furthermore, the voltage difference between no load and full load is the impedance voltage drop (Z_{Vd}) and is the resultant of two components: resistance voltage and reactance voltage drops. This relationship for the two transformers is shown in **Figure 9.101**.

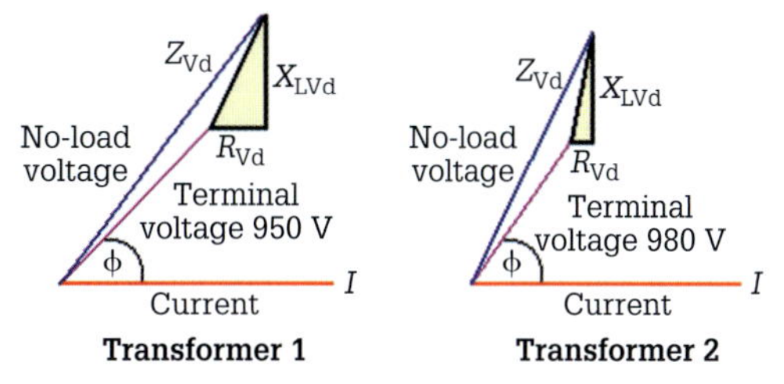

FIGURE 9.101 Voltage drops

The resistance voltage drop depends only on the ohmic resistance of the winding conductors and is in phase with the current. However, the reactance voltage drop depends on the magnetic leakage between the high-voltage winding and the low-voltage winding causing the reactance voltage to drop 90° out of phase with the current. Therefore, the current delivered to the load by the two transformers is numerically equal to the voltage drop divided by the impedance of the transformer windings. The current lags the transformer terminal voltage by a phase angle determined by the ratio of reactance to resistance.

This phase angle (Φ) is not the same for the two transformers and the two currents are not in phase with each other. Therefore, because of their phase difference, the sum of the two currents is greater than the current delivered to the load. The difference in impedance controls the division of current between transformers connected in parallel.

The transformer with lower percentage impedance takes more than its proper share of the load. This implies that it is important to ensure that when two transformers connect in parallel they also have the same percentage of impedance. The result is that each transformer delivers current to its rated kVA capacity. Finally, any electrical difference between transformers connected in parallel could lead to equipment overloads, wasted energy and operational instability.

Single-phase transformer paralleling

The following conditions must exist for single-phase transformers to operate satisfactorily in parallel.

First, the connection diagrams must be identical. There is a simple test illustrated in **Figure 9.102** to check this. In using this diagram, if the voltmeter reads approximately twice the expected secondary voltage, then the connections of either the primary or secondary must change. However, if the voltmeter reads zero volts, all is okay, and the paralleling done.

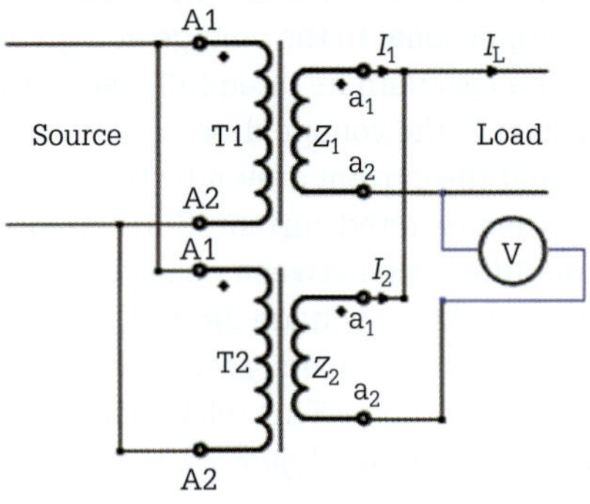

FIGURE 9.102 Parallel single-phase transformers

Second, voltage ratios must be the same. If voltage ratios are not the same, circulating currents will flow in their secondary windings with no or light load and the division of load is unequal. The percentage impedance, including primary and secondary lead impedance to each transformer, should be practically equal. Therefore, if the impedances are equal and the turns ratios are identical, then the paralleled transformers divide the load currents into proportion according to their kVA ratings. However, if the percentage impedances are different, the transformer with lower percentage impedance acquires more than its correct share of the load.

Load sharing in parallel-connected transformers

Two possible considerations for load sharing of parallel-connected transformers are:

1 different impedances but same turns ratio
2 same or different impedances and different turns ratios.

Different impedances but same turns ratio

The load current or kVA is inversely proportional to the transformer impedances. When the turns ratios and voltages are equal, as shown in **Figure 9.103**, no circulating current exists. However, the method shown in **Figure 9.103** is not exact if the transformer resistance and impedance values are not in the same ratio, but is accurate enough for practical purposes. **Example 9.14** shows how it is possible to determine each transformer's share of the load based on percentage impedance values.

EXAMPLE 9.14

A 3 kVA transformer is to be connected in parallel with an existing 4 kVA transformer to supply a load of 5 kVA. Both transformers are rated at 400/230 V with Transformer 1 having an impedance of 2% and Transformer 2 having an impedance of 2.5%. Determine the suitability of both transformers to handle the paralleling requirements when supplying a three-phase resistive load.

Step 1: Determine the load shared by each transformer.

$$VA_{T1} = Load\ VA \times \frac{Z\%_{T2}}{Z\%_{T1} + Z\%_{T2}}$$
$$= 5000 \times \frac{2.5}{2 + 2.5}$$
$$= \mathbf{2778\ VA}$$
$$VA_{T2} = Load\ VA \times \frac{Z\%_{T1}}{Z\%_{T1} + Z\%_{T2}}$$
$$= 5000 \times \frac{2}{2 + 2.5}$$
$$= \mathbf{2222\ VA}$$

Step 2: Determine the division of the load current between the two transformers.

$$I_{T1} = \frac{VA_{T1}}{\sqrt{3}\ V_{Line}}$$
$$= \frac{2778}{\sqrt{3} \times 400}$$
$$= \mathbf{4.01\ A}$$
$$I_{T2} = \frac{VA_{T2}}{\sqrt{3}\ V_{Line}}$$
$$= \frac{2222}{\sqrt{3} \times 400}$$
$$= \mathbf{3.21\ A}$$

This arrangement is suitable.

EXERCISE 9.13

Two 10 kVA transformers are to be connected in parallel to supply a load of 8 kVA. Both transformers are rated at 400/230 V with Transformer 1 having an impedance of 1.8% and Transformer 2 having an impedance of 2%. Determine the suitability of both transformers to handle the paralleling requirements when supplying a three-phase resistive load.

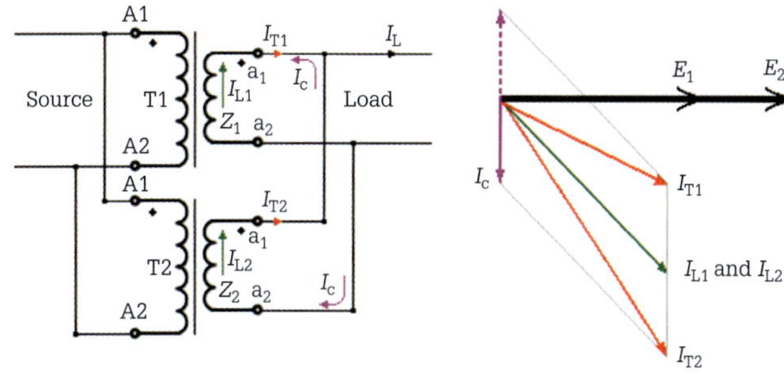

FIGURE 9.103 Parallel single-phase transformers with different turns ratios

Same or different impedances and different turns ratios

If the paralleled transformers have different turns ratios, a circulating current flows due to the difference in terminal voltage. The terminal voltages E_1 and E_2 and the load current cause the circulating current to lag the primary current by approximately 90°. Phasor addition is used to add the value of the circulating current to the load current as shown in **Figure 9.103**.

If the transformers have dissimilar impedances as well as unlike turns ratios, then the distribution of current between them is affected. Furthermore, the value of I_C allowed for constant operation depends upon the resultant loading of the two transformers. Note that the allocation of the load current between the two transformers does not depend on the value of I_C.

REVIEW QUESTIONS

1. Draw a diagram showing a single-phase transformer with two secondary windings connected in parallel.
2. Two-phase transformer windings consist of two single-phase circuits. By how many electrical degrees are the two windings displaced?
3. Name three different types of inter-phase, three-phase transformer connections.
4. Draw a connection diagram showing a three-phase star–star (Y–Y) connected transformer.
5. Draw a connection diagram showing a three-phase star–delta (Y–Δ) connected transformer.
6. What is an application for the delta–star (Δ–Y) connected transformer?
7. Name an advantage of the open delta transformer connection.
8. What type of transformer has a main winding and a teaser winding?
9. What does the vector notation Yy6 mean?
10. State three effects of incorrect connections with transformers.
11. Explain the term 'polarity'.
12. What polarity is the Australian standard with single-phase power transformers?
13. When is transformer polarity unimportant?
14. State the four conditions required when paralleling three-phase transformers.
15. What is the result when two transformers whose phase sequence is different are in parallel across a load?
16. What effect does a difference in turns ratio have when connecting two transformers in parallel?
17. What electrical parameter controls the division of current between transformers?
18. State one reason for paralleling transformers.

9.11 Industry standards and *Wiring Rules* requirements

Many power transformers connect to high voltage supplies. As such it is important to have a good understanding of industry requirements when undertaking high voltage work.

High voltage isolation

High voltage refers to a voltage source in excess of 1000 V a.c. or 1500 V d.c. Before working with high voltages, electrical workers need to have knowledge of the following publications that are applicable for Australia and its states and territories, as well as countries such as New Zealand, Papua New Guinea and the Pacific Islands in which high-voltage work is being conducted:

- *Electrical Safety Act*
- *Electrical Safety Regulations*
- *Electrical Codes of Practice*
- *Electricity entity procedures for safe access to high-voltage electrical apparatus.*

The information detailed in these publications forms a crucial part of providing a safe system of work for workers accessing a high-voltage system. There are many safe systems of work employed by electrical entities. An access permit as illustrated in **Figure 9.104** allows work on or near high-voltage electrical apparatus and a test permit (see **Figure 9.105**) is required for testing the electrical apparatus.

High voltage clearance certificate – access permit

- ✓ Scope of permit
- ✓ The issue of the access permit by the HV switching operator including the operator's instructions
- ✓ Signed receipt by the access permit recipient to acknowledge conditions of the permit
- ✓ Test for dead indicating the location of the test
- ✓ Signature acknowledgement by all workers involved with regard to conditions of the permit and another signature for concluding access
- ✓ Location where earth was connected (signed and dated) and removed (signed and dated)
- ✓ Surrender of access permit by recipient
- ✓ Cancellation of permit by HV switching operator

FIGURE 9.104 Access permit

High voltage clearance certificate – test permit

- ✓ Details of test, including hazard control measures to be undertaken
- ✓ The issue of the access permit by the HV switching operator including the operator's instructions
- ✓ Signed receipt by the access permit recipient to acknowledge conditions of the permit
- ✓ Signature acknowledgement by all workers involved with regard to conditions of the permit and another signature for concluding access
- ✓ Working earth schedule with direct supervision by the access recipient
- ✓ Surrender of access permit by recipient
- ✓ Cancellation of permit by HV switching operator

FIGURE 9.105 Test permit

A switching sheet should also accompany an access permit. A switching sheet, shown in **Figure 9.106**, is an official document that has an exclusive identifying number recording operations of switching in a systematic process.

HIGH VOLTAGE SWITCHING SHEET										Record Number
Section A: SCOPE OF SWITCHING BY WRITER AND CHECKER							Switching sheet			445679
List of lines and apparatus rendered isolated and earthed upon issue of access permit										
Switch writer		Name				Sign		Date		
Switch checker		Name				Sign		Date		
Section B: SEQUENCE OF OPERATIONS										
Op No	Substation area	Voltage	Apparatus	Identification	Tag #	Operation	Operating instruction #	Sw Op	Sw Asst	Time

FIGURE 9.106 Switching sheet

These permits established by the electrical entity allow any person required to be within the exclusion zone to gain access to live parts of high-voltage apparatus for any reason. The issue of an access permit to a person should be within the limits of the access permit as prescribed by the issuing officer.

Before any permit is issued, the person (HV switching operator) who is responsible for implementing the switching sheet requirements must ensure that the high-voltage apparatus isolates all possible sources of supply. Danger tags should be applied to all points of the isolation. Furthermore, locking devices or control systems must be installed. Proof that the apparatus is de-energised, earthed (using an earthing switch or using portable earthing equipment) and short circuited by connecting operational earths, must be available.

Once this has occurred, the access permit work area must be appropriately defined. After these procedures have been finalised an access or test permit may be obtained from the issuing authority.

Operational earthing

'Operational earths' are earths and short circuits of an approved type and size which are implemented as an operation in a switching sheet.

A portable earthing device (PED) is portable earthing equipment of an approved size used to earth parts of a high-voltage circuit where an earthing switch is not available. An earthing switch occurs on high-voltage switchgear for protection of electrical workers during maintenance and overhauls. The switch earths live parts.

Working earths are earths and short circuits used by a worker at the work area (within the boundary of the issued access permit) to provide protection against any potential electrical hazard that may occur.

A three-phase earthing and shorting device is shown in **Figure 9.107**.

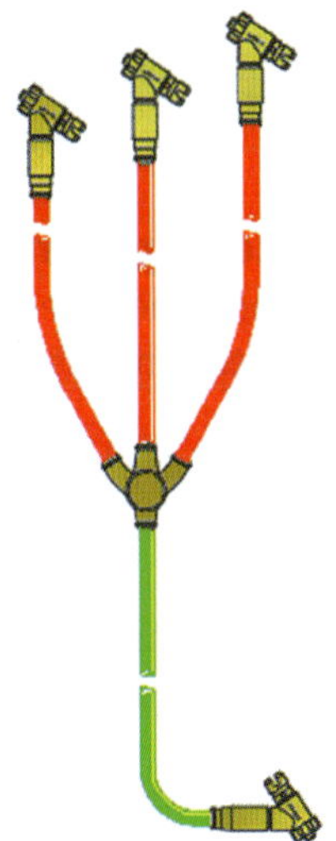

FIGURE 9.107 Three-phase earthing and shorting device

Authorised workers

The entity that is the access authority has procedures in place to ensure that the permits:

- are issued by an authorised person
- are received by an authorised trained (high-voltage) electrical worker
- provide clearance to carry out specified work or testing on specific high-voltage electrical apparatus only
- record the isolation points operated to isolate the high-voltage electrical apparatus
- record the number and location of operational earths and working earths installed.

The recipient of a permit must ensure that the final 'test for dead' happens prior to any work done under the access permit. In addition, if a test for dead with a test instrument is not possible, then another approved method must be used.

An effective point of isolation for high voltage can be a visible break such as racking out a circuit breaker truck or another visual confirmation of isolation. Each point of isolation requires an HV lock box locked and tagged with an HV access tag.

A high-voltage conductor has adequate earthing when earthed by means of an earth truck, portable earthing or integral earth switch, locked into place with an HV lock box lock.

A high-voltage conductor under an access permit is not 'adequately earthed' when failure to achieve or maintain effective isolation from all sources of supply renders the conductor susceptible to becoming live.

Every point of earthing must have an access tag attached to it. Where there is no installed earth point, an approved metal earthing stake must be driven at least 600 mm into the ground. When placing earth leads, the connection at the earth point occurs first, and all electrical workers should keep clear of the earth leads.

All high-voltage electrical workers require training in the use of rescue equipment, and recognising signs and symptoms of electric shock, heart fibrillation, electric burns and first aid. These persons should also be able to identify hazards and any associated risks prior to working on any high-voltage apparatus.

A person working under the authority of an access permit is responsible for carrying out that work under the conditions contained in the work permit. The person must be satisfied that the necessary safety precautions have been taken.

Additional safety procedures

Additional safety measures may include:

- signage
- safe approach distances
- personal protective equipment (PPE)
- surrender of access permit
- step and touch potential clearances.

Signage

Signage is a very important element in providing a safe system of work in high-voltage circumstances. A work area sign as shown in **Figure 9.108** must be used to define the entrance to a high-voltage work area.

FIGURE 9.108 Work area sign

Methods used to indicate the border between live electrical apparatus and the electrical apparatus on which it is safe to perform work under an access permit include insulating screens, yellow tape barriers or rope.

Figure 9.109 shows a permit to work tag. Note that the tag has an area to write the identification number that designates the permit to work authority document.

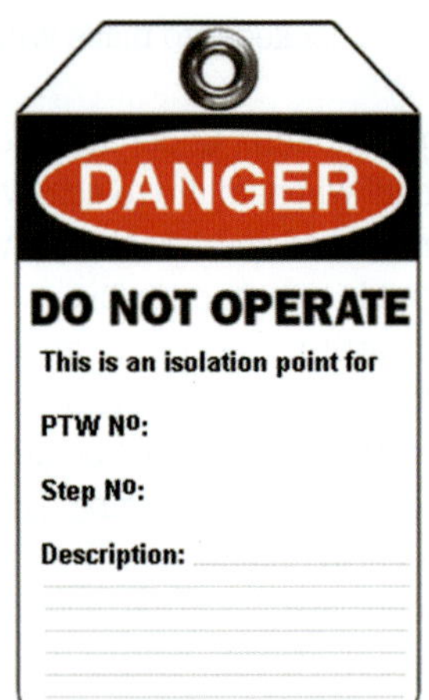

FIGURE 9.109 Permit to work tag

The permit to work tag must:

- be attached to any electrical apparatus, which when in service provides continuity between electrical apparatus and the associated earthing equipment
- be installed so as to be visible to a person attempting to operate the electrical apparatus.

Live high voltage signs as illustrated in **Figure 9.110** placed at various points in the work area show that there is equipment or conductors regarded as live from which workers and other persons need to maintain an exclusion zone.

FIGURE 9.110 Live high voltage signs

High voltage testing signs as shown in **Figure 9.111** indicate that electrical apparatus is under test and that workers and other persons need to maintain the required exclusion zone.

FIGURE 9.111 High voltage testing sign

Safe approach distances

Safe approach distances are areas around electrical apparatus into which no part of a person, mobile plant, equipment or object (other than approved insulated objects) may intrude.

Personal protective equipment (PPE)

Every electrical worker who is listed on the access permit as being allowed to be in close proximity to high-voltage apparatus must wear approved PPE as a control measure against electric shock and arc flash.

> **SWITCH ON**
>
> **Safe approach distances**
>
> Safe approach limits for workers are prescribed by Electricity Regulations.
>
> Always refer to these.
>
> **Note:** These distances may vary across Australia and New Zealand; recourse to the entity in charge of the high-voltage apparatus is necessary.

Surrender of access permit

The recipient may surrender the access permit when the work becomes finalised. An access permit requires surrendering when:

- all working earths have been removed
- the high-voltage electrical workers have all signed off the access permit
- the recipient has visually inspected the worksite to check that no unsafe condition exists.

Step and touch potential

Step potential is the voltage occurring between the feet of a person standing near an energised earthed object. It is the voltage difference between a person's feet (normally one metre apart) caused by the dissipation of a fault current entering the general mass of earth.

Touch potential is defined as the touch voltage between a person's outstretched hand touching an energised object and their foot. A person's maximum reach is one metre. **Figure 9.112** illustrates these potentials.

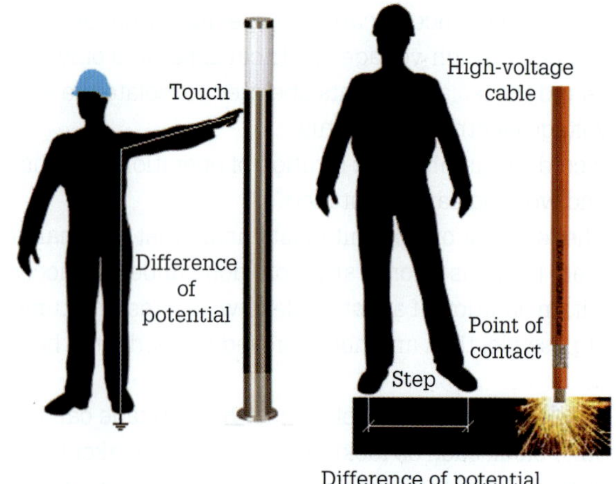

FIGURE 9.112 Touch and step potential

Wiring Rules requirements for transformers

In the Australian and New Zealand standard AS/NZS 3000: 2018 *Wiring Rules* relating to transformers, Section 4.14 must be consulted in order to apply the mandatory requirements. There are other standards applicable to transformers such as the AS/NZS 3108 series and the AS/NZS 61558 series.

Clause 2.5 of AS/NZS 3000:2018 discusses actions to avert danger because of faults between live parts of the electrical installation and circuits supplied at higher voltages. For transformers this action requires adequate insulation, screening or separation of windings.

The main section on transformers is 4.14 of AS/NZS 3000:2018. In this section Clauses 4.14.2 to 4.14.5 provide installation guidance on individual transformers. The electrical areas discussed include the secondary circuit, control and protection, isolating transformers, other transformers, autotransformers and step-up transformers.

Clause 7.4.2 of AS/NZS 3000:2018 discusses the requirements for a transformer as the feed supplying a separated circuit. Clause 7.4.8.1 looks at the testing requirements for a transformer supplying a separated circuit.

The following AS/NZS 3000: 2018 *Wiring Rules* clauses are relevant to transformers:

- Transformers, 4.14
- Autotransformers, 4.14.4
- Control of, 4.14.2.2
- High voltage, 7.6.1
- In separated supply, 7.4.2
- Insulation for, 2.7.2
- Isolating, 4.14.3.1, 7.4.2
- Oil-filled, 4.16.1
- Protection of, 4.14.2.2
- Secondary circuit, 4.14.2
- Step up, 4.14.5.

REVIEW QUESTIONS

1. State the value for high voltage.
2. When is a test permit required?
3. What other document would accompany an access permit?
4. What should be applied to all points of the high-voltage isolation?
5. Describe the function of a portable earthing device (PED).
6. What training must all high-voltage electrical workers receive?
7. Beside permits and training, state another very important element in providing a safe system of work in high-voltage circumstances.
8. Define the term 'safe approach distances'.
9. Describe the term 'touch potential'.
10. Describe the term 'step potential'.

9.12 Transformer nameplates and specifications

Transformers are rated in kilovolt-amperes (kVA), although there are other rating designations. The rating indicates the maximum current that a transformer can deliver without overheating.

In addition, because the transformer is the supply input to a load, the manufacturer provides a power rating. However, the power rating refers to a secondary maximum voltage and current-delivering capacity. (The manufacturer has no control over the power factor of the load the user chooses to connect to the transformer.)

Figure 9.113 shows a nameplate with the kVA rating, serial number and winding connections of the transformer. Note that nameplates in service can be difficult to read.

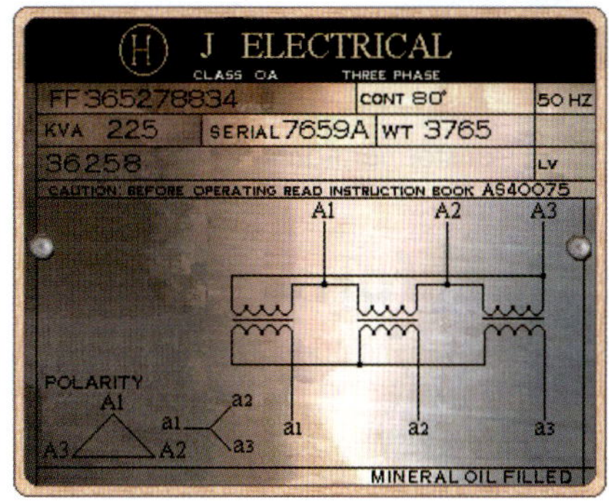

FIGURE 9.113 Transformer nameplate

EXAMPLE 9.15

Determine a transformer's kVA rating for the following conditions. The load is single-phase lighting using fluorescent lamps. Each fixture needs 1.3 A, 230 V, 50 Hz, at a power factor of 0.8. The installation requires 56 fixtures.

$$kVA\ rating = VI = 230 \times (1.3 \times 56)$$

$$= \mathbf{16.74\ kVA}$$

A suitable transformer is 18 kVA, since it has some capacity for future additions.

EXERCISE 9.14

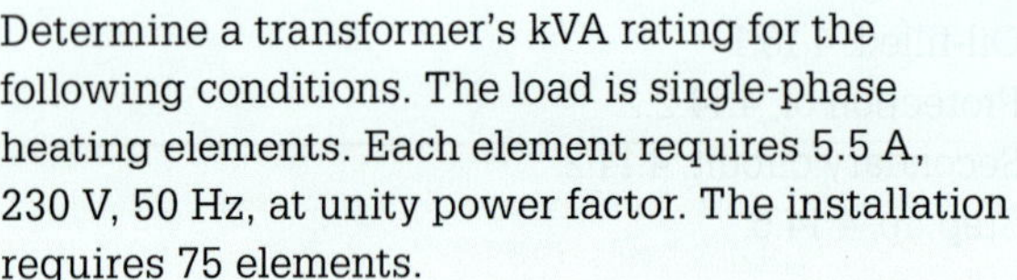

Determine a transformer's kVA rating for the following conditions. The load is single-phase heating elements. Each element requires 5.5 A, 230 V, 50 Hz, at unity power factor. The installation requires 75 elements.

Nameplates are made of brass or stainless steel and are fixed visibly and permanently to the transformer. All pertinent information is engraved. Nameplates may also contain important data items as shown in **Figure 9.114**.

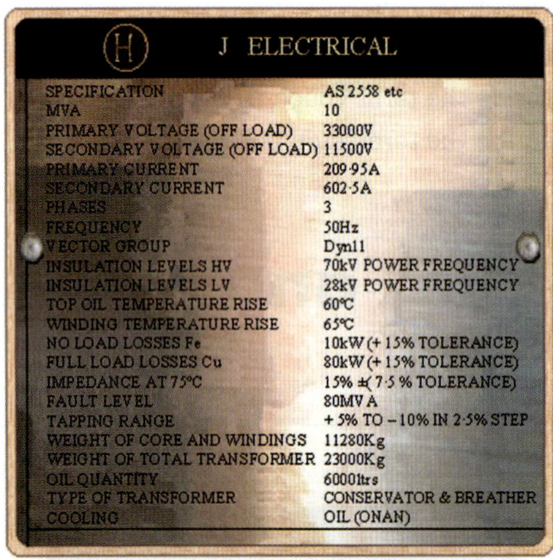

FIGURE 9.114 Transformer nameplate

Basic transformer tests

Basic transformer tests include:

- proving insulation
- testing continuity
- identifying windings.

Proving insulation

The transformer must have an insulation test between windings and each winding to earth using either 500 V or 1000 V d.c.

SWITCH ON

Safety precautions

For insulation measurements to be significant, tests should be made immediately after shutdown while the transformer is still at its operating temperature. For this reason, test errors as a result of moisture condensation on the windings do not occur.

Test transformers at or above their rated voltage to be certain that there are no leakage paths to ground. The specified test voltage, which can range from 100 V to 10 000 V, is applied for one minute. The minimum insulation resistance allowed is 1 MΩ. **Figure 9.115** illustrates this test.

FIGURE 9.115 Proving insulation – winding to winding and winding to earth

Continuity test

The continuity test checks that each winding is electrically continuous from one lead start end to the finish lead end. An 'out of range' reading (e.g. one on some digital meters) indicates an open circuit within the winding. A continuity test can be carried out by using an ohmmeter as shown in **Figure 9.116**.

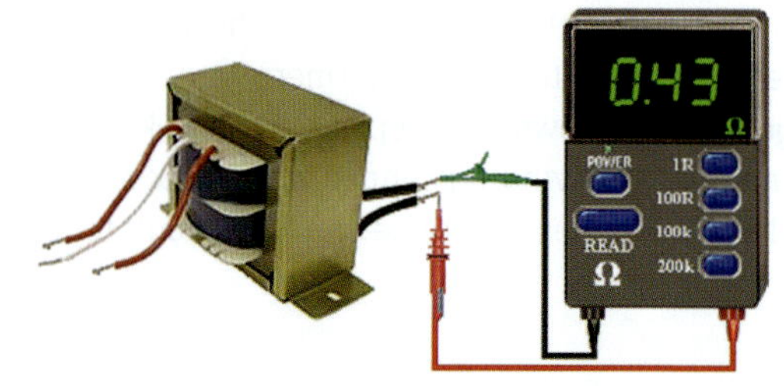

FIGURE 9.116 Continuity

Winding identification test

An ohmmeter can help you determine which termination leads connect to which windings, by looking for continuity. The primary and secondary windings are identified by their resistances. Sometimes you can determine the order of tappings by their resistance. In general, the thicker wires or terminals are for the low-voltage windings. The following provides an outline of a simple test to determine the tapping order.

1 Test the transformer for insulation resistance.
2 Identify the primary and secondary windings by their resistances.
3 Attach an autotransformer (variac) and a low, full-scale deflection ammeter to one of the secondary windings as shown in **Figure 9.117**.

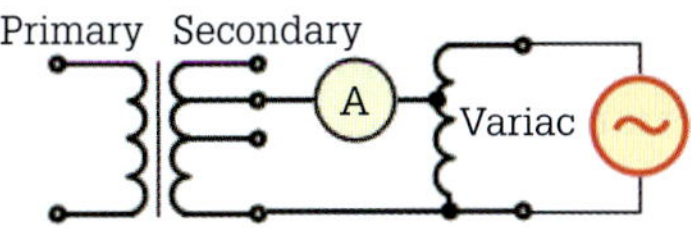

FIGURE 9.117 Winding identification

4 Set the autotransformer to zero on the dial and switch it on. Observe the ammeter, and then **slowly** and **carefully** adjust the autotransformer output current. The current should rise slowly, and **then begin to increase rapidly as the core approaches saturation**.
5 At this point, stop adjusting the autotransformer.
   Note: It may require several attempts to find the point at which the current is just beginning to grow rapidly.
6 Measure and record the voltages across the various leads or terminals of the transformer. You should be able to identify and arrange them in their correct hierarchy.

A transformer-turns-ratio (TTR) instrument as shown in **Figure 9.118** can be easily connected to measure the ratio – just connect the leads, push the test switch and the ratio displays in a few seconds. The transformer-turns-ratio meter is a rugged field and factory instrument that performs fully automatic measurements of voltage ratio and turns ratio.

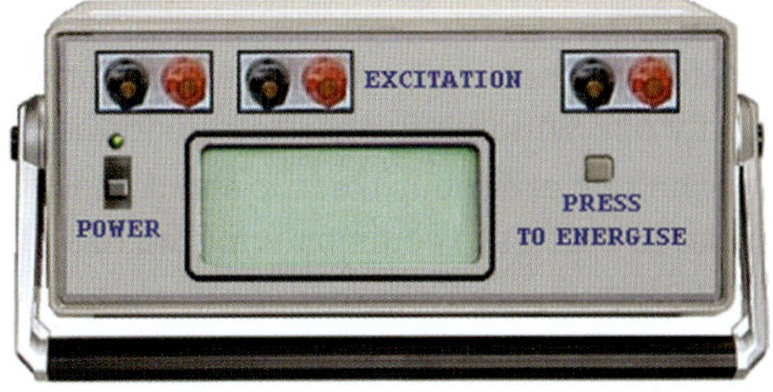

FIGURE 9.118 Transformer-turns-ratio meter

Induced rms voltage

The following three factors determine the magnitude of the induced electromotive force in the winding of a transformer:

- number of turns in the specific winding
- frequency of the supply
- strength of the electromagnetic flux in the core.

These three factors possess an interrelationship as an equation to find the rms voltage induced in either the primary or secondary winding of a transformer:

$$V = 4.44Nf\Phi_{max}$$

where V = the rms value of the induced voltage in volts (V)

N = the number of conductor turns on the coil

f = the frequency of the supply energy in hertz (Hz)

Φ_{max} = the maximum value of the electromagnetic flux in webers (Wb)

EXAMPLE 9.16

A single-phase 50 Hz transformer has 1000 turns on the primary winding and 500 turns on the secondary winding. The net electromagnetic flux established in the core is 1.1 mWb. Calculate the emf induced in:

a the primary winding

$$V = 4.44Nf\Phi_{max}$$
$$= 4.44 \times 1000 \times 50 \times 0.0011$$
$$= \mathbf{244.2\ volts}$$

b the secondary winding

$$V = 4.44Nf\Phi_{max}$$
$$= 4.44 \times 500 \times 50 \times 0.0011$$
$$= \mathbf{122.1\ volts}$$

EXERCISE 9.15

A single-phase 50 Hz transformer has 1200 turns on the primary winding and 350 turns on the secondary winding. The net electromagnetic flux established in the core is 1.5 mWb. Calculate the emf induced in:

a the primary winding
b the secondary winding.

REVIEW QUESTIONS

1 How are transformers rated?
2 What does the transformer rating state?
3 What material is used for transformer nameplates?
4 How are nameplates fixed to a transformer?
5 Name the three basic transformer tests.
6 Which winding of a step-down transformer has the higher resistance and why?
7 A single-phase 50 Hz transformer has 1500 turns on the primary winding and 750 turns on the secondary winding. The net electromagnetic flux established in the core is 1.2 mWb. Calculate the emf induced in:
   a the primary winding
   b the secondary winding.

CHAPTER REVIEW

9.1 Transformer operation

- A transformer consists primarily of a soft-iron core, a primary winding and a secondary winding.
- The winding connected to the energising circuit is the 'primary winding' and that connected to the load is the 'secondary winding'.
- A basic transformer consists of two coupled coils in close proximity to each other, wound on a ferromagnetic substance and sharing a common electromagnetic circuit.
- Coupling measures the effect of the position of the individual coils.

9.2 Transformer core construction, laminations and insulation

- The major components of a transformer are the core, the windings and insulation.
- Toroidal transformers use tape-wound circular ferromagnetic cores.
- A characteristic of the shell-type transformer is a short electromagnetic circuit consisting of two parallel paths.
- There are four winding types: layer, pancake, bobbin and toroidal.
- Lamination design affects transformer size and volume.
- Materials used for transformer laminations include laser-scribed cold-rolled grain-oriented silicon steel, amorphous magnetic metal, cold-rolled grain-oriented silicon steel, cold-rolled non-grain-oriented silicon steel and amorphous steel.

9.3 Application of transformers

- Transformers find use in a variety of applications, ranging from impedance matching to various forms of step-up and step-down potential (voltage) transformers.
- The magnethermic effect of transformer action has heating applications.
- Saturable core reactors control the voltage and current for high-power loads.

9.4 Voltage transformers

- The great variety of transformers makes it necessary to classify them under four general classes: power, volt-ampere, custom and instrument.
- Power transformers have the highest power rating and continuous voltage performance rating of all classes of transformers.
- Volt-ampere transformers have ratings below 300 VA.
- An isolation or linear transformer is a specialty transformer because it has a 1:1 primary to secondary winding turns ratio.

9.5 Transmission and distribution transformers

- Transformers have their high- and low-voltage leads brought out through bushings.
- Rapid pressure rise relays detect abnormal internal tank pressure. A surge diverter redirects any increase of voltage above nominal value to ground by acting as a short circuit.
- Tap changers enable the adjustment for voltage sag in a feeder or main, or change the secondary voltage to the varying requirements of a particular load.
- The conservator allows the dielectric coolant to flow into and out of the reservoir when the coolant in the main tank expands and contracts.
- A Buchholz relay is a gas-and oil-actuated relay device that monitors the presence of gases.
- Temperature indicators should show the actual temperature of the hot spot in the coil windings.

9.6 Autotransformers and instrument transformers

- A transformer that enables voltage transformation with only one winding is an autotransformer.
- Due to the risk of serious electric shock, it is essential to earth the common winding of a step-up autotransformer.
- Applications of autotransformers include speed control of induction motors, lighting control in theatres, hotels and photographic studios, starting polyphase induction motors, power supplies, industrial process and heating control, control of rectifiers in electroplating, and supply voltage adjustment.
- Instrument transformers act as isolating transformers.

- A current transformer is used with an ammeter to measure current. A potential transformer is used with a voltmeter to measure potential difference.
- The polarity marks of a current transformer indicate that when a primary current enters at the polarity mark (P1) of the primary, a current in phase with the primary current and proportional to it in magnitude exits the polarity terminal of the secondary (S1).
- A potential transformer is a step-down transformer.
- A potential transformer can be protected from a secondary short circuit by incorporating fuses into the secondary circuits.

9.7 Transformer losses and efficiency

- Practical transformers have losses – these are iron losses and copper losses.
- Maximum operating efficiency of a transformer occurs when the iron losses are equal to the copper losses.
- 'All-day efficiency' is the ratio of the total energy output of the transformer during a 24-hour period to the total energy input for the same time interval.
- The open-circuit test measures the iron loss of a transformer while the short-circuit test measures the copper loss of a transformer.
- Calculations concerning voltages and currents in a practical transformer use a phasor diagram.

9.8 Transformer cooling

- Transformers can be air natural cooled (AN), air forced cooled (AF), oil natural cooled (ON), oil natural air natural cooled (ONAN), fan cooled (ONAF) and direct water cooled (ONWF).
- The most important factor that determines the life and effective operation of a transformer is the coolant.
- The three principal causes of gas generation are ageing, thermal faults and electrical faults.

9.9 Voltage regulation and percentage impedance

- The percentage impedance of a transformer is the voltage drop on full load due to the winding resistance and leakage reactance expressed as a percentage of the rated voltage.
- The voltage regulation is the term used to measure the performance of the transformer when it is operating between no-load and full-load conditions.
- Harmonics is the name given to contaminating waveforms that are sinusoidal in shape but are multiples of the fundamental waveform.
- Triplen harmonics affect delta–star-connected three-phase transformers.
- Current harmonics can also distort the voltage waveform and cause voltage harmonics.
- Harmonic currents have a major effect on transformer load losses.

9.10 Parallel operation of transformers

- Two-phase transformer windings consist of two single-phase circuits whose voltages are 90 electrical degrees apart.
- The most common connections are star–star (Y–Y), delta–delta (Δ–Δ), star–delta (Y–Δ), delta–star (Δ–Y), open delta (V–V); or Scott (T–T).
- The advantage of the open-delta connection is that it takes two single-phase transformers to produce a three-phase output.
- Polarity refers to the instantaneous voltage direction obtained from the primary winding in relation to the secondary winding.
- In order to place three-phase transformers in parallel, it is necessary to know the relative phase sequence of the three high-voltage leads and the three low-voltage leads.
- There are four conditions – phase sequence, vector grouping, turns ratio and percentage impedance – that must be satisfied for effective parallel operation.
- For single-phase transformers to operate satisfactorily in parallel, connection diagrams must be identical, voltage ratios must be the same and percentage impedance, including primary and secondary leads to each transformer, should be practically equal.

9.11 Industry standards and *Wiring Rules* requirements

- High voltage refers to a voltage source in excess of 1000 V a.c. or 1500 V d.c.
- The permit to work tag has an area to write the identification number which designates the permit-to-work authority document.
- Live high-voltage signs must be placed at various points around the work area to indicate that there is equipment or conductors that should be regarded as live from which workers and other persons need to maintain an exclusion zone.
- High-voltage testing signs indicate that electrical apparatus is under test and that workers and other persons need to maintain the required exclusion zone.
- Every electrical worker who is listed on the access permit as being allowed to be in close proximity to high-voltage apparatus must wear PPE as a control measure against electric shock and arc flash.
- Step potential is the step voltage between the feet of a person standing near an energised earthed object.
- Touch potential is the touch voltage between an extended hand touching the energised object and the feet of a person in contact with the object.

9.12 Transformer nameplates and specifications

- Transformers are rated in kilovolt-amperes (kVA), which indicates the maximum current that a transformer can deliver without overheating.
- Nameplates in service can be difficult to read.
- Nameplates are made of brass or stainless steel and are fixed visibly and permanently to the transformer.
- Basic transformer tests include proving insulation, testing continuity and identifying windings.

TRIAL EXAM

For Chapter 9 knowledge assessment, please complete the following trial exam.

1 Name the electrical device that provides a physical isolation between a supply and a load while maintaining electrical continuity.
  a an inductor
  b a capacitor
  c a transformer
  d a switch

2 The basic transformer consists of:
  a a primary coil and a secondary coil
  b two coupled coils wound on a ferromagnetic substance
  c a coil wound on a silicon steel former
  d steel laminations and a coil

3 Name the current that supplies the magnetomotive force which produces the transformer electromagnetic flux.
  a load current
  b primary current
  c secondary current
  d exciting current

4 The induced voltage across the secondary coil is:
  a voltage of mutual induction
  b electromagnetic voltage
  c inductive voltage
  d electrostatic voltage

5 The primary coil, when the secondary coil is open-circuited, draws:
  a full-load current
  b a small magnetising current only
  c no current
  d an equilibrium current

6 For a maximum mutual induction to take place:
  a the primary coil should be at right angles to the secondary coil
  b the secondary coil should be at right angles to the primary coil
  c both coils must be parallel to each other
  d both coils must be separated by 180 electrical degrees

7 Besides transforming voltage and current, a transformer can be used to:
  a modify power
  b prevent isolation of the load from the primary feed
  c change frequency
  d impedance match the primary to the load

8 What category of transformer has a rating not exceeding 300 VA?
  a volt-ampere transformer
  b distribution transformer
  c autotransformer
  d power transformer

9 Which of the following transformer types has the highest power rating?
  a volt-ampere transformer
  b power transformer
  c ferroresonant transformer
  d autotransformer

10 Which of the following transformer types does not produce a true sine wave but a square wave?
  a current transformer
  b high-reactance (leakage) transformer
  c constant voltage/ferroresonant transformer
  d autotransformer

11 Which of the following transformer types has a 1:1 primary to secondary winding turns ratio?
  a constant voltage transformer
  b step-down transformer
  c high-reactance transformer
  d isolation transformer

12 Which of the following equations correctly states the transformation ratio?
  a $\frac{V_1}{V_2} = \frac{N_2}{N_1} = \frac{I_1}{I_2}$
  b $\frac{V_1}{V_2} = \frac{N_1}{N_2} = \frac{I_1}{I_2}$
  c $\frac{V_2}{V_1} = \frac{N_2}{N_1} = \frac{I_1}{I_2}$
  d $\frac{V_1}{V_2} = \frac{N_2}{N_1} = \frac{I_2}{I_1}$

13 A 400 V (V_1) step-down autotransformer has 1600 turns. If a tapping occurs at 840 turns, determine the output voltage (V_2) at this point.
  a 210 V
  b 115 V
  c 460 V
  d 50 V

14 A 230 V (V_1) step-down autotransformer has 1000 turns. If a tapping occurs at 500 turns, determine the supply current (I_1) at this point when the load demand is 8 A.
  a 0.2 A
  b 4 A
  c 8 A
  d 2.67 A

15 The magnethermic effect of transformer action is used for:
  a controlling the voltage and current for high-power loads
  b electromagnetic increases
  c frequency changes
  d heating applications

16 What creates air gaps in the core that in turn have the effect of increasing the magnetising current because air has a much lower permeability than steel?
a lamination design
b burrs on laminations
c an oxide coating
d soft, low-remanence magnetic steel

17 The various transformer core types fall within one or other of which three general types:
a toroidal type, shell type and core type
b toroidal type, shell type and E type
c toroidal type, C type and E type
d E and I type, core type and C type

18 Which type of transformer core provides an almost perfect magnetic circuit?
a core type
b toroidal type
c C type
d shell type

19 The transformer construction that has the primary and secondary windings both placed on the centre leg of the transformer core is:
a E and I type
b core type
c shell type
d E type

20 What do the types layer, pancake, bobbin and toroidal refer to?
a cooling types
b core types
c winding types
d insulation types

21 State the most significant single factor that has the greatest effect on transformer size and volume.
a conductor design
b winding design
c cooling design
d lamination design

22 Name the transformer lamination stamping in **Figure 9.119**.

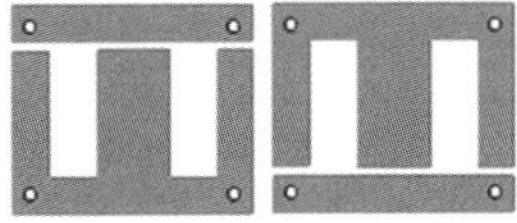

FIGURE 9.119 Lamination stamping for Question 22

a C stampings
b U stampings
c L stampings
d E and I stampings

23 How can a decrease in magnetic flux density, which reduces the iron loss and magnetising current, be achieved in small VA transformers?
a by increasing the primary turns
b by using a coupling coefficient factor of 0.5
c by increasing the secondary turns
d by widening the steel leg that passes through the coils

24 State the purpose of the surface insulation on a transformer lamination.
a to provide secondary insulation for the windings
b to establish high no-load losses
c to decrease the effects of eddy currents
d to electrically couple the individual coils

25 What type of material is more suited to C-core laminations?
a cold-rolled non-grain-oriented silicon steel
b amorphous alloy
c cold-rolled grain-oriented silicon steel
d refractory metals

26 A transformer has 10 000 turns on the primary winding and 3000 turns on the secondary winding. If a 230 V 50 Hz supply connects to the primary winding and a resistive load of 100 Ω connects across the terminals of the secondary, calculate the secondary current.
a 6.9 A
b 0.23 A
c 23 A
d 0.69 A

27 What transformer rating informs us of the maximum current that a transformer can deliver to a load without overheating?
a kVA rating
b secondary voltage rating
c power factor rating
d load rating

28 When should a transformer test be used to prove insulation integrity on a transformer?
a while the transformer is in service
b after the load is isolated with the transformer connected on no-load
c after shutdown when the transformer has cooled
d immediately after shutdown while the transformer is still at its operating temperature

29 A single-phase 50 Hz transformer has 1600 turns on the primary winding and 800 turns on the secondary winding. The net electromagnetic flux established in the core is 2.1 mWb. Calculate the emf induced in the secondary winding.
a 0.3972 V
b 0.74592 V
c 372.96 V
d 745.92 V

30 Which of the following test instruments can help in determining which transformer termination leads connect to which winding?
a turns ratio instrument
b voltmeter
c ohmmeter
d ammeter

31 Why is knowledge of the impedance of a transformer important?
a it provides the transformer rating
b it determines the maximum value of current that will flow under fault conditions
c it enables the determination of the maximum connected load
d it determines the voltage regulation of a transformer

32 Name the term used to give a measure of the performance of the transformer when it is operating between full-load and no-load conditions?
a kVA rating
b load control
c voltage regulation
d percentage impedance

33 A 20 kVA 11 000:400 V distribution transformer has a no-load voltage of 425 V at a power factor of 0.9 lagging. Calculate the voltage regulation.
a 2.5% at 0.9 lag
b 5.88% at 0.9 lag
c 175% at 0.9 lag
d 1.27% at 0.9 lag

34 A 230 V transformer primary fluctuating from –5% to +10% (218.5 V to 253 V) would cause a 60 V rated secondary to fluctuate from:
a 54 V to 63 V
b 63 V to 66 V
c 66 V to 72 V
d 57 V to 66 V

35 What transformer losses are independent of the power used by the transformer?
a iron
b copper
c cooling
d load

36 What is the full-load efficiency of a 1000 kVA transformer having an iron loss of 2.5 kW and a full-load copper loss of 4.5 kW?
a 98.2%
b 99.3%
c 100%
d 97%

37 Name the efficiency parameter that is important when a transformer carries little or no load during large time intervals of a 24-hour period.
a full-load efficiency
b load efficiency
c all-day efficiency
d maximum operating efficiency

38 The open-circuit test is used to measure:
a the secondary current of a transformer
b the load voltage of a transformer
c the I^2R loss of a transformer
d the iron losses of a transformer

39 What factor accelerates the chemical action naturally occurring within insulating materials of a transformer?
a heat
b load
c convection currents
d magnetising force

40 Which of the following transformer cooling methods does not use a coolant?
a AN-type transformers
b ON-type transformers
c ONAN-type transformers
d ONAF-transformers

41 When considering paint for a transformer housing, which heat transference medium does paint affect?
a convection
b radiation
c conduction
d reflection

42 A desirable characteristic in transformer cooling media is:
a high freezing point
b sludge
c low viscosity
d excellent emulsion ability

43 State the purpose of inhibitors used with transformer oils.
a to lower the temperature of the oil
b to improve convection currents
c to develop sludge
d to slow down the oxidation of the oil

44 What can occur within transformers because of ageing, thermal faults and/or electrical faults?
a gas generation
b low oil viscosity
c improved dielectric constant
d thermal stability

45 Name the type of transformer test that evaluates the oil's insulating strength.
a dissolved gas in oil
b dielectric strength
c acidity
d interfacial tension

46 Which of the following is the type of transformer ancillary equipment that offers a degree of protection against lightning and other high-voltage disturbances?
a breather
b pressure relief device
c bushings
d conservator

47 Name the transformer tap to which the rated quantities (voltage, current, power) are related.
a centre tap
b low-voltage tap
c principal tap
d high-voltage tap

48 The function of a Buchholz relay is to:
a monitor the presence of gases
b change the turns ratio of a transformer
c monitor the temperature of the transformer windings
d enable changing of the transformer oil

49 Tap changers are used on a transformer to:
a compensate for variations in the supply voltage
b remove oil sludge
c provide addition points where oil can be added to the transformer tank
d enable gas to be released from the transformer

50 A conservator is used on a transformer to:
a protect the windings from excessive heat
b allow the oil to expand with heat
c isolate the transformer windings from the degrading influence of oxygen and moisture
d maintain the purity of the transformer bushings

51 A desiccant's function is to:
a sustain a state of dryness
b increase moisture
c lower the viscosity of the oil
d filter oil

52 A transformer with only one wound winding and a ratio of 300:5 is usually a:
a distribution transformer
b potential transformer
c current transformer
d autotransformer

53 What type of device is usually associated with a current transformer?
a a resistor
b a fuse
c a voltmeter
d a make-before-break switch

54 An example of a step-down transformer is:
a potential transformer
b current transformer
c autotransformer
d step-up transformer

55 Two-phase transformer windings consist of two single-phase circuits whose voltages are:
a in phase
b 90° degrees apart
c 120° leading
d 30° lagging

56 The vector group notation Dy11 means:
a high current and the tapping marking
b impedance transformer
c star, star with no phase shift
d delta, star with 330° phase shift

57 Which of the following terms only applies to single-phase transformers?
a current
b parallel
c polarity
d potential

58 The Australian standard with single-phase power transformer terminations is:
a positive with negative polarity
b neutral polarity
c additive polarity
d subtractive polarity

59 The positions of the relative connections for a three-phase transformer when viewed from the high-voltage side of the transformer are:
a NABC
b ABCN
c BCNA
d CNAB

60 State the four conditions that must be satisfied for effective paralleling of three-phase transformers.
a percentage impedance, type of oil, vector grouping and load current
b phase sequence, vector grouping, turns ratio and percentage impedance
c type of oil, secondary voltage, load current and power factor
d turns ratio, load current, phase sequence and secondary voltage

61 What condition may be indicated when a transformer experiences higher than normal temperatures, voltage distortion and high crest factor?
a leakage current
b lower hysteresis loss
c current harmonics
d thickening of the oil

62 What document is necessary to obtain before work on or near high-voltage electrical apparatus is started?
a switching sheet
b operational earth performance
c access permit
d test permit

63 The voltage between a person's outstretched hand touching an energised object and their foot is known as:
a the step voltage
b the open-circuit voltage
c touch voltage
d the short-circuit voltage

64 The potential difference between a person's outstretched legs is referred to as:
a main earth voltage
b fault voltage
c step voltage
d touch voltage

65 Prior to an access permit being granted, the high-voltage apparatus must be:
a energised
b identified
c isolated and earthed
d tagged

66 A high voltage transformer has a voltage of 400 V on the secondary winding, and secondary winding turns of 100. If the number of turns on the primary windings are 2750, what will be the applied primary voltage?

67 A single-phase transformer has 1500 A of current flowing in the primary winding. The primary winding has 6600 turns and the secondary winding has 1100 turns. Neglecting losses, calculate the current flowing in the secondary winding.

68 A 25 MVA transformer operating at full-load has an efficiency of 98% at 0.9 lag power factor. Calculate the total losses at this load for the transformer.

69 Two transformers A and B are required to share a load current of 1500 A. The percentage of impedance corresponding to full-load currents is 4.6% for transformer A and 4.2% for transformer B. Calculate the load current carried by transformer B.

70 A potential transformer with a ratio of 100:1 has 66 volts across the secondary winding of the transformer. Determine the voltage in the primary winding.

Single-phase input d.c. power supplies

This chapter provides electrotechnology workers with essential knowledge and skills that relate to basic electronic devices and circuits used for single-phase power supplies. You will also gain a knowledge of how common semi-conductor diodes work by examining the structure of atoms.

LEARNING OBJECTIVES

Atomic structure
- Explain the concept of the atom

Power supply operating principles
- Outline the operating principles of single-phase power supplies

d.c. rectification circuits
- Explain the operation of single-phase rectifier circuits

Filter circuits
- Explain the operation of power supply filter circuits

Zener diode shunt regulator
- Outline the operation of the Zener diode regulator

Three-terminal IC voltage regulator
- Outline the operation of the three-terminal IC regulator

d.c. power supply testing and fault finding
- Describe common d.c. power supply testing and fault-finding procedures

Safe working practices
- Outline safe work procedures for d.c. power supplies

10.1 Atomic structure

Atoms

An atom is the smallest particle that an element can reduce to and still possess the properties of that element. In addition, the number of neutrons and protons in the nucleus of an atom determines the atom's chemical and physical characteristics. The number of protons in an atom, the atomic number, determines the element to which an atom belongs, such as copper. All elements have multiple isotopes which differ from each other only in the number of neutrons in the nucleus. Isotopes of the same element have very similar chemical and physical properties. Refer to **Figure 10.1**, which shows a simplified model of a copper atom.

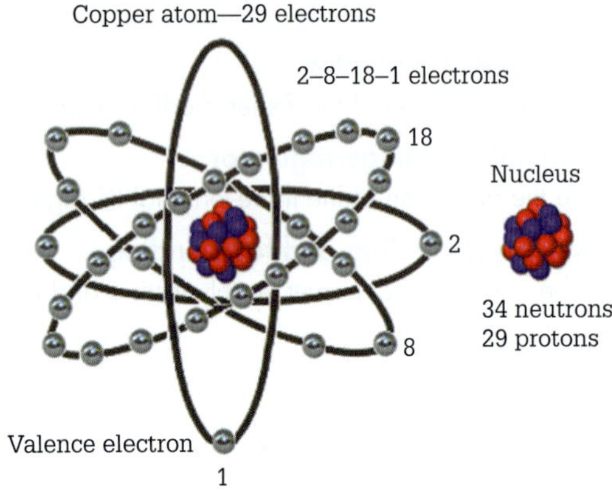

FIGURE 10.1 Copper atom

The protons and neutrons are located in the centre, or nucleus, of the atom, and the electrons travel around the nucleus in orbits within an electron cloud. Atoms consist of three types of subatomic particles as shown in **Figure 10.2**.

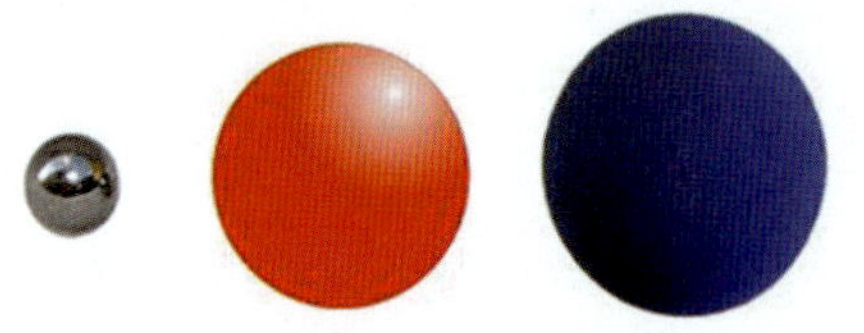

FIGURE 10.2 Electron, proton and neutron

Electrons are the smallest and lightest of the subatomic particles and possess a negative charge. Protons are substantially heavier than electrons and possess a positive charge. Neutrons have a similar mass to protons but have no electrical charge. The positive charge of a proton is equal but opposite to the negative charge of an electron. The number of protons and electrons in an atom tend to be equal; hence the equal and opposite positive and negative charges cancel each other out, rendering the atom itself electrically neutral. The charges on an electron and proton are called electrostatic charges. Particles that hold an electric charge obey the law of electrical charges as illustrated in **Figure 10.3**.

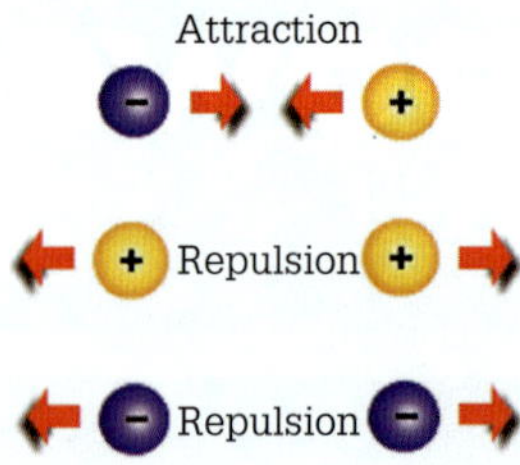

FIGURE 10.3 Law of electric charges

SWITCH ON

If an atom loses or gains an electron, the atom becomes electrically charged. Consequently, if an atom loses an electron; it becomes a positively charged atom and is a positive ion, or 'cation'. By contrast, if an atom gains an electron, it becomes a negatively charged atom, and is called a negative ion, or 'anion'.

Bohr's model of the atom

Niels Bohr was a Danish physicist who developed an early and now obsolete model of what he believed was the structure of an atom. Bohr described the atom as having a positively charged central core or nucleus. In addition, a series of numbered orbits (shells, bands or energy levels) which house negatively charged electrons, circled the nucleus. Furthermore, the orbit next to the nucleus is called the K shell or energy level, and then, working away from the nucleus, the L, M, N and O shells respectively. The force of attraction between the nucleus and the electrons decreases with increasing distance from the nucleus.

Electron orbits can contain only a certain number of electrons. For example, the K orbit can have up to two electrons, the L orbit up to eight; the M orbit can hold up to 18 electrons and the N orbit up to 32 and so on. **Figure 10.4** shows the KLM orbits of the Bohr model of an atom.

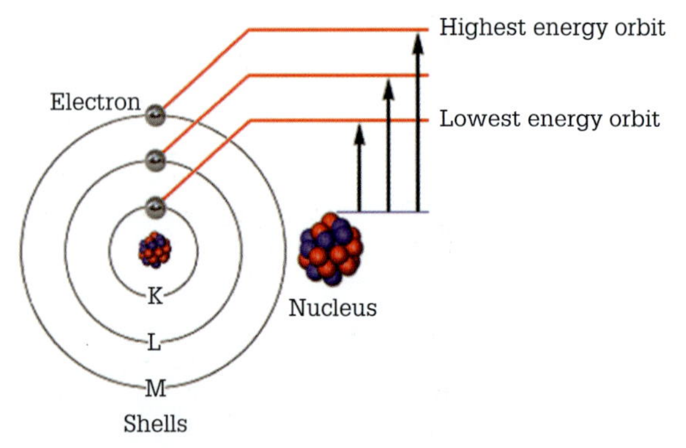

FIGURE 10.4 Atom with three shells: K, L and M

However, the outermost orbit of an atom cannot include more than eight electrons. The outer orbit or shell is the valence shell. The valence shell determines the chemical bonding and electrical qualities of the material. Consequently, the number of valence electrons for insulators is greater than four, for semiconductors it is four, and for conductors three or fewer. Furthermore, electricity develops when valence electrons leave their atom, becoming free electrons.

Free electrons

Each electron within a given orbit has a definite energy level. The more energy an electron acquires, the higher its orbit. If a valence electron absorbs sufficient energy, it attempts to move to a higher orbit. However, in the process of doing so, it can escape the attraction of the atom's nucleus and become a free electron. These electrons can exist by themselves outside the atom, and it is these free electrons which are responsible for most electrical and electronic phenomena.

Quantum mechanics

Quantum mechanics or quantum theory is the study of matter and energy acting as both a particle and as a wave, with the focus being on the atomic and subatomic level. A model of the atom that is more consistent with the principles of quantum mechanics is shown in **Figure 10.5**.

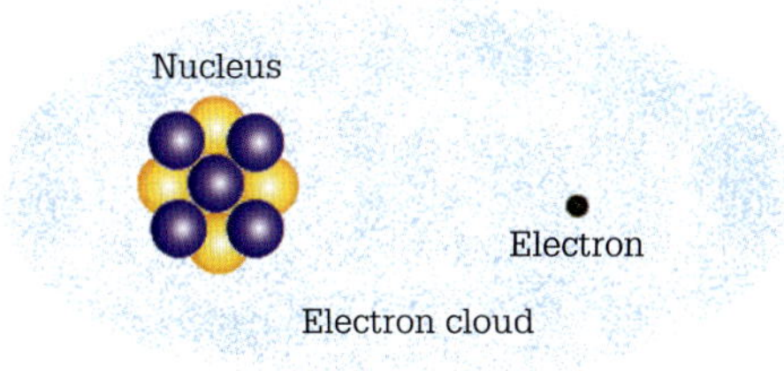

FIGURE 10.5 Nucleus within an electron cloud with an electron

The nucleus

According to quantum theory, a nucleus is a spinning conglomerate of two types of subatomic particles, protons and neutrons, existing within a cloud of negative charge containing electrons exhibiting both particle and wave phenomena. In the nucleus, there exists binding energy between neutrons and protons called the strong nuclear force, which holds the nucleus together.

The electron cloud

With quantum theory, electrons and other particles are considered to have both particle and wave properties. The electron cloud represents the 'probability distributions' for the electrons because it is not possible to precisely know the position or path an electron takes. The area of probabilities an electron of a particular energy level occupies in the cloud is called the electron orbital.

Orbitals

Orbitals exist at certain distances from the nucleus, and they differ from each other in energy, size, angular momentum and magnetic properties. Consequently, electrons in separate orbitals have different energy levels. These energy levels can be close to one another and are described as being so many electron volts (eV) apart. The potential energy states for electrons are arranged in bands. Within each band, the levels are very close together but between each band, there may be energy gaps as shown in **Figure 10.6**.

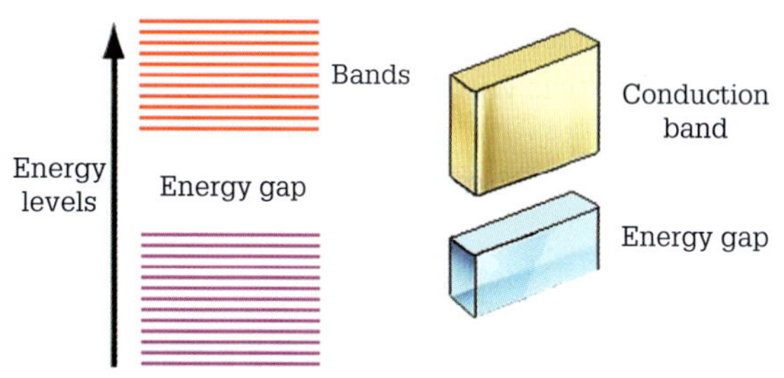

FIGURE 10.6 Energy states

When an atom gains or loses energy because of the movement of an electron, the energy change can cause other electrons to transfer from one orbital or energy state to another. In fact, electrons seek to occupy the lowest energy state. However, greater external energy allows electrons to move to higher and more energetic orbitals. The outer electrons within the electron cloud are easier to move to a higher orbital than electrons in lower orbitals.

In the nomenclature regarding orbitals, the lowest orbital that is closest to the nucleus is given a quantum number of 1. Higher quantum numbers mean that an electron has gained higher energy in order to occupy that orbital. The name given to the states or energy bands are ground – the lowest state; the higher states are excited states with two of these being called valence and conduction states. Orbitals form energy bands likened to shells into which electrons group; within each shell are subshells. When electrons enter energy bands they complete the shell first then each subshell one at a time with one electron.

Spin

Electrons have another intrinsic property called 'spin', or angular momentum. Spin enables electrons to occupy the same orbital as long as their spins are different. Therefore, other electrons complete each subshell by having an opposite spin to the first electron. Once completed, the electrons move on to the next shell. Finally, when an electron can gain a sufficiently high energy state, it may be able to break away from the atom as a free electron and become part of an electric current passing through a substance.

Energy bands

The theory of energy bands is fundamentally important in classifying materials as conductors, insulators or semiconductors. Two bands are of importance: the valence band and conduction band.

The valence band is composed of a series of energy levels containing valence electrons. However, above the valence band is another band called the forbidden band or energy gap. Electrons do not stay in this band but may travel back and forth through it. The conduction band is located above the forbidden band. Electrons in the valence band are more tightly bound to the atom than the electrons in the conduction band. However, when given sufficient energy, electrons in the valence band can move into the conduction band. Consequently, when the valence band electrons have enough energy to pass through the forbidden band into the conduction band as free electrons, an electric current is possible.

REVIEW QUESTIONS

1 What particles comprise an atom?
2 Which particle within an atom is electrically negative?
3 Describe Bohr's model of the atom.
4 According to quantum mechanics, electrons have two types of properties. What are these types?
5 What is the consequence of an atom losing or gaining an electron?

10.2 Power supply operating principles

Power supply function

The basic function of a single-phase d.c. power supply is to convert the 230 V, 50 Hz a.c. voltage available at a standard socket outlet into an extra-low d.c. voltage (not exceeding 120 V ripple-free d.c.). It is one of the most common electronic circuits. The d.c. voltage produced by the power supply powers numerous electronic circuits, such as televisions and portable music players, and is used for charging the battery in your mobile phone.

A simple, single-phase d.c. power supply contains a transformer, rectifier, filter and voltage regulator (as shown in **Figure 10.7** and **Table 10.1**).

TABLE 10.1 The components of a d.c. power supply

Components	Purpose
Transformer	Reduces the mains a.c. voltage to a value suited to electronic equipment.
Rectifier	Converts the a.c. voltage from the transformer into a pulsating d.c. voltage.
Filter	Reduces fluctuations in the raw d.c. voltage output from the rectifier to produce a smoother d.c. voltage.
Voltage regulator	Maintains a constant output voltage irrespective of changes in input voltages or load current.
Load	A circuit or device for which the d.c. power supply is producing the d.c. voltage and load current.

The output of a d.c. power supply may be constant voltage (CV) or constant current (CC). When delivering constant voltage, the power supply maintains a steady output voltage regardless of any changes to the load; this means that the current will vary. So, for example, when

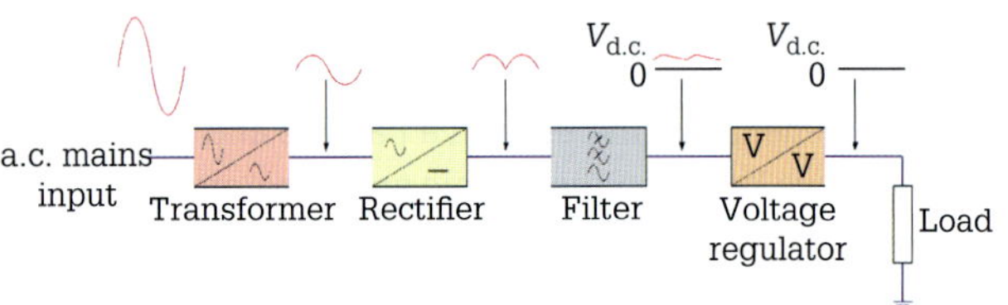

FIGURE 10.7 Block diagram of a d.c. power supply

providing a constant voltage of 10 V to a 100 Ω load, the current is 100 mA but if the load is changed to 10 Ω, the current increases to 1 A.

When delivering constant current, the power supply maintains a steady output current regardless of any changes to the load; this means that the voltage will vary. So, for example, when providing a constant current of 500 mA to a 100 Ω load, the voltage is 50 V but if the load is changed to 10 Ω, the voltage decreases to 5 V.

P-type and N-type semiconductors

Materials classed as semiconductors are not good electrical conductors and have limited use in their intrinsic (pure) state. This is due to the limited number of free electrons in the conduction band and holes in the valence band. Two common semiconductor materials are silicon and germanium. It is possible to improve the electrical conductivity of semiconductor materials through the controlled addition of impurities to the intrinsic semiconductor material. The process of introducing impurities is doping and its purpose is to increase the number of current carriers (either electrons or holes). Depending on the type of impurity added, either an N-type or P-type semiconductor is produced.

N-type semiconductor

Introducing pentavalent impurity atoms to intrinsic silicon increases the number of conduction band electrons to produce an N-type semiconductor. Pentavalent atoms such as arsenic (As), phosphorus (P), bismuth (Bi) and antimony (Sb) have five valence electrons. The added pentavalent atom forms covalent bonds with four adjacent silicon atoms leaving one extra electron. This electron becomes a conduction electron, as it is not attached to any atom. As the pentavalent atom releases an electron, it is a donor atom. The number of conduction electrons is controlled by the number of impurity atoms added to the intrinsic semiconductor material. The doping process does not leave a hole in the valence band as it produces more than enough electrons to fill the valence band.

In N-type material the majority of the current carriers are electrons. The negative charge of these free electrons is balanced by the positive charge of the immobile ions, resulting in an overall neutral charge. There are a number of holes produced when the electron–hole pairs are thermally generated. The addition of pentavalent atoms does not produce these holes, and as such, holes are minority current carriers.

P-type semiconductor

Introducing trivalent impurity atoms to intrinsic silicon increases the number of holes to produce a P-type semiconductor. Trivalent atoms such as boron (B), indium (In) and gallium (Ga) have three valence electrons. The added trivalent atom forms covalent bonds with three adjacent silicon atoms, leaving a hole as another electron is required to form a fourth covalent bond. As the trivalent atom can accept an electron, it is referred to as an acceptor atom. The number of holes is controlled by the number of impurity atoms added to the intrinsic semiconductor material. The doping process does not produce a conduction (free) electron.

In P-type material the majority of the current carriers are holes. The P-type material also exhibits overall neutral charge. There are a number of free electrons produced when the electron–hole pairs are thermally generated. The addition of trivalent atoms does not produce these free electrons and, as such, electrons are minority current carriers.

Diodes

The diode is a simple, two-terminal semiconductor device, which is formed when a portion of intrinsic (pure) material such as silicon is doped (impurities are added) with a trivalent impurity, and another portion is doped with a pentavalent impurity. This process creates a P–N junction between the P-type and N-type regions, as shown in **Figure 10.8**.

The two P- and N-type materials contain majority and minority current carriers. Holes are the majority carriers in the P-type material and are the minority carriers in the N-type material. Electrons are the majority carriers in the N-type material and the minority carriers in the P-type material. When a diode is created there is an interaction between the P- and N-type regions. The electrons are attracted to the holes in the P region, while holes are attracted to the electrons in the N region. The diffusion of holes and electrons across the junction in order to recombine causes positive ions (charge) to be generated by the electrons leaving the N-type material. Negative ions (charge) are generated when the holes leave the P-type material. As the recombination continues, the area on either side of the junction becomes depleted of free electrons and this forms a depletion or barrier zone (as illustrated in **Figure 10.9**) between the P-type and N-type material. This means that no more holes or electrons can recombine. The positive and negative ions have formed a barrier potential across the depletion zone.

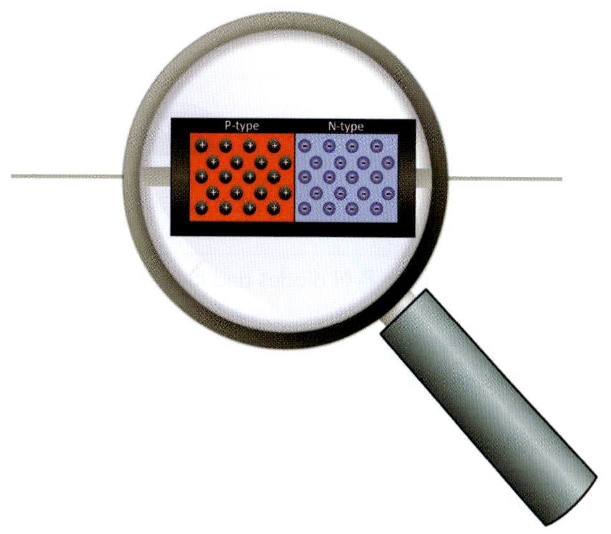

FIGURE 10.8 P–N junction

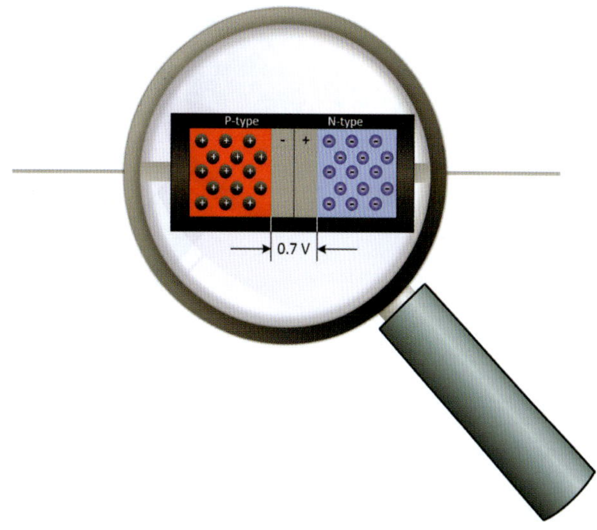

FIGURE 10.9 Depletion zone showing barrier potential

For a diode to conduct, the barrier potential requires overcoming in order to allow current across the P–N junction. This potential exists at all times and for a silicon P–N junction diode is 0.7 V. A diode only conducts when forward biased (see below). Silicon P–N rectifier diodes and the diode standard circuit symbol are shown in **Figure 10.10**.

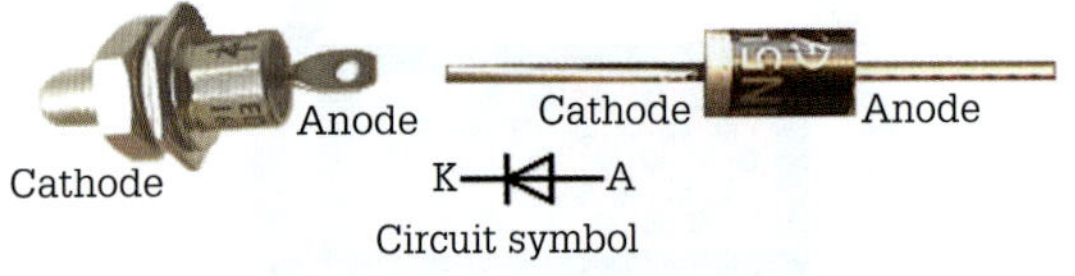

FIGURE 10.10 Silicon P–N diodes and their circuit symbol

> **SWITCH ON**
>
> The two terminals of the diode are the anode (connected to the P-type material) and the cathode (connected to the N-type material). The significance of a P–N junction diode is its ability to conduct current (maximum forward current) in only one direction and to prevent current flow in the other direction.

Forward bias

If a positive potential, which is greater than the barrier potential of 0.7 volts, is applied to the connection of the P-type material and a negative potential to the connection of the N-type material, then current can flow. In this arrangement the diode is said to be 'forward biased' and is depicted in **Figure 10.11**.

FIGURE 10.11 Forward bias condition

At the P–N junction, the forward biasing results in a thinner and less resistive depletion region. If the applied voltage is large enough, the depletion region's resistance becomes negligible. In silicon, this occurs at about 0.7 volts forward bias. From 0 to 0.7 volts, there is still considerable resistance due to the depletion region. Above 0.7 volts, the depletion region's resistance is very small and current flows virtually unimpeded.

Reverse bias

If a positive potential greater than the barrier potential of 0.7 volts is applied to the connection of the N-type material and a negative potential to the connection of the P-type material, then current cannot flow. In this arrangement the diode is said to be 'reverse biased' and is depicted in **Figure 10.12**.

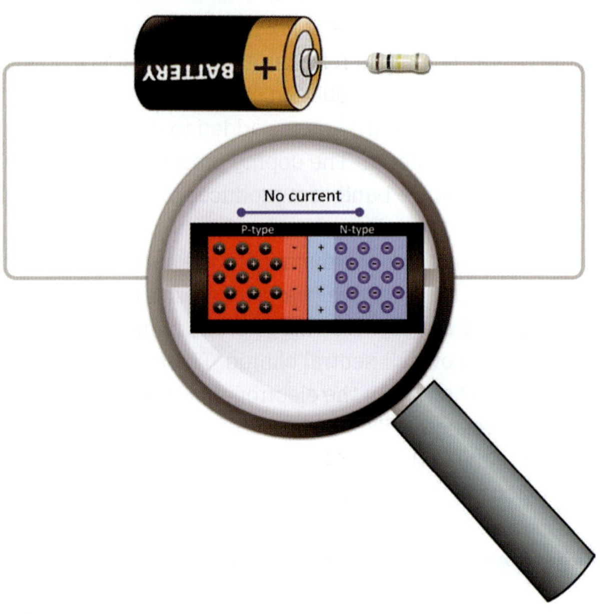

FIGURE 10.12 Reverse bias condition

At the P–N junction, the reverse biasing creates a thicker, more resistive depletion region. If the applied voltage becomes larger, the depletion region becomes thicker and more resistive.

In reality, some current will still flow through this resistance, but the resistance is so high that the current may be considered to be zero. As the applied reverse bias voltage becomes larger, the current flow will saturate at a constant but very small value.

Diode volt–ampere characteristic

Figure 10.13 details the volt-ampere characteristic for a low-power rectifier diode. This characteristic shows:

- Forward bias condition where both diode potential (V_f) and diode current (I_f) are positive
- Reverse bias condition where both diode potential (V_r) and diode current (I_r) are negative
- The knee voltage (V_{knee}), which is 0.7 volts for a silicon diode and 0.3 volts for a germanium diode. Silicon semiconductor is commonly used for power diode applications and germanium semiconductor is commonly used for signal diode applications.
- Reverse breakdown voltage or peak inverse voltage, which is the negative potential that will cause the diode to break down and carry current in the reverse direction. This point is also referred to as the avalanche point, as on the graph the line is almost a vertical fall (avalanche).
- Leakage current, which is an extremely small current flowing when the diode is connected reverse bias.

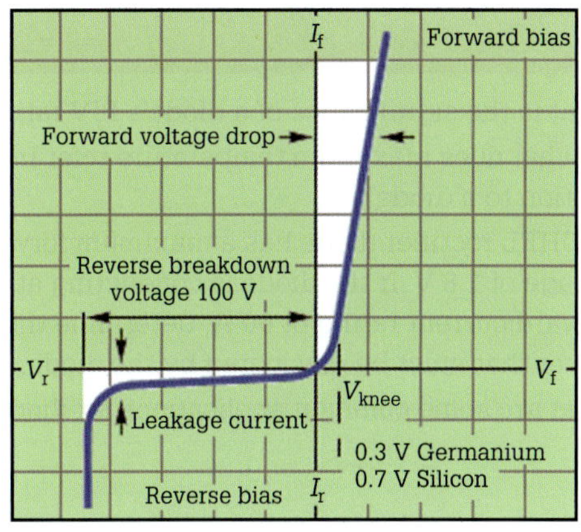

FIGURE 10.13 Diode volt–ampere characteristic

Forward volt drop or barrier potential is the potential difference measured across the P–N junction when the diode is operating in forward bias. As stated earlier, there is a small resistance associated with the junction and as such, an increase in forward current will result in an increased forward voltage drop.

All diodes have a peak inverse voltage (PIV) or Maximum Reverse Voltage (V_{RRM}) allowable between anode and cathode. This voltage rating specifies the maximum voltage a diode can withstand in the reverse-biased direction before breakdown (the diode is forced to conduct in the reverse direction and is irreparably damaged – short circuit). One hundred volts is shown for the diode volt–ampere characteristic shown in **Figure 10.13**. This rating is usually less than the avalanche breakdown level on the reverse bias characteristic curve. Typical values of PIV range from a few volts to thousands of volts and their parameters must be met when replacing a diode in service. A diode must be able to withstand the maximum reverse potential it will likely encounter in service. The peak inverse voltage is a function of diode construction, so manufacturers make diodes that have specific PIV values. The peak inverse voltage is an important parameter and has application for rectifying diodes in a.c. circuits.

Leakage current

When in reverse bias, there is an extremely small current in the order of µA, and this leakage current is dependent on temperature. That is, an increase in temperature causes the junction temperature to rise, creating an increase in leakage current. Importantly, an increase in temperature can also reduce the point at which avalanche occurs.

Maximum forward current

The maximum forward current ($I_{f(max)}$) is the maximum current the diode can carry when forward biased without sustaining damage. The load connected to the diode determines this forward current drawn from the power supply through the diode. When the diode is conducting in the forward bias condition, it has a very small forward resistance across the P–N junction and so power is dissipated across the P–N junction in the form of heat.

If the maximum forward value is exceeded, more heat will be generated across the junction and the diode will fail due to thermal stress. When operating diodes near their maximum current ratings, additional cooling such as a heat sink is necessary to dissipate the heat produced by the diode.

Power rating

High-power diodes have a maximum power dissipation rating (P_R). This rating is the maximum power dissipation of the diode when it is forward biased. As the resistance of the depletion region is non-linear, to find the power dissipated by the diode, we must multiply the voltage drop across it (the barrier potential) by the current flowing through it (forward current):

$$P_R = V_F \times I_F$$

EXAMPLE 10.1

A high-power diode when forward biased has a barrier potential of 0.7 V. If the maximum forward current that the diode can handle continuously is 8 A, determine the power rating requirement of the diode.

$$P_R = V_F \times I_F$$
$$= 0.7 \times 8$$
$$= \mathbf{5.6\ W}$$

EXERCISE 10.1

a An IN4007 rectifier diode has a maximum forward voltage of 0.8 V. If the diode is conducting at a forward current rating of 2.5 A, determine the power that must be dissipated by the diode.

b A 70HFL rectifier diode has a maximum forward voltage of 1.8 V. If the diode is conducting at a forward current rating of 40 A, determine the power that must be dissipated by the diode.

Maximum operating temperature

This temperature rating refers to the junction temperature (T_J) of the diode expressed in degrees Celsius per Watt (°C/W), and is related to maximum power dissipation. It is the maximum temperature allowable before the construction of the diode weakens. For a 12F (R) power diode this temperature could be as high as 144 °C. The maximum operating temperature for a 70HFL rectifier diode is 125 °C.

Applications

Applications for diodes in the power industry are battery chargers, convertors, power supplies and machine tool controls. They are also used as protection diodes with relays to protect relay circuitry from the inductive voltage spike that occurs when a relay coil is switched off.

REVIEW QUESTIONS

1 Draw a block diagram of a power supply.
2 What is meant by constant current when referring to the output of a power supply?
3 In which block of a power supply are you likely to find a P–N junction diode?
4 What type of impurity is used to produce an N-type semiconductor?
5 Draw the circuit symbol for a P–N junction diode.
6 What is the typical barrier potential for a silicon P–N junction diode?
7 What is the significance of a diode's PIV rating?
8 To what does maximum temperature refer in relation to a diode?
9 A 70HFL rectifier diode has a maximum forward voltage of 1.8 V. If the diode is conducting at a forward current rating of 60 A, determine the power that must be dissipated by the diode.
10 What are some common applications for diodes?

10.3 d.c. rectification circuits

Rectification is the process of converting an a.c. voltage waveform to a pulsating unidirectional d.c. waveform by means of diodes or silicon-controlled rectifiers. **Figure 10.14** shows the rectification process.

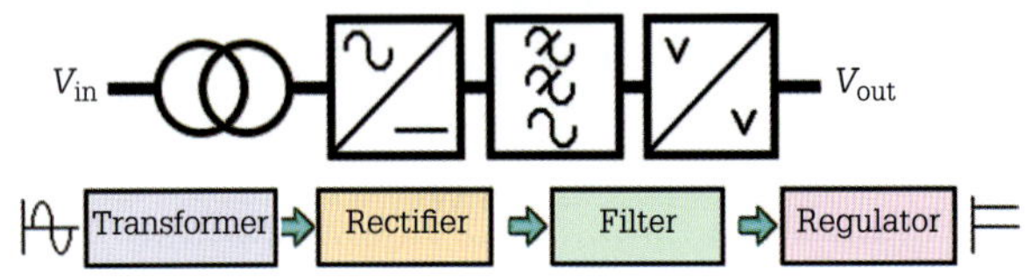

FIGURE 10.14 Operation of the rectification process

Half-wave rectifier

The simplest form of rectification is the half-wave rectifier, which incorporates only a single rectifier diode. In half-wave rectification, only the positive half of the input waveform can pass through the rectifier while the negative half-cycle is blocked. A half-wave rectifier circuit is shown in **Figure 10.15**.

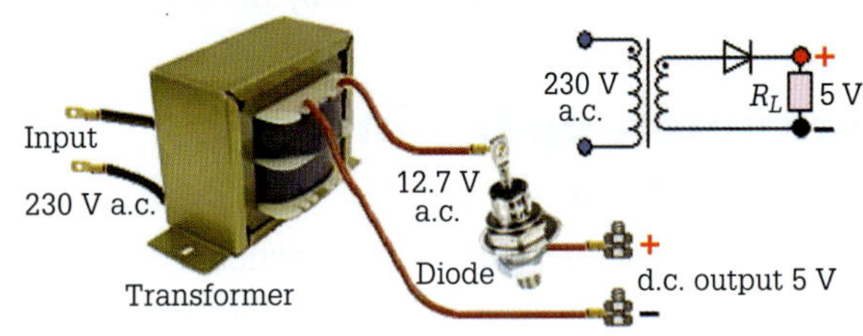

FIGURE 10.15 Half-wave rectifier circuit

Figure 10.16 shows the waveform of the rectified voltage output. The half-cycle for which the diode is forward biased is determined by the connections to the diode.

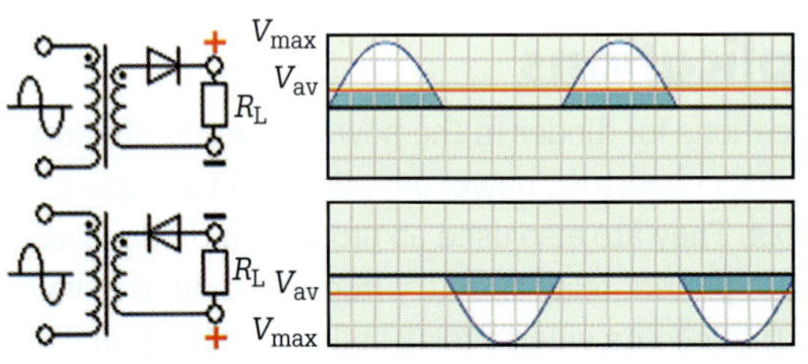

FIGURE 10.16 Reversing the diode

Average d.c. output voltage

The average or mean value for a half sine wave is 0.637 of the maximum or peak value. For a half-wave rectifier, however, only every second half-wave is used, resulting in an average value for the complete wave of 0.318 (0.637/2) of the maximum value. As we normally refer to rms values of voltage and current, the average value of a complete sine wave is 0.45 of the rms value (0.637 / 0.707). So, the average d.c. voltage for a half-wave rectifier is less than half the average value of one cycle of output sine wave. The actual load voltage is reduced even further because of the forward voltage drop across the diode (0.7 V). **Example 10.2** shows how we get the same load voltage value using either maximum or rms values of input voltage.

$V_{av} = 0.318\ V_{max}$ $\quad V_{av} = 0.45\ V_{rms}$ $\quad V_{d.c.} = V_{av} - V_{diode}$

Note that the load connected across a half-wave rectifier only receives current for 50% of the a.c. input cycle. This is because the diode is reverse biased for 50% of the cycle. The current is unidirectional but discontinuous.

EXAMPLE 10.2

A half-wave rectifier is connected to a 40 V rms a.c. supply. Calculate the d.c. load voltage.

$$V_{max} = 1.414 \times V_{rms} = 1.414 \times 40 = \mathbf{56.56\ V}$$

$$V_{av} = V_{max} \times 0.318 = 56.56 \times 0.318 = \mathbf{18\ V}$$

or

$$V_{av} = V_{rms} \times 0.45 = 40 \times 0.45 = \mathbf{18\ V}$$

$$V_{d.c.} = V_{av} - V_{diode} = 18 - 0.7 = \mathbf{17.3\ V}$$

EXERCISE 10.2

a Calculate the d.c. load voltage of a half-wave rectifier when the rms output of the transformer is 21 V if the diode has a forward voltage of 0.7 V.

b A battery charger which uses a half-wave rectifier circuit containing a BYY53 diode is able to deliver 18 A d.c. to a load. What is the d.c. load voltage if the diode has a forward voltage of 0.9 V when supplied with an rms voltage of 42 V?

SWITCH ON

The main advantage of a single-phase half-wave rectifier is its simplicity. However, it is rarely used in industry because the circuit makes ineffectual use of a transformer and the quality of the rectified voltage is poor due to the high ripple output. Ripple is a measure of how much of the sine wave remains on the output voltage. The higher the ripple, the less smooth the output voltage.

Peak inverse voltage

When the diode is not conducting, during the reverse half-cycle, the potential difference across the diode is the maximum value of the sine wave as shown in **Figure 10.17**. As shown earlier, the maximum value of a sine wave is 1.414 $\left(\sqrt{2}\right)$ times the rms value.

$$\text{PIV} = \sqrt{2}\,V_{rms} = V_{max}$$

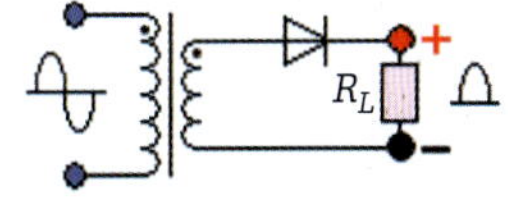

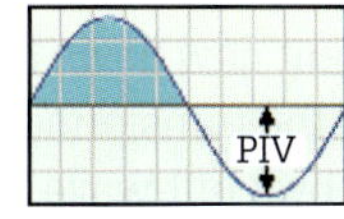

FIGURE 10.17 Peak inverse voltage

It is necessary to increase the PIV rating of a diode used in half-wave rectification when charging a battery. This is because the battery voltage is in series with the maximum reverse voltage of the supply.

Full-wave rectification

Full-wave rectification allows a pulsating unidirectional voltage to appear across a load during each cycle of the input waveform. Either a centre-tapped secondary transformer with diodes or a transformer with a bridge rectifier can be used to provide full-wave rectification. Full-wave rectification uses both halves of the input wave.

Full-wave centre-tapped rectifier

A rectifier circuit is illustrated in **Figure 10.18**. In this rectifier circuit only half of the secondary voltage is used during each cycle of the input waveform. During the first half of one cycle of input we will assume that lead L_1 is positive compared to lead L_2.

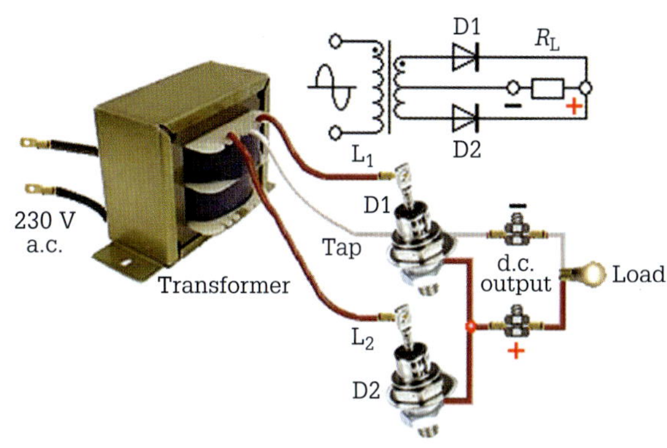

FIGURE 10.18 Full-wave, centre-tapped rectifier

This means that diode D1 will be forward biased and D2 will be reverse biased. Load current is then able to flow through D1 with the load returning through the tap lead of the transformer.

During the second half-cycle leads L_1 and L_2 will reverse polarity. This means that diode D2 is forward biased and D1 is reverse biased. Load current is now able to flow through D2 with the load returning to the tap lead of the transformer.

Note that the current flow through the load is in the same direction. The waveform produced by a full-wave, centre-tapped rectifier is shown in **Figure 10.19**.

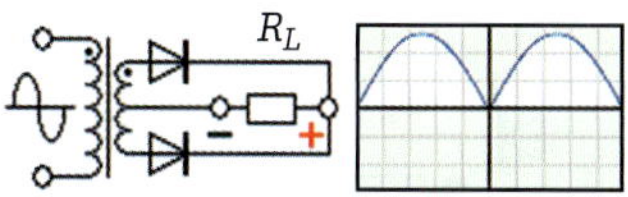

FIGURE 10.19 Centre-tapped rectified waveform

A disadvantage of the two-diode full-wave rectifier is the requirement of a centre-tapped transformer. In addition, the quality of the rectified voltage is low because of very high ripple content.

Average d.c. output voltage

The average d.c. output voltage for a centre-tapped full-wave rectifier is the average value of one cycle of output. This is because the output voltage is applied during each half-cycle. The average or mean value for a half sine wave is 0.637 of the maximum or peak value.

$$V_{av} = 0.637\,V_{max} \qquad V_{av} = 0.9\,V_{rms} \qquad V_{d.c.} = V_{av} - V_{diode}$$

EXAMPLE 10.3

A centre-tapped full-wave rectifier is connected to a 100 V rms supply. Calculate the d.c. load voltage.

$$\begin{aligned} V_{max} &= 1.414 \times V_{rms} \\ &= 1.414 \times 100 \\ &= \mathbf{141.4\ V} \end{aligned}$$

»

$V_{av} = V_{max} \times 0.637$ or $V_{av} = V_{rms} \times 0.9$

$= 141.4 \times 0.637$ or $= 100 \times 0.9$

$= \mathbf{90\ V}$ $= \mathbf{90\ V}$

$$V_{d.c.} = V_{av} - V_{diode}$$
$$= 90 - 0.7$$
$$= \mathbf{89.3\ V}$$

EXERCISE 10.3

a Determine the d.c. load voltage for a centre-tapped full-wave rectifier connected to an 18 V rms feed when the forward voltage of the diode is 0.7 V.

b A centre-tapped full-wave d.c. power supply uses SKR45 type diodes. If the forward voltage of each diode is 1.2 V at 25 °C when delivering 30 A to the load, determine the d.c. load voltage when the transformer feed is 36 V rms.

Peak inverse voltage

The PIV for the diodes of a centre-tapped full-wave rectifier must be twice the V_{max} value as shown in **Figure 10.20**. This is because the diode sees the voltage across both leads of the transformer and not between one lead and the centre tap.

$$PIV = 2 \times V_{max}$$

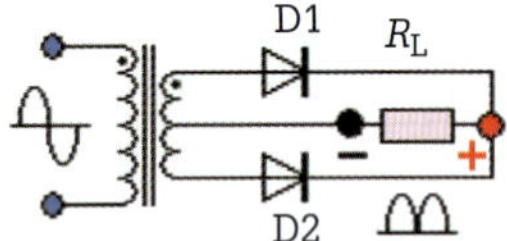

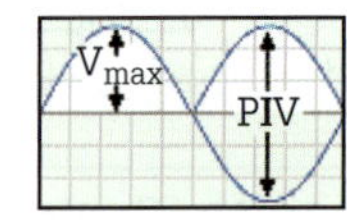

FIGURE 10.20 PIV for a centre-tapped full-wave rectifier

EXAMPLE 10.4

The rms transformer output voltage of a power supply using a centre-tapped full-wave rectifier circuit is 75 V. Determine the PIV that occurs across each diode.

$$V_{max} = 1.414 \times V_{rms}$$
$$= 1.414 \times 75$$
$$= \mathbf{106.05\ V}$$

$$PIV = 2 \times V_{max}$$
$$= 2 \times 106.05$$
$$= \mathbf{212.1\ V}$$

EXERCISE 10.4

Determine the PIV for the diodes of a full wave rectifier circuit using a 32 V rms (16V-0V-16V) 10 A transformer.

Full-wave bridge rectifier

A circuit using a bridge rectifier does not need a centre-tapped transformer but it does need an extra two diodes. In comparison with the centre-tapped transformer, for the same output voltage, the bridge transformer requires only one winding having a voltage output equivalent to the voltage between one lead and the tap on the centre-tapped transformer. A full-wave bridge rectifier circuit is shown in **Figure 10.21**.

Note: The four diodes are manufactured into a single package.

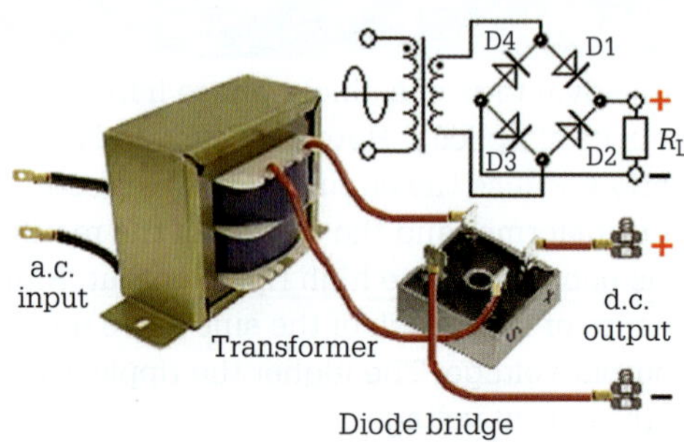

FIGURE 10.21 Full-wave bridge rectifier

For one-half of an a.c. input cycle the transformer has an instantaneous positive secondary terminal with respect to the other secondary terminal. This means that diodes D2 and D4 are reverse biased. Load current is able to flow through diode D1 and the load and returns to the transformer through diode D3.

During the next half-cycle diodes D2 and D4 are forward biased and diodes D1 and D3 are reverse biased. Load current now flows through diode D2 to the load, and returns to the transformer through diode D4. The waveform produced by a full-wave bridge rectifier is shown in **Figure 10.22**.

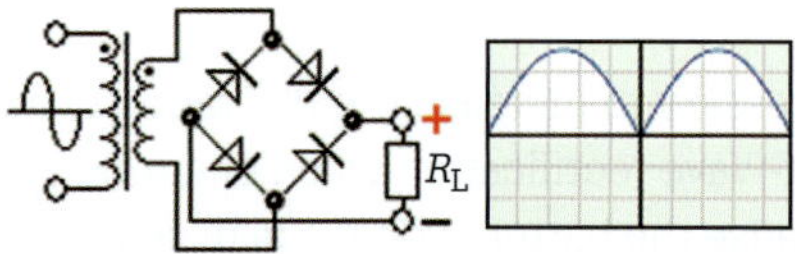

FIGURE 10.22 Bridge rectified waveform

SWITCH ON

The main advantage of the single-phase full-wave rectifier is its more efficient use of the transformer and diodes than in the half-wave rectifier.

For high values of direct output voltage, the use of a bridge rectifier is advantageous as it has a lower PIV than the centre-tapped full-wave rectifier circuit (see **Example 10.5**). However, the quality of the rectified voltage is poor because of very high ripple content. The additional pair of diodes leads to extra voltage drop in the circuit, which is the disadvantage of the bridge rectifier.

Average d.c. output voltage

The average d.c. output voltage for a full-wave rectifier is the average value of one cycle of output. This is because the output voltage is applied during each half-cycle. The average or mean value for a half sine wave is 0.637 of the maximum or peak value.

$V_{av} = 0.637\ V_{max}$ $\quad V_{av} = 0.9\ V_{rms}$ $\quad V_{d.c.} = V_{av} - (2 \times V_{diode})$

EXAMPLE 10.5

A full-wave bridge rectifier is connected to a 100 V rms supply. Calculate the d.c. load output if the forward voltage drop per diode is 0.7 V.

$$V_{max} = 1.414 \times V_{rms} = 1.414 \times 100 = \mathbf{141.4\ V}$$

$$V_{av} = V_{max} \times 0.637 = 141.4 \times 0.637 = \mathbf{90\ V} \quad \text{or} \quad V_{av} = V_{rms} \times 0.9 = 100 \times 0.9 = \mathbf{90\ V}$$

$$V_{d.c.} = V_{av} - (2 \times V_{diode}) = 90 - (2 \times 0.7) = \mathbf{88.6\ V}$$

EXERCISE 10.5

A 40 A bridge rectifier has a single-phase transformer feed of 28 V rms. Determine the average d.c. output voltage if the maximum forward voltage drop per diode element is 0.75 V.

Peak inverse voltage

The PIV for the diodes of a full-wave rectifier equals the secondary maximum voltage:

$$\text{PIV} = V_{max}$$

EXAMPLE 10.6

A 12.6 V rms single-phase 3 A transformer feeds an NTE5312 bridge rectifier rated at 100 V 8 A. Determine the PIV for each diode element.

$$V_{max} = 1.414 \times V_{rms} = 1.414 \times 12.6 = \mathbf{17.8164\ V}$$

$$\text{PIV} = V_{max} = \mathbf{17.8164\ V}$$

EXERCISE 10.6

An SKB52 power bridge rectifier supplies power to a d.c. motor. If each element has a forward voltage drop of 0.8 V, calculate the PIV when the feed to the bridge is 120 V rms.

In practice, it is a good idea to select a diode having a PIV circuit rating greater than 1.414 $(\sqrt{2})$ times the maximum rms voltage of the transformer secondary to allow for a reasonable safety margin. Fluctuations in supply voltage due to spikes on the power line can cause increased voltage across the diode. The IN40xx general-purpose rectifier diodes have V_{RRM} ratings from 50 V to 1000 V.

Ripple frequency and ripple voltage

The d.c. output waveform of rectified direct current consists of a series of unidirectional pulses. The pulses contain the peak variations in the d.c. output, called ripple voltage (V_R) as shown in **Figure 10.22**. These d.c. pulses occur at regular intervals, and this is known as ripple frequency (f_R). The ripple frequency is associated with the a.c. supply frequency and the rectifier design.

For a single-phase half-wave rectifier:

$$f_R = f_{supply} \text{ and } V_R = V_{max} = 1.414\ V_{a.c.}$$

The single-phase half-wave rectifier supplies 50 pulses per second when the a.c. supply frequency is 50 Hz.

For a single-phase full-wave centre-tap rectifier:

$$f_R = 2f_{supply} \text{ and } V_R = 1.414\ V_{a.c.}$$

For a single-phase full-wave bridge rectifier:

$$f_R = 2f_{supply} \text{ and } V_R = 1.414\ V_{a.c.}$$

The single-phase full-wave centre-tap and bridge rectifiers supply 100 pulses per second when the a.c. supply frequency is 50 Hz.

Controlled rectification

With controlled rectification, variable d.c. output voltages ranging from zero to maximum value are obtainable. The control is achieved by replacing the diodes with silicon-controlled rectifiers (SCRs). A full-wave, centre-tapped controlled rectifier circuit is shown in **Figure 10.23**.

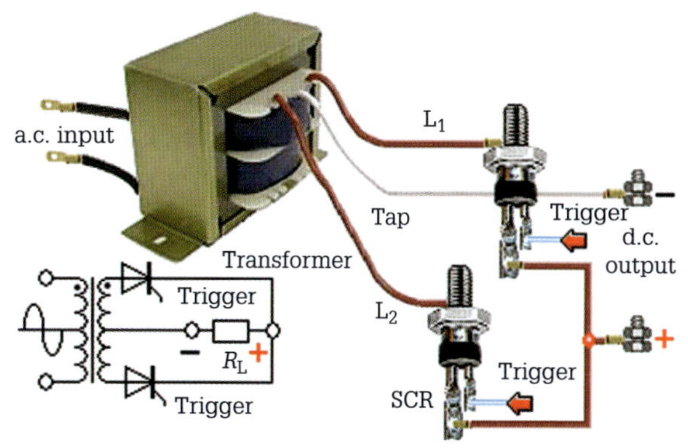

FIGURE 10.23 Controlled rectifier circuit

Direct current output from the rectifier is controlled by varying the duration of the conduction period by varying the point (phase control) along the a.c. input waveform at which the trigger (gate signal) is applied to the SCR as shown in **Figure 10.24**.

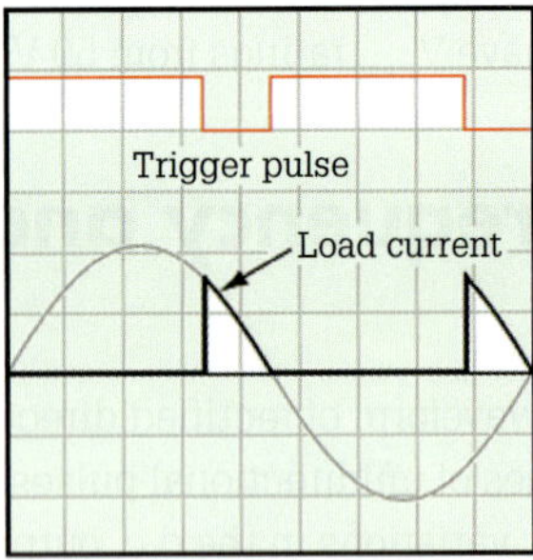

FIGURE 10.24 Controlled d.c. output waveform

Phase angle control offers voltage and current limitation or management, but this is associated with high levels of radio frequency interference (RFI) due to noise associated with the zero crossing commutation (process of turning off the SCR) and harmonic generation. Once conducting, the phase-controlled SCR can only be turned off by reducing the current flowing through it to zero. The SCR automatically switches off as the cycle passes through zero.

SWITCH ON

Controlled rectifiers have a wide range of applications, from small rectifiers to large three-phase, high-voltage d.c. transmission systems. They are used for many kinds of motor drives, traction equipment and controlled power supplies.

REVIEW QUESTIONS

1. Using a circuit diagram, explain how a half-wave rectifier operates.
2. A half-wave rectifier is connected to a 15 V rms a.c. supply. Calculate the d.c. load voltage if the forward voltage drop across the diode is 0.75 V.
3. A centre-tapped full-wave rectifier is connected to a 24 V rms a.c. supply (24V-0V-24V). Calculate the d.c. load voltage if the forward voltage drop across the diode is 0.7 V.
4. What does the PIV for the diodes of a centre-tapped full-wave rectifier equal?
5. Draw the circuit diagram for a full-wave bridge rectifier.
6. A single-phase full-wave bridge rectifier is connected to 20 V rms a.c. supply. Calculate the d.c. load voltage if the forward voltage drop across a diode is 0.7 V.
7. What is the ripple frequency of the output waveform from a single-phase half-wave rectifier when connected to a 14.8 V 50 Hz supply?
8. What is the advantage of controlled rectification?
9. With controlled rectification, what is a disadvantage of using phase angle control?
10. Where is controlled rectification used?

10.4 Filter circuits

The output of the single-phase rectifiers examined previously has a very distinct ripple. In each of the single-phase rectifiers, the voltage drops to zero every half-cycle of the input wave. If this high ripple wave was applied to an amplifier, an intolerable hum of 100 Hz in the case of a full-wave rectifier would be present over the desired sound signal. The high ripple component would have similar adverse effects on other electronic equipment. It is desirable to remove the ripple component completely and have a smooth output voltage like that supplied from a battery. A circuit that adjusts a pulsating d.c. output into a steady d.c. output similar to a battery is known as a filter.

Simple capacitor filter

In this circuit, a single capacitor, usually an electrolytic is connected across the rectifier output and in parallel with the load to achieve the desired filtering. This type of filter uses the ability of a capacitor to charge and to discharge under certain conditions. The filtering action of a simple capacitor filter when used with a half-wave rectifier circuit is illustrated in **Figure 10.25**.

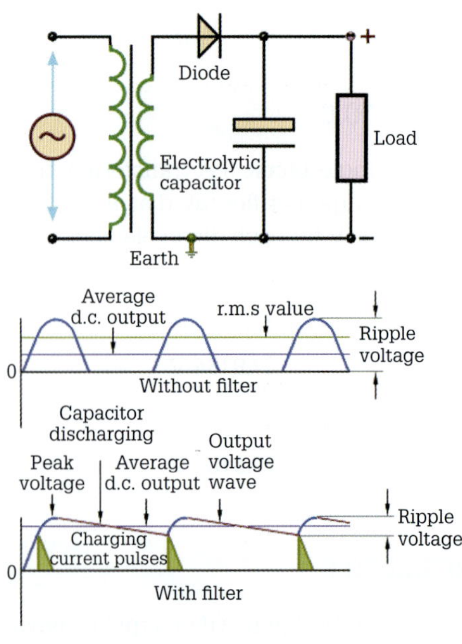

FIGURE 10.25 Half-wave rectifier with simple capacitor filter

As you can see in **Figure 10.25**, the capacitor, which is placed in parallel with the load, will charge to the peak voltage value of the first peak. While this is occurring, current from the transformer flows through the diode to supply the load. When the output voltage wave passes its peak value at 90° the capacitor is then able to discharge slowly through the load. We saw in **Chapter 3** that the voltage decay depends on two factors:

$$\tau = RC$$

Where

τ = circuit time constant in seconds

R = circuit resistance in ohms

C = circuit capacitance in farads

As the filter capacitor connects directly to the rectifier it charges very quickly and the value of its capacitance is selected so that it discharges more slowly. When the next positive half-cycle occurs, the capacitor again begins to charge when the output voltage of the rectifier exceeds the voltage across the capacitor. Once the output voltage wave again passes its peak value at 90° the capacitor discharges through the load. Operating in this manner the capacitor assists in filling in the gaps between the voltage peaks as **Figure 10.25** shows. In **Figure 10.25** the blue portion of the voltage waveform shows when the rectifier provides current to the load, while the red portion shows when the capacitor provides current to the load. The result of using a capacitor filter means that the d.c. output voltage never falls to zero.

The diode in a filter circuit performs two functions:

1. to recharge the capacitor (green pulses in lower wave shown in **Figure 10.25**)
2. to provide the output voltage and load current while the capacitor is charging.

The function of the capacitor is to supply output voltage and the load current.

The value for the capacitor must be carefully selected because if it discharges too quickly, as is the case when load current is high, the ripple voltage across the load will be high. Adding a carefully selected filter capacitor can result in output voltages that are up to 2.5 times the unfiltered voltage. You are probably thinking: how can you get a voltage greater than 100% (2.5 × 45% = 112.5%) – remember that the peak value, the value to which the capacitor is charging, is 141% of the rms value.

For example, a 20 V peak a.c. output from the transformer will have a 14.14 V rms value:

45% of 14.14 V = 6.36 V d.c. output − 0.7 V drop across the diode voltage = 5.66 V

With a capacitor filter the average d.c. voltage output is:

5.66 V × 2.5 = 14.16 V d.c. output

This is less than the peak value of 20 V but significantly higher than the unfiltered voltage of 5.66 V d.c.

Simple capacitor filtering of a full-wave rectifier

When the capacitor is supplied from a full-wave or bridge rectifier, as shown in **Figure 10.26**, it will charge in a similar manner to that of the half-wave filter.

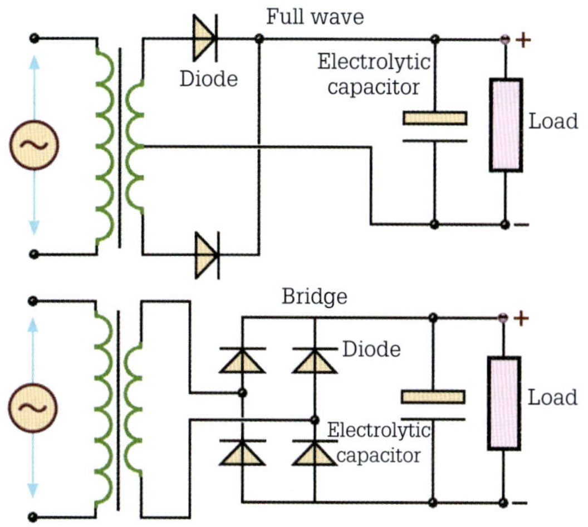

FIGURE 10.26 Simple capacitor filtering a full-wave and a bridge rectifier

The difference is that the capacitor recharges twice as often because the number of voltage peaks will have doubled as illustrated in **Figure 10.27**. In comparison with the half-wave filter, the d.c. average voltage increases slightly towards peak voltage and the ripple voltage reduces. The reduced ripple voltage is because of the shorter discharge time before the capacitor is charged by another pulse of current.

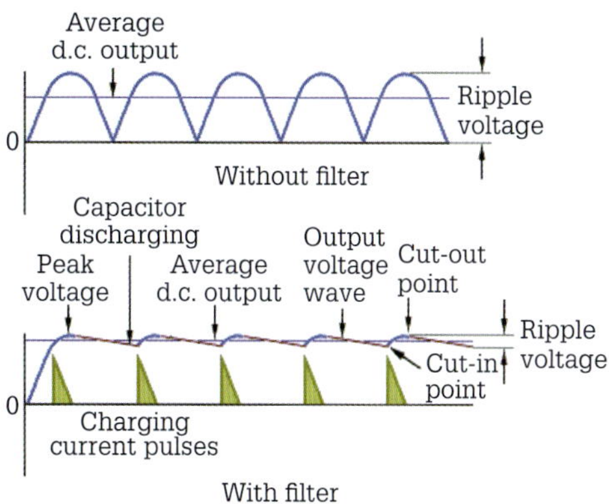

FIGURE 10.27 Full-wave and bridge output waveforms without filter and with filter

As capacitors tend to oppose a change in voltage, a larger-sized capacitor will tend to reduce the size of the ripple voltage for a given load. Equally, this means the opposite will happen should the load decrease. This occurs because ripple voltage is inversely proportional to load resistance. Effective and simple capacitive filtering occurs when the ripple voltage is at a minimum. Capacitive filtering is enhanced when the output from the filter is fed into a voltage regulator.

Simple inductive filter

We learned in **Chapters 4** and **5** that an inductor has the electrical property of inductance, which is defined in Lenz's law as the ability to oppose a change in current. A basic inductor consists of a number of turns of wire around a former, which may be air or ferromagnetic material. At the instant of switch on, an inductive circuit will delay the establishment of current, and once current flows any change in current (either increasing or decreasing) will be delayed.

When an inductor is connected into the output circuit of a single-phase full-wave rectifier as shown in **Figure 10.28**, its effect is to oppose the increase in current from zero to maximum and oppose the decrease in current from maximum to zero.

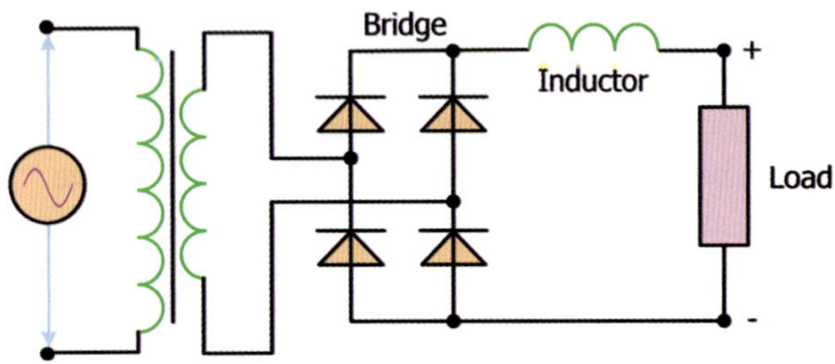

FIGURE 10.28 Full-wave bridge rectifier with inductive filter

As the inductor opposes changing current, its effect as a filter is most evident in high-power rectifiers and as such does not find much application in single-phase, low power, power supplies. Inductors suited to high-power applications are comparatively cheaper than capacitors. **Figure 10.29** shows the filtering action of an inductor on the output from a single-phase, full-wave rectifier.

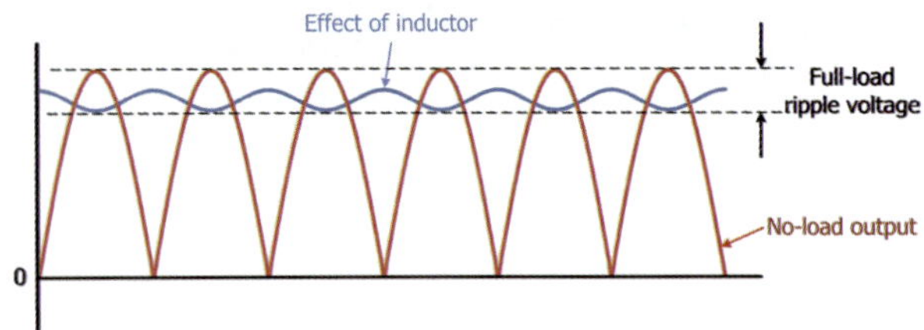

FIGURE 10.29 Filtering action of an inductor on a full-wave bridge rectifier

Low-pass active filters

The output voltage from rectifiers consists of a unidirectional pulsating waveform which also contains an a.c. voltage component. In order to achieve a steady d.c. voltage similar to that of a battery voltage, which is pure d.c., filtering circuits are necessary. A filter allows some frequencies of an output signal to pass more easily than others do. The effect of filtering on a rectified waveform is shown in **Figure 10.30**.

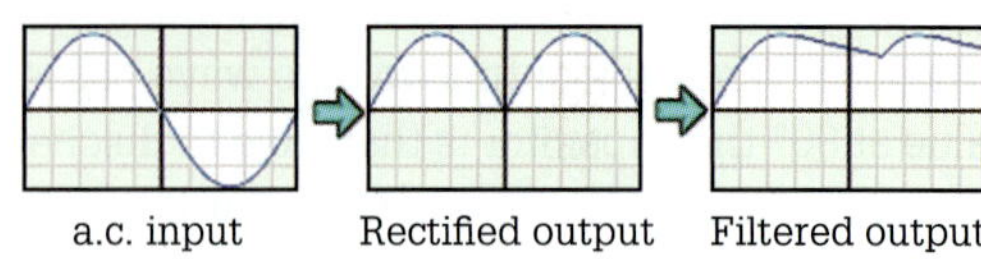

FIGURE 10.30 Filtering a rectified waveform

Active low-pass filters are used in low-frequency, low-power applications to clean up remnant waveform frequencies (ripple) caused by the rectification process. Active filters use active devices, such as transistors or integrated circuits, in addition to passive components, such as resistors, inductors and capacitors. In order to function, active filters require an additional source of power other than that of the rectified output to filter the waveform. A low-pass filter is designed to pass low frequencies but attenuate (block) high frequencies above its cut-off frequency.

As a practical example of a low-pass filter, when very loud music is being played in a room, people in an adjacent room will hear only the bass components of the music. This is because the walls and the air gap between them, act as a low-pass filter where only the lower frequencies pass through and the higher frequencies are prevented from passing.

A simple circuit that acts as a low-pass filter with passive devices consists of a resistor in series with the load and a capacitor connected in parallel with the load, as shown in **Figure 10.31**. **Chapter 5** introduced the electrical property of reactance, which opposes electric current in a circuit. The capacitive component of reactance, called capacitive reactance is inversely proportional to frequency. When connected as a shunt or parallel connected component, as shown in **Figure 10.31**, the capacitor has a high reactance at low frequencies, which blocks low-frequency waves, causing them to go through the load. At higher frequencies, the reactance is lower and the capacitor allows these frequencies to pass through, which prevents them from reaching the load as they are shunted past the load.

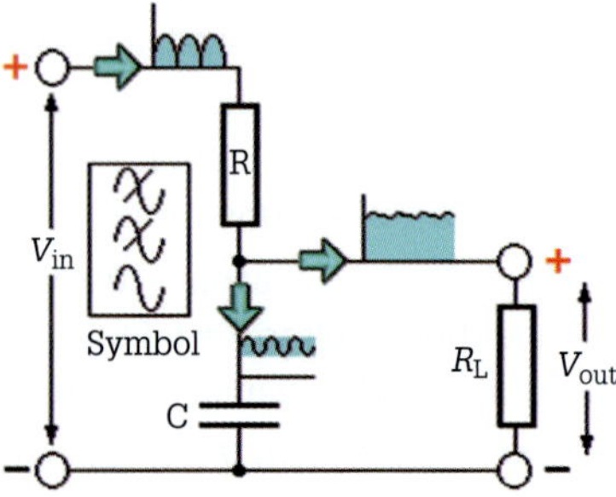

FIGURE 10.31 A simple low-pass filter circuit and its symbol

The active low-pass filter is usually an integrated semiconductor which combines an operational amplifier (Op Amp) and passive components into one package. An illustration of an Op Amp is shown in **Figure 10.32**.

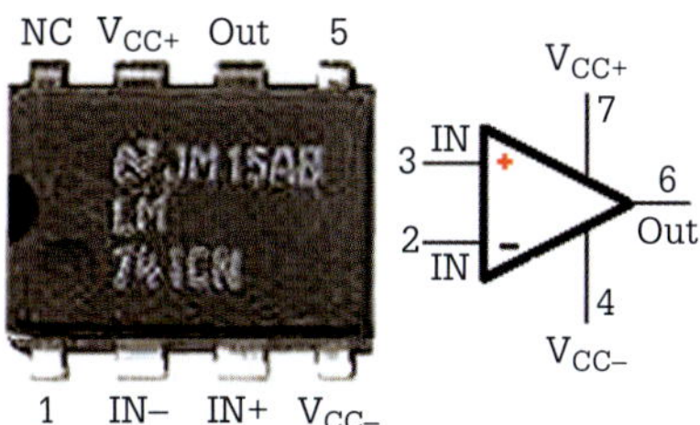

FIGURE 10.32 Operational amplifier and pin diagram

There are two types of low-pass active filters. A first-order filter will reduce the amplitude of the high-frequency ripple by half. A second-order filter will reduce that same high frequency by half again. The difference between the two types is the number of resistors and capacitors connected in the circuit as shown in **Figure 10.33**.

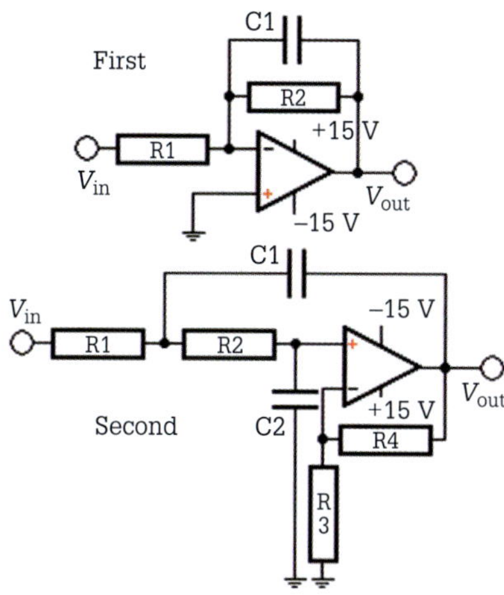

FIGURE 10.33 First- and second-order low-pass active filters

REVIEW QUESTIONS

1. What is the function of a filter circuit?
2. How is a simple capacitor filter connected in a circuit?
3. Why is it necessary to carefully select the value of a filter capacitor?
4. Why does an inductor filter have most effect on the heavily-loaded output voltage from a rectifier?
5. What is the purpose of an active low-pass filter?
6. What components generally make up an active low-pass filter?
7. What happens to the ripple voltage if the load resistance connected to a capacitor-filtered power supply increases?

10.5 Zener diode shunt regulator

Zener diode

A Zener diode is a particular type of diode that will conduct when forward biased like a conventional P–N junction diode, and also when reverse biased after a specific potential exists across the cathode and anode. This effect is achieved by having a highly doped P–N junction.

A conventional P–N junction diode will conduct from cathode to anode when the reverse bias breakdown voltage is exceeded. In this condition, the diode may suffer permanent damage due to overheating. A Zener diode displays similar properties, except that its design allows it to have a reduced breakdown voltage, which is called the Zener voltage. A reverse biased Zener diode displays a controlled breakdown and allows the current to keep the potential across the reverse biased Zener diode close to the Zener breakdown voltage. As an example, a 5.6 V Zener diode has a reverse potential of 5.6 V across a wide range of reverse currents. The Zener diode is ideal for applications such as a voltage regulator for low-current applications.

Illustrations of a Zener diode and its circuit symbol are shown in **Figure 10.34**. Note that the symbol is similar to the junction diode except that the cathode has a different characteristic.

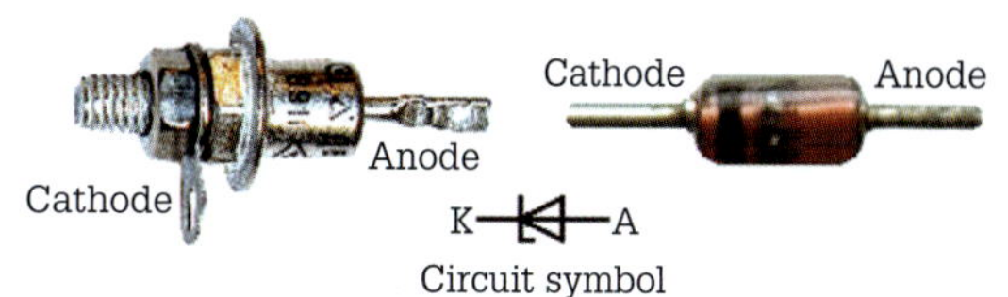

FIGURE 10.34 Zener diode and circuit symbol

Reverse-bias mode

In reverse-bias mode, Zener diodes do not conduct until the applied voltage reaches or exceeds the breakdown voltage, at which point the Zener diode is able to conduct significant current, and in doing so will limit the voltage dropped across it to that breakdown voltage point. So long as the power dissipated by this reverse current (I_{Zmax}) does not exceed the diode's thermal limits, the Zener diode will not be affected. The maximum current that the Zener diode can conduct is limited by the power rating (from 400 mW up to 50 W) of the diode.

$$P_{max} = I_{Zmax} \times V_Z$$

EXAMPLE 10.7

A BZD27-C75 (75 V) Zener diode, connected across a 100 V input, has a maximum current of 10 mA flowing through it. Calculate the power dissipated by the Zener diode.

$$P_{max} = I_{Zmax} V_Z$$
$$= (10 \times 10^{-3}) \times 75$$
$$= \mathbf{750\ mW}$$

EXERCISE 10.7

A 1N5380B 120 V, 5.0 W Zener diode voltage regulator has 40 mA flowing through it. Calculate the power dissipated by the Zener diode.

Breakdown voltage

The breakdown voltage value is determined at the manufacturing stage with respect to the amount of doping required. When the voltage applied across the Zener reaches the breakdown point, the voltage across it remains constant while the current it conducts sharply increases. A Zener diode conduction characteristic is shown in **Figure 10.35**.

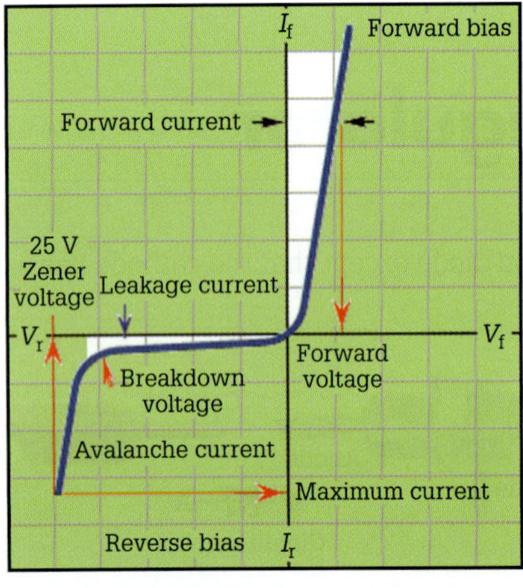

FIGURE 10.35 Zener diode conduction characteristic

Zener diode shunt regulator

Due to their breakdown characteristics, Zener diodes are widely used to generate a reference voltage with varying d.c. voltages at a specific Zener voltage (from 2.4 V up to 270 V). In this way we can regulate the output voltage, from a varying d.c. input voltage, to a constant voltage. This application of a Zener diode is known as a voltage regulator. A circuit diagram of a voltage regulator is shown in **Figure 10.36**.

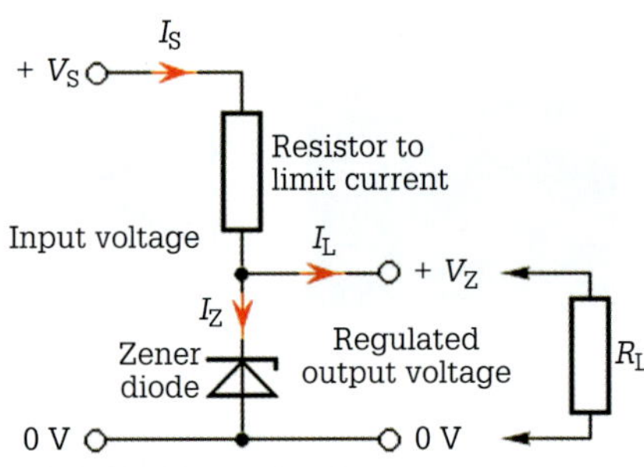

FIGURE 10.36 Zener voltage regulator

In this circuit the Zener diode is in parallel with the load. The Zener diode is used to shunt some current around the load. With the Zener diode placed in the circuit so that it is reverse biased, it will start to conduct once the Zener voltage is reached. At this instant in time the voltage across the Zener diode will remain the same. Because the load is in parallel with the Zener diode its supply voltage is the Zener voltage. Because the Zener voltage does not vary, the load voltage does not change. The load voltage supply is now stabilised or regulated. For effective voltage regulation, the practical operating limits for a Zener shunt regulator are in the range of 10% to 80%.

In the circuit of **Figure 10.36**, the Zener diode can be viewed as a variable resistor.

The total (supply) current (I_S) is:

$$I_S = I_Z + I_L$$

The voltage across the limit resistor (R) is:

$$V_{limit} = (I_Z + I_L) \times R_{limit}$$

As the load voltage across R_L is to be kept constant, the current (I_Z) through the Zener diode can be varied (working current range is from 0.1 A to 0.8 A) so that any voltage above the required load voltage is dropped across the limit resistor. If the voltage across the load increases, the Zener diode reduces its resistance, allowing more current (I_S) to flow through the limit resistor, which causes a higher voltage drop across it. Likewise, if the load voltage decreases, I_S will decrease, allowing the voltage across the limit resistor to decrease. This leaves more voltage for the load.

A main characteristic of a regulator circuit is its ability to maintain a constant voltage as the load demand changes. How well a voltage regulator can maintain a constant output voltage from no-load to full load is known as voltage regulation, which is normally expressed as a percentage. The percentage voltage regulation for voltage regulators can be determined by applying the following equation:

$$V_{reg} = \frac{V_{NL} - V_L}{V_L}$$

where V_{NL} is the no-load voltage and V_L is the loaded voltage. Voltage regulators should have a regulation of less than 5%.

A diagram of a Zener diode shunt regulator circuit is shown in **Figure 10.37**.

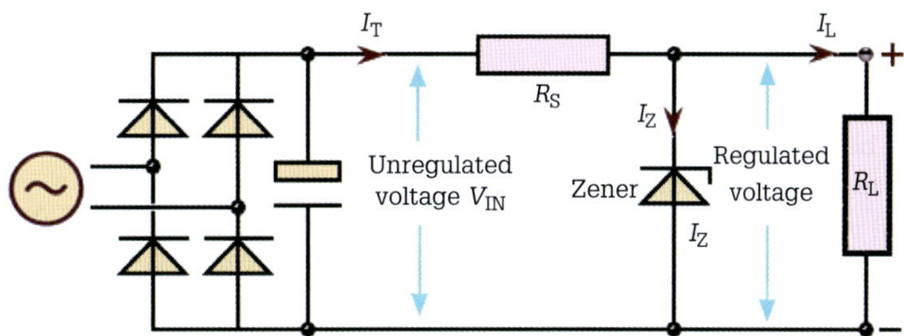

FIGURE 10.37 Zener diode shunt regulator

EXAMPLE 10.8

A 1N5356B 19 V Zener diode provides a load voltage of 18.6 V. Calculate the percentage voltage regulation.

$$V_{reg} = \frac{V_{NL} - V_L}{V_L} = \frac{19 - 18.6}{18.6} = \mathbf{0.0215 \text{ or } 2.15\%}$$

EXERCISE 10.8

A 1N5388B 200 V Zener diode provides a load voltage of 196 V. Calculate the percentage voltage regulation.

EXAMPLE 10.9

1. The Zener diode (IN4738) of **Figure 10.38** has the following electrical characteristics as per the manufacturer's data sheet:

   d.c. power dissipation at 50 °C = 1 W (maximum allowable Zener diode power dissipation)

   $V_z = 8.2$ V at $I_z = 31.0$ mA

   $I_{ZK} = 0.5$ mA

   $R_z = 4.5\ \Omega$

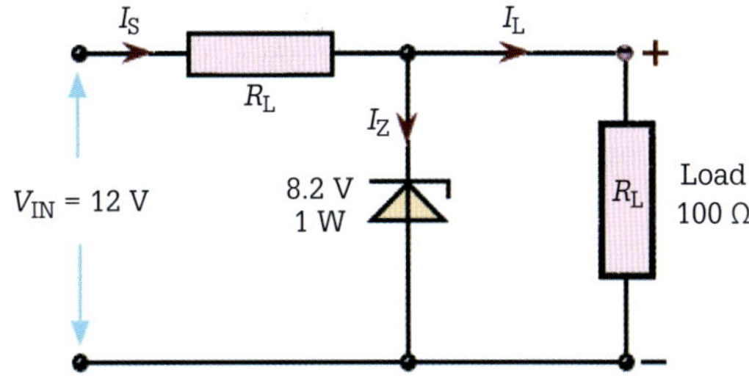

FIGURE 10.38 Zener shunt regulator

Using **Figure 10.38**, determine the following electrical parameters in (a) to (f):

a The load voltage

As $V_L = V_z$ and $V_z = 8.2$ V, then $V_L = \mathbf{8.2\ V}$

b The load current

Using Ohm's law:

$$I_L = \frac{V_L}{R_L} = \frac{8.2}{100} = 0.082\text{A or } \mathbf{82\ mA}$$

c The maximum current the Zener diode can handle I_{ZM}, can be found by:

$$I_{ZM} = \frac{P_{max}}{V_Z} = \frac{1}{8.2} = \mathbf{122\ mA}$$

d The Zener diode's practical operating limits

For effective voltage regulation, the practical operating limits for a Zener shunt regulator are in the range of 10% to 80%, so

$$I_{ZM} = \mathbf{12.2\ mA\ to\ 97.6\ mA}$$

e The value of the series resistor

The current through the series resistor:

$$I_S = I_Z + I_L = 31 + 82 = \mathbf{113\ mA}$$

(**Note:** obtain I_z from manufacturer's specifications)

Using Ohm's law:

$$R_S = \frac{V_S - V_Z}{I_S} = \frac{12 - 8.2}{0.113} = \mathbf{33.6\ \Omega}$$

When designing a circuit you need to round calculated resistor values to a preferred value. Normally this would be the value closest to the one calculated for the series you are using. Because the series resistor (R_S) is a current-limiting resistor, using a value smaller than calculated could cause an overload. Using the E24 (5% tolerance) preferred series, a 39 Ω resistor is recommended.

»

f Resistor power rating.

$$P_{RS} = I_S^2 R_S$$
$$= 0.113^2 \times 39$$
$$= \mathbf{0.498\ W}$$

A 0.5 W resistor would be just acceptable; a preferred option would be to use a 1 W resistor.

2. What would be the load voltage when load current I_L in the circuit of **Figure 10.38** varies from 15 mA to 75 mA? Also, calculate the voltage regulation of the regulator.

For 75.0 mA:

$$I_L = 75.0\ \text{mA}$$
$$I_S = I_{ZM} - I_L = 122 - 75 = 47.0\ \text{mA}$$

Deviation from 31 mA (I_Z) = 47 − 31 = −16.0 mA

V_{drop} across the Zener = $I_Z \times R_Z$ = −16 × 4.5 = −72.0 mV

$$V_L = V_Z + I_Z R_Z = 8.2 + (-0.072) = 8.128\ \text{V}$$

For 15.0 mA:

$$I_L = 15.0\ \text{mA}$$
$$I_S = I_{ZM} - I_L = 122 - 15 = 107.0\ \text{mA}$$

Deviation from 31 mA (I_Z) = 107 − 31 = 76.0 mA

V_{drop} across the Zener = $I_Z \times R_Z$ = 76 × 4.5 = 342.0 mV

$$V_L = V_Z + I_Z R_Z = 8.2 + (0.342) = 8.542\ \text{V}$$

% regulation:

$$\%\ \text{regulation} = \frac{V_{max} - V_{min}}{V_{min}} \times 100$$
$$= \frac{8.542 - 8.128}{8.128} \times 100$$
$$= \mathbf{5.09\%}$$

For some applications, a change in load voltage of 5.09% may be acceptable.

EXERCISE 10.9

The electrical characteristics for a 1N5359B Zener diode are as follows:

d.c. power dissipation @ 25 °C = 5 W maximum

$$V_Z = 24\ \text{V at } I_Z = 50\ \text{mA}$$
$$I_{ZK} = 1\ \text{mA}$$

If the unregulated feed (V_{IN}) to the regulator is 30 V d.c. and the connected resistive load is 110 ohms calculate the following electrical parameters:

a the load voltage
b the load current
c the maximum current the Zener diode can handle
d the Zener diode's practical operating limits
e the value of the series resistor
f the power rating of the series resistor.

REVIEW QUESTIONS

1 What is a Zener diode?
2 What is a common application of a Zener diode?
3 State what happens when the reverse breakdown voltage of a Zener diode is reached.
4 What potential would you expect to measure across a reverse-biased 6.8 V Zener diode?
5 For effective voltage regulation, the practical operating limits for a Zener shunt regulator need to be within what range?
6 A BZD27-C75 (75 V) Zener diode, connected across an 85 V input, has a maximum current of 20 mA flowing through it. Calculate the power dissipated by the Zener diode.
7 To what does voltage regulation refer?
8 The no-load output voltage from a d.c. power supply is 14.2 V and drops to 13.6 V at full-load. What is the percentage voltage regulation for this power supply?

10.6 Three-terminal IC voltage regulator

Three-terminal voltage regulators are used to provide a fixed or variable level of d.c. operating voltage regardless of changes in the input voltage or changes in load conditions.

The three terminals are input, ground (or common) and output. A typical three-terminal regulator, its circuit symbol and internal circuits are shown in **Figure 10.39**.

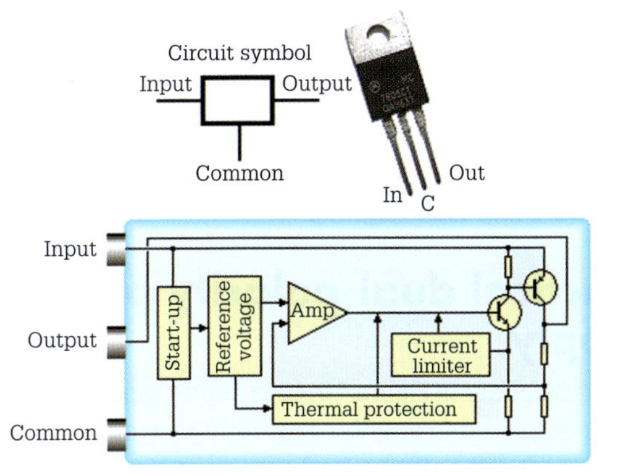

FIGURE 10.39 Three-terminal regulator and basic internal circuit

Operation

While it looks like a transistor, the three-terminal regulator is a complex monolithic integrated circuit (IC). A rectified filtered voltage or a varying voltage is used as the input and a fixed d.c. voltage of either positive or negative polarity is obtained at the output. However, the input voltage must be the same polarity as the regulator's rated output polarity. The most common part numbers for three-terminal regulators start with the numbers 78 or 79 and finish with two digits indicating the output voltage. The number 78 represents positive voltage and 79 represents negative voltage. For example, a positive 12 V regulator (7812) must have a positive input voltage and a negative 12 V regulator (7912) must have a negative input voltage. The circuit diagram in **Figure 10.40** represents a typical use of a positive voltage regulator as a power supply.

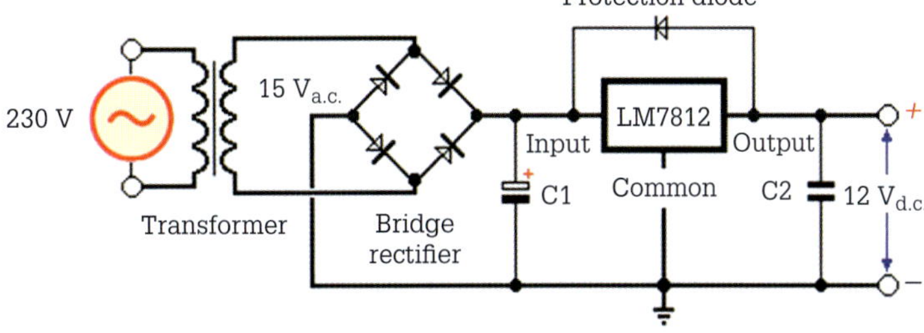

FIGURE 10.40 Three-terminal regulator power supply

The circuit operates in the following manner. The transformer steps down the 230 V a.c. primary voltage to 15 V a.c. secondary. A diode bridge rectifies the 15 V a.c. secondary into a direct current supply feeding into the voltage regulator. Capacitor C1 filters any voltage spikes that could be induced in the input leads connected to the regulator and the smoothed voltage is then applied to the V_{INPUT} terminal of the regulator.

The regulated 12 V d.c., or reference voltage, is obtained from the V_{OUTPUT} terminal.

Capacitor C2 is used to assist the regulator response to possible voltage transients (back emf) around the output from inductive loads. An input short-circuit protection diode is used to safeguard the regulator from a reverse input polarity connection. Also, three-terminal voltage regulators have on-chip circuitry that will automatically shut down the regulator if its rated current is exceeded (called safe area protection) or overheating occurs (junction temperature reaches 150 °C). The voltage across the regulator is called the drop-out voltage and is the difference between the output and the input voltages. Other applications of the three-terminal IC regulators are as constant- or variable-current output devices as illustrated in **Figure 10.41**.

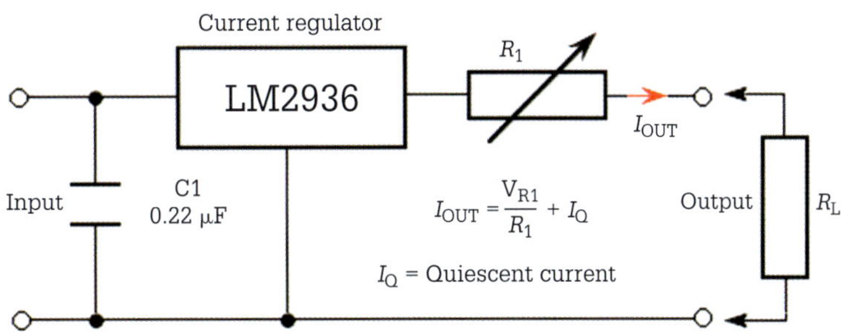

FIGURE 10.41 Variable-current circuit

The LM2936 regulator of **Figure 10.41** holds the reference voltage constant across resistor R_1. Because the voltage is constant the current will vary in direct proportion to changes in resistance. Therefore, at a specific value of resistance the current through R_1 will be constant. Since the current through R_1 also flows through the load resistance R_L, resistor R_1 may be used to control the load current. Regulators can get quite hot and require a heatsink. Without a heatsink, three-terminal regulators can dissipate about 2 W depending upon the package. A simple calculation of the voltage drop-out multiplied by the current drawn will give the power to be dissipated.

Dual-polarity power supply

The dual-polarity power supply is fundamentally two power supplies in one. These power supplies have a positive output rail (V+) and a negative output rail (V–) which are referenced to a common rail.

The three-terminal regulators of a dual power supply must be used in positive and negative pairs but they do not require matching voltage: for example, you can use a +12 V and a –5 V regulator together.

Practical dual-polarity power supply

The transformer of **Figure 10.42** is centre tapped with a power output rated at 9 VA. This rating means that the transformer can deliver a load current of 230 mA for the +/– 15 V output without overheating (0.23 A × 36 V = 8.28 VA). If the transformer was rated by output rms current as 1 A rms then that value would need to be divided by 1.2 to obtain the value of current that could be supplied (1 A rms divided by 1.2 = 830 mA output).

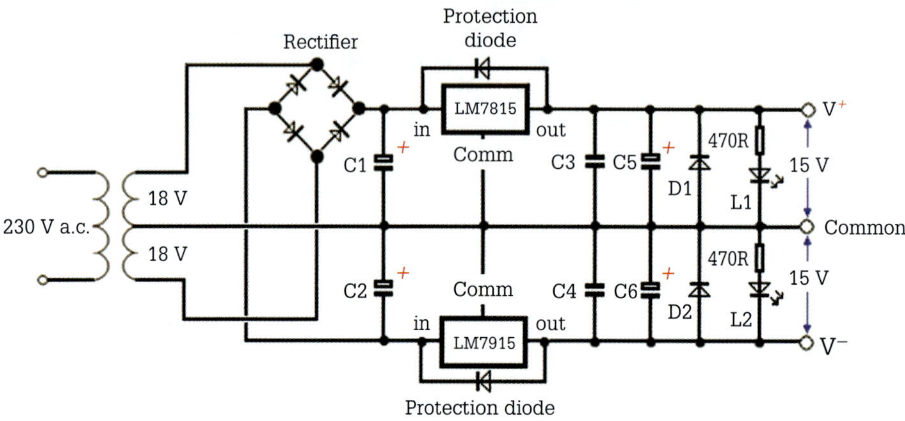

FIGURE 10.42 Dual-polarity power supply circuit

The circuit uses a 4 A full-wave bridge rectifier with a minimum peak inverse voltage (PIV) of twice the maximum or peak voltage (2 × 25.45 V = 50.9 V). Regarding safety in a practical circuit the PIV should be at least three to four times that of the transformer's secondary voltage – in our circuit 101.8 V. The current rating of the bridge should be twice the current that will be drawn by the load. In our example this is 460 mA.

The filter capacitors (C1 and C2) smooth out the a.c. ripple in the output of the rectifier. The amount of ripple that is passed on is determined by the value of the filter capacitor used. The higher the value of capacitance, the smaller is the ripple that is passed on. The working voltage of the capacitor must be greater than the peak output voltage of the centre tapped-transformer (1.414 × 18 V = 25.45 V). Suitable capacitors would be 2200 μF 35 V electrolytic.

The input voltage across the regulator must always be higher than the regulator's output voltage by at least 3 V in order for it to work. The regulators are 7815 and 7915, which provide a +/– 15 V output. These regulators will not need heatsinks because the regulators will be dissipating less than their capacity of 2 W each.

$$P = VI = 6\text{ V (drop-out)} \times 0.23\text{ A} = \mathbf{1.38\ W}$$

Capacitors C5 and C6 assist the regulator to react to sudden changes in load current and also prevent uncontrolled oscillations. These capacitors improve the circuit stability. Suitable capacitors for our circuit would be 100 nF 35 V electrolytics.

Capacitors C3 and C4 provide high-frequency decoupling across the output which keeps the impedance low at high frequencies. Suitable capacitors would be 100 nF of the monoblock variety.

Two LEDs indicate when the regulated output is on-line. The 470 R resistors act as current limiters for the LEDs.

Diodes D1 and D2 (IN4004 types) provide protection against any back emf which may exist across the output terminals when the power supply feeds inductive loads.

Voltage transients

Voltage regulators cannot act to regulate the output voltage if a transient over-voltage occurs across the primary of the transformer. To suppress the transient situation the power supply transformer is protected, in addition to over-current protection (fuse), by a voltage-dependent resistor (VDR) or varistor that is connected across the primary winding. The VDR clamps the high-voltage transient, allowing only the normal mains supply (230 V) across the input of the power supply.

REVIEW QUESTIONS

1. What is the purpose of a three-terminal IC voltage regulator?
2. Name the three terminals of a three-terminal regulator.
3. When not mounted on a heatsink, a three-terminal regulator is able to dissipate a maximum power of ______.
4. A three-terminal IC voltage regulator is identified as 7915. What is the significance of the '79' and the '15'?
5. What would be the complementary three-terminal IC voltage regulator for the one in question 4 in order to produce a dual-polarity power supply?
6. Why would a three-terminal IC voltage regulator be mounted on a heatsink?
7. Why is a reverse-biased diode connected across the input and output terminals of a three-terminal IC voltage regulator?
8. What input voltage would a 7805 regulator require for satisfactory operation?

10.7 d.c. power supply testing and fault finding

Fault-finding is a common form of problem solving. Electricians diagnose faulty systems and take direct, corrective action to eliminate any faults in order to return the systems to their normal states.

SWITCH ON

Before undertaking any testing or fault finding activities on d.c. power supplies, it is essential to undertake and document a risk assessment. Section 10.8 provides guidance on undertaking a risk assessment.

It is possible to consider the d.c. power supply as a number of functional blocks as shown in **Figure 10.43**. By examining the power supply in blocks rather than the whole, it is possible to narrow down the search for the faulty component or device. This process begins by taking measurements to identify the faulty block and then analysing the circuit within that block to locate the faulty component.

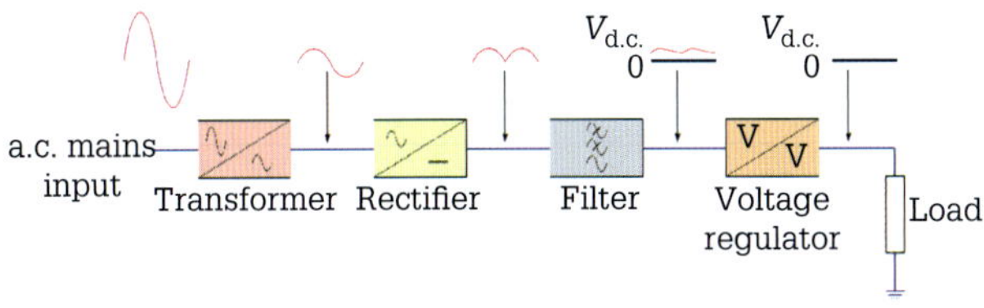

FIGURE 10.43 Block diagram of d.c. power supply

Faults

The range of faults that could exist in a single-phase d.c. power supply includes:

- absence of mains power
- absence of supply power to the load
- short-circuit faults (including transformer windings, diodes, capacitors, resistors, chokes and IC regulators).
- open-circuit faults (including transformer windings, diodes, capacitors, resistors, chokes and IC regulators)
- low-insulation resistance
- high-earth resistance
- high-resistance joints (terminal connections)
- contact deterioration (can be caused by oxidation).

An important test tool for detecting most of the faults listed above is the oscilloscope. An insulation tester and a digital multimeter are also required for determining insulation faults and high-resistance respectively.

Loose terminal connections resulting in high-resistance joints can overheat as they can produce high power ($P = I^2R$) over a small cross-sectional area for a long period. If the power produced is high enough, the connection will radiate heat. Radiated heat can ignite combustible gases coming off insulation in contact with the termination, resulting in fire.

When fault-finding, always use visual, sound, smell and heat senses where appropriate, observe timing of events, and any departure from normal operation. Verify symptoms through direct observation.

Fault-finding process

For the purpose of electrical safety, fault-finding should be undertaken on de-energised equipment. If de-energising is not possible, it is necessary to implement additional hazard controls before commencing work. In order to carry out fault-finding tasks, use only approved test instruments with the correct installation category and current test and calibration status. If working on live equipment the following additional requirements are necessary:

1. Prior to the commencement of fault-finding on energised plant, the electrician should remove all jewellery and metal objects.
2. Always use insulating mats for working on conductive surfaces.
3. Always use insulating barriers or covers over exposed live parts if there is potential to reach them during the fault-finding process. Exposed parts are any terminal, connection, conductor or electrical parts that a standard test finger can touch. If it is determined that there are no exposed parts, perform testing on energised parts with the use of an approved meter using insulating gloves only.
4. The electrician performing the testing must wear the following PPE:
   - insulating gloves rated to the highest expected voltage applicable for the task
   - face shield
   - flame-retardant clothing that is covering the full body, arms and legs.

A faulty single-phase d.c. power supply is one where the input is within design parameters but the output is outside of design parameters. For example, **Figure 10.44** represents an operational d.c. power supply as a single block where both the input and output voltages are within design parameters. **Figure 10.45** represents a faulty d.c. power supply as a single block where the input is within design parameters but the output is outside design parameters; that is, it is faulty.

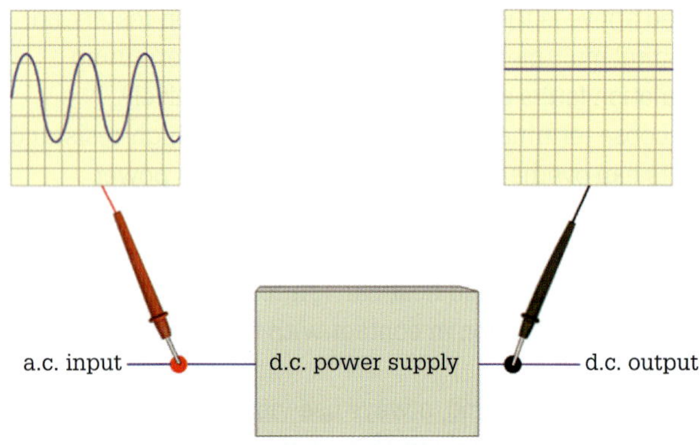

FIGURE 10.44 Correct d.c. voltage output

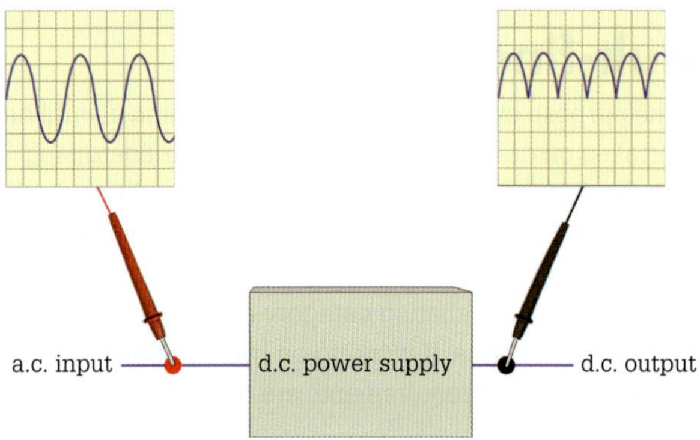

FIGURE 10.45 Incorrect d.c. voltage output

Location of all faults is diagnosed using logical, systematic analysis, supported by observation, measurements and testing. Finally, confirm repair of the fault by operational testing.

The methods used to decide which block is faulty are (1) half-split and (2) systematic.

Half-split method of fault-finding

The logic behind this fault-finding approach is pure. With every step in the fault-finding process, the electrician eliminates half of the functional blocks as the cause of the problem. In applying the half-split method, the first action might be as simple as taking a voltage measurement with an oscilloscope. For example, a lower than specified d.c. output voltage occurs from the power supply.

The block diagram of **Figure 10.43** shows four functional blocks (transformer, full-wave rectifier, capacitor filter and voltage regulator) and expected voltage waveforms at five points. Consequently, these points can be identified as test points (TP) 1 to 5. Since there are four blocks, it is possible to divide the unit in half. Test each half and based on the results obtained, determine which half is not functioning correctly. Split the faulty half in half again and repeat until the faulty block or component is identified.

In our example we know that we have an output voltage but it is not a correct value. We can start the fault-finding by checking the voltage waveform halfway (between blocks two and three, at test point 3 in **Figure 10.46**). This allows us to determine whether the fault exists in the front or rear half of the circuit.

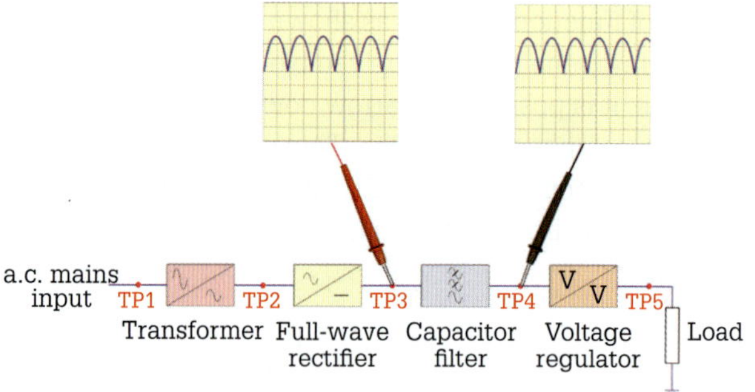

FIGURE 10.46 Example of applying the half-split method to locate a faulty capacitor filter

Test point 3 (TP3) indicates a full-wave rectified voltage, which shows that the transformer and rectifier are operating as expected. The next step in this fault-finding process is to split the remainder of the functional blocks in half, in this case between blocks three and four (test point 4). This measurement indicates that the capacitor filter is malfunctioning, due to the full-wave rectified voltage being observed at TP4. If the capacitor filter was operational, we would observe a smooth d.c. voltage at TP4. This fault would tend to indicate an open-circuit filter capacitor. A shorted filter capacitor would most likely cause the fuse to open circuit. A short anywhere in the power supply is very difficult to isolate, as any over-current protection will operate when a short, particularly to earth, develops.

This example highlights the effectiveness of the half-split method of fault-finding. It took two measurements to identify the fault to the open-circuit capacitor filter. If we started the fault-finding process at the power supply input, it would take four measurements to locate the fault. However, if we started at the output, it would take three measurements as we need to take a measurement at TP3 to confirm correct operation at this point.

Although we still have to find out what is wrong with this part of the circuit at least we identified the possible fault.

The next steps would involve removing the capacitor from the unit and testing it on its own and then replacing as necessary. The cause of the fault may be a weak solder join resulting in an open circuit.

Next you would confirm that the fault no longer exists through operational testing of the power supply.

Systematic method

The input-to-output and output-to-input methods are examples of a systematic approach to fault-finding. With this approach to fault-finding, the electrician should have an up-to-date circuit diagram and all the necessary test equipment.

The electrician then has to describe the fault precisely. This point is most significant. It is useless trying to locate a fault that has an unclear definition. The symptoms require accurate noting. For example, consider a single-phase d.c. power supply experiencing an absence of output power. Before checking the output of the power supply's transformer, the electrician would:

- Check that the main switch or circuit breaker is in the on position.
- Check that the power supply fuse is intact.

If the power supply has no functionality after the above checks, the secondary voltage of the power supply transformer should be the first in a series of checks.

Skill in fault diagnosis involves the identification of the faults that arise in a single-phase d.c. power supply, and a quick and accurate location of the cause of the fault. It is important to identify all of the signs and indications that can designate a fault and its cause, as well as implementing a suitable method of collecting this information.

Fault-finding d.c. power supplies

When fault-finding in the field, verify symptoms through direct observation of the faulty equipment if energised. Use your visual, sound and smell senses to locate the area where the fault may exist.

If available, check circuit diagrams of the power supply as these provide the tool by which the electrician can obtain information about its electrical process. Knowledge of how to read such diagrams is essential to anyone who is required to fault-find power supplies. There are two methods used in fault-finding. One relies on direct knowledge of the machine or electrical equipment and the other is a logical approach to locating the fault. The logical approach involves the gathering of information and then analysing this information to establish the most likely cause of the fault. Location of electrical faults uses logical, systematic analysis, supported by observation, measurements and testing.

Primary fault-finding techniques apply to every situation and occupation. Identification of the fault is essential to solving the electrical condition. Often the inexperienced electrician mistakes one or more of the symptoms for the problem. Solving the symptom normally just postpones the problem to a later date, by which time the problem may have caused further problems.

For example, a fuse in a power supply open circuits and the electrician directly obtains a replacement fuse and inserts it into the fuse holder. There are several factors that could cause the fuse to blow, conditional on the intricacy of the power supply circuit. Excess current caused the fuse to open circuit.

Excess current has causes such as overload on the load; short circuit between the output terminals, earth fault on the output. In addition, a short circuit in the supply's transformer, earth fault in the load, voltage spike, and so on, also cause excess current. If the electrician does not fault-find the power supply circuit prior to replacing the fuse and restoring power, further problems could develop.

It is not uncommon for a power supply to develop a number of small faults and continue to function at a reduced level of operational capability. Then, one last little fault occurs and the whole power supply fails. Finding and correcting the last fault will not necessarily restore the operational capability of the power supply. All the other electrical faults must be identified and corrected before the power supply is restored to full operational capability.

The following text examines some common faults in various single-phase d.c. power supply circuits.

Half-wave rectifier

Figure 10.47 depicts a half-wave filtered rectifier with an open-circuit diode. This fault will result in zero voltage at the output as shown. An open-circuit diode opens the current path from the secondary winding of the transformer to the filter and load resistor, resulting in zero load current.

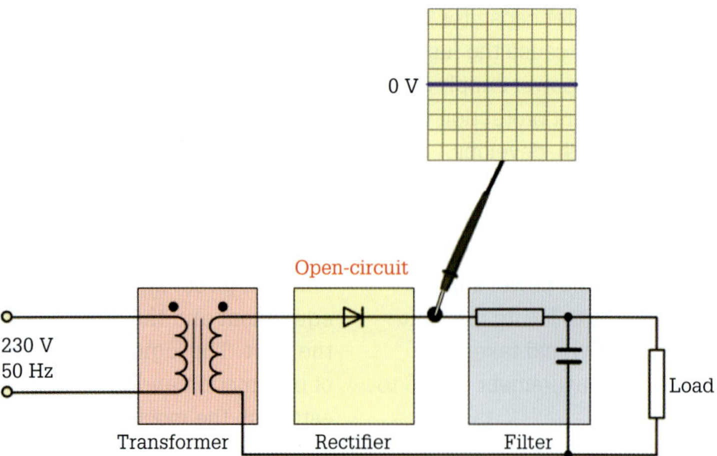

FIGURE 10.47 Effect of an open-circuit diode in a single-phase, half-wave rectifier

Other faults producing zero output voltages from a half-wave filtered rectifier are an open-circuit transformer winding, an open-circuit fuse, or absence of input supply voltage.

Centre-tapped rectifier

Figure 10.48 shows a single-phase, full-wave, centre-tapped, filtered rectifier. If one of the diodes develops an open circuit, the output voltage will have a larger than average ripple voltage at half the expected frequency (50 Hz rather than 100 Hz).

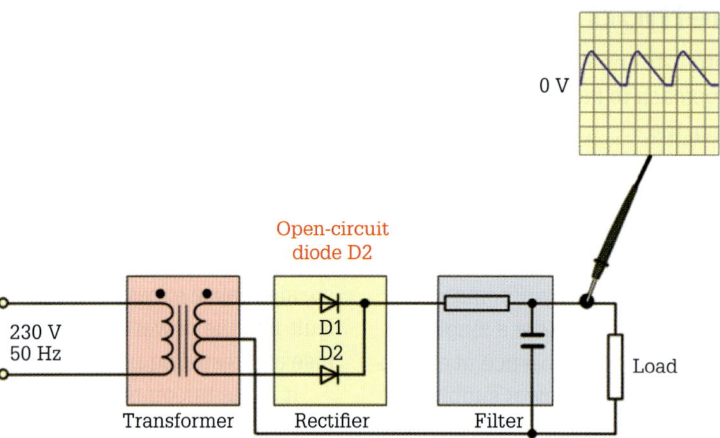

FIGURE 10.48 Effect of an open-circuit diode in a single-phase, centre-tapped rectifier

An open circuit in one-half of the transformer secondary winding will also result in this symptom.

An open circuit occurring in one of the diodes of **Figure 10.48** causes current to flow in R_L once every half-cycle of the input voltage. This current will flow through the operational diode. Current will not flow in R_L during the alternate half-cycle of the input voltage due to the open path caused by the open-circuit diode. This fault results in half-wave rectification, as shown in **Figure 10.48**, which produces the higher ripple voltage at half the expected frequency.

Bridge rectifier

An open circuit occurring in one of the diodes of a full-wave bridge rectifier as shown in **Figure 10.49** causes current to flow in R_L once every half-cycle of the input voltage. This current will flow through the operational diodes. Current will not flow in R_L during the alternate half-cycle of the input voltage due to the open path caused by the open-circuit diode. This fault results in half-wave rectification, as shown in **Figure 10.49**, which again produces the higher ripple voltage at half the expected frequency.

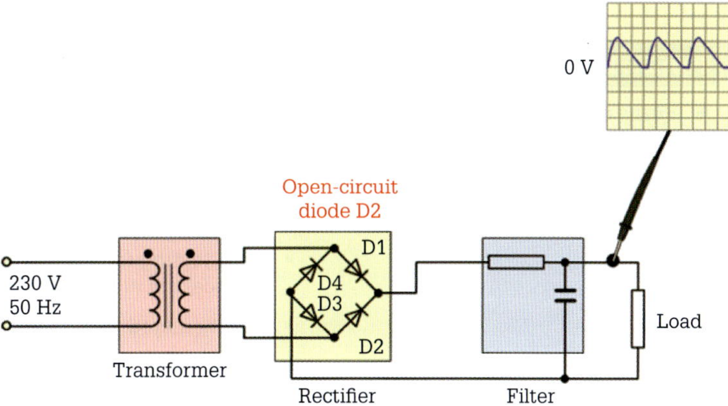

FIGURE 10.49 Effect of an open-circuit diode in a single-phase, bridge rectifier

Faulty transformer

An open circuit primary or secondary winding of a power supply transformer results in the power supply being unable to produce output power.

A partially shorted primary winding (less likely than an open-circuit winding) produces an increased rectifier output voltage due to the effective increase in turns ratio of the transformer resulting in higher secondary output voltage. A partially shorted secondary winding produces a decreased rectifier output voltage, due to the decrease in turns ratio of the transformer, resulting in lower secondary output voltage.

REVIEW QUESTIONS

1. What effect does an open-circuit diode have on the output voltage of a single-phase full-wave bridge rectifier?
2. What effect does an open-circuit diode have on the output voltage of a single-phase full-wave, centre-tapped rectifier?
3. What effect does an open-circuit primary or secondary winding of a power supply transformer have on the output voltage of a single-phase full-wave bridge rectifier?
4. Name two examples of a systematic approach to fault-finding.
5. What is an important test tool for detecting most of the faults that occur in d.c. power supplies?
6. What may be produced by loose terminal connections in d.c. power supplies?
7. What constitutes a faulty single-phase d.c. power supply?

10.8 Safe working practices

Risk assessment

Prior to engaging in fault-finding on a power supply, it is necessary to undertake a risk assessment. This assessment should:

1. Check the condition of PPE such as fire-rated safety clothing, safety footwear and safety glasses.
2. Check/secure area adjacent to work.
3. Check whether an assistant is needed.
4. Consider risks and hazards in the work area (see **Table 10.2**). Are relevant permits in place?
5. Be familiar with the replaced component before carrying out any work.
6. Check for availability of diagrams of the power supply.

On completion of a job:

1. Tidy up, remove and dispose of redundant materials.
2. Sign off any relevant permits.
3. Sign off the job on completion.

TABLE 10.2 Risk assessment

Possible hazards	Possible harm to people	Control measures to prevent harm or reduce hazards
• Eye injury • Electrocution • Electric shock • Explosion	• Burns or fatalities arising from contact with a live conductor. • Eye injury arising from explosion.	• All electrical fault-finding is to be carried out by qualified tradespeople. • Check faults in a planned and concise manner. • Isolate the supply as soon as is practically possible. • Start testing from the source of supply and work by a process of elimination. • Use insulated tools and 'in-service' test equipment.

Safety precautions when working on d.c. power supplies

When working on or testing d.c. power supplies, it is important to be careful. Whatever type of power supply you are testing, whether it be simple or complex, it is paramount to use safe work practices. Working with electrical equipment involves risks that should never be taken lightly. There are a number of safety procedures to follow in order to avoid personal injury, possible damage to equipment or danger of fire.

General safety

Before working on any electronics, consider following these basic safety precautions to help reduce any hazards.

- Disconnect and isolate the power supply on which you are testing or repairing from the power source. Never assume that this is the case. Test and test again with a suitable voltage presence indicator to confirm isolation.
- Remove fuses and replace them only after isolating power to the power supply.
- Do not reconnect power to a power supply until work is complete and checked.
- Ensure that all electronic equipment is properly earthed.
- Replace damaged components rather than attempting inappropriate repairs. As an example, replace damaged cables rather than repairing with insulating tape.
- Use the correct repair and maintenance tools.
- Reinstate covers after removing them to reduce the risk of electric shock.
- Have safety equipment such as a fire extinguisher, basic first aid kit and a mobile phone nearby.

Personal safety

It is important to ensure that you are safe when working on electronic circuits. Here are some personal safety precautions to keep in mind:

- Ensure that the work area is clean, dry, tidy and well ventilated.
- Do not wear loose clothing when working.
- Remove any metallic jewellery such as watches, rings and bracelets from your body.
- Do not handle hot components with bare hands.
- Wear non-conductive footwear when working on power supplies.
- Safely discharge capacitors by using an appropriate load – do not short the terminals.
- Use test equipment rated to Cat III as a minimum, this includes test leads.
- Wear eye protection.

Electric shock

One of the major hazards when working with d.c. power supplies is electric shock. To minimise the risk, you should follow a few safety precautions, including:

- Read any safety procedures that came with the equipment you are about to test or repair.
- Check all wires for solid connections.
- Ensure that all parts of the power supply are securely fixed to prevent accidents.
- Keep water and other liquids away from the power supply under repair or test.
- Check for signs of wear, defects and fraying on cables, cords and connectors.
- Wear insulating safety rubber gloves and footwear when testing and repairing power supplies.

Have a look at https://www.safeworkaustralia.gov.au/safety-topic/hazards/electrical-safety for relevant information.

REVIEW QUESTIONS

1. What should be undertaken prior to engaging in fault-finding on a power supply?
2. Name four possible hazards associated with working on a power supply.
3. What should be removed from the body prior to repairing a power supply?
4. What should be reinstated on a power supply being serviced to reduce the risk of electric shock?
5. When repairing a power supply, in what condition should the work area be?

CHAPTER REVIEW

10.1 Atomic structure

- An atom is the tiniest particle that an element can be condensed to and still possess the properties of that element.
- The atom contains three types of subatomic particles that concern electricity: protons, neutrons and electrons.
- Quantum mechanics or quantum theory concerns matter acting as both a particle and as a wave.

10.2 Power supply operating principles

- A d.c. power supply comprises a transformer, a rectifier, a filter and a voltage regulator.
- Power rectifiers use silicon diodes.
- The diode is a simple P–N junction semiconductor.
- The two terminals of the diode are the anode and the cathode.
- N-type material is formed by doping intrinsic semiconductor material with pentavalent atoms.
- P-type material is formed by doping intrinsic semiconductor material with trivalent atoms.
- A diode only allows current to flow through it in one direction.
- Between the P and N layers of a diode is a depletion layer.
- Before a diode can conduct it is necessary to exceed the barrier potential, which is 0.7 V for a silicon diode.

10.3 d.c. rectification circuits

- In half-wave rectification, only the positive half of the input waveform can pass through the rectifier while the negative half-cycle is blocked.
- $V_{d.c.} = V_{av} = V_{max} \times 0.318 = V_{rms} \times 0.45$
- The peak inverse voltage (PIV) for the diode of a half-wave rectifier equals the secondary maximum voltage.
- The average d.c. output voltage for a full-wave rectifier is the average value of one cycle of output.
- Full-wave rectification allows a pulsating unidirectional voltage to appear across a load during each cycle of the input waveform.
- $V_{d.c.} = V_{av} = V_{max} \times 0.637 = V_{rms} \times 0.9$
- The PIV for the diodes of a centre-tapped full-wave rectifier must be twice the V_{max} value.
- A bridge rectifier does not need a centre-tapped transformer, but it does need an extra two diodes.
- The average d.c. output voltage for a full-wave rectifier is the average value of one cycle of output.
- The PIV for the diodes of a full-wave rectifier equals the secondary maximum voltage.
- The d.c. output waveform of rectified direct current consists of a series of unidirectional pulses.

10.4 Filter circuits

- A circuit that adapts a pulsating d.c. output into a steady d.c. output similar to a battery is known as a filter.
- In a simple capacitor filter circuit, a single capacitor, usually an electrolytic one, is connected across the rectifier output and in parallel with the load to achieve the desired filtering.
- Capacitive filtering is enhanced when the output from the filter feeds into a voltage regulator.
- Inductive filtering is best suited to high-power applications.
- A low-pass filter passes low frequencies but attenuates high frequencies.
- There are two types of low-pass active filters: first-order filters and second-order filters.

10.5 Zener diode shunt regulator

- Voltage regulators are essential for maintaining constant load voltage.
- Voltage regulation is a ratio of the change in voltage between no load and full load.
- A Zener diode operates in the reverse-biased mode.
- A Zener diode shunt regulator comprises a Zener diode in parallel with the load and a resistor in series with the incoming supply.

10.6 Three-terminal IC voltage regulator

- Three-terminal voltage regulators are used to provide a fixed or variable level of d.c. operating voltage regardless of changes in the input voltage or changes in load conditions.
- The three terminals are input, ground (or common) and output.
- The dual-polarity power supply is fundamentally two power supplies in one.
- To suppress transient voltages, the power supply transformer is protected, in addition to over-current protection (fuse), by a voltage-dependent resistor (VDR) or varistor that is connected across the primary winding.

10.7 d.c. power supply testing and fault finding

- It is important to conduct a risk assessment prior to engaging in fault-finding processes.
- When fault-finding always use visual, sound, smell and heat senses where appropriate, observe timing of occurrences, and any departure from normal operation.
- Knowledge of how to read circuit diagrams is essential to anyone who is required to fault-find power supplies.
- It is not uncommon for a power supply to develop a number of small faults and continue to function at a reduced level of operational capability.
- The logic behind the half-split fault-finding approach is pure. With every step in the fault-finding process, the electrician eliminates half the functional blocks as the cause of the problem.

10.8 Safe working practices

- Prior to engaging in fault-finding on a power supply, it is necessary to undertake a risk assessment.
- Check faults in a planned and concise manner.
- Isolate the supply as soon as is practically possible.
- Start testing from the source of supply and work by a process of elimination.
- Use insulated tools and 'in-service' test equipment.

TRIAL EXAM

For Chapter 10 knowledge assessment, please complete the following trial exam.

1 N-type semiconductor material is produced by adding an impurity which has:
  a 2 valence electrons
  b 3 valence electrons
  c 4 valence electrons
  d 5 valence electrons

2 A diode only conducts when:
  a the barrier voltage increases
  b reverse biased
  c forward biased
  d the PIV falls

3 A silicon power diode used as a rectifier would have a typical forward (barrier) potential of:
  a 0.3 volts
  b 0.7 volts
  c 10 volts
  d 400 volts

4 The minimum number of diodes required for a single-phase half-wave rectifier is:
  a 1
  b 2
  c 3
  d 4

5 Converting a.c. into pulsating d.c. is called:
  a filtering
  b power suppling
  c voltage regulation
  d rectification

6 The minimum number of diodes required for a single-phase centre-tap rectifier is:
  a 1
  b 2
  c 3
  d 4

7 The minimum number of diodes required for a single-phase full-wave bridge rectifier is:
  a 2
  b 3
  c 4
  d 6

8 A simple filter consists of:
  a a Zener and a capacitor
  b a capacitor
  c a resistor and a regulator
  d a resistor

9 When a capacitor is used as a filter it is connected:
  a in parallel with the load
  b in series with the load
  c in parallel with the rectifier diodes
  d in parallel with the rectifier input

10 A voltage regulator is used to:
  a keep the load current constant
  b keep the load voltage constant
  c keep the supply voltage constant
  d keep the supply current constant

11 A device that is always connected in reverse bias is a:
  a Zener
  b capacitor
  c diode
  d resistor

12 Three-terminal voltage regulators are connected:
  a before the rectifier
  b after the load
  c between the filter and the load
  d between the rectifier and the filter

13 What component used in power supplies is suitable for filtering when supplying high-current loads?
  a Zener
  b diode
  c capacitor
  d inductor

14 The symbol for a Zener diode in Figure 10.50 is:

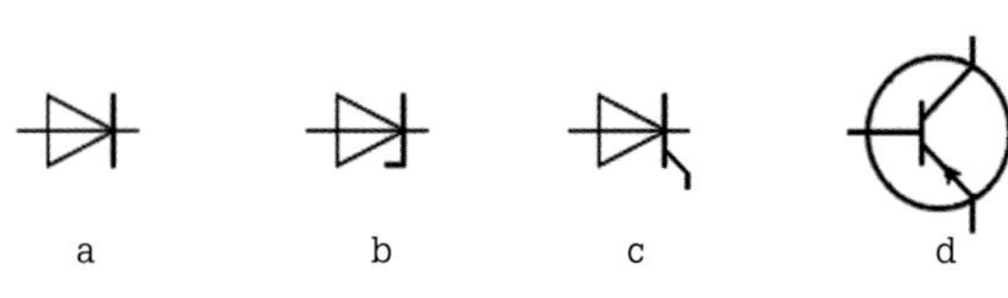

FIGURE 10.50 Problem 14

15 Which semiconductor device looks like a transistor but is a complex integrated circuit?
  a three-terminal regulator
  b triac
  c SCR
  d NPN transistor

16 A rectifier converts an a.c. voltage waveform to a:
  a sinusoidal waveform
  b pulsating unidirectional d.c. waveform
  c displacement waveform
  d lissajous waveform

17 The d.c. voltage output from a single-phase full-wave bridge rectifier is:
  a 1.414 of the rms voltage plus 1.4 V d.c.
  b 0.637 of the rms voltage plus 0.7 V d.c.
  c 0.45 of the rms input voltage minus 0.7 V d.c.
  d 0.9 of the rms input voltage minus 1.4 V d.c.

18 For the circuit shown in Figure 10.51 determine the d.c. load voltage and the load current.
  a 19.3 V and 1.07 A
  b 8.3 V and 461 mA
  c 27.58 V and 1.59 A
  d 11.7 V and 65 mA

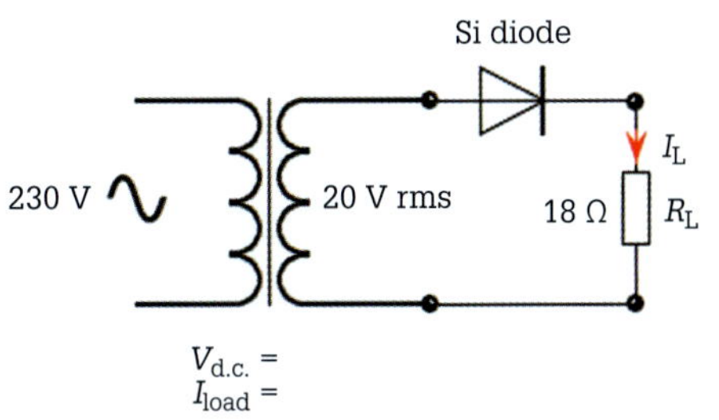

FIGURE 10.51 Circuit for Problem 18

19 The waveform in **Figure 10.52** represents:
- a single-phase full-wave rectification
- b single-phase half-wave rectification
- c three-phase full-wave rectification
- d three-phase half-wave rectification

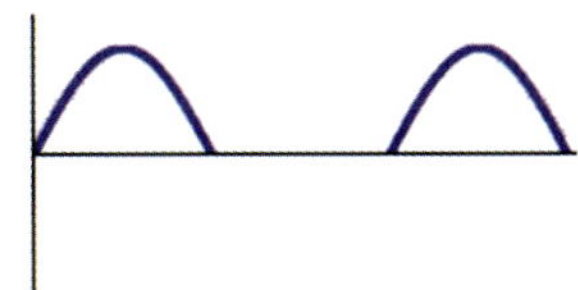

FIGURE 10.52 Waveform for Problem 19

20 For the circuit shown in **Figure 10.53** determine the d.c. load voltage and the load current.
- a 99.3 V and 2.483 A
- b 22.5 V and 562.5 mA
- c 44.3 V and 1.11 A
- d 89.3 V and 2.25 A

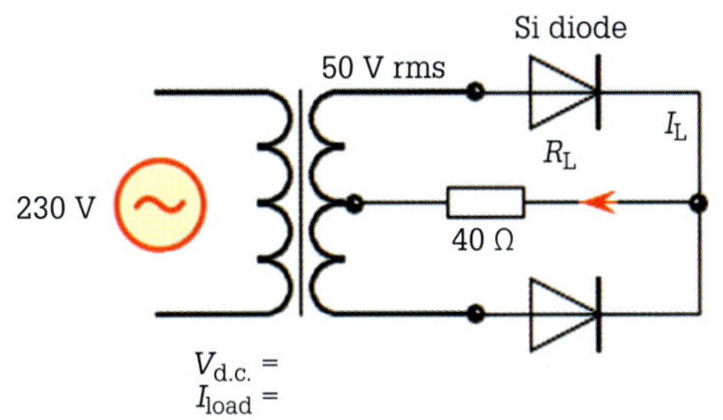

FIGURE 10.53 Circuit for Problem 20

21 The circuit shown in **Figure 10.54** is identified as a:
- a half-wave rectifier circuit
- b two-terminal regulator circuit
- c full-wave centre-tap rectifier circuit
- d full-wave bridge rectifier circuit

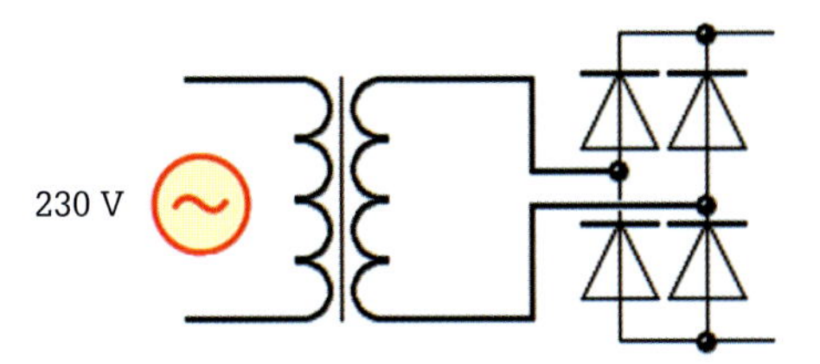

FIGURE 10.54 Circuit for Problem 21

22 For a single-phase half-wave rectifier, the ripple frequency is:
- a the same as the frequency of the supply
- b twice the frequency of the supply
- c half the frequency of the supply
- d d.c. has no frequency

23 The ripple voltage of a single-phase half-wave rectifier is equal to:
- a 0.45 × the d.c. output ($V_{d.c.}$)
- b 0.707 × the peak a.c. input (V_p)
- c 1.414 × the rms a.c. input ($V_{a.c.}$)
- d 0.9 × the d.c. output ($V_{d.c.}$)

24 If the output from a power supply has a higher than normal ripple voltage, the usual fault is:
- a an open-circuit filter capacitor
- b a short-circuit filter capacitor
- c an open-circuit rectifier diode
- d a short-circuit rectifier diode

25 If the output waveform from a full-wave rectifier is the same as that from a half-wave rectifier the most likely fault would be:
- a an open-circuit diode
- b a short-circuit diode
- c an open-circuit filter capacitor
- d a short-circuit filter capacitor

26 What should be undertaken prior to engaging in fault-finding on a power supply?
- a risk assessment
- b cost analysis
- c internet search for similar problems
- d a tool list audit

27 What should be reinstated on a power supply being serviced to reduce the risk of electric shock?
- a fuses
- b source power
- c covers
- d load bank

11 Apply environmentally sustainable procedures

This chapter provides electrotechnology workers with knowledge of environmentally sustainable procedures in the energy sector. In addition, it includes identifying and applying sustainable methods of work practice that minimise energy and material usage and applying energy reduction strategies in the energy sector workplace. This chapter provides underpinning knowledge for the unit UEERE0001 from the UEE training package.

LEARNING OBJECTIVES

Notions of sustainable energy

- Explain the notion of sustainable energy
- Describe the causes and consequences of the greenhouse effect
- Discuss sustainable work practice

Overview of sustainable energy technologies

- Describe geothermal energy
- Describe wave energy and tidal energy
- Describe energy storage technologies

Selection of energy-reducing control systems

- Describe an efficient and effective lighting system
- Name types of energy-efficient bulbs
- Outline how home automation is implemented for energy control and monitoring

11.1 Notions of sustainable energy

It is important to have an understanding of terminology used when discussing sustainability. The following section will define sustainable energy and its guiding principles.

Definition of sustainable energy

Sustainable energy has a minimal impact on the environment and health of all organisms on Earth. Sustainable energy does not harm the effective functioning of ecosystems and using sustainable energy sources and practices will provide existing and future generations with a continuing supply of energy.

Sources of sustainable energy include solar photovoltaic (PV), solar thermal, wind, hydrogen fuel cells, geothermal, small-scale (mini and pico) hydroelectric, tides and waves.

Principles of sustainable energy

Existing data published about sustainable energy articulates several principles that support the sustainable energy drive. These principles are:

- **Equitable access by all persons to sustainable energy resources and development:** to drastically reduce and ultimately eliminate dependence on unsustainable forms of energy.
- **Poverty eradication:** to provide sustainable energy resources to promote the development of poorer countries and populations.
- **Global security:** to provide consistent and reliable sources of energy.
- **Climate protection:** to reduce emissions of greenhouse gases.
- **Environmental and social protection:** to reduce air-, water-, and ground-borne pollutants, which can affect the health of people and the environment.
- **Technological innovation and dissemination:** to promote the accelerated development and dissemination of sustainable energy industries and practices.

The Earth's atmosphere

The atmosphere is composed of different gases. It surrounds the Earth and provides air pressure that protects all organisms by blocking or reducing harmful radiation from the sun and space. The atmosphere is densest at sea level and gradually becomes thinner with increasing altitude. The typical composition of air at sea level includes (by volume) the following gases:

- Nitrogen 78%
- Oxygen 21%
- Argon 0.9%
- Water vapour 0–4%
- Carbon dioxide 0.04%
- Methane 0.00017%
- Nitrous oxide 0.00003%
- Ozone 0.000004%

The concentration of water vapour in the atmosphere is highly variable both geographically and seasonally. Water vapour is a greenhouse gas that absorbs and radiates infrared/heat energy in the atmosphere and helps to maintain a surface temperature range that is suitable for living organisms. Water vapour can undergo phase changes, and form clouds, rain, sleet and snow, making it a significant influence on weather and climates.

Carbon dioxide is another important greenhouse gas responsible for regulating heat in the atmosphere. Carbon dioxide is also necessary for photosynthesis, which involves the conversion of carbon dioxide and water vapour into a sugar called glucose using energy from sunlight. Oxygen is produced as a waste product of this process. The main source of carbon dioxide emissions are from human activity, especially the burning of fossil fuels.

The greenhouse effect

The greenhouse effect is a natural process that assists in maintaining the Earth's temperature and the stability of its climate. The surface of the Earth absorbs radiation from the sun (visible light, infrared and ultraviolet) and re-emits that energy primarily as infrared radiation. When the infrared radiation encounters a greenhouse gas (GHG) in the atmosphere such as carbon dioxide, water vapour, methane or nitrous oxide, the gas absorbs the energy. The GHG then emits the absorbed infrared radiation, which may be absorbed by another GHG, the surface, or lost to space. This repeated absorption and emission process acts like a blanket to slow the loss of heat from the surface of the Earth into space, reducing the difference between day and night temperatures and maintaining a higher overall surface temperature. Refer to **Figure 11.1**.

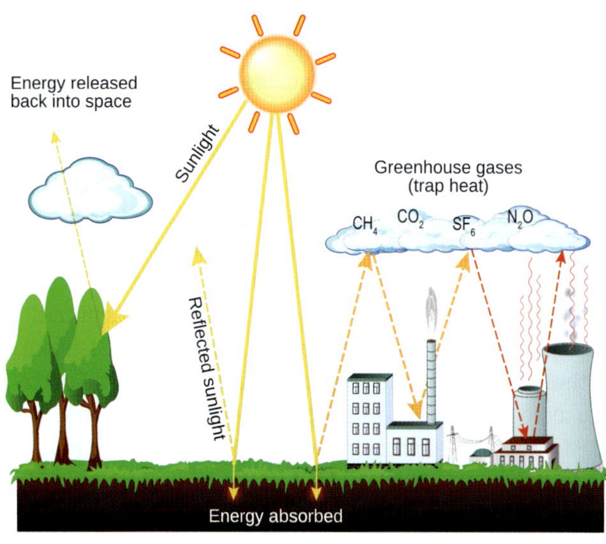

Source: Shutterstock.com/Designua

FIGURE 11.1 The greenhouse effect

In addition to the natural greenhouse effect, the Earth is also warmed by the enhanced greenhouse effect, which is a result of GHG emissions from human activities and is a significant contribution to human-induced climate change. Estimates indicate that Australia contributes approximately 1.15% of global GHG emissions (IEA, 2020). However, Australia's emissions per person (per capita) are among the highest in the world. Fossil fuels provide the vast majority of Australia's energy needs and our industries are energy intensive. The widely separated and sprawling cities around Australia also mean that transport use is high. To reduce the impact of climate change, Australia needs to reduce GHG emissions, especially carbon dioxide, and its reliance on fossil fuels.

Greenhouse gases

In the atmosphere, the main GHG that contributes to the greenhouse effect is water vapour. However, water vapour is not responsible for climate change due to the natural water cycle with the excess water vapour being easily removed through precipitation. For climate change, the main GHG in terms of importance and abundance is carbon dioxide (CO_2). Consequently, other GHGs can be compared in terms of their carbon dioxide equivalent (CO_2–e). The carbon dioxide equivalent of a GHG is found by multiplying the mass of the GHG by its global warming potential (GWP), which in turn is determined from the heat absorbed by that GHG relative to carbon dioxide over a particular length of time. By definition, carbon dioxide has a carbon dioxide equivalent and global warming potential of 1.

Carbon dioxide is the primary GHG contributing to climate change through fossil fuel burning, agriculture and other human activity. Almost half of the carbon dioxide emitted by humanity is absorbed by the oceans, which results in increased acidity of the oceans and affects marine life. Some additional carbon dioxide is removed by plants and soils; however, deforestation and land-use changes are leading to higher emissions. Approximately half of the carbon dioxide emitted remains in the atmosphere and may have a lifetime of hundreds to a thousand years before being absorbed and sequestered.

Methane is a potent GHG and the second most important GHG after carbon dioxide. Approximately half of the methane that is emitted is produced naturally from wetlands, forest fires, certain animals and oceans. The remainder of emitted methane is produced from human activity including the keeping of ruminant livestock, such as cattle and sheep; production and transportation of natural gas, oil and coal; decomposing organic matter in landfills; wastewater treatment; and some agricultural practices, such as rice farming. Methane is less persistent in the atmosphere than carbon dioxide, however, methane is a more potent GHG with a global warming potential of 84 over a 20-year timeframe and 28 over a 100-year timeframe.

The main sources of anthropogenic nitrous oxide emissions are from agricultural use of nitrogen fertilisers, soil disturbance, burning of grasslands and animal waste. Although the amount of nitrous oxide in the atmosphere is much lower than carbon dioxide and methane, its global warming potential of 265 over a 100-year timeframe is far higher, making nitrous oxide a very potent greenhouse gas. Nitrous oxide is also more persistent in the atmosphere with a lifetime of more than 100 years.

Ozone contains three atoms of oxygen (O_3) and in the ozone layer is responsible for absorbing a significant amount of UV radiation from the sun, especially UV-C and UV-B. Ozone is also a minor GHG due to its absorption of some infrared radiation. While the presence of ozone in the ozone layer is important for the survival of life, ozone closer to the Earth's surface is a harmful pollutant. Apart from the natural production of ozone from lightning, ground level ozone is not directly emitted by human activities but instead arises from reactions between other atmospheric pollutants released by human activities.

Although not a gas, aerosols/particulates in the atmosphere affect the greenhouse effect. Aerosols are very fine particles that are produced from diverse sources including bushfires, emissions from volcanoes, and urban and industrial pollution. Aerosols affect the sun's light as it passes through the air. The effects include scattering (which causes cooling) and absorption (which causes warming). While the presence of aerosols in the atmosphere are reducing the impact of the greenhouse effect from GHGs, aerosols can be a significant hazard to human health. The release of aerosols is an issue that affects all countries because of the natural movement of the atmosphere. Some aerosols in the atmosphere have been shown to aggravate asthma and other lung diseases and contribute to millions of premature deaths worldwide.

There are also a number of synthetic greenhouse gases in the atmosphere created from industrial processes, such as hydrofluorocarbons, perfluorocarbons and sulphur hexafluoride gases.

Contributing factors to the emission of greenhouse gases

Contributing factors to the excess emission of GHGs from human activities include:

- fossil fuel burning, especially for electricity generation, transportation and heating
- industrial emissions
- deforestation/land-use change
- livestock and other agricultural practices, such as the production and use of fertilisers
- waste disposal
- emission of other synthetic GHGs.

Factors such as increasing world population, economic growth and technological development all affect energy usage and the production of GHGs. GHG emissions from the existing electricity supply can be substantially reduced and Australia can do this by utilising renewable energy that has low GHG intensity.

Measures to reduce the emission of greenhouse gases

Everyday measures to reduce GHG emissions are intended to reduce energy and resource consumption. Such measures include:

- saving energy (using energy efficient lighting, reducing heater and air conditioner use, green energy suppliers)
- using less water (having a shower rather than a bath, using hot water with a lower temperature)
- reducing waste (recycling and reusing products, using less energy intensive products)
- reducing meat consumption and food wastage

Significant measures at a state or national level include:

- emissions trading – decreases GHG emissions by financially incentivising polluters to reduce emissions
- regulatory approaches – use of performance standards and planning approvals to guide industries towards lower GHG emissions
- investment – funding of projects, technology and research that reduce GHG emissions.

Sustainable work practice

A definition of sustainability can be found in the 1987 United Nations 'Report of the World Commission on Environment and Development' (the *Brundtland Report*). Sustainability is the ability to 'ensure that development meets the needs of the present without compromising the ability of future generations to meet their own needs' (UN, 1987).

According to this report, sustainability has both environmental and social dimensions. Consequently, sustainability is a concept that integrates environmental management, social justice and economic growth.

Definition of sustainable work practice

Sustainable work practices are human activities that sustain the use of raw materials and regenerate or recycle natural resources. Sustainable work practices should improve the skill and employability of workers by establishing work policies that give priority to time, qualifications, task development, working conditions and variable work hours, rather than a monetary salary or wage.

Sustainable work practice is a conscious, empowering, communally organised, needs-centred activity that encompasses social and environmental factors.

For an individual worker, sustainable work practices are those that target personal responsibility:

- understanding, and compliance with, environmental standards, policies and regulation requirements together with knowledge of the environmental impact of actions
- establishing personal and workplace competence in resource use, waste reduction and improvement of work processes to reduce their environmental impact.

Practical application of sustainable work practices for society should:

- reduce GHG emissions (fossil fuels play a dominant role in Australia's primary energy consumption)
- develop better renewable energy systems and transport systems
- engage in sustainable use of land
- facilitate economic development so it can proceed in a sustainable way
- encourage sustainable forestry and vegetation management by planting species suitable for the long term and fast-growing species for human use
- engage in biodiversity by implementing strategies to protect the variety of life under present and future climate conditions
- make certain that food production is not vulnerable by identifying new crops or crop cultivars suitable for use as the climate changes
- develop suitable response actions for coastal management under climate change and associated sea level rise
- identify water management strategies for water resources so that they remain suitable for future use.

Effects of neglecting sustainable work practice

Neglect can be defined as any passive or active omission by humanity that constitutes a failure to ensure that the fundamental needs and protection of ecosystems in their care are adequately and appropriately met.

Viewing incidents of neglect (oil spills, forest destruction, land degradation, GHG emissions) as singular events without probable co-occurrence may lead to the belief that the neglect experienced by the environment is 'low risk', deserving 'no further action'. Assessment of the accumulated harm of incidents of neglect enables early and ongoing intervention to reduce the risk of harm to the environment in the future.

Neglect is always unending in the sense that it is a cumulative force (accumulated harm) and has a significant impact on the environment. Neglecting sustainable work practices results in the entrenchment of poverty, worldwide injustice and environmental damage.

People must start to live sustainably in order to hand on a liveable Earth to future generations. The extraordinary rate of technological innovation has made it possible to take advantage of the supply of natural capital such as living systems, minerals, oil and water. Humankind uses natural capital in its industrial and economic systems to provide the resources and services that humanity depends on.

International and national greenhouse imperatives

The United Nations Framework Convention on Climate Change (UNFCCC or FCCC) is an international environmental treaty produced at the United Nations Conference on Environment and Development held in Brazil in 1992. The treaty became formalised with the goal of achieving stabilisation of GHG concentrations in the atmosphere at a level that would minimise anthropogenic contribution to dangerous climate change.

The UNFCCC (1992) notes 'the largest share of historical and current global emissions has originated in developed countries' and that 'Parties should protect the climate system … on the basis of equity and in accordance with their common but differentiated responsibilities and respective capabilities. Accordingly, the developed countries should take the lead in combating climate change.'

Kyoto Protocol

The Kyoto Protocol was an international agreement that arose from the UNFCCC, which sets goals for developed countries to reduce their GHG emissions of carbon dioxide and five other GHGs: methane (CH_4), nitrous oxide (N_2O), hydrofluorocarbons (HFCs), perfluorocarbons (PFCs) and sulphur hexafluoride (SF_6). Developed countries were able to participate in emissions trading with other countries if they were unable to sufficiently reduce emissions of these gases. The treaty was adopted in Kyoto, Japan, in December 1997.

Many countries had no limitations under the Kyoto Protocol because it was decided that developed countries, due to their carbon-based industries, accounted for the majority of the carbon dioxide build-up in the atmosphere that has occurred since 1880. Additionally, developing countries are least able to implement strategies to reduce GHG emissions and mitigate the effects of climate change. The UNFCCC put forward the principle that developed countries should provide capital and technology to developing countries to manage climate change.

In 2008, Australia contributed 1.32% of the world's GHGs. While this appears small compared to China (23%), the USA (18%) and the European Union (14%), this value does not account for different population sizes. If you look at per capita emissions for the same period, Australians were the world's worst emitters of GHGs, emitting about 27–28 tonnes per person. By contrast, the global average is about 6 tonnes per person, with an average of about 14 tonnes per person in other developed countries.

Under the Kyoto Protocol, countries were required to determine their GHG emissions in a particular year, commonly 1990, to be used as a baseline for emissions reduction. There were two commitment periods, 2008–2012 and 2013–2020, with different emissions reduction targets for each period. Countries that ratified the protocol were required to show through documentation that their emissions have not exceeded their target in the relevant commitment period. As one example, the Australian Government's Greenhouse and Energy Data Officer (GEDO) publishes greenhouse gas emissions and energy consumption data received from registered corporations under the *National Greenhouse and Energy Reporting Act 2007*.

Australia ratified the Kyoto Protocol in November 2007, and the agreement took effect in March 2008. Emissions were aggregated from the following sectors: stationary energy (electricity generation and petroleum refining) and transport; fugitive emissions (from the extraction, transportation and handling of fossil fuels); industrial processes (metals, minerals, chemicals, pulp and paper, and food and beverages); waste (solid waste disposal on land) and agriculture.

For Australia, annual GHG emissions in the period 2008–2012 had to be less than 8% above 1990 levels. Over this period, emissions were on average 3% above 1990 levels each year. For the second commitment period of 2013–2020, the Australian Government pledged to cut emissions by at least 5% compared to 2000 levels or 0.5% compared to 1990 levels (Keywood MD, Emmerson KM, Hibberd MF, 2016). As part of Australia's efforts to reduce GHG emissions, a carbon pricing mechanism was implemented between 2012 and 2014 under the *Clean Energy Act 2011* and covered approximately 60% of annual emissions (Commonwealth of Australia (Clean Energy Regulator), 2015). While the scheme had some impact on emissions reduction, it was repealed in 2014.

Paris Agreement

The Paris Agreement is a successor to the Kyoto Protocol for climate action after 2020 that was adopted in Paris, France, in December 2015 at the Conference of the Parties (COP21) as part of the UNFCCC. The primary aim of the Paris Agreement is to limit global warming to well below 2 °C above pre-industrial levels and preferably keep global warming below 1.5 °C. Countries would also aim to reach peak GHG emissions as soon as possible (United Nations, 2015).

The Paris Agreement established targets for several outcomes including national mitigation of and adaptation to climate change and supporting commitments to help developing countries implement the agreement, which will be reviewed every five years. This agreement is underpinned by a rules-based system that assesses whether countries are meeting their targets.

Australia supports the Paris Agreement, which it ratified on 9 November 2016. Australia's target under the Paris Agreement is to reduce emissions by 26–28% below 2005 levels by 2030 (Department of the Environment and Energy, n.d.). This emissions reduction target, which was determined in 2015 and reaffirmed in 2020, is lower than many other developed countries. Additionally, Australia has yet to commit to any targets beyond 2030.

In the period 2019–2020, seven of the top ten GHG emitters in Australia were electricity generators, mostly using coal-fired generation systems. The following three top GHG emitters were oil and gas producers. These top ten GHG emitters accounted for almost half of all direct emissions reported under the *National Greenhouse and Energy Reporting Act 2007*. Following on from these ten are more electricity generators, oil and gas producers, mining corporations, and metal smelting and refining corporations (Commonwealth of Australia (Clean Energy Regulator), 2021).

When undertaking any emissions reduction or climate action, concerns are often raised regarding the impact such actions will have on economic growth, jobs and international trade. While such concerns are important, it is inevitable that all countries will need to transition to a low-carbon economy, and countries should weigh the benefits and costs of climate action and the increasing effects of climate change on the economy and society.

Economic benefits of sustainable initiatives

In the developed world, a sustainability change is taking place – from a high-carbon, high-pollution, energy wasteful, and ecologically damaging economy, to a clean economy that is low-carbon, low-pollution, energy efficient, and ecologically conscious. The economic benefits of the clean economy focus on six areas: environment, people, resources, energy, water and waste.

Areas where sustainable initiatives are emerging for the clean economy:

- **Agriculture:** land management, forestry practices, carbon sinks, natural pesticides, phosphate-free soil enrichment, reduction in source water use, local food initiatives
- **Air:** emission control, monitoring and compliance
- **Alternative energy development:** wind, solar, hydro, wave, biofuels, geothermal, battery, nuclear
- **Education outcomes:** training, research, relevant skilling
- **Energy efficiency:** lighting, building design, insulation
- **Energy infrastructure:** management, transmission, communication
- **Transportation:** vehicles, fuels
- **Water and grey water:** rainwater harvesting, desalination treatment, water conservation, water management and wastewater treatment
- **Manufacturing and industrial:** efficiency, low energy impact, packaging, production and industrial ecology
- **Recycling and waste:** recycling and waste treatment.

As sustainable enterprises develop and interact with each other as producers, distributors and customers, a clean economy can develop over time, providing redesigned forms of employment and business opportunities. Sustainability solutions that unite improved environmental performance and economic benefits are the means to successful climate action strategies.

REVIEW QUESTIONS

1. State the six principles supporting the drive for sustainable energy.
2. Name the four most abundant gases that comprise the Earth's atmosphere.
3. What is the greenhouse effect?
4. Name three gases that are contributing to climate change.
5. Which GHG is able to readily undergo phase changes at the Earth's surface?
6. To which gas are other GHGs compared with respect to their warming effect?
7. What effect is the emission of carbon dioxide having on the oceans?
8. Name two activities that result in nitrous oxide emissions into the atmosphere.
9. State the purpose of ozone in the ozone layer.
10. Why are aerosols an issue that concerns all countries?
11. State two everyday measures to reduce GHG emissions and energy usage.
12. Describe one major measure to reduce GHG emissions and energy usage.
13. List five contributing factors to the emission of GHGs.
14. Provide a definition for 'sustainability'.
15. Define 'sustainable work practices'.
16. Why did many countries not have a GHG emission limit under the Kyoto Protocol?
17. What is the primary aim of the Paris Agreement on climate change?

11.2 Overview of sustainable energy technologies

Electricity will be increasingly generated from renewable sources as the world transitions to a low-carbon economy and away from fossil fuels. In many regions, solar (photovoltaic) and wind (on-shore and off-shore) will likely contribute to the majority of renewable electricity generation. However, other sustainable energy technologies, such as geothermal, wave and tidal energy, may increase the diversity of electricity generation sources where the appropriate renewable sources exist. Another area of sustainable energy technology is energy storage, such as batteries, pumped hydro and hydrogen gas, which is expected to grow as the proportion of renewable energy in electricity generation increases.

Geothermal energy

Geothermal energy is heat that originates within the Earth. This heat comes mainly from the radioactive decay of elements such as thorium, uranium and potassium within the Earth. Due to the size of the Earth and long half-lives of these radioactive elements, geothermal energy is considered a renewable source of energy. This heat can be used to warm building interiors or generate electricity. Geothermal energy has an output profile similar to non-renewable sources such as coal and oil as it produces base-load electricity that is manageable, predictable and dependable.

In one type of geothermal power plant, water released from aquifers at high temperatures converts liquid isopentane into a high-pressure gas, which in turn drives a turbine to generate power (see **Figure 11.2**). A geothermal power station rated at 120 kW is located in Birdsville on the Great Artesian Basin in Queensland.

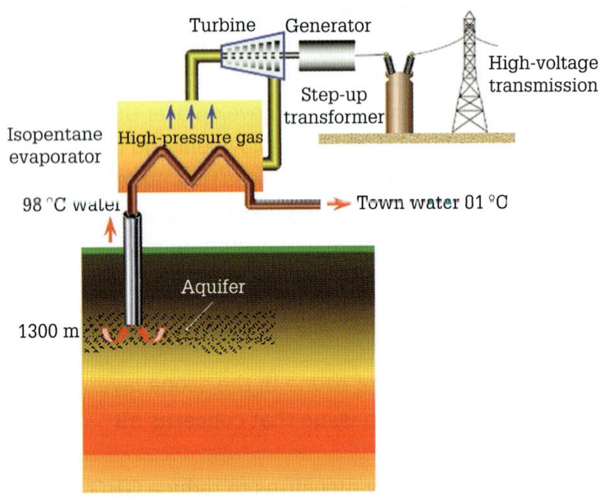

FIGURE 11.2 Geothermal energy from an aquifer

Note that in various locations around the world, steam produced underground is used to directly drive the turbines.

Hot-dry-rock technology, illustrated in **Figure 11.3**, has been developed to harness the geothermal energy found in zones of hot, dry rock, particularly granite. For example, water entering into the zone is heated and returns to the surface as hot water or steam and is used to generate electricity.

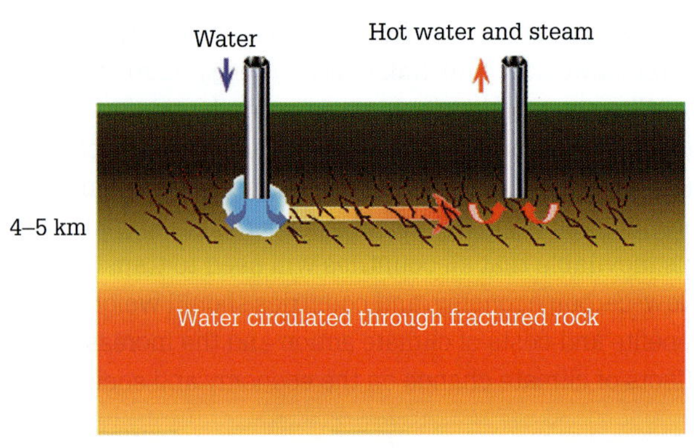

FIGURE 11.3 Hot-dry-rock energy conversion processes

Disadvantages of geothermal energy

Disadvantages of geothermal energy include the following:

- In general, geothermal energy is renewable; however, individual sites could be exhausted of geothermal heat for significant periods.
- Hazardous minerals may be produced which are difficult to dispose of safely.
- The zone of fractured rock may calcify due to salts in the water or experience structural collapse, thereby making the site unproductive.
- Long-term use of the water could damage underground aquifers.

Wave energy and tidal energy

Wave energy (**Figure 11.4**) and tidal energy (**Figure 11.5**) are probably least likely to be significant contributors of base energy in Australia because Australia does not have many suitable sites. Tidal energy possibilities in Australia

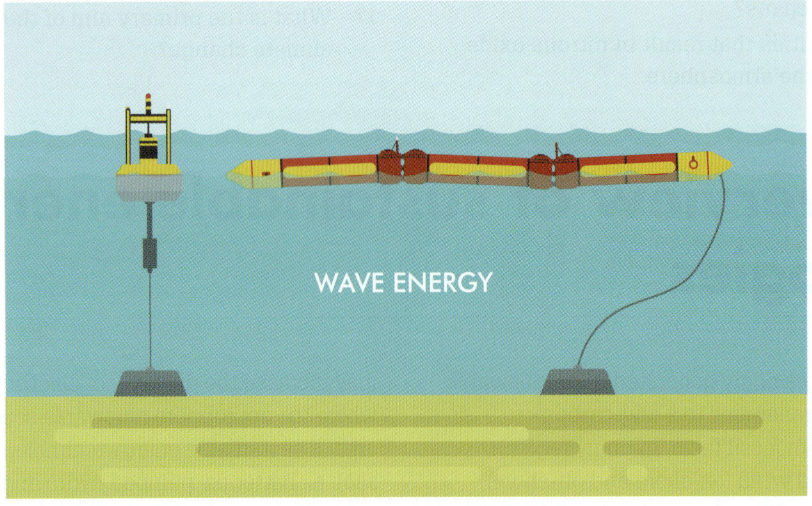

Source: Shutterstock.com/Oleksandr Derevianko

FIGURE 11.4 Wave generator

are best in northern Western Australia and are fair in Bass Strait, Queensland and South Australia. The possibility of wave energy exists on the westerly exposed coasts in Tasmania, South Australia, Victoria and Western Australia.

Source: Shutterstock.com/Alex Mit

FIGURE 11.5 Tidal generator

Tidal energy

Nearly all of Australia's potentially useful tidal energy sources occur in the Kimberley district in the north-west of Western Australia, where tidal ranges vary from 7 m to 12 m.

The first type of tidal energy is the kinetic energy in the water current between the incoming and outgoing tide (in and out movement). The second type is the potential energy in the rising and falling of the water between high and low tides (up and down movement).

Capture of the kinetic energy of the oceanic currents or river currents occurs when water passes through the turbine intake to generate electricity. Tidal turbines, as shown in **Figure 11.5**, function well where tides run at 2–2.5 m/s.

To capture potential energy, a dam called a barrage is built across estuaries. Consequently, when the tide ebbs and flows, water flows through tunnels in the barrage to drive a bank of turbines. The amount of electricity created depends on the head of the tide and the volume of water moving through the barrage tunnels. To be an efficient tidal energy source, the head (the height between the high and low tides) must be more than 4 m. A small head means that vast volumes of water must flow through the turbines to generate significant amounts of electricity. Tidal-generated coastal currents are predictable – the tide comes in and out every 12 hours. Tides develop maximum velocity four times every day. It is believed that 30 per cent of the total tidal energy can be extracted without harmful environmental effects.

Disadvantages of tidal energy

The reduced flushing action of the water system could affect water quality, sediment movement and shoreline vegetation, such as mangroves. Marine animals could have their movements restricted, possibly affecting their eating and mating habits and their nurseries. Bird life that feeds and depends on marine species, and the environment, could be endangered. Furthermore, trying to exclude fish from the turbines is a problem.

Since the electricity developed by tidal power plants is cyclical and varies during the day (two high tides and two low tides), the power generated cannot easily be used as base-load electricity. However, tidal turbines if situated around the Australian coastline could smooth out the peaks and troughs in base generation.

Wave power

Salter's duck is a device that generates electricity by bobbing up and down with the waves. Wave motion turns the duck's generator in a constant direction thereby generating electricity. Other devices can also produce usable energy from sea swell.

Energy storage

Electricity generated from solar and wind varies, depending on the weather conditions and time of day. This means there will be times when there is excess electricity being generated and times where generation is insufficient for the current demand. As renewable energy sources supply a greater proportion of electricity generation, storage of excess energy for later use will become increasingly critical.

Batteries have been receiving considerable attention as a storage option with the growing development of renewable energy projects. Several large-scale battery projects were announced in 2020 and more than 595 MW of capacity is under construction across Australia. Smaller batteries are also increasingly being installed in residential and commercial buildings for storing excess electricity generated from solar panels. While most batteries being constructed or considered use lithium chemistries, other types are also being developed with different properties, especially for stationary power applications.

Another type of large-scale energy storage being considered is pumped hydro, which operates in the exact opposite manner to a conventional hydroelectric plant by using electricity to pump water into a higher water storage reservoir. When the stored potential energy is needed, the water is allowed to flow back down through turbines to generate electricity like a conventional hydroelectric plant. Pumped hydro sites require particular geography and so will often utilise existing hydroelectric sites, such as the 2000 MW/350 000 MWh pumped hydro project being constructed at the Snowy Mountains hydroelectric complex (Clean Energy Council, n.d.).

Excess electricity could also be used to produce hydrogen gas. The hydrogen could be stored on site, transported for use in Australia or exported to international markets. When energy is required, the hydrogen gas can be combusted or used in a fuel cell. Additionally, the hydrogen gas could be used in industrial processes as a green source of hydrogen if it was produced from renewable energy.

REVIEW QUESTIONS

1 What is geothermal energy?
2 Describe hot-dry-rock technology.
3 Describe an advantage of geothermal energy.
4 What two renewable forms of energy are probably least likely to be major contributors of base energy in Australia?
5 What type of renewable energy uses a barrage?
6 State a disadvantage of tidal energy.
7 Why is energy storage becoming increasingly important for renewable energy?
8 Describe the basic principle of pumped hydro as a way to store energy.

11.3 Selection of energy-reducing control systems

Lighting is an area where significant savings in greenhouse gas emissions can be made and this starts with how we control building illumination.

Basic lighting systems

Lighting systems produce around 12% of GHG emissions from the domestic sector and around 25% of emissions from the commercial sector. Therefore, lighting is an area where emission reduction and energy savings can be made by phasing out inefficient lighting systems and replacing them with energy-efficient alternatives.

In simple terms, every lighting system consists of three elements: a lamp, a luminaire fitting and a control system. Each element is important and offers opportunities for improved performance, energy savings and the reduction of emissions.

An efficient and effective lighting system should follow these rules:

- make use of natural light
- provide a high level of visual comfort and prevent glare
- provide the correct luminous efficacy for the task
- provide controls for reduced energy usage
- use light-emitting diode systems as the primary light source
- use LED systems for commercial high-bay lighting applications
- use high-intensity discharge (HID) lighting systems wherever an intense point source of light is required, particularly pulse-start metal halide. Ceramic metal halide lighting systems are a good choice in applications where colour quality is critical.

Note: Efficacy is the efficiency of the lamp at converting electrical energy into light, measured in lumens per watt (lighting output divided by its power consumption). Note that a lumen is a measure of the total amount of light that a lamp emits.

Types of energy-efficient bulbs

Types of energy-efficient bulbs include linear fluorescent lamps, compact fluorescent lamps (CFLs) and LEDs. All these bulbs use less energy to produce light than incandescent lamps. Using less energy means reductions in peak demand, tariff charges and GHG emissions of a lighting system.

Compact fluorescent lamps

The CFL as illustrated in **Figure 11.6** is a miniaturised linear fluorescent tube with an integral ballast (either iron or electronic) in a standard Edison, Swan or pin base (requires a dedicated luminaire). CFLs have largely replaced incandescent lamps but are themselves being replaced by LEDs.

FIGURE 11.6 Compact fluorescent lamps

Self-ballasted (the purpose of the ballast is to produce high-voltage pulses to start the lamp) CFLs combine a lamp, electronic ballast and base in a single sealed unit. The entire unit is discarded when the lamp or ballast burns out. The electronic ballast used with most CFLs raises the frequency of the current going to the lamp, which eliminates fluorescent lamp flicker.

Some disadvantages and advantages of CFLs are shown in **Table 11.1**.

TABLE 11.1 The disadvantages and advantages of CFLs

Disadvantages	Advantages
Elongated or circular shaped CFLs may result in a reduced optimal lighting pattern.	They are rated from 10 000 to 12 000 hours of service life.
They are sensitive to frequent on and off cycling.	They provide excellent colour rendering by the use of rare-earth phosphors that produce bright, vibrant colours.

»

TABLE 11.1 (Continued)

Disadvantages	Advantages
They contain small amounts of mercury (≤ 5 mg per lamp). This metal may be released if the bulb breaks or during disposal.	Most CFLs have reliable starting that is flicker-free and instant. Depending on the wattage, CFLs have a lumen range varying between 450 lm and 2780 lm.

Light-emitting diodes

LEDs, as illustrated in **Figure 11.7**, are solid-state electronic devices that produce light. LEDs are naturally narrow-band sources, and the colour of the light produced depends on the construction materials. LEDs have the ability to vary the colour, create sparkle and aim the light precisely. They offer several advantages over primary light sources, including long life, low power consumption, low heat generation, no infrared or ultraviolet radiation, as well as being shock and vibration resistant.

FIGURE 11.7 Light-emitting diodes (LEDs)

LED lighting finds application in the domestic, industrial and commercial fields. For example, in the signage industry, retail accent lighting (due to their small size and the directional nature of the light output), traffic signals and exit signs. The advantages of LEDs include:

- 230 V a.c. LED bulbs are available (1 watt Cree XR-E LEDs provides 270 lm light output for the cool white condition – 6000 K)
- long life of over 100 000 hours
- practically no maintenance or routine replacement
- more environmentally friendly than other lighting options – no disposal of glass or toxic substances
- brightness and directional control
- very wide operating ambient temperature range.

Lighting control

Since the simplest way to reduce energy consumption by lighting systems – turning lights off when not needed, is also the most unreliable, other methods to modulate the output of lighting systems also exist. Other manual-type energy-saving controls include timed switches (clock timed) and delayed time switches (pneumatic).

A more sophisticated form of energy usage control is the energy management system (EMS). An EMS performs the same function as a timed switch but has additional features (microprocessor based). A typical EMS function is a sweep mode that automatically cycles lights on or off, one section or floor at a time, signalling to occupants the shutting down of the lights. However, occupants can override the shutdown in their area by operating a local switch or by phoning in a code to the EMS.

The most effective variety of automatic lighting control bases its operation on whether the occupant is present to make use of the illumination. Known as occupancy sensors, these sensors are very useful in reducing lighting energy consumption. The two most common types are passive infrared sensors (PIR), which require a direct line of sight to the moving, warm occupant (changes in infrared heat levels are detected); and ultrasonic sensors (these listen for a change in the frequency of reflected sound), which detect any movement, occupant or otherwise (curtains, blinds, air-conditioner cycling). **Figure 11.8** shows a PIR and an ultrasonic sensor.

FIGURE 11.8 Passive infrared sensor and an ultrasonic sensor

Home automation (smart home)

It was estimated that in 2015 the average Australian household had nine connected devices. Three years later this increased to 17 and in 2022 the prediction puts it at 37. Many of these devices will form part of what is referred to as the internet of things or IoT. The IoT allows for everyday household appliances to be made 'smart' through internet connectivity or by connecting to other smart devices via the home wi-fi router. An automated or smart home can have just a couple of connected devices or many connected devices.

In the early days of this technology, IoT devices were limited to basic commands such as turning on and off, setting temperatures, alarms and the like. The technology was further hampered by only being available in a few proprietary products. This meant it was necessary to interface with each device through different apps, which for many did not work with both Apple iOS and Android.

Currently, almost any appliance made by a big name brand has some form of built-in 'smart' technology. This, combined with the proliferation of smartphones, tablets and smart home assistants, which are used to control these products, has made IoT devices more accessible to the average person. This has culminated in making a smart home largely plug and play.

Essentially, there are three main ways to use most smart home devices:

1. apps
2. smart assistants
3. automation.

Apps

While there are many and varied apps, most products support both iOS and Android devices. Most manufacturers use one app to control their range of appliances but these proprietary apps do not play nicely with appliances from other brands. The Google Home app is an exception that provides control of multiple devices, albeit with a reduced suite of features.

Smart assistants

Smart home assistants such as Alexa, Google Assistant and Siri can form the central hub of an automated home. **Figure 11.9** shows two popular home assistants. They make it possible to control many devices from one location using simple voice commands. Android phones have Google Assistant pre-loaded and iPhones have Siri. Once set up, controlling a device is as simple as speaking a command such as, 'Alexa, play Jeff's playlist from Spotify' or 'Hey Google, dim the lights in the theatre'.

Automation

Some apps and devices can communicate with each other without human interaction. As an example, it is possible to set the coffee maker to start making coffee and illuminating the kitchen when the morning alarm activates. Another possibility is to have the air conditioning activate when your smartwatch gets within 2 km of home.

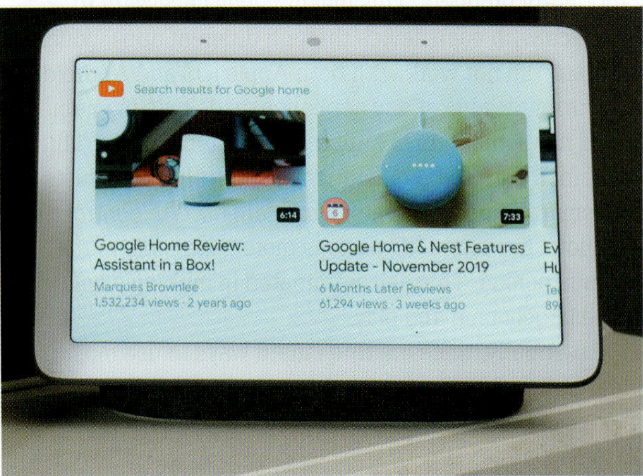

Source: Shutterstock.com/Juan Ci; Shutterstock.com/Vantage_DS

FIGURE 11.9 Smart assistants

Benefits of home automation

Home automation finds use in:

- **Lighting** – smart lights are able to be turned on and off, dimmed, or even change colour through an app or smart assistant. With these, it is also possible to simulate an occupied house when you are away, or create pre-set configurations such as preferred lighting conditions when watching movies.
- **General appliances** – appliances such as refrigerators, ovens and washing machines have smart functionality such as alerting your phone when the washing is finished or when the roast is cooked. Smart refrigerators have a touchscreen on the front that connects to the internet to facilitate online shopping. Smart switches allow devices to be turned on or off using simple voice commands or from a smartphone.
- **Entertainment** – music and television-related smart home devices facilitate the grouping of smart speakers and wi-fi speakers, which allows for being able to select and play music, pause, stop or skip tracks via a smartphone app or through voice commands.
- **Home security** – smart cameras connect to a wi-fi network and are controlled through an app or smart assistant. When combined with a smart lock, it is possible to view who is at the front door and unlock it without leaving your chair. This also makes it possible to give someone access when you are away from home – use an app to see who is requesting access and unlock the door.
- **Sustainability** – some devices make it possible to monitor water and electricity usage.
- **Accessibility** – home automation and smart home assistants have enabled less-abled and less mobile people to use voice commands to control a range of devices – from control of the television to unlocking the front door to let a visitor in without needing to get up.

Privacy concerns

While home automation and smart home assistants are an exciting prospect, they do open up one's life to one or more companies in order to take advantage of the technology. Unfortunately, one's privacy and IoT devices do not go hand in hand, particularly when these 'smart' devices start to learn your daily habits. Having a smart home generally means there is a database somewhere in the ether that is collecting and collating all your commands, habits and interests, which is a prospect that may not sit well for some.

Other considerations are the fact that every connected device added to a home is a potential security risk. Cyber criminals can gain access to your home network and stories of home security being hacked are plentiful. As with most things, protect yourself by having complex passwords, a robust firewall and reliable internet security.

Lighting maintenance

Maintenance is vital to lighting systems and lamp efficiency. Illumination levels decrease over time because of ageing lamps and dirty fixtures and lenses, together with dust covering reflective surfaces. These factors can reduce total illumination by 50 per cent. The following basic maintenance suggestions can help prevent this:

- clean lamps and fixtures every 6–12 months
- replace discoloured glass or plastic lenses
- clean or repaint working areas every two to three years. Dirt collects on surfaces and reduces the amount of light they reflect.
- consider group relamping – replacing all the lamps in a lighting system at once. Grouping saves labour, keeps illumination high and avoids stressing any ballast with failing lamps.

REVIEW QUESTIONS

1 List the three elements that make up a lighting system.
2 List three rules that an efficient and effective lighting system should follow.
3 What does the term 'efficacy' mean?
4 List two disadvantages of CFLs.
5 What are the advantages that LEDs offer over basic light sources?
6 What is the simplest way to reduce the amount of energy consumed by lighting systems?
7 Name three control devices used to minimise energy usage.
8 Describe one typical EMS function related to minimising energy consumption.
9 Name three main ways to use smart home devices.
10 Why do illumination levels decrease over time?

CHAPTER REVIEW

11.1 Notions of sustainable energy

- Use of sustainable energy minimises harmful effects on the environment and health of all organisms on Earth.
- Sources of sustainable energy include solar photovoltaic (PV), solar thermal, wind, hydrogen fuel cells, geothermal, small-scale (mini and pico) hydroelectric, tides and waves.
- The greenhouse effect is a natural process that assists in maintaining the Earth's temperature and the stability of its climate.
- The enhanced greenhouse effect is a result of GHG emissions from human activities and is a significant contributor to human-induced climate change.
- Contributing factors to the excess emission of GHGs from human activities include fossil fuel burning, industrial emissions, deforestation/land-use change, agriculture and waste disposal.
- Everyday measures to reduce GHG emissions are intended to reduce energy and resource consumption.
- Sustainability is a concept that integrates environmental management, social justice and economic growth.
- The Kyoto Protocol was an international agreement that arose from the UNFCCC, which set goals for developed countries to reduce their emissions of carbon dioxide and five other GHGs.
- The Paris Agreement is a successor to the Kyoto Protocol for climate action after 2020 that aims to limit global warming to well below 2 °C above pre-industrial levels and preferably keep global warming below 1.5 °C.

11.2 Overview of sustainable energy technologies

- Geothermal energy is heat that originates from radioactive decay of elements within the Earth.
- Methods to harness geothermal energy include using water released from aquifers at a high temperature to convert liquid isopentane into a high-pressure gas; or passing water through zones of hot, dry rock to produce hot water or steam; which is used to drive a turbine to generate electricity.
- Wave energy and tidal energy are probably least likely to be major contributors of base energy in Australia.
- As renewable energy sources supply a greater proportion of electricity generation, storage of excess energy for later use will be become increasingly critical.
- Types of energy storage being developed or considered are batteries, pumped hydro and hydrogen gas.

11.3 Selection of energy-reducing control systems

- Lighting systems produce around 12% of GHG emissions from the domestic sector and around 25% of emissions from the commercial sector.
- Every lighting system consists of three elements: a lamp, a luminaire fitting and a control system.
- Efficacy is the efficiency of the lamp at converting electrical energy into light, measured in lumens per watt (lighting output divided by its power consumption).

- Types of energy-efficient bulbs include linear fluorescent lamps, CFLs and LEDs.
- The CFL is a miniaturised linear fluorescent tube with integral ballast (either iron or electronic) in a standard Edison, Swan or pin base.
- CFLs combine a lamp, electronic ballast and base in a single sealed unit.
- LEDs are solid-state electronic devices that produce light.
- LEDs have the ability to vary the colour, create sparkle and aim the light precisely.
- A more sophisticated form of energy usage control is the EMS. An EMS performs the same function as a timed switch but has additional features.
- Essentially there are three main ways to use most smart home devices: apps, smart assistants and automation.
- Home automation finds use in: lighting, general appliances, entertainment, home security, sustainability and accessibility.
- Maintenance is vital to lighting systems and lamp efficiency.

TRIAL EXAM

For Chapter 11 knowledge assessment, please complete the following trial exam.

1 State one principle for supporting the drive for sustainable energy.
 a sustainable energy resource control
 b limitation of sustainable energy industries
 c poverty eradication
 d to promote GHG emissions

2 Name the environmental component that protects all organisms by blocking or reducing harmful radiation from the sun and space.
 a precipitation
 b atmosphere
 c lithosphere
 d clouds

3 The natural process that assists in maintaining the Earth's temperature and the stability of its climate is:
 a the greenhouse effect
 b the trade winds
 c the Earth's seasons
 d the chlorophyll effect

4 The Earth is kept warmer by GHGs absorbing and emitting which form of radiation?
 a gamma
 b infrared
 c microwave
 d visible

5 Which of the following gases is the most plentiful GHG that contributes to climate change?
 a water vapour
 b nitrous oxide
 c methane
 d carbon dioxide

6 Which pollutant reduces the greenhouse effect but aggravates asthma and other lung diseases and contributes to millions of premature deaths worldwide?
 a perfluorocarbons
 b nitrous oxide
 c sulphur hexafluoride
 d aerosols/particulates

7 A contributing factor to the excess emission of GHGs is:
 a land-use change
 b biofuels
 c tree planting
 d geothermal energy

8 A major measure used to reduce GHGs is:
 a emissions trading
 b fossil fuel burning
 c deforestation
 d fertiliser use

9 Sustainability was first mentioned in:
 a the Kyoto Protocol
 b the *Brundtland Report*
 c the UNFCCC
 d the *National Greenhouse and Energy Reporting Act 2007*

10 State the concept that integrates environmental management, social justice and economic growth.
 a depletion
 b sustainability
 c attenuation
 d emissions trading

11 State the human endeavour which is a conscious, empowering, communally organised, needs-centred activity that encompasses social and environmental factors.
 a systemic risk
 b consumerism
 c corporate governance
 d sustainable work practice

12 For an individual worker, sustainable work practices should target:
 a self-esteem and confidence
 b personal responsibility
 c behaviour modification
 d performance goals

13 The definition of 'any passive or active omission by humanity' is:
 a reduction
 b neglect
 c attenuation
 d abandonment

14 What is the effect of neglect on the environment?
  a accumulated harm
  b nurturing
  c appreciation
  d adaptation

15 What agreement established five-yearly reviews of targets including national mitigation, adaptation and support commitments, underpinned by a rules-based system to assess whether countries are meeting their targets?
  a UNFCCC
  b the *Brundtland Report*
  c Kyoto Protocol
  d Paris Agreement

16 Which thermal renewable energy source originates from radioactive decay?
  a solar thermal
  b tides
  c geothermal
  d solar PV

17 What will be needed as the proportion of electricity generated from renewable sources increases?
  a coal-fired power stations
  b energy storage
  c reduced GHG emissions
  d lower electricity demand

18 What three elements make up a lighting system?
  a lamp, luminaire fitting and control system
  b domestic conductors, switch and lamp
  c luminaire fitting, circuit breaker and conductors
  d control system, conductors and circuit breaker

19 Which rule should an efficient and effective lighting system follow?
  a use the lowest luminous efficacy lamp
  b avoid providing lighting controls
  c use incandescent bulbs
  d make use of natural light

20 State how climate change strategies in Australia will be most successful.
  a improved environmental performance and economic benefits
  b increasing coal usage to provide wealth
  c minimising public participation
  d focusing only on the short-term economic benefits

REFERENCES

Clean Energy Council (n.d.), *Energy storage*, https://www.cleanenergycouncil.org.au/resources/technologies/energy-storage, viewed 8 September 2021.

Commonwealth of Australia (Clean Energy Regulator) (2015), *2014–2015 Annual Report*, http://www.cleanenergyregulator.gov.au/DocumentAssets/Documents/Annual%20report%202014-15.pdf, viewed 8 September 2021, p. 40.

Commonwealth of Australia (Clean Energy Regulator) (2021), *Corporate emissions and energy data 2019-20*, http://www.cleanenergyregulator.gov.au/NGER/National%20greenhouse%20and%20energy%20reporting%20data/Corporate%20emissions%20and%20energy%20data/corporate-emissions-and-energy-data-2019-20, viewed 9 September 2021.

Department of the Environment and Energy (n.d.), *International activities*, http://www.environment.gov.au/climate-change/government/international, viewed 19 April 2018.

International Energy Agency (IEA) (2020), *Key World Energy Statistics 2020*, https://iea.blob.core.windows.net/assets/1b7781df-5c93-492a-acd6-01fc90388b0f/Key_World_Energy_Statistics_2020.pdf, viewed 27 July 2021, p. 60.

Keywood MD, Emmerson KM, Hibberd MF (2016), *Climate: Kyoto Protocol targets*, https://soe.environment.gov.au/theme/climate/topic/2016/kyoto-protocol-targets, viewed 8 September 2021.

United Nations (1987), *Report of the World Commission on Environment and Development*, http://www.exteriores.gob.es/Portal/es/PoliticaExteriorCooperacion/Desarrollosostenible/Documents/Informe%20Brundtland%20(En%20inglés).pdf, viewed 19 April 2018, p. 15.

United Nations (2015), *Paris Agreement*, https://unfccc.int/files/essential_background/convention/application/pdf/english_paris_agreement.pdf, viewed 9 September 2021, pp. 3-4.

United Nations Framework Convention on Climate Change (UNFCCC) (1992), https://unfccc.int/resource/docs/convkp/conveng.pdf, viewed 19 April 2018.

12 Cabling for telecommunication services

This chapter provides electrotechnology workers with essential knowledge and skills with respect to regulations that apply to telecommunications installations. Workers in the field of telecommunications cabling will also gain knowledge and skills in reading and interpreting telecommunications cabling plans and installing various telecommunications cables, outlets and associated equipment. This chapter covers the essential knowledge and associated skills specified for the unit of competency UEEDV0005 from the UEE training package.

LEARNING OBJECTIVES

Cabling Provider Rules
- Outline the content of the Cabling Provider Rules and explain the responsibilities of the parties involved in their implementation

General installation requirements
- Summarise the general installation requirements for telecommunications cabling and cabling products

Customer interfaces
- Describe common methods for interfacing customer cabling with the telecommunications network

Cable distribution devices
- Explain the purpose of cable distribution devices and outline the requirements pertaining to their installation

Network boundaries
- Delineate the network boundary for various cabling installations
- List the various technologies used in Australia's national broadband network

Cable identification
- Explain the methods to document information about cabling installations

Telecommunication cable types
- Outline the construction and transmission characteristics of common telecommunication cables

Indoor cabling
- Outline the properties of indoor cabling and summarise the requirements relating to their installation

Underground cabling
- Outline the properties of underground cabling and summarise the requirements relating to their installation

Aerial cabling
- Outline the properties of aerial cabling and summarise the requirements relating to their installation

Earthing protection
- List the various telecommunications earthing systems and outline the regulatory requirements that apply

Earthing concepts
- Outline methods used to test and calculate the resistance of a telecommunications earthing system

Surge suppression
- Explain causes of surges on telecommunications systems and the methods employed to minimise their effect

Cable shielding and interference
- Explain the requirement for cabling shielding and associated earthing practice

Miscellaneous regulations
- Summarise miscellaneous regulations that apply to telecommunications installation work

Installation and termination requirements
- Outline various installation points during building construction
- List common cable support systems and state their application

Cable installation
- Explain the installation procedures for telecommunications cables
- Explain the procedure for safely hauling telecommunications cables

Cable termination
- Outline hazards and safety precautions experienced in telecommunications cabling work
- List the various connection techniques used in telecommunications cabling

End-to-end testing
- List the various cable tests and outline telecommunications testing and fault finding procedures

Hazards
- Outline the hazards involved in the installation and maintenance of switching systems

12.1 Cabling Provider Rules

Anyone installing telecommunications cabling that may connect to the telecommunications network, either domestic or commercial, must comply with the Cabling Provider Rules (CPRs). The CPRs are legislative documents that outline mandatory requirements for the installation of telecommunications cabling and provide mechanisms for auditing and self-regulation.

Cabling Provider Rules (CPRs)

The cabling industry was transformed with the introduction of Cabling Provider Rules (CPRs). This process also included the streamlining of mandatory educational requirements and the removal of barriers to entry into the cabling sector in an effort to increase competition and benefit the community overall.

CPRs are based on an industry-run national registration scheme and are designed to promote industry self-regulation in line with government policy. CPRs continue to provide for appropriate community safeguards, including the protection of the health and safety of end-users, other cabling providers and carrier personnel, and the preservation of network integrity (ACMA, 2016a, p. 20).

Failing to comply with the CPRs may result in the Australian Communications and Media Authority (ACMA) issuing a fine to the cabling provider for non-compliance. In the first instance the ACMA will generally advise the cabling provider of areas of non-compliance and the actions necessary to ensure compliance. The cabling provider may also receive a written warning of non-compliance. Subsequent non-compliance, and certain breaches of the *Telecommunications Act 1997*, may result in the ACMA issuing the cabling provider with a telecommunications infringement notice. This notice details the nature of non-compliance, the amount of the penalty, and advises the maximum penalty imposed by a court. See http://acma.gov.au for the latest information regarding penalties.

To determine the effectiveness of self-regulation an ACMA-contracted auditor may contact a cabling provider and arrange to meet with them and examine some of their work. Auditors are authorised by the ACMA for the purpose of monitoring compliance with the CPRs.

The ACMA conducts investigations into written complaints by the public about unsatisfactory cabling work. These enforce compliance with the rules. The ACMA has the power to prosecute a cabling provider who does not comply with the rules.

SWITCH ON

Summary of the Cabling Provider Rules

The ACMA introduced Cabling Provider Rules on 3 October 2000. These CPRs are updated from time to time so it is important for cabling providers to keep up to date. The current Telecommunications Cabling Provider Rules were published in 2014. As this is a legislative document it is not an easy read, but in summary the 10 CPRs include requirements for:

1 Cabling work in the telecommunications, fire, security and data industries to be undertaken by a registered cabling provider.
2 All cabling providers to obtain an Open, Restricted or Lift registration, depending on the scope of cabling work the cabling provider is performing.
3 All cabling work complies with the *Wiring Rules*.*
4 Telecommunications cabling to be adequately separated or segregated from electrical cabling.
5 Cabling providers to only install cables, cabling product, and customer equipment that complies with the Telecommunications Labelling Notice (TLN).
6 Cabling providers to provide customers with a TCA1 sign-off form at the completion of each job.
7 Registered cabling providers to directly supervise the work of unregistered cable installers.
8 Registered cabling providers to be responsible for the work of unregistered cable installers.
9 Cabling providers to provide all reasonable cooperation and assistance to ACMA inspectors.
10 Cabling providers to notify their registrar of any change of contact within 21 days (Australian Government, 2014).

*The *Wiring Rules* is a standard specific to the telecommunications industry and details the mandatory or recommended requirements for working with telecommunications cabling.

Industry regulators

The current Australian regulator for broadcasting, internet, radio communications and telecommunications is the ACMA. The ACMA is a statutory authority within the federal government portfolio of Communications and the Arts.

The ACMA's responsibilities include:

- promoting self-regulation and competition in the telecommunications industry, while protecting consumers and other users from network hazards and unscrupulous operators
- fostering an environment in which electronic media respects community standards and responds to audience and user needs
- managing access to the radio-frequency spectrum, including the broadcasting services bands
- representing Australia's communications and broadcasting interests internationally (ACMA, 2016b).

Another important body is the Communications Alliance (CA), which offers a forum for the telecommunications industry in Australia to make clear and productive contributions to policy development. The CA oversees a number of committees and working groups that work on specific areas of telecommunications. One of these is the Customer Equipment and Cabling Reference Panel, which identifies and reviews issues related to customer equipment and customer cabling that are likely to affect the telecommunications industry. This panel proactively recommends suitable action for the continued efficient operation of the telecommunications industry, which includes the review of current standards, codes, guidelines, and other related documentation.

Labelling of telecommunications products

All customer equipment and customer cabling installed in Australia must comply with the Telecommunications Labelling Notice (TLN). Basically, this notice lets cabling providers know that the cabling product was tested and is compliant with Australian Standards for use on the public switched telephone network (PSTN).

Under the requirements of the TLN, in order to supply a product to the Australian market, a supplier, which includes a manufacturer, importer or their authorised agent, must apply a label to the product to indicate that the product complies with mandatory technical standards. To ensure the legitimacy of labelling, a supplier must register on the national database as a 'responsible supplier' before fixing a compliance label to a product.

There are two forms of labelling identified in the TLN:

- compliance label, which is identified with the Regulatory Compliance Mark (RCM)
- non-compliance label (ACMA, 2016c).

Compliance label

The Regulatory Compliance Mark (RCM), as shown in **Figure 12.1**, is the compliance label used to indicate regulatory compliance of telecommunications products.

Source: © Commonwealth of Australia (Australian Communications and Media Authority) 2018. Material reproduced with permission of www.acma.gov.au.

FIGURE 12.1 RCM compliance mark

A compliance label is a permanent and easy-to-read label that indicates a product complies with the relevant regulatory requirements, which includes any applicable ACMA standards. It is not permissible for a compliance label to be attached to a non-compliant product. In order to display the compliance label, the supplier must hold the documentation relevant to the product. For products that have a built-in display, the compliance label may be shown electronically on the display.

Compliant products may have been labelled with an A-Tick compliance mark prior to 1 March 2016, when the RCM was introduced. It is not a requirement for these products to be re-labelled with the RCM.

The ACMA has no regulatory responsibility for the RCM used under the Electrical Equipment Safety System (EESS), which the Electrical Regulatory Authorities Council (ERAC) administers (ACMA, 2018a).

Non-compliance label

A non-compliance label fixed to a product indicates that the product does not comply with the applicable technical standards or that the product has not been tested against them. It is necessary to satisfy all other regulatory arrangements.

The non-compliance label includes a statement that the product does not comply with each applicable technical standard. The statement must be printed on both the outside of the product's packaging and in the documentation supplied with the product.

If you install a product that carries a non-compliance label you will be held liable for repairs and also possible litigation from failure to comply with the Telecommunications Act. You may also be subject to a fine from the ACMA.

Figure 12.2 is a checklist that shows the steps for compliance under the 'Labelling Notice'. Refer to https://www.acma.gov.au/5-steps-suppliers.

Source: Material reproduced with permission of www.acma.gov.au.

FIGURE 12.2 Steps to compliance

It is the supplier's responsibility to ensure that a compliance label is correctly applied to each product before it is supplied to the market (ACMA, 2016c).

Electronic labelling of products

If a product has a built-in display then the compliance label may be displayed electronically rather than on the surface of the product. The TLN does not prescribe how electronic labels should be displayed.

The ACMA suggests that electronic labels may be displayed:

1 during the device's power-up sequence
2 under the device's system information page
3 under the device's help menu (ACMA, 2016c).

Registration as a cabling provider

Under the CPRs, cabling providers must register with registrars appointed by the ACMA. Cabling providers must meet certain competencies determined by the ACMA in order to register. The ACMA's competencies form the basis for training programs developed by industry skills councils (ISCs).

The ACMA cabling arrangements ensure cabling providers have the necessary skills to perform specialised cabling work for the current and emerging customer cabling environment.

Cabling providers undertaking broadband, structured, optical fibre or coaxial cabling work must have the training competencies relevant to the specialised cabling work. All cabling providers who install those specialised cables must attain the appropriate competencies.

Figure 12.3 shows the steps required to become a registered cabling provider as outlined by the ACMA. Educational requirements for Restricted and Open customer cabling training typically include:

- work health and safety (WHS)
- the regulatory framework, which covers the telecommunications *Wiring Rules*
- basic telephony
- cabling installation practices for domestic work
- basic electrical fundamentals and principles
- workplace experience or practice.

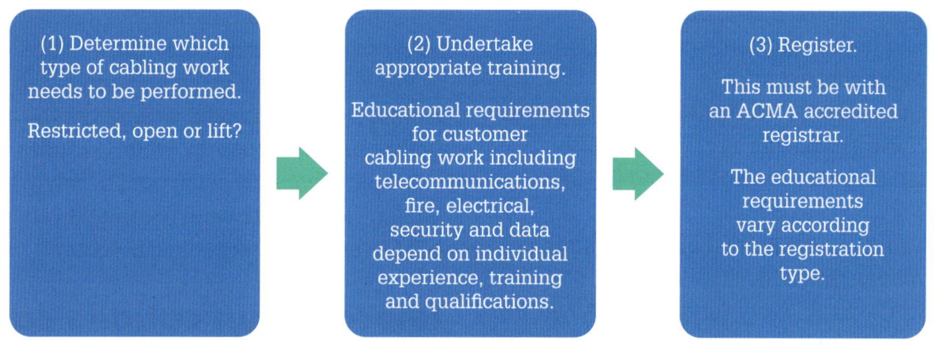

FIGURE 12.3 The three main steps to becoming a registered cabling provider (ACMA)

Persons applying for an open registration need additional training in cabling installation practices for commercial work and customer premises equipment (CPE).

Registration involves signing a declaration confirming that you will cable in accordance with the CPRs. These declarations are subject to audit and also confirm that the applicant has the relevant competencies. Applicants are required to keep evidence of their competencies, which may be statements of attainment or qualifications from registered training organisations.

There are separate competency requirements for Open, Restricted and Lift registrations. These competencies relate to health and safety and network integrity issues. The cabling provider must also demonstrate that they have acquired sufficient experience relevant to the class of registration.

SWITCH ON

Refer to the ACMA website http://acma.gov.au and see cabler registration for the latest information.

»

For more on the customer cabling rules and requirements visit https://www.acma.gov.au/cabling-provider-rules

The ACMA has information about types of cabling work and becoming a registered cabling provider, which you can view at https://www.acma.gov.au/types-cabling-registration

A list of Telecommunications Industry Training Advisory Board (TITAB) registered assessors is available from the TITAB Cabler Registry or the Communications and Information Technology Training Limited (CITT) website at http://citt.com.au

A list of RTOs is available from the Australian Government Department of Industry Skills website at http://training.gov.au

Types of cabling work

The CPRs recognise three types of cabling work, which are summarised in **Table 12.1**.

TABLE 12.1 Types of cabling work

Classification of work	Description of work
Open	All types of cabling work from simple cabling in residential premises to complex cabling in multi-storey buildings.
Restricted	Where the network boundary is a simple socket or a network termination device, which is typically found in residential premises and small businesses and does not include large commercial buildings.
Lift	Telecommunications cabling in lifts and lift wells.

Cabling registrars

After attaining the relevant competency requirements for the desired registration, contact one of the five ACMA accredited registrars to arrange registration. Each registrar offers different services so you need to choose one that suits you. Each also has different fees, so contact the registrars directly to obtain these details. ACMA accredited registrars are listed below (alphabetically):

- Australian Cabler Registration Service (ACRS) Website: http://acrs.com.au
- Australian Security Industry Association Limited (ASIAL) Website: http://www.asial.com.au
- BICSI Registered Cablers Australia Pty Ltd (BRCA) Website: http://www.brca.com.au
- Fire Protection Association Australia (FPA Australia) Website: http://www.fpaa.com.au
- TITAB Australia Cabler Registry Services (TITAB ACRS) Website: http://www.titab.com.au

Supervision

Registered cabling providers who are supervising an unregistered cable installer have to provide supervision of all unqualified cable installers' cabling work. As such, the registered cabling provider must accept full responsibility for all the work done and ensure that it fully complies with the *Wiring Rules*. This is a requirement of Section 4.1 of the CPRs (see Telecommunications Cabling Provider Rules 2014 – F2014L01684 at http://www.comlaw.gov.au). For the purpose of this rule, supervision unambiguously means supervision shall be in the presence of the unqualified worker and not simply on the same site as that unqualified worker.

Inspection of cabling work

The telecommunications carriers are responsible for the public telephone network. They also have the right to inspect cabling work. They may inspect work to ensure that it satisfies the industry regulator's standards for network integrity, personal safety and proper network functioning.

Only the carrier has the right to run network connection jumpers on a main distribution frame (campus distributor or building distributor) or NTD (network termination device) as this marks the demarcation point of the carrier's telecommunications service. A registered cabling provider cannot do this without written approval from the carrier. If a carrier tags a service for use by a customer and notifies the cabling provider to complete the work, then that is the written consent that is required for the cabling provider to work beyond what is deemed the network boundary (see section 13.13 of AS/CA S009: 2020).

ACMA-appointed auditor

An auditor is authorised by the ACMA to inspect cabling providers' records for the purpose of monitoring compliance with these rules. An audit may involve the appointed auditor obtaining telecommunications cabling advice (TCA) forms from a registrar for a number of randomly selected cabling providers. The auditor will review the documents to ensure compliance with the CPRs and then advise the registrar with a decision on whether a cabling provider achieved, or failed to achieve, compliance.

ACMA-appointed inspector

An inspector will be directed by the ACMA to undertake an inspection following a complaint or a random inspection at a construction or building site. An inspector has the power to enter a work site without notice. An inspector will be following up on a complaint submitted by a customer or cabling provider and will work under the guidance of the Telecommunications Act, which includes the CPRs.

See https://www.acma.gov.au/how-we-regulate-cabling-industry for more information about how the ACMA regulates the cabling industry.

Telecommunications cabling advice

All registered cabling providers must complete a Telecommunications Cabling Advice form (TCA1). You must complete and sign a TCA1 form on completion of your cabling job. The ACMA provides guidance on completing these forms, which you can find at https://www.acma.gov.au/cabling-advice-forms. When filling out the form the cabling provider should:

- complete all sections of the form
- print legibly in black ink and
- describe the work clearly (specify the type of work and where it is located, such as room, floor, section, department, building).

The cabling provider must provide a copy of the form to the customer and employer (if applicable). They must keep all copies of TCA1 forms for at least 12 months and these must be made available to ACMA inspectors or auditors on request. Cabling providers are responsible for keeping their TCA1 forms. To avoid additional paperwork, it is acceptable to incorporate the information of the TCA1 form into an existing invoice or other business documentation.

A cabling provider may use an optional form (TCA2) to alert a customer or building manager of any non-compliant cable installations that are outside the contracted scope of work.

Figures 12.4 (A) and **(B)** show completed examples of the Telecommunications Cabling Advice forms.

Telecommunications customer cabling compliance form (TCA1 form)

Copies required for customer, cabler and employer (if applicable)

Instructions for completion

Requirements

> The registered cabling provider (cabler) who performed or supervised the work described in this form must complete this upon finishing the work (except for certain exemptions).

> Cablers must retain a copy of this form for at least 12 months and give a copy to the customer/employer.

> Where proposed works may be compromised by existing cabling, a TCA2 form should be completed.

Enquiries

> For advice on completing this form, please go to the ACMA website at acma.gov.au.

> Technical enquiries about cabling should be directed to:

Email: info@acma.gov.au
Telephone: 1300 850 115

Registered cabling provider

Name

SURNAME	Bear
GIVEN NAMES	Edward

Address

12 Lochinvar Circuit
West Wyoming
POSTCODE 3289

Contact details

WORK (03)	8855 5588
MOBILE	0444 111 222
EMAIL	Edward.Bear@Bearcomm.com.au

Registration number	Expiry date
B300444QLD	21/12/2024

Employer (IF APPLICABLE)

Name of company

Bear Communications

Address

1/22 Enterprise Way
East Wyoming POSTCODE 3288

Contact details

WORK (03)	8866 8866
MOBILE	0444 111 222
EMAIL	Info@Bearcomm.com.au

Description of work (INCLUDING ANY SUPERVISION)

Installation of additional distributor (10B/3) from FD10B located on North Wing 10th Floor. Extension of CSS services (211 - 220) to new offices. PVC trunking provided for new Services, which were terminated on modular 8X8 jacks.

Customer details

Name

Custom Entertainment

Address

22 Solent Circuit
Tilbury POSTCODE 3344

Contact details

HOME (03)	3455 1111
MOBILE	
EMAIL	Info@customentertainment.com.au

Certification

I hereby certify the cabling work described in this advice complies with the Wiring Rules (AS/CA S009:2020 Installation Requirements for Customer Cabling or its replacement).

SIGNATURE	E Bear
DATE	22nd February 2022

PRINT FULL NAME Edward Bear

Source: Australian Communications and Media Authority © Commonwealth of Australia.

FIGURE 12.4(A) Sample Telecommunications Cabling Advice form – TCA1

Telecommunications cabling advice (TCA2)

This form is an optional addition to the TCA1 Telecommunications Cabling Advice form. It may be used by registered cablers to alert the customer or building manager of any non-compliant cable installations that are outside the contracted scope of work.

☐ Pre works advice
☑ Post works advice

Outstanding matters

While undertaking the contracted cabling work, the following issues have been identified with the pre-existing cable installation, which may require your attention:

Issue	Rating
☐ Inadequate separation of communications and electrical cabling	☐ Urgent safety hazard ☐ Attention required – non-urgent ☐ Long term – low safety risk
☐ Inappropriate or inadequate support provided to cables	☐ Urgent safety hazard ☐ Attention required – non-urgent ☐ Long term – low safety risk
☐ Cables not secured or fixed	☐ Urgent safety hazard ☐ Attention required – non-urgent ☐ Long term – low safety risk
☐ Non-compliant cabling product used	☐ Urgent safety hazard ☐ Attention required – non-urgent ☐ Long term – low safety risk
☐ Non-compliant customer equipment installed	☐ Urgent safety hazard ☐ Attention required – non-urgent ☐ Long term – low safety risk
☐ Non-compliant earthing	☐ Urgent safety hazard ☐ Attention required – non-urgent ☐ Long term – low safety risk
☐ Wrong colour conduit used	☐ Urgent safety hazard ☐ Attention required – non-urgent ☐ Long term – low safety risk
☑ Records are missing or out of date	☐ Urgent safety hazard ☑ Attention required – non-urgent ☐ Long term – low safety risk
☐ Pre-existing cables are worn or frayed	☐ Urgent safety hazard ☐ Attention required – non-urgent ☐ Long term – low safety risk
☐ Pre-existing cabling is not compliant (other)	☐ Urgent safety hazard ☐ Attention required – non-urgent ☐ Long term – low safety risk

Source: Australian Communications and Media Authority © Commonwealth of Australia.

FIGURE 12.4(B) Sample Telecommunications Cabling Advice form – TCA2

The TCA1 must include minor work such as the installation of parallel sockets in a domestic installation. The ACMA does not require a cabling provider to complete a cabling advice for any of the following activities:

- replacement of sockets (other than network boundary sockets), detectors (for fire and security alarms) or other minor cabling equipment for maintenance purposes
- running, transposing or removing jumpers from distribution frames
- marking, replacing and upgrading cabling records
- all testing and transmission measurement activities.

SWITCH ON

AS/CA S009: 2020 'Installation Requirements for Customer Cabling' (*Wiring Rules*)

All electrical installations must comply with AS/NZS 3000:2018 *Wiring Rules*. Likewise, all telecommunications installations must comply with AS/CA S009: 2020, which are the wiring rules for telecommunications installations. This Standard, along with others in the series, is available as a free download from http://www.commsalliance.com.au.

REVIEW QUESTIONS

1. Which is the body responsible for regulating telecommunications (including communications cabling) and radio communications in Australia?
2. What industry body offers a forum for the telecommunications industry in Australia to make contributions to policy development?
3. What is the primary function of the telecommunications labelling notice?
4. Outline the purpose of electromagnetic compatibility regulation.
5. What type of cabling work can the holder of an Open registration undertake?
6. What must an applicant do to undertake broadband, structured, optical fibre or coaxial work?
7. With what code or document must cabling installation work comply?
8. At completion of cabling work it is necessary for the registered cabling provider to complete a cabling advice form (TCA1). What is the purpose of this form and who receives a copy?
9. What does supervision of an unregistered or unlicensed cable installer entail?
10. What is the consequence of installing a non-compliant product?
11. A key requirement of the CPRs is that telecommunications cabling must be separated from what other cabling?
12. Is it necessary to complete a TCA1 form if the cabling work only involved updating cabling records?

12.2 General installation requirements

Basic installation requirements include:

- segregation
- cable joints
- safety and integrity
- cable terminations
- conduits
- power feeding.

Segregation

The segregation of telecommunications cabling from hazardous services, such as cables carrying 230 volts, is necessary to minimise the risk of a dangerous situation being present on the telecommunications service. Such situations can arise:

- should a nail, screw, or other metal object simultaneously pierce the insulation of both electrical and telecommunications cables resulting in a dangerous voltage being present on the communications service
- should a telecommunications cable come in contact with a non-electrical hazardous service, such as temperatures above 60 °C, or pipes carrying corrosive or flammable liquids or gases, resulting in damage to the telecommunications cable and interfering with proper functionality
- damaging the sheaths of both a telecommunications cable and electrical cable through poor installation technique resulting in simultaneous burning of telecommunications and electrical cables greatly reducing the effectiveness of the cable insulation of both services.

Segregation of telecommunications cabling from power cabling is important for the general safety of customers, cabling providers and carrier staff. Inadequate segregation results in incidents every year of hazardous voltages on telecommunications lines and equipment. Each situation is potentially life threatening. Widespread use of open modular telecommunications outlets in domestic premises, coupled with the increased use of 'smart' televisions, may increase this risk.

Both AS/CA S009: 2020 and AS/NZS 3000: 2018 require segregation of power and telecommunications cables and their terminations by appropriate means. Insulated and sheathed (double-insulated) electrical cables near or in physical contact with insulated and sheathed telecommunications conductors do not in themselves represent a hazard in normal circumstances. The hazard comes from external influences such as accidental penetration of both cables by nails, screws, drills and the like. Another possible hazard is from cable insulation damage due to vermin, heat, abrasion, mechanical stresses, or over-voltage surges from power system faults or lightning strikes.

There may be no obvious effect on an electrical circuit if the active conductor of a power cable not protected by a residual current device (RCD) comes in electrical contact with a metallic conductor in a telecommunications cable. This could happen because of damage to both layers of cable insulation (penetration by a nail or screw). There will, however, be a catastrophic and dangerous effect on the telecommunications circuit. This may include the presence of 230/400 V power circuits wherever the telecommunications circuit terminates within the customer's premises (telecommunications outlets) and wherever the circuit terminates within the carrier's network (street pillars, joints, exchange and so on). This could result in a fatality or serious injury to an occupant within the customer premises, or to an employee of the carrier. The risk to customers is particularly high in the case of coaxial cable as the shield of the cable usually connects to a coaxial connector with exposed metal parts. If the shield of the cable becomes live, the customer may be at risk of electric shock when plugging or unplugging video appliances.

Cable joints

Joints may be required in either copper or optical fibre cables.

Copper conductors

The method for joining twisted pair communications cable having copper conductors is similar to joining electrical cables. While it is permitted to make joins by twisting and soldering, it is common practice to use a commercially available insulation displacement connector (IDC) as shown in **Figure 12.5**. The 3M company introduced the original Scotchlok UR Connector over 30 years ago. IDCs can connect wires having a range of conductor diameter and an insulation diameter up to 2.08 mm.

The UR2 (red cap) connector is for general butt splicing and accepts a range of conductor sizes from 0.4 mm to 0.9 mm in any combination, which means it is suitable for butt connecting a 0.4 mm conductor to a 0.9 mm conductor. The incorporation of a water-blocking compound ensures that cable joints made with the IDC will not degrade due to the absorption of moisture into the joint. While dry versions of IDCs are available, it is the moisture-resistant connectors (MRCs) that are most prevalent as any moisture in a cable joint can result in an unacceptable noise level.

The UY2 (yellow cap) connector is physically smaller than the UR2 connector. It suits smaller splice bundles

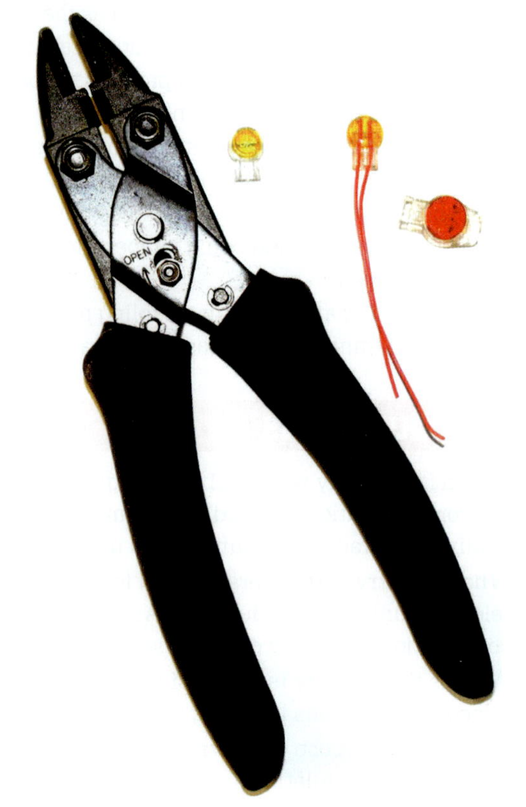

FIGURE 12.5 Common insulation displacement connectors and tooling for twisted pair

such as is the case when effecting encapsulated underground cable joins.

It is essential to use the correct crimping plier, such as that shown in **Figure 12.5**, to ensure an effective joint. After making the joint it is necessary to ensure compliance with AS/CA S009: 2020, which requires that 'Cable joints must be suitably fabricated, enclosed, positioned, and supported to prevent the ingress of dust or moisture'.

Optical fibres

Optical fibre cables may be joined by fusing or mechanically splicing. Fusion splicing involves using a machine to align the two prepared fibre ends and fusing them together under intense heat generated from an arc drawn between two electrodes. Mechanical splices clamp the two fibres end-to-end. The mechanical splice incorporates an index matching gel, which is matched to the refractive index of the fibre to ensure effective light transmission from one fibre to the other. A mechanical splice is shown in **Figure 12.6**.

FIGURE 12.6 Mechanical splice for optical fibres

Fusion splicing is common for outside installations, while mechanical splicing is more common for indoor applications. Again, the joins must be suitably supported and enclosed to prevent the ingress of dust and moisture. Special splice racks are available for this purpose.

Safety and integrity

The cabling provider must ensure that the installation of all customer cabling complies with AS/CA S009: 2020 and the generally accepted principles of safe and sound practice. It is important to be aware of practices that may breach another code or standard. For example, securing customer cabling to water pipes may impede plumbing work.

Some important aspects of AS/CA S009: 2020 pertaining to safety and integrity are summarised in **Table 12.2**.

TABLE 12.2 Requirements for safety and integrity

Clause	Summary of requirement
5.2	Cable and other cabling equipment be installed in accordance with the manufacturer's instructions. This is particularly important for minimum bend radius, hauling-tension, cable tie pressure and the like.
5.3	All customer cabling and customer equipment to comply with the Telecommunications Labelling Notice.
5.4	Requires protection against damage for all parts of an installation.

Other areas of concern are:

- earth potential rise
- optical fibre and coaxial cabling
- explosive atmospheres
- catenary cabling.

Earth potential rise

In the multiple earth neutral (MEN) system of earthing, the neutral conductor joins the general mass of earth at numerous points in the power system. This allows any fault currents to flow back to the supply through the earth. As the earth has some resistance, a current flowing through the earth will cause a potential difference between two points in the ground as shown in **Figure 12.7**. This is known as earth potential rise (EPR).

The point where the active conductor touches the ground will rise in potential to be equal to the supply voltage. The rise in earth potential gets smaller as you move away from this point. Back at the supply, the earth potential remains at zero volts.

Earth potential rise only happens in areas where fault currents may flow through the earth. Clause 6.1 of AS/CA S009: 2020 states that telecommunications equipment and accessories should not be placed in areas where the earth potential rise exceeds 430 V a.c. The only time it is necessary to consider earth potential rise is where the installation is in close proximity to high-voltage (HV) equipment. In this case the power supply authority must be consulted as to the extent of the hazard zone.

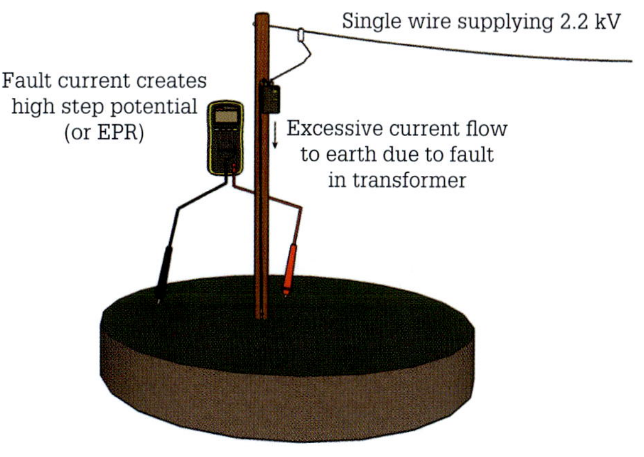

FIGURE 12.7 Earth potential rise

It is acceptable for plastic-sheathed cables installed in plastic conduit to pass through the EPR hazard zone. This also includes the installation of associated pits, access holes and draw-boxes. It is dangerous to install customer equipment, surge suppression devices, earth electrodes, associated earthing equipment, metallic cable supports (conduit, tray and the like) and cable connection devices within the EPR hazard zone. Appendix H to AS/CA S009: 2020 recommends a minimum distance of 15 m from any HV power transformer; including single-wire earth return (SWER) systems.

It may be necessary to install special isolating links and HV isolation units for telecommunications installations within an HV installation.

Optical fibre and coaxial cabling

Optical fibre cables are exempt from many of the separation requirements detailed in AS/CA S009: 2020. However, installation of optical fibre cables should not pose a threat to cabling providers from electrical hazards when working on customer cabling.

Optical fibre systems must comply with the applicable requirements of AS/CA S008, AS/NZS IEC 60825.1 and AS/NZS IEC 60825.2, paying particular attention to the following requirements:

- Optical fibre cables are to carry markings to distinguish them from cables carrying other services.
- There must be appropriate labelling of all optical fibre access points where disconnecting fibres may result in laser radiation exceeding Class 1 being emitted.
- Optical fibre cables are not to damage or obscure any manufacturer warning label.
- Personal protective equipment (PPE) must be used, including but not limited to, eye protection and disposable gloves when terminating or joining optical fibres.
- Removal of all fibre particles, hazardous solvents or chemicals from the site and safe disposal of all 'sharps' must be carried out.

Coaxial cable is still a popular cabling medium and is subject to the requirements of Clause 11.2 of AS/CA S009: 2020. This clause prohibits a telecommunications circuit from connecting to the outer conductor of a coaxial cable that may be touched by the end-user, unless the circuit meets the requirements of a safety extra-low-voltage (SELV) circuit or the outer conductor connects permanently to a protective earth.

Explosive atmospheres

Similar to the requirements for electrical installations, telecommunications installations must be classified in accordance with:

- AS/NZS 60079.10.1: 2009 *Explosive atmospheres – Classification of areas – Explosive gas atmospheres*
- AS/NZS 60079.10.2: 2016 *Explosive atmospheres – Classification of areas – Combustible dust atmospheres.*

Section 7.1 of AS/CA S009: 2020 details the requirements for installing telecommunications cable and equipment in explosive atmosphere environments. Clause 7.1.3.2 calls on AS/NZS 60079.14 when selecting and installing equipment, which includes cabling and connecting hardware, for use in hazardous areas.

Catenary cabling

A catenary support system comprises a wire rope suspended between points under tension to which customer cabling attaches. This may be an indoor or outdoor cabling system. The catenary support can bond to the protective earthing system of the building. It is therefore necessary to separate telecommunications cables, supported on a non-integral catenary support system, from other services.

Cable terminations

In order to ensure the safety of all users of telecommunications services and the health and wellbeing of those working on telecommunications circuits, AS/CA S009: 2020 has a number of requirements pertaining to the termination of telecommunications services. Generally this involves the separation of telecommunications service terminations from other hazardous services by either appropriate physical distance or appropriate barriers. Some important aspects of AS/CA S009: 2020 pertaining to safety and integrity of cable terminations are summarised in **Table 12.3**.

TABLE 12.3 Requirements for safety and integrity

Clause	Summary of requirement
9.1.2	Shared enclosures are permitted for low voltage power and telecommunications terminations provided that accidental access to the LV conductors and terminations by telecommunications workers is prevented by means of a physical barrier or obstruction.
9.1.3.1	It is not permissible to terminate telecommunications conductors in the same enclosure or building cavity as the conductors and terminations of high voltage conductors and terminations.
9.1.3.2	It is necessary to separate enclosed conductors and terminations of telecommunications cables from enclosed conductors and terminations of high voltage cables.
9.2	Separate conductors and terminations of telecommunications cables from non-electrical hazardous services.

Source: Adapted from Standards Australia 2020, AS/CA S009: 2020 'Installation requirements for customer cabling (Wiring Rules)'.

Figures 12.8 to **12.10** show the required separation between telecommunications terminations and those of LV electrical services.

FIGURE 12.8 Separation between LV and telecommunications terminations in common enclosure (note the cover over the power cable terminations on the left)

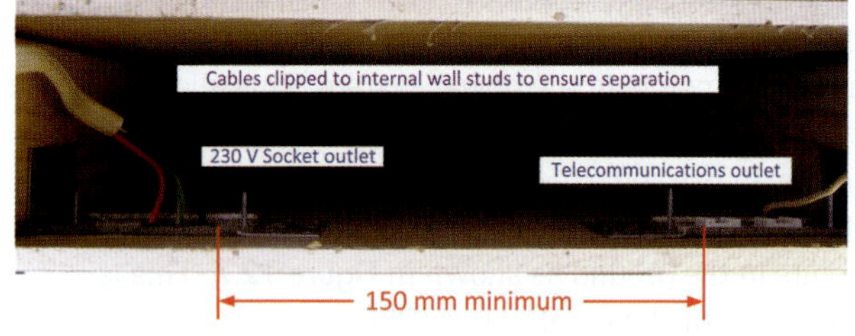

FIGURE 12.9 Separation between LV and telecommunications terminations in common portion of a wall

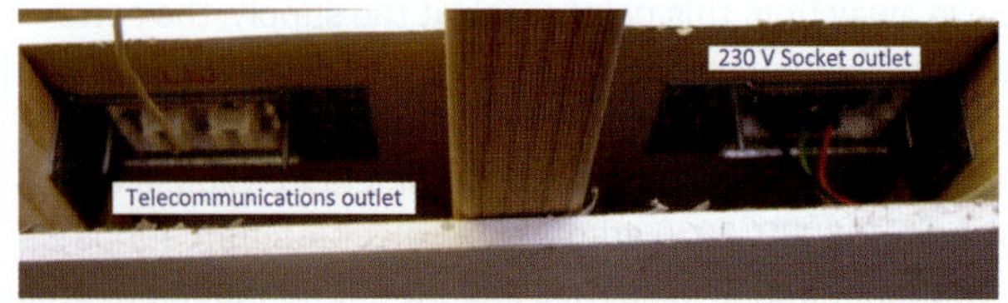

FIGURE 12.10 Separation between LV and telecommunications terminations in separate portions of a wall (attaching the supports to opposite studs facilitates separation when cables pass through the top plate)

LV telecommunications circuits

Telecommunications systems have progressed beyond the original 48 V d.c. (telecommunications network voltage – TNV) circuits of the past to include remote power feeding of active telecommunications devices. As such, telecommunications lines may have voltage of up to 300 V d.c. on them. These are known as LV telecommunications circuits (AS/CA S009: 2020) and are not permitted to share a cable sheath with ELV (extra-low voltage), SELV (safety extra-low voltage), TNV (telecommunications network voltage) or limited-current circuits.

It is a requirement of AS/CA S009: 2020 for ELV circuit terminations to be separated from SELV, TNV or limited-current circuit terminations by a minimum distance of 150 mm. An alternative is to place a rigidly fixed barrier of durable insulating material or earthed metal between the two systems.

Conduits

It is not permissible to install customer cables in a conduit or pipe of a colour specified for use with hazardous services (refer to AS/CA S009: 2020, Clause 8.3). Colours for hazardous services are summarised in **Table 12.4**.

TABLE 12.4 Colour codes for hazardous services

Colour		Hazardous service
	Orange	Electrical
	Yellow or Yellow-ochre	Fuel, process, toxic or medical gases
	Silver-grey	Steam
	Brown	Flammable and combustible liquids
	Violet	Acids and alkalis
	Light Blue	Compressed air

Source: Adapted from Standards Australia 2013, AS/CA S009:2020 'Installation requirements for customer cabling (Wiring Rules)'.

Clause 8.3 of AS/CA S009: 2020 allows for the following exceptions:

- conduits that are used solely to support telecommunications cables and are fully encased in concrete providing that the ends of the conduit are identified white in a durable manner
- sub-ducting of conduits that do not contain high voltage cables with compliant non-conductive conduit that encloses the telecommunications cables
- non-conductive telecommunications cables, such as optical fibre cables, may be installed within conduits identified by a prohibited colour provided that the cable is identified at each access point with a warning label such as 'may contain a hazardous light source'.

Clause 8.3.2 of AS/CA S009: 2020 expressly prohibits a cabling provider from accessing a conduit of a colour specified for non-telecommunications services.

It is a further requirement to ensure total segregation between electrically conductive telecommunications cables and high voltage cables; this requirement precludes sub-ducting services containing high voltage cables.

Power feeding

On occasion a carriage service provider may install equipment in customer premises that derives its electrical power from the telecommunications line. This may cause the line voltage at the customer premises to exceed the limits for TNV.

An example of remote power feeding used in data networks is known as Power Over Ethernet (POE). In this application power from a network switch, which has POE ports, will distribute operational electrical power from the switch into the customer's cable, and power the remote device. Having remote power feeding eliminates the need for power sources such as plug packs at the location of the remote device.

Some examples of POE devices are:

- access control systems
- fire and alarm detection devices
- intercoms
- IP cameras
- IP telephones
- wireless access points

Cable used for any low-voltage telecommunications circuit must be clearly identified at any access point and must be appropriately separated from other services and other telecommunications circuits. Cable identified with a red sheath should be used to identify fire-detection and fire-alarm circuits.

To protect installations from the heating effect of electric current, Clause 20.4 of AS/CA S009: 2020 requires protection of telecommunications cabling for circuits that provide power to equipment other than that derived from a carriage service. Such protection can be by way of circuit breakers, fuses or current limiting.

REVIEW QUESTIONS

1 Why is it necessary to ensure segregation of telecommunications cabling from hazardous services, such as cables carrying 230 volts?

2 What external influences pose a hazard from insulated and sheathed electrical cables near or in physical contact with insulated and sheathed telecommunications conductors?

3 What is a common method for jointing copper telecommunications cables?

»

4 Name two common methods for jointing optical fibre telecommunications cables.
5 With what must customer cabling installation work comply?
6 What is earth potential rise?
7 What precaution is necessary when installing optical fibre telecommunications cables?
8 What is a catenary support system?
9 By what minimum distance is it necessary to separate the conductors and terminations of telecommunications cables from the uninsulated and single-insulated conductors and terminations of LV cables installed in the common portion of a wall?
10 It is not permissible to install customer cables in a conduit or pipe of a colour specified for use with certain services. What are these services?
11 What is a common example of remote power feeding in telecommunications circuits?

12.3 Customer interfaces

A customer interface generally refers to the point at which customer cabling connects to a carrier's network, creating a demarcation point known as the network boundary.

A distributor is a connection device that provides for cross-connection of cables using jumpers or patch cords (AS/CA S009: 2020, Clause 4.2.34). A connection device that does not contain jumpers or patch cords is not a distributor. In a manner similar to how switchboards distribute circuits in an electrical installation, distributors dispense communications circuits in an installation. The distributor located at the network boundary is the main distribution frame (MDF).

The carrier will supply and install the necessary network lead-in cables to the MDF and associated termination modules (blocks) on the MDF. Termination of lead-in cables is the sole responsibility of the carrier. A cabling provider must supply and install the main distribution frame along with sufficient termination modules to terminate the customer's cabling. These termination modules may be of any permitted type. **Figure 12.11** shows a 10-pair MDF, while **Figure 12.12** shows larger MDFs using ADC Krone and '110'-style cable termination systems.

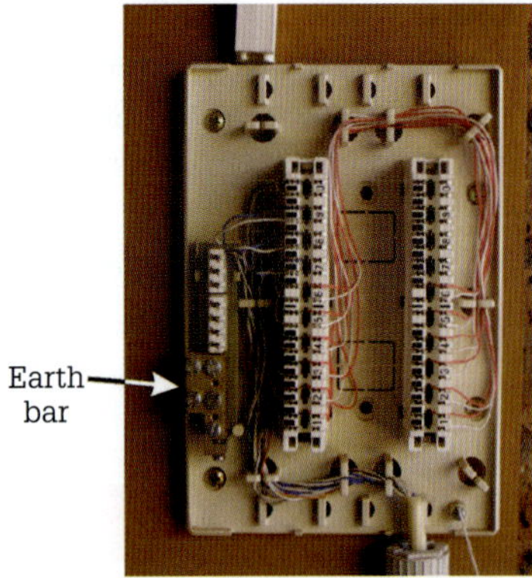

FIGURE 12.11 A 10-pair MDF

Note the large earth bar on the left. This accommodates earthing for surge suppression modules. The cover is also deeper than the similar floor distributor.

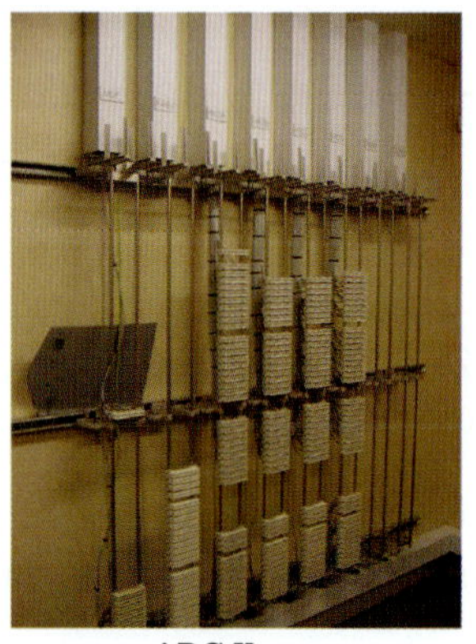
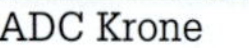
ADC Krone

110 style

FIGURE 12.12 Larger MDFs

Physical properties

In large installations requiring many telephones, the carrier will make enough exchange lines available to satisfy the customers' needs. Installing distributors simplifies the distribution of exchange line services throughout an installation.

The main distribution frame may provide:

- termination facilities for the carrier's cable
- a distribution point for part of the installation
- a testing position
- a telecommunications reference conductor distribution point
- provision for surge protection equipment.

Distributors may be one piece or assembled from a number of parts: a rack, back-mount and termination modules. Back-mount-type distributors (as shown in **Figure 12.12**) differ from one-piece distributors (typically 10 pair similar to that shown in **Figure 12.11**) in that:

- They generally allow the cable to enter from behind the termination modules. Installation of the termination modules occurs after fixing the cable in position.
- The formed cable structure resembles the body of a tree, the branches being groups of cable pairs that will terminate on each termination module (typically 10 pair).

General regulatory requirements

Cross-connections should match or exceed the category rating of the installed cabling system. For example, if the installed cabling is Class D (Cat. 5), then the cross-connections must be at least equal to Class D; it is not acceptable to use normal telephone cross-connections. (Class D was previously referred to as Cat. 5e, or enhanced, to signify the higher bandwidth capability over the previous Cat. 5 cable. However, a Class D system as defined in AS/NZS 11801.1: 2019 *Information technology – Generic cabling for customer premises – General requirements*, which replaced AS/NZS 3080: 2013, is referred to in the industry as Cat. 5.)

Clause 12.3 of AS/CA S009: 2020 requires the cabling provider to supply sufficient cabling records relating to the cabling to enable cables and cross-connections to be correctly identified and connected. The cabling records must be legible and updatable. The records should also include details of any cables external to the boundaries of the premises that connect to the distributor for which the records relate. The records must clearly identify any LV telecommunications circuit.

Distributors located outdoors must have a minimum rating of IPX3 and installation must not degrade this rating. As with electrical enclosures, cable entry holes in distributors must be free of sharp edges or burrs. An alternative to this is to fit an insulating grommet to the hole. If the enclosure is electrically conductive, it must facilitate connection to the protective earth. The enclosure itself must be free of exposed sharp edges.

Main distribution frame

The key definitions and requirements that determine whether a connection device qualifies as an MDF are as follows:

- An MDF is a distributor used to connect a carrier's twisted pair lead-in cabling (AS/CA S009: 2020, Clauses 4.2.67 and 13.1).
- An MDF must be within or attached to a building (AS/CA S009: 2020, Clauses 13.3 and J.3(4)(a)(i)).
- A building is a substantial structure intended to protect persons, animals, vehicles and the like from the weather (AS/CA S009: 2020, Clause 4.2.4), and so a fence, pole, pedestal or similar structure does not qualify as a building.

Location of an MDF

The location of an MDF must be in or on a building ('a substantial structure with a roof and walls'). Another requirement is for the MDF to be close to the electrical switchboard to facilitate earthing and equipotential bonding where required. If an MDF is installed within a building, the position must be free from the ingress of dust and moisture and not subject to damp and humid conditions. Furthermore, the MDF must attach securely to a permanent building element (AS/CA S009: 2020, Clause 13.3).

It is important to consider what constitutes the network boundary for premises serviced by a main distribution frame (see AS/CA S009: 2020 Clause J.3(4)(a)(i)). If the location of the MDF is not in the end-user's building, it is not the network boundary for that end-user's line, in which case the network boundary will default to the end-user's first socket.

In multi-tenancy occupancies, the MDF should be in a common property area. A carrier may not agree to connect to an MDF located in an individual tenancy if it supplies cables to other tenancies, as this could create future access problems for the carrier, cabling providers and service providers, and inconvenience to the tenant with the MDF in their tenancy.

SWITCH ON

Clearances for an MDF

When installing an MDF, provide the minimum clearances specified in Appendix D of AS/CA S009: 2020 around the MDF.

Cable terminations and cross-connections

The cable terminations on the carrier side form part of the carrier's network and the termination modules must be the carrier's standard type so that the carrier can maintain them. The carrier supplies and installs the termination modules for the lead-in cabling. One exception is an ADC Krone 10-pair MDF, which is a legacy from the pre-deregulation era (pre-1989) and comes as a complete assembly including the lead-in termination module. A cabling provider can only install this MDF for a carrier whose standard termination is ADC Krone LSA Plus (e.g. Telstra). Just because the cabling provider supplies the termination modules for the carrier side as part of the MDF does not mean the cabling provider can terminate the lead-in cable on it – this is still the responsibility of the carrier.

The termination system on the customer side can be any type permitted by the regulations; that is, it satisfies compliance requirements. However, a carrier may not be able to terminate a cross-connection on the termination module for the customer side if a suitable termination tool is not available, or if the carrier is unsure where to connect the cross-connection. In such cases, the carrier may choose to 'tag' the customer end of the cross-connect or to 'tag' the carrier-side pair for the end-user's cabling provider or service provider to effect the final connection.

Marking of MDF verticals

It is usual to refer to the columns of termination modules in a distributor as 'verticals'. These are designated alphabetically from left to right, starting with the letter A and omitting the letters I and O as these may be confused with the numerals 1 and 0. For very large distributors (more than 24 verticals) the designations after Z are AA, AB and so on. The range of terminations within each vertical is numbered in ascending order from 1, which is in the bottom left of the vertical. In this way a particular termination is designated by its vertical, and its position within that vertical, for example A101 – vertical A, termination 101. Traditionally, lead-in cabling terminates on the A vertical and the customer cabling terminates from the B vertical onwards. This has led to use of the common terms, A side (carrier side) and B side (customer side). However, a carrier's lead-in cable can terminate on any convenient vertical. A carrier's cable can terminate on intermediate verticals in an MDF and customer cables can terminate on the A vertical. In any case, the MDF and the records must clearly indicate this arrangement. The terminations may be allocated according to the size of the lead-in cable(s) where the MDF has only one vertical. **Figure 12.13** shows the numbering of a distributor with two verticals.

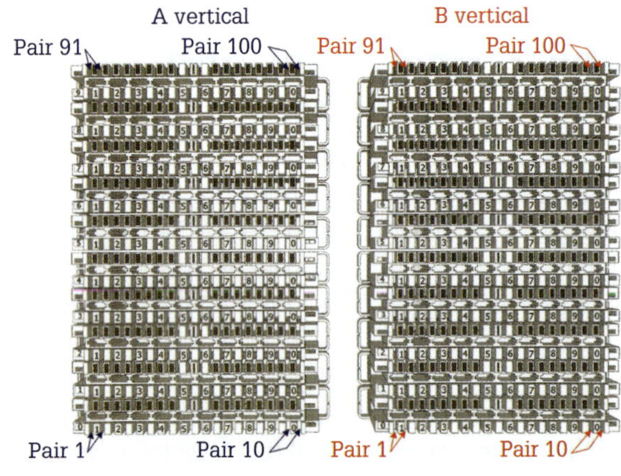

FIGURE 12.13 Distributor labelling

Although the lead-in cable is normally terminated on the bottom modules, this is not essential provided the records are appropriately marked. A carrier may not agree to connect a lead-in cable to an MDF that is not clearly marked in accordance with AS/CA S009: 2020 (Clauses 12.3 and 13.12 apply).

Network termination device

The only device away from a building that is a defined network boundary is a network termination device (NTD), which is provided by a carrier at the carrier's discretion and is marked at manufacture as an NTD.

If the device is permanently marked at manufacture 'network termination device' or 'NTD', then it is an NTD. Once installed it forms part of the carrier's network and requires testing and approval by the carrier. Other markings such as 'network interface device' or 'telephone network interface' do not qualify as valid NTDs. Likewise, devices of similar appearance to an NTD but not marked as an NTD at manufacture are not NTDs. **Figure 12.14** shows a true NTD.

FIGURE 12.14 Network termination device (the carrier's cable is terminated behind the cover)

AS/CA S009: 2020 poses similar restrictions to main distribution frames on network termination devices (see Clauses 14.1 to 14.4). A cabling provider can terminate customer cabling to the customer side of the NTD only and apart from testing will not access the carrier side of the NTD.

REVIEW QUESTIONS

1. Describe a distributor.
2. Who supplies and installs the necessary network lead-in cables to a main distribution frame?
3. Who is responsible for the supply and installation of an MDF?
4. What style of terminating modules are used on the customer's side of an MDF?
5. Outline the requirements for the location of an MDF installed within a building.
6. What identifying letters are *not* used to identify distributor verticals?
7. On which vertical within the MDF can the carrier's lead-in cable terminate?
8. What minimum IP rating is required for distributors located outdoors?
9. When can a cabling provider install cross-connects at the network boundary?
10. What network boundary device is permitted to be installed away from the building?

12.4 Cable distribution devices

In large installations requiring many telephones, the carrier will provide sufficient exchange lines to satisfy the customers' needs. Installing distributors simplifies the distribution of exchange line services throughout an installation. Distributors are referred to by function; hence we have main distribution frame (MDF), building distributor (BD) and floor distributor (FD). **Figure 12.15** shows the hierarchical arrangement of distributors within an installation (often referred to as a campus).

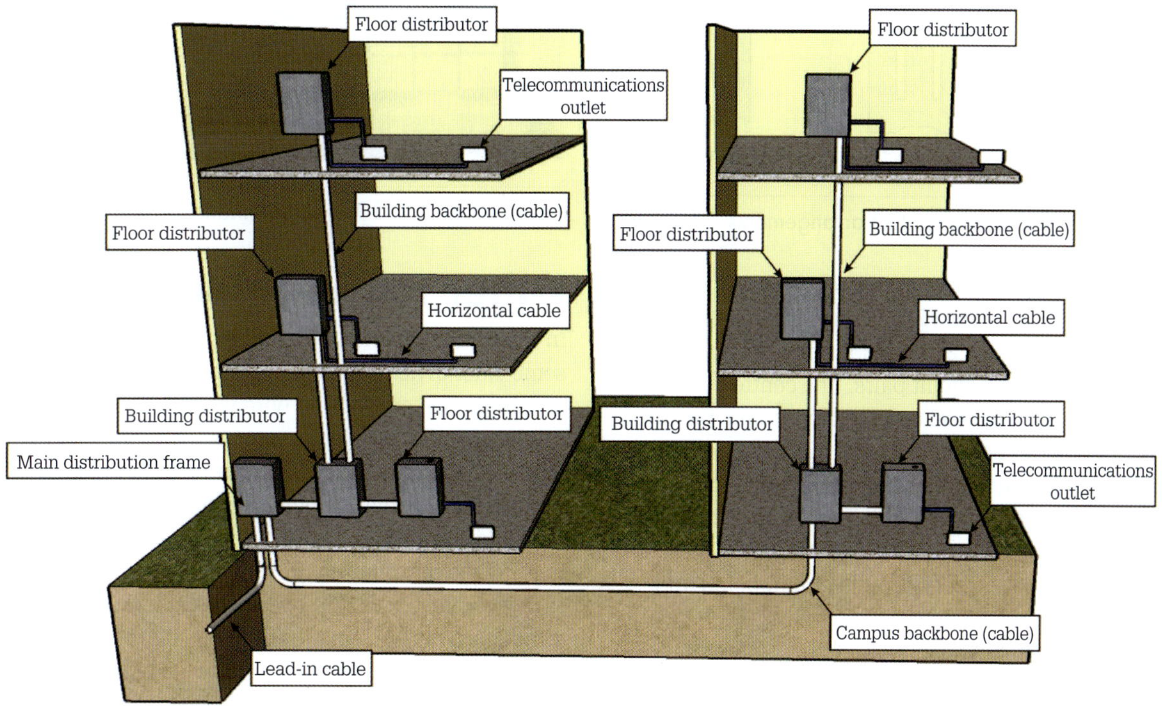

FIGURE 12.15 Distributor hierarchy

The carrier terminates their wiring at the main distribution frame, which is located at the network boundary. This distributor provides:

- termination facilities for the carrier's cable
- a distribution point for part of the installation
- a testing position
- a telecommunications reference conductor (TRC) distribution point
- provision for surge protection equipment.

Distributors may be one piece or assembled from a number of parts. When a distributor is assembled from a number of parts it is referred to as a back-mount frame.

The most common back-mount distributors are surface-mount frames and island, or floor-mount, frames. The features offered by these include:

- provision for the installation of surge suppression devices
- cable entry from the top or bottom of the frame
- cable tie positions at the entry points
- jumper rings or guides to assist in supporting jumper wires
- provision for a connection to earth.

ADC Krone 10-pair termination modules

The ADC Krone 10-pair termination modules are widely used in telecommunications distributors. **Figure 12.29** (see Section 12.6 Cable identification) shows a typical 100-pair ADC Krone distributor. There are three different arrangements for the terminating modules: disconnection, connection and switching.

Regardless of the arrangement, all modules operate on the same principle. The wires of the incoming (feeder) circuit terminate on the unnumbered side of the module and the wires of the outgoing (cross-connect) circuit terminate on the numbered side of the module. Each wire of a pair straddles a number on the module. Internal contacts provide a path between the incoming and outgoing circuits.

The **disconnection** modules utilise a normally closed two-piece contact. The image on the left in **Figure 12.16** depicts this arrangement. Inserting a disconnect plug into a module disconnects the circuit. This feature allows for the insertion of a test cord into a module, to test a wire pair in both directions (inward and outward from the insertion point). The centre contact provides a useful point for connecting monitoring equipment or over-voltage protection devices.

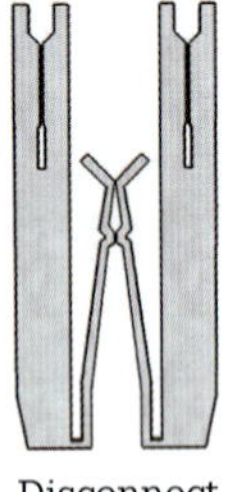

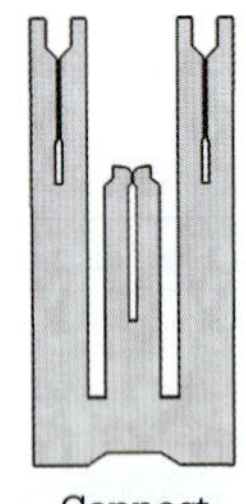

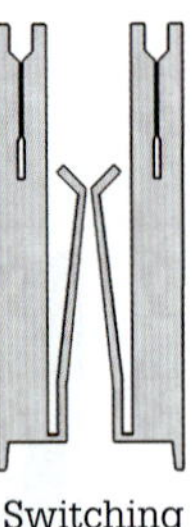

FIGURE 12.16 ADC Krone contact arrangements

The **connection** module utilises a one-piece contact. The centre image in **Figure 12.16** depicts this arrangement. With this contact there is a continuous circuit between the incoming and outgoing circuit pairs. The centre contact point allows for easy parallel testing, monitoring, patching and over-voltage protection.

Switching contacts are usually a normally open two-piece contact. The image on the right in **Figure 12.16** depicts this arrangement. With this contact arrangement there is usually isolation between the unnumbered and numbered sides of the module. This allows for high-density termination and patching facilities.

A fourth variant is the earthing module, which has all the contacts electrically joined internally.

The functional unit of the ADC Krone LSA-Plus contact consists of the contact slot, with flexible contact tags arranged at 45° to the axis of the wire. Pressing the wire into the contact slot with the insertion tool cuts the insulation and pushes the wire between the flexible contact tags. This effectively creates two gas-tight contact points. Clamping ribs tightly grip the wire on both sides of the contact point. This ensures that the connection remains unbroken by mechanical shocks.

'110'-style wiring block

Another popular termination system is the '110'-style wiring block (and its variants), which utilises a quick clip connection. A '110'-style system is shown in **Figure 12.17**. This quick clip connection forms a gas-tight connection that prevents corrosion between the clip and the conductor. The fixed cabling terminates in the base of the '110' wiring blocks while the cross-connects terminate in the top of the connecting block. This method is different from the Krone system where incoming circuits terminate at the top of the wiring module and outgoing circuits originate at the bottom of the module. The '110' system provides high-density terminations.

FIGURE 12.17 '110'-style termination system

Other termination systems

In telecommunications work, the installer may come across some older-style termination systems. These include screw terminals, solder tabs and wire wrap terminals.

Some screw terminals, such as those shown in **Figure 12.18**, utilise an insulation-crushing method of termination. In this system, a spring washer crushes the insulation and makes contact with the conductor of the wire. When terminating under a screw it is important that the wire follows the body of the screw in a clockwise direction. This will ensure that the wire is not forced out from under the head of the screw when tightened. This system is still similar to the modern system in that it facilitates cross-connecting incoming and outgoing cables.

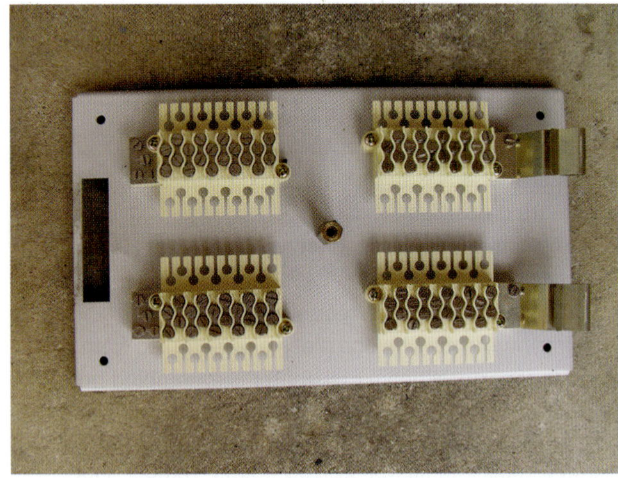

FIGURE 12.18 Screw-type termination system

Note the difficulty many cable installers have in terminating on the solder tab frame.

Solder tab distribution frames, distribution boxes and link mounting frames were introduced over 50 years ago for use as main distribution frames and floor distributors. The cable installer may need to install cross-connects on these frames. As poor termination techniques can result in unacceptable noise on the telephone service, it is important to terminate wires correctly.

The solder tab terminal frame has a pin passing through an insulating block. The incoming wires are soldered to one side and the outgoing wires are soldered to the other side. Again, this style of termination system comprises a number of verticals. As shown in **Figure 12.19**, two grooves are made in each pin. This facilitates parallel connections as one wire can be soldered into each groove.

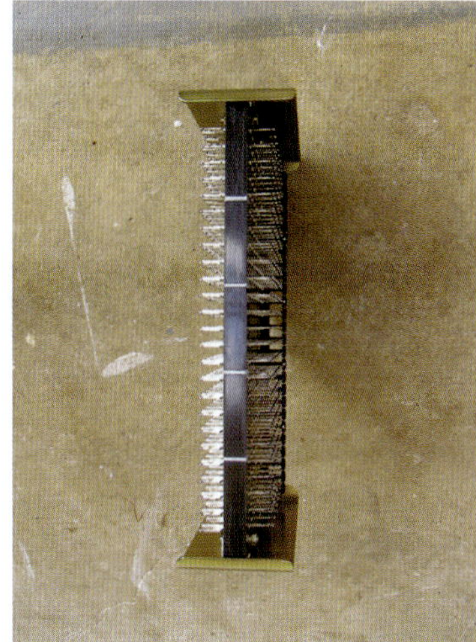

FIGURE 12.19 Solder-tab-type termination system

It is important to remember that the low thermal mass of the telephone wire compared to that of the terminal makes it very easy to burn the wire insulation. Noxious fumes are emitted from burning PVC insulation. **Figure 12.20** shows the accepted method for terminating on solder tab frames.

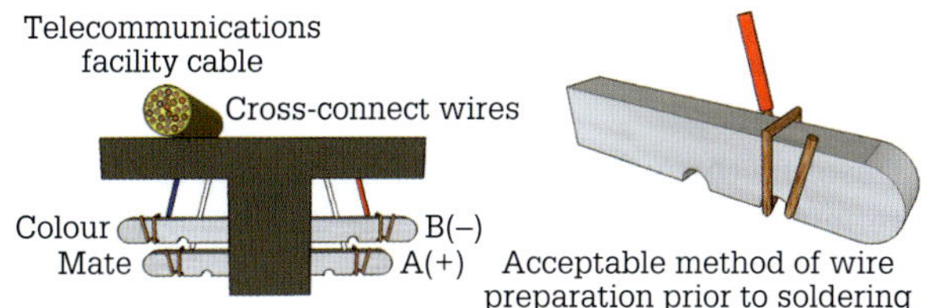

FIGURE 12.20 Termination method for solder tab frames

SWITCH ON

Hazards and safety precautions

Observe all WHS requirements when making solder connections.

Frames using the wire wrap method of termination are used as main distribution frames and floor distributors and predate the common insulation displacement methods used currently. Wire wrap terminals are also common as external distributors as part of Telstra's external plant (see **Figure 12.21**).

FIGURE 12.21 Wire wrap as part of Telstra's external plant

The wire wrap terminals comprise small, square binding posts. A special wire-wrapping tool (right image in **Figure 12.22**) is used to wrap the wire around the post. It is usually necessary to remove some wire insulation. The wire is prepared as necessary and placed in a hole near the edge of the wire-wrap tool. The tool is placed over the post (the post is inserted in the centre axial hole in the tool) and rotated until the required number of wire wraps is made. As the wire is wrapped around the post, the corners of the post cut into the wire producing a cold weld; 1½ turns of insulated wire may be made around the post. This allows for some slight movement of the wire without causing damage to the connection.

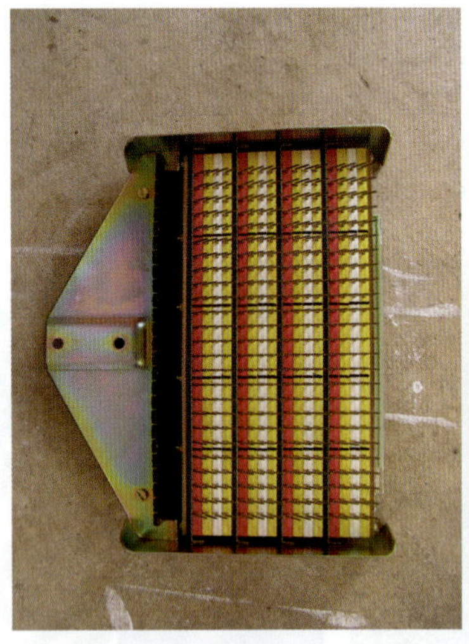

FIGURE 12.22 Wire-wrap-type termination system with wire-wrap tool shown on the right

Technical requirements for cable distribution devices

Clause 5.4 of AS/CA S009: 2020 requires the following of cable distribution devices:

- Cable entry holes must be free of sharp edges or burrs or have an insulating grommet fitted.
- An enclosure must be free of exposed sharp edges.
- Outdoor enclosures must offer a minimum degree of protection to IPX3.
- The main distribution frame must be capable of being locked.
- The main distribution frame must be capable of accommodating the standard terminating modules used by the carrier.
- A minimum of 30 mm clearance is maintained between the carrier's standard termination modules and the inside face of the front cover or door of the MDF to allow the installation of surge suppression devices.

REVIEW QUESTIONS

1. How are distributors referenced?
2. What provisions are afforded by the main distribution frame?
3. What name describes a distributor that is assembled from a number of parts?
4. What are three different arrangements for the terminating modules used in the ADC Krone distributor?
5. What advantage does the '110-style' termination system afford?
6. Describe the insulation-crushing method of termination.
7. In which direction is the wire formed under a screw connection?
8. What may be the result of poor termination techniques for a telephone service?

12.5 Network boundaries

The network boundary is defined in the *Telecommunications Act 1997*, which essentially sets the demarcation point or boundary of a telecommunications carrier's network.

The network boundary is that point of a carrier's physical infrastructure for determining whether cabling or equipment is deemed to be 'customer cabling' or 'customer equipment' for the purpose of technical regulation (by ACMA). Beyond the network boundary, customer cabling and customer equipment has to comply with ACMA technical standards and the customer cabling must be installed, connected or repaired in accordance with the Cabling Provider Rules.

Another important point in a telecommunications network is the service delivery point. This is a point in customer cabling where the telecommunications service provider will supply a carriage service to a customer under the published terms and conditions for the supply of that service, which may be beyond the network boundary. Remember that the network boundary sets a demarcation for who can install and maintain cabling and equipment.

In the days of the 'plain old telephone service' (POTS) the network boundary was clear; either the first telephone outlet or the main distribution frame and then later the network termination device. Now with Australia's national broadband network, this demarcation point is not so well defined. **Table 12.5** summarises the location of the network boundary for various NBN technologies.

TABLE 12.5 Summary of network boundaries for NBN installations

Technology	Port	User Network Interface (UNI)	Premises	Network Boundary
NBN FTTP	Analogue telephone	UNI-V1	All types	UNI
	Ethernet	UNI-D1	All types	UNI
NBN FTTN	VDSL2	UNI-DSL	Single dwelling	TO or NTD
NBN FTTB	VDSL2	UNI-DSL	Multi-dwelling	Customer side of MDF
NBN FTTC	Ethernet	UNI-D1	Single dwelling	TO or NTD

»

TABLE 12.5 (Continued)

Technology	Port	User Network Interface (UNI)	Premises	Network Boundary
NBN HFC	Ethernet	UNI-D1	All types	UNI
NBN Wireless	Ethernet	UNI-D1	All types	UNI

Source: Table J1 from Australian Standards AS/CA S009:2020 - Installation Requirements for Customer Cabling (Wiring Rules), © Communications Alliance Ltd 2020.

An overview of network infrastructure

Australia's new national network employs a range of broadband technologies broadly grouped as fixed line connections and wireless and satellite connections to deliver internet connections to subscribers.

Fixed line technologies

Table 12.6 summarises the distinguishing features of the various fixed line technologies used to get the national broadband to premises (NBN Co., 2018).

TABLE 12.6 Summary of fixed line technologies

Technology	Description	Installation requirements
Fibre to the Premises (FTTP)	An optical fibre runs directly from a fibre node to the premises	An nbn™ access network device, which requires mains power, is installed in the premises. See Figure 12.23.
Fibre to the Node (FTTN)	The existing copper phone and internet network from a nearby fibre node is used to make the final part of the connection to the nbn™ access network. Data speed is comparable to that of ADSL2+	A VDSL modem, which requires power to operate, is installed in the premises. See Figure 12.24.
Fibre to the Building (FTTB)	An optical fibre cable runs to a fibre node installed in the building's communications room, from there, existing cabling is used to connect each apartment.	Fibre node is installed in the form of a secure cabinet in the building's communications room. See Figure 12.25.
Hybrid Fibre Coaxial (HFC)	Existing 'pay TV' or cable network is used to extend an HFC line from a fibre node to the premises.	An nbn™ access network device, which requires mains power, is installed in the premises. See Figure 12.26.

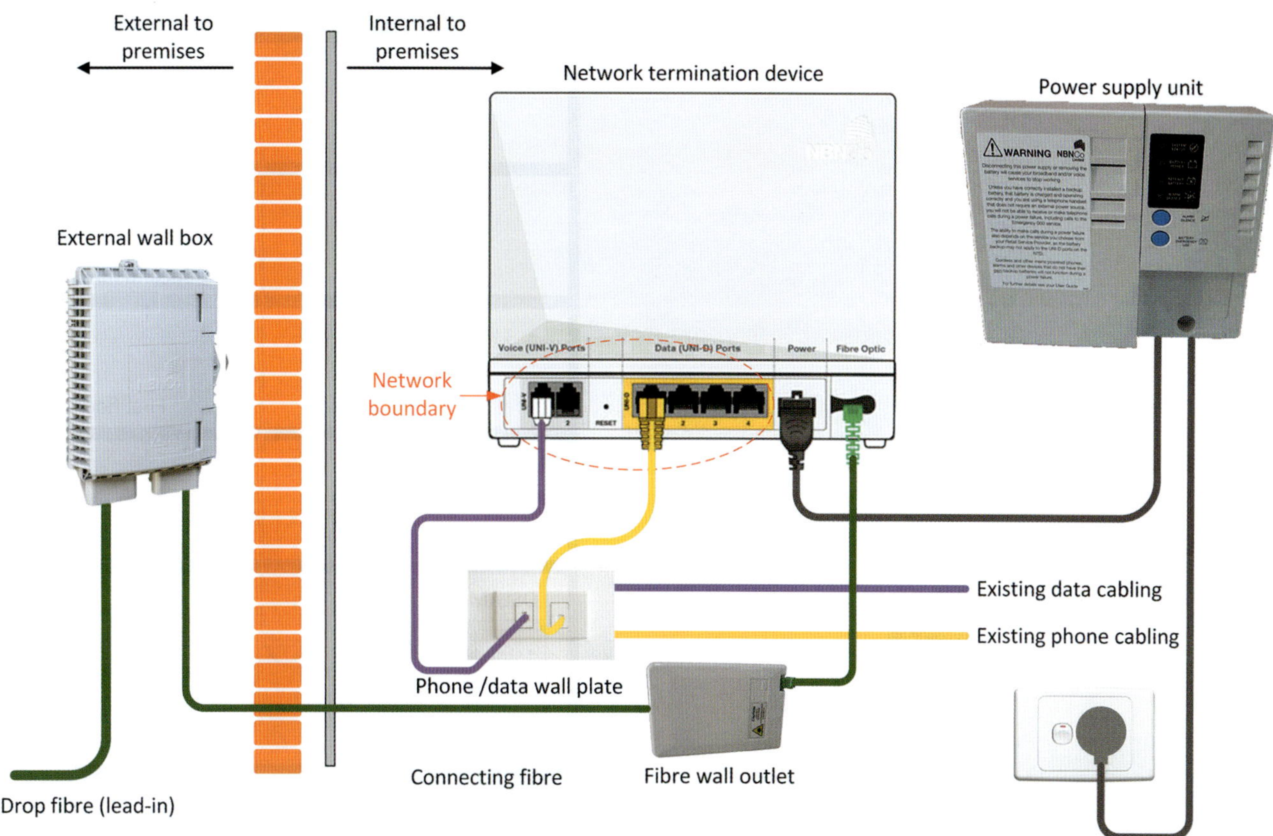

Sources: [L to R] Shutterstock.com/fotobycam; Shutterstock.com/Fluid Shutter; Image by Bidgee/Wikimedia.CC BY-SA 3.0 AU; iStock/Getty Images Plus/Juan_Gomez; Image by Bidgee/Wikimedia. CC BY-SA 3.0 AU.

FIGURE 12.23 The FTTP connection

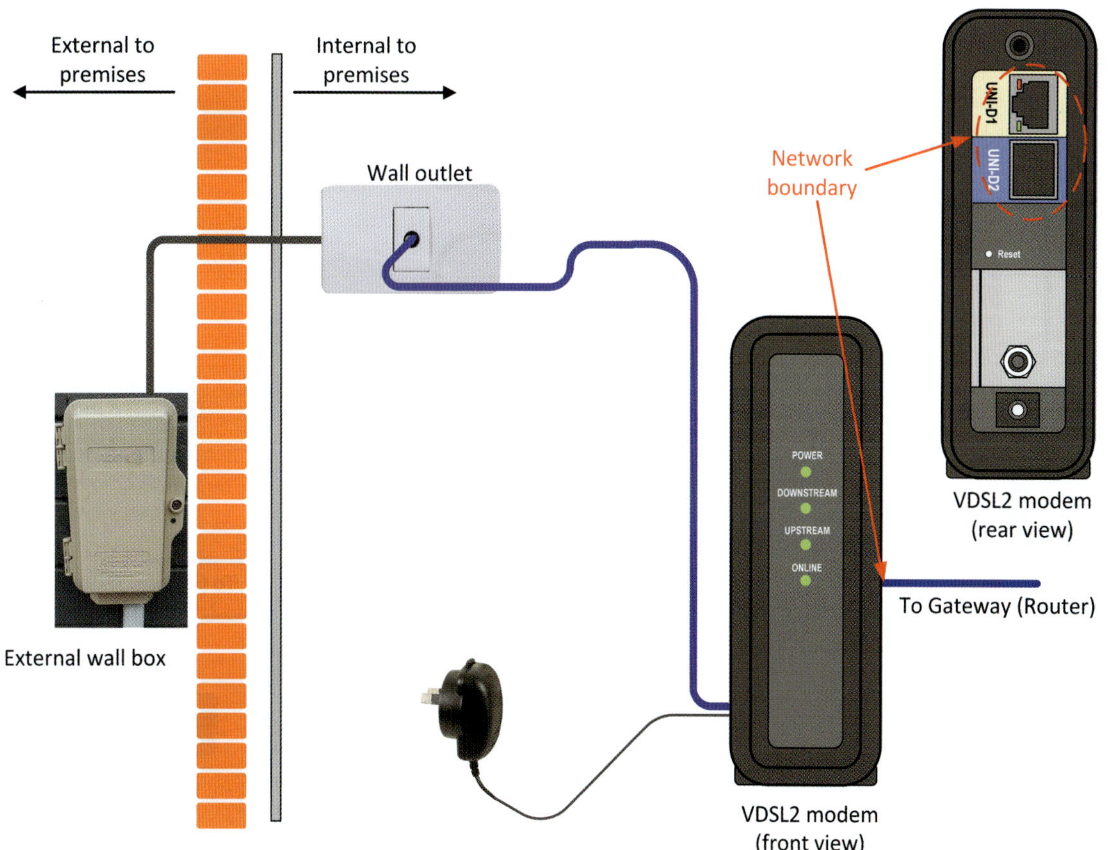

Sources: [L to R] Shutterstock.com/hansbui; Shutterstock.com/aoya.

FIGURE 12.24 The FTTN connection

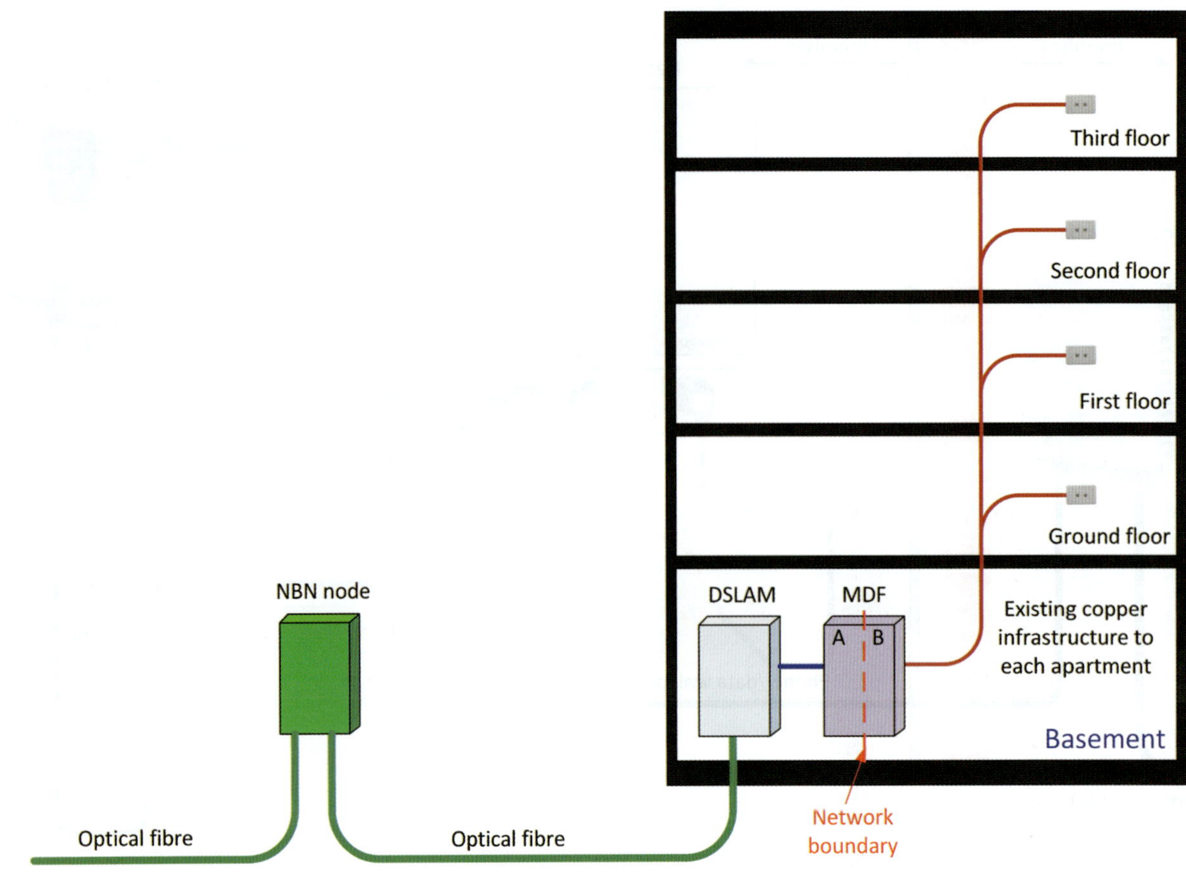

FIGURE 12.25 The FTTB connection

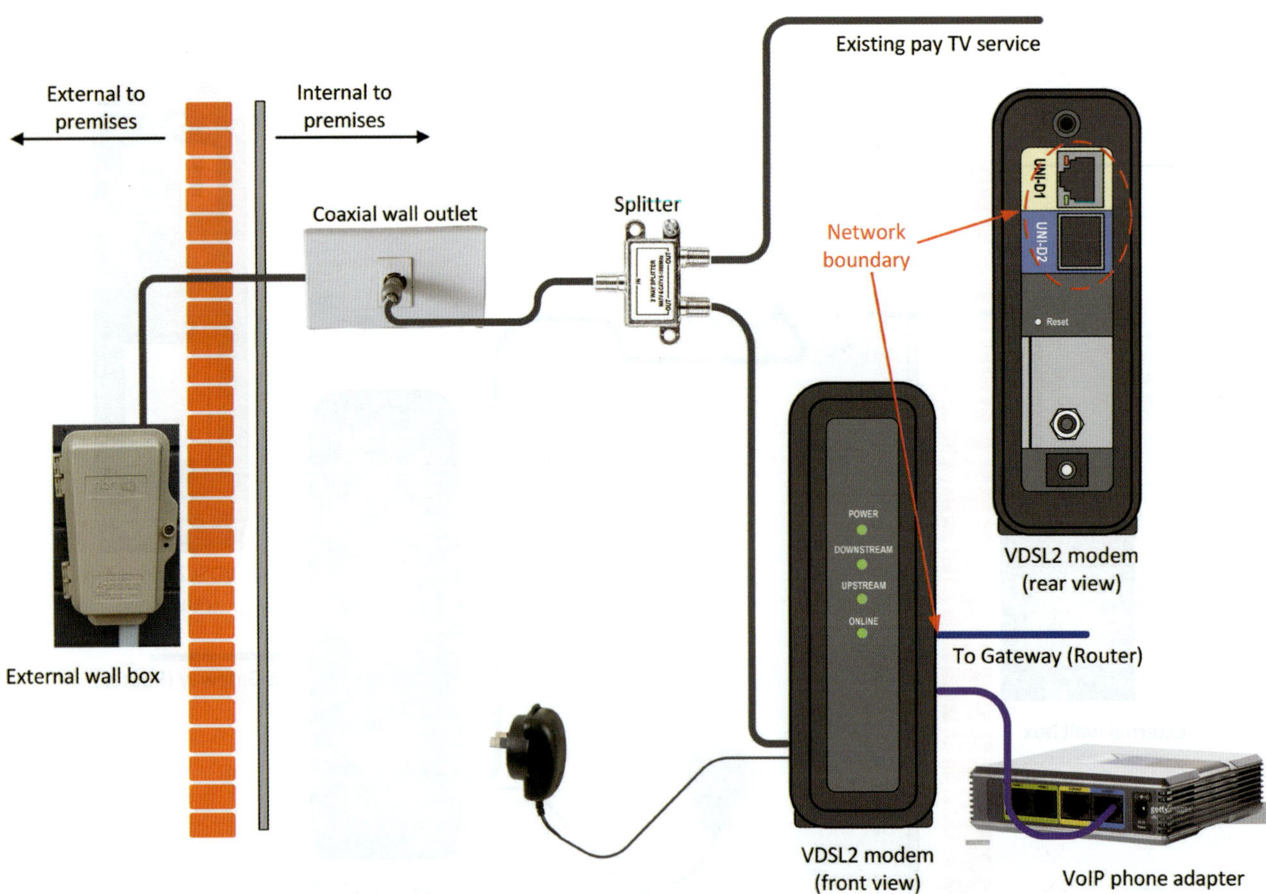

Sources: [L to R] Shutterstock.com/hansbui; Shutterstock.com/mikumistock; Shutterstock.com/Sanit Ratsameephot; iStock.com/vtls.

FIGURE 12.26 The HFC connection

Wireless and satellite technologies

Table 12.7 summarises the distinguishing features of the wireless and satellite technologies used to get the national broadband to premises (NBN Co., 2018).

TABLE 12.7 Summary of wireless and satellite technologies

Technology	Description	Installation requirements
Fixed wireless	Data transmitted as radio signals from a transmission tower, located up to 14 km away, to an nbn™ outdoor antenna at the premises.	Outdoor antenna and nbn™ connection box, which requires power to operate, at the point where the cable from the nbn™ outdoor antenna enters the premises. See Figure 12.27.
Satellite	Data transmitted as radio signals from a Sky Muster™ satellite	Roof mounted satellite dish and nbn™ supplied modem, which requires power to operate, at the point where the cable from the satellite dish enters the premises. See Figure 12.28.

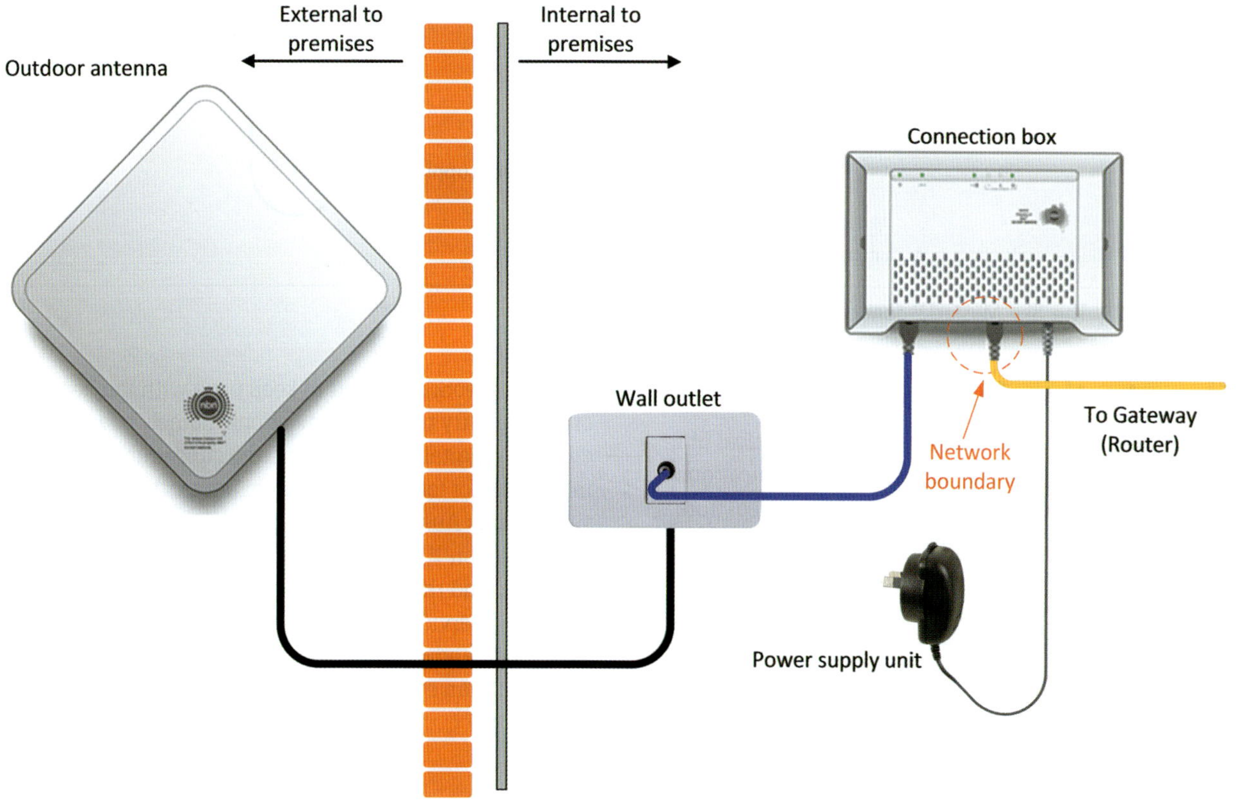

Source: Shutterstock.com/aoya.

FIGURE 12.27 The fixed wireless connection

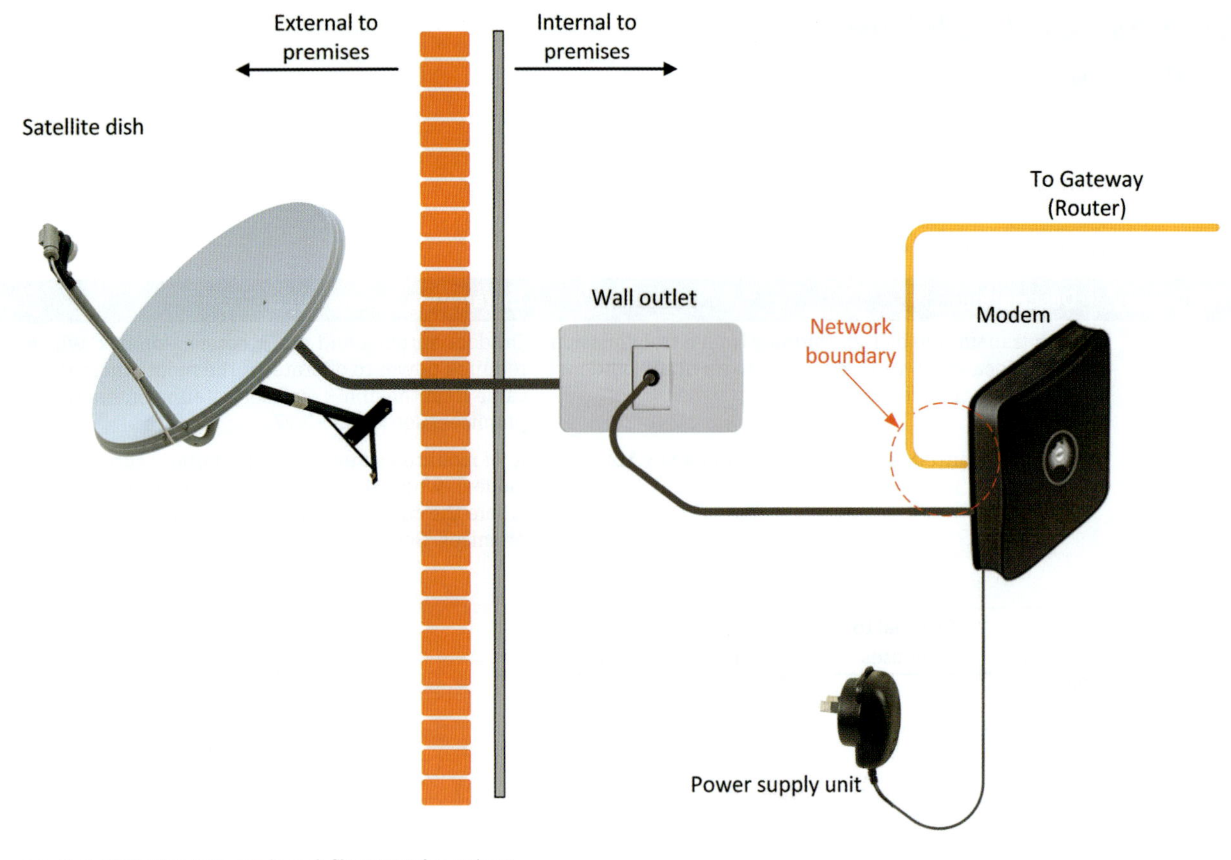

Sources: [L to R] Shutterstock.com/a_v_d; Shutterstock.com/aoya.

FIGURE 12.28 The Sky Muster™ connection

Fibre-to-the-Curb (FTTC)

Fibre-to-the-Curb is a technology where the optical fibre network extends all the way to the telecommunications pit located outside customer premises. The optical fibre connection terminates at a small device known as a distribution point unit (DPU) that then uses the relatively short length of existing copper telephone service to deliver broadband services to the premises. The DPUs are reverse-powered from the end-user premises and should have a lower power requirement per household.

The move to a national broadband network

The need for an improved broadband network infrastructure was recognised when Australia's existing copper network struggled to keep pace with the demands of 21st century telecommunications, namely a fast-changing digital landscape and the subsequent increase in demand for fast and reliable connectivity. **Table 12.8** summarises Australia's internet connection speeds from dial up to nbn™. With most enterprises having an online presence, and the continued evolution of music and video streaming services as well as online gaming, the need for increased bandwidth resulting in faster download speeds is greater than ever before.

TABLE 12.8 Summary of internet connection speed

Decade	Technology	Connection speed download / upload	Cost base
1990s	Dial up	0.056 / 0.006 Mbps	Connection time
Early 2000s	ADSL	1.5 / 0.256 Mbps	Downloads
	Cable	2.9 / 0.256 Mbps	
Late 2000s	ADSL 2+	24 / 0.8 Mbps	Downloads
	Cable	100 / 2 Mbps	

»

Decade	Technology	Connection speed download / upload	Cost base
Mid 2010s	nbn™	12 / 1 Mbps	Speed and downloads
		25 / 5 Mbps	
		50 / 20 Mbps	
		100 / 40 Mbps	

Decommissioning of the copper network

The copper telephone network will be decommissioned in an area 18 months after the national broadband connection is ready for service. Furthermore, new connections must be made to the optical fibre network and not the copper network.

REVIEW QUESTIONS

1 What is the name of the point in a telecommunications network where the service provider will supply a carriage service to a customer?
2 A single domestic residence is served by nbn FTTP technology. What point delineates the network boundary?
3 A multiple domestic residence is served by nbn FTTB technology. What point delineates the network boundary if the premises has a DSLAM and MDF?
4 What is the maximum transmission distance for nbn fixed wireless technology?
5 What existing cabling infrastructure is utilised in the nbn HFC technology?
6 Which nbn technology employs a DPU, which is reverse-powered from the end-user premises?

12.6 Cable identification

Administration is a fundamental part of cable installation. It means accurate identification and record keeping of all components that comprise the cabling system, including pathways, distributors and other spaces housing the cabling system. It is important to promptly update cabling records on completion of work to reflect any changes. Some administration systems are computer based.

Associated documentation may include:

- building drawings detailing cable routes
- location and identification of telecommunications outlets
- distributor construction and layout
- test and certification results
- connectivity mapping.

Distributor records

Distributor records provide a way of documenting the origin of an incoming or backbone cable, and where outgoing (quad) cables go within a particular area serviced by that distributor. Distributor records also have provision for documenting the cross-connection details within that particular distributor.

There are two sections to a distributor record. One section deals with the incoming cabling and the other deals with the outgoing cabling. There is no standard method for completing distributor records. However, regardless of the record book or secure cable distribution record system used, you should complete as many details as possible.

Distributor records must be updatable. For written records entries must be made in pencil so as to allow service and cross-connect details to be changed following relocation of services and the like.

The distributor in the following example consists of two 11-way, 100-pair verticals. One vertical is vertical A and the other is vertical B. For clarity, the example shows only the first 20 pairs of wires. **Figure 12.29** shows the physical layout of the cross-connects, while **Figures 12.30** and **12.31** show the distributor cross-connect records.

FIGURE 12.29 100-pair distributor

TELECOMMUNICATIONS CABLING RECORD

Location: 123 Sea St Meadowfields .... Floor: Gnd... Distributor No: FD-01A... Vertical: .A

Cable Origin		Termination Pair Number	Service Details (eg Telephone number or extension port for telephone services)	Jumper To		Other Details (eg user name or workstation location)
Distributor ID	Vertical Pair			Vertical Frame	Pair	
20 pair underground cable from BD vertical C41 – C60	C60	20	9942 3640	B	12	
	C59	19	9942 3639	B	11	
	C58	18	9942 3638	B	10	
	C57	17	9942 3637	B	9	
	C56	16	9942 3636	B	6	Fax
	C55	15	9942 3635	B	4	
	C54	14	9942 3634	B	5	
	C53	13	9942 3633	B	7	
	C52	12	9942 3632	B	8	
	C51	11	9942 3631	B	3	
	C50	0	9942 3630	B	20	
	C49	9	9942 3629	B	14	
	C48	8	9942 3628	B	15	Fax
	C47	7	9942 3627	B	13	
	C46	6	9942 3626	B	19	
	C45	5	9942 3625	B	17	
	C44	4	9942 3624	B	18	
	C43	3	9942 3623	B	16	
	C42	2	9942 3622	B	2	
	C41	1	9942 3621	B	1	ADSL

Notes: 1. Write lightly with black pencil 2. Complete all relevant particulars

FIGURE 12.30 Basic record for the A vertical – complete all details

TELECOMMUNICATIONS CABLING RECORD

Location: 123 Sea St Meadowfields .... Floor: Gnd... Distributor No: FD-01A... Vertical: .B.

Cable Origin		Termination Pair Number	Service Details (eg Telephone number or extension port for telephone services)	Jumper To		Other Details (eg user name or workstation location)
Distributor ID	Vertical Pair			Vertical Frame	Pair	
		20	9942 3630	A	10	Room 112
		19	9942 3626	A	6	Room 111
		18	9942 3624	A	4	Room 110
		17	9942 3625	A	5	Room 109
		16	9942 3623	A	3	Room 108
		15	9942 3628	A	8	Room 107
		14	9942 3629	A	9	Room 106
		13	9942 3627	A	7	Room 105
		12	9942 3640	A	20	Reception
		11	9942 3639	A	19	Store
		0	9942 3638	A	18	Technician lab
		9	9942 3637	A	17	Technician lab
		8	9942 3632	A	12	Room 104
		7	9942 3633	A	13	Room 103
		6	9942 3636	A	16	Room 102
		5	9942 3634	A	14	Room 101
		4	9942 3635	A	15	Manager
		3	9942 3631	A	11	Electronics lab
		2	9942 3622	A	2	Radio lab
		1	9942 3621	A	1	Staff room

Notes: 1. Write lightly with black pencil 2. Complete all relevant particulars

FIGURE 12.31 Basic record for the B vertical – complete all details

Note that only the lower two blocks on the B vertical are terminated. The remaining blocks are spare.

After terminating the incoming cabling, enter these details in the cross-connect record book. This involves:

- entering the *vertical designation*. In this case it is A
- entering the *origin* of the cable in columns 1 and 2 (*cable origin*)
- entering the *termination pair numbers* in column 3, starting with pair 1 at the bottom
- completing the details of the *service details* in column 4 (*telephone number or extension port for telephone services*). These correspond to those shown in the network boundary distributor record book.

At this stage it is not known what cross-connection is necessary. Complete these details later.

After terminating the telecommunications outlet cabling from the floor distributor it is possible to enter these details in the cross-connect record book. This involves the following process:

- enter the *vertical designation*. In this case it is B
- enter the *termination pair numbers* in column 3, starting with pair 1 at the bottom
- enter the *destination* of the cables in column 7 (*Other details*)
- complete the details of the required *exchange line services* in column 4. Enter these details after completing the cross-connect details if there is no requirement for specific exchange lines to terminate in specific locations
- compare the entries for the A and B verticals. In the A vertical 'Jumper to' column (**Figure 12.30**) progressively write the B vertical pair having the same exchange line number. In the example, the exchange line number for wire pair A1 corresponds to that of wire pair B1. Repeat this process for all wire pairs on the A vertical
- repeat this process for the B vertical
- complete all records before installing cross-connects.

After documenting the required cross-connections it is possible to accurately make the physical cross-connects.

Cabling plans

Construction of commercial buildings is complex, needing much information and planning beforehand as well as the involvement of many people. It is essential for all these people to work together harmoniously if the building is to be functional and comfortable.

The building construction progresses in stages. First is the preparation of the ground, followed by the foundation and the basic structure. Next are the floors, walls, roofing and cladding. Throughout these stages it is necessary to provide and connect various services to cater for the needs of the people using the building. Some of these services require planning and installation at various stages of construction. Some, like telecommunications and low-voltage electrical, also require segregation.

For a structure to be built and for correct inclusion of all these services there must be plans detailing how and where to install the services. It is also necessary to show their relationship with the building as a whole.

To locate cables already installed within customer premises, you must be able to read and interpret plans. For a telecommunications cable installer providing services in a building, interpretation of plans and practical application of their information are most important. Most buildings have a standard layout of details on a plan. This means that people using these plans need to know where to look and what to look for. See *Electrotechnology Practice* for a full treatment of architectural drawings.

Cabling symbols

Cabling symbols pictorially represent equipment used in an installation. The cable installer needs to read plans and record information using common symbols in the course of their work. It is important, therefore, to be able to recognise and draw the symbols that represent various equipment components in the workplace.

Standard cabling symbols ensure that everyone involved in the telecommunications industry understands exactly what a plan details and exactly what is necessary for a particular location in an installation. Standard cabling symbols ensure that anyone who knows how to read and interpret these symbols can read and use plans and reference material. There is a requirement that symbols be drawn to a minimum size on all plans to ensure they are clear and legible.

AS/NZS 3085.1: 2004 *Telecommunications Installations – Administration of Communications Cabling Systems – Basic Requirements* details the cabling symbols for telecommunications use. The various tables of symbols provide information about the use, minimum size, orientation and position of fittings and equipment. They give references to the Australian Standard referring directly to fittings or equipment.

Drawings

Outdoor cabling drawings

An external cabling drawing should show all cabling and distributing arrangements between interconnected buildings in an installation. The cabling provider needs to read external cabling drawings and record information using common symbols in the course of their work. It is important, therefore, to be able to recognise and interpret external cabling drawings related to the workplace. **Figure 12.32** shows some symbols for a common external telecommunications plant.

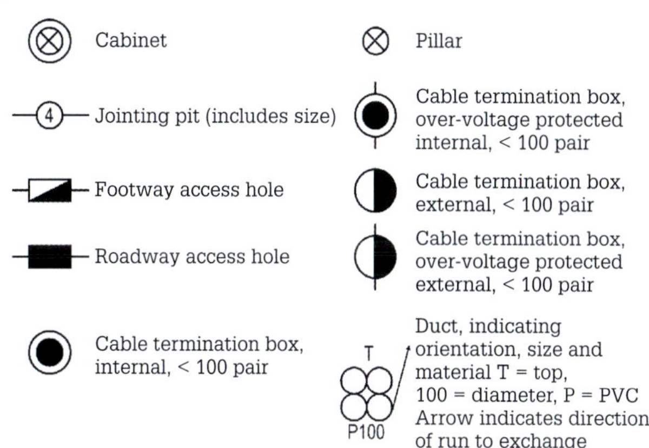

FIGURE 12.32 Common outdoor cabling symbols

External cabling drawings provide a considerable amount of information as **Figure 12.33** shows. Information recorded on an external cabling drawing should include:

- scale of drawing
- orientation and location of buildings

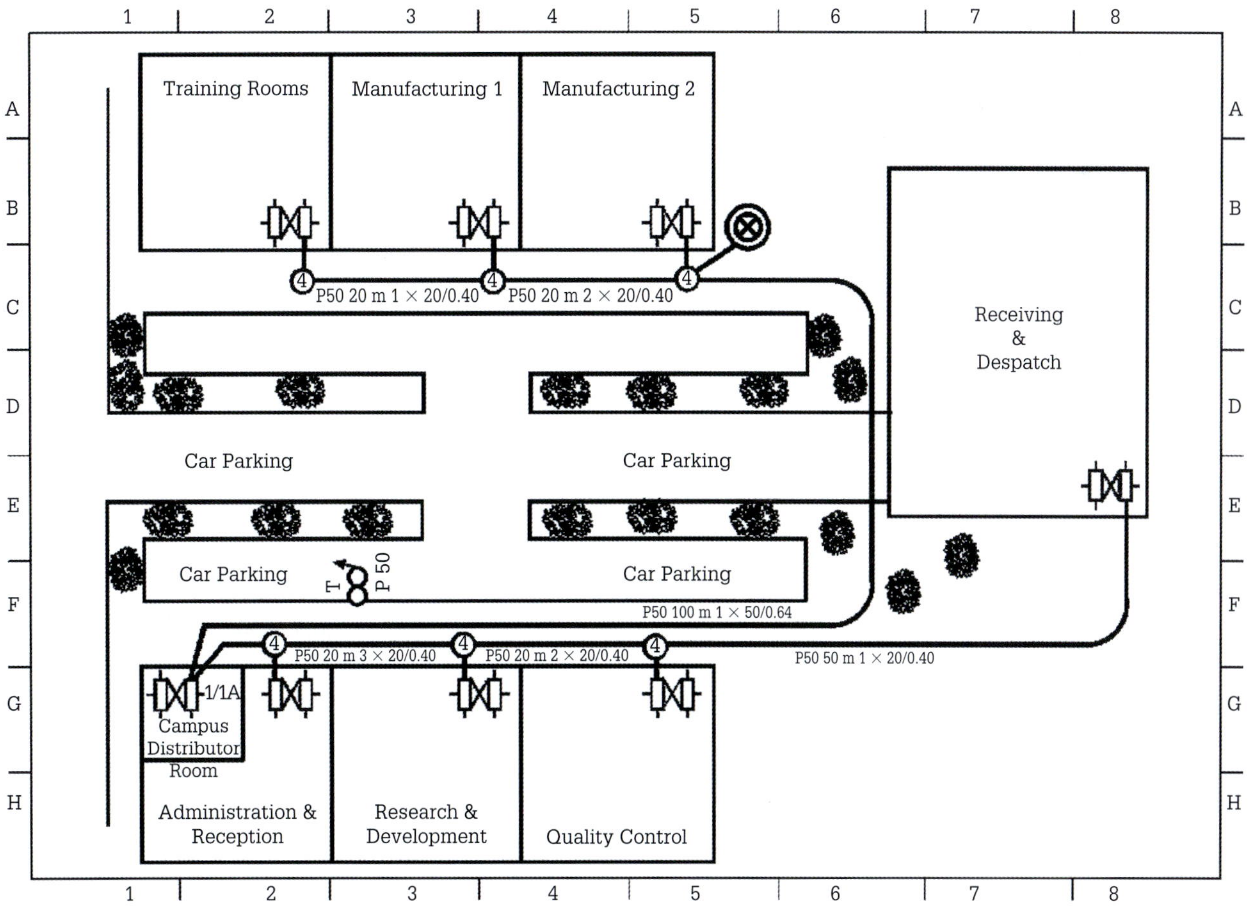

FIGURE 12.33 Sample outdoor cabling plan

- building description or number
- campus distributor
- each building distributor
- pathways, including depth or height and description of capacity of such pathways
- cables (including types)
- pits and other infrastructure, including:
  - cable identification number
  - cable type, size (number of pairs), conductor diameter and length
  - general location and size of pits (where standard-type pits are used), type and size of conduits and distances from the centre of each pit or access hole
  - general location of cabinets and pillars (external distributors).

Indoor cabling drawings

An indoor cabling drawing should show all cabling arrangements, outlets and equipment within an installation. The cable installer needs to read internal cabling drawings and record information using common symbols in the course of their work. Therefore, it is important to be able to recognise and interpret internal cabling drawings related to the workplace. **Figure 12.34** shows some symbols for a common internal telecommunications plant.

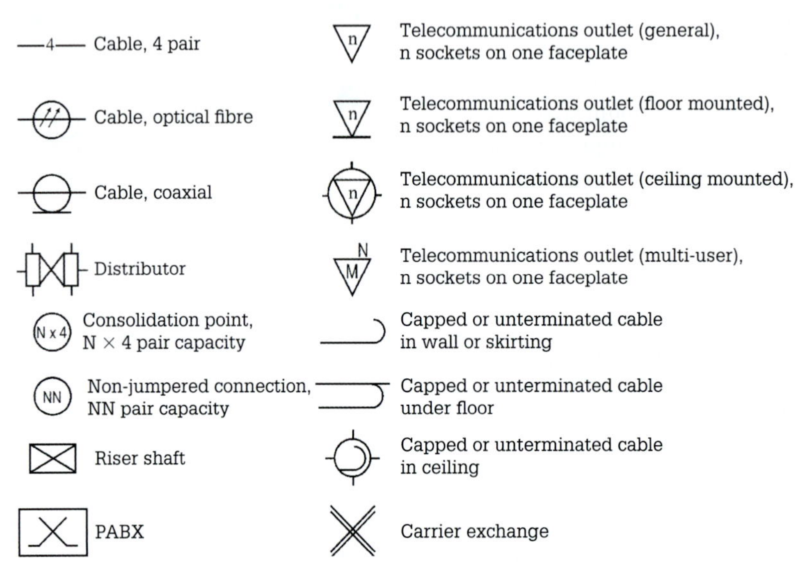

FIGURE 12.34 Common indoor cabling symbols

Floor layout drawings

Floor layout drawings, similar to **Figure 12.35**, should include the following:

- all main structural features such as stairs, lift wells and other permanent fixtures
- orientation and location of the building (on ground floor or street level drawing only)
- location of riser shafts, major pathways, and location and designation of distributors
- location of cables – capped or unterminated, left in walls, skirtings or ceilings
- numbering of telecommunications outlets – every outlet should have a unique number
- location of connection to building protective earth.

Backbone cabling drawings

Backbone cabling drawings should include the following:

- public network interface cable indicating the number of pairs and wire diameter
- cross-connect including, where appropriate, designation of panels and pair (or fibre) capacity of each
- indoor cables with size expressed in number of pairs and category.

Identifiers for distributors should be shown, for example, as:

- 1A (i.e. the A distributor on the first floor)

Identifiers for telecommunication outlets should be shown, for example, as:

- B3A (i.e. the A distributor on the third floor below ground floor).
- Level – Grid Reference – Originating Distributor – TO number.
- The first two identifiers are optional. The telecommunications outlet identifier could be, for example, 2-B12-2B-56.

The backbone cabling diagram (see **Figure 12.36**) provides details of distributor and cable configurations. This information is useful when quoting for a job and when ordering cable and equipment.

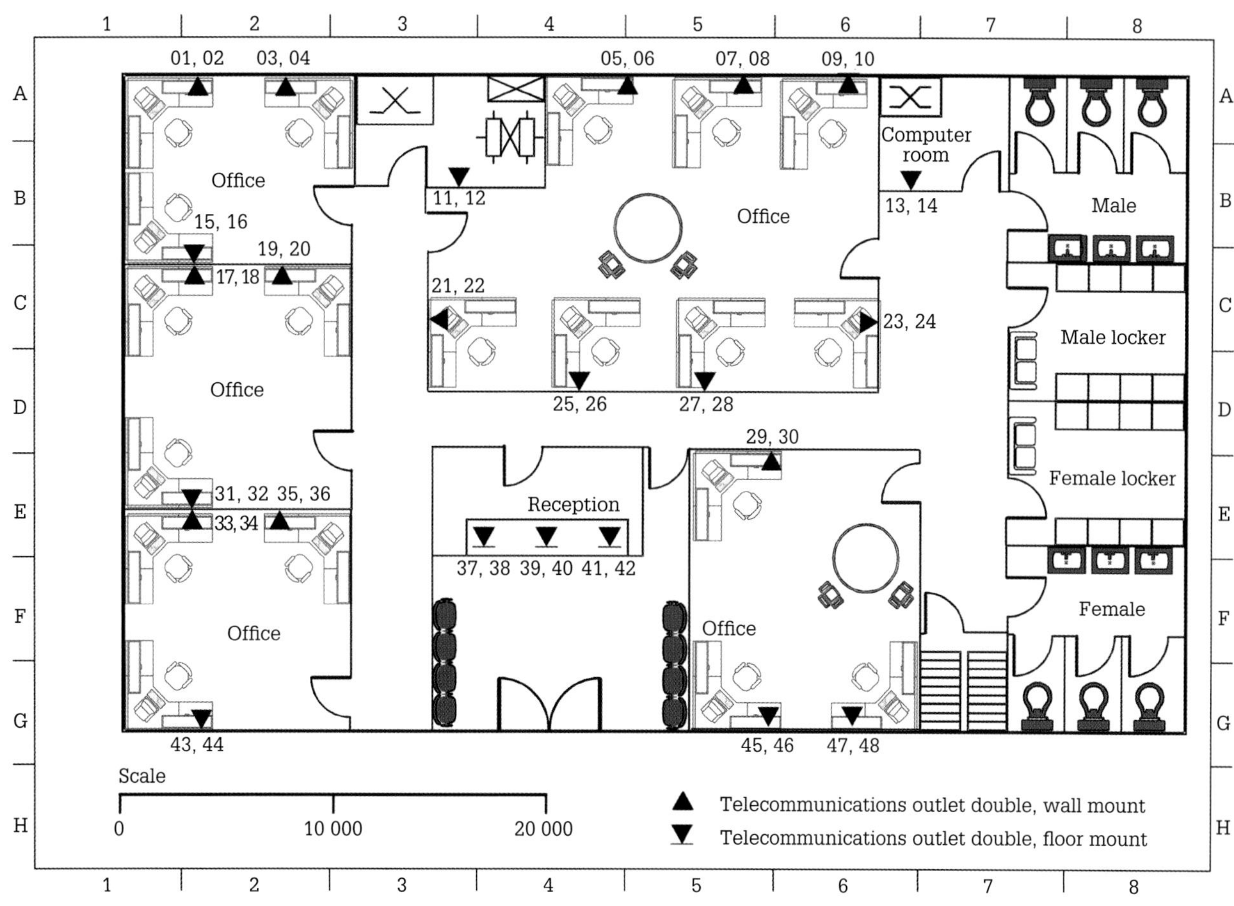

FIGURE 12.35 Sample indoor layout drawing

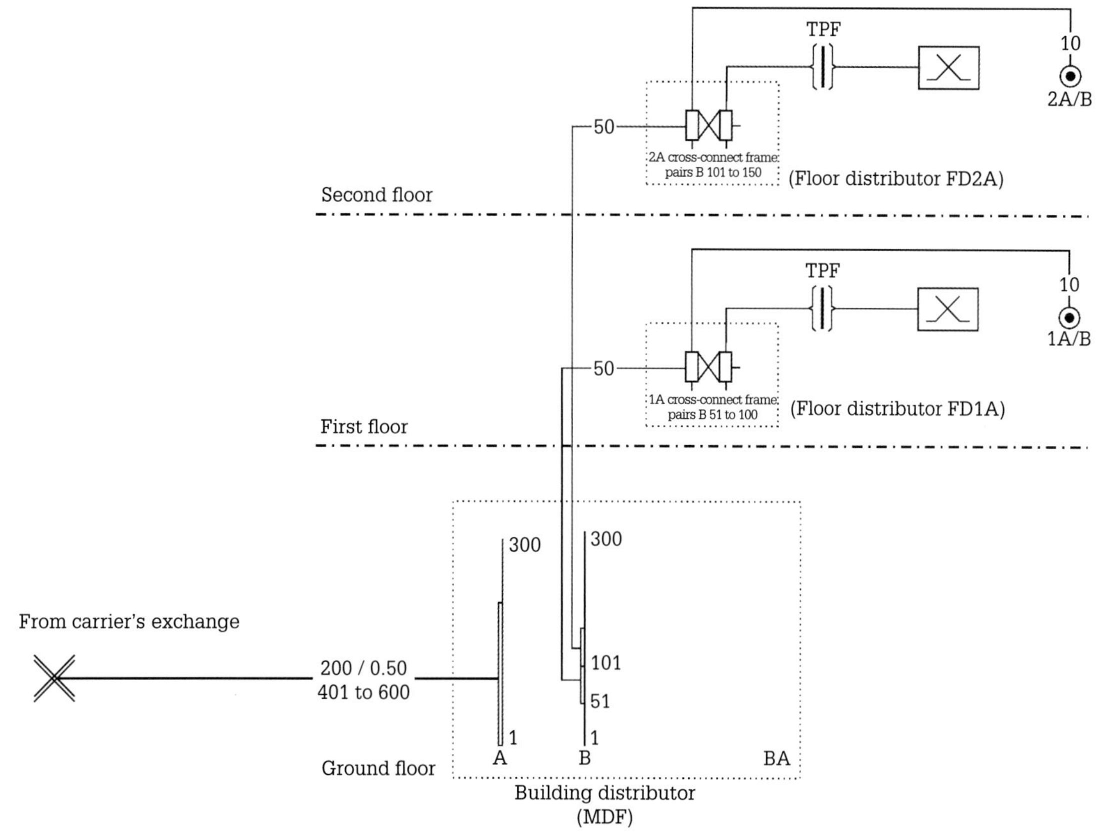

FIGURE 12.36 Sample backbone cabling diagram

REVIEW QUESTIONS

1. Why is cabling administration considered a fundamental part of cable installation?
2. What is provided by distributor records?
3. What are cabling symbols?
4. What should an external cabling drawing show?
5. Draw the outdoor cabling symbol for a cabinet.
6. Draw the outdoor cabling symbol for a roadway access hole.
7. What should an internal cabling drawing show?
8. Draw the indoor cabling symbol for a PABX.
9. Draw the indoor cabling symbol for a coaxial cable.
10. What should the backbone cabling drawing show?

12.7 Telecommunication cable types

Before looking at the types of cables used to convey telephone signals, it is important to consider the electrical characteristics of these cables.

Telecommunications line characteristics

The simplest form of two-wire line consists of bare conductors suspended on insulators at the top of poles (an open line). The wires must not touch each other, as this would result in a short circuit and interrupt the communications.

Another type of two-wire line consists of conductors insulated from each other in a cable, which also has an outer cover of insulation called a sheath. PVC is commonly used as both wire insulation and sheath. The two insulated conductors are twisted together along the length of the cable to form what is known as a twisted pair. One advantage of twisting the wires as pairs is that it helps in identifying which two wires belong together. This is an advantage when you consider that a 100-pair indoor-grade cable has, for example, 20 individual white wires.

Often it is necessary to install many two-wire lines between the same two places. It is more convenient to provide this by making a cable with a number of twisted wire pairs. Sometimes the twists are in pairs and sometimes they are in fours (or quads). To identify the various wires within the cable, each wire has a colouring on the insulating material surrounding it, in accordance with a standard colour code for cable pair identification.

Electrical properties

The conductors have some opposition to electric current (resistance). Furthermore, as no insulating material is perfect, the insulation used to separate the two conductors of a pair will allow a very small current flow between the two conductors.

There are a number of electrical factors that cause the loss of signal along the length of the cable. Cables exhibit the electrical properties of resistance, inductance, and capacitance. Each of these contribute to signal loss in varying degrees. Resistance causes a gradual drop in voltage along the length of the cable. The effect of this is more significant over long lengths of cable and low signal levels.

Inductance is that property of the cable that opposes changes in current. Inductance can have a transformer effect and result in coupling of signals from different wires, which is known as cross-talk. As the signal levels are generally low current, this is not of major concern for an undamaged cable, which is used in the correct application. Inductive reactance, which is an a.c. property resulting in signal attenuation, can be determined from the relationship:

$$X_L = 2\pi fL$$

where X_L is the inductive reactance of the wire pair in ohms, f is the base frequency of the transmitted signal in hertz, and L is the inductance of the wire pair in henrys.

So the inductive reactance of the conductor increases proportionally with frequency (f). This has a more pronounced effect on data cables that carry frequencies in the order of GHz.

Capacitance is that property of the cable that allows it to store electrical energy. Capacitance can have the effect of providing a low impedance path between adjacent conductors, causing a coupling of signals from different wires, which is known as cross-talk. Again this is not of major concern for an undamaged cable, which is used in the correct application. Capacitive reactance, which is an a.c. property can be determined from the relationship:

$$X_c = \frac{1}{2\pi fC}$$

where X_c is the capacitive reactance of the wire pair in ohms, f is the base frequency of the transmitted signal in hertz, and C is the capacitance of the wire pair in farads.

So the capacitive reactance of the conductor decreases proportionally with frequency (f). This has a more pronounced effect on data cables that carry frequencies in the order of GHz as it has the effect of 'bleeding' signals from their intended path.

An electrical signal propagating along a wire also creates a small, circular magnetic field around the wire. If this signal is alternating current, the magnetic fields will be continually building up and collapsing. If two wires are in close proximity, the magnetic field from one wire will pass through the adjacent wire inducing a small voltage (transformer action). In the case of twisted pair media, the paired conductors are carrying equal and opposite signal energy (balanced pair operation), so the two magnetic fields cancel each other.

Balanced pair operation means that the 'a' and 'b' wires of the pair are carrying the same magnitude of current but in opposite directions as current is supplied from the source along one wire, through the destination device, and returns to the source through the other wire. Any electrical disturbance will affect both wires equally due to their close proximity because of the twisting of the pair. If wire 'a' was at 0 V and wire 'b' was at 15 V, the differential voltage is 15 V (15 – 0). If an electrical disturbance induced +8 V on each wire, then wire 'a' is at 8 V and wire 'b' is at 23 V, the differential voltage is 15 V (23 – 8).

Types of copper telecommunications cable

Demand for high-speed telecommunications is increasing worldwide. As such, copper, optical fibre and wireless systems are being used to meet this demand. Recent developments in digital signal processing technology have extended the bandwidth and hence the usefulness of copper cabling. For the telephone network, asymmetric digital subscriber line (ADSL) and fibre to the curb (FTTC) technologies have breathed new life into ageing infrastructure by providing high data speeds over existing twisted pair cables. However, copper wire is still a common transmission path for telecommunications circuits.

Copper conductor cables provide a cost-effective cabling solution for voice and data communications networks. There are two main types of copper cabling media: unshielded twisted pair (UTP) and overall braid-shielded twisted pair (S/UTP).

Indoor telephone cable

Indoor telephone cable usually has a cream-coloured PVC sheath over PVC-insulated wires. It may be either unshielded twisted pair (UTP) or overall foil-shielded twisted pair (F/UTP) construction. This cable is suitable for installation in the roof void, in wall cavities and in the floor void. It is not, however, suitable for use in underground situations, or in locations that expose it to direct sunlight or the effects of weather or moisture. Indoor cable typically has 0.5 mm diameter copper conductors. **Figure 12.37** shows a 10-pair unshielded indoor telephone cable.

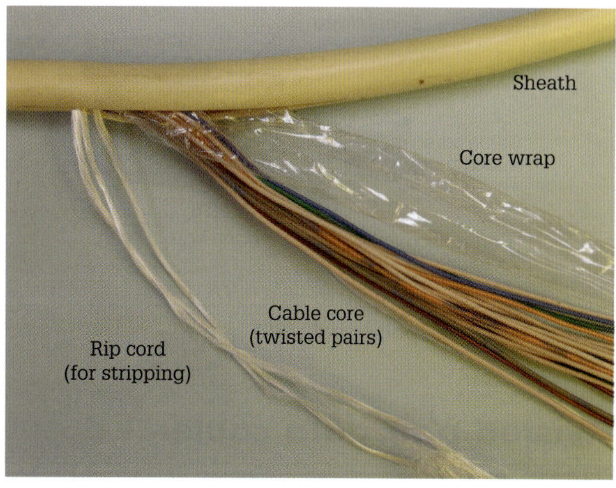

FIGURE 12.37 Ten-pair indoor cable

Outdoor cable

Outdoor cable is categorised into two groups: aerial cable and underground cable. Underground cable usually has polyethylene conductor insulation (solid or foam) and a polyethylene sheath. The core of an underground cable may incorporate water blocking such as a lapped metal tape, grease or gel compound (silicone based). It is usual to call the gel compound 'jelly' and the cable 'filled cable'. Underground cables sometimes have an additional outer jacket of nylon to protect them from termites. Underground cable typically has 0.64 mm diameter copper conductors (0.4 mm and 0.9 mm are also available). **Figure 12.38** shows the typical construction of a jacketed underground cable.

FIGURE 12.38 Fifty-pair underground cable (even with some jelly filling intact it easily separates into 10-pair groupings)

Aerial cable also has a polyethylene sheath and polyethylene conductor insulation. Aerial cables often have a strong support wire, contained within the same jacket, but outside of the screen, known as an integral bearer wire as shown in **Figure 12.39**. This bearer wire may be steel or hard-drawn copper. The core of an aerial cable may also incorporate water blocking such as a lapped metal tape, grease or gel. Aerial cable typically has 0.64 mm diameter copper conductors (0.4 mm and 0.9 mm are also available). The typical construction of a 10-pair aerial cable is shown in **Figure 12.39**.

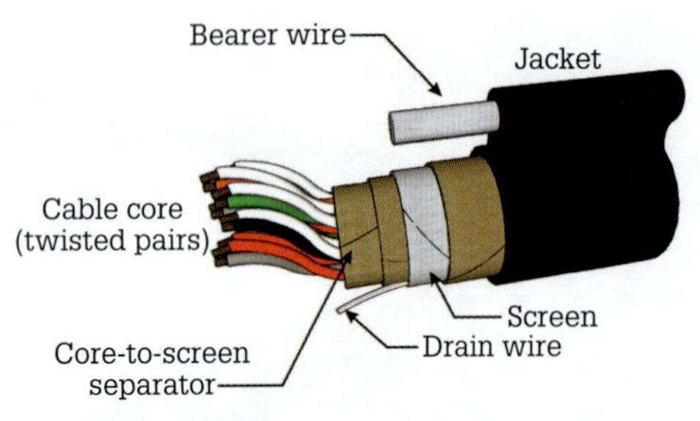

FIGURE 12.39 Ten-pair aerial cable

Twisted pair data cable

The telephone cables pictured in **Figures 12.38** and **12.39** are of unscreened (UTP) or screened construction (F/UTP in the case of the aerial cable). For data applications, the absence of shielding gives UTP cables a high degree of flexibility as well as rugged durability. In cabling environments where electrical noise (EMI) is present, the use of F/UTP cable may be required. The F/UTP cable is slightly larger and less flexible than a comparable UTP cable. UTP cables are found in many data networks and telephone systems.

With the introduction of new cables using individual foil-shielded pairs and overall shield constructions, there has been much confusion regarding the proper terminology used to describe these different designs. AS/NZS 11801: 2019 establishes a systematic naming convention, as shown in **Figure 12.40**.

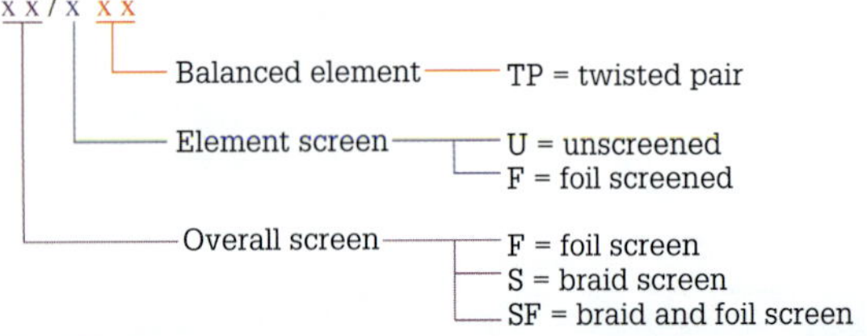

FIGURE 12.40 Cable naming convention

UTP cable

UTP cable constructions feature unshielded twisted pairs enclosed within an overall thermoplastic jacket as shown in **Figure 12.41**. As cable manufacturers produce cables in advance of any standards, they tend to be identified by a category rating, hence Cat. 5, Cat. 6 and Cat. 7. The problem is that when the standard is released, performance requirements exceed the original capabilities of the cable, which itself has advanced. This gives rise to nomenclature such as Cat. 5e (e for *e*nhanced). While this terminology ultimately finishes up as trade jargon, the standard classifications seem to be rarely used in industry. So a cable that satisfies the minimum requirements for Class D applications (as specified in AS/NZS 11801: 2019, up to 100 MHz) is marketed as Cat. 5 but has been branded as Cat. 5e to distinguish it from the original lower performance cable that preceded the Standard.

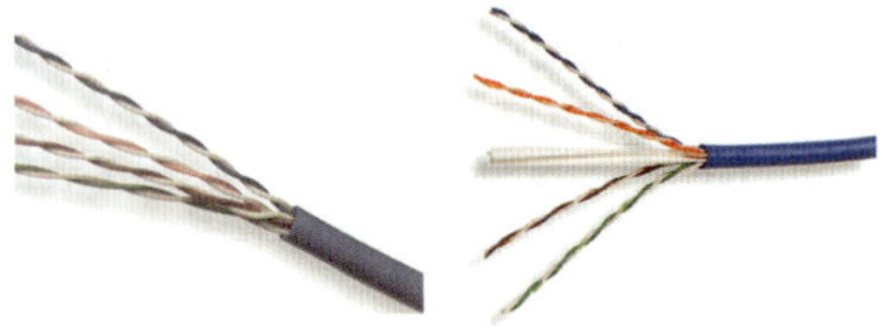

FIGURE 12.41 UTP data cable – for Class D application (left) and Class E application (right)

Note the spline in the centre of the cable for Class E application, which ensures each pair remains physically separated from the others. The inclusion of this spline increases the separation between pairs and maintains pair geometry resulting in lower cross-talk. This in turn affords the cable greater bandwidth, meaning it can carry data at faster speeds than a comparable cable without the spline.

F/UTP cable

F/UTP cable constructions feature unshielded twisted pairs surrounded by an overall conductive mylar-backed aluminium foil shield enclosed within an inclusive thermoplastic jacket as shown in **Figure 12.42**. F/UTP cables were previously known as 'ScTP', 'screened' or 'FTP'.

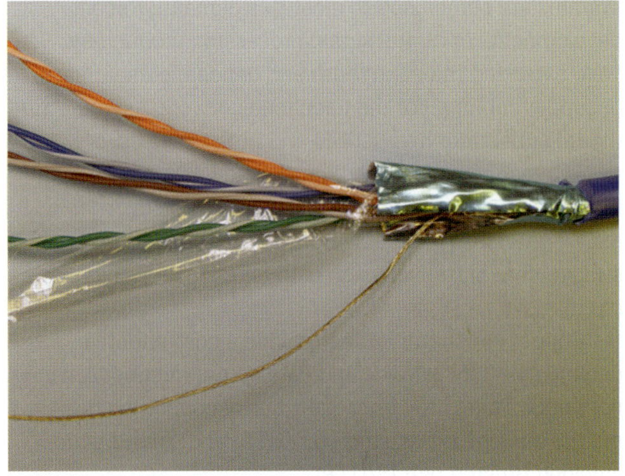

FIGURE 12.42 F/UTP data cable

To comply with AS/CA S008: 2020, a drain wire is in continuous contact with the foil shield. The drain wire affords easy connection to earth and provides electrical continuity in case the cable is bent too tightly causing the lapped tape foil shield to separate and break. Cable manufacturers refer to this cable construction as an overall foil shield. A shielded twisted pair can operate at higher data rates due to its higher immunity from noise. Shielded cables are more difficult to terminate than UTP cables, as good earthing of the shield is essential for effective operation. A potential problem is the possibility of creating circulating currents in the shield if both ends of the shield connect to earth.

S/FTP cable

S/FTP cable constructions feature individually foil-shielded twisted pairs surrounded by an overall braid enclosed within an inclusive thermoplastic jacket as shown in **Figure 12.43**. S/FTP cables used to be referred to as 'PiMF', 'STP', 'SSTP' or 'fully shielded'. The purpose of shielding the individual wire pairs is to reduce the inductive and capacitive coupling between adjacent wire pairs. These cables are harder to terminate than overall shielded twisted pair cables due to the increased number of drain wires and shields needing connection to earth.

FIGURE 12.43 S/FTP data cable

Types of optical telecommunications cables

An optical fibre is a glass or plastic fibre designed to guide light along its length and is used to convey end-to-end digital signals. Optical fibres allow transmission of data over longer distances and at higher data rates than other forms of cabling medium. One advantage of optical fibre is its low signal loss and its relative immunity to electrical and magnetic interference.

Optical fibre construction

The optical fibre comprises two concentric layers called the core and the cladding. The central core is the part of the fibre that carries the light. The cladding that surrounds the core provides the difference in refractive index that allows total internal reflection of light through the core. The cladding behaves as a mirror, reflecting light back into the core (total internal reflection). The refractive index of the cladding is less than 1% lower than that of the core. For example, if the core of a particular optical fibre had a refractive index of 1.47, then the associated cladding would have a refractive index of 1.46. The manufacturing process of the fibre is tightly controlled to ensure that large differences in the refractive index do not occur.

Optical fibres have an additional coating around the cladding of one or more layers of polymer to provide mechanical protection for the fibre. This coating, called the buffer, has no optical properties affecting the propagation of light through the fibre. It acts as a shock absorber. **Figure 12.44** shows the simplified construction of single-mode optical fibre (SMOF) and multi-mode optical fibre (MMOF) cables. Core diameters for single-mode fibres range from 5 μm to 10 μm, with a standardised cladding diameter of 125 μm. Core diameters for multi-mode fibres range from 50 μm to 100 μm, with a standardised cladding diameter of 125 μm.

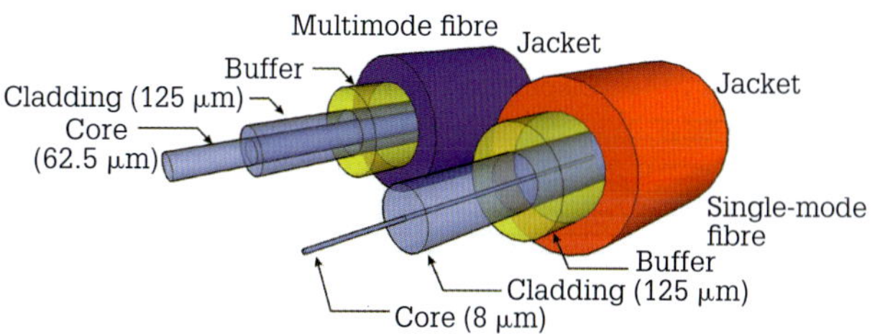

FIGURE 12.44 Construction of optical fibre cables

Classification of fibres

Optical fibres are classified by:

- material from which they are made
- modes of propagation.

Classification by material

Glass fibres have a glass core and glass cladding. The glass used in their construction is ultra-pure, ultra-transparent silicon dioxide or fused quartz. Impurities are added to the pure glass to achieve the desired refractive index.

Plastic-clad silica (PCS) fibres comprise a glass core and a plastic cladding inside an outer jacket. They do not perform as well as pure glass fibres as they have higher signal attenuation and lower bandwidth. They find use in industrial, medical, and sensing applications where a larger fibre core is advantageous.

Plastic fibres have a plastic core and plastic cladding inside an outer jacket. Plastic fibres have very low production tolerances and equally low performance. They are suited to low-bandwidth, short-distance applications. Their immunity to electromagnetic radiation and high flexibility make them the ideal cable to connect digital data streams between audio-visual equipment.

Classification by propagation

Optical fibre has two modes of propagation: multimode and single mode. These perform differently in terms of attenuation and signal spread. When a light wave is guided along an optical fibre, it exhibits specific modes. For a given optical fibre, the modes that exist depend on the dimensions of the cable and the variation in refractive indices. These modes include multimode step-index, multimode graded-index, and single mode step-index, which refer to the transition between the core and cladding. Step-index has a sharp transition, while graded-index has a gradual transition.

Figure 12.45 shows the various modes of light propagation through optical fibres.

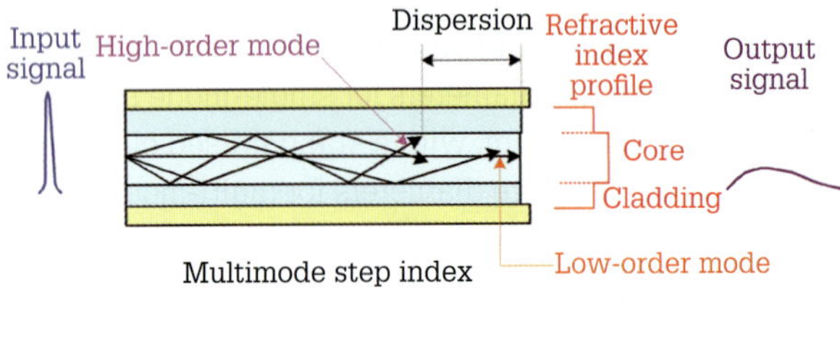

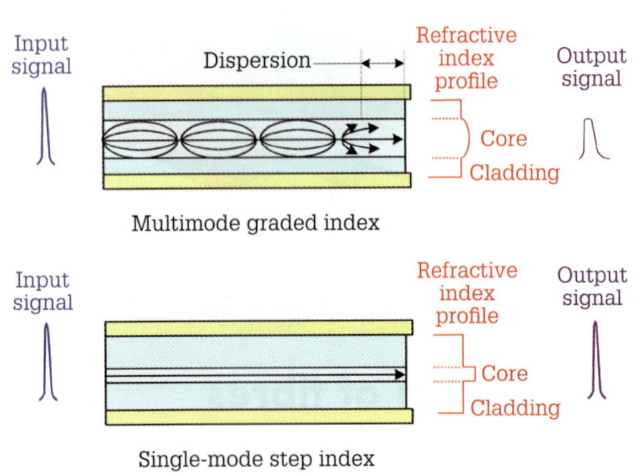

FIGURE 12.45 Optical fibre modes of propagation

Optical fibre cable

Optical fibres are packaged into cables, which contain:

- optical fibre
- buffer
- strength member
- jacket.

Buffer

When packaged as a cable, an additional buffer is added by the cable manufacturer. This may be of the loose-buffer or tight-buffer type. The **loose buffer** has a hard plastic tube with an inside diameter several times greater than that of the fibre. One or more fibres lie within the buffer tube. The tube isolates the fibre from the rest of the cable and the mechanical forces to which the cable may be subjected. The buffer becomes the load-bearing member. The fibre remains relatively unaffected by cable expansion and contraction with changes in temperature. A fibre has a lower temperature coefficient than most cable elements, meaning that it expands and contracts less with changes in temperature. Typically, some excess fibre is in the tube – the fibre in the tube is slightly longer than the tube itself. This allows the cable to expand and contract without stressing the fibre. Cables comprising loose-tube construction may incorporate water-blocking components such as gel or other water blocking compounds, or armouring. Loose-tube cables provide the best protection in outdoor environments.

The **tight buffer** has a plastic directly applied over the fibre coating. This construction provides better crush and impact resistance. However, it does not protect the fibre as well from the stresses of temperature variations. As the plastic expands and contracts at a different rate to the fibre, contractions caused by variations in temperature can result in loss-producing micro-bends. Tight-buffered cables are often referred to as premise or distribution cables and are frequently used in indoor applications where the temperature is more constant and environment less hostile than outdoors.

Another advantage of the tight buffer is that it is more flexible and allows tighter turn radii. This advantage can make tight-tube buffers useful for indoor applications where the ability to make tight turns inside walls is desired.

Strength members

Strength members add mechanical strength to the fibre. During and after installation, the strength members handle the tensile stresses applied to the cable so that the fibre is not damaged. The most common strength members are made from Kevlar aramid yarn, steel or fibreglass epoxy rods. Kevlar is most commonly used when individual fibres are placed within their own jackets. Steel and fibreglass members find use in multi-fibre cables. Steel offers better strength than fibreglass, but it is sometimes undesirable when one wishes to maintain an all-dielectric cable. Steel, for example, attracts lightning, whereas fibreglass does not. *Cables must only be secured by or drawn in by the strength members.*

Jacket

The **jacket** provides protection from the effects of abrasion, oil, ozone, acids, alkalis, solvents and the like. The choice of jacket material depends on the degree of resistance required for the different influences as well as cost. PVC is a common material for the jacket of indoor cables, while various forms of polyethylene and nylon find use in outdoor cables.

Optical fibre cables are made to suit indoor and outdoor purposes:

- Outdoor cables are generally stronger and stiffer to better suit the harsh environments in which they are installed.
- Indoor cables are built for flexibility and are generally made of a halogen-free insulation.

Outdoor cables

For outdoor applications there are mainly three types of cables used in both underground and aerial situations in Australia. They are slotted core, loose tube and indoor/outdoor riser cable.

Slotted-core and loose-tube cables are a 250 µm buffered fibre (i.e. 125 µm diameter fibre covered with a buffer material, making the diameter up to 250 µm). They are filled with a gel to prevent the ingress of moisture and covered with an outer black polythene sheath. A hard jacket of nylon can also be provided for termite protection. The versatility of these two types of cables ensures that they can be ploughed into the ground, pulled into ducts or suspended on poles.

The slotted-core structure (see **Figure 12.46**) is a core containing a central strength member which has been extruded over with plastic such that slots (hence its name) or grooves are formed in the periphery of the core. The core serves as the support and the strain-relief element.

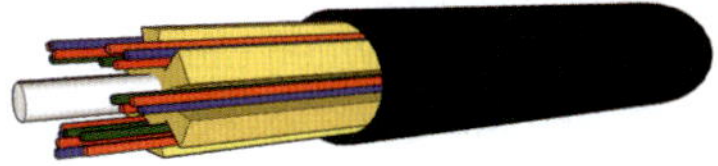

FIGURE 12.46 Slotted-core underground optical fibre cable

Each slot forms a continuous channel along the length of the cable and the optical fibres are placed loosely within the confines of the slot. For improved bending performance, each slot is formed either into a continuous helix or made to oscillate around the cable axis.

The loose-tube structure (see **Figure 12.47**) is made up of a number of elements stranded around a strength member. Each element is formed by extruding a loose-fitting tube over a single optical fibre or multiple numbers of optical fibres. For moisture protection, a filling gel is also injected during extrusion. The inner wall of the tube is smooth to ensure that the fibres can move freely. The tube's dimension is specially selected so there is enough room for the optical fibres to move in both the longitudinal and radial directions.

FIGURE 12.47 Loose-tube underground optical fibre cable

Indoor cables

Optical fibre cables for use within a building are generally sheathed with flame-retardant PVC or non-halogen material. The mechanical characteristic of the cable tends to be less stringent (not stiff) and the emphasis is more on flexibility and easy access to the optical fibres. The central strength member can be a fibreglass rod but this reduces the flexibility of the cable.

The preferred cable type is one where a number of individually jacketed fibres are enclosed in the same PVC sheath stranded around a central strength member.

The strengthened tight-buffered fibre is called a **unit cord cable** (see **Figure 12.48**). Each fibre unit can be directly terminated into a connector. This style of cable is also commonly referred to as a **break-out-style cable** or a **heavy-duty riser cable**. This type of cable can run vertically in a building, the cable being supported by its internal strength members. It can also be drawn into conduits or placed on cable trays as backbone cable.

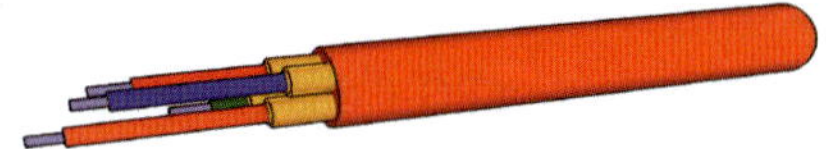

FIGURE 12.48 Cord indoor optical fibre cable

Connectors

Connectors can be fitted directly to each fibre in the cable. This is done by stripping away some buffer material and gluing the exposed fibre along with some unstripped buffer into the connector. The aramid strength member is crimped to the connector case for greater mechanical protection.

Some indoor applications require additional flexibility and less stringent mechanical requirements, such as horizontal runs within an office floor. In these situations the light-duty riser cable (see **Figure 12.49**) is preferred. Light-duty riser cable consists of a number of tightly buffered fibres bunched together, not individually sheathed, but surrounded by a Kevlar strength member and then sheathed with fire-retardant PVC.

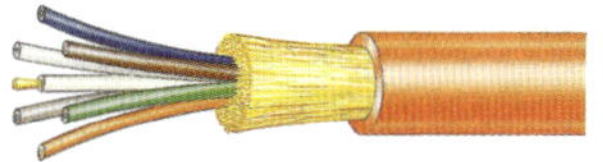

FIGURE 12.49 Riser optical fibre cable

Connectors can also be fitted directly to each fibre of this cable, by stripping away some buffer material and gluing the fibre directly into the connector. Special 900 µm rubber boots are usually used to provide additional protection to the buffered fibre as it leaves the connector ferrule.

Patch panel cable is very flexible and robust as aramid strength members are used in its construction. Connectors can be fitted using the same method as for the unit cord cable. The outer jacket is made in 2.5 mm and 4 mm diameter sizes to suit connector boots.

Connectors for optical fibre cable

Optical fibre connectors connect the fibres to optical transceivers and cross-connect points. Optical fibre connectors mate with other connectors via a through adaptor (sleeve or bulkhead). The adaptor allows two connectors to lock into it, which ensures the necessary physical contact and alignment of the connector ferrules. This minimises connector loss or attenuation.

AS/NZS 11801: 2019 recognises the SC connector as the preferred connector for optical fibres in Australia. A duplex SC (SC-D) connector is used at the telecommunications outlet. The term SC comes from 'subscriber connector', which describes its original telecommunications use. The basic SC connector consists of a plug assembly containing the ferrule. The plug mates into a connector housing. The SC connector has a pull-proof design. This means that a slight pull on the cable will not interrupt the optical contact, but a forceful pull would interrupt the optical contact. A duplex SC connector is shown in **Figure 12.50**.

FIGURE 12.50 Optical fibre connectors

Connectors exclusively for plastic optical fibres are cheap and easy to apply. It is possible to trim the plastic with a hot sharp knife, which alleviates the need for polishing as is required with glass fibres. The fibre is held in the connector through mechanical means such as barbs that grip the fibre cladding. One major connector style for plastic fibres is the digital audio connector as shown in **Figure 12.50**.

Another popular connector is the LC type. It is a small-form connector, which allows a duplex connection in the same space as a simplex SC connector. The LC adaptor has a latch similar to a modular plug. To install an LC connector it is necessary to align the latch with the latch-way of the adaptor and push the connector until the latch engages. Another popular connector for multimode networks is the ST connector, which has a round footprint with a barrel that is pushed forward onto the adaptor and rotated 90° to lock it in place.

Termination methods for fitting connectors

There are three common termination methods suitable for fitting SC, LC and ST connectors in the field. These are:

- no-epoxy/no-polish
- epoxy and polish
- pigtail splicing.

The **no-epoxy/no-polish** method is the fastest and simplest termination of the three. It utilises a factory-polished connector incorporating a factory-cleaved internal fibre and mechanical fibre splice. To install this termination it is necessary to strip the fibre to the glass, use a precision fibre cleaver to cleave the fibre, insert the prepared fibre into the connector and activate the connector by mechanical means.

With the **epoxy and polish** method, the installer prepares the fibre by stripping the fibre to the glass, injecting an epoxy adhesive into the connector ferrule using a syringe, applying an activator or hardener on the bare fibre, and inserting the fibre through the connector ferrule so that it protrudes beyond the ferrule by a couple of millimetres. This method does not require an external heat source to activate the epoxy; however, when the epoxy is pre-injected into the ferrule, it is necessary to heat the entire connector to activate the epoxy. The fibre is secured within the ferrule when the epoxy hardens. It is then necessary to remove the excess fibre by scribing and then polishing the end face of the connector ferrule. A polishing puck holds the connector to ensure it remains perpendicular to the lapping film during the polishing process. It takes a great deal of practice to become proficient at manually polishing fibre connectors.

The **pigtail splicing** method involves fusion splicing a short factory-made, normally unbuffered, optical fibre having a pre-installed optical connector, to the installed fibre cable. It is usual to route the pigtails into a splice housing and into a splice tray or accessory within the housing to provide mechanical support.

Coaxial cable

The construction of a coaxial cable is essentially two conductors having a common axis. The centre conductor is either solid or stranded. The dielectric surrounds the centre conductor and maintains the separation between the centre conductor and the shield. The shield surrounds the dielectric and is usually of single- or double-layered braid construction. The outer sheath protects against mechanical damage and is usually made of PVC. For high-temperature applications the sheath is usually made of Teflon FEP (FEP is fluorinated ethylene propylene), which affords more flexibility than PTFE (a common insulant for power cables).

Coaxial cable is available in a variety of different diameters and material types. Cable manufacturers and governments (for military specification) determine the size of coaxial cable and hence its code. The cable code gives no indication of its attenuation or quality.

It is common to refer to different types of coaxial cable using terminology such as RG-59 (with 75 Ω characteristic impedance) and RG-58 (with 50 Ω characteristic impedance).

The RG prefix designates government-specified cables. For example, an RG-58C/U cable equates to:

- R = radio
- G = government
- 58 = 58th cable approved
- C = revision
- U = universal approval.

It is possible to purchase an RG-58C/U cable with quality insulation for premises cabling or as patch cords or even as aerial cable. The code number gives no indication of usage.

Figure 12.51 shows an RG-6 quad shield cable and **Figure 12.52** shows common jacks and connectors.

FIGURE 12.51 RG-6 quad shield coaxial cable

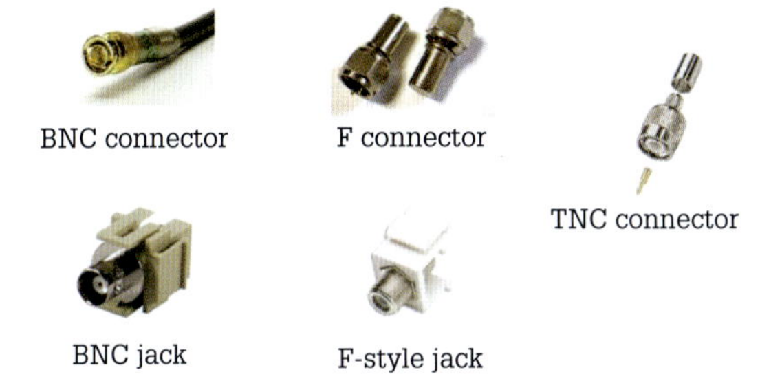

FIGURE 12.52 Coaxial cable connectors

Characteristic impedance

The various coaxial cables have different characteristic impedances and bandwidth limitations. It is therefore important to select a cable that matches the impedance of the receiving and transmitting devices. Furthermore, the cable must be capable of operating at the required frequency. For example, RG-174/U has a much higher attenuation than RG-214/U although they are both 50 Ω cables. **Table 12.9** gives the operational specifications for some coaxial cables.

The benefits of a coaxial cable over twisted pair are shown in **Table 12.10**.

Broadband and baseband transmission

Broadband in telecommunications is a term that refers to a signalling method that includes a wide range of frequencies. This frequency range may be divided into a number of channels or frequency bands. If a particular medium has a large bandwidth it has the capability of carrying more information than a medium having a narrow bandwidth. For example, a very narrow band radio signal can carry telemetry signals; a broader band can carry speech; a still broader band is needed to carry music without losing the high audio frequencies required for realistic sound reproduction.

TABLE 12.9 Operational specifications for some coaxial cables

Cable	Z_0	Attenuation (dB/100 m)					Volts
type	Ohms	50	100	200	400	1	rms
		MHz	MHz	MHz	MHz	GHz	
RG-8/U	52	5.2	7.2	10.5	15.4	29.2	5000
RG-11/U	75	4.3	6.6	9.5	13.8	23.3	5000
RG-59B/U	75	8.5	11.5	16.1	23.3	39.4	2300
RG-174/U	50	34.4	45.9	62.3	91.9	150.9	1000
RG-214/U	50	5.2	7.2	10.5	15.4	29.2	5000
RG-6 quad	75	5.3	8.5	10	12.5	21	1000

TABLE 12.10 Coaxial cables in comparison to twisted pair cables

Advantages	Disadvantages
▪ higher bandwidth ▪ greater immunity from noise ▪ minimum EMI radiation ▪ high electromagnetic compatibility (EMC) ▪ not affected by environmental EMI or RFI (radio frequency interference)	▪ larger medium ▪ heavier medium ▪ less flexible medium ▪ more difficult and costly to terminate ▪ requires a good earthing system for the shield

For telecommunications, multiple channels of data are transmitted over a single communications medium, typically using some form of frequency or wave division multiplexing. This is the case with asymmetric digital subscriber line (ADSL) where the twisted pair telephone cable (subscriber line) carries a broadband signal. This signal comprises a low-frequency band for telephone, and higher frequency bands for downstream data (data downloaded by user) having a high transmission rate and upstream data having a lower transmission rate. Each signal occupies a particular frequency band.

A baseband transmission sends one signal using a medium's full bandwidth, as in 100BASE-T Ethernet, which is a common Class D cabling application where one twisted pair carries data away from a computer (transmit pair) and another pair carries data to the computer (receive pair). A baseband signal is considered to include frequencies from near 0 Hz up to the highest frequency in the signal having significant power.

Any of the cabling media mentioned previously, that is, twisted pair, coaxial or optical fibre, is capable of carrying broadband and baseband transmissions. As optical fibre has a much higher bandwidth than twisted pair, it is capable of transmitting a larger quantity of data over a given period of time.

REVIEW QUESTIONS

1. Describe open line construction.
2. Name the electrical properties of a telecommunications cable that contribute to signal loss.
3. What does the term 'balanced pair operation' mean?
4. What has extended the bandwidth and hence the usefulness of copper cabling?
5. What effect does increasing the frequency of a signal have on transmission properties of a twisted pair wire?
6. What is the constructional characteristic of UTP cable?
7. What is the constructional characteristic of S/FTP cable?
8. What is an optical fibre?
9. What are the two modes of signal propagation for optical fibres?
10. Describe the single mode step-index feature.
11. What are the main components that make up an optical fibre cable?
12. What is the preferred connector type for optical fibres?
13. What is the characteristic of a coaxial cable designated RG-6/U?
14. What advantages do coaxial cables have over twisted pair cables?
15. What is the difference between broadband and baseband transmission?

12.8 Indoor cabling

Two insulated copper conductors form the basis of the telephone service to the customer's premises. Twisting insulated wires together forms a twisted pair. This twisted pair is the pathway for signals between the customer's telephone and the carrier's exchange.

Twisting two insulated wires together forms a pair and twisting four wires together forms a quad (quad cable). The length of twist varies so that no two adjacent pairs or quads have the same length or twist. This reduces induction, or cross-talk, between pairs. **Figure 12.53** shows the construction of a simple twisted pair wire.

FIGURE 12.53 Twisted pair (jumper or cross-connect) wire

Telecommunications services use various styles of twisted pair cables. The twisted pair wire shown in **Figure 12.53** is a cross-connect wire. A cable may consist of one pair or hundreds of pairs inside a sheath. A colour code distinguishes the individual pairs. In cables with larger pair counts, a binding tape identifies the individual binder groups.

Indoor cable usually has a cream-coloured PVC sheath over PVC-insulated wires. Cable construction may be either unshielded twisted pair (UTP) or overall foil-shielded twisted pair (FTP). Either type of indoor cable is suitable for installation in the roof void, in wall cavities, and in the floor void. It is not, however, suitable for use in underground situations, locations that expose it to direct sunlight or to the effects of weather or moisture. Indoor cable typically has 0.5 mm diameter copper conductors.

It is common to use a quad cable to connect to individual telephones or other telecommunications equipment. A quad cable allows for connection of a two-line or a four-wire digital service. **Figure 12.54** shows a quad indoor cable (notice how all four wires twist together).

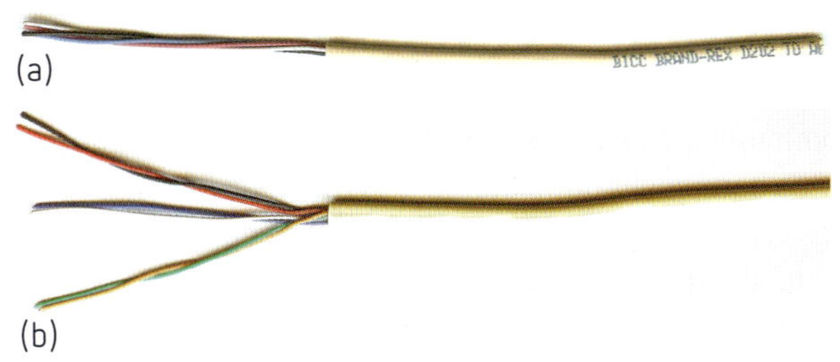

FIGURE 12.54 Indoor cable (a) quad cable and (b) three pair

SWITCH ON

Flammability

It is important to minimise the spread of fire in an installation. As such, the installation of any communications cabling must comply with the requirements of the Building Code of Australia. As with electrical cables, suitable fire stopping is required where a communications cable passes through a fire-isolating wall, floor or riser.

Separation

As was the case for terminations of communications cabling, it is important to maintain separation between communications cables and those of circuits supplying low-voltage and high-voltage power. The main reason for maintaining separation from low-voltage cables is to minimise the risk of accidental penetration of both cables by nails, screws, drills and the like. In this case, separation is easily achieved by maintaining a minimum air separation of 50 mm. Where both communications and low-voltage power cables are installed lying flat on the ceiling, additional protection measures are needed as cables from both systems are free to move. Cables from both systems need to be clipped in position to maintain the required 50 mm separation. Another separation measure includes installing a barrier of durable insulating material or earthed metal between the two cables (installing the communications cable(s) in conduit satisfies this requirement). **Figure 12.55** shows permitted separation measures between a communications cable having metallic conductors and a low-voltage power cable. A stud, nogging, joist or rafter of timber or metal can also serve as the required separation (AS/CA S009: 2020, Clause 16.3). **Figure 12.10** on page 516 shows a timber wall stud providing the required separation.

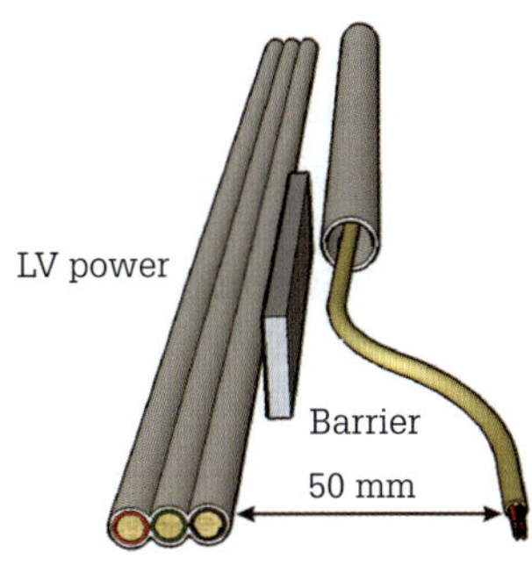

FIGURE 12.55 Separation between LV and telecommunications surface cabling

Similar requirements apply for cables in trunking. Cable trunking is an enclosure with a removable cover to facilitate the installation of cables after installation of the trunking. For example, skirting trunking has separate compartments for communications and power. If under-carpet cabling is employed, the 50 mm separation requirement still applies. This is difficult to achieve where communications cable crosses power cable, and so in this case the communications cable must be uppermost, the crossing must be at right angles and an earthed metal barrier must be placed between the two cables. **Figure 12.56** shows the required separation where an under-carpet communications cable crosses an under-carpet LV power cable.

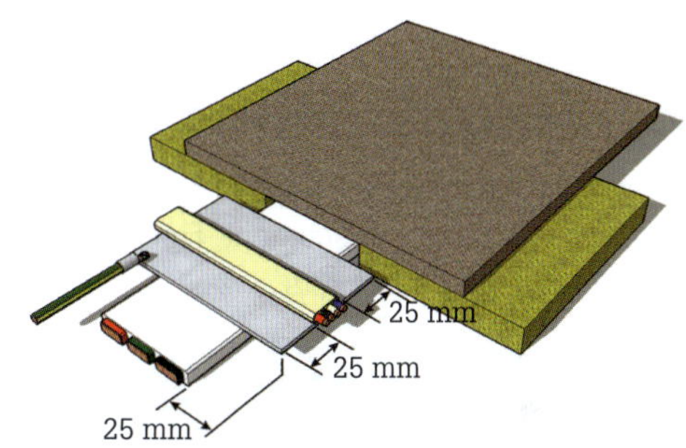

FIGURE 12.56 Crossing of under-carpet LV power and telecommunications cables

High-voltage circuits pose a risk in the form of low-frequency induction. As such, Clause 16.4 of AS/CA S009: 2020 requires telecommunications cables be separated by a minimum distance of 450 mm from single-core high-voltage cables irrespective of any interposing barriers. For multi-core high-voltage cables the minimum separation is reduced to 300 mm owing to reduced electromagnetic emissions. This distance is further reduced to 150 mm where there is an interposing barrier of durable insulating material or earthed metal, providing the distance around the barrier is not less than 175 mm when measured sheath to sheath. **Figure 12.57** shows the required separation between a communications cable having conductive elements and HV power cables.

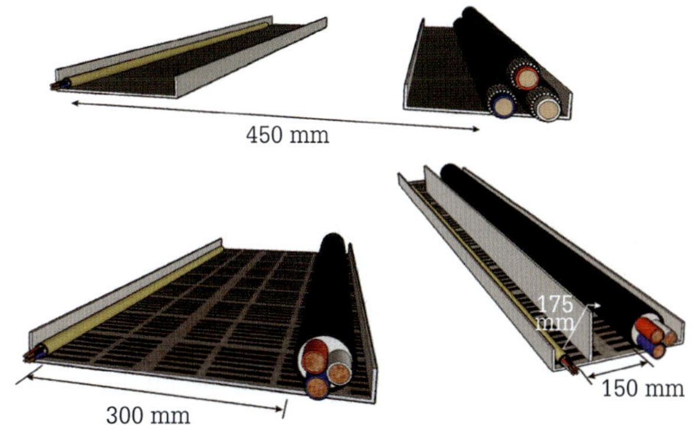

FIGURE 12.57 Separation between HV and telecommunications surface cabling

Installation

In areas exposed to elevated temperatures (> 60 °C), such as lift or hoist shafts, the use of non-flame-propagating conduit and fittings is required (AS/CA S009: 2020, Clause 16.5). Also, in a lift or hoist shaft it is not always possible to attain the required 50 mm separation between communications cabling and low-voltage power cabling. In this situation, compliant cable is required or a compliant line-isolation device is fitted to each end of the communications circuit. Similar measures apply to travelling cables (AS/CA S009: 2020, Clause 16.6).

While it is acceptable for flexible equipment connecting cords to run on the surface of the floor, fixed wiring requires suitable protection (AS/CA S009: 2020, Clause 16.7).

Telecommunications outlets

The '600 Series' of sockets was once commonplace but is now only used for a direct replacement. **Figure 12.58** shows the range of these types of sockets.

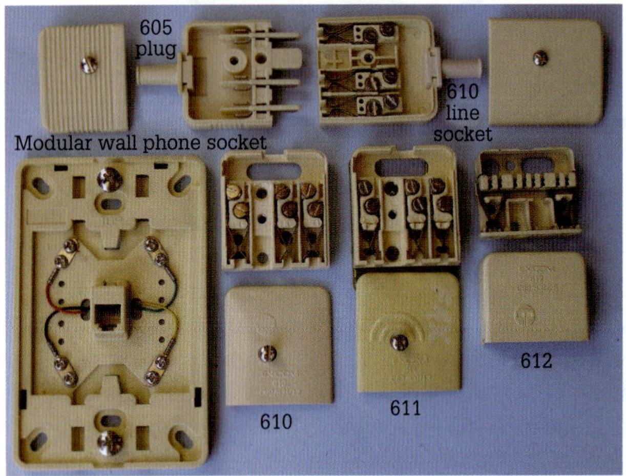

FIGURE 12.58 Configuration of various telephone sockets

These sockets include:

- **610 socket:** The main use for this socket is connecting to older-style desk phones. Contacts 3 and 4 make contact when the plug is withdrawn from the socket. Often a metal link joins contacts 2 and 3 – this is only necessary to provide the bell circuit for old telephones; it has no functional purpose for modern telephones.
- **611 socket:** The main use for this socket is connecting burglar alarm systems, computer modems and fax machines incorporating line switching circuits to isolate 'downstream' wiring and equipment. The original 611 socket has two semi-circular bars profiled on the cover for identification. Contacts 1 and 2, 3 and 4, and 5 and 6 normally make contact when the plug is withdrawn from the socket; it is possible to wedge the contacts open by turning the cam between the contacts 180°.
- **612 socket:** This is similar to the 610 socket but uses insulation displacement for connecting wires. This socket is now largely obsolete.

Modular connectors

Modular connectors are preferred for connecting communications equipment. Modern wall phones and office equipment often connect to this type of connector. The types of connector in this family include:

- modular 6 position, 4 contact (6P4C)
- modular 6 position, 6 contact (6P6C)
- modular 8 position, 8 contact (8P8C).

Modular plugs are often incorrectly referred to as RJ11, RJ12 and RJ45. Use of these terms can result in confusion as the RJ designations refer to very specific wiring arrangements. These arrangements are covered in the Universal Service Ordering Codes (USOC). The name 'RJ' means registered jack. It is possible to wire each jack style (e.g. 6 position) for different RJ arrangements, including:

- RJ11C for one pair
- RJ14C for two pair
- RJ25C for three pair.

There are two variations of the modular plug to accommodate either flat or round cable. The lower face of the cable entry on the rear of the plug body can be either flat or round. The round type should always be used when terminating round cables to ensure that the jacket and conductors are not crushed during termination. The flat type must be used to ensure adequate strain relief for flat cable.

The insulation displacement contact varies to accommodate either solid conductors or stranded conductors. Plugs for solid wires have a three-pronged contact that forks over the solid conductor. The contacts for stranded conductors have two spikes that penetrate between the strands. Failure to use the correct contact type will result in unreliable terminations.

The **6 position modular plug** is most commonly used for line cords in voice applications and is often referred to as RJ11 or RJ12. The 6 position polycarbonate body can be loaded only with the centre two contacts (6 position, 2 contacts, or 6P2C) or the centre four contacts (6 position, 4 contacts, or 6P4C). When fully loaded with all six contacts it is a 6 position, 6 contact (6P6C) modular plug. For basic two-wire voice applications, the centre two contacts are used and the plug should be correctly described as 6P2C (6 position, 2 contact). Where four wires are required for key telephone stations or two lines, the centre four contacts are used and the plug should be correctly described as 6P4C. In this application the two centre contacts connect to the first wire pair and the contact either side connects to the second wire pair.

The **4 position modular plug** (4P4C) is narrower than the 6 position modular plug, and is usually only used on coiled telephone handset cords.

The **8 position modular plug** is the standard for data and voice applications. This arrangement is usually referred to as RJ45. The plug body is usually supplied with all 8 gold contacts loaded and should be correctly referred to as 8P8C. For basic voice line cords, the plug may be loaded with only the centre two or four gold contacts and is referred to as 8P2C or 8P4C respectively.

For data applications, the most widely used specific wiring arrangement for the 8P8C modular plug in Australia is known as EIA/TIA 568A. Some common modular plugs and the EIA/TIA 568A termination arrangement are shown in **Figure 12.59**.

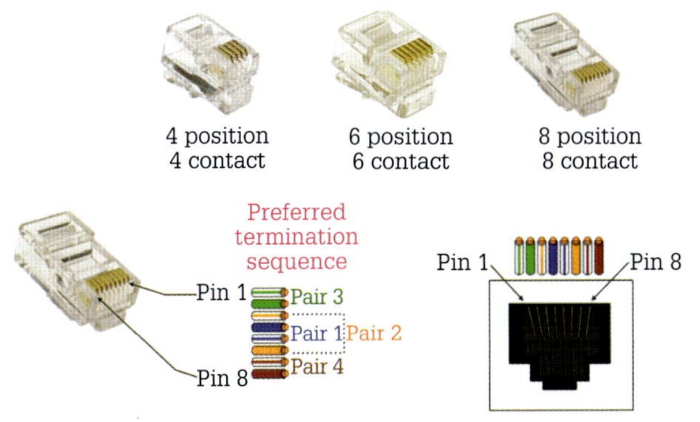

FIGURE 12.59 Configuration of modular plugs and jacks (sockets)

The basic set-up of a modular 6 position 6 contact jack for single-line telecommunications use has line – (blue) to 3 and line + (white) to 4.

To comply with AS/CA S009: 2020, 8 position modular and similar sockets in locations frequented by small children should be installed out of reach or have mechanical protection to prevent small children touching the exposed contacts (Clause 15.2). Furthermore, in damp areas such as bathrooms, laundries and the like, the location of the telecommunications outlet must be outside the damp area restricted zones delineated in AS/NZS 3000: 2018 (and Appendix A of AS/CA S009: 2020) and installed in a manner that minimises moisture collecting inside. Customer equipment installed in the damp area restricted zone must be rated at IPX7 for a bathroom and IPX6 for a shower room (AS/CA S009: 2020, Clause 7.2.3).

REVIEW QUESTIONS

1. What type of telecommunications cable has four wires twisted together?
2. What type of telecommunications cable has a cream-coloured PVC sheath over PVC-insulated wires?
3. What is the typical conductor diameter for indoor telephone cable?
4. What minimum separation is required between indoor telecommunications cable and LV power cables?
5. What are the requirements where under-carpet telecommunications cable crosses an LV power cable?
6. What is the requirement for conduit and fittings used to enclose telecommunications cables when installed in areas exposed to elevated temperatures (> 60 °C)?
7. What minimum separation is required between indoor telecommunications cable and single-core HV power cables without an interposing barrier?
8. What are the preferred connectors for connecting communications equipment?
9. A modular plug is identified as 4P4C. What does this mean?
10. What is a common application for a 6P4C modular plug?

12.9 Underground cabling

Underground cable usually has polyethylene conductor insulation (solid or foam) and a polyethylene sheath. The core of an underground cable may incorporate a water-blocking compound or tape. Underground cables sometimes employ an additional outer jacket of nylon to protect the cable from termites. Underground cable typically has 0.64 mm diameter copper conductors (0.4 mm and 0.9 mm are also available). **Figure 12.60** depicts the typical construction of a quad underground cable. Twisted pair telecommunications cables must comply with the requirements of AS/CA S008: 2010.

FIGURE 12.60 Quad underground cable

Pillars, pits and access holes

Pillars or cabinets located in public areas must be locked to prevent access by unauthorised persons.

It is a requirement to clearly identify any pits or access holes. This is achieved by having a descriptive caption such as 'COMMUNICATIONS' cast into the cover during manufacture. The only exception is where another service passes through the communications pit or access hole. It is important to consider possible dangers associated with pits. These include confined spaces, syringes, snakes and spiders. It is necessary to take precautions to strengthen or protect pits installed in driveways (AS/CA S009: 2020, Clause 18.1). Precautions should be taken to prevent fluids and gases entering a building through conduits.

Installation requirements

Underground communications cable installed under a public footway or roadway must be enclosed in compliant conduit or have a white marker tape laid 100 mm above the unenclosed cable. Similar measures apply for communications cable installed underground on private property; an alternative is to install markers at regular intervals to mark the cable route. Conduit complying with AS/CA S008:2020 is suitable for enclosing underground telecommunications cables. Non-metallic conduit is marked 'COMMUNICATIONS' and is either white in colour or contains a white stripe.

The installation of underground communications cables is similar to underground low-voltage cables. The depth of laying for cables installed under a public roadway or footway is 450 mm. In other areas the minimum depth is 300 mm unless the soil conditions preclude this depth, in which case additional measures are needed to protect the cable (AS/CA S009: 2020, Clause 18.6).

Where an underground communications cable crosses another service, the separation is in accordance with the utility concerned. AS/CA S009: 2020 details required clearances for telecommunications cables from low-voltage and high-voltage power cables (see Table 4 of AS/CA S009: 2020). Where a communications cable crosses a power cable, the communications cable should be enclosed in conduit and cross above the power cable. Additional protection is required where an unprotected power cable must cross above the communications cable.

Communications cables may be installed in either an exclusive trench or a shared trench. An exclusive trench is one where the telecommunications cable is the only service installed in that particular trench. This means multiple services are **not together** in a trench. A shared trench is one where the telecommunications cable is installed with other services in that particular trench. This means multiple services **are together** in a trench. **Figure 12.61** provides a pictorial representation of common installations from Table 4 of AS/CA S009: 2020, which should be referred to for the required separation distances.

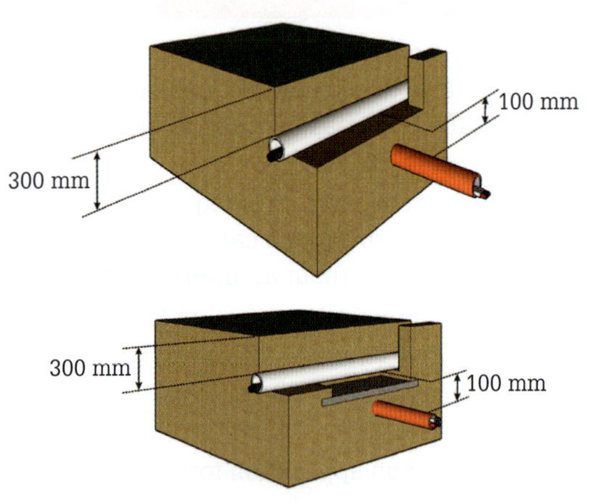

Exclusive trench crossing protected LV power

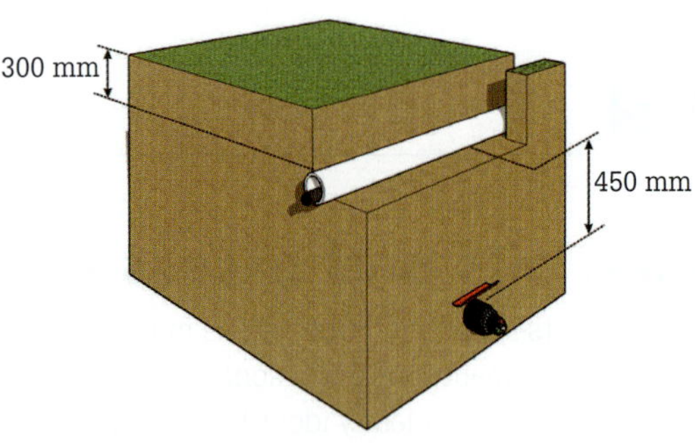

Exclusive trench crossing unprotected HV power

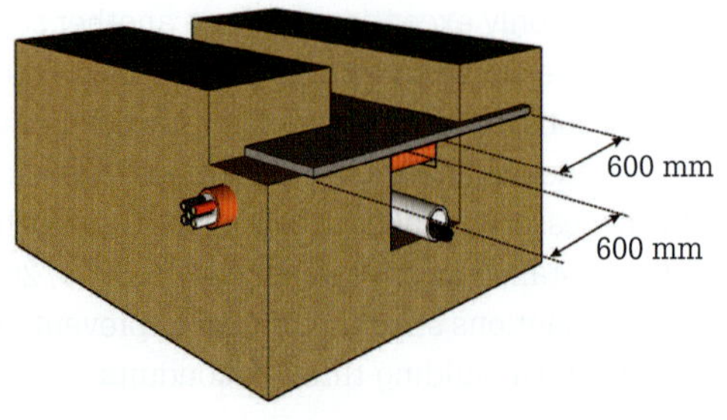

Exclusive trench with power uppermost

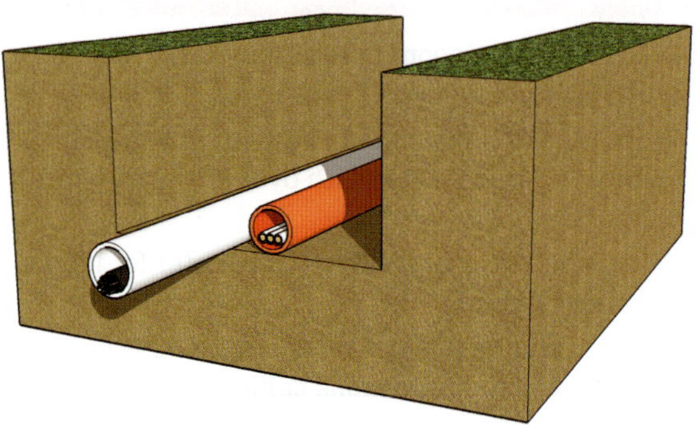

Shared trench

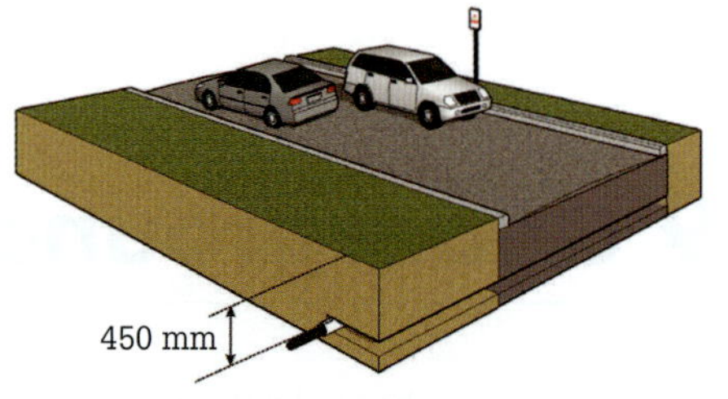

Increased depth under roadway

FIGURE 12.61 Pictorial representation of common installations from Table 4 of AS/CA S009: 2020

REVIEW QUESTIONS

1 Describe the construction of underground telecommunications cable.
2 What is the range of conductor sizes for underground telecommunications cables?
3 How are pits or access holes used for telecommunications services identified?
4 Particular requirements exist for pillars or cabinets located in public areas. What are these requirements?
5 How are underground non-metallic communications conduits distinguished from conduits used for other services?
6 If customer cables enclosed in white communications conduit share an underground trench with LV cables, which are enclosed in HDUPVC conduit, how can the required segregation be achieved?
7 What is an exclusive trench?
8 Who determines the required separation if a telecommunications cable crosses another service?
9 What is the requirement where a telecommunications cable crosses a power cable?
10 How can the required segregation be achieved where an unprotected power cable must cross above a telecommunications cable?

12.10 Aerial cabling

Aerial cable employs polyethylene insulation for both the sheath and the conductor insulation. Aerial cables often incorporate a strong support wire known as an 'integral bearer wire'. This bearer wire may be steel or hard-drawn copper. The core of an aerial cable may also incorporate a water-blocking compound. Aerial cable typically has 0.64 mm diameter copper conductors (0.4 mm and 0.9 mm are also available). **Figure 12.62** depicts the typical construction of a quad aerial lead-in cable.

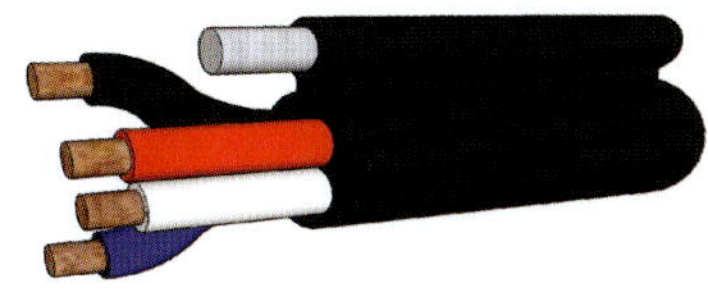

FIGURE 12.62 Quad aerial cable

Bearer wires must be insulated from any conductive pole or structure carrying an aerial power line (AS/CA S009: 2020, Clause 19.5.8). Furthermore, they must be insulated or shrouded to prevent accidental contact by a power-line worker and must not connect to earth. The latter is important to minimise the chance of a lightning strike.

Poles and supports

Poles and other supporting structures for aerial communications cables must be fit for purpose and adequately anchored to support the load, allowing for a suitable safety factor. As is the case for power poles, supports for aerial communications cables should be inspected regularly for deterioration.

Clause 19.1.2 of AS/CA S009: 2020 specifies the minimum distances between power and communications for parallel pole routes. Bear in mind that a distance of 50 m is specified for aerial power lines exceeding 330 kV.

Clearances

Clause 19.2 of AS/CA S009: 2020 specifies the minimum clearance for aerial customer cables. In summary these are:

- 2.7 m over private property not traversed by road vehicles
- 3.5 m over any residential driveway
- 4.9 m over any commercial or industrial driveway or private roadway.

For public roadways and footpaths the clearance is specified by the relevant authority but should not be less than 4.9 m. **Figure 12.63** shows the clearances necessary at a pole shared by low-voltage power and communications cables having conductive elements.

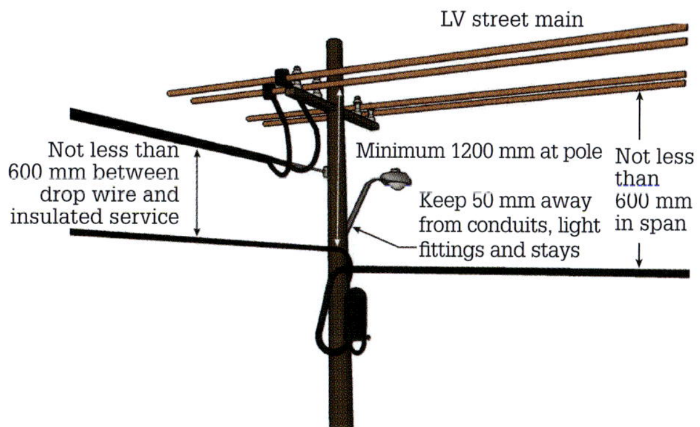

FIGURE 12.63 Clearances at shared pole

Installation requirements

Aerial communications cable must be suitable for installation in direct sunlight and satisfy the requirements of AS/CA S008: 2010. Any support for the bearer wire or catenary must be capable of supporting the cable under extreme weather conditions and tighten under increasing load. High-wind areas and areas subject to ice and snow can cause an appreciable increase in the weight a support needs to bear.

To avoid unnecessary duplication of cable supports, under certain circumstances it is possible for aerial communications cables to attach to poles carrying power cables (joint use). Clause 19.5 of AS/CA S009: 2020 details these requirements. Essentially, the following conditions must be satisfied:

- The owner of the pole agrees.
- The power cables do not exceed 66 kV.
- Attachment to poles carrying high voltage not exceeding 66 kV requires a non-conductive structure. Low-voltage power lines are supported below the high-voltage power lines, or the attachment is for a crossing only.
- Attachment to a pole or structure carrying a high-voltage transformer is not permitted.

Where it is necessary for an aerial communications cable to cross aerial power cables the following applies:

- The power cables do not exceed 330 kV.
- The communications cable crosses below the aerial power cable. In rare situations the communications cable may cross above the power cables but the span of the communications and its height above the power line must be such that the communications cable clears the power line by at least 5 m should the communications cable break at either end. A further requirement is that the owner of the power cable agrees with the arrangement.
- Separation distances comply with Table 3 of AS/CA S009: 2020.
- Crossing is as far as practical from mid-span of the power cable.

REVIEW QUESTIONS

1. What insulation does aerial cable use?
2. State the function of an 'integral bearer wire'.
3. What is the range of wire diameters for aerial cables?
4. What are the requirements for poles and other supporting structures for aerial communications cables?
5. How should 'integral bearer wires' terminate?
6. When installing aerial customer cables over any residential driveway, what is the minimum required ground clearance?
7. When installing aerial customer cables over any public roadway or footway, what is the minimum required ground clearance?
8. What factors can cause an appreciable increase in the weight a support needs to bear?
9. If an aerial customer cable attaches to a pole, how can the required segregation from an insulated service lead be achieved?
10. If an aerial customer cable attaches to a pole, how can the required segregation from light fittings, conduits and stays be achieved?

12.11 Earthing protection

There are four forms of earthing system recognised in AS/CA S009: 2020. These are:

1. Communications earthing system (CES) is used for both protective and functional purposes, the conductors of which are identified by green/yellow insulation.
2. Telecommunications reference conductor (TRC) provides a low-noise telecommunications earthing system for functional purposes, the conductors of which are identified by violet insulation.
3. ELV d.c. power supply distributes earth to communications equipment through the positive or negative ELV d.c. supply conductor.
4. d.c. earth return provides a dedicated telecommunications earth electrode to discharge continuous d.c. current to earth, the conductors of which are coloured violet.

General requirements

Where customer equipment requires connection to a functional earth, unless specified otherwise, it may connect to any of the recognised earthing systems. Customer equipment manufactured prior to 1998 can connect to any of the recognised earthing systems. Customer equipment manufactured after 1997 must connect to the specified earthing system.

Where customer equipment or cabling requires connection to a protective earth it can connect to:

- the communications earthing system
- the communications earth terminal
- the bar, terminal or back-mount at the distributor where equipotential bond is made
- the electrical earth electrode via an equipotential bonding conductor as appropriate

- the earthing system via a protective earthing conductor
- in the absence of an electrical system, directly to an earth electrode complying with AS/CA S009: 2020 ensuring the resistance to earth does not exceed 30 Ω (10 Ω is preferred) and equipotential bonding to structural building elements or services such as a water pipe is also made.

Earth cables are multi-stranded copper conductor, single-core PVC-insulated cables having a working voltage of 0.6/1.0 kV. It is usual to refer to the size of the earth cable based on its cross-sectional area in mm^2. Multi-pair cable is not generally used for earthing.

Earth bars and terminals

Clause 20.9.1 of AS/CA S009: 2020 requires that earth bars and terminals in the telecommunications earthing system be capable of terminating cables of at least 6 mm^2 in cross-sectional area. Alternatively, they should be capable of terminating cable lugs designed to terminate 6 mm^2 cables. Earth bars or links must have adequate capacity for the installation.

When terminating earth cables it is necessary to remove only sufficient insulation to secure all the conductor strands in the termination. It is also necessary to consolidate the bare conductor strands by twisting or other suitable means before termination. Consolidation **cannot** be by soldering.

It is necessary to effect crimp or compression connections so that a suitable ferrule securely retains the conductors, and to use the appropriate tool for this purpose. It is **not** permissible to use spade-type connectors. It is allowable to solder cable lugs after crimping to exclude moisture.

Cable joints

It is necessary to maintain electrical continuity through joints in telecommunications earthing system cables. Joints can be soldered, clamped, tunnel-type or crimped. Joints in telecommunications earthing system cables must be insulated or enclosed to prevent the entry of dust and moisture.

Equipotential bonding

Equipotential bonding minimises risk associated with the voltage differences between the accessible metallic parts of electrical equipment and the accessible metallic parts **not** associated with the electrical installation. Clause 20.11 of AS/CA S009: 2020 covers the requirements for equipotential bonding.

Communications bonding conductors have green/yellow coloured insulation and a minimum cross-sectional area of 6 mm^2. The communications bonding conductor must be as short and direct as practicable, and it must be labelled at the electrical installation end and, where it is not readily identifiable, at the telecommunications earthing system end with a descriptive caption such as 'communications bonding conductor'. The resistance of the bonding conductor cannot exceed 0.5 Ω.

Communications earth terminal

The communications earth terminal (CET) must be installed in a convenient and readily accessible location; it must not be located within the electrical switchboard. It must have a descriptive caption such as 'communications earth terminal' (AS/CA S009: 2020, Clause 20.11.3).

The connection of the bonding conductor to the electrical system must comply with the requirements of AS/NZS 3000: 2018 and must be protected against corrosion. A cabling provider must **not** reduce the effectiveness of the electrical earthing system.

Only licensed electrical workers can cut and rearrange the main or sub-main earthing system of the electrical installation and make connections to the protective earth inside the switchboard.

Earthing systems for telecommunications

There are three earthing systems recognised in AS/CA S009: 2020:

- communications earth system
- telecommunications reference conductor
- d.c. earth return systems.

Communications earth system

The communications earth system (CES) is a system of earthing using common elements to provide earthing facilities for electrical and communications equipment within premises. This includes protective and functional earthing for telecommunications. Section 20.12 of AS/CA S009: 2020 details the requirements for the communications earth system.

In the CES, the electrical protective earth is extended to a communications earth terminal (CET) using a minimum 6 mm^2 green/yellow coloured conductor. This occurs at the relevant floor or section of the electrical installation. From the CET a 2.5 mm^2 or 6 mm^2 (in the case of over-voltage protection requirements), green/yellow coloured conductor extends the floor distributor. There is a requirement for the resistance of the CES cable not to exceed 1 Ω when measured between the bond at the earthing system of the electrical installation and the earth bar or terminal at any NTD, distributor or customer switching system.

A solid electrical connection to the earthing system of an electrical installation will ensure any over-voltage conditions or surges will have a low resistance path to earth to allow effective operation.

Figure 12.64 details the arrangement of the communications earth system.

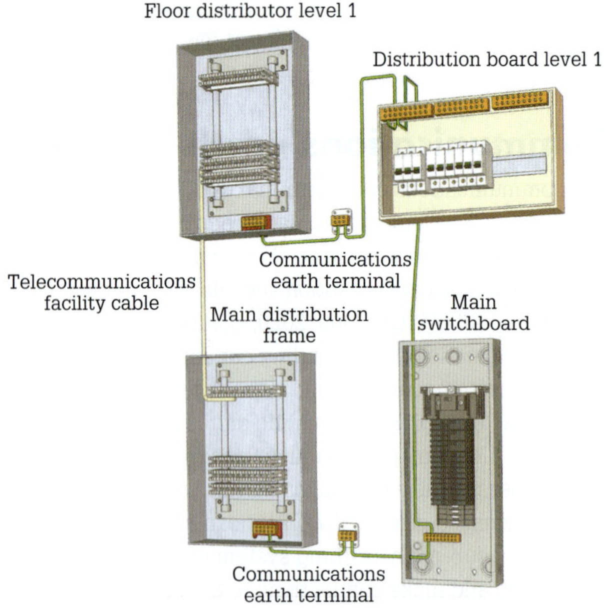

FIGURE 12.64 Communications earth system

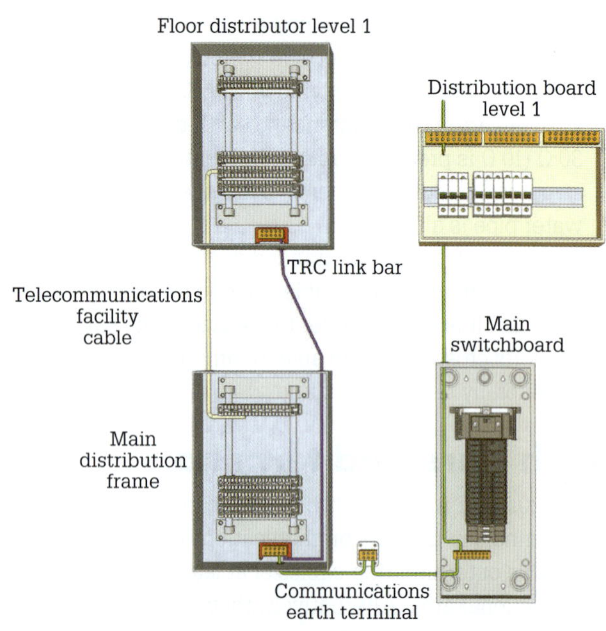

FIGURE 12.65 Preferred bonding method

Telecommunications reference conductor system

Should a communications earth system prove to suffer from electrical noise due to its connection to the power system earth, it might be necessary to install a telecommunications reference conductor (TRC) system. The TRC is a low-noise functional earthing system often supplied to ensure equipment reliability and provide for functional testing. The TRC does not provide protection, such as a protective earthing conductor and is designated by the cable insulation colour of violet. Requirements for the TRC are detailed in Section 20.13 of AS/CA S009: 2020 with cable sizes designated in Table 7 of that standard.

The TRC bonds to the protective earth of the electrical installation via a communications earth terminal as **Figure 12.65** shows, and is referred to as the 'preferred bonding method'. In the TRC system, the electrical protective earth is extended to a CET using a minimum 6 mm² green/yellow coloured conductor. This occurs at the main electrical switchboard for the installation. From the CET a 6 mm² green/yellow coloured conductor extends the earth bar in the MDF. There is a requirement for the resistance of the TRC system to not exceed 5 Ω when measured between the TRC link bar at the MDF and the TRC link bar at any other distributor or customer access equipment (CAE).

Differential earth clamp

If noise is a problem on the TRC it is possible to equipotential bond the telecommunications reference conductor to the earthing system of the electrical installation via a differential earth clamp. A differential clamp is a device that electrically connects two earthing systems under over-voltage conditions but remains electrically disconnected under normal operating conditions. It is necessary to install a telecommunications functional earth electrode to ensure functional earthing of the telecommunications system. **Figure 12.66** shows the arrangement for the differential earth clamp.

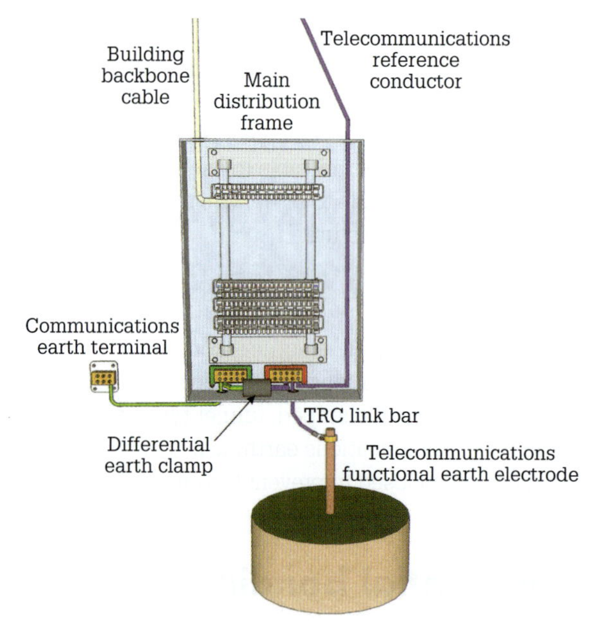

FIGURE 12.66 Differential earthing of the TRC

It is important to ensure that any over-voltage protection devices installed between a telecommunications line and earth have the earth connection made at the earth bar and not the TRC link bar. This will ensure that the differential earth clamp does not carry this surge current.

When using a differential earth clamp it is a requirement of clause 20.13.9.2 that all TRC link bars must be enclosed or otherwise installed to prevent access by end-users and have a warning label indicating that a hazardous voltage may be present.

Telecommunications functional earth electrode

An installation may include a telecommunications functional earth electrode (TFEE) to limit the magnitude of direct current flowing in the communications bonding conductor and hence the electrical system main earth conductor. When installed, a TFEE must comply with the requirements of AS/NZS 3000: 2018. The relevant clauses describe the type and size of suitable materials as well as the method for installing earth electrodes.

There are a number of items that are suitable for use as a telecommunications functional earth electrode, including a galvanised star picket (AS/CA S009: 2020, Clause 20.13.8.2). Installation requirements for the telecommunications functional earth electrode are similar to those for an electrical earth electrode: exposed to the weather, external to the building and separated from metallic enclosures of other underground services. The telecommunications functional earth electrode must be permanently labelled 'telecommunications electrode'. The minimum size conductor that connects the TFEE to the TRC link bar at the nominated distributor is 4 mm^2 and it has violet coloured insulation.

Separate building containing an electrical switchboard

A TRC system installed in a separate building having an electrical switchboard must bond to the earthing system of the electrical installation within **that** building in accordance with AS/CA S009: 2020, Section 20.13. This connection must **not** be to the TRC system of the main building.

Careful consideration is necessary when installing a TRC system in outbuildings to ensure that there are no circulating currents between the electrical and telecommunications earthing systems as this could result in dangerous voltage gradients. **Figure 12.67** details the bonding arrangement for a TRC within a separate building having an electrical switchboard.

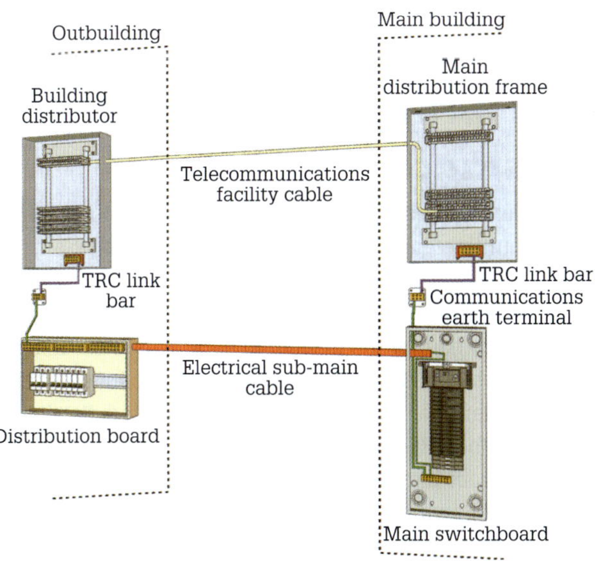

FIGURE 12.67 Bonding arrangement for a TRC within a separate building having an electrical switchboard

Separate building without an electrical switchboard

Where a separate building **without** an electrical switchboard requires an earth reference it is possible to feed a TRC to the separate building via a violet cable or cable pair, provided end-user access to the TRC in the separate building is prevented by effective means. **Figure 12.68** shows this arrangement.

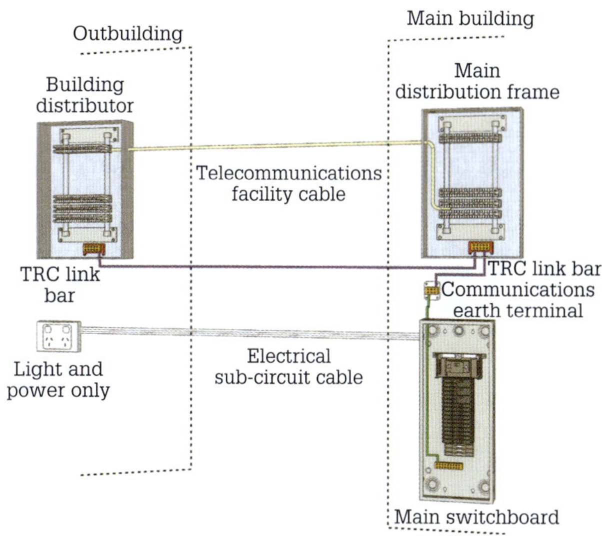

FIGURE 12.68 Bonding arrangement for a TRC within a separate building without an electrical switchboard

d.c. earth return systems

Telecommunication installations operating at extra low voltage (ELV) d.c. and located in restricted access locations may utilise d.c. earth return paths in accordance with AS/NZS 3015. In this system, the earth may be distributed to equipment using either the positive or negative conductor of the d.c. supply.

This system of earthing requires adequate labelling and the installation of circuit breakers or other current-limiting devices that protect the circuit to ensure the safety of the cabling provider and general public.

A d.c. earth return circuit may be required where continuous d.c. current will be discharged to earth. Such circuits require the installation of a dedicated earth electrode to prevent damage to the electrical earthing system. Earthing conductors used for this purpose are a minimum of 4 mm^2, have violet insulation and are labelled at each termination point such as 'd.c. functional earth'. Refer to clause 20.16 of AS/CA S009: 2020 for details of the d.c. earth return system.

REVIEW QUESTIONS

1 What are the forms of earthing recognised under AS/CA S009: 2020?
2 Where customer equipment or cabling requires connection to a protective earth, to where can it connect?
3 Earth bars and terminals in a telecommunications system must be capable of terminating cables of what size?
4 What methods are suitable for consolidating bare conductor strands of telecommunication earth cables?
5 What is the purpose of equipotential bonding?
6 What is the size, colour and maximum resistance of the equipotential bonding conductor?
7 Where must the communications earth terminal (CET) be installed?
8 What is a differential earth clamp?
9 The telecommunications reference conductor requires equipotential bonding to what other component?
10 What is the colour and minimum conductor size for the cable that connects the TFEE to the TRC link bar?

12.12 Earthing concepts

Apart from functional earthing, there are a number of other telecommunications system components that require an earth connection. Some of these are discussed below.

Earthing of cable shields

Depending on operational requirements it is not necessary to earth cable shields. It is not acceptable to bond cable shields to the TRC. Where required, a cable shield must connect to earth at a point in the protective earthing conductor of the electrical installation or to any metal bonded to earth in accordance with AS/NZS 3000: 2018.

Where a shielded cable connects between equipment in separate buildings having separate electrical installations **only one** end bonds to earth and the other end is insulated from any earth reference or connects to earth via a differential earth clamp.

Earthing of metallic frames

Metallic frames, back-mounts, enclosures, trays, conduits and ducts must **not** bond to the TRC. They connect to a compliant earth reference through a green/yellow insulated copper cable with a minimum cross-sectional area of 2.5 mm^2. It is not mandatory to earth metallic frames, back-mounts, enclosures, trays, conduits and ducts (AS/CA S009: 2020, Clause 20.19).

Earthing of surge suppression devices

Surge suppression devices for protection of telecommunications line conductors must connect to a compliant earth reference through a green/yellow insulated copper cable with a minimum cross-sectional area of 6 mm^2 when used to protect end-users. A green/yellow insulated copper cable with a minimum cross-sectional area of 2.5 mm^2 is used to protect equipment. The earth cable connecting the surge suppression device to earth should not be longer than 1.5 m (AS/CA S009: 2020, Clause 20.20). **Figure 12.69** shows the earthing arrangement for surge arrestors installed at the network boundary.

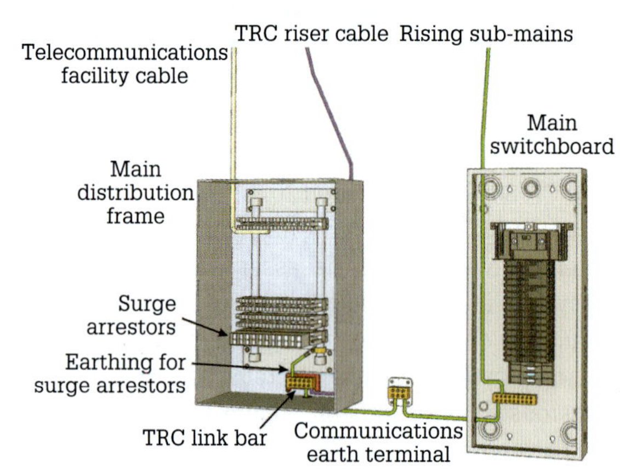

FIGURE 12.69 Earthing of surge suppression devices via back-mount

Measuring TRC system resistance

To measure the resistance of the TRC system you use a digital multimeter set to the ohms range. The steps are then:

1 Select a floor distributor riser cable pair to use as a reference pair for the measurements. This is an unused pair in the building backbone cable that connects the particular floor distributor to the designated distributor (where the TRC cable originates).
2 Disconnect the TRC, under test, from the distributor or customer switching system earth bar.
3 Connect each wire of the reference pair to the TRC connection at the distributor or customer switching system under test. Commercially available plugs make it easy to connect the wire pair to the TRC link bar.

4 At the designated distributor (the distributor where the TRC originates), measure and document the loop resistance of the reference pair. As an example, in **Figure 12.70** the resistance of the reference pair (R_1) is 33 Ω.

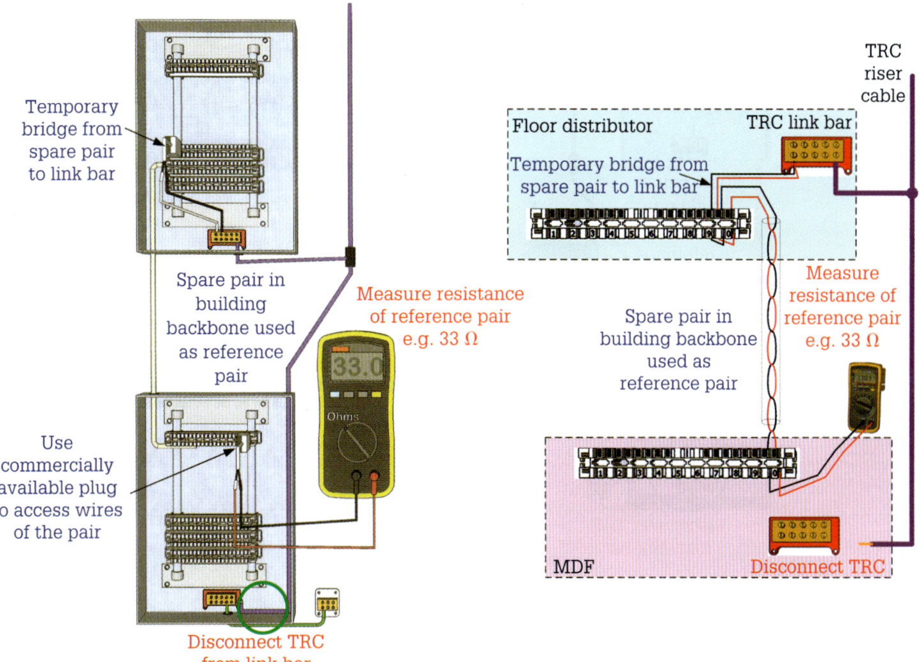

FIGURE 12.70 Measuring the resistance of the reference pair

5 Measure and document the resistance of the A-wire of the reference pair and the main TRC cable. As an example, in **Figure 12.71** the resistance of the A-wire in the reference pair and the main TRC cable (R_2) is 19.5 Ω.

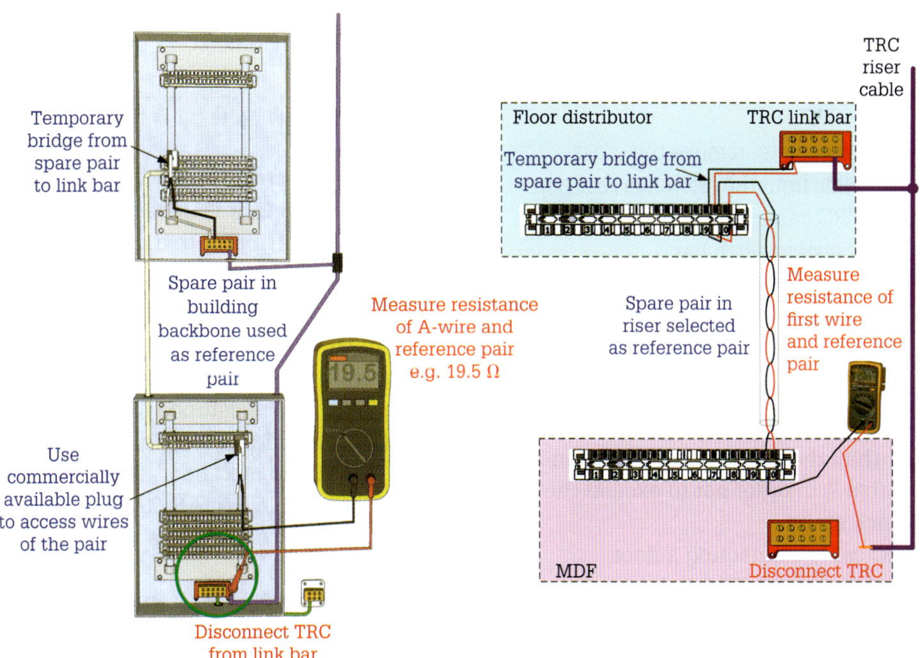

FIGURE 12.71 Measuring the resistance of the A-wire in reference pair and TRC

6 Measure and document the resistance of the B-wire of the reference pair and the main TRC cable. As an example, in **Figure 12.72** the resistance of the B-wire in the reference pair and the main TRC cable (R_3) is 18.9 Ω.

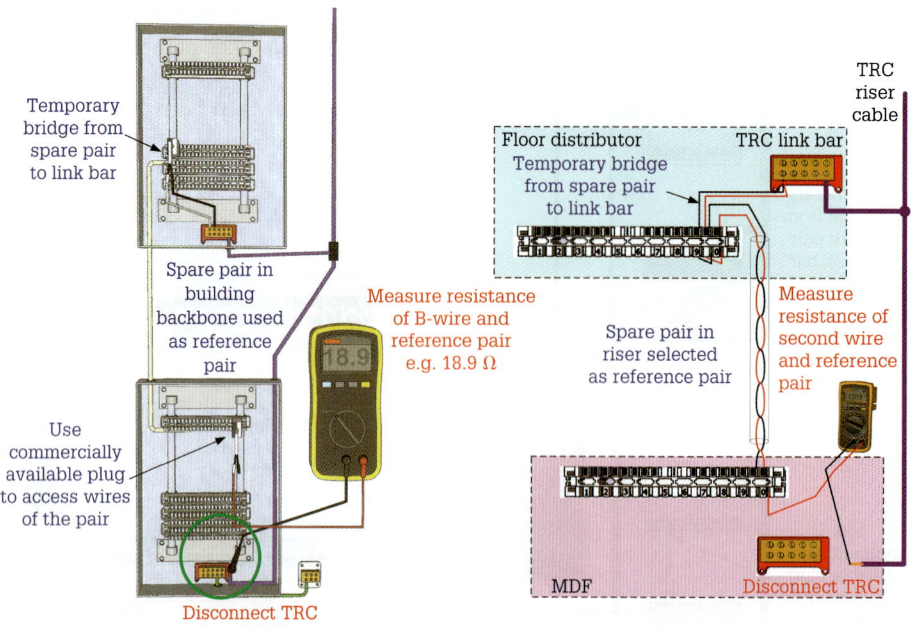

FIGURE 12.72 Measuring the resistance of the B-wire in reference pair and TRC

7 Substitute the values for R_1, R_2 and R_3 in the following equation to calculate the value of the main TRC resistance.

$$R_{TRC} = \frac{R_2 + R_3 - R_1}{2}$$

In this example R_1 = 33 Ω, R_2 = 19.5 Ω and R_3 = 18.9 Ω:

$$R_{TRC} = \frac{R_2 + R_3 - R_1}{2} = \frac{19.5 + 18.9 - 33}{2} = \mathbf{2.7\ \Omega}$$

8 When the testing is complete, remove the cross-connects and test leads from the reference test pair.

9 Reinstall the main TRC cable in the distributor or customer switching system earth bar.

EXERCISE 12.1

Calculating resistance of the TRC system

In some installations it may be necessary to install a TRC that is larger than the minimum specified in AS/CA S009: 2020, Table 7, in order not to exceed the maximum resistance limits.

A test on a TRC cable connecting a floor distributor TRC link bar to the TRC link bar at the designated distributor yielded the following resistance measurements:

- Resistance of the reference pair (R_1) = 15 Ω
- Resistance of the main TRC and A leg (R_2) = 7.8 Ω
- Resistance of the main TRC and B leg (R_3) = 8.2 Ω

Calculate the resistance of the TRC between the nominated distributor TRC link bar and this floor distributor.

Extending the TRC cable

Often it is necessary to extend the TRC to another distributor. It is important to know how much further to extend the TRC. Table 7 in AS/CA S009: 2020 provides the resistance of various copper conductors. The following example demonstrates how to calculate the allowable continuation of the TRC so as not to exceed the maximum permissible resistance.

EXAMPLE 12.1

The resistance of the TRC to a floor distributor measures 1.9 Ω. It is necessary to extend the TRC from this distributor to a new distributor. How much further can the TRC extend to the other floor distributor using 4 mm² copper cable?

SOLUTION:

The maximum allowable resistance is 5 Ω (Clause 20.13.11). The resistance to the existing floor distributor is 1.9 Ω. This allows an additional 3.1 Ω (5 – 1.9) to the next floor distributor.

Using Table 6 in AS/CA S009: 2020 as a guide, 4 mm² copper cable has a resistance of 4.61 Ω per kilometre (R_C), so:

$$length = \frac{R_{available}}{R_C} \times 1000 = \frac{3.1}{4.61} \times 1000 = \mathbf{672\ metres}$$

It is possible to extend the TRC cable another 672 m.

EXERCISE 12.2

a The resistance of the TRC to a floor distributor measures 3.8 Ω. It is necessary to extend the TRC from this distributor to a new distributor. How much further can the TRC extend to the other floor distributor using 6 mm² copper cable?

b A test on a TRC cable connecting a local distributor to a floor distributor yielded a resistance of 1.8 Ω. Resistance of the TRC between the designated distributor and the floor distributor is 1.5 Ω. How much further is it possible to move the local distributor using 2.5 mm² TRC cable?

REVIEW QUESTIONS

1 What is the colour and minimum conductor size for a cable connecting metallic frames, backmounts, enclosures, trays, conduits and ducts to a protective earth?

2 What is the minimum size and colour of the cable connecting surge suppression devices to earth?

3 Is it necessary to earth cable shields?

4 A shielded cable connects between equipment in separate buildings having separate electrical installations. Describe the earthing arrangement for this cable.

5 What is the maximum length of the earth cable connecting the surge suppression device to earth?

6 A test on a TRC cable connecting a local distributor to the designated distributor yielded the following resistance measurements:
- Resistance of the reference pair = 1.7 Ω
- Resistance of the main TRC and A leg = 2 Ω
- Resistance of the main TRC and B leg = 2.3 Ω

Does the cable between the designated distributor and the local distributor satisfy the resistance requirements of AS/CA S009: 2020?

7 A test on a TRC cable connecting a local distributor to the designated distributor yielded the following resistance measurements:
- Resistance of the reference pair = 14 Ω
- Resistance of the main TRC and A leg = 8.3 Ω
- Resistance of the main TRC and B leg = 8.5 Ω

How much further is it possible to move the local distributor using the existing 4 mm² TRC cable?

12.13 Surge suppression

The purpose of a telecommunications network is to allow data exchange (analogue or digital) between numerous subscribers. Different elements combine to form the network and each element is subject to various disturbances. Transmission lines are the most susceptible element due to their length and remoteness.

Disturbances affect the lines somewhere along their length and then migrate to the ends of the lines. This can have significant effects on telecommunications equipment. The effects may range from minor noise or disruption of the service due to induction, to severe damage of cable and equipment and injury to users caused by insulation failure and high induced voltages. Note the warnings in all telephone directories to avoid using telecommunications equipment during thunderstorms.

Telecommunications cables, like power cables, are prone to over-voltage from external and internal sources.

Protection techniques

Techniques for protecting telecommunications equipment and personnel from over-voltage include the following:
- **Intercepting** is where a lightning arrestor (or rod) attracts the lightning to itself rather than to the telecommunications network. This method is common in the telecommunications network but not in individual installations.
- **Clamping** or **clipping** is a way of limiting the amplitude of the over-voltage surge.
- **Shunting** provides a path to earth for the over-voltage surge.
- **Interrupting** opens the circuit for the duration of the over-voltage surge.
- **Isolating** provides a barrier between hostile environments and sensitive telecommunications circuits through the use of opto-isolators or transformers.

Modern line interface equipment usually has a low clamping factor. *Where the clamping factor is the ratio of normal operating voltage to the maximum voltage that the device can withstand*, a single protection device generally will not be able to give enough surge suppression to ensure that the over-voltage reaching the interface is within this limit. To protect equipment adequately, it is therefore necessary to provide two stages of over-voltage protection.

Effect of surges

Most communications circuits employ a balanced-pair line configuration. In balanced-pair operation, the two wires typically carry equal and opposite signals, known as differential mode. The two signals are then combined by addition at the destination. The common-mode noise (signal induced onto each wire of the pair) from the two wires cancel each other in this addition. The difference in potential between the two wires remains the same at the receiver as shown in **Figure 12.73**.

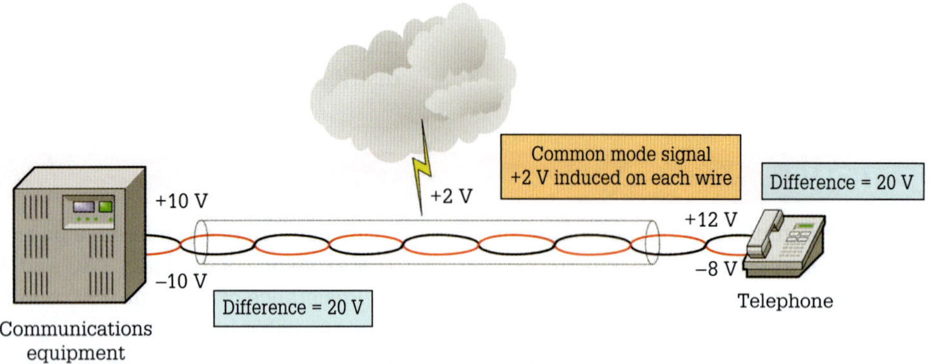

FIGURE 12.73 Effect of common-mode signal on differential-mode transmission

The line signals appear across the pair in differential mode, but surges induced on to balanced-pair circuits are generally common mode. Additional to over-voltage protection devices suitable for use on power circuits, longitudinal chokes are employed to protect telecommunications circuits. A balanced longitudinal choke placed in the line has the effect of passing the wanted (differential-mode) signal while presenting high impedance to common-mode signals such as induced surges. The longitudinal choke is placed in series with each wire and effectively blocks any induced surge voltage. When used in conjunction with other protection devices (e.g. gas arrestor), longitudinal chokes provide effective protection against induced surges of both long and short duration. **Figure 12.74** shows such an arrangement.

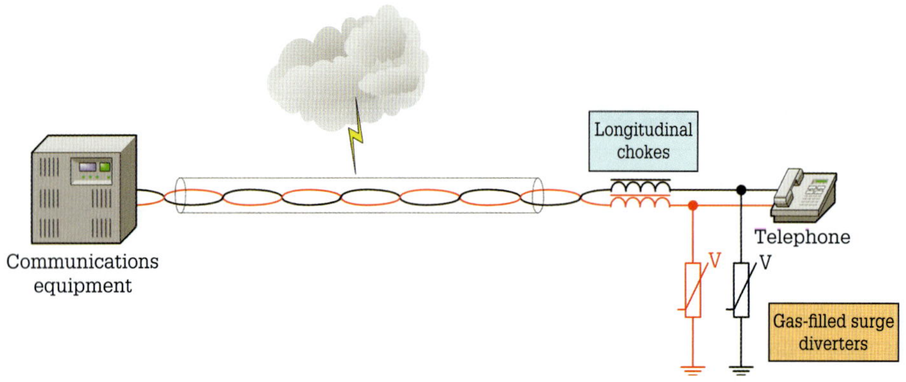

FIGURE 12.74 Protecting a communication circuit from the effects of over-voltages

Gas-filled surge arrestors

At telecommunications cabling installations where high-energy surges are not commonplace, gas-filled surge arrestors are the main form of primary protection. A gas-filled surge arrestor is a gas discharge device that conducts when the voltage between its terminals exceeds a predetermined value. Common conducting voltages for gas-filled surge arrestors used in telecommunications circuits include 90 V, 230 V, 350 V and 500 V, with a typical tolerance of ±20%. Gas-filled surge arrestors have a certain surge-current capability and are generally available as 5 kA/5 A, 10 kA/10 A and 20 kA/20 A. The rating refers to the peak surge current and continuous rms sinusoidal current-carrying capabilities of the arrestor under explicit test conditions. In most underground telecommunications cabling environments, 5 kA/5 A and 10 kA/10 A rated arrestors are suitable. The 20 kA/20 A type arrestors are more suited to protecting long runs of aerial cables.

Arrestors with a 350 V conducting value are common in telephone circuits. The telephone ring voltage is about 75 V rms (106 V peak). When this value is superimposed on the 48 V d.c. value of the telephone line it results in a voltage up to 154 V peak. This means that the 350 V arrestor provides sufficient margin above the normal line conditions, which in turn means it is unlikely to conduct to earth during normal operation, taking into account that the conducting voltage of 350 V is nominal and may actually conduct at values less than this. For data circuits, it is more common to use 230 V arrestors as the normal line-operating voltages are not subject to the additional ring voltage.

Three-electrode gas-filled arrestors, such as those shown in **Figure 12.75**, are common for protection of telecommunications circuits. As most long-distance telecommunications circuits use balanced pair transmission, it is preferable to use a symmetric protection arrangement to provide simultaneous protection on both legs of the pair. A three-electrode arrestor, is essentially a pair of two-electrode devices in a single package. It has two circuit terminals and a centre ground terminal, all connected to a common gas chamber. The premise is that if either side conducts, the discharge spreads rapidly to the other side. This will effectively maintain circuit balance and rapidly quench the surge on both legs of the circuit.

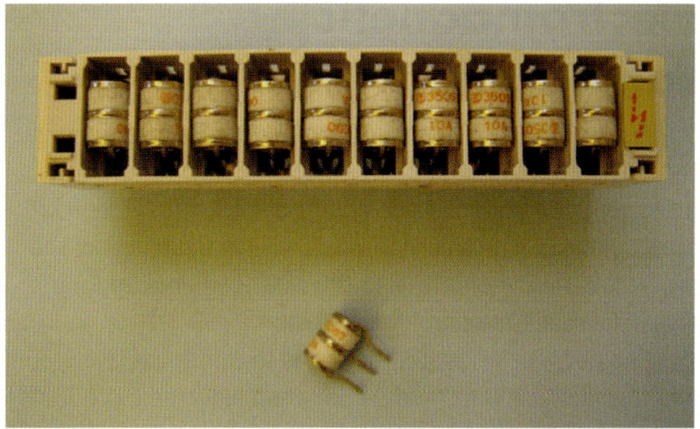

FIGURE 12.75 Three-electrode gas-filled arrestors

Regulatory requirements

AS/CA S009: 2020 requires the cabling provider to assess the need for surge protection (see clause 10.1), in accordance with AS 4262.1: 1995 *Telecommunications Overvoltages – Protection of Persons*, where the customer cabling comprises copper twisted pair and the network boundary is not at that particular building. For example, in a campus arrangement, the network boundary would be located in one building but other buildings supplied beyond the network boundary would be susceptible to over-voltages and require over-voltage protection.

It is the cabling provider's responsibility to install surge suppression where required if there is a high risk of injury as assessed in accordance with AS 4262.1: 1995. Surge suppression devices connecting between telecommunications line conductors and earth must comply with AS/NZS 4117: 1999 *Surge Protective Devices for Telecommunication Applications*. When located within a main distribution frame (MDF) either a Class 1 or Class 3 device is suitable. In other locations, a Class 1 device is suitable. Class 1 devices have a nominal conducting voltage between 500 V and 600 V; Class 3 devices have a nominal conducting voltage of 230 V.

Surge suppression devices may connect to earth through the back-mount on which the wiring modules terminate. This is standard practice in most terminating systems and it requires the back-mount on which the protected wiring terminates to connect to the protective earth. It is possible for a registered telecommunications cable installer to make this connection directly to the building distributor earth bar or terminal if the earth bar or terminal connects to the protective earth through the bonding conductor.

Alternatively, a licensed electrician may make the connection from the back-mount in the building distributor to the protective earth in the nearest electrical switchboard. In other locations in the installation, such as floor distributors, the back-mounts containing surge suppressors must bond to the protective earth at the nearest electrical switchboard. Only a licensed electrician may make this connection at the electrical switchboard.

REVIEW QUESTIONS

1 A common technique used in over-voltage protection is 'shunting'. What does this mean?
2 A technique used in over-voltage protection is 'isolating'. What does this mean?
3 Describe the function of a longitudinal choke.
4 Class 3 surge suppression devices located in the MDF, installed in the customer cabling, and connected between telecommunications line conductors and earth, must have a nominal firing voltage greater than what value?
5 What type of surge arrestor is common for protection of telecommunications circuits?
6 To where must back-mounts containing surge suppressors in locations such as floor distributors bond?

12.14 Cable shielding and interference

Telecommunications cables may be installed in electrically 'noisy' environments such as near high voltage power lines, near devices that produce alternating magnetic fields, or near other cables and equipment containing signals operating at radio frequencies (>20 kHz). The general term used to describe this electrical noise is electromagnetic interference (EMI). If a cable is not protected from this noise, the noise may combine with the signal on the metallic conductor telecommunications cable resulting in a corrupted signal. It is important to note that optical fibres are immune from this noise emanating from electrical sources, so this discussion pertains to telecommunications cables having metallic conductors.

Cable shielding

The use of effective physical separation and/or using a telecommunications cable that has a continuous shield or screen that connects to earth may reduce or totally eliminate this noise from the telecommunications circuit. Shielding also reduces electromagnetic radiation from the telecommunications cable.

There is a tendency in recent times for power cables to be constructed so that they are electromagnetic compatible (EMC) to minimise noise generation which affects many other systems like radio and data communication. The magnetic fields produced by multi-core three-phase HV power cables supplying balanced loads do not generally induce hazardous energy into nearby telecommunications cables as the magnetic fields produced by the currents in the balanced phase conductors tend to cancel each other out and the resultant EMI will be negligible.

Shielded telecommunications cables (F/UTP construction) are not as prone to the effects of EMI on the data transmitted as unshielded cables. To further reduce the effect of inter-pair EMI or cross talk and coupling, the telecommunication cables may have the individual pairs shielded (S/FTP construction).

Types of cable shield

There is a lot of engineering involved in cable shielding, which may be braided wire, spiral design, or metal-coated Mylar or foil shield. The shielding can mitigate noise by 85% to 100%, depending on the configuration. Using a thin layer of Mylar or aluminium foil eliminates the gaps that occur with braided designs. The foil attaches to a polyester backing to provide total coverage. Foil shielding is susceptible to damage in high-flex applications, so spiral or braided designs are better suited to these applications.

Braided shielding comprises a mesh of bare or tinned copper wires that are woven together, which assists in terminating on a soldered or crimped connector. The braiding creates small gaps of coverage resulting in a shielding performance of about 90%. This rating should be sufficient if the cable is not subject to movement or flexing.

In a particularly noisy environment, a telecommunications cable may employ multiple layers of shielding using both the braided and foil designs. In these environments it may be best to opt for S/FTP cable.

Earthing of cable shields

Certain shield designs allow for termination by crimping or soldering. The connector must also offer similar ratings as the cable for effective shielding. It is important to earth the shield at only one end of the cable to eliminate the potential for noise loops and circulating earth currents due to potential difference between the two local earths.

Clause 20.18 of AS/CA S009: 2020 covers the earthing requirements of cable shields, which allows for connection to a point connected to the protective earth or any metallic part that connects to the earthing system of the electrical installation in accordance with AS/NZS 3000. It is a further requirement that where an earthed shielded cable connects separate buildings or structures it needs to be insulated from any earth reference within the building or structure or only connect to earth through a differential earth clamp having a minimum firing voltage of 400 V d.c.

Electromagnetic compatibility (EMC)

To deal with the problem of electromagnetic interference (EMI), electromagnetic compatibility (EMC) regulatory arrangements under the *Radiocommunications Act 1992* are in place. All products that fall within the scope of the regulation must comply with the arrangements and must show the relevant compliance mark.

The purpose of the regulation is to minimise electromagnetic interference between electronic products which may diminish the performance of electrical products or disrupt essential communications (ACMA, 2018b). This is increasingly important with the greater use of electronic systems and digital technology in the commercial and domestic environments so that all users can access the radiofrequency spectrum without risk of interference.

The EMC regulatory arrangements introduce technical limits for emissions from electrical/electronic products and communications services. Accountability lies with Australian suppliers responsible for placing the products on the market. To establish compliance with the regulatory arrangements, suppliers must demonstrate that products satisfy relevant standards before being available for supply in Australia.

Electromagnetic energy (EME)

The ACMA has regulations in place to limit human exposure to electromagnetic energy (EME) emitted by radiocommunications transmitters. Portable transmitters with integral antennae, such as mobile phones and hand-held two-way radios, are regulated at point of supply. Responsibility for compliance lies with the manufacturer or importer. Other transmitters, such as mobile phone base stations and television broadcast towers, are regulated through licence conditions. Responsibility for compliance lies with the licensee.

A number of other terms used interchangeably with EME include EMR (electromagnetic radiation) and EMF (electromagnetic fields). EME generally refers to the radiofrequency portion of the electromagnetic spectrum.

Low-frequency induction

It is possible for 50 Hz currents to be induced in the metallic conductors of communications cables where they run parallel to high-voltage power lines carrying low-frequency (50 Hz) currents. This is particularly true for cable lengths greater than 200 m under power-system fault conditions, where the high-voltage cable can be carrying very high current. While the fault condition may be of short duration (less than 2 s), it can pose a real threat to telecommunications workers and equipment that attaches to the telecommunications line. Low-voltage a.c. power lines (less than 1000 volts) do not generally cause concern for low-frequency induction due to the decreased energy level. The voltage induced in the telephone line from the power cables happens in a manner similar to the operation of a step-down transformer. A higher potential on the primary circuit results in a higher potential on the secondary circuit. Hence, low-voltage a.c. power lines pose a threat for direct contact, but less of a threat for dangerous voltage being induced on the telephone line.

It is important to maintain spatial separation between conductive telecommunications cables and HV power cables to ensure that the possibility of induced voltage on the communications cables does not exceed 430 V a.c. If it is not possible to maintain spatial separation, an engineered design is needed to ensure the safety of the network and the network users. Additionally, the relevant carrier must be advised of the proposed arrangement prior to commencement of the installation.

The magnitude of induced voltage is dependent on:

- the magnitude of the fault current on the HV line
- the length of the cable segment parallel with the HV line
- the distance between the communications cable and the HV line
- the presence of shielding conductors or environmental shielding such as metallic pipes between the communications cable and the HV line.

Personnel and telecommunications equipment can be protected from the effects of low-voltage induction by:

- using shielded (or screened) communications cable or a communications cable having a metallic moisture barrier. In both cases the shield or moisture barrier must be electrically continuous for the entire length of the cable and be suitably connected to earth
- installing an interposing shielding barrier parallel to and in close proximity to the communications cable; again the barrier must be suitably earthed
- installing gas-filled surge suppression devices to each pair in the communications cable.

REVIEW QUESTIONS

1. Name three cabling environments that may be sources of electrical 'noise' on telecommunications cables.
2. What are two measures employed to reduce noise on a metallic conductor telecommunications cable?
3. How are three-phase multi-core cables electromagnetic compatible (EMC)?
4. Which construction of metallic conductor telecommunications cable affords the best noise immunity?
5. Why is earthing of cable shields at both ends inadvisable?
6. Under what condition is it possible for 50 Hz currents to be induced in metallic conductors of telecommunications cables?

12.15 Miscellaneous regulations

Miscellaneous regulations include those dealing with heritage sites and buildings, cable installation, pollution, and hazardous areas.

Heritage sites and buildings

The term 'heritage' may be broadly defined as 'the possessions, traditions or conditions that have been passed from one generation to another'. There are many buildings, places and relics in Australia that are considered to be part of Australia's heritage. These sites are historically significant and considered worthy of preservation for future generations. They may be of cultural, architectural or archaeological significance and as such are part of our heritage.

For the telecommunications cable installer, this means that cabling in any heritage building needs special consideration to preserve the existing character of the building. Legislation has been passed in this country to protect such heritage sites for future generations and to ensure coexistence with current occupants. Before examining the installation requirements, it is useful to have a general overview of the heritage laws and the government departments involved.

Australian Heritage Commission

The Australian Heritage Commission is a federal government body. The main purpose of the commission is to compile a register of heritage sites. This national inventory may be used by other government bodies or any other interested parties in determining the significance of a particular site. When that is done, steps may then be taken to ensure the conservation of the site.

State heritage listings

State governments each have their own heritage Act, administered by a separate department, which deals with the heritage sites in their state. It is at this level that conservation orders may be placed on sites. These conservation orders may prohibit a range of activities on the site and work may not start without approval of the relevant authority.

Local government heritage listings

Local government bodies may list heritage sites through planning. For example, a local government may rezone an area considered to be historically significant so that all existing buildings remain and are preserved in their original condition. An example of this is The Rocks area in Sydney. Individual buildings may also be affected if considered of great significance. Building works cannot proceed without approval from local government, so that heritage sites are protected.

Cable installation

Cabling installation work may require referencing regulations that pertain to heritage-listed buildings, fire regulations, joint use agreements, traffic control, and reinstatement works.

Installation in heritage buildings

Cabling work in a heritage-listed building is usually specialised work and should only proceed on the advice of a specialist in the area, such as an architect or builder. Permission may need to be obtained from the relevant authorities before any cabling work can take place. In many instances builders or architects will get this approval. In some situations it may be the responsibility of the cable installer. When it comes to the installation of services, the areas of most concern to the cable installer are concealing wiring within the installation and locating equipment in areas that are unobtrusive and aesthetically compatible with the building.

In many situations a licensed cable installer may be asked to install telecommunications cabling and equipment in areas where the existing finishes are to stay intact. It is therefore essential that the wiring methods used conceal the wiring. Installation methods should cause little or no harm to the existing structure and finishes. Where damage has occurred, repairs should be made to return the finish to its original condition.

Wherever possible, cabling in these structures must be done in roof spaces, underfloor areas and cavity walls. Due to the nature of many old buildings, wall cavities are often solid and this can make cabling difficult. It may be necessary to go to extreme lengths to conceal wiring. As an example, if wiring was required to a point on a solid wall, skirting boards could be removed, walls behind the boards chased and cabling installed in conduits or ducts behind the skirting. Cabling from floor to ceiling in such cases may be installed within a cupboard, or a boxed section may be constructed using a style to match the existing finish. Putting equipment in dedicated rooms or in cupboards will not affect the aesthetics of the building.

Fire regulations and cable installations

One of the most important aspects of the Building Code of Australia is fire prevention and safety. It is essential that a cable installer be aware of how the installation of cables and support systems may affect the fire safety of a building. These are:

- The way in which cables are installed in a building may affect the safety of the building. Penetrations in fire-rated walls, floors and ceilings may dramatically affect the fire safety of a building.
- The materials used in a cable system (cables, conduits and other cabling products) may be combustible and provide fuel for a fire or produce toxic fumes when burnt.

Joint-use agreements

Joint-use agreements are usually arranged between telecommunications carriers and other utility providers. Both the 'sharing of trenches' agreement and the 'joint use of poles' agreement relate to public services. Note that in some circumstances these and other agreements may be binding on private cable installers. Many other utility providers may have similar agreements for sharing of easements or trenches.

Traffic control regulations

When installing cables in public places, it may be necessary to block or close a road for trenching purposes. Before performing any cabling work involving the control or redirection of traffic, it is important to obtain permission from the appropriate authorities. Traffic control regulations vary from state to state. In some Australian states police are ultimately responsible for traffic control; in other states separate government departments are responsible for road closures and work on public roads.

The best place to start investigations is usually with the local police. If they are not directly responsible, they will be able to direct you to the department with responsibility. When getting permission to close a road totally or partially for any reason it is important to obtain details on the signalling and control measures to be used. Again, these requirements will vary from state to state.

The traffic load of the road will also influence the control procedures to be used. For example, permission to close a very busy road is not likely to be granted. It may only be possible to close half the road at one time. Work may also have to proceed out of peak hours so as to cause minimum disruption. Each situation must be taken on its own merit as local conditions and regulations will apply.

Requirements for reinstatement

Whenever cabling work is done on private property, it is regarded as 'best practice' to leave the job site in an 'as found' condition. This is of particular importance when it comes to installing cable and equipment in external locations, and applies to cabling in and on buildings.

When cabling in or on a building, it is essential to replace any building components that are removed for cabling purposes and to repair and make good any damage to paintwork or surfaces. This is particularly important where penetrations are made in building structures. It is vital that any penetrations into buildings are made both water and vermin resistant to prevent any future problems.

In external locations, particularly where cable trenching is involved, leaving the area as found may be more difficult. One major problem resulting from trenching is sinking of backfilled areas, leaving the trench lower than the surrounding ground level. This problem results from the trench being backfilled and finished to existing ground level without compacting the fill. After a period of time or heavy rain, the trench will sink, leaving a depression. There are two simple ways of overcoming this:

1. compacting of backfill material using a compacter
2. finishing the trench above ground level to allow for sinking.

If a compacter is used to compress backfill during the filling-in stage, there is no need to allow for future compaction. Using this method, trenches are filled in stages (150 mm to 300 mm of backfill) and compacted at each stage. The trench is then finished to existing ground level. This method is recommended wherever a trench crosses a roadway or vehicular path. If the spoil removed from the trench is not suitable for compacting, additional material must be used and the original spoil removed.

If trenches are not compacted during the backfill stage, it is essential to overfill the trench, leaving a small mound of spoil to compact with time. This is not a preferred method, but may be used in areas where appearance is not important, such as open paddocks, unused areas and some garden beds.

Care must also be taken when excavating so that minimal damage is inflicted on the surrounding area. In grassed areas turf and topsoil should be removed carefully and placed to one side of the trench ready for replacement. The excavated dirt should then be placed on the other side of the trench. In garden areas plants should be carefully removed and set aside ready for replanting. The trench route should be selected to avoid large trees, as trenching may weaken the root system and cause the trees to fall in a storm.

It is also important when trenching to avoid other services. Any damage resulting from hitting other services could be hazardous and costly.

'Dial before you dig'

Before starting any trenching on private or public property it is a good idea to determine the exact location of all existing services before excavation. This will make trenching safer and save considerable time, money and resources by preventing damage to existing services.

There are several ways to determine the location of services prior to trenching. 'Dial before you dig' is one solution to this problem. It simply involves telephoning the 'Dial before you dig' service on 1100 or visiting their website at http://1100.com.au. This is a one-stop referral service where one phone call will result in details on all services.

It is always good practice to check locations with all utilities before beginning excavation.

Pollution

Pollution encompasses noise, visual and waste.

Noise pollution

Any noise that is offensive may be regarded as noise pollution, and some state legislation enforces the reduction or elimination of the source of noise. These laws will vary from state to state, and in some areas it is left to the local government authority to enforce noise abatement laws.

Noise abatement is of interest to the telecommunications cable installer as many activities involved with cabling produce noise. For example, trenching using a backhoe or trenching machine in a suburban garden could result in offensive noise. It is the responsibility of the cable installer to determine the noise abatement requirements for each installation and to comply with these regulations. Usually, restrictions will apply to maximum noise levels and hours during which noisy work may take place.

Visual pollution

The visual environment in which we live may also be protected by pollution control legislation. This legislation may be of particular importance in areas of natural or cultural heritage. As far as the telecommunications industry is concerned, the running of aerial cables most frequently comes under the scrutiny of these laws. It is fair to say that this legislation may also vary between states and local government authorities. Local governments may zone certain areas to protect the visual environment and maintain the character of the area under their jurisdiction. It is again important for the telecommunications installer to be familiar with these requirements.

Waste pollution

One other area of concern for cable installers is the disposal of waste cabling products. Most products are safe for normal disposal, but some need special attention. Any chemicals should be disposed of in accordance with manufacturers' recommendations, including items such as solvents, cleaners or filling from filled cables (jelly). Other materials such as lead from lead-sheathed cables should be recycled to prevent pollution.

Hazardous areas

Wherever flammable or combustible materials are used there is a real danger of a fire, or even an explosion. In the past, areas of main concern have been underground mines and sites containing potentially flammable dust, such as flour and saw mills. In modern society, other areas of industry, such as the chemical industry, pose a real threat to safety.

Hazardous areas are not just restricted to heavy industry. Dry cleaners, petrol service stations and spray painting shops are also hazardous areas containing flammable and explosive materials. One factor common to these areas is the use of electricity, both as a power source and as a means for communication.

To prevent any electrical equipment from becoming a source of ignition in a hazardous area, precautions must be taken in the design, construction and installation of the equipment.

The explosive atmosphere in a hazardous area may result from the presence of a flammable liquid, vapour or gas, or from the presence of combustible dust. It is important to note that not all areas of a particular installation will be regarded as hazardous, and it is normal practice for industry to minimise the extent of hazardous areas through good design. In situations where only portions of an installation are regarded as hazardous, the telecommunications cable installer can simply locate all equipment and cable outside the hazardous areas and treat the installation as any other. If materials have to be installed within restricted areas, special precautions must be taken.

There are two main areas defined as hazardous when it comes to government legislation. Underground coal mines are clearly covered by state legislation (see below). Other hazardous areas are covered by miscellaneous state legislation with various departments administering this legislation. Apart from these government bodies, there are a number of other groups which may have interests or responsibilities when it comes to hazardous areas other than coal mines.

A golden rule for any electrical or telecommunications installation is to avoid hazardous areas. Wherever possible, all equipment, cabling and joints should be placed outside a hazardous area.

If cable must pass through a hazardous area, the cable must be suitably enclosed (depending on the class and zone) and sealed at the entry and exit points. If terminations or equipment are to be installed in a hazardous area they must be suitably enclosed in an enclosure rated to suit the area.

The cable installer needs to know that installations in these areas are specialised and require a properly engineered solution to comply with the relevant codes.

Mining installations are covered by state legislation and any installation in a mine must comply with this legislation. There are special installation requirements for many areas in mines, such as hazardous areas. Underground mines have more stringent requirements than open-cut mines. This is largely because of the hazardous gases that build up in the mine or are trapped in the mine.

Underground coal mines must only use equipment approved by the Chief Inspector of Coal Mines (CICM) or equivalent.

Confined spaces

Confined spaces may be broadly defined as areas in which the rate of air being consumed is greater than the air being replaced. A confined space can generally be defined in one of four ways:

1 **Contents** The presence of toxic or highly flammable gases may make a workplace a confined space.
2 **Work activity** Power tools, welding and thermal cutting may produce noise and toxic fumes in a small area.
3 **Location** Pits in the ground and low work areas may contain heavy gases that can replace oxygen.
4 **Construction** Tanks, pipelines and similar enclosed structures may provide a confined space.

Accidents in confined spaces are often fatal, as air becomes toxic or oxygen is depleted. These accidents usually result from a failure to identify the hazardous situation. Many problems can be alleviated by ensuring adequate ventilation in confined spaces. Testing of the air before entering the confined space will also prevent many accidents. Personal protective equipment such as respirators and protective clothing may also be used for work in confined spaces.

REVIEW QUESTIONS

1 What is meant by the term 'heritage'?
2 What is the purpose of conservation orders?
3 How may local government bodies list heritage sites?
4 Why is it important for a cable installer to have an understanding of fire regulations?
5 What is a joint-use agreement?
6 What precautions would you take when installing cables in public places?
7 What is considered best practice when cabling on private property?
8 What is meant by 'Dial before you dig'?
9 What forms may pollution take?
10 What is a confined space?

12.16 Installation and termination requirements

While cables used for both domestic and non-domestic installations may be the same, the installation and termination methods may vary. The following looks at some of these differences.

Cabling in domestic buildings

Before installing additional services to an existing domestic dwelling, it is important to consider the location of existing services. These services include:

- electrical power cables, which may be within any wall (internal or external). They may run vertically to a light switch or power outlet or they may run horizontally between power outlets. Exercise *extreme care* when drilling through walls to avoid electric shock
- water pipes, including hot water, cold water and waste pipes. These are usually in the external walls, although they may be within internal walls. Water pipes may run vertically or horizontally
- gas pipes, which usually run through the roof void or floor void and then drop or rise vertically to the appliance. There are usually fewer gas pipes than water pipes
- existing telephone cables, which usually run through the roof void or floor void and then drop or rise vertically to the telephone outlet. Telephone cables generally run down the wall from directly above the outlet socket. Look for wall phones on internal walls and locate any skirting sockets and avoid drilling close to them
- coaxial cables for television, which usually run through the roof void or floor void and then drop or rise vertically to the television outlet.

SWITCH ON

Any one or a combination of building services may be within a wall that you intend to drill through. Be very careful in dealing with these potential hazards.

Locating services in dwellings

In existing dwellings it can be difficult to look inside a wall, but before drilling into a cavity wall you could look from above. Use a torch to see if there is any likelihood of damaging a service in the wall. Be careful not to use more force on the drill than is necessary. Ease the feed on the drill if you feel it starting to break the interior surface. Remove the drill when through and use a torch to see if there are any cables or pipes in the wall. You might have to redrill the hole in a more suitable place.

There are devices to help you find services within walls. These may be magnetic or electronic. One device is the Stud Finder. It detects the presence of nails in the wall. The problem with this device is that while the nails are usually in the studs they may also be in a nogging. Additionally, the device does not detect non-ferrous metals, or copper or plastic pipes.

It is possible to locate electrical power cables using a volt stick. However, this device has low sensitivity and may not detect live power cables through brick or between stud and brick exterior walls (brick veneer construction).

Cabling in existing commercial buildings

The first step in adding cabling to any existing commercial building is to establish the exact location of all proposed additional services. A detailed site inspection should then follow to establish the building structure and the location of existing services. These services would include:

- electrical power cables, usually supported and transported between floors by a dedicated riser, while skirting trunking, internal walls and ceiling spaces, or conduits within slabs transport and support electrical cables from the distribution boards to service outlets
- plumbing services, which may be in dedicated risers or ceiling or floor spaces
- air-conditioning and ventilation services in risers, false ceiling spaces and plant rooms
- gas services, usually in risers and ceiling spaces
- fire services in dedicated risers, ceiling spaces and plant rooms
- building automation services, which may share telecommunications risers and, in some cases, the same sheath as telecommunications conductors
- telecommunications services in risers, ceiling spaces, walls and skirting trunking.

Remember the regulations when installing additional telecommunications services close to other services. It is necessary to maintain segregation from other services. Usually it is possible to make use of the following areas when installing these additional services.

Existing false ceiling spaces or voids

These areas may have conduits or cable trays containing existing telecommunications services. Install the new cabling with the existing services wherever possible. If access to the ceiling space is not available, then another cable route is necessary. It is possible to install the new cables in the ceiling void if the space is accessible and clear of existing services. Do not install the new cables directly onto the ceiling or tie them to the ceiling hangers. A cable support system must be installed to accommodate the new cables.

In multi-storeyed buildings you can run cables in the false ceiling space to access equipment on the floor above. You may have to drill through the floor slab to do this. If so, be careful not to damage existing services within the concrete slab. If damage does occur to services such as electrical conduits, then a qualified tradesperson must make prompt repairs.

Existing floor spaces and voids

Treat these areas the same as for ceiling spaces. Avoid harmful and hazardous services and utilise all existing cabling systems where possible. Again, when drilling through a concrete slab, exercise care to ensure damage does not occur to any existing services embedded in the concrete. Many offices have false floors, often called computer floors.

Walls and skirtings

Walls in existing buildings are usually difficult to get into. In constructions with steel-frame partition walls it may be possible to gain access through the top plate of the wall vertically down to the required location. The only way to access such walls horizontally is through skirting trunking.

Skirting trunking is usually on all walls of a building to allow for flexibility with services. Installation of conduits is normally done during construction to link service cupboards and risers to the trunking. These conduits normally have ample spare capacity to allow for future installations. Use this spare capacity when installing additional services.

Skirting trunking has separate compartments for electrical and telecommunications services, and it is essential to maintain this segregation.

Cabling in new buildings

When cabling in new commercial premises it is advantageous to install the cable systems progressively to match the progress of building construction. The telecommunications cable installer should arrange with the builder to be on site during the stages when the installation of conduits, tray systems and equipment is the easiest. It may be possible to wire an entire building after its construction is complete by utilising risers and skirting trunking. However, experience has proved this not to be advisable.

A telecommunications system in large commercial premises would consist of the following components:

- Cable runway goes from building entry point to the main distribution frame. The carrier supplies and installs the lead-in cable. The cable runway may be a conduit, cable tray or other approved cabling system. Installation of this system is usually in a basement or sub-floor area; it may also be a conduit in a slab or a cable tray in a ceiling space.
- The main distribution frame consists of back-mounts suitable for terminating the lead-in cable and capable of accepting the carrier's termination modules (supplied by the carrier). This distributor will also contain back-mounts and termination modules for the customer cabling. It is usually situated in a dedicated cupboard forming part of a riser or in a dedicated room.
- Backbone cabling rises from the main distribution frame to floor distributors on each level or in each section of the building. Installation of this cabling may be on a tray in a dedicated riser, in conduits or in ceiling spaces. The location of the floor distributors may be in dedicated cupboards forming part of the riser or recessed into walls in open locations.
- Backbone cabling from the floor distributors usually runs in conduits in the slab or through false ceiling spaces in conduits or on a tray or tied to a catenary. Location of the local distributors is usually throughout the floor, evenly distributed and close to skirting trunking.
- Skirting trunking usually runs around all walls, and access to local distributors is usually via conduits. It is normally the responsibility of the electrician to install the skirting trunking.

It is important to realise that as building construction proceeds it is the responsibility of the cable installer to install all necessary components before being built out. **Built out** is the term used when building construction work prevents the installation of equipment. For example, if you need to install a conduit in a slab, but the slab is already poured, then you have been built out. This can occur in many other situations. The critical stages for installing cabling systems are as follows.

Conduits in slabs

Installation of conduits in slabs must occur directly after laying of the first layer of steelwork and before the installation of the second layer. Locations for conduit 'turn ups' are marked in the formwork and then the conduits are installed and tied with wire to the reinforcing steelwork. It is important to seal the ends of the conduits to prevent the entry of foreign objects that might block the conduit. It is also vital to be on site during the concrete pour in case there is damage to a conduit.

Cable systems in ceilings

Installation of cable trays, catenary wires and conduits in false ceiling spaces must happen before installation of the ceiling hangers and tiles. Although it is possible to complete this work afterwards, it is much easier and quicker to complete it before installation of the ceiling components.

Distributed cabling

Installation of distributed cabling should occur before painting and finishing of the building. This will prevent damage to already completed surfaces. Remember that each building is different. Give careful consideration to the most suitable cabling methods *before* cabling starts.

Fire stop systems

When wiring systems pass through a fire-rated wall, floor or ceiling, it is important that the penetration does not result in a reduction to the original fire rating. It is usually necessary to install a proprietary fire stop system.

The Building Code of Australia (BCA) outlines responsibilities regarding fire stopping of service penetrations. Essentially, the BCA says that the penetration should not reduce the required Fire Resistance Level (FRL) of the penetrating element itself, such as the fire rated wall, floor or ceiling.

The FRL is a grading period in minutes, determined in accordance with AS 1530.4:2005 for structural adequacy, integrity and insulation. For example, FRL = 120/120/60 means that structural adequacy is maintained for 120 minutes; the integrity of the structure is maintained for 120 minutes; and the insulation is maintained for 60 minutes.

The choice of a specific fire stop system will depend on the specific application and any additional requirements such as whether the penetration is temporary or permanent, whether re-routing is expected, the required movement capabilities, environmental conditions, exposure to oils and grease and budgetary constraints.

When dealing with fire stop systems the terms 'intumescent material' or 'intumescence' arise. An intumescent material is a substance that expands as a result of exposure to heat, thereby increasing in volume and decreasing in density. Intumescent materials are common in passive fire protection systems.

Examples of fire stop systems suitable for maintaining the FRL where cables pass through a fire-rated wall, floor or ceiling include:

- pillows
- mortars
- coated panels
- sealants/mastics
- putties
- RTV silicone foams
- blocks/bricks
- boards.

Pillows

Pillows are fire-retardant fabric bags that contain high-temperature granulated rockwool. These pillows offer high flexibility in securing penetrations through fire-rated construction elements against the passage of smoke and the spread of fire. The pillows are packed tightly into the opening and around the services, as **Figure 12.76** shows. Pillows are easily removed to permit the installation of new services. Pillows are available in several sizes to allow easy installation in service penetrations.

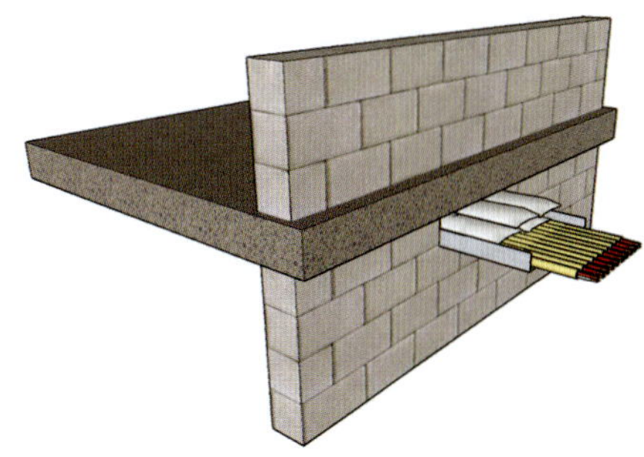

FIGURE 12.76 Fire stopping a penetration using pillows

Mortars

These are generally made from a gypsum- or cement-based powder blended with inorganic lightweight fillers, composite reinforcement and chemical modifiers. Adding water to the mortar forms a lightweight, low-slump mix for installation around pipe or cable penetrations in walls and floors. Mortars are suitable for backfilling and panel forming in wall and floor openings where it is necessary to maintain a specific FRL. The advantage of mortar is that it forms easily around multiple service penetrations and has an equivalent fire rating to concrete. Working with mortar is also easy as it is readily cut, drilled and shaped with conventional wood-working tools.

Coated panels

Coated panels are suitable for fire stopping mechanical and electrical services where they pass through fire-rated walls and floors. Typically the coated panel is made by spraying high-density rock fibreboard with a fire-rated ablative coating. The coated panels are simply cut and friction fitted between the services and the edges of the structure as shown in **Figure 12.77**. Plastic pipes may require sleeving with fire protection collars prior to installation of the coated panel system. All joints in the barrier are pointed in with an intumescent sealant to form a monolithic layer.

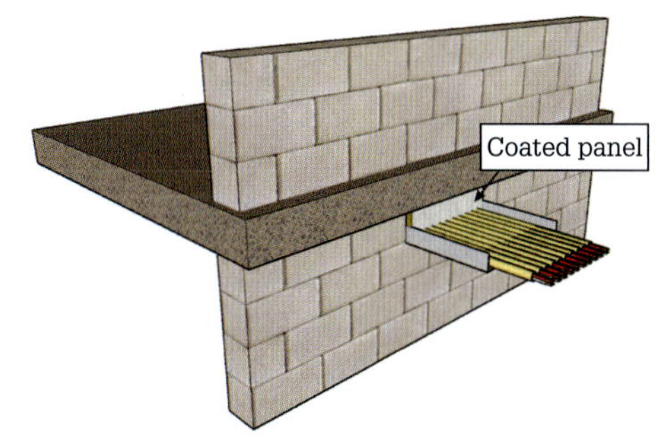

FIGURE 12.77 Fire stopping a penetration using a coated panel

In plasterboard partitions, a coated panel system must be fitted to both sides of the wall to fire protect the studs. In floor applications, the coated panel requires protection from physical damage by foot traffic. A plywood or chequer-plate platform will suffice.

Sealants/mastics

Fire-resisting silicone sealants are a one-part alkoxy cure silicone offering excellent unprimed adhesion to most building surfaces. They are designed for use in areas where heat and fire resistance is required. Fire-resisting silicone sealants are extruded from a cartridge that is loaded into a standard sealant gun. The depth of the joint will depend on the gap to be filled and the required fire rating.

All surfaces must be thoroughly clean and free of bond-breaking contaminants prior to application of the sealant. No priming is required for most construction substrates; however, it is recommended that a small area be tested on substrates.

Putties

Putties are single-pack materials which are similar to some sealants/mastics. The difference is that putties are capable of being formed and installed directly by hand.

RTV silicone foams

Room-temperature vulcanising (RTV) foams are two-component silicone materials which, when mixed together, cause the materials to foam and increase in volume.

Blocks/bricks

Blocks and bricks are made from bonded vermiculite, modified rubber, polyurethane or flexible intumescent material. They come in many shapes and sizes, but are typically shaped like conventional bricks, and are stacked in the penetrating opening in a similar way to laying conventional bricks, but without the need for mortar.

Boards

There are many types of fire-resistant board materials including fibre-reinforced autoclaved calcium silicate, vermiculite-bonded sodium, and potassium silicate. The board is cut to fit around electrical services penetrating the opening. Boards are typically used where large penetrations are present and often require the use of additional fire stop products.

Cable support systems

Cable support systems (including conduits, trays, ducts, ladders and cable distribution equipment) require fixing to or supporting on surfaces made of different materials. In addition, the building material may vary in thickness and composition; it may be soft or hard, damp or wet. The fixing may have to be permanent or it may be such that removal or replacement of the fixture is possible. It is also necessary to consider other factors such as mechanical strength and the aesthetics of the fixing method.

Fixing devices

As well as having a working knowledge of building materials and their assembly, it is essential to have an awareness of methods and devices for fixing equipment to structural and architectural components. The cable installer must be able to fix equipment to a structure without impairing the structural integrity of a building member. Furthermore, this fixing must not affect or damage other services and must not detract from the appearance of the structure. However, perhaps the most important consideration is safety. The fixing must not put the installer or others in the vicinity at risk.

A fastener is a general term used to describe a component that connects two items together, while a fixing more specifically refers to a component that attaches a removable item to a permanent structure. A variety of fixing devices are used when installing telecommunications infrastructure:

- masonry screws
- plastic expansion plugs
- expanding sleeve anchors
- chemical anchors
- toggle anchors
- bolts and screws.

Fixing requirements

Fixing devices (fasteners or anchors) are suitable for securing a variety of components (fixtures) to walls, ceilings and other structural components.

Under the Work Health and Safety Regulations (WHS) installers must:

- Ensure the health and safety of themselves, their employees and other workers at the installation site during the installation process.
- Ensure that once installed the equipment is not a safety or health hazard when properly used and maintained.

The process of fixing equipment to various surfaces involves the use of powered and hand-operated tools, ladders, trestles and lifting equipment. It may involve the use of explosive-powered tools or various chemicals, possibly in confined spaces. It is therefore essential to take proper safety measures.

These measures include but are not limited to the personal protective equipment (PPE) listed in **Table 12.11**.

TABLE 12.11 Personal protective equipment

Equipment	Environmental hazard	Most common high risk situations
Goggles, safety glasses or a full visor Eye protection must be worn Source: Shutterstock.com/lindaks	There is a chance of particles becoming airborne and damaging your eyes.	▪ using power drills in masonry or metal ▪ using power impact drills in masonry ▪ using cold chisels and the like on masonry surfaces ▪ when others in the vicinity are using the above tools
Hearing protection, such as earplugs or earmuffs Hearing protection must be worn Source: Shutterstock.com/lindaks	Loud, sustained noise is likely.	▪ near machinery, such as compressors and portable generators ▪ while using power tools ▪ while using hand tools, such as hammers and chisels
Breathing respirators or face filters Wear dust mask Source: Shutterstock.com/lindaks	Where any airborne particles are likely to affect your breathing.	▪ dust from the work surface or surrounding work areas ▪ fumes from painting or cleaning/degreasing operations

Cable enclosures

AS/CA S009: 2020, Clause 4.2.20, defines an enclosure as a housing or covering for cables or equipment providing an appropriate degree of protection against external influences or end-user contact with hazardous voltages, extra-low voltage (ELV) or telecommunications network voltage (TNV).

The main support systems suitable for telecommunications cables include:

- conduit
- cable tray
- cable ladder
- free-standing trunking
- in-floor trunking
- skirting trunking
- partition trunking
- column trunking
- in-floor duct
- catenary system.

AS/CA S009: 2020, Clause 8.1, requires telecommunications cabling to be supported or secured at suitable intervals to ensure the safe passage of persons, to maintain separation from hazardous services and to comply with the instructions provided by the cable manufacturer.

AS/CA S009: 2020, Clause 8.6, requires the removal of sharp edges from the cable-bearing surface of conduits, trays and trunking.

Conduit

Where particular products exist for use specifically with telecommunications installations, they must comply with the requirements of AS/CA S008:2010. In particular, underground non-metallic conduit must be white in colour.

An alternative is for the conduit to contain an indelible, durable continuous white stripe. Furthermore, non-metallic underground cable must be legibly and durably marked 'COMMUNICATIONS' at regular intervals (≤1 m). See Clause 5.3 of AS/CA S008: 2010 for specific requirements.

REVIEW QUESTIONS

1 Before installing additional services to an existing residential dwelling, it is important to consider the location of existing services. What are some of these services?
2 What is a stud finder and what is its use?
3 Before installing additional services to an existing commercial building, it is important to consider the location of existing services. What are some of these services?
4 What may be contained within existing false ceiling spaces or voids to assist the installation of new customer cabling?
5 What components would a telecommunications system in large commercial premises comprise?
6 What is meant by the term 'built out'?
7 What remediation action is required where a wiring system passes through a fire-rated wall, floor or ceiling?
8 What is meant by a fire resistance level of 60/60/60?
9 What are four fire stop systems suitable for maintaining the fire resistance level where cables pass through a fire-rated wall?
10 What is a typical application for 'coated panels'?
11 What is the most important consideration when installing a fixing?
12 What is a fixing?
13 What are some common fixing devices for fixing telecommunications infrastructure to a structure?
14 What are installers required to do under the Work Health and Safety Regulations when installing fixing devices?
15 What are common support systems suitable for telecommunications cables?
16 What personal protection equipment (PPE) should be worn when installing fixing devices?
17 How is non-metallic underground telecommunications conduit distinguished from electrical conduit?

12.17 Cable installation

Installation of cables may involve unrolling, pulling, lifting, dragging, bending, tying and crimping of the cable. During this time the cable is at risk of damage due to the ingress of moisture and air as the ends are open and unterminated.

All communication cables are relatively easy to damage and this damage is most likely to occur during installation.

Only the worst type of damage such as a tear in the outer sheath or sheared conductors will be obvious to the installer. More often than not, the damage caused by mishandling the cable will not become apparent until after the cable has been put into service. The harm caused to the cable during installation can show up as degradation of cable performance or early breakdown of the cable.

The amount of tensile stress, shear stress and torsional stress a cable will absorb without deviating from the manufacturer's performance specifications varies with the type of cable and even with the manufacturer. The same type of cable, from two different manufacturers, may differ in how much stress the cable can withstand without causing permanent damage.

Damage during installation

Cables can incur damage from the following:

- pulling the cable too hard (excessive hauling tension)
- allowing the cable to stretch under its own weight
- bending the cable too sharply around corners
- kinking the cable
- twisting the cable
- dragging the cable over rough or sharp surfaces
- crushing the cable
- water entering the cable.

Pulling the cable too hard can cause permanent localised elongation of the outer sheath, the screening material, the individual core insulation and the conductor or fibre-optic filament itself. Stretching of the outer sheath can cause its premature breakdown, reducing its effectiveness as a moisture, air and vermin barrier. Stretching of insulation around the individual cores of a cable reduces the effective thickness of the insulating material leading to signal leakage (as cross-talk) or possible insulation breakdown between cores (low insulation resistance); all of which add to decreased signal quality. Stretching of screening material can reduce its effectiveness as a shield thereby increasing the possibility of cross-talk and outside EM interference. Stretching of individual cores reduces the cross-sectional area of a conductor, increasing its resistance and leading to a reduction in signal power. Optical fibre cables, which transmit data through glass fibres using light, are particularly susceptible to any change in cross-sectional area. At worst, stretching of cables may lead to a break in one or more cores leading to total signal loss.

Allowing a cable to stretch under its own weight can occur, for example, when installing cable in riser ducts of multi-storeyed buildings or where a long, horizontal span of cable is temporarily unsupported during installation. This can result in similar problems as with 'pulling the cable too hard'.

Bending the cable too sharply causes stretching of the cable components near the outer radius of the bend. It also causes compression of the components near the inner radius of the bend. Stretching of components causes an effective reduction in cross-sectional area with similar problems as with 'pulling the cable too hard'. Compression of components causes an effective increase in cross-sectional area and this in turn affects the performance limits of a cable. The centre of the cable remains unstressed. It is important therefore to observe the minimum bend radii specified by the cable manufacturer. AS/CA S009: 2020 provides useful guidelines in the absence of manufacturer's data.

Kinking a cable usually occurs if the cable is incorrectly rolled off the drum and a loop is left in the slack cable. When the cable is tightened a kink forms. Kinking will cause similar problems to bending the cable too sharply.

Twisting a cable usually occurs if the cable is incorrectly rolled off the drum. Twisting the cable may distort the twist ratio for which the cable was designed, which may result in it operating outside its design parameters.

Dragging a cable over rough or sharp surfaces can cause the outer sheathing material to abrade, increasing the possibility of moisture or air ingress.

Crushing a cable generally occurs through negligence in allowing heavy objects to rest on the cable for extended periods of time, causing cable cores, insulation, screening and sheathing to be permanently distorted or even partially severed. Once again, this will lead to a degradation of cable performance.

To assist installers and system designers, cable manufacturers publish data, usually in cable catalogues, of important cable parameters such as maximum pulling tension, minimum bending radius, crush resistance and weight. Most cable manufacturers give the maximum tensile strength of a cable in newtons (N). To convert newtons force to kilograms, use the following relationship:

$$F = ma$$

where F = force in newtons

m = mass in kg

a = acceleration due to gravity (9.8 ms^{-2})

EXAMPLE 12.2

Consider the following cable:

Pairs count:	10
Overall diameter:	14 mm
Minimum bend radius:	120 mm
Maximum pull tension:	740 N
Cable mass:	20 kg/100 m
Conductor resistance:	3.8 Ω/100 m
Maximum current:	2.8 A

a The cable is lowered down a service duct in a multi-storeyed building. The cable jams after dispensing 60 m. To free the jam an 85 kg person swings on the end of the cable. Does this force exceed the manufacturer's recommended tensile strength (pull tension)?

It is necessary to add the dispensed cable weight to the weight of the person.

$$F = m \times a$$
$$= \left(\left(\frac{20 \times 60}{100}\right) + 85\right) \times 9.81$$
$$= \mathbf{951.6\,N}$$

This exceeds the maximum tensile strength of 740 N.

b The cable is lowered down a service duct in a multi-storeyed building. What is the maximum length of unsupported cable that will not exceed the manufacturer's recommended tensile strength (pull tension)?

It is necessary to find the mass that will result in a 740 N force.

$$F = m \times a$$
$$\text{so } m = \frac{F}{a}$$
$$= \frac{740}{9.81}$$
$$= 75.4 \text{ kg}$$
$$length = \frac{m}{cable\ weight} \times 100$$
$$= \frac{75.4}{20} \times 100$$
$$= \mathbf{377\ m}$$

The × 100 is performed as the cable mass is expressed in kg/100 m.

Note that sometimes the letter g is used to represent acceleration due to gravity instead of the letter a.

Figure 12.78 shows where force is exerted on a cable by applying weight to the end of a cable during drawing-in.

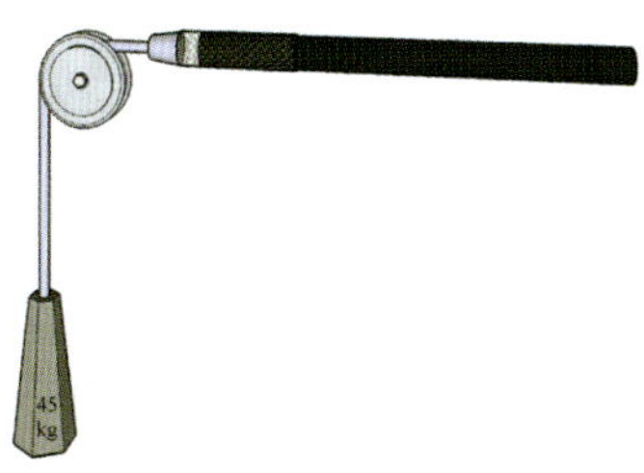

FIGURE 12.78 Effect of weight on a cable

Damage during termination

Nicks or cuts to the cable sheath facilitate the ingress of moisture or air into the cable. This may significantly degrade the performance of the cable.

Nicks or cuts to the wire insulation can result in atmospheric contamination of the conductor. This may cause the resistance of the conductor to increase and will decrease the insulation resistance. A decrease in insulation resistance can cause an increase in the effect of cross-talk.

Nicks in the conductor cause impedance anomalies, and these cause the signal to reflect, which is particularly noticeable on data circuits.

Burning the wire insulation may occur when soldering the conductor. This causes a decrease in insulation resistance, resulting in an increase in cross-talk. Excessive heat may melt the conductors.

Cable dispensers

Cables are usually in one of four packages:

- plastic bag
- box (usually 305 m or 500 m)
- spool (usually 100 m)
- drum (100 m, 500 m or 1000 m).

All cable packages allow the easy dispensing of cable. The box and plastic bag have the advantages of being easy to carry and not rolling away. However, the cable will tangle easily in a plastic bag dispenser if you do not take care when storing.

Cable usually has length markings printed onto the cable sheath. These are useful in determining the amount of cable dispensed for a particular cable run. For example, if the length marking at the start is 1017 m and after installation the length marking is 957 m, then the total cable dispensed is the difference between the two readings (60 m).

The **plastic bag** dispenser lets the installer see how much cable is left and has the added advantage of being easy to carry. Take care when preparing for initial use by carefully cutting the retaining ties and locating the inner end of the cable for easy dispensing.

The **box dispenser** is ready to use. Simply remove the restraining tape and pull the cable end.

Cable in lengths of 100 m, 305 m or 500 m usually comes on a **spool** made of lightweight sheet metal, cardboard or plastic. Take care when dispensing cable from a spool to ensure that the cable does not kink or twist when being unrolled, and that too much cable does not unroll at once. Usually the spool is set up on a small stand or roller. It is important to secure the end of the cable left on the spool to prevent unravelling.

Long lengths of cable and cables of large diameter often come on wooden **drums** ranging in weight from 100 kg to several tonnes. Use extreme care with cable drums as you could injure yourself through lifting incorrectly or through the drum moving. It is usual to mount drums on an axle supported on a substantial frame. This arrangement may be portable or on the back of a truck or trailer. You need mechanical lifting devices when manoeuvring heavy cable drums.

Cable hauling

To haul cables, particularly larger-diameter cables, through conduits, along cable tray or ladder routes, or along aerial routes using a draw line, requires considerable force. It is essential that the method of securing the cable to a draw line is reliable. Moreover, it must not place undue stress on the cable. To aid in the fastening of cables to draw lines a range of hauling eyes, cable grips and crimping kits are available.

Securing the cable

There are a number of options available to secure the telecommunications cable to a draw line.

Hauling eyes

Hauling eyes are crimped-on or clamped-on cable-end sleeves terminating in an 'eyelet' (for attaching the towline). They are available in a range of sizes and designs to suit different types of cables.

An appropriate hauling eye for copper conductor telephone cable features a central ridged spike. Install the hauling eye by driving it into the end of the cable, locating the spike through the centre cores. It is then necessary to crimp the outer sleeve onto the outer layer of the cable. This type of hauling eye is typically available in a range of sizes to suit cables with external diameters of between 19 mm and 92 mm. There is a range of tools available to help attach the device to the cable end. These include hand boring tools and driving tools. Hand boring tools are suitable for making a pilot hole to help centre the pulling eye. Driving tools fit over the hauling eye and disperse the force when driving the eye, minimising risk of damage to the eye itself.

Crimping

A hydraulic crimping die provides a convenient way of crimping eyelet fittings to cables. It is available with a kit of crimping dies for accommodating cables from 14 mm to 92 mm outside diameter. This power unit consists of a portable electrically driven hydraulic pump with pressure gauge and foot control switch.

Cable grips

Cable grips minimise damage to the cable end and do not need any special tools for fitting. The grip consists of a wire mesh terminated in an eyelet. Pulling force on the eyelet makes the mesh grip the cable jacket to prevent slippage. Cable grip devices are suitable for any cable type, including copper telephone and data cables, optical fibre cables, bare power cables, rope cables and even cable bundles. Cable grips come in a variety of sizes, mesh configurations and eyelet types to suit most cable types and sizes.

Manufacturers provide data on approximate breaking strength, so with careful selection of the type of grip you will not exceed the maximum allowable tensile strength of the cable. In other words, the cable grip acts as a mechanical 'fuse', causing it to break if the applied force exceeds its maximum tensile strength. Although cable grips are reusable, they are subject to much abrasion and other mechanical damage, which reduces their maximum pulling force. Inspect the cable grip closely after each use and replace the device if necessary.

An 'open tubular double eye' configuration has an open end for pushing the grip along the cable when it is not possible to use the end of the cable for hauling. This type of eye suits applications such as hauling a cable along a catenary. One loop is secured to the catenary and the other loop is for hauling. The effective strength of the grip reduces by half when hauling with only one eye.

An 'open tubular offset eye' configuration is primarily used for pulling slack during final placement of cable in exchange buildings and at termination points such as pole tops where the end of the cable is unavailable for hauling.

Where you have to haul multiple cables at the same time through the same pathway, it is preferable for ease of haulage to stagger the attachment points of the cables. Some of the devices are typically available in 2, 3, 4 or 5 leg configurations as either chain or cable.

Swivels

A swivel forms a link between the draw line and the cable. The swivel has a clevis at each end with a removable pivot pin or bolt that passes through the eyelet or loop of the cable-hauling device. It prevents the cable from twisting during hauling.

A breakaway swivel allows the cable to separate if the hauling tension exceeds the load rating of the swivel. It incorporates a pin that fractures at a predesigned rating. Replacement pins of varying load capacities range from 667 N to about 8000 N. Breakaway swivels are suitable for hauling stress-sensitive optical fibre cables.

For rod systems, an attachment can screw directly to the rod end-piece, allowing hauling of cables directly by a rod system.

Guiding the cable

Hauling a cable involves transferring the cable from a drum or reel into the prepared pathway. A cable drum or reel may be as small as 250 mm in diameter and weigh a few kilograms or as large as 1600 mm in diameter and weigh a tonne, depending on the size, type and length of cable. The broad steps involved are:

1 Set up the feeder drum at one end of the pathway.
2 Set up cable guides and intermediate haulage devices.
3 Set up hauling devices at the other end of the pathway.

For smooth unrolling of the cable, the drum or reel needs to have an axle suspended in such a way as to allow unobstructed rotation of the drum or reel. With small reels this may be a simple device, set up onsite, consisting of a length of pipe supported between a pair of trestles and secured to the trestles with conduit saddles.

For larger drums, a purpose-built cable stand is often easier and safer to use. **Figure 12.79** shows one such device consisting of a pair of jacks with adjustable axle heights. This allows the cable stand to accommodate cable drums up to 1950 mm in diameter and weighing up to 3000 kg.

FIGURE 12.79 Jack stand

A portable cable stand, such as shown in **Figure 12.80**, facilitates the dispensing of cable from smaller cable drums. This particular stand comes as one unit and features adjustable axle height and flotation wheels for mobility.

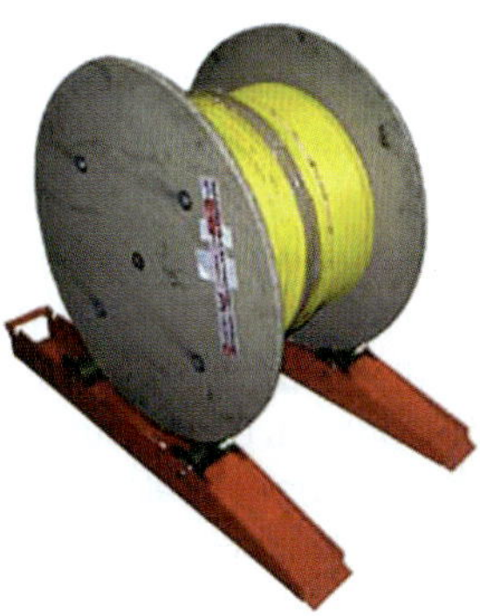

FIGURE 12.80 Portable cable stand

As mentioned earlier, cables can be damaged during installation by stretching, twisting, kinking, abrasion, crushing and exposure to moisture or air. To prevent damage to the cable along the intended route, keep it away from abrasive surfaces and sharp edges. It needs guiding around changes of direction either by the intrinsic pathway design or by other means. It needs to flow smoothly through the pathway with no sudden changes of tension and no twisting.

Conduit guides

Most of the damage cables suffer when pulled into conduit occurs when feeding the cable into the conduit entry and, though not so often, when drawing the cable out of the conduit exit. There are a number of devices to minimise the risk of damage at these points.

A 'split lock conduit guide' clamps to the outside of the conduit entry, allowing the drawing of the cable over the roller. This comes in sizes to suit conduits of between 50 mm and 150 mm in diameter.

Mechanised hauling

Mechanised drawing of a cable overcomes the inadequacies of hand pulling, provides a means of controlling or limiting the stresses placed on the cable and avoids damage to the cable.

Hauling a cable through a pathway by hand may be suitable for small-diameter cables or short cable runs. One person or a number of people can grip the draw line or cable itself and pull. A mechanised device may be needed, however, to haul longer, thicker and heavier cables through a pathway or to monitor closely the tension on the cable during hauling.

A basic cable puller (or winch) has the following main components:

- a motor which can be either electric or hydraulic
- a drive train consisting of a speed-reduction gearbox
- a capstan for reeling in the draw line.

If the cable winch has a hydraulic motor, it requires a source of hydraulic power. This can take the form of a matched portable hydraulic power unit, gas or diesel driven, or it can be an onsite source of hydraulic power such as a backhoe, bobcat or truck. The advantage of using a hydraulically driven hauling device over an electrically driven unit is that it allows stepless variation in speed, from zero to maximum, without loss of torque at low speeds. The speed controller usually consists of a foot-operated guarded pedal. The maximum pulling speed of a hydraulically driven unit is about 20 m a minute.

Electrically driven units, although more convenient, usually have a choice of only high or low speeds. However, electrically driven units can connect directly to a mains supply or to a portable generator. The maximum pulling speed of an electrically driven unit is about 12 m a minute.

All motorised hauling units, whether hydraulic or electric, have a gauge that displays the pulling force. The gauge may work on hydraulic pressure or electric current, both of which increase with increasing torque.

REVIEW QUESTIONS

1. What type of cable damage may be obvious to the installer?
2. How can cables sustain damage during installation?
3. What is the likely effect of 'pulling copper telecommunications cable too hard' during installation?
4. What is the likely effect of bending a copper telecommunications cable too sharply?
5. To assist installers and system designers, cable manufacturers publish data, usually in cable catalogues, of important cable parameters. What are some of these parameters?
6. A cable having a mass of 18 kg per 100 m is lowered down a service duct in a multi-storey building. The cable jams after dispensing 40 m. To free the jam a 90 kg person swings on the end of the cable. Does this force exceed the manufacturer's recommended tensile strength (pull tension) of 600 N?
7. What damage may a cable sustain during termination?
8. What are three types of cable dispenser?
9. What is a hauling eye?
10. What is a cable grip when used in cable hauling?
11. What is a swivel when used in cable hauling?
12. Name the three main components of a cable puller (winch).
13. What are the broad steps involved in dispensing a cable from a cable drum?
14. What is the function of a conduit guide?

12.18 Cable termination

When preparing an indoor cable for termination it is important to keep the individual pairs together within their binder group of 10 pairs. Telecommunications cables are normally terminated in 10-pair groups. This is particularly true of the ADC Krone 10-pair modules, which are common throughout the telecommunications industry.

Preparing indoor cable for termination on an ADC Krone frame

1. Remove 50 mm of sheath from a 100-pair telecommunications cable and cable tie all the pairs together.
2. Remove about 600 mm of additional outer sheathing by drawing on the cotton stripping string.
3. Progressively withdraw each wire pair. *Be careful to keep pairs together.* After drawing each pair from the main tree it is useful to 'crank' it (see **Figure 12.81**). Cranking refers to grasping the end of the wire pair between the thumb and forefinger of each hand, leaving a gap of about 50 mm. A Z shape or crank handle is formed in the wires, which are then rotated in a cranking fashion.
4. Progressively identify each 10-pair group of wires and lace them to form a tree (see **Figure 12.82**), with a branch at about 25 mm intervals. Use the cable colour code in **Figure 12.83** as a guide.

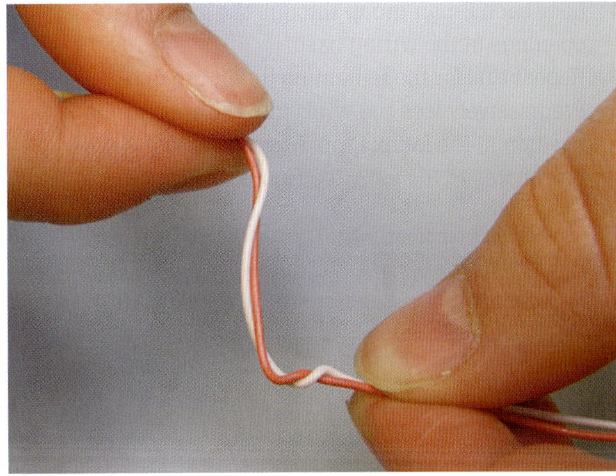

FIGURE 12.81 'Cranking' pairs (left-hander shown)

FIGURE 12.82 Attaching the wire tree to the frame

5 Fix the cable to the distribution frame back-mount so that the cable sheath is approximately 30 mm into the frame.

6 Take the first 10-pair group and leave enough wire to form a goose neck at the rear of the termination module (approximately 200 mm). Lay this horizontally behind the termination module and position to allow for easy removal of the module.

7 Clip the first (bottom-most) termination module to one of the back-mount rods.

8 Pass the goose-neck tail through the cable entry guide at the rear of the module and secure it to the cable tie point.

9 Fan out the wires and place them in sequence through the wire fanning strip on the top of the module; that is, wire 1 of pair 1 through the left-most fanning strip, wire 2 of pair 1 to the adjacent fanning strip and so on.

Note: The numbers on the termination module indicate the bottom of the module.

Pair No	Wire A	Wire B	Pair No	Wire A	Wire B
1	White	Blue	51	Black	Orange / White
2	White	Orange	52	Black	Orange / Green
3	White	Green	53	Black	Orange / Brown
4	White	Brown	54	Black	Orange / Grey
5	White	Grey	55	Black	Green / White
6	White	Blue / White	56	Black	Green / Brown
7	White	Blue / Orange	57	Black	Green / Grey
8	White	Blue / Green	58	Black	Brown / White
9	White	Blue / Brown	59	Black	Brown / Grey
10	White	Blue / Grey	60	Black	Grey / White
11	White	Orange / White	61	Violet	Blue
12	White	Orange / Green	62	Violet	Orange
13	White	Orange / Brown	63	Violet	Green
14	White	Orange / Grey	64	Violet	Brown
15	White	Green / White	65	Violet	Grey
16	White	Green / Brown	66	Violet	Blue / White
17	White	Green / Grey	67	Violet	Blue / Orange
18	White	Brown / White	68	Violet	Blue / Green
19	White	Brown / Grey	69	Violet	Blue / Brown
20	White	Grey / White	70	Violet	Blue / Grey
21	Yellow	Blue	71	Violet	Orange / White
22	Yellow	Orange	72	Violet	Orange / Green
23	Yellow	Green	73	Violet	Orange / Brown
24	Yellow	Brown	74	Violet	Orange / Grey
25	Yellow	Grey	75	Violet	Green / White
26	Yellow	Blue / White	76	Violet	Green / Brown
27	Yellow	Blue / Orange	77	Violet	Green / Grey
28	Yellow	Blue / Green	78	Violet	Brown / White
29	Yellow	Blue / Brown	79	Violet	Brown / Grey
30	Yellow	Blue / Grey	80	Violet	Grey / White
31	Yellow	Orange / White	81	Red	Blue
32	Yellow	Orange / Green	82	Red	Orange
33	Yellow	Orange / Brown	83	Red	Green
34	Yellow	Orange / Grey	84	Red	Brown
35	Yellow	Green / White	85	Red	Grey
36	Yellow	Green / Brown	86	Red	Blue / White
37	Yellow	Green / Grey	87	Red	Blue / Orange
38	Yellow	Brown / White	88	Red	Blue / Green
39	Yellow	Brown / Grey	89	Red	Blue / Brown
40	Yellow	Grey / White	90	Red	Blue / Grey
41	Black	Blue	91	Red	Orange / White
42	Black	Orange	92	Red	Orange / Green
43	Black	Green	93	Red	Orange / Brown
44	Black	Brown	94	Red	Orange / Grey
45	Black	Grey	95	Red	Green / White
46	Black	Blue / White	96	Red	Green / Brown
47	Black	Blue / Orange	97	Red	Green / Grey
48	Black	Blue / Green	98	Red	Brown / White
49	Black	Blue / Brown	99	Red	Brown / Grey
50	Black	Blue / Grey	100	Red	Grey / White

FIGURE 12.83 Colour code for 100-pair indoor cable

10 Locate each wire in the appropriate contact slot ready for termination; that is, wire 1 of pair 1 to the left-most contact slot, wire 2 of pair 1 to the adjacent contact slot and so on. Be certain to leave approximately 50 mm of excess wire in the termination to avoid fouling the contacts with short off-cuts. See **Figure 12.84**.

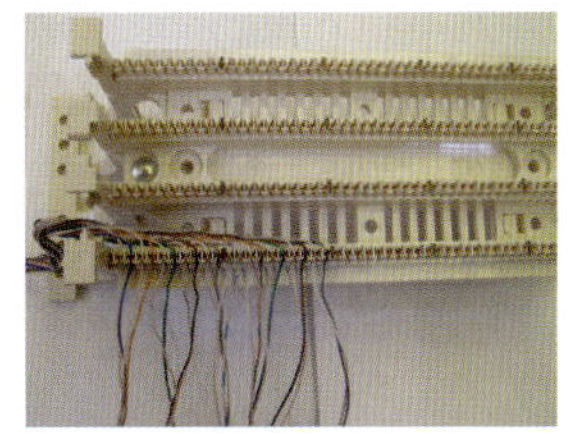

FIGURE 12.84 Fanning wires into module

11 Clip the other end of the termination module to the other back-mount rod.

12 Insert the termination tool over the wire, being careful to align the tool in the slot. Push the tool forward until you hear a click. This indicates correct termination and cutting of the wire. Repeat this step for the remaining wires in the 10-pair group. See **Figure 12.85**.

FIGURE 12.85 Terminating wires into module

13 Progressively install the termination modules on the frame and terminate each 10-pair group of wires.

14 Complete the frame designation scheme by placing the numbered plugs into the appropriate holes in the termination modules.

15 Insert the numbers 0 to 9 progressively into the bottom left-hand side of the module, starting at the bottom-most module.

16 Insert the numbers 10 to 100 progressively into the top right-hand side of the module, starting at the bottom-most module. These indicate the pair number sequences.

Preparing indoor cable for termination on a '110'-style frame

This involves the following procedure.

1 Install the '110' wiring blocks as appropriate. Next install cable hangers above and below each '110' wiring block. The cable hangers assist in proper cable management.

2 Strip the cable and prepare the 25-pair groupings in a similar way to that outlined previously in section 12.15.

3 Progressively identify each 5-pair group of wires and lace them to form a tree, with a branch at each 5-pair position in the wiring block. Use the cable colour code in **Figure 12.83** as a guide.

4 Lace each 5-pair group through the appropriate openings in the wiring base.

5 Repeat this for the remaining 5-pair groups.

6 Progressively lace the wires into the appropriate position in the index strip. (Remember the A-wire terminates to the left.) See **Figure 12.86**.

Viewed from front

Viewed from rear

FIGURE 12.86 Fanning wires into index strip

7 Visually inspect the cable termination at this point to eliminate any mis-wires or reversals.

8 Seat the conductors and trim off excess wire with an impact tool. See **Figure 12.87**.

FIGURE 12.87 Seating wires into index strip

9 Carefully position the connecting block over the wiring base. The blue marking must be to the left side of the block. See **Figure 12.88**.

FIGURE 12.88 Placing biscuit on index strip

10 Seat the connecting block using a 5-pair impact tool. See **Figure 12.89**.

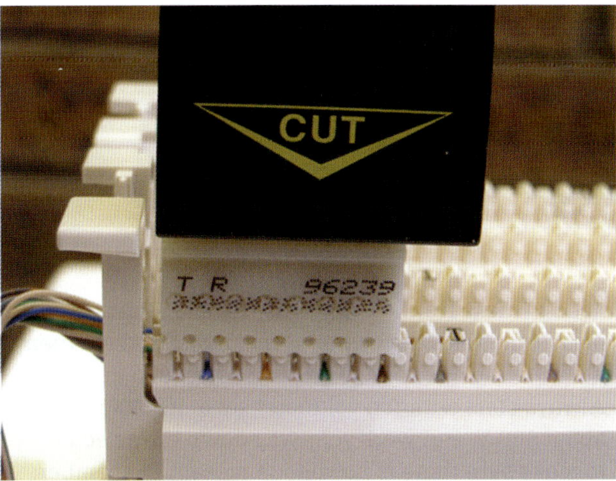

FIGURE 12.89 Installing connecting block on index strip

11 Label the circuits.
12 Slide the corresponding designation strip into the holder and snap this onto the wiring block.
13 Cross-connects terminate in the top of the connecting blocks. The cross-connects exit the wiring blocks to the left and right.
14 The connecting blocks *only* accept 22 to 26 AWG (American Wire Gauge) cross-connect wire of 1.6 mm overall diameter.
15 Position the cross-connects in the top of the connecting block.
16 Seat the conductors and trim off the excess wire using an impact tool. Set the impact tool to low impact when terminating cross-connects.

Preparing outdoor cable for termination

This involves the following procedure:

1 Remove 50 mm of sheath from the 50-pair telecommunications cable and cable tie all the pairs together.
2 Remove about 500 mm of additional outer sheathing.
3 Progressively withdraw each wire pair. *Be careful to keep pairs together.* After drawing each pair from the main tree it is useful to 'crank' them.
4 Progressively identify each 10-pair group of wires and lace them to form a tree, with a branch at about 25 mm intervals. Use the cable colour code in **Figure 12.90** as a guide.

Pair No.	Whipping or Binder	Wire A	Wire B
1	Blue	White	Blue
2		White	Orange
3		White	Green
4		White	Brown
5		White	Grey
6		Red	Blue
7		Red	Orange
8		Red	Green
9		Red	Brown
10		Red	Grey
11	Orange	White	Blue
12		White	Orange
13		White	Green
14		White	Brown
15		White	Grey
16		Red	Blue
17		Red	Orange
18		Red	Green
19		Red	Brown
20		Red	Grey
21	Green	White	Blue
22		White	Orange
23		White	Green
24		White	Brown
25		White	Grey
26		Red	Blue
27		Red	Orange
28		Red	Green
29		Red	Brown
30		Red	Grey
31	Brown	White	Blue
32		White	Orange
33		White	Green
34		White	Brown
35		White	Grey
36		Red	Blue
37		Red	Orange
38		Red	Green
39		Red	Brown
40		Red	Grey
41	Grey	White	Blue
42		White	Orange
43		White	Green
44		White	Brown
45		White	Grey
46		Red	Blue
47		Red	Orange
48		Red	Green
49		Red	Brown
50		Red	Grey

Pair No.	Whipping or Binder	Wire A	Wire B
51	Blue / White	White	Blue
52		White	Orange
53		White	Green
54		White	Brown
55		White	Grey
56		Red	Blue
57		Red	Orange
58		Red	Green
59		Red	Brown
60		Red	Grey
61	Orange / White	White	Blue
62		White	Orange
63		White	Green
64		White	Brown
65		White	Grey
66		Red	Blue
67		Red	Orange
68		Red	Green
69		Red	Brown
70		Red	Grey
71	Green / White	White	Blue
72		White	Orange
73		White	Green
74		White	Brown
75		White	Grey
76		Red	Blue
77		Red	Orange
78		Red	Green
79		Red	Brown
80		Red	Grey
81	Brown / White	White	Blue
82		White	Orange
83		White	Green
84		White	Brown
85		White	Grey
86		Red	Blue
87		Red	Orange
88		Red	Green
89		Red	Brown
90		Red	Grey
91	Grey / White	White	Blue
92		White	Orange
93		White	Green
94		White	Brown
95		White	Grey
96		Red	Blue
97		Red	Orange
98		Red	Green
99		Red	Brown
100		Red	Grey

FIGURE 12.90 Colour code for 100-pair outdoor cable

When terminating telecommunications cables, it is important to follow the manufacturer's instructions.

While many termination systems utilise the insulation displacement connection (IDC) system, there are subtle variations. You cannot generally use one manufacturer's termination tool on another manufacturer's termination system.

Preparing filled cables

The preparation of filled cables is done in much the same way as for any other cable but with one exception. It is necessary to carefully remove the filling material around the wire pairs prior to terminating. After removing the sheath from the cable and separating the wire pairs, remove the filling material using a solvent such as isopropyl alcohol. At all times wear rubber gloves, and eye protection is advisable. It is important to have good cross-ventilation when cleaning the cable.

SWITCH ON

Hazards and safety precautions

The filling material in these cables may cause allergic reactions, so it is important to take care during the preparation. It is also important to dispose of any rags used to wipe down the cable in sealed plastic bags. At the end of the task, dispose of gloves in the same way. Immediately remove any filling compound from the worksite to prevent future contamination. Wash your hands after completing the work and clean all tools.

REVIEW QUESTIONS

1 What is the recommended sequence of termination for the whipping of a 100-pair outdoor telecommunications cable?
2 What is the colour of the A-wire of pair 50 in a 100-pair indoor metallic conductor telecommunications cable?
3 What is the colour of the B-wire of pair 68 in a 100-pair outdoor metallic conductor telecommunications cable?
4 What is the colour of the whipping of binder unit 8 in a 100-pair outdoor metallic conductor cable?
5 What is the colour of the A- and B-wire of pair 57 in a 100-pair outdoor metallic conductor telecommunications cable?

12.19 End-to-end testing

Businesses using telecommunications networks tend to rely heavily on these networks to conduct their day-to-day business. Inevitably, one day when a user picks up the telephone it will not work. It is possible to eliminate many failures by performing simple but thorough tests on the cabling system after installation.

There is a large range of sophisticated test equipment available for testing metallic conductor cabling systems. However, it is possible to discover many cable faults by using simple testing techniques with low-cost test instruments. For example:

- End-to-end continuity testing is simple and effective for detecting wiring errors and shorted or open conductors.
- Measuring the resistance of a conductor and comparing with the manufacturer's specifications.
- Measuring the resistance of the cable insulation by connecting an insulation resistance tester to the isolated conductors.

SWITCH ON

Safety note

Before doing any test on a telecommunications cable you must determine its state. To do this you will need to measure the voltage on the line. Start by measuring the a.c. voltage; if this indicates zero then measure the d.c. voltage – this too needs to be zero. Always start with the voltmeter set at a voltage of about 500 V, and work your way down the selector to make sure there is no voltage present. Remember: *never assume you already know the state of any cable.*

Equivalent circuit

Before looking at cable faults and the methods to detect them, it is worth looking at the equivalent circuit of a metallic conductor wire pair. As **Figure 12.91** shows, a wire pair comprises: series resistance, which is attributable to the wire diameter; series inductance, which is inherent in any conductor and increased slightly by the twisting of the wire; mutual capacitance, which occurs when two parallel conductors are separated by insulation; and the insulation resistance, which is attributable to the dielectric strength of the wire insulation.

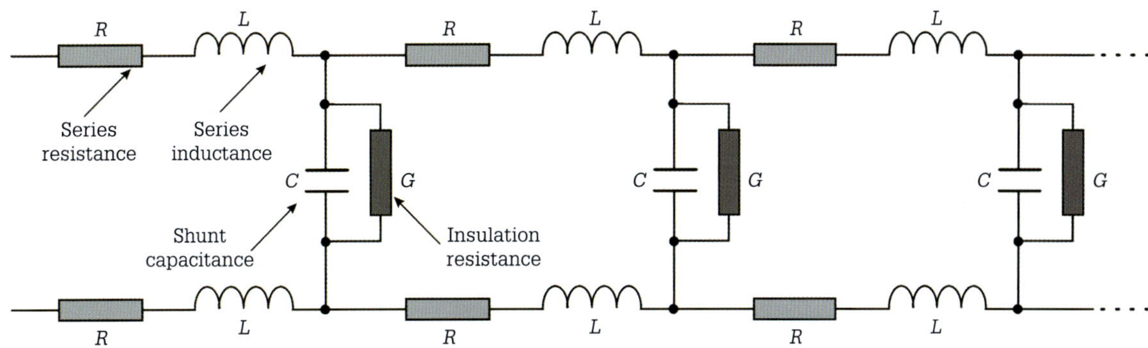

FIGURE 12.91 Equivalent circuit of a wire pair

The cable representation in **Figure 12.91** is a basic building block. As the length of the wire pair increases, more building blocks are added end-for-end. From this we can determine that, for increased length, conductor resistance increases, conductor inductance increases, insulation resistance decreases and mutual capacitance increases.

Resistance measurement

Resistance measurement gives an indication of basic cable performance. An ohmmeter can locate high-resistance faults or identify cables. When the far end of the wire pair is open-circuited, a normal wire pair will show a resistance reading of infinity (∞) between the two conductors and between each conductor and earth. It is, in most cases, extremely difficult to place a meter probe at either end of a single conductor of a wire pair due to the large distance involved. Instead it is more practical to place a loop across the conductors at the far end and measure the loop resistance (A-wire + B-wire) as **Figure 12.92** shows.

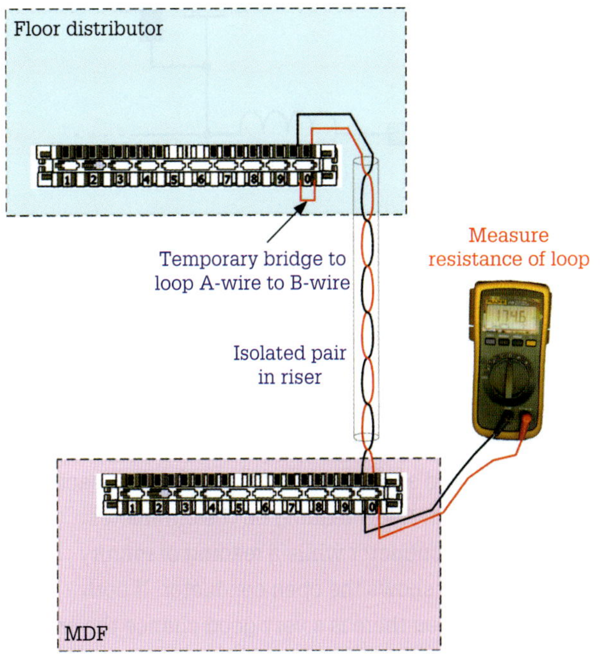

FIGURE 12.92 Principle of loop resistance test

SWITCH ON

Safety note

You must be certain that the wire pair is not in service for this procedure. The effect of placing a short on a live service could be dangerous.

With a loop across the far end you can measure the resistance between the A-wire and B-wire of the service. This reading is called the loop resistance, which is the sum of the A-wire, the B-wire, contact resistance and the meter leads. Using the cable manufacturer's data you can work out the health of a wire pair. The manufacturer may specify conductor resistance in Ω/km or the loop resistance in Ω/km. In either case it is possible to work out the expected loop resistance of the installed cable.

If the conductor resistance is specified use the equation:

$$R_{LOOP} = \frac{R_{CONDUCTOR} \times 2 \times \text{distance}}{1000}$$

where R_{LOOP} = the loop resistance in Ω

$R_{CONDUCTOR}$ = the conductor resistance in Ω/km

distance = conductor length in metres

It is necessary to double the distance (2 × distance) to account for both the A-wire and the B-wire.

If the loop resistance is specified use the equation:

$$R_{LOOP} = \frac{R_{LOOP\ SPECIFIED} \times \text{distance}}{1000}$$

where R_{LOOP} = the loop resistance in Ω

$R_{LOOP\ SPECIFIED}$ = the specified loop resistance in Ω/km

distance = conductor length in metres

If you use an analogue ohmmeter for the loop resistance you may notice that at first it reads a low resistance and then gradually moves down scale to settle on its reading. This is particularly noticeable when you use the Ω × 200 k range to measure the loop resistance on a long cable. It is due to the capacitive effect of the wire pair. A capacitor exists when insulation separates two conductors (a wire pair). An uncharged capacitor draws a large current from the charging source (ohmmeter in this case) and so appears as a short circuit momentarily; as the charging current decreases the resistance appears to increase.

If you ever notice this then you are probably testing a long length of cable. Using a capacitance meter will give an accurate measurement of the capacitance that you can use to determine the length of the cable.

EXAMPLE 12.3

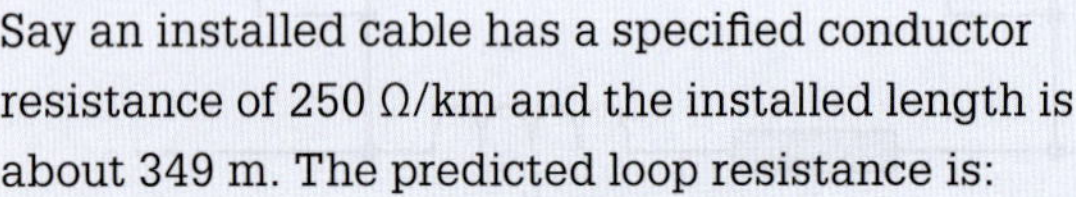

Say an installed cable has a specified conductor resistance of 250 Ω/km and the installed length is about 349 m. The predicted loop resistance is:

$$R_{LOOP} = \frac{R_{SPECIFIED} \times \text{distance} \times 2}{1000}$$
$$= \frac{250 \times 349 \times 2}{1000}$$
$$= \mathbf{174.5\,\Omega}$$

EXERCISE 12.3

An installed cable has a specified conductor resistance of 150 Ω/km and the installed length is about 259 m. What is the predicted loop resistance?

Detecting cable faults

You could find any of a number of faults on cable pairs, including:

- open-circuit pair
- short-circuit pair
- split pair
- reversed pair
- crossed pair.

The first two faults are relatively easy to detect. The others are slightly more complicated to diagnose. Additionally, to complicate fault diagnosis, a fault may exist in isolation or in any combination. In all cases, before starting any test, make sure the pair you are testing is not in service.

Open-circuit pair

An open-circuit pair exists when one or both of the conductors in the pair are not electrically continuous. This fault will generally appear as a loss of telecommunications signal at either end of the pair. To test for an open-circuit, loop the far end of the conductors as in **Figure 12.92** and measure the resistance across the open end of the pair. An open-circuit pair will show on the ohmmeter as infinity (∞) – make sure you have the ohmmeter set to maximum ohms so that you can accurately measure an open circuit. You could have a poorly terminated conductor, creating a high-resistance contact that will appear as an infinity reading on a low-ohms range. **Figure 12.93** shows an open circuit in the equivalent circuit of a wire pair.

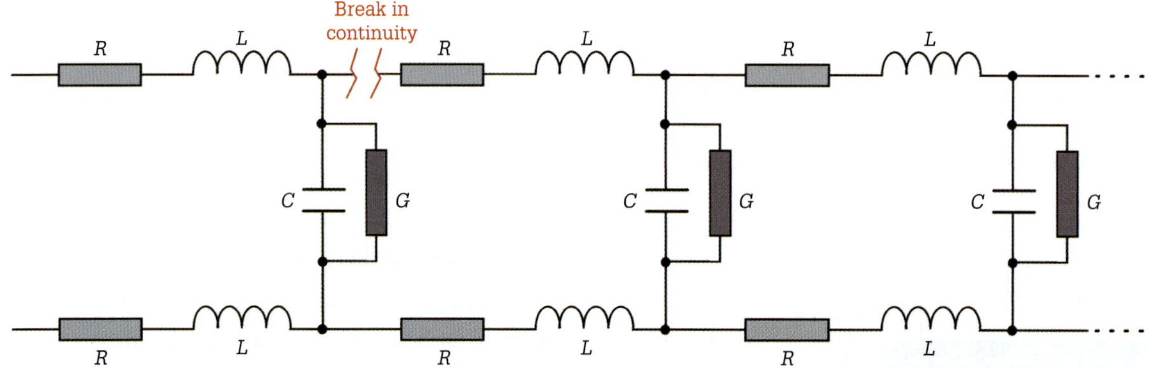

FIGURE 12.93 Effect of open circuit on a wire pair

As you can see in the equivalent circuit in **Figure 12.93**, the only available current path is through the insulation resistance as the capacitor 'blocks' the direct current from the ohmmeter. The insulation resistance of the wires is normally an open circuit.

Possible causes of open circuits in a cable include:

- wires connected to wrong pins at connector or terminating modules
- faulty connections
- cables routed to the wrong location
- wires broken by stress at connections
- damaged connector
- cuts or breaks in cable.

Special test equipment is needed to locate the open circuit accurately. However, to determine if one or both conductors are open you can loop both conductors to earth at the far end and measure the resistance of each conductor to earth. A low reading on the ohmmeter shows the good conductor while a reading of infinity (see **Figure 12.94**) shows the open conductor. If both conductors are open there is a very good chance the entire cable is damaged.

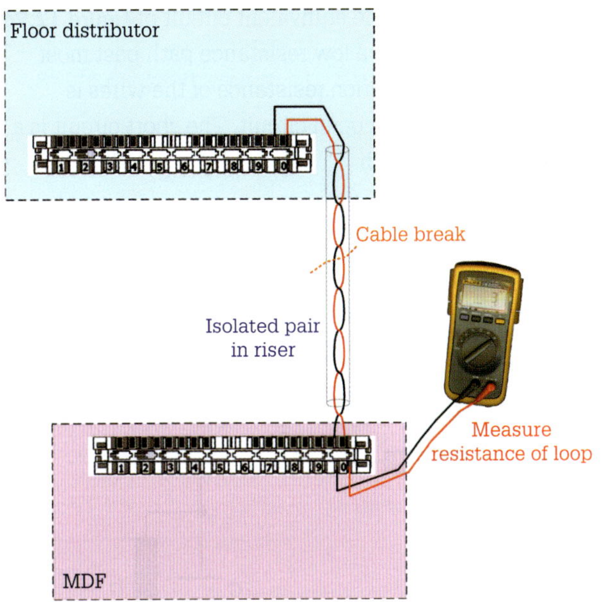

FIGURE 12.94 Testing for distance to open circuit

Cable capacitance

As the equivalent circuit for a wire pair (see **Figure 12.91**) shows, a wire pair has capacitance, which will increase as the length of the cable increases. You can use this effect to estimate the length of a cable. To do this, connect a capacitance meter across the A- and B- wires of an isolated cable pair and read the capacitance.

$$C_{LOOP} = \frac{C_{MUTUAL} \times \text{distance}}{1000}$$

where C_{LOOP} = the loop capacitance (usually in pF)

C_{MUTUAL} = the specified mutual capacitance (usually in pF)

distance = conductor length in metres

EXAMPLE 12.4

Say an installed cable has a specified mutual capacitance of 150 pF/km and the installed length is about 349 m. The predicted loop capacitance is:

$$C_{LOOP} = \frac{C_{MUTUAL} \times \text{distance}}{1000} = \frac{150 \times 349}{1000} = \mathbf{52.35\,pF}$$

EXERCISE 12.4

An installed cable has a specified mutual capacitance of 150 pF/km and the installed length is about 250 m. What is the predicted loop capacitance?

Short-circuit pair

A short-circuit will have the conductors of the A-wire and the B-wire touching somewhere along their length. This type of fault appears as a loss of telecommunications signal on the pair. To test for a short circuit, do not place a loop at the far end as previously described. Instead measure the resistance across the conductors of the wire pair as **Figure 12.95** shows. A short circuit may not measure zero ohms but it will read much less than infinity. Make sure the ohmmeter is set to the low ohms range.

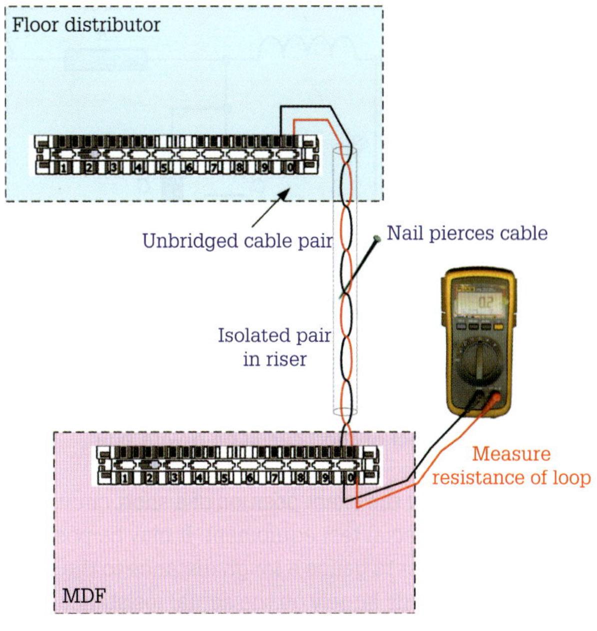

FIGURE 12.95 Testing for a short circuit

By using the measurement of cable capacitance it is possible to determine the approximate distance to a cable break. Calculate the length of a cable based on mutual capacitance from this equation:

$$\text{distance} = \frac{C_{LOOP} \times 1000}{C_{MUTUAL}}$$

EXAMPLE 12.5

A test for an open circuit measured 89 pF from one end and 118 pF from the other end. The mutual capacitance of the wire pair in good condition is 210 pF. The manufacturer specifies a mutual capacitance of 300 pF/km. The installed length of the cable is 700 m.

Adding 89 pF and 118 pF gives 207 pF, which is near enough to 210 pF for a distance approximation. So:

$$\text{distance}_1 = \frac{C_{LOOP_1} \times 1000}{C_{MUTUAL}} = \frac{89 \times 1000}{300} = \mathbf{296.7\ m}$$

$$\text{distance}_2 = \frac{C_{LOOP_2} \times 1000}{C_{MUTUAL}} = \frac{118 \times 1000}{300} = \mathbf{393.3\,m}$$

EXERCISE 12.5

A test for an open circuit measured 65 pF from one end and 98 pF from the other end. The mutual capacitance of the wire pair in good condition is 165 pF. The manufacturer specifies a mutual capacitance of 300 pF/km. The installed length of the cable is 550 m. Calculate the distance to the open circuit from each end of the cable.

As you can see in the equivalent circuit of **Figure 12.96**, a short circuit provides a low resistance path past most of the cable. The insulation resistance of the wires is normally considered an open circuit. The short circuit is a very low resistance path.

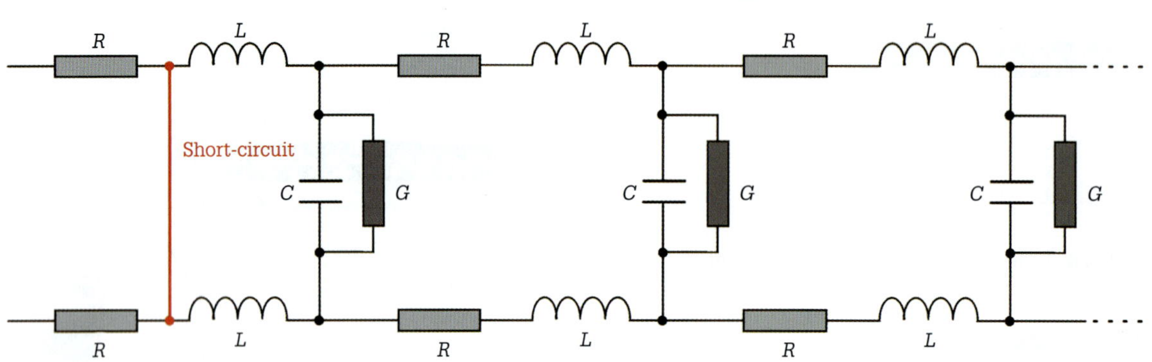

FIGURE 12.96 Effect of short circuit on a wire pair

It is hard to find the exact position of a short circuit in a cable without special test equipment. It may be possible to use an ohmmeter to gauge a rough distance to the short-circuit but it depends largely on the actual resistance of the short (most are not 0 Ω). To do this, measure the unlooped resistance of the pair from either end. If the sum of the two readings equals the loop resistance of the cable pair in good condition, then the fault is a true short circuit and you can rearrange the loop resistance equation to determine the distance to the fault.

This gives us a good point to start looking for the cable fault.

$$\text{distance}_1 = \frac{R_{\text{LOOP}_1} \times 1000}{R_{\text{SPECIFIED}} \times 2}$$

Possible causes of short circuits in a cable include:

- wires connected to wrong pins at connector or termination modules
- conductive material caught between pins at a connection
- damage to cable insulation.

EXAMPLE 12.6

A test for a short circuit measured 19 Ω from one end and 34.5 Ω from the other end. The loop resistance of the wire pair in good condition is 53 Ω. The manufacturer specifies a conductor resistance of 94.5 Ω /km. The installed length of the cable is 280 m.

Adding 19 Ω and 34.5 Ω gives 53.5 Ω – near enough to 53 Ω for a distance approximation. So:

$$\begin{aligned}\text{distance}_1 &= \frac{R_{\text{LOOP}_1} \times 1000}{R_{\text{SPECIFIED}} \times 2}\\ &= \frac{19 \times 1000}{94.5 \times 2}\\ &= \mathbf{100.5\ m}\end{aligned}$$

$$\begin{aligned}\text{distance}_2 &= \frac{R_{\text{LOOP}_2} \times 1000}{R_{\text{SPECIFIED}} \times 2}\\ &= \frac{34.5 \times 1000}{94.5 \times 2}\\ &= \mathbf{182.5\ m}\end{aligned}$$

> **EXERCISE 12.6**
>
> A test for a short circuit measured 15 Ω from one end and 18 Ω from the other end. The loop resistance of the wire pair in good condition is 33 Ω. The manufacturer specifies a conductor resistance of 94.5 Ω /km. The installed length of the cable is 175 m.
>
> Calculate the distance to the fault from each end of the cable.

Intermediate cable testing

Intermediate cable testing involves the use of devices used specifically to test data network cabling. They can perform a number of tests as well as monitor active networks. They are also suitable for testing telecommunications cabling for more complex faults than a multimeter can reveal.

Cable meters, such as Fluke's Cable IQ, can test:

- multi-pair wires (up to 4 pair)
- coaxial cable
- end-to-end connectivity
- cabling faults (open, short, crossed, reversed and split pairs)
- distance to fault
- cable length.

An important function of a cable meter is a graphical display that is able to indicate the wire map of an installed cable. The wire map essentially shows on its display which connector pin (1–8) at one end of the cable connects with which connector pin (1–8) at the other end of the cable. The cable meter tests and displays the wire connections between the near and far ends of a cable on all four pairs. If shielded cable is used the tester also tests the continuity of the shield. The wire map test detects and reports opens, shorts, crossed pairs, split pairs and reversed pairs. The tester will display the wire map at completion of the test.

Figure 12.97 shows the testing arrangement: a wire map adaptor connects into the jack at the far end of the cable while the cable tester connects to the near end of the cable (at a patch panel or distributor). In this case the Cable IQ™ shows that no faults are detected.

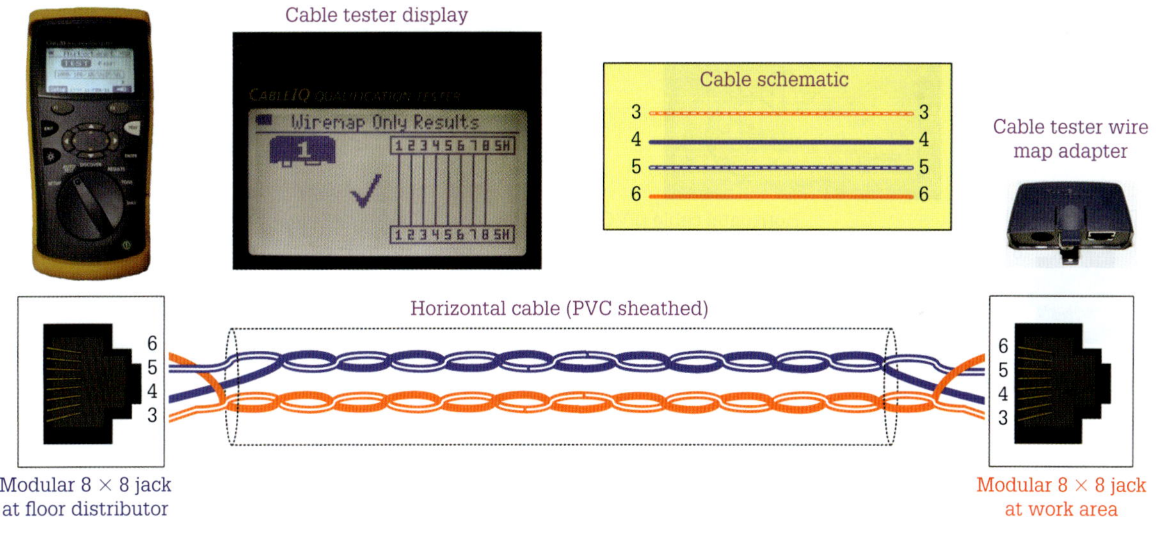

FIGURE 12.97 Using a cable meter to test cabling – good cable (only two pairs are shown for clarity)

Open circuits

Figure 12.98 shows the results of a test on cable having an open circuit on one wire. Note that the display indicates there is an open circuit on pin 5, which is the A-wire of pair 1. Note the display for the open-circuit wire is at the near end (closest to the cable tester).

Possible causes of open circuits in a cable include:

- wires connected to wrong pins at connector or termination modules
- faulty connections
- cables routed to the wrong location
- wires broken by stress at connections
- damaged connector
- cuts or breaks in cable.

Short circuits

Figure 12.99 shows the results of a test on cable having a short circuit on one pair. Note that the display indicates that there is a short circuit on pins 4 and 5, which is pair 1.

Possible causes of short circuits in a cable include:

- wires connected to wrong pins at connector or termination modules
- conductive material caught between pins at a connection
- damage to cable insulation.

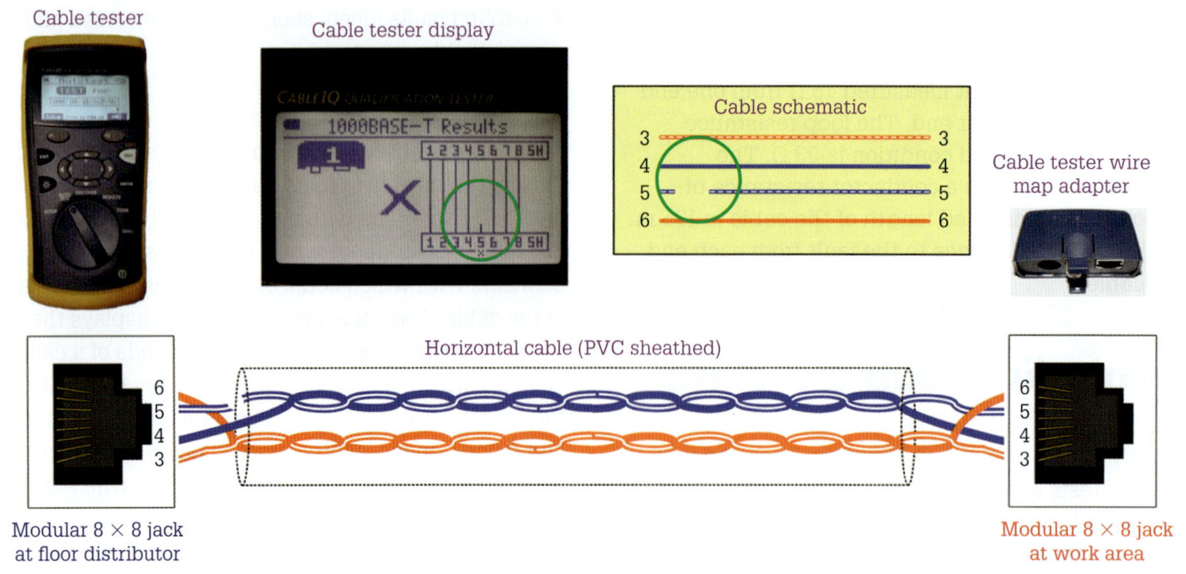

FIGURE 12.98 Using a cable meter to test cabling – open circuit on pin 5 (only two pairs are shown for clarity)

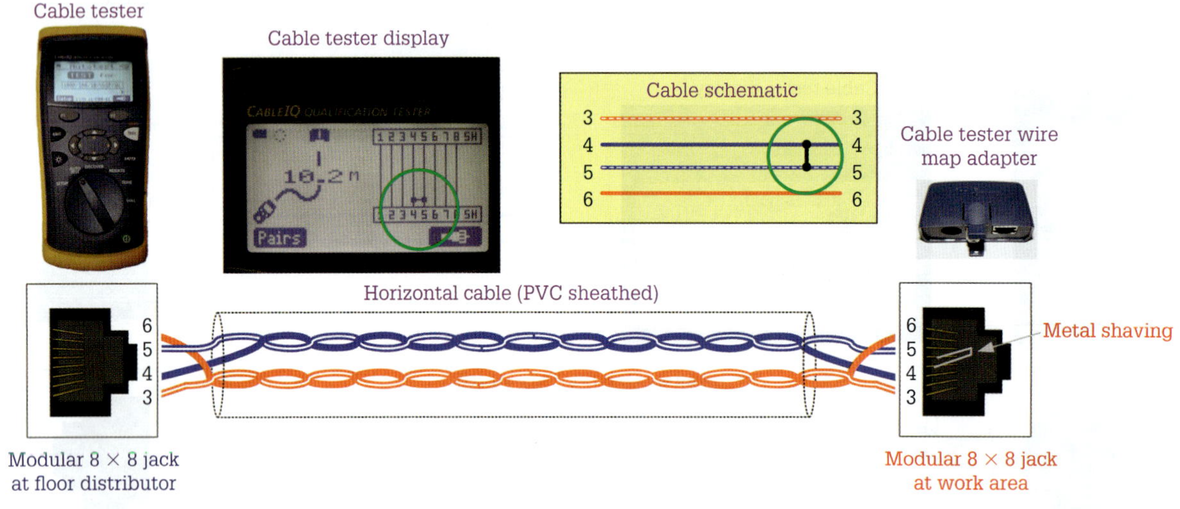

FIGURE 12.99 Using a cable meter to test cabling – short circuit on pins 4 and 5 (only two pairs are shown for clarity)

Split pairs

A split pair occurs when one wire from a cable pair twists together and terminates with a wire from a different cable pair. Split pairs most frequently result from mis-wires at termination modules and cable connectors. For example, the A-wire of pair 1 is paired with the B-wire of pair 2; also the B-wire of pair 1 is paired with the A-wire of pair 2. **Figure 12.100** shows an example of split-pair wiring. You will notice the pin-to-pin connections are correct but the pairs twisted together do not make a complete circuit (pins 4 and 5 should be pair 1; also pins 3 and 6 should be pair 2).

Split pairs cause cross-talk as the signals in the twisted pairs come from different circuits. In telephony circuits, you may notice excessive cross-talk using a test telephone. Detection of split-pair wiring needs sophisticated test equipment.

Possible causes of split pairs in a cable include wires connected to wrong pins at connector or termination modules.

Reversed pairs

Reversed-pair wiring occurs, for example, when the A-wire terminates in the correct position at one end but terminates in the position for the B-wire at the other end and vice versa. This situation will have little or no effect on a simple telephone service but is important on systems requiring correct polarity. **Figure 12.101** shows typical reversed-pair wiring.

You can detect a reversed pair by connecting one leg of the pair to earth and testing each leg for a connection to earth at the other end. For example, if you connect the A-wire to earth at one end you should see earth on the A-wire at the other end. If the B-wire is showing a connection to earth then the pair is reversed.

Possible causes of reversed pairs in a cable include wires connected to wrong pins at connector or termination modules.

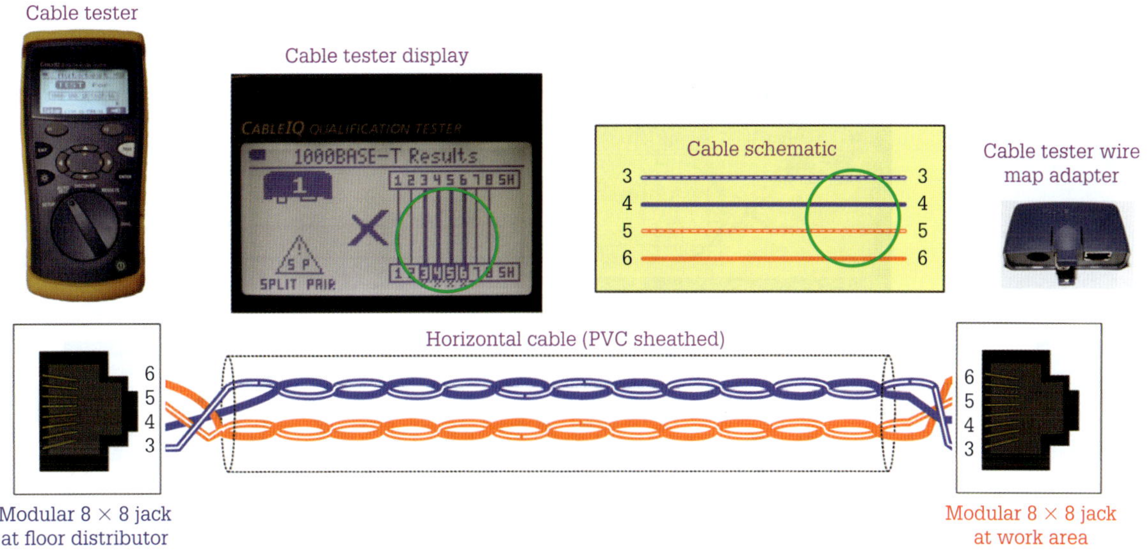

FIGURE 12.100 Using a cable meter to test cabling – split pairs 1 and 2 (only two pairs are shown for clarity)

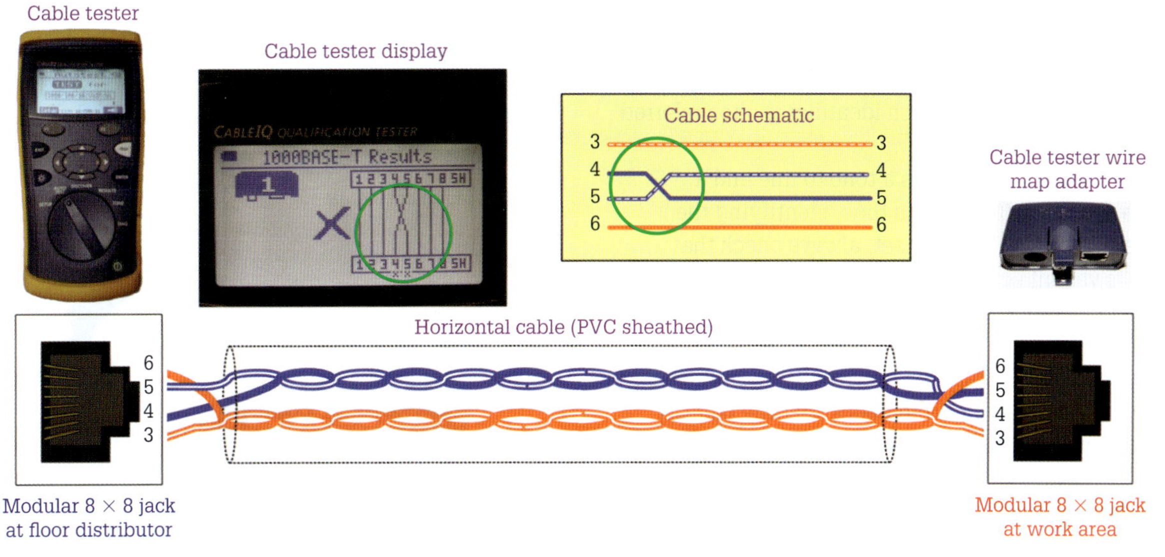

FIGURE 12.101 Using a cable meter to test cabling – reversed pair 1 on pins 4 and 5 (only two pairs are shown for clarity)

Crossed pairs

Crossed-pair wiring occurs, for example, when the A-wire and B-wire of pair 1 terminate in the correct position at one end but terminate in the position of the A-wire and B-wire of pair 2 at the other end and vice versa. This may be noticeable on a simple telephone service – it will mean a customer will not have their correct telephone service number. **Figure 12.102** shows an example of crossed pair-wiring.

Possible causes of crossed pairs in a cable include wires connected to wrong pins at connector or termination modules.

Other cable faults and tests

Earth fault

An earth fault happens when a conductor of a pair accidentally comes in contact with the system earth. This could impair a telephony service but it may not entirely fail. In telephony circuits an earth fault usually appears as a loud hum over the speech signal. This means that you can use a field test telephone to identify an earth fault by listening to the noise on a circuit.

To confirm an earth fault you can isolate the circuit and measure the resistance between each conductor of the pair and earth. You should notice a very low reading between one of the conductors and earth. To locate the fault, you can use the method described in the previous topic for finding a short circuit.

Foreign battery

Another problem on wire media is that of foreign battery. This is where a voltage from another source is present on a telecommunications circuit. Measuring the voltage across the A-wire and B-wire of an isolated pair will detect a foreign battery. Finding its source is more difficult.

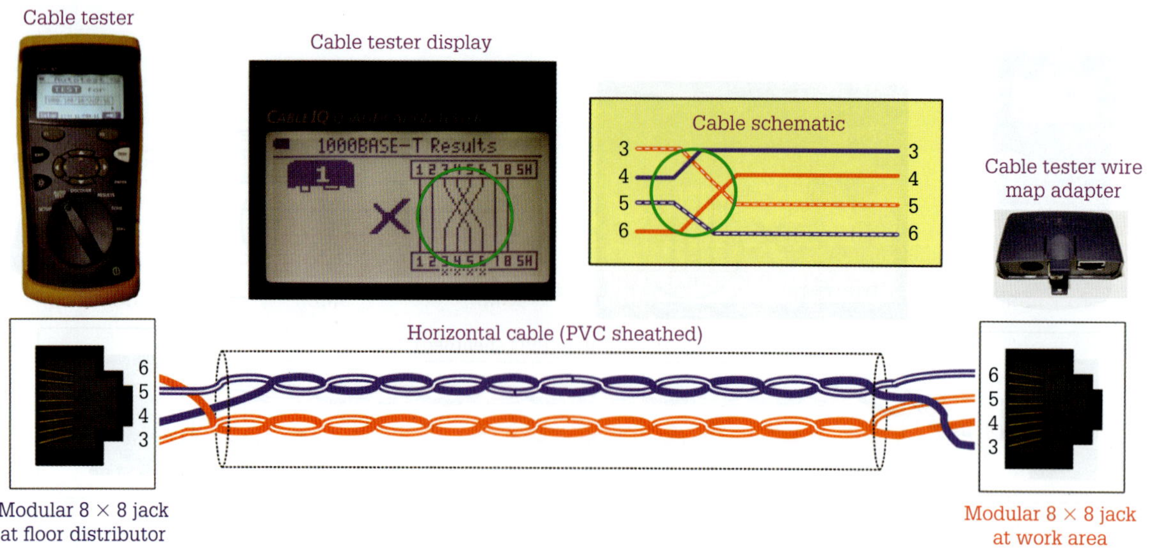

FIGURE 12.102 Using a cable meter to test cabling – crossed pairs 1 and 2 (only two pairs are shown for clarity)

Tone search identifier

A useful item of test equipment is the 'cable pair identification set', incorporating a polarity tester and audio tone generator. The tone search identifier is often referred to as an F set.

The audio oscillator sends a tone to line and a separate probe 'sniffs out' this tone, identifying the cable pair. When using the F set, always check that the pair is vacant or that a working telephone service is idle before connecting the oscillator to the line. Use a permitted connector to connect to the cable pair and do not pierce the insulation. Connect the instrument to the idle telephone service and check that the polarity indicator lamp is on. In this condition the circuit is looped to provide a balanced line. Reverse the connection of the test leads if the indicator lamp is not on. F sets can identify the A-leg (+ earth) and the B-leg (exchange battery negative). **Figure 12.103** shows the principle of operation of the F set.

On short cable distances, the signal may be so strong that it links adjacent cable pairs. In this situation the probe seems to identify several pairs. To overcome this you can de-tune the probe by using a metal probe to touch the conductor terminations in the module and then hold the probe close to your side. In this way the probe will sound only when you are in contact with a conductor of the particular pair. The signal should stop if you short both the A- and B-wires in the pair. This indicates continuity of the pair.

Some cable test sets can detect and demodulate high-speed signals in digital systems. Demodulation is where a particular data signal is separated from the combined signal. This lets you identify working circuits that may not give an indication on a test telephone or other simple test instrument. Accidental interference with one of these circuits could result in the loss of a data link or a large number of telephone circuits.

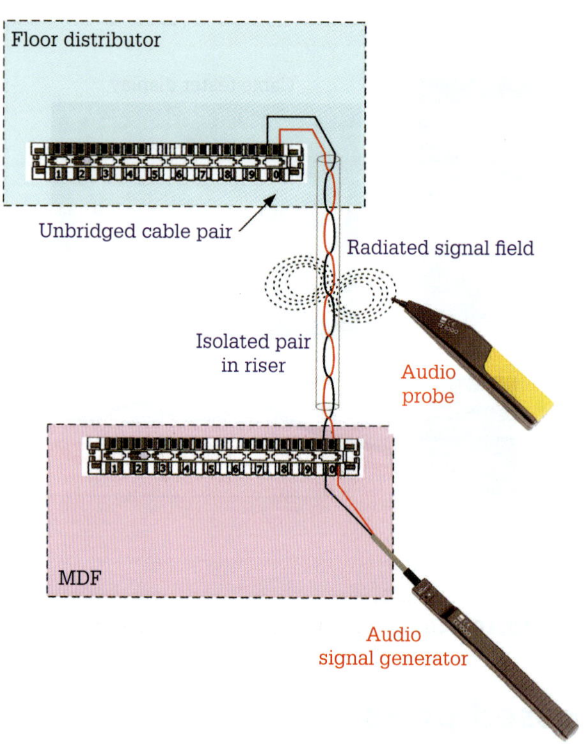

FIGURE 12.103 Tone search identification

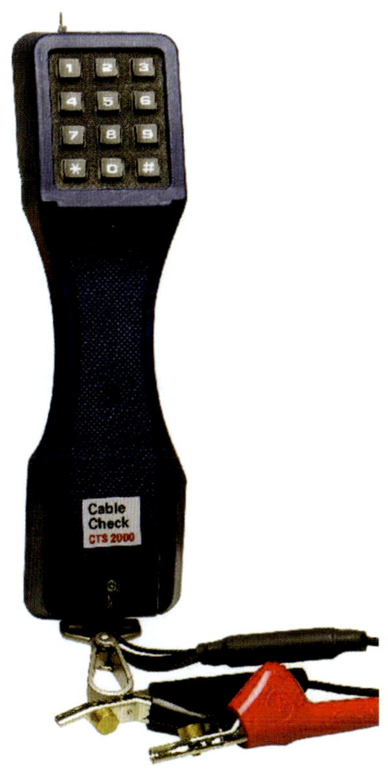

FIGURE 12.104 Basic field test telephone

Field test telephone

The field test telephone is a special telephone capable of performing basic tests on telephone lines. In its simplest form the field test telephone allows you to connect across a wire pair to test polarity, listen for noise or dial a PABX for test functions and the like. A basic field test telephone is shown in **Figure 12.104**.

SWITCH ON

It is illegal to listen in on telephone conversations. You must make every effort to ensure the service is idle before connecting a field test telephone.

REVIEW QUESTIONS

1 Why is it important to test installed communications cables for the presence of voltage?
2 What reading indicates cable continuity on an ohmmeter?
3 Why is it necessary to test for continuity in a terminated communication cable?
4 What cabling term indicates a break in a conductor?
5 What instrument would you use to test for a foreign battery on a cable?
6 How would you use a multimeter to detect a short circuit in an installed communications cable?
7 How would you use a multimeter to detect an open circuit in an installed communications cable?
8 An open-circuit test on a cable gave 18.5 Ω from one end and 25.2 Ω from the other end. The loop resistance of the wire pair in good condition is 43.5 Ω. The manufacturer specifies a conductor resistance of 94.5 Ω/km. The installed length of the cable is 230 m. What is the fault on the cable? What is the distance to this fault?
9 A test for an open circuit measured 55 pF from one end and 110 pF from the other end. The mutual capacitance of the wire pair in good condition is 165 pF. The manufacturer specifies a mutual capacitance of 300 pF/km. The installed length of the cable is 550 m. Calculate the distance to the open circuit from each end of the cable.
10 How can you use the capacitive component of a cable to estimate its length?

12.20 Hazards

Printed circuit board assemblies (PBAs) in modern electronic equipment are precision assemblies with sensitive components. These components (and, indeed, the PBA) demand proper handling both mechanically and electrically. A PBA is usually called a card.

Static electricity

You may recall one of your earliest experiments at school – rubbing a plastic ruler on a piece of cloth and using it to attract pieces of paper or making someone's hair stand on end. You experience the same phenomenon when you walk across a carpeted floor and turn on the television. Sometimes you hear the *crack*, other times you just feel the sensation of the shock. These are examples of the effects of static electricity.

While the effect of this static electricity is often unpleasant for us, the effect it has on the sophisticated electronic devices in modern equipment can be damaging. Discharging this static energy into assemblies containing solid-state electronic devices can be disastrous – particularly if these devices are metal oxide semiconductors (MOS devices for short).

Why MOS devices are static sensitive

To understand why MOS devices suffer damage from static electricity you need to know about their construction. It may be useful to think of the basic MOS device as a relay having an extremely small distance between the contacts. Static discharge across these imaginary contacts will cause arcing, which in turn will destroy the insulation, leaving a permanent circuit across these contacts. In reality, the MOS device is a transistor and the integrated circuit (IC) of today contains many hundreds of thousands of these devices. Damage to one transistor could render the entire IC useless.

Handling printed circuit board assemblies

It is important to remember that customer access equipment contains a considerable number of MOS and other static-sensitive components. To reduce the chance of premature failure of a component due to static discharge, take notice of the following steps.

Electrical considerations

- Ensure power is disconnected from the customer access equipment when removing or inserting PBAs.
- Bring your body to the same potential as the equipment on which you intend to work. Do this by connecting an antistatic strap to your wrist or ankle and to the frame of the equipment. This strap should stay in place as long as you are working on the equipment. If you are working on PBAs at a bench you will need to connect the antistatic strap to the bench surface.
- Use antistatic floor mats.
- Handle PBAs by the edge or by the handles (in much the same way you would handle a photographic negative). Do not touch any of the tracks on a PBA. This may cause corrosion of the copper tracks.
- Do not touch any component on a PBA as they are physically delicate. Finger pressure could be enough to fracture a component.
- It is important to protect PBAs containing memory from sunlight and ultraviolet light.

Mechanical considerations

Most PBAs of today use surface mount devices (SMDs). This is a construction technique that facilitates high-density designs through the use of miniature components that attach directly to the surface of the PBA. This manufacturing technique leaves the PBA prone to damage through fracturing of the bond between a component and the PBA.

In customer access equipment (such as a PABX) the PBA connects to the main equipment board through a multi-pin connector. These connectors are usually at the rear of the PBA for easy replacement of the card. The main equipment usually contains several cards arranged on a rack to ensure proper alignment of card and connector.

When inserting cards into the main equipment, ensure that they slide freely in the guide. Push them firmly home but *never* use force. Many cards employ a 'locking' mechanism to facilitate proper seating in the connector.

Things to watch for:

- Build-up of dust and debris may require cleaning with an appropriate circuit-board cleaner.
- Do not touch circuit tracks with your bare skin – this may leave a chemical film or residue on the tracks that can cause corrosion of the conductive material.
- Do not force cards into the main equipment.
- Flexing the card could fracture components or tracks on the circuit board.

REVIEW QUESTIONS

1. What is the main static-sensitive component contained in customer access equipment?
2. To what electrical component is the basic MOS device analogous?
3. In customer access equipment, what is generally referred to as a card?
4. What is the advantage of using SMDs on printed circuit boards?
5. What is required when inserting cards into the main equipment?
6. What could result from flexing a circuit card?
7. What method is suitable to bring your body to the same potential as the equipment on which you intend to work?
8. What should you do before removing or inserting circuit boards into customer access equipment?
9. Why should you not touch circuit tracks with your bare skin?
10. What method is suitable for removing a build-up of dust and debris on circuit boards?

CHAPTER REVIEW

12.1 Cabling Provider Rules

- The Australian Communications and Media Authority (ACMA) is responsible for the regulation of broadcasting, radio communications, telecommunications and online content.
- The Communications Alliance (CA) has responsibility for coordinating technical standards.
- The ACMA has the power to prosecute a cabling provider who does not comply with the cabling rules.
- The Cabling Provider Rules (CPRs) form the regulatory framework for performance of cabling work.
- Cabling registration covers the Open, Restricted and Lift categories.
- All people undertaking cabling work must register under CPRs in order to work legally.
- Cable, cabling product and customer equipment must comply with requirements in the Labelling Notice.

12.2 General installation requirements

- Inadequate segregation of telecommunications cabling from power cabling results in incidents of hazardous voltages on telecommunications lines and equipment. Each situation is potentially life threatening.
- Copper conductor communications cable is joined by twisting and soldering or using a commercially available insulation displacement connector.
- The cabling provider must ensure that installation of all customer cabling complies with AS/CA S009: 2020 and the generally accepted principles of safe and sound practice.
- Telecommunications equipment and accessories should not be positioned in areas where the earth potential rise exceeds 430 V a.c.
- Optical fibre cables are exempt from many of the separation requirements detailed in AS/CA S009: 2020.
- It is necessary to separate the enclosed conductors and terminations of telecommunications cables from the enclosed conductors and terminations of HV cables.
- It is not permissible to install customer cables in a conduit or pipe of a colour specified for use with hazardous services.

12.3 Customer interfaces

- A distributor is a connection device that provides for cross-connection of cables using jumpers or patch cords.
- Installing distributors simplifies the distribution of exchange line services throughout an installation.
- The cabling provider must supply sufficient cabling records relating to the cabling to enable cables and cross-connections to be correctly identified and connected. The cabling records must be legible and updatable.
- The verticals are marked alphabetically from left to right, starting with the letter A and omitting the letters I and O. The range of terminations within each vertical is numbered in ascending order from 1, which is in the bottom left of the vertical.
- The only device away from a building that is a defined network boundary is a network termination device (NTD).

12.4 Cable distribution devices

- The use of distributors simplifies the provisioning of telecommunications services in large installations.
- Distributors are referred to by their function: main distribution frame, building distributor and floor distributor.
- Terminations may be insulation displacement, wire wrap, solder or screw.
- Observe all WHS requirements when making solder connections.

12.5 Network boundaries

- The network boundary is defined in the *Telecommunications Act 1997*, which essentially sets the demarcation point or boundary of a telecommunications carrier's network.
- With Australia's national broadband network, network boundary is not so well defined.
- Demands of twenty-first century telecommunications meant that Australia's existing copper network was struggling to keep pace with a fast-changing digital landscape.
- The nbn™ network uses a range of broadband technologies including fixed line, which incorporates FTTP, FTTN, FTTB and HFC, as well as fixed wireless and Sky Muster™ satellite for rural Australia.
- Fibre-to-the-Curb is an emerging technology where optical fibre cable is installed to the telecommunications pit located outside customer premises.

12.6 Cable identification

- Administration is the accurate identification and record keeping of all components that comprise the cabling system, including pathways, distributors and other spaces.
- Distributor records provide a way of documenting the origin of a cable and where cables go to at a particular distributor.
- Distributor records have provision for showing the cross-connects within a particular distributor.
- Cabling symbols show pictorially equipment used in an installation.
- Standard cabling symbols ensure universal understanding of what a plan details and exactly what is necessary for a particular location in an installation.
- External cabling drawings show all cabling and distributing arrangements between interconnected buildings.
- Internal cabling drawings show all cabling arrangements, outlets and equipment within an installation.

12.7 Telecommunication cable types

- The most common transmission path for telecommunications circuits is copper wire.
- Twisting wire pairs reduces cross-talk.
- Indoor cable usually has a cream-coloured PVC sheath over PVC-insulated wires.
- Outdoor cable usually has polyethylene conductor insulation and a polyethylene sheath.
- Coaxial cable comprises two conductors having a common axis and separated by a dielectric.
- Baseband transmission is where the entire bandwidth of the transmission medium is used for a signal.
- Broadband transmission uses multiplexing to divide the bandwidth of a single transmission medium into multiple channels of information.

12.8 Indoor cabling

- The minimum separation between LV power and telecommunications cables is 50 mm.
- Separation may include installing a barrier of durable insulating material or earthed metal between LV power and telecommunications cables.
- For high-voltage circuits a minimum separation of 450 mm is required from single-core cables and telecommunications cables irrespective of any interposing barriers.
- For multi-core high-voltage cables the minimum cable separation is reduced to 300 mm, which can be further reduced to 150 mm where there is an interposing barrier of durable insulating material or earthed metal, providing the distance around the barrier is not less than 175 mm when measured sheath to sheath.
- In areas of elevated temperatures (greater than 60 °C) the use of non-flame-propagating conduit and fittings is required.

12.9 Underground cabling

- Underground cable usually has polyethylene conductor insulation and a polyethylene sheath.
- The core of an underground cable may incorporate a water-blocking compound (silicone-based jelly) or tape.
- Pillars or cabinets located in public areas must be locked.
- Any pit or access hole must be legibly and permanently labelled on its cover.
- The minimum depth of laying underground communications cables is 300 mm unless under a roadway, where it is 450 mm.
- Telecommunications cables must be separated from LV and HV cables.

12.10 Aerial cabling

- Aerial cable employs polyethylene sheath and conductor insulation.
- Bearer wires must be insulated from any conductive pole or structure carrying an aerial power line.
- It is necessary to maintain separation between aerial telecommunications and aerial power cables.
- Aerial communications cable must be suitable for installation in direct sunlight (UV stabilised).

12.11 Earthing protection

- There are four forms of earthing system: communications earthing system (CES), telecommunications reference conductor (TRC), ELV d.c. power supply and d.c. earth return.
- Telecommunications earthing systems must comply with the requirements of AS/CA S009: 2020.
- Only one direct connection of the TRC to the building earth system is permissible under the regulations.
- The TRC is exclusive to telecommunications services.
- An installation can include a telecommunications functional earth electrode to limit the magnitude of direct current flowing in the communications bonding current and hence the electrical system main earth conductor.

12.12 Earthing concepts

- Metallic frames, back-mounts, enclosures, trays, conduits and ducts connect to a compliant earth reference through a green/yellow insulated copper cable.
- Surge suppression devices for protection of telecommunication line conductors must connect to a compliant earth reference through a green/yellow insulated copper cable.
- It is important to measure the resistance of the telecommunications earthing system to ensure that it is within acceptable limits to ensure proper functionality.

12.13 Surge suppression

- Techniques for protecting telecommunications equipment and personnel from over-voltage include intercepting, clamping or clipping, shunting, interrupting, and isolating.
- When located within an MDF, surge suppression devices may be either Class 1 or Class 3. In other locations, a Class 1 device is suitable.

12.14 Cable shielding and interference

- The use of effective physical separation and/or using a telecommunications cable that has a continuous shield or screen that connects to earth may reduce or totally eliminate noise from the telecommunications circuit.
- Earth the shield at only one end of the cable to eliminate the potential for noise loops and circulating earth currents due to potential difference between the two local earths.

12.15 Miscellaneous regulations

- The term 'heritage' may be broadly defined as 'the possessions, traditions or conditions that have been passed from one generation to another'.
- State governments each have their own heritage Act, administered by a separate department, which deals with the heritage sites in their state.
- Whenever cabling work is done on private property, it is regarded as 'best practice' to leave the job site in an 'as found' condition.

- Before starting any trenching on private or public property it is a good idea to determine the exact location of all existing services before excavation.
- Pollution may be visual, noise, or waste. A tidy worksite is also a safe worksite.
- A confined space is any area where air is consumed at a greater rate than it is replaced.

12.16 Installation and termination requirements

- Before installing additional services to an existing dwelling, it is important to consider the location of existing services: electrical cables, water pipes, gas pipes, coaxial television cables and telephone cables.
- Use a stud finder and a volt stick, or similar, to assist in the location of services in a wall.
- Install new telecommunications cable in designated cabling pathways.
- The Building Code of Australia (BCA) outlines responsibilities regarding fire stopping of service penetrations.
- Any penetration should not reduce the required fire resistance level (FRL) of the penetrating element.
- Most generic forms of fire stop systems can provide an FRL rating.
- The choice of fire stop system depends on specific application and additional requirements.
- Cable support systems and cable distribution equipment require fixing to, or supporting on, surfaces made of different materials, which dictate the type of fixing.
- Installers must observe all relevant WHS requirements when installing fixings.
- Wearing of PPE may be required when installing fixings.
- A fastener generally connects together two items, while a fixing attaches a removable item to a permanent structure.
- The main support systems suitable for telecommunications cables include conduit, cable tray, cable ladder, free-standing trunking, in-floor trunking, skirting trunking, partition trunking, column trunking, catenary and in-floor duct.
- When installed underground, conduit enclosing telecommunications cables must be white in colour.
- An enclosure is a housing or covering for cables or equipment providing an appropriate degree of protection against external influences or end-user contact with hazardous voltages, ELV or TNV (AS/CA S008: 2020, Clause 4.2.20).
- AS/CA S008: 2020 details the constructional requirements for cable distribution devices.

12.17 Cable installation

- Telecommunications cables can suffer damage from kinking, stretching, burning, crushing, bending too sharply and nicking.
- Adherence to WHS requirements is paramount during cable installation.
- Dispensers for telecommunications cables include drums, spools, bags and boxes.
- Considerable force is needed to haul large-diameter cables through conduits, along cable tray or ladder routes or along aerial routes using a draw line.
- It is essential that the method of securing a cable to a draw line is totally reliable.
- A range of hauling eyes, cable grips and crimping kits are available to aid in the fastening of cables to draw lines.
- A swivel forms a link between the draw line and the cable, which prevents the cable from twisting during hauling.

12.18 Cable termination

- A variety of termination systems are available for telecommunications cables.
- Standard colour codes exist for indoor and outdoor telecommunications cables.
- It is important to observe all WHS requirements when terminating filled cables.
- Many termination systems utilise insulation displacement connections. However, you cannot generally use one manufacturer's termination tool on another manufacturer's termination system.

12.19 End-to-end testing

- Before making any test on a telecommunications cable make sure it is isolated.
- A wire pair comprises: series resistance, series inductance, mutual capacitance and insulation resistance.
- Cable faults include: open-circuit pair, short-circuit pair, split pair, reversed pair and crossed pair.
- Loop resistance provides an indication as to the length of an installed cable or the distance to a short.
- Mutual capacitance also provides an indication as to the length of an installed cable or the distance to a break.
- An earth fault is where a conductor of a pair is in contact with the system earth.
- Foreign battery is where a voltage from another source is present on a telecommunications circuit.
- The tone search identifier uses an audio oscillator to send a tone to line and a separate probe to detect this tone, thereby identifying the cable pair.

12.20 Hazards

- Printed circuit board assemblies (PBAs) in modern electronic equipment are precision assemblies with sensitive components.
- PBAs can be damaged from mechanical, chemical and electrical hazards.
- The MOS device is a transistor and the integrated circuit (IC) of today contains many hundreds of thousands of these devices. Damage to one transistor could render the entire IC useless.

TRIAL EXAM

For Chapter 12 knowledge assessment, please complete the following trial exam.

1 Who is the body responsible for regulating telecommunications and radio communications in Australia?
 a Australian Competition and Consumers' Commission
 b Australian Communications Industry Forum
 c Australian Communications and Media Authority
 d Communications Alliance

2 What could happen if a registered cabling provider fails to comply with the cabling provider rules?
 a nothing as the CPRs are only a guide
 b the cabling registrar may disconnect the cabling or equipment from the line
 c cabling registrar may change the conditions of the registration
 d the ACMA may issue a fine to the cabler

3 A telecommunications labelling notice is
 a a document supplied to the customer on completion of cabling work
 b a label that indicates compliance of equipment
 c a series of records that are associated with a distributor
 d notice that a cabler's installation has been selected for audit by the ACMA

4 At completion of cabling work it is necessary for the registered cabling provider to complete a cabling advice form (TCA1). On completion, the form is:
 a forwarded to the ACMA and copies kept for at least one year
 b forwarded to the cabling registrar and copies kept for at least three years and made available to the ACMA and auditors upon request
 c given to employer or customer and copies kept for at least one year and made available to the ACMA and auditors upon request
 d given to employer or customer and copies kept for at least three years and made available to the ACMA and auditors upon request

5 Joints in internal customer cables must be:
 a wholly concealed in a suitable enclosure or wall, floor or ceiling space
 b suitably insulated and protected to prevent injury to customers
 c suitably concealed from public view and protected from interference or damage
 d suitably manufactured, enclosed, positioned and supported to prevent the ingress of dust or moisture

6 A cabling provider must check with the power supply authority about the extent of the EPR hazard in proposed locations where the installation is near:
 a a 230 V service line
 b an HV transformer
 c an LV transformer
 d an installation supplied at 430 V a.c.

7 When working with optical fibre telecommunications cables the cabling provider must ensure:
 a fibre particles and hazardous solvents are removed from site at completion of work
 b manufacturer's labels are removed subsequent to installation
 c optical fibre cables carry identical markings to those of hazardous electrical services
 d all access points where excess fibres may emit laser light exceeding AEL Class 1 are unsecured

8 What is the required minimum separation between the terminations of telecommunications cables and low voltage (LV) terminations in a shared enclosure?
 a the required separation is 150 mm or a permanent rigidly fixed barrier of durable insulating material or earthed metal
 b the required separation is 300 mm or a permanent rigidly fixed barrier of durable insulating material
 c LV terminations cannot share a common enclosure with those of telecommunications conductors and terminations
 d the location of the MDF should be within its own room remote from the electrical switchboard

9 Distributors on an external wall must satisfy the minimum degree of protection to:
 a IPX2
 b IPX3
 c IPX4
 d IPX8

10 Once carrier authorisation has been granted, when can a cabling provider install cross-connects at the network boundary?
 a under the direct supervision of the carrier's technician
 b with written permission from the customer
 c when authorised by the regulatory authority or under the direct supervision of an authorised cabling provider
 d generally a cabling provider cannot install cross-connects at the network boundary

11 The network boundary in a residential premises with no more than two lines in the lead in cable is usually the
 a main distribution frame
 b network termination device
 c wireless router
 d connection box

12 Verticals in an MDF are labelled in alphabetical order from left to right omitting the letters
 a A and Z
 b A and B
 c I and O
 d X, Y and Z

13 Distributor records must be
 a written in ink
 b kept safe in a lockable cabinet
 c legible and updatable
 d stored electronically

14 If a building is only supplied with 230 V a.c. single phase power or 400 V a.c. three phase power, then
a earth potential rise must be assessed
b the main distribution frame must be equipotential bonded to protective earth
c ACMA must be advised of the need to assess the extent of earth potential rise
d there is no need to consider earth potential rise

15 Which of the following types of cable are generally exempt from electrical separation requirements of AS/CA S009: 2020?
a Cat. 6 UTP cables
b Cat. 6 S/FTP cables
c optical fibre cables
d coaxial cables

16 Underground telecommunications conduit is identified by the colour:
a white
b orange
c brown
d yellow

17 Cable shields may connect to:
a earth at one end
b earth at both ends
c a remote power feeding circuit
d the telecommunications reference conductor

18 When hauling or drawing cables though an underground conduit or duct, care should be taken to:
a prevent damaging the cable by monitoring the hauling tension
b use winches that are electrically operated only
c haul using the outside sheath of the cable only
d haul using the conductors of the cable only

19 Which is the preferred connector for optical fibre cables in Australia?
a SC
b LC
c ST
d BNC

20 Where an external customer cable in an exclusive trench runs parallel with LV cables, the minimum separation between the two systems is:
a 100 mm where power cables are installed under a protective covering to AS/NZS 3000, or 300 mm without such a protective covering
b 300 mm where power cables are installed under a protective covering to AS/NZS 3000, or 450 mm without such a protective covering
c 300 mm irrespective of whether the power cables are installed under a protective covering
d 450 mm irrespective of whether the power cables are installed under a protective covering

21 Where customer cabling is in a trench together with pipes supplying services for public utilities such as gas and water, the customer cabling:
a must be uppermost in the trench
b should be alongside, not above or below, the other services
c must be separated in accordance with the utility concerned
d must incorporate shielding in accordance with AS/CA S008: 2020

22 Where customer cabling is in a trench together with LV power cables and the power cables are not installed in conduit, separation must be such that:
a the cables are side by side
b the telecommunications cables are above the electrical cables
c the telecommunications cables are below the electrical cables
d each service can be accessed without disturbing the other

23 The minimum height of aerial customer cables above a private roadway is:
a 2.4 m
b 2.7 m
c 3.5 m
d 4.9 m

24 The minimum separation between aerial customer cables and LV power main conductors in-span is:
a 450 mm
b 600 mm
c 1.2 m
d 1.8 m

25 What minimum size cable must an earth bar or terminal be capable of accepting?
a link bar must be capable of accepting 6 mm^2 cable
b link bar must be capable of accepting at least 6 mm^2 cable and cable lugs
c link bar must be capable of accepting 6 mm^2 cable or cable lugs of 16 mm^2 and smaller
d link bar must be capable of accepting 16 mm^2 cable or cable lugs of 16 mm^2 and smaller

26 The colour of the B-wire of pair 8 in a 10-pair indoor metallic conductor telecommunications cable is:
a blue/grey
b white/blue
c red/grey
d blue/green

27 The industry standard AS/CA S009: 2020 recognises four forms of earthing system. The one that provides both protective and functional earthing is:
a communications earthing system
b telecommunications reference conductor
c ELV d.c. power
d d.c. earth return

28 The minimum size and colour of the equipotential bonding conductor is
a 2.5 mm² green/yellow
b 4 mm² green/yellow
c 6 mm² green/yellow
d 2.5 mm² violet
e 4 mm² violet
f 6 mm² violet

29 The maximum resistance of the communications earth system cable between the bond at the earthing system of the electrical installation and the earth bar or terminal at any NTD, distributor or customer switching system is:
a 0.5 Ω
b 1 Ω
c 2 Ω
d 5 Ω

30 The minimum size and colour of the cable connecting surge suppression devices to earth for devices intended to protect end users is:
a 2.5 mm² green/yellow
b 4 mm² green/yellow
c 6 mm² green/yellow
d 2.5 mm² violet
e 4 mm² violet
f 6 mm² violet

31 It is essential to update cabling records prior to completion of work to reflect any changes to be made.
a True
b False

32 A test on a TRC cable connecting a local distributor to the MDF yielded the following resistance measurements:
- Resistance of the reference pair = 27.3 Ω
- Resistance of the main TRC and A-leg = 14 Ω
- Resistance of the main TRC and B-leg = 14.3 Ω

Does the cable between the MDF and the local distributor satisfy the resistance requirements of AS/CA S009?

33 A test on a TRC cable connecting a local distributor to the MDF yielded the following resistance measurements:
- Resistance of the reference pair = 27.5 Ω
- Resistance of the main TRC and A-leg = 14 Ω
- Resistance of the main TRC and B-leg = 14.5 Ω

How much further is it possible to move the local distributor using the existing 2.5 mm² TRC cable?

34 Draw the symbol for a single ceiling-mounted TO.
35 Draw the symbol for a roadway access hole.
36 Draw the symbol for an outdoor pillar.
37 Draw the symbol for a PABX.
38 Draw the symbol for a distributor.

Task

Your task is to complete the cross-connect records for a 20-pair floor distributor.

Installation details

A 20-pair indoor grade cable, from the building distributor, supplies the floor distributor. This cable originates on the B vertical of the building distributor (B 151 to B 170). The exchange line services are 9999 1220 to 9999 1239. The B vertical of the floor distributor distributes customer cabling for 20 shops in a larger shopping centre. Enter suitable details in the records for the shops. The directory for the tenants appears in the table below.

TENANTS LIST

9999 1220	Centre management	Shop 1
9999 1221	Hilda's Boutique	Shop 2
9999 1222	Hill's Flowers	Shop 3
9999 1223	Formal Hire	Shop 4
9999 1224	Formal Hire	Shop 5
9999 1225	Sven's Jewellers	Shop 6
9999 1226	Mr Whippy	Shop 7
9999 1227	Pics & Frames	Shop 8
9999 1228	Specs R Us	Shop 9
9999 1229	Cards & Gifts	Shop 10
9999 1230	Beck's Pets	Shop 11
9999 1231	Big John's	Shop 12
9999 1232	Liquor Land	Shop 13
9999 1233	Faye's Frocks	Shop 14
9999 1234	Mick's Meats	Shop 15
9999 1235	Little Deli	Shop 16
9999 1236	Vacant	Shop 17
9999 1237	Vacant	Shop 18
9999 1238	The Little Shop	Shop 19
9999 1239	Kateez	Shop 20

Procedure

The details for the floor cabling are shown on the record for the building distributor B vertical (shown following).

Using the Tenants List above, complete the cross-connect records for the floor distributor A and B verticals.

VERTICAL: A					
CABLE DISTRIBUTION RECORDS			NOTES: 1. WRITE LIGHTLY WITH BLACK PENCIL 2. COMPLETE ALL RELEVANT PARTICULARS		
CABLE DETAILS	PAIR NO.	SERVICE NUMBER OR PORT	NAME OR OTHER PARTICULARS OF SERVICE	JUMPERED TO	
				VERTICAL	PAIR
	0				
	9				
	8				
	7				
	6				
	5				
	4				
	3				
	2				
	1				
	0				
	9				
	8				
	7				
	6				
	5				
	4				
	3				
	2				
	1				

VERTICAL: B					
CABLE DISTRIBUTION RECORDS			NOTES: 1. WRITE LIGHTLY WITH BLACK PENCIL 2. COMPLETE ALL RELEVANT PARTICULARS		
CABLE DETAILS	PAIR NO.	SERVICE NUMBER OR PORT	NAME OR OTHER PARTICULARS OF SERVICE	JUMPERED TO	
				VERTICAL	PAIR
	0		Sven's Jewellers		
	9		Pics & Frames		
	8		Beck's Pets		
	7		Hilda's Boutique		
	6		Big John's		
	5		Mr Whippy		
	4		Formal Hire		
	3		Faye's Frocks		
	2				
	1		Liquor Land		
	0				
	9		Centre Management		
	8		Mick's Meats		
	7		Kateez		
	6		Specs R Us		
	5		The Little Shop		
	4		Hill's Flowers		
	3		Little Deli		
	2		Cards & Gifts		
	1		Formal Hire		

REFERENCES

Australian Communications and Media Authority (ACMA) (2016a), *ACMA cabling provider rules. Pathways to cabling registration: November 2016 update*, viewed 29 May 2018, https://www.acma.gov.au/-/media/Technical-Regulation-Development/Information/pdf/Pathways_to_cabler_registration.pdf?la=en, p. 20.

Australian Communications and Media Authority (ACMA) (2016b), *Regulatory responsibility*, viewed 29 May 2018, https://www.acma.gov.au/theACMA/About/Corporate/Responsibilities/regulation-responsibilities-acma

Australian Communications and Media Authority (ACMA) (2016c), *Labelling – telecommunications products*, viewed 29 May 2018, https://acma.gov.au/Industry/Suppliers/Regulatory-arrangements/Telecommunications-customer-equipment-and-cabling/labelling—telecommunications-products

Australian Communications and Media Authority (ACMA) (2018a), *EMC labelling requirements*, viewed 29 May 2018, https://www.acma.gov.au/theACMA/emc-labelling-requirements

Australian Communications and Media Authority (ACMA) (2018b), *EMC regulatory arrangements*, viewed 29 May 2018, https://www.acma.gov.au/Industry/Suppliers/Regulatory-arrangements/EMC-Electromagnetic-compatibility/emc-regulatory-arrangements

Australian Government (2014), *Telecommunications Cabling Provider Rules 2014*, viewed 29 May 2018, https://www.legislation.gov.au/Details/F2014L01684

NBN Co. (2018), nbn™ Fixed Line connections, viewed 1 June 2018, https://www.nbnco.com.au/residential/learn/network-technology.html

Answers to the exercises

Chapter 1

Exercise 1.1
a 88
b 45
c 10

Exercise 1.2
a 4
b 1
c 4
d 4
e 3

Exercise 1.3
a 1240
b 600
c 1760 or 1770
d 500
e 4131.0
f 290 or 300

Exercise 1.4
a 27 000
b 0.00861
c 85.37
d 65 674 000
e 0.04008
f 583 000
g 5

Exercise 1.5
a 1.2845×10^4
b 2.345×10^6
c 1.0×10^{-2}
d 4.5×10^{-5}

Exercise 1.6
a 15.845×10^3
b 2.365×10^6
c 50×10^{-3}
d 65×10^{-6}

Exercise 1.7
a 16 807
b 1728
c 0.003086

Exercise 1.8
a 2^7
b 4^5
c 6^9
d 3^4
e 5^2
f 6^{-2}
g 4^6
h 6^5
i 3^6
j 4^2

Exercise 1.9
a $R_1 = R_T - (R_2 + R_3)$ or $R_1 = R_T - R_2 - R_3$
b $V_A = E_G - V_T$
c $L = \dfrac{F}{Bi}$
d $\rho = \dfrac{Ra}{l}$
e $f = \dfrac{X_L}{2\pi L}$
f $T = \dfrac{60P}{2\pi N}$
g $V = IR$
h $V = \sqrt{PR}$
i $R_B = \dfrac{R_A \times R}{R_X}$
j $f = \dfrac{NP}{120}$

Exercise 1.10
a $10x$
b $2x$
c $10a$
d $-4a$
e $-3x$
f $-1x + 4$
g $2a - 14$
h $2x + 5y$

Exercise 1.11
a $9a - 10b + 4c - 3d$
b $4x + 2y + 3z$
c $2ab - 4a + 12b - 17$
d $2x^2 + 11x + 16$
e $2n + 2$
f $6a - 4x + 6$
g -8

Exercise 1.12
a $24a^4b$
b $16x^2y^5$

Exercise 1.13
a x^6
b m^2n^6
c $10u^{-3}v^{-3}$

Exercise 1.14
$\dfrac{12b}{10} + \dfrac{b}{10} = \dfrac{12b + b}{10} = \dfrac{13b}{10}$

Exercise 1.15
a $2\sqrt{24}$
b $5\sqrt{10}$

Exercise 1.16

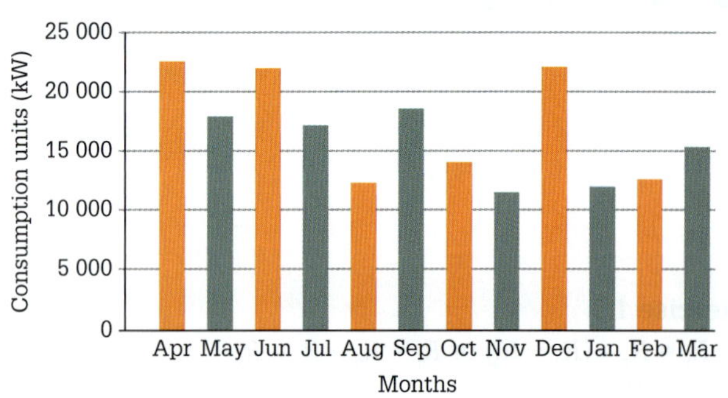

Exercise 1.17

a 5.0 m

b 189.7 m

Exercise 1.18

a $c = 32.31$ Sin A = 0.37 Cos A = 0.93
Tan A = 0.40 Sin B = 0.93 Cos B = 0.37 Tan B = 2.50

b $c = 49.24$ Sin A = 0.91 Cos A = 0.41
Tan A = 2.25 Sin B = 0.41 Cos B = 0.91 Tan B = 0.44

c $c = 29.15$ Sin A = 0.86 Cos A = 0.51
Tan A = 1.67 Sin B = 0.51 Cos B = 0.86 Tan B = 0.60

Chapter 2

Exercise 2.1

a 1200 rpm

b 6000 rpm

c 375 rpm

d 4 poles

e 500 rpm

Exercise 2.2

a 4 A

b 192 V

c 25 Ω

Exercise 2.3

Resistor A = 1.6 Ω
Resistor B = 0.8 Ω

Exercise 2.4

a 6 N

b 19.6 N

c 2.45 kN

Exercise 2.5

26 Nm

Exercise 2.6

9.87 kW

Exercise 2.7

a 1495 W

b 56.25 W

c 36 W

Exercise 2.8

a 90.9%

b 20 kW

Exercise 2.9

a 960 Ω

b 94 Ω

Exercise 2.10

55.4 V

Exercise 2.11

69.2 W

Chapter 3

Exercise 3.1

a 4.5 Ω

b 26.7 m

c 0.653 Ω

d 10 mm^2

Exercise 3.2

a 2.3 Ω

b 0.5 mm^2

c 0.137 Ω

Exercise 3.3

a 8.88 Ω

b 25 Ω

c 179 Ω

Exercise 3.4

a 42.8 Ω

b 101 Ω

Exercise 3.5

a 85.0 Ω

b 1.00 A

c 1.00 A, 1.00 A, 500 mA, 500 mA

d 18.0 V, 33.0 V, 34.0 V, 34.0 V

e 18.0 W, 33.0 W, 34.0 W, 34.0 W

Exercise 3.6

a
$$R_{Multiplier} = \frac{E-(I_{FSD} \times R_{meter})}{I_{FSD}} = \frac{50-\left(\left(200 \times 10^{-6}\right) \times \left(5 \times 10^{3}\right)\right)}{\left(200 \times 10^{-6}\right)} = 245\,\text{k}\Omega$$

b
$$R_{Shunt} = \frac{I_{FSD} \times R_{meter}}{I - I_{FSD}} = \frac{\left(200 \times 10^{-6}\right) \times \left(5 \times 10^{3}\right)}{\left(500 \times 10^{-3}\right) - \left(200 \times 10^{-6}\right)} = 2.000\,\Omega$$

c
$$sensitivity = \frac{1}{I_{FSD}} = \frac{1}{200 \times 10^{-6}} = 5\ \text{k}\Omega/\text{V}$$

Exercise 3.7

a 5000 Ω

b 38.1 Ω

Exercise 3.8

265 pF

Exercise 3.9

230 pF

Exercise 3.10

80 kV/mm

Exercise 3.11

2 J

Exercise 3.12

a 10 s
b 50 s
c 0 A as steady state assumed after 5 time constants

Exercise 3.13

a 16.6 μF
b 403 μF
c 200 μF
d 160 μF

Chapter 4

Exercise 4.1

a i 500 At ii 1600 At
b 240 At
c 167 mA
d 25 000 At
e i Fm is halved
  ii Fm is doubled
  iii Fm is doubled as current is proportional to potential difference
  iv Fm is doubled
  v Fm is quartered

Exercise 4.2

a i 785×10^{-9} Wb ii 1.26×10^{-6} Wb
b 133 Wb
c 625 μA
d i 3.33 Wb ii 13.33 Wb

Exercise 4.3

a 1.5 T
b 200 mm^2

Exercise 4.4

a 15.9×10^6 At/Wb
b 248×10^3 At/Wb
c 0.5 mm
d 4.77 cm^2
e i halves ii doubles

Exercise 4.5

a 3000 At/m
b 133 mm
c i halves ii doubles
  iii increases proportionally

Exercise 4.6

b i 6 At/m ii 6 At/m iii 30 At/m
c i 1.0 T ii 2.6 T iii 2.97 T

Exercise 4.7

b i 1.0 ii 3.8 iii 3.15
c i 200 At/m
  ii 400 At/m and 480 At/m
  iii 310 At/m and 900 At/m

Exercise 4.8

a 18.7 mV
b 20 ms^{-1}
c 12 mT

Exercise 4.9

a −60 V
b 60 mH

Exercise 4.10

a 500 mH
b 104.7 μH remember $\mu_0 = 4\pi \times 10^{-7}$

Exercise 4.11

1 a 80 ms b 400 ms c 8.0 A
2 5 J

Chapter 5

Exercise 5.1

a
$$V_{Reg} = \frac{V_{NL} - V_{Rated}}{V_{Rated}} = \frac{475 - 452}{452} = \mathbf{0.0509 \text{ or } 5.09\%}$$

b
$$V_{Reg} = \frac{V_{NL} - V_{Rated}}{V_{Rated}}$$
$$\therefore V_{NL} = (V_{Reg} \times V_{Rated}) + V_{Rated} = (0.045 \times 448) + 448 = \mathbf{468.16 \text{ V}}$$

Exercise 5.2

a 262.75 V
b 200 V

Exercise 5.3

a 232.2 V
b 302.8 V

Exercise 5.4

a 541.7 V
b 8
c 480

Exercise 5.5

3.28 N

Exercise 5.6

a 0.075 Nm (0.0375 N produced by each coil side)
b 1.67 A

Exercise 5.7

a 79.58 Nm

b 8.17 kW

Exercise 5.8

a 0.833 or 83.3%

b 12 kW

c 1.75 kW

Chapter 6

Exercise 6.1

a 200 mm

b 28.72 mm

c 13 cm

Exercise 6.2

a 85 mm

b 147 mm

Exercise 6.3

a 30.3°

b 42.8°

c 44.4°

Exercise 6.4

Sin 330°	Cos 230°	Tan 345°	Sin 135°	Cos 120°	Tan 210°
−0.5	−0.64278	−0.26795	0.7071	−0.5	0.57735

Exercise 6.5

a 103.92 V

b 41.04 V

c −103.92 V

d −60 V

Exercise 6.6

a 40 ms

b 10 ms

c 1 ms

d 50 us

Exercise 6.7

a 250 Hz

b 10 Hz

c 50 kHz

d 2 kHz

Exercise 6.8

Angle	i (A)
0	0.00
15	0.52
30	1.00
45	1.41
60	1.73
75	1.93
90	2.00
105	1.93
120	1.73
135	1.41
150	1.00
165	0.52
180	0.00
195	−0.52
210	−1.00
225	−1.41
240	−1.73
255	−1.93
270	−2.00
285	−1.93
300	−1.73
315	−1.41
330	−1.00
345	−0.52
360	0.00

Exercise 6.9

a 216.58 V

b 251.18 V

Exercise 6.10

a 240.5 V

b 141 A

Exercise 6.11

a 0.8726 rads

b 3.490 rads

c 5.5846 rads

d 1.571 rads

e 2.6178 rads

Exercise 6.12

120°

Exercise 6.13

a **i**

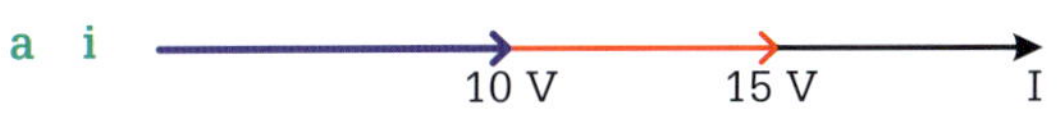

ii

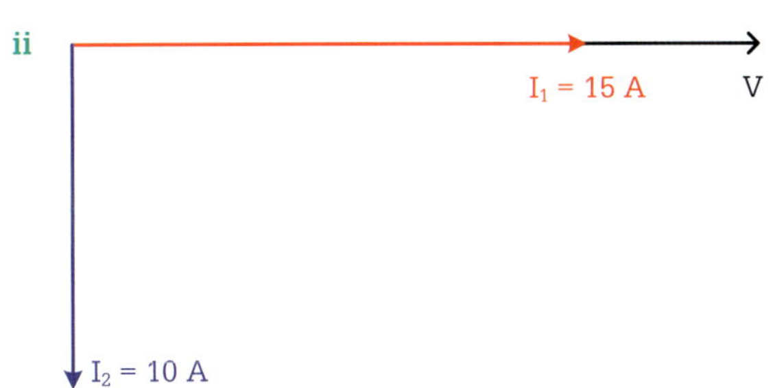

iii

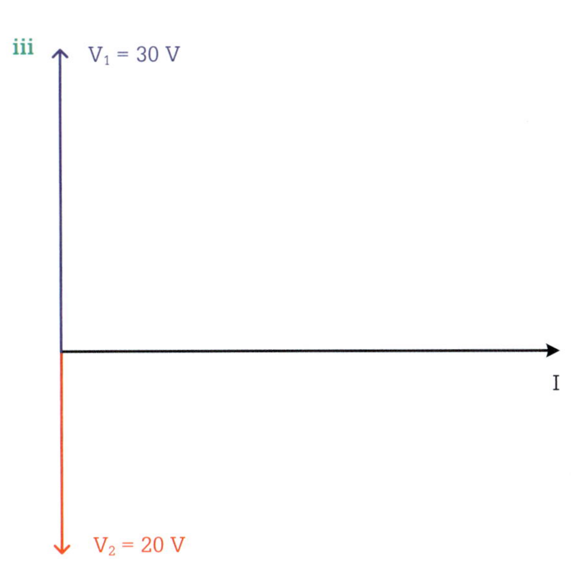

iv

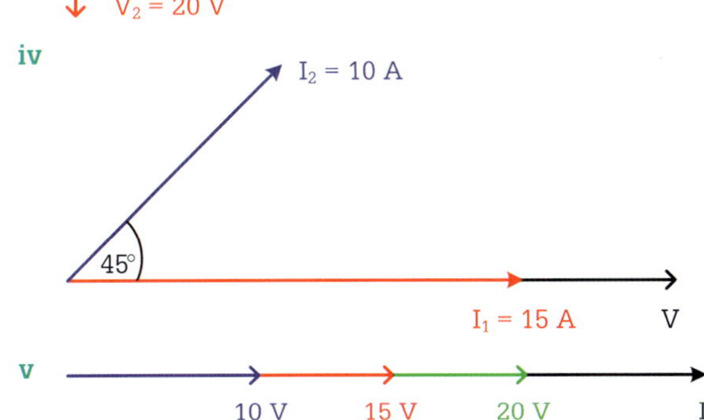

v

10 V 15 V 20 V I

b Two voltages, 8 V and 12 V in phase with current
Current I_1 is in phase with voltage V and current I_2 leads voltage by 90°
Voltage V_2 lags Current I by 90° and Voltage V_1 leads current by 90°
Current of 12 A is in phase with Voltage and current of 10 A leads voltage by 35°
Voltages of 8, 12, and 20 V in phase with current

Exercise 6.14

a **i**

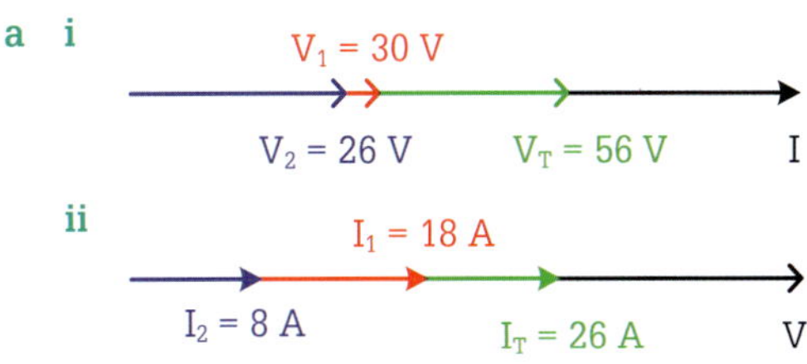

iii

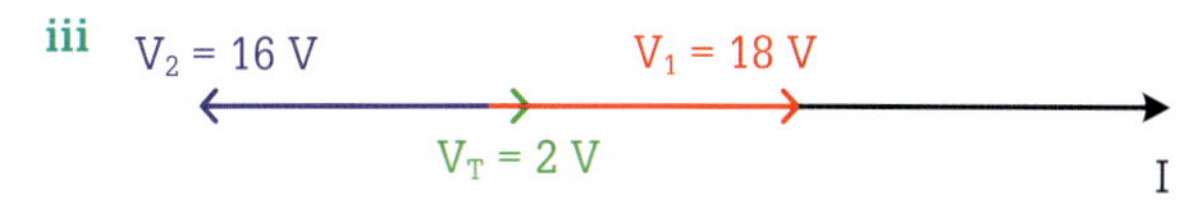

iv

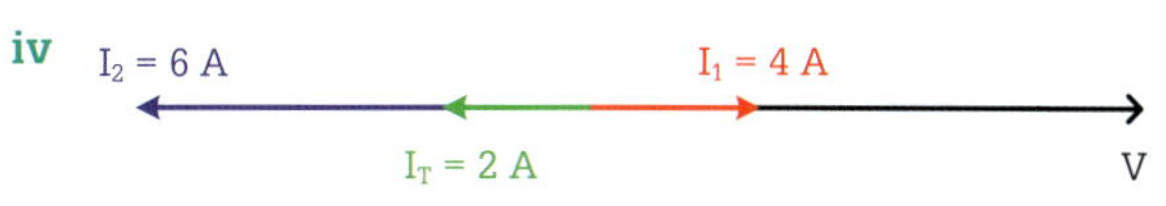

b **i**

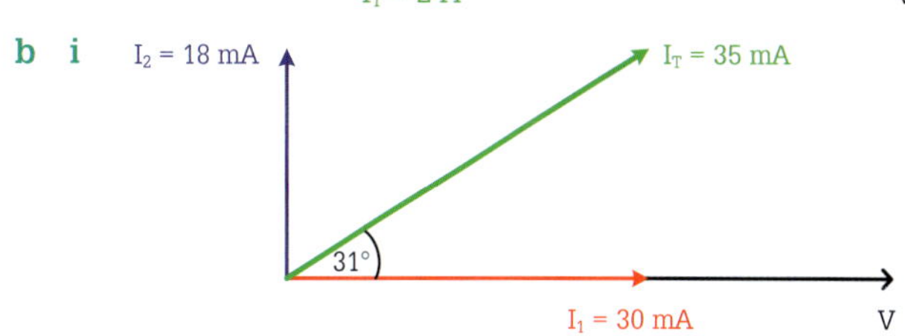

ii

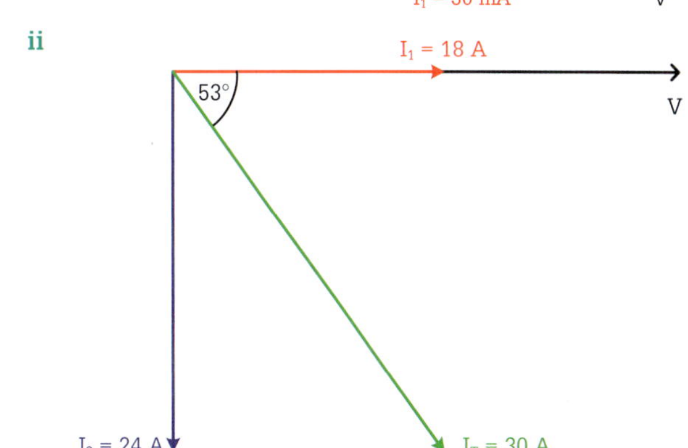

Exercise 6.15

a **i** **ii**

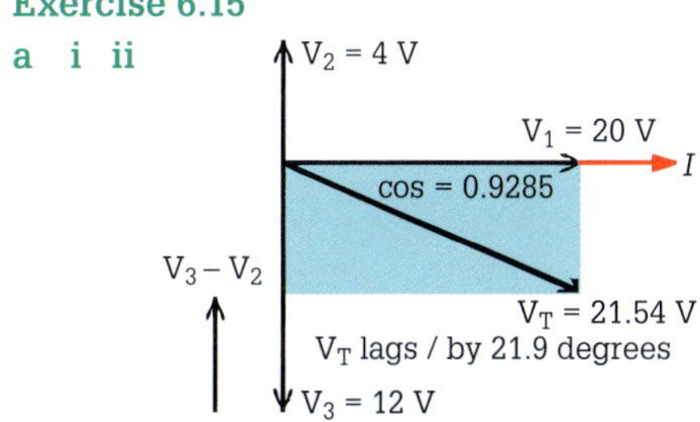

b **i**

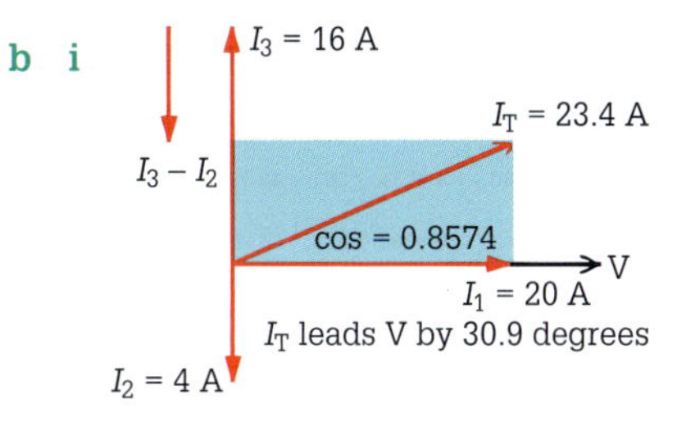

ii 12 A

Exercise 6.16

a R2 = 400 Ω, V1 = 16V
b 600 Ω
c **i** 500 Ω **ii** 5 V **iii** 0.5 W
d **i** 40 V **ii** 50 mW **iii** 56 kΩ
e **i** 45.4 Ω **ii** 833 mW **iii** 37 mA **iv** 2.2 W
f **i** 11 mA **ii** 545.5 Ω **iii** 8.5 V **iv** 180 mW
g 31.3 A
h **i** 22.4 V **ii** 3133 Ω
i **i** 14.3 mA **ii** 7.83 V
iii 8.8 mA **iv** 10.6 V

Exercise 6.17

a V_1 = 6.25 V V_2 = 12.5 V V_T = 18.75 V

6.25 V 12.5 V I = 25 mA

b **i, ii** $I_1 = 1.5$ A $I_2 = 0.5$ A

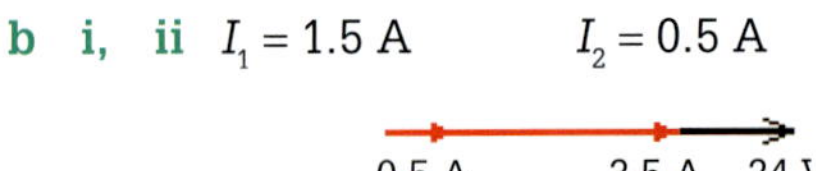

c **i** 43.65 Ω **ii** 1.746 V **iii** 113 mA p–p

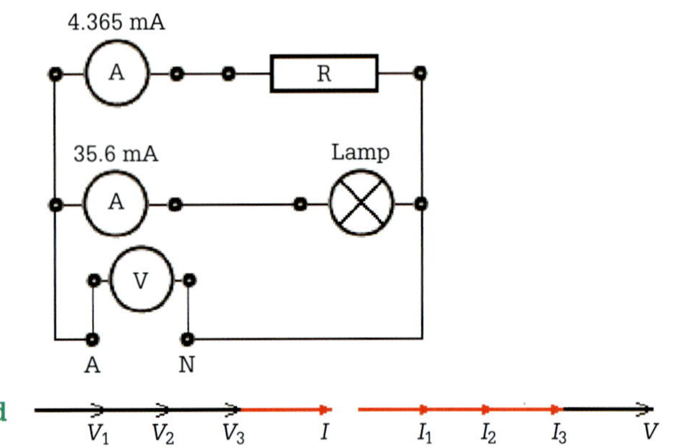

d V_1 V_2 V_3 I I_1 I_2 I_3 V

e

f $R_1 = 16\,\Omega$, $R_2 = 24\,\Omega$ and Total power = 160 W

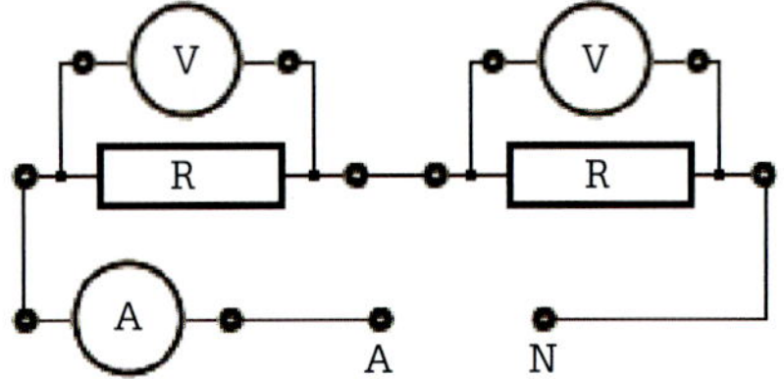

Exercise 6.18

a

angle	sin	e	i	*P*	angle	*P*	angle	*P*	angle	*P*
0	0.0000	0	0	0	100	38.7939	200	4.6791	300	30.0000
20	0.3420	3.4202	1.3681	4.6791	120	30.0000	220	16.5270	320	16.5270
40	0.6428	6.4279	2.5712	16.5270	140	16.5270	240	30.0000	340	4.6791
60	0.8660	8.6603	3.4641	30.0000	160	4.6791	260	38.7939	360	0.0000
80	0.9848	9.8481	3.9392	38.7939	180	0.0000	280	38.7939		

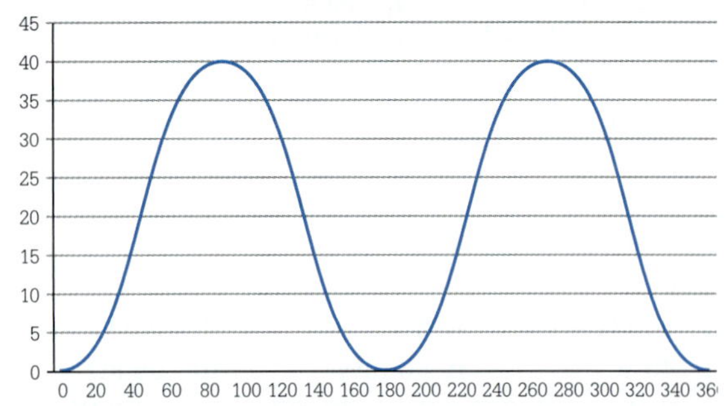

b

angle	sin	i	*P*	angle	*P*	angle	*P*	angle	*P*
0	0.0000	0	0.000	100	9.698	200	1.170	300	7.500
20	0.3420	0.0342	1.170	120	7.500	220	4.132	320	4.132
40	0.6428	0.0643	4.132	140	4.132	240	7.500	340	1.170
60	0.8660	0.0866	7.500	160	1.170	260	9.698	360	0.000
80	0.9848	0.0985	9.698	180	0.000	280	9.698		

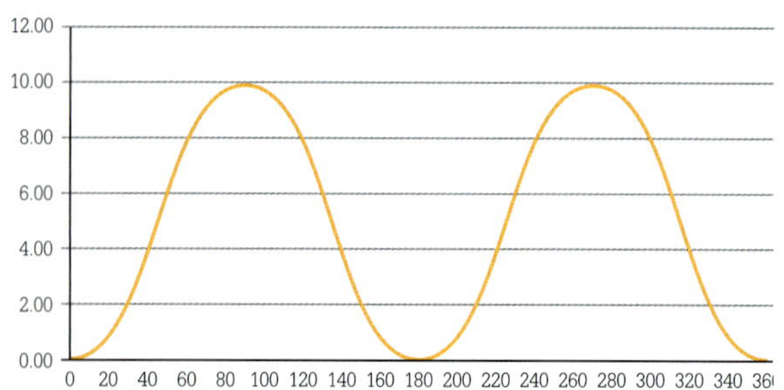

Exercise 6.19

a 47.12 Ω, 78.54 Ω

b 31.42 Ω

c 6.366 H

d 159 Hz

e 1885 Ω, 2827 Ω

f 306 mA

g 3.39 H

h 60 kHz

i 27.3 mA

j 36.2 V

Exercise 6.20

a

		Frequency (Hz)										
Inductor		0	100	200	300	400	500	600	700	800	900	1000
100	mH	0.00	62.83	125.66	188.50	251.33	314.16	376.99	439.82	502.65	565.49	628.32
200	mH	0.00	125.66	251.33	376.99	502.65	628.32	753.98	879.65	1005.31	1130.97	1256.64

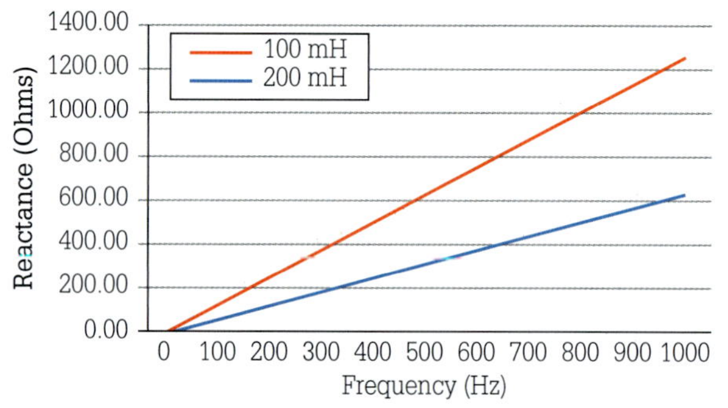

280 Ω @ 445 Hz for 100 mH and 223 Hz for 200 mH

b

		Frequency (Hz)											
Inductor		40	100	200	300	400	500	600	700	800	900	1000	
100	mH	1592	637	318	212	159	127	106	91	80	71	64	mA
200	mH	796	318	159	106	80	64	53	45	40	35	32	mA

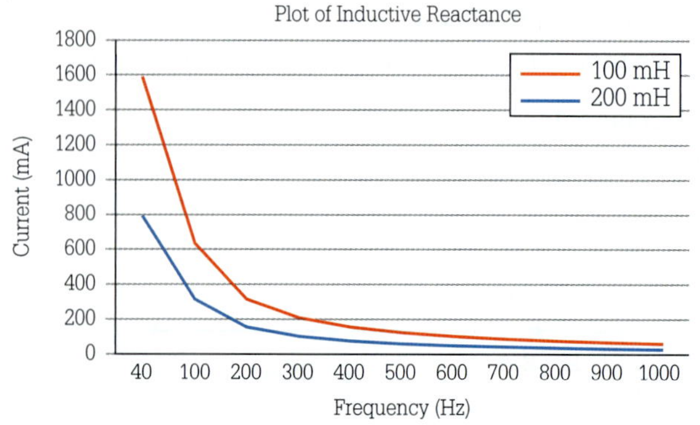

@ 600 Hz current through 100 mH = 106 mA and through 200 mH = 53 mA

Exercise 6.21

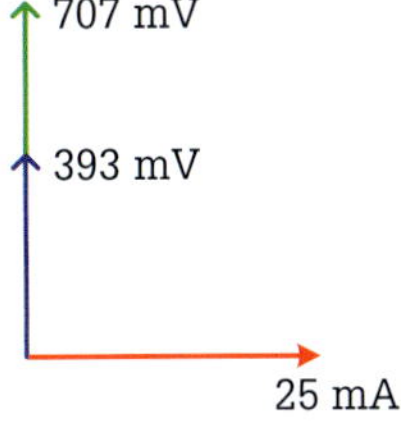

Exercise 6.22

a

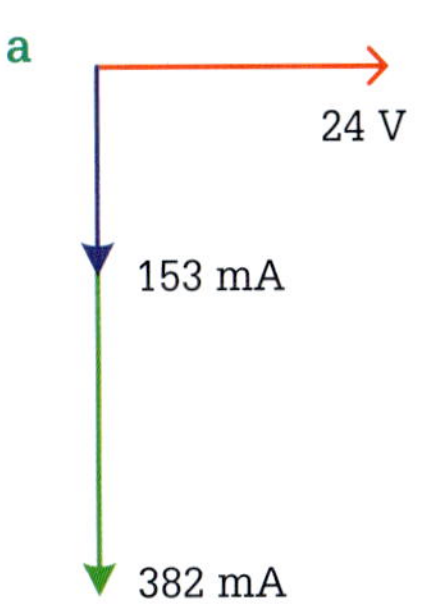

b i 47.5 mA

ii

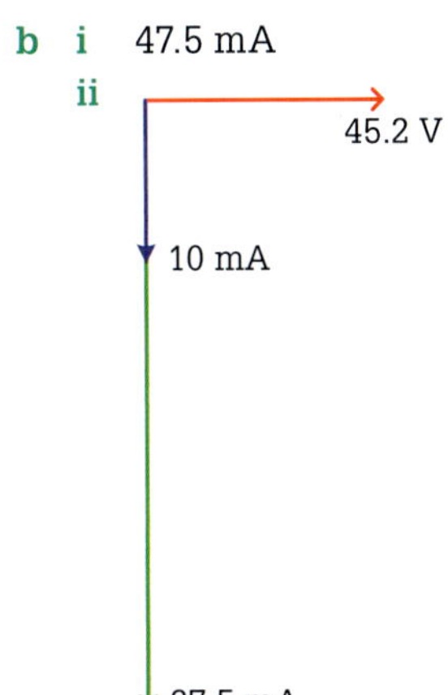

c

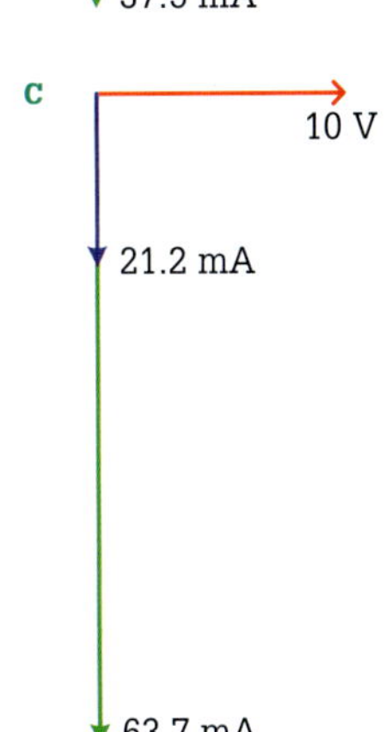

Exercise 6.23

Table results

angle	sin	I	e	P	angle	P	angle	P	angle	P
0	0	0	0							
15	0.259	0.259	0.518	0.134	105	1.866	195	0.134	285	1.866
30	0.5	0.5	1.0	0.5	120	1.5	210	0.5	300	1.5
45	0.707	0.707	1.414	0.575	135	0.575	225	0.575	315	0.575
60	0.866	0.866	1.732	1.5	150	0.5	240	1.5	330	0.5
75	0.966	0.966	1.932	1.866	165	0.134	255	1.866	345	0.134
90	1.0	1.0	2.0	2.0	180	0	270	2.0	360	0

Exercise 6.24

1 31.83 Ω

2 250 Hz

3 42.4 µF

Exercise 6.25

a 2.67 kΩ

b 50.3 mA

c 165 mA

Exercise 6.26

a i V1 = 16.5 V V2 = 7.5 V

ii

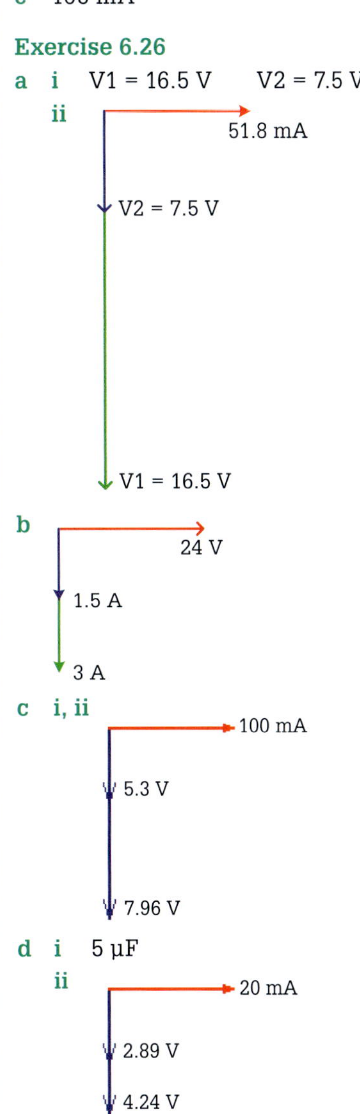

Exercise 6.27

angle	sin	*e*	i	*P*
0	0.0000	0	1.00	0
10	0.1736	0.3473	0.98	0.3420
20	0.3420	0.6840	0.94	0.6428
30	0.5000	1.0000	0.87	0.8660
40	0.6428	1.2856	0.77	0.9848
50	0.7660	1.5321	0.64	0.9848
60	0.8660	1.7321	0.50	0.8660
70	0.9397	1.8794	0.34	0.6428
80	0.9848	1.9696	0.17	0.3420
90	1.0000	2.0000	0.00	0.0000
100	0.9848	1.9696	−0.17	−0.3420
110	0.9397	1.8794	−0.34	−0.6428
120	0.8660	1.7321	−0.50	−0.8660
130	0.7660	1.5321	−0.64	−0.9848
140	0.6428	1.2856	−0.77	−0.9848
150	0.5000	1.0000	−0.87	−0.8660
160	0.3420	0.6840	−0.94	−0.6428
170	0.1736	0.3473	−0.98	−0.3420
180	0.0000	0.0000	−1.00	0.0000
190	−0.1736	−0.3473	−0.98	0.3420
200	−0.3420	−0.6840	−0.94	0.6428
210	−0.5000	−1.0000	−0.87	0.8660
220	−0.6428	−1.2856	−0.77	0.9848
230	−0.7660	−1.5321	−0.64	0.9848
240	−0.8660	−1.7321	−0.50	0.8660
250	−0.9397	−1.8794	−0.34	0.6428
260	−0.9848	−1.9696	−0.17	0.3420
270	−1.0000	−2.0000	0.00	0.0000
280	−0.9848	−1.9696	0.17	−0.3420
290	−0.9397	−1.8794	0.34	−0.6428
300	−0.8660	−1.7321	0.50	−0.8660
310	−0.7660	−1.5321	0.64	−0.9848
320	−0.6428	−1.2856	0.77	−0.9848
330	−0.5000	−1.0000	0.87	−0.8660
340	−0.3420	−0.6840	0.94	−0.6428
350	−0.1736	−0.3473	0.98	−0.3420
360	0.0000	0.0000	1.00	0.0000

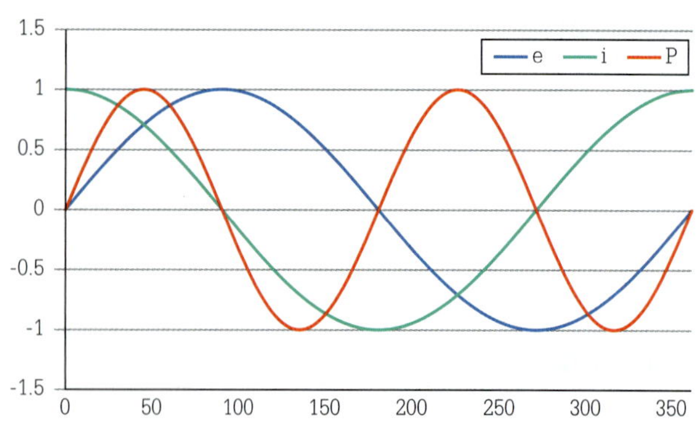

Exercise 6.28

a 43.0 V
b 40.25 V

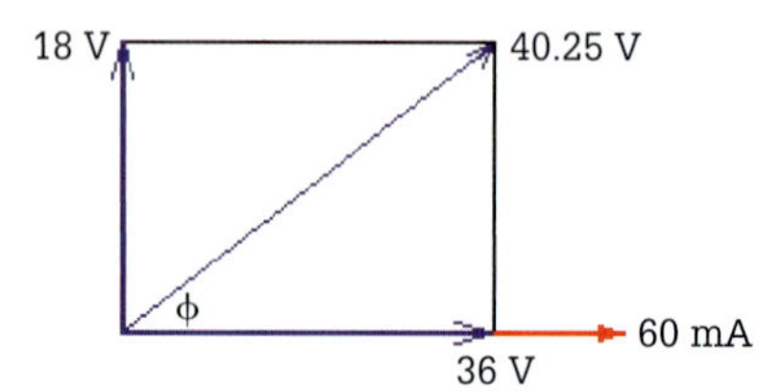

c

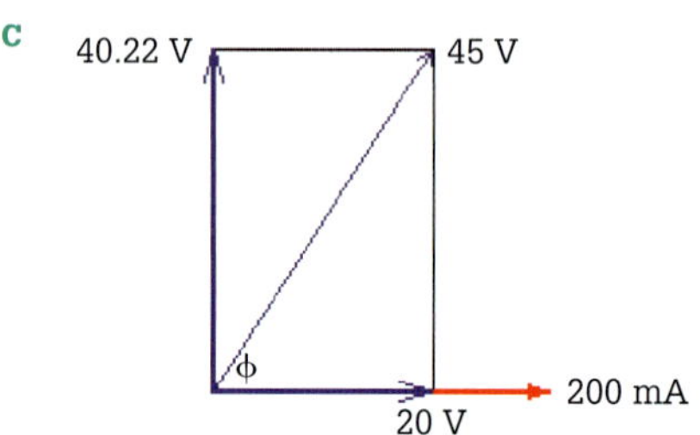

d 158.1 Ω
e 22.9 V
f 127 V
g 2.52 kΩ
h i 103 Ω ii 160.6 Ω
i 27.5 mH as R = 1 kΩ, Z = 2 kΩ, then XL = 1.73 kΩ
j 5.5 mH

Exercise 6.29

a 71.6° lagging
b 81° lagging
c 63.6° lagging

Exercise 6.30

a 0.5 lagging
b 0.33 lagging
c 18 W
d 0.907 lagging

Exercise 6.31

a [61 V]

b

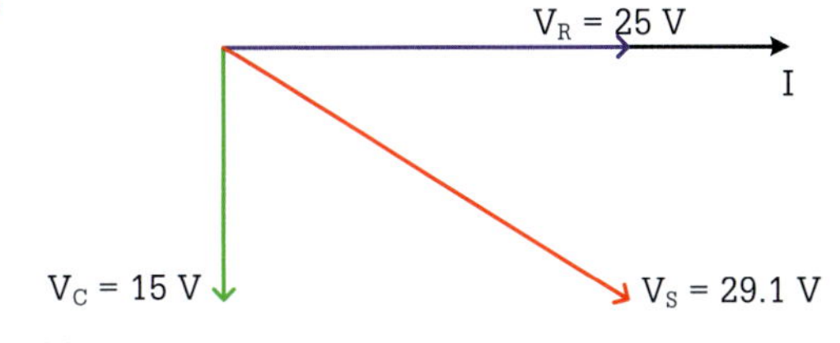

c 207 V

d

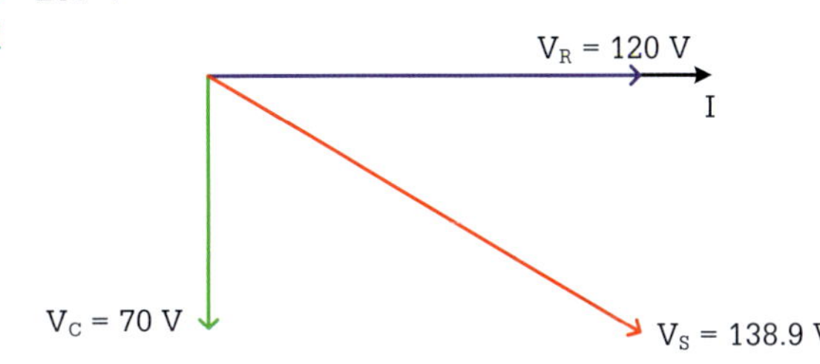

e 417.6 Ω

f 376 Ω

g 23.0 Ω, 22.4 Ω

h 38.7 Ω

i 1.84 µF

Exercise 6.32

a 17.7° leading

b 26.6° leading

c 12.5° leading

Exercise 6.33

a 44.7 V

b 123.3 V

Exercise 6.34

a 53.85 Ω

b 101.2 Ω

c 245 Ω

Exercise 6.35

a 26.6° leading

b 7.75° leading

c 2078 Ω

d 90 W

e pf = 0.7071 leading, P = 1.18 W

Exercise 6.36

a 49.2 A

b 360 mA

c 347 mH

Exercise 6.37

a 31°

b 39.1°

c 45°

d i 0.954 ii 45 W

Exercise 6.38

a 8 A

b Ic = 1.16A, Ir = 2.3A, It = 2.58A

c 84.2 Ω

Exercise 6.39

a 21.8° leading

b 45° leading

Exercise 6.40

a 3.6 A leading

b 4.9 A lagging

c 340 mA

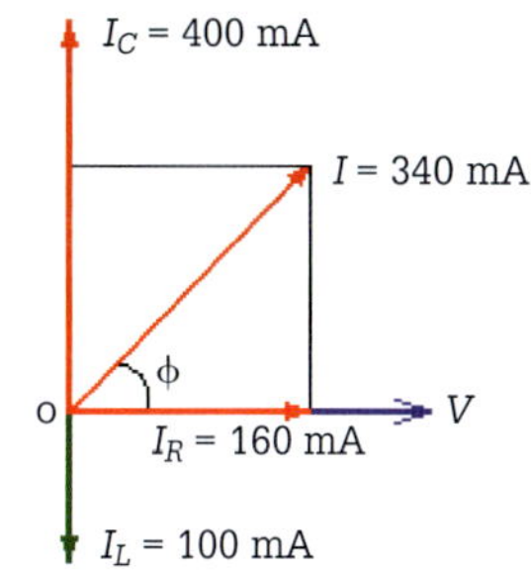

Exercise 6.41

a 0.894 leading

b 0.928 leading

c 0.832 lagging

Exercise 6.42

a i 1458 W ii 3177 VAr iii 3450 VA

b i 352 W ii 295.7 VAr iii 460 VA

Exercise 6.43

134 kVAr

Exercise 6.44

0.725 lagging

Exercise 6.45

1

Graph table	Reactance (Ω)		
	$f(H_z)$	(X_L)	X_C
	100	12.57	79.6
	200	25.14	39.8
	300	37.7	26.5
	400	50.3	19.9
	500	62.8	15.9
	600	75.4	13.26

a 250 Hz b 250–600 Hz c 0–250 Hz

2 a 250 Hz

b 31.83 Hz, 1 A

c i 199 Ω ii 158.3 mH

d 8.11 µF

e 726 mH @ 59 Hz

Exercise 6.46

1 306.3 Hz

2 562.7 Hz

Exercise 6.47

Graph table	f (H_z)	X_L	I_L	X_C	I_C	$I_L - I_C$
	100	50.3	0.31	796	0.02	0.29
	200	100.5	0.16	398	0.04	0.12
	300	151	0.106	265	0.06	0.046
	400	201	0.08	199	0.08	0
	500	251	0.064	159	0.1	0.036
	600	302	0.053	133	0.12	0.067

a 400 Hz b 0–399 Hz c 401–600 Hz

Chapter 7

Exercise 7.1
2200

Exercise 7.2
1 2200 V
2 12.7 kV

Exercise 7.3
From AS/NZS 3008.1.1 Table 30, 16 mm² has a reactance of 0.0861 Ω/km

From AS/NZS 3008.1.1 Table 35, 16 mm² has a resistance of 1.40 Ω/km at 75°C, which is the operating temperature of V90 cables.

$$R_C = \frac{length \times Resistance / km}{1000} = \frac{45 \times 1.4}{1000} = 0.063\ \Omega$$

$$X_C = \frac{length \times Reactance / km}{1000} = \frac{45 \times 0.0861}{1000} = 0.00387\ \Omega$$

$$Z_C = \sqrt{R_C^2 + X_C^2} = \sqrt{0.063^2 + 0.00387^2} = 0.0631\ \Omega$$

$$V_d = IZ_C = 55 \times 0.0631 = \mathbf{3.47\ V}$$

Exercise 7.4
a 381 V
b 381 V
c 25.4 A
d 44 A
e 44 A

Exercise 7.5
a 250 V
b 144 V
c 14.4 A
d 14.4 A
e 8.31 A

Exercise 7.6
13.51 kW

Exercise 7.7
a 124.7 kVA
b 106 kW

Exercise 7.8
0.866

Exercise 7.9
From AS/NZS 3008.1.1 Table 30, 10 mm² has a reactance of 0.0906 Ω/km

From AS/NZS 3008.1.1 Table 35, 10 mm² has a resistance of 2.23 Ω/km at 75°C, which is the operating temperature of V90 cables.

$$R_{PH} = \frac{length \times Resistance / km}{1000} = \frac{30 \times 2.23}{1000} = 0.0669\ \Omega$$

$$X_{PH} = \frac{length \times Reactance / km}{1000} = \frac{30 \times 0.0906}{1000} = 0.00272\ \Omega$$

$$Z_{PH} = \sqrt{{R_{PH}}^2 + {X_{PH}}^2} = \sqrt{0.0669^2 + 0.00272^2} = 0.06696\ \Omega$$

Impedance of the circuit comprises $Z_{PH} + Z_{PH} + Z_{Fault}$

$$Z_{Circuit} = Z_{PH} + Z_{PH} + Z_{Fault} = 0.06696 + 0.06696 + 0.5 = 0.6339\ \Omega$$

The short-circuit current, I_{sc}, can be determined as follows:

$$I_{sc} = \frac{V}{Z_{circuit}} = \frac{230}{0.6339} = \mathbf{363\ A}$$

Chapter 8

Exercise 8.1
a 1000 rpm
b 500 rpm
c 3000 rpm

Exercise 8.2
a 120 rpm, 4%
b 50 rpm, 5%

Exercise 8.3
a 11.9 Nm
b 45.2 Nm

Exercise 8.4
a 7.9 kW
b 3.58 kW

Exercise 8.5

a 0.903 or 90.3%

b 0.906 or 90.6%

c 0.785 or 78.5%

Exercise 8.6

a 0.938

b 9.04 kW

Exercise 8.7

a A 160 A aM fuse link can support 819 A for over 6 s.

b A 32 A aM fuse link can support 195 A for over 6 s.

Exercise 8.8

a 72 A for three seconds at a ratio of 3.8 with an I_N of 18.95 A, therefore a 20 A Type C CB is suitable.

b 67.5 A for seven seconds at a ratio of 2.75 with an I_N of 24.5 A, therefore a 25 A Type C CB is suitable.

Exercise 8.9

152.8 Nm

Exercise 8.10

83.1 kW

Exercise 8.11

0.95 or 95%

Exercise 8.12

17.3 A

Exercise 8.13

1.2 kV

Exercise 8.14

a
$$\begin{aligned} V_{Reg} &= \frac{V_{NL} - V_{Rated}}{V_{Rated}} \\ &= \frac{465 - 448}{448} \\ &= \mathbf{0.03795 \text{ or } 3.79\%} \end{aligned}$$

b
$$\begin{aligned} V_{Reg} &= \frac{V_{NL} - V_{Rated}}{V_{Rated}} \\ &= \frac{465 - 478}{478} \\ &= \mathbf{-0.0272 \text{ or } -2.27\%} \end{aligned}$$

Exercise 8.15

a 7.5 kW

b 2

c 13.1 A

d No

e One

f $P = \sqrt{3}\ V_L I_L\ pf = \sqrt{3} \times 400 \times 13.1 \times 0.89 = 8077\,W$

g 0.928 or 92.8%

h specified efficiency is lower than calculated by 2.6%

i 6203 and 6203

j Yes

Chapter 9

Exercise 9.1

a 24 V

b 1.5 A

c 156.5 mA

Exercise 9.2

V_1 at the principal tap = **22 000 V**

V_1 at the +2.5% tap = **22 500 V**

V_1 at the −2.5% tap = **21 450 V**

V_1 at the −5% tap = **20 900 V**

V_1 at the −7.5% tap = **20 350** V

Exercise 9.3

a 4:1

b 75 V

c 5 A

d 1.63 A

e 3.37 A

Exercise 9.4

a 460 V

b 30 A

Exercise 9.5

$$\begin{aligned} \text{burden} &= I^2 Z \\ &= 1^2 \times 0.15 \\ &= \mathbf{0.15\ V\ A} \end{aligned}$$

Exercise 9.6

a
$$\begin{aligned} I_{Line} &= \frac{kVA}{\sqrt{3} \times V_{Line}} \\ &= \frac{1500 \times 10^3}{\sqrt{3} \times 11\,000} \\ &= \mathbf{78.7\ A} \end{aligned}$$

Current transformers with 100:5 ratios would be suitable.

The line-to-line voltage of 11 000 V requires that an 11 000:230 or a 50:1 ratio would be suitable for the two potential transformers.

b
$$\begin{aligned} I_{Line} &= \frac{kVA}{\sqrt{3} \times V_{Line}} \\ &= \frac{2500 \times 10^3}{\sqrt{3} \times 11\,000} \\ &= \mathbf{131.2\ A} \end{aligned}$$

A current transformer with a 200:5 ratio would be suitable.

The line-to-neutral voltage is: **6351 V**

Therefore, a 50:1 ratio would be suitable for the two potential transformers.

Exercise 9.7

$$\begin{aligned} \eta &= \frac{P_{out}}{P_{out} + P_{loss}} \\ &= \frac{1000}{1000 + (3.5 + 9.5)} \\ &= \mathbf{0.987 \text{ or } 98.7\%} \end{aligned}$$

Exercise 9.8

No. of hours	Loading in kW	Power factor	kVA loading	P_{Cu} FL = 10
12	850	0.85	1000 kVA	10
7	650	0.75	867 kVA	7.52
5	200	0.9	222 kVA	0.493

Total copper loss in 24 hours

$$= (12 \times 10) + (7 \times 7.52) + (5 \times 0.493) = 175.1 \text{ kWh.}$$

Total iron loss over 24 hours

$$= 24 \times 8 = 192 \text{ kWh.}$$

Transformer output over the 24 hours

$$= (12 \times 850) + (7 \times 650) + (5 \times 200) = 15750 \text{ kWh}$$

$$\therefore \text{all day} = \frac{15\,750}{15\,750 + 175.1 + 192}$$

$$= \mathbf{0.977 \text{ or } 97.7\%}$$

Exercise 9.9

The losses before this occurred were:

- full-load copper loss was 18 W per transformer at full load
- for the 45% loaded unit, copper loss $= 0.45^2 \times 18.0 = 3.645$ W
- for the 25% loaded unit, copper loss $= 0.25^2 \times 18.0 = 1.125$ W
- full-load iron loss was 5 W per transformer at full load
- total losses = 3.645 + 1.125 + 5 + 5 = **14.77 W**.

With one transformer only on load:

- iron loss = 5 W
- copper loss $= (0.45 + 0.25)^2 \times 18 = 8.82$ W
- total loss = 5 + 8.82 = **13.82 W**.

Over a 12-month period of 260 working days, 24 hours per day:

Original loss:

$$\text{loss}_{orig} = \frac{260 \times 24 \times 14.77}{1000}$$

$$= 92.2 \text{ kWh}$$

New loss:

$$loss_{orig} = \frac{260 \times 24 \times 13.82}{1000}$$

$$= 86.2 \text{ kWh}$$

Saving = 92.2 – 86.2 = **6.0 kWh**, which is a 6.5% saving in energy loss.

Exercise 9.10

a 0.0336 or 3.36% @ 0.95 lagging pf

b 0.0476 or 4.76% @ 0.89 lagging pf

Exercise 9.11

a

$$I_{Full-load} = \frac{VA}{V_{Line}}$$

$$= \frac{50\,000}{230}$$

$$= 217.4 \text{ A}$$

The fault current is calculated taking into account the percentage impedance:

$$I_{Fault} = I_{Full-load} \times \frac{100}{Z\%}$$

$$= 217.4 \times \frac{100}{5}$$

$$= \mathbf{4348 \text{ A}}$$

The circuit breaker or fuse would have a minimum fault current capacity of 4348 A at 230 V.

b

$$I_{Full-load} = \frac{VA}{\sqrt{3} \times V_{Line}}$$

$$= \frac{1\,000\,000}{\sqrt{3} \times 400}$$

$$= 1443 \text{ A}$$

$$I_{Fault} = I_{Full-load} = \frac{100}{Z\%}$$

$$= 1443 \times \frac{100}{4.5}$$

$$= \mathbf{32\,067 \text{ A}}$$

The circuit breaker or fuse would have a minimum fault current capacity of 32 067 A at 400 V.

Exercise 9.12

$$\text{I}_{\text{THD}} = \frac{\sqrt{\text{I}_5^{\,2} + \text{I}_7^{\,2}}}{\text{I}_1}$$

$$= \frac{\sqrt{35^2 + 18^2}}{250}$$

$$= \mathbf{15.7\% \text{ THD}}$$

Exercise 9.13

Step 1: Determine the load shared by each transformer.

$$VA_{T1} = Load\ VA \times \frac{Z\%_{T2}}{Z\%_{T1} + Z\%_{T2}}$$

$$= 8000 \times \frac{2}{1.8 + 2}$$

$$= \mathbf{4210 \text{ VA}}$$

$$VA_{T2} = Load\ VA \times \frac{Z\%_{T1}}{Z\%_{T1} + Z\%_{T2}}$$

$$= 8000 \times \frac{1.8}{1.8 + 2}$$

$$= \mathbf{3790 \text{ VA}}$$

Step 2: Determine the division of the load current between the two transformers.

$$I_{T1} = \frac{VA_{T1}}{\sqrt{3}\,V_{Line}}$$

$$= \frac{4210}{\sqrt{3} \times 400}$$

$$= \mathbf{6.08 \text{ A}}$$

$$I_{T2} = \frac{VA_{T2}}{\sqrt{3}\,V_{Line}}$$

$$= \frac{3790}{\sqrt{3} \times 400}$$

$$= \mathbf{5.47 \text{ A}}$$

This arrangement is suitable.

Exercise 9.14

$kVA\ rating = VI$

$= 230 \times (5.5 \times 75)$

$= \mathbf{94.875\ kVA}$

A suitable transformer is 100 kVA, and it has some capacity for future additions.

Exercise 9.15

a $V = 4.44 Nf\Phi_{max}$

$= 4.44 \times 1200 \times 50 \times 0.0015$

$= \mathbf{399.6\ volts}$

b $V = 4.44 Nf\Phi_{max}$

$= 4.44 \times 350 \times 50 \times 0.0015$

$= \mathbf{116.5\ volts}$

Chapter 10

Exercise 10.1

a 2 W

b 72 W

Exercise 10.2

a 8.75 V

b 18.0 V

Exercise 10.3

a 15.5 V

b 31.2 V

Exercise 10.4

45.2 V

Exercise 10.5

23.7 V

Exercise 10.6

169.7 V

Exercise 10.7

4.8 W

Exercise 10.8

2%

Exercise 10.9

a 24 V

b 218 mA

c 208 mA

d 10% to 80% = 20.8 mA to 166.4 mA

e 22.4 Ω A E22 or an E24 can be used

f 1.58 W using an E22 resistor

Chapter 12

Exercise 12.1

0.5 Ω

Exercise 12.2

a 389 metres

b 229 metres

Exercise 12.3

$$R_{Loop} = \frac{R_{Specified} \times \text{distance} \times 2}{1000} = \frac{150 \times 259 \times 2}{1000} = \mathbf{77.7\,\Omega}$$

Exercise 12.4

$$C_{Loop} = \frac{C_{Mutual} \times \text{distance}}{1000} = \frac{150 \times 250}{1000} = \mathbf{32.5\,pF}$$

Exercise 12.5

$$Distance_1 = \frac{C_{Loop_1} \times 1000}{C_{Mutual}} = \frac{65 \times 1000}{300} = 217\ \text{metres}$$

$$Distance_2 = \frac{C_{Loop_2} \times 1000}{C_{Mutual}} = \frac{98 \times 1000}{300} = 327\ \text{metres}$$

Exercise 12.6

$$Distance_1 = \frac{R_{Loop_1} \times 1000}{R_{Specified} \times 2} = \frac{15 \times 1000}{94.5 \times 2} = 79\ \text{metres}$$

$$Distance_2 = \frac{R_{Loop_2} \times 1000}{R_{Specified} \times 2} = \frac{18 \times 1000}{94.5 \times 2} = 95\ \text{metres}$$

Provide solutions to routine electro-technology problems

APPENDIX

This appendix provides maths extension material, which builds on the concepts discussed in Chapter 1 Solving electrotechnology problems. This chapter provides underpinning knowledge for the unit UEECD0038 from the UEE Training Package.

LEARNING OBJECTIVES

Setting out problems
- Follow a clear procedure for setting out electrotechnology problems

Significant digits
- Express values to a specific number of significant digits

Scientific and engineering notation
- Express values using scientific and engineering notation

Transposition
- Transpose equations to change the subject of the equation and solve

Graphs
- Express quantities in graphical form

Setting out problems

If you follow a clear procedure, similar to that presented here, you will demonstrate to your teacher that you have an obvious understanding of the problem. All answers must be noticeably discernible from the working, which must be complete. If you adopt this method your teacher will be able to follow your progress through the problem and identify any areas requiring further attention. Another important consideration is that marks may be awarded for the steps involved in arriving at a solution to the problem; so even if you do not calculate the correct answer, you can still receive marks for your working.

Adopt a procedural approach

Consider the following sample problem.

Sample question

A series circuit consists of three resistors. The resistors are 5 Ω, 10 Ω and 15 Ω respectively. The combination is connected to a 120 volt d.c. supply.

Calculate:

a the total resistance

b current drawn from the supply.

Step 1: Draw the circuit diagram

Step 1 shows your ability to relate descriptive information to an accurate diagram. Make sure you show all relevant information on your diagram. Use the information presented in the problem to produce a diagram. **Figure A.1** shows what this looks like for the sample problem.

Step 2: Set out knowns and unknowns

Step 2 allows you to summarise the information you already know and what it is you need to calculate (the unknowns).

Step 3: Perform the first calculation

In Step 3, make sure you indicate clearly which part of the problem you are solving. In this example part (a) is solved. Remember it is important to show all steps that are necessary to arrive at the answer. In this example, the steps are equation, substitute 'real values' in equation and

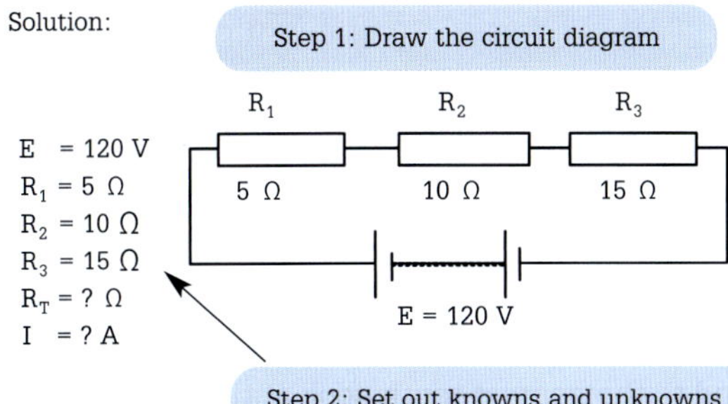

Step 3: Perform first calculation

(a) $R_T = R_1 + R_2 + R_3$
$= 5 + 10 + 15$
$= \underline{\underline{30\ \Omega}}$

Highlight or underline answer

Step 4: Perform next calculation

(b) $I = E / R_T$
$= 120 / 30$
$= \underline{\underline{4A}}$

FIGURE A.1

solve the equation. Not only does this demonstrate that you clearly know what you are doing, it provides a blow-by-blow account for later revision. Underline or highlight your answer.

Step 4: Perform remaining calculations

In Step 4, perform the remaining calculations following the same procedure outlined in Step 3. Indicate which part of the problem you are solving. In this example part (b) is solved. Remember to underline or highlight your answer.

SWITCH ON

Explanation

You may not realise that when an expression is written as 160/40 or alternatively as $\frac{160}{40}$ it means 160 divided by 40 and you use the ÷ key on your calculator. Figure A.2 shows this and the result obtained.

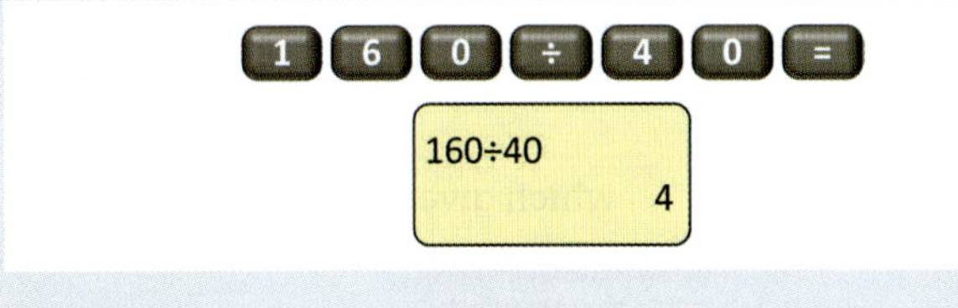

FIGURE A.2

Rounding

Before we proceed any further let's examine exactly what is meant by a number. Each digit in the number has a specific weighting. For example, in the number 415:

4 is in the hundreds place and carries a weight of
4×100 (= 400)
1 is in the tens place and carries a weight of
1×10 (= 10)
5 is in the units place and carries a weight of
5×1 (= 5)

Another way to look at this is:

Hundreds	Tens	Units
4	1	5

If we wanted to round off 415 to the nearest hundred, the answer is 400. You probably guessed that before you read this far. It's okay to guess because it helps you make an educated judgement as to the correctness of your answer. In this case you know that 415 is closer to 400 than it is to 500, which is the next hundred value. Let's examine the mathematical processes involved.

EXAMPLE A.1

To round to the nearest hundred, look at the number immediately to the right of the hundreds value. In this case it is 15.

Compare this value to half of what you are rounding to (half of 100 is 50).

If the value to the right of the hundred is less than 50, replace it with 00.

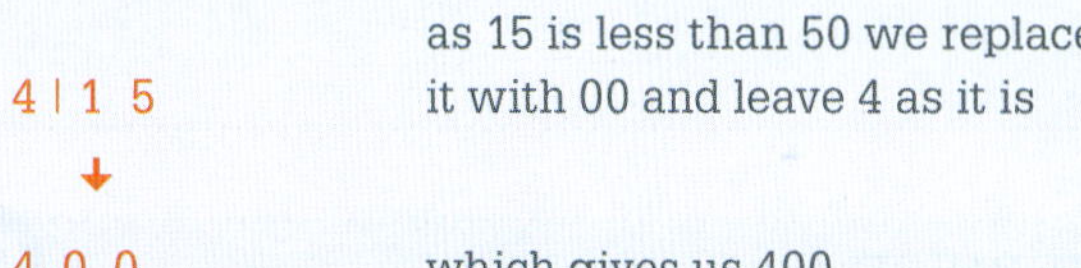

If the value to the right of the hundred is more than 50, replace it with 00 and increase the hundred value by 1.

EXAMPLE A.2

So, to round 660 to the nearest hundred:

6 | 6 0 — as 60 is more than 50 we replace it with 00 and increase 6 by 1

↓

7 0 0 — which gives us 700

If the value you were rounding was 650, then you could round it either up to 700 or down to 600. Either answer is correct.

Let's try these again, but this time we will round off to the nearest ten.

EXAMPLE A.3

If we wanted to round off 412 to the nearest ten:

To round to the nearest ten, look at the number immediately to the right of the ten value, in this case 2.

Compare this value to half of what we are rounding to (half of 10 is 5).

If the value to the right of the ten is less than 5, replace it with 0.

4 1 | 2 — as 2 is less than 5 we replace it with 0 and leave 4 and 1 as they are

↓

4 1 0 — which gives us 410

EXAMPLE A.4

So to round 415 to the nearest ten:

4 1 | 5 — as 5 is neither more than 5 or less than 5 we can replace it with 0 and increase 41 by 1 or leave it as it is

↓

4 1 0 — which gives us 410

OR

4 2 0 — which gives us 420

EXERCISE A.1

a Round 118 to the nearest hundred.
b Round 248 to the nearest ten.
c Round 67480 to the nearest hundred.
d Round 34430 to the nearest thousand.
e Round 33967 to the nearest ten.

Rounding decimals

A similar process to that outlined above applies to rounding of decimals. Again, each digit in the number has a specific weighting. For example, in the number 234.83:

2 is in the hundreds place and carries a weight of 2×100 (= 200)

3 is in the tens place and carries a weight of 3×10 (= 30)

4 is in the units place and carries a weight of 4×1 (= 4)

8 is in the tenths place and carries a weight of 8×0.1 (= 0.8)

3 is in the hundredths place and carries a weight of 3×0.01 (= 0.03)

Another way to look at this is:

Hundreds	Tens	Units	Tenths	Hundredths
2	3	4	8	3

If we wanted to round off 234.83 to the nearest tenth:

EXAMPLE A.5

To round to the nearest tenth, look at the number immediately to the right of the tenths value, in this case 3.

Compare this value to half of what we are rounding to (half of 10 is 5).

If the value to the right of the tenth is less than 5, ignore it.

2 3 4 . 8 | 3 — as 3 is less than 5 we ignore it and leave 8 as it is

↓

2 3 4 . 8 — which gives us 234.8 (ignore any zero after the 8)

If the value to the right of the tenth is more than 5, ignore it and increase the tenths value by 1.

EXAMPLE A.6

If we wanted to round off 253.18 to the nearest tenth:

2 5 3 . 1 | 8 — as 8 is more than 5 we ignore it but increase the 1 to 2

↓

2 5 3 . 2 — which gives us 253.2 (ignore any zero after the 2)

If increasing the tenths value by 1 results in 10, then we replace the 9 with 0 and increase the units value by 1.

EXAMPLE A.7

If we wanted to round off 18.98 to the nearest tenth:

1 8 . 9 | 8 — as 8 is more than 5 we ignore it but increase the 9 to 0 (10) and add 1 to the units (8 + 1 = 9)

↓

1 9 . 0 — which gives us 19.0 (ignore any trailing zeros)

Note: In **Example A.7**, the answer is expressed as 19.0 and not just 19. Showing the zero after the decimal point indicates the degree of precision used.

EXERCISE A.2

a Round 1.118 to the nearest hundredth.
b Round 2.48 to the nearest tenth.
c Round 6.7482 to the nearest thousandth.
d Round 34.439 to the nearest hundredth.
e Round 339.67 to the nearest tenth.

REVIEW QUESTIONS

1 Round 1598 to the nearest hundred.
2 Round 846.9 to the nearest ten.
3 Round 9.8 to the nearest unit.
4 Round 19.33 to the nearest unit.
5 Round 10.57 to the nearest tenth.
6 Round 33.24 to the nearest tenth.
7 Round 183 to the nearest ten.
8 Round 1857 to the nearest ten.
9 Round 86.433 to the nearest hundredth.
10 Round 15.737 to the nearest hundredth.

Significant digits

If we consider a mobile telephone number comprising the digits 0 4 4 4 1 1 1 2 2 2, every digit represented is significant. We need to dial all 10 digits in order to connect with that particular mobile phone. If we ignored the last digit, the call would not connect.

Consider now a car that is nine years, seven months and 22 days old; this is also equivalent to 9.6452 years. In either case, we don't really care about the months and days – they are insignificant. Essentially, we would be considering a 10-year-old car.

Consider another example: the televised day five of the test cricket between Australia and Sri Lanka stated that the attendance was 45 000 people. A spectator at the match said the crowd was 45 396, as displayed on the scoreboard late in the afternoon. There appears to be some discrepancy. However, the figures from both sources are correct – just expressed to different degrees of precision. The actual crowd attendance may have been 45 396 but when we round to the nearest thousand we get an attendance of 45 000. We have expressed the attendance correct to two significant digits that is, 45 thousand.

The first non-zero digit, when read from left to right, is the first significant digit. In the crowd attendance of 45 396, 4 is the first significant digit. When we round the attendance value to two significant digits, we have two non-zero digits on the left, which are followed by a number of zeros. As in this case the third digit is less than five, we replace it with zero along with all subsequent (or following) digits.

Consider now expressing the crowd attendance of 45 396 to:	
1 significant digit	50 000
2 significant digits	45 000
3 significant digits	45 400
4 significant digits	45 400
5 significant digits	45 396

The precision of the answer depends on the number of significant digits. The answer is more precise when expressed to a higher number of significant digits.

Rules for significant digits

Consider what happens when we express the number 53.810 756 to one significant digit, three significant digits and six significant digits.

EXAMPLE A.8

Express 53.810 756 to one significant digit.

5 | 3 . 8 1 0 7 5 6 — as 3 is less than 5 we replace it with 0 and ignore all following digits

↓

5 0 — which gives us 50 (ignore any trailing zeros)

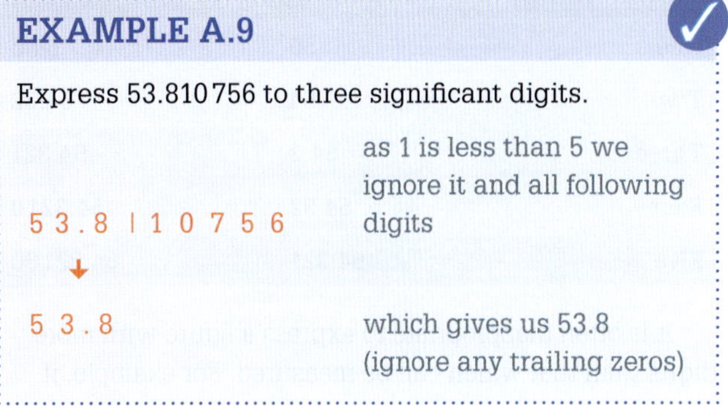

EXAMPLE A.10

Express 53.810756 to six significant digits.

5 3 . 8 1 0 7 | 5 6	as the 7th digit is 5 we need to look at the next digit, which is 6. This forces us to round up, so the 7 becomes 8, and we ignore all following digits
↓	
5 3 . 8 1 0 8	which gives us 53.8108 (ignore any trailing zeros)

EXERCISE A.3

a Express 11.8 to two significant digits.
b Express 2.408 to three significant digits.
c Express 674 802 to four significant digits.
d Express 34430 to two significant digits.
e Express 33967 to five significant digits.

Arithmetic precision

The precision of a value describes the number of digits used to express that value. In an engineering environment this is the number of significant digits or, less commonly, the number of decimal places (the number of digits after the decimal point). The latter is useful in financial and engineering applications where the number of digits in the fractional part has particular importance. In either case, the term 'precision' describes the position at which an inexact result is rounded.

As an example, the number 54.321 can be expressed with various numbers of significant digits or decimal places. If insufficient precision is available, then the number is rounded in some manner to fit the available precision. The following table shows the results for various total precisions and decimal places, with the results rounded to the nearest number.

Consider now expressing the number 54.321 to:		
Level of precision	**Rounded to significant digits**	**Rounded to decimal places**
One	50	54.3
Two	54	54.32
Three	54.3	54.321
Four	54.32	54.3210
Five	54.321	54.32100

It is often inappropriate to express a figure with more digits than that which can be measured. For example, if a set of scales measures to the nearest gram and gives a reading of 4.321 kg, it would create false precision if it were expressed as 4.321000 kg.

When rounding to a specific number of significant digits, there are a few general rules to follow (n represents the number of significant digits):

1. If the digit immediately to the right of the nth significant digit is greater than 5, the nth number is rounded up.
2. If the digit immediately to the right of the nth significant digit is less than 5, the nth number remains unchanged.
3. If the digit immediately to the right of the nth significant digit is 5 and there are non-zero digits after the 5, the number is rounded up.
4. If the digit immediately to the right of the nth significant digit is 5 and there are no subsequent non-zero digits, there are two commonly used conventions. In 'common rounding', such a digit is always rounded up; in 'unbiased rounding' (also known as 'round-to-even'), it is rounded in whichever direction leaves the nth digit even. For instance, under unbiased rounding, 61.5 would be rounded up to 62, but 64.5 would be rounded down to 64.

Consider what happens when we express the number 0.004 675 607 to one significant digit, two significant digits and four significant digits.

EXAMPLE A.11A

Express 0.004675607 to one significant digit.

0 . 0 0 4 | 6 7 5 6 0 7	as 6 is more than 5 we increase the first significant digit by 1 and ignore all following digits.
↓	
0 . 0 0 5	Remember the zeros immediately following the decimal points are not significant. This gives us 0.005 (ignore trailing zeros).

EXAMPLE A.11B

Express 0.004675607 to two significant digits.

0 . 0 0 4 6 | 7 5 6 0 7	as 7 is more than 5 we increase the second significant digit by 1 and ignore all following digits.
↓	
0 . 0 0 4 7	Remember the zeros immediately following the decimal point are not significant. This gives us 0.0047 (ignore trailing zeros).

EXAMPLE A.11C

Express 0.004 675 607 to four significant digits.

0.0 0 4 6 7 5 | 6 0 7 — as 6 is more than 5 we increase the fourth significant digit by 1 and ignore all following digits.

↓

0.0 0 4 6 7 6 — Remember the zeros immediately following the decimal point are not significant. This gives us 0.004676 (ignore trailing zeros).

EXERCISE A.4

a Express 0.118 to one significant digit.
b Express 0.02408 to three significant digits.
c Express 0.674 802 to four significant digits.
d Express 0.034 430 to two significant digits.
e Express 0.003 967 to five significant digits.

REVIEW QUESTIONS

1 Express 3098 to two significant digits.
2 Express 644.4 to three significant digits.
3 Express 9.38 to two significant digits.
4 Express 29.83 to three significant digits.
5 Express 0.0657 to two significant digits.
6 Express 0.003 374 to three significant digits.
7 Express 1.4302 to three significant digits.
8 Express 10.57 to three significant digits.
9 Express 36.734 to four significant digits.
10 Express 0.000 563 764 to four significant digits.

Scientific and engineering notation

In the field of electrotechnology we are often called upon to work with numbers that are very large and very small. Two ways to express these numbers in a more readable form are scientific notation, which is sometimes referred to as standard form, and engineering notation, which is a slight variation of standard form.

Scientific notation

Scientific notation is an easier way of writing very large and very small numbers. A number written in scientific notation has several properties that make it very useful to us. The basic form is shown in the following SWITCH ON feature:

SWITCH ON

Value = $N \times 10^x$ where N = any real number, referred to as the significand; it is in the form of a whole number between 1 and 9, a decimal point and a string of numbers of any length; x is a whole number, referred to as the exponent; it may be positive or negative.

Whether the exponent is positive or negative depends on whether the decimal point is moved to the right or to the left. Moving the decimal point to the right makes the exponent negative; moving it to the left makes the exponent positive.

When written in the form $N \times 10^x$, the choice of exponent x is such that the absolute value of N remains at least one but less than ten.

Scientific notation is a very convenient way to write large or small numbers and do calculations with them. It also quickly conveys two properties of a measurement that are useful to scientists – significant digits and order of magnitude.

Examples of use

- An electron has a mass of about 0.00000000000000000000000000000091093826 kg. In scientific notation this is written $9.1093826 \times 10^{-31}$ kg.
- The Earth has a mass of about 5 973 600 000 000 000 000 000 000 kg. In scientific notation this is written 5.9736×10^{24} kg.
- The circumference of the Earth is about 40 000 000 m. In scientific notation this is written 4×10^7 m.

Scientific notation also enables simple order of magnitude comparisons. Magnitude simply means how large a value is. A proton's mass is 0.000 000 000 000 000 000 000 000 001 672 6 kg. Writing this in scientific notation as 1.6726×10^{-27} kg makes it easier to compare it to the mass of the electron given above. The order of magnitude of the ratio of the masses can be obtained by comparing the exponents rather than counting all the leading zeros. In this case, '–27' for the proton is larger than '–31' for the electron. Therefore, the proton is four orders of magnitude (about 10 000 times) more massive than the electron. This is obtained from the difference between '–27' and '–31'.

Scientific notation also avoids misunderstandings due to local differences in certain quantifiers, such as 'billion', which might indicate either 10^9 or 10^{12}.

Using a calculator to solve problems in scientific notation

The scientific calculator makes it easy to solve seemingly complex equations, provided you know how to drive it. One common error students make is in the interpretation of $\times 10^x$. A single button on the calculator performs this function. This button is marked either [$\times 10^x$] or [EXP]. In either case simply enter the significand followed by the [$\times 10^x$] button followed by the value of the exponent. If the exponent is negative, *do not* use the standard minus button on the calculator, *use* the button specifically made for this function [(–)].

Try the following for yourself. Your calculator may not show the answer in scientific notation, so you will need to perform the conversions outlined above.

EXERCISE A.5

a Use a scientific calculator to solve $(1.86 \times 10^{-4}) \times (7.53 \times 10^{-3})$
b Use a scientific calculator to solve $(1.59 \times 10^{4}) \times (3.79 \times 10^{3})$
c Use a scientific calculator to solve $(2.53 \times 10^{-4}) \div (7.86 \times 10^{-3})$
d Use a scientific calculator to solve $(2.53 \times 10^{3}) \div (3.86 \times 10^{4})$

Engineering notation

Engineering notation is similar to scientific notation inasmuch as $Value = N \times 10^x$ but differs by restricting the exponent x to multiples of 3. As such the value of N can be equal to or larger than 10 (ranging from 1 to 1000 but not including 1000). Numbers in this form are easily read out using magnitude (or metric) prefixes like mega (where $x = 6$), kilo (where $x = 3$), milli (where $x = -3$), micro (where $x = -6$) or nano (where $x = -9$). For example, 12.5×10^{-9} m can be expressed as 12.5 nm.

Table A.1 summarises some common metric prefixes used in electrotechnology.

TABLE A.1 Magnitude prefixes for electrotechnology

Prefix	Symbol	Engineering notation	Decimal weighting
Giga	G	$\times 10^9$	× 1 000 000 000
Mega	M	$\times 10^6$	× 1 000 000
kilo	k	$\times 10^3$	× 1000
milli	m	$\times 10^{-3}$	$\times \frac{1}{1000}$
micro	μ	$\times 10^{-6}$	$\times \frac{1}{1\,000\,000}$
nano	n	$\times 10^{-9}$	$\times \frac{1}{1\,000\,000\,000}$
pico	p	$\times 10^{-12}$	$\times \frac{1}{1\,000\,000\,000\,000}$

The following examples show conversion of values based on their metric prefix to the value using engineering notation.

EXAMPLE A.12

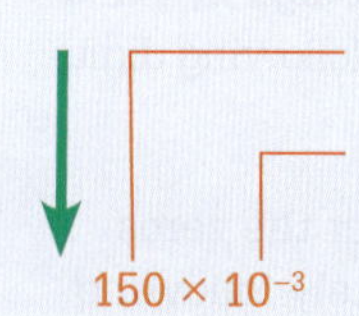

To write 150 milli-ampere in engineering notation: express the significand, in this case 150 then, referring to **Table A.1**, express the exponent, in this case $\times 10^{-3}$

EXAMPLE A.13

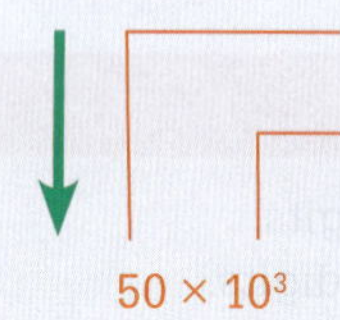

To write 50 kilovolts in engineering notation: express the significand, in this case 50 then, referring to **Table A.1**, express the exponent, in this case $\times 10^{3}$

EXAMPLE A.14

455 × 10⁶ ohms → 455 mega-ohms

To write 455×10^6 ohms using magnitude prefix: write the significand, in this case 455 then, referring to **Table A.1**, express the exponent as its prefix, in this case $\times 10^6$ has the prefix **mega**

EXAMPLE A.15

15 × 10⁻⁹ amperes → 15 nano-ampere

To write 15×10^{-9} amperes using magnitude prefix: write the significand, in this case 15 then, referring to **Table A.1**, express the exponent as its prefix, in this case $\times 10^{-9}$ has the prefix **nano**

EXERCISE A.6

a Express 33 kV in engineering notation.
b Express 660 kΩ in engineering notation.
c Express 33 mA in engineering notation.
d Express 45 µA in engineering notation.
e Express 4.5×10^3 ampere using the appropriate metric prefix.
f Express 68×10^6 ohms using the appropriate metric prefix.
g Express 105×10^{-6} ampere using the appropriate metric prefix.
h Express 270×10^{-3} volts using the appropriate metric prefix.
i Express 330×10^3 ohms using the appropriate metric prefix.
j Express 270×10^9 joules using the appropriate metric prefix.

Using a calculator to solve problems in engineering notation

Using a calculator to solve problems having values expressed in engineering notation is much the same as solving problems having values expressed in scientific notation. The only difference is that you need to convert the significand and magnitude prefix into the form $N \times 10^x$ using the steps outlined above.

EXERCISE A.7

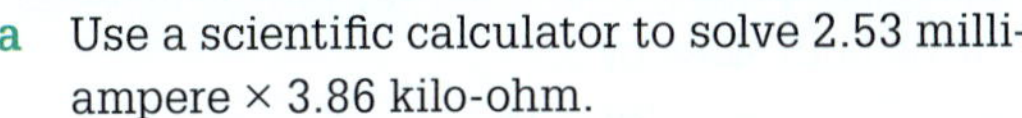

a Use a scientific calculator to solve 2.53 milli-ampere × 3.86 kilo-ohm.

b Use a scientific calculator to solve 8.76 kilovolt ÷ 2.14 milli-ampere.

The answer above is a bit 'messy'. In electrotechnology we don't need numbers expressed to the above level of precision; all we need is an expectation of what reading we are likely to get on a meter or how much current a cable is likely to carry. In these cases, it is usual to express the answer correct to three significant digits.

Furthermore, we express the answer using magnitude prefixes. The calculator has a handy button to assist in this. This is the ENG [engineering] button. Look at the following example and try it for yourself.

EXERCISE A.8

Use a scientific calculator to solve 2.76 kilovolt ÷ 8.14 milli-ampere.

$2.76 \times 10^3 \div 8.14 \times 10^{-3} =$
339.0663391×10^3

Answer: 339×10^3 ohms or 339 kilo-ohm (expressed to three significant digits)

REVIEW QUESTIONS

1 Express 4098 in scientific notation.
2 Express 455.4 in scientific notation.
3 Express 289.31 in scientific notation.
4 Express 49.83 in scientific notation.
5 Express 0.0752 in scientific notation.
6 Express 0.005324 in scientific notation.
7 Express 1.4302 in scientific notation.
8 Express 320.57 in scientific notation.
9 Express 36.534 in scientific notation.
10 Express 0.000563764 in scientific notation.
11 The prefix micro is used when a unit is multiplied by what exponent?
12 Giga is the prefix used when a unit is multiplied by what exponent?
13 What is the prefix for a unit multiplied by 10^{-9}?
14 What is the prefix for a unit multiplied by 10^6?

Express your answers to the following correct to three significant digits.

15 Express 33000 ohms as kilo-ohms.
16 Express 133000 volts as kilovolts.
17 Express 2.8 kilo-amperes as amperes.
18 Express 66 kilovolts as volts.
19 Express 0.8 volt as millivolts.
20 Express 0.45 ampere as milli-amperes.
21 Express 0.0065 watt as milliwatts.
22 Express 3850 micro-amperes as milli-amperes.
23 Express 0.00000675 volt as microvolts.
24 Express 0.00000865 watt as microwatts.
25 Express 0.00000052 ampere as micro-amperes.
26 Express 554000000 micro-ohms as ohms.
27 Express 18000000 watts as megawatts.
28 Express 265000000 ohms as giga-ohms.
29 Express 220 volts as milli-volts.
30 Express 0.0558 mega-ampere as kilo-ampere.
31 Express 62500000 micro-volt as volt.
32 Express 6450 microvolts as millivolts.
33 Express 37500 amperes as kilo-amperes.
34 Express 58200 microwatts as milliwatts.
35 Express 5270 milli-amperes as amperes.
36 Express 0.0543 millivolt as microvolt.
37 Express 0.0665 milli-ampere as micro-ampere.
38 Express 2240 milli-ohms as ohms.
39 Express 2800 watts as kilowatts.
40 Express 0.145 milliwatt as microwatt.

Transposition

Let's look at how we can rearrange equations to change the subject or, in other words, transpose the equation. You probably do this on many occasions without realising it. Consider the following example:

Your family has two cars. You filled one car with petrol and your sister filled the other. Due to strong sibling rivalry, you want to know who bought the cheapest fuel, but each car was filled with a different amount of fuel. So it is necessary to compare the cost per litre but neither of you can remember the cost per litre. You put 61 litres in the tank at a total cost of $82.35. Your sister put 45 litres in the tank at a total cost of $57.60. The total dollar amount is calculated at the pump using the equation:

$$\text{total cost} = \text{cost per litre} \times \text{litres}$$

In order to determine the cost per litre, you divide the total cost by the number of litres. So, you bought petrol at $1.35 per litre and your sister paid $1.28 per litre. This is an example of transposing the equation, total cost = cost per litre × litres, to find cost per litre.

All equations have two sides separated by an equals (=) sign. Let's call these the left-hand side (LHS) and the right-hand side (RHS). For example, in the equation below $3x + 8$ is on the LHS of the equation and 23 is on the RHS of the equation:

$$3x + 8 = 23$$

$$\text{LHS} = \text{RHS}$$

It may be useful to think of the equals sign (=) as being a balance scale. For the scale to balance, the LHS must be of the same weight as the RHS. Clearly, in the picture below the scale is not balanced as LHS ≠ RHS (where ≠ means does not equal).

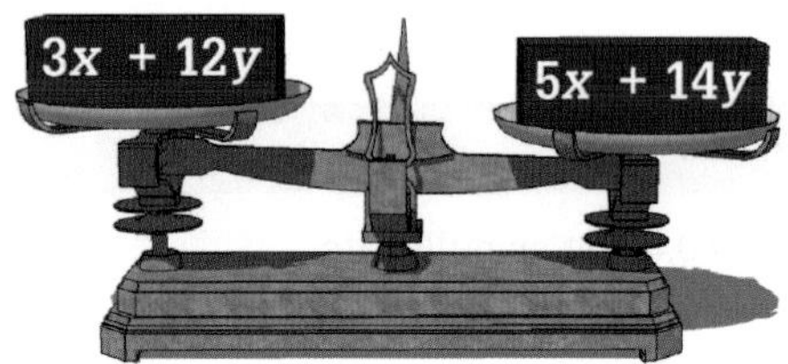

There are many ways to transpose equations. One method is to do the same thing to both sides of the equation. The aim is to bring like terms together and isolate the unknown quantity.

So, to find out what x is equal to in the equation $3x + 8 = 23$, we need to move the numbers around to be left with x = some finite value. Moving numbers methodically from one side of the equals sign to the other side is known as transposition by opposites. In our equation the 8 is begging to be moved from the LHS to the RHS. To do this it is necessary to subtract 8 from the LHS and the RHS, so:

$$3x + 8 - 8 = 23 - 8$$

We can simplify this to:

$$3x + 0 = 15$$

As it is not needed to show the zero, our equation becomes:

$$3x = 15$$

As we are multiplying x by 3, the opposite is dividing. We now divide both sides by 3:

$$\frac{3x}{3} = \frac{15}{3}$$

This simplifies to:

$$1x = 5$$

As it is not necessary to show the 1, then:

$$x = 5$$

An important final step is to put this value back into the equation, so:

$$(3 \times 5) + 8 = ?$$

$$15 + 8 = 23$$

We can safely say our transposition is correct.

SWITCH ON

Transposition: Two basic rules

1. Do the same thing to both sides of the equals sign (=).
2. To move a value across the = sign, do the opposite function.

Equations involving addition

	$V_T = V_1 + V_2$	Transpose to find V_1
Step 1	$+V_T = V_1 + V_2$	Make the subject positive. That is, $+V_T$. Both V_1 and V_2 are positive, though there is no sign written before V_1.
Step 2	$+V_T - V_2 = V_1 + V_2 - V_2$	As this equation involves addition, the opposite is subtraction. To find V_1 subtract V_2 from both sides.
Step 3	$+V_T - V_2 = V_1 + 0$	As the result of subtracting anything from itself is 0, $V_2 - V_2 = 0$.
Step 4	$+V_T - V_2 = V_1$	It is not necessary to show the 0.
Step 5	$V_1 = V_T - V_2$	Tidy up by showing the subject on the left and remove + from the front of V_T.

	$R_T = R_1 + R_2$	Transpose to find R_2
Step 1	$+R_T = +R_1 + R_2$	Show R_T and R_1 as positive.
Step 2	$+R_T - R_1 = +R_1 + R_2 - R_1$	To leave R_2 on the RHS, subtract R_1 from RHS and LHS.
Step 3	$+R_T - R_1 = +R_2 + 0$	As the result of subtracting anything from itself is 0, $+R_1 - R_1 = 0$
Step 4	$+R_T - R_1 = +R_2$	It is not necessary to show the 0.
Step 5	$R_2 = R_T - R_1$	Tidy up by showing the subject on the left and removing the redundant +.
Practice	Transpose to find R_1.	Your answer should be $R_1 = R_T - R_2$.

Equations involving subtraction

	$P_{Loss} = P_{In} - P_{Out}$	Transpose to find P_{In}
Step 1	$+P_{Loss} = P_{In} - P_{Out}$	Make the subject positive. That is, $+ P_{Loss}$. P_{In} is positive, though there is no sign written before it.
Step 2	$+P_{Loss} + P_{Out} = P_{In} - P_{Out} + P_{Out}$	As this equation involves subtraction, the opposite is addition. To find P_{In} add P_{Out} to both sides.
Step 3	$+P_{Loss} + P_{Out} = P_{In} + 0$	As the result of adding anything to its negative self is 0, $- P_{Out} + P_{Out} = 0$
Step 4	$+P_{Loss} + P_{Out} = P_{In}$	It is not necessary to show the 0.
Step 5	$P_{In} = P_{Loss} + P_{Out}$	Tidy up by showing the subject on the left and remove + from the front of P_{Loss}.

	$I_1 = I_T - I_2$	Transpose to find I_T
Step 1	$+I_1 = +I_T - I_2$	Show I_T and I_1 as positive.
Step 2	$+I_2 + I_1 = +I_T - I_2 + I_2$	To leave I_T on the RHS, add I_2 to RHS and LHS.
Step 3	$+I_2 + I_1 = +I_T + 0$	As the result of adding a positive to the same negative value is 0, $- I_2 + I_2 = 0$.
Step 4	$+I_2 + I_1 = + I_T$	It is not necessary to show the 0.
Step 5	$I_T = I_2 + I_1$	Tidy up by showing the subject on the left and remove the redundant +.
Practice:	If $C_2 = C_T - C_1$, find C_T.	Your answer should be $C_T = C_1 + C_2$.

Equations involving multiplication

	$V = I \times R$	Transpose to find R
Step 1	$+V = I \times R$	Make the subject positive. Expand to show the multiplication. As this equation involves multiplication, the opposite is division.
Step 2	$\frac{+V}{+I} = \frac{I \times R}{I}$	To find R divide both sides by I.
Step 3	$\frac{+V}{+I} = 1 \times R$	As the result of dividing anything by itself is 1, $I/I = 1$.
Step 4	$\frac{+V}{+I} = R$	It is not necessary to show the 1.
Step 5	$R = \frac{V}{I}$	Tidy up by showing the subject on the left and remove + from the front of V and I.

	$P = E \times I$	Transpose to find I
Step 1	$\frac{P}{E} = \frac{E \times I}{E}$	To leave I on the RHS, divide RHS and LHS by E.
Step 2	$\frac{P}{E} = 1 \times I$	As the result of dividing anything by itself is 1, $E/E = 1$.
Step 3	$\frac{P}{E} = I$	It is not necessary to show the 1.
Step 4	$I = \frac{P}{E}$	Tidy up by showing the subject on the left.
Practice:	Transpose to find E.	Your answer should be $E = \frac{P}{I}$

Equations involving division

	$a=\frac{b}{c}$	Transpose to find b
Step 1	$a\times c=\frac{b}{c}\times c$	As this equation involves division, the opposite is multiplication. To find b multiply both sides by c.
Step 2	$a\times c=\frac{b}{\cancel{c}}\times\cancel{c}$	As the result of dividing anything by itself is 1, $c/c = 1$.
Step 3	$a\times c=b$	It is not necessary to show the × 1.
Step 4	$b = ac$	Tidy up by showing the subject on the left.

	$I=\frac{E}{R}$	Transpose to find E
Step 1	$R\times I=\frac{E}{R}\times R$	To leave E on the RHS, multiply RHS and LHS by R.
Step 2	$R\times I=E\times 1$	As the result of dividing anything by itself is 1, $R/R = 1$.
Step 3	$R\times I=E$	It is not necessary to show the × 1.
Step 4	$E=R\times I$	Tidy up by showing the subject on the left.
Practice:	If $a=\frac{F}{m}$ find F	Your answer should be $F = m \times a$.

Equations involving squares

	$c^2 = 2ab$	Transpose to find c
Step 1	$\sqrt{c^2}=\sqrt{2ab}$	As this equation involves squaring, the opposite is square rooting. To find c, take the square root of both sides.
Step 2	$\sqrt{c^2}=\sqrt{2ab}$	As the result of taking the square root of the square of a value equals the value, $\sqrt{c^2}$ equals c.
Step 3	$c=\sqrt{2ab}$	

	$I^2=\frac{P}{R}$	Transpose to find I
Step 1	$\sqrt{I^2}=\sqrt{\frac{P}{R}}$	To leave I on the LHS, take the square root of RHS and LHS.
Step 2	$I=\sqrt{\frac{P}{R}}$	As the result of square rooting a square leaves the original value, $\sqrt{I^2}=I$.
Practice:	If $V^2=P\times R$, find V.	Your answer should be $V=\sqrt{P\times R}$

Equations involving square roots

	$\sqrt{c}=4ab$	Transpose to find c
Step 1	$\left(\sqrt{c}\right)^2=(4ab)^2$	As this equation involves a square root, the opposite is squaring. To find c, take the square of both sides. We need to square $4ab$.
Step 2	$c=(4ab)^2$	As the result of taking the square of the square root of a value equals the value, $\left(\sqrt{c}\right)^2$ equals c.

»

	$\sqrt{Q} = 4\pi \times Z \times T$	**Transpose to find Q**
Step 1	$\sqrt{Q} = (4\pi \times Z \times T)$	Treat the RHS as a single group by enclosing in brackets.
Step 2	$(\sqrt{Q})^2 = (4\pi \times Z \times T)^2$	To leave Q on the LHS, square the RHS and LHS.
Step 3	$Q = (4\pi \times Z \times T)^2$	As the result of squaring a square root leaves the original value, $(\sqrt{Q})^2 = Q$
Practice:	If $\sqrt{Z} = P \times Q$, find Z.	Your answer should be $Z = (P \times Q)^2$.

The examples shown so far cover the basic rules of transposition. They are deliberately kept to a single operation to help you build your confidence. The transposition of complex equations involves the sequential application of the steps outlined above. In order to execute these correctly you need to have an understanding of the order of operations. The mathematical order of operations is similar to how we construct coherent sentences. For example, one would say that 'James is 180 centimetres tall' rather than '180 centimetres tall is James'. We construct the sentence following a particular set of rules.

REVIEW QUESTIONS

Transpose the following equations.

1. $R_T = R_1 + R_2 + R_3$ Find R_3.
2. $V_T = E_G - V_A$ Find E_G.
3. $F = B\,I\,I$ Find B.
4. $R = \frac{\rho l}{a}$ Find a.
5. $X_L = 2\pi fL$ Find f.
6. $P = \frac{2\pi NT}{60}$ Find T.
7. $C = \frac{\varepsilon a}{d}$ Find ε.
8. $P = \frac{V^2}{R}$ Find V.
9. $R_x = \frac{R_A \times R}{R_B}$ Find R_A.
10. $N = \frac{120f}{P}$ Find f.

Graphs

Consider the following graph:

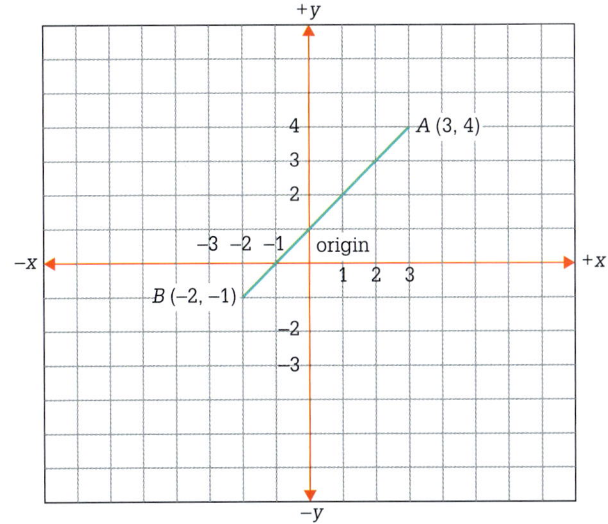

An **ordered pair** is any two numbers where the order is important. For example, in the adjacent graph (3, 4) is an ordered pair. It gives the location of point *A*. Point *A* is located 3 units along the *x*-axis and 4 units along the *y*-axis. Point *B* is located 2 units in the negative direction along the *x*-axis and 1 unit in the negative direction along the *y*-axis. In other words, in an ordered pair the first number gives the number of units along the *x*-axis and the second number gives the number of units along the *y*-axis.

A **relation** is a set of ordered pairs. These are usually set out in a table:

x	−2	−1	0	1	2
y	−1	0	1	2	3

The **domain** of the relation is {−2, −1, 0, 1, 2}, which are the first-named numbers.

The **range** of the relation is {−1, 0, 1, 2, 3}, which are the second-named numbers.

The **rule** or **equation** is how the numbers are related to each other. In the above relation, we can write the rule as $y = x + 1$. This is a linear relationship.

The x-**intercept** is where the graph cuts the x-axis.

The y-**intercept** is where the graph cuts the y-axis.

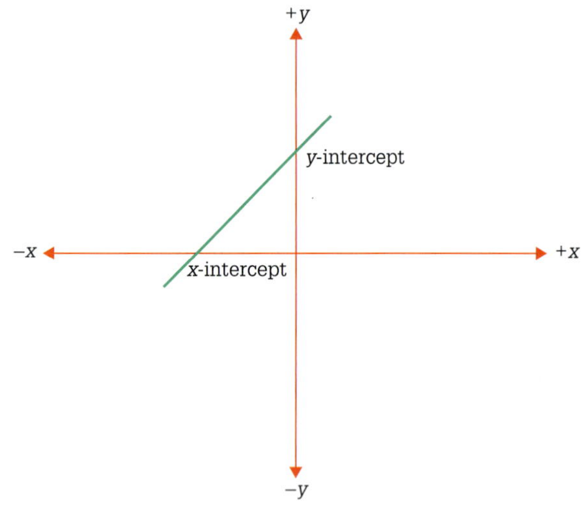

The **gradient** is the slope of a straight line.

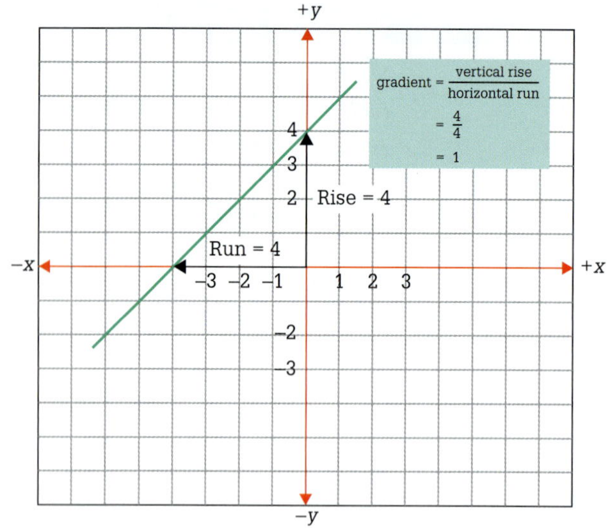

It is common to use the **run** as the x-axis and the **rise** as the y-axis. You may like to think of these as the x-axis representing the 'cause', such as different voltages applied to a circuit. Then the different voltages will cause a particular current to flow – it has an 'effect', which is plotted along the y-axis.

REVIEW QUESTIONS

1 On graph paper with a 5 mm grid, plot the following to a scale of 10 mm = 10 volt and 10 mm = 0.5 A:

x-axis	Label = Voltage	Unit = Volts	10	30	45	70	100
y-axis	Label = Current	Unit = A	0.50	1.50	2.25	3.50	5.00

2 What is the gradient of the line?

3 Use your graph to predict the value of current when the voltage is 50 volts.

4 Write the equation for the graph in terms of x and y.

5 What is the range of the plotted graph?

Useful equations

Note: The following equations use symbols detailed in AS 1046, part 1. There are alternative recognised symbols in use. Transposition of equations will be necessary to solve problems.

$Q = It$	$v = \frac{s}{t}$	$a = \frac{\Delta v}{t}$
$F = ma$	$W = Fs$	$W = mgh$
$W = Pt$	$\eta = \frac{P_{Out}}{P_{In}}$	$I = \frac{V}{R}$
$P = VI$	$P = I^2 R$	$P = \frac{V^2}{R}$
$R_2 = \frac{R_1 A_1 l_2}{A_2 l_1}$	$R_2 = R_1(1 + \alpha(T_2 - T_1))$	$R = \frac{\rho l}{a}$
$R_T = R_1 + R_2 + R_3$	$V_T = V_1 + V_2 + V_3$	$\frac{1}{R_T} = \frac{1}{R_1} + \frac{1}{R_2} + \frac{1}{R_3}$

»

$I_T = I_1 + I_2 + I_3$	$V_2 = V_T \frac{R_2}{R_1+R_2}$	$I_2 = I_T \frac{R_1}{R_1+R_2}$
$R_X = \frac{R_A R}{R_B}$	$C = \frac{Q}{V}$	$\tau = RC$
$\frac{1}{C_T} = \frac{1}{C_1} + \frac{1}{C_2} + \frac{1}{C_3}$	$C_T = C_1 + C_2 + C_3$	$C = \frac{A\varepsilon_0\varepsilon_r}{d}$
$F_m = IN$	$H = \frac{F_m}{l}$	$B = \frac{\Phi}{A}$
$\Phi = \frac{F_m}{R_m}$	$R_m = \frac{l}{\mu_0\mu_r A}$	$V = N\frac{\Delta\Phi}{\Delta t}$
$e = Blv$	$L = \frac{\mu_0\mu_r AN^2}{l}$	$L = N\frac{\Delta\Phi}{\Delta I}$
$V = L\frac{\Delta I}{\Delta t}$	$\tau = \frac{L}{R}$	$F = BIl$
$T = Fr$	$E_g = \frac{\Phi zNP}{60\,a}$	$P = \frac{2\pi\, NT}{60}$
$\tau = \frac{1}{f}$	$f = \frac{NP}{120}$	$V = 0.707V_{max}$
$I = 0.707I_{max}$	$V_{ave} = 0.637V_{max}$	$I_{ave} = 0.637I_{max}$
$V = V_{max}\sin\theta$	$i = I_{max}\sin\theta$	$I = \frac{V}{Z}$
$Z = \sqrt{R^2 + (X_L - X_C)^2}$	$X_L = 2\pi fL$	$X_C = \frac{1}{2\pi fC}$
$\cos\theta = \frac{P}{S}$	$\cos\theta = \frac{R}{Z}$	$S = \sqrt{P^2 + Q^2}$
$S = VI$	$P = VI\cos\theta$	$Q = VI\sin\theta$
$f_0 = \frac{1}{2\pi\sqrt{LC}}$	$V_L = \sqrt{3}V_p$	$I_L = \sqrt{3}I_p$
$S = \sqrt{3}V_L I_L$	$P = \sqrt{3}V_L I_L\cos\theta$	$Q = \sqrt{3}V_L I_L\sin\theta$
$\tan\phi = \sqrt{3}\left(\frac{W_2 - W_1}{W_2 + W_1}\right)$	$Q = mC\Delta t$	
$V' = 4.4\Phi fN$	$\frac{V_1}{V_2} = \frac{N_1}{N_2}$	$\frac{I_1}{I_2} = \frac{N_1}{N_2}$
$N_s = \frac{120f}{P}$	$slip = \frac{N_s - N}{N_s}$	$f_r = \frac{s\% \times f}{100}$
$V_{reg}\% = \frac{(V_{NL} - V_{FL})}{V_{FL}} \times \frac{100}{1}$		$I_{st} = \frac{1}{3} \times I_{DOL}$
$T_{st} = \frac{1}{3} \times T_{DOL}$	$I_{st} = \frac{V_{ST}}{V} \times I_{DOL}$	$T_{ST} = \left(\frac{V_{ST}}{V}\right) \times T_{DOL}$
$I_{motor_{st}} = \frac{\%Tap}{100} \times I_{DOL}$	$I_{Line_{st}} = \left(\frac{\%Tap}{100}\right)^2 \times I_{DOL}$	$E = \frac{\Phi_v}{I}$
$E = \frac{I}{d^2}$	$n_v = \frac{\Phi_v}{p}$	$V_L = 0.45V_{ac}$
$V_L = 0.9V_{ac}$	$V_L = 1.17V_{phase}$	$V_L = 1.35V_{line}$
$PRV = \sqrt{2}V_{ac}$	$PRV = 2\sqrt{2}V_{ac}$	$PRV = 2.45V_{ac}$
$V_{ripple} = \sqrt{2}V_{ac}$	$V_{ripple} = 0.707V_{phase}$	$V_{ripple} = 0.1895V_{line}$

Periodic table of elements

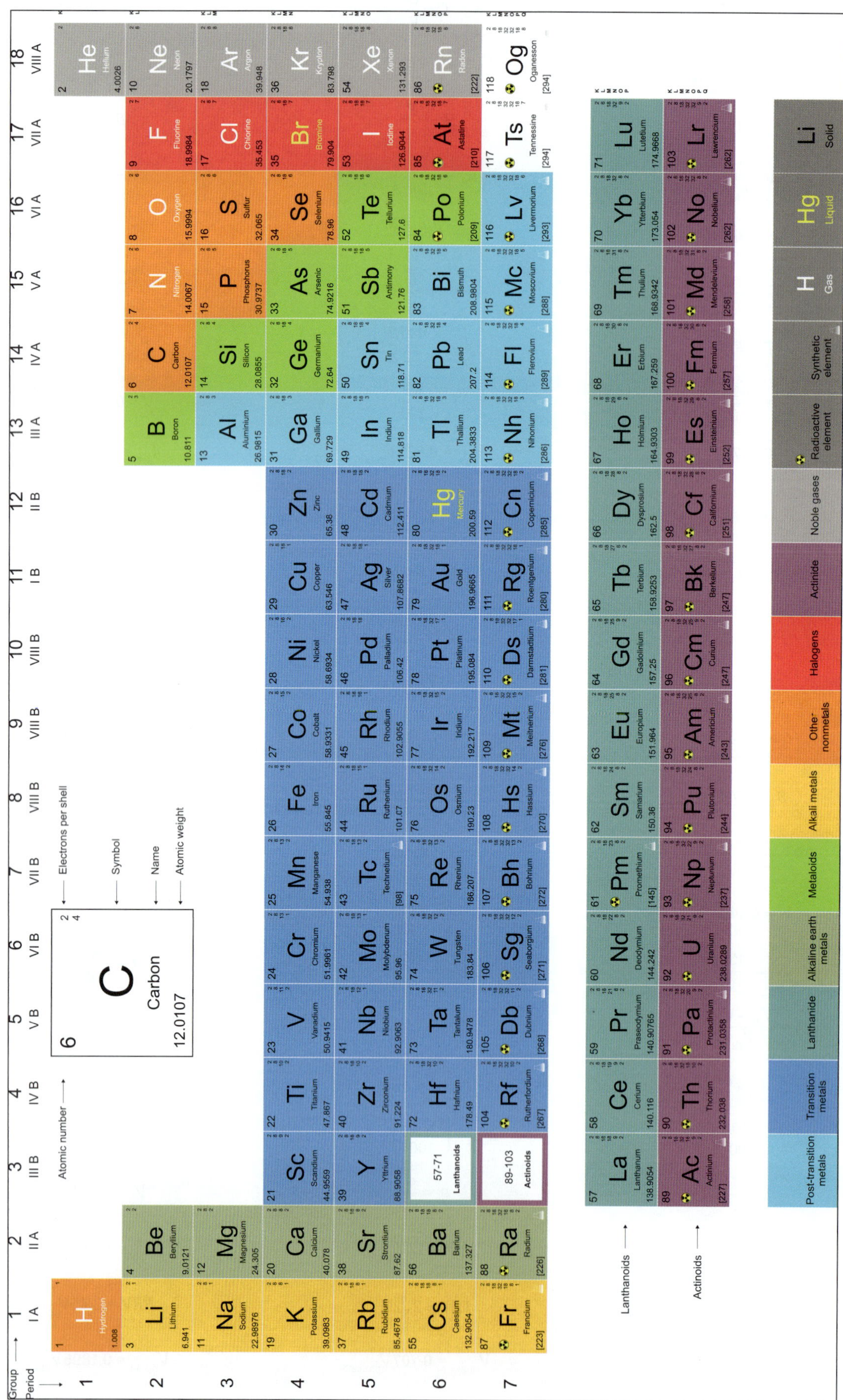

Source: Shutterstock.com/concept w

Index